T0189003

Lecture Notes in Computer Science 12261

More information about this series at http://www.springer.com/series/7412

Anne L. Martel · Purang Abolmaesumi ·
Danail Stoyanov · Diana Mateus ·
Maria A. Zuluaga · S. Kevin Zhou ·
Daniel Racoceanu · Leo Joskowicz (Eds.)

Medical Image Computing and Computer Assisted Intervention – MICCAI 2020

23rd International Conference
Lima, Peru, October 4–8, 2020
Proceedings, Part I

Editors

Anne L. Martel
University of Toronto
Toronto, ON, Canada

Purang Abolmaesumi
The University of British Columbia
Vancouver, BC, Canada

Danail Stoyanov
University College London
London, UK

Diana Mateus
École Centrale de Nantes
Nantes, France

Maria A. Zuluaga
EURECOM
Biot, France

S. Kevin Zhou
Chinese Academy of Sciences
Beijing, China

Daniel Racoceanu
Sorbonne University
Paris, France

Leo Joskowicz
The Hebrew University of Jerusalem
Jerusalem, Israel

ISSN 0302-9743 ISSN 1611-3349 (electronic)
Lecture Notes in Computer Science
ISBN 978-3-030-59709-2 ISBN 978-3-030-59710-8 (eBook)
https://doi.org/10.1007/978-3-030-59710-8

LNCS Sublibrary: SL6 – Image Processing, Computer Vision, Pattern Recognition, and Graphics

This Springer imprint is published by the registered company Springer Nature Switzerland AG
The registered company address is: Gewerbestrasse 11, 6330 Cham, Switzerland

Preface

The 23rd International Conference on Medical Image Computing and Computer-Assisted Intervention (MICCAI 2020) was held this year under the most unusual circumstances, due to the COVID-19 pandemic disrupting our lives in ways that were unimaginable at the start of the new decade. MICCAI 2020 was scheduled to be held in Lima, Peru, and would have been the first MICCAI meeting in Latin America. However, with the pandemic, the conference and its program had to be redesigned to deal with realities of the "new normal", where virtual presence rather than physical interactions among attendees, was necessary to comply with global transmission control measures. The conference was held through a virtual conference management platform, consisting of the main scientific program in addition to featuring 25 workshops, 8 tutorials, and 24 challenges during October 4–8, 2020. In order to keep a part of the original spirit of MICCAI 2020, SIPAIM 2020 was held as an adjacent LatAm conference dedicated to medical information management and imaging, held during October 3–4, 2020.

The proceedings of MICCAI 2020 showcase papers contributed by the authors to the main conference, which are organized in seven volumes of *Lecture Notes in Computer Science* (LNCS) books. These papers were selected after a thorough double-blind peer-review process. We followed the example set by past MICCAI meetings, using Microsoft's Conference Managing Toolkit (CMT) for paper submission and peer reviews, with support from the Toronto Paper Matching System (TPMS) to partially automate paper assignment to area chairs and reviewers.

The conference submission deadline had to be extended by two weeks to account for the disruption COVID-19 caused on the worldwide scientific community. From 2,953 original intentions to submit, 1,876 full submissions were received, which were reduced to 1,809 submissions following an initial quality check by the program chairs. Of those, 61% were self-declared by authors as Medical Image Computing (MIC), 6% as Computer Assisted Intervention (CAI), and 32% as both MIC and CAI. Following a broad call to the community for self-nomination of volunteers and a thorough review by the program chairs, considering criteria such as balance across research areas, geographical distribution, and gender, the MICCAI 2020 Program Committee comprised 82 area chairs, with 46% from North America, 28% from Europe, 19% from Asia/Pacific/Middle East, 4% from Latin America, and 1% from Australia. We invested significant effort in recruiting more women to the Program Committee, following the conference's emphasis on equity, inclusion, and diversity. This resulted in 26% female area chairs. Each area chair was assigned about 23 manuscripts, with suggested potential reviewers using TPMS scoring and self-declared research areas, while domain conflicts were automatically considered by CMT. Following a final revision and prioritization of reviewers by area chairs in terms of their expertise related to each paper, over 1,426 invited reviewers were asked to bid for the papers for which they had been suggested. Final reviewer allocations via CMT took account of reviewer bidding, prioritization of area chairs, and TPMS scores, leading to allocating about 4 papers per reviewer. Following an initial double-blind review phase by reviewers, area chairs provided a meta-review summarizing key points of reviews and a recommendation for each paper. The program chairs then evaluated the reviews and their scores, along with the recommendation from the area chairs, to directly accept 241 papers (13%) and reject 828 papers (46%); the remainder of the papers were sent for rebuttal by the authors. During the rebuttal phase, two additional area chairs were assigned to each paper using the CMT and TPMS scores while accounting for domain conflicts. The three area chairs then independently scored each paper to accept or reject, based on the reviews, rebuttal, and manuscript, resulting in clear paper decisions using majority voting. This process resulted in the acceptance of a further 301 papers for an overall acceptance rate of 30%. A virtual Program Committee meeting was held on July 10, 2020, to confirm the final results and collect feedback of the peer-review process.

For the MICCAI 2020 proceedings, 542 accepted papers have been organized into seven volumes as follows:

- Part I, LNCS Volume 12261: Machine Learning Methodologies
- Part II, LNCS Volume 12262: Image Reconstruction and Machine Learning
- Part III, LNCS Volume 12263: Computer Aided Intervention, Ultrasound and Image Registration
- Part IV, LNCS Volume 12264: Segmentation and Shape Analysis
- Part V, LNCS Volume 12265: Biological, Optical and Microscopic Image Analysis
- Part VI, LNCS Volume 12266: Clinical Applications
- Part VII, LNCS Volume 12267: Neurological Imaging and PET

For the main conference, the traditional emphasis on poster presentations was maintained; each author uploaded a brief pre-recorded presentation and a graphical abstract onto a web platform and was allocated a personal virtual live session in which they talked directly to the attendees. It was also possible to post questions online allowing asynchronous conversations – essential to overcome the challenges of a global conference spanning many time zones. The traditional oral sessions, which typically included a small proportion of the papers, were replaced with 90 "mini" sessions where all of the authors were clustered into groups of 5 to 7 related papers; a live virtual session allowed the authors and attendees to discuss the papers in a panel format.

We would like to sincerely thank everyone who contributed to the success of MICCAI 2020 and the quality of its proceedings under the most unusual circumstances of a global pandemic. First and foremost, we thank all authors for submitting and presenting their high-quality work that made MICCAI 2020 a greatly enjoyable and successful scientific meeting. We are also especially grateful to all members of the Program Committee and reviewers for their dedicated effort and insightful feedback throughout the entire paper selection process. We would like to particularly thank the MICCAI society for support, insightful comments, and continuous engagement with organizing the conference. Special thanks go to Kitty Wong, who oversaw the entire process of paper submission, reviews, and preparation of conference proceedings. Without her, we would have not functioned effectively. Given the "new normal", none of the workshops, tutorials, and challenges would have been feasible without the true leadership of the satellite events organizing team led by Mauricio Reyes: Erik Meijering (workshops), Carlos Alberola-López (tutorials), and Lena Maier-Hein (challenges). Behind the scenes, MICCAI secretarial personnel, Janette Wallace and Johanne Langford, kept a close eye on logistics and budgets, while Mehmet Eldegez and his team at Dekon Congress and Tourism led the professional conference organization, working tightly with the virtual platform team. We also thank our sponsors for financial support and engagement with conference attendees through the virtual platform. Special thanks goes to Veronika Cheplygina for continuous engagement with various social media platforms before and throughout the conference to publicize the conference. We would also like to express our gratitude to Shelley Wallace for helping us in Marketing MICCAI 2020, especially during the last phase of the virtual conference organization.

The selection process for Young Investigator Awards was managed by a team of senior MICCAI investigators, led by Julia Schnabel. In addition, MICCAI 2020 offered free registration to the top 50 ranked papers at the conference whose primary authors were students. Priority was given to low-income regions and Latin American students. Further support was provided by the National Institutes of Health (support granted for MICCAI 2020) and the National Science Foundation (support granted to MICCAI 2019 and continued for MICCAI 2020) which sponsored another 52 awards for USA-based students to attend the conference. We would like to thank Marius Linguraru and Antonion Porras, for their leadership in regards to the NIH sponsorship for 2020, and Dinggang Shen and Tianming Liu, MICCAI 2019 general chairs, for keeping an active bridge and engagement with MICCAI 2020.

Marius Linguraru and Antonion Porras were also leading the young investigators early career development program, including a very active mentorship which we do hope, will significantly catalize young and briliant careers of future leaders of our scientific community. In link with SIPAIM (thanks to Jorge Brieva, Marius Linguraru, and Natasha Lepore for their support), we also initiated a Startup Village initiative, which, we hope, will be able to bring in promissing private initiatives in the areas of MICCAI. As a part of SIPAIM 2020, we note also the presence of a workshop for Peruvian clinicians. We would like to thank Benjaming Castañeda and Renato Gandolfi for this initiative.

MICCAI 2020 invested significant efforts to tightly engage the industry stakeholders in our field throughout its planning and organization. These efforts were led by Parvin Mousavi, and ensured that all sponsoring industry partners could connect with the conference attendees through the conference's virtual platform before and during the meeting. We would like to thank the sponsorship team and the contributions of Gustavo Carneiro, Benjamín Castañeda, Ignacio Larrabide, Marius Linguraru, Yanwu Xu, and Kevin Zhou.

We look forward to seeing you at MICCAI 2021.

October 2020

Anne L. Martel
Purang Abolmaesumi
Danail Stoyanov
Diana Mateus
Maria A. Zuluaga
S. Kevin Zhou
Daniel Racoceanu
Leo Joskowicz

Organization

General Chairs

Daniel Racoceanu	Sorbonne Université, Brain Institute, France
Leo Joskowicz	The Hebrew University of Jerusalem, Israel

Program Committee Chairs

Anne L. Martel	University of Toronto, Canada
Purang Abolmaesumi	The University of British Columbia, Canada
Danail Stoyanov	University College London, UK
Diana Mateus	Ecole Centrale de Nantes, LS2N, France
Maria A. Zuluaga	Eurecom, France
S. Kevin Zhou	Chinese Academy of Sciences, China

Keynote Speaker Chair

Rene Vidal	The John Hopkins University, USA

Satellite Events Chair

Mauricio Reyes	University of Bern, Switzerland

Workshop Team

Erik Meijering (Chair)	The University of New South Wales, Australia
Li Cheng	University of Alberta, Canada
Pamela Guevara	University of Concepción, Chile
Bennett Landman	Vanderbilt University, USA
Tammy Riklin Raviv	Ben-Gurion University of the Negev, Israel
Virginie Uhlmann	EMBL, European Bioinformatics Institute, UK

Tutorial Team

Carlos Alberola-López (Chair)	Universidad de Valladolid, Spain
Clarisa Sánchez	Radboud University Medical Center, The Netherlands
Demian Wassermann	Inria Saclay Île-de-France, France

Challenges Team

Lena Maier-Hein (Chair) — German Cancer Research Center, Germany
Annette Kopp-Schneider — German Cancer Research Center, Germany
Michal Kozubek — Masaryk University, Czech Republic
Annika Reinke — German Cancer Research Center, Germany

Sponsorship Team

Parvin Mousavi (Chair) — Queen's University, Canada
Marius Linguraru — Children's National Institute, USA
Gustavo Carneiro — The University of Adelaide, Australia
Yanwu Xu — Baidu Inc., China
Ignacio Larrabide — National Scientific and Technical Research Council, Argentina
S. Kevin Zhou — Chinese Academy of Sciences, China
Benjamín Castañeda — Pontifical Catholic University of Peru, Peru

Local and Regional Chairs

Benjamín Castañeda — Pontifical Catholic University of Peru, Peru
Natasha Lepore — University of Southern California, USA

Social Media Chair

Veronika Cheplygina — Eindhoven University of Technology, The Netherlands

Young Investigators Early Career Development Program Chairs

Marius Linguraru — Children's National Institute, USA
Antonio Porras — Children's National Institute, USA

Student Board Liaison Chair

Gabriel Jimenez — Pontifical Catholic University of Peru, Peru

Submission Platform Manager

Kitty Wong — The MICCAI Society, Canada

Conference Management

DEKON Group
Pathable Inc.

Program Committee

Ehsan Adeli	Stanford University, USA
Shadi Albarqouni	ETH Zurich, Switzerland
Pablo Arbelaez	Universidad de los Andes, Colombia
Ulas Bagci	University of Central Florida, USA
Adrien Bartoli	Université Clermont Auvergne, France
Hrvoje Bogunovic	Medical University of Vienna, Austria
Weidong Cai	The University of Sydney, Australia
Chao Chen	Stony Brook University, USA
Elvis Chen	Robarts Research Institute, Canada
Stanley Durrleman	Inria, France
Boris Escalante-Ramírez	National Autonomous University of Mexico, Mexico
Pascal Fallavollita	University of Ottawa, Canada
Enzo Ferrante	CONICET, Universidad Nacional del Litoral, Argentina
Stamatia Giannarou	Imperial College London, UK
Orcun Goksel	ETH Zurich, Switzerland
Alberto Gomez	King's College London, UK
Miguel Angel González Ballester	Universitat Pompeu Fabra, Spain
Ilker Hacihaliloglu	Rutgers University, USA
Yi Hong	University of Georgia, USA
Yipeng Hu	University College London, UK
Heng Huang	University of Pittsburgh and JD Finance America Corporation, USA
Juan Eugenio Iglesias	University College London, UK
Madhura Ingalhalikar	Symbiosis Center for Medical Image Analysis, India
Pierre Jannin	Université de Rennes, France
Samuel Kadoury	Ecole Polytechnique de Montreal, Canada
Bernhard Kainz	Imperial College London, UK
Marta Kersten-Oertel	Concordia University, Canada
Andrew King	King's College London, UK
Ignacio Larrabide	CONICET, Argentina
Gang Li	University of North Carolina at Chapel Hill, USA
Jianming Liang	Arizona State University, USA
Hongen Liao	Tsinghua University, China
Rui Liao	Siemens Healthineers, USA
Feng Lin	Nanyang Technological University, China
Mingxia Liu	University of North Carolina at Chapel Hill, USA
Jiebo Luo	University of Rochester, USA
Xiongbiao Luo	Xiamen University, China
Andreas Maier	FAU Erlangen-Nuremberg, Germany
Stephen McKenna	University of Dundee, UK
Bjoern Menze	Technische Universität München, Germany
Mehdi Moradi	IBM Research, USA

Mentorship Program (Mentors)

Ehsan Adeli	Stanford University, USA
Stephen Aylward	Kitware, USA
Hrvoje Bogunovic	Medical University of Vienna, Austria
Li Cheng	University of Alberta, Canada
Marleen de Bruijne	University of Copenhagen, Denmark
Caroline Essert	University of Strasbourg, France
Gabor Fichtinger	Queen's University, Canada
Stamatia Giannarou	Imperial College London, UK
Juan Eugenio Iglesias Gonzalez	University College London, UK
Bernhard Kainz	Imperial College London, UK
Shuo Li	Western University, Canada
Jianming Liang	Arizona State University, USA
Rui Liao	Siemens Healthineers, USA
Feng Lin	Nanyang Technological University, China
Marius George Linguraru	Children's National Hospital, George Washington University, USA
Tianming Liu	University of Georgia, USA
Xiongbiao Luo	Xiamen University, China
Dong Ni	Shenzhen University, China
Wiro Niessen	Erasmus MC - University Medical Center Rotterdam, The Netherlands
Terry Peters	Western University, Canada
Antonio R. Porras	University of Colorado, USA
Daniel Racoceanu	Sorbonne University, France
Islem Rekik	Istanbul Technical University, Turkey
Nicola Rieke	NVIDIA, USA
Julia Schnabel	King's College London, UK
Ruby Shamir	Novocure, Switzerland
Stefanie Speidel	National Center for Tumor Diseases Dresden, Germany
Martin Styner	University of North Carolina at Chapel Hill, USA
Xiaoying Tang	Southern University of Science and Technology, China
Pallavi Tiwari	Case Western Reserve University, USA
Jocelyne Troccaz	CNRS, Grenoble Alpes University, France
Pierre Jannin	INSERM, Université de Rennes, France
Archana Venkataraman	Johns Hopkins University, USA
Linwei Wang	Rochester Institute of Technology, USA
Guorong Wu	University of North Carolina at Chapel Hill, USA
Li Xiao	Chinese Academy of Science, China
Ziyue Xu	NVIDIA, USA
Bochuan Zheng	China West Normal University, China
Guoyan Zheng	Shanghai Jiao Tong University, China
S. Kevin Zhou	Chinese Academy of Sciences, China
Maria A. Zuluaga	EURECOM, France

Additional Reviewers

Alaa Eldin Abdelaal
Ahmed Abdulkadir
Clement Abi Nader
Mazdak Abulnaga
Ganesh Adluru
Iman Aganj
Priya Aggarwal
Sahar Ahmad
Seyed-Ahmad Ahmadi
Euijoon Ahn
Alireza Akhondi-asl
Mohamed Akrout
Dawood Al Chanti
Ibraheem Al-Dhamari
Navid Alemi Koohbanani
Hanan Alghamdi
Hassan Alhajj
Hazrat Ali
Sharib Ali
Omar Al-Kadi
Maximilian Allan
Felix Ambellan
Mina Amiri
Sameer Antani
Luigi Antelmi
Michela Antonelli
Jacob Antunes
Saeed Anwar
Fernando Arambula
Ignacio Arganda-Carreras
Mohammad Ali Armin
John Ashburner
Md Ashikuzzaman
Shahab Aslani
Mehdi Astaraki
Angélica Atehortúa
Gowtham Atluri
Kamran Avanaki
Angelica Aviles-Rivero
Suyash Awate
Dogu Baran Aydogan
Qinle Ba
Morteza Babaie
Hyeon-Min Bae
Woong Bae
Wenjia Bai
Ujjwal Baid
Spyridon Bakas
Yaël Balbastre
Marcin Balicki
Fabian Balsiger
Abhirup Banerjee
Sreya Banerjee
Sophia Bano
Shunxing Bao
Adrian Barbu
Cher Bass
John S. H. Baxter
Amirhossein Bayat
Sharareh Bayat
Neslihan Bayramoglu
Bahareh Behboodi
Delaram Behnami
Mikhail Belyaev
Oualid Benkarim
Aicha BenTaieb
Camilo Bermudez
Giulia Bertò
Hadrien Bertrand
Julián Betancur
Michael Beyeler
Parmeet Bhatia
Chetan Bhole
Suvrat Bhooshan
Chitresh Bhushan
Lei Bi
Cheng Bian
Gui-Bin Bian
Sangeeta Biswas
Stefano B. Blumberg
Janusz Bobulski
Sebastian Bodenstedt
Ester Bonmati
Bhushan Borotikar
Jiri Borovec
Ilaria Boscolo Galazzo

Alexandre Bousse
Nicolas Boutry
Behzad Bozorgtabar
Nadia Brancati
Christopher Bridge
Esther Bron
Rupert Brooks
Qirong Bu
Tim-Oliver Buchholz
Duc Toan Bui
Qasim Bukhari
Ninon Burgos
Nikolay Burlutskiy
Russell Butler
Michał Byra
Hongmin Cai
Yunliang Cai
Sema Candemir
Bing Cao
Qing Cao
Shilei Cao
Tian Cao
Weiguo Cao
Yankun Cao
Aaron Carass
Heike Carolus
Adrià Casamitjana
Suheyla Cetin Karayumak
Ahmad Chaddad
Krishna Chaitanya
Jayasree Chakraborty
Tapabrata Chakraborty
Sylvie Chambon
Ming-Ching Chang
Violeta Chang
Simon Chatelin
Sudhanya Chatterjee
Christos Chatzichristos
Rizwan Chaudhry
Antong Chen
Cameron Po-Hsuan Chen
Chang Chen
Chao Chen
Chen Chen
Cheng Chen
Dongdong Chen
Fang Chen
Geng Chen
Hao Chen
Jianan Chen
Jianxu Chen
Jia-Wei Chen
Jie Chen
Junxiang Chen
Li Chen
Liang Chen
Pingjun Chen
Qiang Chen
Shuai Chen
Tianhua Chen
Tingting Chen
Xi Chen
Xiaoran Chen
Xin Chen
Yuanyuan Chen
Yuhua Chen
Yukun Chen
Zhineng Chen
Zhixiang Chen
Erkang Cheng
Jun Cheng
Li Cheng
Xuelian Cheng
Yuan Cheng
Veronika Cheplygina
Hyungjoo Cho
Jaegul Choo
Aritra Chowdhury
Stergios Christodoulidis
Ai Wern Chung
Pietro Antonio Cicalese
Özgün Çiçek
Robert Cierniak
Matthew Clarkson
Dana Cobzas
Jaume Coll-Font
Alessia Colonna
Marc Combalia
Olivier Commowick
Sonia Contreras Ortiz
Pierre-Henri Conze
Timothy Cootes

Luca Corinzia
Teresa Correia
Pierrick Coupé
Jeffrey Craley
Arun C. S. Kumar
Hui Cui
Jianan Cui
Zhiming Cui
Kathleen Curran
Haixing Dai
Xiaoliang Dai
Ker Dai Fei Elmer
Adrian Dalca
Abhijit Das
Neda Davoudi
Laura Daza
Sandro De Zanet
Charles Delahunt
Herve Delingette
Beatrice Demiray
Yang Deng
Hrishikesh Deshpande
Christian Desrosiers
Neel Dey
Xinghao Ding
Zhipeng Ding
Konstantin Dmitriev
Jose Dolz
Ines Domingues
Juan Pedro Dominguez-Morales
Hao Dong
Mengjin Dong
Nanqing Dong
Qinglin Dong
Suyu Dong
Sven Dorkenwald
Qi Dou
P. K. Douglas
Simon Drouin
Karen Drukker
Niharika D'Souza
Lei Du
Shaoyi Du
Xuefeng Du
Dingna Duan
Nicolas Duchateau
James Duncan
Jared Dunnmon
Luc Duong
Nicha Dvornek
Dmitry V. Dylov
Oleh Dzyubachyk
Mehran Ebrahimi
Philip Edwards
Alexander Effland
Jan Egger
Alma Eguizabal
Gudmundur Einarsson
Ahmed Elazab
Mohammed S. M. Elbaz
Shireen Elhabian
Ahmed Eltanboly
Sandy Engelhardt
Ertunc Erdil
Marius Erdt
Floris Ernst
Mohammad Eslami
Nazila Esmaeili
Marco Esposito
Oscar Esteban
Jingfan Fan
Xin Fan
Yonghui Fan
Chaowei Fang
Xi Fang
Mohsen Farzi
Johannes Fauser
Andrey Fedorov
Hamid Fehri
Lina Felsner
Jun Feng
Ruibin Feng
Xinyang Feng
Yifan Feng
Yuan Feng
Henrique Fernandes
Ricardo Ferrari
Jean Feydy
Lucas Fidon
Lukas Fischer
Antonio Foncubierta-Rodríguez
Germain Forestier

Reza Forghani
Nils Daniel Forkert
Jean-Rassaire Fouefack
Tatiana Fountoukidou
Aina Frau-Pascual
Moti Freiman
Sarah Frisken
Huazhu Fu
Xueyang Fu
Wolfgang Fuhl
Isabel Funke
Philipp Fürnstahl
Pedro Furtado
Ryo Furukawa
Elies Fuster-Garcia
Youssef Gahi
Jin Kyu Gahm
Laurent Gajny
Rohan Gala
Harshala Gammulle
Yu Gan
Cong Gao
Dongxu Gao
Fei Gao
Feng Gao
Linlin Gao
Mingchen Gao
Siyuan Gao
Xin Gao
Xinpei Gao
Yixin Gao
Yue Gao
Zhifan Gao
Sara Garbarino
Alfonso Gastelum-Strozzi
Romane Gauriau
Srishti Gautam
Bao Ge
Rongjun Ge
Zongyuan Ge
Sairam Geethanath
Yasmeen George
Samuel Gerber
Guido Gerig
Nils Gessert
Olivier Gevaert
Muhammad Usman Ghani
Sandesh Ghimire
Sayan Ghosal
Gabriel Girard
Ben Glocker
Evgin Goceri
Michael Goetz
Arnold Gomez
Kuang Gong
Mingming Gong
Yuanhao Gong
German Gonzalez
Sharath Gopal
Karthik Gopinath
Pietro Gori
Maged Goubran
Sobhan Goudarzi
Baran Gözcü
Benedikt Graf
Mark Graham
Bertrand Granado
Alejandro Granados
Robert Grupp
Christina Gsaxner
Lin Gu
Shi Gu
Yun Gu
Ricardo Guerrero
Houssem-Eddine Gueziri
Dazhou Guo
Hengtao Guo
Jixiang Guo
Pengfei Guo
Yanrong Guo
Yi Guo
Yong Guo
Yulan Guo
Yuyu Guo
Krati Gupta
Vikash Gupta
Praveen Gurunath Bharathi
Prashnna Gyawali
Stathis Hadjidemetriou
Omid Haji Maghsoudi
Justin Haldar
Mohammad Hamghalam

Bing Han
Hu Han
Liang Han
Xiaoguang Han
Xu Han
Zhi Han
Zhongyi Han
Jonny Hancox
Christian Hansen
Xiaoke Hao
Rabia Haq
Michael Hardisty
Stefan Harrer
Adam Harrison
S. M. Kamrul Hasan
Hoda Sadat Hashemi
Nobuhiko Hata
Andreas Hauptmann
Mohammad Havaei
Huiguang He
Junjun He
Kelei He
Tiancheng He
Xuming He
Yuting He
Mattias Heinrich
Stefan Heldmann
Nicholas Heller
Alessa Hering
Monica Hernandez
Estefania Hernandez-Martin
Carlos Hernandez-Matas
Javier Herrera-Vega
Kilian Hett
Tsung-Ying Ho
Nico Hoffmann
Matthew Holden
Song Hong
Sungmin Hong
Yoonmi Hong
Corné Hoogendoorn
Antal Horváth
Belayat Hossain
Le Hou
Ai-Ling Hsu
Po-Ya Hsu
Tai-Chiu Hsung
Pengwei Hu
Shunbo Hu
Xiaoling Hu
Xiaowei Hu
Yan Hu
Zhenhong Hu
Jia-Hong Huang
Junzhou Huang
Kevin Huang
Qiaoying Huang
Weilin Huang
Xiaolei Huang
Yawen Huang
Yongxiang Huang
Yue Huang
Yufang Huang
Zhi Huang
Arnaud Huaulmé
Henkjan Huisman
Xing Huo
Yuankai Huo
Sarfaraz Hussein
Jana Hutter
Khoi Huynh
Seong Jae Hwang
Emmanuel Iarussi
Ilknur Icke
Kay Igwe
Alfredo Illanes
Abdullah-Al-Zubaer Imran
Ismail Irmakci
Samra Irshad
Benjamin Irving
Mobarakol Islam
Mohammad Shafkat Islam
Vamsi Ithapu
Koichi Ito
Hayato Itoh
Oleksandra Ivashchenko
Yuji Iwahori
Shruti Jadon
Mohammad Jafari
Mostafa Jahanifar
Andras Jakab
Amir Jamaludin

Won-Dong Jang
Vincent Jaouen
Uditha Jarayathne
Ronnachai Jaroensri
Golara Javadi
Rohit Jena
Todd Jensen
Won-Ki Jeong
Zexuan Ji
Haozhe Jia
Jue Jiang
Tingting Jiang
Weixiong Jiang
Xi Jiang
Xiang Jiang
Jianbo Jiao
Zhicheng Jiao
Amelia Jiménez-Sánchez
Dakai Jin
Taisong Jin
Yueming Jin
Ze Jin
Bin Jing
Yaqub Jonmohamadi
Anand Joshi
Shantanu Joshi
Christoph Jud
Florian Jug
Yohan Jun
Alain Jungo
Abdolrahim Kadkhodamohammadi
Ali Kafaei Zad Tehrani
Dagmar Kainmueller
Siva Teja Kakileti
John Kalafut
Konstantinos Kamnitsas
Michael C. Kampffmeyer
Qingbo Kang
Neerav Karani
Davood Karimi
Satyananda Kashyap
Alexander Katzmann
Prabhjot Kaur
Anees Kazi
Erwan Kerrien
Hoel Kervadec
Ashkan Khakzar
Fahmi Khalifa
Nadieh Khalili
Siavash Khallaghi
Farzad Khalvati
Hassan Khan
Bishesh Khanal
Pulkit Khandelwal
Maksym Kholiavchenko
Meenakshi Khosla
Naji Khosravan
Seyed Mostafa Kia
Ron Kikinis
Daeseung Kim
Geena Kim
Hak Gu Kim
Heejong Kim
Hosung Kim
Hyo-Eun Kim
Jinman Kim
Jinyoung Kim
Mansu Kim
Minjeong Kim
Seong Tae Kim
Won Hwa Kim
Young-Ho Kim
Atilla Kiraly
Yoshiro Kitamura
Takayuki Kitasaka
Sabrina Kletz
Tobias Klinder
Kranthi Kolli
Satoshi Kondo
Bin Kong
Jun Kong
Tomasz Konopczynski
Ender Konukoglu
Bongjin Koo
Kivanc Kose
Anna Kreshuk
AnithaPriya Krishnan
Pavitra Krishnaswamy
Frithjof Kruggel
Alexander Krull
Elizabeth Krupinski
Hulin Kuang

Serife Kucur
David Kügler
Arjan Kuijper
Jan Kukacka
Nilima Kulkarni
Abhay Kumar
Ashnil Kumar
Kuldeep Kumar
Neeraj Kumar
Nitin Kumar
Manuela Kunz
Holger Kunze
Tahsin Kurc
Thomas Kurmann
Yoshihiro Kuroda
Jin Tae Kwak
Yongchan Kwon
Aymen Laadhari
Dmitrii Lachinov
Alexander Ladikos
Alain Lalande
Rodney Lalonde
Tryphon Lambrou
Hengrong Lan
Catherine Laporte
Carole Lartizien
Bianca Lassen-Schmidt
Andras Lasso
Ngan Le
Leo Lebrat
Changhwan Lee
Eung-Joo Lee
Hyekyoung Lee
Jong-Hwan Lee
Jungbeom Lee
Matthew Lee
Sangmin Lee
Soochahn Lee
Stefan Leger
Étienne Léger
Baiying Lei
Andreas Leibetseder
Rogers Jeffrey Leo John
Juan Leon
Wee Kheng Leow
Annan Li
Bo Li
Chongyi Li
Haohan Li
Hongming Li
Hongwei Li
Huiqi Li
Jian Li
Jianning Li
Jiayun Li
Junhua Li
Lincan Li
Mengzhang Li
Ming Li
Qing Li
Quanzheng Li
Shulong Li
Shuyu Li
Weikai Li
Wenyuan Li
Xiang Li
Xiaomeng Li
Xiaoxiao Li
Xin Li
Xiuli Li
Yang Li (Beihang University)
Yang Li (Northeast Electric Power University)
Yi Li
Yuexiang Li
Zeju Li
Zhang Li
Zhen Li
Zhiyuan Li
Zhjin Li
Zhongyu Li
Chunfeng Lian
Gongbo Liang
Libin Liang
Shanshan Liang
Yudong Liang
Haofu Liao
Ruizhi Liao
Gilbert Lim
Baihan Lin
Hongxiang Lin
Huei-Yung Lin

Jianyu Lin
C. Lindner
Geert Litjens
Bin Liu
Chang Liu
Dongnan Liu
Feng Liu
Hangfan Liu
Jianfei Liu
Jin Liu
Jingya Liu
Jingyu Liu
Kai Liu
Kefei Liu
Lihao Liu
Luyan Liu
Mengting Liu
Na Liu
Peng Liu
Ping Liu
Quande Liu
Qun Liu
Shengfeng Liu
Shuangjun Liu
Sidong Liu
Siqi Liu
Siyuan Liu
Tianrui Liu
Xianglong Liu
Xinyang Liu
Yan Liu
Yuan Liu
Yuhang Liu
Andrea Loddo
Herve Lombaert
Marco Lorenzi
Jian Lou
Nicolas Loy Rodas
Allen Lu
Donghuan Lu
Huanxiang Lu
Jiwen Lu
Le Lu
Weijia Lu
Xiankai Lu
Yao Lu
Yongyi Lu
Yueh-Hsun Lu
Christian Lucas
Oeslle Lucena
Imanol Luengo
Ronald Lui
Gongning Luo
Jie Luo
Ma Luo
Marcel Luthi
Khoa Luu
Bin Lv
Jinglei Lv
Ilwoo Lyu
Qing Lyu
Sharath M. S.
Andy J. Ma
Chunwei Ma
Da Ma
Hua Ma
Jingting Ma
Kai Ma
Lei Ma
Wenao Ma
Yuexin Ma
Amirreza Mahbod
Sara Mahdavi
Mohammed Mahmoud
Gabriel Maicas
Klaus H. Maier-Hein
Sokratis Makrogiannis
Bilal Malik
Anand Malpani
Ilja Manakov
Matteo Mancini
Efthymios Maneas
Tommaso Mansi
Brett Marinelli
Razvan Marinescu
Pablo Márquez Neila
Carsten Marr
Yassine Marrakchi
Fabio Martinez
Antonio Martinez-Torteya
Andre Mastmeyer
Dimitrios Mavroeidis

Jamie McClelland
Verónica Medina Bañuelos
Raghav Mehta
Sachin Mehta
Liye Mei
Raphael Meier
Qier Meng
Qingjie Meng
Yu Meng
Martin Menten
Odyssée Merveille
Pablo Mesejo
Liang Mi
Shun Miao
Stijn Michielse
Mikhail Milchenko
Hyun-Seok Min
Zhe Min
Tadashi Miyamoto
Aryan Mobiny
Irina Mocanu
Sara Moccia
Omid Mohareri
Hassan Mohy-ud-Din
Muthu Rama Krishnan Mookiah
Rodrigo Moreno
Lia Morra
Agata Mosinska
Saman Motamed
Mohammad Hamed Mozaffari
Anirban Mukhopadhyay
Henning Müller
Balamurali Murugesan
Cosmas Mwikirize
Andriy Myronenko
Saad Nadeem
Ahmed Naglah
Vivek Natarajan
Vishwesh Nath
Rodrigo Nava
Fernando Navarro
Lydia Neary-Zajiczek
Peter Neher
Dominik Neumann
Gia Ngo
Hannes Nickisch
Dong Nie
Jingxin Nie
Weizhi Nie
Aditya Nigam
Xia Ning
Zhenyuan Ning
Sijie Niu
Tianye Niu
Alexey Novikov
Jorge Novo
Chinedu Nwoye
Mohammad Obeid
Masahiro Oda
Thomas O'Donnell
Benjamin Odry
Steffen Oeltze-Jafra
Ayşe Oktay
Hugo Oliveira
Marcelo Oliveira
Sara Oliveira
Arnau Oliver
Sahin Olut
Jimena Olveres
John Onofrey
Eliza Orasanu
Felipe Orihuela-Espina
José Orlando
Marcos Ortega
Sarah Ostadabbas
Yoshito Otake
Sebastian Otalora
Cheng Ouyang
Jiahong Ouyang
Cristina Oyarzun Laura
Michal Ozery-Flato
Krittin Pachtrachai
Johannes Paetzold
Jin Pan
Yongsheng Pan
Prashant Pandey
Joao Papa
Giorgos Papanastasiou
Constantin Pape
Nripesh Parajuli
Hyunjin Park
Sanghyun Park

Seyoun Park
Angshuman Paul
Christian Payer
Chengtao Peng
Jialin Peng
Liying Peng
Tingying Peng
Yifan Peng
Tobias Penzkofer
Antonio Pepe
Oscar Perdomo
Jose-Antonio Pérez-Carrasco
Fernando Pérez-García
Jorge Perez-Gonzalez
Skand Peri
Loic Peter
Jorg Peters
Jens Petersen
Caroline Petitjean
Micha Pfeiffer
Dzung Pham
Renzo Phellan
Ashish Phophalia
Mark Pickering
Kilian Pohl
Iulia Popescu
Karteek Popuri
Tiziano Portenier
Alison Pouch
Arash Pourtaherian
Prateek Prasanna
Alexander Preuhs
Raphael Prevost
Juan Prieto
Viswanath P. S.
Sergi Pujades
Kumaradevan Punithakumar
Elodie Puybareau
Haikun Qi
Huan Qi
Xin Qi
Buyue Qian
Zhen Qian
Yan Qiang
Yuchuan Qiao
Zhi Qiao
Chen Qin
Wenjian Qin
Yanguo Qin
Wu Qiu
Hui Qu
Kha Gia Quach
Prashanth R.
Pradeep Reddy Raamana
Jagath Rajapakse
Kashif Rajpoot
Jhonata Ramos
Andrik Rampun
Parnesh Raniga
Nagulan Ratnarajah
Richard Rau
Mehul Raval
Keerthi Sravan Ravi
Daniele Ravì
Harish RaviPrakash
Rohith Reddy
Markus Rempfler
Xuhua Ren
Yinhao Ren
Yudan Ren
Anne-Marie Rickmann
Brandalyn Riedel
Leticia Rittner
Robert Robinson
Jessica Rodgers
Robert Rohling
Lukasz Roszkowiak
Karsten Roth
José Rouco
Su Ruan
Daniel Rueckert
Mirabela Rusu
Erica Rutter
Jaime S. Cardoso
Mohammad Sabokrou
Monjoy Saha
Pramit Saha
Dushyant Sahoo
Pranjal Sahu
Wojciech Samek
Juan A. Sánchez-Margallo
Robin Sandkuehler

Rodrigo Santa Cruz
Gianmarco Santini
Anil Kumar Sao
Mhd Hasan Sarhan
Duygu Sarikaya
Imari Sato
Olivier Saut
Mattia Savardi
Ramasamy Savitha
Fabien Scalzo
Nico Scherf
Alexander Schlaefer
Philipp Schleer
Leopold Schmetterer
Julia Schnabel
Klaus Schoeffmann
Peter Schueffler
Andreas Schuh
Thomas Schultz
Michael Schwier
Michael Sdika
Suman Sedai
Raghavendra Selvan
Sourya Sengupta
Youngho Seo
Lama Seoud
Ana Sequeira
Saeed Seyyedi
Giorgos Sfikas
Sobhan Shafiei
Reuben Shamir
Shayan Shams
Hongming Shan
Yeqin Shao
Harshita Sharma
Gregory Sharp
Mohamed Shehata
Haocheng Shen
Mali Shen
Yiqiu Shen
Zhengyang Shen
Luyao Shi
Xiaoshuang Shi
Yemin Shi
Yonghong Shi
Saurabh Shigwan
Hoo-Chang Shin
Suprosanna Shit
Yucheng Shu
Nadya Shusharina
Alberto Signoroni
Carlos A. Silva
Wilson Silva
Praveer Singh
Ramandeep Singh
Rohit Singla
Sumedha Singla
Ayushi Sinha
Rajath Soans
Hessam Sokooti
Jaemin Son
Ming Song
Tianyu Song
Yang Song
Youyi Song
Aristeidis Sotiras
Arcot Sowmya
Rachel Sparks
Bella Specktor
William Speier
Ziga Spiclin
Dominik Spinczyk
Chetan Srinidhi
Vinkle Srivastav
Lawrence Staib
Peter Steinbach
Darko Stern
Joshua Stough
Justin Strait
Robin Strand
Martin Styner
Hai Su
Pan Su
Yun-Hsuan Su
Vaishnavi Subramanian
Gérard Subsol
Carole Sudre
Yao Sui
Avan Suinesiaputra
Jeremias Sulam
Shipra Suman
Jian Sun

Liang Sun
Tao Sun
Kyung Sung
Chiranjib Sur
Yannick Suter
Raphael Sznitman
Solale Tabarestani
Fatemeh Taheri Dezaki
Roger Tam
José Tamez-Peña
Chaowei Tan
Jiaxing Tan
Hao Tang
Sheng Tang
Thomas Tang
Xiongfeng Tang
Zhenyu Tang
Mickael Tardy
Eu Wern Teh
Antonio Tejero-de-Pablos
Paul Thienphrapa
Stephen Thompson
Felix Thomsen
Jiang Tian
Yun Tian
Aleksei Tiulpin
Hamid Tizhoosh
Matthew Toews
Oguzhan Topsakal
Jordina Torrents
Sylvie Treuillet
Jocelyne Troccaz
Emanuele Trucco
Vinh Truong Hoang
Chialing Tsai
Andru Putra Twinanda
Norimichi Ukita
Eranga Ukwatta
Mathias Unberath
Tamas Ungi
Martin Urschler
Verena Uslar
Fatmatulzehra Uslu
Régis Vaillant
Jeya Maria Jose Valanarasu
Marta Vallejo
Fons van der Sommen
Gijs van Tulder
Kimberlin van Wijnen
Yogatheesan Varatharajah
Marta Varela
Thomas Varsavsky
Francisco Vasconcelos
S. Swaroop Vedula
Sanketh Vedula
Harini Veeraraghavan
Gonzalo Vegas Sanchez-Ferrero
Anant Vemuri
Gopalkrishna Veni
Ruchika Verma
Ujjwal Verma
Pedro Vieira
Juan Pedro Vigueras Guillen
Pierre-Frederic Villard
Athanasios Vlontzos
Wolf-Dieter Vogl
Ingmar Voigt
Eugene Vorontsov
Bo Wang
Cheng Wang
Chengjia Wang
Chunliang Wang
Dadong Wang
Guotai Wang
Haifeng Wang
Hongkai Wang
Hongyu Wang
Hua Wang
Huan Wang
Jun Wang
Kuanquan Wang
Kun Wang
Lei Wang
Li Wang
Liansheng Wang
Manning Wang
Ruixuan Wang
Shanshan Wang
Shujun Wang
Shuo Wang
Tianchen Wang
Tongxin Wang

Wenzhe Wang
Xi Wang
Xiangxue Wang
Yalin Wang
Yan Wang (Sichuan University)
Yan Wang (Johns Hopkins University)
Yaping Wang
Yi Wang
Yirui Wang
Yuanjun Wang
Yun Wang
Zeyi Wang
Zhangyang Wang
Simon Warfield
Jonathan Weber
Jürgen Weese
Donglai Wei
Dongming Wei
Zhen Wei
Martin Weigert
Michael Wels
Junhao Wen
Matthias Wilms
Stefan Winzeck
Adam Wittek
Marek Wodzinski
Jelmer Wolterink
Ken C. L. Wong
Jonghye Woo
Chongruo Wu
Dijia Wu
Ji Wu
Jian Wu (Tsinghua University)
Jian Wu (Zhejiang University)
Jie Ying Wu
Junyan Wu
Minjie Wu
Pengxiang Wu
Xi Wu
Xia Wu
Xiyin Wu
Ye Wu
Yicheng Wu
Yifan Wu
Zhengwang Wu
Tobias Wuerfl
Pengcheng Xi
James Xia
Siyu Xia
Yingda Xia
Yong Xia
Lei Xiang
Deqiang Xiao
Li Xiao (Tulane University)
Li Xiao (Chinese Academy of Science)
Yuting Xiao
Hongtao Xie
Jianyang Xie
Lingxi Xie
Long Xie
Xueqian Xie
Yiting Xie
Yuan Xie
Yutong Xie
Fangxu Xing
Fuyong Xing
Tao Xiong
Chenchu Xu
Hongming Xu
Jiaofeng Xu
Kele Xu
Lisheng Xu
Min Xu
Rui Xu
Xiaowei Xu
Yanwu Xu
Yongchao Xu
Zhenghua Xu
Cheng Xue
Jie Xue
Wufeng Xue
Yuan Xue
Faridah Yahya
Chenggang Yan
Ke Yan
Weizheng Yan
Yu Yan
Yuguang Yan
Zhennan Yan
Changchun Yang
Chao-Han Huck Yang
Dong Yang

Fan Yang (IIAI)
Fan Yang (Temple University)
Feng Yang
Ge Yang
Guang Yang
Heran Yang
Hongxu Yang
Huijuan Yang
Jiancheng Yang
Jie Yang
Junlin Yang
Lin Yang
Xiao Yang
Xiaohui Yang
Xin Yang
Yan Yang
Yujiu Yang
Dongren Yao
Jianhua Yao
Jiawen Yao
Li Yao
Chuyang Ye
Huihui Ye
Menglong Ye
Xujiong Ye
Andy W. K. Yeung
Jingru Yi
Jirong Yi
Xin Yi
Yi Yin
Shihui Ying
Youngjin Yoo
Chenyu You
Sahar Yousefi
Hanchao Yu
Jinhua Yu
Kai Yu
Lequan Yu
Qi Yu
Yang Yu
Zhen Yu
Pengyu Yuan
Yixuan Yuan
Paul Yushkevich
Ghada Zamzmi
Dong Zeng
Guodong Zeng
Oliver Zettinig
Zhiwei Zhai
Kun Zhan
Baochang Zhang
Chaoyi Zhang
Daoqiang Zhang
Dongqing Zhang
Fan Zhang (Yale University)
Fan Zhang (Harvard Medical School)
Guangming Zhang
Han Zhang
Hang Zhang
Haopeng Zhang
Heye Zhang
Huahong Zhang
Jianpeng Zhang
Jinao Zhang
Jingqing Zhang
Jinwei Zhang
Jiong Zhang
Jun Zhang
Le Zhang
Lei Zhang
Lichi Zhang
Lin Zhang
Ling Zhang
Lu Zhang
Miaomiao Zhang
Ning Zhang
Pengfei Zhang
Pengyue Zhang
Qiang Zhang
Rongzhao Zhang
Ru-Yuan Zhang
Shanzhuo Zhang
Shu Zhang
Tong Zhang
Wei Zhang
Weiwei Zhang
Wenlu Zhang
Xiaoyun Zhang
Xin Zhang
Ya Zhang
Yanbo Zhang
Yanfu Zhang

Yi Zhang
Yifan Zhang
Yizhe Zhang
Yongqin Zhang
You Zhang
Youshan Zhang
Yu Zhang
Yue Zhang
Yulun Zhang
Yunyan Zhang
Yuyao Zhang
Zijing Zhang
Can Zhao
Changchen Zhao
Fenqiang Zhao
Gangming Zhao
Haifeng Zhao
He Zhao
Jun Zhao
Li Zhao
Qingyu Zhao
Rongchang Zhao
Shen Zhao
Tengda Zhao
Tianyi Zhao
Wei Zhao
Xuandong Zhao
Yitian Zhao
Yiyuan Zhao
Yu Zhao
Yuan-Xing Zhao
Yue Zhao
Zixu Zhao
Ziyuan Zhao
Xingjian Zhen
Hao Zheng
Jiannan Zheng
Kang Zheng
Yalin Zheng
Yushan Zheng
Jia-Xing Zhong
Zichun Zhong
Haoyin Zhou
Kang Zhou
Sanping Zhou
Tao Zhou
Wenjin Zhou
Xiao-Hu Zhou
Xiao-Yun Zhou
Yanning Zhou
Yi Zhou (IIAI)
Yi Zhou (University of Utah)
Yuyin Zhou
Zhen Zhou
Zongwei Zhou
Dajiang Zhu
Dongxiao Zhu
Hancan Zhu
Lei Zhu
Qikui Zhu
Weifang Zhu
Wentao Zhu
Xiaofeng Zhu
Xinliang Zhu
Yingying Zhu
Yuemin Zhu
Zhe Zhu
Zhuotun Zhu
Xiahai Zhuang
Aneeq Zia
Veronika Zimmer
David Zimmerer
Lilla Zöllei
Yukai Zou
Gerald Zwettler
Reyer Zwiggelaa

Contents – Part I

Machine Learning Methodologies

Machine Learning Methodologies

Attention, Suggestion and Annotation: A Deep Active Learning Framework for Biomedical Image Segmentation

Haohan Li[1(✉)] and Zhaozheng Yin[2]

[1] Department of Computer Science, Missouri University of Science and Technology, Rolla, MO 65409, USA
hl87c@umsystem.edu
[2] AI Institute, Department of Biomedical Informatics, Department of Computer Science, Department of Applied Mathematics and Statistics (Affiliated), Stony Brook University, Stony Brook, NY 11794, USA
zyin@cs.stonybrook.edu

Abstract. Despite the great success, deep learning based segmentation methods still face a critical obstacle: the difficulty in acquiring sufficient training data due to high annotation costs. In this paper, we propose a deep active learning framework that combines the attention gated fully convolutional network (ag-FCN) and the distribution discrepancy based active learning algorithm (dd-AL) to significantly reduce the annotation effort by iteratively annotating the most informative samples to train the ag-FCN for the better segmentation performance. Our framework is evaluated on 2015 MICCAI Gland Segmentaion dataset and 2017 MICCAI 6-month infant brain MRI Segmentation dataset. Experiment results show that our framework can achieve state-of-the-art segmentation performance by using only a portion of the training data.

1 Introduction

Automated image segmentation is a cornerstone of many image analysis applications. Recently, thanks to their representation power and generalization capability, deep learning models have achieved superior performance in many image segmentation tasks [1,2]. However, despite the success, deep learning based segmentation still faces a critical hindrance: the difficulty in acquiring sufficient training data due to the high annotation cost. In biomedical image segmentation, this hindrance is even more severe for the reason that: (1) Only experts can provide precise annotations for biomedical image segmentation tasks, making crowd-computing quite difficult; (2) Biomedical images from high-throughput experiments contain big data of images, which require extensive workforces to provide pixel-level annotations; (3) Due to the dramatic variations in biomedical images (e.g., different imaging modalities and specimens), deep learning models need specific sets of training data to achieve good segmentation performances,

A. L. Martel et al. (Eds.): MICCAI 2020, LNCS 12261, pp. 3–13, 2020.
https://doi.org/10.1007/978-3-030-59710-8_1

rather than using a general training dataset and transfer learning techniques to solve all kinds of segmentation tasks. Due to these reasons, a real-world biomedical image segmentation project may require thousands of annotation hours from multiple domain experts. Thus, there is a great need to develop annotation suggestion algorithms that can assist human annotators by suggesting the most informative data for annotation to accomplish the task with less human efforts.

1.1 Related Works

Despite the great success of deep learning in image segmentation tasks [3,4], deep learning based segmentation algorithms still face a critical difficulty in acquiring sufficient training data due to the high annotation cost. To alleviate the annotation burden in image segmentation tasks, weakly supervised segmentation algorithms [5–7] have been proposed. However, how to select representative data samples for annotation is overlooked. To address this problem, active learning [8] can be utilized as an annotation suggestion to query informative samples for annotation. As shown in [9], by using active learning, good performance can be achieved using significantly less training data in natural scene image segmentation. However, this method is based on the pre-trained region proposal model and pre-trained image descriptor network, which cannot be easily acquired in the biomedical image field due to the large variations in various biomedical applications. A progressively trained active learning framework is proposed in [10], but it only focuses on the uncertainty and the representativeness of suggested samples in the unlabeled set and ignores the rarity of suggested samples in the labeled set, which can easily incur serious redundancy in the labeled set.

1.2 Our Proposal and Contribution

In this work, we propose a deep active learning framework, combining a new deep learning model and a new active learning algorithm, which iteratively suggests the most informative annotation samples to improve the model's segmentation performance progressively.

Although the motivation seems to be straightforward, it is challenging to design a framework that can perfectly integrate a deep learning model into an active learning process due to the following challenges: (1) The deep learning model should have a good generalization capability so that it can produce reasonable results when little training data are available in the active learning process; (2) The deep learning model should perform well when using the entire training set so that it can provide a good upper-bound of the performance for the active learning framework; (3) The active learning algorithm should be able to make judicious annotation suggestions based on the limited information provided by a not-well-trained deep learning model in the early training stage. To overcome these three challenges, we design a deep active learning framework with two major components: (1) the attention gated Fully Convolutional Network (**ag-FCN**) and (2) the distribution discrepancy based active learning algorithm (**dd-AL**):

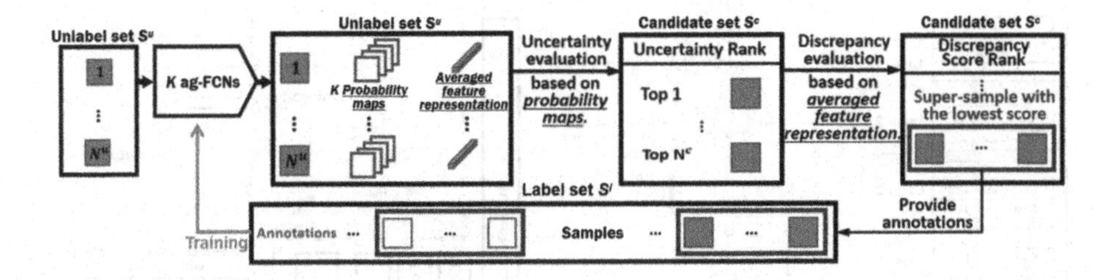

Fig. 1. The workflow of our deep active learning framework.

- **Attention:** For the first and second challenges, we design a novel ag-FCN that uses attention gate units (AGUs) to automatically highlight salient features of the target content for accurate pixel-wise predictions. In addition, both of the ag-FCN and the AGU are built using bottleneck designs to significantly reduce the number of network parameters while maintaining the same number of feature channels at the end of each residual module. This design ensures the good generality of the proposed ag-FCN.
- **Suggestion and Annotation:** For the third challenge, we design the dd-AL to achieve the final goal of the iterative annotation suggestion: decreasing the distribution discrepancy between the labeled set and the unlabeled set[1]. If the discrepancy between these two sets is small enough, which means their distributions are similar enough, the classifier trained on the labeled set can achieve similar performance compared to the classifier trained on the entire training dataset with all samples annotated. Therefore, besides the uncertainty metric, dd-AL also evaluates each unlabeled sample's effectiveness in decreasing the distribution discrepancy between the labeled set and the unlabeled set after we annotate it, which is further evaluated by the representativeness and rarity metrics.

2 Method

Figure 1 shows the workflow of our deep active learning framework. In each annotation suggestion stage, first we pass each unlabeled sample through K ag-FCNs to obtain its K segmentation probability maps and the corresponding averaged feature representation. Then, dd-AL selects the most informative unlabeled samples based on their uncertainties to the currently-trained ag-FCNs and effectiveness in decreasing the data distribution discrepancy between the labeled and unlabeled set. Finally, the small set of suggested samples are annotated for fine-tuning the ag-FCNs. We conduct this annotation suggestion process iteratively until satisfied.

[1] In this paper, the labeled set and unlabeled set refer to the labeled and unlabeled portions of a training dataset, respectively.

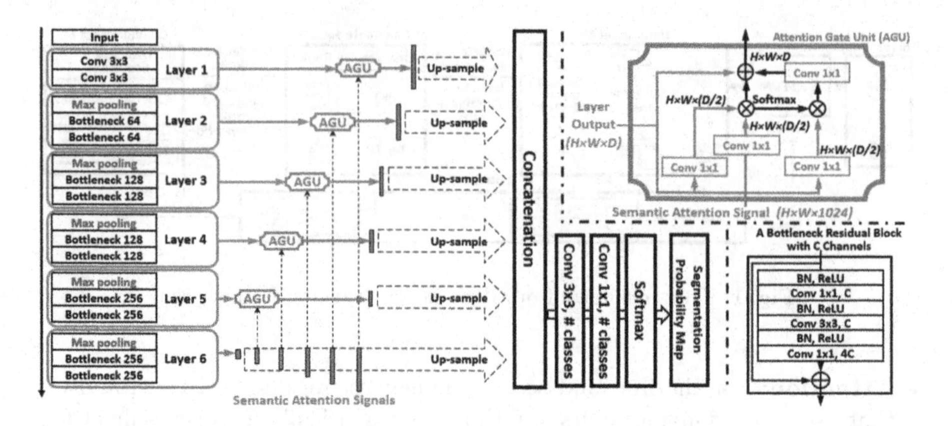

Fig. 2. The architecture of the semantic attention guided fully convolutional network.

2.1 Attention Gated Fully Convolutional Network

Based on recent advances of deep neural networks such as the fully convolutional network (FCN, [3]) and non-local network [11], we propose an attention gated fully convolutional network (ag-FCN) that can not only conduct accurate image segmentation but also be suitable for active learning. Compared with the original FCN, our ag-FCN, shown in Fig. 2, has three main improvements:

Attention Gate Units: We propose the Attention Gate Unit (AGU) to fuse the high-level semantic features to low- and mid-level features. AGU exploits the high-level semantic information as soft-attentions that lead low- and mid-level features to focus on target areas and highlight the feature activations that are relevant to the target instance. Hence, AGU ensures that the ag-FCN can conduct accurate segmentation on object instances with high variabilities.

Feature Fusion Strategy: Compared with the conventional skip-connections that progressively merge low-level features to the up-sampling process of high-level features [3], the feature fusion strategy in the ag-FCN considers each layer's attentive features (with semantic attention) as an up-sampling seed. All seeds will be progressively up-sampled to the input image size, and then be concatenated for generating smooth segmentation results.

Bottleneck Residual Modules: In ag-FCN, we replace most convolutional layers by bottleneck residual modules to significantly reduce the number of parameters while maintaining the same receptive field size and feature channels at the end of each module. This design reduces the training cost with less parameters (i.e., suitable for the iterative active learning) and maintains ag-FCN's generalization capability.

These three improvements of our ag-FCN are essential when combining deep neural networks and active learning. By using our AGUs and feature fusion strategy, the ag-FCN can achieve state-of-the-art segmentation performance using all training data, which provides a good upper-bound performance for our framework. By using the bottleneck residual blocks, the ag-FCN can have good generalization capability even when very little training data are available, and facilitate the iterative active learning with small sets of network parameters.

2.2 Distribution Discrepancy Based Active Learning Algorithm

In general, our distribution discrepancy based active learning algorithm (dd-AL) suggests samples for annotation based on two criteria: (1) the uncertainty to the segmentation network and (2) the effectiveness in decreasing the distribution discrepancy between the labeled set and unlabeled set. Since parallelly evaluating these two criteria of each unlabeled sample is computational expensive, dd-AL conducts the annotation suggestion process in two sequential steps. As shown in Fig. 1, first, dd-AL selects N^c samples with the highest uncertainty scores from the unlabeled set as candidate samples. Secondly, among these N^c candidate samples, dd-AL selects a subset of them that have the highest effectiveness in decreasing the distribution discrepancy between the labeled and unlabeled set.

Evaluating a Sample's Uncertainty: In the first step of dd-AL, to evaluate the uncertainty of each unlabeled sample, we adopt the bootstrapping strategy that trains K ag-FCNs, each of which only uses a subset of the suggested data for training in each annotation suggestion stage, and calculates the disagreement among these K models. Specifically, in each annotation suggestion stage, for each unlabeled sample s^u whose spatial dimension is $h \times w$, we first use K ag-FCNs to generate K segmentation probability maps of s^u. Then, we compute an uncertainty score $u_k^{s^u}$ of the k-th ($k \in [1, K]$) segmentation probability map of s^u by using the Best-versus-Second-Best (BvSB) strategy:

$$u_k^{s^u} = \frac{1}{h \times w} \sum_{i=1}^{h \times w} (1 - \left|p_{k,i}^{best} - p_{k,i}^{second}\right|), \tag{1}$$

where $p_{k,i}^{best}$ and $p_{k,i}^{second}$ denote the probability values of the most-possible class and second-possible class of the i-th pixel on s^u, respectively, predicted by the k-th ag-FCN. $(1 - \left|p_{k,i}^{best} - p_{k,i}^{second}\right|)$ denotes the pixel-wise BvSB score, where a larger score indicates more uncertainty. In Eq. 1, the uncertainty score of s^u estimated by the k-th ag-FCN is the average of the BvSB scores of all pixels in this image. We compute the final uncertainty score of s^u by averaging the uncertainty scores predicted by all K ag-FCNs:

$$u_{final}^{s^u} = \frac{1}{K} \sum_{k=1}^{K} u_k^{s^u}. \tag{2}$$

Then, we rank all the unlabeled samples based on their final uncertainty scores and select the top N^c samples with the highest uncertainty scores as the candidate set S^c for the second step of dd-AL.

Evaluating a Sample's Effectiveness in Decreasing Discrepancy: In the second step of dd-AL, we aim to annotate a subset of the candidate set S^c that can achieve the smallest distribution discrepancy between the labeled set and unlabeled set after the annotation. After several annotation suggestion stages, if the distributions of the labeled set and unlabeled set are similar enough, the classifier trained on the labeled set can achieve similar performance compared to the classifier trained on the entire dataset with all samples annotated.

In each annotation suggestion stage, we define S^l as the labeled set with N^l samples and S^u as the unlabeled set with N^u samples. We use the i-th candidate sample s_i^c in S^c, where $i \in [1, N^c]$, as a reference data point to estimate the data distributions of the unlabeled set S^u and the labeled set S^l, and compute a distribution discrepancy score d_i^c that represents the distribution discrepancy between S^u and S^l after annotating s_i^c:

$$d_i^c = \frac{1}{N^l+1} \sum_{j=1}^{N^l+1} Sim(s_i^c, s_j^l) - \frac{1}{N^u-1} \sum_{j=1}^{N^u-1} Sim(s_i^c, s_j^u). \quad (3)$$

In Eq. 3, the first term represents the data distribution of the labeled set S^l estimated by s_i^c, where $Sim(s_i^c, s_j^l)$ represents the *cosine similarity* between s_i^c and the j-th sample s_j^l in the labeled set S^l in the high-dimensional feature space[2]. The second term in Eq. 3 represents the data distribution of the unlabeled set S^u estimated by s_i^c, where $Sim(s_i^c, s_j^u)$ represents the *cosine similarity* between s_i^c and the j-th sample s_j^u of the unlabeled set S^u in the high-dimensional feature space. After we compute the distribution discrepancy scores for all candidate samples in S^c, the candidate sample with the lowest score can be chosen as the most informative sample for annotation.

To accelerate the annotation suggestion process, we prefer to suggest multiple samples for the annotation in each stage instead of suggesting one sample at a time. However, directly ranking the candidate samples in an ascending order based on their distribution discrepancy scores and suggesting the top ones is inaccurate. Since the distribution discrepancy of the labeled and unlabeled sets is computed based on annotating one sample at a time.

To address this problem, we propose the idea of super-sample s^{super}, which is a m-combination of the candidate set S^c with N^c samples. In total, there are $\binom{N^c}{m}$ possible super-samples that can be generated from S^c. The feature representation of each super-sample is the average of the feature representations of the m samples within it. Thus, we can rewrite the distribution discrepancy score computation in Eq. 3 into a super-sample version as:

$$d_q^{super} = \frac{1}{N^l+m} \sum_{j=1}^{N^l+m} Sim(s_q^{super}, s_j^l) - \frac{1}{N^u-m} \sum_{j=1}^{N^u-m} Sim(s_q^{super}, s_j^u), \quad (4)$$

[2] The encoding part of each ag-FCN can be utilized as a feature extractor. Given an input image to K ag-FCNs, the average of outputs of Layer 6 in these ag-FCNs can be viewed as a high-dimensional feature representation of the input image.

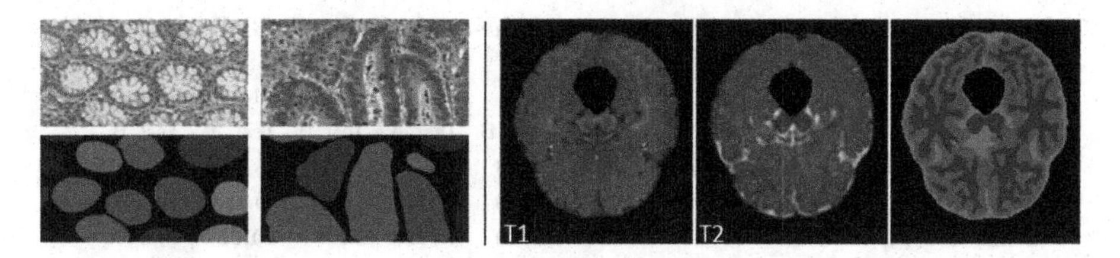

Fig. 3. Some qualitative results of our framework on GlaS dataset (left) and iSeg dataset (right, pink: Cerebrospinal Fluid; purple: White Matter; green: Gray Matter) using only 50% training data. (Color figure online)

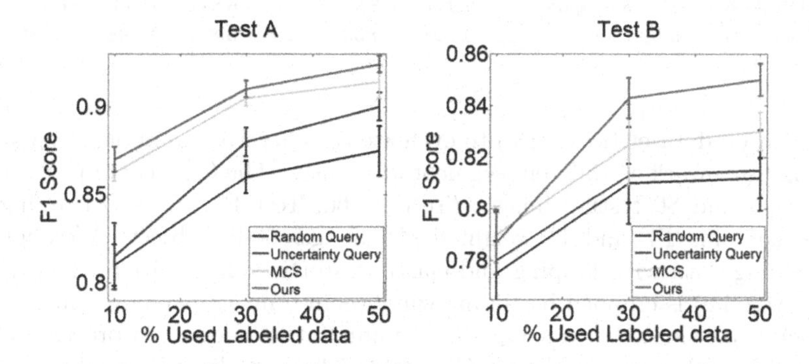

Fig. 4. Comparison using limited training data of GlaS dataset. FCN-MCS [10] is an active learning algorithm only considering uncertainty and representativeness.

where d_q^{super} denotes the distribution discrepancy score of the q-th super-sample s_q^{super} in the candidate set S^c. Then, the super-sample with the lowest distribution discrepancy score will be suggested, where the m samples within this super-sample will be the final suggested samples in this annotation suggestion stage. Finally, these samples will be annotated for fine-tuning the ag-FCNs.

The suggestion is to find the super-sample with the lowest distribution discrepancy score in Eq. 4. In other words, dd-AL aims to suggest samples that can minimize the first term in Eq. 4, which is equivalent to minimizing the similarity between suggested samples and the labeled set S^l. Therefore, the proposed dd-AL ensures the **high rarity** of suggested samples in the labeled set. Also, in Eq. 4, dd-AL aims to suggest samples that can maximize the second term, which is equivalent to maximizing the similarity between suggested samples and the unlabeled set S^u. Therefore, the proposed dd-AL can also ensure the **high representativeness** of suggested samples regarding to the unlabeled set.

3 Experiment

Dataset. As the same as [10,14–16], we use the 2015 MICCAI gland segmentation dataset (GlaS, [12]) and the training set of 2017 MICCAI infant brain

Table 1. Comparing accuracies with state-of-the-art methods on GlaS dataset [12].

Method	F1 score		ObjectDice		ObjectHausdorf	
	Test A	Test B	Test A	Test B	Test A	Test B
CUMedNet [15]	0.912	0.716	0.897	0.781	45.418	160.347
TCC-MSFCN [16]	0.914	0.850	0.913	0.858	39.848	93.244
MILD-Net [14]	0.920	0.820	0.918	0.836	**39.390**	103.070
FCN-MCS [10]	0.921	0.855	0.904	0.858	44.736	96.976
Ours (full training data)	**0.938**	**0.866**	**0.929**	**0.870**	40.548	**93.103**
FCN-MCS [10] (50% training data)	0.913	0.832	0.901	0.836	–	–
ag-FCN-MCS (50% training data)	0.915	0.842	0.911	0.850	43.363	95.796
FCN-dd-AL (50% training data)	0.920	0.836	0.912	0.844	43.143	96.131
Ours (50% training data)	**0.924**	**0.850**	**0.918**	**0.866**	**41.988**	**95.331**

segmentation dataset (iSeg, [13]) to evaluate the effectiveness of our deep active learning framework on different segmentation tasks. The GlaS contains 85 training images and 80 testing images (Test A: 60; Test B: 20). The training set of iSeg contains T1- and T2-weighted MR images of 10 subjects. We augment the training data with flipping and elastic distortion. In addition, the original image (volume) is cropped by sliding windows into image patches (cubes), each of which is considered as a sample in the annotation suggestion process. There are 27200 samples generated from GlaS and 16380 samples generated from iSeg.

Implementation Details. In our experiments, we train 3 ag-FCNs ($K = 3$). For each annotation stage, we select top 16 uncertain samples in the first step of dd-AL ($N^c = 16$), and our super-sample size is 12 ($m = 12$) in the second step of dd-AL. At the end of each stage, ag-FCNs will be fine-tuned with all available labeled data.

Experiments on GlaS. First, we compared our ag-FCNs using all training data with state-of-the-art methods. As shown in Table 1, our ag-FCNs achieve very competitive segmentation performances (best in five columns), which shows the effectiveness of our ag-FCN in producing accurate pixel-wise predictions. Secondly, to validate our entire framework (ag-FCNs and dd-AL), we simulate the annotation suggestion process by only providing the suggested samples and their annotations to the ag-FCNs for training. For fair comparison, we follow [10] to consider the annotation cost as the number of annotated pixels and set the annotation cost budget as 10%, 30% and 50% of the overall labeled pixels. Note, though the suggested samples are generated from the original image data by using data augmentation techniques, the annotation cost budget is based on the annotation of the original image data. Our framework is compared with (1) Random Query: randomly selecting samples; (2) Uncertainty Query: suggesting samples only considering uncertainties; (3) FCN-MCS, an active learning algorithm that only considers the uncertainty and representativeness proposed in [10]. We follow [10] to randomly divide the GlaS training set into ten folds,

Table 2. Analyzing the effect of changing the super-sample size on GlaS Test A.

Super-sample size	w/o	4	6	8	10	**12**	14
F1-score	0.918	0.920	0.920	0.919	0.923	**0.924**	0.919
Total training time	13.3 h	8.2 h	7.1 h	6.3 h	5.1 h	**4.4 h**	2.5 h

Table 3. Comparing computation cost with FCN-MCS [10], the current best annotation suggestion algorithm on GlaS dataset. Note, though our GPUs are different from [10], V100 is only 2.9x faster than P100 on training typical deep learning benchmarks on average [17]. So, the lead cause of reducing the computation cost is our newly-designed framework, not the GPU hardware.

Methods	Time cost per stage	# of stages	Total training time	GPU
FCN-MCS [10]	10 min	2000	333.3 h	4 Tesla P100
Ours	20 s	800	4.4 h	4 Tesla V100

each of which is used as the initial training data for one experiment. The average results are reported. As shown in Fig. 4, our framework is consistently better than the other three query methods. Thirdly, we conduct ablation studies on our framework (ag-FCN and dd-AL) by replacing our dd-AL by the active learning algorithm MCS proposed in [10] (shown as ag-FCN-MCS in Table 1) and replacing our ag-FCN by the FCN proposed in [10] (shown as FCN-dd-AL in Table 1). As shown in Table 1, both ag-FCN-MCS and FCN-dd-AL outperform the deep active learning framework FCN-MCS proposed in [10], and our framework obtains the best performance among all the four methods using 50% training data, which reveals that the boosted performance of our framework is due to both our ag-FCN and dd-AL. Fourthly, we study the effect of changing the size of the super-sample. As shown in Table 2, compared with our framework without using super-samples, our framework using super-sample size 12 largely improves the training time by 9 h and the F1-score by 0.6% on the GlaS Test A, which validates the effectiveness of our super-sample version of the distribution discrepancy score computation. Fifthly, in addition to outperforming the current best annotation suggestion algorithm [10] on biomedical image segmentation in terms of accuracy (Table 1), our framework is more efficient (Table 3).

Experiments on iSeg. We also extend our ag-FCN into the 3D version (3D-ag-FCN)[3] and test our framework (3D-ag-FCN and dd-AL) on the training set of iSeg using 10-fold cross-validation (9 subjects for training, 1 subject for testing, repeat 10 times). As shown in Table 4, our framework still achieves competitive performances even using only 50% training data. Figure 3 shows some qualitative examples on the two datasets.

[3] We replace all 2D operations with 3D operations (e.g., 2D conv → 3D conv, etc.).

Table 4. Comparison with state-of-the-art on iSeg dataset [13]. (DICE: Dice Coefficient; MHD: Modified Hausdorff Distance; ASD: Average Surface Distance).

Method	White matter			Gray matter			Cerebrospinal fluid		
	DICE	MHD	ASD	DICE	MHD	ASD	DICE	MHD	ASD
3D-Unet [18]	0.896	5.39	0.44	0.907	5.90	0.38	0.944	13.86	0.15
VoxResNet [19]	0.899	**5.20**	0.44	0.906	5.48	0.38	0.943	13.93	0.15
3D-SDenseNet [20]	0.910	5.92	**0.39**	0.916	5.75	**0.34**	0.949	13.64	0.13
Ours (full training data)	**0.928**	5.21	0.40	**0.921**	**5.44**	0.35	**0.960**	**13.26**	**0.12**
Ours (50% training data)	0.920	5.33	0.44	0.914	5.61	0.37	0.951	13.58	0.17

4 Conclusion

To significantly alleviate the burden of manual labeling in the biomedical image segmentation task, we propose a deep active learning framework that consists of: (1) an attention gated fully convolutional network (ag-FCN) that achieves state-of-the-art segmentation performances when using the full training data and (2) a distribution discrepancy based active learning algorithm that progressively suggests informative samples to train the ag-FCNs. Our framework achieves state-of-the-art segmentation performance by only using a portion of the annotated training data on two MICCAI challenges.

References

1. Kamnitsas, K., Ledig, C., et al.: Efficient multi-scale 3D CNN with fully connected CRF for accurate brain lesion segmentation. Med. Image Anal. **36**, 61–78 (2017)
2. Liao, F., Liang, M., et al.: Evaluate the malignancy of pulmonary nodules using the 3-D deep leaky noisy-or network. IEEE Trans. Neural Netw. Learn. Syst. **30**(11), 3484–3495 (2019)
3. Long, J., Shelhamer, E., et al.: Fully convolutional networks for semantic segmentation. In: Proceedings of the IEEE Conference on Computer Vision and Pattern Recognition, pp. 3431–3440 (2015)
4. He, K., Gkioxari, G., et al.: Mask R-CNN. In: Proceedings of the IEEE International Conference on Computer Vision, pp. 2961–2969 (2017)
5. Papandreou, G., Chen, L.C., et al.: Weakly-and semi-supervised learning of a DCNN for semantic image segmentation. arXiv, arXiv preprint arXiv:1502.02734 (2015)
6. Xiao, H., Wei, Y., et al.: Transferable semi-supervised semantic segmentation. In: AAAI Conference on Artificial Intelligence (2018)
7. Hong, S., Noh, H., et al.: Decoupled deep neural network for semi-supervised semantic segmentation. In: Advances in Neural Information Processing Systems, pp. 1495–1503 (2015)
8. Settles, B.: Active learning literature survey. University of Wisconsin-Madison Department of Computer Sciences (2009)
9. Dutt Jain, S., Grauman, K.: Active image segmentation propagation. In: Proceedings of the IEEE Conference on Computer Vision and Pattern Recognition, pp. 2864–2873 (2016)

10. Yang, L., Zhang, Y., et al.: Suggestive annotation: a deep active learning framework for biomedical image segmentation. In: International Conference on Medical Image Computing and Computer-Assisted Intervention, pp. 399–407 (2017)
11. Wang, X., Girshick, R., et al.: Non-local neural networks. In: Proceedings of the IEEE Conference on Computer Vision and Pattern Recognition, pp. 7794–7803 (2018)
12. Sirinukunwattana, K., Pluim, J.P., et al.: Gland segmentation in colon histology images: the glas challenge contest. Med. Image Anal. **35**, 489–502 (2017)
13. Wang, L., Nie, D., et al.: Benchmark on automatic six-month-old infant brain segmentation algorithms: the iSeg-2017 challenge. IEEE Trans. Med. Imaging **38**(9), 2219–2230 (2019)
14. Graham, S., Chen, H., et al.: MILD-Net: minimal information loss dilated network for gland instance segmentation in colon histology images. Med. Image Anal. **52**, 199–211 (2019)
15. Graham, S., Chen, H., et al.: DCAN: deep contour-aware networks for accurate gland segmentation. In: Proceedings of the IEEE Conference on Computer Vision and Pattern Recognition, pp. 2487–2496 (2016)
16. Ding, H., Pan, Z., et al.: Multi-scale fully convolutional network for gland segmentation using three-class classification. Neurocomputing **380**, 150–161 (2020)
17. TESLA V100 Performance Guide. https://www.nvidia.com/content/dam/en-zz/Solutions/Data-Center/tesla-product-literature/v100-application-performance-guide.pdf
18. Çiçek, Ö., Abdulkadir, A., et al.: 3D U-Net: learning dense volumetric segmentation from sparse annotation. In: International Conference on Medical Image Computing and Computer-Assisted Intervention, pp. 424–432 (2016)
19. Chen, H., Dou, Q., et al.: VoxResNet: deep voxelwise residual networks for brain segmentation from 3D MR images. NeuroImage **170**, 446–455 (2018)
20. Bui, T.D., Shin, J., et al.: Skip-connected 3D DenseNet for volumetric infant brain MRI segmentation. Biomed. Signal Process. Control **54**, 101613 (2019)

Scribble2Label: Scribble-Supervised Cell Segmentation via Self-generating Pseudo-Labels with Consistency

Hyeonsoo Lee[1] and Won-Ki Jeong[2(✉)]

[1] School of Electrical and Computer Engineering,
Ulsan National Institute of Science and Technology, Ulsan, South Korea
hslee1@unist.ac.kr

[2] College of Informatics, Department of Computer Science and Engineering,
Korea University, Seoul, South Korea
wkjeong@korea.ac.kr

Abstract. Segmentation is a fundamental process in microscopic cell image analysis. With the advent of recent advances in deep learning, more accurate and high-throughput cell segmentation has become feasible. However, most existing deep learning-based cell segmentation algorithms require fully annotated ground-truth cell labels, which are time-consuming and labor-intensive to generate. In this paper, we introduce `Scribble2Label`, a novel weakly-supervised cell segmentation framework that exploits only a handful of scribble annotations without full segmentation labels. The core idea is to combine pseudo-labeling and label filtering to generate reliable labels from weak supervision. For this, we leverage the consistency of predictions by iteratively averaging the predictions to improve pseudo labels. We demonstrate the performance of `Scribble2Label` by comparing it to several state-of-the-art cell segmentation methods with various cell image modalities, including bright-field, fluorescence, and electron microscopy. We also show that our method performs robustly across different levels of scribble details, which confirms that only a few scribble annotations are required in real-use cases.

Keywords: Cell segmentation · Weakly-supervised learning · Scribble annotation

1 Introduction

Micro- to nano-scale microscopy images are commonly used for cellular-level biological image analysis. In cell image analysis, segmentation serves as a crucial task to extract the morphology of the cellular structures. Conventional cell segmentation methods are mostly grounded in model-based and energy minimization methods, such as Watershed [20], Chan-Vese with the edge model [10], and gradient vector flow [7]. The recent success of deep learning has gained much attention in many image processing and computer vision tasks. A common approach to achieve highly-accurate segmentation performance is to train deep neural networks using ground-truth labels [1,19,21]. However, generating a

A. L. Martel et al. (Eds.): MICCAI 2020, LNCS 12261, pp. 14–23, 2020.
https://doi.org/10.1007/978-3-030-59710-8_2

sufficient number of ground-truth labels is time-consuming and labor-intensive, which is becoming a major bottleneck in the segmentation process. Additionally, manually generated segmentation labels are prone to errors due to the difficulty in drawing pixel-level accurate region masks.

To address such problems, weakly-supervised cell segmentation methods using point annotation have recently been proposed [12,13,22]. Yoo et al. [22] and Qu et al. [13] introduced methods that generate coarse labels only from point annotations using a Voronoi diagram. Further, Nishimura et al. [12] proposed a point detection network in which output is used for cell instance segmentation. Even though point annotation is much easier to generate compared to full region masks, the existing work requires point annotations for the entire dataset – for example, there are around 22,000 nuclei in 30 images of the MoNuSeg dataset [5]. Moreover, the performance of the work mentioned above is highly sensitive to the point location, i.e., the point should be close to the center of the cell.

Recently, weakly-supervised learning using scribble annotations, i.e., scribble-supervised learning, has actively been studied in image segmentation as a promising direction for lessening the burden of manually generating training labels. Scribble-supervised learning exploits scribble labels and regularized networks with standard segmentation techniques (e.g., graph-cut [8], Dense Conditional Random Field [DenseCRF] [3,16]) or additional model parameters (e.g., boundary prediction [17] and adversarial training [18]). The existing scribble-supervised methods have demonstrated the possibility to reduce manual efforts in generating training labels, but their adaptation in cell segmentation has not been explored yet.

In this paper, we propose a novel weakly-supervised cell segmentation method that is highly accurate and robust with only a handful of manual annotations. Our method, `Scribble2Label`, uses scribble annotations as in conventional scribble-supervised learning methods, but we propose the combination of pseudo-labeling and label-filtering to progressively generate full training labels from a few scribble annotations. By doing this, we can effectively remove noise in pseudo labels and improve prediction accuracy. The main contributions of our work are as follows.

- We introduce a novel iterative segmentation network training process (Fig. 1) that generates training labels automatically via weak-supervision using only a small set of manual scribbles, which significantly reduces the manual effort in generating training labels.
- We propose a novel idea of combining pseudo-labeling with label filtering, exploiting consistency to generate reliable training labels, which results in a highly accurate and robust performance. We demonstrate that our method consistently outperforms the existing state-of-the-art methods across various cell image modalities and different levels of scribble details.
- Unlike existing scribble-supervised segmentation methods, our method is an end-to-end scheme that does not require any additional model parameters or external segmentation methods (e.g, Graph-cut, DenseCRF) during training.

To the best of our knowledge, this is the first scribble-supervised segmentation method applied to the cell segmentation problem in various microscopy images.

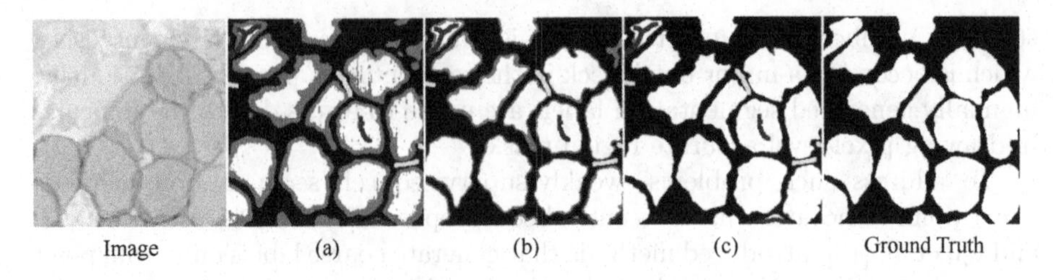

Fig. 1. An example of iterative refinement of pseudo labels during training. Blue and yellow: scribbles for cells and background, respectively (Ω_s); red: the pixels below the consistency threshold τ, which will be ignored when calculating the unscribbled pixel loss ($\mathcal{L}_{up}$); white and black: cell or background pixels over τ (Ω_g). (a) – (c) represent the filtered pseudo-labels from the predictions over the iterations (with Intersection over Union [IoU] score): (a): 7th (0.5992), (b): 20th (0.8306), and (c): 100th (0.9230). The actual scribble thickness used in our experiment was 1 pixel, but it is widened to 5 pixels in this figure for better visualization. (Color figure online)

2 Method

In this section, we describe the proposed segmentation method in detail. The input sources for our method are the image x and the user-given scribbles s (see Fig. 2). Here, the given scribbles are *labeled* pixels (denoted as blue and yellow for the foreground and background, respectively), and the rest of the pixels are *unlabeled* pixels (denoted as black). For labeled (scribbled) pixels, a standard cross-entropy loss is applied. For unlabeled (unscribbled) pixels, our network automatically generates reliable labels using the exponential moving average of the predictions during training. Training our model consists of two stages. The first stage is initialization (i.e., a warm-up stage) by training the model using only the scribbled pixel loss ($\mathcal{L}_{sp}$). Once the model is initially trained via the warm-up stage, the prediction is iteratively refined by both scribbled and unscribbled losses ($\mathcal{L}_{sp}$ and $\mathcal{L}_{up}$). Figure 2 illustrates the overview of the proposed system.

2.1 Warm-Up Stage

At the beginning, we only have a small set of user-drawn scribbles for input training data. During the first few iterations (warm-up stage), we train the model only using the given scribbles, and generate the average of predictions which can be used in the following stage (Sect. 2.2). Here, the given scribbles is a subset of the corresponding mask annotation. By ignoring unscribbled pixels, the proposed network is trained with cross entropy loss as follows:

$$\mathcal{L}_{sp}(x, s) = -\frac{1}{|\Omega_s|} \sum_{j \in \Omega_s} [s_j \log(f(x; \theta_i)) + (1 - s_j) \log(1 - f(x; \theta_i))], \quad (1)$$

where x is an input image, s is a scribble annotation, and Ω_s is a set of scribbled pixels. $f(x; \theta_i)$ is the model's prediction at iteration i. This warm-up stage continues until we reach the warm-up Epoch E_W.

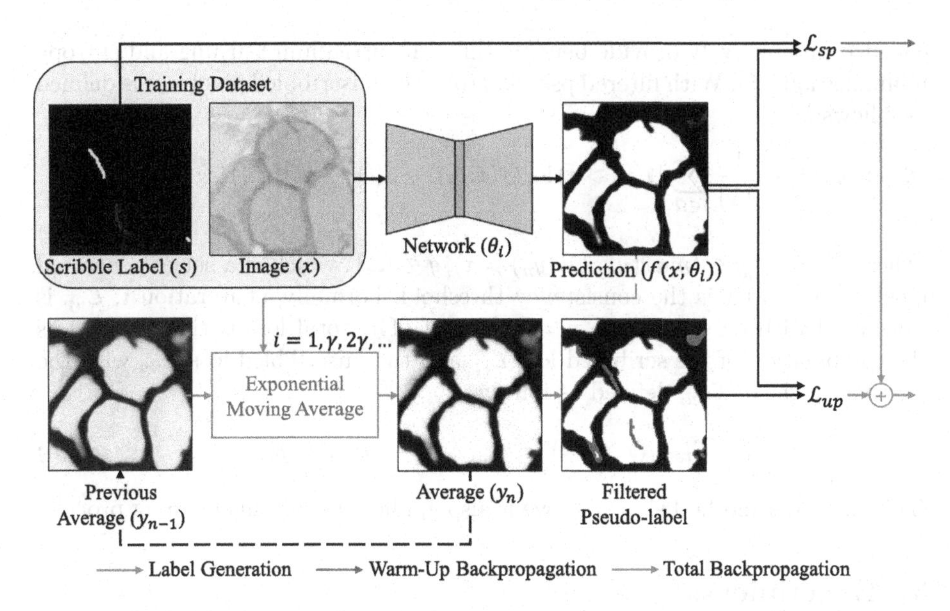

Fig. 2. The overview of the proposed method (`Scribble2Label`). The pseudo-label is generated from the average of predictions. Following, $\mathcal{L}_{sp}$ is calculated with the scribble annotation, and $\mathcal{L}_{up}$ is calculated with the filtered pseudo-label. The prediction ensemble process occurs every γ epochs, where γ is the ensemble interval. n represents how many times the predictions are averaged. (Color figure online)

Moreover, we periodically calculate the exponential moving average (EMA) of the predictions over the training process: $y_n = \alpha f(x;\theta_i) + (1-\alpha)y_{n-1}$ where α is the EMA weight, y is the average of predictions, $y_0 = f(x;\theta_1)$, and n is how many times the predictions are averaged. This process is called a prediction ensemble [11]. Note that, since we use data augmentation for training, the segmentation prediction is not consistent for the same input image. Our solution for this problem is splitting the training process into training and ensemble steps. In the ensemble phase, an un-augmented image is used for the input to the network, and EMA is applied to that predictions. Moreover, in the scribble-supervised setting, we cannot ensemble the predictions when the best model is found, as in [11], because the given label is not fully annotated. To achieve the valuable ensemble and reduce computational costs, the predictions are averaged every γ epochs, where γ is the ensemble interval.

2.2 Learning with a Self-generated Pseudo-Label

The average of the predictions can be obtained after the warm-up stage. This can be used for generating a reliable pseudo-label of unscirbbled pixels. For filtering the pseudo-label, the average is used. The pixels with consistently the same result are one-hot encoded and used as a label for unscribbled pixels with standard cross entropy. Using only reliable pixels and making these one-hot encoded progressively provide benefits through curriculum learning and entropy minimization [15]. With filtered pseudo-label, the unscribbled pixel loss is defined as follows:

$$\mathcal{L}_{up}(x, y_n) = -\frac{1}{|\Omega_g|}\sum_{j\in\Omega_g}[\mathbb{1}(y_n > \tau)\log(f(x;\theta_i)) + \mathbb{1}((1-y_n) > \tau))\log(1-f(x;\theta_i))], \tag{2}$$

where $\Omega_g = \{g|g \in (max(y_n, 1-y_n) > \tau), g \notin \Omega_s\}$, which is a set of generated label pixels, and τ is the consistency threshold. Formally, at iteration i, $\mathcal{L}_{up}$ is calculated with (x, y_n), where $n = \lfloor i/\gamma \rfloor + 1$. The total loss is then defined as the combination of the scribbled loss $\mathcal{L}_{sp}$ and the unscribbled loss $\mathcal{L}_{up}$ with the relative weight of $\mathcal{L}_{up}$, defined as follows:

$$\mathcal{L}_{total}(x, s, y_n) = \mathcal{L}_{sp}(x, s) + \lambda\mathcal{L}_{up}(x, y_n) \tag{3}$$

Note the EMA method shown above is also applied during this training process.

3 Experiments

3.1 Datasets

We demonstrated the efficacy of our method using three different cell image datasets. The first set, MoNuSeg [5], consists of 30 1000×1000 histopathology images acquired from multiple sites covering diverse nuclear appearances. We conducted a 10-fold cross-validation for the MoNuSeg dataset. BBBC038v1 [2], the second data set, which is known as Data Science Bowl 2018, is a set of nuclei 2D images. We used the stage 1 training dataset, which is fully annotated, and further divided it into three main types, including 542 fluorescence (DSB-Fluo) images of various sizes, 108 320×256 histopathology images (DSB-Histo), and 16 bright-field 1000×1000 (DSB-BF) images. Each dataset is split into training, validation, and test sets, with ratios of 60%, 20%, and 20%, respectively. EM is an internally collected serial-section electron microscopy image dataset of a larval zebrafish. We used three sub-volumes of either $512 \times 512 \times 512$ or $512 \times 512 \times 256$ in size. The size of the testing volume was $512 \times 512 \times 512$. The scribbles of MoNuSeg and DSBs were manually drawn by referencing the full segmentation labels. To ensure the convenience of scribbles, we annotate images up to 256×256 within 1 min, 512×512 within 2 min, and 1024×1024 within 4 min. For the EM dataset, the scribble annotation was generated by a scribble generation algorithm in [18] with a 10% ratio.

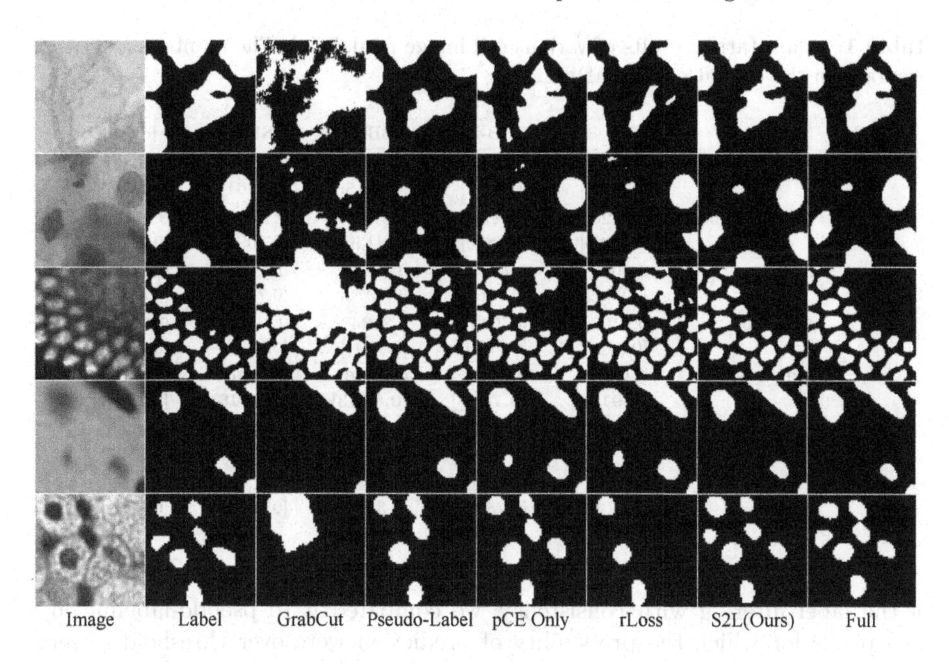

Fig. 3. Qualitative results comparison. From the top to the bottom, EM, DSB-BF [2], DSB-Fluo, DSB-Histo, and MoNuSeg [5] are shown.

3.2 Implementation Details

Our baseline network was U-Net [14] with the ResNet-50 [4] encoder. For comparison with [13] in histopathology experiments (MoNuSeg, DSB-Histo), we used ResNet-34 for the encoder. The network was initialized with pre-trained parameters, and RAdam [9] was used for all experiments. To regularize the network, we used conventional data augmentation methods, such as cropping, flipping, rotation, shifting, scaling, brightness change, and contrast changes.

The hyper-parameters used for our model are as follows: Consistency Threshold $\tau = 0.8$; EMA Alpha $\alpha = 0.2$; Ensemble Momentum $\gamma = 5$; $\mathcal{L}_{up}$'s weight $\lambda = 0.5$; and warm-up epoch $E_W = 100$. For the MoNuSeg dataset (which is much noisier than other datasets), we used $\tau = 0.95$ and $\alpha = 0.1$ to cope with noisy labels.

3.3 Results

We evaluated the performance of semantic segmentation using the intersection over union (IoU) and the performance of instance segmentation using mean Dice-coefficient (mDice) used in [12].

Comparison with Other Methods: We compared our method to the network trained with full segmentation annotation, scribble annotation (pCE Only) [16], and the segmentation proposal from Grab-Cut [8]. To demonstrate the efficacy

Table 1. Quantitative results of various cell image modalities. The numbers represent accuracy in the format of IoU[mDice].

Label	Method	EM	DSB-BF	DSB-Fluo	DSB-Histo	MoNuSeg
Scribble	GrabCut [8]	0.5288 [0.6066]	0.7328 [0.7207]	0.8019 [0.7815]	0.6969 [0.5961]	0.1534 [0.0703]
	Pseudo-Label [6]	0.9126 [0.9096]	0.6177 [0.6826]	0.8109 [0.8136]	0.7888 [0.7096]	0.6113 [0.5607]
	pCE Only [16]	0.9000 [0.9032]	0.7954 [0.7351]	0.8293 [0.8375]	0.7804 [0.7173]	0.6319 [0.5766]
	rLoss [16]	0.9108 [0.9100]	0.7993 [0.7280]	0.8334 [0.8394]	0.7873 [0.7177]	0.6337 [0.5789]
	S2L(Ours)	**0.9208 [0.9167]**	**0.8236 [0.7663]**	**0.8426 [0.8443]**	**0.7970 [0.7246]**	**0.6408 [0.5811]**
Point	*Qu*[13]	–	–	–	0.5544 [0.7204]	0.6099 [0.7127]
Full	Full	0.9298 [0.9149]	0.8774 [0.7879]	0.8688 [0.8390]	0.8134 [0.7014]	0.7014 [0.6677]

of the label filtering with consistency, we compared it to pseudo-labeling [6]. The pixels for which the probability of prediction were over threshold τ were assigned to be a pseudo-label, where τ was same as our method setting. Our method was also compared to Regularized Loss (rLoss) [16], which integrates the DenseCRF into the loss function. The hyper-parameters of rLoss are $\sigma_{XY} = 100$ and $\sigma_{RGB} = 15$ (Fig. 3).

Table 1 shows the quantitative comparison of our method with several representative methods. Overall, our method outperformed all methods on both IoU and mDice quality metrics. We observed that our method achieved even higher mDice accuracy compared to the full method (i.e., trained using full segmentation labels) on EM, DSB-BF, and DSB-Histo datasets. Note also that MoNuSeg dataset contains many small cluttering cells, which are challenge to separate individually. However, our method showed outstanding instance segmentation results in this case, too.

Grab-Cut's [8] segmentation proposal and the pseudo-label [6] were erroneous. Thus, training with these erroneous segmentation labels impairs the performance of the method. Qu et al.'s method [13] performed well for instance-level segmentation on MoNuSeg dataset, however, it performed worse on DSB-histo dataset. Because [13] used a clustering label that has circular shape cell label, it was hard to segment the non-circular cell. Learning with pCE [16] showed stable results on various datasets. However, due to learning using only scribbles, the method failed to correctly predict boundary accurately as in our method. rLoss [16] outperformed most of the previous methods, but our method generally showed better results. We also observed that leveraging consistency by averaging predictions is crucial to generate robust pseudo-labels. Scribble2Label's results also confirm that using pseudo label together with scribbles is effective to generate accurate boundaries, comparable to the ground-truth segmentation label.

Table 2. Quantitative results using various amounts of scribbles. DSB-Fluo [2] was used for the evaluation. The numbers represent accuracy in the format of IoU[mDice].

Method	10%	30%	50%	100%	Manual
GrabCut [8]	0.7131 [0.7274]	0.8153 [0.7917]	0.8244 [0.8005]	0.8331 [0.8163]	0.8019 [0.7815]
Pseudo-label [6]	0.7920 [0.8086]	0.7984 [0.8236]	0.8316 [0.8392]	0.8283 [0.8251]	0.8109 [0.8136]
pCE Only [16]	0.7996 [0.8136]	0.8180 [0.8251]	0.8189 [0.8204]	0.8098 [0.8263]	0.8293 [0.8375]
rLoss [16]	0.8159 [0.8181]	0.8251 [0.8216]	0.8327 [0.8260]	0.8318 [0.8369]	0.8334 [0.8394]
S2L (Ours)	**0.8274 [0.8188]**	**0.8539 [0.8407]**	**0.8497 [0.8406]**	**0.8588 [0.8443]**	**0.8426 [0.8443]**
Full	0.8688 [0.8390]				

Effect of Amount of Scribble Annotations: To demonstrate the robustness of our method over various levels of scribble details, we conducted an experiment using scribbles automatically generated using a similar method by Wu et al. [18] (i.e., foreground and background regions are skeletonized and sampled). The target dataset was DSB-Fluo, and various amounts of scribbles, i.e., 10%, 30%, 50%, and 100% of the skeleton pixels extracted from the full segmentation labels (masks), are automatically generated. Table 2 summarizes the results with different levels of scribble details. Our method `Scribble2Label` generated stable results in both the semantic metric and instance metric from sparse scribbles to abundant scribbles.

The segmentation proposal from Grab-Cut [8] and the pseudo-lable [6] were noisy in settings lacking annotations, which resulted in degrading the performance. rLoss [16] performed better than the other methods, but it sometimes failed to generate correct segmentation results especially when the background is complex (causing confusion with cells). Our method showed very robust results over various scribble amounts. Note that our method performs comparable to using full segmentation masks only with 30% of skeleton pixels.

4 Conclusion

In this paper, we introduced `Scribble2Label`, a simple but effective scribble-supervised learning method that combines pseudo-labeling and label-filtering with consistency. Unlike the existing methods, `Scribble2Label` demonstrates highly-accurate segmentation performance on various datasets and at different levels of scribble detail without extra segmentation processes or additional model parameters. We envision that our method can effectively avoid time-consuming and labor-intensive manual label generation, which is a major bottleneck in image segmentation. In the future, we plan to extend our method in more general problem settings other than cell segmentation, including semantic and instance segmentation in images and videos. Developing automatic label generation for the segmentation of more complicated biological features, such as tumor regions in histopathology images and mitochondria in nano-scale cell images, is another interesting future research direction.

Acknowledgements. This work was partially supported by the Bio & Medical Technology Development Program of the National Research Foundation of Korea (NRF) funded by the Ministry of Science and ICT (NRF-2015M3A9A7029725, NRF-2019M3E5D2A01063819), and the Korea Health Technology R&D Project through the Korea Health Industry Development Institute (KHIDI), funded by the Ministry of Health & Welfare, Republic of Korea (HI18C0316).

References

1. Arbelle, A., Raviv, T.R.: Microscopy cell segmentation via convolutional LSTM networks. In: 2019 IEEE 16th International Symposium on Biomedical Imaging (ISBI 2019), pp. 1008–1012. IEEE (2019)
2. Caicedo, J.C., et al.: Nucleus segmentation across imaging experiments: the 2018 data science bowl. Nat. Methods **16**(12), 1247–1253 (2019)
3. Can, Y.B., Chaitanya, K., Mustafa, B., Koch, L.M., Konukoglu, E., Baumgartner, C.F.: Learning to segment medical images with scribble-supervision alone. In: Stoyanov, D., et al. (eds.) DLMIA/ML-CDS -2018. LNCS, vol. 11045, pp. 236–244. Springer, Cham (2018). https://doi.org/10.1007/978-3-030-00889-5_27
4. He, K., Zhang, X., Ren, S., Sun, J.: Deep residual learning for image recognition. In: Proceedings of the IEEE Conference on Computer Vision and Pattern Recognition, pp. 770–778 (2016)
5. Kumar, N., Verma, R., Sharma, S., Bhargava, S., Vahadane, A., Sethi, A.: A dataset and a technique for generalized nuclear segmentation for computational pathology. IEEE Trans. Med. Imaging **36**(7), 1550–1560 (2017)
6. Lee, D.H.: Pseudo-label: the simple and efficient semi-supervised learning method for deep neural networks. In: Workshop on Challenges in Representation Learning, ICML, vol. 3, p. 2 (2013)
7. Li, C., Liu, J., Fox, M.D.: Segmentation of edge preserving gradient vector flow: an approach toward automatically initializing and splitting of snakes. In: 2005 IEEE Computer Society Conference on Computer Vision and Pattern Recognition (CVPR 2005), vol. 1, pp. 162–167. IEEE (2005)
8. Lin, D., Dai, J., Jia, J., He, K., Sun, J.: Scribblesup: scribble-supervised convolutional networks for semantic segmentation. In: Proceedings of the IEEE Conference on Computer Vision and Pattern Recognition, pp. 3159–3167 (2016)
9. Liu, L., et al.: On the variance of the adaptive learning rate and beyond. arXiv preprint arXiv:1908.03265 (2019)
10. Maška, M., Daněk, O., Garasa, S., Rouzaut, A., Muñoz-Barrutia, A., Ortiz-de Solorzano, C.: Segmentation and shape tracking of whole fluorescent cells based on the Chan-Vese model. IEEE Trans. Med. Imaging **32**(6), 995–1006 (2013)
11. Nguyen, D.T., Mummadi, C.K., Ngo, T.P.N., Nguyen, T.H.P., Beggel, L., Brox, T.: Self: learning to filter noisy labels with self-ensembling. arXiv preprint arXiv:1910.01842 (2019)

12. Nishimura, K., Ker, D.F.E., Bise, R.: Weakly supervised cell instance segmentation by propagating from detection response. In: Shen, D., et al. (eds.) MICCAI 2019. LNCS, vol. 11764, pp. 649–657. Springer, Cham (2019). https://doi.org/10.1007/978-3-030-32239-7_72
13. Qu, H., et al.: Weakly supervised deep nuclei segmentation using points annotation in histopathology images. In: International Conference on Medical Imaging with Deep Learning, pp. 390–400 (2019)
14. Ronneberger, O., Fischer, P., Brox, T.: U-Net: convolutional networks for biomedical image segmentation. In: Navab, N., Hornegger, J., Wells, W.M., Frangi, A.F. (eds.) MICCAI 2015. LNCS, vol. 9351, pp. 234–241. Springer, Cham (2015). https://doi.org/10.1007/978-3-319-24574-4_28
15. Sohn, K., et al.: Fixmatch: simplifying semi-supervised learning with consistency and confidence. arXiv preprint arXiv:2001.07685 (2020)
16. Tang, M., Perazzi, F., Djelouah, A., Ben Ayed, I., Schroers, C., Boykov, Y.: On regularized losses for weakly-supervised cnn segmentation. In: Proceedings of the European Conference on Computer Vision (ECCV), pp. 507–522 (2018)
17. Wang, B., et al.: Boundary perception guidance: a scribble-supervised semantic segmentation approach. In: Proceedings of the 28th International Joint Conference on Artificial Intelligence, pp. 3663–3669. AAAI Press (2019)
18. Wu, W., Qi, H., Rong, Z., Liu, L., Su, H.: Scribble-supervised segmentation of aerial building footprints using adversarial learning. IEEE Access **6**, 58898–58911 (2018)
19. Xing, F., Xie, Y., Yang, L.: An automatic learning-based framework for robust nucleus segmentation. IEEE Trans. Med. Imaging **35**(2), 550–566 (2015)
20. Yang, X., Li, H., Zhou, X.: Nuclei segmentation using marker-controlled watershed, tracking using mean-shift, and Kalman filter in time-lapse microscopy. IEEE Trans. Circ. Syst. I Regul. Pap. **53**(11), 2405–2414 (2006)
21. Yi, J., et al.: Multi-scale cell instance segmentation with keypoint graph based bounding boxes. In: Shen, D., et al. (eds.) MICCAI 2019. LNCS, vol. 11764, pp. 369–377. Springer, Cham (2019). https://doi.org/10.1007/978-3-030-32239-7_41
22. Yoo, I., Yoo, D., Paeng, K.: PseudoEdgeNet: nuclei segmentation only with point annotations. MICCAI 2019. LNCS, vol. 11764, pp. 731–739. Springer, Cham (2019). https://doi.org/10.1007/978-3-030-32239-7_81

Are Fast Labeling Methods Reliable? A Case Study of Computer-Aided Expert Annotations on Microscopy Slides

Christian Marzahl[1,2](✉), Christof A. Bertram[3], Marc Aubreville[1], Anne Petrick[3], Kristina Weiler[4], Agnes C. Gläsel[4], Marco Fragoso[3], Sophie Merz[3], Florian Bartenschlager[3], Judith Hoppe[3], Alina Langenhagen[3], Anne-Katherine Jasensky[5], Jörn Voigt[2], Robert Klopfleisch[3], and Andreas Maier[1]

[1] Pattern Recognition Lab, Department of Computer Science, Friedrich-Alexander-Universität Erlangen-Nürnberg, Erlangen, Germany
christian.marzahl@gmail.com

[2] Research and Development, EUROIMMUN Medizinische Labordiagnostika AG Lübeck, Lübeck, Germany

[3] Institute of Veterinary Pathology, Freie Universität Berlin, Berlin, Germany

[4] Department of Veterinary Clinical Sciences, Clinical Pathology and Clinical Pathophysiology, Justus-Liebig-Universität Giessen, Giessen, Germany

[5] Laboklin GmbH und Co. KG, Bad Kissingen, Germany

Abstract. Deep-learning-based pipelines have shown the potential to revolutionalize microscopy image diagnostics by providing visual augmentations and evaluations to a trained pathology expert. However, to match human performance, the methods rely on the availability of vast amounts of high-quality labeled data, which poses a significant challenge. To circumvent this, augmented labeling methods, also known as expert-algorithm-collaboration, have recently become popular. However, potential biases introduced by this operation mode and their effects for training deep neuronal networks are not entirely understood. This work aims to shed light on some of the effects by providing a case study for three pathologically relevant diagnostic settings. Ten trained pathology experts performed a labeling tasks first without and later with computer-generated augmentation. To investigate different biasing effects, we intentionally introduced errors to the augmentation. In total, the pathology experts annotated 26,015 cells on 1,200 images in this novel annotation study. Backed by this extensive data set, we found that the concordance of multiple experts was significantly increased in the computer-aided setting, versus the unaided annotation. However, a significant percentage of the deliberately introduced false labels was not identified by the experts.

Keywords: Expert-algorithm collaboration · Computer-aided labelling · Microscopy · Pathology

A. L. Martel et al. (Eds.): MICCAI 2020, LNCS 12261, pp. 24–32, 2020.
https://doi.org/10.1007/978-3-030-59710-8_3

1 Introduction

The field of computer vision strongly relies on the availability of high quality, expert-labelled image data sets to develop, train, test and validate algorithms. The availability of such data sets is frequently the limiting factor for research and industrial projects alike. This is especially true for the medical field, where expert resources are restricted due to the high need for trained experts in clinical diagnostics. Consequently, the generation of high-quality and high-quantity data sets is limited and there is a growing need for a highly efficient labeling processes. To explore the potential of reducing the expert annotation effort while maintaining expert-level quality, we reviewed a method called expert-algorithm collaboration in three types of data sets. In this approach, experts manually correct labels pre-computed typically by a machine learning-based algorithm. While there are numerous successful applications of crowd-sourcing in the medical field [15] crowd-algorithm collaboration or expert-algorithm collaboration has been applied rarely and only in order to solve highly specific tasks. Some examples are: Maier-Hein et al. [9] on endoscopic images, Ganz et al. [6] on MR-based cortical surface delineations or Marzahl et al. [13] on a pulmonary haemorrhage cytology data set. There is no previous research regarding the best way to apply expert-algorithm collaboration, or its challenges and limitations. Therefore, this study aims to investigate the following questions: First, is the expert-algorithm collaboration performance independent from the type of task? Second, can the findings in [13] regarding the performance gains for some specific tasks also be applied to different types of medical data sets? Finally, is there a bias towards accepting pre-computed annotations? To test our hypothesis, ten medical experts participated in our expert-algorithm collaboration study on a mitotic figure detection, asthma diagnosis, and pulmonary haemorrhage cytology data set.

2 Materials and Methods

For our experiments, we selected three different types of medical object detection data sets. First, as an example of a less challenging classification task, we chose the microscopic diagnosis of equine asthma. Equine asthma is diagnosed by counting five types of cells (eosinophils, mast cell, neutrophils, macrophages, lymphocytes) and calculating their relative occurrence. The cells are visually mostly distinguishable from each other due to their distinct morphology. The data set [3] was created by digitisation of six May-Grunwald Giemsa-stained cytocentrifugated equine bronchoalveolar lavage fluids. Second, equine exercise-induced pulmonary haemorrhage (EIPH) is diagnosed by counting hemosiderin-laden macrophages, which can be divided into five groups based on their hemosiderin content according to Golde et al. [7]. In contrast to equine asthma, the cells are stained with Prussian Blue or Turnbull's Blue in order to clearly visualise the iron pigments contained in the hemosiderin. The grading task, however, is particularly challenging because the hemosiderin absorption is a continuous process which is mapped to a discrete grading system. The last task we considered was the detection of rare events on microscopy images with high inconsistency of classification. As such, the identification of mitotic figures (i.e., cells undergoing cell division) is a important example used in the vast majority of tumor grading schemes, such as in canine cutaneous mast cell tumors [8], and known to have high inter-rater disagreement [4,14]. Due to the rareness of mitotic figures in histopathological specimens, this represents a challenging task with high demands on concentration and expert knowledge, and is thus a very suitable candidate for visual augmentation in a decision support system in a clinical diagnostic setting. Whole slide images (WSI) containing 21 tumor cases with mixed grade were selected from the training set of a publicly available data set of canine cutaneous mast cell tumor [2], which represents hematoxylin and eosin stained specimen with variable mitotic density at a resolution of $0.25\frac{\mu m}{\mathrm{px}}$. Experts were requested to annotate mitotic figures they identified. To quantify the quality of the experts' annotations, we calculated the mean intersection over union (mIoU).

2.1 Patch Selection

For labelling and training the data set was subdivided into patches. The patch selection process aims to represent the variability of the data set as carefully as possible while providing a realistic workflow of integrating pre-computed annotations into clinical routine. Therefore, we selected twenty patches for EIPH, Asthma and mitotic figures: For EIPH, we used twenty algorithmically chosen patches which had to fulfill the following criteria: each patch had to cover as many grades as possible, the two staining types had to be represented equally, at least one patch was sampled from each publicly available WSI [11] and as recommended for grading by Golde et al. [7] around 300 hemosiderophages were visible on all patches combined. The twenty asthma patches were selected on the condition that all cell types are represented approximately equally. Additionally, the patches had to have the high visual quality. From the WSI of the mitotic figure data set, a board-certified pathologist selected a region of interest spanning 10 high power fields (10 HPF, total area = $2.37\,\mathrm{mm}^2$), as done in a diagnostic setting. This restriction of area was performed in order to reduce known selection bias for this study [1].

2.2 Label Generation

We used the labels provided as ground truth from the respective data sets, and we deliberately included modifications to these labels to investigate the effects thereof and potential biases in the resulting data set generated by the experts annotating in an expert-algorithm collaborative setting. Matching the tasks, we introduced different annotation errors:

Equine Asthma Data Set. For the equine asthma data set, we randomly changed the cell type of 15 cells on five of the images. Additionally, on a separate set of five images we removed the annotation of one cell, thus introducing a missing annotation.

EIPH Scoring. For the EIPH grading task, we also introduced missing cells on five of the images. On a distinct set of five further images, we increased the grade of each of the cells by one. Finally, five images contained in total ten standard artifacts that could be generated by a typical object detection pipeline, such as false annotations (where no relevant cell was present), or multiple annotations.

Mitotic Figures. For the mitotic figure identification task, we removed 20 % of all mitotic figures (resulting in 149 missing objects) present in the data set and added the same amount of false mitotic figures from the non-mitotic figure annotations of the data set. To further understand the effects introduced by this, the mitotic figures were categorized by a CNN-based classifier w.r.t. their model score of being a mitotic figure. We did this in order to provide easy to spot false annotations, hard to distinguish candidates, easily recognizable mitotic figures (as per model score), and hard to identify objects. We grouped the mitotic figures accordingly as 0: easy (n = 49), 1: medium (n = 50) and 2: difficult (n = 50) for the fake mitotic figure task. The cutoff thresholds were chosen at $p_0 \leq 0.2$, $0.2 < p_1 \leq 0.4$ and $p_2 > 0.4$, respectively, where p is the final model score in a binary classification experiment. For the deleted true mitotic figures, we also performed a selection according to their group, where group 0 represented the hard to spot mitotic figures (n = 49), 1 the medium level (n = 59) and 2 the easy to spot mitotic figures (n = 50), as given by the model's score. To define the groups, we randomly selected according to the thresholds $p_0 \leq 0.33$, $0.33 < p_1 \leq 0.66$ and $p_2 > 0.6$.

While participants were informed that the proposed labels would be computer-generated, they were unaware about the distinctively introduced false labels in the three datasets. This allowed us to investigate the effects of error propagation introduced by our systematic labeling flaws.

Voted Ground Truth. The ground truth taken from the three respective data sets was created by pathology experts annotating the data sets. This process is highly subjective, and the quality of the ground truth annotations is hard to quantify if there is no non-subjective measurable alternative. To minimise the effects of any biases introduced by the imported ground truth, we defined a second ground truth called voted ground truth. The voted ground truth was calculated retrospectively by majority vote from the results of the participating pathologist on the unaided annotation mode. For mitotic figures, the new ground truth is defined by three out of ten experts agreeing on a mitotic figure due to the high variability of results, which is a known challenge in mitotic figure detection [5,10,17].

The participating experts comprised two board-certified anatomical pathologists, five trainee anatomical pathologists and three clinical pathologists.

2.3 Labelling Platform

We used the open-source platform EXACT [12] to host our experiments. Multiple pre-integrated features of the EXACT platform made it an ideal candidate for our study. These included: open source, team and user management, image set creation and annotation upload in combination with a screening mode for large images and a web client.

2.4 Label Experiment Design

We designed our experiments with the intent to assess the effect of computer-aided annotation methods on medical experts. For that purpose, we created three modes on the EXACT server for each of the three types of data sets (EIPH, Mitotic figures, Asthma). The first mode is the training mode which can be used to get familiar with the user interface and the EXACT features. The training was supported by providing YouTube videos[1] describing the user interface as well as the tasks. The second mode is the annotation mode, here the experts performed all annotations without algorithmic support. Finally, there is the expert-algorithm mode where the participants were asked to enhance the generated labels and the artificial flaws. As recommended by "Validating Whole Slide Imaging for Diagnostic Purposes in Pathology" guidelines from Pantanowitz et al. [16], the participants had a two weeks break between the annotation and the expert-algorithm mode to reduce potential bias of a memory effects (Fig. 2).

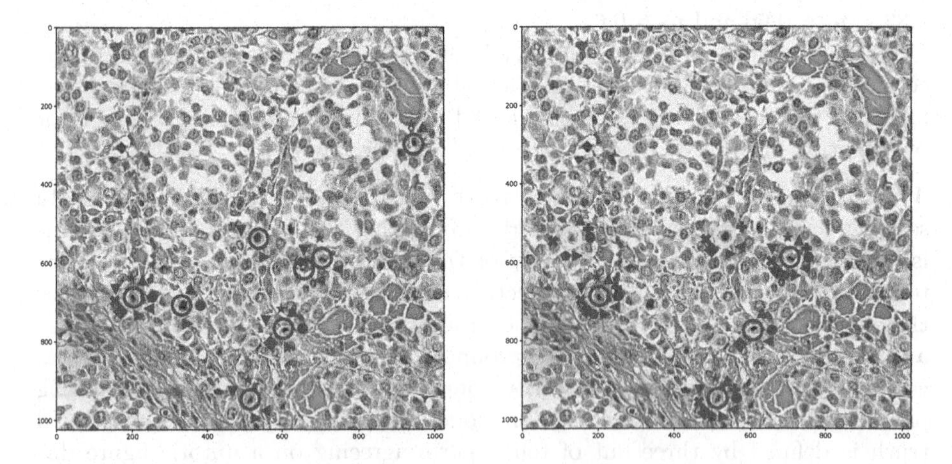

Fig. 1. Examplary annotation results without (left) and with (right) algorithmic support. Green circles represent ground truth mitotic figures, red false-positives made by experts and yellow fake (artificially introduced) mitotic figures. A symbol at the corresponding annotation represents each expert. (Color figure online)

[1] https://youtu.be/XG05RqDM9c4.

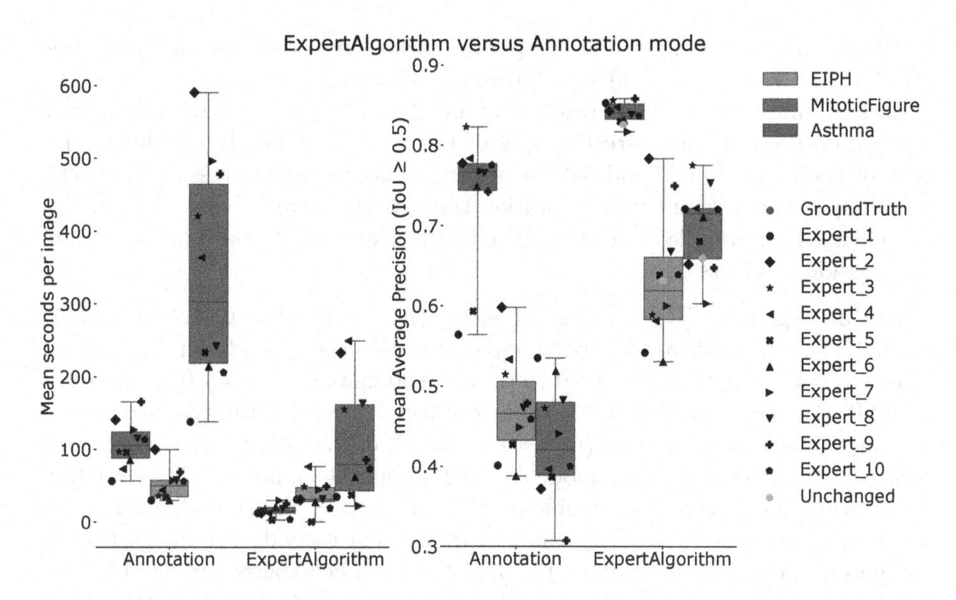

Fig. 2. Comparison of annotation and expert-algorithm collaboration mode. Left panel shows the mean number of seconds per image, while the right panel shows the mAP ($IoU \geq .5$) for each expert on the original ground truth from the three respective data sets. Additionally, we present the mAP if all computer-aided annotations were accepted unmodified (Unchanged).

3 Results

In total, ten experts created 26,015 annotations on 1,200 images. The annotation accuracy was calculated by the mean Average Precision (mAP) with a intersection over union (IoU) greater or equal 0.5.

Asthma: The computer-aided mode led to a significantly increased mAP of μ=0.84 (min = 0.82, max = 0.86, $\sigma = 0.01$) with the ground truth [F(1, 19) = 81.61, p<0.01], compared to the annotation mode (μ = 0.73, min = 0.56, max = 0.82, $\sigma = 0.08$) while also decreasing the average annotation time from μ = 106 (min = 56, max = 164, σ = 30) to μ = 15 (min = 3, max = 29, σ = 11) [F(1, 19) = 7.17, p = 0.01] seconds. Both ground truth variants resulted in indistinguishable performance metrics. The experts found and corrected 78% of the artificially falsely classified cells, 78% of the deleted cells and 67% of the non maximum suppression artefacts in the expert-algorithm mode. In comparison, without pre-annotation of the cells the experts correctly classified 86% of the changed classes, 84% of the deleted cells and 75% of the non maximum suppression artefacts.

EIPH: The annotation mode (μ = 0.47, min = 0.39, max = 0.60, σ = 0.06) shows significantly smaller mAP with the ground truth [F(1, 19)=42.04, p < 0.01], compared to the computer-aided mode (μ = 0.59, min = 0.53, max = 0.67,

$\sigma = 0.05$) in terms of mAP and comparable results in terms of annotation time $\mu = 51$ (min = 29, max = 99, $\sigma = 20$) to $\mu = 45$ (min = 27, max = 76, $\sigma = 20$) s. Both ground truth variants resulted in indistinguishable performance metrics. The experts found and corrected 57% of the artificially falsely classified cells, 10% of the deleted cells and 54% of the non-maximum suppression artefacts. For comparison: without pre-annotation the experts correctly classified 59% of the changed classes, 28% of the deleted cells and 60% of the non maximum suppression artefacts.

Mitotic Figures: The annotation mode ($\mu = 0.43$, min = 0.31, max = 0.54, $\sigma = 0.07$) shows significantly smaller agreement with the ground truth, compared to the computer-aided mode $\mu = 0.70$ (min = 0.60, max = 0.77, $\sigma = 0.05$) in terms of mAP [$F(1, 19) = 94.71$, $p < 0.01$]. For the voted ground truth, the mAP in the computer-aided mode $\mu = 0.63$ (min = 0.54, max = 0.69, $\sigma = 0.03$) also increased in comparison to the annotation mode ($\mu = 0.48$, min = 0.35, max = 0.56, $\sigma = 0.06$). Annotation time decreased significantly from $\mu = 338$ (min = 137, max = 590, $\sigma = 144$) to $\mu = 111$ (min = 222, max = 248, $\sigma = 78$) seconds per image for the computer-aided mode [$F(1, 19) = 17.73$, $p < 0.01$]. The experts corrected 18% of the artificially removed grade zero cells, 26% of the grade one cells and 41% of the grade two cells. They did not remove 71% of the grade zero fake mitosis, 77% of grade one and 84% of grade two. In comparison, without pre-annotation, the experts correctly annotated 26% grade zero, 43% grade one and 66% grade two mitotic figures which were artificially deleted in the computer-aided mode. Furthermore, the experts annotated 14% of the grade zero fake mitotic figures, 17% of grade one and 26% of grade two. According to Kiupel et al. [8] seven mitotic figures per ten high power fields is the threshold for grading tumours. In the annotation mode, for six cases experts over-estimated the mitotic figure count, while in 15 cases experts under-estimated the mitotic figure count compared to nine over-estimations and 13 under-estimations in the computer-aided mode compared to the imported ground truth.

The analysis code is available online[2], together with the anonymised participant contributions. The image set is accessible online at https://exact.cs.fau.de/ with the user name: "StudyJan2020" and the password "Alba2020". The study remains online for further contributions upon reasonable written request.

4 Discussion and Outlook

Our study shows that computer-assisted annotations can lead to a significant improvement regarding the annotation accuracy while also reducing the annotation time, which was a consistent finding in all three experiments. In detail, however, there were differences. The mitotic figure data set benefited from the computer-aided mode the most with an increase in mAP of 27%, which is likely related to mitotic figures being rare and ambiguous to classify. Even more than

[2] https://github.com/DeepPathology/Results-Exact-Study.

the two other tasks, accurate annotation of mitotic figures in histological images probably heavily relies on experience and routine. It cannot be excluded that the lack of lengthy experience of some experts for this specific task has resulted in higher uncertainty for the annotation mode and higher benefits for the expert-algorithm collaboration mode. Regardless, even two board-certified anatomical pathologists heavily benefited from the agorithm collaboration, underlining the general high level of difficulty in identification and classification of mitotic figures. Microscopic asthma diagnosis, as a rather simple and unambiguous task, showed least benefits in terms of accuracy, but the processing speed was increased by a factor of five, which can be attributed to the fact that pathologists were able to check the results at a glance. For EIPH, the picture is slightly different. Between the annotation mode and the computer-aided annotated mode no significant time reduction was measurable. We attribute this to the fact that the EIPH grading is more of an estimation that cannot be easily surveyed. Nevertheless, also for EIPH grading the computer-aided annotated mode showed significantly increased concordance with the ground truth. Overall, we observed that the variability of results in the computer-aided mode was reduced, resulting in higher comparability and reproducibility of results, which is highly desirable in medicine. Furthermore, we were able to show that it was more likely that the experts overlooked artificially inserted errors (see Fig. 1). The generalisability of the computer-aided annotation mode was further underlined by the observation that also for the independently retrospectively calculated voted ground truth the annotation accuracy increased for mitotic figure detection. Nonetheless, the overlooked artificially inserted errors are a particularly critical observation, as it shows that besides all the advantages of speed and accuracy, the quality of the computer-aided annotation is crucial for the result and should be of the highest possible standard.

Acknowledgements. We thank all contributors for making this work possible. CAB gratefully acknowledges financial support received from the Dres. Jutta & Georg Bruns-Stiftung für innovative Veterinärmedizin.

References

1. Aubreville, M., et al.: Field of interest prediction for computer-aided mitotic count. arXiv.org 0(1902.05414) (2019)
2. Bertram, C.A., Aubreville, M., Marzahl, C., Maier, A., Klopfleisch, R.: A large-scale dataset for mitotic figure assessment on whole slide images of canine cutaneous mast cell tumor. Sci. Data **6**(274), 1–9 (2019)
3. Bertram, C.A., Dietert, K., Pieper, L., Erickson, N.A., Barton, A.K., Klopfleisch, R.: Effects of on-slide fixation on the cell quality of cytocentrifuged equine bronchioalveolar lavage fluid. VET CLIN PATH **47**(3), 513–519 (2018). https://doi.org/10.1111/vcp.12623
4. Bertram, C.A., et al.: Computerized calculation of mitotic count distribution in canine cutaneous mast cell tumor sections: mitotic count is area dependent. Vet. Pathol. **57**(2), 214–226 (2020). https://doi.org/10.1177/0300985819890686. pMID: 31808382

5. Bertram, C.A., et al.: Computerized calculation of mitotic count distribution in canine cutaneous mast cell tumor sections: mitotic count is area dependent. Vet. Pathol. **57**(2), 214–226 (2020)
6. Ganz, M., Kondermann, D., Andrulis, J., Knudsen, G.M., Maier-Hein, L.: Crowd-sourcing for error detection in cortical surface delineations. Int. J. Comput. Assist. Radiol. Surg. **12**(1), 161–166 (2016). https://doi.org/10.1007/s11548-016-1445-9
7. Golde, D.W., et al.: Occult pulmonary haemorrhage in leukaemia. Br. Med. J. **2**(5964), 166–168 (1975)
8. Kiupel, M., et al.: Proposal of a 2-tier histologic grading system for canine cutaneous mast cell tumors to more accurately predict biological behavior. Vet. Pathol. **48**(1), 147–155 (2011). https://doi.org/10.1177/0300985810386469
9. Maier-Hein, L., et al.: Crowd-algorithm collaboration for large-scale endoscopic image annotation with confidence. In: Ourselin, S., Joskowicz, L., Sabuncu, M.R., Unal, G., Wells, W. (eds.) MICCAI 2016. LNCS, vol. 9901, pp. 616–623. Springer, Cham (2016). https://doi.org/10.1007/978-3-319-46723-8_71
10. Malon, C., et al.: Mitotic figure recognition: agreement among pathologists and computerized detector. Anal. Cell. Pathol. **35**(2), 97–100 (2012)
11. Marzahl, C., et al.: Deep learning-based quantification of pulmonary hemosiderophages in cytology slides (2019)
12. Marzahl, C., et al.: EXACT: a collaboration toolset for algorithm-aided annotation of almost everything (2020)
13. Marzahl, C., et al.: Is crowd-algorithm collaboration an advanced alternative to crowd-sourcing on cytology slides? In: Tolxdorff, T., Deserno, T., Handels, H., Maier, A., Maier-Hein, K., Palm, C. (eds.) Bildverarbeitung für die Medizin 2020. Informatik aktuell, pp. 26–31. Springer, Wiesbaden (2020). https://doi.org/10.1007/978-3-658-29267-6_5
14. Meyer, J.S., et al.: Breast carcinoma malignancy grading by Bloom-Richardson system vs proliferation index: reproducibility of grade and advantages of proliferation index. Mod. Pathol. **18**(8), 1067–1078 (2005). https://doi.org/10.1038/modpathol.3800388
15. Ørting, S., et al.: A survey of crowdsourcing in medical image analysis. arXiv preprint arXiv:1902.09159 (2019)
16. Pantanowitz, L., et al.: Validating whole slide imaging for diagnostic purposes in pathology: guideline from the college of american pathologists pathology and laboratory quality center. Arch. Pathol. Lab. Med. **137**(12), 1710–1722 (2013)
17. Veta, M., et al.: Assessment of algorithms for mitosis detection in breast cancer histopathology images. Med. Imag Anal. **20**(1), 237–248 (2015)

Deep Reinforcement Active Learning for Medical Image Classification

Jingwen Wang[1], Yuguang Yan[2], Yubing Zhang[1](✉), Guiping Cao[1], Ming Yang[1], and Michael K. Ng[2]

[1] CVTE Research, Guangzhou, China
{wangjingwen7003,zhangyubing,caoguiping,yangming}@cvte.com
[2] The University of Hong Kong, Hong Kong, China
{ygyan,mng}@maths.hku.hk

Abstract. In this paper, we propose a deep reinforcement learning algorithm for active learning on medical image data. Although deep learning has achieved great success on medical image processing, it relies on a large number of labeled data for training, which is expensive and time-consuming. Active learning, which follows a strategy to select and annotate informative samples, is an effective approach to alleviate this issue. However, most existing methods of active learning adopt a hand-design strategy, which cannot handle the dynamic procedure of classifier training. To address this issue, we model the procedure of active learning as a Markov decision process, and propose a deep reinforcement learning algorithm to learn a dynamic policy for active learning. To achieve this, we employ the actor-critic approach, and apply the deep deterministic policy gradient algorithm to train the model. We conduct experiments on two kinds of medical image data sets, and the results demonstrate that our method is able to learn better strategy compared with the existing hand-design ones.

Keywords: Active learning · Deep reinforcement learning · Medical image classification.

1 Introduction

In the last decades, benefiting from the powerful ability of representation learning, deep learning has achieved great success on object recognition, natural image understanding, and medical image analysis [7,13]. Nevertheless, existing methods heavily rely on a large of high-quality labeled data, which is expensive and

J. Wang and Y. Yan—are the co-first authors. Y. Zhang—is the corresponding author.
This work was supported by HKRGC GRF 12306616, 12200317, 12300218, 12300519, and 17201020.

Electronic supplementary material The online version of this chapter (https://doi.org/10.1007/978-3-030-59710-8_4) contains supplementary material, which is available to authorized users.

A. L. Martel et al. (Eds.): MICCAI 2020, LNCS 12261, pp. 33–42, 2020.
https://doi.org/10.1007/978-3-030-59710-8_4

time-consuming. This issue becomes even severer in medical image analysis, since it requires experienced experts to annotate medical images.

Active learning is an effective approach to address the problem of labeled data scarcity [11,16]. In the iterative procedure of active learning, some informative unlabeled samples are selected and annotated. After that, the classifier is trained with the help of new labeled data and is expected to achieve better performance than before. Most existing methods of active learning prefer to select and annotate samples with high uncertainty, since these samples can provide more information and are usually difficult to be classified correctly. Motivated by this, some methods based on uncertainty are proposed, such as least confidence [11], margin sampling [10], entropy [12], *etc.* However, these methods rely on fixed hand-design strategies, which are not able to handle the dynamic procedure of model training. As the model changes, the predefined strategy may be inappropriate to select the most informative samples.

In this paper, to address the above issue, we propose a new active learning algorithm named **D**eep **R**einforcement **L**earning for **A**ctive learning (DRLA) for medical image classification. Rather than adopting a hand-design data selection strategy, we seek to learn a dynamic policy to select samples for annotation. To this end, we model the procedure of active learning as a Markov decision process, and apply deep reinforcement learning [8,14] to learn a data selection strategy, which takes the state of the classifier into consideration, thus can obtain a better strategy compared with hand-design ones. Specifically, we employ the actor-critic approach [3] to generate and judge decisions of data selection, and apply the deep deterministic policy gradient algorithm (DDPG) [6] to train the model. We conduct experiments on two kinds of medical image data sets to evaluate the performance of our proposed method.

1.1 Related Works

Active learning aims to select and annotate informative samples for improving the performance [4,11,16]. The most common strategy is based on data uncertainty. In [11], the least confidence method is proposed to select the samples whose probabilities of the most probable classes are still low. The margin sampling method [10] calculates the margin between the first and second most probable classes for each sample, and selects the samples with small margin values. The entropy method measures the uncertainty of each sample based on the entropy of the predicted class label probabilities [12]. In [17], a fusion strategy is proposed to combine the above methods. In [19], a deep active learning method based on uncertainty and similarity information is proposed for biomedical image segmentation.

Reinforcement learning is a classic algorithm in artificial intelligence [14]. Thanks to the great progress of deep neural network, deep reinforcement learning has shown powerful ability in learning policy for decision problems. Deep Q-learning extends traditional Q-learning by leveraging deep neural network to learn a Q-value function [8]. After that, deep deterministic policy gradient algorithm (DDPG) is proposed to adapt deep Q-learning to handle the continuous action space [6]. In [2], the Twin Delayed Deep Deterministic policy gradient algorithm (TD3) further extends DDPG by maintaining a pair of critics along with a single actor, which obtains better performance and efficiency.

2 Methodology

2.1 Overview

Figure 1 illustrates the main idea of our proposed method DRLA. A classifier network for disease diagnosis is trained on labeled training data, including samples with label in advance (*i.e.*, (X_l, Y_l)), and samples which are selected and annotated in the procedure of active learning. During the learning process, an actor network is devoted to selecting the most informative samples from unlabeled training data (*i.e.*, X_u) according to the current state and a learned policy. After that, an annotator is responsible to annotate the selected samples. As a result, we have more and more labeled training data to update the classifier gradually. Last but not the least, a critic network is trained to evaluate if the selection of the actor network is effective to improve the performance of the classifier. By employing a deep reinforcement learning approach to train the actor network and the critic network, we can select and annotate the most informative samples that are beneficial for training an effective classifier, and further improve the classification performance.

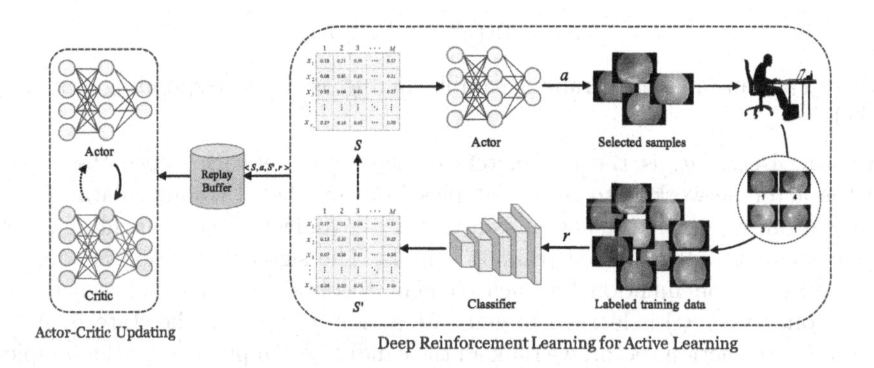

Fig. 1. The illustration of our proposed method DRLA.

2.2 Classifier Training

Let the parameters of the classifier network be θ_d. At the beginning of the learning procedure, we can use labeled training data to pretrain the classifier. After that, we apply deep reinforcement active learning to select and annotate samples, and further train the classifier to enhance the performance. Given the labeled training samples $X_l = \{x_i\}_{i=1}^{n_l}$ with corresponding labels $Y_l = \{y_i\}_{i=1}^{n_l}$, where n_l is the number of labeled data. The classifier network is trained by minimizing the cross-entropy loss, which is defined as

$$\mathcal{L}_{ce} = -\sum_{i=1}^{n_l}\sum_{j=1}^{M} \mathbb{I}(y_i = j) \log \Pr(y_i = j \mid x_i; \theta_d), \tag{1}$$

where M is the number of classes, $\mathbb{I}(\cdot)$ is the indicator function, and $\mathbb{I}(y_i = j)$ judge if the label of the i-th sample is j or not. $\Pr(y_i = j \mid x_i; \theta_d)$ is the softmax output of the classifier given x_i for the j-th class, which indicates the probability of x_i belonging to label j obtained from the classifier with parameters θ_d.

2.3 Deep Reinforcement Active Learning

In this part, we propose a new active learning method named DRLA to learn a policy, which guides the actor network to select samples for annotation. In the following, we discuss our proposed method in detail.

State. In order to select the samples which are the beneficial for improving the classification performance, the prediction of the current classifier should be taken into consideration. Motivated by this, we design the state $S \in (0,1]^{n_u \times M}$ as a matrix including all the predicted values for unlabeled training samples X_u, where n_u is the number of unlabeled training samples, and M is the number of the classes. Mathematically, the (i,j)-th element of S is defined as

$$S_{ij} = \Pr(y_i^u = j \mid x_i^u; \theta_d), \tag{2}$$

where x_i^u is an unlabeled training sample, and y_i^u is the corresponding unknown label.

Action. Define θ_a is the parameters of the actor network. Since the target of the actor network is to select samples from unlabeled training data set for annotation, we define the action as a vector $a \in (0,1)^{n_u}$, each element of which corresponds to an unlabeled training sample. The sigmoid function is adopted as the activation function for each element to obtain a value between 0 and 1. A policy $\pi(S;\theta_a)$ is learned to generate action a based on the state S. After obtaining the action vector, we rank all the candidate samples except the samples which are selected already in descending order, and select the first n_s samples with the highest values for annotation. The selected samples and the labels provided by the annotator are denoted as $\{(x_i^s, y_i^s)\}_{i=1}^{n_s}$, and the augmented labeled training data are denoted as $(X_l, Y_l) := (X_l, Y_l) \cup \{(x_i^s, y_i^s)\}_{i=1}^{n_s}$.

State Transition. After the selected samples are annotated and added into the labeled training data, we can update the classifier with the augmented training data set (X_l, Y_l) by minimizing the cross-entropy loss in Eq. (1). After that, we can use the new classifier to obtain the new state matrix S' based on Eq. (2).

Reward. In order to enhance the performance of the classifier, we propose to make the actor concentrate more on those samples which are highly possible to be misclassified by the classifier. To achieve this, we design a novel reward function by considering the predicted values and true labels obtained from the annotator. In specific, for the selected sample x_i^s, define k_i as the true label obtained from the annotator, and $\hat{k}_i$ as the predicted label obtained from the classifier, *i.e.*, $\hat{k}_i = \max_j \Pr(y_i^s = j \mid x_i^s; \theta_d)$. The reward is defined as

$$r(S, a) = \frac{1}{n_s} \sum_{i=1}^{n_s} \Pr(y_i^s = \hat{k}_i \mid x_i^s; \theta_d) - \Pr(y_i^s = k_i \mid x_i^s; \theta_d). \tag{3}$$

If the sample x_i^s is classified correctly, then $k_i = \hat{k}_i$, and $\Pr(y_i^s = \hat{k}_i \mid x_i^s; \theta_d) - \Pr(y_i^s = k_i \mid x_i^s; \theta_d) = 0$. On the other hand, a high reward indicates that the selected samples are classified incorrectly. This implies that these samples with the wrong predictions should be paid more attention by the classifier. Therefore, these samples are encouraged to be selected by the actor network.

At state S, reinforcement learning aims to maximize the expected reward in the future, which is defined as a Q-value function. Similar to Q-learning in traditional reinforcement learning, the Q-value function is used to evaluate the state-action pair (S, a), and is represented by the Bellman equation as $Q(S, a; \theta_c) = \mathbb{E}\big[\gamma Q(S', \pi(S'; \theta_a); \theta_c) + r(S, a)\big]$, where γ is the delay parameter. Here we adopt a critic network with parameters θ_c to approximate the Q-value function. Inspired by deep Q-Learning [8], we aim to learn a greedy policy for the actor by solving the following problem

$$\max_{\theta_a} Q(S, \pi(S; \theta_a); \theta_c). \tag{4}$$

We define $\tilde{Q}(S, a; \theta_c) = \gamma Q(S', \pi(S'; \theta_a); \theta_c) + r(S, a)$, and train the critic network by solving the following problem

$$\min_{\theta_c} \big(\tilde{Q}(S, a; \theta_c) - Q(S, a; \theta_c)\big)^2. \tag{5}$$

Training with Target Networks. In order to stabilize the training of the actor and critic networks, we follow [6] to employ a separate target network to calculate $\tilde{Q}(S, a; \theta_c)$. According to Problem (5), $\tilde{Q}(S, a; \theta_c)$ depends on the new state S', the actor to output action $\pi(S'; \theta_a)$, and the critic to evaluate $(S', \pi(S', \theta_a))$. We adopt a separate target actor network parameterized by $\theta_{a'}$ and a separate target critic network parameterized by $\theta_{c'}$ to calculate $\tilde{Q}(S, a; \theta_c)$. As a result, we rewrite Eq. (5) as

$$\min_{\theta_c} \big(\gamma Q'(S', \pi'(S'; \theta_{a'}); \theta_{c'}) + r(S, a) - Q(S, a; \theta_c)\big)^2, \tag{6}$$

where $\pi'(\cdot; \theta_{a'})$ is the target policy estimated by the target actor, and $Q'(\cdot, \cdot; \theta_{c'})$ is the function of the target critic. This problem can be optimized by the deep deterministic policy gradient algorithm (DDPG) [6].

At the last of each epoch, the target actor and critic are updated by

$$\theta_{a'} := \lambda \theta_a + (1 - \lambda)\theta_{a'}, \quad \theta_{c'} := \lambda \theta_c + (1 - \lambda)\theta_{c'}, \tag{7}$$

where $\lambda \in (0, 1)$ is a trade-off parameter.

To train the actor and critic in the mini-batch paradigm, we use a replay buffer to store samples $\{(S, a, S', r)\}$. As a result, we can uniformly select training samples from the replay buffer to update the actor and critic networks [6].

Algorithm 1 summarizes our proposed method.

Algorithm 1. Deep Reinforcement Learning for Active learning (DRLA)

Input: Labeled training data (X_l, Y_l), unlabeled training data X_u.
Initialize: Pretrain the classifier to get $f(\cdot; \theta_d)$ based on labeled training data (X_l, Y_l).
1: **for** each epoch **do**
2: Compute the state S according to Eq. (2).
3: Select n_s unlabeled training sample $\{x_i^s\}_{i=1}^{n_s}$ based on the actor $a = \pi(S; \theta_a)$.
4: Annotate $\{x_i^s\}_{i=1}^{n_s}$ to get $\{(x_i^s, y_i^s)\}_{i=1}^{n_s}$.
5: Update the classifier parameters θ_d using $(X_l, Y_l) := (X_l, Y_l) \cup \{(x_i^s, y_i^s)\}_{i=1}^{n_s}$.
6: Calculate the state S' based on Eq. (2), and the reward r based on Eq. (3).
7: Save the sample (S, a, S', r) into the replay buffer.
8: **for** each training sample for actor and critic **do**
9: Update the critic by optimizing Problem (6).
10: Update the actor by optimizing Problem (4).
11: Update the target actor and critic according to Eq. (7).
12: **end for**
13: **end for**

3 Experiments

3.1 Data Sets and Evaluation Metrics

- **chestCT**[1] is a Computed Tomography (CT) data set for lung disease detection with four kinds of diseases, including pulmonary nodule, pulmonary cord, arteriosclerosis and calcification of lymph node. It contains 1,470 CT scans. We randomly pick up scans from them to construct training and testing data sets. After that, we take the regions with the ground-truth label to obtain samples, and then randomly pick up 3,500 samples as training samples and 3,500 samples as testing samples.
- The **Retinopathy** data set[2] contains 35,126 fundus images collected by different devices from different environments. Each fundus image is rated from 0 to 4 according to the presence and degree of diabetic retinopathy (DR), *i.e.*, no DR, mild, moderate, severe, and proliferative DR. We randomly pick up 2,230 images as training data and 2,230 images as testing data.

[1] https://tianchi.aliyun.com/competition/entrance/231724/introduction.
[2] https://www.kaggle.com/c/diabetic-retinopathy-detection/.

We adopt two performance metrics, *i.e.*, Macro F1 and Micro F1 scores, for evaluation. Taking the j-th class as the positive label while the other classes as the negative label, we can define TP_j, TN_j, FP_j and FN_j as the numbers of true positive, true negative, false positive and false negative, respectively. The two evaluation metrics are defined as Macro F1 $= \frac{1}{M}\sum_{j=1}^{M}\frac{2\cdot TP_j}{2\cdot TP_j+FN_j+FP_j}$, Micro F1 $= \frac{2\cdot\sum_{j=1}^{M}TP_j}{2\cdot\sum_{j=1}^{M}TP_j+\sum_{j=1}^{M}FN_j+\sum_{j=1}^{M}FP_j}$.

3.2 Experimental Settings

In the experiments, we compare our method with several active learning methods, including random selection (RANDOM), least confidence (LC) [11], margin sampling (MS) [10], entropy (EN) [12], and FUSION [17]. The FUSION method combines the three above mentioned criteria, *i.e.*, LC, MS and EN. In specific, FUSION selects top $\frac{K}{2}$ samples according to LC, MS and EN, respectively. After that, FUSION removes the replicate ones from the $\frac{3K}{2}$ samples, and randomly selects K samples from them to annotate. We also conduct a method named "ALL", which takes all the training data as labeled ones to train the model.

For the chestCT data set, we randomly select 5% samples of each class from the training set to initialize the network, and the rest are for the incremental learning process. In each epoch, we randomly select 5 samples from the unlabeled training set to annotate, and then add them into the labeled training data to update the classifier.

For the Retinopathy data set, we randomly select 10% images of each class from the training set to initialize the network, and the rest are for the incremental learning process. In each epoch, we randomly select 1 sample from the unlabeled training set to annotate, and then add them into the labeled training data to update the classifier.

All the methods are implemented on the PyTorch platform [9]. We use ResNet-50 [5] with 3D convolutional layers [15] as the architecture for chestCT, and ResNet-50 pretrained on ImageNet data set [1] to initialize the classifier for the Retinopathy data set. We adopt the SGD optimizer with the learning rate 0.0001 to train the classifier network. The actor network and the critic network have the same architecture, which consists of three fully connected layers. Both of them are trained using the Adam optimizer with the learning rate 0.001. We set the delay factor as $\gamma = 0.99$, the trade-off parameter as $\lambda = 0.005$, and the batch size as 16. For the DDPG algorithm, we adopt the noise exploration mechanism used in [18].

3.3 Results and Discussion

Tables 1 and 2 present the results on chestCT. Our proposed method DRLA outperforms the compared active learning methods. This demonstrates that compared with the hand-design strategies used in the compared methods, DRLA is able to learn a more effective strategy to select informative samples for improving the performance. Besides, to achieve 0.70 F1 scores, DRLA only needs around 40% labeled training data, while the other active learning method require around 68% labeled training data. This indicates that DRLA can reduce the need of labeled data.

Table 1. Macro F1 results on the chestCT data set.

Percentage of training samples	26%	40%	54%	68%	82%	100%
ALL	–	–	–	–	–	0.7494
RANDOM	0.6530	0.6994	0.7014	0.7101	0.7125	0.7176
LC	0.6813	0.7193	0.7235	0.7383	0.7396	0.7389
MS	0.6431	0.6902	0.7007	0.7173	0.7182	0.7207
EN	0.6730	0.7194	0.7287	0.7353	0.7330	0.7348
FUSION	0.6733	0.6845	0.7112	0.7198	0.7190	0.7192
DRLA	**0.6869**	**0.7362**	**0.7419**	**0.7451**	**0.7481**	**0.7525**

Table 2. Micro F1 results on the chestCT data set.

Percentage of training samples	26%	40%	54%	68%	82%	100%
ALL	–	–	–	–	–	0.7462
RANDOM	0.6574	0.7023	0.7029	0.7109	0.7131	0.7180
LC	0.6829	0.7197	0.7231	0.7363	0.7369	0.7363
MS	0.6449	0.6926	0.7011	0.7143	0.7169	0.7180
EN	0.6857	0.7200	0.7280	0.7329	0.7311	0.7343
FUSION	0.6914	0.6891	0.7137	0.7197	0.7191	0.7203
DRLA	**0.6940**	**0.7343**	**0.7409**	**0.7451**	**0.7480**	**0.7537**

Figure 2 shows the results of learning procedures on the two data sets, respectively. We observe that as the number of labeled training data increases, all the active learning methods can obtain better performance. Besides, after receiving 20% labeled training data, our proposed method DRLA consistently achieves the best or highly comparable performance compared with the other active learning methods. This further verifies that DRLA is able to select informative samples to improve the classification performance.

More experimental results could be found in Supplementary Material.

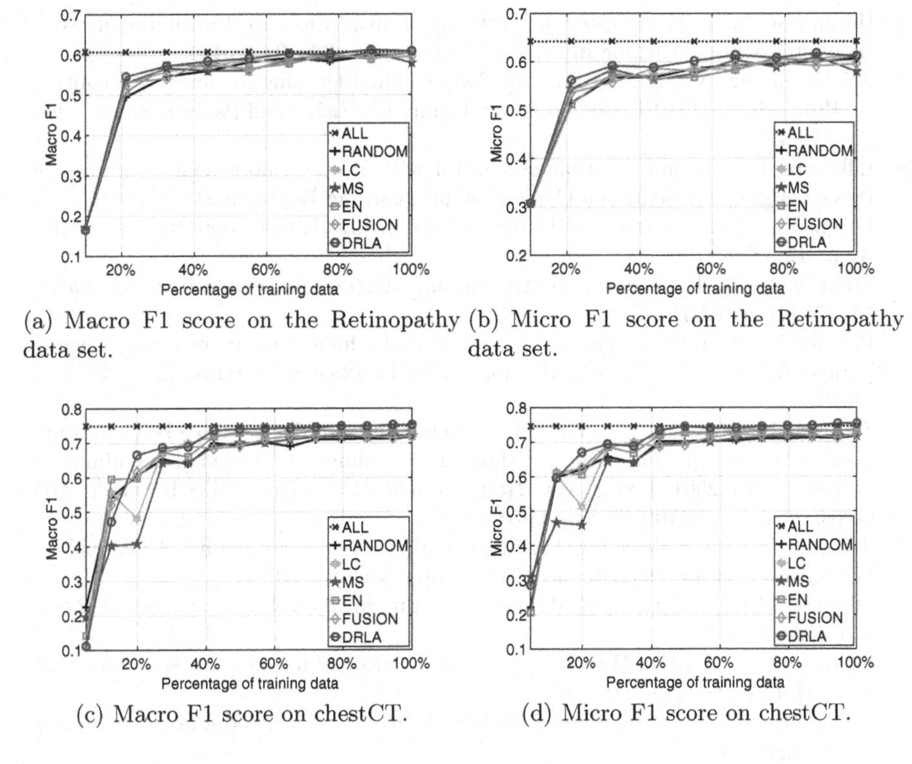

(a) Macro F1 score on the Retinopathy data set.

(b) Micro F1 score on the Retinopathy data set.

(c) Macro F1 score on chestCT.

(d) Micro F1 score on chestCT.

Fig. 2. Macro F1 and micro F1 results on the data sets.

4 Conclusion

In this paper, we propose a deep reinforcement active learning algorithm for medical image classification. To learn a dynamic strategy for active learning, we apply deep reinforcement learning to learn a policy to select samples for annotation, and employ deep deterministic policy gradient algorithm under the actor-critic paradigm to train the model. We conduct experiments on two medical image data sets to demonstrate the effectiveness of the proposed method.

References

1. Deng, J., Dong, W., Socher, R., Li, L.J., Li, K., Fei-Fei, L.: ImageNet: a large-scale hierarchical image database. In: Proceedings of IEEE Conference on Computer Vision and Pattern Recognition, pp. 248–255 (2009)
2. Fujimoto, S., Hoof, H., Meger, D.: Addressing function approximation error in actor-critic methods. In: International Conference on Machine Learning, pp. 1587–1596 (2018)
3. Haarnoja, T., Zhou, A., Abbeel, P., Levine, S.: Soft actor-critic: off-policy maximum entropy deep reinforcement learning with a stochastic actor. In: Proceedings of International Conference on Machine Learning, pp. 1861–1870 (2018)

4. Hatamizadeh, A., et al.: Deep active lesion segmentation. In: International Workshop on Machine Learning in Medical Imaging, pp. 98–105 (2019)
5. He, K., Zhang, X., Ren, S., Sun, J.: Deep residual learning for image recognition. In: Proceedings of IEEE Conference on Computer Vision and Pattern Recognition, pp. 770–778 (2016)
6. Lillicrap, T.P., et al.: Continuous control with deep reinforcement learning. In: Proceedings of International Conference on Learning Representations (2015)
7. Litjens, G., et al.: A survey on deep learning in medical image analysis. Med. Image Anal. **42**, 60–88 (2017)
8. Mnih, V., et al.: Human-level control through deep reinforcement learning. Nature **518**(7540), 529–533 (2015)
9. Paszke, A., et al.: PyTorch: an imperative style, high-performance deep learning library. In: Advances in Neural Information Processing Systems, pp. 8024–8035 (2019)
10. Scheffer, T., Decomain, C., Wrobel, S.: Active hidden Markov models for information extraction. In: Hoffmann, F., Hand, D.J., Adams, N., Fisher, D., Guimaraes, G. (eds.) IDA 2001. LNCS, vol. 2189, pp. 309–318. Springer, Heidelberg (2001). https://doi.org/10.1007/3-540-44816-0_31
11. Settles, B.: Active learning literature survey. Technical report, University of Wisconsin-Madison Department of Computer Sciences (2009)
12. Shannon, C.E.: A mathematical theory of communication. Bell Syst. Tech. J. **27**(3), 379–423 (1948)
13. Shen, D., Wu, G., Suk, H.I.: Deep learning in medical image analysis. Annu. Rev. Biomed. Eng. **19**, 221–248 (2017)
14. Sutton, R.S., Barto, A.G.: Reinforcement Learning: An Introduction. MIT Press, Cambridge (2018)
15. Tran, D., Bourdev, L., Fergus, R., Torresani, L., Paluri, M.: Learning spatiotemporal features with 3D convolutional networks. In: Proceedings of IEEE International Conference on Computer Vision, pp. 4489–4497 (2015)
16. Tuia, D., Volpi, M., Copa, L., Kanevski, M., Munoz-Mari, J.: A survey of active learning algorithms for supervised remote sensing image classification. IEEE J. Sel. Top. Signal Process. **5**(3), 606–617 (2011)
17. Wang, K., Zhang, D., Li, Y., Zhang, R., Lin, L.: Cost-effective active learning for deep image classification. IEEE Trans. Circ. Syst. Video Technol. **27**(12), 2591–2600 (2016)
18. Wawrzynski, P.: Control policy with autocorrelated noise in reinforcement learning for robotics. Int. J. Mach. Learn. Comput. **5**(2), 91 (2015)
19. Yang, L., Zhang, Y., Chen, J., Zhang, S., Chen, D.Z.: Suggestive annotation: a deep active learning framework for biomedical image segmentation. In: Descoteaux, M., Maier-Hein, L., Franz, A., Jannin, P., Collins, D.L., Duchesne, S. (eds.) MICCAI 2017. LNCS, vol. 10435, pp. 399–407. Springer, Cham (2017). https://doi.org/10.1007/978-3-319-66179-7_46

An Effective Data Refinement Approach for Upper Gastrointestinal Anatomy Recognition

Li Quan(✉), Yan Li, Xiaoyi Chen, and Ni Zhang

NEC Laboratories, China, Beijing, China
{quan_li,li-yan,chen_xiaoyi,zhangni_nlc}@nec.cn

Abstract. Accurate recognition of anatomy sites is important for evaluating the quality of esophagogastroduodenoscopy (EGD) examinations. However, because some anatomy sites have similar appearances and anatomical landmarks are lacking, gastric-anatomy image annotations are less than accurate. The annotations by doctors with various experience levels vary widely. Deep learning–based systems trained on these noisy annotations have poor recognition performance. In this work, we propose a novel data refinement approach to alleviate the problem of noisy annotations and improve the upper gastrointestinal anatomy recognition performance. In essence, we introduce a new uncertainty inference module for deep convolutional neural networks (CNNs) and leverage Bayesian uncertainty estimates to select possibly noisy data. In addition, we employ an ensemble of semi-supervised learning to rectify noisy labels and produce refined training data. We validate the proposed approach via controlled experiments on CIFAR-10, in which the noise rate is adjusted and noisy data are made known. It shows much improvement on classification accuracy using the refined dataset, and outperforms state-of-the-art robust training methods, e.g., MentorNet and Co-teaching. An evaluation of the upper gastrointestinal anatomy recognition task proves that our proposed method effectively improves the recognition accuracy for real, noisy clinical data. The proposed data refinement approach reduces the human effort needed to filter out and manually rectify noisy annotations. It can also be applied to wider scenarios where accurate expert labeling is expensive.

Keywords: Bayesian uncertainty estimates · Semi-supervised learning · Noisy label · Medical annotation · Data refinement

1 Introduction

A good-quality esophagogastroduodenoscopy (EGD) examination is essential for early gastric-cancer diagnoses. Because of their operating habits, many inexperienced doctors tend to miss certain mucosal areas or fail to observe physiological anatomical structures during examinations, resulting in missed diagnoses [1]. A poor-quality examination incompletely covers key anatomy sites that are prone to having lesions. Development of an artificial intelligence-assisted system to supervise and evaluate the examination coverage is a promising approach to ensure the examination quality.

A. L. Martel et al. (Eds.): MICCAI 2020, LNCS 12261, pp. 43–52, 2020.
https://doi.org/10.1007/978-3-030-59710-8_5

In this study, we train a convolutional neural network (CNN) to recognize 30 key anatomy sites in the upper digestive tract, from the esophagus to the stomach and to the duodenum. If all of the sites can be detected during a doctor's examination process, it ensures a high-quality EGD examination. However, owing to the absence of clear anatomical landmarks, gastric anatomical sites—especially the anterior/posterior walls and the greater/lesser curvature in the lower, middle, and upper body of the stomach—can easily be incorrectly annotated. The annotations of doctors with various experience levels vary widely.

Training with noisy annotations can cause learning biases in deep models and make them overfit to the noise, leading to poor generalization performance on clean data [2]. Training a robust deep neural network with noisy labels has attracted considerable research attention [3–6]. Prior studies assume the knowledge of a small clean dataset. One study proposes to distill the knowledge from clean labels to train a better model from noisy labels [3]. Another study trained a label-cleaning network, using a small set of clean labels to reduce the noise in large-scale noisy labels [4]. These methods have limited use in practice, especially in medical-diagnosis applications, as a clean label set cannot be easily obtained because of large variations in annotation among doctors.

The latest research follows a learning-to-learn paradigm for the robust learning of noisy labels. MentorNet [5] pre-trains an extra teacher network to filter out noisy data to enable its student network to learn robustly from noisy labels. Co-teaching [6] improves on MentorNet by simultaneously training two deep neural networks, and having them select possibly clean data from every mini-batch to teach each other; this method achieves state-of-the-art performance. Li et al. [7] proposed to simulate actual training by generating noisy synthetic labels such that the model did not overfit to specific noises.

All of these methods reduce the side effects of noisy labels by adjusting their weights during training. Noisy labels are judged based on the measured training loss. These methods of treating noisy labels during the training process do not support the possibility that noisy labels can be rectified, by either manual efforts or algorithmic-based methods, to improve the model performance.

In this work, we propose a more flexible two-stage data-refining process to improve the model's performance on noisy labels. Our hypothesis is that if noisy labels can be selected and then re-labeled with possibly correct labels, the model can achieve better performance using the refined data, compared with training directly on the noisy labels.

Notably, unlike the aforementioned robust training methods, we propose to leverage Bayesian uncertainty estimates, instead of the training loss, to select possibly noisy labels. The uncertainty measures how confidently a model makes its prediction, based on given inputs. The higher the model uncertainty is, the more likely it is to be noisy label. We employ semi-supervised learning to rectify the noisy labels and provide possibly correct labels. This is different from conventional semi-supervised learning methods, where unlabeled data are leveraged to produce additional signals to improve the performance of a model trained on a limited amount of labeled data.

We validate the effectiveness of our proposal via controlled experiments on CIFAR-10, in which the noise rate is adjusted and noisy data are made known, as well as in a real application of upper gastrointestinal anatomy recognition. This shows that the proposed data-refining process significantly improves the classification accuracy and outperforms the state-of-the-art (SOTA) robust training methods, e.g., MentorNet and Co-teaching. Moreover, the two-stage data-refining process is flexible and reduces the human effort needed to filter and manually rectify noisy annotations. It can also be applied to wider scenarios where accurate expert labeling is expensive.

2 Our Methods

As illustrated in Fig. 1, our two-stage data-refining process first selects possibly noisy labels, based on Bayesian uncertainty estimates. Unlike previous works [8, 9] that use MC-Dropout for uncertainty estimates, we employ a more sophisticated variational Bayesian (VB) approximate inference, using CNNs to obtain better uncertainty estimates. After training a Bayesian CNN, we evaluate all of the training samples and obtain their uncertainty estimates. The higher the uncertainty is, the more likely the label is to be noisy.

In the second stage, we remove the labels of possibly noisy data and employ semi-supervised learning to assign them possibly correct labels. Specifically, the noisy-label data are treated as unlabeled training data in a virtual adversarial training (VAT) model. The VAT model assigns new labels to the noisy data, after learning from possibly not-noisy data. An ensemble of multiple VAT models is employed to deal with the unknown noise rate and to guarantee more plausible re-labeling results. This two-stage data-refining process outputs a cleaned training dataset, which should improve the performance. The following subsections explain key methods in each stage in more detail.

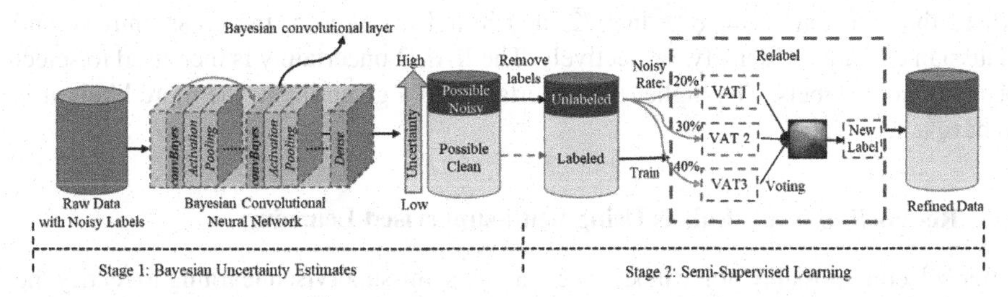

Fig. 1. Our proposal of two-stage data refinement approach. The first stage leverages Bayesian uncertainty estimates to select possibly noisy labels. The second stage employs semi-supervised learning to rectify those noisy labels.

2.1 Selecting Noisy Labels Using Bayesian Uncertainty Estimates

The Bayesian approach provides a theoretical framework for modeling uncertainty. Uncertainty measures a model's confidence in its prediction for any given input. Previous research indicates that uncertainty outputs directly correlate with the difficulty level of the input samples [8]. In this work, we propose a new viewpoint to leverage Bayesian uncertainty estimates to identify possibly noisy labeled data.

We employ a more sophisticated variational Bayesian approximate inference in the CNN layers to obtain better uncertainty estimates, as the commonly used MC-Dropout method severely underestimates the uncertainty [10]. As illustrated in Fig. 1, we introduce layer-wise stochasticity to the CNN layers to form a Bayesian convolutional layer ("*convBayes*"). Following a local re-parameterization mechanism [11], we define convolutional kernel weights $w \sim N\left(\mu, \lambda\mu^2\right)$, and thus obtain the output of each convolutional layer:

$$Y = \frac{1}{K}\sum\nolimits_{i=1}^{K}\left(x \odot \mu + \varepsilon\sqrt{x^2 \odot \left(\lambda\mu^2\right)}\right), \tag{1}$$

where x is the layer input, $\odot$ represents the convolution operation, and ε is random noise that follows a normal distribution. The learnable parameters are μ and λ. The feature response (Y) of each layer is the average of K-time sampling (K = 20 is used in this work) and follows a Gaussian distribution: $\mathrm{Y} \sim \mathrm{N}\left(x \odot \mu, \sqrt{x^2 \odot \left(\lambda\mu^2\right)}\right)$.

With the estimated mean $\gamma = x \odot \mu$ and variance $\sigma = \sqrt{x^2 \odot \left(\lambda\mu^2\right)}$, we calculate the uncertainty for each input sample using the Bayesian Active Learning by Disagreement (BALD) metric [12]. BALD is a mutual information metric and has been previously used in active learning for CNNs [13]. We calculate the BALD uncertainty metric using Eq. (2) [12]:

$$U_{\mathrm{BALD}} = \mathrm{h}\left(\rho\left(\frac{\gamma}{\sqrt{\sigma^2+1}}\right)\right) - \left(\frac{C}{\sqrt{\sigma^2+C^2}}\right)\exp\left(-\frac{\gamma^2}{2\left(\sigma^2+C^2\right)}\right), \tag{2}$$

where the constant $C = \sqrt{(\pi \ln 2)/2}$, and h and ρ represent the cross entropy and Gaussian cumulative density, respectively. The BALD uncertainty is then used to select possibly noisy labels. The higher the uncertainty of a given input, the more likely it is to have a noisy label.

2.2 Re-labeling Noisy Labels Using Semi-supervised Learning

After selecting possibly noisy labels, we employ semi-supervised learning to rectify the noisy labels and produce a refined dataset. The latest survey [14] shows that the state-of-the-art semi-supervised learning methods can achieve superior classification accuracy with a limited amount of labeled data; they even outperform supervised learning on benchmark datasets, e.g., CIFAR-10. Our method removes possibly noisy labels and treats them as unlabeled data in semi-supervised learning. This is different from the conventional way of semi-supervised learning in which unlabeled data are leveraged to provide additional signals to improve the model performance, instead of re-labeling.

Among SOTA semi-supervised learning methods, we choose VAT [15] as a base model to rectify possibly noisy labels, based on learning from possibly clean labeled data. VAT defines an effective regularization method, based on measuring the local smoothness of the conditional label distribution, given inputs. It is consistent with the label-propagation technique [16] that smooths the model around the input data and extrapolates the labels of the unlabeled data.

An ensemble of multiple VAT models is employed to guarantee more plausible re-labeling results. It is usually unknown in practice how much noise is present in a dataset. Hence, we train multiple VAT models that use different ratios of labeled (possibly clean) and unlabeled (possibly noisy) data. The proportion of possibly noisy data is decided by setting a threshold for uncertainty estimates. Each VAT model produces a prediction result for all unlabeled data. The possibly correct label is decided by voting from multiple VAT models, thereby avoiding a biased prediction by a single model.

3 Validation Experiments on CIFAR-10

The effectiveness of the proposed two-stage data refinement method depends on the following: 1) That Bayesian uncertainty estimates can correctly select possibly noisy labels, and 2) that semi-supervised learning can re-label correctly to rectify noisy labels. We separately validated the effects of the two aspects via controlled experiments on CIFAR-10, in which the noise rate is adjusted and noisy data are made known.

The original CIFAR-10 is a clean dataset consisting of 50,000 training images and 10,000 testing images from ten classes. Following the symmetry-flipping method [6], we prepared a noisy CIFAR-10 training set by assigning all training images a 20% possibility to flip their labels to any other class label of equivalent chance. The testing set remains clean. Table 1 summarizes the distributions of clean and noisy labels of our noisy CIFAR-10.

Table 1. Distributions of clean and noisy labels of noisy CIFAR-10

Noisy rate	Training		Testing
	Clean labels	Noisy labels	Clean labels
0 (Clean)	50000	0	10000
0.2	39849	10151	10000

We chose ResNet50 [17] as our baseline model and trained using Noisy CIFAR-10 (denoted as the N_{ori} dataset). The Noisy CIFAR-10 dataset was also used as a benchmark for two SOTA robust training methods, i.e., MentorNet and Co-teaching. We used their original code implementations[1,2] for model training and testing.

To evaluate the effect of the proposed Bayesian uncertainty estimates on selecting noisy labels, we constructed a Bayesian ResNet50 model (introduced in Sect. 2.1) and obtained an uncertainty estimate for every training image. The higher the uncertainty value is, the more likely it is to be a noisy label. We sorted all of the training images based on their uncertainty, selected 20% of the possibly noisy data (denoted as N_{ui}), and left 80% as a clean subset (denoted as C_{ui}). As a comparison, we also produced another noisy subset and clean subset, based on the training loss (denoted as N_{loss} and

[1] https://github.com/google/mentornet.

[2] https://github.com/bhanML/Co-teaching.

C_{loss}, respectively), because SOTA robust training methods [5, 6, 18] commonly use the training loss to select noisy labels. The baseline ResNet50 model was then trained using the C_{ui} and C_{loss} datasets respectively for a performance comparison.

To evaluate the effect of re-labeling by semi-supervised learning, we removed the labels of the possibly noisy data, selected based on uncertainty estimates (i.e., N_{ui}), and treated them as unlabeled data to train together with the clean subsets C_{ui}. A single VAT model was employed to re-label the 20% possibly noisy data and produce a refined dataset (denoted as $N_{\text{ui}}^{R} \cup C_{\text{ui}}$). In a similar manner, another refined dataset was processed, based on the training loss (denoted as $N_{\text{loss}}^{R} \cup C_{\text{loss}}$). The baseline ResNet50 model was then trained on the two refined datasets for a performance comparison.

Table 2 summarizes the test accuracy of all aforementioned trained models. It shows that our proposed two-stage data-refining process significantly improves the accuracy of the noisy CIFAR-10, i.e., from 78.34% to 84.36%, and outperforms the SOTA robust training methods, MentorNet and Co-teaching. The empirical results validate our assumption that, if the noisy labels could be selected and re-labeled with possibly correct labels, a better performance could be achieved on the cleaned data, compared with that obtained by training directly on noisy labels.

In addition, this proves that leveraging Bayesian uncertainty estimates to select possibly noisy labels more effectively improves the accuracy than the conventional method based on the training loss, i.e., 80.87% vs. 82.90%. Moreover, employing semi-supervised learning to rectify noisy labels is effective, as the accuracy improves for both models, based on uncertainty estimates and training loss.

Table 2. Comparison of test accuracy on noisy CIFAR-10

Methods	Dataset	20% noisy rate
Baseline model	N_{ori}	78.34%
MentorNet	N_{ori}	80.76%
Co-teaching	N_{ori}	82.32%
Loss (baseline model)	C_{loss}	80.87%
UI (baseline model) (ours)	C_{ui}	82.90%
Loss + VAT	$N_{\text{loss}}^{R} \cup C_{\text{loss}}$	83.05%
UI + VAT (ours)	$N_{\text{ui}}^{R} \cup C_{\text{ui}}$	**84.36%**

4 Evaluation in Upper Gastrointestinal Anatomy Recognition

A good-quality EGD examination is essential for early gastric-cancer diagnoses. We developed a deep learning–based system to recognize 30 key anatomy sites in the upper digestive tract, from the esophagus to the stomach and to the duodenum, to ensure a good-quality EGD examination, if all of the sites can be detected during a doctor's examination process. Because of the absence of clear anatomical landmarks, some annotations of gastric anatomical sites are very inaccurate. The annotations of doctors with various experience levels vary widely.

4.1 Gastroscopy Dataset and Noisy Annotations

In total, 37,423 anatomy images of 30 key sites in the upper gastrointestinal tract were captured during the EGD examination of 744 patients using the Olympus Evis Lucera Elite CV 290. The images were then annotated by a team of 16 expert doctors, who had two to ten years of experience in EGD examination. It was found that in the lower, middle, and upper body of the stomach, the images of the anterior wall (A), greater curvature (G), lesser curvature (L), and posterior wall (P) looked very similar, and it was difficult for doctors to reach a consensus when labeling these anatomy sites.

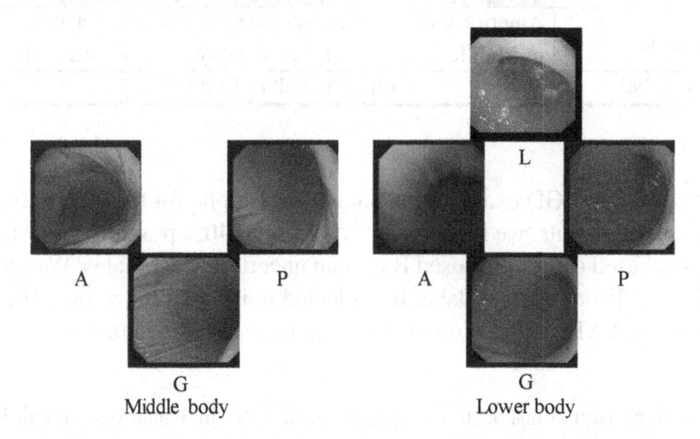

Fig. 2. Examples of gastric anatomy sites with similar appearance

Figure 2 exemplifies some anatomy sites that share a similar appearance and are difficult to distinguish. Table 3 presents some noisy annotations, in which the given label was annotated by a doctor, and the true label was obtained by a consensus from multiple expert doctors. It appears that anatomy sites with similar appearances are easily mistaken for each other. In our experiments, the doctors annotated those images individually, without leveraging the temporal relations of the images. This is in accordance with the examination practice in which doctors do not follow the same route when going through anatomy sites. Leveraging temporal information may not help alleviate the problem of noisy annotations.

4.2 Experiments and Results

We used ResNet50 as the baseline model to classify 30 key anatomy sites of the upper digestive tract. We split 37,423 anatomy images into a training set and a testing set with a 9:1 ratio, based on the patients' IDs. The annotations of the testing set were cleaned up by asking multiple expert doctors to reach a consensus. The training set remained noisy, as each image was annotated by only one doctor. Because it is hard to quantify the

Table 3. Examples of noisy annotation on upper gastrointestinal anatomy images

Anatomy Images	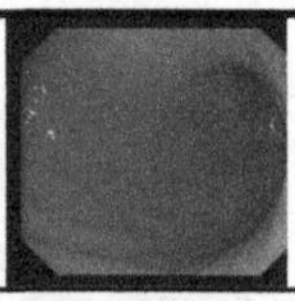	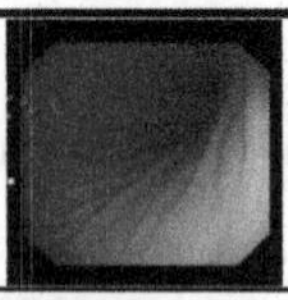	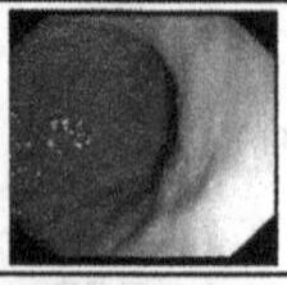
True Label	Greater Curvature of Lower body	Posterior Wall of Middle body	Posterior Wall of Lower body
Given Label	Greater Curvature of Lower body		
Anatomy Images	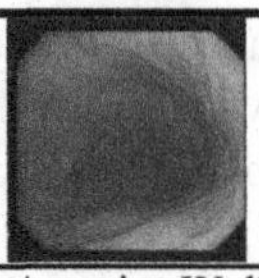	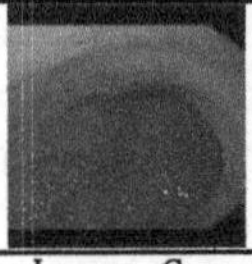	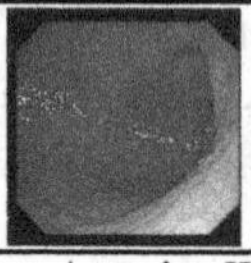
True Label	Anterior Wall of Middle body	Lesser Curvature of Lower body	Anterior Wall of Lower body
Given Label	Anterior Wall of Middle body		

actual noise rate of an EGD examination dataset, as is done for the Noisy CIFAR-10 in the controlled experiments, we selected 20%, 30%, and 40% possibly noisy labels from the training set, based on our proposed Bayesian uncertainty estimates. We trained three corresponding VAT models to re-label the selected noisy data by voting. The choice of using one or more VATs depends on the noise condition of the datasets.

Table 4. Comparison of test accuracy of upper gastrointestinal anatomy recognition task

Methods	Dataset	Test accuracy
Baseline model	N_{ori}	68.36%
Co-teaching	N_{ori}	69.71%
UI + Ensemble(ours)	$N_{\mathrm{ui}}^{R} \cup C_{\mathrm{ui}}$	74.27%

We conducted a trial test by asking several doctors to annotate the same 3,000 images; a 30% average noise rate with a 10% fluctuation was estimated for the EGD examination dataset. As a result, we employed three VATs to correspond to the 20%, 30%, and 40% noise rates. Table 4 shows that our proposed two-stage data-refining process leads to the best recognition accuracy. It is much better than training directly on noisy data (N_{ori}), and outperforms the robust training method Co-teaching by a large margin.

5 Conclusion

Tackling noisy annotations is a prerequisite for AI-assisted medical diagnosis applications. We developed an effective two-stage data refinement approach to solve the problem of training with noisy labels. This technique leveraged Bayesian uncertainty inferences and semi-supervised learning to select and rectify possibly noisy labels. Our proposal was validated by controlled experiments on the CIFAR-10 dataset and real applications in upper gastrointestinal anatomy recognition. The accuracy of our method was greatly improved and it outperformed the robust SOTA training methods, e.g., MentorNet and Co-teaching.

Notably, the two-stage data-refining process is flexible and can be used for downstream tasks, e.g., active learning. It reduces the human effort needed to filter out and manually rectify noisy annotations and can be applied to wider scenarios where accurate expert labeling is expensive. Further work involves investigating the ensemble effect of multiple SOTA semi-supervised learning methods for further accuracy improvement. We will also investigate the idea of directly combining the data refinement procedure with a classification network to achieve end-to-end model training.

References

1. Zhu, R., Zhang, R., Xue, D.: Lesion detection of endoscopy images based on convolutional neural network features. In: 2015 8th International Congress on Image and Signal Processing, pp. 372–376 (2015)
2. Zhang, C., Recht, B., Bengio, S., Hardt, M., Vinyals, O.: Understanding deep learning requires rethinking generalization. In: 5th International Conference on Learning Representations (2017)
3. Li, Y., Yang, J., Song, Y., Cao, L., Luo, J., Li, L.J.: Learning from noisy labels with distillation. In: International Conference on Computer Vision, pp. 1928–1936 (2017)
4. Veit, A., Alldrin, N., Chechik, G., Krasin, I., Gupta, A., Belongie, S.: Learning from noisy large-scale datasets with minimal supervision. In: 30th IEEE Conference on Computer Vision and Pattern Recognition, pp. 6575–6583 (2017)
5. Jiang, L., Zhou, Z., Leung, T., Li, L.J., Fei-Fei, L.: MentorNet: learning data-driven curriculum for very deep neural networks on corrupted labels. In: 35th International Conference on Machine Learning, pp. 2304–2313 (2018)
6. Han, B., et al.: Co-teaching: robust training of deep neural networks with extremely noisy labels. In: Advances in Neural Information Processing Systems, pp. 8527–8537 (2018)
7. Li, J., Wong, Y., Zhao, Q., Kankanhalli, M.S.: Learning to learn from noisy labeled data. In: Proceedings of the IEEE Computer Society Conference on Computer Vision and Pattern Recognition, pp. 5051–5059 (2019)
8. Khan, S., Hayat, M., Zamir, S.W., Shen, J., Shao, L.: Striking the right balance with uncertainty. In: Proceedings of the IEEE Computer Society Conference on Computer Vision and Pattern Recognition, pp. 103–112 (2019)
9. Kendall, A., Badrinarayanan, V., Cipolla, R.: Bayesian SegNet: model uncertainty in deep convolutional encoder-decoder architectures for scene understanding. In: British Machine Vision Conference (2017)
10. Gal, Y., Ghahramani, Z.: Dropout as a Bayesian approximation: representing model uncertainty in deep learning. In: 33rd International Conference on Machine Learning, pp. 1050–1059 (2016)
11. Kingma, D.P., Salimans, T., Welling, M.: Variational dropout and the local reparameterization trick. In: Advances in Neural Information Processing Systems, pp. 2575–2583 (2015)
12. Houlsby, N., Husz, F., Ghahramani, Z., Lengyel, M.: Bayesian active learning for classification and preference learning. Comput. Sci. 10–13 (2011)

13. Gal, Y., Islam, R., Ghahramani, Z.: Deep Bayesian active learning with image data. In: 34th International Conference on Machine Learning, pp. 1183–1192 (2017)
14. Schmarje, L., Santarossa, M., Schr, S., Koch, R.: A survey on semi-, self-and unsupervised techniques in image classification similarities, differences & combinations. arXiv:2002.08721 (2020)
15. Miyato, T., Maeda, S.I., Koyama, M., Ishii, S.: Virtual adversarial training: a regularization method for supervised and semi-supervised learning. IEEE Trans. Pattern Anal. Mach. Intell. **41**, 1979–1993 (2019)
16. Zhu, X., Ghahramani, Z.: Learning from labeled and unlabeled data with label propagation. Sch. Comput Science Carnegie Mellon University Technical report CMUCALD02107 (2002)
17. He, K., Zhang, X., Ren, S., Sun, J.: Deep residual learning for image recognition. In: Proceedings of the IEEE Computer Society Conference on Computer Vision and Pattern Recognition, pp. 770–778 (2016)
18. Zhang, Z., Sabuncu, M.R.: Generalized cross entropy loss for training deep neural networks with noisy labels. In: Neural Information Processing Systems, pp. 8792–8802 (2018)

Synthetic Sample Selection via Reinforcement Learning

Jiarong Ye[1], Yuan Xue[1], L. Rodney Long[2], Sameer Antani[2], Zhiyun Xue[2], Keith C. Cheng[3], and Xiaolei Huang[1(✉)]

[1] College of Information Sciences and Technology, The Pennsylvania State University, University Park, PA, USA
sharon.x.huang@psu.edu

[2] National Library of Medicine, National Institutes of Health, Bethesda, MD, USA

[3] College of Medicine, The Pennsylvania State University, Hershey, PA, USA

Abstract. Synthesizing realistic medical images provides a feasible solution to the shortage of training data in deep learning based medical image recognition systems. However, the quality control of synthetic images for data augmentation purposes is under-investigated, and some of the generated images are not realistic and may contain misleading features that distort data distribution when mixed with real images. Thus, the effectiveness of those synthetic images in medical image recognition systems cannot be guaranteed when they are being added randomly without quality assurance. In this work, we propose a reinforcement learning (RL) based synthetic sample selection method that learns to choose synthetic images containing reliable and informative features. A transformer based controller is trained via proximal policy optimization (PPO) using the validation classification accuracy as the reward. The selected images are mixed with the original training data for improved training of image recognition systems. To validate our method, we take the pathology image recognition as an example and conduct extensive experiments on two histopathology image datasets. In experiments on a cervical dataset and a lymph node dataset, the image classification performance is improved by 8.1% and 2.3%, respectively, when utilizing high-quality synthetic images selected by our RL framework. Our proposed synthetic sample selection method is general and has great potential to boost the performance of various medical image recognition systems given limited annotation.

1 Introduction

The success of deep learning in vision tasks heavily relies on large scale datasets with high quality annotations [4,9]. However, in the domain of medical images,

J. Ye and Y. Xue—These authors contributed equally to this work.

Electronic supplementary material The online version of this chapter (https://doi.org/10.1007/978-3-030-59710-8_6) contains supplementary material, which is available to authorized users.

A. L. Martel et al. (Eds.): MICCAI 2020, LNCS 12261, pp. 53–63, 2020.
https://doi.org/10.1007/978-3-030-59710-8_6

large scale datasets, especially with expert annotations, are often unavailable due to the high cost of annotation and privacy concerns. To mitigate the data insufficiency in medical image tasks, one feasible solution is to manually augment the original dataset. However, random augmentation [16] including random spatial and intensity augmentation are difficult to generalize and sometimes generate unrealistic images. Recently, pioneering works [1,2,5,6,20] have been done using synthetic data generated by Generative Adversarial Network (GAN) models to augment medical training datasets with limited annotation. While achieving promising results, previous works mainly focus on advancing the image synthesis models and often neglect the quality assessment of generated images in terms of their effectiveness in downstream tasks such as improved image recognition. Unlike real images, synthetic images are with unequal qualities and many of them may contain features that skew the data distribution. Thus, directly expanding the original dataset with synthetic images cannot guarantee performance improvement on various downstream tasks such as image recognition and classification.

In this work, we aim at complementing the medical image synthesis with synthetic sample selection, *i.e.*, how to select high quality synthetic samples to improve medical image recognition systems. The first step of our method is synthesizing images for selection. For small training sets, advanced images synthesis models include conditional GAN (cGAN) [10] which generates fake images from class label and noise vector, and cycleGAN [21] which translates existing images into new synthetic images. We focus on the cGAN setting and design a new cGAN model for histopathology image synthesis. For synthetic sample selection, a recent work [18] has attempted to use distance to class centroids in the feature space as a criterion to select synthetic images; however, their approach is not learning based and lacks generality.

Considering that the selection of synthetic images can be modeled as a binary decision making problem, the decision-making criteria can be either handcrafted or learned. Compared to using a handcrafted metric such as that in [18], an agent trained with a learning algorithm can automatically make decision based on a more comprehensive pool of learned features and classification performance gains, thus is more likely to achieve superior performances in downstream tasks. Intuitively, such selection mechanism can be framed as a model-free, policy-based reinforcement learning (RL) process, in which a controller determines the actions applied to the candidate synthetic images and keeps updating based on the reward of downstream task performance until convergence. Therefore, we propose a reinforcement learning based selection model to automatically choose the most representative images with highest quality. The backbone of our selection model is a combination of ResNet34 [7] model for feature extraction and a transformer [14] model for sample selection. We use the validation classification accuracy as the reward of RL training and use the state-of-the-art proximal policy optimization (PPO) [13] to update the model weights.

To the best of our knowledge, our work is the first to systematically investigate the synthetic sample selection problem for medical images. Our proposed method is validated on two histopathology datasets [15,18] with limited number of annotations. Experimental results show that our method substantially outperform all previous methods including traditional augmentation, direct synthetic augmentation with GAN-generated images, and selective augmentation based on feature distances. With RL as the learning engine, our proposed synthetic sample selection method provides a principled approach to a critical missing link in medical image synthetic augmentation. Moreover, comparison between selected and discarded samples can help us gain deeper understanding about which types of feature encode semantically-relevant information.

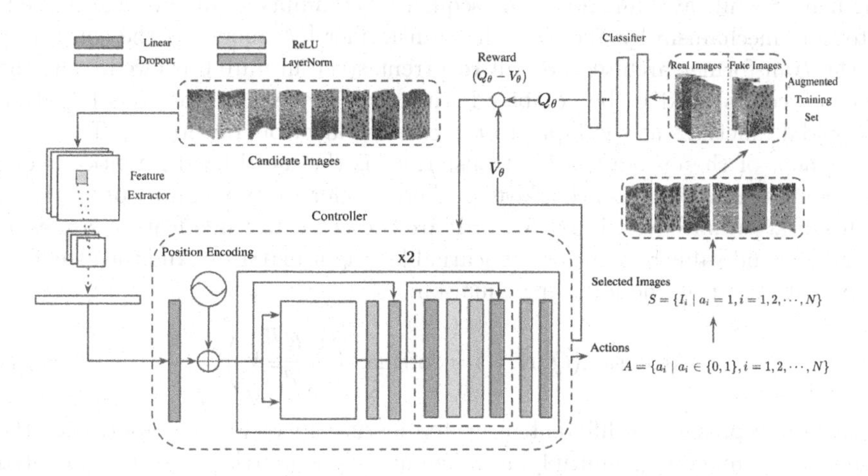

Fig. 1. The architecture of our proposed selective synthetic augmentation framework based on reinforcement learning.

2 Methodology

The detailed architecture of our proposed framework is shown in Fig. 1. The candidate pool of synthetic images is generated with HistoGAN designed for synthesizing histopathology patches (details are provided in supplementary materials). The controller which is a transformer model [14] plays a vital role of deciding whether to select a synthetic sample or not. In each training iteration, it takes the feature vectors extracted from a ResNet34 [7] model trained on the original training images as the input, then outputs the binary action (select or discard) for each candidate synthetic image. After sample selection, we train the classifier on the expanded dataset and take the maximum validation accuracy of last five epochs [22] as the the reward to update the policy. To improve the stability of policy update and avoid severe fluctuation caused by parameter change, we adopt Proximal Policy Optimization (PPO) [13] as the policy gradient method in our synthetic sample selection model. Next, we will introduce the detailed design of the controller and policy gradient method, which are two main components in our proposed framework.

2.1 Controller

The rationale behind the choice of the controller is based on the feature dependencies among candidate images. We hypothesize that the order of augmentation is not entirely independent since the late adds are subject to a constraint to differentiate with the early adds to assure diversity of the entire augmented training set. In order to address the potential relation existing between images augmented while avoiding an strong sequential assumption, we leverage the self-attention mechanism by adopting the transformer [14] model as the controller. In the transformer architecture, all recurrent structure are eschewed, thus the feature vectors need to be combined with their embedded positions based on sinusoidal functions as the input to the encoder layer of transformer. The main component of the encoder inside transformer is the multi-head attention block that consists of n self-attention layers, where n refers to the number of heads. In each self-attention layer, input features are projected to three feature spaces as query, key and value by multiplying learnable weight matrices. And the attention map is obtained by the following equation:

$$\text{Attention}(Q, K, V) = \text{softmax}\left(\frac{QK^T}{\sqrt{d_k}}V\right) . \tag{1}$$

Each head represents a different projected feature space for the input, with the same input embedding multiplying different weight matrix, then concatenated at the end to generate the final attention map as:

$$\text{MultiHead}(Q, K, V) = [\text{head}_1; \text{head}_2]W^O , \tag{2}$$

where $\text{head}_i = \text{Attention}(QW_i^Q, KW_i^K, VW_i^V)$. The learnable weight matrices are denoted as: $W^Q \in \mathbb{R}^{d_{\text{input}} \times d_k}$, $W^K \in \mathbb{R}^{d_{\text{input}} \times d_k}$, $W^V \in \mathbb{R}^{d_{\text{input}} \times d_v}$, $W^O \in \mathbb{R}^{hd_v \times d_{\text{input}}}$, here h represents the number of heads. Then after applying the attention map, the context vector is fed into the feed forward layer as follows:

$$F(x) = \max(0, xW_1 + b_1)W_2 + b_2 . \tag{3}$$

Skip connections are adopted in the process to combine learned features from both the higher and lower abstract levels. Considering the task of the controller is to output a binary action for each input feature vector, as shown in Fig. 1, the decoder for transformer is a linear layer used as the policy network. All in all, applying the transformer as the controller for our reinforcement learning based image selection framework is advantageous for its self-attention mechanism to capture the dependencies among input feature vectors. We perform comprehensive ablation study over the choice of controller and results are presented in Sect. 3.2.

2.2 Policy Gradient Method

An effective while efficient policy gradient method is crucial for the entire reinforcement learning process to optimally leverage the reward as the feedback to the controller. Proximal Policy Optimization (PPO) [13] has become a popular choice of policy gradient algorithm, due to its computational efficiency and satisfactory performance compared with previous algorithms such as TRPO [12]. In our framework we use the PPO to stabilize the RL training process. To reduce complexity while maintaining comparable performance, in the algorithm of PPO, the KL convergence constraint enforced in TRPO on the size of policy update is replaced with clipped probability ratio between the current and the previous policy within a small interval around 1. At time step t, let A_θ be the advantage function, the objective function is as follows:

$$\mathcal{L}(\theta) = \mathbb{E}\left[\min(\gamma_\theta(t)A_\theta(s_t, a_t), \text{clip}(\gamma_\theta(t), 1-\epsilon, 1+\epsilon)A_\theta(s_t, a_t))\right] \ , \quad (4)$$

where $A_\theta(s_t, a_t) = Q_\theta(s_t, a_t) - V_\theta(s_t, a_t)$. As part of the transformer output, $V_\theta(s_t, a_t)$ is a learned state-value taken off as the baseline from the q-value to lower the variation of the rewards along the training process. π refers to the probability of actions. $Q_\theta(s_t, a_t)$ is the q-value at time t defined as the smooth version of the max validation accuracy among the last 5 epochs in the classification task. Since our target tasks are trained on limited number of data, in order to get a robust estimation of reward changing pattern, we apply the Exponential Moving Average (EMA) [8] algorithm to smooth the curve of original reward. With smoothing, the final reward at time t is:

$$\hat{Q}_\theta(s_t, a_t) = \begin{cases} \mathrm{Q}_\theta(s_t, a_t), & t = 1 \\ \alpha\hat{Q}_\theta(s_{t-1}, a_{t-1}) + (1-\alpha)\mathrm{Q}_\theta(s_t, a_t), & t > 1 \end{cases} . \quad (5)$$

Based on the idea of importance sampling, the weight assigned to the current policy also depends on older policies. The probability ratio between previous and current policies $\gamma_\theta(t)$ is defined by:

$$\gamma_\theta(t) = \frac{\pi_\theta(a_t \mid s_t)}{\pi_\theta(a_{t-1} \mid s_{t-1})} \ , \quad (6)$$

where $a_t \in \mathbb{R}^{N \times 2}$, N refers to the number of synthetic images in the candidate pool. If at time step t, $a_i(t) = 0, i \in \{1, 2, \cdots N\}$, then the candidate i is discarded, otherwise it is added to the original training set for further steps. We also compare PPO with the classic REINFORCE [17] algorithm in experiments.

3 Experiments

We conduct comprehensive experiments on two histopathology datasets. The first dataset is a cervical histopathology dataset where all images are annotated by the same pathologist. The data processing follows [18], and results in patches with a unified size of 256×128 pixels. Compared with the dataset used in [18], we include more data for more comprehensive experiments. In total, there are 1,284 Normal, 410 CIN1, 481 CIN2, 472 CIN3 patches. Examples of the images can be found in Fig. 2. We randomly split the dataset, by patients, into training,

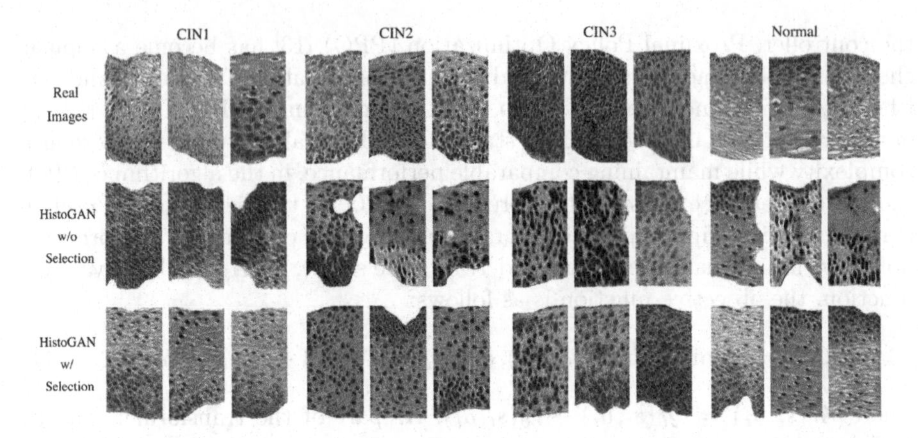

Fig. 2. Examples of real images and synthetic images generated by HistoGAN trained on cervical histopathology dataset and selected with our RL-based framework. Zoom in for better view.

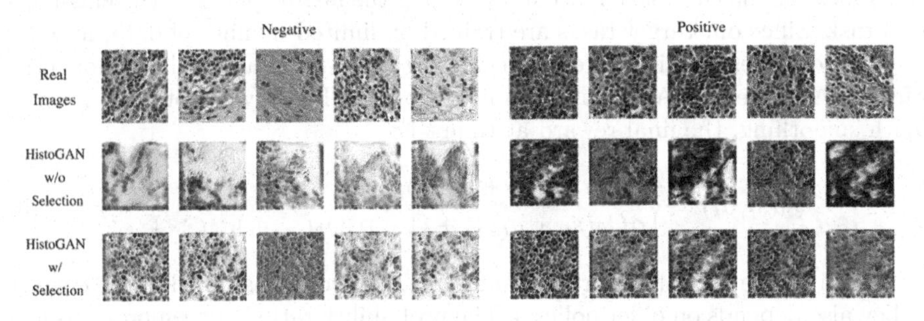

Fig. 3. Examples of real images and synthetic images generated by our HistoGAN model trained on 3% of PCam dataset and selected with our framework.

validation, and testing sets, with ratio 7:1:2 and keep the ratio of image classes almost the same among different sets. All evaluations and comparisons reported in this section are carried out on the test set.

To further prove the generality of our proposed method, we also conduct experiments on the public PatchCamelyon (PCam) benchmark [15], which consists of patches extracted from histopathologic scans of lymph node sections with unified size of 96×96 pixels. To simulate the scenario of limited training data, we use the randomly selected 3% of the training set in PCam, which results 3,931 negative and 2,757 positive patches, to train the HistoGAN model and the baseline classifier. All trained models are evaluated on the full testing set. Example results are illustrated in Fig. 3.

3.1 Implementation Details

As illustrated in Fig. 1, the candidate pool contains 1,024 HistoGAN-generated images for each class (CIN1, CIN2, CIN3, NORMAL) of the cervical

Table 1. Classification results of baseline and augmentation models on the cervical dataset. We reimplemented [18] and a metric learning model with triplet loss [11] using the same pool of synthetic images generated by HistoGAN for fair comparison.

	Accuracy	AUC	Sensitivity	Specificity
Baseline model [7]	.754 ± .012	.836 ± .008	.589 ± .017	.892 ± .005
+ Traditional augmentation	.766 ± .013	.844 ± .009	.623 ± .029	.891 ± .006
+ GAN augmentation	.787 ± .005	.858 ± .003	.690 ± .014	.909 ± .003
+ Metric learning (Triplet loss) [11]	.798 ± .016	.865 ± .010	.678 ± .048	.909 ± .013
+ Selective augmentation (Centroid Distance) [18]*	.808 ± .005	.872 ± .004	.639 ± .015	.912 ± .006
+ Selective augmentation (Transformer-PPO, ours)	**.835 ± .007**	**.890 ± .005**	**.747 ± .013**	**.936 ± .003**

histopathology dataset, and 2,048 images for each class (Negative, Positive) of the PCam dataset. These candidates are fed into a ResNet34 model pretrained on the original dataset and the feature vectors of the fully-connected layer are extracted as input for the controller. The entire set of input is first sorted by the cosine distances to the corresponding centroids and divided into 8 batches, with each containing 128 images from each class for histopathology dataset, and 256 images from each class for PCam dataset. The controller consists of 2 encoder layers, in which the multi-head attention block contains 2 heads. It outputs a binary action vector A_θ for the further selection of the augmented training set, as well as a value V_θ used in the calculation of reward for the policy gradient method. The classifier is also a ResNet34 model of the same structure as the feature extractor. Similar to [22], we adopt the max accuracy obtained from testing on the validation set of the last 5 epochs in the classification task. To further stabilize the training reward, we use the EMA-smoothed max validation accuracy as the reward with $\alpha = 0.8$. In the policy gradient algorithm PPO, the policy function π is obtained from the softmax layer of the ResNet34 model, after the policy network, *i.e.*, a linear layer with output dimension set to the number of classes in the corresponding dataset. ϵ in Eq. 4 is set to 0.2, providing the upper and lower bound for the ratio of policy functions at current time step t and previous time step $t-1$. In all synthetic augmentation methods, the augmentation ratio is set to 0.5 to be consistent with [18]. In our RL based method, the number of selected images are determined by the controller. The learning rate for the reinforcement learning framework training is $2.5e-04$, and for the attention mechanism in the ablation study in Table 3 is $1e-04$.

3.2 Result Analysis

In Fig. 2 and Fig. 3, we show qualitative results of synthetic images on cervical and lymph node datasets generated by HistoGAN, and images selected by our proposed synthetic sample selection model. From visual results, images selected by our method clearly contain more realistic features than images before selection. With HistoGAN generated images as candidates, we compare our RL based synthetic sample selection with traditional augmentation method [16] and other synthetic augmentation methods in Table 1 and 3. The Traditional Augmentation includes horizontal flipping and color jittering of original training data; GAN Augmentation refers to randomly adding HistoGAN generated images to the original training set; In the Metric Learning method, we adopt the Triplet Loss [11] which minimizes the intra-class differences and maximizes the inter-class differences between samples. We also compare with the current state-of-the-art sample selection method [18] which ranks samples based on distance to class centroids in feature space. We report quantitative evaluation scores of all methods using the accuracy, area under the ROC curve (AUC), sensitivity and specificity. All models are run for 5 rounds with random initialization for fair comparison. The mean and standard deviation results of the 5 runs are reported. For fair comparison, we use the same backbone ResNet34 classifier with same hyperparameters setting in all experiments and use the same candidate pool generated by HistoGAN for sample selection to ensure that differences only come from the augmented training set.

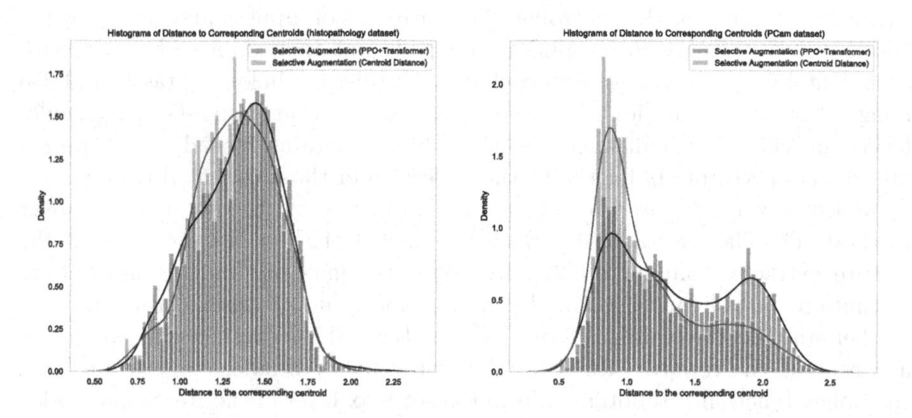

Fig. 4. Comparison between histograms of the distances to the corresponding centroid of each class in the cervical histopathology dataset and PCam dataset.

From Table 1 and 2, one can observe that compared with previous augmentation and sample selection methods, our proposed selective augmentation delivers superior result on both datasets using the same candidate synthetic image pool. While achieving markedly better performances than the second best method [18] in all metrics, we further analyse how images selected by our learning based method differ from those selected by the previous handcrafted sample selection method [18]. We compare the distribution of selected images over the feature distance to corresponding class centroids. As shown in Fig. 4, both methods discard samples with too small or too large centroid distance as they are either too similar to original training data, or are outliers. Especially on PCam dataset,

Table 2. Classification results of baseline and augmentation models on the PCam dataset.

	Accuracy	AUC	Sensitivity	Specificity
Baseline model [7]	.853 ± .003	.902 ± .002	.815 ± .008	.877 ± .009
+ Traditional augmentation	.860 ± .005	.907 ± .003	.823 ± .015	.885 ± .017
+ GAN augmentation	.859 ± .001	.906 ± .001	.822 ± .014	.884 ± .011
+ Metric learning (Triplet loss) [11]	.864 ± .004	.910 ± .003	.830 ± .012	.887 ± .008
+ Selective augmentation (Centroid distance) [18]*	.868 ± .002	.912 ± .002	.835 ± .010	.890 ± .006
+ Selective augmentation (Transformer-PPO, ours)	**.876 ± .001**	**.917 ± .001**	**.846 ± .010**	**.895 ± .005**

[18] tends to select samples with relatively lower centroid distances. Meanwhile, our controller learns to select samples distributed evenly on different centroid distances and achieves promising results. We hypothesizes that a learning based method selects samples in a more general way than handcrafted method, and the histogram analysis further proves that handcrafted method based on a single metric may not be able to find an optimal pattern for sample selection.

To validate our choice of the Transformer and PPO, we perform ablation study on the cervical dataset and report results in Table 3. For the controller, we experiment with GRU [3] and GRU with attention [19] (GRU-Attn); For the policy gradient algorithm, we compare PPO with the classic REINFORCE [17]. Compared with other controller and policy gradient algorithms, our full model with transformer and PPO achieves best performances in all metrics, which justifies our choices.

Table 3. Ablation study of our proposed reinforcement learning framework for synthetic images selection on the cervical histopathology dataset.

	Accuracy	AUC	Sensitivity	Specificity
Selective augmentation (GRU-REINFORCE)	.789 ± .011	.859 ± .007	.687 ± .014	.908 ± .005
Selective augmentation (GRU-Attn-REINFORCE)	.804 ± .019	.869 ± .012	.674 ± .039	.914 ± .010
Selective augmentation (Transformer-REINFORCE)	.812 ± .008	.875 ± .005	.724 ± .022	.920 ± .006
Selective Augmentation (GRU-PPO)	.792 ± .017	.862 ± .012	.701 ± .039	.912 ± .010
Selective augmentation (GRU-Attn-PPO)	.811 ± .014	.874 ± .010	.751 ± .034	.919 ± .007
Selective augmentation (Transformer-PPO)	**.835 ± .007**	**.890 ± .005**	**.747 ± .013**	**.936 ± .003**

4 Conclusions

In this paper, we propose a reinforcement learning based synthetic sample selection method. Compared with previous methods using handcrafted selection metrics, our proposed method achieves state-of-the-art results on two histopathology datasets. In future works, we expect to extend our method to more medical image recognition tasks where annotations are limited. We also plan to investigate the usage of our RL based method for sample selection toward other purposes such as active learning for annotation.

Acknowledgements. This work was supported in part by the Intramural Research Program of the National Library of Medicine and the National Institutes of Health. We gratefully acknowledge the help with expert annotations from Dr. Rosemary Zuna, M.D., of the University of Oklahoma Health Sciences Center, and the work of Dr. Joe Stanley of Missouri University of Science and Technology that made the histopathology data collection possible.

References

1. Bowles, C., et al.: GAN augmentation: augmenting training data using generative adversarial networks. arXiv preprint arXiv:1810.10863 (2018)
2. Chaitanya, K., Karani, N., Baumgartner, C.F., Becker, A., Donati, O., Konukoglu, E.: Semi-supervised and task-driven data augmentation. In: Chung, A.C.S., Gee, J.C., Yushkevich, P.A., Bao, S. (eds.) IPMI 2019. LNCS, vol. 11492, pp. 29–41. Springer, Cham (2019). https://doi.org/10.1007/978-3-030-20351-1_3
3. Chung, J., Gulcehre, C., Cho, K., Bengio, Y.: Empirical evaluation of gated recurrent neural networks on sequence modeling. arXiv preprint arXiv:1412.3555 (2014)
4. Deng, J., Dong, W., Socher, R., Li, L.J., Li, K., Fei-Fei, L.: ImageNet: a large-scale hierarchical image database. In: 2009 IEEE Conference on Computer Vision and Pattern Recognition, pp. 248–255. IEEE (2009)
5. Frid-Adar, M., Diamant, I., Klang, E., Amitai, M., Goldberger, J., Greenspan, H.: GAN-based synthetic medical image augmentation for increased CNN performance in liver lesion classification. Neurocomputing **321**, 321–331 (2018)
6. Gupta, A., Venkatesh, S., Chopra, S., Ledig, C.: Generative image translation for data augmentation of bone lesion pathology. In: International Conference on Medical Imaging with Deep Learning, pp. 225–235 (2019)
7. He, K., Zhang, X., Ren, S., Sun, J.: Deep residual learning for image recognition. In: Proceedings of the IEEE Conference on Computer Vision and Pattern Recognition, pp. 770–778 (2016)
8. Hunter, J.S.: The exponentially weighted moving average. J. Qual. Technol. **18**(4), 203–210 (1986)
9. Lin, T.-Y., et al.: Microsoft COCO: common objects in context. In: Fleet, D., Pajdla, T., Schiele, B., Tuytelaars, T. (eds.) ECCV 2014. LNCS, vol. 8693, pp. 740–755. Springer, Cham (2014). https://doi.org/10.1007/978-3-319-10602-1_48
10. Mirza, M., Osindero, S.: Conditional generative adversarial nets. arXiv preprint arXiv:1411.1784 (2014)
11. Schroff, F., Kalenichenko, D., Philbin, J.: FaceNet: a unified embedding for face recognition and clustering. In: Proceedings of the IEEE Conference on Computer Vision and Pattern Recognition, pp. 815–823 (2015)

12. Schulman, J., Levine, S., Abbeel, P., Jordan, M., Moritz, P.: Trust region policy optimization. In: International Conference on Machine Learning, pp. 1889–1897 (2015)
13. Schulman, J., Wolski, F., Dhariwal, P., Radford, A., Klimov, O.: Proximal policy optimization algorithms. arXiv preprint arXiv:1707.06347 (2017)
14. Vaswani, A., et al.: Attention is all you need. In: Advances in Neural Information Processing Systems, pp. 5998–6008 (2017)
15. Veeling, B.S., Linmans, J., Winkens, J., Cohen, T., Welling, M.: Rotation equivariant CNNs for digital pathology. In: Frangi, A.F., Schnabel, J.A., Davatzikos, C., Alberola-López, C., Fichtinger, G. (eds.) MICCAI 2018. LNCS, vol. 11071, pp. 210–218. Springer, Cham (2018). https://doi.org/10.1007/978-3-030-00934-2_24
16. Wang, J., Perez, L.: The effectiveness of data augmentation in image classification using deep learning. Convolutional Neural Netw. Vis. Recognit. 11 (2017)
17. Williams, R.J.: Simple statistical gradient-following algorithms for connectionist reinforcement learning. Mach. Learn. **8**(3–4), 229–256 (1992)
18. Xue, Y., et al.: Synthetic augmentation and feature-based filtering for improved cervical histopathology image classification. In: Shen, D., et al. (eds.) MICCAI 2019. LNCS, vol. 11764, pp. 387–396. Springer, Cham (2019). https://doi.org/10.1007/978-3-030-32239-7_43
19. Yang, Z., Yang, D., Dyer, C., He, X., Smola, A., Hovy, E.: Hierarchical attention networks for document classification. In: Proceedings of the 2016 Conference of the North American Chapter of the Association for Computational Linguistics: Human Language Technologies, pp. 1480–1489 (2016)
20. Zhao, A., Balakrishnan, G., Durand, F., Guttag, J.V., Dalca, A.V.: Data augmentation using learned transformations for one-shot medical image segmentation. In: Proceedings of the IEEE Conference on Computer Vision and Pattern Recognition, pp. 8543–8553 (2019)
21. Zhu, J.Y., Park, T., Isola, P., Efros, A.A.: Unpaired image-to-image translation using cycle-consistent adversarial networks. In: Proceedings of the IEEE International Conference on Computer Vision, pp. 2223–2232 (2017)
22. Zoph, B., Le, Q.V.: Neural architecture search with reinforcement learning. arXiv preprint arXiv:1611.01578 (2016)

Dual-Level Selective Transfer Learning for Intrahepatic Cholangiocarcinoma Segmentation in Non-enhanced Abdominal CT

Wenzhe Wang[1], Qingyu Song[1], Jiarong Zhou[2], Ruiwei Feng[1], Tingting Chen[1], Wenhao Ge[2], Danny Z. Chen[3], S. Kevin Zhou[4,5], Weilin Wang[2(✉)], and Jian Wu[1(✉)]

[1] College of Computer Science and Technology, Zhejiang University, Zhejiang, China
wujian2000@zju.edu.cn

[2] Department of Hepatobiliary and Pancreatic Surgery, The Second Affiliated Hospital, Zhejiang University School of Medicine, Zhejiang, China
wam@zju.edu.cn

[3] Department of Computer Science and Engineering, University of Notre Dame, Notre Dame, USA

[4] Institute of Computing Technology, Chinese Academy of Sciences, Beijing, China

[5] Peng Cheng Laboratory, Shenzhen, China

Abstract. Automatic and accurate Intrahepatic Cholangiocarcinoma (ICC) segmentation in non-enhanced abdominal CT images can provide significant assistance for clinical decision making. While deep neural networks offer an effective tool for ICC segmentation, collecting large amounts of annotated data for deep network training may not be practical for this kind of applications. To this end, transfer learning approaches utilize abundant data from similar tasks and transfer the prior-learned knowledge to achieve better results. In this paper, we propose a novel Dual-level Selective Transfer Learning (DSTL) model for ICC segmentation, which selects similar information at global and local levels from a source dataset and produces transfer learning using the selected hierarchical information. Besides the basic segmentation networks, our DSTL model is composed of a global information selection network (GISNet) and a local information selection network (LISNet). The GISNet is utilized to output weights for global information selection and to mitigate the gap between the source and target tasks. The LISNet outputs weights for local information selection. Experimental results show that our DSTL model achieves superior ICC segmentation performance and outperforms the original and image selection based transfer learning and joint training strategies. To the best of our knowledge, this is the first method for ICC segmentation in non-enhanced abdominal CT.

Keywords: Tumor segmentation · Intrahepatic Cholangiocarcinoma · Selective transfer learning

W. Wang, Q. Song and J. Zhou—The first three authors contributed equally.

A. L. Martel et al. (Eds.): MICCAI 2020, LNCS 12261, pp. 64–73, 2020.
https://doi.org/10.1007/978-3-030-59710-8_7

1 Introduction

Intrahepatic Cholangiocarcinoma (ICC) is the second most popular primary malignant liver tumors, and about 15% of liver cancers are estimated to be ICC [9]. Automatic and accurate segmentation of ICC in non-enhanced abdominal Computer Tomography (CT) images can provide significant assistance for the diagnosis, preoperative planning, and prognosis, and at the same time avoid the high-risk, time-consuming, and expensive issues caused by contrast agents. For Computer-Aided Diagnosis (CAD) of ICC in CT, a texture-based method was proposed [7] for benign/malignant tumor segmentation and classification, but it did not classify tumors into detailed classes. To the best of our knowledge, no method was known for ICC segmentation in non-enhanced abdominal CT.

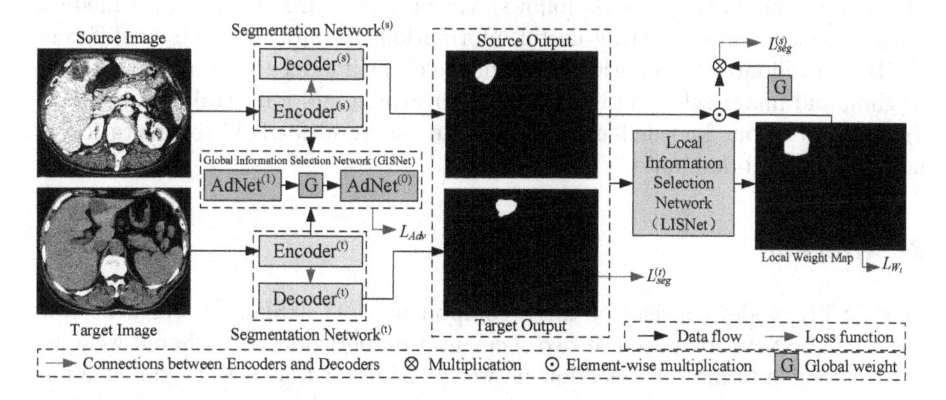

Fig. 1. Our proposed Dual-level Selective Transfer Learning (DSTL) model for ICC segmentation in non-enhanced abdominal CT images, which selects similar information at global and local levels from a source dataset and performs transfer learning using the selected hierarchical information. We use the same color for modules in the same networks. Note that the two segmentation networks share their weights, and AdNet^1 and AdNet^0 are two adversarial learning modules (best viewed in color) (Color figure online).

With the rapid development of deep learning techniques, especially Convolutional Neural Networks (CNNs), medical image segmentation has been largely improved [5,8,10]. To achieve accurate segmentation results, datasets with large amounts of annotated samples are commonly used for deep learning model training. However, for real-world applications like ICC segmentation, annotated data collection may be difficult due to multiple reasons (e.g., the incidence of diseases and difficulties of image acquisition). Under such circumstances, public datasets from related tasks (i.e., source tasks) are commonly utilized to perform transfer learning for target task model training [11]. A straightforward way to conduct transfer learning is to fine-tune pre-trained models on target tasks [4,11]. These methods learn or un-learn from holistic images within source datasets. However, not all information in source datasets is useful [12]. It is more beneficial to select helpful data for transfer learning. To deal with this problem, 2D image selection strategies [1,3] were proposed to mitigate the gap between the source and target tasks. But for 3D medical image segmentation tasks, selection of both global and local information should be considered at the same time.

This paper aims to develop Dual-level (i.e., both global and local levels) Selective Transfer Learning (DSTL) for ICC segmentation in non-enhanced abdominal CT. Liver segmentation produced at the same time can help improve the ICC segmentation performance. We leverage a public liver tumor segmentation dataset containing portal venous phase CT [2] as our source dataset. We utilize a Siamese Segmentation Network (SSNet) for ICC and liver feature learning, and global and local information selection networks (GISNet and LISNet) for hierarchical information selection. The architecture of our model is shown in Fig. 1. Our main contributions are as follows. We propose a DSTL model for medical image segmentation, selecting useful information in source datasets at different levels. We validate the superior performance of our model and compare it to the original and image selection based transfer learning and joint training methods. To the best of our knowledge, this is the first method for ICC segmentation in non-enhanced abdominal CT.

2 Method

Our DSTL model consists of three components: (1) a Siamese Segmentation Network (SSNet) composed of two segmentation networks with shared weights; (2) a global information selection network (GISNet) that selects global information and mitigates the gap between the source and target tasks; (3) a local information selection network (LISNet) that selects semantic information. The components are trained in an end-to-end manner and are described as follows.

2.1 Siamese Segmentation Network

Under our problem setting, the learning difficulty lies in the information gap between the source and target datasets. To compare the similarity of information in the source and target datasets, a Siamese structure is utilized.

For medical image segmentation, U-Net and its variants [5,10] have been commonly adopted because of their straightforward and successful architectures. This kind of networks utilizes an encoder to capture context information and a decoder to enable precise localization. Without loss of generality, we utilize the original 2D U-Net [10] in this work as our base network for semantic information learning. We divide CT images into slices as the inputs of 2D U-Nets. Formally, let $(X^{(s)}, Y^{(s)}) \in D^{(s)}$ and $(X^{(t)}, Y^{(t)}) \in D^{(t)}$ denote CT slices and their corresponding annotations in the source and target datasets, respectively. As shown in Fig. 1, $(X^{(s)}, Y^{(s)})$ and $(X^{(t)}, Y^{(t)})$ are fed to two weight-sharing U-Nets for semantic information learning synchronously. Features belonging to the two tasks can be learned jointly. We utilize $\mathcal{F}_{Seg}(\cdot)$ to denote the function of the segmentation networks. The normal Cross-Entropy loss is utilized for the target task training, which is defined as follows:

$$L_{Seg}^{(t)}(\mathcal{F}_{Seg}) = -\sum Y^{(t)} \log \mathcal{F}_{Seg}(X^{(t)}) \quad (1)$$

The segmentation loss utilized for the source task training is defined later as it involves information selection.

2.2 Global Information Selection Network

Following the ideas of [6,13], we apply the adversarial learning strategy to improve the expressiveness of U-Nets. Our global information selection network (GISNet) contains two adversarial learning modules (i.e., AdNet$^{(1)}$ and AdNet$^{(0)}$) that select global information and mitigate the information gap between the source and target tasks, respectively.

The encoders of the two U-Nets encode CT slices from the source and target datasets into high-level representations. The first adversarial module (i.e., AdNet$^{(1)}$) takes the representations as input. The output of the optimum parameters of the module gives the probability of the representations coming from the source task. AdNet$^{(1)}$ is a binary classifier with all the source data labeled as 1 and all the target data labeled as 0. The loss $L_{Adv}^{(1)}$ is defined as follows:

$$L_{Adv}^{(1)}(\mathcal{F}_A^{(1)}, \mathcal{F}_E) = \mathbb{E}_{D^{(s)}}[\log \mathcal{F}_A^{(1)}(\mathcal{F}_E(X^{(s)}))] + \mathbb{E}_{D^{(t)}}[\log(1 - \mathcal{F}_A^{(1)}(\mathcal{F}_E(X^{(t)})))] \quad (2)$$

where $\mathcal{F}_A^{(1)}(\cdot)$ is the function of AdNet$^{(1)}$ and $\mathcal{F}_E(\cdot)$ is the function of the encoders. $D^{(s)}$ and $D^{(t)}$ are the source and target distributions.

Since the output of AdNet$^{(1)}$ denotes the probability of the representations coming from the source task, it can be utilized as a weight for global information selection. The intuition for the weighting scheme is that if the activation of AdNet$^{(1)}$ is large, then the samples can be almost perfectly discriminated from the target task. And this is not suitable for transfer learning. The weight of the global information selection M_g is defined as follows:

$$M_g = \frac{1 - \mathcal{F}_A^{(1)}(\mathcal{F}_E(X^{(s)}))}{\mathbb{E}[1 - \mathcal{F}_A^{(1)}(\mathcal{F}_E(X^{(s)}))]} \quad (3)$$

so that the expectation of M_g is equal to 1.

On the other hand, we would like to mitigate the information gap between the two tasks. However, if we directly apply the weight M_g to the same adversarial module, then the theoretical results of the min-max game will not be reducing the Jensen-Shannon divergence between the two tasks [13]. That is to say, the gradients of AdNet$^{(1)}$ are learned on the unweighted source and target samples which would not be a good indicator for mitigating the information gap between the two tasks. Hence, we resolve this issue by introducing another adversarial module, AdNet$^{(0)}$, for comparing the weighted source and original target data. After all, it is AdNet$^{(0)}$ that plays the min-max game with the two encoders to drive the representations of the source samples close to the distribution of the target ones. The encoders and the adversarial module are alternately trained using the adversarial loss $L_{Adv}^{(0)}$ so that the encoders can generate task-invariant features that can fool the adversarial module. The objective function of AdNet$^{(0)}$ is defined as follows:

$$\min_{\mathcal{F}_E} \max_{\mathcal{F}_A^{(0)}} L_{Adv}^{(0)}(\mathcal{F}_A^{(0)}, \mathcal{F}_E) = \\ \mathbb{E}_{D^{(s)}}[M_g \log \mathcal{F}_A^{(0)}(\mathcal{F}_E(X^{(s)}))] + \mathbb{E}_{D^{(t)}}[\log(1 - \mathcal{F}_A^{(0)}(\mathcal{F}_E(X^{(t)})))] \tag{4}$$

where $\mathcal{F}_A^{(0)}(\cdot)$ denotes the function of AdNet$^{(0)}$. Both of AdNet$^{(1)}$ and AdNet$^{(0)}$ are composed of one max-pooling layer to down-sample representations, and two groups of fully-connected and LeakyReLU layers and a final fully-connected layer for adversarial learning. Our GISNet is composed of AdNet$^{(1)}$ and AdNet$^{(0)}$, and the total adversarial loss L_{adv} is defined as the sum of $L_{Adv}^{(1)}$ and $L_{Adv}^{(0)}$.

2.3 Local Information Selection Network

We propose a local information selection network (LISNet) for local-level similar information selection. LISNet is an attention network that assigns more attention to the source image regions that are similar to the regions in the target ones. A straightforward way for information selection is to compare the segmentation outputs of the source and target slices (i.e., $\mathcal{F}_{Seg}(X^{(s)})$ and $\mathcal{F}_{Seg}(X^{(t)})$), where similarly structured regions can be found directly. Our attention network assigns more attention to those pixels with similar structures.

LISNet takes the outputs of both the U-Nets discussed in Sect. 2.1 (i.e., $\mathcal{F}_{Seg}(X^{(s)})$ and $\mathcal{F}_{Seg}(X^{(t)})$) as input, which have a size of 512 × 512. Formally, we define the ground-truth labels, Y_l, of LISNet in the following way: The ground-truth label of $\mathcal{F}_{Seg}(X^{(s)})$ (i.e., $Y_l^{(s)}$) is an all-zero map. We set the ground-truth label of $\mathcal{F}_{Seg}(X^{(t)})$ (i.e., $Y_l^{(t)}$) equal to the label of ICC in $Y^{(t)}$. This makes LISNet concentrate on the structure of ICC in the target dataset and ignore other tissues. When $\mathcal{F}_{Seg}(X^{(s)})$ is fed to LISNet, the network can then find structures similar to ICC, and the regions thus found are selected for transfer learning. The objective function of LISNet is defined as follows:

$$L_{LIS}(\mathcal{F}_L) = -\sum Y_l^{(s)} \log \mathcal{F}_L(\mathcal{F}_{Seg}(X^{(s)})) - \sum Y_l^{(t)} \log \mathcal{F}_L(\mathcal{F}_{Seg}(X^{(t)})) \tag{5}$$

where $\mathcal{F}_L(\cdot)$ denotes the function of LISNet. We use only $\mathcal{F}_L(\mathcal{F}_{Seg}(X^{(s)}))$ to generate the local weight map.

The output of LISNet has the same size as its input. The network is composed of four groups of 'Conv2d-BN-LeakyReLU' layers, one Conv2d layer, and an up-sample layer. The kernel size, stride, and padding of the first four Conv2d layers are set to 3, 2, and 1, respectively. And these of the final Conv2d layer are set to 3, 0, and 1. The parameter of LeakyReLU is set to 0.01. Since the first four

Conv2d layers down-sample features to 1/16 of the size of the input, we up-sample them to 512 × 512 utilizing the nearest up-sampling. After acquiring the up-sampled outputs, we conduct thresholding to generate binary maps. Since using a fixed value as the threshold cannot always adapt to the current score range, an adaptive thresholding operation is utilized to acquire the weight map of local information selection M_l, which is defined as follows:

$$M_l = I(\mathcal{F}_L(\mathcal{F}_{Seg}(X^{(s)})) > mean(\mathcal{F}_L(\mathcal{F}_{Seg}(X^{(s)})))) \quad (6)$$

where $I(\cdot)$ is an indicator function. The adaptive mean value of $\mathcal{F}_L(\mathcal{F}_{Seg}(X^{(s)}))$ is utilized as its threshold. The final attention map takes into account the local and global information for similarity-based transfer learning. The segmentation loss of the source task $L_{Seg}^{(s)}$ is defined as a weighted Cross-Entropy loss:

$$L_{Seg}^{(s)}(\mathcal{F}_{Seg}) = -\sum M_g Y^{(s)}(M_l \odot \log \mathcal{F}_{Seg}(X^{(s)})) \quad (7)$$

where $\odot$ denotes the element-wise multiplication. The total segmentation loss L_{Seg} is defined as the sum of $L_{Seg}^{(s)}$ and $L_{Seg}^{(t)}$. The total loss of our model is defined as follows:

$$L = L_{Seg} + L_{Adv} + L_{LIS} \quad (8)$$

3 Experiments

3.1 Datasets and Experimental Setup

We use the training set of the public LiTS challenge 2017 [2] as our source dataset, which contains 130 portal venous phase CT images with ternary annotations (i.e, liver, tumor, and background). For our target dataset, we collected 25 non-enhanced abdominal CT images from a local hospital, each of which contains at least one ICC and was annotated by three experienced radiologists. The resolution of the slices in both the source and target datasets is 512 × 512.

For image pre-processing, we use the same window width and window level for the source and target datasets. We take both the source and target datasets for model training, and only the target dataset for validation. Note that there is no overlap between the training and validation images we use. For the training of our DSTL model, we randomly select one CT image from each of the source and target datasets, respectively, and feed the slices in the selected image pair to the segmentation networks. Since the purpose of this paper is to evaluate the effectiveness of our selective transfer learning model DSTL, rather than proposing a carefully designed segmentation model for ICC segmentation, we do not utilize data augmentation even though it may help improve the segmentation results. For all experiments in this work, we divide the target dataset into five folds for cross-validation. All the networks are trained for 200 epochs using the Adam optimizer with a learning rate gradually decreasing from 1e-4. The experiments are conducted on an Nvidia TITAN XP GPU.

3.2 Results

We compare our DSTL model with several baseline approaches and conduct ablation study to evaluate the components in our DSTL model. We consider two metrics: Dice-Sørensen Coefficient (*DSC*) and Jaccard Index (*JI*), which are defined as follows:

$$DSC(Y^{(t)}, \mathcal{F}_{Seg}(X^{(t)})) = \frac{2 \times |Y^{(t)} \cap \mathcal{F}_{Seg}(X^{(t)})|}{|Y^{(t)}| + |\mathcal{F}_{Seg}(X^{(t)})|},$$
$$JI(Y^{(t)}, \mathcal{F}_{Seg}(X^{(t)})) = \frac{|Y^{(t)} \cap \mathcal{F}_{Seg}(X^{(t)})|}{|Y^{(t)} \cup \mathcal{F}_{Seg}(X^{(t)})|} \quad (9)$$

The mean and standard deviation values of *DSC* and *JI* are used to evaluate the performance of our model. Table 1 shows the segmentation results.

Four baseline approaches are used for comparison. First, we directly train a U-Net with only the target dataset and denote it as 'Direct Training' in Table 1. Second, we train a U-Net with data in both the source and target datasets without information selection and denote it as 'Joint Training'. Third, a U-Net is pre-trained with the source dataset and fine-tuned with the target one, which is denoted as 'Target Fine-tuning'. Finally, the image selection-based approach [3] is re-implemented for the ICC segmentation task. For fair comparison, all the four baseline approaches utilize the same experimental setups as SSNet.

As shown in Table 1, the performance of 'Joint Training' is worse than 'Direct Training'. This is because of the information gap between the source and target datasets (the gap confuses the segmentation network during training). At the same time, the performance of 'Target Fine-tuning' is slightly worse than 'Direct Training'. This may also be due to the information gap, and the current experimental setups are not sufficient for transfer learning. We further train the network in 'Target Fine-tuning' for more epochs, and the results catch up with and finally surpass those of 'Direct Training'. When an image selection based method [3] is applied, segmentation results are improved. Our DSTL model achieves results of 80.55 ± 4.21% *DSC*, and 67.64 ± 5.94% *JI*, which are the best compared to all the baseline approaches. Compared to 'Target Fine-tuning', even though our DSTL model contains more parameters, the extra parameters are utilized for information selection rather than ICC and liver feature learning. Parameters utilized for ICC and liver feature learning are the same (i.e., the parameters in U-Net), and our DSTL uses less information to achieve better results. This suggests that our DSTL can select useful information and exclude useless information for segmentation. Our DSTL and the method in [3] are both information selection approaches, and parameters used for ICC and liver feature learning are also the same. Our DSTL can achieve better results, which suggests that global and local information selection is more suitable than image selection approaches for medical image segmentation tasks. Figure 2 shows some examples of ICC segmentation results for comparison (Fig. 3).

Table 1. Experimental results of ICC segmentation in non-enhanced abdominal CT. All the results are reported based on five-fold cross-validation.

Method	Mean *DSC* ± SD (%)	Mean *JI* ± SD (%)
Direct training	76.10 ± 6.59	61.88 ± 8.84
Joint training	74.90 ± 7.16	60.39 ± 9.09
Target fine-tuning	76.01 ± 7.86	61.25 ± 9.95
Re-implemented version of [3]	76.86 ± 6.47	62.19 ± 8.62
DSTL w/o GISNet	79.12 ± 6.26	65.89 ± 8.41
DSTL w/o LISNet	77.48 ± 6.06	63.65 ± 8.24
DSTL	**80.55 ± 4.21**	**67.64 ± 5.94**

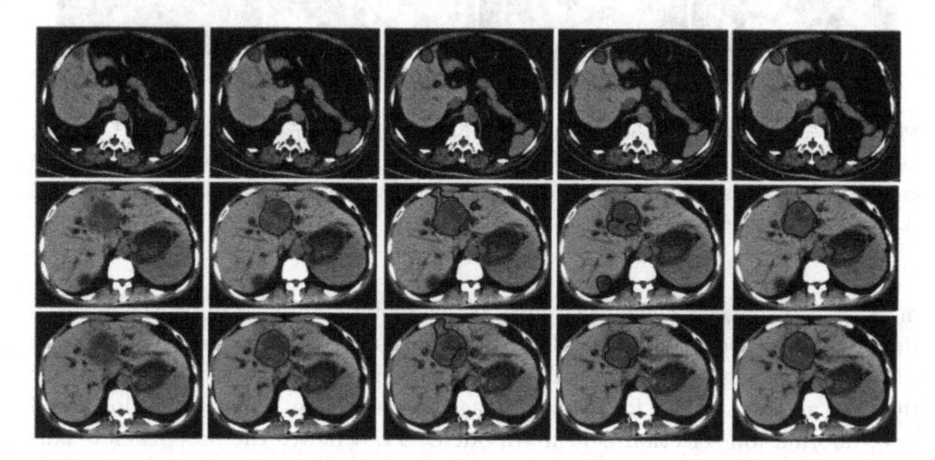

Fig. 2. Examples of ICC segmentation results for comparison. From left to right: (1) Target slices; (2) ground truth labels; (3) results of 'Target Fine-tuning'; (4) results of the re-implemented method [3]; (5) results of our DSTL model (best viewed in color) (Color figure online).

3.3 Ablation Study

We conduct ablation study to evaluate the effect of the components in our DSTL model. Our model contains three main components, which are SSNet, GISNet, and LISNet. Note that SSNet is composed of two weight-sharing segmentation networks in a Siamese structure. When only SSNet is utilized for segmentation, our model degenerates to 'Joint Training' in Table 1. To evaluate the effectiveness of GISNet and LISNet, we train DSTL without these two networks, respectively, and show the experimental results in Table 1.

When only one AdNet (AdNet$^{(1)}$) is utilized in our GISNet, the mean DSC of our model is only 75.62%, which is even worse than 'Direct Training'. This suggests the necessity of introducing two adversarial modules (i.e., AdNet$^{(1)}$ and AdNet$^{(0)}$). As shown in Table 1, when only global or local level information selection is utilized for segmentation, our model already outperforms the method

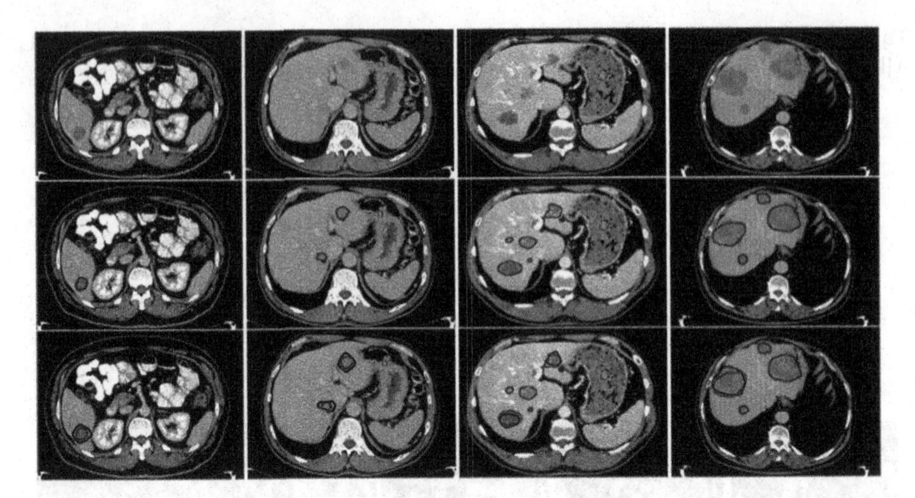

Fig. 3. Examples of local-level similar information selection results. From top to down: (1) Source slices; (2) ground truth labels; (3) results of LISNet. Green curves denote ground truth labels and red curves denote the results of LISNet (best viewed in color) (Color figure online).

in [3], which suggests the effectiveness of the global and local information selection networks. Specifically, 'DSTL w/o GISNet' slightly outperforms 'DSTL w/o LISNet'. This suggests that local-level information selection is more important than the global level one for medical image segmentation. When both global and local information selection is utilized, our DSTL achieves the best results. This suggests that the combination of hierarchical information selection is better than using only one level of information selection.

4 Conclusions

We have developed a new Dual-level Selective Transfer Learning (DSTL) model for ICC segmentation in non-enhanced abdominal CT images. Our model selects global and local level similar information from a similar task (the source task) and conducts transfer learning using the selected information to improve the segmentation results. Our model contains an SSNet for ICC and liver feature learning, a GISNet to extract global information and at the same time to mitigate the information gap between the source and target datasets, and a LISNet for local information selection. This is the first method for ICC segmentation in non-enhanced abdominal CT. Our DSTL model selects useful information in source datasets and achieves superior segmentation performance.

Acknowledgement. The research of J. Wu was partially supported by ZUEF under grants No. K17-511120-017 and No. K17-518051-021, NSFC under grant No. 61672453, and the National Key R&D Program of China under grant No. 2018AAA0102100, No. 2019YFC0118802, and No. 2019YFB1404802. The research of D.Z. Chen was partially supported by NSF Grant CCF-1617735.

References

1. Azizpour, H., Sharif Razavian, A., et al.: From generic to specific deep representations for visual recognition. In: CVPR Workshops, pp. 36–45 (2015)
2. Bilic, P., Christ, P.F., et al.: The liver tumor segmentation benchmark (LiTS). arXiv preprint arXiv:1901.04056 (2019)
3. Ge, W., Yu, Y.: Borrowing treasures from the wealthy: deep transfer learning through selective joint fine-tuning. In: CVPR, pp. 1086–1095 (2017)
4. Hong, S., Oh, J., et al.: Learning transferrable knowledge for semantic segmentation with deep convolutional neural network. In: CVPR, pp. 3204–3212 (2016)
5. Isensee, F., Petersen, J., et al.: nnU-Net: self-adapting framework for U-Net-based medical image segmentation. arXiv preprint arXiv:1809.10486 (2018)
6. Li, C., Wand, M.: Precomputed real-time texture synthesis with markovian generative adversarial networks. In: Leibe, B., Matas, J., Sebe, N., Welling, M. (eds.) ECCV 2016. LNCS, vol. 9907, pp. 702–716. Springer, Cham (2016). https://doi.org/10.1007/978-3-319-46487-9_43
7. Mala, K., Sadasivam, V.: Wavelet based texture analysis of liver tumor from computed tomography images for characterization using linear vector quantization neural network. In: ICACC, pp. 267–270 (2006)
8. Milletari, F., Navab, N., et al.: V-Net: fully convolutional neural networks for volumetric medical image segmentation. In: 3DV, pp. 565–571 (2016)
9. Nakanuma, Y., Sripa, B., et al.: Intrahepatic cholangiocarcinoma. World Health Organization classification of tumours: pathology and genetics of tumours of the digestive system, pp. 173–180 (2000)
10. Ronneberger, O., Fischer, P., Brox, T.: U-Net: convolutional networks for biomedical image segmentation. In: Navab, N., Hornegger, J., Wells, W.M., Frangi, A.F. (eds.) MICCAI 2015. LNCS, vol. 9351, pp. 234–241. Springer, Cham (2015). https://doi.org/10.1007/978-3-319-24574-4_28
11. Shang, H., et al.: Leveraging other datasets for medical imaging classification: evaluation of transfer, multi-task and semi-supervised learning. In: Shen, D., et al. (eds.) MICCAI 2019. LNCS, vol. 11768, pp. 431–439. Springer, Cham (2019). https://doi.org/10.1007/978-3-030-32254-0_48
12. Sun, R., Zhu, X., et al.: Not all areas are equal: transfer learning for semantic segmentation via hierarchical region selection. In: CVPR, pp. 4360–4369 (2019)
13. Zhang, J., Ding, Z., et al.: Importance weighted adversarial nets for partial domain adaptation. In: CVPR, pp. 8156–8164 (2018)

BiO-Net: Learning Recurrent Bi-directional Connections for Encoder-Decoder Architecture

Tiange Xiang[1], Chaoyi Zhang[1], Dongnan Liu[1], Yang Song[2], Heng Huang[3,4], and Weidong Cai[1](✉)

[1] School of Computer Science, University of Sydney, Sydney, Australia
{txia7609,dliu5812}@uni.sydney.edu.au
{chaoyi.zhang,tom.cai}@sydney.edu.au

[2] School of Computer Science and Engineering, University of New South Wales, Sydney, Australia
yang.song1@unsw.edu.au

[3] Electrical and Computer Engineering, University of Pittsburgh, Pittsburgh, USA
henghuanghh@gmail.com

[4] JD Finance America Corporation, Mountain View, CA, USA

Abstract. U-Net has become one of the state-of-the-art deep learning-based approaches for modern computer vision tasks such as semantic segmentation, super resolution, image denoising, and inpainting. Previous extensions of U-Net have focused mainly on the modification of its existing building blocks or the development of new functional modules for performance gains. As a result, these variants usually lead to an unneglectable increase in model complexity. To tackle this issue in such U-Net variants, in this paper, we present a novel **Bi**-directional **O**-shape network (BiO-Net) that reuses the building blocks in a recurrent manner without introducing any extra parameters. Our proposed bi-directional skip connections can be directly adopted into any encoder-decoder architecture to further enhance its capabilities in various task domains. We evaluated our method on various medical image analysis tasks and the results show that our BiO-Net significantly outperforms the vanilla U-Net as well as other state-of-the-art methods. Our code is available at https://github.com/tiangexiang/BiO-Net.

Keywords: Semantic segmentation · Bi-directional connections · Recursive neural networks

1 Introduction

Deep learning based approaches have recently prevailed in assisting medical image analysis, such as whole slide image classification [27], brain lesion

Electronic supplementary material The online version of this chapter (https://doi.org/10.1007/978-3-030-59710-8_8) contains supplementary material, which is available to authorized users.

A. L. Martel et al. (Eds.): MICCAI 2020, LNCS 12261, pp. 74–84, 2020.
https://doi.org/10.1007/978-3-030-59710-8_8

segmentation [26], and medical image synthesis [10]. U-Net [22], as one of the most popular deep learning based models, has demonstrated its impressive representation capability in numerous medical image computing studies. U-Net introduces skip connections that aggregate the feature representations across multiple semantic scales and helps prevent information loss.

U-Net Variants. Recent works were proposed to extend the U-Net structure with varying module design and network construction, illustrating its potentials on various visual analysis tasks. V-Net [17] applies U-Net on higher dimension voxels and keeps the vanilla internal structures. W-Net [25] modifies U-Net to tackle unsupervised segmentation problem by concatenating two U-Nets via an autoencoder style model. Compared to U-Net, M-Net [16] appends different scales of input features to different levels, thus multi-level visual details can be captured by a series of downsampling and upsampling layers. Recently, U-Net++ [28] adopts nested and dense skip connections to represent the fine-grained object details more effectively. Moreover, attention U-Net [20] uses extra branches to adaptively apply attention mechanism on the fusion of skipped and decoded features. However, these proposals may involve additional building blocks, which lead to a greater number of network parameters and thus an increased GPU memory. Unlike above variants, our BiO-Net improves the performance of U-Net via a novel feature reusing mechanism where it builds bi-directional connections between the encoder and decoder to make inference from a recursive manner.

Recurrent Convolutional Networks. Using recurrent convolution to iteratively refine the features extracted at different times has been demonstrated to be feasible and effective for many computer vision problems [1,7,8,24]. Guo *et al.* [7] proposed to reuse residual blocks in ResNet so that available parameters would be fully utilized and model size could be reduced significantly. Such a mechanism also benefits the evolution of U-Net. As a result, Wang *et al.* [24] proposed R-U-Net, which recurrently connects multiple paired encoders and decoders of U-Net to enhance its discrimination power for semantic segmentation, though, extra learnable blocks are introduced as a trade-off. BiO-Net distinguishes R-U-Net from its design of backward skip connections, where latent features in every decoding levels are reused, enabling more intermediate information aggregations with gradients preserved among temporal steps. R2U-Net [1] adopts a similar approach that only recurses the last building block at each level of refinement. By contrast, our method learnt recurrent connections in the existing encoder and decoder rather than recursing the same level blocks without the involvement of refined decoded features.

To this end, we propose a recurrent U-Net with an bi-directional O-Shape inference trajectory (BiO-Net) that maps the decoded features back to the encoder through the backward skip connections, and recurses between the encoder and the decoder. Compared to previous works, our approach achieves a better feature refinement, as multiple encoding and decoding processes are triggered in our BiO-Net. We applied our BiO-Net to perform semantic segmentation on nuclei segmentation task and EM membrane segmentation task and our results show that the proposed BiO-Net outperforms other U-Net variants, including the recurrent counterparts and many state-of-the-art approaches. Experiments on super resolution tasks also demonstrate the significance of our BiO-Net applied to different scenarios.

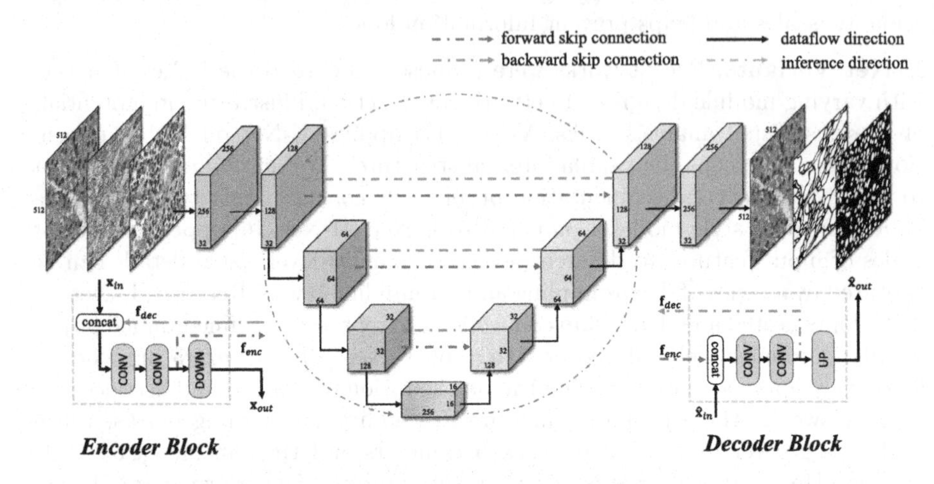

Fig. 1. Overview of our BiO-Net architecture. The network inferences recurrently in an O-shape manner. `CONV` represents the sequence of convolution, non-linearity and batch-norm layers. `DOWN` stands for downsampling (implemented by a `CONV` followed by max-pooling), while `UP` denotes upsampling (archived by transpose convolution).

2 Methods

As shown in Fig. 1, BiO-Net adopts the same network architecture as U-Net, without any dependencies on extra functional blocks but with paired bi-directional connections. It achieves better performance as t increases with no extra trainable parameters introduced during its unrolling process. Moreover, our method is not restricted to U-Net and can be integrated into other encoder-decoder architectures for various visual analysis tasks.

2.1 Recurrent Bi-directional Skip Connections

The main uniqueness of our BiO-Net model is the introduction of bi-directional skip connections, which facilitate the encoders to process the semantic features in the decoders and vice versa.

Forward Skip Connections. Forward skip connections linking encoders and decoders at the same level can preserve the encoded low-level visual features $\mathbf{f}_{enc}$, with their gradients well-preserved [9,22]. Hence, the l-th decoder block could fuse $\mathbf{f}_{enc}$ with its input $\hat{\mathbf{x}}_{in}$ generated from lower blocks, and propagate them through the decoding convolutions DEC to generate $\mathbf{f}_{dec}$, which will be further restored to higher resolutions via UP block. This process can be defined as:

$$\mathbf{f}_{dec} = \mathtt{DEC}([\mathbf{f}_{enc}, \hat{\mathbf{x}}_{in}]), \tag{1}$$

where concatenation is employed as our fusion mechanism $[\cdot]$. The index notation (l-th) for encoders and decoders is omitted in this paper for simplicity purpose.

Backward Skip Connections. With the help of our novel backward skip connections, which pass the decoded high-level semantic features $\mathbf{f}_{dec}$ from the decoders to the encoders, our encoder can now combine $\mathbf{f}_{dec}$ with its original input $\mathbf{x}_{in}$ produced by previous blocks, and therefore achieves flexible aggregations of low-level visual features and high-level semantic features. Similar to the decoding path enhanced by forward skip connections, our encoding process can thus be formulated with its encoding convolutions ENC as:

$$\mathbf{f}_{enc} = \mathtt{ENC}([\mathbf{f}_{dec}, \mathbf{x}_{in}]). \tag{2}$$

The DOWN block feeds $\mathbf{f}_{enc}$ to subsequent encoders for deeper feature extraction.

Recursive Inferences. The above bi-directional skip connections create an O-shaped inference route for encoder-decoder architectures. Noticeably, this O-shaped inference route can be recursed multiple times to receive immediate performance gains, and more importantly, this recursive propagation policy would not introduce any extra trainable parameters. Hence, the outputs of encoders and decoders equipped with our proposed O-shaped connections can be demonstrated as follows, in terms of their current inference iteration i:

$$\begin{aligned} \mathbf{x}^i_{out} &= \mathtt{DOWN}(\mathtt{ENC}([\mathtt{DEC}([\mathbf{f}^{i-1}_{enc}, \hat{\mathbf{x}}^{i-1}_{in}]), \mathbf{x}^i_{in}])), \\ \hat{\mathbf{x}}^i_{out} &= \mathtt{UP}(\mathtt{DEC}([\mathtt{ENC}([\mathbf{f}^i_{dec}, \mathbf{x}^i_{in}]), \hat{\mathbf{x}}^i_{in}])). \end{aligned} \tag{3}$$

Compared to the vanilla U-Net, our BiO-Net takes both encoded and decoded features into consideration and performs refinement according to features from the previous iterations.

2.2 BiO-Net Architecture

We use the plain convolutional layers, batch normalization [11] layers, and ReLU [18] layers in the network architecture. No batch normalization layer is reused.

The input image is first fed into a sequence of three convolutional blocks to extract low-level features. Note that there is no backward skip connection attached to the first stage block and, hence, the parameters in the first stage block will not be reused when recursing. The extracted features are then sent to a cascade of encode blocks that utilize max pooling for feature downsampling. During the encoding stage, the parameters are reused and the blocks are recursed through the paired forward and backward connections as shown in

Fig. 1. After the encoding phase, an intermediate stage of which contains convolutional blocks that are used to further refine the encoded features. Then, the features are subsequently passed into a series of decode blocks that recover encoded details using convolutional transpose operations. During the decoding stage, our proposed backward skip connections preserve retrieved features by concatenating them with the features from the same level encoders as depicted in Sect. 2.1. The recursion begins with the output generated from the last convolutional block in the decoding stage. After recursing the encoding and decoding stages, the updated output will be fed into the last stage block corresponding to the first stage blocks. Similarly, the last stage blocks will not be involved in the recurrence.

We define several symbols for better indication: 't' represents the total recurrence time; '$\times n$' represents the expansion multiplier times to all hidden output channel numbers; 'w' represents the number of backward skip connections used from the deepest encoding level; '`INT`' represents stacking decoded features from each iteration and feeding them into the last stage block as a whole and 'l' represents the most encoding depth. Details can be seen in Fig. 2.

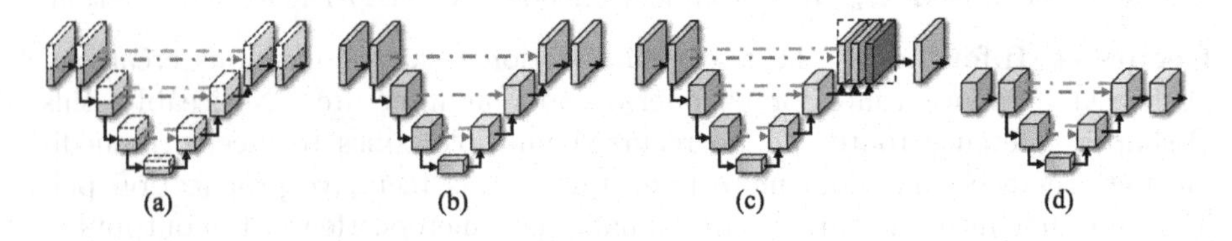

Fig. 2. Visualizations of setups in the ablation experiments. (a) BiO-Net with $\times 0.5$. (b) BiO-Net with $w = 2$. (c) BiO-Net with $t = 3$ and `INT`. (d) BiO-Net with $l = 3$.

3 Experiments

Datasets. Our method was evaluated on three common digital pathology image analysis tasks: nuclei segmentation, EM membrane segmentation, and image super resolution on a total of four different datasets. Two publicly available datasets, namely MoNuSeg [13] and TNBC [19], were selected to evaluate our method for nuclei semantic segmentation. The MoNuSeg dataset consists of a 30-image training set and a 14-image testing set, with images of size 1000^2 sampled from different whole slide images of multiple organs. We extract 512^2 patches from 4 corners of each image, which enlarges the dataset by 4 times. TNBC is comprised of 50 histopathology images of size 512^2 without any specific testing set. Both datasets include pixel-level annotation for the nuclei semantic segmentation problem. The second task we evaluated is EM membrane segmentation, where the piriform cortex EM dataset of the mice collected from [15], which contains four stacks of EM images with the slice image sizes of 255^2,

512^2, 512^2, and 256^2, respectively. Image super resolution is the last task we evaluated our method on, the dataset was constructed from a whole slide image collected by MICCAI15 CBTC. We sampled 2900 patches of size 512^2 at $40\times$ magnification level, with the 9 : 1 split ratio for the training and testing set, respectively.

Implementation Details. We used Adam [12] optimizer with an initial learning rate of 0.01 and a decay rate of 0.00003 to minimize cross entropy loss in segmentation tasks and mean square error in super resolution tasks. The training dataset was augmented by applying random rotation (within the range [$-15°$, $+15°$]), random shifting (in both x- and y-directions; within the range of [-5%, 5%]), random shearing, random zooming (within the range [0, 0.2]), and random flipping (both horizontally and vertically). The batch size is set to 2 in both training and testing phases. Unless explicitly specified, our BiO-Net is constructed with an encoding depth of 4 and a backward skip connection built at each stage of the network. Our BiO-Net was trained by 300 epochs in all experiments, which were conducted on a single NVIDIA GeForce GTX 1080 GPU with Keras. Given the GPU limitation, we explore the performance improvement to its maximum possible temporal step at $t = 3$.

Table 1. Comparison of segmentation methods on MoNuSeg testing set and TNBC.

	MoNuSeg		TNBC			
Methods	IoU	DICE	IoU	DICE	#params	Model size
U-Net [22] **w.** ResNet-18 [9]	0.684	0.810	0.459	0.603	15.56 M	62.9 MB
U-Net++ [28] **w.** ResNet-18 [9]	0.683	0.811	0.526	0.652	18.27 M	74.0 MB
U-Net++ [28] **w.** ResNet-50 [9]	0.695	0.818	0.542	0.674	37.70 M	151.9 MB
Micro-Net [21]	0.696	0.819	0.544	0.701	14.26 M	57.4 MB
Naylor *et al.* [19]	0.690	0.816	0.482	0.623	36.63 M	146.7 MB
M-Net [16]	0.686	0.813	0.450	0.569	0.6 M	2.7 MB
Att U-Net [20]	0.678	0.810	0.581	0.717	33.04 M	133.2 MB
R2U-Net, $t = 2$ [1]	0.678	0.807	0.532	0.650	37.02 M	149.2 MB
R2U-Net, $t = 3$ [1]	0.683	0.815	0.590	0.711	37.02 M	149.2 MB
LinkNet [3]	0.625	0.767	0.535	0.682	11.54 M	139.4 MB
BiO-LinkNet [3], $t = 2$ (Ours)	0.621	0.766	0.541	0.690	11.54 M	139.4 MB
BiO-LinkNet [3], $t = 3$ (Ours)	0.634	0.774	0.571	0.716	11.54 M	139.4 MB
BiO-Net, $t = 1$ (ours)	0.680	0.803	0.456	0.608	15.0 M	60.6 MB
BiO-Net, $t = 2$ (ours)	0.694	0.816	0.548	0.693	15.0 M	60.6 MB
BiO-Net, $t = 3$ (ours)	0.700	0.821	0.618	0.751	15.0 M	60.6 MB
BiO-Net, $t = 3$, INT (Ours)	**0.704**	**0.825**	**0.651**	**0.780**	15.0 M	60.6 MB

3.1 Semantic Segmentation

Nuclei Segmentation. In this task, our method is compared to the baseline U-Net [22] and other state-of-the-art methods [1,16,19–21,28]. Following [6], models were trained on the MoNuSeg training set only and evaluated on the MoNuSeg testing set and TNBC dataset. Dice coefficient (DICE) and Intersection over Union (IoU) are evaluated. As shown in Table 1, our results are better than others on the MoNuSeg testing set. Our results on the TNBC dataset are also higher than the others by a large margin, which demonstrates a strong generalization ability. Additionally, compared with other extensions of U-Net, and, our BiO-Net is more memory efficient. Qualitative comparison of our method and the recurrent counterpart R2U-Net [1] is shown in Fig. 3. It can be seen that our model segments nuclei more accurately as the inference time increases. In our experiments, BiO-Net infers a batch of two predictions in 35, 52, and 70 ms when $t=1$, $t=2$, and $t=3$, respectively. Further evaluation of incorporating our method into another encoder-decoder architecture LinkNet [3], which is also shown in the table. Our BiO-LinkNet adds the skipped features element-wisely and, hence, shares the same number of parameters as the vanilla LinkNet.

Table 2. Ablative results. The parameters are defined as depicted in Sect. 2.2. IoU(DICE), number of parameters, and, model size are reported.

	MoNuSeg			TNBC				
	$t=1$	$t=2$	$t=3$	$t=1$	$t=2$	$t=3$	#params	Model size
×1.25	0.685 (0.813)	0.698 (0.819)	0.695 (0.817)	0.490 (0.637)	0.557 (0.697)	0.623 (0.758)	23.5 M	94.3 MB
×1.0	0.680 (0.803)	0.694 (0.816)	0.700 (0.821)	0.456 (0.608)	0.548 (0.693)	0.618 (0.751)	15.0 M	60.6 MB
×0.75	0.676 (0.800)	0.678 (0.805)	0.691 (0.815)	0.516 (0.661)	0.571 (0.710)	0.598 (0.738)	8.5 M	34.3 MB
×0.5	0.668 (0.792)	0.680 (0.806)	0.691 (0.814)	0.491 (0.644)	0.543 (0.679)	0.611 (0.742)	3.8 M	15.8 MB
×0.25	0.667 (0.791)	0.678 (0.804)	0.677 (0.804)	0.524 (0.674)	0.535 (0.678)	0.575 (0.710)	0.9 M	4.0 MB
$w=3$	0.680 (0.803)	0.694 (0.817)	0.688 (0.814)	0.456 (0.608)	0.510 (0.656)	0.620 (0.757)	15.0 M	60.3 MB
$w=2$	0.680 (0.803)	0.672 (0.801)	0.686 (0.813)	0.456 (0.608)	0.527 (0.679)	0.601 (0.742)	14.9 M	60.1 MB
INT	0.680 (0.803)	0.689 (0.812)	**0.704 (0.825)**	0.456 (0.608)	0.588 (0.728)	**0.651(0.780)**	15.0 M	60.6 MB
$l=3$	0.681 (0.806)	0.679 (0.805)	0.689 (0.812)	0.613 (0.742)	0.594 (0.733)	0.615 (0.741)	3.8 M	15.4 MB
$l=2$	0.690 (0.810)	0.695 (0.817)	0.697 (0.818)	0.596 (0.734)	0.647 (0.775)	0.596 (0.735)	0.9 M	4.0 MB

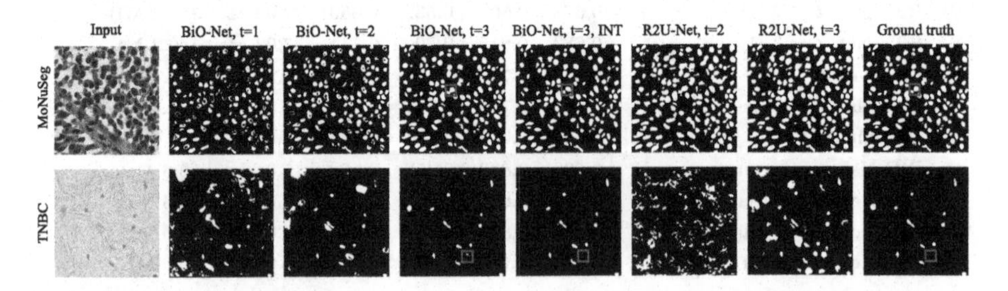

Fig. 3. Qualitative comparison between our models and the R2U-Net [1] on the MoNuSeg testing set and TNBC. Red boundary boxes indicate the effects of integrating features from each iteration at the last stage block. (Color figure online)

Table 2 demonstrates our ablation study by varying the setups as defined in Fig. 2. The results show that recursing through the encoder and the decoder with the proposed bi-directional skip connections improves network performances generally. Integrating decoded features from all inference recurrences yields state-of-the-art performances in both datasets. Furthermore, we find that when there are insufficient parameters in the network, increasing inference recurrence has little improvement or even makes the results worse. It is also interesting to observe that when constructing BiO-Net with shallower encoding depth, our models perform better on the two datasets than those with deeper encoding depth.

EM Membrane Segmentation. We further evaluated our method by segmenting Mouse Piriform Cortex EM images [15], where the models are trained on stack1 and stack2, and validated on stack4 and stack3. The results were evaluated by Rand F-score [2]. As shown in Table 3, our method demonstrates better segmentation results with the proposed bi-directional O-shaped skip connections.

Table 3. Comparison of different U-Net variants in EM membrane segmentation.

Variants	Stack3	Stack4	#params	Ours	Stack3	Stack4	#params
U-Net	0.939	0.821	15.56 M	BiO-Net, t = 1	0.941	0.827	15.0 M
Att U-Net	0.937	0.833	33.04M	BiO-Net, t = 2	0.955	0.871	15.0 M
U-Net++	0.940	0.844	18.27 M	BiO-Net, t = 3	**0.958**	**0.887**	15.0 M

3.2 Super Resolution

In addition to segmentation tasks, we are also interested in experimenting with our BiO-Net on a significantly different task: image super resolution, which has been studied actively [4,23]. In the super resolution task, low-resolution (down-sampled) images are used as inputs to train the networks toward their original high-resolution ground truth, which can assist medical imaging analysis by

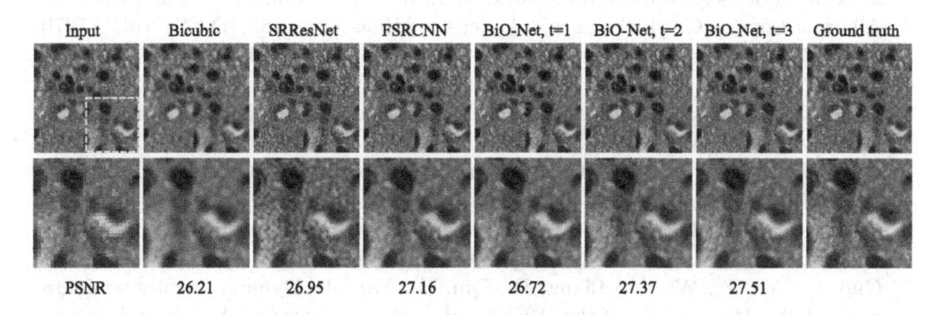

Fig. 4. Comparison of different methods in super resolution task. PSNR scores of the methods over the entire testing set are reported as well. The second row is projected from the yellow boundary box to have a closer view on the super resolution achieved in higher resolution. (Color figure online)

recovering missing details and generating high-resolution histopathology images based on the low-resolution ones. Two state-of-the-art methods, FSRCNN [5] and SRResNet [14], are adopted to compare with our BiO-Net. The qualitative results along with the Peak Signal to Noise Ratio (PSNR) score over the entire testing set are shown in Fig. 4. It can be seen that, our method outperforms the state-of-the-art methods by a safe margin, which validates the feasibility of applying our BiO-Net on different visual tasks.

4 Conclusion

In this paper, we introduced a novel recurrent variant of U-Net, named BiO-Net. BiO-Net is a compact substitute of U-Net with better performance and no extra trainable parameters, which utilizes paired forward and backward skip connections to compose a complex relation between the encoder and decoder. The model can be recursed to reuse the parameters during training and inference. Extensive experiments on semantic segmentation and super-resolution tasks indicate the effectiveness of our proposed model, which outperforms the U-Net and its extension methods without introducing auxiliary parameters.

References

1. Alom, M.Z., Yakopcic, C., Taha, T.M., Asari, V.K.: Nuclei segmentation with recurrent residual convolutional neural networks based U-Net (R2U-Net). In: IEEE National Aerospace and Electronics Conference, pp. 228–233. IEEE (2018)
2. Arganda-Carreras, I., et al.: Crowdsourcing the creation of image segmentation algorithms for connectomics. Front. Neuroanat. **9**, 142 (2015)
3. Chaurasia, A., Culurciello, E.: LinkNet: exploiting encoder representations for efficient semantic segmentation. In: IEEE Visual Communications and Image Processing (VCIP), pp. 1–4. IEEE (2017)
4. Chen, Y., Shi, F., Christodoulou, A.G., Xie, Y., Zhou, Z., Li, D.: Efficient and accurate mri super-resolution using a generative adversarial network and 3D multi-level densely connected network. In: Frangi, A., Schnabel, J., Davatzikos, C., Alberola-López, C., Fichtinger, G. (eds.) MICCAI 2018. LNCS, vol. 11070, pp. 91–99. Springer, Heidelberg (2018). https://doi.org/10.1007/978-3-030-00928-1_11
5. Dong, C., Loy, C.C., Tang, X.: Accelerating the super-resolution convolutional neural network. In: Leibe, B., Matas, J., Sebe, N., Welling, M. (eds.) ECCV 2016. LNCS, vol. 9906, pp. 391–407. Springer, Heidelberg (2016). https://doi.org/10.1007/978-3-319-46475-6_25
6. Graham, S., et al.: Hover-net: Simultaneous segmentation and classification of nuclei in multi-tissue histology images. Med. Image Anal. (MIA) **58**, 101563 (2019)
7. Guo, Q., Yu, Z., Wu, Y., Liang, D., Qin, H., Yan, J.: Dynamic recursive neural network. In: Proceedings of the IEEE Conference on Computer Vision and Pattern Recognition (CVPR), pp. 5147–5156 (2019)
8. Han, W., Chang, S., Liu, D., Yu, M., Witbrock, M., Huang, T.S.: Image super-resolution via dual-state recurrent networks. In: Proceedings of the IEEE Conference on Computer Vision and Pattern Recognition (CVPR), pp. 1654–1663 (2018)

9. He, K., Zhang, X., Ren, S., Sun, J.: Deep residual learning for image recognition. In: Proceedings of the IEEE Conference on Computer Vision and Pattern Recognition (CVPR), pp. 770–778 (2016)
10. Hou, L., Agarwal, A., Samaras, D., Kurc, T.M., Gupta, R.R., Saltz, J.H.: Robust histopathology image analysis: to label or to synthesize? In: Proceedings of the IEEE Conference on Computer Vision and Pattern Recognition (CVPR), pp. 8533–8542 (2019)
11. Ioffe, S., Szegedy, C.: Batch normalization: accelerating deep network training by reducing internal covariate shift. In: International Conference on Machine Learning (ICML), pp. 448–456 (2015)
12. Kingma, D.P., Ba, J.: Adam: a method for stochastic optimization. In: International Conference on Learning Representations (ICLR) (2015)
13. Kumar, N., Verma, R., Sharma, S., Bhargava, S., Vahadane, A., Sethi, A.: A dataset and a technique for generalized nuclear segmentation for computational pathology. IEEE Trans. Med. Imaging (TMI) **36**(7), 1550–1560 (2017)
14. Ledig, C., et al.: Photo-realistic single image super-resolution using a generative adversarial network. In: Proceedings of the IEEE Conference on Computer Vision and Pattern Recognition (CVPR), pp. 4681–4690 (2017)
15. Lee, K., Zlateski, A., Ashwin, V., Seung, H.S.: Recursive training of 2D–3D convolutional networks for neuronal boundary prediction. In: Advances in Neural Information Processing Systems (NeurIPS), pp. 3573–3581 (2015)
16. Mehta, R., Sivaswamy, J.: M-net: A convolutional neural network for deep brain structure segmentation. In: 14th International Symposium on Biomedical Imaging (ISBI), pp. 437–440. IEEE (2017)
17. Milletari, F., Navab, N., Ahmadi, S.A.: V-net: fully convolutional neural networks for volumetric medical image segmentation. In: 4th International Conference on 3D Vision (3DV), pp. 565–571. IEEE (2016)
18. Nair, V., Hinton, G.E.: Rectified linear units improve restricted boltzmann machines. In: Proceedings of the 27th International Conference on Machine Learning (ICML), pp. 807–814 (2010)
19. Naylor, P., Laé, M., Reyal, F., Walter, T.: Segmentation of nuclei in histopathology images by deep regression of the distance map. IEEE Trans. Med. Imaging (TMI) **38**(2), 448–459 (2018)
20. Oktay, O., et al.: Attention U-net: learning where to look for the pancreas. In: 1st Conference on Medical Imaging with Deep Learning (MIDL) (2018)
21. Raza, S.E.A., et al.: Micro-net: a unified model for segmentation of various objects in microscopy images. Med. Image Anal. (MIA) **52**, 160–173 (2019)
22. Ronneberger, O., Fischer, P., Brox, T.: U-Net: convolutional networks for biomedical image segmentation. In: Navab, N., Hornegger, J., Wells, W.M., Frangi, A.F. (eds.) MICCAI 2015. LNCS, vol. 9351, pp. 234–241. Springer, Cham (2015). https://doi.org/10.1007/978-3-319-24574-4_28
23. Sui, Y., Afacan, O., Gholipour, A., Warfield, S.K.: Isotropic MRI super-resolution reconstruction with multi-scale gradient field prior. In: Shen, D., et al. (eds.) MICCAI 2019. LNCS, vol. 11766, pp. 3–11. Springer, Cham (2019). https://doi.org/10.1007/978-3-030-32248-9_1
24. Wang, W., Yu, K., Hugonot, J., Fua, P., Salzmann, M.: Recurrent U-net for resource-constrained segmentation. In: The IEEE International Conference on Computer Vision (ICCV) (2019)
25. Xia, X., Kulis, B.: W-net: a deep model for fully unsupervised image segmentation. arXiv preprint arXiv:1711.08506 (2017)

26. Zhang, C., et al.: Ms-GAN: GAN-based semantic segmentation of multiple sclerosis lesions in brain magnetic resonance imaging. In: 2018 Digital Image Computing: Techniques and Applications (DICTA), pp. 1–8. IEEE (2018)
27. Zhang, C., Song, Y., Zhang, D., Liu, S., Chen, M., Cai, W.: Whole slide image classification via iterative patch labelling. In: 25th IEEE International Conference on Image Processing (ICIP), pp. 1408–1412. IEEE (2018)
28. Zhou, Z., Rahman Siddiquee, M.M., Tajbakhsh, N., Liang, J.: UNet++: a nested U-Net architecture for medical image segmentation. In: Stoyanov, D., et al. (eds.) DLMIA/ML-CDS -2018. LNCS, vol. 11045, pp. 3–11. Springer, Cham (2018). https://doi.org/10.1007/978-3-030-00889-5_1

Constrain Latent Space for Schizophrenia Classification via Dual Space Mapping Net

Weiyang Shi[1,2], Kaibin Xu[1], Ming Song[1,2], Lingzhong Fan[1,2,4], and Tianzi Jiang[1,2,3,4,5,6](✉)

[1] Brainnetome Center and National Laboratory of Pattern Recognition, Institute of Automation, Chinese Academy of Sciences, Beijing, China
jiangtz@nlpr.ia.ac.cn
[2] School of Artificial Intelligence, University of Chinese Academy of Sciences, Beijing, China
[3] Center for Excellence in Brain Science and Intelligence Technology, Chinese Academy of Sciences, Shanghai, China
[4] Innovation Academy for Artificial Intelligence, Chinese Academy of Sciences, Beijing, China
[5] Key Laboratory for Neuro Information of Ministry of Education, School of Life Science and Technology, University of Electronic Science and Technology of China, Chengdu, China
[6] Queensland Brain Institute, University of Queensland, Brisbane, Australia

Abstract. Mining potential biomarkers of schizophrenia (SCZ) while performing classification is essential for the research of SCZ. However, most related studies only perform a simple binary classification with high-dimensional neuroimaging features that ignore individual's unique clinical symptoms. And the biomarkers mined in this way are more susceptible to confounding factors such as demographic factors. To address these questions, we propose a novel end-to-end framework, named Dual Spaces Mapping Net (DSM-Net), to map the neuroimaging features and clinical symptoms to a shared decoupled latent space, so that constrain the latent space into a solution space associated with detailed symptoms of SCZ. Briefly, taking functional connectivity patterns and the Positive and Negative Syndrome Scale (PANSS) scores as input views, DSM-Net maps the inputs to a shared decoupled latent space which is more discriminative. Besides, with an invertible space mapping sub-network, DSM-Net transforms multi-view learning into multi-task learning and provides regression of PANSS scores as an extra benefit. We evaluate the proposed DSM-Net with multi-site data of SCZ in the leave-one-site-out cross validation setting and experimental results illustrate the effectiveness of DSM-Net in classification, regression performance, and unearthing neuroimaging biomarkers with individual specificity, population commonality and less effect of confusions.

Keywords: Schizophrenia · Clinical symptoms · Multi-view learning

A. L. Martel et al. (Eds.): MICCAI 2020, LNCS 12261, pp. 85–94, 2020.
https://doi.org/10.1007/978-3-030-59710-8_9

1 Introduction

Schizophrenia (SCZ) is one of the most disabling psychiatric disorders worldwide. Nevertheless, the complex etiologic and biological underpinnings of SCZ have not been completely understood. In addition, the current diagnosis still relies on consultation performed by qualified psychiatrists, lacking objective biomarkers.

To shed light on the pathogenesis of SCZ and further establish objective diagnostic criteria, large-scale related studies have been done, falling broadly into statistics based group-level and classification based individual-level researches. As group-level studies cannot do inference at the individual level, increasing researches shift attention to individual-level analysis, mainly based on machine learning algorithms [6,12]. Almost all of these studies regard the patients with SCZ and healthy controls as two homogeneous classes and perform simple binary classification. In fact, there are some commonly used clinical scales that provide crucial information in clinical practice. For example, the Positive and Negative Syndrome Scale (PANSS) [5], a well-established assessment of SCZ psychopathology, is commonly used to assess individual symptoms such as delusions and hallucinations. However, this detailed information has often been ignored in previous neuroimage-based classification studies [7,12]. This neglect hinders the adoption of these methods in clinical practice.

Given the current understanding of mental illness as abnormalities in functional connectomics [8], functional connectivity, the pairwise Pearson's correlation between pre-defined brain areas, extracted from resting-state functional magnetic resonance imaging (rs-fMRI) is widely used in the studies of SCZ [6,12]. It was hoped that fMRI would lead to the development of imaging-based biomarkers and objective diagnosis. However, due to the high-dimensional property of functional connectivity and the complex etiology of SCZ, the biomarkers discovered based on simple binary classification are more susceptible to confounding factors such as demographic or other unidentified irrelevant factors. Besides, the analysis of biomarkers mined in this way in previous studies was performed at group level to a certain degree that only considered consistent features among patients. Focusing on the common biomarkers hinders further exploration of the heterogeneity of SCZ and precision psychiatry.

Taking all of the above challenges into account, we introduce PANSS as a constraint view to constrain the solution space. Specifically, we take functional connectivity and PANSS as two input views and map these two feature spaces to a shared decoupled latent space via the proposed Dual Space Mapping Net (DSM-Net). Based on invertible blocks, the multi-view learning task can further be formed as a general multi-task learning process with functional connectivity as the only input feature sets so as to provide the extra regression of PANSS scores. Thus, we can optimize the classifier of SCZ, the regressor of PANSS scores, and the decoupling constraint of latent space simultaneously. Applying DSM-Net to a large-scale multi-site SCZ dataset in the leave-one-site-out cross validation setting, we improved the classification accuracy to 84.33% and provided regression of the PANSS scores which is important for clinical practice but often overlooked in similar studies. Besides, with the good interpretability and clinical symptom constraint, DSM-Net mined potential biomarkers with group commonality and individual specificity which are meaningful for precision medicine.

The contribution of this paper is summarized as follows: (1) Within the proposed DSM-Net, the dual space mapping sub-networks mapped functional connectivity and PANSS to a shared decoupled space which is more discriminative and contains more relevant information about SCZ; (2) Benefit from the invertible mapping sub-network, we transformed multi-view learning into a general form of multi-task learning and additionally obtained the clinically meaningful regression of PANSS; (3) Considering individual's unique symptoms, the proposed DSM-Net unearthed neuroimaging biomarkers with individual specificity, population commonality, and less effect of confusion.

2 Method

In order to overcome the shortcomings of ignoring detailed symptoms and the existence influence of confounding factors caused by simple binary classification with high dimensional input, we constrain the solution space with the proposed novel multi-view learning framework DSM-Net, as shown in Fig. 1(a). Denote $X = [\boldsymbol{x_1}, \boldsymbol{x_2}, ..., \boldsymbol{x_n}]^T \in \mathbb{R}^{n \times d}$ as the functional connectivity matrix for all subjects, where d is the feature dimension of functional connectivity for each sample and n is the number of subjects. Similarly, we denote $S = [\boldsymbol{s_1}, \boldsymbol{s_2}, ..., \boldsymbol{s_n}]^T \in \mathbb{R}^{n \times m}$ and $Y = [y_1, y_2, ..., y_n]^T \in \mathbb{R}^n$ as the associated PANSS score matrix and label vector respectively, where m is the dimension of PANSS scores for each sample, $y_i \in \{0, 1\}$ is the associated class label (i.e., healthy control or SCZ), and $\boldsymbol{s_i}$ is the PANSS score vector of the i-th subject.

Within DSM-Net, we try to map X in space $\mathcal{X}$ to Z in the shared latent space $\mathcal{Z}$ with the constraint that bidirectional mapping can be realized between space $\mathcal{S}$ and $\mathcal{Z}$. And then perform the classification task in this latent space $\mathcal{Z}$. In particular, inspired by soft CCA [2], we additionally introduce a decorrelation loss to further constrain space $\mathcal{Z}$ as a decoupled space which is theoretically possible to obtain more information capacity. In the rest of this section, we will cover the details of the proposed DSM-Net.

2.1 Invertible Block Based Dual Space Mapping

Due to the interpretability [13] and reversibility of invertible block, we adopt invertible block as the main component of DSM-Net. Invertible block proposed by Gomez *et al.* [4] can fully recover the input from the output, as illustrated in Fig. 1(b). In the forward computation, invertible block divides the input $\boldsymbol{x} \in \mathbb{R}^d$ into two parts: $\boldsymbol{x}^{(1)} \in \mathbb{R}^{d/2}$ and $\boldsymbol{x}^{(2)} \in \mathbb{R}^{d/2}$ of the same size. With input $\boldsymbol{x}$ and output $\boldsymbol{o}$, the forward and backward computation formed as using Eq. 1. In order to ensure that the input $\boldsymbol{x}$ can be recovered from the output $\boldsymbol{o}$, the dimension of $\boldsymbol{x}$ and $\boldsymbol{o}$ must be the same. As there are no additional constraints, F and G can be any arbitrary function.

$$\begin{cases} \boldsymbol{o}^{(2)} = \boldsymbol{x}^{(2)} + F(\boldsymbol{x}^{(1)}) \\ \boldsymbol{o}^{(1)} = \boldsymbol{x}^{(1)} + G(\boldsymbol{o}^{(2)}) \end{cases} \quad \begin{cases} \boldsymbol{x}^{(1)} = \boldsymbol{o}^{(1)} - G(\boldsymbol{o}^{(2)}) \\ \boldsymbol{x}^{(2)} = \boldsymbol{o}^{(2)} - F(\boldsymbol{x}^{(1)}) \end{cases} \tag{1}$$

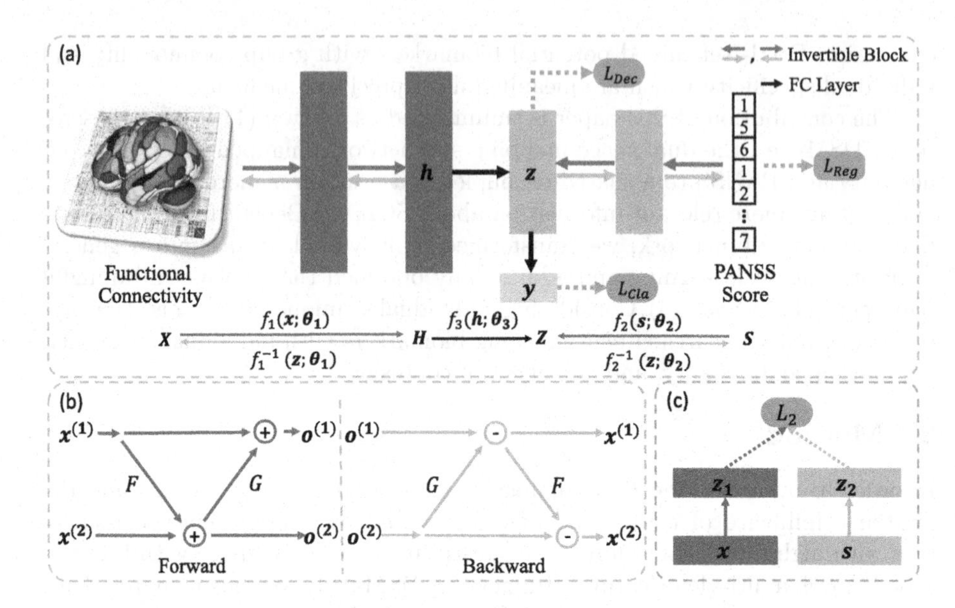

Fig. 1. Overview of the proposed dual space mapping scheme. (a) Our proposed DSM-Net architecture. Both $f_1(\cdot)$ and $f_2(\cdot)$ are composed of two invertible blocks in series, $f_3(\cdot)$ is a linear projection transformation, and θ_1, θ_2 and θ_3 are their parameters. By concatenating $f_1(\cdot)$ and $f_3(\cdot)$, we can map X to Z in the latent space $\mathcal{Z}$ and S is mapped to this share latent space by $f_2(\cdot)$. Then we can perform the SCZ classification task on the learned latent space $\mathcal{Z}$. The network is optimized by the combination of constraints of space $\mathcal{Z}$ and classification objective. (b) Invertible Block. (c) A typical form of multi-view learning.

As shown in Fig. 1(a), DSM-Net consists of four parts: two space mapping sub-networks $f_1(\cdot)$ and $f_2(\cdot)$, a linear projection transformation function $f_3(\cdot)$, and a linear classifier $g(\cdot)$, where $f_1(\cdot)$ and $f_2(\cdot)$ are both made up by two invertible blocks. Because $f_1(\cdot)$ is invertible ($H = f_1(X), X = f_1^{-1}(H)$), H and X in the same dimension contain the same amount of information. Similarly, this relationship also exists between Z and S as $Z = f_2(S), S = f_2^{-1}(Z)$. Therefore, H contains much more information compared to Z since H contains all information that can be inferred from fMRI but only schizophrenia-related information is stored by Z. In order to align these two mapped spaces, a fully-connection (FC) layer is used to form a projection transformation function, $Z = f_3(H)$. This dimensionality reduction can effectively eliminate confounding variables as only the information related to Z in H is retained in this process. And the SCZ classification task can be formed as

$$\hat{y}_i = g(f_3(f_1(\boldsymbol{x}_i))). \tag{2}$$

Transformation from Multi-view to Multi-task Learning. For a general multi-view learning task, the usual approach is to optimize the $\mathcal{L}_2$ loss between

$\boldsymbol{z_1}$ and $\boldsymbol{z_2}$ mapped from different views where $\boldsymbol{z_1} = f_3(f_1(\boldsymbol{x}))$ and $\boldsymbol{z_2} = f_2(\boldsymbol{s})$ as shown in Fig. 1(c). Because there is a one-to-one mapping between the input S and the associated output Z_2 as $f_2(\cdot)$ is invertible, the constraints of output space Z_2 can be transformed into the input space S intuitively. Specifically, because $\boldsymbol{z_2} = \boldsymbol{z_1} + \varepsilon_1$ and $\boldsymbol{s} + \varepsilon_2 = f_2^{-1}(\boldsymbol{z_1})$, the space mapping alignment loss $\mathcal{L}_2$ which minimizes ε_1 can be softly replaced by a mean square error (MSE) regression loss $\mathcal{L}_{Reg}$ of $\boldsymbol{s}_i$ to minimize ε_2 for the i-th subject. In this way, the general multi-view learning degenerates to a general form of multi-task learning with a single input view, bringing the predicted PANSS scores as an extra benefit, formed as Eq. 3.

$$\hat{\boldsymbol{s}}_i = f_2^{-1}(f_3(f_1(\boldsymbol{x}_i))) \tag{3}$$

Mining Biomarkers. The discovery of biomarkers is very important for deepening the understanding of SCZ. To dig this valuable information, we combine multiple linear transformations and invertible blocks. Specifically, we combine the projection transformation $f_3(\cdot)$ and the classifier $g(\cdot)$ into a linear classifier, since they are both linear function and are connected in series, with the weight vector written as $\boldsymbol{w}$ and the bias written as b:

$$\hat{\boldsymbol{y}}_i = g(f_3(f_1(\boldsymbol{x}_i))) = g(f_3(\boldsymbol{h}_i)) = \langle \boldsymbol{w}, \boldsymbol{h}_i \rangle + b. \tag{4}$$

Absorbing the spirit of the interpretation method for invertible block based networks [13], we next calculate the mapped data $\boldsymbol{h}_i$ using $\boldsymbol{h}_i = f_1(\boldsymbol{x}_i)$ and calculate its projection onto the decision boundary as

$$\boldsymbol{h}_i^{'} = \boldsymbol{h}_i - \langle \frac{\boldsymbol{w}}{\|\boldsymbol{w}\|_2}, \boldsymbol{h}_i \rangle \frac{\boldsymbol{w}}{\|\boldsymbol{w}\|_2} - b\frac{\boldsymbol{w}}{\|\boldsymbol{w}\|_2}. \tag{5}$$

Then map the projection $\boldsymbol{h}_i^{'}$ to the original input space using $\boldsymbol{x}_i^{'} = f_1^{-1}(\boldsymbol{h}_i^{'})$. Finally, we use Eq. 6, where $\otimes$ is element-wise product, to define the importance of each feature of $\boldsymbol{x}_i$ in this classification task.

$$\boldsymbol{importance} = |\nabla g(f_3(f_1(\boldsymbol{x}_i))) \otimes (\boldsymbol{x}_i - \boldsymbol{x}_i^{'})| \tag{6}$$

2.2 Objective Function

Decoupling Constraint. To make full use of the low-dimensional latent space $\mathcal{Z}$, we further introduce a decoupling loss as a constraint. As $Z \in \mathbb{R}^{n \times k}$ and the rank of a matrix reflects its maximal linearly independent dimension, if the rank of Z is equal to its dimension, the learned shared latent space $\mathcal{Z}$ will obtain the maximum representation capacity under the constraint of a given number of dimensions. Based on the fact that if the two vectors are orthogonal, then they must be linearly independent, and at the same time, for the sake of computational efficiency, we introduce a semi-orthogonal constraint [2] instead of maximizing the kernel norm to practice this idea. To this end, we first normalize Z to unit variance, denoted as Z_{norm}. And then calculate its covariance matrix Σ

with Eq. 7. $\mathcal{L}_{Dec}$ penalizes the off-diagonal elements of Σ to decorrelate features in the latent space.

$$\Sigma = Z_{norm}^T Z_{norm} \tag{7}$$

To adapt the mini-batch gradient descent algorithm, $\mathcal{L}_{Dec}$ can be formed as

$$\mathcal{L}_{Dec} = \sum_{i \neq j} |\hat{\Sigma}_{i,j}|, \tag{8}$$

where $\hat{\Sigma}$ is a running average of Σ over batch.

Combination of Multiple Loss Functions. In order to improve the classification accuracy of SCZ and obtain potential biomarkers with group commonality and individual specificity simultaneously, the proposed DSM-Net is optimized by the combination of classification objective and constraints of latent space as follows:

$$\mathcal{L} = \mathcal{L}_{Cla} + \lambda \mathcal{L}_{Reg} + \gamma \mathcal{L}_{Dec}, \tag{9}$$

where $\mathcal{L}_{Cla}$ is the cross-entropy loss between y_i and $\hat{\boldsymbol{y}}_i$, $\mathcal{L}_{Reg}$ is the MSE regression loss between $\boldsymbol{s}_i$ and $\hat{\boldsymbol{s}}_i$, and $\mathcal{L}_{Dec}$ is the decoupling constraint of learned latent space. λ and γ are both set to 0.05 during training.

3 Experiments

3.1 Materials and Preprocessing

In this work, we report on rs-fMRI collected from seven sites with three types of scanners (three sites with 3.0 T Siemens Trio Tim Scanner, one site with 3.0 T Siemens Verio Scanner and the other three sites with 3.0 T Signa HDx GE Scanner). The 1061 participants used in this work, including 531 patients with SCZ (27.80 ± 7.13 years, 276 males) and 530 healthy controls (28.43 ± 7.21 years, 270 males), were recruited with the same recruitment criterion. The SCZ patients diagnosed by experienced psychiatrists were evaluated based on the Structured Clinical Interview for DSM disorders (SCID) and fulfilled the DSM-IV-TR criteria for SCZ. The symptom severity of the patients was measured with the PANSS assessment.

The scan parameters of rs-fMRI are as follows: TR = 2000 ms, TE = 30 ms, flip angle = 90°, voxel size = $3.44 \times 3.44 \times 4$ mm^3 and FOV = 220×220 mm^2. The preprocessing pipeline performed with BRANT [11] includes six steps: (1) discarding the first 10 timepoints; (2) slice timing correction, head motion correction, and registration; (3) resampling to $3 \times 3 \times 3$ mm^3 voxel size; (4) regressing out averaged tissue time series and motion-related temporal covariates; (5) 0.01 Hz–0.08 Hz band-pass filtering and smoothing with a $6 \times 6 \times 6$ mm^3 Gaussian kernel; (6) extracting mean time series of each brain nodes defined by the Human Brainnetome Atlas [3] and calculating functional connectivity. After that, we get the functional connectivity as one input view of length $246 \times 245/2 = 30135$, and take the vectorized PANSS (including 30 items) as the other input view. Each score of PANSS is rated on a scale of 1 to 7. For the healthy controls, we set all their PANSS scores to 1, indicating no corresponding symptoms.

3.2 Results of Classification and Regression

Each space mapping sub-network has two invertible blocks and the F and G in each block are FC-ReLU-FC-ReLU. For $f_1(\cdot)$, the first FC layer maps the half of the input vector from the original dimension ($\lceil 30135/2 \rceil = 15068$) to 1024 and the second FC layer maps the 1024 dimensional vector to 15068. The first and second FC layers of $f_2(\cdot)$ map PANSS score vector from 15 ($30/2 = 15$) to 64 and 15 in series. The model is trained with Adam optimizer for 100 epochs with learning rate $\eta = 1.0 \times 10^{-3}$ and decay rate set to 0.5 every 10 epochs. And all methods are validated by leave-one-site-out cross validation. We compare DSM-Net with the Invertible-Net which is most related to our work, and a general multi-task DNN model which replaces all invertible blocks of DSM-Net with FC layers at the expense of a certain degree of interpretability. As shown in Table 1, in addition to the classification task, DSM-Net has learned to predict PANSS scores. Besides, compared with Invertible-Net, DSM-Net without decoupling constraint has no obvious improvement on the classification metrics except for a significant difference in the number of potential biomarkers which we will show in Sect. 3.3. We deem that introducing PANSS scores only constrains the solution space to exclude some irrelevant information without introducing additional information and does not change the representation capability of the baseline network. So, the classification metrics are close. However, after introducing the decoupling constraint on the learned latent space, which increases the representation capability, the classification accuracy has been improved to 84.33%, while the other metrics have been improved to varying degrees. Moreover, the decorrelation constraint reduces the regression error of the PANSS scores, i.e. MSE and normalized mean square error (NMSE), considerably.

Table 1. Leave-one-site-out cross-validation performance of classification of SCZ and regression of PANSS scores.

Method	Classification				Regression	
	Accuracy	Sensitivity	Specificity	F1 Score	MSE	NMSE
Invertible-Net [13]	82.78 ± 3.11	82.70 ± 5.51	83.53 ± 5.67	82.56 ± 4.05	–	–
Multi-task DNN	83.43 ± 2.96	83.98 ± 5.49	83.60 ± 5.09	83.33 ± 3.84	2.03 ± 0.44	0.35 ± 0.03
DSM-Net w/o $\mathcal{L}_{Dec}$	82.72 ± 3.36	82.80 ± 5.00	83.53 ± 6.21	82.52 ± 4.35	1.81 ± 0.30	0.32 ± 0.03
DSM-Net (ours)	$\mathbf{84.33} \pm 2.49$	$\mathbf{84.99} \pm 4.89$	$\mathbf{84.02} \pm 4.87$	$\mathbf{84.20} \pm 3.68$	$\mathbf{1.52} \pm 0.25$	$\mathbf{0.27} \pm 0.02$

3.3 Biomarker Discovery

As the DNN model with high-dimensional input lacks good interpretability, we compare the potential biomarkers mined by the other three models. According to the *importance* of each connection, we select the top 50 connections of each individual and take their union to form a potential biomarker collection for each site. The average number of connections in the collections of seven sites is 99.14 without introducing PANSS as a constraint view. However, using PANSS as a constraint view reduces the average number of potential biomarkers to 62.14. The details are shown in Fig. 2(a). In terms of reducing the number of mined potential biomarkers while improving the performance of classification and regression tasks as shown in Table 1, DSM-Net effectively reduces the number of mined confounding variables.

Then we define a common index of each potential biomarker according to the frequency they appear. In detail, the common index of connection i is formed as $c_i = n_i/N$, where n_i represents the number of occurrences of connection i, N represents the number of participants in this site and $c_i \in [0, 1]$. If c_i is closer to 1, the corresponding functional connection is more common among patients; on the contrary, if c_i is closer to 0, the connection is more specific and the possibility of being a confounding variable is greater. Take one site as an example, there are 143 participants in this site and the potential biomarker collection of this site has 61 connections shown as Fig. 2(b). The more common connections ($c_i \geq 0.95$, shown with solid lines) are mostly among the frontal lobe, temporal lobe, and subcutaneous nuclei. And the more specific connections ($c_i < 0.95$, shown with dashed lines) are mostly related to the parietal lobe, insular lobe, and limbic lobe. In particular, we found that the more common connections are widely reported [1,10], whereas the more specific connections are reported relatively infrequently within group-level studies. And this is in line with the characteristic of group-level studies that tend to ignore individual-specific markers. Moreover, we mapped the mined biomarkers to Yeo 7-Network Solution Networks [9] and found that the connections are clearly concentrated in the limbic network and subcutaneous nuclei.

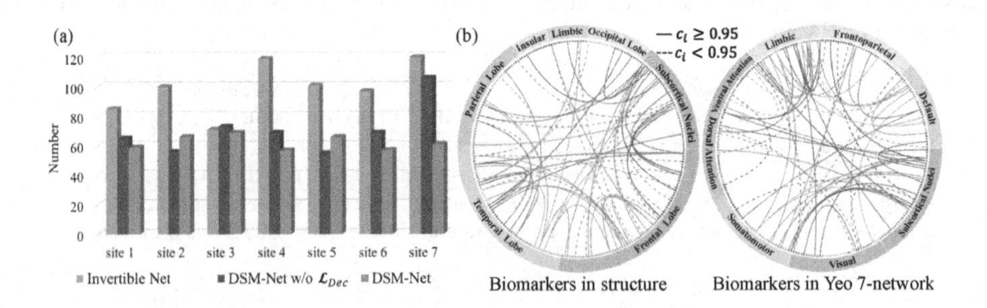

Fig. 2. Result of biomarker discovery. (a) The number of potential connections in the collection for each site; (b) Mined potential biomarkers of site 7.

4 Conclusion

Although PANSS provides individual-level symptoms and reflects clinical heterogeneity of SCZ patients, most previous studies ignored this information and only performed simple binary classification task, which resulted in the inexhaustible extraction of related information and the vulnerability to confounding factors. To address these questions, we propose a novel multi-view learning framework DSM-Net with good interpretability to classify SCZ. The DSM-Net extracts a more detailed representation of SCZ through mapping functional connectivity features and PANSS scores to shared latent space. And the decoupling constraint on the solution space further improves the representation capability of the proposed framework. We empirically show that the DSM-Net can effectively improve the classification accuracy and reduce the probability of mining confounding biomarkers. Besides, the framework unifies multi-view and multi-task learning to a certain extent. By transforming multi-view learning into multi-task learning, the DSM-Net can generate predicted PANSS scores as an extra benefit. It should be noted that the regression of PANSS scores for each patient is important for improving the credibility of the model and valuable for clinical practice but often overlooked in classification-based studies. Moreover, with the good interpretability of the DSM-Net, we defined common or individual-specific potential biomarkers driven by the adopted data which is desirable to precision psychiatry. It is hoped that this framework and the results are meaningful to precision psychiatry and may provide a new research perspective for the study of SCZ or other mental illness at the individual level.

Acknowledgements. This work was partly supported by the Natural Science Foundation of China (Grant Nos. 31620103905, 31870984); Science Frontier Program of the Chinese Academy of Sciences (grant No. QYZDJ-SSW-SMC019).

References

1. Buchsbaum, M.S.: The frontal lobes, basal ganglia, and temporal lobes as sites for schizophrenia. Schizophr. Bull. **16**(3), 379–389 (1990)
2. Chang, X., Xiang, T., Hospedales, T.M.: Scalable and effective deep CCA via soft decorrelation. In: Proceedings of the IEEE Conference on Computer Vision and Pattern Recognition, pp. 1488–1497 (2018)
3. Fan, L., et al.: The human brainnetome atlas: a new brain atlas based on connectional architecture. Cereb. Cortex **26**(8), 3508–3526 (2016)
4. Gomez, A.N., Ren, M., Urtasun, R., Grosse, R.B.: The reversible residual network: backpropagation without storing activations. In: Advances in Neural Information Processing Systems, pp. 2214–2224 (2017)
5. Kay, S.R., Fiszbein, A., Opler, L.A.: The positive and negative syndrome scale (PANSS) for schizophrenia. Schizophr. Bull. **13**(2), 261–276 (1987)
6. Lei, D., et al.: Detecting schizophrenia at the level of the individual: relative diagnostic value of whole-brain images, connectome-wide functional connectivity and graph-based metrics. Psychol. Med. 1–10 (2019)
7. Nieuwenhuis, M., van Haren, N.E., Pol, H.E.H., Cahn, W., Kahn, R.S., Schnack, H.G.: Classification of schizophrenia patients and healthy controls from structural mri scans in two large independent samples. Neuroimage **61**(3), 606–612 (2012)
8. Rubinov, M., Bullmore, E.: Fledgling pathoconnectomics of psychiatric disorders. Trends Cogn. Sci. **17**(12), 641–647 (2013)

9. Thomas Yeo, B., et al.: The organization of the human cerebral cortex estimated by intrinsic functional connectivity. J. Neurophysiol. **106**(3), 1125–1165 (2011)
10. Weinberger, D.R.: Schizophrenia and the frontal lobe. Trends Neurosci. **11**(8), 367–370 (1988)
11. Xu, K., Liu, Y., Zhan, Y., Ren, J., Jiang, T.: BRANT: a versatile and extendable resting-state fMRI toolkit. Front. Neuroinf. **12**, 52 (2018)
12. Yan, W., et al.: Discriminating schizophrenia using recurrent neural network applied on time courses of multi-site fMRI data. EBioMedicine **47**, 543–552 (2019)
13. Zhuang, J., Dvornek, N.C., Li, X., Ventola, P., Duncan, J.S.: Invertible network for classification and biomarker selection for ASD. In: Shen, D., et al. (eds.) MICCAI 2019. LNCS, vol. 11766, pp. 700–708. Springer, Cham (2019). https://doi.org/10.1007/978-3-030-32248-9_78

Have You Forgotten? A Method to Assess if Machine Learning Models Have Forgotten Data

Xiao Liu[1(✉)] and Sotirios A. Tsaftaris[1,2]

[1] School of Engineering, University of Edinburgh, Edinburgh EH9 3FB, UK
{Xiao.Liu,S.Tsaftaris}@ed.ac.uk
[2] The Alan Turing Institute, London, UK

Abstract. In the era of deep learning, aggregation of data from several sources is a common approach to ensuring data diversity. Let us consider a scenario where several providers contribute data to a consortium for the joint development of a classification model (hereafter the target model), but, now one of the providers decides to leave. This provider requests that their data (hereafter the query dataset) be removed from the databases but also that the model 'forgets' their data. In this paper, for the first time, we want to address the challenging question of whether data have been forgotten by a model. We assume knowledge of the query dataset and the distribution of a model's output. We establish statistical methods that compare the target's outputs with outputs of models trained with different datasets. We evaluate our approach on several benchmark datasets (MNIST, CIFAR-10 and SVHN) and on a cardiac pathology diagnosis task using data from the Automated Cardiac Diagnosis Challenge (ACDC). We hope to encourage studies on what information a model retains and inspire extensions in more complex settings.

Keywords: Privacy · Statistical measure · Kolmogorov-Smirnov

1 Introduction

Deep learning requires lots of, diverse, data and in healthcare likely we will need to enlist several data providers (e.g. hospital authorities, trusts, insurance providers who own hospitals, etc) to ensure such data diversity. To develop models, data from several providers must thus be either centrally aggregated or leveraged within a decentralized federated learning scheme that does not require the central aggregation of data (e.g. [1,13,14,17,19]). Thus, several providers will contribute data for the development of a deep learning model e.g. to solve a simple classification task of normal vs. disease. Suddenly, one of the providers

Electronic supplementary material The online version of this chapter (https://doi.org/10.1007/978-3-030-59710-8_10) contains supplementary material, which is available to authorized users.

A. L. Martel et al. (Eds.): MICCAI 2020, LNCS 12261, pp. 95–105, 2020.
https://doi.org/10.1007/978-3-030-59710-8_10

decides to leave and asks for the data to be deleted but more critically that *the model 'forgets' the data.*

Let us now assume that the model has *indeed* not used the data and has 'forgotten' them (we will not care how herein) but we want to *verify* this. In other words, and as illustrated in Fig. 1, our problem is to assess whether the model has not used the data under question in the training set.

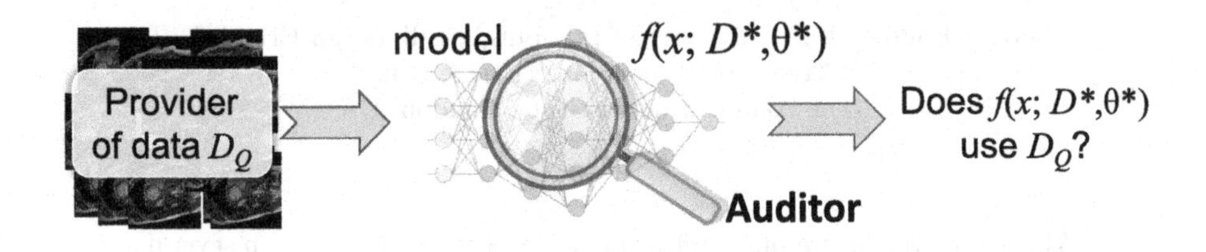

Fig. 1. Our goal. We aim to develop an approach that an 'auditor' can use to ascertain whether a model $f(x; D^*; \theta)$ has used data D_Q during training.

We consider that an *auditor* will have access to the query dataset D_Q and the model $f(\mathbf{x}; D^*, \theta^*)$, (trained on data D^*) to render a decision whether $f()$ retains information about D_Q. We assume that the training dataset D^* is *unknown* but both D^* and D_Q are sampled from a *known* domain $\mathbb{D}$ for a given *known task.*[1] We emphasize that we are *not* proposing a method that forgets data, as for example the one in [8]. We believe that assessing whether data have been forgotten is an important task that should come first (because without means to verify claims, anyone can claim that data have been forgotten).

Related work is limited. In fact, our communities (machine learning or medical image analysis) have not considered this problem yet in this form. The closest problem in machine learning is the *Membership Inference Attack* (MIA) [4,5,16,18,20]. While these works borrow concepts from MIA, as we detail in the related work section, MIA sets different hypotheses that have considerable implications in returning false positives when data sources overlap. Our major novelty is setting appropriate test hypotheses that calibrate for data overlap.

Our approach builds on several key ideas. First, inspired by [18], we adopt the Kolmogorov-Smirnov (K-S) distance to compare statistically the similarity between the distribution of a model $f()$'s outputs and several purposely constructed reference distributions. To allow our approach to operate in a black-box setting (without precise knowledge of $f()$, we construct 'shadow' models (inspired by [20]) to approximate the target model $f()$; however, we train the shadow models on D_Q but also on another dataset D_C sampled from domain $\mathbb{D}$ which does not overlap (in element-wise sense) D_Q.

[1] Knowledge of the task (e.g. detect presence of a pathology in cardiac MRI images), implies knowledge of the domain $\mathbb{D}$ (e.g. the space of cardiac MRI images). Without this assumption, D^* can be anything, rendering the problem intractable.

Contributions:

1. To introduce a new problem in data privacy and retention to our community.
2. To offer a solution that can be used to detect whether a model has forgotten specific data including the challenging aspect when data sources may overlap.
3. Experiments in known image classification benchmark datasets to verify the effectiveness of our approach. Experiments on the pathology classification component of the ACDC simulating a healthcare-inspired scenario.

2 Related Work

The purpose of *Membership Inference Attack* is to learn which data were used to train a model. The hypothesis is that a model *has* used some data and the goal is to find which part of the query data are in the training set. Here we briefly review key inspirational approaches.[2]

In [20] they train a model f_{attack} that infers whether some data $\mathbf{D}$ were used by f. f_{attack} accepts as inputs the decisions of f (the softmax outputs), $\mathbf{D}$ and the ground-truth class. Their premise is that machine learning models often behave "*differently on the data that they were trained on vs. data that they 'see' for the first time*". Thus, the task reduces to training this attack model. They rely on being able to generate data that resemble $\mathbf{D}$ (or not) and train several, different, 'shadow' models that mimic f to obtain outputs that can be used to train f_{attack}. We rely here on the same premise and use the idea of training shadow models of $f()$ to enable gray-box inference on $f()$ using only its outputs.

In [18], in the context of keeping machine learning competitions fair, using the same premise as [20], they propose a statistical approach to compare model outputs, and eschew the need for f_{attack}. They use Kolmogorov-Smirnov (K-S) distance to measure the statistical similarity between emissions of the classifier layer of a network between query data and a reference dataset to see whether models have used validation data to train with. However, their approach assumes that both query (validation set) and reference datasets, which in their context is the testing set, are known *a priori*. We adopt (see Sect. 3) the K-S distance as a measure of distribution similarity but different from [18] we construct reference distributions to also calibrate for information overlap.

Why our Problem is Not the Same as MIA: An MIA algorithm, by design, assumes that query data *may* have been used to train a model. Whereas, we care to ascertain if a model has *not* used the data. These are different hypotheses which appear complementary but due to data source overlap have considerable implications on defining false positives and true negatives.[3] When there is no

[2] We do not cover here the different task of making models and data more private, by means e.g. of *differential privacy*. We point readers to surveys such as [9,10] and a recent (but not the only) example application in healthcare [14].

[3] In fact, overlap will frequently occur in the real world. For example, datasets collected by different vendors can overlap if they collaborate with a same hospital or if a patient has visited several hospitals. Our method has been designed to address this challenging aspect of overlap.

information overlap between datasets, a MIA algorithm will return the right decision/answer. However, when D_Q and D^* statistically overlap (i.e. their manifolds overlap), for a sample both in D_Q or D^*, a MIA algorithm will return a false positive even when the model was trained on D^* alone. In other words, MIA *cannot* tell apart data overlap. Since overlap between D_Q and D^* is likely and *a priori* the auditor does not know it, the problem we aim to address has more *stringent* requirements than MIA. To address these requirements we assume that we can sample another dataset D_C from $\mathbb{D}$ (that we can control not to at least sample-wise overlap with D_Q) to train shadow models on D_C.

Algorithm 1. The proposed method.

Input: the target model $f(\mathbf{x}; D^*, \theta^*)$, query dataset D_Q and dataset domain $\mathbb{D}$.
Output: answer to hypothesis if $f(\mathbf{x}; D^*, \theta^*)$ is trained with D_Q.
Step 1. Train $\widetilde{f}(\mathbf{x}; D_Q, \theta_Q)$ with the same design of $f(\mathbf{x}; D^*, \theta^*)$.
Step 2a. Sample the calibration dataset D_C from domain $\mathbb{D}$ but without overlapping (in sample-wise sense) D_Q.
Step 2b. Train $\widetilde{f}(\mathbf{x}; D_C, \theta_C)$ with the same design of $f(\mathbf{x}; D^*, \theta^*)$.
Step 3. Find $r(T_{D_Q|D_Q})$, $r(T_{D_Q|D^*})$ and $r(T_{D_Q|D_C})$ (Eq. 1).
Step 4. Find $KS(r(T_{D_Q|D_Q}), r(T_{D_Q|D^*}))$, $KS(r(T_{D_Q|D_Q}), r(T_{D_Q|D_C}))$ (Eqs. 2, 3).
Step 5. Find $\rho = \frac{KS(r(T_{D_Q|D_Q}), r(T_{D_Q|D^*}))}{KS(r(T_{D_Q|D_Q}), r(T_{D_Q|D_C}))}$ (Eq. 4).
return If $\rho \geq 1$, the target model has forgotten D_Q. Otherwise ($\rho < 1$), has not.

---

3 Proposed Method

We measure similarity of distributions to infer whether a model is trained with the query dataset D_Q, i.e. if D^* has D_Q or not. Note that this is not trivial since D^* is *unknown*. In addition, the possible overlap between D^* and D_Q introduces more challenges as we outlined above. To address both, we introduce another dataset D_C sampled from domain $\mathbb{D}$ but without overlapping (in element-wise sense) D_Q. The method is summarized in Algorithm 1 with steps detailed below.

Notation: We will consider $\mathbf{x}$, a tensor, the input to the model e.g. an image. We will denote the *target* model as $f(\mathbf{x}; D^*, \theta^*)$ that is trained on dataset D^* parametrised by θ^*. Similarly, we define *query* $\widetilde{f}(\mathbf{x}; D_Q, \theta_Q)$ and *calibration* models $\widetilde{f}(\mathbf{x}; D_C, \theta_C)$, where both models share model design with the target model. If D_Q has N samples ($N \gg 1$), we use $y^{\{i\}}, i = \{1, \cdots, N\}$, a scalar, to denote the ground-truth label of data $\mathbf{x}^{\{i\}}$ of D_Q. We denote with $\mathbf{t}^{\{i\}}, i = \{1, \cdots, N\}$, a M-dimensional vector, the *output* of a model with input $\mathbf{x}^{\{i\}}$. Hence, $\mathbf{t}^{\{i\}}[y^{\{i\}}]$ denotes the output of the model $\mathbf{t}^{\{i\}}$ for the ground-truth class $y^{\{i\}}$ (confidence score). We further define the $N \times 1$ vector $\mathbf{c}_{D_Q|D^*}$, which contains the values of $\mathbf{t}^{\{i\}}[y^{\{i\}}]$, $i = \{1, \cdots, N\}$ in an increasing order. We define as $T_{D_Q|D^*}$ the $N \times M$ matrix that contains all outputs of the target model $f(\mathbf{x}; D^*, \theta^*)$ that is tested with D_Q. Similarly, we define $T_{D_Q|D_Q}$ and $T_{D_Q|D_C}$ for the outputs of $\widetilde{f}(\mathbf{x}; D_Q, \theta_Q)$ and $\widetilde{f}(\mathbf{x}; D_C, \theta_C)$, respectively, when both are tested with D_Q.

3.1 Assumptions on $f()$

We follow a gray-box scenario: access to the output of the model $f(\mathbf{x}; D^*, \theta^*)$ before any thresholds and knowledge of design (e.g. it is a neural network) but not of the parameters (e.g. weights) θ^*. (Model providers typically provide the best (un-thresholded) value as surrogate of (un)certainty).

3.2 Kolmogorov-Smirnov Distance Between Distributions

We denote as $r(T_{D_Q|D^*})$ (a $N \times 1$ vector) the empirical cumulative distribution (cdf) of the output of the target model when tested with D_Q. The nth element $r_n(T_{D_Q|D^*})$ is defined as:

$$r_n(T_{D_Q|D^*}) = \frac{1}{N} \sum_{i=1}^{N} I_{[-\infty, \mathbf{c}^{\{n\}}_{D_Q|D^*}]}(\mathbf{c}^{\{i\}}_{D_Q|D^*}), \tag{1}$$

where $I_{[-\infty, \mathbf{c}^{\{n\}}_{D_Q|D^*}]}(\mathbf{c}^{\{i\}}_{D_Q|D^*})$ the indicator function, is equal to 1 if $\mathbf{c}^{\{i\}}_{D_Q|D^*} \leq \mathbf{c}^{\{n\}}_{D_Q|D^*}$ and equal to 0 otherwise.

Our next step is to create a proper empirical distribution as reference to compare against $r(T_{D_Q|D^*})$. Motivated by [20], we propose to train a query model $\widetilde{f}(\mathbf{x}; D_Q, \theta_Q)$ with the same model design as the target model. We then obtain (as previously) the output cumulative distribution $r(T_{D_Q|D_Q})$. We propose to use $r(T_{D_Q|D_Q})$ as the reference distribution of dataset D_Q. Hence, measuring the similarity between $r(T_{D_Q|D^*})$ with $r(T_{D_Q|D_Q})$ can inform on the relationship between D_Q and D^*, which is extensively explored in the field of dataset bias [21]. Following [18], we use Kolmogorov–Smirnov (K-S) distance as a measure of the similarity between the two empirical distributions.

K-S distance was first used in [6] to compare a sample with a specific reference distribution or to compare two samples. For our purpose we will peruse the two-sample K-S distance, which is given by:

$$KS(r(T_{D_Q|D_Q}), r(T_{D_Q|D^*})) = sup|r(T_{D_Q|D_Q}) - r(T_{D_Q|D^*})|_1, \tag{2}$$

where *sup* denotes the largest value. K-S distance $\in [0, 1]$ with lower values pointing to greater similarity between $r(T_{D_Q|D^*})$ and $r(T_{D_Q|D_Q})$.

If D^* contains D_Q, the K-S distance between $r(T_{D_Q|D_Q})$ and $r(T_{D_Q|D^*})$ will be very small and ≈ 0, i.e. $KS(r(T_{D_Q|D_Q}), r(T_{D_Q|D^*})) \approx 0$. On the contrary, if D^* has no samples from D_Q, then the value of $KS(r(T_{D_Q|D_Q}), r(T_{D_Q|D^*}))$ depends on the statistical overlap of D^* and D_Q. However, this overlap is challenging to measure because the training dataset D^* is assumed unknown (and we cannot approximate statistical overlap with element-wise comparisons).

3.3 Calibrating for Data Overlap

We assume the training data D^* of model $f(\mathbf{x}; D^*, \theta^*)$ are sampled from a domain $\mathbb{D}$. To calibrate against overlap between D^* and D_Q, we sample a *calibration* dataset D_C from $\mathbb{D}$ but ensure that no samples in D_Q are included in D_C (by sample-wise comparisons) and train the calibration model $\widetilde{f}(\mathbf{x}; D_C, \theta_C)$. Inference on model $\widetilde{f}(\mathbf{x}; D_C, \theta_C)$ with D_Q, the output empirical cumulative distribution $r(T_{D_Q|D_C})$ can be calculated in a similar fashion as Eq. 1. To compare the similarity between D_Q and D_C, we calculate:

$$KS(r(T_{D_Q|D_Q}), r(T_{D_Q|D_C})) = sup|r(T_{D_Q|D_Q}) - r(T_{D_Q|D_C})|_1. \tag{3}$$

We argue that Eq. 3 can be used to calibrate the overlap and inform if D^* contains D_Q.[4] We discuss this for two scenarios:

1. D^* Does Not Include any Samples of D_Q. Since D_C and D_Q do not overlap in a element-wise sense by construction, then data of D^* should have higher probability of statistical overlap with D_C. In other words, the calibration model can be used to mimic the target model. By sampling more data from domain $\mathbb{D}$ as D_C, there will be more statistical overlap between D_C and D_Q. Hence, the statistical overlap between a large-size dataset D_C and D_Q can be used to approximate the (maximum) possible overlap between D^* and D_Q i.e. $KS(r(T_{D_Q|D_Q}), r(T_{D_Q|D^*})) \geq KS(r(T_{D_Q|D_Q}), r(T_{D_Q|D_C}))$.

2. D^* Contains Samples of D_Q. In this case, samples in D_Q that are also in D^* will introduce statistical overlap between D_Q and D^*, which is well demonstrated in [18]. Based on our experiments, we find that the overlap between D_Q and D^* is consistently less than that of D_Q and D_C i.e. $KS(r(T_{D_Q|D_Q}), r(T_{D_Q|D^*})) < KS(r(T_{D_Q|D_Q}), r(T_{D_Q|D_C}))$.

Thus, we propose to use $KS(r(T_{D_Q|D_Q}), r(T_{D_Q|D_C}))$ as a data-driven indicator for detecting if D^* contains D_Q. To quantify how much the target model has forgotten about D_Q, we capture the two inequalities above in the ratio:

$$\rho = \frac{KS(r(T_{D_Q|D_Q}), r(T_{D_Q|D^*}))}{KS(r(T_{D_Q|D_Q}), r(T_{D_Q|D_C}))}. \tag{4}$$

In the first scenario i.e. D^* does not include any samples of D_Q, the inequality $KS(r(T_{D_Q|D_Q}), r(T_{D_Q|D^*})) \geq KS(r(T_{D_Q|D_Q}), r(T_{D_Q|D_C}))$ translates to $\rho \geq 1$. Instead, in the second scenario, $\rho < 1$. Thus, if $\rho \geq 1$, this implies the target model has forgotten D_Q. On the contrary, if $\rho < 1$, implies the target model has not forgotten D_Q.

[4] As mentioned previously we cannot measure statistical overlap between D^* and D_C (or D_Q) since D^* is unknown. Furthermore, statistical overlap in high-dimensional spaces is not trivial [2,7], and hence we want to avoid it.

4 Experiments

We first perform experiments on benchmark datasets such as MNIST [12], SVHN [15] and CIFAR-10 [11] to verify the effectiveness of the proposed method. Then, we test our method on the ACDC dataset using the pathology detection component of the challenge [3]. All classifiers in our experiments are well-trained (adequate training accuracy) such that the output statistics well approximate the input statistics. All model designs are included as Supplementary Material.

4.1 Benchmark Datasets - MNIST, SVHN and CIFAR-10

Benchmark Datasets: MNIST contains 60,000 images of 10 digits with image size 28×28. Similar to MNIST, SVHN has over 600,000 digit images obtained from house numbers in Google Street view images. The image size of SVHN is 32×32. Since both datasets are for the task of digit recognition/classification we consider to belong to the same domain $\mathbb{D}$. We use CIFAR-10 as an *out of domain* $\mathbb{D}$ dataset to validate the method. CIFAR-10 has 60,000 images (size 32×32) of 10-class objects (airplane, bird, etc). To train our models, we preprocess images of all three datasets to gray-scale and rescale to size 28×28.

Table 1. Benchmark datasets results. The query dataset D_Q is MNIST. The training dataset D^* has different variants as listed below. The K-S distances are calculated with Eq. 2. D_C for defining ρ is SVHN. Bold number in the table is the threshold.

Training dataset (D^*)	MNIST	%SVHN			SVHN + %MNIST			CIFAR-10
		50%	75%	100%	10%	50%	100%	
$KS(r(t_{D_Q \mid D_Q}), r(t_{D_Q \mid D^*}))$	0.049	0.669	0.652	**0.596**	0.056	0.039	0.029	0.957
ρ	0.082	1.122	1.094	1.000	0.094	0.065	0.049	1.606

Results and Discussion: We consider MNIST as D_Q. Experimental results are listed in Table 1. We first verify if our method can detect the simple case where the target model has been fully trained with D_Q. According to our findings (first column of Table 1) as expected $\rho = 0.082 \ll 1$. Other cases where we assume D^* is drawn only from D_C (SVHN) give $\rho > 1$. In the challenging setting when the target model was trained with a mix of data from SVHN and parts of MNIST, $\rho \sim 0$. We draw attention to the case of SVHN+%100MNIST. $\rho = 0.049$ indicates that the target definitely has not forgotten the query dataset.

For exploration purposes, we consider the extreme case of D^* not in the domain of D_Q. We train a target model with CIFAR-10 which has little statistical information overlap with MNIST. Hence, we would expect our method to return a $\rho > 1$, which agrees with the result $\rho = 1.606$ (highest in the table).

Next we assess the scenarios that D^* does not include any samples of D_Q. We consider SVHN as D_C and train target models with D^*, subsets of SVHN.

The results ρ of 50% and 75% SVHN are 1.122 and 1.094 that both higher than $\rho = 1$. Higher ρ points to lower information overlap between D_Q and D^*. Hence, the large-size D_C i.e. %100SVHN has the largest statistical overlap with D_Q.

To check the sensitivity of our method when D^* contains part of D_Q, we train two models with SVHN and part of MNIST (10% and 50%). We observe that both have $\rho < 1$. Specifically, including only 10% of MNIST causes $(1-0.094) \div 1 = 90.6\%$ drop of ρ value. This suggests that if the training dataset contains part of D_Q, our method can still accurately detect these cases.

Some uncertainty may arise if ρ is hovering around 1. However, we did not observe such cases when D^* contains samples from D_Q in our extensive experiments. We advise to run multiple runs of experiments to statistically eliminate such uncertainty if needed.

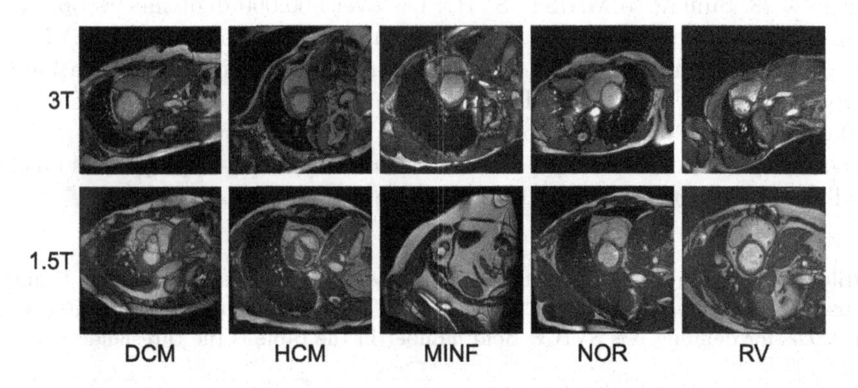

Fig. 2. Example images of ACDC 1.5T and ACDC 3T datasets. DCM: dilated cardiomyopathy. HCM: hypertrophic cardiomyopathy. MINF: myocardial infarction. NOR: normal subjects. RV: abnormal right ventricle.

4.2 Medical Dataset - ACDC

ACDC Dataset: Automated Cardiac Diagnosis Challenge (ACDC) dataset is composed of cardiac images of 150 different patients that are acquired with two MRI scanners of different magnetic strengths, i.e. 1.5T and 3.0T. For each patient, ACDC provides 4-D data (weight, height, diastolic phase instants, systolic phase instants). To simulate several data providers we split the original training data of ACDC into two datasets: we consider the images at 1.5 T field strength as one source and those at 3 T as the other. Overall, ACDC 1.5T has 17,408 images and ACDC 3T has 7,936 images (Examples are shown in Fig. 2.). For both datasets, four classes of pathology are provided and one for normal subjects (for a total of 5 classes). We randomly crop all images to 128 × 128 during training.

Results and Discussion: We consider ACDC 3T as D_Q and assume the domain $\mathbb{D}$ spans the space of cine images of the heart in 1.5 T or 3T strength. ACDC 1.5T is the calibration dataset D_C. Results are shown in Table 2.

We first perform experiments to verify if our method can correctly detect whether D_Q is in D^*. Similar to benchmark datasets, all models trained with ACDC 3T or part of ACDC 3T have $\rho < 1$. Other models trained without any data from ACDC 3T have $\rho \geq 1$. Hence, for ACDC dataset, our method returns correct answers in all experiments.

According to %ACDC 1.5T results, a large-size dataset D_C (%100 ACDC 1.5T) achieves lower K-S distance compared to 50% and 75% ACDC 1.5T. This supports our discussion of the scenario when D^* does not include any samples of D_Q. Note that when mixing all data of ACDC 1.5T and only 10% of 3T, the K-S distance drops $(1 - 0.750) \div 1 = 25\%$. It suggests even if part of D_Q is in D^*, the proposed method is still sensitive to detect such case.

Table 2. ACDC dataset results. The query dataset D_Q is ACDC 3T. The training dataset D^* has different variants as listed below. The K-S distances are calculated with Eq. 2. D_C for defining ρ is ACDC 1.5T. Bold number in the table is the threshold.

Training dataset (D^*)	ACDC 3T	%ACDC 1.5T			ACDC 1.5T + % 3T		
		50%	75%	100%	10%	50%	100%
$KS(r(T_{D_Q\|D_Q}), r(T_{D_Q\|D^*}))$	0.036	0.885	0.793	**0.772**	0.579	0.529	0.103
ρ	0.047	1.146	1.027	1.000	0.750	0.685	0.133

5 Conclusion

We introduced an approach that uses Kolmogorov-Smirnov (K-S) distance to detect if a model has used/forgotten a query dataset. Using the K-S distance we can obtain statistics about the output distribution of a target model without knowing the weights of the model. Since the model's training data are unknown we train new (shadow) models with the query dataset and another calibration dataset. By comparing the K-S values we can ascertain if the training data contain data from the query dataset even for the difficult case where data sources can overlap. We showed experiments in classical classification benchmarks but also classification problems in medical image analysis. We did not explore the effect of the query dataset's sample size but the cumulative distribution remains a robust measure even in small sample sizes. Finally, extensions to segmentation or regression tasks and further assessment whether differential-privacy techniques help protect data remain as future work.

Acknowledgment. This work was supported by the University of Edinburgh by a PhD studentship. This work was partially supported by the Alan Turing Institute under the EPSRC grant EP/N510129/1. S.A. Tsaftaris acknowledges the support of the Royal

Academy of Engineering and the Research Chairs and Senior Research Fellowships scheme and the [in part] support of the Industrial Centre for AI Research in digital Diagnostics (iCAIRD) which is funded by Innovate UK on behalf of UK Research and Innovation (UKRI) [project number: 104690] (https://icaird.com/).

References

1. Barillot, C., et al.: Federating distributed and heterogeneous information sources in neuroimaging: the neurobase project. Stud. Health Technol. Inf. **120**, 3 (2006)
2. Belghazi, M.I., et al.: Mutual information neural estimation. In: Dy, J., Krause, A. (eds.) Proceedings of the 35th International Conference on Machine Learning. Proceedings of Machine Learning Research, vol. 80, pp. 531–540. PMLR, Stockholmsmässan, Stockholm Sweden, 10–15 July 2018
3. Bernard, O., et al.: Deep learning techniques for automatic MRI cardiac multi-structures segmentation and diagnosis: is the problem solved? IEEE Trans. Med. Imaging **37**(11), 2514–2525 (2018)
4. Carlini, N., Liu, C., Erlingsson, U., Kos, J., Song, D.: The secret sharer: evaluating and testing unintended memorization in neural networks. In: Proceedings of the 28th USENIX Conference on Security Symposium, SEC 2019, pp. 267–284. USENIX Association, Berkeley, CA, USA (2019)
5. Cherubin, G., Chatzikokolakis, K., Palamidessi, C.: F-BLEAU: fast Black-box Leakage Estimation, February 2019. http://arxiv.org/abs/1902.01350
6. Feller, W.: On the Kolmogorov-Smirnov limit theorems for empirical distributions. In: Schilling, R., Vondraček, Z., Woyczyński, W. (eds.) Selected Papers I, pp. 735–749. Springer, Cham (2015). https://doi.org/10.1007/978-3-319-16859-3_38
7. Glazer, A., Lindenbaum, M., Markovitch, S.: Learning high-density regions for a generalized Kolmogorov-Smirnov test in high-dimensional data. In: NIPS (2012)
8. Golatkar, A., Achille, A., Soatto, S.: Eternal sunshine of the spotless net: selective forgetting in deep networks (2019)
9. Gong, M., Xie, Y., Pan, K., Feng, K., Qin, A.K.: A survey on differentially private machine learning [review article]. IEEE Comput. Intell. Mag. **15**(2), 49–64 (2020)
10. Ji, Z., Lipton, Z.C., Elkan, C.: Differential privacy and machine learning: a survey and review (2014)
11. Krizhevsky, A., Hinton, G., et al.: Learning multiple layers of features from tiny images (2009)
12. LeCun, Y., Bottou, L., Bengio, Y., Haffner, P.: Gradient-based learning applied to document recognition. Proc. IEEE **86**(11), 2278–2324 (1998)
13. Li, T., Sahu, A.K., Talwalkar, A., Smith, V.: Federated learning: challenges, methods, and future directions. IEEE Signal Process. Mag. **37**(3), 50–60 (2020)
14. Li, W., et al.: Privacy-preserving federated brain tumour segmentation. In: Suk, H.I., Liu, M., Yan, P., Lian, C. (eds.) MLMI 2019. LNCS, vol. 11861, pp. 133–141. Springer, Cham (2019). https://doi.org/10.1007/978-3-030-32692-0_16
15. Netzer, Y., Wang, T., Coates, A., Bissacco, A., Wu, B., Ng, A.Y.: Reading digits in natural images with unsupervised feature learning (2011)
16. Pyrgelis, A., Troncoso, C., De Cristofaro, E.: Under the hood of membership inference attacks on aggregate location time-series, February 2019. http://arxiv.org/abs/1902.07456
17. Roy, A.G., Siddiqui, S., Pölsterl, S., Navab, N., Wachinger, C.: BrainTorrent: a peer-to-peer environment for decentralized federated learning, May 2019. http://arxiv.org/abs/1905.06731

18. Sablayrolles, A., Douze, M., Schmid, C., Jégou, H.: D\'ej\'a Vu: an empirical evaluation of the memorization properties of ConvNets, September 2018. http://arxiv.org/abs/1809.06396
19. Sheller, M.J., Reina, G.A., Edwards, B., Martin, J., Bakas, S.: Multi-institutional deep learning modeling without sharing patient data: a feasibility study on brain tumor segmentation. In: Crimi, A., Bakas, S., Kuijf, H., Keyvan, F., Reyes, M., van Walsum, T. (eds.) BrainLes 2018. LNCS, vol. 11383, pp. 92–104. Springer, Cham (2019). https://doi.org/10.1007/978-3-030-11723-8_9
20. Shokri, R., Stronati, M., Song, C., Shmatikov, V.: Membership inference attacks against machine learning models. In: 2017 IEEE Symposium on Security and Privacy (SP), pp. 3–18, May 2017. https://doi.org/10.1109/SP.2017.41
21. Torralba, A., Efros, A.: Unbiased look at dataset bias. In: Proceedings of the 2011 IEEE Conference on Computer Vision and Pattern Recognition, pp. 1521–1528 (2011)

Learning and Exploiting Interclass Visual Correlations for Medical Image Classification

Dong Wei(✉), Shilei Cao, Kai Ma, and Yefeng Zheng

Tencent Jarvis Lab, Shenzhen, China
{donwei,eliasslcao,kylekma,yefengzheng}@tencent.com

Abstract. Deep neural network-based medical image classifications often use "hard" labels for training, where the probability of the correct category is 1 and those of others are 0. However, these hard targets can drive the networks over-confident about their predictions and prone to overfit the training data, affecting model generalization and adaptation. Studies have shown that label smoothing and softening can improve classification performance. Nevertheless, existing approaches are either non-data-driven or limited in applicability. In this paper, we present the Class-Correlation Learning Network (CCL-Net) to learn interclass visual correlations from given training data, and produce soft labels to help with classification tasks. Instead of letting the network directly learn the desired correlations, we propose to learn them implicitly via distance metric learning of class-specific embeddings with a lightweight plugin CCL block. An intuitive loss based on a geometrical explanation of correlation is designed for bolstering learning of the interclass correlations. We further present end-to-end training of the proposed CCL block as a plugin head together with the classification backbone while generating soft labels on the fly. Our experimental results on the International Skin Imaging Collaboration 2018 dataset demonstrate effective learning of the interclass correlations from training data, as well as consistent improvements in performance upon several widely used modern network structures with the CCL block.

Keywords: Computer-aided diagnosis · Soft label · Deep metric learning

1 Introduction

Computer-aided diagnosis (CAD) has important applications in medical image analysis, such as disease diagnosis and grading [4,5]. Benefiting from the progress of deep learning techniques, automated CAD methods have advanced remarkably

Electronic supplementary material The online version of this chapter (https://doi.org/10.1007/978-3-030-59710-8_11) contains supplementary material, which is available to authorized users.

A. L. Martel et al. (Eds.): MICCAI 2020, LNCS 12261, pp. 106–115, 2020.
https://doi.org/10.1007/978-3-030-59710-8_11

in recent years, and are now dominated by learning-based classification methods using deep neural networks [9,11,15]. Notably, these methods mostly use "hard" labels as their targets for learning, where the probability for the correct category is 1 and those for others are 0. However, these hard targets may adversely affect model generalization and adaption as the networks become over-confident about their predictions when trained to produce extreme values of 0 or 1, and prone to overfit the training data [17].

Studies showed that label smoothing regularization (LSR) can improve classification performance to a limited extent [13,17], although the smooth labels obtained by uniformly/proportionally distributing probabilities do not represent genuine interclass relations in most circumstances. Gao *et al.* [6] proposed to convert the image label to a discrete label distribution and use deep convolutional networks to learn from ground truth label distributions. However, the ground truth distributions were constructed based on empirically defined rules, instead of data-driven. Chen *et al.* [1] modeled the interlabel dependencies as a correlation matrix for graph convolutional network based multilabel classification, by mining pairwise cooccurrence patterns of labels. Although data-driven, this approach is not applicable to single-label scenarios in which labels do not occur together, restricting its application. Arguably, softened labels reflecting real data characteristics improve upon the hard labels by more effectively utilizing available training data, as the samples from one specific category can help with training of similar categories [1,6]. For medical images, there exist rich interclass visual correlations, which have a great potential yet largely remain unexploited for this purpose.

In this paper, we present the Class-Correlation Learning Network (CCL-Net) to learn interclass visual correlations from given training data, and utilize the learned correlations to produce soft labels to help with the classification tasks (Fig. 1). Instead of directly letting the network learn the desired correlations, we propose implicit learning of the correlations via distance metric learning [16,20] of class-specific embeddings [12] with a lightweight plugin CCL block. A new loss is designed for bolstering learning of the interclass correlations. We further present end-to-end training of the proposed CCL block together with the classification backbone, while enhancing classification ability of the latter with soft labels generated on the fly. In summary, our contributions are three folds:

- We propose a CCL block for data-driven interclass correlation learning and label softening, based on distance metric learning of class embeddings. This block is conceptually simple, lightweight, and can be readily plugged as a CCL head into any mainstream backbone network for classification.
- We design an intuitive new loss based on a geometrical explanation of correlation to help with the CCL, and present integrated end-to-end training of the plugged CCL head together with the backbone network.
- We conduct thorough experiments on the International Skin Imaging Collaboration (ISIC) 2018 dataset. Results demonstrate effective data-driven CCL, and consistent performance improvements upon widely used modern network structures utilizing the learned soft label distributions.

2 Method

Preliminaries. Before presenting our proposed CCL block, let us first review the generic deep learning pipeline for single-label multiclass classification problems as preliminaries. As shown in Fig. 1(b), an input x first goes through a feature extracting function f_1 parameterized as a network with parameters θ_1, producing a feature vector $\boldsymbol{f}$ for classification: $\boldsymbol{f} = f_1(x|\theta_1)$, where $\boldsymbol{f} \in \mathbb{R}^{n_1 \times 1}$. Next, $\boldsymbol{f}$ is fed to a generalized fully-connected (fc) layer f_c (parameterized with θ_{fc}) followed by the softmax function, obtaining the predicted classification probability distribution $\boldsymbol{q}^{\text{cls}}$ for x: $\boldsymbol{q}^{\text{cls}} = \text{softmax}(f_c(\boldsymbol{f}|\theta_{\text{fc}}))$, where $\boldsymbol{q}^{\text{cls}} = [q_1^{\text{cls}}, q_2^{\text{cls}}, \ldots, q_K^{\text{cls}}]^T$, K is the number of classes, and $\sum_k q_k^{\text{cls}} = 1$. To supervise the training of f_1 and f_c (and the learning of θ_1 and θ_{fc} correspondingly), for each training sample x, a label $y \in \{1, 2, \ldots, K\}$ is given. This label is then converted to the one-hot distribution [17]: $\boldsymbol{y} = [\delta_{1,y}, \delta_{2,y}, \ldots, \delta_{K,y}]^T$, where $\delta_{k,y}$ is the Dirac delta function, which equals 1 if $k = y$ and 0 otherwise. After that, the cross entropy loss l_{CE} can be used to compute the loss between $\boldsymbol{q}^{\text{cls}}$ and $\boldsymbol{y}$: $l_{\text{CE}}(\boldsymbol{q}^{\text{cls}}, \boldsymbol{y}) = -\sum_k \delta_{k,y} \log q_k^{\text{cls}}$. Lastly, the loss is optimized by an optimizer (e.g., Adam [10]), and θ_1 and θ_{fc} are updated by gradient descent algorithms.

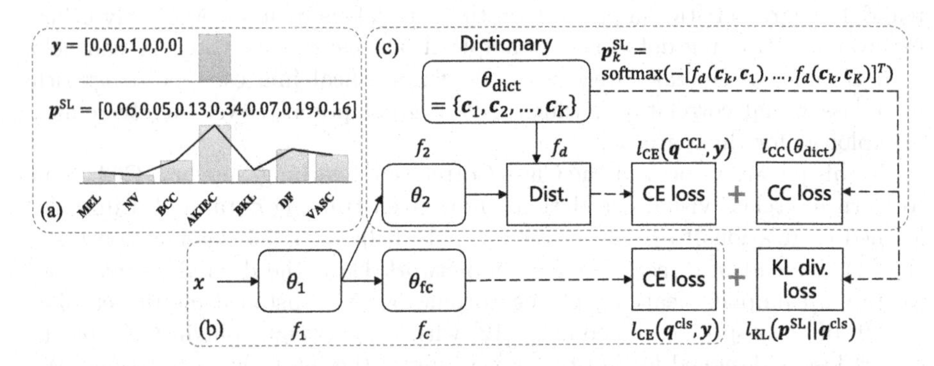

Fig. 1. Diagram of the proposed CCL-Net. (a) Soft label distributions ($\boldsymbol{p}^{\text{SL}}$) are learned from given training data and hard labels $\boldsymbol{y}$. (b) Generic deep learning pipeline for classification. (c) Structure of the proposed CCL block. This lightweight block can be plugged into any classification backbone network as a head and trained end-to-end together, and boost performance by providing additional supervision signals with $\boldsymbol{p}^{\text{SL}}$.

As mentioned earlier, the "hard" labels may adversely affect model generalization as the networks become over-confident about their predictions. In this sense, LSR [17] is a popular technique to make the network less confident by smoothing the one-hot label distribution $\boldsymbol{y}$ to become $\boldsymbol{p}^{\text{LSR}} = [p_{1,y}^{\text{LSR}}, \ldots, p_{K,y}^{\text{LSR}}]^T$, where $p_{k,y}^{\text{LSR}} = (1-\epsilon)\delta_{k,y} + \epsilon u(k)$, ϵ is a weight, and $u(k)$ is a distribution over class labels for which the uniform [13] or *a priori* [13,17] distributions are proposed. Then, the cross entropy is computed with $\boldsymbol{p}^{\text{LSR}}$ instead of $\boldsymbol{y}$.

Learning Interclass Visual Correlations for Label Softening. In most circumstances, LSR cannot reflect the genuine interclass relations underlying the given training data. Intuitively, the probability redistribution should be biased towards visually similar classes, so that samples from these classes can boost training of each other. For this purpose, we propose to learn the underlying visual correlations among classes from the training data and produce soft label distributions that more authentically reflect intrinsic data properties. Other than learning the desired correlations directly, we learn them implicitly by learning interrelated yet discriminative class embeddings via distance metric learning. Both the concepts of feature embeddings and deep metric learning have proven useful in the literature (e.g., [12,16,20]). To the best of our knowledge, however, combining them for data-driven learning of interclass visual correlations and label softening has not been done before.

The structure of the CCL block is shown in Fig. 1(c), which consists of a lightweight embedding function f_2 (parameterized with θ_2), a dictionary θ_{dict}, a distance metric function f_d, and two loss functions. Given the feature vector $\boldsymbol{f}$ extracted by f_1, the embedding function f_2 projects $\boldsymbol{f}$ into the embedding space: $\boldsymbol{e} = f_2(\boldsymbol{f}|\theta_2)$, where $\boldsymbol{e} \in \mathbb{R}^{n_2 \times 1}$. The dictionary maintains all the class-specific embeddings: $\theta_{\text{dict}} = \{\boldsymbol{c}_k\}$. Using f_d, the distance between the input embedding and every class embedding can be calculated by $d_k = f_d(\boldsymbol{e}, \boldsymbol{c}_k)$. In this work, we use $f_d(\boldsymbol{e}_1, \boldsymbol{e}_2) = \left\| \frac{\boldsymbol{e}_1}{\|\boldsymbol{e}_1\|} - \frac{\boldsymbol{e}_2}{\|\boldsymbol{e}_2\|} \right\|^2$, where $\|\cdot\|$ is the L2 norm. Let $\boldsymbol{d} = [d_1, d_2, \ldots, d_K]^T$, we can predict another classification probability distribution $\boldsymbol{q}^{\text{CCL}}$ based on the distance metric: $\boldsymbol{q}^{\text{CCL}} = \text{softmax}(-\boldsymbol{d})$, and a cross entropy loss $l_{\text{CE}}(\boldsymbol{q}^{\text{CCL}}, \boldsymbol{y})$ can be computed accordingly. To enforce interrelations among the class embeddings, we innovatively propose the class correlation loss l_{CC}:

$$l_{\text{CC}}(\theta_{\text{dict}}) = 1/K^2 \sum_{k_1=1}^{K} \sum_{k_2=1}^{K} \left| f_d(\boldsymbol{c}_{k_1}, \boldsymbol{c}_{k_2}) - b \right|_+ \quad (1)$$

where $|\cdot|_+$ is the Rectified Linear Unit (ReLU), and b is a margin parameter. Intuitively, l_{CC} enforces the class embeddings to be no further than a distance b from each other in the embedding space, encouraging correlations among them. Other than attempting to tune the exact value of b, we resort to the geometrical meaning of correlation between two vectors: if the angle between two vectors is smaller than 90°, they are considered correlated. Or equivalently for L2-normed vectors, if the Euclidean distance between them is smaller than $\sqrt{2}$, they are considered correlated. Hence, we set $b = (\sqrt{2})^2$. Then, the total loss function for the CCL block is defined as $\mathcal{L}_{\text{CCL}} = l_{\text{CE}}(\boldsymbol{q}^{\text{CCL}}, \boldsymbol{y}) + \alpha_{\text{CC}} l_{\text{CC}}(\theta_{\text{dict}})$, where α_{CC} is a weight. Thus, θ_2 and θ_{dict} are updated by optimizing $\mathcal{L}_{\text{CCL}}$.

After training, we define the soft label distribution $\boldsymbol{p}_k^{\text{SL}}$ for class k as:

$$\boldsymbol{p}_k^{\text{SL}} = \text{softmax}(-[f_d(\boldsymbol{c}_1, \boldsymbol{c}_k), \ldots, f_d(\boldsymbol{c}_K, \boldsymbol{c}_k)]^T) = [p_{1,k}^{\text{SL}}, \ldots, p_{K,k}^{\text{SL}}]^T. \quad (2)$$

Algorithm 1. End-to-end training of the proposed CCL-Net.

Input: Training images $\{x\}$ and labels $\{y\}$
Output: Learned network parameters $\{\theta_1, \theta_{\text{fc}}, \theta_2, \theta_{\text{dict}}\}$
1: Initialize $\theta_1, \theta_{\text{fc}}, \theta_2, \theta_{\text{dict}}$
2: **for** number of training epochs **do**
3: **for** number of minibatches **do**
4: Compute soft label distributions $\{\boldsymbol{p}^{\text{SL}}\}$ from θ_{dict}
5: Sample minibatch of m images $\{x^{(i)}|i \in \{1,\ldots,m\}\}$, compute $\{\boldsymbol{f}^{(i)}\}$
6: Update θ_1 and θ_{fc} by stochastic gradient descending: $\nabla_{\{\theta_1,\theta_{\text{fc}}\}} \frac{1}{m}\sum_{i=1}^{m} \mathcal{L}_{\text{cls}}$
7: Update θ_2 and θ_{dict} by stochastic gradient descending: $\nabla_{\{\theta_2,\theta_{\text{dict}}\}} \frac{1}{m}\sum_{i=1}^{m} \mathcal{L}_{\text{CCL}}$
8: **end for**
9: **end for**

It is worth noting that by this definition of soft label distributions, the correct class always has the largest probability, which is a desired property especially at the start of training. Next, we describe our end-to-end training scheme of the CCL block together with a backbone classification network.

Integrated End-to-End Training with Classification Backbone. As a lightweight module, the proposed CCL block can be plugged into any mainstream classification backbone network—as long as a feature vector can be pooled for an input image—and trained together in an end-to-end manner. To utilize the learned soft label distributions, we introduce a Kullback-Leibler divergence (KL div.) loss $l_{\text{KL}}(\boldsymbol{p}^{\text{SL}}||\boldsymbol{q}^{\text{cls}}) = \sum_k p_k^{\text{SL}} \log(p_k^{\text{SL}}/q_k^{\text{cls}})$ in the backbone network (Fig. 1), and the total loss function for classification becomes

$$\mathcal{L}_{\text{cls}} = l_{\text{CE}}(\boldsymbol{q}^{\text{cls}}, \boldsymbol{y}) + l_{\text{KL}}(\boldsymbol{p}^{\text{SL}}||\boldsymbol{q}^{\text{cls}}). \quad (3)$$

Consider that l_{CC} tries to keep the class embeddings within a certain distance of each other, it is somehow adversarial to the goal of the backbone network which tries to push them away from each other as much as possible. In such cases, alternative training schemes are usually employed [7] and we follow this way. Briefly, in each training iteration, the backbone network is firstly updated with the CCL head frozen, and then it is frozen to update the CCL head; more details about the training scheme are provided in Algorithm 1. After training, the prediction is made according to $\boldsymbol{q}^{\text{cls}}$: $\hat{y} = \text{argmax}_k(q_k^{\text{cls}})$.

During training, we notice that the learned soft label distributions sometimes start to linger around a fixed value for the correct classes and distribute about evenly across other classes after certain epochs; or in other words, approximately collapse into LSR with $u(k)$ = uniform distribution (the proof is provided in the supplementary material). This is because of the strong capability of deep neural networks in fitting any data [19]. As l_{CC} forces the class embeddings to be no further than b from each other in the embedding space, a network of sufficient capacity has the potential to make them exactly the distance b away from each other (which is overfitting) when well trained. To prevent this from happening, we use the average correct-class probability $\bar{p}_{k,k}^{\text{SL}} = 1/K \sum_k p_{k,k}^{\text{SL}}$ as a measure of the total softness of the label set (the lower the softer, as the correct classes distribute more probabilities to other classes), and consider that the CCL head has converged if $\bar{p}^{\text{SL}}_{k,k}$ does not drop for 10 consecutive epochs. In such case, θ_{dict} is frozen, whereas the rest of the CCL-Net keeps updating.

3 Experiments

Dataset and Evaluation Metrics. The ISIC 2018 dataset is provided by the Skin Lesion Analysis Toward Melanoma Detection 2018 challenge [2], for prediction of seven disease categories with dermoscopic images, including: melanoma (MEL), melanocytic nevus (NV), basal cell carcinoma (BCC), actinic keratosis/Bowen's disease (AKIEC), benign keratosis (BKL), dermatofibroma (DF), and vascular lesion (VASC) (example images are provided in Fig. 2). It comprises 10,015 dermoscopic images, including 1,113 MEL, 6,705 NV, 514 BCC, 327 AKIEC, 1,099 BKL, 115 DF, and 142 VASC images. We randomly split the data into a training and a validation set (80:20 split) while keeping the original interclass ratios, and report evaluation results on the validation set. The employed evaluation metrics include accuracy, Cohen's kappa coefficient [3], and unweighed means of F1 score and Jaccard similarity coefficient.

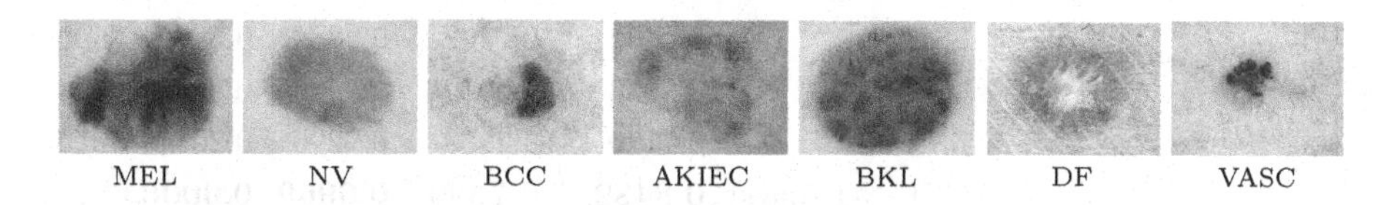

Fig. 2. Official example images of the seven diseases in the ISIC 2018 disease classification dataset [2].

Implementation. In this work, we use commonly adopted, straightforward training schemes to demonstrate effectiveness of the proposed CCL-Net in data-driven learning of interclass visual correlations and improving classification performance, rather than sophisticated training strategies or heavy ensembles. Specifically, we adopt the stochastic gradient descent optimizer with a momentum of 0.9, weight decay of 10^{-4}, and the backbone learning rate initialized to 0.1 for all experiments. The learning rate for the CCL head (denoted by lr_{CCL}) is initialized to 0.0005 for all experiments, except for when we study the impact of varying lr_{CCL}. Following [8], we multiply the initial learning rates by 0.1 twice during training such that the learning process can saturate at a higher limit.[1] A minibatch of 128 images is used. The input images are resized to have a short side of 256 pixels while maintaining original aspect ratios. Online data augmentations

[1] The exact learning-rate-changing epochs as well as total number of training epochs vary for different backbones due to different network capacities.

including random cropping, horizontal and vertical flipping, and color jittering are employed for training. For testing, a single central crop of size 224×224 pixels is used as input. Gradient clipping is employed for stable training. α_{CC} is set to 10 empirically. The experiments are implemented using the PyTorch package. A singe Tesla P40 GPU is used for model training and testing. For f_2, we use three fc layers of width 1024, 1024, and 512 (i.e., $n_2 = 512$), with batch normalization and ReLU in between.

Table 1. Experimental results on the ISIC18 dataset, including comparisons with the baseline networks and LSR [17]. Higher is better for all evaluation metrics.

Backbone	Method	Accuracy	F1 score	Kappa	Jaccard
ResNet-18 [8]	Baseline	0.8347	0.7073	0.6775	0.5655
	LSR-u1	0.8422	0.7220	0.6886	0.5839
	LSR-u5	0.8437	0.7211	0.6908	0.5837
	LSR-a1	0.7883	0.6046	0.5016	0.4547
	LSR-a5	0.7029	0.2020	0.1530	0.1470
	CCL-Net (ours)	**0.8502**	**0.7227**	**0.6986**	**0.5842**
EfficientNet-B0 [18]	Baseline	0.8333	0.7190	0.6696	0.5728
	LSR-u1	0.8382	0.7014	0.6736	0.5573
	LSR-u5	0.8432	0.7262	0.6884	0.5852
	LSR-a1	0.8038	0.6542	0.5526	0.5058
	LSR-a5	0.7189	0.2968	0.2295	0.2031
	CCL-Net (ours)	**0.8482**	**0.7390**	**0.6969**	**0.6006**
MobileNetV2 [14]	Baseline	0.8308	0.6922	0.6637	0.5524
	LSR-u1	0.8248	0.6791	0.6547	0.5400
	LSR-u5	0.8253	0.6604	0.6539	0.5281
	LSR-a1	0.8068	0.6432	0.5631	0.4922
	LSR-a5	0.7114	0.2306	0.2037	0.1655
	CCL-Net (ours)	**0.8342**	**0.7050**	**0.6718**	**0.5648**

*LSR settings: -u1: $u(k)$ = uniform, $\epsilon = 0.1$; -u5: $u(k)$ = uniform, $\epsilon = 0.5228$; -a1: $u(k)$ = *a priori*, $\epsilon = 0.1$; -a5: $u(k)$ = *a priori*, $\epsilon = 0.5228$.

Comparison with Baselines and LSR. We quantitatively compare our proposed CCL-Net with various baseline networks using the same backbones. Specifically, we experiment with three widely used backbone networks: ResNet-18 [8], MobileNetV2 [14], and EfficientNet-B0 [18]. In addition, we compare our CCL-Net with the popular LSR [17] with different combinations of ϵ and $u(k)$: $\epsilon \in \{0.1, 0.5228\}$ and $u(k) \in \{\text{uniform}, \textit{a priori}\}$, resulting in a total of four settings (we compare with $\epsilon = 0.5228$ since for the specific problem, the learned soft label distributions would eventually approximate uniform LSR with this ϵ value, if the class embeddings θ_{dict} are not frozen after convergence). The results are charted in Table 1. As we can see, the proposed CCL-Net achieves the best performances on all evaluation metrics for all backbone networks, including Cohen's kappa [3] which is more appropriate for imbalanced data than accuracy. These results demonstrate effectiveness of utilizing the learned soft label distributions in improving classification performance of the backbone networks. We also note that moderate improvements are achieved by the LSR settings with $u(k)$ = uniform on two of the three backbone networks, indicating effects of this simple strategy. Nonetheless, these improvements are outweighed by those achieved by our CCL-Net. In addition to the superior performances to LSR, another advantage of the CCL-Net is that it can intuitively reflect intrinsic interclass correlations underlying the given training data, at a minimal extra overhead. Lastly, it is worth noting that the LSR settings with $u(k)$ = *a priori* decrease all evaluation metrics from the baseline performances, suggesting inappropriateness of using LSR with *a priori* distributions for significantly imbalanced data.

Table 2. Properties of the CCL by varying the learning rate of the CCL head. ResNet-18 [8] backbone is used.

lr_{CCL}	0.0001	0.0005	0.001	0.005	0.01	0.05
Epochs of converge	Never	137	76	23	12	5
$\bar{p}^{\text{SL}}_{k,k}$	0.48	0.41	0.38	0.35	0.33	0.30
Accuracy	0.8422	**0.8502**	0.8452	0.8442	0.8422	0.8417
Kappa	0.6863	**0.6986**	0.6972	0.6902	0.6856	0.6884

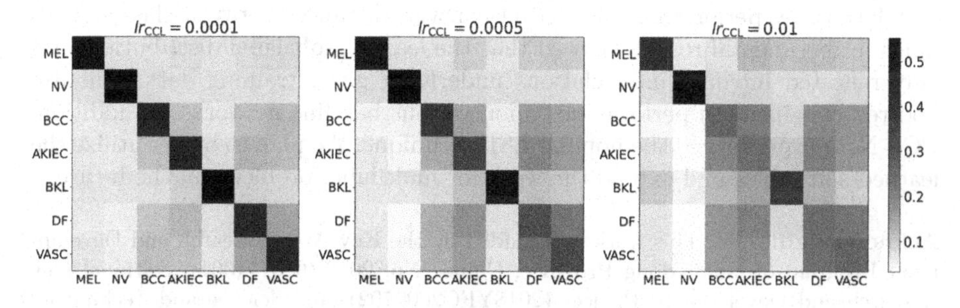

Fig. 3. Visualization of the learned soft label distributions using different lr_{CCL}, where each row of a matrix represents the soft label distribution of a class.

Analysis of Interclass Correlations Learned with CCL-Net. Next, we investigate properties of the learned interclass visual correlations by the proposed CCL-Net, by varying the value of lr_{CCL}. Specifically, we examine the epochs and label softness when the CCL head converges, as well as the final accuracies and kappa coefficients. Besides $\bar{p}^{\mathrm{SL}}_{k,k}$, the overall softness of the set of soft label distributions can also be intuitively perceived by visualizing all $\{\boldsymbol{p}^{\mathrm{SL}}_k\}$ together as a correlation matrix. Note that this matrix does not have to be symmetric, since the softmax operation is separately conducted for each class. Table 2 presents the results, and Fig. 3 shows three correlation matrices using different lr_{CCL}. Interestingly, we can observe that as lr_{CCL} increases, the CCL head converges faster with higher softness. The same trend can be observed in Fig. 3, where the class probabilities become more spread with the increase of lr_{CCL}. Notably, when $lr_{\mathrm{CCL}} = 0.0001$, the CCL head does not converge in given epochs and the resulting label distributions are not as soft. This indicates that when lr_{CCL} is too small, the CCL head cannot effectively learn the interclass correlations. Meanwhile, the best performance is achieved when $lr_{\mathrm{CCL}} = 0.0005$ in terms of both accuracy and kappa, instead of other higher values. This may suggest that very quick convergence may also be suboptimal, probably because the prematurely frozen class embeddings are learned from the less representative feature vectors in the early stage of training. In summary, lr_{CCL} is a crucial parameter for the CCL-Net, though it is not difficult to tune based on our experience.

4 Conclusion

In this work, we presented CCL-Net for data-driven interclass visual correlation learning and label softening. Rather than directly learning the desired correlations, CCL-Net implicitly learns them via distance-based metric learning of class-specific embeddings, and constructs soft label distributions from learned correlations by performing softmax on pairwise distances between class embeddings. Experimental results showed that the learned soft label distributions not only reflected intrinsic interrelations underlying given training data, but also boosted classification performance upon various baseline networks. In addition, CCL-Net outperformed the popular LSR technique. We plan to better utilize the learned soft labels and extend the work for multilabel problems in the future.

Acknowledgments. This work was funded by the Key Area Research and Development Program of Guangdong Province, China (No. 2018B010111001), National Key Research and Development Project (2018YFC2000702), and Science and Technology Program of Shenzhen, China (No. ZDSYS201802021814180).

References

1. Chen, Z.M., Wei, X.S., Wang, P., Guo, Y.: Multi-label image recognition with graph convolutional networks. In: Proceedings of the IEEE Conference on Computer Vision and Pattern Recognition, pp. 5177–5186 (2019)
2. Codella, N., et al.: Skin lesion analysis toward melanoma detection 2018: a challenge hosted by the International Skin Imaging Collaboration (ISIC). arXiv preprint arXiv:1902.03368 (2019)
3. Cohen, J.: A coefficient of agreement for nominal scales. Educ. Psychol. Measur. **20**(1), 37–46 (1960)
4. Doi, K.: Current status and future potential of computer-aided diagnosis in medical imaging. Br. J. Radiol. **78**(suppl_1), s3–s19 (2005)
5. Doi, K.: Computer-aided diagnosis in medical imaging: historical review, current status and future potential. Comput. Med. Imaging Graph. **31**(4–5), 198–211 (2007)
6. Gao, B.B., Xing, C., Xie, C.W., Wu, J., Geng, X.: Deep label distribution learning with label ambiguity. IEEE Trans. Image Process. **26**(6), 2825–2838 (2017)
7. Goodfellow, I., et al.: Generative adversarial nets. In: Advances in Neural Information Processing Systems, pp. 2672–2680 (2014)
8. He, K., Zhang, X., Ren, S., Sun, J.: Deep residual learning for image recognition. In: Proceedings of the IEEE Conference on Computer Vision and Pattern Recognition, pp. 770–778 (2016)
9. Ker, J., Wang, L., Rao, J., Lim, T.: Deep learning applications in medical image analysis. IEEE Access **6**, 9375–9389 (2017)
10. Kingma, D.P., Ba, J.: Adam: a method for stochastic optimization. arXiv preprint arXiv:1412.6980 (2014)
11. Litjens, G., et al.: A survey on deep learning in medical image analysis. Med. Image Anal. **42**, 60–88 (2017)
12. Mikolov, T., Sutskever, I., Chen, K., Corrado, G.S., Dean, J.: Distributed representations of words and phrases and their compositionality. In: Advances in Neural Information Processing Systems, pp. 3111–3119 (2013)
13. Müller, R., Kornblith, S., Hinton, G.E.: When does label smoothing help? In: Advances in Neural Information Processing Systems, pp. 4696–4705 (2019)
14. Sandler, M., Howard, A., Zhu, M., Zhmoginov, A., Chen, L.C.: MobileNetV2: inverted residuals and linear bottlenecks. In: Proceedings of the IEEE Conference on Computer Vision and Pattern Recognition, pp. 4510–4520 (2018)
15. Shen, D., Wu, G., Suk, H.I.: Deep learning in medical image analysis. Annu. Rev. Biomed. Eng. **19**, 221–248 (2017)
16. Sohn, K.: Improved deep metric learning with multi-class N-pair loss objective. In: Advances in Neural Information Processing Systems, pp. 1857–1865 (2016)
17. Szegedy, C., Vanhoucke, V., Ioffe, S., Shlens, J., Wojna, Z.: Rethinking the Inception architecture for computer vision. In: Proceedings of the IEEE Conference on Computer Vision and Pattern Recognition, pp. 2818–2826 (2016)
18. Tan, M., Le, Q.V.: EfficientNet: rethinking model scaling for convolutional neural networks. arXiv preprint arXiv:1905.11946 (2019)
19. Zhang, C., Bengio, S., Hardt, M., Recht, B., Vinyals, O.: Understanding deep learning requires rethinking generalization. arXiv preprint arXiv:1611.03530 (2016)
20. Zhe, X., Chen, S., Yan, H.: Directional statistics-based deep metric learning for image classification and retrieval. Pattern Recogn. **93**, 113–123 (2019)

Feature Preserving Smoothing Provides Simple and Effective Data Augmentation for Medical Image Segmentation

Rasha Sheikh and Thomas Schultz(✉)

University of Bonn, Bonn, Germany
{rasha,schultz}@cs.uni-bonn.de

Abstract. CNNs represent the current state of the art for image classification, as well as for image segmentation. Recent work suggests that CNNs for image classification suffer from a bias towards texture, and that reducing it can increase the network's accuracy. We hypothesize that CNNs for medical image segmentation might suffer from a similar bias. We propose to reduce it by augmenting the training data with feature preserving smoothing, which reduces noise and high-frequency textural features, while preserving semantically meaningful boundaries. Experiments on multiple medical image segmentation tasks confirm that, especially when limited training data is available or a domain shift is involved, feature preserving smoothing can indeed serve as a simple and effective augmentation technique.

1 Introduction

Image segmentation is a key problem in medical image analysis. Convolutional neural networks (CNNs) often achieve high accuracy, but only if trained on a sufficient amount of annotated data. In medical applications, the number of images that are available for a given task is often limited, and the time of experts who can provide reliable labels is often scarce and expensive. Therefore, data augmentation is widely used to increase the ability to generalize from limited training data, and it is often indispensable for achieving state-of-the-art results.

Augmentation generates artificial virtual training samples by applying certain transformations to the original training images. Many of these transformations reflect variations that are expected in test images. Examples are geometric transformations such as image flipping, rotations, translations, elastic deformations [19], or cropping, but also intensity or color space transformations, which can be used to simulate changes in illumination or acquisition device characteristics [5,22]. More complex augmentation has been performed via data-driven generative models that either generate images for augmentation directly [6,21], or generate spatial and appearance transformations [7,25]. However, implementing and training these approaches requires a relatively high effort.

In this work, we demonstrate that feature preserving smoothing provides a novel, simple, and effective data augmentation approach for CNN-based semantic segmentation. In particular, we demonstrate the benefit of augmenting the

A. L. Martel et al. (Eds.): MICCAI 2020, LNCS 12261, pp. 116–126, 2020.
https://doi.org/10.1007/978-3-030-59710-8_12

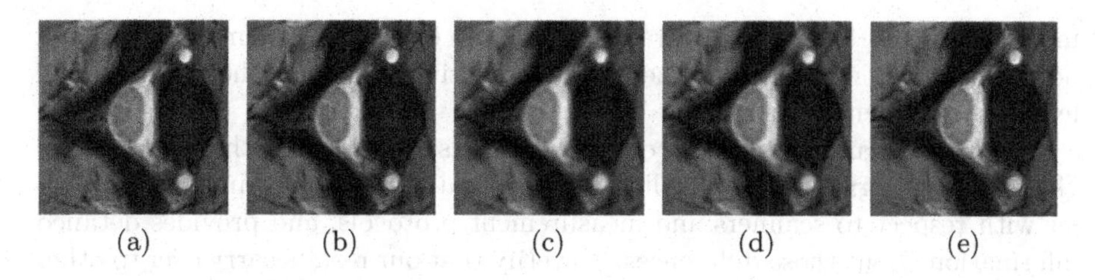

Fig. 1. Sample image (a) and the results of smoothing with Total Variation (b), Gaussian (c), bilateral (d), and guided filters (e).

original training images with copies that have been processed with total variation (TV) based denoising [20]. As shown in Fig. 1(b), TV regularization creates piecewise constant images, in which high frequency noise and textural features are removed, but sharp outlines of larger regions are preserved.

Unlike the above-mentioned augmentation techniques, ours does not attempt to generate realistic training samples. Rather, it is inspired by the use of neural style transfer for augmentation. Neural style transfer combines two images with the goal of preserving the semantic contents of one, but the style of the other [8]. It is often used for artistic purposes, but has also been found to be effective as an augmentation technique when training CNNs for image classification [12]. To explain this, Geirhos et al. [9] perform experiments for which they generated images with conflicting shape and texture cues. When interpreting such images, ImageNet-trained CNNs were found to favor texture, while humans favor shape. Since shape is more robust than texture against many image distortions, this might be one factor that allows human vision to generalize better than CNNs.

Our work is based on the hypothesis that CNNs for image segmentation suffer from a similar bias towards texture, and will generalize better when increasing the relative impact of shape. Augmenting with TV regularized images should contribute to this, since it preserves shapes, but smoothes out high frequency textural features. In a similar spirit, Zhang et al. [24] use superpixelization for augmentation, and Ma et al. [15] use lossy image compression. Both argue that the respective transformations discard information that is less relevant to human perception, and thus might prevent the CNN from relying on features that are irrelevant for human interpretation. However, Ma et al. only evaluate JPEG augmentation in one specialized application, the segmentation of sheep in natural images. As part of our experiments, we verify that their idea, which has a similar motivation as ours, carries over to medical image segmentation.

2 Materials and Methods

2.1 Selection of Datasets

We selected datasets that allow us to test whether feature preserving smoothing would be an effective data augmentation technique when dealing with limited training data in medical image segmentation. Moreover, we hypothesized that this augmentation might reduce the drop in segmentation accuracy that is frequently associated with domain shifts, such as changes in scanners. If differences in noise levels or textural appearance contribute to those problems, increasing a network's robustness towards them by augmenting with smoothed data should lead to better generalization.

As the primary dataset for our experiments, we selected the Spinal Cord Grey Matter Segmentation Challenge [18], because it includes images that differ with respect to scanners and measurement protocols, and provides detailed information about those differences. To verify that our results carry over to other medical image segmentation tasks, we additionally present experiments on the well-known Brain Tumor Segmentation Challenge [3,4,16], as well as the White Matter Hyperintensity Segmentation Challenge [13].

2.2 CNN Architecture

We selected the U-Net architecture [19] for our experiments, since it is widely used for medical image segmentation. In particular, our model is based on a variant by Perone et al. [17], who train it with the Adam optimizer, dropout for regularization, and the dice loss

$$\text{dice} = \frac{2\sum pg}{\sum p^2 + \sum g^2}, \tag{1}$$

where p is the predicted probability map and g is the ground-truth mask. Our model is trained to learn 2D segmentation masks from 2D slices. The initial learning rate is 0.001, the dropout rate is 0.5, betas for the Adam optimizer are 0.9 and 0.999, and we train the model for 50, 30, and 100 epochs for the SCGM, BraTS, and WMH datasets respectively. The number of epochs is chosen using cross-validation, the others are the default settings of the framework we use.

2.3 Feature-Preserving Smoothing

We mainly focus on Total Variation based denoising [20], which we consider to be a natural match for augmentation in image segmentation problems due to its piecewise constant, segmentation-like output. The TV regularized version of an n-dimensional image $f : D \subset \mathbb{R}^n \to \mathbb{R}$, with smoothing parameter α, can be defined as the function $u : D \to \mathbb{R}$ that minimizes

$$E(u; \alpha, f) := \int_D \left(\frac{1}{2} (u - f)^2 + \alpha \|\nabla u\| \right) dV, \tag{2}$$

where the integration is performed over the n-dimensional image domain D. Numerically, we find u by introducing an artificial time parameter $t \in [0, \infty)$, setting $u(\mathbf{x}, t = 0) = f$, and evolving it under the Total Variation flow [1]

$$\frac{\partial u}{\partial t} = \operatorname{div} \left(\frac{\nabla_{\mathbf{x}} u}{\|\nabla_{\mathbf{x}} u\|} \right), \tag{3}$$

using an additive operator splitting scheme [23]. The resulting image $u(\mathbf{x}, t)$ at time t approximates a TV regularized version of f with parameter $\alpha = t$. We apply TV smoothing to all images in the training set, and train on the union of both sets, randomly shuffling all images before each epoch. We did not observe a clear difference between this basic strategy and a stratified sampling which ensured that half of the images in each batch were TV smoothed.

For comparison, we also consider two alternative feature preserving filters, the bilateral filter [2] and the guided filter [10], as well as standard, non feature preserving Gaussian smoothing. We hypothesized that feature preserving filters other than TV regularization might also be effective for augmentation, even though maybe not as much, since they do not create a segmentation-like output to the same extent as TV. We expected that Gaussian smoothing would not be helpful, because it does not preserve sharp edges, and therefore does not provide clear shape cues to the network. Moreover, it can be expressed using a simple convolution which, if useful, could easily be learned by the CNN itself.

3 Experiments

3.1 Spinal Cord Grey Matter

This challenge dataset consists of spinal cord MR images of healthy subjects acquired from four different sites with different scanners and acquisition parameters. The task of the challenge is to segment the grey matter in those images. The publicly available training data includes MR images of 10 subjects per site for a total of 40 subjects. Each MR image was annotated by four raters. A sample image from site 2 and its four masks are shown in Fig. 2.

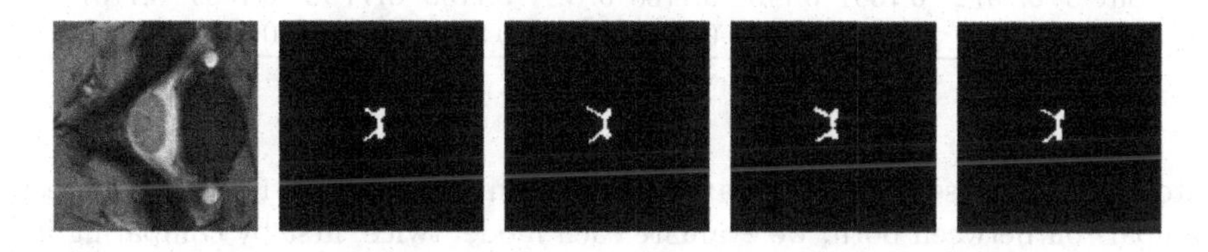

Fig. 2. Spinal cord image and segmentation masks from Site 2.

Setup. The data was split into 80% training and 20% test sets. To evaluate the effect of TV augmentation under domain shift, we train on data from only one site (Montreal) and report results on the test sets from all sites. Since the four sites have different slice resolutions (0.5 mm, 0.5 mm, 0.25 mm, 0.3 mm), we resample images to the highest resolution. We also standardize intensities.

We consider it unusual that each image in this dataset has annotations from four raters. In many other cases, only a single annotation would be available per training image, due to the high cost of creating annotations. We expected that training with multiple annotations per image would provide an additional regularization. To investigate how it interacts with data augmentation, we conducted two sets of experiments. In the first, we train using annotations from the first

Table 1. Dice scores for the model trained on Site 2 using annotations from the first rater, and evaluated by comparing the predictions to annotations made by the same rater (top rows) and annotations made by all raters (bottom rows).

	Original	TV	Flip	Rotate	Elastic	Gauss	Bilateral	Guided	JPEG
Site 1	0.5930	**0.7332**	0.5148	0.4842	0.6484	0.4670	0.5530	0.6050	0.5674
Site 2	0.8337	**0.8610**	0.8254	0.8289	0.8258	0.8341	0.8356	0.8262	0.8567
Site 3	0.5950	0.6466	0.5952	0.6260	0.6260	0.6397	0.6323	0.6207	**0.6605**
Site 4	0.7978	**0.8395**	0.7896	0.8011	0.7960	0.8021	0.8006	0.7909	0.8233
Site 1	0.5695	**0.7044**	0.4934	0.4599	0.6344	0.4533	0.5392	0.5714	0.5366
Site 2	0.8185	**0.8483**	0.8129	0.8158	0.8218	0.8209	0.8245	0.8104	0.8435
Site 3	0.6707	0.7475	0.6664	0.7012	0.7109	0.7182	0.7090	0.6960	**0.7582**
Site 4	0.7776	**0.8184**	0.7751	0.7798	0.7843	0.7813	0.7840	0.7708	0.8046

Table 2. Dice scores for the model trained on Site 2 using annotations from all raters, and evaluated by comparing the predictions to annotations made by the first rater (top rows) and annotations from all raters (bottom rows).

	Original	TV	Flip	Rotate	Elastic	Gauss	Bilateral	Guided	JPEG
Site 1	0.7936	**0.8254**	0.8227	0.7875	0.7823	0.8238	0.8084	0.8057	0.8236
Site 2	0.8710	**0.8819**	0.8771	0.8729	0.8453	0.8758	0.8786	0.8756	0.8674
Site 3	0.6507	0.6511	0.6582	0.6632	0.6501	0.6620	**0.6680**	0.6546	0.6585
Site 4	0.8480	0.8572	0.8535	0.8486	0.8487	0.8583	0.8544	0.8562	**0.8610**
Site 1	0.7827	**0.8303**	0.8083	0.7617	0.7754	0.8096	0.8000	0.7912	0.8048
Site 2	0.8786	**0.8956**	0.8869	0.8750	0.8554	0.8866	0.8846	0.8796	0.8790
Site 3	0.7672	0.7651	0.7734	0.7768	0.7617	0.7763	**0.7790**	0.7689	0.7761
Site 4	0.8430	**0.8562**	0.8510	0.8424	0.8445	0.8549	0.8497	0.8494	0.8561

rater only. In the second, we use annotations from all raters. To facilitate a direct comparison between both, we evaluate each model twice, first by comparing its predictions to annotations from rater one, second by using those from all raters.

Other Augmentation Techniques. We compare TV augmentation to random flipping, rotation with a random angle between [−10, 10] degrees, and elastic deformations. We also compare against augmentation with Gaussian, bilateral, and guided filters, as well as JPEG compression [15]. We chose filter parameters visually, so that they result in a smoothed image while not distorting the gray matter shape. Examples are shown in Fig. 1.

Results. Table 1 shows the average dice score of each site's test subjects when training using annotations from the first rater. The top rows compare the predictions to annotations made by the same rater, the bottom rows show the average dice across all four raters.

In Table 2, we train using annotations from all four raters. At training time, each input image is replicated four times with a different mask for each input.

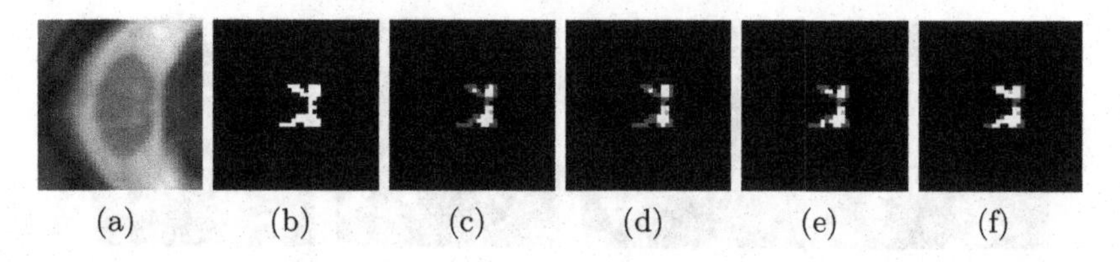

Fig. 3. Sample image from Site 1 (a), its ground truth (b), the prediction of a model trained with no augmentation (c), or with bilateral (d), jpeg (e), and TV augmentation (f), respectively. White, red, and blue colors indicate TP, FN, and FP, respectively. Quantitative performance is shown in Table 1. (Color figure online)

We again evaluate the results by comparing the predictions to the annotations provided by the first rater (top) and annotations from all raters (bottom).

Discussion. TV augmentation improved results almost always, by a substantial margin in some of the cases that involved a domain shift. We show qualitative results of a challenging image from Site 1 in Fig. 3. Most of the time, TV performed better than any other augmentation technique. In the few cases where it did not, the conceptually similar bilateral or JPEG augmentation worked best. As expected, augmentation with Gaussian smoothing did not lead to competitive results. For Site 3, we observed that TV sometimes smoothes out very fine details in the spinal cord structure. This might explain why JPEG augmentation produced slightly higher dice scores than TV on that site.

Traditional augmentation techniques such as flipping, rotation, and elastic deformation did not show a clear benefit in this specific task. Even though it is possible to combine them with TV augmentation, our results suggest that it is unlikely to benefit this particular task. It is thus left for future work.

Comparing Table 2 to Table 1, we can see the largest benefits when annotations by only one rater are available at training time. This agrees with our intuition that repeated annotations provide a form of regularization. Despite this, we continue seeing a benefit from data augmentation.

Impact of Smoothing Parameter and Multi Scale Augmentation. The effect of the TV smoothing parameter α that was discussed in Sect. 2.3 is illustrated in Fig. 4. Previous experiments used $\alpha = 5$. Table 3 studies how sensitive TV augmentation is with respect to this parameter. Again, training was on Site 2 using annotations from the first rater, and evaluation compared the predictions to annotations made by the same rater (left) and annotations made by all raters (right). All scales perform quite well, and combining different ones, as shown in the last column, yields the best dice score in nearly all cases.

3.2 BraTS 2019

This challenge data consists of brain MRI scans with high- and low-grade gliomas, collected at different institutions. The publicly available training data includes 3D scans of 335 subjects.

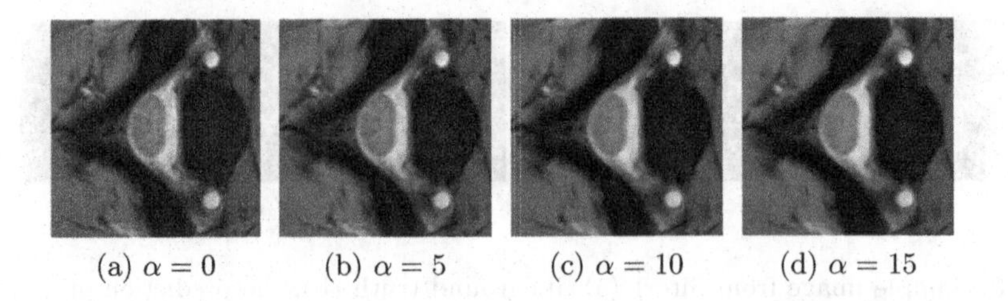

(a) $\alpha = 0$ (b) $\alpha = 5$ (c) $\alpha = 10$ (d) $\alpha = 15$

Fig. 4. Sample image and the results of different TV smoothing parameters.

Table 3. Results with varying smoothing parameters suggest that TV augmentation is robust to its choice, and combining multiple scales is a feasible strategy. Dice scores on the left are from the same rater, on the right from all raters.

	$\alpha = 5$	$\alpha = 10$	$\alpha = 15$	Combined
Site 1	0.7332	0.7423	0.7573	**0.8187**
Site 2	0.8610	0.8653	0.8599	**0.8757**
Site 3	0.6466	**0.6581**	0.6579	0.6487
Site 4	0.8395	0.8457	0.8428	**0.8525**

	$\alpha = 5$	$\alpha = 10$	$\alpha = 15$	Combined
Site 1	0.7044	0.7130	0.7305	**0.7848**
Site 2	0.8483	0.8530	0.8536	**0.8749**
Site 3	0.7475	0.7576	0.7574	**0.7633**
Site 4	0.8184	0.8217	0.8215	**0.8378**

Setup. We again aimed for an evaluation that would separately assess the benefit with or without a domain shift. Although there was no explicit mapping of individual subjects to their respective institutions, we could identify four groups based on the challenge data description and the filenames: Brats2013 (30 subjects), CBICA (129 subjects), TCGA (167 subjects), and TMC (9 subjects).

We split each group's data into 80% training and 20% test sets. We train on data from only one group (TMC) and report results on the test sets from all groups. An example image with the ground truth and the smoothed augmentation is shown in Fig. 5.

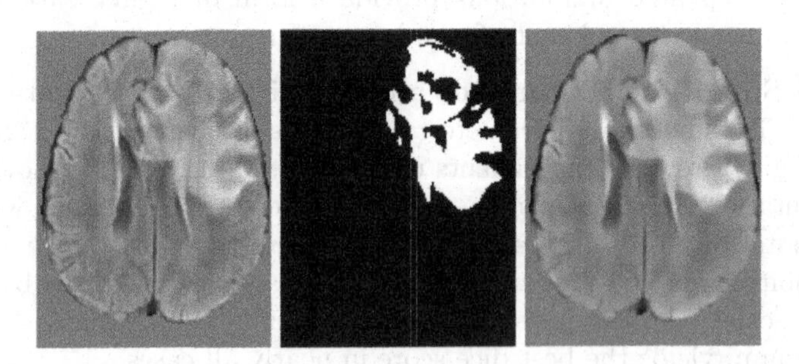

Fig. 5. BraTS (TMC) training image, ground truth, and the TV smoothed image.

As input we use the FLAIR modality and train the model to learn the Whole Tumor label. We follow the preprocessing steps of [11] and standardize intensities to zero mean and unit variance.

Results. As shown in Table 4, TV augmentation increased segmentation accuracy in three out of the four groups. The largest benefit was observed in the group Br13. This involved a domain shift, as the model was only trained on TMC. On the group where the performance slightly decreases (TCGA), we found that the predicted segmentation sometimes misses some of the finer tumor details. Such fine structures might be more difficult to discern after TV smoothing.

Table 4. Dice Score on held-out test sets of different subsets of BraTS 19.

	Br13	CBICA	TCGA	TMC
Original	0.8088	0.7933	**0.7929**	0.7894
TV smoothed	**0.8570**	**0.8152**	0.7885	**0.8233**

3.3 White Matter Hyperintensity

The publicly available data from the WMH challenge is acquired with different scanners from 3 institutions. 2D FLAIR and T1 MR images and segmentation masks are provided for 60 subjects, 20 from each institution.

Setup. The data is split into 80% training and 20% test sets. We combined the training data from all three sites, and report results on the test sets we held out ourselves, as well as on the official test sets, by submitting to the challenge website. To avoid creating a large number of submissions, we do not investigate the impact of training on one compared to all sites in this case. We follow the preprocessing steps of [14] and standardize intensities. An example image with the ground truth and TV smoothed augmentation is shown in Fig. 6.

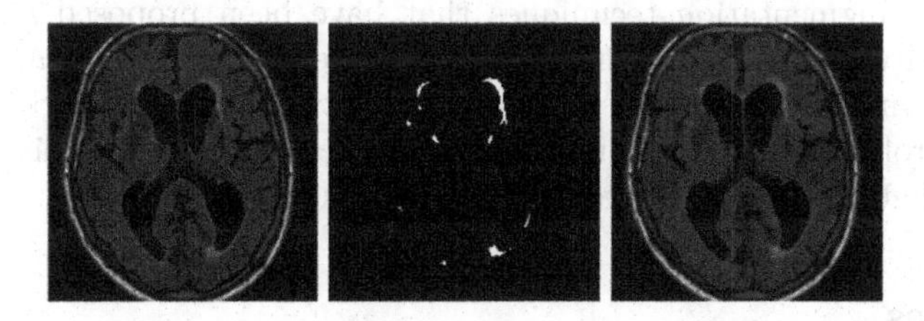

Fig. 6. WMH site 1 training image, ground-truth, and the TV smoothed image.

Results. Table 5 shows that TV augmentation improved segmentation accuracy on the held-out data in all three sites. Table 6 shows a clear improvement also on the official test set. The evaluation criteria in this challenge are Dice Score (DSC), Hausdorff distance (H95), Average Volume Difference (AVD), Recall for individual lesions, and F1 score for individual lesions.

Table 5. Dice Score on held-out test sets of WMH challenge

	Site 1	Site 2	Site 3
Original	0.6879	0.8348	0.7170
TV smoothed	**0.7329**	**0.8480**	**0.7685**

Table 6. Results on the official test set of the challenge. Higher values for DSC, Recall, F1 are better, and lower values for H95, AVD are better.

	DSC	H95	AVD	Recall	F1
Original	0.74	9.05	29.73	0.65	0.64
TV smoothed	**0.77**	**7.42**	**24.97**	**0.76**	**0.67**

Discussion. Results on both the held-out test set and the official test set show that TV smoothing improves the segmentation performance, in some cases by a substantial margin. TV-smoothing performs better with respect to all evaluation criteria in the detailed official results. The most pronounced improvement is in the number of lesions that the model detects (recall).

4 Conclusion

Our results indicate a clear benefit from using feature preserving smoothing for data augmentation when training CNN-based medical image segmentation on limited data. Advantages were especially pronounced when using TV smoothing, which creates a piecewise constant, segmentation-like output. TV augmentation also helped when a domain shift between training and test data was involved. Consequently, we propose that TV smoothing can be used as a relatively simple and inexpensive data augmentation method for medical image segmentation.

In the future, we hope to better characterize the exact conditions under which the different augmentation techniques that have been proposed for semantic segmentation work well, and when it makes sense to combine them. We expect that factors such as the nature of differences between training and test images will play a role, as well as characteristics of the images (e.g., noise), and the structures that should be segmented (e.g., presence of fine details).

References

1. Andreu, F., et al.: Minimizing total variation flow. Diff. Integral Eq. **14**(3), 321–360 (2001)
2. Aurich, V., Weule, J.: Non-linear Gaussian filters performing edge preserving diffusion. In: Sagerer, G., Posch, S., Kummert, F. (eds.) Mustererkennung. Informatik Aktuell, pp. 538–545. Springer, Heidelberg (1995). https://doi.org/10.1007/978-3-642-79980-8_63
3. Bakas, S., et al.: Advancing the cancer genome atlas glioma MRI collections with expert segmentation labels and radiomic features. Sci. Data **4**, 170117 (2017)

4. Bakas, S., et al.: Identifying the best machine learning algorithms for brain tumor segmentation, progression assessment, and overall survival prediction in the BRATS challenge. Technical report 1811.02629, arXiv (2018)
5. Billot, B., Greve, D., Van Leemput, K., Fischl, B., Iglesias, J.E., Dalca, A.V.: A learning strategy for contrast-agnostic MRI segmentation. Technical report 2003.01995, arXiv (2020)
6. Bowles, C., et al.: GAN augmentation: augmenting training data using generative adversarial networks. Technical report 1810.10863, arXiv (2018)
7. Chaitanya, K., Karani, N., Baumgartner, C.F., Becker, A., Donati, O., Konukoglu, E.: Semi-supervised and task-driven data augmentation. In: Chung, A.C.S., Gee, J.C., Yushkevich, P.A., Bao, S. (eds.) IPMI 2019. LNCS, vol. 11492, pp. 29–41. Springer, Cham (2019). https://doi.org/10.1007/978-3-030-20351-1_3
8. Gatys, L.A., Ecker, A.S., Bethge, M.: Image style transfer using convolutional neural networks. In: IEEE Conference on Computer Vision and Pattern Recognition (CVPR), pp. 2414–2423 (2016)
9. Geirhos, R., Rubisch, P., Michaelis, C., Bethge, M., Wichmann, F.A., Brendel, W.: ImageNet-trained CNNs are biased towards texture; increasing shape bias improves accuracy and robustness. In: International Conference on Learning Representations (ICLR) (2019)
10. He, K., Sun, J., Tang, X.: Guided image filtering. In: Daniilidis, K., Maragos, P., Paragios, N. (eds.) ECCV 2010. LNCS, vol. 6311, pp. 1–14. Springer, Heidelberg (2010). https://doi.org/10.1007/978-3-642-15549-9_1
11. Isensee, F., Kickingereder, P., Wick, W., Bendszus, M., Maier-Hein, K.H.: Brain tumor segmentation and radiomics survival prediction: contribution to the BRATS 2017 challenge. In: Crimi, A., Bakas, S., Kuijf, H., Menze, B., Reyes, M. (eds.) BrainLes 2017. LNCS, vol. 10670, pp. 287–297. Springer, Cham (2018). https://doi.org/10.1007/978-3-319-75238-9_25
12. Jackson, P.T.G., Abarghouei, A.A., Bonner, S., Breckon, T.P., Obara, B.: Style augmentation: Data augmentation via style randomization. In: CVPR Deep Vision Workshop, pp. 83–92 (2019)
13. Kuijf, H.J., et al.: Standardized assessment of automatic segmentation of white matter hyperintensities and results of the WMH segmentation challenge. IEEE Trans. Med. Imaging **38**(11), 2556–2568 (2019)
14. Li, H., et al.: Fully convolutional network ensembles for white matter hyperintensities segmentation in MR images. NeuroImage **183**, 650–665 (2018)
15. Ma, R., Tao, P., Tang, H.: Optimizing data augmentation for semantic segmentation on small-scale dataset. In: Proceedings of International Conference on Control and Computer Vision (ICCCV), pp. 77–81 (2019)
16. Menze, B.H., et al.: The multimodal brain tumor image segmentation benchmark (BRATS). IEEE Trans. Med. Imaging **34**(10), 1993–2024 (2014)
17. Perone, C.S., Ballester, P., Barros, R.C., Cohen-Adad, J.: Unsupervised domain adaptation for medical imaging segmentation with self-ensembling. NeuroImage **194**, 1–11 (2019)
18. Prados, F., et al.: Spinal cord grey matter segmentation challenge. NeuroImage **152**, 312–329 (2017)
19. Ronneberger, O., Fischer, P., Brox, T.: U-Net: convolutional networks for biomedical image segmentation. In: Navab, N., Hornegger, J., Wells, W.M., Frangi, A.F. (eds.) MICCAI 2015. LNCS, vol. 9351, pp. 234–241. Springer, Cham (2015). https://doi.org/10.1007/978-3-319-24574-4_28
20. Rudin, L.I., Osher, S., Fatemi, E.: Nonlinear total variation based noise removal algorithms. Physica D **60**(1), 259–268 (1992)

21. Sandfort, V., Yan, K., Pickhardt, P.J., Summers, R.M.: Data augmentation using generative adversarial networks (CycleGAN) to improve generalizability in CT segmentation tasks. Sci. Rep. **9**(1), 16884 (2019)
22. Shorten, C., Khoshgoftaar, T.M.: A survey on image data augmentation for deep learning. J. Big Data **6**(1), 60 (2019)
23. Weickert, J., Romeny, B.T.H., Viergever, M.: Efficient and reliable schemes for nonlinear diffusion filtering. IEEE Trans. Image Process. **7**(3), 398–410 (1998)
24. Zhang, Y., et al.: SPDA: superpixel-based data augmentation for biomedical image segmentation. In: International Conference on Medical Imaging with Deep Learning (MIDL). Proceedings of Machine Learning Research, vol. 102, pp. 572–587 (2019)
25. Zhao, A., Balakrishnan, G., Durand, F., Guttag, J.V., Dalca, A.V.: Data augmentation using learned transformations for one-shot medical image segmentation. In: Proceedings of IEEE Conference on Computer Vision and Pattern Recognition (CVPR), pp. 8543–8553 (2019)

Deep kNN for Medical Image Classification

Jiaxin Zhuang[1], Jiabin Cai[1], Ruixuan Wang[1(✉)], Jianguo Zhang[2(✉)], and Wei-Shi Zheng[1,3,4]

[1] School of Data and Computer Science, Sun Yat-sen University, Guangzhou, China
wangrui5@mail.sysu.edu.cn

[2] Department of Computer Science and Engineering, Southern University of Science and Technology, Shenzhen, China
zhangjg@sustech.edu.cn

[3] Key Laboratory of Machine Intelligence and Advanced Computing, MOE, Guangzhou, China

[4] Pazhou Lab, Guangzhou, China

Abstract. Human-level diagnostic performance from intelligent systems often depends on large set of training data. However, the amount of available data for model training may be limited for part of diseases, which would cause the widely adopted deep learning models not generalizing well. One alternative simple approach to small class prediction is the traditional k-nearest neighbor (kNN). However, due to the non-parametric characteristics of kNN, it is difficult to combine the kNN classification into the learning of feature extractor. This paper proposes an end-to-end learning strategy to unify the kNN classification and the feature extraction procedure. The basic idea is to enforce that each training sample and its K nearest neighbors belong to the same class during learning the feature extractor. Experiments on multiple small-class and class-imbalanced medical image datasets showed that the proposed deep kNN outperforms both kNN and other strong classifiers.

Keywords: Small class · Deep kNN · Intelligent diagnosis

1 Introduction

With recent advance particularly in deep learning, intelligent diagnosis has shown human-level performance for various diseases [1,2]. However, in many intelligent diagnosis tasks, the amount of available data for model training is often limited for some or all diseases. Small training data often leads to the over-fitting issue, i.e., the trained model does not generalize well to new data. While the issue can be often alleviated by fine-tuning a model which was originally trained in another task [3], such transfer learning may not work well if the

Electronic supplementary material The online version of this chapter (https://doi.org/10.1007/978-3-030-59710-8_13) contains supplementary material, which is available to authorized users.

A. L. Martel et al. (Eds.): MICCAI 2020, LNCS 12261, pp. 127–136, 2020.
https://doi.org/10.1007/978-3-030-59710-8_13

image domain in the current task is far from that of the original task. Recently developed meta-learning techniques for few-shot learning problems seem to provide a plausible solution to the small-class classification tasks [4]. However, these techniques often presume the access to a large number of additional small classes for model training, which is impractical in the tasks of intelligent diagnosis.

Another simple but often effective approach is the k-nearest neighbor (kNN) classification [5,7], where feature extraction and/or dimensionality reduction are often performed to obtain a concise feature vector to represent the original image [6,8,9]. However, because the feature extraction procedure is often predetermined and independent of the classification task of interest, the extracted features may not be discriminative enough for the classification task. To alleviate this issue, metric learning can be applied to transform the extracted features into a new feature space where the transformed features become more discriminative, such as neighborhood component analysis (NCA) [10] and large margin nearest neighbor methods [11]. The metric learning methods depend on the originally extracted features which may already omit certain features helpful for the classification task. To learn to extract discriminative features directly from original image data, recently a triplet loss function was proposed to train a convolutional neural network (CNN) through which an image will be transformed to a low-dimensional but discriminative feature vector [12,13]. However, the process of CNN training is independent of the latter kNN classification step. It would be desirable to unify feature extraction and kNN classification into a single step as done in current CNN classifiers. However, searching for K nearest neighbours for each training data is a non-differentiable operation, and therefore it is not easy to combine the neighbor search into the training of a feature extractor.

This paper proposes an end-to-end learning strategy to unify the kNN classification and the feature extraction process, particularly for classification of small classes. The basic idea is to enforce that each training image and its K nearest neighbors belong to the same class during learning feature extractor. By unifying the feature extraction and the kNN classification procedure, a better feature extractor can be learned specifically for the kNN classifier and the task of interest. Comprehensive evaluations on multiple medical image datasets showed that the proposed approach, called deep kNN, outperforms various kNNs and even CNN classifiers particularly for small class prediction. Compared to the traditional triplet loss, the proposed novel loss function can help train the feature extractor much faster, as confirmed in experiments. What's more, the proposed deep kNN is independent of network architectures, and therefore can be directly combined with any existing convolutional or fully connected neural network architectures.

2 Deep kNN

A traditional kNN classifier does not include feature extraction, i.e., the feature extraction is done separately from the k-nearest neighbor search. Since the feature extraction does not consider any task-specific information, extracted features could be not discriminative enough for the kNN classification. It would be ideal to learn to extract features specific to the classification task of interest. However, due to the non-parametric characteristics of the nearest neighbor search process, so far it is not clear how to combine feature learning and the k-nearest neighbor search into a unified process for the kNN classifier.

2.1 Problem Formulation

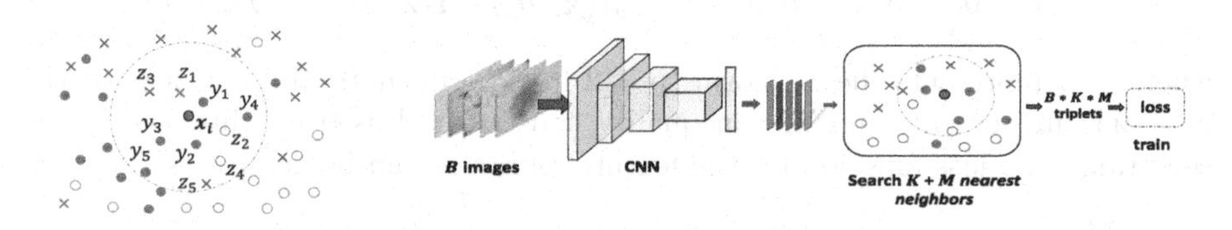

Fig. 1. Demonstration of neighbor search (left) and deep kNN training (right). Left: for each image $\mathbf{x}_i$, K nearest neighbors $\{\mathbf{y}_k^{t,i}\}$ of the same class and M nearest neighbors $\{\mathbf{z}_m^{t,i}\}$ from all other classes are searched to generate $K \cdot M$ triplets. Here $K = 5$, $M = 5$, $\mathbf{y}_k^{t,i}$ and $\mathbf{z}_m^{t,i}$ are simplified to $\mathbf{y}_k$ and $\mathbf{z}_m$. Right: during training, K within-class and M cross-class nearest neighbors are searched within the mini-batch for each image.

We propose a deep learning strategy to naturally unify the two steps of kNN classifier, where the feature extractor is represented by a convolutional or fully connected neural network whose output of the last layer is a feature vector representing the input image. The intuitive idea behind the strategy is to train a feature extractor based on which any training image and its k-nearest neighbors are forced to belong to the same class. In this way, the k-nearest neighbor search procedure is naturally incorporated into the process of training the feature extractor. If such a feature extractor can be trained, we would expect any test image and its K nearest neighbors in the training set is probably from the same class. The challenge is how to formulate this idea for feature extractor training.

We propose a novel triplet loss to solve this challenge. At the beginning stage of training a feature extractor, since the parameters of the feature extractor are initially either randomly set or from a pre-trained model based on a public dataset (e.g., ImageNet), it is not surprising that the distributions of different classes of images would be interleaved in the feature space. That means, among the several nearest neighbors of any specific training image, one or more neighbors may not be from the same class of the training image. To make any image and its nearest neighbors belong to the same class, the feature extractor needs to be updated such that the distance between the image and any of its nearest neighbors is closer than the distance between the image and any image of other classes in the feature space. Formally, denote the feature extractor after the $(t-1)^{th}$ training iteration by $\mathbf{f}(\cdot; \boldsymbol{\theta}_{t-1})$, where $\boldsymbol{\theta}$ represents the parameters of the feature extractor to be learned. Also denote the i-th training image by $\mathbf{x}_i$, its K nearest neighbors of the same class after the $(t-1)^{th}$ training iteration by $\{\mathbf{y}_k^{t,i}, k=1,\ldots,K\}$, and its M nearest neighbors from all the other classes after the $(t-1)^{th}$ training iteration by $\{\mathbf{z}_m^{t,i}, m=1,\ldots,M\}$ (Fig. 1, Left). Then, after the tth training iteration, for each training image $\mathbf{x}_i$, we expect its K nearest neighbors of the same class based on the updated feature extractor $\mathbf{f}(\cdot;\boldsymbol{\theta}_t)$ are closer to the training image $\mathbf{x}_i$ compared to the distance between $\mathbf{x}_i$ and any image of the other classes (particularly the M nearest neighbors from the other classes), i.e., the following inequality is expected to be satisfied,

$$\|\mathbf{f}(\mathbf{x}_i;\boldsymbol{\theta}_t)-\mathbf{f}(\mathbf{y}_k^{t,i};\boldsymbol{\theta}_t)\|+\alpha<\|\mathbf{f}(\mathbf{x}_i;\boldsymbol{\theta}_t)-\mathbf{f}(\mathbf{z}_m^{t,i};\boldsymbol{\theta}_t)\|,\ \forall k,m \tag{1}$$

where $\|\cdot\|$ represents the L_p norm ($p=2$ in experiments), and α is a positive constant further enforcing the inequality constraint. Based on this inequality constraint, the loss function for the feature extractor can be defined as

$$l(\mathbf{x}_i,\mathbf{y}_k^{t,i},\mathbf{z}_m^{t,i};\boldsymbol{\theta}_t)=[\|\mathbf{f}(\mathbf{x}_i;\boldsymbol{\theta}_t)-\mathbf{f}(\mathbf{y}_k^{t,i};\boldsymbol{\theta}_t)\|+\alpha-\|\mathbf{f}(\mathbf{x}_i;\boldsymbol{\theta}_t)-\mathbf{f}(\mathbf{z}_m^{t,i};\boldsymbol{\theta}_t)\|]_+ \tag{2}$$

where $[d]_+=\max(0,d)$, such that the loss is 0 when the inequality constraint is satisfied, and becomes larger when the constraint is further from being satisfied. Considering all N images in the training set, the loss function becomes

$$L(\boldsymbol{\theta}_t)=\frac{1}{NKM}\sum_{i=1}^{N}\sum_{k=1}^{K}\sum_{m=1}^{M}l(\mathbf{x}_i,\mathbf{y}_k^{t,i},\mathbf{z}_m^{t,i};\boldsymbol{\theta}_t). \tag{3}$$

2.2 Neighbor Search During Training

Since feature extractor is updated over training iterations, part of (or the whole) K nearest neighbors for one specific training image based on the feature extractor at previous iteration could become no longer the K nearest neighbors at current iteration. Therefore, after updating the feature extractor by minimizing $L(\boldsymbol{\theta}_t)$ at the tth iteration, for each training image, its K nearest neighbors of the same class and M nearest neighbors from the other classes will be searched (based on Euclidean distance here) and updated again before updating the feature extractor in next iteration. In this way, the k-nearest neighbor classification performance on the training dataset will be naturally evaluated and gradually improved during feature extractor training. Therefore, the proposed training strategy unifies the feature extractor and the kNN classification, thus called *deep kNN*. Note during the training process, the non-parametric k-nearest neighbor search plays the role of providing training data at each iteration, and the search process is not involved in the derivative of extractor parameters. Therefore, the difficulty in differentiating the non-parametric process is naturally circumvented.

In analogy to the stochastic gradient descent (SGD) widely adopted for training deep learning models, a similar SGD method can be used here to update the feature extractor. In this case, the feature extractor will be updated once per mini-batch of training images. However, this would cause the update (search) of K nearest neighbors for each image in the next mini-batch set much more computationally expensive, because all training images need to be fed into the feature extractor after each mini-batch training to find the K nearest neighbors for each image in the new mini-batch set. To alleviate this issue, similar to the sampling strategy for minimizing the triplet loss in related work [12,14], the $(K+M)$ nearest neighbors for each image may be searched just within the mini-batch (Fig. 1, Right). To guarantee that enough number of within- and cross-class nearest neighbors exist in each mini-batch, stratified sampling was adopted for mini-batch generation (see Experimental setup). Experiments showed such within-batch nearest neighbor search did not downgrade deep kNN performance compared to searching for nearest neighbors from the whole dataset.

2.3 Comparison with Traditional Triplet Loss

While the proposed triplet loss has a similar form compared to the traditional triplet loss [12,13], there exists a few significant differences between them. Traditionally, triplets are formed often by the *furthest* within-class pairs and the nearest cross-class pairs, while the proposed triplet is formed by the *nearest* within-class pairs and nearest cross-class pairs. The objective of traditional triplet loss is to enforce all pair-wise distances within a class are smaller than the distances between any data of the class and all data of other classes. In comparison, the proposed method only requires any data and its K nearest neighbors of the same class are closer than the distance between the data and other classes of data. That means, traditional triplet loss is used to train a feature extractor based on which the distribution of each class of data is clustered more compactly, while the proposed triplet loss is used to help separate distributions of different classes apart from each other such that the K nearest neighbors of any data belong to the same class as that of the data, without requiring the distribution of each class to be compactly clustered (Supplementary Figure S1(a)(b)). Such a difference also indicates that the feature extractor based on the proposed triple loss can be trained more easily (i.e., faster convergence) than that based on traditional triplet loss. Last but important, the proposed triplet loss can be used to train a deep kNN by combining the two steps of kNN classification, while the traditional triplet loss cannot. As shown below, deep kNN performs much better than traditional two-step kNN classifiers.

3 Experiment

Experimental Setup: Experiments were extensively carried out on two skin image datasets, SD-198 and Skin-7, and one Pneumonia X-ray dataset [1], each including small classes and/or different level of class imbalance (Table 1). For each dataset, images were resized into 256×256 and randomly cropped to 224×224 pixels for training, followed by a random horizontal flip. The similar operation was performed for testing, except that only one cropped image

[1] https://www.kaggle.com/c/rsna-pneumonia-detection-challenge.

Table 1. Dataset statistics. More ticks denote more imbalanced.

Dataset	#Class	#Train	#Test	ImageSize	SmallestClass	LargestClass	Imbalance
Pneumonia	3	21345	5339	1024 × 1024	6012	11821	✓
SD-198 [15]	198	5,206	1377	[260, 4752]	9	60	✓✓
Skin-7 [16]	7	8005	2005	600 × 450	115	6705	✓✓✓

was generated from the center region of each test image. During training a deep kNN, the batch size B varied a bit for different datasets due to the varying scales of datasets, set to 96 on Skin-7 and pneumonia dataset, and 90 on SD-198 dataset. To guarantee K nearest neighbors of the same class for each image in a batch particularly on the SD-198 dataset, the stratified sampling was adopted, i.e., forming batches by randomly sampling 9 classes and then randomly sampling 10 images for each of the 9 class. During training, the SGD optimizer was used, with the initial learning rate 0.01. Learning rate decayed by 0.5 for every 20 epochs on Skin-7 dataset and pneumonia dataset, and every 30 epochs on SD-198 dataset.

During testing, all test and training images were fed into the trained feature extractor only once to get the corresponding feature vectors. For each test image, its k-nearest neighbors within the training dataset were found in the feature space. The class label of the test image was then determined by the majority class in those neighbors. For simplicity, K denotes the number of nearest neighbors used in training while K_p denotes the number of nearest neighbors used in testing. Overall accuracy (Acc), mean class recall over all classes (MCR) and the recall on the smallest class (RS) were used for measurement.

Effectiveness of Deep kNN: To demonstrate effectiveness of the proposed deep kNN, our method is compared to several baseline methods. Two baseline kNN classifiers, *kNN (VGG19)* and *kNN (ResNet)*, were respectively built on the feature extractor part of *VGG19* [17] and *ResNet50* [18] pretrained on ImageNet. The other baseline kNN *Triplet (ResNet50)* was built on the feature extractor trained with the traditional triplet loss [13], using the pretrained ResNet50 on ImageNet as the backbone. The same ResNet50 backbone was trained with the proposed deep kNN learning strategy. Note that similar amount of effort was put into tuning each baseline method. Table 2 showed that the proposed deep kNN (rows 4, 8) outperforms all the three baseline kNN classifiers (rows 1–3, 5–7). It is reasonable that both the baseline *kNN (VGG19)* and *kNN (ResNet)* were outperformed by the *Triplet (ResNet)* and the proposed deep kNN, because the latter were trained using the specific data of the task of interest. However, it is surprising that the deep kNN also outperforms the *Triplet (ResNet)*, especially considering that the distribution of the *training* data based on the traditional triplet loss are more compactly clustered than that based on the proposed triplet loss (Supplementary Figure S1(a)(b)). Detailed inspection shows that the distribution of the test dataset based on the deep kNN is more compactly clustered than that based on the traditional triplet loss (Supplementary Figure S1(c)(d)),

Table 2. Comparison between deep kNN and various baselines on multiple datasets. $K = 5$ and $M = 5$. Similar findings were obtained with other values for K and M. For each model, multiple runs were performed with similar performance observed.

Datasets	Skin-7			Pneumonia			SD-198		
Methods	Acc	MCR	RS	Acc	MCR	RS	Acc	MCR	RS
kNN (VGG19, $K_p = 1$) [17]	71.37	42.03	37.93	52.67	52.36	43.06	24.49	24.37	20.83
kNN (ResNet50, $K_p = 1$) [18]	75.56	49.92	55.17	52.37	51.24	38.90	27.62	28.38	29.16
Triplet (ResNet50, $K_p = 1$) [12]	82.91	64.21	63.21	67.84	66.32	65.21	59.96	60.05	47.31
Deep kNN (ours, $K_p = 1$)	**88.2**	**77.4**	**77.0**	**71.0**	**69.3**	**67.3**	**64.5**	**64.1**	**48.3**
kNN (VGG19, $K_p = 5$) [17]	70.72	30.77	10.34	55.42	55.71	44.97	15.12	13.28	4.17
kNN (ResNet50, $K_p = 5$) [18]	74.41	37.53	34.48	55.52	54.66	39.65	19.11	17.45	12.50
Triplet (ResNet50, $K_p = 5$) [12]	84.29	68.31	67.34	70.00	68.73	66.32	60.17	60.02	47.21
Deep kNN (ours, $K_p = 5$)	**89.1**	**78.9**	**77.3**	**71.1**	**69.4**	**69.0**	**65.1**	**64.3**	**48.3**
Weighted-CE (ResNet50) [19]	88.02	80.21	76.60	71.12	69.11	70.05	61.90	62.40	47.67
Deep kNN* (ours, $K_p = 5$)	**90.3**	**81.0**	**80.4**	**71.6**	**71.1**	**70.9**	**66.4**	**66.4**	**51.5**

suggesting the feature extractor trained by the proposed triplet loss is more generalizable to unknown data than that by the traditional triplet loss.

We also compared deep kNN with CNN classifier of the same ResNet50 backbone. Standard deviations over multiple runs are within the range [0.5%, 1.2%] for accuracy, mean class recall (MCR), and the recall on the smallest class in all experiments. The overall performance (Acc, MCR) of deep kNN (Table 2, rows 4, 8) is close to that of CNN classifier (Table 2, 2nd last row, using class-weighted cross entropy to handle data imbalance) on Skin-7 and Pneumonia. Importantly, on the small-class dataset SD-198 and the small class (column RS) of Skin-7 (no small class on Pneumonia), deep kNN clearly outperforms the CNN classifier, confirming that deep kNN works better particularly for small class prediction. CNN classifier could become overfitting on small classes, while deep kNN could largely alleviate this issue by increasing the number of training data (triplets) on small classes. Also interestingly, after fine-tuning the trained CNN classifiers (with the last fully connected layer removed) with the proposed triplet loss, the resulting new deep kNN (Table 2, last row) further improved the performance.

Deep kNN with MLP: In some scenarios, only feature vectors were originally available in dataset. In this case, the deep kNN learning strategy still works, not based on a CNN structure but on a multilayer perceptron (MLP) structure. To demonstrate effectiveness of deep kNN under this condition, features vector of each image in Skin-7 was extracted from output of the last convolutional layer of a pre-trained ResNet50, and then with the vectors as original data, three-layer MLPs (with batch normalization and ReLU activation) were trained respectively based the proposed triplet loss (*MLP+deep-kNN* in Table 3), the traditional triplet loss (*MLP+triplets*), and the cross-entropy loss (*MLP+CE*). The neighborhood component analysis (NCA) and large margin nearest neighbor (LMNN) methods were also evaluated with the feature vectors as input (note NCA and LMNN were not used as baselines in Table 2 because two-dimensional

Table 3. Performance comparison when MLP was used as backbone.

Methods	kNN-basic	NCA	LMNN	MLP+triplets	MLP+CE	MLP+deep-kNN
Acc	64.53	75.06	81.85	77.13	78.85	**85.09**
MCR	34.49	36.39	61.78	62.22	63.11	**66.08**
RS	10.34	27.59	62.07	63.43	64.21	**67.30**

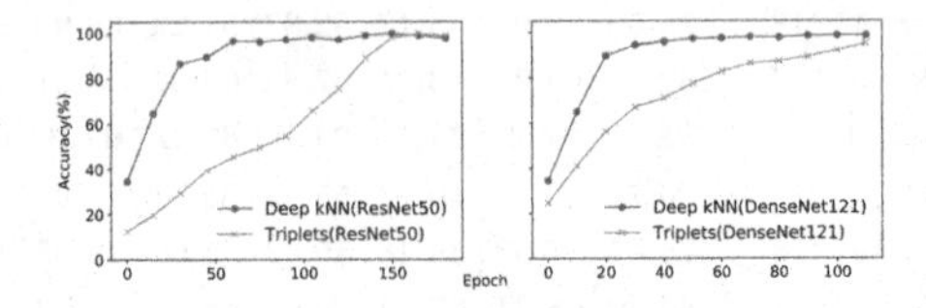

Fig. 2. Training process of feature extractor with the proposed (blue) and traditional triplet loss (orange). Left: ResNet50 on SD-198; Right: DenseNet121 on Skin-7. (Color figure online)

image data cannot be used as input to the two methods). Table 3 shows that all the linear (*NCA*, *LMNN*) and non-linear (*MLP+triplets*, *MLP+CE*, *MLP+deep-kNN*) transformations of the initial feature vectors would help improve the performance of kNN classification compared to the basic kNN (*kNN-basic*). Among them, deep kNN outperforms all others, again confirming the effectiveness of the proposed learning strategy for kNN classification.

Speed of Convergence: We compared the effects of the proposed triplet loss with the traditional triplet loss on the convergence of the optimization using different backbone structures and datasets. It was observed that the proposed triplet loss takes much fewer epochs than the traditional loss to reach the same level of training accuracy (Fig. 2), faster in convergence regardless of model structures and datasets. Note that the deep kNN learning strategy only requires that each sample and its K nearest neighbours of the same class are close enough, while the traditional triplet loss requires all samples of the same class should be closer than the cross-class samples. The stronger constraints in the traditional triple loss is probably the key cause to the slower convergence.

Flexibility with Architecture. Deep kNN is independent of choice of model architectures. To show this, we test deep kNN with four widely used CNNs. For each backbone, the network was trained with cross-entropy loss on the dataset of task of interest, and the output layer was then removed to get the task-specific feature extractor. Such feature extractor was fixed and then a traditional kNN was used to predict each test image, resulting in a strong baseline kNN (because such baseline kNN is based on a task-specific feature extractor). With each backbone network, the strong baseline kNN was compared to the corresponding deep kNN model. Table 4 shows deep kNNs with different backbones perform slightly differently, but all outperforming the corresponding strong baseline kNNs, demonstrating the deep kNN learning strategy is robust to structures

Table 4. Performance of the deep kNN with different CNN backbones on the SD-198 dataset. Similar findings were obtained on other datasets. $K=5$, $M=5$ and $K_p=5$.

Models	VGG19		ResNet50		Dense121		SE-ResNet50	
	kNN	deep kNN	kNN	deep kNN	kNN	deep kNN	kNN	deep kNN
Acc	59.16	**61.51**	60.17	**65.12**	60.54	**64.02**	61.85	**62.79**
MCR	56.33	**61.93**	57.35	**64.34**	60.12	**65.11**	59.32	**62.27**
RS	43.12	**45.43**	46.01	**48.31**	50.42	**52.21**	50.01	**53.12**

of feature extractors. Additional evaluations shows the deep kNN is also robust to the hyper parameters K, M, and K_p (Supplementary Tables S3 and S4).

4 Conclusion

In this paper, we introduced a novel deep learning approach, called deep kNN, which for the first time unifies the feature extraction and the kNN classification procedure. Experiments showed that keep kNN not only performs better than traditional two-step kNN classifiers, but also better than CNN classifiers particularly on small class prediction. Therefore, deep kNN provides a new approach to intelligent diagnosis of diseases with limited training data.

Acknowledgement. This work is supported in part by the National Key Research and Development Program (grant No. 2018YFC1315402), the Guangdong Key Research and Development Program (grant No. 2019B020228001), the National Natural Science Foundation of China (grant No. U1811461), and the Guangzhou Science and Technology Program (grant No. 201904010260).

References

1. Esteva, A., et al.: Dermatologist-level classification of skin cancer with deep neural networks. Nature **542**(7639), 115–118 (2017)
2. Litjens, G., et al.: A survey on deep learning in medical image analysis. Med. Image Anal. **42**, 60–88 (2017)
3. He, K., Girshick, R., Dollr, P.: Rethinking imageNet pre-training. In: CVPR, pp. 4918–4927(2019)
4. Li, A., et al.: Large-scale few-shot learning: Knowledge transfer with class hierarchy. In: CVPR, pp. 7212–7220 (2019)
5. Bingham, E., Mannila, H.: Random projection in dimensionality reduction: applications to image and text data. In: KDD, pp. 245–250 (2001)
6. Turk, M., Pentland, A.: Face recognition using eigenfaces. In: CVPR, pp. 586–587 (1991)
7. Zhang, H., et al.: SVM-KNN: discriminative nearest neighbor classification for visual category recognition. In: CVPR, pp. 2126–2136 (2006)
8. Koniusz, P., et al.: Higher-order occurrence pooling for bags-of-words: visual concept detection. IEEE Trans. Pattern Anal. Mach. Intell. **39**(2), 313–326 (2016)

9. Yosinski, J., et al.: How transferable are features in deep neural networks? In: NeurIPS, pp. 3320–3328 (2014)
10. Goldberger, J., et al.: Neighbourhood components analysis. In NeurIPS, pp. 513–520 (2005)
11. Weinberger, K.Q., Saul, L.K.: Distance metric learning for large margin nearest neighbor classification. J. Mach. Learn. Res. **10**(2), 207–244 (2009)
12. Hermans, A., Beyer, L., Leibe, B.: In defense of the triplet loss for person re-identification. arXiv:1703.07737 (2017)
13. Schroff, F., Kalenichenko, D., Philbin, J.: FaceNet: a unified embedding for face recognition and clustering. In: CVPR, pp. 815–823 (2015)
14. Gordo, A., et al.: End-to-end learning of deep visual representations for image retrieval. IJCV **124**(2), 237–254 (2017)
15. Sun, X., Yang, J., Sun, M., Wang, K.: A benchmark for automatic visual classification of clinical skin disease images. In: Leibe, B., Matas, J., Sebe, N., Welling, M. (eds.) ECCV 2016. LNCS, vol. 9910. Springer, Cham (2016). https://doi.org/10.1007/978-3-319-46466-4_13
16. Codella, N.C.F., et al.: Skin lesion analysis toward melanoma detection: a challenge at the 2017 international symposium on biomedical imaging (ISBI), hosted by the international skin imaging collaboration (ISIC). In: ISBI, pp. 168–172 (2018)
17. Simonyan, K., Zisserman, A.: Very deep convolutional networks for large-scale image recognition. arXiv:1409.1556 (2014)
18. He, K., et al. Deep residual learning for image recognition. In: CVPR, pp. 770–778 (2016)
19. Zhuang, J., et al.: Care: class attention to regions of lesion for classification on imbalanced data. In: MIDL, pp. 588–597 (2019)

Learning Semantics-Enriched Representation via Self-discovery, Self-classification, and Self-restoration

Fatemeh Haghighi[1], Mohammad Reza Hosseinzadeh Taher[1], Zongwei Zhou[1], Michael B. Gotway[2], and Jianming Liang[1](✉)

[1] Arizona State University, Tempe, AZ 85281, USA
{fhaghigh,mhossei2,zongweiz,jianming.liang}@asu.edu
[2] Mayo Clinic, Scottsdale, AZ 85259, USA
Gotway.Michael@mayo.edu

Abstract. Medical images are naturally associated with rich semantics about the human anatomy, reflected in an abundance of recurring anatomical patterns, offering unique potential to foster deep semantic representation learning and yield semantically more powerful models for different medical applications. But how exactly such strong yet free semantics embedded in medical images can be harnessed for self-supervised learning remains largely unexplored. To this end, we train deep models to learn semantically enriched visual representation by self-discovery, self-classification, and self-restoration of the anatomy underneath medical images, resulting in a semantics-enriched, general-purpose, pre-trained 3D model, named Semantic Genesis. We examine our Semantic Genesis with all the publicly-available pre-trained models, by either self-supervision or fully supervision, on the six distinct target tasks, covering both classification and segmentation in various medical modalities (*i.e.,* CT, MRI, and X-ray). Our extensive experiments demonstrate that Semantic Genesis significantly exceeds all of its 3D counterparts as well as the *de facto* ImageNet-based transfer learning in 2D. This performance is attributed to our novel self-supervised learning framework, encouraging deep models to learn compelling semantic representation from abundant anatomical patterns resulting from consistent anatomies embedded in medical images. Code and pre-trained Semantic Genesis are available at https://github.com/JLiangLab/SemanticGenesis.

Keywords: Self-supervised learning · Transfer learning · 3D model pre-training

1 Introduction

Self-supervised learning methods aim to learn general image representation from unlabeled data; naturally, a crucial question in self-supervised learning is how to

Electronic supplementary material The online version of this chapter (https://doi.org/10.1007/978-3-030-59710-8_14) contains supplementary material, which is available to authorized users.

A. L. Martel et al. (Eds.): MICCAI 2020, LNCS 12261, pp. 137–147, 2020.
https://doi.org/10.1007/978-3-030-59710-8_14

"extract" proper supervision signals from the unlabeled data directly. In large part, self-supervised learning approaches involve predicting some hidden properties of the data, such as colorization [16,17], jigsaw [15,18], and rotation [11,13]. However, most of the prominent methods were derived in the context of natural images, without considering the unique properties of medical images.

In medical imaging, it is required to follow protocols for defined clinical purposes, therefore generating images of similar anatomies across patients and yielding recurrent anatomical patterns across images (see Fig. 1a). These recurring patterns are associated with rich semantic knowledge about the human body, thereby offering great potential to foster deep semantic representation learning and produce more powerful models for various medical applications. However, it remains an unanswered question: *How to exploit the deep semantics associated with recurrent anatomical patterns embedded in medical images to enrich representation learning?*

To answer this question, we present a novel self-supervised learning framework, which enables the capture of semantics-enriched representation from unlabeled medical image data, resulting in a set of powerful pre-trained models. We call our pre-trained models **Semantic Genesis**, because they represent a significant advancement from Models Genesis [25] by introducing two *novel* components: self-discovery and self-classification of the anatomy underneath medical images (detailed in Sect. 2). Specifically, our *unique* self-classification branch, with a small computational overhead, compels the model to learn semantics from consistent and recurring anatomical patterns discovered during the self-discovery phase, while Models Genesis learns representation from random subvolumes with no semantics as no semantics can be discovered from random subvolumes. By explicitly employing the strong yet free semantic supervision signals, Semantic Genesis distinguishes itself from all other existing works, including colorization of colonoscopy images [20], context restoration [9], Rubik's cube recovery [26], and predicting anatomical positions within MR images [4].

As evident in Sect. 4, our extensive experiments demonstrate that (1) learning semantics through our two innovations significantly enriches existing self-supervised learning approaches [9,19,25], boosting target tasks performance dramatically (see Fig. 2); (2) Semantic Genesis provides more generic and transferable feature representations in comparison to not only its self-supervised learning counterparts, but also (fully) supervised pre-trained 3D models (see Table 2); and Semantic Genesis significantly surpasses any 2D approaches (see Fig. 3).

This performance is ascribed to the semantics derived from the consistent and recurrent anatomical patterns, that not only can be automatically discovered from medical images but can also serve as strong yet free supervision signals for deep models to learn more semantically enriched representation automatically via self-supervision.

2 Semantic Genesis

Figure 1 presents our self-supervised learning framework, which enables training Semantic Genesis from scratch on unlabeled medical images. Semantic Genesis

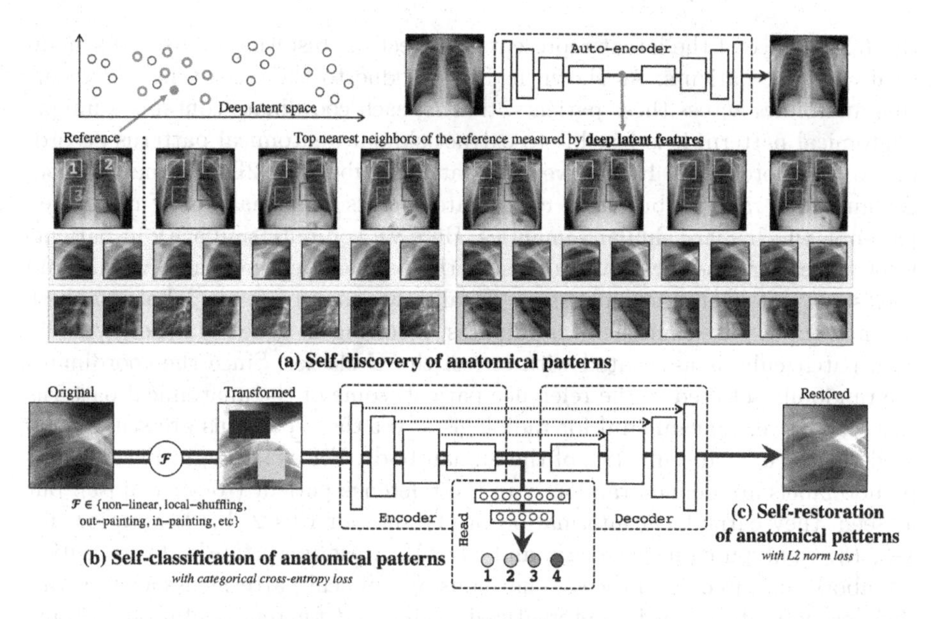

Fig. 1. Our self-supervised learning framework consists of (a) self-discovery, (b) self-classification, and (c) self-restoration of anatomical patterns, resulting in semantics-enriched pre-trained models—Semantic Genesis—an encoder-decoder structure with skip connections in between and a classification head at the end of the encoder. Given a random reference patient, we find similar patients based on deep latent features, crop anatomical patterns from random yet fixed coordinates, and assign pseudo labels to the crops according to their coordinates. For simplicity and clarity, we illustrate our idea with four coordinates in X-ray images as an example. The input to the model is a transformed anatomical pattern crop, and the model is trained to classify the pseudo label and to recover the original crop Thereby, the model aims to acquire semantics-enriched representation, producing more powerful application-specific target models.

is conceptually simple: an encoder-decoder structure with skip connections in between and a classification head at the end of the encoder. The objective for the model is to learn different sets of semantics-enriched representation from multiple perspectives. In doing so, our proposed framework consists of three important components: 1) self-discovery of anatomical patterns from similar patients; 2) self-classification of the patterns; and 3) self-restoration of the transformed patterns. Specifically, once the self-discovered anatomical pattern set is built, we jointly train the classification and restoration branches together in the model.

1) Self-discovery of anatomical patterns: We begin by building a set of anatomical patterns from medical images, as illustrated in Fig. 1a. To extract deep features of each (whole) patient scan, we first train an auto-encoder network with training data, which learns an identical mapping from scan to itself. Once trained, the latent representation vector from the auto-encoder can be used as an indicator of each patient. We randomly anchor one patient as a reference and search for its nearest neighbors through the entire dataset by computing the $L2$ distance of the latent representation vectors, resulting in a set of semantically similar patients. As shown in Fig. 1a, due to the consistent and recurring anatomies across these patients, that is, each coordinate contains a unique anatomical pattern, it is feasible to extract similar anatomical patterns according to the coordinates. Hence, we crop patches/cubes (for 2D/3D images) from C number of random but fixed coordinates across this small set of discovered patients, which share similar semantics. Here we compute similarity in patient-level rather than pattern-level to ensure the balance between the diversity and consistency of anatomical patterns. Finally, we assign pseudo labels to these patches/cubes based on their coordinates, resulting in a new dataset, wherein each patch/cube is associated with one of the C classes. Since the coordinates are randomly selected in the reference patient, some of the anatomical patterns may not be very meaningful for radiologists, yet these patterns are still associated with rich local semantics of the human body. For example, in Fig. 1a, four pseudo labels are defined randomly in the reference patient (top-left most), but as seen, they carry local information of (1) anterior ribs 2–4, (2) anterior ribs 1–3, (3) right pulmonary artery, and (4) LV. Most importantly, by repeating the above self-discovery process, enormous anatomical patterns associated with their pseudo labels can be automatically generated for representation learning in the following stages (refer to Appendix Sect. A).

2) Self-classification of anatomical patterns: After self-discovery of a set of anatomical patterns, we formulate the representation learning as a C-way multi-class classification task. The goal is to encourage models to learn from the recurrent anatomical patterns across patient images, fostering a deep semantically enriched representation. As illustrated in Fig. 1b, the classification branch encodes the input anatomical pattern into a latent space, followed by a sequence of fully-connected (*fc*) layers, and predicts the pseudo label associated with the pattern. To classify the anatomical patterns, we adopt categorical cross-entropy loss function: $\mathcal{L}_{cls} = -\frac{1}{N}\sum_{b=1}^{N}\sum_{c=1}^{C}\mathcal{Y}_{bc}\log\mathcal{P}_{bc}$, where N denotes the batch size; C denotes the number of classes; $\mathcal{Y}$ and $\mathcal{P}$ represent the ground truth (one-hot pseudo label vector) and the prediction, respectively.

3) Self-restoration of anatomical patterns: The objective of self-restoration is for the model to learn different sets of visual representation by recovering original anatomical patterns from the transformed ones. We adopt the transformations proposed in Models Genesis [25], *i.e.*, non-linear, local-shuffling, out-painting, and in-painting (refer to Appendix Sect. B). As shown in Fig. 1c, the restoration branch encodes the input transformed anatomical pattern into a latent space and decodes back to the original resolution, with an aim to recover the original anatomical pattern from the transformed one. To let Semantic Genesis restore the transformed anatomical patterns, we compute $L2$ distance between original pattern and reconstructed pattern as loss function: $\mathcal{L}_{rec} = \frac{1}{N}\sum_{i=1}^{N}\|\mathcal{X}_i - \mathcal{X}'_i\|_2$, where N, $\mathcal{X}$ and $\mathcal{X}'$ denote the batch size, ground truth (original anatomical pattern) and reconstructed prediction, respectively.

Formally, during training, we define a multi-task loss function on each transformed anatomical pattern as $\mathcal{L} = \lambda_{cls}\mathcal{L}_{cls} + \lambda_{rec}\mathcal{L}_{rec}$, where λ_{cls} and λ_{rec}

Table 1. We evaluate the learned representation by fine-tuning it for six publicly-available medical imaging applications including 3D and 2D image classification and segmentation tasks, across diseases, organs, datasets, and modalities.

Code[a]	Object	Modality	Dataset	Application
NCC	Lung nodule	CT	LUNA-2016 [21]	Nodule false positive reduction
NCS	Lung nodule	CT	LIDC-IDRI [3]	Lung nodule segmentation
LCS	Liver	CT	LiTS-2017 [6]	Liver segmentation
BMS	Brain tumor	MRI	BraTS2018 [5]	Brain Tumor Segmentation
DXC	Chest diseases	X-ray	ChestX-Ray14 [23]	Fourteen chest diseases classification
PXS	Pneumothorax	X-ray	SIIM-ACR-2019 [1]	Pneumothorax Segmentation

[a] The first letter denotes the object of interest ("N" for lung nodule, "L" for liver, etc.); the second letter denotes the modality ("C" for CT, "X" for X-ray, "M" for MRI); the last letter denotes the task ("C" for classification, "S" for segmentation).

regulate the weights of classification and reconstruction losses, respectively. Our definition of $\mathcal{L}_{cls}$ allows the model to learn more semantically enriched representation. The definition of $\mathcal{L}_{rec}$ encourages the model to learn from multiple perspectives by restoring original images from varying image deformations. Once trained, the encoder alone can be fine-tuned for target *classification* tasks; while the encoder and decoder together can be fine-tuned for target *segmentation* tasks to fully utilize the advantages of the pre-trained models on the target tasks.

3 Experiments

Pre-training Semantic Genesis: Our Semantic Genesis 3D and 2D are self-supervised pre-trained from 623 CT scans in LUNA-2016 [21] (same as the publicly released Models Genesis) and 75,708 X-ray images from ChestX-ray14 [22] datasets, respectively. Although Semantic Genesis is trained from only unlabeled images, we do not use all the images in those datasets to avoid test-image leaks between proxy and target tasks. In the self-discovery process, we select top K most similar cases with the reference patient, according to the deep features computed from the pre-trained auto-encoder. To strike a balance between diversity and consistency of the anatomical patterns, we empirically set K to 200/1000 for 3D/2D pre-training based on the dataset size. We set C to 44/100 for 3D/2D images so that the anatomical patterns can largely cover the entire image while avoiding too much overlap with each other. For each random coordinate, we extract multi-resolution cubes/patches, then resize them all to $64 \times 64 \times 32$ and 224×224 for 3D and 2D, respectively; finally, we assign C pseudo labels to the cubes/patches based on their coordinates. For more details in implementation and meta-parameters, please refer to our publicly released code.

Baselines and Implementation: Table 1 summarizes the target tasks and datasets. Since most self-supervised learning methods are initially proposed in 2D, we have extended two most representative ones [9,19] into their 3D version

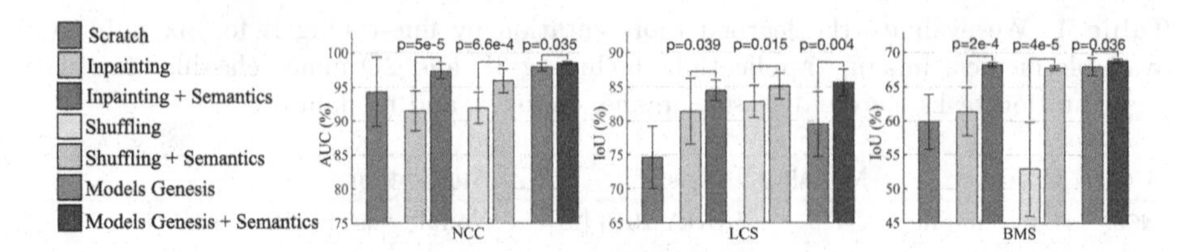

Fig. 2. With and without semantics-enriched representation in the self-supervised learning approaches contrast a substantial ($p < 0.05$) performance difference on target classification and segmentation tasks. By introducing self-discovery and self-classification, we enhance semantics in three most recent self-supervised learning advances (*i.e.*, image in-painting [19], patch-shuffling [9], and Models Genesis [25]).

for a fair comparison. Also, we compare Semantic Genesis with Rubik's cube [26], the most recent multi-task self-supervised learning method for 3D medical imaging. In addition, we have examined publicly available pre-trained models for 3D transfer learning in medical imaging, including NiftyNet [12], MedicalNet [10], Models Genesis [25], and Inflated 3D (I3D) [8] that has been successfully transferred to 3D lung nodule detection [2], as well as ImageNet models, the most influential weights initialization in 2D target tasks. 3D U-Net[1]/U-Net[2] architectures used in 3D/2D applications, have been modified by appending fully-connected layers to end of the encoders. In proxy tasks, we set $\lambda_{rec} = 1$ and $\lambda_{cls} = 0.01$. Adam with a learning rate of 0.001 is used for optimization. We first train classification branch for 20 epochs, then jointly train the entire model for both classification and restoration tasks. For CT target tasks, we investigate the capability of both 3D volume-based solutions and 2D slice-based solutions, where the 2D representation is obtained by extracting axial slices from volumetric datasets. For all applications, we run each method 10 times on the target task and report the average, standard deviation, and further present statistical analyses based on independent two-sample t-test.

4 Results

Learning semantics enriches existing self-supervised learning approaches: Our proposed self-supervised learning scheme should be considered as an *add-on*, which can be added to and boost existing self-supervised learning methods. Our results in Fig. 2 indicate that by simply incorporating the anatomical patterns with representation learning, the semantics-enriched models consistently outperform each and every existing self-supervised learning method [9,19,25]. Specifically, the semantics-enriched representation learning achieves performance gains by 5%, 3%, and 1% in NCC, compared with the original in-painting, patch-shuffling, and Models Genesis, respectively; and the performance improved by 3%, 2%, and 6% in LCS and 6%, 14%, and 1% in BMS. We

[1] 3D U-Net: github.com/ellisdg/3DUnetCNN.

[2] Segmentation Models: github.com/qubvel/segmentation_models.

Table 2. Semantic Genesis outperforms learning 3D models from scratch, three competing publicly available (fully) supervised pre-trained 3D models, and four self-supervised learning approaches in four target tasks. For every target task, we report the mean and standard deviation (mean ± s.d.) across ten trials and further perform independent two sample t-test between the best (bolded) vs. others and highlighted in italic when they are not statistically significantly different at $p = 0.05$ level.

Pre-training	Initialization	NCC (AUC%)	LCS (IoU%)	NCS (IoU%)	BMS[a] (IoU%)
	Random	94.25 ± 5.07	74.60 ± 4.57	74.05 ± 1.97	59.87 ± 4.04
Supervised	NiftyNet [12]	94.14 ± 4.57	83.23 ± 1.05	52.98 ± 2.05	60.78 ± 1.60
	MedicalNet [10]	95.80 ± 0.51	83.32 ± 0.85	75.68 ± 0.32	66.09 ± 1.35
	I3D [8]	98.26 ± 0.27	70.65 ± 4.26	71.31 ± 0.37	67.83 ± 0.75
Self-supervised	Autoencoder	88.43 ± 10.25	78.16 ± 2.07	75.10 ± 0.91	56.36 ± 5.32
	In-painting [19]	91.46 ± 2.97	81.36 ± 4.83	75.86 ± 0.26	61.38 ± 3.84
	Patch-shuffling [9]	91.93 ± 2.32	82.82 ± 2.35	75.74 ± 0.51	52.95 ± 6.92
	Rubik's Cube [26]	95.56 ± 1.57	76.07 ± 0.20	70.37 ± 1.13	62.75 ± 1.93
	Self-restoration [25]	98.07 ± 0.59	78.78 ± 3.11	**77.41 ± 0.40**	67.96 ± 1.29
	Self-classification	97.41 ± 0.32	83.61 ± 2.19	76.23 ± 0.42	66.02 ± 0.83
	Semantic Genesis 3D	**98.47 ± 0.22**	**85.60 ± 1.94**	*77.24 ± 0.68*	**68.80 ± 0.30**

[a] Models Genesis used only *synthetic* images of BraTS-2013, however we examine *real* and only MR Flair images for segmenting brain tumors, so the results are not submitted to BraTS-2018.

conclude that our proposed self-supervised learning scheme, by autonomously discovering and classifying anatomical patterns, learns a unique and complementary visual representation in comparison with that of an image restoration task. Thereby, due to this combination, the models are enforced to learn from multiple perspectives, especially from the consistent and recurring anatomical structure, resulting in more powerful image representation.

Semantic Genesis 3D provides more generic and transferable representations in comparison to publicly available pre-trained 3D models: We have compared our Semantic Genesis 3D with the competitive publicly available pre-trained models, applied to four distinct 3D target medical applications. Our statistical analysis in Table 2 suggests three major results. Firstly, compared to learning 3D models from scratch, fine-tuning from Semantic Genesis offers performance gains by at least 3%, while also yielding more stable performances in all four applications. Secondly, fine-tuning models from Semantic Genesis achieves significantly higher performances than those fine-tuned from other self-supervised approaches, in all four distinct 3D medical applications, *i.e.*, NCC, LCS, NCS, and BMS. In particular, Semantic Genesis surpasses Models Genesis, the state-of-the-art 3D pre-trained models created by image restoration based self-supervised learning, in three applications (*i.e.*, NCC, LCS, and BMS), and offers equivalent performance in NCS. Finally, even though our Semantic Genesis learns representation without using any human annotation, we still have examined it with 3D models pre-trained from full supervision, *i.e.*, MedicalNet, NiftyNet, and I3D. Without any bells and whistles, Semantic Genesis outperforms supervised pre-trained models in all four target tasks. Our results evidence that in contrast

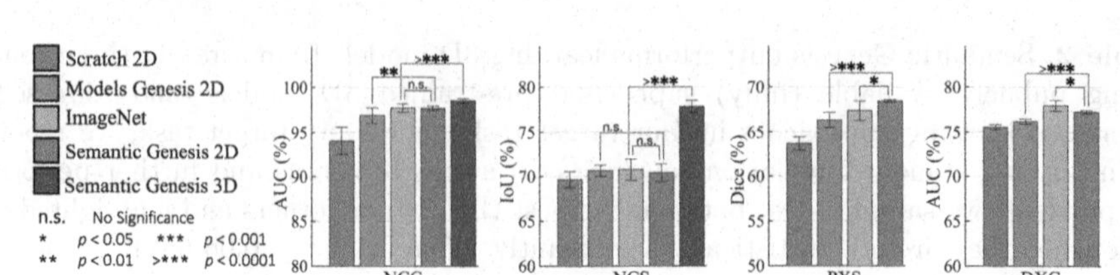

Fig. 3. To solve target tasks in 3D medical modality (NCC and NCS), 3D approaches empowered by Semantic Genesis 3D, significantly outperform any 2D slice-based approaches, including the state-of-the-art ImageNet models, confirming that 3D medical problems demand 3D solutions. For target tasks in 2D modality (PXS and DXC), Semantic Genesis 2D outperforms Models Genesis 2D and, noticeably, yields higher performance than ImageNet in PXS.

to other baselines, which show fluctuation in different applications, Semantic Genesis is consistently capable of generalizing well in all tasks even when the domain distance between source and target datasets is large (*i.e.*, LCS and BMS tasks). Conversely, Semantic Genesis benefits explicitly from the deep semantic features enriched by self-discovering and self-classifying anatomical patterns embedded in medical images, and thus contrasts with any other existing 3D models pre-trained by either self-supervision or full supervision.

Semantic Genesis 3D significantly surpasses any 2D approaches: To address the problem of limited annotation in volumetric medical imaging, one can reformulate and solve 3D imaging tasks in 2D [25]. However, this approach may lose rich 3D anatomical information and inevitably compromise the performance. Evidenced by Fig. 3 (NCC and NCS), Semantic Genesis 3D outperforms all 2D solutions, including ImageNet models as well as downgraded Semantic Genesis 2D and Models Genesis 2D, demonstrating that 3D problems in medical imaging demand 3D solutions. Moreover, as an ablation study, we examine our Semantic Genesis 2D with Models Genesis 2D (self-supervised) and ImageNet models (fully supervised) in four target tasks, covering classification and segmentation in CT and X-ray. Referring to Fig. 3, Semantic Genesis 2D: 1) significantly surpasses training from scratch and Models Genesis 2D in all four and three applications, respectively; 2) outperforms ImageNet model in PXS and achieves the performance equivalent to ImageNet in NCC and NCS, which is a significant achievement because to date, all self-supervised approaches lag behind fully supervised training [7,14,24].

Self-classification and self-restoration, lead to complementary representation: In theory, our Semantic Genesis benefits from two sources: pattern classification and pattern restoration, so we further conduct an ablation study to investigate the effect of each isolated training scheme. Referring to Table 2, the combined training scheme (Semantic Genesis 3D) consistently offers significantly higher and more stable performance compared to each of the isolated training schemes (self-restoration and self-classification) in NCS, LCS, and BMS. Moreover, self-restoration and self-classification reveal better performances in four target applications, alternatingly. We attribute their complementary results to the different visual representations that they have captured from each isolated pre-training scheme, leading to different behaviors in different target applications. These complementary representations, in turn, confirm the importance of the unification of self-classification and self-restoration in our Semantic Genesis and its significance for medical imaging.

5 Conclusion

A key contribution of ours is designing a self-supervised learning framework that not only allows deep models to learn common visual representation from image data directly, but also leverages semantics-enriched representation from the consistent and recurrent anatomical patterns, one of a broad set of unique properties that medical imaging has to offer. Our extensive results demonstrate that Semantic Genesis is superior to publicly available 3D models pre-trained by either self-supervision or even full supervision, as well as ImageNet-based transfer learning in 2D. We attribute this outstanding results to the compelling deep semantics learned from abundant anatomical patterns resulted form consistent anatomies naturally embedded in medical images.

Acknowledgments. This research has been supported partially by ASU and Mayo Clinic through a Seed Grant and an Innovation Grant, and partially by the NIH under Award Number R01HL128785. The content is solely the responsibility of the authors and does not necessarily represent the official views of the NIH. This work has utilized the GPUs provided partially by the ASU Research Computing and partially by the Extreme Science and Engineering Discovery Environment (XSEDE) funded by the National Science Foundation (NSF) under grant number ACI-1548562. We thank Zuwei Guo for implementing Rubik's cube and evaluating MedicalNet, M. M. Rahman Siddiquee for examining NiftyNet, and Jiaxuan Pang for evaluating I3D. The content of this paper is covered by patents pending.

References

1. Siim-acr pneumothorax segmentation (2019). https://www.kaggle.com/c/siim-acr-pneumothorax-segmentation/
2. Ardila, D., et al.: End-to-end lung cancer screening with three-dimensional deep learning on low-dose chest computed tomography. Nat. Med. **25**(6), 954–961 (2019)
3. Armato III, S.G., et al.: The lung image database consortium (LIDC) and image database resource initiative (IDRI): a completed reference database of lung nodules on CT scans. Med. Phys. **38**(2), 915–931 (2011)
4. Bai, W., et al.: Self-supervised learning for cardiac MR image segmentation by anatomical position prediction. In: Shen, D., et al. (eds.) MICCAI 2019. LNCS, vol. 11765, pp. 541–549. Springer, Cham (2019). https://doi.org/10.1007/978-3-030-32245-8_60
5. Bakas, S., et al.: Identifying the best machine learning algorithms for brain tumor segmentation, progression assessment, and overall survival prediction in the brats challenge. arXiv preprint arXiv:1811.02629 (2018)

6. Bilic, P., et al.: The liver tumor segmentation benchmark (LITS). arXiv preprint arXiv:1901.04056 (2019)
7. Caron, M., Bojanowski, P., Mairal, J., Joulin, A.: Unsupervised pre-training of image features on non-curated data. In: Proceedings of the IEEE International Conference on Computer Vision, pp. 2959–2968 (2019)
8. Carreira, J., Zisserman, A.: Quo vadis, action recognition? A new model and the kinetics dataset. In: Proceedings of the IEEE Conference on Computer Vision and Pattern Recognition, pp. 6299–6308 (2017)
9. Chen, L., Bentley, P., Mori, K., Misawa, K., Fujiwara, M., Rueckert, D.: Self-supervised learning for medical image analysis using image context restoration. Med. Image Anal. **58**, 101539 (2019)
10. Chen, S., Ma, K., Zheng, Y.: Med3D: transfer learning for 3D medical image analysis. arXiv preprint arXiv:1904.00625 (2019)
11. Feng, Z., Xu, C., Tao, D.: Self-supervised representation learning by rotation feature decoupling. In: Proceedings of the IEEE Conference on Computer Vision and Pattern Recognition, pp. 10364–10374 (2019)
12. Gibson, E., et al.: NiftyNet: a deep-learning platform for medical imaging. Comput. Methods Programs Biomed. **158**, 113–122 (2018)
13. Gidaris, S., Singh, P., Komodakis, N.: Unsupervised representation learning by predicting image rotations. arXiv preprint arXiv:1803.07728 (2018)
14. Hendrycks, D., Mazeika, M., Kadavath, S., Song, D.: Using self-supervised learning can improve model robustness and uncertainty. In: Advances in Neural Information Processing Systems, pp. 15637–15648 (2019)
15. Kim, D., Cho, D., Yoo, D., Kweon, I.S.: Learning image representations by completing damaged jigsaw puzzles. In: 2018 IEEE Winter Conference on Applications of Computer Vision (WACV), pp. 793–802. IEEE (2018)
16. Larsson, G., Maire, M., Shakhnarovich, G.: Learning representations for automatic colorization. In: Leibe, B., Matas, J., Sebe, N., Welling, M. (eds.) ECCV 2016. LNCS, vol. 9908, pp. 577–593. Springer, Cham (2016). https://doi.org/10.1007/978-3-319-46493-0_35
17. Larsson, G., Maire, M., Shakhnarovich, G.: Colorization as a proxy task for visual understanding. In: Proceedings of the IEEE Conference on Computer Vision and Pattern Recognition, pp. 6874–6883 (2017)
18. Noroozi, M., Favaro, P.: Unsupervised learning of visual representations by solving jigsaw puzzles. In: Leibe, B., Matas, J., Sebe, N., Welling, M. (eds.) ECCV 2016. LNCS, vol. 9910, pp. 69–84. Springer, Cham (2016). https://doi.org/10.1007/978-3-319-46466-4_5
19. Pathak, D., Krahenbuhl, P., Donahue, J., Darrell, T., Efros, A.A.: Context encoders: feature learning by inpainting. In: Proceedings of the IEEE Conference on Computer Vision and Pattern Recognition, pp. 2536–2544 (2016)
20. Ross, T., et al.: Exploiting the potential of unlabeled endoscopic video data with self-supervised learning. Int. J. Comput. Assist. Radiol. Surg. **13**(6), 925–933 (2018). https://doi.org/10.1007/s11548-018-1772-0
21. Setio, A.A.A., et al.: Validation, comparison, and combination of algorithms for automatic detection of pulmonary nodules in computed tomography images: the luna16 challenge. Med. Image Anal. **42**, 1–13 (2017)
22. Wang, H., et al.: Comparison of machine learning methods for classifying mediastinal lymph node metastasis of non-small cell lung cancer from 18 F-FDG PET/CT images. EJNMMI Res. **7**(1), 11 (2017)

23. Wang, X., Peng, Y., Lu, L., Lu, Z., Bagheri, M., Summers, R.M.: Chestx-ray8: hospital-scale chest x-ray database and benchmarks on weakly-supervised classification and localization of common thorax diseases. In: Proceedings of the IEEE Conference on Computer Vision and Pattern Recognition, pp. 2097–2106 (2017)
24. Zhang, L., Qi, G.J., Wang, L., Luo, J.: AET vs. AED: unsupervised representation learning by auto-encoding transformations rather than data. In: Proceedings of the IEEE Conference on Computer Vision and Pattern Recognition, pp. 2547–2555 (2019)
25. Zhou, Z., et al.: Models genesis: generic autodidactic models for 3D medical image analysis. In: Shen, D., et al. (eds.) MICCAI 2019. LNCS, vol. 11767, pp. 384–393. Springer, Cham (2019). https://doi.org/10.1007/978-3-030-32251-9_42
26. Zhuang, X., Li, Y., Hu, Y., Ma, K., Yang, Y., Zheng, Y.: Self-supervised feature learning for 3D medical images by playing a Rubik's cube. In: Shen, D., et al. (eds.) MICCAI 2019. LNCS, vol. 11767, pp. 420–428. Springer, Cham (2019). https://doi.org/10.1007/978-3-030-32251-9_46

DECAPS: Detail-Oriented Capsule Networks

Aryan Mobiny(✉), Pengyu Yuan, Pietro Antonio Cicalese, and Hien Van Nguyen

Department of Electrical and Computer Engineering, University of Houston, Houston, USA
amobiny@uh.edu

Abstract. Capsule Networks (CapsNets) have demonstrated to be a promising alternative to Convolutional Neural Networks (CNNs). However, they often fall short of state-of-the-art accuracies on large-scale high-dimensional datasets. We propose a Detail-Oriented Capsule Network (DECAPS) that combines the strength of CapsNets with several novel techniques to boost its classification accuracies. First, DECAPS uses an Inverted Dynamic Routing (IDR) mechanism to group lower-level capsules into heads before sending them to higher-level capsules. This strategy enables capsules to selectively attend to small but informative details within the data which may be lost during pooling operations in CNNs. Second, DECAPS employs a Peekaboo training procedure, which encourages the network to focus on fine-grained information through a second-level attention scheme. Finally, the distillation process improves the robustness of DECAPS by averaging over the original and attended image region predictions. We provide extensive experiments on the CheXpert and RSNA Pneumonia datasets to validate the effectiveness of DECAPS. Our networks achieve state-of-the-art accuracies not only in classification (increasing the average area under ROC curves from 87.24% to 92.82% on the CheXpert dataset) but also in the weakly-supervised localization of diseased areas (increasing average precision from 41.7% to 80% for the RSNA Pneumonia detection dataset).

Keywords: Capsule network · Chest radiography · Pneumonia

1 Introduction

Convolutional neural networks (CNNs) have achieved state-of-the-art performance in many computer vision tasks due to their ability to capture complex representations of the desired target concept [4,11,14,16,17,19]. These architectures are composed of a sequence of convolutional and pooling layers, with

Electronic supplementary material The online version of this chapter (https://doi.org/10.1007/978-3-030-59710-8_15) contains supplementary material, which is available to authorized users.

A. L. Martel et al. (Eds.): MICCAI 2020, LNCS 12261, pp. 148–158, 2020.
https://doi.org/10.1007/978-3-030-59710-8_15

max-pooling being popularized in the literature due to its positive effect on performance. The max-pooling operation allows CNNs to achieve some translational invariance (meaning they can identify the existence of entities regardless of their spatial location) by attending only to the most active neuron. However, this operation has been criticized for destroying spatial information which can be valuable for classification purposes. Capsule networks (CapsNets) aim to address this fundamental problem of max pooling and has become a promising alternative to CNNs [18]. Previous works have demonstrated that CapsNets possess multiple desirable properties; they are able to generalize with fewer training examples, and are significantly more robust to adversarial attacks and noisy artifacts [5,15]. CapsNets utilize view-point invariant transformations that learn to encode the relative relationships (including location, scale, and orientation) between sub-components and the whole object, using a dynamic routing mechanism to determine how information should be categorized. This effectively gives CapsNets the ability to produce interpretable hierarchical parsing of each desired scene. By looking at the paths of the activations, we can navigate the hierarchy of the parts and know exactly the parts of an object. This property has prompted several research groups to develop new capsule designs and routing algorithms [3,5,10,13].

In recent years, CapsNets have been widely adopted and used in various medical image analysis tasks [2,6,9]. Jiao *et al.* used CapsNet for the diagnosis of mild cognitive impairment [8]. Mobiny *et al.* proposed Fast CapsNet which exploits novel techniques to improve the inference time and prediction performance of CapsNets for the lung nodule classification in 3D CT scans [15]. Lalonde *et al.* proposed SegCaps to expand capsule networks to segment pathological lungs from low dose CT scans [12].

In medical image analysis, identifying the affected area and attending to small details is critical to diagnostic accuracy. In this paper, we propose a novel version of CapsNets called Detail-Oriented Capsule Networks (DECAPS) which simulate this process by attending to details within areas that are relevant to a given task while suppressing noisy information outside of the region of interest (or ROI). We can effectively describe our architecture as having both a coarse and fine-grained stage. First, the architecture groups capsules into submodules named capsule heads, each of which is trained to extract particular visual features from the input image. It then employs an inverted routing mechanism in which groups of lower-level capsules compete with each other to characterize the task-specific regions in the image, generating coarse level predictions. The ROIs are then cropped and used in the fine-grained prediction scheme, where the model learns to interpret high-resolution representations of areas of interest. The two predictions are then combined and generate a detail-oriented output which improves performance. We will make DECAPS implementation publicly available.

2 Background on Capsule Networks

A CapsNet is composed of a cascade of capsule layers, each of which contains multiple capsules. A capsule is the basic unit of CapsNets and is defined as a

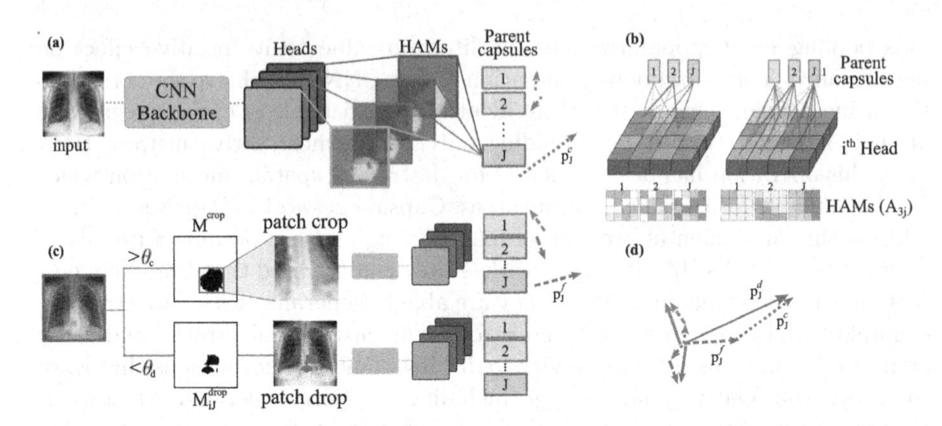

Fig. 1. **(a)**: DECAPS architecture. Head activation maps (HAMs) are presented for the Jth class. **(b)**: Dynamic (left) vs. Inverted dynamic routing (right). Inverted dynamic routing places the competition between children capsules of a head yielding discriminative and localized HAMs. **(c)**: Peekaboo training. **(d)**: The distillation process to fine-tune the coarse-grained prediction (p^c) using the fine-grained prediction (p^f).

group of neurons whose output forms a *pose* vector. This is in contrast to traditional deep networks which use neurons as their basic unit. Let Ω_L denote the sets of capsules in layer L. Each capsule $i \in \Omega_L$ has a pose vector p_i^L. The length (the norm or magnitude) of the pose vector encodes the probability that an object of interest is present, while its direction represents the object's pose information, such as location, size, and orientation. The i-th capsule in Ω_L propagates its information to j-th capsule in Ω_{L+1} through a linear transformation $\mathrm{v}_{ij}^L = \mathrm{W}_{ij}^L \mathrm{p}_i^L$, where v_{ij}^L is called a *vote* vector. The pose vector of capsule $j \in \Omega_{L+1}$ is a convex combination of all the votes from child capsules: $\mathrm{p}_j^{(L+1)} = \sum_i r_{ij} \mathrm{v}_{ij}^L$, where r_{ij} are routing coefficients and $\sum_i r_{ij} = 1$. These coefficients are determined by the dynamic routing algorithm [18] which iteratively increases the routing coefficients r_{ij} if the corresponding voting vector v_{ij}^L is similar to p_j^{L+1} and vice versa. Dynamic routing ensures that the output of each child capsule gets sent to proper parent capsules. Through this process, the network gradually constructs a transformation matrix for each capsule pair to encode the corresponding part-whole relationship and retains geometric information of the input data.

Notations: Throughout the paper, r, r, R, **R** represent a scalar, a vector, a 2D matrix, and a tensor (i.e. a higher dimensional matrix; usually a 3D matrix of capsule activations), respectively. Note that multiplying a transformation matrix and a tensor of poses is equivalent to applying the transformation to each pose.

3 Detail-Oriented Capsule Network

In the original CapsNet [18], the vote of each child capsule contributes directly to the pose of all parent capsules. This ultimately has a negative effect on the quality of the final prediction, due to the noisy votes derived from non-descriptive areas. DECAPS utilize a modified architecture, loss, and routing mechanism that favors votes from ROIs, thus improving the quality of the inputs being routed to the parent capsules. Inspired by the Transformers architecture [22], we group capsules within a grid and call them *Capsule Heads* (see Fig. 1 (a)). One can think of a capsule head as being a grid of capsules that routes information independently of the other heads. Ideally, each capsule head is responsible for detecting a particular visual feature in the input image. To accomplish this, each head shares a transformation matrix W_{ij}^{L} between all capsules for each output class. This contrasts against the original architecture which uses one transformation matrix per capsule, per class. This reduces the required number of trainable parameters by an order of head size (i.e. the number of capsules within a head, 26×26 in our proposed DECAPS); this allows DECAPS to properly scale to large, high-dimensional input images. Additionally, we use the head activation regularization loss (explained in Sect. 3.2) to force the capsules within a head to seek the same semantic concept for each diagnostic task.

Let $\mathbf{P}_i^L \in \mathbb{R}^{h_L \times w_L \times d_L}$ denote the pose matrix of the capsules of the ith head where h_L and w_L represent the height and width of the head respectively, and d_L is the capsule dimension (i.e. the number of hidden units grouped together to make capsules in layer L). Note that i is the child capsule index in the original CapsNet, but is changed to represent the head index in our architecture. We want the length of the pose vector of each capsule to represent the probability of existence for a given entity of interest in the current input. The capsule outputs are passed through a nonlinear squash function [18] to ensure that the length of the pose vectors is normalized between zero and one. Then we say that $\mathbf{V}_{ij}^L = \mathrm{W}_{ij}^L \mathbf{P}_i^L$ is the votes from the capsules of the ith head to the jth parent capsule. To preserve the capsule's location, we perform Coordinate Addition: at each position within a capsule head, the capsule's relative coordinates (row and column) are added to the final two entries of the vote vector [5]. Once we have generated the votes, the routing mechanism determines how information should flow to generate each parent's pose vector.

3.1 Inverted Dynamic Routing

Dynamic routing [18] is a *bottom-up* approach which forces higher-level capsules to compete with each other to collect lower-level capsule votes. We propose an inverted dynamic routing (IDR) technique which implements a *top-down* approach, effectively forcing lower-level capsules to compete for the attention of higher-level capsules (see Fig. 1 (b)). During each iteration of the routing procedure, we use a softmax function to force the routing coefficients between all capsules of a single head and a single parent capsule to sum to one (see Algorithm 1). The pose of the jth parent capsule, p_j^{L+1}, is then set to the squashed weighted-sum over all votes from the earlier layer (line 6 in Algorithm 1). Given the vote map computed as $\mathbf{V}_{ij}^L = \mathrm{W}_{ij}^L \mathbf{P}_i^L \in \mathbb{R}^{h_L \times w_L \times d_{L+1}}$, the proposed algorithm generates a routing map $\mathrm{R}_{ij} \in \mathbb{R}^{h_L \times w_L}$ from each capsule head to each output class.

Algorithm 1. Inverted Dynamic Routing (IDR). Note that i and j are the indices of capsule heads in layer L and $L+1$ respectively.

1: **procedure** IDR($\mathbf{V}_{ij}^{L}, n_{\text{iter}}$) ▷ given the votes and number of routing iterations
2: $\mathrm{R}_{ij}^{\text{pre}} \leftarrow 0, \ \forall j$ ▷ initialize the routing coefficients
3: **for** n_{iter} iterations **do**
4: $\mathrm{R}_{ij} \leftarrow \texttt{softmax}(\mathrm{R}_{ij}^{\text{pre}})$ ▷ softmax among capsules in head i
5: $\tilde{\mathbf{A}}_{ij} \leftarrow \mathrm{R}_{ij} \odot \mathbf{V}_{ij}^{L}$ ▷ $\odot$ is the Hadamard product
6: $\mathrm{p}_{j}^{L+1} \leftarrow \texttt{squash}(\sum_{i}\sum_{xy}\tilde{\mathbf{A}}_{ij})$ ▷ $\sum_{xy}$ is the sum over spatial locations
7: $\mathrm{R}_{ij}^{\text{pre}} \leftarrow \mathrm{R}_{ij}^{\text{pre}} + \mathrm{p}_{j}^{(L+1)}.\mathbf{V}_{ij}^{L}$
8: **end for**
9: $\mathrm{A}_{ij} \leftarrow \texttt{length}(\tilde{\mathbf{A}}_{ij})$ ▷ `length` computes Eq. 1
10: **return** $\mathrm{p}_{j}^{L+1}, \mathrm{A}_{ij}$
11: **end procedure**

The voting map describes the children capsules' votes for the parent capsule's pose. The routing map depicts the weights of the children capsules according to their agreements with parent capsules, with winners having the highest r_{ij}. We combine these maps to generate head activation maps (or HAMs) following

$$\mathrm{A}_{ij} = \left(\sum\nolimits_{d}\tilde{\mathbf{A}}_{ij}^{2}\right)^{1/2}, \quad \text{where} \quad \tilde{\mathbf{A}}_{ij} = \mathrm{R}_{ij} \odot \mathbf{V}_{ij}^{L} \tag{1}$$

where A_{ij} is the HAM from the ith head to the jth parent, and $\sum_{d}$ is the sum over d_{L+1} channels along the third dimension of $\mathbf{V}_{ij}^{L}$. A_{ij} highlights the informative regions within an input image corresponding to the jth class, captured by the ith head. IDR returns as many activation maps as the number of capsule heads per output class (see Fig. 1 (a)). Class-specific activation maps are the natural output of the proposed framework, unlike CNNs which require the use of additional modules, such as channel grouping, to cluster spatially-correlated patterns [24]. We utilize the activation maps to generate ROIs when an object is detected. This effectively yields a model capable of weakly-supervised localization which is trained end-to-end; we train on the images with categorical annotations and predict both the category and the *location* (i.e. mask or bounding box) for each test image. This framework is thus able to simultaneously generate multiple ROIs within the same image that are associated with different medical conditions.

3.2 Loss Function

The loss function we define is the sum of two terms and is described as follows:

Margin Loss: We use margin loss to enforce the activation vectors of the top-level capsule j to have a large magnitude if and only if the object of the corresponding class exists in the image [18]. The total margin loss is the sum of the losses for each output capsule as given by

$$L_{\text{margin}} = \sum\nolimits_{j}[T_j \max(0, m^{+} - \|\mathrm{p}_j\|)^2 + \lambda(1-T_j)\max(0, \|\mathrm{p}_j\| - m^{-})^2] \tag{2}$$

where $T_j = 1$ when class j is present (else $T_j = 0$). Minimizing this loss forces $\|\mathrm{p}_j\|$ of the correct class to be higher than m^+, and those of the wrong classes to be lower than m^-. In our experiments, we set $m^+ = 0.9$, $m^- = 0.1$, and $\lambda = 0.5$.

Head Activation Regularization: We propose a regularization loss function to supervise the head activation learning process. We want each head activation map A_{ij} to capture a unique semantic concept of the jth output category. Inspired by center loss [23], we define a feature template $\mathrm{t}_{ij} \in \mathbb{R}^{d_{L+1}}$ for the ith semantic concept of the jth output category. We compute the semantic features $\mathrm{f}_{ij} \in \mathbb{R}^{d_{L+1}}$ using the information routed from the ith head to the jth output category. While the magnitude of f_{ij} represents the presence of the desired semantic concept, the orientation captures the instantiation parameters (i.e. pose, deformation, texture, etc.). We, therefore, want to regularize the orientation of f_{ij} for a given capsule head to guarantee that it is capturing the same semantic concept among all training images. Each value of t_{ij} is initialized to zero and is updated using a moving average as

$$\mathrm{t}_{ij} \leftarrow \mathrm{t}_{ij} + \gamma(\hat{\mathrm{f}}_{ij} - \hat{\mathrm{t}}_{ij}), \quad \text{where } \mathrm{f}_{ij} = 1/n_i \sum\nolimits_{xy} \tilde{\mathbf{A}}_{ij}, \tag{3}$$

where $\hat{\mathrm{f}}_{ij}$ and $\hat{\mathrm{t}}_{ij}$ are the normalized vectors, γ is the update step, which we set to 10^{-4}, while n_i represents the number of capsules in head i. To accomplish this, we penalize the network when the orientation of a head's features f_{ij} deviate from the template t_{ij} following

$$L_{\mathrm{HAR}} = \frac{1}{IJ} \sum\nolimits_i \sum\nolimits_j (1 - \mathtt{cosine}(\mathrm{f}_{ij}, \mathrm{t}_{ij})) \tag{4}$$

where I and J represent the total number of child and parent capsules.

3.3 Peekaboo: The Activation-Guided Training

To further promote DECAPS to focus on fine-grained details, we propose the Peekaboo strategy for capsule networks. Our strategy boosts the performance of DECAPS by forcing the network to look at all relevant parts for a given category, not just the most discriminative parts [20]. Instead of hiding random image patches, we use the HAMs to guide the network's attention process. For each training image, we randomly select an activation map A_{ij} for each recognized category. Each map is then normalized in the range $[0, 1]$ to get the normalized HAM $\mathrm{A}^*_{ij} \in \mathbb{R}^{h_L \times w_L}$. We then enter a two step process: patch cropping, which extracts a fine-grained representation of the ROI to learn how to encode details, and patch dropping, which encourages the network to attend to multiple ROIs. In patch cropping, a mask $\mathrm{M}^{crop}_{ij} \in \mathbb{R}^{h_L \times w_L}$ is obtained by setting all elements of A^*_{ij} which are less than a cropping threshold $\theta_c \in [0, 1]$ to 0, and 1 otherwise. We then find the smallest bounding box which covers the entire ROI, and crop it from the raw image (Fig. 1 (c)). It is then upsampled and fed into the network to generate a detailed description of the ROI. During the patch dropping procedure, M^{drop}_{ij} is used to remove the ROI from the raw image by using a dropping threshold $\theta_d \in [0, 1]$. The new patch-dropped image is then fed to the network for prediction. This encourages the network to train capsule heads to attend to multiple discriminative semantic patterns. At test time, we first input the whole image to obtain the coarse prediction vectors (p_j^c for the jth class) and the HAMs A_{ij} from all capsule heads. We then average all maps across the heads, crop and upsample the ROIs, and feed the regions to the network to obtain the fine-grained prediction vectors (p_j^f). The final prediction p_j^d, referred to as distillation (Fig. 1 (d)), is the average of the p_j^c and p_j^f.

4 Experiments and Results

Implementation Details: In our experiments, we use Inception-v3 as the backbone and take the *Mix6e* layer output as the CNN feature maps. We then compress the feature maps using 1×1 convolutional kernels to generate 256 maps. We split the maps into four capsule heads, each of which includes a grid of 64-dimensional capsules. These heads employ the described inverted routing procedure to route into the final 16-dimensional class capsules. The best performance was achieved using 3 routing iterations, $\theta_c = 0.5$, and $\theta_d = 0.3$. The network is trained using the Adam optimizer with $\beta_1 = 0.5$, $\beta_2 = 0.999$ and a learning rate of 10^{-4} which is fixed for the duration of training.

Datasets: We use two datasets to evaluate the performance of the proposed DECAPS architecture. The CheXpert [7] radiography dataset is used for the detection of 5 selected pathologies, namely Atelectasis, Cardiomegaly, Consolidation, Edema, and Pleural effusion (see Table 1 of [7] for more information on the data distribution). We also report our results on the RSNA Pneumonia detection data which includes bounding box annotations of the affected regions in the images [1]. It is important to note that our approach only uses the category labels (not the bounding boxes) for the localization of pneumonia localization.

Evaluation Metrics: We use the area under ROC curve (AUC) to report the prediction accuracy on the CheXpert dataset. We use the mean intersection over union (mIoU) to evaluate the localization accuracy of the model on the RSNA dataset. We also compute the average precision (AP) at different IoU thresholds. At each threshold, a true positive (TP) is counted when a predicted object matches a ground truth object with an IoU above the threshold. A false

Table 1. Prediction performance of models trained on the CheXpert dataset. For each model, average result is reported over the best 10 trained model checkpoints.

	Cardiomeg.	Edema	Consolid.	Atelectasis	Pleural Eff.	mAUC (%)
Inception-v3 [21]	0.841(±0.052)	0.876(±0.055)	0.891(±0.044)	0.833(±0.032)	0.921(±0.038)	87.24
DenseNet121 [7]	0.832(±0.047)	0.941(±0.031)	0.899(±0.037)	0.858(±0.042)	0.934(±0.027)	89.28
DenseNet121+HaS [20]	0.849(±0.041)	0.940(±0.055)	0.904(±0.039)	0.867(±0.050)	0.938(±0.024)	89.96
CapsNet [18]	0.835(±0.033)	0.915(±0.038)	0.890(±0.035)	0.845(±0.031)	0.949(±0.033)	88.68
DECAPS	0.852(±0.048)	0.935(±0.039)	0.897(±0.028)	0.865(±0.045)	0.946(±0.022)	89.90
DECAPS+Peekaboo	**0.895(±0.044)**	**0.972(±0.027)**	**0.913(±0.033)**	**0.883(±0.029)**	**0.978(±0.019)**	**92.82**

positive (FP) indicates a predicted object had no associated ground truth object. A false negative (FN) indicates a ground truth object had no associated predicted object. We then calculated AP as TP/(TP+FP+FN) over all test samples [1].

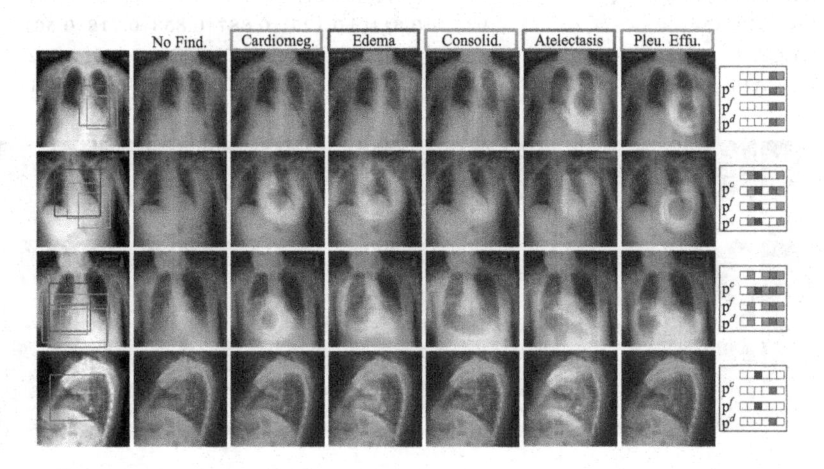

Fig. 2. Qualitative results on the CheXpert dataset. True pathologies (T), coarse (p^c), fine-grained (p^f), and distilled (p^d) predictions are presented for each case.

Results on CheXpert Dataset: The quantitative classification results are summarized in Table 1. We compare our results with Inception-v3 (which is the backbone used in our framework), DenseNet121 (the best performing baseline CNN according to [7]), a DenseNet121 model trained with the Hide-and-Seek (HaS) strategy [20] to boost the weakly-supervised localization of the model, and the vanilla CapsNet with the same backbone. The vanilla DECAPS architecture yielded significantly higher classification accuracies than the baseline networks and achieves performance on par with DenseNet121+HaS. Adding the proposed Peekaboo method to our framework significantly improves the prediction and localization performance of the model. Examples are shown in Fig. 2. Each HAM is activated when the model detects a visual representation associated with the pathology of interest. It highlights the ROI which will be cropped and passed through the fine-grained prediction stage to then generate the distilled prediction. The first row in Fig. 2 shows an example with accurate classifications, while the second and third rows show samples that benefit from fine-grained prediction and distillation (atelectasis and edema are correctly removed). The fourth row shows a failure case that is diagnosed as Edema, but predicted as Atelectasis (see more examples and ablation study in the supplementary material section).

Results on RSNA Dataset: We compare our qualitative results with a weakly-supervised localization approach [20], as well as a supervised (Faster RCNN [17]) detection method as shown in Table 2. The prediction and localization metrics are computed at two levels for the DECAPS model: level-1 refers to the coarse

Table 2. Test prediction accuracy (%), mean intersection over union (mIoU), and average precision (AP) over various IoU thresholds for RSNA Pneumonia detection.

	Unsupervised	%Acc	mIoU	$AP_{0.3}$	$AP_{0.4}$	$AP_{0.5}$	$AP_{0.6}$
Inception-v3+HaS	✓	87.14	0.314(±0.321)	0.417	0.370	0.241	0.194
Faster-RCNN		92.77	**0.611(±0.125)**	**0.887**	**0.853**	**0.718**	**0.561**
DECAPS (level-1)	✓	86.25	0.401(±0.176)	0.642	0.537	0.460	0.322
DECAPS (level-2)	✓	**94.02**	0.509(±0.130)	0.800	0.771	0.594	0.471

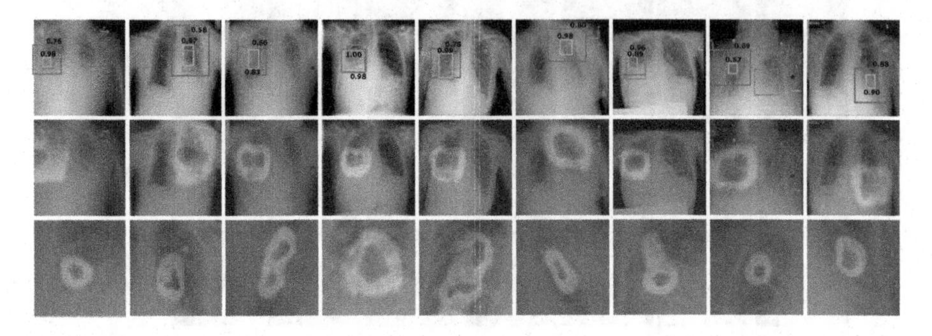

Fig. 3. Qualitative results on the RSNA dataset. **Top:** Input images with ground truth (green), level-1 (red), and level-2 (orange) bounding boxes. **Middle** Course-grained (level-1) activation maps. **Bottom:** Fine-grained (level-2) activation maps. (Color figure online)

prediction on the whole image while level-2 refers to the localization result of the fine-grained prediction (i.e. the ROI within the cropped region, examples shown in Fig. 3). We observe that the fine-grained prediction stage significantly improves the weakly-supervised localization performance over the coarse prediction stage and the baseline weakly supervised method (Inception-v3+HaS). We also note that the prediction accuracy of the fine-grained prediction stage exceeds the supervised Faster-RCNN prediction diagnosis, while localization accuracy is lower. We hypothesize that this is due to the coarse nature of the ground truth bounding boxes which also capture superfluous information from other tissues.

5 Conclusions

In this work, we present a novel network architecture, called DECAPS, that combines the strength of CapsNets with detail-oriented mechanisms. DECAPS is first applied to the whole image to extract global context and generate saliency maps that provide coarse localization of possible findings. This is analogous to a radiologist roughly scanning through the entire image to obtain a holistic view. It then focuses in the informative regions to extract fine-grained visual details from the ROIs. Finally, it employs a distillation process that aggregates information from both global context and local details to generate the final prediction. DECAPS achieves the highest accuracies on CheXpert and RSNA Pneumonia datasets. Despite being trained with only image-level labels, DECAPS are able to accurately localize the region of interests which enhances the model's interpretability. We expect our method to be widely applicable to image detection and recognition tasks, especially for medical image analysis tasks where small details significantly change the diagnostic outcomes.

Acknowledgments. This research was supported by the National Science Foundation (NSF-IIS 1910973).

References

1. RSNA pneumonia detection challenge. https://www.kaggle.com/c/rsna-pneumonia-detection-challenge/overview/evaluation. Accessed 15 Mar 2020
2. Afshar, P., et al.: 3D-MCN: a 3D multi-scale capsule network for lung nodule malignancy prediction. Sci. Rep. **10**(1), 1–11 (2020)
3. Ahmed, K., Torresani, L.: Star-caps: capsule networks with straight-through attentive routing. In: Advances in Neural Information Processing Systems, pp. 9098–9107 (2019)
4. He, K., Zhang, X., Ren, S., Sun, J.: Deep residual learning for image recognition. In: Proceedings of the IEEE Conference on Computer Vision and Pattern Recognition, pp. 770–778 (2016)
5. Hinton, G.E., Sabour, S., Frosst, N.: Matrix capsules with EM routing. In: International Conference on Learning Representations (2018)
6. Iesmantas, T., Alzbutas, R.: Convolutional capsule network for classification of breast cancer histology images. In: Campilho, A., Karray, F., ter Haar Romeny, B. (eds.) ICIAR 2018. LNCS, vol. 10882, pp. 853–860. Springer, Cham (2018). https://doi.org/10.1007/978-3-319-93000-8_97
7. Irvin, J., et al.: ChexPert: a large chest radiograph dataset with uncertainty labels and expert comparison. In: Proceedings of the AAAI Conference on Artificial Intelligence, vol. 33, pp. 590–597 (2019)
8. Jiao, Z., et al.: Dynamic routing capsule networks for mild cognitive impairment diagnosis. In: Shen, D., et al. (eds.) MICCAI 2019. LNCS, vol. 11767, pp. 620–628. Springer, Cham (2019). https://doi.org/10.1007/978-3-030-32251-9_68
9. Jiménez-Sánchez, A., Albarqouni, S., Mateus, D.: Capsule networks against medical imaging data challenges. In: Stoyanov, D., et al. (eds.) LABELS/CVII/STENT -2018. LNCS, vol. 11043, pp. 150–160. Springer, Cham (2018). https://doi.org/10.1007/978-3-030-01364-6_17
10. Kosiorek, A., Sabour, S., Teh, Y.W., Hinton, G.E.: Stacked capsule autoencoders. In: Advances in Neural Information Processing Systems, pp. 15486–15496 (2019)
11. Krizhevsky, A., Sutskever, I., Hinton, G.E.: ImageNet classification with deep convolutional neural networks. In: Advances in Neural Information Processing Systems, pp. 1097–1105 (2012)
12. LaLonde, R., Bagci, U.: Capsules for object segmentation. arXiv preprint arXiv:1804.04241 (2018)
13. Mobiny, A., Lu, H., Nguyen, H.V., Roysam, B., Varadarajan, N.: Automated classification of apoptosis in phase contrast microscopy using capsule network. IEEE Trans. Med. Imaging **39**(1), 1–10 (2019)
14. Mobiny, A., Singh, A., Van Nguyen, H.: Risk-aware machine learning classifier for skin lesion diagnosis. J. Clin. Med. **8**(8), 1241 (2019)

15. Mobiny, A., Van Nguyen, H.: Fast CapsNet for lung cancer screening. In: Frangi, A.F., Schnabel, J.A., Davatzikos, C., Alberola-López, C., Fichtinger, G. (eds.) MICCAI 2018. LNCS, vol. 11071, pp. 741–749. Springer, Cham (2018). https://doi.org/10.1007/978-3-030-00934-2_82
16. Ravindran, A.S., Mobiny, A., Cruz-Garza, J.G., Paek, A., Kopteva, A., Vidal, J.L.C.: Assaying neural activity of children during video game play in public spaces: a deep learning approach. J. Neural Eng. **16**(3), 036028 (2019)
17. Ren, S., He, K., Girshick, R., Sun, J.: Faster R-CNN: towards real-time object detection with region proposal networks. In: Advances in Neural Information Processing Systems, pp. 91–99 (2015)
18. Sabour, S., Frosst, N., Hinton, G.E.: Dynamic routing between capsules. In: Advances in Neural Information Processing Systems (2017)
19. Shahraki, F.F., Prasad, S.: Graph convolutional neural networks for hyperspectral data classification. In: 2018 IEEE Global Conference on Signal and Information Processing (GlobalSIP), pp. 968–972. IEEE (2018)
20. Singh, K.K., Lee, Y.J.: Hide-and-seek: forcing a network to be meticulous for weakly-supervised object and action localization. In: 2017 IEEE International Conference on Computer Vision (ICCV), pp. 3544–3553. IEEE (2017)
21. Szegedy, C., Vanhoucke, V., Ioffe, S., Shlens, J., Wojna, Z.: Rethinking the inception architecture for computer vision. In: Proceedings of the IEEE Conference on Computer Vision and Pattern Recognition, pp. 2818–2826 (2016)
22. Vaswani, A., et al.: Attention is all you need. In: Advances in Neural Information Processing Systems, pp. 5998–6008 (2017)
23. Wen, Y., Zhang, K., Li, Z., Qiao, Yu.: A discriminative feature learning approach for deep face recognition. In: Leibe, B., Matas, J., Sebe, N., Welling, M. (eds.) ECCV 2016. LNCS, vol. 9911, pp. 499–515. Springer, Cham (2016). https://doi.org/10.1007/978-3-319-46478-7_31
24. Zheng, H., Fu, J., Mei, T., Luo, J.: Learning multi-attention convolutional neural network for fine-grained image recognition. In: Proceedings of the IEEE International Conference on Computer Vision, pp. 5209–5217 (2017)

Federated Simulation for Medical Imaging

Daiqing Li[1(✉)], Amlan Kar[1,2,3], Nishant Ravikumar[4,5], Alejandro F. Frangi[4,5,6], and Sanja Fidler[1,2,3]

[1] NVIDIA, Toronto, Canada
lidaiqing2016@gmail.com
[2] Vector Institute, Toronto, Canada
[3] Department of Computer Science, University of Toronto, Toronto, Canada
[4] CISTIB Centre for Computational Imaging and Simulation Technologies in Biomedicine, School of Computing, University of Leeds, Leeds, UK
[5] LICAMM Leeds Institute of Cardiovascular and Metabolic Medicine, School of Medicine, University of Leeds, Leeds, UK
[6] Department of Cardiovascular Sciences, and Department of Electrical Engineering, ESAT/PSI, KU Leuven, Leuven, Belgium

Abstract. Labelling data is expensive and time consuming especially for domains such as medical imaging that contain volumetric imaging data and require expert knowledge. Exploiting a larger pool of labeled data available across multiple centers, such as in federated learning, has also seen limited success since current deep learning approaches do not generalize well to images acquired with scanners from different manufacturers. We aim to address these problems in a common, learning-based image simulation framework which we refer to as *Federated Simulation*. We introduce a physics-driven generative approach that consists of two learnable neural modules: 1) a module that synthesizes 3D cardiac shapes along with their materials, and 2) a CT simulator that renders these into realistic 3D CT Volumes, with annotations. Since the model of geometry and material is disentangled from the imaging sensor, it can effectively be trained across multiple medical centers. We show that our data synthesis framework improves the downstream segmentation performance on several datasets.

Keywords: CT synthesis · Cardiac segmentation · Federated learning

1 Introduction

High quality pixel-level annotations, necessary for training fully supervised segmentation approaches, are prohibitively expensive to source for medical imaging data especially in the context of 3D volumes. This is due to the high dimensionality of the data and the complexity of the task of identifying tissue boundaries and manually delineating the object(s) of interest. Furthermore, identifying regions of interests often requires expert knowledge.

Electronic supplementary material The online version of this chapter (https://doi.org/10.1007/978-3-030-59710-8_16) contains supplementary material, which is available to authorized users.

A. L. Martel et al. (Eds.): MICCAI 2020, LNCS 12261, pp. 159–168, 2020.
https://doi.org/10.1007/978-3-030-59710-8_16

An appealing alternative to labeling data, is to *synthesize* it [5,13,16]. Generating synthetic data sets to learn from, by simulating medical images, has been proposed in several previous studies. This includes simulation of cardiac CTs [13] from mesh-based parametric representations of cardiac shape. Among these, Unberath et al. [14] were the first to use deep learning to estimate X-ray scatter and simulate digitally reconstructed radiographs (DRRs) from annotated CT volumes. [16] employed unpaired image-to-image style transfer to further improve the similarity between real X-ray images and the DRRs. These methods focus on synthesizing DRRs or CTs conditioned on (given) annotations, but cannot produce novel shapes and annotations from the data distribution. Furthermore, these methods work at the pixel-level and need to be re-trained per modality. We, instead, break the synthesis process into both, synthesizing novel 3D shapes and materials, and physical sensor simulation allowing us to generate multiple imaging modalities along with their annotations. We additionally aim to effectively exploit pools of data available across different acquisition sensors with as little as annotations as possible. We refer to this learning-based imaging simulation framework as 'Federated Simulation'.

We introduce a physics-driven generative approach that consists of two modules: 1) a module that synthesizes 3D cardiac shapes along with their materials, and 2) a CT simulator that renders these into realistic CT volumes. Both are implemented as learnable neural network modules enabling us to simulate realistic cardiac CTs. Since the model of geometry and material is disentangled from the imaging sensor, it can effectively be trained across different centers in a privacy preserving manner. Once trained, our model can synthesize a virtually infinite amount of data in a desired imaging modality. By design, our approach also produces ground-truth labels along with the CTs, enabling training of downstream machine learning models. We showcase our data simulation framework to outperform the traditional federated learning approaches in our use case.

2 Methodology

We aim to learn a generative model S_θ, parametrized using neural networks, to synthesize CT volumes and their corresponding labels (in our case voxel segmentation labels). Here, θ are learnable parameters of our model that we learn from a given CT dataset D with few annotated and several unlabeled volumes. We wish to learn S_θ such that it captures the *essence* of the dataset D, and can generate new realistic samples from its distribution. These samples are then used as an auxiliary labeled dataset for training downstream machine learning models (in our case a 3D segmentation neural network).

We introduce our generative model in Sect. 2.1 and explain how we learn it for a single site in Sect. 2.2. Finally, in Sect. 2.3 we propose how to implement the algorithm in a federated setting across multiple data sites.

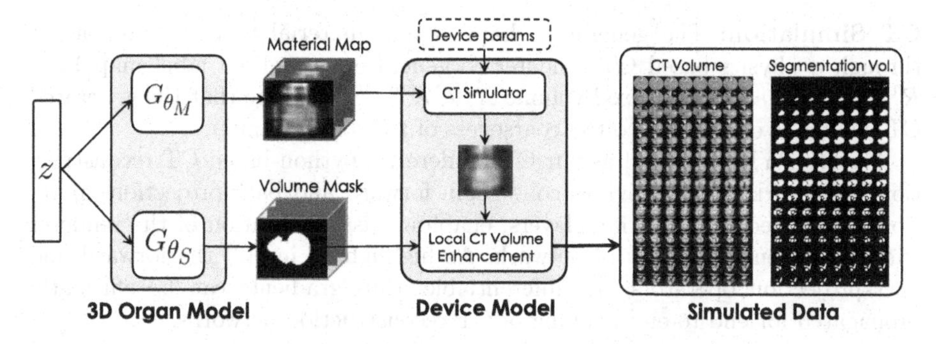

Fig. 1. Our Generative Model: We sample a latent vector from a normal distribution, and pass it through two neural networks to produce the organ's *shape* and a *material map*. These are then input to a differentiable CT renderer to produce a CT volume. A conditional GAN is then used to further improve the realism of the volume.

2.1 Generative Model

Our generative model (Fig. 1) generates an organ shape and a material map, both independent from an imaging device, that are then passed through a CT simulator to generate a synthetic CT volume with labels. Additionally, we *enhance* the generated CT volume using a conditional Generative Adversarial Network (GAN) [8] to further improve on realism. In the federated simulation setting, different sites jointly learn a global shape and material model, while each site maintains a site-specific GAN. Federated setting is discussed in details in Sect. 2.3.

Shape/Material Generation: From a latent vector $z \in \mathcal{Z}$, we aim to learn to generate a 3D organ shape and material properties around the shape. To ensure tractable learning, we constrain generated shapes to be physically plausible and provide control over the shape using a reduced set of parameters through a Statistical Shape Model (SSM). Thus, the shape parameter $\tau_S = G_{\theta_S}(z) \in \mathcal{R}^{21}$ operates on the SSM in order to generate a mesh of the organ. Along with a vector of SSM weights (in $\mathcal{R}^{14}$), we also generate a rigid transformation to be applied to the mesh (in $\mathcal{R}^7$, 3 rotation, 3 translation and 1 scale). In our case, the SSM is a parametric representation of the whole heart and its associated great vessel trunks (pulmonary artery and aorta), including seven regions, namely, the blood pool and myocardium of the left ventricle, the right ventricle, the left and right atria, and the vessels. We estimated our SSM using PCA [10] (see supplementary material for details). We obtain the organ's mesh and convert it to a volumetric (voxel) representation for the CT simulator as S = voxelize($B(\tau_S)$), with B being our PCA basis.

From the same latent vector z, we also generate a coarse material voxel map $\tau_M = G_{\theta_M}(z) \in \mathcal{R}^{16\times16\times16}$. The material map is a combined representation of the voxel-wise tissue-properties, energy-dependent linear attenuation coefficient and material density at each point in the object. See supplementary material for implementation details of G_{θ_S} and G_{θ_M}.

CT Simulation: The generated shape S and material map τ_M are passed through a physics-based CT renderer to generate a voxelized label map $Y_z \in \mathcal{R}^{128\times128\times128}$ and a CT voxel volume $\tilde{X}_z \in \mathcal{R}^{16\times16\times16}$. Note that the generated CT volume is coarse due to the coarseness of the material map.

We use PYRO-NN [12] as our CT renderer, a python-based CT reconstruction framework which provides cone-beam forward and back-projection operations embedded as Tensorflow layers, enabling easy integration of the renderer within our generative neural network. As highlighted in [12] the forward and back-projection operators are differentiable, thus gradients can be efficiently propagated for end-to-end training of CT reconstruction networks.

CT Volume Enhancement: CT-Images are dependent on factors such as the scanning machine and acquisition protocol at a site, which are not all modelled by our CT simulator. Additionally, our generated CT volume is a coarse representation. We thus use a GAN to both enhance and translate our simulated CT slices to look similar to target images, bridging the gap between simulation and real data. The generated coarse CT volume $\tilde{X}_z$ and label map Y_z are used to generate the final high resolution synthetic CT slices. Specifically, we utilize GauGAN [8] (G_{GAN}) to take a slice k of $\tilde{X}_z$, Y_z and slice index k as input and produce a final CT slice image $X_{z,k}$, with label $Y_{z,k}$. We choose GauGAN since it is designed to respect semantic shape input, which is the label map in our case. Particularly, 1) we take the kth slice of $\tilde{X}_z$ and trilinearly upsample it to $\mathcal{R}^{128\times128}$, 2) take the kth slice of Y_z and 3) create a 128×128 slice with a constant value of $\frac{k}{128}$ and concatenate them together as the input to G_{GAN}.

Complete generative process can be written succinctly as,

$$\begin{aligned} S_z &= \text{SSM}(G_{\theta_S}(z)) \\ \tilde{X}_z, Y_z &= \text{CT}_{\text{sim}}(S_z, G_{\theta_M}(z)) \\ X_{z,k} &= G_{\text{GAN}}(\tilde{X}_{z,k}, Y_{z,k}, k) \quad \forall\, k \end{aligned}$$

where S_z represents the shape obtained from the SSM with parameters $G_{\theta_S}(z)$.

2.2 Learning

We train our generative model using the Generative Latent Optimization (GLO) [2] framework in two stages: pre-training and unsupervised training. First, we pre-train the model using the labeled training subset, and then fine-tune the model in a semi-supervised fashion using the rest of the unlabelled training data. We first introduce the GLO framework and then describe our training stages.

GLO [2]**:** GLO is a technique for learning generative networks using only reconstruction losses. In our case, every volume in the training set is coupled with a particular latent vector z^i (initialized from a unit normal distribution), which is simultaneously learnt along with a generation process that transforms z^i into a volume. Learning is done by optimizing reconstruction losses between the generated volume and the corresponding ground truth volume. The set of learnt latent

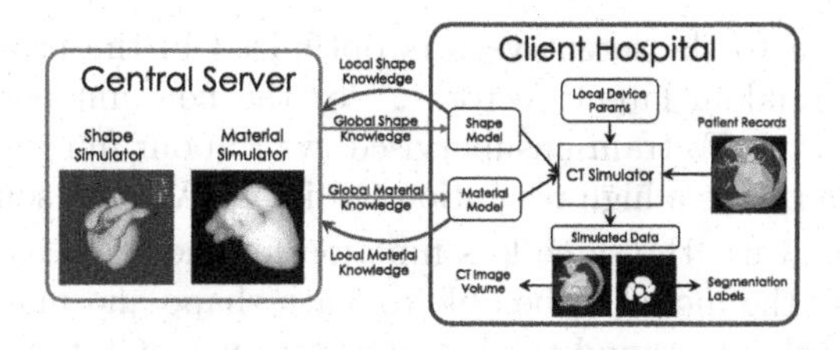

Fig. 2. Federated Simulation: Central server trains a generative model of 3D organ shape and material, which is deployed to centers. Clients train local device models that mimic their sensors. Clients send gradients w.r.t. shape and material back to server.

vectors $\{z^i\}$ can be fit to a parametric distribution (*e.g.* multivariate normal), which is sampled from to generate new volumes. We choose GLO due to its stability in training while enjoying visual-appealing samples property of GANs. We also found that GLO has better sample quality in small-data regime comparing to Meta-Sim [5].

Pre-training: We use the training CT volumes which have ground truth annotations to pretrain the parameter generation module (G_{θ_S} and G_{θ_M}). We use the mean Intersection-Over-Union (mIoU) metric to learn G_{θ_S} (in our case, generated Y_z^i and ground-truth Y^i are binary), and a combination of mean-square-error and perceptual loss [4] using all VGG-19 [11] layer features for G_{θ_M}.

$$L_{\text{IoU}}(\theta_S) = 1 - \frac{Y_z^i \odot Y^i}{Y_z^i + Y^i - Y_z^i \odot Y^i}$$
$$L(\theta_M) = \left\| \tilde{X}_z^i - \tilde{X}^i \right\|_2^2 + L_{\text{perc}}(\tilde{X}_z^i, \tilde{X}^i)$$

where $\tilde{X}^i \in \mathcal{R}^{16\times16\times16}$ is the trilinearly downsampled real CT volume X^i , $\odot$ represents the hadamard product and other operations are element-wise for matrices. L_{perc} is implemented following [8]. Backpropagation through the SSM is done by calculating the gradients using finite-differences on the low-dimensional SSM parameters.

After pre-training G_{θ_S} and G_{θ_M}, we pre-train our conditional GAN G_{GAN}. Specifically, in this phase, we use a slice k of the ground truth labels Y^i instead of the generated labels Y_z^i, and the generated material $\tilde{X}_z^i$ as input to the GAN, and supervise it with the associated real CT Volume X^i, using the same loss functions as [8]. Using the ground truth label ensures that the input label and the output image have the same exact shape, resulting in the GAN learning to generate a high-resolution CT that respects the input shape.

Semi-supervised Learning: We utilize the unlabelled data (which is typically more widely available) in the training set to improve our simulation. The abundance of unlabelled data and the cost of annotation makes this a compelling proposition. In this stage, we alternate between training on supervised data (explained above), and training on unsupervised data. We first fit a multivariate normal distribution to the latent vectors optimized in the pre-training phase, and sample new random latent vectors z^i for the new unlabelled data-points from this distribution. To train unsupervised, we run our full generative process from every z^i to generate a high resolution CT image $X^i_{z,k}$ for some random slice index k from z^i and use the same loss function for the GAN as above to train. This phase adapts the model to be able to learn shape and material properties for data points without ground-truth segmentation annotations. Note that we freeze the GAN discriminator in this phase, and observe that it strongly improves training stability.

2.3 Federated Simulation

Medical data (CTs in our case) is usually available at multiple sites, each with their specific acquisition parameters and privacy concerns, which makes both, consolidating data and training on consolidated data (domain adaptation) difficult. With our learning-based simulation, we demonstrate *federated simulation*, where we learn our generative model with data from multiple hospitals in a federated fashion. Because of our disentanglement of shape and material from the CT process, we are in an advantageous position where we can learn shape and material parameters across multiple sites, and *render* generated shapes and materials from this global distribution into CT volumes from a particular hospital's distribution through the CT simulator and the site-specific (local) enhancement GAN. This mitigates both issues of data consolidation and domain adaptation.

Figure 2 depicts this process. We learn one G_{θ_S} and G_{θ_M} to model the distribution of shapes and materials across sites, and learn a site-specific G_{GAN}. In every case, we take one step update from each site, accumulate gradients and run a step of gradient descent at the server, and broadcast the updated weights θ_S and θ_M back to the sites. Note that this is a simple federated setting and serves to demonstrate our method; but a real deployment would require additional engineering considerations. See supplementary material for details.

3 Experiments

We validate our method on three Cardiac-CT datasets on both the single hospital train/test and our federated scenario. Additional ablation studies are conducted to validate our design choices. We split each dataset into train, validation and test subsets and experiment with different sizes of labels made available in the training set. All experiments are done using *five-fold cross validation*, and we report the mean and std. deviation of the test set performance.

Datasets: We will refer to our datasets as **CT20**, **CT34LC** and **CT34MC**. CT20 data was collected from healthy adults at the Shanghai Shugang Hospital, China, and provided as part of the MM-WHS challenge [17]. The CT34LC and CT34MC data [15] were collected from congenital heart disease patients, whose ages ranged from 1 month-21 years. We split it into two equal-sized datasets

Table 1. Dataset split sizes

Dataset	Train	Val	Test
CT20 [17]	12	4	4
CT34LC [15]	20	7	7
CT34MC[15]	20	7	7

Table 2. Ablation studies

Method	Perf
Rand Shape + GAN	67.07
+ Alpha blend	68.80
+ Poisson blend	76.44
Ours Fix-Mat	**78.92**

based on pathological differences. Table 1 details the split sizes for each of the three datasets. Large difference in age demographics and the presence/absence of pathology correspond to substantial variations in cardiac shape pose a significant challenge for generalization in learning.

Evaluation Metric: To evaluate the performance of our generated dataset, we train a 3D-Unet [3] for binary segmentation of the heart region and measure its performance on the respective test set. Such task-based performance evaluation of generative models has been proposed in [1], and we adopt it here. For every experiment with synthetically generated data, we evaluate by first pre-training on the synthetic data and then fine-tuning on the available real data. We note that semi-supervised training techniques [6] could be used for the segmentation model to make use of unlabelled CT-volumes in all cases, which we omit here.

Single Site Simulation: We first evaluate our model independently per dataset. We compare against training a supervised model using the subset of training dataset with labels (**Lower Bound**) and training the supervised model on the full training dataset (**Upper Bound**) using extra training labels that our method does not have access to. We also compare three variants of our model, 1) instead of predicting a material map, using a fixed atlas of attenuation coefficients [14] (**Ours Fix-Mat**) (in this setting, the material map and the rendering is already high-resolution), 2) using only our pre-trained method (**Ours Pre**) and 3) using our full method with semi-supervised training (**Ours-Full**). We also evaluate with different amounts of labels available from the training set. 20 synthetic volumes are generated when using our generative model to synthesize data, and all 3D-Unet training was done with a batch size of one. Table 3 summarizes these results, and shows the effect of learning material parameters, as well as utilizing unlabelled data to learn the simulator. We see that our method beats the lower bound across all datasets, sometimes even beating the upper bound, meaning that using our synthesized data improves performance more than using all annotations from the training set. Some generated samples per dataset are shown in Fig. 3. In the figure, we also show that our method generates novel samples by showing the nearest neighbour sample (computed on the whole CT-Volume) from the training set.

Table 3. Quantitative Results of training a Unet-3D binary segmentation model on our generated data on three datasets. We see that data generated by our methods (with access to a small subset of training labels) outperforms baselines on both the single site and federated simulation case. It sometimes outperforms the upper bound of using the full training set as well. Method with highest mean is in bold.

		CT20		CT34LC		CT34MC	
		Label Sz. 4	Label Sz. 8	Label Sz. 6	Label Sz. 12	Label Sz. 6	Label Sz. 12
Single site simulation	Lower Bound	87.65 ± 2.20	86.33 ± 2.09	85.25 ± 4.57	87.27 ± 2.52	85.32 ± 1.50	84.65 ± 2.23
	Ours-Fix-Mat	87.87 ± 4.28	88.32 ± 5.11	86.33 ± 1.69	88.32 ± 1.81	83.13 ± 2.19	84.91 ± 1.97
	Ours-Pre	87.78 ± 4.28	91.35 ± 1.55	**87.39 ± 2.34**	88.01 ± 1.30	**85.92 ± 1.37**	84.91 ± 1.35
	Ours-Full	**88.95 ± 2.97**	**91.39 ± 1.27**	85.91 ± 3.61	**88.98 ± 1.86**	84.79 ± 1.73	**85.13 ± 2.31**
	Upper Bound	89.81 ± 2.50		88.76 ± 1.82		85.75 ± 2.43	
Fede. Sim	Direct-FL	91.65 ± 2.69	92.45 ± 1.52	89.34 ± 1.27	90.17 ± 1.40	87.11 ± 1.55	87.29 ± 1.84
	Ours-Sim-FL	90.71 ± 1.62	**93.45 ± 1.21**	88.33 ± 2.10	**90.19 ± 2.21**	**87.15 ± 1.66**	**87.40 ± 1.71**

Federated Simulation: Next, we evaluate our method in the Federated setting. Specifically, we simulate the federated setting by using our three datasets as three different sites. We compare against training the segmentation network directly in a federated setting [7] (**Direct-FL**). In the federated simulation scenario (**Ours-Sim-FL**), each hospital generates data by sampling shapes and materials from the shared model and using their local GAN (see Fig. 2). We argue that this can reduce domain-adaptation issues in sharing data (different devices, protocols etc.), since shapes and materials from other hospitals' distributions can be *rendered* under another hospital's conditions using the CT Simulator and the site-specific GAN. The federated learning baseline (Direct-FL) averages three gradients from the sites, and thus runs at an effective batch size of three. Therefore, for fair reporting, all other 3D-Unet training also uses a batch size of three. Table 3 presents these results. For all methods, we used the combination of lower label sizes (4, 6, 6) of all datasets for one experiment, and all the higher label sizes (8, 12, 12) for another. Every model was fine-tuned on the particular site's labelled training data before reporting. We see that using federated-simulation performs comparably or slightly worse with the federated baseline across all datasets, showing that we can indeed learn to simulate and share a simulator across sites instead of sharing/working on real data samples, which comes with significant privacy concerns.

Performance on Out-of-Distribution Samples: We simulate the case where a patient from one hospital A goes to hospital B, by running the GAN trained for hospital B on the patient's GT segmentation mask and downsampled CT image. An ideal segmentation network would perform well on this out-of-distribution sample. See supplementary material for results.

Ablation Studies: We ablate our choice of learning shape parameters for the SSM in Table 2. These experiments use are performed in the **Ours-Fix-Mat** setting. Specifically, we compare with randomly generating shapes (instead of learning) from the SSM, alpha blending the heart (foreground) from the CT_{sim} output with the background from the GAN output, poisson blending [9] in the

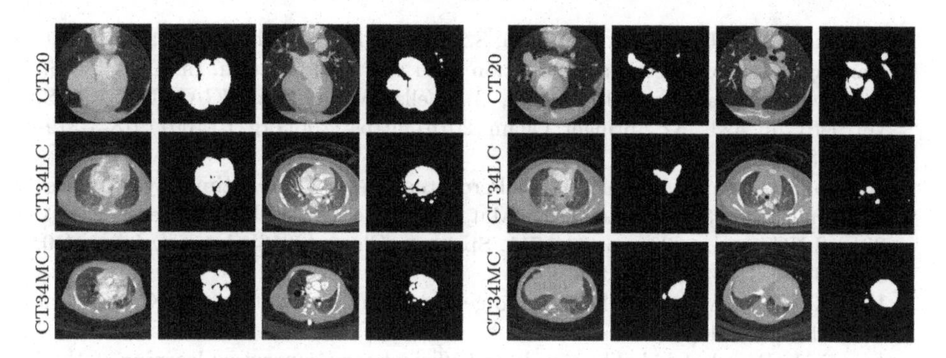

Fig. 3. Qualitative Results: Two sets of examples per dataset. First two columns show random samples (one slice) from our full model on each of the datasets. Last two columns show nearest neighbour (same slice) from the training set. Our model can generate plausible yet novel data samples with annotations.

same case and our method of learning the shape and using the full output of the GAN. These results show the efficacy of learning the shape parameters, and using the conditional GAN to generate the final enhanced output. All experiments here train *only* on synthetic data and are trained on the CT20 dataset (with 4 training labels).

4 Conclusion

In this paper, we introduced a generative model for synthesizing labeled cardiac CT volumes that mimic real world data. Our model abstracts modeling of the shape and material away from the imaging sensor, which enables it to learn in a federated setting, within a framework we call federated simulation. We show that using data generated by our method in both single-site and federated settings improves performance of a downstream 3D segmentation network. Our method currently is using a SSM to parameterize the shape which has limited representation ability. In the future work, we aim to explore a more flexible shape representation and extend the current framework to generate and learn from multiple image sensors (MR, CT etc.).

References

1. Bass, C., et al.: Image synthesis with a convolutional capsule generative adversarial network. In: International Conference on Medical Imaging with Deep Learning, pp. 39–62 (2019)
2. Bojanowski, P., Joulin, A., Lopez-Paz, D., Szlam, A.: Optimizing the latent space of generative networks. arXiv preprint arXiv:1707.05776 (2017)

3. Çiçek, Ö., Abdulkadir, A., Lienkamp, S.S., Brox, T., Ronneberger, O.: 3D U-Net: learning dense volumetric segmentation from sparse annotation. In: Ourselin, S., Joskowicz, L., Sabuncu, M.R., Unal, G., Wells, W. (eds.) MICCAI 2016. LNCS, vol. 9901, pp. 424–432. Springer, Cham (2016). https://doi.org/10.1007/978-3-319-46723-8_49
4. Johnson, J., Alahi, A., Fei-Fei, L.: Perceptual losses for real-time style transfer and super-resolution. In: Leibe, B., Matas, J., Sebe, N., Welling, M. (eds.) ECCV 2016. LNCS, vol. 9906, pp. 694–711. Springer, Cham (2016). https://doi.org/10.1007/978-3-319-46475-6_43
5. Kar, A., et al.: Meta-sim: learning to generate synthetic datasets. arXiv preprint arXiv:1904.11621 (2019)
6. Lee, D.H.: Pseudo-label: The simple and efficient semi-supervised learning method for deep neural networks
7. McMahan, H.B., Moore, E., Ramage, D., Hampson, S., et al.: Communication-efficient learning of deep networks from decentralized data. arXiv preprint arXiv:1602.05629 (2016)
8. Park, T., Liu, M.Y., Wang, T.C., Zhu, J.Y.: Semantic image synthesis with spatially-adaptive normalization. In: Proceedings of the IEEE Conference on Computer Vision and Pattern Recognition, pp. 2337–2346 (2019)
9. Pérez, P., Gangnet, M., Blake, A.: Poisson image editing. In: ACM SIGGRAPH 2003 Papers, pp. 313–318 (2003)
10. Ravikumar, N., Gooya, A., Frangi, A.F., Taylor, Z.A.: Generalised coherent point drift for group-wise registration of multi-dimensional point sets. In: Descoteaux, M., Maier-Hein, L., Franz, A., Jannin, P., Collins, D.L., Duchesne, S. (eds.) MICCAI 2017. LNCS, vol. 10433, pp. 309–316. Springer, Cham (2017). https://doi.org/10.1007/978-3-319-66182-7_36
11. Simonyan, K., Zisserman, A.: Very deep convolutional networks for large-scale image recognition. arXiv preprint arXiv:1409.1556 (2014)
12. Syben, C., Michen, M., Stimpel, B., Seitz, S., Ploner, S., Maier, A.K.: PYRO-NN: python reconstruction operators in neural networks. Med. Phys. **46**(11), 5110–5115 (2019)
13. Unberath, M., Maier, A., Fleischmann, D., Hornegger, J., Fahrig, R.: Open-source 4D statistical shape model of the heart for x-ray projection imaging. In: 2015 IEEE 12th International Symposium on Biomedical Imaging (ISBI), pp. 739–742. IEEE (2015)
14. Unberath, M., et al.: DeepDRR – a catalyst for machine learning in fluoroscopy-guided procedures. In: Frangi, A.F., Schnabel, J.A., Davatzikos, C., Alberola-López, C., Fichtinger, G. (eds.) MICCAI 2018. LNCS, vol. 11073, pp. 98–106. Springer, Cham (2018). https://doi.org/10.1007/978-3-030-00937-3_12
15. Xu, X., et al.: Whole heart and great vessel segmentation in congenital heart disease using deep neural networks and graph matching. In: Shen, D., et al. (eds.) MICCAI 2019. LNCS, vol. 11765, pp. 477–485. Springer, Cham (2019). https://doi.org/10.1007/978-3-030-32245-8_53
16. Zhang, Y., Miao, S., Mansi, T., Liao, R.: Unsupervised x-ray image segmentation with task driven generative adversarial networks. Med. Image Anal. **62**, 101664 (2020)
17. Zhuang, X., Shen, J.: Multi-scale patch and multi-modality atlases for whole heart segmentation of MRI. Med. Image Anal. **31**, 77–87 (2016)

Continual Learning of New Diseases with Dual Distillation and Ensemble Strategy

Zhuoyun Li[1,2], Changhong Zhong[1,2], Ruixuan Wang[1,2](✉), and Wei-Shi Zheng[1,2,3]

[1] School of Data and Computer Science, Sun Yat-sen University, Guangzhou, China
wangruix5@mail.sysu.edu.cn
[2] Key Laboratory of Machine Intelligence and Advanced Computing, MOE, Guangzhou, China
[3] Pazhou Lab, Guangzhou, China

Abstract. Most intelligent diagnosis systems are developed for one or a few specific diseases, while medical specialists can diagnose all diseases of certain organ or tissue. Since it is often difficult to collect data of all diseases, it would be desirable if an intelligent system can initially diagnose a few diseases, and then continually learn to diagnose more and more diseases with coming data of these new classes in the future. However, current intelligent systems are characterised by catastrophic forgetting of old knowledge when learning new classes. In this paper, we propose a new continual learning framework to alleviate this issue by simultaneously distilling both old knowledge and recently learned new knowledge and by ensembling the class-specific knowledge from the previous classifier and the learned new classifier. Experiments showed that the proposed method outperforms state-of-the-art methods on multiple medical and natural image datasets.

Keywords: Continual learning · Distillation · Ensemble

1 Introduction

Deep learning has been a common tool for medical image analysis in recent years and achieved human-level performance in diagnosis of various diseases [2,5,18]. Most intelligent diagnosis systems would be fixed once developed and deployed. However, it is difficult to collect training data for all diseases of certain organ or tissue. As a result, every current intelligent system can diagnose just one or a few diseases, unable to diagnose all diseases as medical specialists do. Therefore, it would be desirable to enable an intelligent system to continually learn to diagnose more and more diseases, finally becoming a human-like specialist. Such a continual or lifelong learning process often presumes that the intelligent

Z. Li and C. Zhong—The authors contribute equally to this paper.

A. L. Martel et al. (Eds.): MICCAI 2020, LNCS 12261, pp. 169–178, 2020.
https://doi.org/10.1007/978-3-030-59710-8_17

systems can access little or no old data of previously learned diseases due to various factors (e.g., privacy, data not shared across institutes). Unlike humans who can learn new knowledge without forgetting old knowledge, current intelligent classification systems are characterised by catastrophic forgetting of old classes when learning new classes [7,8,13]. This makes it very challenging to develop an intelligent system which can continually learn to diagnose new diseases without sacrificing diagnosis performance on previously learned old diseases [3].

Typically, there always exists a trade-off between learning new knowledge and keeping previously acquired knowledge about old classes. To mitigate the forgetting of previously learned knowledge, one way is to find system units (e.g., kernels in convolutional neural networks) which are crucial for old knowledge, and then try to keep parameters of such units unchanged when learning new knowledge [6,14,15,17]. Although this can help the system keep old knowledge to some extent, it would cause more and more difficulty in continually learning new knowledge, particularly when most kernels in the convolutional neural network (CNNs) become crucial for keeping old knowledge. To solve this dilemma, researchers tried to dynamically add new components to the existing system specifically for new knowledge [1,12,20,24,25]. However, this approach often assumes that the intelligent system can discriminate between old and new classes in advance, which is impractical in an intelligent diagnosis system where old and new diseases need to be diagnosed together. Inspired by the human learning process which often replays memory of old knowledge during learning new knowledge, another approach is to either store a small subset of original exemplars [4,10,11,16,19] or regenerate synthetic data [23,26] for each old class, which has been shown to help reduce catastrophic forgetting more effectively in continual learning of new classes. However, synthetic data may not faithfully contain fine-grained discriminative features for each old class, particularly when applied to different diseases part of which may look similar to each other, and imbalanced small subset of exemplars for each old class compared to relatively large set of data for new classes often makes the intelligent system biased to recently learned (relatively new) class during prediction [22].

Here we propose a more effective framework to keep old knowledge during learning new classes, with a better way to simulate the idea of memory replay. In this framework, an *expert* classifier only for new classes is first trained by fine-tuning from the *original* classifier of old classes, and then the knowledge from both classifiers are distilled to teach a new classifier to recognize both old and new classes. The fine-tuning from the *original* classifier to the *expert* classifier would make the *expert* classifier largely contain old knowledge while learning new classes, therefore distillation from the *expert* classifier to the new classifier would additionally help keep knowledge of old classes. Besides the dual distillation, an ensemble strategy is applied to help keep old knowledge by combining the information of old classes extracted from both the *original* classifier and the new classifier. The proposed approach shows state-of-the-art performance in continual learning on two skin disease datasets and one natural image dataset.

To our best knowledge, this is the first time to explore *continual learning* of intelligent diagnosis on relatively large number of (i.e., 40) diseases.

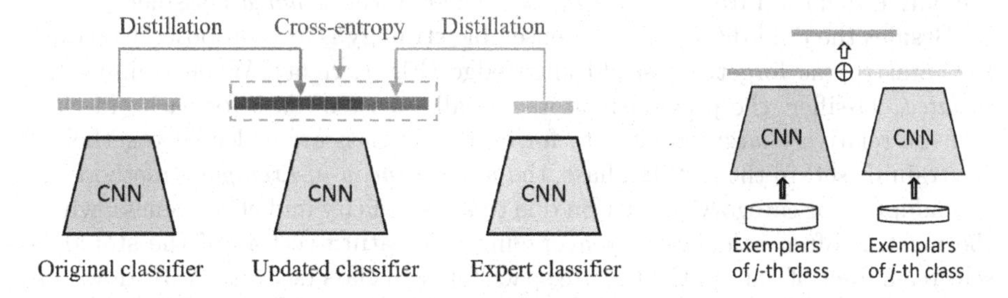

Fig. 1. The proposed continual learning framework. Left: distillation of old and new knowledge from the *original* classifier and the *expert* classifier to the *updated* classifier. Right: ensemble (denote by $\oplus$ above) of mean feature vectors for the j-th old class. (Color figure online)

2 Continual Learning

The objective is to make an intelligent diagnosis system continually learn to diagnose new classes of diseases while keeping its diagnosis performance on old classes not reduced largely. One presumption of the continual learning problem is that most, if not all, training data of the old classes are not available when continually learning new classes. The key challenge is to reduce the catastrophic forgetting of old knowledge when continually learning new classes each time. Considering that keeping small subset of data for each old class in continual learning has been confirmed crucial to potentially reduce the catastrophic forgetting of old knowledge in previous studies [19], here we also assume that a small subset of data for each old class is available during continual learning.

2.1 Overview

In this study, we apply a dual distillation strategy to effectively reduce the catastrophic forgetting of old knowledge. The continual learning framework is demonstrated in Fig. 1, with the diagnosis system for the old classes represented by the *original* CNN classifier (Fig. 1, blue), and the system after continual learning of a set of new classes represented by the *updated* CNN classifier (Fig. 1, orange). Different from existing approaches [4,19], an *expert* CNN classifier composed of the feature extractor part of the *original* CNN classifier and a new fully connected layer is trained specifically for classification of the new classes (Fig. 1, green), and then the knowledge from both the *original* classifier and the *expert* classifier is distilled to train the *updated* classifier (Fig. 1, left). Since the *expert* classifier is fine-tuned from the *original* classifier, it is expected that knowledge of the old classes would be partly kept in the *expert* classifier, and therefore distillation from the *expert* classifier to the *updated* classifier would distill not only knowledge of new classes, but also partial knowledge of old classes, thus helping the *updated* classifier keep more knowledge of old classes compared to the only distillation from the *original* classifier to the *updated* classifier.

Besides the dual distillation, an ensemble strategy is also proposed to reduce the catastrophic forgetting of old knowledge (Fig. 1, right). When training the *updated* classifier, the previously stored small subset of data for each old class and the relatively large set of data for each new class are collected together as the training set. In the testing phase, the *nearest-mean-of-exemplars* method [19] was adopted for category prediction due to its simplicity and effectiveness, where the mean is often calculated by averaging the feature vectors of the stored or selected subset of data in the feature space for each class (see Sect. 2.3). However, the imbalanced training set could lead to the bias toward the new classes when predicting the category of any test data by the *updated* classifier. To mitigate the prediction bias, for each old class, the mean feature vectors from the *original* classifier and the *updated* classifier were ensembled (i.e., averaged here) for prediction. Such ensemble strategy makes the mean feature vector of each old class deviated less from the mean feature vector previously generated with the *original* classifier, thus helping the knowledge representation of each old class changed less. Such ensemble was shown to reduce the catastrophic forgetting of old knowledge (Sect. 3.4). It is worth noting that the *updated* classifier would become the *original* classifier in next-round continual learning.

2.2 Dual Distillation

The dual distillation is for the training of the *updated* classifier. Formally, let $D = \{(\mathbf{x}_i, \mathbf{y}_i), i = 1, \ldots, N\}$ denote the collection of previously stored small subset of data for each old class and the whole set of data for each new class, with $\mathbf{x}_i$ representing an image and the one-hot vector $\mathbf{y}_i$ representing the corresponding class label. For each image $\mathbf{x}_i$, denote by $\mathbf{z}_i = [z_{i1}, z_{i2}, \ldots, z_{is}]^\mathsf{T}$ the logit output (just before performing the softmax operation) of the *original* classifier, where s is the number of old classes. Similarly, denote by $\hat{\mathbf{z}}_i = [\hat{z}_{i1}, \hat{z}_{i2}, \ldots, \hat{z}_{i,s+t}]^\mathsf{T}$ the logit output of the *updated* classifier, where t is the number of new classes. Then, the distillation of old knowledge from the *original* classifier to the *updated* classifier can be realized by minimization of the distillation loss $\mathcal{L}_o$,

$$\mathcal{L}_o(\boldsymbol{\theta}) = -\frac{1}{N}\sum_{i=1}^{N}\sum_{j=1}^{s} p_{ij} \log \hat{p}_{ij}, \tag{1}$$

where $\boldsymbol{\theta}$ represents the to-be-learned parameters of the *updated* classifier, and p_{ij} and $\hat{p}_{ij}$ are from the modified softmax operation,

$$p_{ij} = \frac{\exp(z_{ij}/T_o)}{\sum_{k=1}^{s} \exp(z_{ik}/T_o)}, \quad \hat{p}_{ij} = \frac{\exp(\hat{z}_{ij}/T_o)}{\sum_{k=1}^{s} \exp(\hat{z}_{ik}/T_o)}, \tag{2}$$

and $T_o \geq 1$ is the distillation parameter. Larger T_o will force the *updated* classifier to learn a more fine-grained separation between different feature vectors $\mathbf{z}_i$'s [9].

Similarly, the distillation of new knowledge from the *expert* classifier to the *updated* classifier can be realized by minimization of the distillation loss $\mathcal{L}_n$,

$$\mathcal{L}_n(\boldsymbol{\theta}) = -\frac{1}{N}\sum_{i=1}^{N}\sum_{j=1}^{t} q_{ij} \log \hat{q}_{ij}, \tag{3}$$

where q_{ij} and $\hat{q}_{ij}$ are respectively from the modified softmax over the logit of the *expert* classifier and the corresponding logit part of the *updated* classifier, with the distillation parameter T_n.

Besides the two distillation losses, the cross-entropy loss $\mathcal{L}_c$ is also applied to help the *updated* classifier discriminate both old and new classes,

$$\mathcal{L}_c(\boldsymbol{\theta}) = -\frac{1}{N}\sum_{i=1}^{N}\sum_{j=1}^{s+t} y_{ij} \log \hat{y}_{ij}, \tag{4}$$

where y_{ij} is the j-th component of the one-hot class label $\mathbf{y}_i$, and the $\hat{y}_{ij}$ is the j-th output of the *updated* classifier for the input image $\mathbf{x}_i$. In combination, the *updated* classifier can be trained by minimizing the overall loss $\mathcal{L}$,

$$\mathcal{L}(\boldsymbol{\theta}) = \mathcal{L}_c(\boldsymbol{\theta}) + \lambda_1 \mathcal{L}_o(\boldsymbol{\theta}) + \lambda_2 \mathcal{L}_n(\boldsymbol{\theta}), \tag{5}$$

where λ_1 and λ_2 is a coefficient constant balancing different loss terms.

2.3 Ensemble of Class Means

Considering that the majority of the training set is from new classes, and often only a small subset of data for each old class is stored, directly using output of the *updated* classifier to predict the category of any test data would be inevitably biased to new classes. As adopted in recent work [19], the nearest-mean-of-exemplars method is also used for category prediction of any test data in this study. The basic idea is to extract and average the feature vectors of all available data for each class based on the output of the penultimate layer of the *updated* classifier, and then any test data is categorized as the class whose mean vector is nearest to the test data in the feature space. That means, for each new class, the class mean is the average of feature vectors over the whole training set of this class based on the output of the penultimate layer of the *updated* classifier. However, for each old class, different from previous work, we propose an ensemble strategy for calculating the class mean. Specifically, suppose $\mathbf{m}_j$ is the previously obtained class mean for the j-th old class (in the previous round of continual learning), and $\hat{\mathbf{m}}_j$ is the average of feature vectors over the stored small subset of data for the j-th class based on output of the penultimate layer of the well-trained *updated* classifier. Then the ensemble class mean of the j-th old class is the average of the two means, i.e., $(\mathbf{m}_j + \hat{\mathbf{m}}_j)/2$. Using the ensemble class mean instead of the updated individual mean $\hat{\mathbf{m}}_j$ for each old class is experimentally shown to improve the prediction performance for old classes.

Before going to the next round of continual learning, due to the limited memory to store exemplar training data for each class, a small subset of training data needs to be selected from the whole training set for each new class, and the previously stored subset of data for each old class needs to be further reduced. Suppose the memory size is K, then only $K/(s+t)$ exemplars would be stored in the memory for each of the $(s+t)$ classes. As in the iCaRL method, the herding selection strategy is applied to generate a sorted list of exemplars for each (new) class, with exemplars approximating the class mean having higher priorities [19]. Thus, the first $K/(s+t)$ exemplars in the sorted list would be selected and stored for each new class. For each old class, the first $K/(s+t)$ exemplars from the already shortened list would be kept, discarding the others in the memory.

Table 1. Statistics of three datasets used in experiments.

Datasets	#Classes	Train set	Test set	New classes per time	Size
CIFAR100	100	50,000	10,000	5, 10, 20	32×32
Skin40	40	2,000	400	2, 5, 10	[420, 1640]
Skin8	8	3,555	705	2	[600, 1024]

3 Experimental Results

3.1 Experimental Setup

Three datasets were used to extensively evaluate the proposed method (Table 1). Among them, the Skin8 dataset is from the classification challenge of dermoscopic images held by ISIC'2019 [21]. The original Skin8 is highly imbalanced between classes. To alleviate the potential effect of imballance in continual learning, 628 images were randomly selected from six classes and all images (fewer than 260) were kept for the other two smaller classes. The dataset Skin40 consists of 40 classes of skin disease images collected from the internet, with 50 training images and 10 test images for each class. In training, each image was randomly cropped within scale range [0.8, 1.0] and then resized to 224×224 pixels.

On each dataset, an *original* CNN classifier was first trained for a small number (e.g., 2, 5) of classes, and then a set of (e.g., 2, 5) new classes' data were provided to train the *expert* classifier and the *updated* classifier, finishing the first-round continual learning. SGD optimizer (batch size 16) was used, with initial learning rate 0.01 and then divided by 10 at the 80th, and 160th epoch respectively. Weight decay (0.00001) and momentum (0.9) were also applied. Each model was trained for up to 200 epochs, with training convergence consistently observed. After training the *updated* classifier each time, the average accuracy over all learned classes so far was calculated. Such a training and evaluation process was repeated in next-round continual learning. For each experiment, the average and standard deviation of accuracy over five runs were reported, each run with a different order of classes to be learned. Unless otherwise mentioned, ResNet-18 was used as backbone, memory size $K = 50$, parameters $T_o = T_n = 4.0$ on Skin40, $T_o = T_n = 2.0$ on Skin8, and $\lambda_1 = 3.0$, $\lambda_2 = 1.0$.

3.2 Performance on Continual Learning of Medical Image Classes

Our method was first evaluated on skin datasets by comparing with state-of-the-art baselines, including iCaRL [19], End-to-End Incremental Learning (End2End) [4], Distillation and Retrospection (DR) [10] and LwF [16]. Similar amount of effort was put into tuning each baseline method. The same memory size was used for all methods except the LwF which does not store old exemplars. In addition, an upper-bound result was also reported (Fig. 2, **black** cross) by training a non-continual classifier with all classes of training data. Figure 2 shows that our method outperforms all the strong baselines when the classifier continually learn either 2, 5, or 10 classes every time on Skin40 (first three subfigures), and 2 classes every time on Skin8 (last one), all confirming the effectiveness of our method. Smaller improvement on Skin8 than on Skin40 may be due to fewer classes and fewer rounds of continual learning on Skin8.

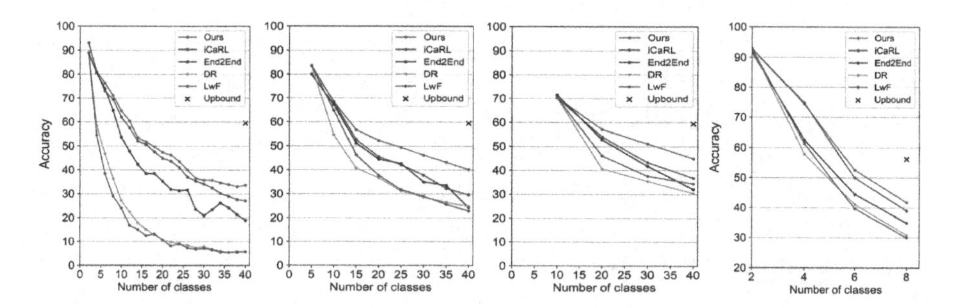

Fig. 2. Performance comparison between the proposed method and baselines on Skin40 (first three) and Skin8 (last). Standard deviation ([0.7, 3.8]) omitted for clearer views.

3.3 Robustness and Generalization of Continual Learning

To show the robustness of our method, multiple CNN backbones were used for continual learning of skin diseases. As shown in Table 2, our method consistently outperforms the strong baselines with all three CNN backbones, no matter whether the classifier continually learn 2, 5, or 10 classes each time on Skin40 (only showing 5-class case). Besides the CNN backbones, different memory sizes ($K = 100, 200$) were also evaluated (not shown due to limited space), all consistently showing our method is superior to the strong baselines.

Table 2. Performance comparison between the proposed method and baselines with different CNN backbones. The average and standard deviation of accuracy at the end of continual learning of 5 classes every time over five runs were reported.

Backbones	ResNet18				AlexNet				VGG19			
Methods	iCaRL	E2E	DR	Ours	iCaRL	E2E	DR	Ours	iCaRL	E2E	DR	Ours
Accuracy	29.6 ± 2.1	24.3 ± 1.4	24.9 ± 3.8	**40.2± 1.6**	29.3 ± 0.4	20.8 ± 1.6	20.6 ± 3.9	**33.0± 0.8**	25.5 ± 2.5	19.3 ± 0.8	20.3 ± 3.8	**28.8± 1.7**

Considering that the baselines were evaluated only on natural image datasets in previous studies, our method was also evaluated on CIFAR100. Consistently, our method showed better performance in continual learning of 5, 10, or 20 new classes each time (Fig. 3), supporting that our method is generalizable to various domains. One interesting observation is that the performance of our method is similar to that of iCaRL at first few rounds of continual learning, and then becomes better at subsequent rounds. This might be because the stored old data is sufficient in first few rounds to help the classifier reduce catastrophic forgetting of old classes, leading to the negligible effect of the *expert* classifier and the ensemble strategy in our method. This is confirmed in Fig. 3 (right), where our method outperforms iCaRL from earlier rounds of continual learning when memory size is smaller ($K = 500$ here versus $K = 2000$ in 2nd subfigure).

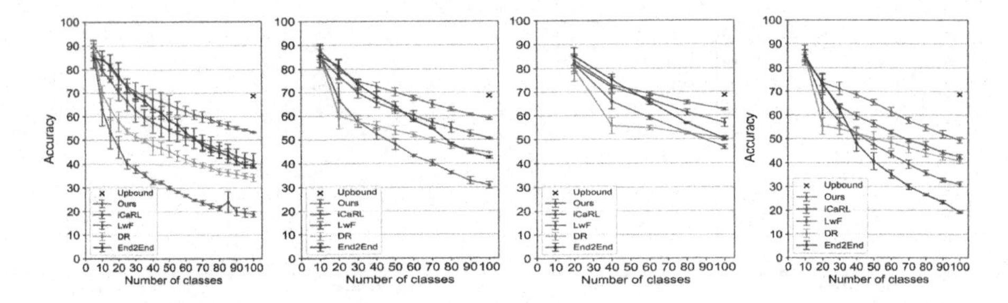

Fig. 3. Performance of continual learning on the CIFAR100 dataset. From left to right: continual learning of 5, 10, and 20 new classes each time with memory size $K = 2000$, and the continual learning of 10 new classes each time with $K = 500$.

3.4 Ablation Study

This section evaluates the role of the *expert* classifier and the ensemble strategy in continual learning. Table 3 shows that although the *expert* classifier trained from scratch each time improve learning performance little (column 3 vs. column 2), the *expert* classifier fine-tuned from the *original* classifier clearly improves more (column 4). Adding the ensemble strategy further improve performance (column 5). It is worth noting the ensemble strategy would help more if memory size becomes smaller (columns 4–5, $K = 2000$ vs. 500 vs. 100).

Table 3. Ablation study on CIFAR100. 'Original+CE': training the *updated* classifier with the cross-entropy loss and the single distillation from the *original* classifier. Final-round learning performance was reported, with 10 new classes learned each time.

K	Original+CE (OCE)	OCE+expert (random)	OCE+expert (finetune)	OCE+expert (finetune)+ensemble
2000	55.7	55.4	57.9	**59.2**
500	44.2	45.6	47.4	**49.3**
100	27.8	31.9	32.7	**34.8**

4 Conclusion

In this study, we proposed a new continual learning framework to help intelligent systems reduce catastrophic forgetting of old knowledge during learning new classes. Experiments showed that dual distillation from two teachers can help teach a new classifier more effectively during continual learning of new diseases, and the ensemble prediction from old and new classifiers further alleviate forgetting old knowledge.

Acknowledgement. This work is supported in part by the National Key Research and Development Program (grant No. 2018YFC1315402), the Guangdong Key Research and Development Program (grant No. 2019B020228001), the National Natural Science Foundation of China (grant No. U1811461), and the Guangzhou Science and Technology Program (grant No. 201904010260).

References

1. Aljundi, R., Chakravarty, P., Tuytelaars, T.: Expert gate: lifelong learning with a network of experts. In: Conference on Computer Vision and Pattern Recognition, pp. 3366–3375 (2017)
2. Ardila, D., et al.: End-to-end lung cancer screening with three-dimensional deep learning on low-dose chest computed tomography. Nat. Med. **25**(6), 954–961 (2019)
3. Baweja, C., Glocker, B., Kamnitsas, K.: Towards continual learning in medical imaging. In: Medical Imaging Meets NIPS Workshop (2018)
4. Castro, F.M., Marín-Jiménez, M.J., Guil, N., Schmid, C., Alahari, K.: End-to-end incremental learning. In: Ferrari, V., Hebert, M., Sminchisescu, C., Weiss, Y. (eds.) ECCV 2018. LNCS, vol. 11216, pp. 241–257. Springer, Cham (2018). https://doi.org/10.1007/978-3-030-01258-8_15
5. Esteva, A., et al.: Dermatologist-level classification of skin cancer with deep neural networks. Nature **542**(7639), 115–118 (2017)
6. Fernando, C., et al.: Pathnet: evolution channels gradient descent in super neural networks. arXiv preprint arXiv:1701.08734 (2017)
7. French, R.M.: Catastrophic forgetting in connectionist networks. Trends Cogn. Sci. **3**(4), 128–135 (1999)
8. Goodfellow, I.J., Mirza, M., Da Xiao, A.C., Bengio, Y.: An empirical investigation of catastrophic forgeting in gradient-based neural networks. In: International Conference on Learning Representations (2014)

9. Hinton, G., Vinyals, O., Dean, J.: Distilling the knowledge in a neural network. In: NIPS Deep Learning and Representation Learning Workshop (2015)
10. Hou, S., Pan, X., Loy, C.C., Wang, Z., Lin, D.: Lifelong learning via progressive distillation and retrospection. In: Ferrari, V., Hebert, M., Sminchisescu, C., Weiss, Y. (eds.) ECCV 2018. LNCS, vol. 11207, pp. 452–467. Springer, Cham (2018). https://doi.org/10.1007/978-3-030-01219-9_27
11. Isele, D., Cosgun, A.: Selective experience replay for lifelong learning. In: AAAI Conference on Artificial Intelligence, pp. 3302–3309 (2018)
12. Karani, N., Chaitanya, K., Baumgartner, C., Konukoglu, E.: A lifelong learning approach to brain MR segmentation across scanners and protocols. In: Frangi, A.F., Schnabel, J.A., Davatzikos, C., Alberola-López, C., Fichtinger, G. (eds.) MICCAI 2018. LNCS, vol. 11070, pp. 476–484. Springer, Cham (2018). https://doi.org/10.1007/978-3-030-00928-1_54
13. Kemker, R., McClure, M., Abitino, A., Hayes, T.L., Kanan, C.: Measuring catastrophic forgetting in neural networks. In: AAAI Conference on Artificial Intelligence, pp. 3390–3398 (2018)
14. Kim, H.E., Kim, S., Lee, J.: Keep and learn: continual learning by constraining the latent space for knowledge preservation in neural networks. In: International Conference on Medical Image Computing and Computer-Assisted Intervention, pp. 520–528 (2018)
15. Kirkpatrick, J., et al.: Overcoming catastrophic forgetting in neural networks. Nat. Acad. Sci. **114**(13), 3521–3526 (2017)
16. Li, Z., Hoiem, D.: Learning without forgetting. IEEE Trans. Pattern Anal. Mach. Intell. **40**(12), 2935–2947 (2017)
17. Mallya, A., Lazebnik, S.: Packnet: adding multiple tasks to a single network by iterative pruning. In: Conference on Computer Vision and Pattern Recognition, pp. 7765–7773 (2018)
18. McKinney, S.M., et al.: International evaluation of an AI system for breast cancer screening. Nature **577**(7788), 89–94 (2020)
19. Rebuffi, S.A., Kolesnikov, A., Sperl, G., Lampert, C.H.: iCaRL: incremental classifier and representation learning. In: Conference on Computer Vision and Pattern Recognition, pp. 2001–2010 (2017)
20. Rusu, A.A., et al.: Progressive neural networks. arXiv preprint arXiv:1606.04671 (2016)
21. Tschandl, P., Rosendahl, C., Kittler, H.: The HAM10000 dataset, a large collection of multi-source dermatoscopic images of common pigmented skin lesions. Sci. Data **5**, 180161 (2018)
22. Wu, Y., et al.: Large scale incremental learning. In: Conference on Computer Vision and Pattern Recognition, pp. 374–382 (2019)
23. Xiang, Y., Fu, Y., Ji, P., Huang, H.: Incremental learning using conditional adversarial networks. In: International Conference on Computer Vision, pp. 6619–6628 (2019)
24. Xu, J., Zhu, Z.: Reinforced continual learning. In: Advances in Neural Information Processing Systems, pp. 899–908 (2018)
25. Yoon, J., Yang, E., Lee, J., Hwang, S.J.: Lifelong learning with dynamically expandable networks. In: International Conference on Learning Representations (2018)
26. Zhai, M., Chen, L., Tung, F., He, J., Nawhal, M., Mori, G.: Lifelong GAN: continual learning for conditional image generation. In: International Conference on Computer Vision, pp. 2759–2768 (2019)

Learning to Segment When Experts Disagree

Le Zhang[1,2(✉)], Ryutaro Tanno[2,4], Kevin Bronik[2], Chen Jin[2], Parashkev Nachev[3], Frederik Barkhof[1,2], Olga Ciccarelli[1], and Daniel C. Alexander[2]

[1] Queen Square Multiple Sclerosis Centre, Department of Neuroinflammation, Queen Square Institute of Neurology, Faculty of Brain Sciences, University College London, London, UK
le.zhang@ucl.ac.uk
[2] Centre for Medical Image Computing, Department of Computer Science, University College London, London, UK
[3] High Dimensional Neurology Group, Queen Square Institute of Neurology, University College London, London, UK
[4] Healthcare Intelligence, Microsoft Research, Cambridge, UK

Abstract. Recent years have seen an increasing use of supervised learning methods for segmentation tasks. However, the predictive performance of these algorithms depend on the quality of labels, especially in medical image domain, where both the annotation cost and inter-observer variability are high. In a typical annotation collection process, different clinical experts provide their estimates of the "true" segmentation labels under the influence of their levels of expertise and biases. Treating these noisy labels blindly as the ground truth can adversely affect the performance of supervised segmentation models. In this work, we present a neural network architecture for jointly learning, from noisy observations alone, both the reliability of individual annotators and the true segmentation label distributions. The separation of the annotators' characteristics and true segmentation label is achieved by encouraging the estimated annotators to be maximally unreliable while achieving high fidelity with the training data. Our method can also be viewed as a translation of STAPLE, an established label aggregation framework proposed in Warfield et al. [1] to the supervised learning paradigm. We demonstrate first on a generic segmentation task using MNIST data and then adapt for usage with MRI scans of *multiple sclerosis* (MS) patients for lesion labelling. Our method shows considerable improvement over the relevant baselines on both datasets in terms of segmentation accuracy and estimation of annotator reliability, particularly when only a single label is available per image. An open-source implementation of our approach can be found at https://github.com/UCLBrain/MSLS.

L. Zhang, R. T—These authors contributed equally.

A. L. Martel et al. (Eds.): MICCAI 2020, LNCS 12261, pp. 179–190, 2020.
https://doi.org/10.1007/978-3-030-59710-8_18

1 Introduction

Segmentation of anatomical structures in medical images is known to suffer from high inter-reader variability [2,3], affecting the performance of downstream supervised machine learning models. For example, accurate identification of multiple sclerosis (MS) lesions in MRIs is difficult even for experienced experts due to variability in lesion location, size, shape and anatomical variability across patients [4]. Further aggravated by differences in levels of expertise, annotations of MS lesions suffer from high annotation variations [5]. Despite the present abundance of medical imaging data thanks to over two decades of digitisation, the world still remains relatively short of access to data with curated labels [6], that is amenable to machine learning, necessitating an intelligent method to learn robustly from such noisy annotations.

To mitigate inter-reader variations, different pre-processing techniques are commonly used to curate annotations by fusing labels from different experts. The most basic yet popular approach is based on the majority vote where the most representative opinion of the experts is treated as the ground truth (GT). A smarter version that accounts for similarity of classes has proven effective in aggregation of brain tumour segmentation labels [2]. A key limitation of such approaches, however, is that all experts are assumed to be equally reliable. Warfield *et al.* [1] proposed a label fusion method, called STAPLE that explicitly models the reliability of individual experts and use such information to "weigh" their opinions in the label aggregation step. After consistent demonstration of its superiority over the standard majority vote preprocessing in multiple applications, STAPLE has become a staple label fusion method in the creation of medical image segmentation datasets e.g., ISLES [7], MSSeg [8], Gleason'19 [9] datasets. Asman *et al.* later extended this approach in [10] by accounting for voxel-wise consensus to address the issue of under-estimation of annotators' reliability. In [11], another extension was proposed in order to model the reliability of annotators across different pixels in images. More recently, within the context of multi-atlas segmentation problems [12] where image registration is used to warp segments from labeled images ("atlases") onto a new scan, STAPLE has been enhanced in multiple ways to encode the information of the underlying images into the label aggregation process. A notable example is STEP proposed in Cardoso *et al.* [13] who designed a strategy to further incorporate the local morphological similarity between atlases and target images, and different extensions of this approach such as [14,15] have since been considered. However, these previous label fusion approaches share a drawback—they critically lack a mechanism to integrate information across different training images, fundamentally limiting the remit of applications to cases where each image receives a reasonable number of annotations from multiple experts, which may be expensive in practice. Moreover, relatively simplistic functions are employed to model the relations between observed noisy annotations, true labels and reliability of experts, which may fail to capture complex characteristics of human experts.

Our Contributions: In this work, we introduce the first instance of an end-to-end supervised segmentation method that jointly estimates, from noisy labels alone, the reliability of multiple human annotators and true segmentation labels. Specifically, we achieve this by translating the long-standing STAPLE framework [1] to the supervised learning setting. The proposed approach (Fig. 1) consists of two coupled CNNs where one estimates the true segmentation probabilities and the other models the characteristics of individual annotators (e.g., prone to over-segmentation) by estimating the pixel-wise confusion matrices on a per image basis. The parameters of our models are global variables that are optimised across different training samples; this enables the model to infer annotators' characteristics and true labels based on correlations between similar image samples even when only a single annotation is available per image. In contrast, this would not be possible with STAPLE [1] and its variants [11,13,14] where the annotators' parameters are estimated on every image separately. Lastly, unlike the previous label fusion methods, as a supervised approach, our method produces a model that can segment test images without needing to acquire labels from annotators or atlases.

For evaluation, we first simulate a diverse range of annotator types on the MNIST dataset by performing morphometric operations with Morpho-MNIST framework [16]. Then we demonstrate the potential in the real-world task of MS lesion segmentation on ISBI 2015 challenge dataset. Experiments on both datasets demonstrate that our method leads to better performance with respect to the widely adopted STAPLE framework and the naive CNNs trained on traditional labels, and is capable of recovering the true label distributions even when there is only one label available per example.

Other Related Works: Our work also relates to a recent strand of methods that aim to generate a set of diverse and plausible segmentation proposals on a given image. Notably, probabilistic U-net [17] and its recent variant, PHiSeg [18] have shown that the aforementioned inter-reader variation in segmentation labels can be modelled with sophisticated forms of probabilistic CNNs. Such approaches, however, fundamentally differ from ours in that variable annotations from many experts in the training data are assumed to be all realistic instances of the true segmentation; we assume, on the other hand, that there is a single true segmentation map of the underlying anatomy, and the variations arise from the characteristics of individual annotators.

We should also note that, in standard supervised classification problems [19–22], several methods have shown promising results in modelling the label noise of multiple annotators and thereby restoring the true label distribution in medical imaging applications and beyond. By contrast, no attention has been paid to the same problem in more complicated, structured prediction tasks – to our knowledge, our work makes the first attempt to address such problem for segmentation where the outputs are high dimensional and structured. In particular, our approach yields the estimate of reliability in every pixel as a function of the input image, crucial in capturing the complex spatial variations in annotators' characteristics and absent in the classification setting.

2 Method

2.1 Problem Set-Up

In this work, we consider the problem of learning a supervised segmentation model from noisy labels acquired from multiple human annotators. Specifically, we consider a scenario where set of images $\{\mathbf{x}_n \in \mathbb{R}^{W \times H \times C}\}_{n=1}^{N}$ (with W, H, C denoting the width, height and channels of the image) are assigned with noisy segmentation labels $\{\tilde{\mathbf{y}}_n^{(r)} \in \mathcal{Y}^{W \times H}\}_{n=1,...,N}^{r \in S(\mathbf{x}_i)}$ from multiple annotators where $\tilde{\mathbf{y}}_n^{(r)}$ denotes the label from annotator $r \in \{1, ..., R\}$ and $S(\mathbf{x}_n)$ denotes the set of all annotators who labelled image $\mathbf{x}_i$ and $\mathcal{Y} = [1, 2, ..., L]$ denotes the set of classes.

Here we assume that every image $\mathbf{x}$ annotated by at least one person i.e., $|S(\mathbf{x})| \geq 1$, and no GT labels $\{\mathbf{y}_n \in \mathcal{Y}^{W \times H}\}_{n=1,...,N}$ are available. The problem of interest here is to *learn the unobserved true segmentation distribution* $p(\mathbf{y} \mid \mathbf{x})$ *from such noisy labelled dataset* $\mathcal{D} = \{\mathbf{x}_n, \tilde{\mathbf{y}}_n^{(r)}\}_{n=1,...,N}^{r \in S(\mathbf{x}_n)}$ i.e., the combination of images, noisy annotations and experts' identities for labels (which label was obtained from whom).

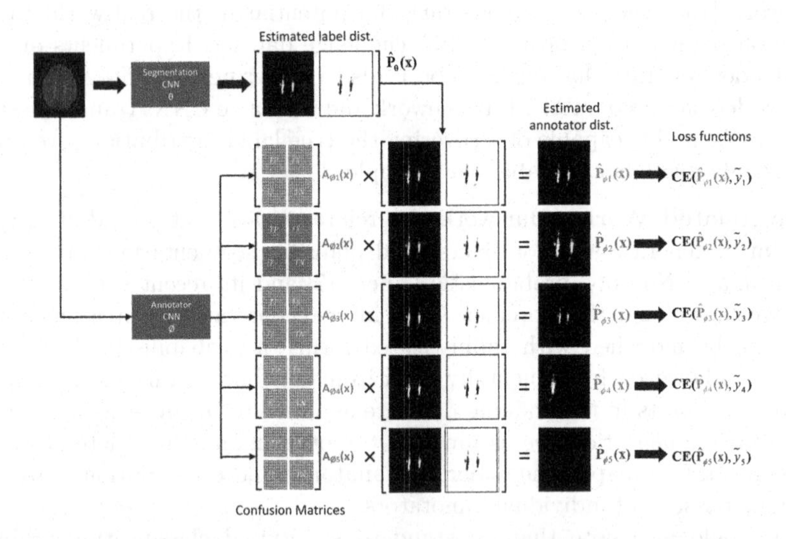

Fig. 1. General schematic of the model in the presence of 5 annotators. The method consists of two components: (1) Segmentation network parametrised by θ that generates an estimate of the GT segmentation probabilities, $p_\theta(\mathbf{x})$ for the given input image $\mathbf{x}$; (2) Annotator network, parametrised by ϕ, that estimates the confusion matrices (CMs) $\{\mathbf{A}_\phi^{(r)}(\mathbf{x})\}_{r=1}^{5}$ of the annotators. The segmentation probabilities of respective annotators $\hat{\mathbf{p}}_\phi^{(r)}(\mathbf{x}) := \mathbf{A}_\phi^{(r)}(\mathbf{x}) \cdot \mathbf{p}_\theta(\mathbf{x})$ are then computed. The model parameters $\{\theta, \phi\}$ are optimized to minimize the sum of five cross-entropy losses between each estimated annotator distribution $\mathbf{p}_\phi^{(r)}(\mathbf{x})$ and the noisy labels $\tilde{\mathbf{y}}^{(r)}$ observed from each annotator.

We also emphasise that *the goal at inference time is to segment a given unlabelled test image* but not to fuse multiple available labels as is typically done in multi-atlas segmentation approaches [12].

2.2 Probabilistic Model and Proposed Architecture

Here we describe the probabilistic model of the observed noisy labels from multiple annotators. We make two key assumptions: (1) annotators are statistically independent, (2) annotations over different pixels are independent given the input image. Under these assumptions, the probability of observing noisy labels $\{\tilde{\mathbf{y}}^{(r)}\}_{r \in S(\mathbf{x})}$ on $\mathbf{x}$ factorises as:

$$p(\{\tilde{\mathbf{y}}^{(r)}\}_{r \in S(\mathbf{x})} \mid \mathbf{x}) = \prod_{r \in S(\mathbf{x})} p(\tilde{\mathbf{y}}^{(r)} \mid \mathbf{x}) = \prod_{r \in S(\mathbf{x})} \prod_{\substack{w \in \{1,...,W\} \\ h \in \{1,...,H\}}} p(\tilde{y}^{(r)}_{wh} \mid \mathbf{x}) \quad (1)$$

where $\tilde{y}^{(r)}_{wh} \in [1, ..., L]$ denotes the $(w, h)^{\text{th}}$ elements of $\tilde{\mathbf{y}}^{(r)} \in \mathcal{Y}^{W \times H}$. Now we rewrite the probability of observing each noisy label on each pixel (w, h) as:

$$p(\tilde{y}^{(r)}_{wh} \mid \mathbf{x}) = \sum_{y_{wh}=1}^{L} p(\tilde{y}^{(r)}_{wh} \mid y_{wh}, \mathbf{x}) \cdot p(y_{wh} \mid \mathbf{x}) \quad (2)$$

where $p(y_{wh} \mid \mathbf{x})$ denotes the GT label distribution over the $(w, h)^{\text{th}}$ pixel in the image $\mathbf{x}$, and $p(\tilde{y}^{(r)}_{wh} \mid y_{wh}, \mathbf{x})$ describes the noisy labelling process by which annotator r corrupts the true segmentation label. In particular, we refer to the $L \times L$ matrix whose each $(i, j)^{\text{th}}$ element is defined by the second term $\mathbf{a}^{(r)}(\mathbf{x}, w, h)_{ij} := p(\tilde{y}^{(r)}_{wh} = i \mid y_{wh} = j, \mathbf{x})$ as the *confusion matrix* (CM) of annotator r at pixel (w, h) in image $\mathbf{x}$.

We introduce a CNN-based architecture which models the different constituents in the above joint probability distribution $p(\{\tilde{\mathbf{y}}^{(r)}\}_{r \in S(\mathbf{x})} \mid \mathbf{x})$ as illustrated in Fig. 1. The model consists of two components: (1) *Segmentation Network*, parametrised by θ, which estimates the GT segmentation probability map, $\hat{\mathbf{p}}_\theta(\mathbf{x}) \in \mathbb{R}^{W \times H \times L}$ whose each $(w, h, i)^{\text{th}}$ element approximates $p(y_{wh} = i \mid \mathbf{x})$; (2) *Annotator Network*, parametrised by ϕ, that generate estimates of the pixel-wise confusion matrices of respective annotators as a function of the input image, $\{\hat{\mathbf{A}}^{(r)}_\phi(\mathbf{x}) \in [0, 1]^{W \times H \times L \times L}\}_{r=1}^{R}$ whose each $(w, h, i, j)^{\text{th}}$ element approximates $p(\tilde{y}^{(r)}_{wh} = i \mid y_{wh} = j, \mathbf{x})$. Each product $\hat{\mathbf{p}}^{(r)}_\phi(\mathbf{x}) := \hat{\mathbf{A}}^{(r)}_\phi(\mathbf{x}) \cdot \hat{\mathbf{p}}_\theta(\mathbf{x})$ represents the estimated segmentation probability map of the corresponding annotator. Note that here "$\cdot$" denotes the element-wise matrix multiplications in the spatial dimensions W, H. At inference time, we use the output of the segmentation network $\hat{\mathbf{p}}_\theta(\mathbf{x})$ to segment test images.

2.3 Learning Spatial Confusion Matrices and True Segmentation

Next, we describe how we jointly optimise the parameters of segmentation network, θ and the parameters of annotator network, ϕ. In short, we minimise the negative log-likelihood of the probabilistic model plus a regularisation term via stochastic gradient descent. A detailed description is provided below.

Given training input $\mathbf{X} = \{\mathbf{x}_n\}_{n=1}^N$ and noisy labels $\tilde{\mathbf{Y}}^{(r)} = \{\tilde{\mathbf{y}}_n^{(r)} : r \in S(\mathbf{x}_n)\}_{n=1}^N$ for $r = 1, ..., R$, we optimaize the parameters $\{\theta, \phi\}$ by minimizing the negative log-likelihood (NLL), $-\log p(\tilde{\mathbf{Y}}^{(1)}, ..., \tilde{\mathbf{Y}}^{(R)}|\mathbf{X})$. From Eqs. (1) and (2), this optimization objective equates to the sum of cross-entropy losses between the observed noisy segmentations and the estimated annotator label distributions:

$$-\log p(\tilde{\mathbf{Y}}^{(1)}, ..., \tilde{\mathbf{Y}}^{(R)}|\mathbf{X}) = \sum_{n=1}^{N}\sum_{r=1}^{R} \mathbb{1}(\tilde{\mathbf{y}}_n^{(r)} \in \mathcal{S}(\mathbf{x}_n)) \cdot \mathrm{CE}(\hat{\mathbf{A}}_\phi^{(r)}(\mathbf{x}) \cdot \hat{\mathbf{p}}_\theta(\mathbf{x}_n),\ \tilde{\mathbf{y}}_n^{(r)}) \tag{3}$$

Minimizing the above encourages each annotator-specific prediction $\hat{\mathbf{p}}^{(r)}(\mathbf{x}) := \hat{\mathbf{A}}_\phi^{(r)} \hat{\mathbf{p}}_\theta(\mathbf{x})$ to be as close as possible to the true noisy label distribution of the annotator $\mathbf{p}^{(r)}(\mathbf{x})$. However, this loss function alone is not capable of separating the annotation noise from the true label distribution; there are many combinations of pairs $\hat{\mathbf{A}}_\phi^{(r)}(\mathbf{x})$ and segmentation model $\hat{\mathbf{p}}_\theta(\mathbf{x})$ such that $\hat{\mathbf{p}}^{(r)}(\mathbf{x})$ perfectly matches the true annotator's distribution $\mathbf{p}^{(r)}(\mathbf{x})$ for any input $\mathbf{x}$ e.g., permutation of rows in the CMs. To combat this problem, inspired by Tanno *et al.* [21], which addressed a similar issue for simple classification, we add the trace of the estimated CMs to the loss in Eq. (3) as a regularisation term. We thus optimize the combined loss:

$$\sum_{n=1}^{N}\sum_{r=1}^{R} \mathbb{1}(\tilde{\mathbf{y}}_n^{(r)} \in \mathcal{S}(\mathbf{x}_i)) \cdot \left[\mathrm{CE}(\hat{\mathbf{A}}_\phi^{(r)}(\mathbf{x}) \cdot \hat{\mathbf{p}}_\theta(\mathbf{x}_n),\ \tilde{\mathbf{y}}_n^{(r)}) + \lambda \cdot \mathrm{tr}(\hat{\mathbf{A}}_\phi^{(r)}(\mathbf{x}_n))\right] \tag{4}$$

where $\mathcal{S}(\mathbf{x}))$ denotes the set of all labels available for image $\mathbf{x}$, and $\mathrm{tr}(\mathbf{A})$ denotes the trace of matrix $\mathbf{A}$. Intuitively, minimising the trace encourages the estimated annotators to be maximally unreliable while minimising the cross entropy ensures fidelity with observed noisy annotators. We minimise this combined loss via stochastic gradient descent to learn both $\{\theta, \phi\}$.

While it still remains unclear whether the theoretical justification for the trace regularisation in Tanno *et al.* [21] holds in this sample-specific setting, we demonstrate empirically that such regularisation consistently improves the performance of both segmentation and the estimation of confusion matrices.

Table 1. Comparison of segmentation accuracy and error of CM estimation for different methods with dense labels (mean ± standard deviation).

Models	MNIST DICE (%) (testing)	MNIST CM estimation (validation)	MSLesion DICE (%) (testing)	MSLesion CM estimation (validation)
Naive CNN on mean labels	38.36 ± 0.41	n/a	46.55 ± 0.53	n/a
Naive CNN on mode labels	62.89 ± 0.63	n/a	47.82 ± 0.76	n/a
Separate CNNs on annotators	70.44 ± 0.65	n/a	46.84 ± 1.24	n/a
STAPLE [1]	78.03 ± 0.29	0.1241 ± 0.0011	55.05 ± 0.53	0.1502 ± 0.0026
Spatial STAPLE [11]	78.96 ± 0.22	0.1195 ± 0.0013	58.37 ± 0.47	0.1483 ± 0.0031
Ours without Trace ($\lambda = 0$)	79.63 ± 0.53	0.1125 ± 0.0037	65.77 ± 0.62	0.1342 ± 0.0053
Our method ($\lambda = 0.001$)	82.02 ± 0.21	0.0979 ± 0.0016	66.23 ± 0.39	0.0956 ± 0.0031
Our method ($\lambda = 0.01$)	82.73 ± 0.21	0.0913 ± 0.0014	66.42 ± 0.37	0.0939 ± 0.0027
Our method ($\lambda = 0.1$)	82.92 ± 0.19	0.0893 ± 0.0009	67.55 ± 0.31	0.0811 ± 0.0024
Our method ($\lambda = 0.7$)	82.97 ± 0.14	0.0887 ± 0.0008	67.58 ± 0.29	0.0805 ± 0.0021
Our method ($\lambda = 0.9$)	82.94 ± 0.18	0.0891 ± 0.0009	67.56 ± 0.33	0.0809 ± 0.0023
Oracle (Ours but with known CMs)	83.29 ± 0.11	0.0238 ± 0.0005	78.86 ± 0.14	0.0415 ± 0.0017

3 Experiments and Analysis

Experimental Settings. In this work, we evaluate our method on two datasets: MNIST segmentation dataset [23] and ISBI 2015 MS lesion segmentation challenge dataset [24]. The MNIST dataset consists of 60,000 training and 10,000 testing examples, all of which are 28 × 28 grayscale images of digits from 0 to 9, and we derive the segmentation labels by thresholding the intensity values at 0.5. The MS dataset is publicly available and comprises 21 3D scans from 5 subjects each with T1w (voxel size = $0.82 \times 0.82 \times 1.17$ mm^3) and FLAIR (voxel size = $0.82 \times 0.82 \times 2.2$ mm^3) MRIs. All scans are split into 10 for training and 11 for testing. We hold out 20% of training images as a validation set for both datasets.

For both datasets, we simulate a group of 5 annotators of disparate characteristics by performing morphological transformations (e.g., thinning, thickening, fractures, etc.) on the ground-truth (GT) segmentation labels, using Morpho-MNIST software [16] (see Fig. 2(a) for examples). In particular, the first annotator provides faithful segmentation ("good-segmentation") with approximate GT, the second tends over-segment ("over-segmentation"), the third tends to

Table 2. Comparison of segmentation accuracy and error of CM estimation for different methods with one label per image (mean ± std). We note that 'Naive CNN' is trained on randomly selected annotations for each image.

Models	MNIST DICE (%) (testing)	MNIST CM estimation (validation)	MSLesion DICE (%) (testing)	MSLesion CM estimation (validation)
Naive CNN	32.79 ± 1.13	n/a	27.41 ± 1.45	n/a
STAPLE [1]	54.07 ± 0.68	0.2617 ± 0.0064	35.74 ± 0.84	0.2833 ± 0.0081
Spatial STAPLE [11]	56.73 ± 0.53	0.2384 ± 0.0061	38.21 ± 0.71	0.2591 ± 0.0074
Ours without Trace ($\lambda = 0$)	74.48 ± 0.37	0.1538 ± 0.0029	54.76 ± 0.66	0.1745 ± 0.0044
Our method ($\lambda = 0.001$)	75.42 ± 0.28	0.1402 ± 0.0015	55.67 ± 0.50	0.1623 ± 0.0028
Our method ($\lambda = 0.01$)	75.93 ± 0.27	0.1394 ± 0.0014	55.81 ± 0.49	0.1581 ± 0.0027
Our method ($\lambda = 0.1$)	76.48 ± 0.25	0.1329 ± 0.0012	56.43 ± 0.47	0.1542 ± 0.0023
Our method ($\lambda = 0.7$)	76.51 ± 0.22	0.1324 ± 0.0011	56.49 ± 0.45	0.1538 ± 0.0022
Our method ($\lambda = 0.9$)	76.49 ± 0.24	0.1326 ± 0.0011	56.45 ± 0.43	0.1540 ± 0.0024

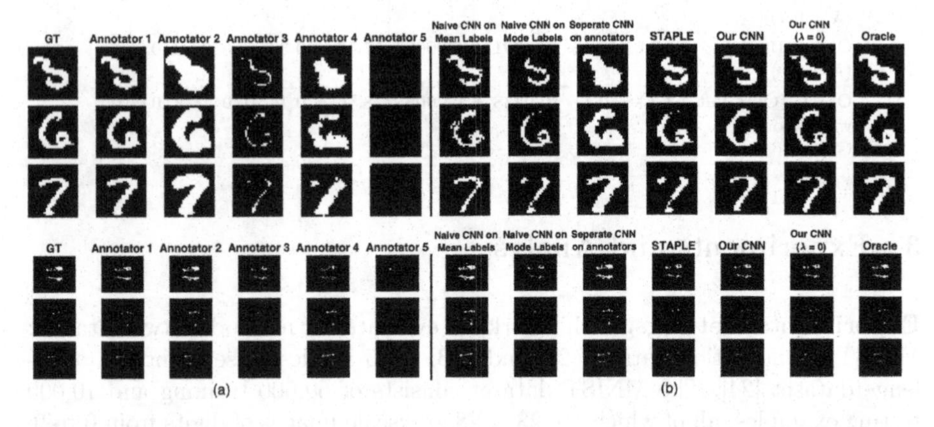

Fig. 2. Visualisation of segmentation labels on two datasets: (a) GT and simulated annotator's segmentations (Annotator 1–5); (b) the predictions from the supervised models.

under-segment ("under-segmentation"), the fourth is prone to the combination of small fractures and over-segmentation ("wrong-segmentation") and the fifth always annotates everything as the background ("blank-segmentation"). We create training data by deriving labels from the simulated annotators. Based on the simulated segmentation annotations, we compute the corresponding CMs of respective annotators for evaluation by comparing against the GT labels. We quantify the segmentation accuracy using the dice similarity coefficient (DICE)

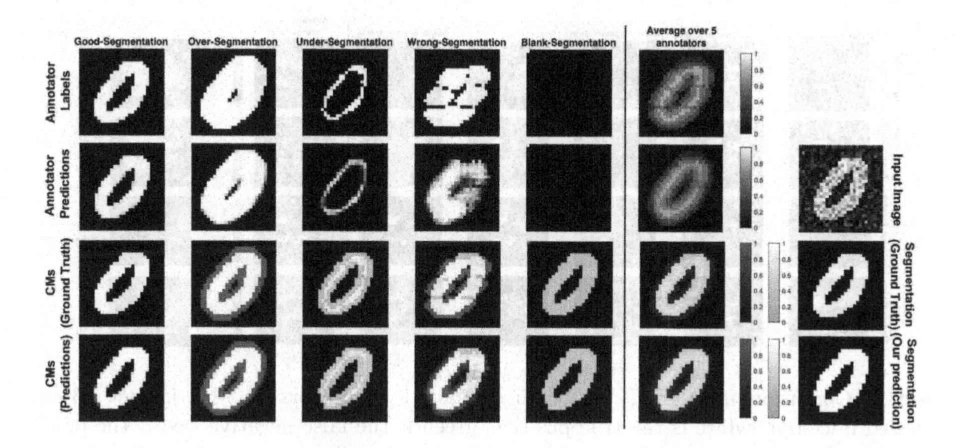

Fig. 3. Confusion matrices (CMs) of 5 simulated annotators on the MNIST dataset (Best viewed in colour: white is the true positive, green indicates the false negative, red is the false positive and black is the true negative). (Color figure online)

and the error of CM estimation by the root-mean-squared-error between each CM and its estimate over the annotators.

We examine the ability of our method to learn the CMs of annotators and the true label distribution. We compared the performance of our method against several baselines, the original STAPLE algorithm [1] and the Spatial STAPLE algorithm [11]. The first baseline is the naive CNN trained on the mean labels and the majority vote labels across the 5 annotators. The second baseline is the separate CNNs trained on 5 annotator labels and evaluate on their mean output. All the baselines and the annotator CNN, the segmentation CNN in our model are implemented with the NicMSlesions architecture described in [25]. We also evaluate on the validation set the effects of regularisation coefficient $\lambda \in \{0, 0.001, 0.01, 0.1, 0.7, 0.9\}$ of the trace-norm in Eq. 4 on the accuracy of segmentation and CM estimation. The "oracle" model is the idealistic scenario where CMs of the annotators are a priori known to the model while "annotators" indicate the average labeling accuracy of each annotator group.

Performance on MNIST Segmentation Dataset. The segmentation results of several examples are given in Fig. 2(b). Overall, models utilizing CMs are more effective than the naive CNN trained on traditional labels. Except the results from oracle model trained on GT, our proposed model achieves a higher dice similarity coefficient than STAPLE and Spatial STAPLE on both of the dense labels and single label (i.e., 1 label per image) scenarios (shown in Table 1 and Table 2). In addition, our model outperforms STAPLE and Spatial STAPLE in terms of CM estimation by a large margin, even removing the trace norm can achieve reasonably high segmentation accuracy and low CM estimation error. Figure 3 illustrates that our method can estimate CMs of the 5 very different annotators. We can see our method clearly capturing the patterns of mistakes for each annotator.

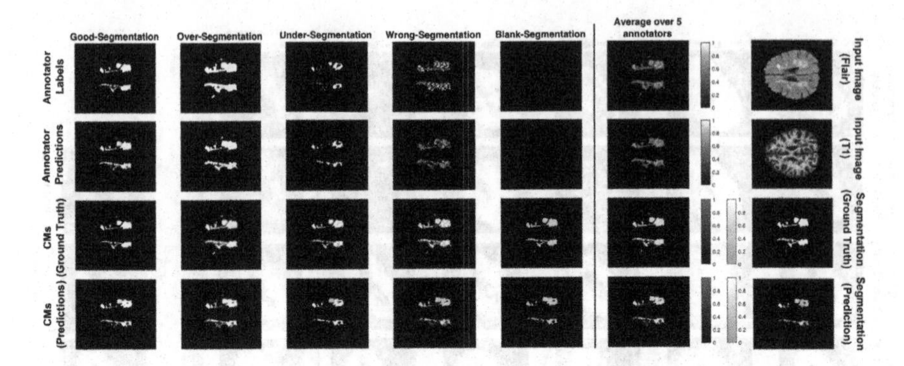

Fig. 4. Confusion matrices (CMs) of 5 simulated annotators on MS dataset (Best viewed in colour: white is the true positive, green is the false negative, red is the false positive and black is the true negative). (Color figure online)

Results on MS Dataset. Table 1 and Table 2 also show a strong correlation between the segmentation accuracy and the error of CM estimation on MS dataset. We observe that our model displays consistently better performance in terms of both segmentation accuracy and estimation of CMs with dense labels and single label. Examples of the different lesion segmentation results are shown in Fig. 2b. Our proposed algorithm shows comparable performance because of the benefits from the pixel information of the image, which provides additional lesion level information. Although the MS lesion is more diverse in shape and size than the MNIST digital data, Fig. 4 shows that our model can still recover the CMs of the 5 different annotators, and presents high segmentation consistency between the GT and our prediction.

4 Conclusion

We introduced the first method for simultaneously recovering the label noise of multiple annotators and the GT label distribution for supervised segmentation problem. We demonstrated this method on the MNIST segmentation dataset and MS lesions dataset respectively. Our method is capable of estimating individual annotators and thereby improving robustness to label noise. Experiments have shown considerable improvement over the common CNNs trained on aggregated labels based on averaging, the majority vote and the widely used STAPLE and Spatial STAPLE framework in terms of both segmentation accuracy and the quality of confusion matrix estimation.

In the future, we plan to extend to accommodate knowledge about the meta-information of annotators (e.g., number of years of experience) and also the non-image data (e.g., genetics) that may influence the pattern of the underlying segmentation label such as lesion appearance. We are also interested in assessing the downstream utility of our approach in active data collection schemes where the segmentation model $\hat{\mathbf{p}}_\theta(\mathbf{x})$ is used to select which samples to annotate ("active learning"), and the annotator models $\{\hat{\mathbf{A}}_{\phi}^{(r)}(\mathbf{x})\}_{r=1}^{R}$ are used to decide which experts to label them ("active labelling").

Acknowledge Funding Sources. EPSRC grants EP/R006032/1, EP/M020533/1, CRUK/EPSRC grant NS/A000069/1, and the NIHR UCLH Biomedical Research Centre all support this work.

References

1. Warfield, S.K., Zou, K.H., Wells, W.M.: Simultaneous truth and performance level estimation (staple): an algorithm for the validation of image segmentation. IEEE Trans. Med. Imaging **23**(7), 903–921 (2004)
2. Menze, B.H., et al.: The multimodal brain tumor image segmentation benchmark (BRATS). IEEE Trans. Med. Imaging **34**(10), 1993–2024 (2014)
3. Joskowicz, L., Cohen, D., Caplan, N., Sosna, J.: Inter-observer variability of manual contour delineation of structures in CT. Eur. Radiol. **29**(3), 1391–1399 (2019)
4. Zhang, H., et al.: Multiple sclerosis lesion segmentation with Tiramisu and 2.5D stacked slices. In: Shen, D., et al. (eds.) MICCAI 2019. LNCS, vol. 11766, pp. 338–346. Springer, Cham (2019). https://doi.org/10.1007/978-3-030-32248-9_38
5. Kats, E., Goldberger, J., Greenspan, H.: A soft STAPLE algorithm combined with anatomical knowledge. In: Shen, D., et al. (eds.) MICCAI 2019. LNCS, vol. 11766, pp. 510–517. Springer, Cham (2019). https://doi.org/10.1007/978-3-030-32248-9_57
6. Harvey, H., Glocker, B.: A standardised approach for preparing imaging data for machine learning tasks in radiology. In: Ranschaert, E.R., Morozov, S., Algra, P.R. (eds.) Artificial Intelligence in Medical Imaging, pp. 61–72. Springer, Cham (2019). https://doi.org/10.1007/978-3-319-94878-2_6
7. Winzeck, S., et al.: Isles 2016 and 2017-benchmarking ischemic stroke lesion outcome prediction based on multispectral MRI. Front. Neurol. **9**, 679 (2018)
8. Commowick, O., et al.: Objective evaluation of multiple sclerosis lesion segmentation using a data management and processing infrastructure. Sci. Rep. **8**(1), 1–17 (2018)
9. Gleason 2019 challenge. https://gleason2019.grand-challenge.org/Home/. Accessed 30 Feb 2020
10. Asman, A.J., Landman, B.A.: Robust statistical label fusion through consensus level, labeler accuracy, and truth estimation (collate). IEEE Trans. Med. Imaging **30**(10), 1779–1794 (2011)
11. Asman, A.J., Landman, B.A.: Formulating spatially varying performance in the statistical fusion framework. IEEE Trans. Med. Imaging **31**(6), 1326–1336 (2012)
12. Iglesias, J.E., Sabuncu, M.R., Van Leemput, K.: A unified framework for cross-modality multi-atlas segmentation of brain MRI. Med. Image Anal. **17**(8), 1181–1191 (2013)
13. Jorge Cardoso, M., et al.: Steps: similarity and truth estimation for propagated segmentations and its application to hippocampal segmentation and brain parcelation. Med. Image Anal. **17**(6), 671–684 (2013)
14. Asman, A.J., Landman, B.A.: Non-local statistical label fusion for multi-atlas segmentation. Med. Image Anal. **17**(2), 194–208 (2013)
15. Akhondi-Asl, A., et al.: A logarithmic opinion pool based staple algorithm for the fusion of segmentations with associated reliability weights. IEEE Trans. Med. Imaging **33**(10), 1997–2009 (2014)

16. Castro, D.C., Tan, J., Kainz, B., Konukoglu, E., Glocker, B.: Morpho-MNIST: quantitative assessment and diagnostics for representation learning. J. Mach. Learn. Res. **20** (2019)
17. Kohl, S., et al.: A probabilistic U-Net for segmentation of ambiguous images. In: Advances in Neural Information Processing Systems, pp. 6965–6975 (2018)
18. Baumgartner, C.F., et al.: PHiSeg: capturing uncertainty in medical image segmentation. In: Shen, D., et al. (eds.) MICCAI 2019. LNCS, vol. 11765, pp. 119–127. Springer, Cham (2019). https://doi.org/10.1007/978-3-030-32245-8_14
19. Raykar, V.C., et al.: Learning from crowds. J. Mach. Learn. Res. **11**, 1297–1322 (2010)
20. Khetan, A., Lipton, Z.C., Anandkumar, A.: Learning from noisy singly-labeled data. arXiv preprint arXiv:1712.04577 (2017)
21. Tanno, R., Saeedi, A., Sankaranarayanan, S., Alexander, D.C., Silberman, N.: Learning from noisy labels by regularized estimation of annotator confusion. arXiv preprint arXiv:1902.03680 (2019)
22. Sudre, C.H., et al.: Let's agree to disagree: learning highly debatable multirater labelling. In: Shen, D., et al. (eds.) MICCAI 2019. LNCS, vol. 11767, pp. 665–673. Springer, Cham (2019). https://doi.org/10.1007/978-3-030-32251-9_73
23. LeCun, Y., Bottou, L., Bengio, Y., Haffner, P., et al.: Gradient-based learning applied to document recognition. Proc. IEEE **86**(11), 2278–2324 (1998)
24. Jesson, A., Arbel, T.: Hierarchical MRF and random forest segmentation of MS lesions and healthy tissues in brain MRI. In: Proceedings of the 2015 Longitudinal Multiple Sclerosis Lesion Segmentation Challenge, pp. 1–2 (2015)
25. Valverde, S., et al.: One-shot domain adaptation in multiple sclerosis lesion segmentation using convolutional neural networks. NeuroImage Clin. p. 101638 (2018)

Deep Disentangled Hashing with Momentum Triplets for Neuroimage Search

Erkun Yang[1], Dongren Yao[1,2], Bing Cao[1], Hao Guan[1], Pew-Thian Yap[1], Dinggang Shen[1], and Mingxia Liu[1](✉)

[1] Department of Radiology and BRIC, University of North Carolina at Chapel Hill, Chapel Hill, NC 27599, USA
mxliu@med.unc.edu

[2] Brainnetome Center and National Laboratory of Pattern Recognition, Institute of Automation, University of Chinese Academy of Sciences, Beijing 100190, China

Abstract. Neuroimaging has been widely used in computer-aided clinical diagnosis and treatment, and the rapid increase of neuroimage repositories introduces great challenges for efficient neuroimage search. Existing image search methods often use triplet loss to capture high-order relationships between samples. However, we find that the traditional triplet loss is difficult to pull positive and negative sample pairs to make their Hamming distance discrepancies larger than a small fixed value. This may reduce the discriminative ability of learned hash code and degrade the performance of image search. To address this issue, in this work, we propose a *deep disentangled momentum hashing* (DDMH) framework for neuroimage search. Specifically, we first investigate the original triplet loss and find that this loss function can be determined by the inner product of hash code pairs. Accordingly, we disentangle hash code norms and hash code directions and analyze the role of each part. By decoupling the loss function from the hash code norm, we propose a unique *disentangled triplet loss*, which can effectively push positive and negative sample pairs by desired Hamming distance discrepancies for hash codes with different lengths. We further develop a *momentum triplet strategy* to address the problem of insufficient triplet samples caused by small batch-size for 3D neuroimages. With the proposed disentangled triplet loss and the momentum triplet strategy, we design an end-to-end trainable deep hashing framework for neuroimage search. Comprehensive empirical evidence on three neuroimage datasets shows that DDMH has better performance in neuroimage search compared to several state-of-the-art methods.

1 Introduction

Neuroimage analysis plays a vital role in modern clinical analysis [1], image-guided surgery [2], and automated disease diagnosis [3]. Nowadays, tremendous amounts of neuroimaging data have been captured and recorded in digital formats. However, interpreting neuroimaging data usually requires extensive practical experience and professional knowledge, and may be affected by inter-observer

A. L. Martel et al. (Eds.): MICCAI 2020, LNCS 12261, pp. 191–201, 2020.
https://doi.org/10.1007/978-3-030-59710-8_19

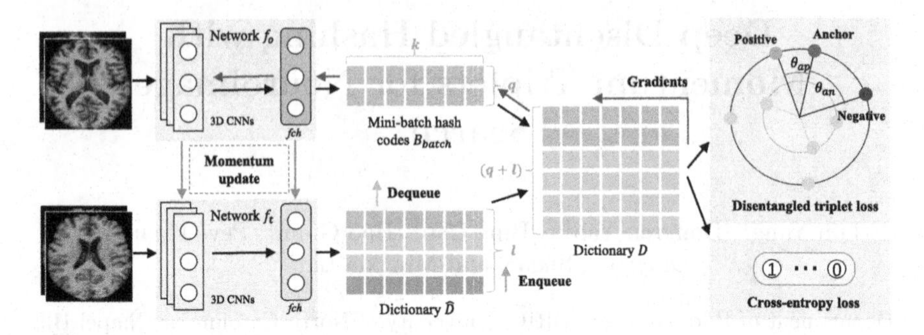

Fig. 1. Proposed deep disentangled momentum hashing (DDMH) framework for neuroimage search, including three key components: (1) a disentangled triplet loss, which improves the traditional triplet loss by decoupling it from the hash code length; (2) a cross-entropy loss, which is used to optimize the hash codes by jointly learning a linear classifier; (3) a momentum triplet strategy, which builds a large and consistent dictionary on-the-fly and enables efficient triplet-based learning even with a small batch-size.

variance [4,5]. To facilitate clinical decision-making, it is important to provide physicians with previous similar cases and corresponding treatment records to form case-based reasoning and evidence-based medicine [6]. In practice, such neuroimage search systems are required to be both scalable and accurate. Due to a sensible balance between search quality and computational cost, hashing-based search methods [7,8], which encode neuroimages as binary embeddings in Hamming space, have recently attracted increasing attention in the field.

According to whether supervisory signals are involved in the learning phase, existing hashing methods can be roughly divided into two groups: 1) *unsupervised hashing* [9–11] and 2) *supervised hashing* [12–18]. Unsupervised methods usually learn hash functions from original feature space to Hamming space by exploiting topological information and data distributions. In contrast, supervised hashing incorporates semantic labels to improve the quality of hash function learning. In this paper, we focus on supervised hashing, as it can take advantage of the supervisory signals and usually outperforms unsupervised methods.

Existing supervised hashing methods usually rely on ranking-based loss functions [15–17,19–21] to preserve similarity between samples, such as contrastive loss and triplet loss. Recent studies [22,23] have shown that the triplet loss encourages samples from the same category to reside on a manifold and generally outperforms the contrastive loss. Even though many triplet-based hashing methods have been proposed [17,19], we find a crucial misspecification problem in the traditional triplet loss, that is, traditional triplet loss is difficult to pull positive and negative pairs to make their Hamming distance discrepancies larger than a small fixed value. This may reduce the discriminative ability of the learned hash code, thereby degrading the performance of triplet loss-based methods.

To this end, we propose a supervised hash model, called *deep disentangled momentum hashing* (DDMH), for neuroimage search. As illustrated in Fig. 1, the proposed DDMH consists of three key components: (1) a disentangled triplet loss, which decouples the traditional triplet loss with hash code norms and can effectively pull positive and negative pairs by a desired Hamming distance discrepancies for hash codes with different lengths; (2) a cross-entropy loss, which is used to optimize the hash codes with a jointly learned linear classifier; (3) a momentum triplet strategy to enable efficient triplet-based learning even with a small batch-size. These three components are seamlessly incorporated into an end-to-end trainable deep hashing framework for neuroimage search. Extensive experiments on three neuroimage datasets demonstrate that the proposed DDMH outperforms several state-of-the-art methods in neuroimage search.

2 Method

2.1 Problem Formulation

Boldface uppercase letters (e.g., $\boldsymbol{A}$) denote matrices, and $\boldsymbol{A}^T$ represents the transpose of a matrix $\boldsymbol{A}$. Also, sign$(\cdot)$ represents the element-wise signum function, defined as

$$sign(x) = \begin{cases} 1, \text{ if } x \geq 0, \\ -1, \text{ if } x < 0. \end{cases} \quad (1)$$

Neuroimage hashing aims to learn hash functions that can map neuroimages into hash codes in Hamming space. Here, hash codes represent binary vectors, and the Hamming space contains a set of binary vectors. Assume that we have N neuroimages in a given dataset $\boldsymbol{X} = \{\boldsymbol{x}_i\}_{i=1}^N$, with each image $\boldsymbol{x}_i$ associated with a label vector $\boldsymbol{l}_i$. We denote $\boldsymbol{B} = \{\boldsymbol{b}_i\}_{i=1}^N$ as the hash codes for $\boldsymbol{X}$, where $\boldsymbol{b}_i \in \{-1, +1\}^k$ represents the binary hash code for the sample $\boldsymbol{x}_i$, k is the hash code length, and b_{ix} indicates the x-th element of $\boldsymbol{b}_i$.

2.2 Deep Disentangled Hash Code Learning

For effective neuroimage search, we aim to learn discriminative binary codes to well preserve the original semantic information, that is, semantically similar neuroimages are expected to have similar hash codes and vice versa. To achieve this goal, we employ a triplet-based ranking loss, which is defined as

$$\mathcal{L}_o(\boldsymbol{b}_a, \boldsymbol{b}_p, \boldsymbol{b}_n) = \log(1 + e^{d_h(\boldsymbol{b}_a, \boldsymbol{b}_p) - d_h(\boldsymbol{b}_a, \boldsymbol{b}_n)}) = \log(1 + e^{-\alpha}), \quad (2)$$

where $\boldsymbol{b}_a$ is the learned hash code for an anchor neuroimage $\boldsymbol{x}_a$, $\boldsymbol{b}_p$ is the binary code for a positive neuroimage $\boldsymbol{x}_p$ that shares the same semantic label with $\boldsymbol{x}_a$, and $\boldsymbol{b}_n$ is the binary code for a negative neuroimage $\boldsymbol{x}_n$ that has a different label from $\boldsymbol{x}_a$. Also, $(\boldsymbol{b}_a, \boldsymbol{b}_p, \boldsymbol{b}_n)$ is a randomly selected triplet sample. $(\boldsymbol{b}_a, \boldsymbol{b}_p)$ represents a positive pair and $(\boldsymbol{b}_a, \boldsymbol{b}_n)$ denotes a negative pair. $d_h(\cdot, \cdot)$ is the Hamming distance function for a pair of neuroimages. $\alpha = d_h(\boldsymbol{b}_a, \boldsymbol{b}_n) - d_h(\boldsymbol{b}_a, \boldsymbol{b}_p)$

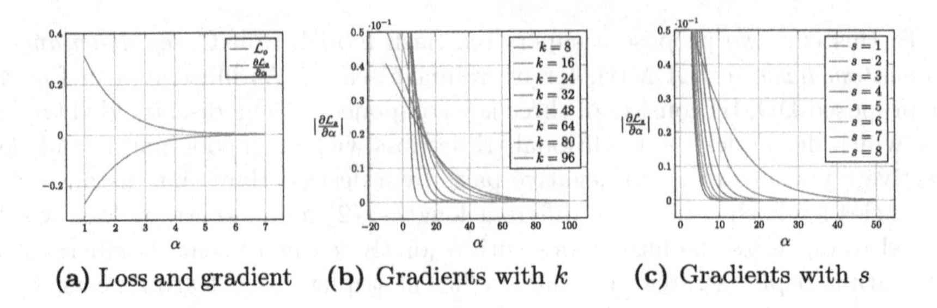

(a) Loss and gradient (b) Gradients with k (c) Gradients with s

Fig. 2. The loss and gradients of conventional triplet loss $\mathcal{L}_o$ w.r.t. α is close to 0 when α is larger than 6, which is ill-specified for hash code learning. By replacing α with $s\beta$, the gradients of our disentangled triplet loss $\mathcal{L}_s$ are more meaningful for different k. We set $s = 3$ in (b) and $k = 16$ in (c).

represents the discrepancy between the corresponding Hamming distances. If there is no ambiguity, we will use $\mathcal{L}_o$ to represent $\mathcal{L}_o(\boldsymbol{b}_a, \boldsymbol{b}_p, \boldsymbol{b}_n)$ in the following.

Since the Hamming distance function $d_h(\boldsymbol{b}_i, \boldsymbol{b}_j) = |b_{ix} \neq b_{jx}, 1 \leq x \leq k|$ requires discrete operations and is non-differentiable, directly optimizing Eq. (2) with back propagation is not feasible. As indicated in [12], there exists a nice relationship between the inner product and the Hamming distance as

$$d_h(\boldsymbol{b}_i, \boldsymbol{b}_j) = \frac{k - \langle \boldsymbol{b}_i, \boldsymbol{b}_j \rangle}{2}. \tag{3}$$

Thus, we can obtain the following

$$\alpha = \frac{k - \langle \boldsymbol{b}_a, \boldsymbol{b}_n \rangle}{2} - \frac{k - \langle \boldsymbol{b}_a, \boldsymbol{b}_p \rangle}{2} = \frac{\langle \boldsymbol{b}_a, \boldsymbol{b}_p \rangle - \langle \boldsymbol{b}_a, \boldsymbol{b}_n \rangle}{2}. \tag{4}$$

Substituting Eq. (4) into Eq. (2), the loss function $\mathcal{L}_o$ can be readily optimized.

Although the loss in Eq. (2) can help preserve the original semantic information, we observe a key misspecification problem when directly applying Eq. (2) to hash code learning. As illustrated in Fig. 2 (a), when the Hamming distance discrepancy α is large than 6, the loss function value of $\mathcal{L}_o$ and its gradients (w.r.t. α) are close to 0. This implies that, no matter how large the hash code length k is, it is difficult to pull the Hamming distance between positive and negative pairs to be greater than 6 by optimizing the objective function $\mathcal{L}_o$. For effective hash code learning, we generally expect that longer hash codes can be more discriminative and the Hamming distance discrepancy α can be increased with the increase of the hash code length. Unfortunately, the triplet loss function in Eq. (2) is misspecified for this goal.

To tackle the above misspecification problem, we investigate the relationship between the loss function $\mathcal{L}_o$ and the hash code length k carefully. From Eq. (2) and Eq. (4), we can see that the loss function $\mathcal{L}_o$ can be determined by the inner product of hash code pairs. Considering that both the norm and the direction of a vector can affect the inner product value, we disentangle these two items as

$$\alpha = \frac{k}{2}(\cos\theta_{ap} - \cos\theta_{an}) = o\beta, \tag{5}$$

where θ_{ap} is the angle between $\boldsymbol{b}_a$ and $\boldsymbol{b}_p$, θ_{an} is the angle between $\boldsymbol{b}_a$ and $\boldsymbol{b}_n$, $o = \frac{k}{2}$, and $\beta = \cos\theta_{ap} - \cos\theta_{an}$. It should be noted that this work performs the disentanglement to deal with the misspecification problem in Eq. (2), which is different from previous studies [24,25] in face verification. Actually, for a given fixed hash code length k, the cosine similarity discrepancy β is proportional to the Hamming distance discrepancy α, while for different hash code length k, α up-weights β with o. By combining Eq. (5) and Eq. (2), one can observe that when the hash code length k is relatively large, the weight o may greatly hinder the optimization. To eliminate this influence, we replace the weight o with a new scale factor s and design a unique *disentangled triplet loss* as follows

$$\mathcal{L}_s(\boldsymbol{b}_a, \boldsymbol{b}_p, \boldsymbol{b}_n) = \log(1 + e^{s(\cos\theta_{an} - \cos\theta_{ap})}) = \log(1 + e^{-\frac{2s}{k}\alpha}). \tag{6}$$

The gradients of $\mathcal{L}_s(\boldsymbol{b}_a, \boldsymbol{b}_p, \boldsymbol{b}_n)$ w.r.t. the Hamming distance discrepancy α with different k and s are illustrated in Fig. 2 (b) and (c), respectively. From these figures we can see that the saturation region can be tuned by selecting different s. That is, by using a proper s, we can decouple the hash code length from the loss function and enable desired Hamming distance discrepancies for hash codes with different lengths, which helps generate discriminative hash codes.

2.3 Overall Objective Function

Following previous studies [26,27], we assume that optimal hash codes can be jointly learned with a classification task, and use a linear classifier $f(\cdot)$ to model the relationship between learned binary codes and semantic labels, defined as

$$f(\boldsymbol{b}_i) = \sigma(\boldsymbol{W}^\top \boldsymbol{b}_i), \tag{7}$$

where σ is the sigmoid function and $\boldsymbol{W}$ is the parameter. A cross-entropy loss is then adopted as

$$\mathcal{L}_c(\boldsymbol{b}_i) = -\boldsymbol{l}_i^\top \ln f(\boldsymbol{b}_i). \tag{8}$$

By combining Eq. (8) and Eq. (6), we can obtain the overall loss function as

$$\mathcal{L} = \sum\nolimits_{(\boldsymbol{b}_a, \boldsymbol{b}_p, \boldsymbol{b}_n) \in \mathcal{T}} \mathcal{L}_s(\boldsymbol{b}_a, \boldsymbol{b}_p, \boldsymbol{b}_n) + \lambda \sum_{i=1}^{N} \mathcal{L}_c(\boldsymbol{b}_i), \tag{9}$$

where $\mathcal{T}$ is the set for all triplets, and λ is a hyper-parameter to balance the contributions of the disentangled triplet loss and the cross-entropy loss in Eq. (9).

2.4 Optimization with Momentum Triplets

A critical challenge for optimizing our triplet-based 3D deep model is the insufficient number of triplets. Actually, to construct a sufficient amount of triplets, triplet-based deep models often require a relatively large batch size to include more samples in a mini-batch. However, the memory requirement and computation time will increase as the batch size increases, which is particularly significant for 3D convolution neural networks (CNNs). Therefore, it is desirable to design a feasible solution to optimize our model in Eq. (9) using 3D neuroimages.

Motivated by [28], we introduce a momentum strategy to supervised hashing and propose *momentum triplets* to address the above insufficient triplet problem. As shown in Fig. 1, we design two 3D CNNs, i.e., f_o and f_t, which share the same architecture. Denote q as the mini-batch size. For each mini-batch entity $\boldsymbol{X}_{batch} = \{\boldsymbol{x}_1, \boldsymbol{x}_2, \cdots, \boldsymbol{x}_q\}$, we calculate their corresponding hash codes $\boldsymbol{B}_{batch} = \{\boldsymbol{b}_1, \boldsymbol{b}_2, \cdots, \boldsymbol{b}_q\}$ by forward propagating them from f_o. We also maintain a l-length dynamic dictionary $\hat{D} = \{\hat{\boldsymbol{b}}_1, \hat{\boldsymbol{b}}_2, \cdots, \hat{\boldsymbol{b}}_l\}$, with each element obtained from the network f_t. In each iteration, the dictionary keys of $\hat{D}$ are updated as a queue, with the current mini-batch enqueued and the oldest mini-batch dequeued. To facilitate the construction of more triplets, for $\boldsymbol{X}_{batch} = \{\boldsymbol{x}_1, \boldsymbol{x}_2, \cdots, \boldsymbol{x}_q\}$, we concatenate its corresponding codes $\boldsymbol{B}_{batch}$ with the dynamic dictionary $\hat{D}$, yielding a new dictionary $D = \{\boldsymbol{b}_1, \boldsymbol{b}_2, \cdots, \boldsymbol{b}_q, \hat{\boldsymbol{b}}_1, \hat{\boldsymbol{b}}_2, \cdots, \hat{\boldsymbol{b}}_l\}$ with the length of $q + l$. The network f_o are then updated with the loss obtained with all triplets from the dictionary D by standard back-propagation. And the network f_t is updated with a momentum strategy via

$$\theta_t = m\theta_t + (1 - m)\theta_o, \tag{10}$$

where θ_o and θ_t are parameters of f_o and f_t, respectively, and $m \in (0, 1]$ is a momentum coefficient. The momentum update strategy for f_t is used to reduce the discrepancy between f_o and f_t and improve the hash code consistency.

2.5 Implementation

The 3D networks for f_o and f_t share the same architecture, which has 12 layers. The first 10 layers are 3D convolutional (Conv) layers (kernel size: $3 \times 3 \times 3$; stride: 1; zero padding). The last 2 layers are fully-connected (FC) layers with 256 and k hidden units, respectively. Batch normalization and ReLU activation are applied to all Conv layers. A 3D max-pooling with stride 2 is also applied to every two Conv layers. For the last FC layer, we use $\text{Tanh}(\cdot)$ to squeeze activation within $[-1, +1]$ to reduce quantization error and also denote it as fully-connected hash (*fch*) layer. We train the network from scratch with Adam [29] by setting the batch size as 2 and the dictionary size l as 10. The learning rate is set to 10^{-3}, λ in Eq. (9) is set to 1, the scale factor s is set to 3, and m is set to 0.999. Assuming the computation cost for one similarity calculation is d, the computation cost to rank all points will be $O(dN)$ and can be further reduced to $O(1)$ with predefined hash look-up tables.

3 Experiments

3.1 Experimental Setup

Materials. We evaluate the proposed DDMH method on three popular benchmark datasets: (1) Alzheimer's Disease Neuroimaging Initiative (ADNI1) [30],

Table 1. The MAP results of eight different methods on three datasets.

Method	ADNI1				ADNI2				AIBL			
	8	16	24	32	8	16	24	32	8	16	24	32
SH	0.403	0.389	0.386	0.384	0.422	0.377	0.375	0.372	0.613	0.611	0.605	0.604
SpH	0.398	0.398	0.400	0.391	0.375	0.384	0.385	0.375	0.599	0.587	0.584	0.578
ITQ	0.391	0.396	0.396	0.393	0.398	0.389	0.384	0.390	0.596	0.582	0.579	0.577
DSH	0.390	0.402	0.398	0.396	0.408	0.387	0.393	0.394	0.596	0.577	0.575	0.576
KSH	0.372	0.384	0.386	0.391	0.385	0.369	0.376	0.374	0.600	0.591	0.588	0.583
DPSH	0.438	0.460	0.447	0.467	0.403	0.413	0.415	0.391	0.702	0.701	0.700	0.684
SSDH	0.421	0.427	0.435	0.422	0.397	0.408	0.406	0.411	0.712	0.710	0.715	0.698
Ours	**0.478**	**0.480**	**0.498**	**0.511**	**0.602**	**0.612**	**0.618**	**0.632**	**0.758**	**0.763**	**0.761**	**0.764**

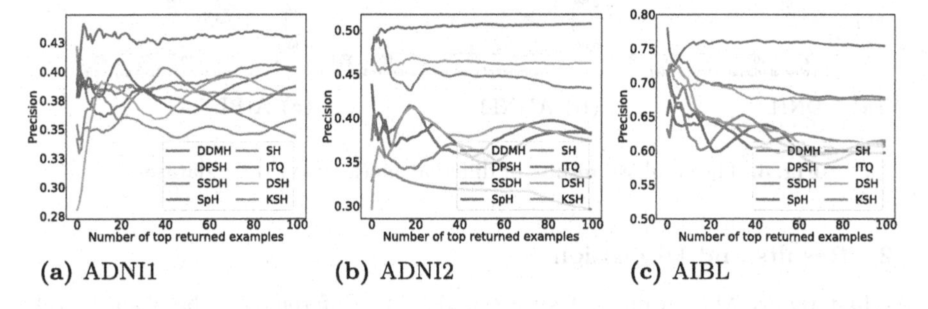

Fig. 3. The topN-precision results of different methods on three datasets.

(2) ANDI2, and (3) Australian Imaging, Biomarkers and Lifestyle dataset (AIBL) [31]. ADNI1 contains 821 subjects with 1.5T T1-weighted structural MRI scans. ADNI2 contains 636 subjects with 3T T1-weighted MRI scans. Besides, AIBL contains 614 subjects with 3T T1-weighted MRI data. Each subject from these three datasets was annotated by a category-level label, *i.e.*, Alzheimer's disease (AD), normal control (NC), or mild cognitive impairment (MCI). For each dataset, we randomly select 10% samples as a query set and others as a retrieval set. All MRIs were pre-processed via skull-stripping, intensity correction, and spatial normalization to Automated Anatomical Labeling (AAL) template.

Evaluation Setting. Our **DDMH** is compared with five state-of-the-art shallow hashing methods, i.e., **SH** [32], **SpH** [33], **ITQ** [34], **DSH** [35], and **KSH** [12], as well as two recent deep hashing methods, i.e., **DPSH** [36] and **SSDH** [37]. Volumes of gray matter tissue inside 90 regions-of-interest (defined in AAL) are used as features for shallow hashing methods, while two deep models and our DDMH employ MRI scans as input. mean of average precision (MAP), topN-precision, and recall@K are used as evaluation metrics.

Table 2. The MAP values of DDMH and its two variants on ADNI1.

Method	$k = 8$	$k = 16$	$k = 24$	$k = 32$
DDMH-C	0.441	0.453	0.456	0.457
DDMH-D	0.462	0.460	0.459	0.466
DDMH	**0.478**	**0.480**	**0.498**	**0.511**

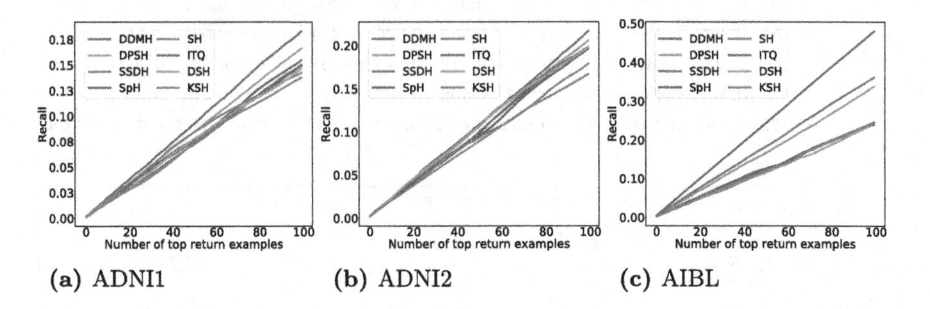

Fig. 4. The recall@k results of different methods on three datasets.

3.2 Results and Discussion

We first report MAP values of all methods with different lengths of hash code (*i.e.*, k) to provide a global evaluation. Then we report the topN-precision and recall@K curves with $k = 16$ for comprehensive contrastive study.

MAP. The MAP results achieved by all methods on the three datasets are reported in Table 1. The results suggest that, compared with shallow hashing methods, deep learning approaches usually yield better performance. This could be attributed to the fact that deep networks enable joint learning of image representations and hash codes directly from the original image. Besides, our DDMH usually outperforms other competing methods by large margins.

TopN-Precision. The topN-precision curves are shown in Fig. 3. This figure suggests that DDMH generally achieves higher precision, which is consistent with the MAP evaluation. It is worth noting that, in the task of neuroimage retrieval, users usually focus on top returned instances. Therefore, it is essential to provide users with top returned instances that are highly relevant to the query. From the results, one can see that DDMH outperforms other methods by a large margin when the number of returned instances is small. This verifies that DDMH can effectively return relevant examples and is suitable for neuroimage search.

Recall@K. The recall@K curves of all methods are reported in Fig. 4. From Fig. 4, one can observe that our DDMH always yields higher recall values for different numbers of returned examples on three datasets, which further proves that DDMH can effectively learn hash codes with high quality.

Ablation Study. We investigate two variants of our DDMH, including (1) **DDMH-C** without using the cross-entropy loss, *i.e.*, $\lambda = 0$ in Eq. (9); and (2)

DDMH-D with the traditional triplet loss *i.e.*, $s = \frac{k}{2}$ in Eq. (6). MAP results of DDMH and its two variants on ADNI1 are reported in Table 2. From the results, we can see that DDMH is consistently superior to its two variants, implying that both the classification loss and the disentangled triplet loss contribute to the performance improvement. In particular, DDMH achieves higher MAP values using longer hash codes, but DDMH-D benefits little from longer hash codes. These results show that compared with the traditional triplet loss (used in DDMH-D), the proposed disentangled loss (used in DDMH) can effectively promote the learning of longer hash codes. On the other hand, from Tables 1 and 2, one can observe that DDMH-D generally outperforms existing state-of-the-art hashing methods on ADNI1. This implies that, with the proposed momentum triplet strategy, DDMH-D enables efficient hashing learning even with the traditional triplet loss function.

4 Conclusion

This paper presents a deep disentangled momentum hashing (DDMH) method for neuroimage search. Specifically, by disentangling the hash code norm and directions, we propose a unique disentangled triplet loss, which can pull positive and negative pairs to have a desired Hamming distance discrepancy with different hash code lengths. We also design a momentum triplet strategy to provide a sufficient number of triplets for model training, even with a small batch-size. Using a 3D CNN as hash functions, DDMH can generate discriminative hash codes by jointly optimizing the disentangled triplet loss and a cross-entropy loss. Extensive experiments on three neuroimage datasets show that our DDMH achieves better performance than several state-of-the-art methods.

Acknowledgments. This work was partly supported by NIH grants (Nos. AG041721, AG053867).

References

1. Graham, R.N., Perriss, R., Scarsbrook, A.F.: DICOM demystified: a review of digital file formats and their use in radiological practice. Clin. Radiol. **60**(11), 1133–1140 (2005)
2. Grimson, W.E.L., Kikinis, R., Jolesz, F.A., Black, P.: Image-guided surgery. Sci. Am. **280**(6), 54–61 (1999)
3. Owais, M., Arsalan, M., Choi, J., Park, K.R.: Effective diagnosis and treatment through content-based medical image retrieval (CBMIR) by using artificial intelligence. J. Clin. Med. **8**(4), 462 (2019)
4. Cheng, B., Liu, M., Shen, D., Li, Z., Zhang, D.: Multi-domain transfer learning for early diagnosis of Alzheimer's disease. Neuroinformatics **15**(2), 115–132 (2017)
5. Liu, M., Zhang, J., Adeli, E., Shen, D.: Landmark-based deep multi-instance learning for brain disease diagnosis. Med. Image Anal. **43**, 157–168 (2018)
6. Holt, A., Bichindaritz, I., Schmidt, R., Perner, P.: Medical applications in case-based reasoning. Knowl. Eng. Rev. **20**(3), 289–292 (2005)

7. Kulis, B., Grauman, K.: Kernelized locality-sensitive hashing for scalable image search. In: ICCV, pp. 2130–2137 (2009)
8. Yang, E., Deng, C., Liu, W., Liu, X., Tao, D., Gao, X.: Pairwise relationship guided deep hashing for cross-modal retrieval. In: AAAI, pp. 1618–1625 (2017)
9. Kulis, B., Grauman, K.: Kernelized locality-sensitive hashing. IEEE Trans. Pattern Anal. Mach. Intell. **34**(6), 1092–1104 (2012)
10. Dai, B., Guo, R., Kumar, S., He, N., Song, L.: Stochastic generative hashing. arXiv preprint arXiv:1701.02815 (2017)
11. Yang, E., Deng, C., Liu, T., Liu, W., Tao, D.: Semantic structure-based unsupervised deep hashing. IJCA **I**, 1064–1070 (2018)
12. Liu, W., Wang, J., Ji, R., Jiang, Y., Chang, S.F.: Supervised hashing with kernels. In: CVPR, pp. 2074–2081 (2012)
13. Gui, J., Liu, T., Sun, Z., Tao, D., Tan, T.: Fast supervised discrete hashing. IEEE Trans. Pattern Anal. Mach. Intell. **40**(2), 490–496 (2017)
14. Yang, E., Deng, C., Li, C., Liu, W., Li, J., Tao, D.: Shared predictive cross-modal deep quantization. IEEE Trans. Neural Netw. Learn. Syst. **29**(11), 5292–5303 (2018)
15. Cao, Y., Long, M., Liu, B., Wang, J., KLiss, M.: Deep Cauchy hashing for hamming space retrieval. In: CVPR, pp. 1229–1237 (2018)
16. Cao, Y., Liu, B., Long, M., Wang, J., KLiss, M.: HashGAN: deep learning to hash with pair conditional Wasserstein GAN. In: CVPR, pp. 1287–1296 (2018)
17. Zhang, R., Lin, L., Zhang, R., Zuo, W., Zhang, L.: Bit-scalable deep hashing with regularized similarity learning for image retrieval and person re-identification. IEEE Trans. Image Process. **24**(12), 4766–4779 (2015)
18. Deng, C., Yang, E., Liu, T., Li, J., Liu, W., Tao, D.: Unsupervised semantic-preserving adversarial hashing for image search. IEEE Trans. Image Process. **28**(8), 4032–4044 (2019)
19. Deng, C., Chen, Z., Liu, X., Gao, X., Tao, D.: Triplet-based deep hashing network for cross-modal retrieval. IEEE Trans. Image Process. **27**(8), 3893–3903 (2018)
20. Chen, L., Honeine, P., Qu, H., Zhao, J., Sun, X.: Correntropy-based robust multi-layer extreme learning machines. Pattern Recogn. **84**, 357–370 (2018)
21. Chen, L., Qu, H., Zhao, J., Chen, B., Principe, J.C.: Efficient and robust deep learning with correntropy-induced loss function. Neural Comput. Appl. **27**(4), 1019–1031 (2016)
22. Schroff, F., Kalenichenko, D., Philbin, J.: FaceNet: a unified embedding for face recognition and clustering. In: CVPR, pp. 815–823 (2015)
23. Hermans, A., Beyer, L., Leibe, B.: In defense of the triplet loss for person re-identification. arXiv preprint arXiv:1703.07737 (2017)
24. Deng, J., Guo, J., Xue, N., Zafeiriou, S.: ArcFace: additive angular margin loss for deep face recognition. In: CVPR, pp. 4690–4699 (2019)
25. Wang, H., et al.: CosFace: large margin cosine loss for deep face recognition. In: CVPR, pp. 5265–5274 (2018)
26. Chen, Z., Cai, R., Lu, J., Feng, J., Zhou, J.: Order-sensitive deep hashing for multimorbidity medical image retrieval. In: Frangi, A., Schnabel, J., Davatzikos, C., Alberola-Lopez, C., Fichtinger, G. (eds.) MICCAI 2018. LNCS, vol. 11070. Springer, Cham (2018). https://doi.org/10.1007/978-3-030-00928-1_70
27. Li, Q., Sun, Z., He, R., Tan, T.: Deep supervised discrete hashing. In: NeurIPS,pp. 2482–2491 (2017)
28. He, K., Fan, H., Wu, Y., Xie, S., Girshick, R.: Momentum contrast for unsupervised visual representation learning. arXiv preprint arXiv:1911.05722 (2019)

29. Kingma, D.P., Ba, J.: Adam: A method for stochastic optimization. CoRR (2014)
30. Jack Jr, C.R., et al.: The Alzheimer's disease neuroimaging initiative (ADNI): MRI methods. J. Magn. Reson. Imaging Offic. J. Int. Soc. Magn. Reson. Med. **27**(4), 685–691 (2008)
31. Ellis, K.A., et al.: The Australian Imaging, Biomarkers and Lifestyle (AIBL) study of aging: methodology and baseline characteristics of 1112 individuals recruited for a longitudinal study of Alzheimer's disease. Int. Psychogeriatr. **21**(4), 672–687 (2009)
32. Weiss, Y., Torralba, A., Fergus, R.: Spectral hashing. In: NeurIPS, pp. 1753–1760 (2009)
33. Heo, J.P., Lee, Y., He, J., Chang, S.F., Yoon, S.E.: Spherical hashing. In: CVPR, pp. 2957–2964 (2012)
34. Gong, Y., Lazebnik, S., Gordo, A., Perronnin, F.: Iterative quantization: a procrustean approach to learning binary codes for large-scale image retrieval. IEEE Trans. Pattern Anal. Mach. Intell. **35**(12), 2916–2929 (2013)
35. Jin, Z., Li, C., Lin, Y., Cai, D.: Density sensitive hashing. IEEE Trans. Cybern. **44**(8), 1362–1371 (2014)
36. Li, W.J., Wang, S., Kang, W.C.: Feature learning based deep supervised hashing with pairwise labels. IJCA **I**, 1711–1717 (2016)
37. Yang, H.F., Lin, K., Chen, C.S.: Supervised learning of semantics-preserving hash via deep convolutional neural networks. IEEE Trans. Pattern Anal. Mach. Intell. **40**(2), 437–451 (2017)

Learning Joint Shape and Appearance Representations with Metamorphic Auto-Encoders

Alexandre Bône(✉), Paul Vernhet, Olivier Colliot, and Stanley Durrleman

ARAMIS Lab, ICM, Inserm U1127, CNRS UMR 7225, Sorbonne University, Inria, Paris, France
{alexandre.bone,paul.vernhet,olivier.colliot, stanley.durrleman}@icm.institute.org

Abstract. Transformation-based methods for shape analysis offer a consistent framework to model the geometrical content of images. Most often relying on diffeomorphic transforms, they lack however the ability to properly handle texture and differing topological content. Conversely, modern deep learning methods offer a very efficient way to analyze image textures. Building on the theory of metamorphoses, which models images as combined intensity-domain and spatial-domain transforms of a prototype, we introduce the "metamorphic" auto-encoding architecture. This class of neural networks is interpreted as a Bayesian generative and hierarchical model, allowing the joint estimation of the network parameters, a representative prototype of the training images, as well as the relative importance between the geometrical and texture contents.

We give arguments for the practical relevance of the learned prototype and Euclidean latent-space metric, achieved thanks to an explicit normalization layer. Finally, the ability of the proposed architecture to learn joint and relevant shape and appearance representations from image collections is illustrated on BraTs 2018 datasets, showing in particular an encouraging step towards personalized numerical simulation of tumors with data-driven models.

Keywords: Numerical brain atlas · Shape analysis · Metamorphosis

1 Introduction

The shape analysis literature offers a number of tools to perform statistical analysis tasks on geometrical objects. At the core of the founding works [12,14] lies the idea to quantify the differences between two shapes thanks to large

A. Bône and P. Vernhet—Equal contributions.

Electronic supplementary material The online version of this chapter (https://doi.org/10.1007/978-3-030-59710-8_20) contains supplementary material, which is available to authorized users.

A. L. Martel et al. (Eds.): MICCAI 2020, LNCS 12261, pp. 202–211, 2020.
https://doi.org/10.1007/978-3-030-59710-8_20

parametric classes of deformations that warp one into the other: once the optimal transformation found, its norm provides a proxy distance metric. Diffeomorphic transformations are for instance widely used for medical image analysis [31], with applications to image registration or atlas building [1,25,32]. However, these transformations are purely geometrical, and cannot account for potential texture (or "appearance") variability in the considered images. In particular, images with differing topological contents cannot be diffeomorphically warped one into the other. This limitation gave birth to the early theoretical work [29] where the proposed "metamorphoses" jointly transform the geometry and the intensity of an image. If some authors built on this idea for brain tumor monitoring [23], sub-cortical brain segmentation [24] or learning generalized principal component analysis models [3], the literature is fairly limited.

Conversely, if modern deep learning architectures are commonly believed to aggregate information up until deep filters able recognize entire shapes [19,21], some recent works [6,8,13] question this so-called shape hypothesis, suggesting that texture (or "appearance") features carry more weight. Beyond the local spatial invariances achieved by max-pooling layers (with receptive fields typically of the order of a few pixels), deep learning methods are agnostic of the data nature: in other words, potential powerful prior knowledge is not taken into account. This agnostic approach presents two disadvantages: the interpretability of the network is often limited, and huge amounts of data might be needed for the network to re-discover already-known data properties.

Most data augmentation techniques are attempts to alleviate this second point, by artificially increasing the data set size according to priors such as invariance to small intensity-domain or affine spatial-domain transformations [9,20,27]. At the cost of a longer training time, the network learns the implicitly-encoded invariance properties. Instead of implicitly suggesting invariances or manipulating data set biases with data augmentation, learning from small data sets and enhanced interpretability can be achieved by designing adapted architectures that explicitly enforce priors [15,28]. Some recent attempts to fill the gap between classical model-based shape analysis tools and data-driven deep learning methods managed to combine the learning flexibility of the former with the theoretical guarantees and interpretability of the latter. Building on the variational auto-encoding architecture introduced in [16], the approaches described in [11,18] learn probabilistic diffeomorphic registration models, which cannot handle varying topology in data, and do not allow the construction of an atlas, i.e. a reference image estimated by groupwise registrations among a training image data set. [7] learns atlas models but is bound to the same topological limitations as the previous ones. In [26], the authors go beyond pure spatial-warping layers and further introduce joint spatial and intensity transformations in auto-encoding networks. This first attempt, still only of its kind to the best of our knowledge, relies on an appearance prediction decoder followed by an ad-hoc warping module specifically developed. Once learned, the latent representations can be manipulated to complete disentangled shape or appearance reconstruction and interpolation. This work is the closest to metamorphosis one, however the regularity of the deformation field is controlled by a loss penalty and thus not ensured by design of the network, furthermore no links are made with the rich theoretical background of metamorphosis.

In this paper, we introduce the metamorphic auto-encoders (MAEs) which relies on the prior that training images can be seen as "metamorphic" transformations of a prototype. We show that thanks to this assumption, estimating MAE architectures is equivalent to learning disentangled shape and appearance low-dimensional representations from imaging data sets in an unsupervised fashion. Thanks to an isometry-enforcing layer inspired by the underlying theory, the learned representations are embedded in a relevant metric space where readily-available Euclidean operations potentially allow to perform image manipulation. This introduced class of neural networks is interpreted as a Bayesian generative and hierarchical model, allowing the joint estimation of the network parameters, a representative prototype of the training images, as well as the relative importance between the geometrical and texture contents. This work is therefore a point of convergence between Bayesian generative statistics, metamorphoses-based shape analysis and variational auto-encoding methods.

Section 2 details the chosen transformation model, from its geometrical interpretation to its practical implementation. Section 2.3 presents the proposed MAE architecture, its Bayesian probabilistic interpretation, and the chosen optimization scheme. Section 3 shows the ability of MAE to learn relevant disentangled shape and appearance representations of imaging data sets and illustrates the potential of allowed post-processing image manipulation applications. Section 4 concludes.

2 Metamorphic Transformation Model

We first detail briefly our theoretical metamorphic transformation model, which builds on different fields of the shape analysis literature [2,29,31]. Illustrated by Fig. 1, its discrete counterpart is then derived to prepare the integration of metamorphoses into neural network architectures. See *Supplementary materials* for a self-contained minimal presentation of the metamorphosis framework.

2.1 Continuous Theory

Let $\Omega \subset \mathbb{R}^d$ with $d \in \{2, 3\}$ be a spatial domain, on which is defined the image $I_0 : \Omega \rightarrow \mathbb{R}$. This image can be deformed into $\phi_1 \star I_0 = I_0 \circ \phi_1$ by any diffeomorphism ϕ_1 of Ω. Similarly to [2], we choose to construct diffeomorphisms by following the flow of static and smooth velocity vector fields $v \in V \subset C_0^\infty(\Omega, \mathbb{R}^d)$ for a unit time period $t \in [0, 1]$, i.e. by integrating:

$$\partial_t \phi_t = v \circ \phi_t \quad \text{from the identity} \quad \phi_0 = \mathrm{Id}_\Omega. \tag{1}$$

We denote $\Phi : v \rightarrow \phi_1$ the mapping which associates the obtained diffeomorphism from the vector field v. Abusing slightly of notations, for any velocity field

v and intensity increment $\delta \in D \subset C_0^\infty(\Omega, \mathbb{R})$, the action of the metamorphosis operator $\Phi(v, \delta)$ on images is given by:

$$\Phi(v, \delta) \star I_0 = \Phi(v) \star (I_0 + \delta) = (I_0 + \delta) \circ \Phi(v). \tag{2}$$

A new metamorphic distance $d_{I_0} \geq 0$ on images can be defined as:

$$d_{I_0}(I, I')^2 = \|v' - v\|_V^2 + \|\delta' - \delta\|_D^2 \text{ where } (v, \delta) = \Phi^{-1}(I) \text{ and } (v', \delta') = \Phi^{-1}(I'). \tag{3}$$

whose norms can be computed by further considering than v and δ live in reproducing kernel Hilbert spaces with kernels K_V and K_D respectively.

2.2 Practical Discrete Case

In practice, the physical domain Ω is discretized into a regular grid g, and the time segment $[0, 1]$ into 2^T time-points, with $T \in \mathbb{N}^*$. We choose K_V and K_D as simple radial Gaussian kernels of respective radiuses $\rho_V > 0$ and $\rho_D > 0$. The g-discretized fields $\underline{v}^*$, $\underline{\delta}^*$ and atlas $\underline{I}_0$ are fed as inputs to the discrete metamorphic transformation module. As illustrated by the right-side of Fig. 1, the discrete velocity field $\underline{v}$ and intensity increment $\underline{\delta}$ are first computed according to the filtering formulae $\underline{v} = \underline{K}_V \cdot \underline{v}^*$ and $\underline{\delta} = \underline{K}_D \cdot \underline{\delta}^*$ where for any grid index k_0:

$$\left[\underline{K}_V\right]_{k_0} = \sum_k \exp\left(\frac{-\|g_k - g_{k_0}\|_{\ell^2}^2}{\rho_V^2}\right) \text{ and } \left[\underline{K}_D\right]_{k_0} = \sum_k \exp\left(\frac{-\|g_k - g_{k_0}\|_{\ell^2}^2}{\rho_D^2}\right) \tag{4}$$

which corresponds to the discrete version of the reproducing Hilbert norm, which writes $\|\underline{v}\|_V = (\underline{v}^*)^\top \cdot \underline{K}_V \cdot \underline{v}^*$ (and similarly on D). As originally described in [2], the integration along the streamlines of v defined by Eq. 1 is discretely carried out with the scaling-and-squaring algorithm which consists in applying T times:

$$\underline{x}_{t+1} = \underline{x}_t + \mathcal{I}(\underline{x}_t - g,\, \underline{x}_t) \qquad \text{from} \quad \underline{x}_0 = g + \underline{v} \,/\, 2^T \tag{5}$$

where $\mathcal{I}(\underline{x}_t - g,\, \underline{x}_t)$ simply denotes the interpolation of the displacement field $\underline{x}_k - g$ at the physical locations $\underline{x}_k$. The metamorphosis of I_0 (i.e. Eq. 2) is finally approximated as:

$$\Phi(v, \delta) \star I_0 \;\approx\; \Phi(\underline{v}, \underline{\delta}) \star \underline{I}_0 \;=\; \mathcal{I}(\underline{I}_0 + \underline{\delta},\, \underline{x}_T) \tag{6}$$

where $\mathcal{I}(\underline{I}_0 + \underline{\delta},\, \underline{x}_T)$ here denotes the interpolation[1] of the intensity values $\underline{I}_0 + \underline{\delta}$ at locations $\underline{x}_T$.

2.3 Variational Formulation for Generative Modeling

Our statistical model is based on the variational framework [16,17], and consists in assuming that the observed images $(I_i)_{i=1}^n$ are hierarchically distributed according to

$$I_i \overset{\text{iid}}{\sim} \mathcal{N}\left\{\Phi\left[S_\sigma(s_i),\, A_\alpha(a_i)\right] \star I_0;\; \epsilon^2\right\} \text{ with } s_i \overset{\text{iid}}{\sim} \mathcal{N}(0, \lambda_s^2) \text{ and } a_i \overset{\text{iid}}{\sim} \mathcal{N}(0, \lambda_a^2) \tag{7}$$

[1] We used bilinear interpolation scheme.

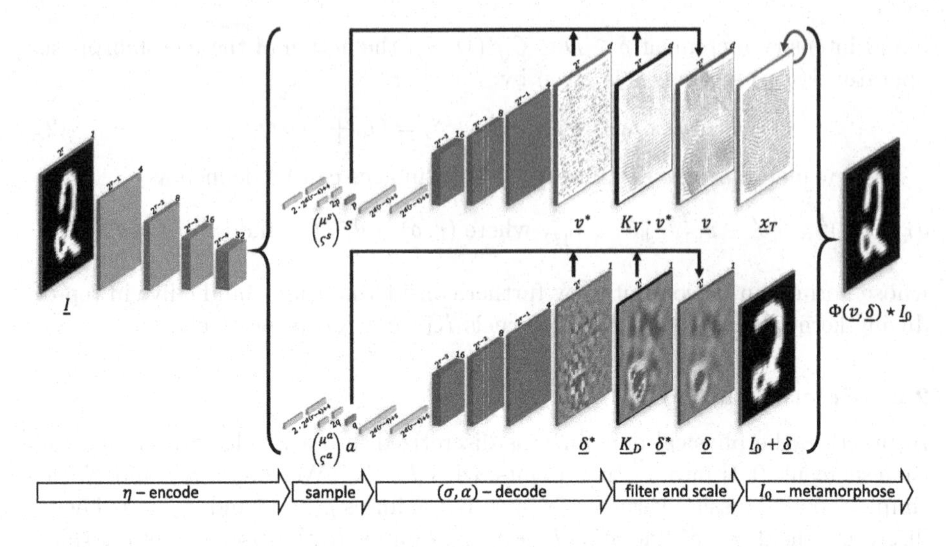

Fig. 1. Architecture of the metamorphic auto-encoder. The input image $\underline{I}$ is encoded by four convolution layers (in green), followed by two parallel pairs of fully-connected layers (in yellow). The encoder outputs are interpreted as characterizing two normal distributions, from which are sampled the latent codes $s \in \mathbb{R}^p$ and $a \in \mathbb{R}^q$. Two parallel decoders successively composed of two fully connected and four deconvolution layers map those latent shape and appearance representations to the velocity field $\underline{v}^*$ and intensity increment $\underline{\delta}^*$ duals. After filtering by the operators $\underline{K}_V$ and $\underline{K}_D$, the obtained vectors are explicitly scaled, enforcing the equality of their Hilbert norm with the Euclidean norm of the corresponding codes s and a (see Eq. 4). The resulting velocity field $\underline{v}$ (see Eq. 5) and intensity increment $\underline{\delta}$ (see Eq. 6) are finally combined to metamorphose the prototype image parameter $\underline{I}_0$. (Color figure online)

where I_0 is the learned atlas, S_σ (respectively A_α) is a σ-parametric (respectively α-parametric) neural network mapping, that associates the velocity field v (respectively the intensity increment δ) to any code $s \in \mathbb{R}^p$ (respectively $a \in \mathbb{R}^q$).

Figure 1 details the architecture of the metamorphic auto-encoder. The first important observation is that we structurally[2] impose the metric equivalence between the metamorphic distance d_{I_0} defined by Eq. 3 and the induced latent norm $d_0(z, z') = \|z - z'\|_{\ell^2}$: the mappings are therefore isometric, i.e. that $\|v\|_V = \|s\|_{\ell^2}$ and $\|\delta\|_D = \|a\|_{\ell^2}$.
Furthermore, we chose tanh activation functions after convolutions (at the exception of the last encoding one) and deconvolutions, and that all decoding layers are chosen without bias : these two last hypotheses ensure the infinite differentiability of the mappings $\underline{S}_\sigma$, $\underline{A}_\alpha$ and $\underline{E}_\eta$.

[2] via an explicit normalization layer.

Lastly, we chose to encode the euclidean difference $I_i - I_0$ as input in our network, which associated with null bias and tanh activations imposes that the null latent-space vector $z = 0$ is mapped to I_0; the prior distribution defined on the random effects $(z_i)_i$ therefore defines I_0 as a statistical average of the observations $(I_i)_i$. In other words, estimating the model parameters θ and I_0 can legitimately be interpreted as computing a Fréchet average of the training data set [25].

3 Experiments

The ability to disentangle shape and appearance becomes necessary if we want to manipulate data where abnormalities, such as tumors, appears at a visible scale. We applied our unsupervised framework to the task of reconstructing brains with tumors, in the spirit of personalized numerical modeling, on BraTs 2018 dataset [4,5,22]. Images have been obtained by selecting an axial section of T1 contrast enhanced 3D brain volumes, pre-processed with standard skull-removing and affine alignment pipelines.

Out first series of experiments compares behavior of metamorphic model to its known diffeomorphic equivalent[7]: Fig. 2 depicts how the flow of transformation for each model works. A comparison with the diffeomorphic equivalent model to ours shows clearly the necessity to disentangle: without the δ module, atlas learned has no proper geometrical features, in addition to a poor reconstruction of brain. Indeed the tumors are obtained through inclusion of saturated points on the learned atlas when diffeomorphism only are used, and even then reconstruction requires high deformations. On the contrary, metamorphisms naturally handle the separation of information, with smoother velocity fields, clear localization of the tumor on the intensity map δ as well as a visually sharp atlas representing a brain without tumors. As a sanity check, simple classification of tumor grade (high vs low) was performed taking the various latent representations learned as features and summarized on Table 1.

Table 1. Average quadratic discriminant analysis (QDA) balanced accuracy classification scores (stratified 10-fold method, chance level 50%).

(%)	Full latent space	Shape latent space	Appearance latent space
Metamorphic	64.0 ± 13.0	64.0 ± 15.0	67 ± 12.0
Diffeomorphic	57.0 ± 16.0	–	–

For a more quantitative understanding, we also applied the Chan Vese level-set segmentation algorithm [10] on δ maps to quantitatively assess the quality of the learned atlas: relatively high Dice scores, up to 0.7 can be attained by segmenting this map and transporting, through v, the obtained mask. Fig. 3 shows

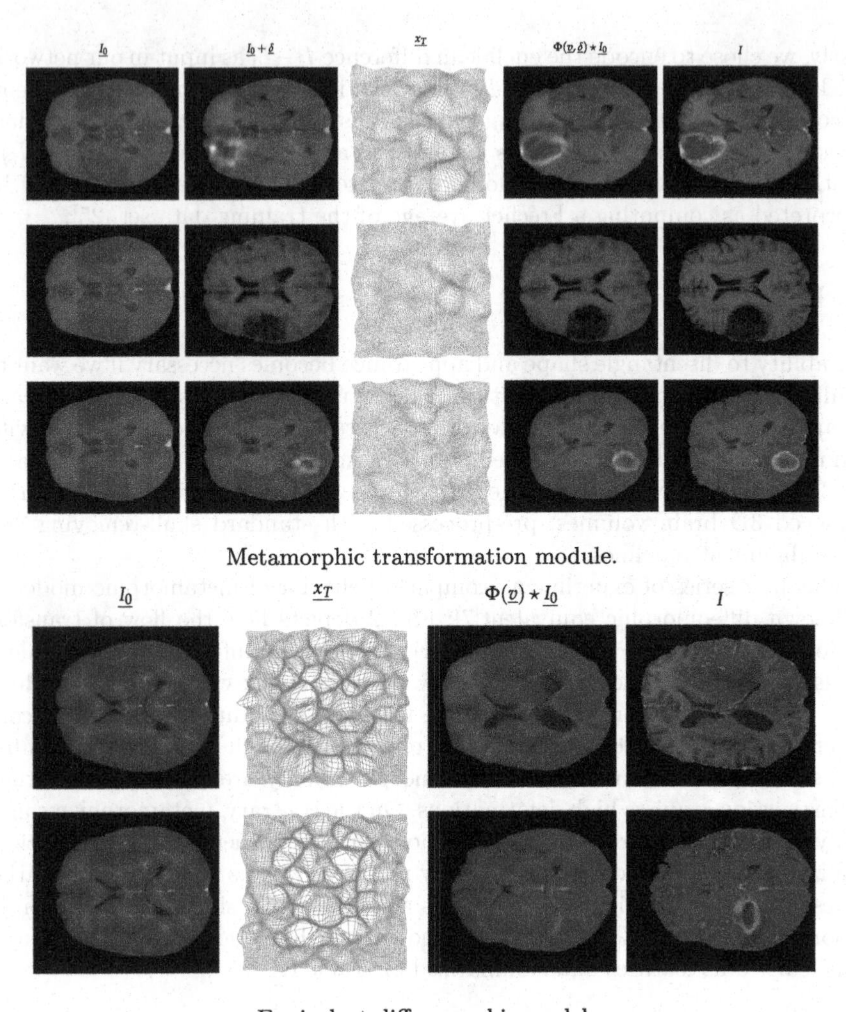

Fig. 2. Comparison of diffeomorphic and metamorphic models

examples of obtained masks. This experiment suggests than intrinsic information of the tumor was mainly grasped by the δ (for the appearance part) and v (for the geometric localization) components of our model, leaving the atlas $\mathcal{I}_0$ to be understood as a control-like Frechet-mean.

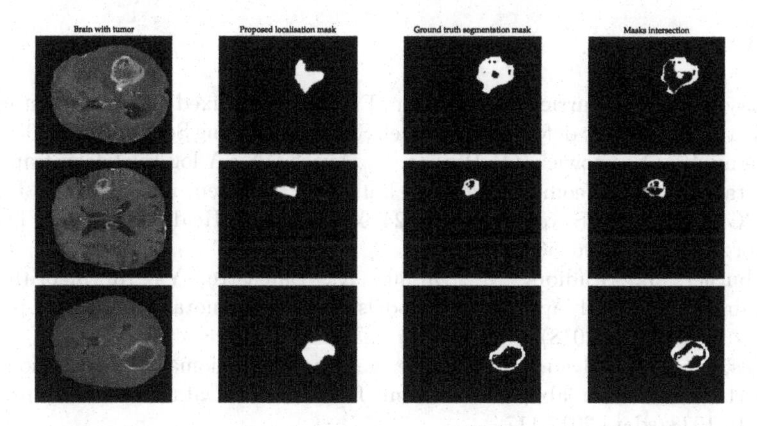

Fig. 3. Unsupervised localisation of tumor using Chan-Vese segmentation algorithm [10] on appearance maps δ (on this figure, $DICE \geq 0.60$), from left to right: input brain, computed segmentation mask, ground-truth tumor mask, intersection of masks

4 Conclusion

Our method may be understood at the crossroad of simulation numerical models, which are based on sound mathematical priors (metamorphosis decomposition of brain variability), and data-driven statistics which may soon enable efficient generative models to emerge, then applicable to personalize individual evolution of pathology.

Untangling the shape and appearance variabilities is of key interest in computational anatomy to better interpret and apprehend the total variability of a collections of organs or anatomical shapes: better disease markers for medical images could for instance be identified, and in the case of tumors improved growth models could be developed.

Qualitative results were presented aiming towards well-behaved models evidence, in particular the segmentation of δ-map which led to very coherent localization of tumors without complex treatment nor specific dataset filtering. To the best of our knowledge, no publicly available code was developed for formal metamorphosis framework, which therefore didn't allow for a direct comparison with our deep-learning method, though a natural advantage is its sampling efficiency and thus potential use in numerical simulation scenarios or for data augmentation.
Lastly, our network was composed in a very simple fashion, and increasing the architecture complexity, in particular through powerful U-net structures [30] will naturally allow to reach stronger reconstruction accuracies and better performances.

Acknowledgments. This work has been partly funded by the European Research Council with grant 678304, European Union's Horizon 2020 research and innovation program with grant 666992, and the program Investissements d'avenir ANR-10-IAIHU-06.

References

1. Allassonnière, S., Durrleman, S., Kuhn, E.: Bayesian mixed effect atlas estimation with a diffeomorphic deformation model. SIAM J. Imaging Sci. **8**, 1367–1395 (2015)
2. Arsigny, V., Commowick, O., Pennec, X., Ayache, N.: A log-Euclidean framework for statistics on diffeomorphisms. In: Larsen, R., Nielsen, M., Sporring, J. (eds.) MICCAI 2006. LNCS, vol. 4190, pp. 924–931. Springer, Heidelberg (2006). https://doi.org/10.1007/11866565_113
3. Ashburner, J., Brudfors, M., Bronik, K., Balbastre, Y.: An algorithm for learning shape and appearance models without annotations. arXiv preprint arXiv:1807.10731 (2018)
4. Bakas, S., et al.: Advancing the cancer genome atlas glioma MRI collections with expert segmentation labels and radiomic features. Sci. Data **4** (2017). https://doi.org/10.1038/sdata.2017.117
5. Bakas, S., et al.: Identifying the best machine learning algorithms for brain tumor segmentation, progression assessment, and overall survival prediction in the BRATS challenge. arXiv e-prints arXiv:1811.02629, November 2018
6. Ballester, P., Araujo, R.M.: On the performance of GoogleNet and AlexNet applied to sketches. In: Thirtieth AAAI Conference on Artificial Intelligence (2016)
7. Bône, A., Louis, M., Colliot, O., Durrleman, S.: Learning low-dimensional representations of shape data sets with diffeomorphic autoencoders. In: Chung, A.C.S., Gee, J.C., Yushkevich, P.A., Bao, S. (eds.) IPMI 2019. LNCS, vol. 11492, pp. 195–207. Springer, Cham (2019). https://doi.org/10.1007/978-3-030-20351-1_15
8. Brendel, W., Bethge, M.: Approximating CNNs with bag-of-local-features models works surprisingly well on ImageNet. arXiv preprint arXiv:1904.00760 (2019)
9. Cireşan, D., Meier, U., Schmidhuber, J.: Multi-column deep neural networks for image classification. arXiv preprint arXiv:1202.2745 (2012)
10. Cohen, R.: The Chan-Vese Algorithm. arXiv e-prints arXiv:1107.2782, July 2011
11. Dalca, A.V., Balakrishnan, G., Guttag, J., Sabuncu, M.R.: Unsupervised learning for fast probabilistic diffeomorphic registration. In: Frangi, A.F., Schnabel, J.A., Davatzikos, C., Alberola-López, C., Fichtinger, G. (eds.) MICCAI 2018. LNCS, vol. 11070, pp. 729–738. Springer, Cham (2018). https://doi.org/10.1007/978-3-030-00928-1_82
12. D'Arcy Wentworth, T.: On growth and form. In: Tyler Bonner, J. (ed.) Abridged. Cambridge University Press (1917)
13. Geirhos, R., Rubisch, P., Michaelis, C., Bethge, M., Wichmann, F.A., Brendel, W.: ImageNet-trained CNNs are biased towards texture; increasing shape bias improves accuracy and robustness. arXiv preprint arXiv:1811.12231 (2018)
14. Grenander, U.: General Pattern Theory-A Mathematical Study of Regular Structures. Clarendon Press, Oxford (1993)
15. Jaderberg, M., Simonyan, K., Zisserman, A., et al.: Spatial transformer networks. In: Advances in Neural Information Processing Systems, pp. 2017–2025 (2015)
16. Kingma, D.P., Welling, M.: Auto-encoding variational Bayes. stat 1050, 10 (2014)
17. Kingma, D.P., Welling, M.: An introduction to variational autoencoders. arXiv e-prints arXiv:1906.02691, June 2019
18. Krebs, J., Delingette, H., Mailhé, B., Ayache, N., Mansi, T.: Learning a probabilistic model for diffeomorphic registration. IEEE Trans. Med. Imaging (2019)
19. Kriegeskorte, N.: Deep neural networks: a new framework for modeling biological vision and brain information processing. Ann. Rev. Vis. Sci. **1**, 417–446 (2015)

20. Krizhevsky, A., Sutskever, I., Hinton, G.E.: ImageNet classification with deep convolutional neural networks. In: Advances in Neural Information Processing Systems, pp. 1097–1105 (2012)
21. LeCun, Y., Bengio, Y., Hinton, G.: Deep learning. Nature **521**(7553), 436 (2015)
22. Menze, B., et al.: The multimodal brain tumor image segmentation benchmark (BRATS). IEEE Trans. Med. Imaging **99** (2014). https://doi.org/10.1109/TMI.2014.2377694
23. Niethammer, M., et al.: Geometric metamorphosis. In: Fichtinger, G., Martel, A., Peters, T. (eds.) MICCAI 2011. LNCS, vol. 6892, pp. 639–646. Springer, Heidelberg (2011). https://doi.org/10.1007/978-3-642-23629-7_78
24. Patenaude, B., Smith, S.M., Kennedy, D.N., Jenkinson, M.: A Bayesian model of shape and appearance for subcortical brain segmentation. Neuroimage **56**(3), 907–922 (2011)
25. Pennec, X.: Intrinsic statistics on Riemannian manifolds: Basic tools for geometric measurements. J. Math. Imaging Vis. **25**(1), 127–154 (2006)
26. Shu, Z., Sahasrabudhe, M., Alp Güler, R., Samaras, D., Paragios, N., Kokkinos, I.: Deforming autoencoders: unsupervised disentangling of shape and appearance. In: Ferrari, V., Hebert, M., Sminchisescu, C., Weiss, Y. (eds.) ECCV 2018. LNCS, vol. 11214, pp. 664–680. Springer, Cham (2018). https://doi.org/10.1007/978-3-030-01249-6_40
27. Simard, P.Y., Steinkraus, D., Platt, J.C., et al.: Best practices for convolutional neural networks applied to visual document analysis. In: ICDAR, vol. 3 (2003)
28. Skafte Detlefsen, N., Freifeld, O., Hauberg, S.: Deep diffeomorphic transformer networks. In: Proceedings of the IEEE Conference on Computer Vision and Pattern Recognition, pp. 4403–4412 (2018)
29. Trouvé, A., Younes, L.: Metamorphoses through lie group action. Found. Comput. Math. **5**(2), 173–198 (2005)
30. Tudosiu, P.D., et al.: Neuromorphologicaly-preserving Volumetric data encoding using VQ-VAE. arXiv e-prints arXiv:2002.05692 (Feb 2020)
31. Younes, L.: Shapes and Diffeomorphisms. Appl. Math. Sci. Springer, Heidelberg (2010). https://books.google.fr/books?id=SdTBtMGgeAUC
32. Zhang, M., Singh, N., Fletcher, P.T.: Bayesian estimation of regularization and atlas building in diffeomorphic image registration. IPMI **23**, 37–48 (2013)

Collaborative Learning of Cross-channel Clinical Attention for Radiotherapy-Related Esophageal Fistula Prediction from CT

Hui Cui[1], Yiyue Xu[2], Wanlong Li[2], Linlin Wang[2(✉)], and Henry Duh[1]

[1] Department of Computer Science and Information Technology, La Trobe University, Melbourne, Australia

[2] Department of Radiation Oncology, Shandong Cancer Hospital and Institute, Shandong First Medical University and Shandong Academy of Medical Sciences, Jinan, China

wanglinlinatjn@163.com

Abstract. Early prognosis of the radiotherapy-related esophageal fistula is of great significance in making personalized stratification and optimal treatment plans for esophageal cancer (EC) patients. The effective fusion of diagnostic consideration guided multi-level radiographic visual descriptors is a challenging task. We propose an end-to-end clinical knowledge enhanced multi-level cross-channel feature extraction and aggregation model. Firstly, clinical attention is represented by contextual CT, segmented tumor and anatomical surroundings from nine views of planes. Then for each view, a Cross-Channel-Atten Network is proposed with CNN blocks for multi-level feature extraction, cross-channel convolution module for multi-domain clinical knowledge embedding at the same feature level, and attentional mechanism for the final adaptive fusion of multi-level cross-domain radiographic features. The experimental results and ablation study on 558 EC patients showed that our model outperformed the other methods in comparison with or without multi-view, multi-domain knowledge, and multi-level attentional features. Visual analysis of attention maps shows that the network learns to focus on tumor and organs of interests, including esophagus, trachea, and mediastinal connective tissues.

Keywords: Esophageal fistula prediction · CT · Cross channel attention

1 Introduction

Globally, esophageal cancer (EC) is the eighth malignant tumor [1]. Esophageal fistula is one of the serious complications which can be life-threatening. 4.3%–24% EC patients were diagnosed with esophageal fistula after chemoradiotherapy [2, 3], and the median survival rate is extremely poor. It is shown by clinical research that the survival rate is only two or three months, due to related infection, massive hemorrhage or unhealed abscess [4–6]. Clinically guided treatment options often circumvent some treatment methods. However, it results in contradictory outcomes that it is not suitable for tumor control . [3]

A. L. Martel et al. (Eds.): MICCAI 2020, LNCS 12261, pp. 212–220, 2020.
https://doi.org/10.1007/978-3-030-59710-8_21

Therefore, the early prognosis of the radiotherapy-related esophageal fistula is of great significance in making patient customized stratification, and optimal treatment plans to improve the treatment outcome.

Cancer imaging and computerized models are promising tools in the quantitative interpretation of tumor metabolism, assessment and monitoring of treatment response. [7] For EC, however, the quantitative modelling of clinical factors and radiographic features to predict the occurrence of the esophageal fistula is underdeveloped in clinical and computer science research. Recently, Xu et al. [3] proposed the first risk prediction model based on logistic regression using clinical, pathological, and serum information. For image-based prediction, a 3D convolutional neural network (CNN) based model [8] using positron emission tomography (PET) is one of the few models developed for EC classification. Jin et al. [9] proposed an esophageal gross tumor segmentation model using PET/CT. Thus, there are great necessities to push data-driven and computerized methods to predict radiotherapy-related esophageal fistula, which may impact future directions in cancer care for EC patients.

Existing computerized methods for image-centric disease or cancer prediction include domain knowledge-based and learning-based approaches. Domain knowledge-based methods are widely used by clinicians and radiologists, which are designed with specific diagnostic consideration or clinical experience. [10–12] Even though most domain knowledge-based methods provide interpretable decision-making process, the major drawback is that the hand-crafted features are explicitly designed for a specific type of disease or cancer which may not be generally applied to others. For learning based models, the mapping between medical images and the target is obtained during the training process, which can be used as feature extractors and predictors for unseen cases. With the recent development of computer vision techniques and the increasing amount of medical imaging scans, learning-based methods demonstrate profound success in computer-assisted diagnosis and prognosis. However, most of those models ignore valuable domain knowledge and clinical attentions. Recently, Xie et al. [10] proposed a clinical knowledge-based collaborative learning model for lung nodule classification. Overall appearance, heterogeneity of voxel values and heterogeneity of lung nodules shapes were extracted to represent domain knowledge in benign and malignant lung nodule classification before feeding into a deep neural network. Liu et al. [12] proposed a multi-branch CNN to adopt basic, context and margin information of thyroid nodule from ultrasound images.

For the afore-mentioned multi-view, multi-domain or multi-branch models, deep features of each view or domain knowledge extracted via convolutional layers were aggregated together at global scale via fully connected layers. The global level aggregation, however, cannot preserve local and lower-level visual descriptors which may also be effective factors in medical image classification and segmentation [14, 15]. Attention mechanism [17] can extract intermediate representations in CNN pipeline and recognize local attentional patterns in images, which improved the performance of various image processing tasks such as image classification [17], rain removal [13], and fetal ultrasound screening [14]. The effective extraction and fusion of diverse clinical attention constrained visual features in a multi-view or multi-branch architecture is still a challenging task.

In this work, we propose an end-to-end clinical knowledge guided multi-level cross-channel radiographic descriptors extraction and aggregation model for esophageal fistula prediction. The first contribution of our model is the representation of clinical attention by contextual, tumor and anatomical surroundings extracted from nine views of planes. Secondly, we propose a new cross-channel attention fusion module which enables the extraction and adaptive integration of multi-channel clinical attention enhanced multi-level visual concepts and global image descriptors. To the best of our knowledge, this is the first CT radiographic feature-based model for esophageal fistula risk prediction in EC patients.

2 Dataset

The experimental dataset consists of 558 esophageal cancer (EC) patients who didn't receive esophageal surgery between 2014 and 2019. There were 186 CT scans of patients who developed esophageal fistula (positive cases) and 372 CT scans of controls (1:2 matched with the diagnosis time of EC, sex, marriage, and race) who did not experience esophageal fistula (negative cases). The CT images were reconstructed using a matrix of 512×512 pixels with a voxel size of $0.8027 \times 0.8027 \times 5\ \text{mm}^3$. The number of slices in thorax CT scans varies across patients and is between 71 and 114. Manual delineations of tumor were performed by an experienced radiation oncologist on CT with reference on PET, Esophagoscopy, or barium meal. The CT voxel size was resampled to $1 \times 1 \times 1\ \text{mm}^3$ for further processing.

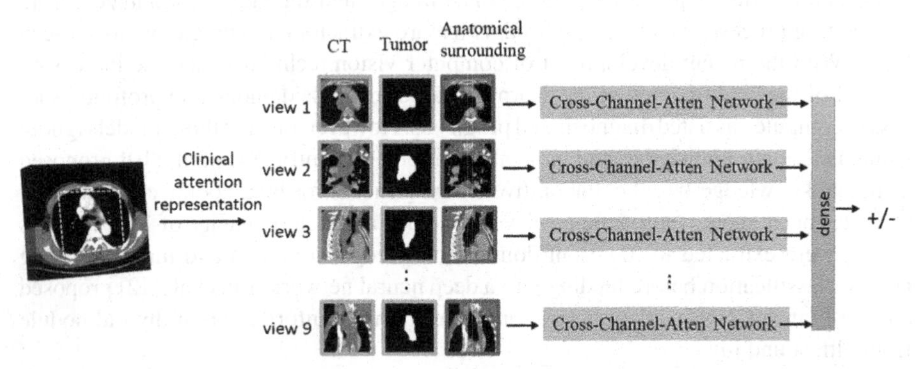

Fig. 1. Overview of the esophageal fistula prediction model using multi-view Cross-Channel-Atten Network. Given CT and segmented tumor as input, patches of clinical attentions including contextual CT, tumor and anatomical surroundings are extracted from 9 view of planes. The outputs of Cross-Channel-Atten Networks are concatenated by a fully connected layer for final classification.

3 Methods

The overview of the proposed model is given in Fig. 1. Given an input CT volume and manual segmentation of esophageal tumor, the major components in our model include clinical attention representation, and Cross-Channel-Atten Networks for the extraction and adaptive fusion of cross-domain knowledge enhanced multi-level visual features.

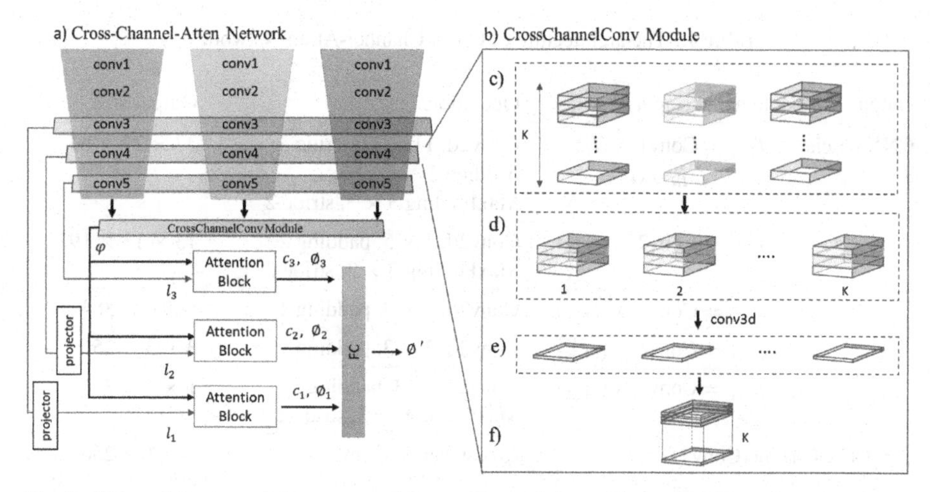

Fig. 2. The architecture of the proposed Cross-Channel-Atten Network. The major components of the network include CNN blocks for multi-level feature extraction, CrossChannelConv module to fuse the same level features from different domain knowledge, and attention block for adaptive fusion.

3.1 Clinical Attention Representation

Esophageal cancers often invade adjacent healthy anatomical structures such as trachea and bronchi. [4] Fistula may be formed during radiotherapy because of the damaging of the esophageal walls and surrounding healthy organs. Accordingly, we extract contextual CT, segmented tumor and anatomical surroundings from 9 views of planes to represent clinical attention as shown in Fig. 1. We define a $200 \times 200 \times 200\ \text{mm}^3$ cube for volume of interest located at the center of manually segmented tumor volume. Its transverse, sagittal, coronal and six diagonal planes are used as 9 views for 2D slices extraction. Then, contextual, tumor and anatomical surrounding patches in each view are extracted. The contextual patch is defined as a 2D slice in a view from the cropped CT cube, which represents the contextual information of the tumor and its neighboring environment. The tumor patch is extracted from the cube of the segmented tumor volume, providing explicit shape and boundary information. To generate anatomical surrounding patch, the pixels inside the tumor is set as zero on contextual patch.

3.2 Cross-channel-Atten Network

Given contextual, tumor and anatomical surrounding patches from nine views, we propose a Cross-Channel-Atten Network for each view as shown in Fig. 2. The network is designed to learn cross-domain knowledge guided multi-level features generated through convolutions and adaptively integrate them for final decision making. The major components of the network include CNN blocks for multi-level feature extraction, cross-channel convolution (CrossChannelConv) module to fuse the same level features from different domain knowledge, and attention block for adaptive fusion. The detailed architecture is given in Table 1.

Table 1. The architecture of Cross-Channel-Atten Network.

Output = module name (input)		Operations	Output size
CNN blocks	$f1\_x$ = Conv1_x (I_x) $x \in \{ct, tu, as\}$	Conv2d, 11 × 11, stride 4, padding 2 MaxPooling 3 × 3, stride 2	27 × 27 × 64
	$f2\_x$ = Conv2_x ($f1\_x$)	Conv2d, 5 × 5, padding 2 MaxPooling 3 × 3, stride 2	13 × 13 × 192
	$f3\_x$ = Conv3_x ($f2\_x$)	Conv2d, 3 × 3, padding 1	6 × 6 × 384
	$f4\_x$ = Conv4_x ($f3\_x$)	Conv2d, 3 × 3, padding 1	3 × 3 × 256
	$f5\_x$ = Conv5_x ($f4\_x$)	Conv2d, 3×3, padding 1 MaxPooling 3×3, stride 2	1 × 1 × 256
φ = CrossChannelConv ($f5\_{ct}$, $f5\_{tu}$, $f5\_{as}$)		CrossChannelConv	1 × 1 × 256
$c_1, \emptyset_1$ = CrossAttenBlock1($f2\_{ct}$, $f2\_{tu}$, $f2\_{as}$, φ)		CrossChannelConv ($f2\_{ct}$, $f2\_{tu}$, $f2\_{as}$)	13 × 13 × 192
		l_1 = projector	3 × 3 × 256
		$c_1, \emptyset_1$ = attention block (l_1, φ)	1 × 1 × 256
$c_2, \emptyset_2$ = CrossAttenBlock2($f3\_{ct}$, $f3\_{tu}$, $f3\_{as}$, φ)		CrossChannelConv ($f3\_{ct}$, $f3\_{tu}$, $f3\_{as}$)	6 × 6 × 384
		l_2 = projector	3 × 3 × 256
		$c_2, \emptyset_2$ = attention block (l_2, φ)	1 × 1 × 256
$c_3, \emptyset_3$ = CrossAttenBlock3($f4\_{ct}$, $f4\_{tu}$, $f4\_{as}$, φ)		l_3 = CrossChannelConv ($f_{4,ct}$, $f_{4,tu}$, $f_{4,as}$)	3 × 3 × 256
		$c_3, \emptyset_3$ = attention block (l_3, φ)	1 × 1 × 256
$\emptyset'$ = FC ($\emptyset_1, \emptyset_2, \emptyset_3$)		Fully connected layers	2

CNN Blocks for Multi-level Visual Feature Extraction. Given the input image I_x and N convolutional blocks, where x can be ct, tu, or as representing a patch of contextual CT, segmented tumor or anatomical surroundings, the outputs of the (N − 3), (N − 2) and (N − 1)-th block are taken as local descriptors representing low-level, high-level and higher-order visual concepts respectively. The output l_N of the N-th block is a global feature descriptor (denoted by φ in Fig. 2 and Table 1). Different backbones can be used for feature extraction; in this work, we adopt the architecture of AlexNet.

CrossChannelConv Module for Cross-domain Knowledge Fusion. The CrossChannelConv module exploits the embedding mechanism proposed by [16] as shown in Fig. 2(b). Given intermediate features fn_{ct}, fn_{tu}, fn_{as}, n $\in [N-3, N]$ from three inputs (shown in Fig. 2(c)), where each of them is of size h × w × K, K is the number of channels, the features of the same channel is firstly stacked together as shown in Fig. 2(d). By such, there are K h × w × 3 features. After reshaping the features to 3 × h × w and applying 3D convolutions with kernel size 3 × 1 × 1, we have K 1 × h × w features (Fig. 2(e)).

During this step, the 3D convolution sums up three modalities using different weights in the output. Finally, the K reshaped h × w features are stacked together, generating an output feature of size h × w × K, as shown in Fig. 2(f). Because of restacking and embedding strategies, the features learnt during the training process are cross-domain clinical attention guided.

Attention Block for Cross-domain Multi-level Feature Embedding. Attention module is used to learn the adaptive weights of different levels of cross fused features. Firstly, each level of cross-domain knowledge guided features is projected to the same dimension as l_N, and denoted by l_i. Secondly, the compatibility score is calculated as [17] by $c_{i,s} = l_{i,s}, l_N$ depending on the alignment between a local descriptor and global descriptor, where $\langle\,\rangle$ denotes dot product, s denotes the spatial location in the feature map l_i. Then $c_{i,s}$ is normalized by a SoftMax operation and a new representation for l_i is obtained by

$$\emptyset_i = \sum\nolimits_{s=1}^{N_{l_i}} a_{i,s} \cdot l_{i,s} \tag{1}$$

where $a_{i,s}$ is the normalized $c_{i,s}$.

$\emptyset_1, \emptyset_2, \emptyset_3$ are concatenated and sent to fully connected layers where are composed of a sequence of Dropout, fully connected layer with 768 units, ReLu, Dropout, fully connected layer with 256 units, ReLu, and fully connected layer with two units.

3.3 Multi-view Aggregation and Training Strategy

To transfer the feature extraction capacity of CNN to Esophageal CT, pre-trained AlexNet on ImageNet is used to initialize the CNN blocks in Cross-modal-Atten network in each view. A fully connected layer concatenates the outputs of these networks for final classification. Fully connected layers are initialized by Xavier algorithm.

Our model is trained using cross-entropy loss, an initial learning rate of 0.01, and SGD as an optimizer with momentum = 0.9 and weight decay = 5e−4. All the input images are resized to 224 × 224, the batch size is 40, and epoch is set as 300. During the five-fold cross-validation process, in each round, all the patients were divided into a training set (150 pairs of positive patients and the controls) and a testing set (36 pairs of positive and controls) randomly. For the training set, data augmentation was performed, including pixel shifting and rotation. As there were imbalanced positive and negative cases, shifting operations of [−10, −5, 0, +5, +10] along x and y-axis and rotations of [−10, +10] degrees were performed for positive cases, resulting in 150 × 25 × 3 = 11250 positive samples. For negative training cases, 300 × 12 × 3 = 10800 negative samples were obtained after shifting operations of [−5, 0, +5, +10] along x and [−5, 0, +5] along y-axis and rotations of [−10, +10] degrees.

4 Experiments and Results

Evaluation Metrics. Evaluation metrics include accuracy, sensitivity, and specificity. Given true positive (TP), false negative (FN), true negative (TN), and false positive (FP) numbers, accuracy, sensitivity and specificity are obtained as $accuracy = (TP + TN)/(TP + FN + TN + FP)$, $sensitivity = TP/(TP + FN)$, $specificity = TN/(TN + FP)$.

Comparison with Other Methods and Ablation Study. To evaluate the contributions of multi-view, multi-domain clinical knowledge and attention modules in the prediction results, the proposed method is compared with the following CNN-based architectures using: 1) multi-view single modal image without attention [18], which was initially designed for shape recognition, 2) multi-domain clinical knowledge from a single view without attention [12] intended for thyroid classification in ultrasound, 3) multi-view multi-domain knowledge without attention [10] proposed for lung nodule classification in CT, and 4) single CT single view VGG with attention module [17]. For models [12] and [17] where multi-view information is not considered, 2D patches from axial view were used as input. For [17, 18] where multi-domain knowledge is not available, contextual CT was used as input.

The comparison results are given in Table 2. As shown, PM achieved the best accuracy, sensitivity and specificity of 0.857, 0.821 and 0.935. The second best was obtained by [10], which indicated that the spatial information in multi-view based methods contributed to the improved performance. Results (p-value < 0.05) proved that the improvement is statistically significant. The ablation study results in Table 3 show that all three major components in our model contributed to the improved prediction results. In particular, given multi-view patches, the effectiveness of CrossChannelConv is demonstrated by the comparison between our model without attention and method [10] in terms of accuracy, sensitivity, and specificity. When multi-view information is not available, CrossChannelConv in our model improved the sensitivity against architecture [12]. Our model without CrossChannelConv proved the contribution of multi-level local visual descriptors in the attention module when compared with methods [10].

Table 2. Comparison with other architectures by accuracy, sensitivity and specificity.

Architecture	Accuracy	Sensitivity	Specificity
Jetley *et al.* 2018 [17]	0.725	0.603	0.812
Su *et al.* 2015 [18]	0.746	0.482	0.881
Liu *et al.* 2019 [12]	0.766	0.604	0.842
Xie *et al.* 2019 [10]	0.810	0.723	0.912
PM	**0.857**	**0.821**	**0.935**

Visual Interpretation of Cross-fused Multi-level Attentions. We draw attention maps to interpret the network's focus during different levels of feature extraction and cross-channel fusion. As shown in Fig. 3, the attention maps atten3, atten2, and atten1 represent

Table 3. Ablation study results. When multi-view is not available, axial view is used as input.

Method			Results		
Multi-view	CrossChannelConv	Attention	Accuracy	Sensitivity	Specificity
×	√	×	0.756	0.614	0.825
×	×	√	0.725	0.603	0.812
×	√	√	0.802	0.684	0.853
√	√	×	0.836	0.756	0.928
√	×	√	0.829	0.807	0.916
√	√	√	**0.857**	**0.821**	**0.935**

the alignments between global descriptor and higher-order, high-level, and low level local visual descriptors, respectively. It is interesting to observe that network learned to pay attention to tumor region in higher-order feature extraction, which corresponds to the conv5 block. During lower level feature extractions in conv4 and conv3, the network fixated on tumor's surrounding organs including esophagus, trachea, and mediastinal connective tissues. These findings show that the system learns to pay attention to organs of interests which may lead to radiotherapy-related esophageal fistula in esophageal cancer patients. It also proved our hypothesis that local and lower-level attentional features are also active factors which would contribute to the improved prediction results.

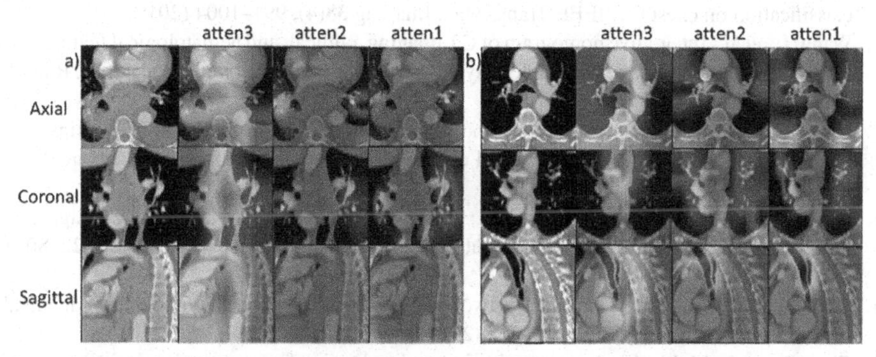

Fig. 3. Two examples of attention map visualization for the interpretation of the network's focuses during different levels of feature extraction and cross-channel fusion. Atten3, atten2, and atten1 denote the attentions corresponding to conv5, conv4 and conv3 blocks.

5 Conclusion

We propose a radiotherapy-related esophageal fistula prediction model which adaptively fuses cross-domain clinical knowledge enhanced multi-level visual features. The experimental results demonstrated improved performance. The interpretable results proved that the local and multi-level cross-fused attentional features are also useful radiographic factors in this task.

References

1. Ferlay, J., et al.: Cancer incidence and mortality worldwide: sources, methods and major patterns in GLOBOCAN 2012. Int. J. Cancer **136**(5), 359–386 (2015)
2. Zhou, Z., et al.: Salvage radiotherapy in patients with local recurrent esophageal cancer after radical radiochemotherapy. J. Radiat. Oncol. **2015**(10), 1–7 (2015)
3. Xu, Y., et al.: Development and validation of a risk prediction model for radiotherapy-related esophageal fistula in esophageal cancer. Radiat. Oncol. **14**, 181 (2019)
4. Zhang, Y., Li, Z., Zhang, W., Chen, W., Song, Y.: Risk factors for esophageal fistula in patients with locally advanced esophageal carcinoma receiving chemoradiotherapy. OncoTargets Therapy **11**, 2311–2317 (2018)
5. Rodriguez, A.N., Diaz-Jimenez, J.P.: Malignant respiratory–digestive fistulas. Curr. Opin. Pulm. Med. **16**(4), 329–333 (2010)
6. Chen, H., Ma, X., Ye, M., Hou, Y., Xie, H., Bai, Y.: Esophageal perforation during or after conformal radiotherapy for esophageal carcinoma. J. Radiat. Res. **55**(5), 940–947 (2014)
7. Bi, W.L., Hosny, A., Schabath, M.B., Giger, M.L., Birkbak, N.J., Mehrtash, A., et al.: Artificial intelligence in cancer imaging: clinical challenges and applications. CA Cancer J. Clin. **69**(2), 127–157 (2019)
8. Yang, C.-K., et al.: Deep convolutional neural network-based positron emission tomography analysis predicts esophageal cancer outcome. J. Clin. Med. **8**(6), 844 (2019)
9. Jin, D., Ho, T.-Y., Harrison, A.P., Xiao, J., Tseng, C.-k., Lu, L.: Accurate esophageal gross tumor volume segmentation in PET/CT using two-stream chained 3D deep network fusion. In: Shen, D., et al. (eds.) MICCAI 2019. LNCS, vol. 11765, pp. 182–191. Springer, Cham (2019). https://doi.org/10.1007/978-3-030-32245-8_21
10. Xie, Y., et al.: Knowledge-based collaborative deep learning for benign-malignant lung nodule classification on chest CT. IEEE Trans. Med. Imaging **38**(4), 991–1004 (2019)
11. Wang, L., et al.: Integrative nomogram of CT imaging, clinical, and hematological features for survival prediction of patients with locally advanced non-small cell lung cancer. Eur. Radiol. **29**(6), 2958–2967 (2019)
12. Liu, T., et al.: Automated detection and classification of thyroid nodules in ultrasound images using clinical-knowledge-guided convolutional neural networks. Med. Image Anal. **58**, 101555 (2019)
13. Hu, X., Fu, C.W., Zhu, L., Heng, P.A.: Depth-attentional features for single-image rain removal. In: IEEE Conference on Computer Vision and Pattern Recognition, pp. 8022–8031 (2019)
14. Schlemper, J., et al.: Attention gated networks: learning to leverage salient regions in medical images. Med. Image Anal. **53**, 197–207 (2019)
15. Chen, T., et al.: Multi-view learning with feature level fusion for cervical dysplasia diagnosis. In: Shen, D., et al. (eds.) MICCAI 2019. LNCS, vol. 11764, pp. 329–338. Springer, Cham (2019). https://doi.org/10.1007/978-3-030-32239-7_37
16. Tseng, K.-L., Lin, Y.-L., Hsu, W., Huang, C.-Y.: Joint sequence learning and cross-modality convolution for 3D biomedical segmentation. In: Conference on Computer Vision and Pattern Recognition, pp. 6393–6400 (2017)
17. Jetley, S., Lord, N.A., Lee, N., Torr, P.H.: Learn to pay attention. In: International Conference on Learning Representations. CoRR: arXiv:1804.02391 (2018)
18. Su, H., Maji, S., Kalogerakis, E., Learned-Miller, E.: Multi-view convolutional neural networks for 3D shape recognition. In: IEEE International Conference on Computer Vision, pp. 945–953 (2015)

Learning Bronchiole-Sensitive Airway Segmentation CNNs by Feature Recalibration and Attention Distillation

Yulei Qin[1,2,4], Hao Zheng[1,2], Yun Gu[1,2(✉)], Xiaolin Huang[1,2], Jie Yang[1,2(✉)], Lihui Wang[3], and Yue-Min Zhu[4]

[1] Institute of Image Processing and Pattern Recognition, Shanghai Jiao Tong University, Shanghai, China
{geron762,jieyang}@sjtu.edu.cn
[2] Institute of Medical Robotics, Shanghai Jiao Tong University, Shanghai, China
[3] Key Laboratory of Intelligent Medical Image Analysis and Precise Diagnosis of Guizhou Province, College of Computer Science and Technology, Guizhou University, Guiyang, China
[4] UdL, INSA Lyon, CREATIS, CNRS UMR 5220, INSERM U1206, Lyon, France

Abstract. Training deep convolutional neural networks (CNNs) for airway segmentation is challenging due to the sparse supervisory signals caused by severe class imbalance between long, thin airways and background. In view of the intricate pattern of tree-like airways, the segmentation model should pay extra attention to the morphology and distribution characteristics of airways. We propose a CNNs-based airway segmentation method that enjoys superior sensitivity to tenuous peripheral bronchioles. We first present a feature recalibration module to make the best use of learned features. Spatial information of features is properly integrated to retain relative priority of activated regions, which benefits the subsequent channel-wise recalibration. Then, attention distillation module is introduced to reinforce the airway-specific representation learning. High-resolution attention maps with fine airway details are passing down from late layers to previous layers iteratively to enrich context knowledge. Extensive experiments demonstrate considerable performance gain brought by the two proposed modules. Compared with state-of-the-art methods, our method extracted much more branches while maintaining competitive overall segmentation performance.

Keywords: Airway segmentation · Recalibration · Distillation

This work was partly supported by National Key R&D Program of China (No. 2019YFB1311503), Committee of Science and Technology, Shanghai, China (No. 19510711200), Shanghai Sailing Program (No. 20YF1420800), National Natural Science Foundation of China (No. 61661010, 61977046), International Research Project METISLAB, Program PHC-Cai Yuanpei 2018 (No. 41400TC), and China Scholarship Council.

A. L. Martel et al. (Eds.): MICCAI 2020, LNCS 12261, pp. 221–231, 2020.
https://doi.org/10.1007/978-3-030-59710-8_22

1 Introduction

Extraction of airways from computed tomography (CT) is critical for quantifying the morphological changes of chest (e.g., bronchial stenosis) and thereby indicating the presence and stage of related diseases. The intrinsic tree-like structure of airways demands great efforts for manual delineation and therefore several automatic airway segmentation methods have been proposed [11,12,19,21]. Most of them employed techniques like thresholding, region growing, and tubular structure enhancing. Since the intensity contrast between airway wall and lumen weakens as airways split into thinner branches, these methods often failed to extract peripheral bronchioles and produced leakage outside wall. Recent progress of convolutional neural networks (CNNs) [4,5,14,17] has spawned research on airway segmentation using CNNs [3,7–9,13,15,18,20,22,25]. 2-D and 2.5-D CNNs [3,22] were adopted to refine the initial coarsely segmented bronchi. 3-D CNNs were developed for direct airway extraction on CT volume in either an optimized tracking way [13,25] or a sliding window way [8]. Wang et al. [20] proposed recurrent convolution and radial distance loss to capture airways' topology. Qin et al. [15] transformed the airway segmentation task into connectivity prediction task for inherent structure comprehension. Graph neural networks [9,18] were also explored to incorporate nodes' neighborhood knowledge.

Despite the improved performance by deep learning, there still remain limitations to be overcome. First, severe class imbalance between airways and background poses a threat to the training of 3-D CNNs. Most of the current CNNs heavily rely on airway ground-truth as supervisory signals. Unlike bulky or spheroid-like organs (e.g., liver and kidney), tree-like airways are thin, tenuous, and divergent. It is difficult to train deep models using such sparse and scattered target labels. Although weighted cross-entropy loss was proposed to focus on positive samples, single source of supervisory signals from deficient airway labels still makes optimization ineffective. Second, the characteristics of airways require the model to utilize both global-scale and local-scale context knowledge to perceive the main body (trachea and main bronchus) and limbs (peripheral bronchi). Previous models used 2 or 3 pooling layers and the coarsest resolution features provide limited long-range context. If more layers are simply piled up, the increased parameters may cause over-fitting due to inadequate training data.

To address these two concerns, we present a CNNs-based airway segmentation method with high sensitivity to peripheral bronchioles. Our contributions are threefold: 1) The recalibration module is proposed to maximally utilize the learned features. On one hand, to avoid over-fitting of deep CNNs, the number of feature channels is reduced to limit model complexity. On the other hand, we do not expect the model's learning capacity to diminish because of such reduction. Under this circumstance, feature recalibration seems to be a reasonable solution. We hypothesize that spatial information of features is indispensable for channel-wise recalibration and should be treated differently from position to position, layer to layer. The average pooling used in [16,26] for spatial knowledge compression may not well capture the location of various airways in different resolution scales. In contrast, we aim at prioritizing information at key positions with learnable weights, which provides appropriate spatial hints to model inter-channel dependency and thereafter improves recalibration. 2) The attention distillation module is introduced to reinforce representation learning of airway structure. Attention maps of different features enable us to potentially reveal the morphology and distribution pattern of airways. Inspired by knowledge distillation [6,24], we refine the attention maps of lower resolution by mimicking those of higher resolution. Finer attention maps (teacher's role) with richer context can cram coarser ones (student's role) with details about airways and lung contours. The model's ability to recognize delicate bronchioles is ameliorated after iteratively focusing on the anatomy of airways. In addition, the distillation itself acts as auxiliary supervision that provides extra supervisory signals to assist training. 3) With extensive experiments on 90 clinical 3-D CT scans, our method achieved superior sensitivity to extraction of thin airways and maintained competitive overall segmentation performance.

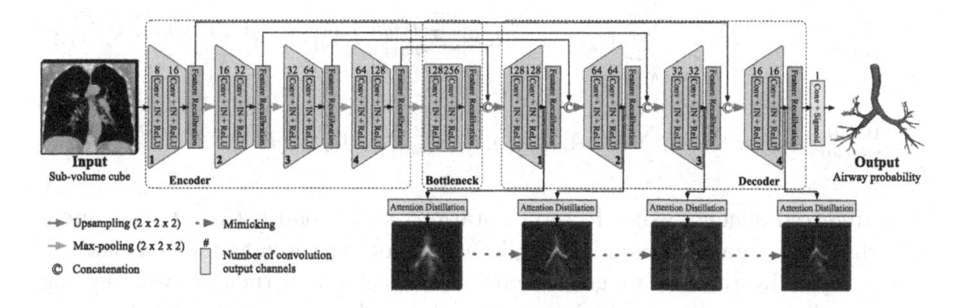

Fig. 1. Overview of the proposed method for airway segmentation.

2 Method

In this section, we present the design of the above-mentioned two modules: feature recalibration and attention distillation. The overview of the proposed method is illustrated in Fig. 1. For airway segmentation, the CNNs-based prediction can be formulated as $P = \mathcal{F}(X)$, where P denotes the predicted probability of airway at each voxel x of the input volume X. Given the airway label Y, the objective is to learn an end-to-end mapping $\mathcal{F}$ that minimizes the difference between P and Y by CNNs. Assuming the segmentation model has M convolution layers in total, we denote the activation output of the m-th convolution feature as $A_m \in R^{C_m \times D_m \times H_m \times W_m}, 1 \leq m \leq M$. The number of its channel, depth, height, and width are respectively denoted as C_m, D_m, H_m, and W_m.

Feature Recalibration: We propose the mapping block $\mathcal{Z}(\cdot)$ that generates a channel descriptor U_m to recalibrate the learned feature A_m. The recalibration by multiplying U_m with A_m unearths and strengthens basis channels that affect most the output decision. Unlike previous recalibration methods [16,26], we integrate spatial information using weighted combination of features along each spatial dimension. Our hypothesis is that different positions may hold different degree of importance both within A_m and across resolution scales (e.g., A_{m-1}, A_{m+1}). The operation like adaptive or global pooling is not spatially discriminating for the finest features (containing thin bronchioles that are easily "erased" by averaging) and the coarsest features (containing mostly thick bronchi). Therefore, we introduce the following spatial integration method $\mathcal{Z}_{spatial}(\cdot)$ that preserves relatively important regions. It can be formulated as:

$$\mathcal{Z}_{spatial}(A_m) = \mathcal{B}(\mathcal{Z}^{Depth}_{spatial}(A_m)) + \mathcal{B}(\mathcal{Z}^{Height}_{spatial}(A_m)) + \mathcal{B}(\mathcal{Z}^{Width}_{spatial}(A_m)), \quad (1)$$

$$\mathcal{Z}^{Depth}_{spatial}(A_m) = \sum_{j=1}^{H_m} h_j \sum_{k=1}^{W_m} w_k A_m[:, :, j, k], \mathcal{Z}^{Depth}_{spatial}(A_m) \in R^{C_m \times D_m \times 1 \times 1}, \quad (2)$$

$$\mathcal{Z}^{Height}_{spatial}(A_m) = \sum_{i=1}^{D_m} d_i \sum_{k=1}^{W_m} w_k A_m[:, i, :, k], \mathcal{Z}^{Height}_{spatial}(A_m) \in R^{C_m \times 1 \times H_m \times 1}, \quad (3)$$

$$\mathcal{Z}^{Width}_{spatial}(A_m) = \sum_{i=1}^{D_m} d_i \sum_{j=1}^{H_m} h_j A_m[:, i, j, :], \mathcal{Z}^{Width}_{spatial}(A_m) \in R^{C_m \times 1 \times 1 \times W_m}, \quad (4)$$

where indexed slicing (using Python notation) and broadcasting $\mathcal{B}(\cdot)$ are performed. The learnable parameters d_i, h_j, w_k denote the weight for each feature slice in depth, height, and weight dimension, respectively. Crucial airway regions are gradually preferred with higher weights during learning. To model the inter-channel dependency, we adopt the excitation technique [16] on the compressed spatial knowledge. Specifically, the channel descriptor is obtained by:

$$U_m = \mathcal{Z}(A_m) = f_2(K_2 * f_1(K_1 * \mathcal{Z}_{spatial}(A_m))), \quad (5)$$

where K_1, K_2 are 3-D kernels of size $1 \times 1 \times 1$ and "$*$" denotes convolution. Convolving with K_1 decreases the channel numbers to C_m/r and that with K_2 recovers back to C_m. The ratio r is the compression factor. $f_1(\cdot)$ and $f_2(\cdot)$ are non-linear activation functions. We choose Rectified Linear Unit (ReLU) as $f_1(\cdot)$ and Sigmoid as $f_2(\cdot)$ in the present study. Multiple channels are recombined through such channel reduction and increment, with informative ones emphasized and redundant ones suppressed. The final recalibrated feature $\hat{A}_m = U_m \odot A_m$ is calculated as element-wise multiplication between U_m and A_m.

Attention Distillation: Recent studies [6,24] on knowledge distillation showed that attention maps serve as valuable knowledge and can be transferred layer-by-layer from teacher networks to student networks. The motivation of our proposed attention distillation is that activation-based attention maps, which guide where to look at, can be distilled and exploited during backward transfer. Without separately setting two different models, later layers play the role of teacher and "impart" such attention to earlier layers in the same model. Besides, to tackle the challenge of insufficient supervisory signals caused by severe class imbalance, the distillation can be viewed as another source of supervision. It produces additional gradient that helps to train deep CNNs for airway segmentation. Specifically, the attention distillation is performed between two consecutive features A_m and A_{m+1}. Firstly, the activation-based attention map is generated by $G_m = \mathcal{G}(A_m), G_m \in R^{1\times D_m\times H_m\times W_m}$. Each voxel's absolute value in G_m reflects the contribution of its correspondence in A_m to the entire segmentation model. One way of constructing the mapping function $\mathcal{G}(\cdot)$ is to compute the statistics of activation values A_m across channel: $G_m = \sum_{c=1}^{C_m}|A_m[c,:,:,:]|^p$. The element-wise operation $|\cdot|^p$ denotes the absolute value raised to the p-th power. More attention is addressed to highly activated regions if $p > 1$. Here, we adopt channel-wise summation $\sum_{c=1}^{C_m}(\cdot)$ instead of maximizing $\max_c(\cdot)$ or averaging $\frac{1}{C_m}\sum_{c=1}^{C_m}(\cdot)$ because it is comparatively less biased. The summation operation retains all implied salient activation information without ignoring non-maximum elements or weakening discriminative elements. Visual comparison in Fig. 2 exhibits that summation with $p > 1$ intensifies most the sensitized airway regions. Secondly, trilinear interpolation $\mathcal{I}(\cdot)$ is performed to ensure that the processed 3-D attention maps share the same dimension. Then, spatial-wise Softmax $\mathcal{S}(\cdot)$ is applied to normalize all elements. Finally, we drive the distilled attention $\hat{G}_m$ closer to $\hat{G}_{m+1}$ by minimizing the following loss function:

$$\mathcal{L}_{distill} = \sum_{m=1}^{M-1} \|\hat{G}_m - \hat{G}_{m+1}\|_F^2, \hat{G}_m = \mathcal{S}(\mathcal{I}(G_m)), \tag{6}$$

where $\|\cdot\|_F^2$ is the squared Frobenius norm. With the current $\hat{G}_m$ iteratively mimicking its successor $\hat{G}_{m+1}$, visual attention guidance is transmitted from the deepest to the shallowest layer. Note that such distillation process does not require extra annotation labor and can work with arbitrary CNNs freely. To ensure that the direction of knowledge distillation is from back to front, we detach the $\hat{G}_{m+1}$ from the computation graph for each m in Eq. 6. Hence, the backward propagating gradients will not change the value of $\hat{G}_{m+1}$.

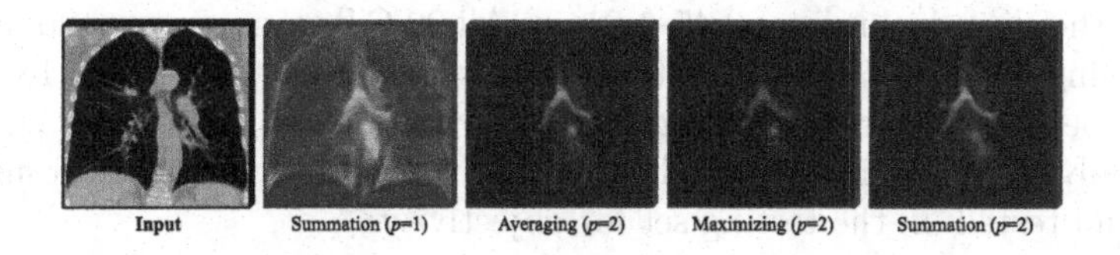

Fig. 2. The difference among mapping functions $\mathcal{G}(\cdot)$ for the lattermost attention map.

Model Design: The proposed method employs U-Net [5] architecture. To enlarge the receptive field of CNNs and facilitate the feature learning of long-range relationship, four pooling layers are used with five resolution scales. At each scale, both encoder and decoder have two convolution layers (kernel size $3 \times 3 \times 3$) followed by instance normalization (IN) and ReLU. Feature recalibration is applied at the end of each resolution scale. Since high-level features in decoder are also of high-resolution, we perform the decoder-side attention distillation to pass down the fine-grained details that are missing in previous low-resolution attention maps. The encoder-side distillation is not favored because low-level features are more local-scale and not strongly related to airway description. Preliminary experiments confirmed that such design is optimum for our task.

Training Loss: To deal with hard samples, we use both the Dice [14] and Focal loss [10] for the segmentation task. Given the binary label $y(x)$ and prediction $p(x)$ of each voxel x in the volume set X, the combined loss is defined as:

$$\mathcal{L}_{seg} = -(\frac{2\sum_{x \in X} p(x)y(x)}{\sum_{x \in X}(p(x) + y(x)) + \epsilon}) - \frac{1}{|X|}(\sum_{x \in X}(1 - p_t(x))^2 \log(p_t(x))), \quad (7)$$

where $p_t(x) = p(x)$ if $y(x) = 1$. Otherwise, $p_t(x) = 1 - p(x)$. Parameter ϵ is used to avoid division by zero. Together with attention distillation loss, the total loss function is given by $\mathcal{L}_{total} = \mathcal{L}_{seg} + \alpha \mathcal{L}_{distill}$, where α balances these two terms.

3 Experiments and Results

Datasets: For evaluation, we conducted airway segmentation experiments on 90 3-D thoracic CT scans collected from two public datasets: 1) 70 scans from LIDC [1]; 2) 20 scans from the training set of EXACT'09 [12]. For the LIDC dataset, 70 CT scans whose slice thickness $\leq$0.625 mm were randomly chosen. For the EXACT'09 dataset, 20 training scans were provided without airway annotation. The airway ground-truth of each CT scan was manually corrected based on the interactive segmentation results of ITK-SNAP [23]. The acquisition and investigation of data were conformed to the principles outlined in the declaration of Helsinki [2]. The axial size of all CT scans was 512×512 pixels with the spatial resolution of 0.5–0.781 mm. The number of slices in each scan varied from 157 to 764 with the slice thickness of 0.45–1.0 mm. All 90 CT scans were randomly split into training set (63 scans), validation set (9 scans), and testing set (18 scans). In the experiments, model training and hyper-parameter fine-tuning were performed only on the training set. The model with the best validation results was chosen and tested on the testing set for objectivity.

Setup: The CT pre-processing included Hounsfield Unit truncation, intensity normalization, and lung field segmentation [15]. In training phase, sub-volume CT cubes of size $64 \times 224 \times 304$ were densely cropped near airways. Random horizontal flipping, shifting, and voxel intensity jittering were applied as on-the-fly data augmentation. The Adam optimizer ($\beta_1 = 0.5, \beta_2 = 0.999$) with the learning rate 2×10^{-4} was used and the training converged after 50 epochs. In validation and testing phases, we performed the sliding window prediction with axial

Table 1. Results (%) of comparison on the testing set (Mean ± Standard deviation).

Method	BD	TD	TPR	FPR	DSC
3-D U-Net [5]	87.2 ± 13.7	73.8 ± 18.7	85.3 ± 10.4	0.021 ± 0.015	91.5 ± 2.9
V-Net [14]	91.0 ± 16.2	81.6 ± 19.5	87.1 ± 13.6	0.024 ± 0.017	92.1 ± 3.6
VoxResNet [4]	88.2 ± 12.6	76.4 ± 13.7	84.3 ± 10.4	0.012 ± 0.009	92.7 ± 3.0
Wang et al. [20]	93.4 ± 8.0	85.6 ± 9.9	88.6 ± 8.8	0.018 ± 0.012	93.5 ± 2.2
Juarez et al. [9]	77.5 ± 20.9	66.0 ± 20.4	77.5 ± 15.5	**0.009 ± 0.009**	87.5 ± 13.2
Qin et al. [15]	91.6 ± 8.3	82.1 ± 10.9	87.2 ± 8.9	0.014 ± 0.009	**93.7 ± 1.9**
Juarez et al. [8]	91.9 ± 9.2	80.7 ± 11.3	86.7 ± 9.1	0.014 ± 0.009	93.6 ± 2.2
Jin et al. [7]	93.1 ± 7.9	84.8 ± 9.9	88.1 ± 8.5	0.017 ± 0.010	93.6 ± 2.0
Our proposed	**96.2 ± 5.8**	**90.7 ± 6.9**	**93.6 ± 5.0**	0.035 ± 0.014	92.5 ± 2.0

Table 2. Results (%) of ablation study on the testing set (Mean ± Standard deviation).

Method	BD	TD	TPR	FPR	DSC
Baseline	91.6 ± 9.2	81.3 ± 11.5	87.2 ± 8.6	**0.014 ± 0.008**	**93.7 ± 1.7**
Baseline + cSE [26]	95.1 ± 6.2	88.5 ± 8.3	92.4 ± 5.5	0.033 ± 0.015	92.3 ± 1.9
Baseline + PE [16]	95.7 ± 5.1	88.4 ± 7.9	92.3 ± 5.9	0.037 ± 0.019	91.8 ± 2.8
Baseline + FR	96.1 ± 5.9	**90.8 ± 7.5**	92.9 ± 5.9	0.034 ± 0.016	92.3 ± 2.3
Baseline + AD	94.9 ± 6.9	88.3 ± 8.2	91.8 ± 6.2	0.029 ± 0.014	92.8 ± 1.4
Our proposed	**96.2 ± 5.8**	90.7 ± 6.9	**93.6 ± 5.0**	0.035 ± 0.014	92.5 ± 2.0

stride of 48. Results were averaged on overlapping margins and thresholded by 0.5 for binarization. No post-processing was involved. For the hyper-parameter setting, we empirically chose $\alpha = 0.1$, $\epsilon = 10^{-7}$, $p = 2$, and $r = 2$. Such setting may not be optimum and elaborate tuning may be conducted in future work.

Metrics: Considering clinical practice, only the largest connected component of segmented airways was kept and five metrics were used: (a) Branches detected (BD), (b) Tree-length detected (TD), (c) True positive rate (TPR), (d) False positive rate (FPR), and (e) Dice coefficient (DSC). We referred to [12] for airway tree-based metric definition (a)–(d) and the trachea region is excluded. For overall segmentation estimation by DSC, trachea is included in computation.

Quantitative Results: Since we adopted U-Net as backbone, comparison experiments were performed with other encoder-decoder CNNs: the original 3-D U-Net [5], its variants V-Net [14], and VoxResNet [4]. We also compared our method with five state-of-the-art methods: Wang et al. [20], Juarez et al. [9], Qin et al. [15], Juarez et al. [8], and Jin et al. [7]. All these methods were re-implemented in PyTorch and Keras by ourselves and fine-tuned on the same dataset. Table 1 shows that our method achieved the highest BD, TD, and TPR with a compelling DSC. The first three metrics directly reflect that our model outperformed the others in the detection of airways, especially distal parts of thin branches. Its superior sensitivity also comes with a side effect of slight performance decline in FPR and DSC. This may be ascribed to: 1) Our model successfully detected some true thin airways that were too indistinct to be annotated properly by experts. When calculating the evaluation metrics, these real branches were counted as false positives and therefore causing higher FPR and lower DSC. 2) A little leakage was produced at bifurcations where the contrast between airway lumen and wall was fairly low. In this situation, the segmentation model was inclined to predict voxels as airway due to the reinforced feature learning of tiny branches. However, it is clinically worthwhile that much more branches were segmented at such a trivial loss. Furthermore, we conducted ablation study to investigate the validity of each component in our method. The model trained without feature recalibration (FR) and attention distillation (AD) was indicated as baseline. Two very recently proposed recalibration methods (cSE [26] and PE [16]) were introduced into our baseline for comparison. As shown in Table 2, FR brings more performance gain than cSE and PE under the same experiment setting, confirming the importance of reasonably integrating spatial knowledge for recalibration. All components (cSE, PE, FR, and AD) increased the baseline model's sensitivity of detecting airways and also produced more false positives. In view of the metrics BD, TD, and TPR, both modules of FR and AD do contribute to the high sensitivity of our model.

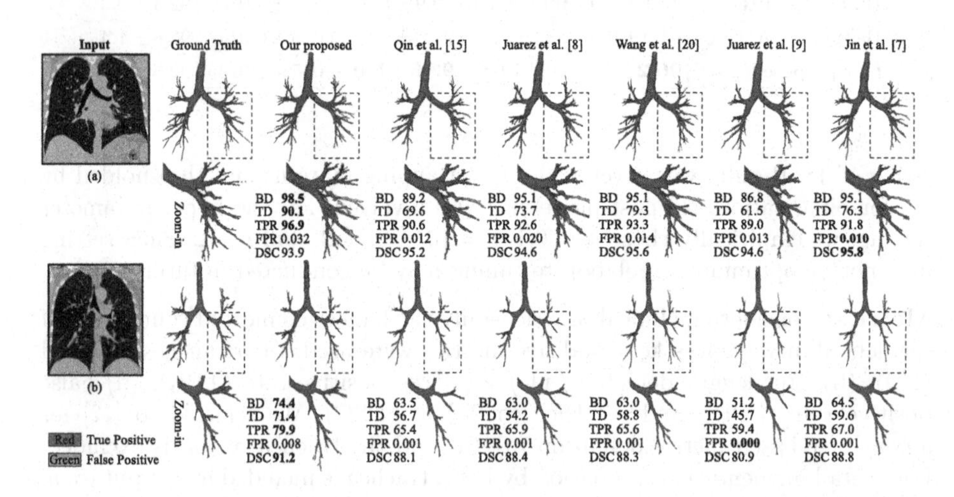

Fig. 3. Rendering of segmentation results of (a) easy and (b) hard testing cases.

Qualitative Results: Segmentation results are shown in Fig. 3 to demonstrate the robustness of our method on both easy and hard cases. In line with Table 1, all methods performed well on extracting thick bronchi. Compared with state-of-the-art methods, more visible tiny branches were reconstructed by the proposed method with high overall segmentation accuracy maintained. In Fig. 3 (a) and (b), some false positives were identified as true airways after retrospective evaluation of labels. These branches were unintentionally neglected due to annotation difficulty. Moreover, to intuitively assess the effect of attention distillation, attention maps from decoder 1–4 are visualized in Fig. 4. After distillation, the activated regions become more distinct and the task-related objects (e.g., airway, lung) are enhanced accordingly. The improved attention on airways and lung boundary explains that the model comprehended more context and therefore achieved higher sensitivity to intricate bronchioles. Another interesting finding is that although the last attention map is not refined in distillation, it still gets polished up because better representation learned at previous layers in turn affects late-layer features.

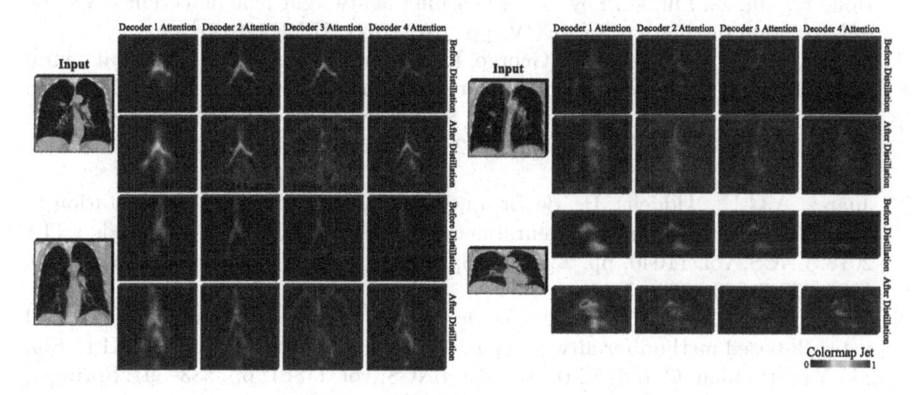

Fig. 4. Pseudo-color rendering of attention maps (decoder 1–4) before and after distillation process. These maps are min-max scaled and rendered with Jet colormap.

4 Conclusion

This paper proposed a highly sensitive method for CNNs-based airway segmentation. With the spatial-aware feature recalibration module and the gradually-reinforced attention distillation module, feature learning of CNNs becomes more effective and relevant to airway perception. Extensive experiments showed that our method detected much more bronchioles while maintaining competitive overall segmentation performance, which corroborates its superior sensitivity over state-of-the-art methods and the validity of the two constituent modules.

References

1. Armato III, S.G., et al.: The lung image database consortium (LIDC) and image database resource initiative (IDRI): a completed reference database of lung nodules on CT scans. Med. Phys. **38**(2), 915–931 (2011)

2. Association, W.M., et al.: World medical association declaration of Helsinki. Ethical principles for medical research involving human subjects. Bull. World Health Organ. **79**(4), 373 (2001)
3. Charbonnier, J.P., et al.: Improving airway segmentation in computed tomography using leak detection with convolutional networks. MedIA **36**, 52–60 (2017)
4. Chen, H., Dou, Q., Yu, L., Qin, J., Heng, P.A.: VoxResNet: deep voxelwise residual networks for brain segmentation from 3D MR images. NeuroImage **170**, 446–455 (2018)
5. Çiçek, Ö., Abdulkadir, A., Lienkamp, S.S., Brox, T., Ronneberger, O.: 3D U-Net: learning dense volumetric segmentation from sparse annotation. In: Ourselin, S., Joskowicz, L., Sabuncu, M.R., Unal, G., Wells, W. (eds.) MICCAI 2016. LNCS, vol. 9901, pp. 424–432. Springer, Cham (2016). https://doi.org/10.1007/978-3-319-46723-8_49
6. Hou, Y., Ma, Z., Liu, C., Loy, C.C.: Learning lightweight lane detection CNNs by self attention distillation. In: ICCV, pp. 1013–1021. IEEE (2019)
7. Jin, D., Xu, Z., Harrison, A.P., George, K., Mollura, D.J.: 3D convolutional neural networks with graph refinement for airway segmentation using incomplete data labels. In: Wang, Q., Shi, Y., Suk, H.I., Suzuki, K. (eds.) MLMI 2017. LNCS, vol. 10541, pp. 141–149. Springer, Cham (2017). https://doi.org/10.1007/978-3-319-67389-9_17
8. Juarez, A.G.U., Tiddens, H., de Bruijne, M.: Automatic airway segmentation in chest CT using convolutional neural networks. In: Stoyanov, D., et al. (eds.) TIA 2018. LNCS, vol. 11040, pp. 238–250. Springer, Cham (2018). https://doi.org/10.1007/978-3-030-00946-5_24
9. Juarez, A.G.U., Selvan, R., Saghir, Z., de Bruijne, M.: A joint 3D unet-graph neural network-based method for airway segmentation from chest CTs. In: Suk, H.I., Liu, M., Yan, P., Lian, C. (eds.) MLMI 2019. LNCS, vol. 11861, pp. 583–591. Springer, Cham (2019). https://doi.org/10.1007/978-3-030-32692-0_67
10. Lin, T.Y., Goyal, P., Girshick, R., He, K., Dollár, P.: Focal loss for dense object detection. In: ICCV, pp. 2980–2988. IEEE (2017)
11. Lo, P., Sporring, J., Ashraf, H., Pedersen, J.J., de Bruijne, M.: Vessel-guided airway tree segmentation: a voxel classification approach. MedIA **14**(4), 527–538 (2010)
12. Lo, P., et al.: Extraction of airways from CT (EXACT 2009). IEEE TMI **31**(11), 2093–2107 (2012)
13. Meng, Q., Roth, H.R., Kitasaka, T., Oda, M., Ueno, J., Mori, K.: Tracking and segmentation of the airways in chest CT using a fully convolutional network. In: MICCAI. pp. 198–207. Springer (2017)
14. Milletari, F., Navab, N., Ahmadi, S.A.: V-Net: Fully convolutional neural networks for volumetric medical image segmentation. In: IC3DV. pp. 565–571. IEEE (2016)
15. Qin, Y., et al.: AirwayNet: a voxel-connectivity aware approach for accurate airway segmentation using convolutional neural networks. In: Shen, D., et al. (eds.) MICCAI 2019. LNCS, vol. 11769, pp. 212–220. Springer, Cham (2019). https://doi.org/10.1007/978-3-030-32226-7_24
16. Rickmann, A.M., Roy, A.G., Sarasua, I., Navab, N., Wachinger, C.: 'Project & Excite' modules for segmentation of volumetric medical scans. In: Shen, D., et al. (eds.) MICCAI 2019. LNCS, vol. 11765, pp. 39–47. Springer, Cham (2019). https://doi.org/10.1007/978-3-030-32245-8_5
17. Ronneberger, O., Fischer, P., Brox, T.: U-Net: Convolutional networks for biomedical image segmentation. In: Navab, N., Hornegger, J., Wells, W., Frangi, A. (eds.) MICCAI 2015. LNCS, vol. 9351, pp. 234–241. Springer, Cham (2015). https://doi.org/10.1007/978-3-319-24574-4_28

18. Selvan, R., Kipf, T., Welling, M., Juarez, A.G.-U., Pedersen, J.H., Petersen, J., de Bruijne, M.: Graph refinement based airway extraction using mean-field networks and graph neural networks. Med. Image Anal. **64**, 101751 (2020). https://doi.org/10.1016/j.media.2020.101751
19. Van Rikxoort, E.M., Baggerman, W., van Ginneken, B.: Automatic segmentation of the airway tree from thoracic CT scans using a multi-threshold approach. In: Proceedings of Second International Workshop on Pulmonary Image Analysis, pp. 341–349 (2009)
20. Wang, C., et al.: Tubular structure segmentation using spatial fully connected network with radial distance loss for 3D medical images. In: Shen, D., et al. (eds.) MICCAI 2019. LNCS, vol. 11769, pp. 348–356. Springer, Cham (2019). https://doi.org/10.1007/978-3-030-32226-7_39
21. Xu, Z., Bagci, U., Foster, B., Mansoor, A., Udupa, J.K., Mollura, D.J.: A hybrid method for airway segmentation and automated measurement of bronchial wall thickness on CT. MedIA **24**(1), 1–17 (2015)
22. Yun, J., et al.: Improvement of fully automated airway segmentation on volumetric computed tomographic images using a 2.5 dimensional convolutional neural net. MedIA **51**, 13–20 (2019)
23. Yushkevich, P.A., et al.: User-guided 3D active contour segmentation of anatomical structures: significantly improved efficiency and reliability. NeuroImage **31**(3), 1116–1128 (2006)
24. Zagoruyko, S., Komodakis, N.: Paying more attention to attention: Improving the performance of convolutional neural networks via attention transfer. In: ICLR. OpenReview.net (2017)
25. Zhao, T., Yin, Z., Wang, J., Gao, D., Chen, Y., Mao, Y.: Bronchus segmentation and classification by neural networks and linear programming. In: Shen, D., et al. (eds.) MICCAI 2019. LNCS, vol. 11769, pp. 230–239. Springer, Cham (2019). https://doi.org/10.1007/978-3-030-32226-7_26
26. Zhu, W., et al.: AnatomyNet: deep learning for fast and fully automated whole-volume segmentation of head and neck anatomy. Med. Phys. **46**(2), 576–589 (2019)

Learning Rich Attention for Pediatric Bone Age Assessment

Chuanbin Liu, Hongtao Xie(✉), Yunyan Yan, Zhendong Mao, and Yongdong Zhang

School of Information Science and Technology, University of Science and Technology of China, Hefei 230026, China
htxie@ustc.edu.cn

Abstract. Bone Age Assessment (BAA) is a challenging clinical practice in pediatrics, which requires rich attention on multiple anatomical Regions of Interest (RoIs). Recently developed deep learning methods address the challenge in BAA with a hard-crop attention mechanism, which segments or detects the discriminative RoIs for meticulous analysis. Great strides have been made, however, these methods face severe requirements on precise RoIs annotation, complex network design and expensive computing expenditure. In this paper, we show it is possible to learn rich attention without the need for complicated network design or precise annotation – a simple module is all it takes. The proposed Rich Attention Network (RA-Net) is composed of a flexible baseline network and a lightweight Rich Attention module (RAm). Taking the feature map from baseline network, the RA module is optimized to generate attention with **discriminability** and **diversity**, thus the deep network can learn rich pattern attention and representation. With this artful design, we enable an end-to-end framework for BAA without RoI annotation. The RA-Net brings significant margin in performance, meanwhile negligible additional overhead in parameter and computation. Extensive experiments verify that our method yields state-of-the-art performance on the public RSNA datasets with mean absolute error (MAE) of 4.10 months.

Keywords: Bone age assessment · Deep learning · Visual attention

1 Introduction

Bone age assessment (BAA) is an important procedure in pediatric radiology [1–3]. Based on the discrepancy between the reading of the bone age and the chronological age, physicians can make accurate diagnoses of endocrinology problems, growth estimation and genetic disorders in children. Generally in clinical practice, radiologists estimate the bone age through elaborate visual inspection of multiple anatomical Regions of Interest (RoIs) to capture the subtle visual

A. L. Martel et al. (Eds.): MICCAI 2020, LNCS 12261, pp. 232–242, 2020.
https://doi.org/10.1007/978-3-030-59710-8_23

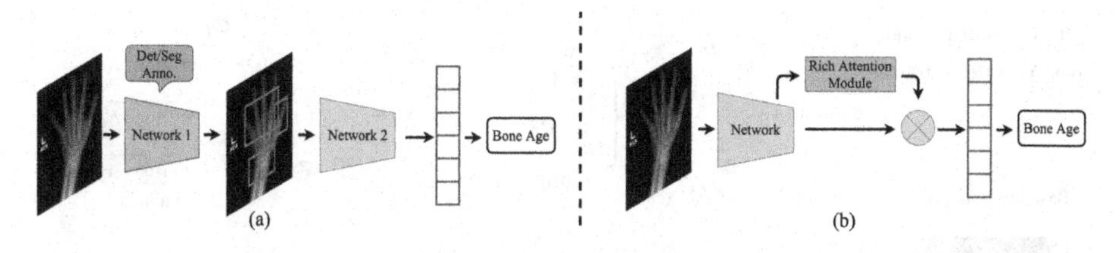

Fig. 1. The comparison between (a) existing SOTA methods and (b) RA-Net. RA-Net is able to learn rich attention without complicated network design or precise annotation.

traits. For example, the TW [3] method is accomplished through local individual analysis on 20 Regions of Interest (RoIs). Consequently, the BAA is a time-consuming and experience-sensitive task. Moreover, due to the complex pattern of bones and discrepant expertise of radiologist, the BAA task suffers from intra-inter observer error and problematic judgments. Although there have been many efforts to push forward computer-assisted system to assist BAA (e.g. BoneXpert [4]), they are still unsatisfactory due to their strict condition requirement and limited pattern recognition ability.

Related Work: Recently, deep learning has become a leading solution in medical image analysis [5,6], as it exhibits superior ability to exploit hierarchical feature representations from big data [7,8]. Drawing the success of deep network, a series of deep learning approaches have been proposed for BAA [9–12].

To accomplish the elaborate inspection on anatomical RoIs, existing state-of-the-art approaches utilize multi-step processing with hard-crop attention mechanism. As shown in Fig. 1(a), these methods firstly segment or detect the discriminative RoIs then corp them out for meticulous analysis. Iglovikov et al. [13] employ U-Net [14] to segment the hands out from radiography, then crop out the carpal bones, metacarpals and proximal phalanges for ensemble regression. Ren et al. [15] use Faster-RCNN [16] to find the hand foreground and apply Inception-V3 for feature extraction and age estimation. Escobar et al. [17] propose a three-stage pipeline with hand detection, hand pose estimation and final age assessment. Ji et al. [18] introduce a foreground segmentation model to remove background and then crop the top-ranking RoIs for assisting BAA. Liu et al. [19] propose an end-to-end structure to extract discriminative bone parts for assembling age assessment. Great strides have been made, however, these methods all face series of requirements: (i) it requires extra precise RoI annotation, which brings expensive and time-consuming labor to domain experts, (ii) it requires extra processing in RoI detection/segmentation, which leads to complex network design and expensive computing expenditure. Consequently, the severe requirements limit the clinical value of these methods.

Contribution: To address the challenging dilemma in bone age assessment, we aim to design a flexible and lightweight attention mechanism for deep learning

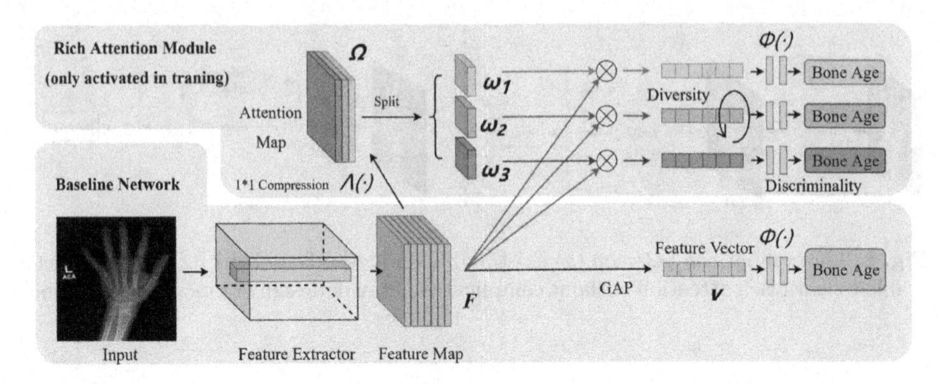

Fig. 2. The framework of RA-Net. The RA-Net is composed of a flexible baseline network and a lightweight Rich Attention module (RAm).

network. In this paper, we propose a novel Rich Attention Network (RA-Net) which can learn rich attention on RoIs automatically without precise annotation and complicated network design. As shown in Fig. 1(b), the proposed RA-Net is composed of a flexible baseline network and a lightweight Rich Attention module(RAm). And we creatively put forward **discriminability** component and **diversity** component to optimize the attention learning. The RA-Net brings significant margin in performance, meanwhile negligible additional overhead in parameter and computation. Extensive experiments verify that our method yields state-of-the-art performance on the public RSNA datasets with mean absolute error (MAE) of 4.10 months.

2 Method

Overall Framework: Figure 2 illustrates the overall framework of RA-Net for bone age assessment. The proposed RA-Net is composed of a flexible baseline network and a lightweight Rich Attention Module (RAm). Taking the final feature map from baseline network, the RAm generates a set of attention maps to indicate the attention pattern, each attention map is then element-wise multiplied to the original feature map to get the attention enhanced feature. The RA module is optimized to generate rich attention with: (i) **discriminability** that forces all enhanced feature maps to provide accurate age assessment, (ii) **diversity** that constraints different enhanced feature maps become mutually exclusive. Correspondingly, the deep network can get optimized to learn rich pattern attention and representation. With this artful design, we enable an end-to-end training for BAA without any precise RoI annotation.

Furthermore, owing to the parameter-sharing and backward-propagation, the feature extractor of baseline network is vested with strong representation learning in rich bone patterns, therefore the RA module can be shelved in testing. Consequently, there is no computation overhead at inference time except the feed-forwarding of standard baseline network, which makes our RA-Net a far more efficient solution for practical usage.

Baseline Network: Our RA-Net is highly flexible and its baseline network can be easily embedded by various convolutional neural networks, including VGG [20], ResNet [21], DenseNet [22]. Given a hand radiograph as input, the baseline network utilizes its feature extractor to extract the feature map $F_{BN} \in R^{H \times W \times N}$, followed by a Global Average Pooling (GAP) operation $g(\cdot)$ to obtain the feature vector $v_{BN} \in R^{1 \times N}$, where

$$v_{BN} = g(F_{BN}). \tag{1}$$

The original output layer is replaced with a fully connected layer with one output unit to estimate the bone age,

$$E_{BN} = \Phi(v_{BN}), \tag{2}$$

where the $\Phi(\cdot)$ represents the fully connected operation. It is worth mentioning that we share the same $\Phi(\cdot)$ for age estimation in both baseline network and RAm. Owing to the parameter-sharing and backward-propagation, the feature extractor can get simultaneously optimized by baseline network and RAm. Therefore the baseline network is vested with strong representation learning in rich bone patterns, even if the RAm is shelved in testing.

Rich Attention Module: The attention and local analysis on multiple anatomical RoIs is the critical step in BAA. In RA-Net, we propose a novel lightweight Rich Attention Module to indicate the locations of attention pattern and enhance the original feature map.

Taking the feature map F_{BN} from baseline network, the RAm generates a set of attention maps $\Omega \in R^{M \times H \times W}$ to indicate the distributions of attention on different M bone patterns, which is formulated as

$$\Omega = \Lambda(F_{BN}), \tag{3}$$

where $\Lambda(\cdot)$ represents a 1×1 $N \rightarrow M$ convolutional operation for attention grouping. Attention maps are then split by channel into M attention maps $\Omega = \{\omega_1, \omega_2, ..., \omega_M\}$ in which the $\omega_k \in R^{1 \times H \times W}$ is expected to reflect the attention map of the k_{th} RoI.

After that, we element-wise multiply feature maps F_{BN} by each attention map ω_k to generate M attention enhanced feature maps F_{ω_k}, as shown in Eq. 4,

$$F_{RA}^k = \omega_k \odot F_{BN} \quad (k = 1, 2, ..., M), \tag{4}$$

where $\odot$ indicates element-wise multiplication for two tensors. Similar with baseline network, we further extract the attention enhanced feature vector v_{RA}^k by Global Average Pooling.

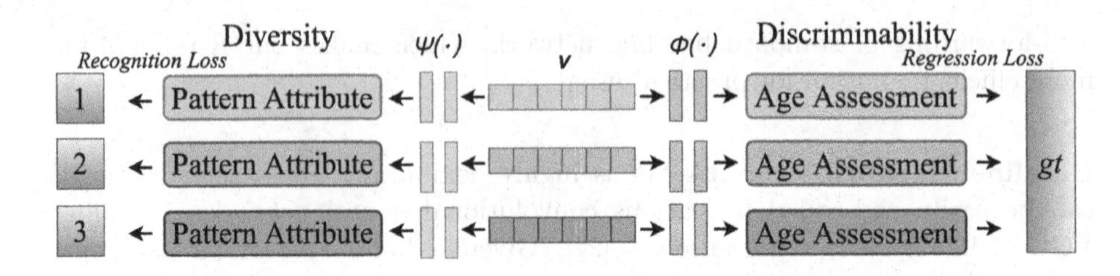

Fig. 3. The discriminability component and diversity component in RAm.

As shown in Fig. 3, each enhanced feature vector v_{RA}^k is utilized to estimate bone age E_{RA}^k by fully connected operation $\Phi(\cdot)$, which shares the same parameters in baseline network.

$$E_{RA}^k = \Phi(v_{RA}^k) = \Phi(g(\omega_k \odot F_{BN})) \quad (k = 1, 2, ..., M). \tag{5}$$

Moreover, each enhanced feature vector v_{ω_k} is operated with another fully connected operation $\Psi(\cdot)$ with M output unit to estimate the probability distributions $\mathbb{P}_{RA_k}$ for the M attributes of the attention pattern.

$$\mathbb{P}_{RA}^k = \Psi(v_{RA}^k) = \Psi(g(\omega_k \odot F_{BN})) \quad (k = 1, 2, ..., M). \tag{6}$$

Multi-task Formulation: The proposed RA-Net is simultaneously optimized by baseline network and RAm. Specifically, we formulate the objective function as a multi-task optimization problem, which is defined as follows:

$$\mathcal{L} = \alpha \mathcal{L}_{BN} + \lambda \mathcal{L}_{RA}. \tag{7}$$

The baseline network is optimized to estimate age assessment E_{BN} close to ground truth age gt, which can be formulated as a regression problem. Correspondingly the loss function $\mathcal{L}_{BN}$ is defined as Eq. 8

$$\mathcal{L}_{BN} = f_{reg}(E_{BN}, gt), \tag{8}$$

where the function f_{reg} is L1 loss function $f_{reg}(x, y) = |x - y|$ in our experiment.

The RAm is optimized to generate rich attention on different patterns. As shown in Fig. 3, we creatively propose two components to enable the optimization: (i) a **discriminability** component that forces all enhanced features to provide accurate age assessment, (ii) and a **diversity** component that constraints different enhanced features become mutually exclusive.

$$\mathcal{L}_{RA} = \beta \mathcal{L}_{dis} + \gamma \mathcal{L}_{div}. \tag{9}$$

Similar with baseline network, the discriminability component is defined as an age regression loss,

$$\mathcal{L}_{dis} = \sum_{k=1}^{M} f_{reg}(E_{RA}^k, gt). \tag{10}$$

Besides, the diverse attention on multiple different bone patterns can further benefit rich representation learning with complementary information. That is: (i) different attention map ω_k with the same input should reflect the attention on different pattern, which makes enhanced features mutually exclusive, (ii) same attention map ω_k according to different input should reflect the attention on the same pattern, which makes the enhanced features semantically consistent. We simply set the attribute of the attention pattern as $\{k\}_{k=1}^{M}$, and define the diversity component as an attribute recognition loss,

$$\mathcal{L}_{div} = \sum_{k=1}^{M} f_{rec}(\mathbb{P}_{RA}^{k}, k). \tag{11}$$

where the function f_{rec} is cross-entropy loss function in our experiment.

During training, we optimize our RA-Net following a multi-task learning strategy and the loss function is defined as below:

$$\mathcal{L} = \alpha\mathcal{L}_{BN} + \beta\mathcal{L}_{dis} + \gamma\mathcal{L}_{div}. \tag{12}$$

During testing, we remove the RAm and only report age assessment from baseline network E_{BN} as the final result.

3 Experiments and Results

Experiment Setup: We evaluate RA-Net on the dataset of RSNA 2017 Pediatric Bone Age Challenge, which consists of 12611 images for training and 200 images for testing. The Mean Absolute Error (MAE) between predicted age E_{BN} and ground-truth age gt on test set is reported as the final evaluation standard.

Our RA-Net is totally implemented with PyTorch [23] and we optimize it separately for male and female. RA-Net is highly flexible and its baseline network can be easily embedded by various convolutional neural networks, such as VGG [20], ResNet [21] and DenseNet [22]. For fair comparison with existing state-of-the-art solutions [11,19], we choose ResNet50 as our baseline network. The hyper-parameters α, β, γ are simply set as 1. Comparative experiments are carried out on the number of attention maps, where M ranges from 2 to 6, and finally we set $M = 4$ as the optimal value. We use SGD to optimize RA-Net, where each mini-batch consists of 32 512×512 images. The initial learning rate is set as 1e-3 and is multiplied by 0.1 after every 50 epochs. Without tips and tricks, out RA-Net can be easily optimized on a Ubuntu workstation with 2 NVIDIA GeForce 1080Ti GPU within 12 h.

Performance Comparison: Table 1 compares RA-Net with other state-of-art methods. RA-Net obtains the best accuracy with MAE of 4.10 months. Generally, the approaches with attention mechanism [13,15,17–19] can achieve better accuracy than that without attention mechanism [9–11]. This clearly states the attention on anatomical RoIs has critical significance for BAA. However, the

Table 1. Comparison results with the state-of-art methods.

Method	Dataset	Attention mech.	Multiple step	MAE (Months)
Human expert [11]	RSNA2017			7.32
Mutasa et al. [9]	Private	×	×	7.64
Spampinato et al. [10]	Private	×	×	9.12
Larson et al. [11]	RSNA2017	×	×	6.24
Iglovikov et al. [13]	RSNA2017	✓	✓	4.97
Ren et al. [15]	RSNA2017	✓	✓	5.20
Escobar et al. [17]	RSNA2017	✓	✓	4.14
Ji et al. [18]	RSNA2017	✓	✓	4.49
Liu et al. [19]	RSNA2017	✓	✓	4.38
Our method	RSNA2017	✓	×	**4.10**

existing methods actualize the attention learning by a hard-crop attention mechanism, which requires extra RoI annotation and image processing. The precise annotation, complex network design and expensive computing expenditure limit the generalization of these methods and make them less influential for clinical application.

Our RA-Net is a highly flexible and lightweight design, which is totally free from extra RoI annotation and multi-step processing. The fully automatic attention learning entrusts RA-Net stronger representation learning ability than the hard-crop attention methods. Correspondingly, RA-Net achieves an appreciable improvement over these methods, and exhibits remarkable value in clinical practice.

Ablation Studies: Ablation studies are first conducted to investigate the margin on model parameters (*w.r.t* Param), computing expenditure (*w.r.t* Flops) and assessment accuracy (MAE). As shown in Table 2, RA-Net brings significant improvement over the original network with same ResNet50 as backbone. At the same time, it costs negligible additional overhead ($\ll 0.1\%$) in parameter and computation during training, as we only introduce one 1×1 convolutional layer $\Lambda(\cdot)$ for attention grouping and one fully connected layer $\Psi(\cdot)$ to estimate the attribute of attention pattern. Furthermore, as the RAm is shelved in testing, there is no additional overhead except the backbone network feedforward during inference, making our RA-Net a highly efficient solution for practical usage.

Moreover, ablation studies are carried out on the number of attention maps M. As shown in Fig. 4, we achieve the best accuracy when $M = 4$, which indicates that visual attention on 4 different pattern is optimal in BAA task. This discovery keeps close consistence to the previous researches [18,19].

Table 2. Ablation studies on the margin.

ResNet50	Param (M)	FLOPs (G)	MAE (months)
Original	23.5101	21.4699	5.83
RA-Net	23.5265	21.4741	4.10
Margin	↑ 0.070%	↑ 0.020%	↓ 1.73

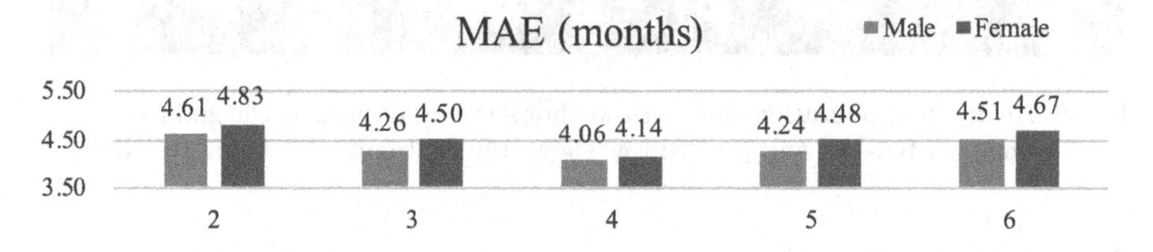

Fig. 4. Ablation studies on the number of attention maps.

Visualization Experiments: We firstly conduct visualization experiments to investigate the generation of attention maps. As shown in Fig. 5, attention maps ω_k with different inputs exhibit similar responses, while attention maps ω_k with different indexes present different distribution. This certifies the RAm can get optimized to generate attention with **discriminability** and **diversity**, thus the deep network can learn rich pattern attention and representation.

Besides, we also conduct visualization experiments to investigate the feature activation heatmap by Grad-CAM [24]. As shown in Fig. 6, the original network exhibits a circumscribed activation on the metacarpal bone part, meanwhile RA-Net generates a wide activation on different bone parts. This demonstrates RA-Net possesses a rich pattern representation learning ability over serval RoIs, which finally brings better assessment performance.

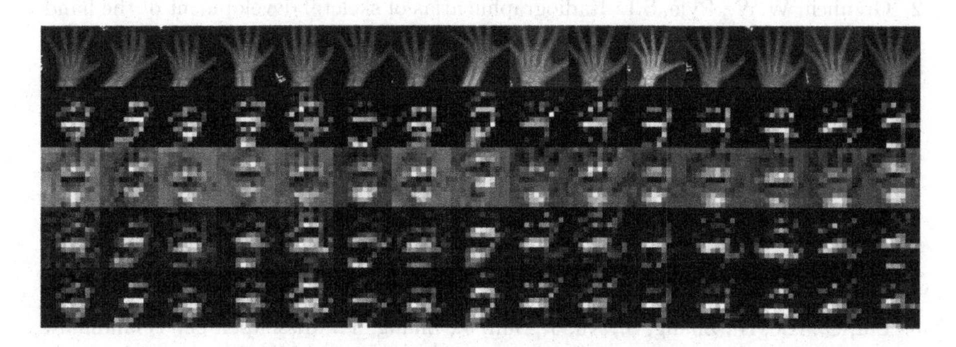

Fig. 5. Attention map visualization. The 1st row shows the input images, and the 2nd to 5th rows show the attention maps ω_1, ω_2, ω_4, ω_4.

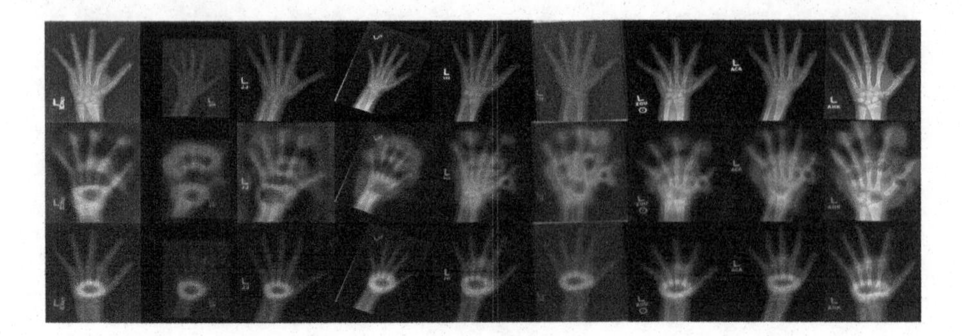

Fig. 6. Heatmap visualization. The 1st row shows the input images, the 2nd row shows the heatmaps of RA-Net and the 3rd row shows the heatmaps of original ResNet50.

4 Conclusion

This work presents RA-Net, a novel flexible and lightweight approach with discriminative and diverse attention for bone age assessment. We show it is possible to learn rich attention without the need of complicated network design or extra RoI annotation. Our method obtains state-of-the-art performance on the public RSNA 2017 datasets with MAE of 4.10 months.

Acknowledgements. This work is supported by the National Nature Science Foundation of China (61976008, U19A2057), the Youth Innovation Promotion Association Chinese Academy of Sciences (2017209). Authors would thank the Chinese Government for protecting her nationals against COVID-19.

References

1. King, D., et al.: Reproducibility of bone ages when performed by radiology registrars: an audit of tanner and whitehouse II versus Greulich and Pyle methods. Br. J. Radiol. **67**(801), 848–851 (1994)
2. Greulich, W.W., Pyle, S.I.: Radiographic atlas of skeletal development of the hand and wrist. Am. J. Med. Sci. **238**(3), 393 (1959)
3. Tanner, J.M., Whitehouse, R., Cameron, N., Marshall, W., Healy, M., Goldstein, H., et al.: Assessment of Skeletal Maturity and Prediction of Adult Height (TW2 Method), vol. 16. Academic Press, London (1975)
4. Thodberg, H.H., Kreiborg, S., Juul, A., Pedersen, K.D.: The BoneXpert method for automated determination of skeletal maturity. IEEE Trans. Med. Imaging **28**(1), 52–66 (2009). https://doi.org/10.1109/TMI.2008.926067
5. Xie, H., Yang, D., Sun, N., Chen, Z., Zhang, Y.: Automated pulmonary nodule detection in CT images using deep convolutional neural networks. Pattern Recognit. **85**, 109–119 (2019). https://doi.org/10.1016/j.patcog.2018.07.031
6. Liu, C., Xie, H., Zhang, S., Xu, J., Sun, J., Zhang, Y.: Misshapen pelvis landmark detection by spatial local correlation mining for diagnosing developmental dysplasia of the hip. In: Shen, D., et al. (eds.) MICCAI 2019. LNCS, vol. 11769, pp. 441–449. Springer, Cham (2019). https://doi.org/10.1007/978-3-030-32226-7_49

7. Wang, Y., Xie, H., Zha, Z.J., Tian, Y., Fu, Z., Zhang, Y.: R-net: arelationship network for efficient and accurate scene text detection. IEEE Trans. Multimedia (2020)
8. Liu, C., Xie, H., Zha, Z., Yu, L., Chen, Z., Zhang, Y.: Bidirectional attention-recognition model for fine-grained object classification. IEEE Trans. Multimedia **22**(7), 1785–1795 (2020)
9. Mutasa, S., Chang, P.D., Ruzal-Shapiro, C., Ayyala, R.: MABAL: a novel deep-learning architecture for machine-assisted bone age labeling. J. Digit. Imaging **31**(4), 513–519 (2018). https://doi.org/10.1007/s10278-018-0053-3
10. Spampinato, C., Palazzo, S., Giordano, D., Aldinucci, M., Leonardi, R.: Deep learning for automated skeletal bone age assessment in X-ray images. Med. Image Anal. **36**, 41–51 (2017). https://doi.org/10.1016/j.media.2016.10.010
11. Larson, D.B., Chen, M.C., Lungren, M.P., Halabi, S.S., Stence, N.V., Langlotz, C.P.: Performance of a deep-learning neural network model in assessing skeletal maturity on pediatric hand radiographs. Radiology **287**(1), 313–322 (2018)
12. Halabi, S.S., et al.: The RSNA pediatric bone age machine learning challenge. Radiology **290**(2), 498–503 (2019). https://doi.org/10.1148/radiol.2018180736
13. Iglovikov, V.I., Rakhlin, A., Kalinin, A.A., Shvets, A.A.: Paediatric bone age assessment using deep convolutional neural networks. In: Stoyanov, D., et al. (eds.) DLMIA/ML-CDS -2018. LNCS, vol. 11045, pp. 300–308. Springer, Cham (2018). https://doi.org/10.1007/978-3-030-00889-5_34
14. Ronneberger, O., Fischer, P., Brox, T.: U-net: convolutional networks for biomedical image segmentation. In: Navab, N., Hornegger, J., Wells, W.M., Frangi, A.F. (eds.) MICCAI 2015. LNCS, vol. 9351, pp. 234–241. Springer, Cham (2015). https://doi.org/10.1007/978-3-319-24574-4_28
15. Ren, X., Li, T., Wang, Q.: Regression convolutional neural network for automated pediatric bone age assessment from hand radiograph. IEEE J. Biomed. Health Inf. **23**(5), 2030–2038 (2018)
16. Ren, S., He, K., Girshick, R., Sun, J.: Faster R-CNN: towards real-time object detection with region proposal networks. In: Advances in Neural Information Processing Systems, pp. 91–99 (2015)
17. Escobar, M., González, C., Torres, F., Daza, L., Triana, G., Arbeláez, P.: Hand pose estimation for pediatric bone age assessment. In: Shen, D., et al. (eds.) MICCAI 2019. LNCS, vol. 11769, pp. 531–539. Springer, Cham (2019). https://doi.org/10.1007/978-3-030-32226-7_59
18. Ji, Y., Chen, H., Lin, D., Wu, X., Lin, D.: PRSNet: part relation and selection network for bone age assessment. In: She, D., et al. (eds.) MICCAI 2019. LNCS, vol. 11769, pp. 413–421. Springer, Cham (2019). https://doi.org/10.1007/978-3-030-32226-7_46
19. Liu, C., Xie, H., Liu, Y., Zha, Z., Lin, F., Zhang, Y.: Extract bone parts without human prior: end-to-end convolutional neural network for pediatric bone age assessment. In: Shen, D., et al. (eds.) MICCAI 2019. LNCS, vol. 11769, pp. 667–675. Springer, Cham (2019). https://doi.org/10.1007/978-3-030-32226-7_74
20. Simonyan, K., Zisserman, A.: Very deep convolutional networks for large-scale image recognition. In: 3rd International Conference on Learning Representations, ICLR 2015, San Diego, CA, USA, May 7–9 2015, Conference Track Proceedings (2015). http://arxiv.org/abs/1409.1556
21. He, K., Zhang, X., Ren, S., Sun, J.: Deep residual learning for image recognition. In: 2016 IEEE Conference on Computer Vision and Pattern Recognition, CVPR 2016, Las Vegas, NV, USA, June 27–30 2016, pp. 770–778 (2016). https://doi.org/10.1109/CVPR.2016.90

22. Huang, G., Liu, Z., Van Der Maaten, L., Weinberger, K.Q.: Densely connected convolutional networks. In: Proceedings of the IEEE Conference on Computer Vision and Pattern Recognition, pp. 4700–4708 (2017)
23. Paszke, A., et al.: Pytorch: an imperative style, high-performance deep learning library. In: Wallach, H.M., Larochelle, H., Beygelzimer, A., d'Alché-Buc, F., Fox, E.B., Garnett, R. (eds.) Advances in Neural Information Processing Systems 32: Annual Conference on Neural Information Processing Systems 2019, NeurIPS 2019, Canada, Vancouver, BC, 8–14 December 2019, pp. 8024–8035 (2019)
24. Selvaraju, R.R., Cogswell, M., Das, A., Vedantam, R., Parikh, D., Batra, D.: Grad-cam: visual explanations from deep networks via gradient-based localization. In: IEEE International Conference on Computer Vision, ICCV 2017, Venice, Italy, 22–29 October 2017, pp. 618–626 (2017). https://doi.org/10.1109/ICCV.2017.74

Weakly Supervised Organ Localization with Attention Maps Regularized by Local Area Reconstruction

Heng Guo, Minfeng Xu, Ying Chi(✉), Lei Zhang, and Xian-Sheng Hua

Alibaba Group, Hangzhou, China
xinyi.cy@alibaba-inc.com

Abstract. Fully supervised methods with numerous dense-labeled training data have achieved accurate localization results for anatomical structures. However, obtaining such a dedicated dataset usually requires clinical expertise and time-consuming annotation process. In this work, we tackle the organ localization problem under the setting of image-level annotations. Previous Class Activation Map (CAM) and its derivatives have proved that discriminative regions of images can be located with basic classification networks. To improve the representative capacity of attention maps generated by CAMs, a novel learning-based Local Area Reconstruction (LAR) method is proposed. Our weakly supervised organ localization network, namely OLNet, can generate high-resolution attention maps that preserve fine-detailed target anatomical structures. Online generated pseudo ground-truth is utilized to impose geometric constraints on attention maps. Extensive experiments on In-house Chest CT Dataset and Kidney Tumor Segmentation Benchmark (KiTS19) show that our approach can provide promising localization results both in saliency map and semantic segmentation perspectives.

Keywords: Organ localization · Local area reconstruction · Attention map.

1 Introduction

Many supervised machine learning approaches have been proposed for automatic localization of anatomical structures from CT or MRI, such as marginal space learning [20], ensemble learning [6,22] and regression forest [5]. These learning-based methods are known to perform well with the presence of numerous training data. Encountering the difficulties of collecting dense-labeled data, many researchers have explored how to utilize image-level annotations. Zhou et al. [21] found that Class Activation Map (CAM) extracted from a classification network with global average pooling layer has remarkable localization ability despite being trained on image-level labels. This gave rise to subsequent great works, such as Grad-CAM [15] and Grad-CAM++ [4]. A series of weakly supervised semantic segmentation approaches have emerged prosperously based on the localization

A. L. Martel et al. (Eds.): MICCAI 2020, LNCS 12261, pp. 243–252, 2020.
https://doi.org/10.1007/978-3-030-59710-8_24

cues provided by CAMs [2]. Although most of these methods have been developed on natural scene images, there are still some related works in the medical image literature. Huang et al. [11] proposed a weakly supervised learning method that leverages Grad-CAM and saliency map in a complementary way to provide accurate evidence localization on pathology images. Feng et al. [7] employed an image classification model to detect slices with lung nodule, and simultaneously learn the discriminative regions from the activation maps of convolutional units at different scales for segmentation. Gondal et al. [8] aimed to learn a representation that can enable localization of discriminative features in a retina image while achieving good classification accuracy simultaneously. And Chan et al. [3] proposed HistoSegNet to semantically segment whole slide pathology images with patch-level annotations.

Few works have been conducted in the literature on applying CAMs to localize target organ from medical volumes. In this work, we propose a weakly supervised Organ Localization Network (OLNet) to tackle this problem. For a 3D medical volume, we extract axial slices as positive and negative samples referring to the upper and lower bounds in coronal plane of the target organ. A binary classifier is trained to predict the positive or negative presence of the organ in each axial slice. Rough localization results could be obtained using methods based on CAMs in our preliminary experiments. Observing that the representative capacity of CAMs is limited due to the interpolation stage, which increases the spatial resolution of attention maps, a novel Local Area Reconstruction (LAR) method is proposed to impose geometric constraints on attention maps generated by CAMs. To provide supervision for LAR, we construct an online pseudo ground-truth generation strategy, which can omit the external object proposals. Under mild assumptions our method can also be regarded as a plug-in to existing CAMs computation methods to improve their localization performance. When post-processed with a fully-connected CRF [13], our regularized high-resolution attention maps can obtain highly precise semantic segmentation results.

OLNet is a novel multi-task network with two-fold capacities: 1) discriminate the presence of the target organ in axial slices extracted from a 3D CT volume; and 2) demonstrate fine-detailed geometric structures of target organ in positive slices with regularized attention maps. To train OLNet, annotators only need to annotate the upper and lower bounds indices of the target organ in a 3D volume, which is promising for reducing their workload by a large margin.

2 Method

In this section, we firstly introduce our weakly supervised solution for organ localization, then we demonstrate the online pseudo Ground-Truth (GT) generation strategy and the LAR method aiming at improving the localizability for classification network.

2.1 Organ Localization via Image-Level Supervision

Unlike previous weakly supervised methods which utilize CAMs out of the shelf, OLNet is constructed on the observation of the imprecise low-resolution attention maps of CAMs. We will introduce the details of OLNet architecture, basic concepts of CAMs and insights of OLNet construction.

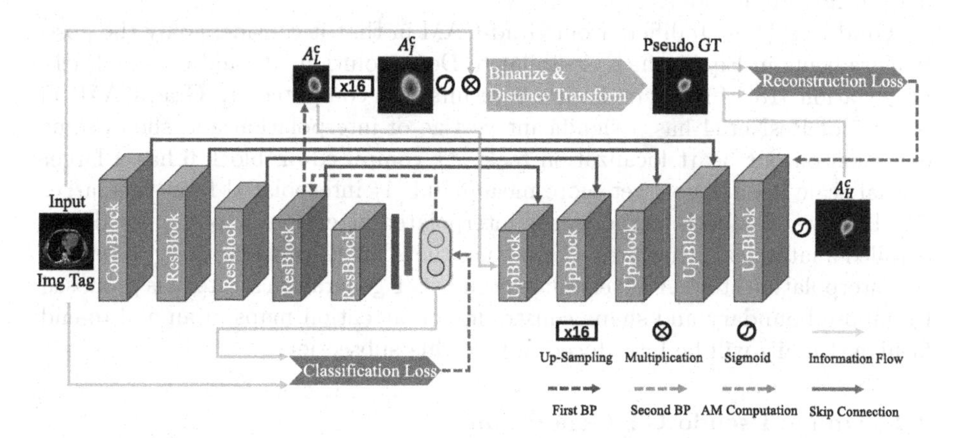

Fig. 1. Illustration of OLNet architecture and training pipeline. OLNet has two back-propagations (BP) which act asynchronously. Low-resolution attention map (AM) is computed in the first BP and the disparities between decoded high-resolution AM and online generated pesudo GT are collected and propagated with classification loss in the second BP. Obvious direct connections are omitted for simplicity.

We use pretrained ResNet34 [9] model, which is a PyTorch implementation consisting of a feature extractor and a classifier, as our classification backbone in the encoder part. The pretrained feature extractor is split sequentially into 8 parts in its namespace. To simplify, we use *ConvBlock* to represent the first four layers and the following four *ResBlocks* to represent the core residual blocks {4, 5, 6, 7} in ResNet34 respectively, as shown in Fig. 1.

With such a typical classification network, CAM [21] and its derivatives [4,15] can compute visual clues which support the discriminative decision. For a given image I, let $f_{l,k,i,j}$ be the activation on pixel (i, j) of feature map k in the l-th layer. For the class c which is expected to be probed, Grad-CAM [15] computes the gradient of score s^c for class c, with respect to activation maps in target layer l. Global average pooling is performed on these back-propagated gradients to obtain the neuron importance weights $w_{l,k}^c$:

$$w_{l,k}^c = \frac{1}{Z} \sum_i \sum_j \frac{\partial s^c}{\partial f_{l,k,i,j}}, \tag{1}$$

then the weight matrix w_l^c and the feature map matrix f_l could be integrated through a *conv2d* operation, in which w_l^c serves as the kernel. A *ReLU* operation is then used to get the attention map in layer l. Previous approaches often extract this attention map from the last convolution layer of a network. Taking ResNet34 for example, there are 5 downsample operations, yielding the output feature map with a shape of $(7, 7)$ in height and width for a $(224, 224)$ input. We denote the attention maps computed on these low-resolution feature maps as A_L^c for class c. Interpolation is used to upgrade the resolution to get the original size attention map, which we denote as A_I^c.

Grad-CAM++ [4] differs from Grad-CAM in that it considers only the positive gradients in Eq. 1, which is similar to Deconvolution [19] and Guided Backpropagation [18]. In our preliminary results, A_I^c computed by Grad-CAM on block 7 of ResNet34 has a significant vestige of interpolation and shows some deviation on our heart localization task. A_L^c computed on block 6 has a larger spatial scale itself, so we get more meaningful A_I^c interpolated from this larger A_L^c. However, for both methods, the interpolated attention maps cannot depict detailed anatomical structures. Can we use the CNN learning capacity to replace the interpolation step? A novel online pseudo GT generation strategy is proposed to impose boundary and shape constraints on attention maps in an end-to-end fashion. Details will be introduced in following subsection.

2.2 Online Pseudo GT Generation

In our experiments, we find that the attention map extracted from block 6 in ResNet34 has a good balance between spatial information and high level semantics, thus we execute the first backpropagation towards block 6 to compute A_L^c as shown with red dotted arrow in Fig. 1.

Given the low-resolution attention map A_L^c, we can directly up-sample it by interpolating to get an amplified A_I^c as previous methods do. Then we can compute a soft mask, which will then be applied on the original input image to get a discriminative local area for class c, denoted by I^{*c}, as in Eq. 2:

$$I^{*c} = S\left(A_I^c\right) \odot I, \tag{2}$$

where $\odot$ denotes element-wise multiplication. $S\left(A_I^c\right)$ is a Sigmoid operation performed on A_I^c to get a soft mask. And σ which is set to 0.3 in our experiments, serves as a threshold, and ω is a scale factor. The formulation is in Eq. 3:

$$S\left(A_I^c\right) = \frac{1}{1 + \exp\left(-\omega\left(A_I^c - \sigma\right)\right)}. \tag{3}$$

The discriminative local area I^{*c} contains abundant anatomical structure information belonging to target organ, which is able to guide the learning process of attention maps. Basic morphological operations are conducted to get the pseudo GT, as defined in Eq. 4:

$$G_p = T(B(I^{*c})), \tag{4}$$

where $B(\cdot)$ denotes binarization with mean value as the threshold and $T(\cdot)$ denotes distance transformation. The whole pseudo GT synthesis pipeline is illustrated at the upper branch in Fig. 1.

2.3 Local Area Reconstruction

Given the pseudo GT G_p, and a decoder network composed of 5 sequential *UpBlocks* as shown in Fig. 1, we can construct a learning-based pipeline from low-resolution attention map A_L^c to a learned high-resolution attention map, denoted by A_H^c. We use *conv2d-conv2d-pixelshuffle* in sequential combination to build the *UpBlock*. Low level features extracted by encoder are concatenated with corresponding decoder upsampled features to complement the morphological information.

For the output of our LAR module, we compute reconstruction loss using $L1$ norm as in Eq. 5. We only consider the reconstruction performance of positive samples.

$$\mathcal{L}_{re} = \mathbb{E}\|A_H^c - G_p\|_1, \tag{5}$$

where $c = 1$. For the classification task, a common cross entropy loss is used to guide the network learning. Thus, our total loss can be defined as:

$$\mathcal{L} = \lambda_1 \mathcal{L}_{ce} + \lambda_2 \mathcal{L}_{re}, \tag{6}$$

where λ_1 and λ_2 control the trade-off between the classification accuracy and the fidelity of spatial information. In our experiments, we set $\lambda_1 = 1$ and $\lambda_2 = 0.01$.

3 Experiments

We perform minmax normalization, random crop, elastic deformation and resize image to (224, 224) on axial slices extracted from 3D medical volumes. Pixelshuffle proposed in [16] is used to alleviate checkerboard artifacts. Weakly supervised localization results are reported from two perspectives: saliency map performance and semantic segmentation performance. Our method is implemented with PyTorch [14] and one NVIDIA graphics card.

3.1 Datasets

In-house Chest CT Dataset (HLoc). We obtain a low-dose chest CT dataset containing 237 volumes collected from 237 subjects by our cooperative hospital. Localizing heart from a chest CT is a prerequisite for subsequent analysis on cardiovascular diseases, thus the dataset is named HLoc meaning heart localization. All of these chest CT volumes have an equal slice thickness of 3 mm. We randomly split the dataset into three parts (200 for training, 7 for validation and 30 for testing). 2D axial slices are extracted from these 3D CT images with positive tag for slices between upper and lower bounds and negative tag otherwise.

Kidney Tumor Segmentation Benchmark (KiTS19). KiTS19 [10] dataset is publicly available and now serves as an ongoing benchmarking resource. To evaluate our weakly supervised localization algorithm on KiTS19, for the 210 volumes with ground-truth, we again randomly split them into three parts (180 for training, 5 for validation and 25 for testing). We merge the kidney and kidney tumor segmentation to get a binary kidney mask. Slices are extracted in the same way as the chest CT dataset.

Table 1. Quantitative results on validation and test sets. All these methods have a ResNet [9] classification backbone except CAM [21], which has a VGG [17] backbone following the original implementation.

Method	Dataset							
	HLoc				KiTS19			
	Valid.		Test		Valid.		Test	
	F_β	F_β^{max}	F_β	F_β^{max}	F_β	F_β^{max}	F_β	F_β^{max}
CAM [21]	0.386	0.410	0.388	0.437	0.261	0.635	0.276	0.628
Grad-CAM [15]	0.542	0.674	0.532	0.698	0.232	0.602	0.275	0.572
Grad-CAM++ [4]	0.473	0.677	0.467	0.717	0.173	0.595	0.175	0.549
OLNet(A_I^c, ours)	0.432	0.692	0.427	0.716	0.206	0.653	0.231	0.670
OLNet(A_H^c, ours)	**0.699**	**0.809**	**0.715**	**0.829**	**0.315**	**0.676**	**0.364**	**0.682**

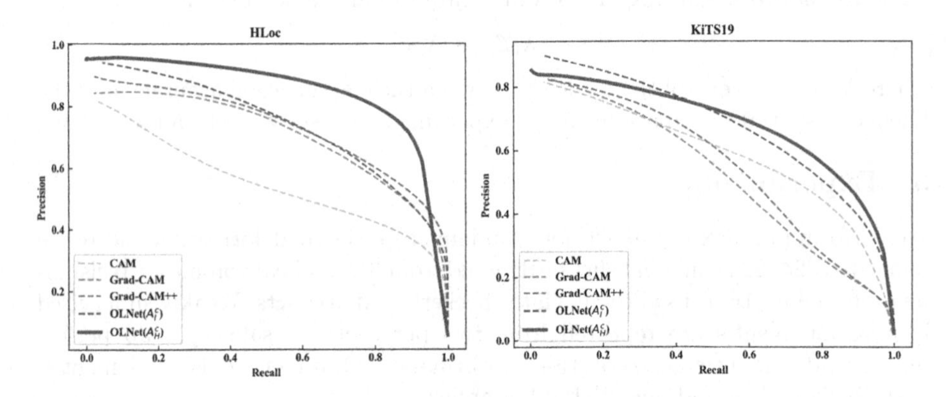

Fig. 2. PR curves of OLNet and previous methods evaluated on HLoc and KiTS19.

3.2 Metrics

We use *Precision-Recall Curves* to evaluate non-binary attention maps. To fulfill this, we binarize the attention map with a threshold sliding from 0 to 255, then compare the binary maps with the reference ground-truth. Beyond PR curves, *F-measure* can give a thorough evaluation taking both precision and recall into consideration as described in [1]. Binary map to be evaluated with *F-measure* is obtained by thresholding the attention map via 2× mean attention value. The maximum F-measure (F_β^{max}) computed from all precision-recall pairs will also be presented. As for semantic segmentation evaluation, *Dice Similarity Coefficient (DSC)* is adopted to provide measurements.

3.3 Regularized Attention Map Localization Performance

To evaluate whether the representative capacity of attention maps have been improved through our learning-based regularization method, we compare our

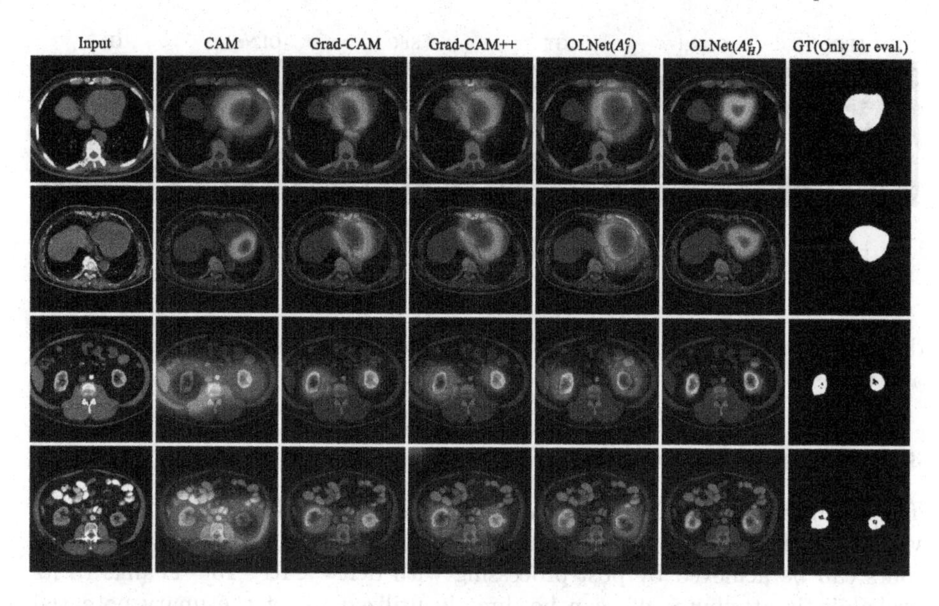

Fig. 3. Qualitative results of different methods. The first two rows for HLoc and the last two rows for KiTS19. Ground-truths (GT) are only used for evaluation. For attention maps, red means high attention value.

approaches with several previous CAMs adapted to heart and kidney localization tasks, both quantitatively and qualitatively.

Heart Localization on HLoc. Quantitative results are summarized in Table 1. The effectiveness of LAR-assisted OLNet is demonstrated by the increased localization accuracy. Among the methods purely using image-level labels, our OLNet achieves the best performance on F_β and F_β^{max}, outperforming previous CAMs by a large margin. In Fig. 2(a), we also observe that the area under OLNet PR curve is significantly larger than other curves. This indicates that the high-resolution attention map generated by OLNet is less sensitive to threshold choosing, i.e., good performance is observed in a large threshold range. Qualitatively, the first two rows of Fig. 3 show that OLNet captures more accurate geometric structures of the target organ.

Kidney Localization on KiTS19. Different from heart localization, kidney localization poses a bigger challenge because of the supposed bimodal distribution of attention maps in positive kidney slices. The last two rows of Fig. 3 present the qualitative results on KiTS19, as can be seen, although all methods can focus the high attention (red area) on the discriminative region, OLNet shows a specific capacity of capturing the accurate geometric contours of target organ with more focused attention values inside target region. And OLNet can deal with this bimodal situation robustly despite the more complicated surroundings compared with heart localization task. With LAR, the learned high-resolution attention map can capture full extent of target region while preserving fine-

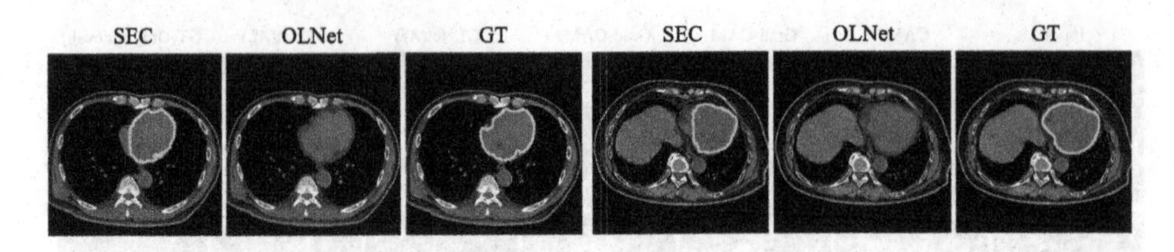

Fig. 4. Qualitative results of SEC and OLNet, two cases are demonstrated.

detailed geometric structures. Quantitative improvements are shown in Table 1 and the PR curves in Fig. 2(b).

3.4 Semantic Segmentation Performance

The resultant high-resolution attention maps can delineate the organ boundary via stratified heatmap values, but more accurate conformation with actual contours can be achieved by post-processing with dense CRF [13]. Thanks to its probabilistic attribute, A_H^c can be directly utilized to set the unary potential for dense CRF, then we can get a binary mask and its contours without much additional efforts. To evaluate the semantic segmentation performance of OLNet, we adapt a baseline Weakly Supervised Semantic Segmentation (WSSS) method SEC [12], which has a dedicated segmentation network, to our HLoc dataset. Mean DSC are shown in Table 2 and qualitative visual results are shown in Fig. 4. Although OLNet was not originally designed for pixel-level segmentation, it can get competitive segmentation results with one-step post-processing. And compared to other WSSS methods, there is no need to prepare localization cues offline for OLNet.

Table 2. Mean DSC of weakly supervised semantic segmentation results.

Method	Dataset	
	HLoc	
	Valid.	Test
SEC [12]	0.790	0.803
OLNet	0.802	0.818

4 Conclusions

In this work, we propose a weakly supervised organ localization approach suitable for 3D medical images. It can learn high-resolution and fine-detailed attention maps with image-level labels. We achieve this through online pseudo ground-truth generation and local area reconstruction, which directly guide the learning process of the attention maps. Quantitative and qualitative results show that our approach can achieve competitive organ localization performance both in saliency map and semantic segmentation perspectives.

Acknowledgements. The author would like to thank Erico Tjoa for proofreading the manuscript.

References

1. Achanta, R., Hemami, S., Estrada, F., Süsstrunk, S.: Frequency-tuned salient region detection. In: Proceedings of IEEE Conference on Computer Vision and Pattern Recognition, pp. 1597–1604 (2009)
2. Chan, L., Hosseini, M.S., Plataniotis, K.N.: A comprehensive analysis of weakly-supervised semantic segmentation in different image domains. arXiv preprint arXiv:1912.11186 (2019)
3. Chan, L., Hosseini, M.S., Rowsell, C., Plataniotis, K.N., Damaskinos, S.: Histoseg-net: semantic segmentation of histological tissue type in whole slide images. In: Proceedings of the IEEE International Conference on Computer Vision, pp. 10662–10671 (2019)
4. Chattopadhay, A., Sarkar, A., Howlader, P., Balasubramanian, V.N.: Grad-cam++: generalized gradient-based visual explanations for deep convolutional networks. In: IEEE Winter Conference on Applications of Computer Vision, pp. 839–847. IEEE (2018)
5. Criminisi, A., et al.: Regression forests for efficient anatomy detection and localization in computed tomography scans. Med. Image Anal. **17**(8), 1293–1303 (2013)
6. Dikmen, M., Zhan, Y., Zhou, X.S.: Joint detection and localization of multiple anatomical landmarks through learning. In: Medical Imaging 2008: Computer-Aided Diagnosis, vol. 6915, p. 691538. International Society for Optics and Photonics (2008)
7. Feng, X., Yang, J., Laine, A.F., Angelini, E.D.: Discriminative localization in CNNs for weakly-supervised segmentation of pulmonary nodules. In: Descoteaux, M., Maier-Hein, L., Franz, A., Jannin, P., Collins, D.L., Duchesne, S. (eds.) MICCAI 2017. LNCS, vol. 10435, pp. 568–576. Springer, Cham (2017). https://doi.org/10.1007/978-3-319-66179-7_65
8. Gondal, W.M., Köhler, J.M., Grzeszick, R., Fink, G.A., Hirsch, M.: Weakly-supervised localization of diabetic retinopathy lesions in retinal fundus images. In: Proceedings of International Conference on Image Processing, pp. 2069–2073. IEEE (2017)
9. He, K., Zhang, X., Ren, S., Sun, J.: Deep residual learning for image recognition. In: Proceedings of IEEE Conference on Computer Vision and Pattern Recognition, pp. 770–778 (2016)
10. Heller, N., et al.: The kits19 challenge data: 300 kidney tumor cases with clinical context, CT semantic segmentations, and surgical outcomes. arXiv preprint arXiv:1904.00445 (2019)
11. Huang, Y., Chung, A.: Celnet: evidence localization for pathology images using weakly supervised learning. arXiv preprint arXiv:1909.07097 (2019)
12. Kolesnikov, A., Lampert, C.H.: Seed, expand and constrain: three principles for weakly-supervised image segmentation. In: Leibe, B., Matas, J., Sebe, N., Welling, M. (eds.) ECCV 2016. LNCS, vol. 9908, pp. 695–711. Springer, Cham (2016). https://doi.org/10.1007/978-3-319-46493-0_42

13. Krähenbühl, P., Koltun, V.: Efficient inference in fully connected CRFs with Gaussian edge potentials. In: Advances in Neural Information Processing Systems, pp. 109–117 (2011)
14. Paszke, A., et al.: Pytorch: an imperative style, high-performance deep learning library. In: Advances in Neural Information Processing Systems, pp. 8024–8035 (2019)
15. Selvaraju, R.R., Cogswell, M., Das, A., Vedantam, R., Parikh, D., Batra, D.: Grad-cam: visual explanations from deep networks via gradient-based localization. In: Proceedings of International Conference on Computer Vision, pp. 618–626 (2017)
16. Shi, W., et al.: Real-time single image and video super-resolution using an efficient sub-pixel convolutional neural network. In: Proceedings of IEEE Conference on Computer Vision and Pattern Recognition, pp. 1874–1883 (2016)
17. Simonyan, K., Zisserman, A.: Very deep convolutional networks for large-scale image recognition. arXiv preprint arXiv:1409.1556 (2014)
18. Springenberg, J.T., Dosovitskiy, A., Brox, T., Riedmiller, M.: Striving for simplicity: the all convolutional net. arXiv preprint arXiv:1412.6806 (2014)
19. Zeiler, M.D., Fergus, R.: Visualizing and understanding convolutional networks. In: Fleet, D., Pajdla, T., Schiele, B., Tuytelaars, T. (eds.) ECCV 2014. LNCS, vol. 8689, pp. 818–833. Springer, Cham (2014). https://doi.org/10.1007/978-3-319-10590-1_53
20. Zheng, Y., Barbu, A., Georgescu, B., Scheuering, M., Comaniciu, D.: Fast automatic heart chamber segmentation from 3D CT data using marginal space learning and steerable features. In: Proceedings of International Conference on Computer Vision, pp. 1–8. IEEE (2007)
21. Zhou, B., Khosla, A., Lapedriza, A., Oliva, A., Torralba, A.: Learning deep features for discriminative localization. In: Proceedings of IEEE Conference on Computer Vision and Pattern Recognition, pp. 2921–2929 (2016)
22. Zhou, X., et al.: Automatic localization of solid organs on 3D CT images by a collaborative majority voting decision based on ensemble learning. Comput. Med. Imaging Graph. **36**(4), 304–313 (2012)

High-Order Attention Networks for Medical Image Segmentation

Fei Ding[1], Gang Yang[1,2](✉), Jun Wu[3], Dayong Ding[4], Jie Xv[5], Gangwei Cheng[6], and Xirong Li[1,2]

[1] AI & Media Computing Lab, School of Information, Renmin University of China, Beijing, China
yanggang@ruc.edu.cn
[2] MOE Key Lab of DEKE, Renmin University of China, Beijing, China
[3] Northwestern Polytechnical University, Xi'an, China
[4] Vistel AI Lab, Visionary Intelligence Ltd., Beijing, China
[5] Beijing Tongren Hospital, Beijing, China
[6] Peking Union Medical College Hospital, Beijing, China

Abstract. Segmentation is a fundamental task in medical image analysis. Current state-of-the-art Convolutional Neural Networks on medical image segmentation capture local context information using fixed-shape receptive fields and feature detectors with position-invariant weights, which limits the robustness to the variance of input, such as medical objects of variant sizes, shapes, and domains. In order to capture global context information, we propose High-order Attention (HA), a novel attention module with adaptive receptive fields and dynamic weights. HA allows each pixel to has its own global attention map that models its relationship to all other pixels. In particular, HA constructs the attention map through graph transduction and thus captures high relevant context information at high-order. Consequently, feature maps at each position are selectively aggregated as a weighted sum of feature maps at all positions. We further embed the proposed HA module into an efficient encoder-decoder structure for medical image segmentation, namely High-order Attention Network (HANet). Extensive experiments are conducted on four benchmark sets for three tasks, i.e., REFUGE and Drishti-GS1 for optic disc/cup segmentation, DRIVE for blood vessel segmentation, and LUNA for lung segmentation. The results justify the effectiveness of the new attention module for medical image segmentation.

Keywords: Segmentation · Receptive field · High-order graph

G. Cheng—This work is supported by the Beijing Natural Science Foundation (No. 4192029, No. 4202033).

Electronic supplementary material The online version of this chapter (https://doi.org/10.1007/978-3-030-59710-8_25) contains supplementary material, which is available to authorized users.

A. L. Martel et al. (Eds.): MICCAI 2020, LNCS 12261, pp. 253–262, 2020.
https://doi.org/10.1007/978-3-030-59710-8_25

1 Introduction

Medical image segmentation provides both qualitative and quantitative evidence for early diagnosis of most diseases. It has been well handled by the powerful methods driven by Convolutional Neural Networks (CNNs). But clinically, there are many factors that affect CNN's performance, such as the object with variant appearance and the distribution difference between testing data and training data which is easily caused by device parameters, clinician's shooting way and so on. Therefore, it is necessary to design a general approach for robust medical image segmentation.

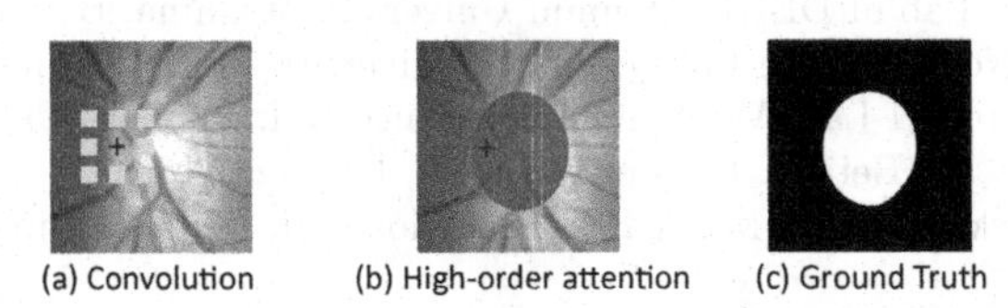

Fig. 1. Illustration of the receptive field in 3×3 convolution and our high-order attention for the pixel marked by "+". The target pixel is updated with the weighted sum of pixels in the receptive field.

Various pioneering fully convolutional network (FCN) approaches have taken into account the contexts to improve the performance of CNNs. The U-shaped networks [7] enable the use of rich context information from the multi-level. Other methods exploit context information from multiple distances by the multi-scale dilated convolutions [4,13] or multi-scale inputs [2]. However, current networks that utilize the fixed-shape receptive fields do not satisfy the variance of input. Moreover, the different objects with similar appearance and the cross-domain input are very common in medical images, which confuse CNNs that utilize feature detectors with position-invariant weights hard to extract discriminative features, leading to false predictions.

To address the above problems, we propose High-order attention (HA), a novel attention module designed from the view of the high-order graph, and embed it to an efficient encoder-decoder structure for medical image segmentation, called High-order Attention Network (HANet). HA aggregates context information using the adaptive receptive field and dynamic weights. Specifically, we first compute the initial attention map for each pixel that models its relationship to all other pixels. Then HA constructs the attention map through graph transduction and thus captures relevant context information at high-order. Finally, pixels at each position are augmented by a weighted sum of context information at all positions. As shown in Fig. 1, in our high order attention mechanism, the receptive field of the target pixel is adapted to the object it belongs to, and the weights of the weighted summation are dynamic. Extensive experiments on four datasets, including REFUGE [6], Drishti-GS1 [9], LUNA[1], and DRIVE

[1] https://www.kaggle.com/kmader/finding-lungs-in-ct-data/data/.

[10] demonstrate the superiority of our approach over existing state-of-the-art methods.

2 High-Order Attention Network

As illustrated in Fig. 2, our high-order Attention (HA) is embedded to an encoder-decoder architecture to capture global context information over local features, i.e., HANet. First, a medical image is encoded by an encoder to form a feature map with the spatial size $(h \times w)$. In the HA module, the feature map $\mathbf{X} \in \mathbb{R}^{c \times h \times w}$ is transformed by two branches. 1) the Dense Relation block and the Attention Propagation block transform $\mathbf{X}$ orderly to generate initial attention map $\mathbf{A}^*$ and high-order attention maps $\{\mathbf{A}^h \mid h \in \{1, 2, ..., n\}\}$ which represents the high relevant neighbors or receptive fileds for each pixel. 2) $\mathbf{X}$ is transformed by a bottleneck layer to produce feature map $\mathbf{H}$. Further, the feature map $\mathbf{H}$ and high-order attention maps $\{\mathbf{A}^h \mid h \in \{1, 2, ..., n\}\}$ are utilized to mix context information of multiple levels to produce new feature map $\mathbf{X}^+$ via the Information Aggregation block. Finally, the feature map $\mathbf{X}^+$ is transformed by a decoder to generate the final segmentation results.

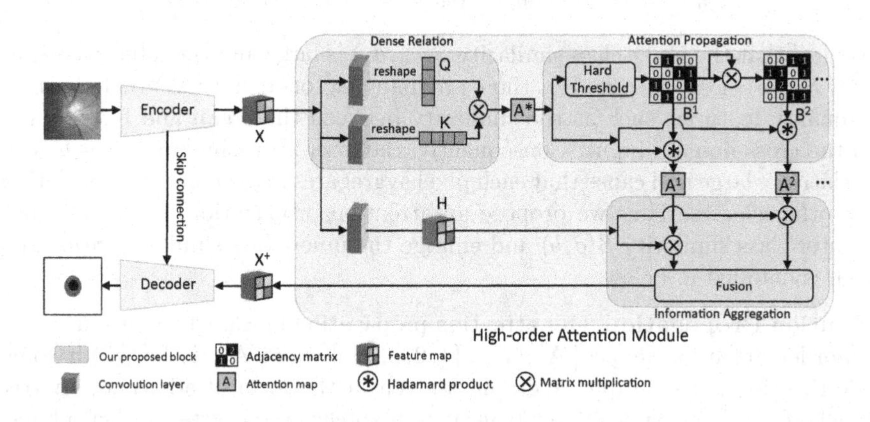

Fig. 2. The overall structure of our proposed High-order Attention network. An input image is passed through an encoder to produce the feature map $\mathbf{X}$. Then $\mathbf{X}$ is fed into the High-order Attention module to reinforce the feature representation by our proposed Dense Relation block, Attention Propagation block, and Information Aggregation block orderly. Finally, the reinforced feature map $\mathbf{X}^+$ is transformed by a decoder to generate the final segmentation results. The corresponding low-level and high-level features are connected by skip connections.

Dense Relation. The dense relation block computes the initial attention map $\mathbf{A}^*$ for each pixel, which models the relationship to all other pixels. We calculate $\mathbf{A}^*$ in a dot-product manner [11,14]. Given the feature map $\mathbf{X}$, we feed it to two parallel 1×1 convolutions to generate two new feature maps with shape $c \times h \times w$. Then they are reshaped to $\mathbb{R}^{(h \times w) \times c}$ and $\mathbb{R}^{c \times (h \times w)}$ respectively, namely $\mathbf{Q}$ and $\mathbf{K}$. The initial attention map $\mathbf{A}^* \in \mathbb{R}^{(h\times w)\times(h\times w)}$ is calculated by matrix multiplication of $\mathbf{Q}$ and $\mathbf{K}$, as follows:

$$\mathbf{A}^* = \alpha\mathbf{QK} \tag{1}$$

where α is a scaling factor to counteract the effect of the numerical explosion. Following previous work [11], we set α as $\frac{1}{\sqrt{c}}$ and c is the channel number of $\mathbf{K}$. Previous work [11] directly used A^* to capture global context information in natural language processing, called Self-attention. But it may not be directly used for robust medical image segmentation. For example, the attention map $A^* = \begin{bmatrix} S(a_0,a_0) & S(a_0,a_1) & S(a_0,b_0) \\ S(a_1,a_0) & S(a_1,a_1) & S(a_1,b_0) \\ S(b_0,a_0) & S(b_0,a_1) & S(b_0,b_0) \end{bmatrix}$ denotes pairwise similarities for the features of three pixels $\{a_0, a_1, b_0\}$, where a and b come from different classes, e.g., $S(a_0, a_1)$ denotes the inner-class similarity and $S(a, b)$ denotes the inter-class similarity. These features are updated by the weighted summation as below:

$$\begin{aligned} a_0^+ &= S(a_0,a_0)\cdot a_0 + S(a_0,a_1)\cdot a_1 + S(a_0,b_0)\cdot b_0 \\ a_1^+ &= S(a_1,a_0)\cdot a_0 + S(a_1,a_1)\cdot a_1 + S(a_1,b_0)\cdot b_0 \\ b_0^+ &= S(b_0,a_0)\cdot a_0 + S(b_0,a_1)\cdot a_1 + S(b_0,b_0)\cdot b_0 \end{aligned} \tag{2}$$

It is ideal that the inner-class similarity $S(a_0, a_1)$ is large and the inter-class similarity $S(a, b)$ closes to zero. But there are many factors confuse CNNs to extract low-quality features, such as the similarity between the organ and background and the cross-domain input. Consequently, the inter-class similarity $S(a, b)$ will be relatively large and cause that each pixel aggregates many context information from other classes. Thus we propose an attention propagation block to reduce the inter-class similarity $S(a, b)$ and enlarge the inner-class similarity $S(a_0, a_1)$, which is detailed in below.

Attention Propagation. Our attention propagation block, which produces the high-order attention maps $\{\mathbf{A}^h \mid h \in \{1, 2, ..., n\}\}$ from $\mathbf{A}^*$, is based on some basic theories of the graph. A graph is given in the form of adjacency matrix $\mathbf{B}^h$, where $\mathbf{B}^h[i, j]$ is a positive integer if vertex i can reach vertex j after h hops, otherwise $\mathbf{B}^h[i, j]$ is zero. And $\mathbf{B}^h$ can be computed by the adjacency matrix $\mathbf{B}^1$ multiplied by itself $h - 1$ times:

$$\mathbf{B}^h = \underbrace{\mathbf{B}^1\mathbf{B}^1 \; ... \; ... \; \mathbf{B}^1}_{h} \tag{3}$$

If $\mathbf{B}^h$ is normalized into a bool accessibility matrix, where we set zero to false and others to true, then as h increases, $\mathbf{B}^h$ tends to be equal to $\mathbf{B}^{h-1}$. At this point, $\mathbf{B}^h$ is the transitive closure[2] of the graph. The transitive closure is the best case for this attention mechanism, where the receptive field of each pixel precisely adapts to the object it belongs to.

[2] https://en.wikipedia.org/wiki/Transitive_closure.

Based on the above graph theories, we propose a novel way to optimize the receptive field and the weights. As presented previously, the element of $\mathbf{A}^*$ represents the correlation between two corresponding features. We consider $\mathbf{A}^*$ as the adjacency matrix of a graph, where an edge means the degree that two nodes belong to the same category. As shown in Fig. 2, given the $\mathbf{A}^*$, we erase those low confidence edges to produce the down-sampled graph $\mathbf{B}^1$ via thresholding. We expect each node to be connected only to the nodes with the same label so that to obtain accurate receptive field. The down-sampled graph $\mathbf{B}^1$ are generated as follows:

$$\mathbf{B}^1[i,j] = \begin{cases} 1 & \mathbf{A}^*[i,j] \geq \delta \\ 0 & \mathbf{A}^*[i,j] < \delta \end{cases} \tag{4}$$

where δ is a threshold. We simply set δ to 0.5 and develop richer useful information through a high-order graph (we set h to 2). The high-order graph $\mathbf{B}^h$ can be obtained with Eq. (3), which indicates the h-degree neighbors of each node. Finally, we obtain the high-order attention maps by the Hadamard product:

$$\mathbf{A}^h[i,j] = \mathbf{A}^*[i,j] \times \mathbf{B}^h[i,j] \tag{5}$$

where h is an integer adjacency power indicating the steps of attention propagation. Thus the attention information of different levels is decoupled into different attention maps via $\mathbf{B}^h$ and more highly relevant neighbors are obtained. The produced high-order attention map $\mathbf{A}^h$ are used to aggregate hierarchical context information in next section.

Information Aggregation. As shown in the lower part of Fig. 2, we aggregate context information of multiple levels to generate $\mathbf{X}^+$ in a weighted summation manner, where we perform matrix multiplication between feature map $\mathbf{H}$ and the normalized high-order attention map $\tilde{\mathbf{A}}^h$, as follows:

$$\mathbf{X}^+ = \Gamma_\theta\left(\Big\Vert_{h=1}^{n} W_\theta^h\left(\mathbf{H}\tilde{\mathbf{A}}^h\right)\right) \tag{6}$$

where $\tilde{\mathbf{A}}^h[i,j] = \frac{exp(\mathbf{A}^h[i,j])}{\sum_j exp(\mathbf{A}^h[i,j])}$, and $\Vert$ denotes channel-wise concatenation. W_θ^h and Γ_θ are 1×1 convolution layers. Finally, $\mathbf{X}^+$ is enriched by the decoder to generate accurate segmentation results. In the decoder, the $\mathbf{X}^+$ is first bilinearly upsampled and then fused with the corresponding low-level features. We adopt a simple yet effective way as [1] that only takes the output of the first block of encoder as low-level features. The output of the decoder is bilinearly upsampled to the same size as the input image.

3 Experiments

3.1 Experimental Setup

Datasets. REFUGE dataset [6] is arranged for the segmentation of optic disc and cup, which consists of 400 training images and 400 validation images.

Table 1. Ablation study on REFUGE validation set. DRB, APB, and IAB represent the dense relation block, attention propagation block, and information aggregation block respectively. The mean inference time per image is obtained on the TITAN Xp GPU with CUDA 8.0.

Method	Encoder	DRB	APB	IAB	mDice	$Dice_d$	$Dice_c$	Time
Baseline	ResNet50				0.9063	0.9513	0.8613	27 ms
HANet w/o APB	ResNet50	✓		✓	0.9104	0.9521	0.8613	38 ms
HANet	ResNet50	✓	✓	✓	0.9261	0.9563	0.8959	38 ms
Baseline	ResNet101				0.9088	0.9574	0.8603	43 ms
HANet w/o APB	ResNet101	✓		✓	0.9124	0.9573	0.8675	52 ms
HANet	ResNet101	✓	✓	✓	0.9302	0.9599	0.9005	52 ms

Drishti-GS1 dataset [9] contains 50 training images and 51 testing images for optic disc/cup segmentation. The annotations are provided in the form of average boundaries. Following all other compared methods, we first localize the disc and then transmit the cropped images into our network in both optic disc/cup segmentation datasets. **LUNA** dataset contains 267 2D CT images of dimensions 224×224 pixels from the Lung Nodule Analysis (LUNA) competition which can be freely downloaded from the website (see footnote 1). It is divided into 214 images for training and 53 images for testing. **DRIVE** dataset [10] is arranged for blood vessel segmentation, where 20 images are used for training and the remaining 20 images for testing. Manual annotations are provided by two experts, and the annotations of the first expert are used as the gold standard. Following the common methods [12,16,17], we orderly extract 48×48 patches for training and testing.

Implementation Details. We build our networks with PyTorch based on the TITAN Xp GPU. The ImageNet [8] pre-trained ResNet50 [5], or ResNet101 [5] is used as our encoder, where we replace the convolution within the last blocks by dilated convolutions [1,18], and the output stride is 8 for blood vessel segmentation and 16 for others. We compute losses via cross-entropy loss function. The Stochastic Gradient Descent with a mini-batch of 8, momentum of 0.9, and weight decay of 5e-4 is used as the optimizer. Initializing the learning rate to 0.01, we set training time to 100 epochs and the learning rate is multiplied by 0.1 if the performance has no improvement within the previous five epochs. The input spatial resolution is 224×224 for blood vessel segmentation and 513×513 for others. We also apply photometric distortion, random scale cropping, left-right flipping, and gaussian blurring during training for data augmentation.

Performance Metrics. For optic disc/cup segmentation, the results are represented by mean dice coefficient (mDice) and mean absolute error of the cup to disc ratio (E_{CDR}), where $Dice_d$ and $Dice_c$ denote dice coefficients of optic disc and cup respectively. The results are represented by accuracy (ACC), mean Intersection over Union (mIoU), and sensitivity (Sen) for lung segmentation, and F1 score (F1), Intersection over Union (IoU), and Specificity (Sp) for blood vessel segmentation.

3.2 Results and Analysis

Ablation Study. To examine the contribution of each component in HANet, we conduct an ablation study as follows. For the baseline, we replace the high-order attention module with a bottleneck layer to fit the input dimension. Given varied combinations of the components, six models are trained. Table 1 summarizes the results. For the HANet which combines all components, it improves the performance remarkably with a small time cost. In particular, the proposed HA module only has 3.29M parameters and its computational complexity is 6.17 GFLOPs. Figure 3 shows that the Attention propagation block (APB) generates more accurate global attention map for marked pixels and thus brings the best segmentation results.

Robustness. We report the results of HANet on Drishti-GS1 dataset [9] to evaluate its robustness on the cross-domain dataset, where our model and compared

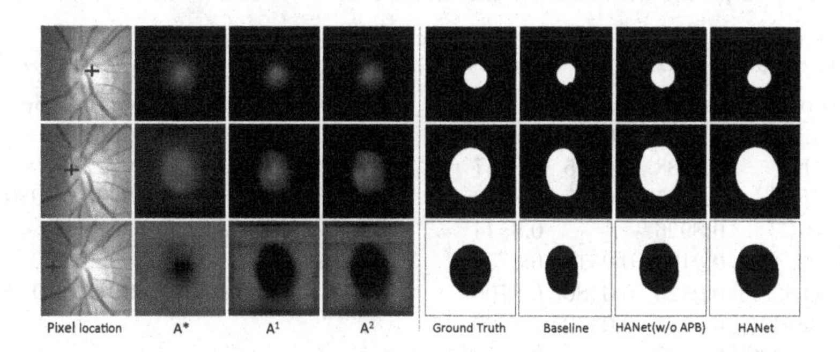

Fig. 3. Visualization of the attention maps and segmentation results on REFUGE validation set. The second to fourth columns show the attention maps (i.e., receptive fields that are marked by hot color) corresponding to the three pixels marked by "+" (i.e., optic disc, optic cup, and background), where the A^1 and A^2 generated by Attention Propagation block (APB) show more accurate receptive field than the initial attention map A^*. Consequently, the segmentation results of HANet are more accurate.

Table 2. The performance on cross-domain dataset of Drishti-GS1. We report the results of HANet which is trained on REFUGE training set and tested on the whole Drishti-GS1 dataset.

Method	mDice	E_{CDR}	$Dice_d$	$Dice_c$
M-Net [2]	0.8515	0.1660	0.9370	0.7660
Ellipse detection [15]	0.8520	0.1590	0.9270	0.7770
Baseline	0.8496	0.1694	0.9324	0.7668
HANet	**0.9117**	**0.1091**	**0.9721**	**0.8513**

methods all are only trained on the REFUGE dataset. Most clinical systems are challenged by the distribution difference between testing data and training data which is easily caused by device parameters, clinician's shooting way and so on. Table 2 shows our network that without any domain adaption technique achieves outstanding robustness on the cross-domain dataset and outperforms the baseline by 6.21% on mDice, by 8.45% on $Dice_c$ (dice coefficients of optic cup). More quantitative results are shown in the supplementary material.

Comparing with State-of-the-Art. We compare our HANet with existing state-of-the-art methods on REFUGE dataset, LUNA dataset, and DRIVE dataset. Table 3 shows our HANet beats state-of-the-art methods on REFUGE dataset. Especially, HANet outperforms the methods leading the REFUGE challenge[3], such as the AIML which combines the results from different networks using the backbone of ResNet50, ResNet101, ResNet-152, and Wide ResNet-38 respectively and achieves the first place on the REFUGE validation set. Table 3 shows the HANet achieves state-of-the-art performance on LUNA dataset, obtaining an mIoU of 98.49%, accuracy of 99.45%, and sensitivity of

Table 3. Comparing with state-of-the-art on REFUGE dataset and LUNA dataset.

(a) Results on REFUGE val. set.

Method	mDice	E_{CDR}	$Dice_c$
AIML [9]	0.9250	0.0376	0.8916
BUCT [9]	0.9188	0.0395	0.8857
CUMED[9]	0.9185	0.0425	0.8848
U-Net [7]	0.8926	-	0.8544
pOSAL [13]	0.9105	0.0510	0.8750
M-Net [2]	0.9120	0.0480	0.8700
Ellipse [15]	0.9125	0.0470	0.8720
ET-Net [20]	0.9221	-	0.8912
Baseline	0.9088	0.0526	0.8603
HANet	**0.9302**	**0.0347**	**0.9005**

(b) Results on LUNA testing set.

Method	mIoU	ACC	Sen
U-Net [7]	0.9479	0.9663	-
M-Net [2]	0.9492	0.9727	-
CE-Net [4]	-	0.9900	0.9800
ET-Net [20]	0.9623	0.9868	-
Baseline	0.9539	0.9829	0.9796
HANet	**0.9849**	**0.9945**	**0.9879**

(c) Results on DRIVE testing set.

Method	ACC	F1	Sen	Sp	IoU
DeepVessel [3]	0.9523	0.7900	0.7603	-	-
MS-NFN [17]	0.9567	-	0.7844	0.9819	-
Vessel-Net [16]	0.9578	-	0.8038	0.9802	-
ET-Net [20]	0.9560	-	-	-	-
DEU-Net [12]	0.9567	0.8270	0.7940	0.9816	-
M-Net [2]	0.9674	-	0.7680	0.9868	0.6726
AG-Net [19]	0.9692	-	0.8100	0.9848	0.6965
Baseline	0.9676	0.7953	0.7420	**0.9885**	0.6601
HANet	**0.9712**	**0.8300**	**0.8297**	0.9843	**0.7094**

[3] https://refuge.grand-challenge.org/Results-ValidationSet_Online/.

98.79%. Table 3 shows HANet achieves the best overall performance on DRIVE dataset, with 82.97% sensitivity, 97.12% accuracy, 83% F1 score, and 70.94% IoU. The above results again justify the effectiveness of HANet.

4 Conclusion

In this paper, we propose a novel High-order Attention module that captures global context information with adaptive receptive filed and dynamic weights, and we further propose the HANet for medical image segmentation. In particular, HA constructs the attention map through graph transduction and thus captures high relevant context information at high-order. Notably, the adaptive receptive filed and dynamic weights are beneficial to the medical objects with variant appearance, and capturing contexts at high-order is robust to the low-quality features. Extensive experiments on four benchmark datasets have demonstrated the superiority of our method compared to other state-of-the-art methods.

References

1. Chen, L.C., Zhu, Y., Papandreou, G., Schroff, F., Adam, H.: Encoder-decoder with atrous separable convolution for semantic image segmentation. In: ECCV, pp. 801–818 (2018)
2. Fu, H., Cheng, J., Xu, Y., Wong, D.W.K., Liu, J., Cao, X.: Joint optic disc and cup segmentation based on multi-label deep network and polar transformation. TMI **37**(7), 1597–1605 (2018)
3. Fu, H., Xu, Y., Lin, S., Wong, D.W.K., Liu, J.: DeepVessel: retinal vessel segmentation via deep learning and conditional random field. In: Ourselin, S., Joskowicz, L., Sabuncu, M.R., Unal, G., Wells, W. (eds.) MICCAI 2016. LNCS, vol. 9901, pp. 132–139. Springer, Cham (2016). https://doi.org/10.1007/978-3-319-46723-8_16
4. Gu, Z., et al.: CE-net: context encoder network for 2D medical image segmentation. TMI **38**(10), 2281–2292 (2019)
5. He, K., Zhang, X., Ren, S., Sun, J.: Deep residual learning for image recognition. In: CVPR, pp. 770–778 (2016)
6. Orlando, J.I., et al.: Refuge challenge: a unified framework for evaluating automated methods for glaucoma assessment from fundus photographs. Med. Image Anal. **59**, 101570 (2020)
7. Ronneberger, O., Fischer, P., Brox, T.: U-net: convolutional networks for biomedical image segmentation. In: Navab, N., Hornegger, J., Wells, W.M., Frangi, A.F. (eds.) MICCAI 2015. LNCS, vol. 9351, pp. 234–241. Springer, Cham (2015). https://doi.org/10.1007/978-3-319-24574-4_28
8. Russakovsky, O., et al.: Imagenet large scale visual recognition challenge. Int. J. Comput. Vision **115**(3), 211–252 (2015)
9. Sivaswamy, J., Krishnadas, S., Joshi, G.D., Jain, M., Tabish, A.U.S.: Drishti-GS: retinal image dataset for optic nerve head (ONH) segmentation. In: ISBI, pp. 53–56. IEEE (2014)
10. Staal, J., Abràmoff, M.D., Niemeijer, M., Viergever, M.A., Van Ginneken, B.: Ridge-based vessel segmentation in color images of the retina. TMI **23**(4), 501–509 (2004)

11. Vaswani, A., et al.: Attention is all you need. In: NIPS, pp. 5998–6008 (2017)
12. Wang, B., Qiu, S., He, H.: Dual encoding U-net for retinal vessel segmentation. In: Shen, D., et al. (eds.) MICCAI 2019. LNCS, vol. 11764, pp. 84–92. Springer, Cham (2019). https://doi.org/10.1007/978-3-030-32239-7_10
13. Wang, S., Yu, L., Yang, X., Fu, C.W., Heng, P.A.: Patch-based output space adversarial learning for joint optic disc and cup segmentation. TMI **38**(11), 2485–2495 (2019)
14. Wang, X., Girshick, R., Gupta, A., He, K.: Non-local neural networks. In: CVPR, pp. 7794–7803 (2018)
15. Wang, Z., Dong, N., Rosario, S.D., Xu, M., Xie, P., Xing, E.P.: Ellipse detection of optic disc-and-cup boundary in fundus images. In: ISBI, pp. 601–604. IEEE (2019)
16. Wu, Y., et al.: Vessel-net: retinal vessel segmentation under multi-path supervision. In: Shen, D., et al. (eds.) MICCAI 2019. LNCS, vol. 11764, pp. 264–272. Springer, Cham (2019). https://doi.org/10.1007/978-3-030-32239-7_30
17. Wu, Y., Xia, Y., Song, Y., Zhang, Y., Cai, W.: Multiscale network followed network model for retinal vessel segmentation. In: Frangi, A.F., Schnabel, J.A., Davatzikos, C., Alberola-López, C., Fichtinger, G. (eds.) MICCAI 2018. LNCS, vol. 11071, pp. 119–126. Springer, Cham (2018). https://doi.org/10.1007/978-3-030-00934-2_14
18. Yu, F., Koltun, V.: Multi-scale context aggregation by dilated convolutions. In: ICLR (2016)
19. Zhang, S., et al.: Attention guided network for retinal image segmentation. In: Shen, D., et al. (eds.) MICCAI 2019. LNCS, vol. 11764, pp. 797–805. Springer, Cham (2019). https://doi.org/10.1007/978-3-030-32239-7_88
20. Zhang, Z., Fu, H., Dai, H., Shen, J., Pang, Y., Shao, L.: ET-net: a generic edge-attention guidance network for medical image segmentation. In: Shen, D., et al. (eds.) MICCAI 2019. LNCS, vol. 11764, pp. 442–450. Springer, Cham (2019). https://doi.org/10.1007/978-3-030-32239-7_49

NAS-SCAM: Neural Architecture Search-Based Spatial and Channel Joint Attention Module for Nuclei Semantic Segmentation and Classification

Zuhao Liu[1], Huan Wang[1], Shaoting Zhang[1,2], Guotai Wang[1(✉)], and Jin Qi[1(✉)]

[1] University of Electronic Science and Technology of China, Chengdu, China
{guotai.wang,jqi}@uestc.edu.cn
[2] SenseTime Research, Shanghai, China

Abstract. The segmentation and classification of different types of nuclei plays an important role in discriminating and diagnosing of the initiation, development, invasion, metastasis and therapeutic response of tumors of various organs. Recently, deep learning method based on attention mechanism has achieved good results in nuclei semantic segmentation. However, the design of attention module architecture relies heavily on the experience of researchers and a large number of experiments. Therefore, in order to avoid this manual design and achieve better performance, we propose a new Neural Architecture Search-based Spatial and Channel joint Attention Module (NAS-SCAM) to obtain better spatial and channel weighting effect. To the best of our knowledge, this is the first time to apply NAS to the attention mechanism. At the same time, we also use synchronous search strategy to search architectures independently for different attention modules in the same network structure. We verify the superiority of our methods over the state-of-the-art attention modules and networks in public dataset of MoNuSAC 2020. We make our code and model available at https://github.com/ZuhaoLiu/NAS-SCAM.

Keywords: Neural architecture search · Nuclei segmentation · Attention mechanism

1 Introduction

The segmentation and classification of different types of nuclei plays a great role in discriminating and diagnosing of the initiation, development, invasion, metastasis and therapeutic response of tumors of various organs [1]. In recent years, deep learning methods are widely applied in nuclei segmentation and classification [2–6]. For example, Kang et al. [4] proposed stacking two U-Nets [7] and Yoo et al. [5] proposed a weakly supervised method for nuclei segmentation.

Z. Liu and H. Wang—Equal contribution.

A. L. Martel et al. (Eds.): MICCAI 2020, LNCS 12261, pp. 263–272, 2020.
https://doi.org/10.1007/978-3-030-59710-8_26

However, as an important step of nuclei instance segmentation, the results of semantic segmentation are greatly impacted by the imbalance of different types of nuclei. Therefore, based on deep neural network, attention module is applied to alleviate class imbalance problem because it can make network automatically enhance important features and suppress unimportant ones.

However, the choice of attention module architecture is based on prior knowledge or limited experiments, which has a lot of disadvantages. Firstly, attention modules with the same architecture may not be applicable to all datasets. However, due to the diversity of optional operations within the attention module, it takes a lot of experimental resources to find the optimal attention module architecture. Secondly, attention modules are often used in different positions of the network. Traditionally, in the same network, these are multiple attention modules with the same architecture. However, due to the influence of convolution and nonlinear operations in network, the spatial and channel information contained in the feature maps of different positions is often different. So, adopting attention modules with different architectures can lead to better use of information in different positions, but this will significantly increase the amount of trail in traditional methods. Therefore, it is time-consuming to try all the experiments to finish attention module selection.

In recent years, with the original intention of simplifying the difficulty of hyper parameter adjustment, Neural Architecture Search (NAS) is widely used in classification [8,9] and semantic segmentation [10]. It can automatically learn the optimal architecture of the network, thus greatly reducing the difficulty of model architecture selection. Therefore, based on NAS, our paper aims to explore the optimal architectures of attention modules. At the same time, we also propose a new searching strategy, i.e., synchronous search strategy, which can search out architectures of different attention modules in the same network.

Our contributions mainly include two folds: (1) We propose a new attention module, i.e., Neural Architecture Search-based Spatial and Channel joint Attention Module (NAS-SCAM), which can efficiently complete the automatic search of architecture to produce better space and channel weighting effect. As far as we know, this is the first application of NAS in the field of attention mechanism, which provides a new development direction for the application of attention module. (2) We propose a synchronous search strategy, which can make our attention module search out different architectures in different positions of the same network, and make the attention module more suitable for certain position of the network, so as to produce a better weighting effect. We have verified our results on the public dataset, MoNuSAC 2020 [1]. Compared with the state-of-the-art attention modules and networks, our method achieves better results in nuclei semantic segmentation and classification.

2 Methods

2.1 NAS-Based Spatial and Channel Joint Attention Module

NAS-SCAM is composed of NAS-based spatial attention module (NAS-SAM) and NAS-based channel attention module (NAS-CAM), which can generate spatial and channel weighting effect, respectively.

The architecture of NAS-SAM is shown in Fig. 1(a). Assuming that the input feature map is $M = [m_1, m_2, ..., m_c]$, which has width W, height H, and channel C. M is transformed into spatial weight map $n \in \mathbb{R}^{H \times W}$ by one or multiple convolutions and nonlinear operations $F_{NS}(.)$ in search space. Through the learning of parameters in search space, n contains the spatial weighting information. Finally, we use the multiplication operation to fuse spatial weight map n into the input feature map M and generate output feature map M' as equation Eq. (1).

$$M' = n \otimes M = [nm_1, nm_2, ..., nm_C] \tag{1}$$

The proper architecture selection of $F_{NS}(.)$ is the key operation and it can make a great difference in the weighting effect. However, because $F_{NS}(.)$ has numerous choices, it is difficult to find the optimal one. So, we propose to select the proper architecture of $F_{NS}(.)$ by using NAS.

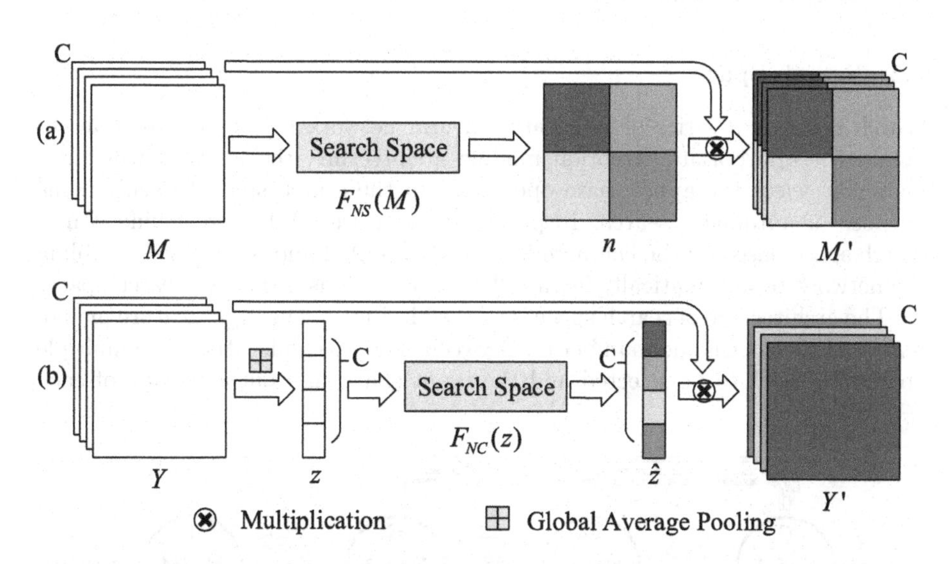

Fig. 1. Illustration of NAS attention modules, (a) is NAS spatial attention module and (b) is NAS channel attention module.

The architecture of NAS-CAM is shown in Fig. 1(b). In order to generate channel weighting effect on the premise of keeping spatial information unchanged. Assuming that the input feature map is $Y = [y_1, y_2, ..., y_c], y_i \in$

$\mathbb{R}^{H \times W}$. We firstly use a global average pooling along the spatial dimension as Eq. (2) and generate vector $z \in \mathbb{R}^{1 \times 1 \times C}$.

$$z_i = \text{Avgpool}(y_i) = \frac{1}{H \times W} \sum_{p=1}^{H} \sum_{q=1}^{W} y_i(p, q) \tag{2}$$

Then, one or multiple convolutions and nonlinear operations $F_{NC}(.)$ in search space are used to generate channel weight vector $\hat{z} \in \mathbb{R}^{1 \times 1 \times C}$ which contains channel-wise weighting information. Then, we use NAS to search for the optimal selection of $F_{NC}(.)$. Finally, the output feature map Y' is generated by recalibrating $\hat{z}$ to Y by Eq. (3).

$$Y' = \hat{z} \otimes Y = [\hat{z}_1 y_1, \hat{z}_2 y_2, ..., \hat{z}_C y_C] \tag{3}$$

NAS-SAM and NAS-CAM can be combined in series or in parallel, which are NAS-SCAM-P and NAS-SCAM-S, respectively. For NAS-SCAM-P, the input feature map is weighted along spatial dimension and channel dimension independently, and then element-wise maximum operation is used to fuse two output feature maps to retain more important weighting information. And for NAS-SCAM-S, the input feature map is weighted in the order of spatial dimension followed by channel dimension.

2.2 Search Space

Search space is a series of alternative neural network structures. In order to obtain the appropriate attention module architecture, the network will automatically select the appropriate operations in the search space to achieve the purpose of automatic search. Inspired by literatures [8,10], we define a new search space based on the characteristics of NAS-SAM and NAS-CAM, enabling the network to automatically learn different operations inside the search space.

The architecture of search space is showed in Fig. 2. The input feature map of search space has channel number C. Between each two nodes, there are multiple operations need to be selected and the input shape and output shape of each

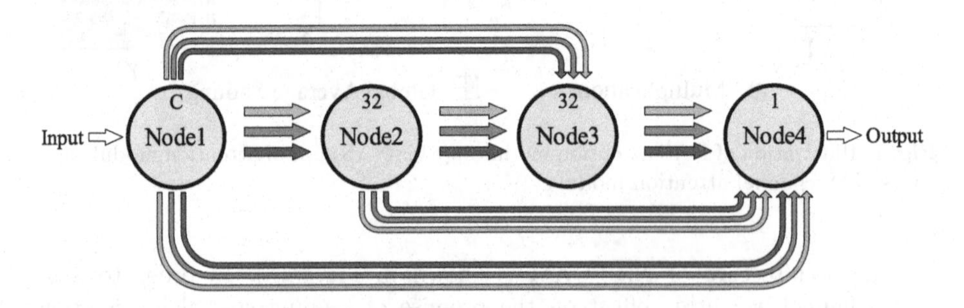

Fig. 2. Architecture of search space.

Table 1. Operations of NAS-SAM and NAS-CAM

NAS-SAM		NAS-CAM	
Zero (No connection)	Conv2D 5	Zero (No connection)	Conv1D 9
Conv2D 1	Atrous Conv2D 3	Conv1D 1	Conv1D 15
Conv2D 3	Atrous Conv2D 5	Conv1D 3	Atrous Conv1D 3
		Conv1D 5	Atrous Conv1D 5

operation are the same except channel dimension. In addition, in order to make attention module have better spatial and channel information learning ability, we set the channel number of Node2 and Node3 to 32. And the number of Node4 is set to 1 to generate weight map along spatial dimension or channel dimension. Every operation in search space is followed by batch normalization to normalize the output value. And activate function following every operation is ReLU except the operations connected to Node4, whose activate function is sigmoid to make output values between 0 and 1.

Because NAS-SAM and NAS-CAM have different architectures, so we choose different operations between each two nodes in their search space. So, the operations of NAS-SAM and NAS-CAM are listed in Table 1. For NAS-SAM, because we need to extract information from the spatial dimension, we use 2D convolutions with different filter sizes to extract information from receptive fields with different sizes. We also use dilated convolutions to increase long-ranged information learning ability. For NAS-SAM, because of the influence of global average pooling, we use 1D convolutions to extract channel information. Furthermore, zero operation is also applied to both NAS-SAM and NAS-CAM to represent no connection between two nodes.

We use continuous relaxation to learn the optimal operation between two nodes. Assuming that operation set between i-th node and j-th node is $O^{(i,j)}(.)$, and $o(.)$ is certain operation in $O^{(i,j)}(.)$. $u_o^{(i,j)}$ is continuous weight coefficient which reflects weight of each $o(.)$ and $x(i)$ is the output of i-th node, $O^{(i,j)}(.)$ is defined as Eq. (4).

$$O^{(i,j)}(x(i)) = \sum_{o \in O^{(i,j)}} u_o^{(i,j)} o(x(i)) \tag{4}$$

Continuous weight coefficient $u_o^{(i,j)}$ is generated from continuous variable $\alpha_o^{(i,j)}$ by softmax function, which is showed in Eq. (5).

$$u_o^{(i,j)} = \frac{\exp(\alpha_o^{(i,j)})}{\sum_{o \in O^{(i,j)}} \exp(\alpha_o^{(i,j)})} \tag{5}$$

Consequently, updating $\alpha_o^{(i,j)}$ through backward propagation can change values of $u_o^{(i,j)}$, and different value of $u_o^{(i,j)}$ represents different importance of each operation. Finally, we choose the operation corresponding to the highest $u_o^{(i,j)}$ as the final operation between i-th node and j-th node.

In both NAS-SAM and NAS-CAM, every node is connected to all previous nodes by summing all output of previous nodes. The output of j-th node $x(j)$ is defined as Eq. (6).

$$x(j) = \sum_{x<j} O^{(x,j)}(x(i)) \tag{6}$$

2.3 Synchronous Search Strategy

In order to search optimal architectures of multiple attention modules in the same network, we propose to use synchronous search strategy to search each attention module independently. Conventionally, after designing an architecture of attention module, the module with the same architecture will be plugged into the end of every down and up sampling block [11,12]. However, the feature maps in different positions of network have large semantic difference because of the convolution and pooling. Therefore, searching different architectures of attention module can make them more suitable for different positions, i.e., down and up sampling blocks in network.

The synchronous search strategy initializes a unique attention module for each down and up sampling block and optimizes them independently. Because the adjustment of attention module architecture is through optimizing continuous variable α, different α in different attention module will have different gradient in the optimization process, so as to be optimized towards different direction and generate more suitable architecture for certain position.

3 Experiments

3.1 Dataset Description

Multi-organ nuclei segmentation and classification dataset in MoNuSAC 2020 is used to validate the performance of the proposed NAS attention module. This dataset contains 209 annotated H&E stained histopathology images and four types of nuclei including epithelial, lymphocytes, macrophages, and neutrophils. We randomly select 120 images as training set, 39 images as validation set and 50 images as testing set. We use Dice similarity coefficient (DSC) as evaluation metric which can calculate the similarity between prediction and ground truth.

3.2 Implementation Details

In order to augment the training data, firstly, we use overlapping crop to generate more training data, and the cropped image size is set to 256×256 and the total number of cropped images is 4803. Secondly, we use a series of augmentation methods including random rotate, random flip, Gaussian blur, median blur and elastic transformation. To normalize the data, we use standard color normalization to preprocess dataset. In addition, it is noticeable that each class containing the same amount of data in each training batch can potentially alleviate the class imbalance problem and achieve better results.

Table 2. DSC scores (%) of different approaches on multi-organ nuclei dataset

Methods	Average	Epithelial	Lymphocyte	Macrophage	Neutrophil
U-Net [7]	56.77 ± 3.40	77.58 ± 3.08	75.40 ± 2.95	27.35 ± 3.46	46.75 ± 4.86
Deeplab V3+ [14]	62.37 ± 3.03	78.91 ± 0.45	76.46 ± 0.85	33.61 ± 11.52	60.53 ± 0.98
CBAM [11]	59.40 ± 3.51	77.51 ± 2.14	77.11 ± 0.88	25.93 ± 6.09	57.07 ± 6.65
scSE [12]	56.86 ± 4.27	79.48 ± 0.46	77.59 ± 0.45	27.94 ± 8.30	42.44 ± 9.55
NAS-SCAM **without** synchronous search strategy					
NAS-SCAM-S	61.22 ± 3.19	78.30 ± 0.79	76.68 ± 0.15	31.39 ± 6.92	58.52 ± 5.45
NAS-SCAM-P	62.80 ± 1.77	79.58 ± 0.36	77.62 ± 0.33	37.75 ± 3.84	56.27 ± 8.44
NAS-SCAM **with** synchronous search strategy					
NAS-SCAM-S	62.47 ± 1.23	78.55 ± 0.78	76.64 ± 1.76	35.42 ± 6.53	59.28 ± 0.15
NAS-SCAM-P	65.01 ± 1.07	80.67 ± 0.66	77.43 ± 1.94	40.85 ± 6.06	61.10 ± 2.72

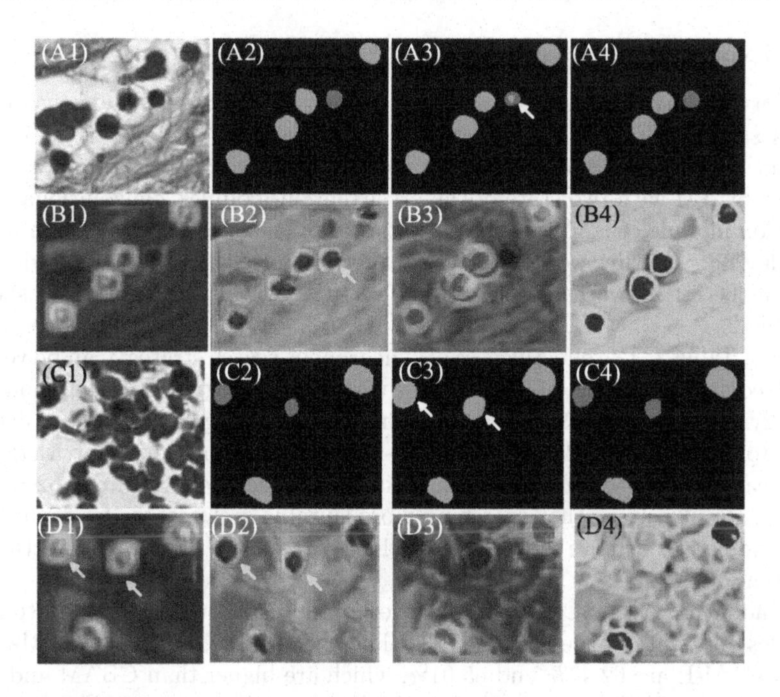

Fig. 3. Visualization of segmentation results. A1 and C1 are original images; A2 and C2 are ground truth; A3 and C3 are segmentation results of CBAM; A4 and C4 are segmentation results of NAS-SCAM-P; B1-2 and D1-2 are attention maps in CBAM; B3-4 and D3-4 are attention maps in NAS-SCAM-P. White arrows highlight the wrong classification in segmentation results and red arrows highlight misjudgment of attention maps.

We implemented our experiments in Python and machine learning framework Pytorch, and two Nvidia RTX 2080 GPU are used. Exponential logarithmic loss described in literature [13] and Adam optimizer are used in optimization process.

The baseline of the networks is U-Net which contains four down sampling and four up sampling blocks. For the networks with attention modules, attention module is plugged into the end of each down or up sampling block.

We use the first-order approximation of bilevel optimizer described in literature [8] to optimize our network. In searching process, the total epoch number is 120, and we only update network parameters in the first 40 epochs. The update of w and update of continuous variable α are implemented alternatively after first 40 epochs. The learning rate is 0.0001 when updating w and 0.001 when updating α. After learned an optimized architecture, we rebuild network based on learned α. In rebuilding process, the total epoch number is 300, and learning rate is 0.0001. We saved the model that performed best in validation set as the model in testing set. All experiments are implemented three times, and mean and deviation of DSC of each class are calculated in each experiment.

3.3 Experimental Results

In order to verify the effectiveness of our proposed NAS-SCAM and synchronous search strategy. Firstly, we compare the performances of NAS-SCAM with or without synchronous search strategy. Secondly, we compare the performances of NAS-SCAM with two state-of-the-art attention modules, convolutional block attention module (CBAM) [11] and spatial-channel squeeze & excitation (scSE) module [12]. Thirdly, we compared the performance of NAS-SCAM with two state-of-the-art networks: U-Net [7] and Deeplab V3+ [14]. Results are showed in Table 2.

From Table 2, The effectiveness of synchronous search strategy can be verified by the comparison between NAS-SCAM with this strategy and that without this strategy. For NAS-SCAM-S, synchronous search strategy can increase average DSC from 61.22% to 62.47%. For NAS-SCAM-P, synchronous search strategy has greater influence on results which increases average DSC from 62.80% to 65.01%. Results prove that searching more suitable architecture for each attention module can achieve better results than searching a unique architecture for all attention modules.

In addition, NAS-SCAM can achieve better results than existing attention modules and state-of-the-art networks. The average DSC of NAS-SCAM-S and NAS-SCAM-P are 62.47% and 65.01%, which are higher than CBAM and scSE whose average DSC are 59.40% and 56.86%. And the result of NAS-SCAM is also better than that of Deeplab V3+ and U-Net, whose average DSC are 62.37% and 56.77%, respectively. Moreover, NAS-SCAM-P can achieve the best results for all type of nuclei except lymphocyte. The results show the effectiveness of NAS-SCAM compared with state-of-the-art attention modules and networks.

3.4 Visualization

From Fig. 3, we visualize the segmentation results and spatial attention maps generated by the last two attention blocks in network from CBAM and NAS-SCAM-P. The red mask is lymphocyte and green mask is neutrophil. It is noted that compared with NAS-SCAM-P, CBAM generates wrong classification results in A3 and C3 as white arrows point out. The misjudgment of attention maps has direct relationship with this result, which is showed in B2 and D1-2, where attention maps generated from CBAM give equal weights to two classes as red arrows point out. But in B3-4 and D3-4, attention maps generated from NAS-SCAM-P give different weights to different classes, so as to generate better results.

4 Conclusion

In this paper, we propose new attention module, NAS-SCAM, which is the first application of NAS in attention mechanism. This provides a new direction for the development of attention mechanism. We also propose a new search strategy, synchronous search strategy, which can make searched architecture of attention module better fit to the network. Our proposed methods can generate better spatial and channel weighting effect which is beneficial for network to distinguish different types of nuclei, so as to achieve better results in nuclei semantic segmentation and classification and can be better applied in clinical diagnosis.

Future work will focus on improving the search strategy and increasing the search space. We are dedicated to add channel number and activation function selection into our search space, and find more effective attention module.

Acknowledgements. This work is supported by Sichuan Jiuzhou electric Group Co. Ltd, Sichuan, Mianyang, 621000, China, and National Natural Science Foundation of China under grant no. 81771921, and Glasgow College, University of Electronic Science and Technology of China.

References

1. Verma, R., Kumar, N., Patil, A., et al.: Multi-organ Nuclei Segmentation and Classification Challenge (2020, unpublished). https://doi.org/10.13140/RG.2.2.12290.02244/1
2. Su, H., Shi, X., Cai, J., Yang, L.: Local and global consistency regularized mean teacher for semi-supervised nuclei classification. In: Shen, D., et al. (eds.) MICCAI 2019. LNCS, vol. 11764, pp. 559–567. Springer, Cham (2019). https://doi.org/10.1007/978-3-030-32239-7_62
3. Saha, M., Chakraborty, C.: Her2Net: a deep framework for semantic segmentation and classification of cell membranes and nuclei in breast cancer evaluation. TIP **27**, 2189–2200 (2018)
4. Kang, Q., Lao, Q., Fevens, T.: Nuclei segmentation in histopathological images using two-stage learning. In: Shen, D., et al. (eds.) MICCAI 2019. LNCS, vol. 11764, pp. 703–711. Springer, Cham (2019). https://doi.org/10.1007/978-3-030-32239-7_78
5. Yoo, I., Yoo, D., Paeng, K.: PseudoEdgeNet: nuclei segmentation only with point annotations. In: Shen, D., et al. (eds.) MICCAI 2019. LNCS, vol. 11764, pp. 731–739. Springer, Cham (2019). https://doi.org/10.1007/978-3-030-32239-7_81

6. Qu, H., Yan, Z., Riedlinger, G.M., De, S., Metaxas, D.N.: Improving nuclei/gland instance segmentation in histopathology images by full resolution neural network and spatial constrained loss. In: Shen, D., et al. (eds.) MICCAI 2019. LNCS, vol. 11764, pp. 378–386. Springer, Cham (2019). https://doi.org/10.1007/978-3-030-32239-7_42
7. Ronneberger, O., Fischer, P., Brox, T.: U-net: convolutional networks for biomedical image segmentation. In: Navab, N., Hornegger, J., Wells, W.M., Frangi, A.F. (eds.) MICCAI 2015. LNCS, vol. 9351, pp. 234–241. Springer, Cham (2015). https://doi.org/10.1007/978-3-319-24574-4_28
8. Liu, H., Simonyan, K., Yang, Y.: Darts: differentiable architecture search. In: International Conference on Learning Representations (ICLR) (2019)
9. Real, E., Aggarwal, A. et al.: Regularized evolution for image classifier architecture search. In: AAAI, vol. 33, no. 01 (2019)
10. Liu, C. et al.: Auto-DeepLab: hierarchical neural architecture search for semantic image segmentation. In: CVPR, pp. 82–92 (2019)
11. Woo, S., Park, J., Lee, J.Y., Kweon, I.N.: CBAM: convolutional block attention module. In: ECCV, pp. 3–19 (2018)
12. Roy, A.G., Navab, N., Wachinger, C.: Concurrent spatial and channel 'squeeze & excitation' in fully convolutional networks. In: Frangi, A.F., Schnabel, J.A., Davatzikos, C., Alberola-López, C., Fichtinger, G. (eds.) MICCAI 2018. LNCS, vol. 11070, pp. 421–429. Springer, Cham (2018). https://doi.org/10.1007/978-3-030-00928-1_48
13. Wong, K.C.L., Moradi, M., Tang, H., Syeda-Mahmood, T.: 3D segmentation with exponential logarithmic loss for highly unbalanced object sizes. In: Frangi, A.F., Schnabel, J.A., Davatzikos, C., Alberola-López, C., Fichtinger, G. (eds.) MICCAI 2018. LNCS, vol. 11072, pp. 612–619. Springer, Cham (2018). https://doi.org/10.1007/978-3-030-00931-1_70
14. Chen, L., Zhu, Y., Papandreou, G., Schroff, F., Adam, H.: Encoder-decoder with atrous separable convolution for semantic image segmentation. In: ECCV, pp. 801–818 (2018)

Scientific Discovery by Generating Counterfactuals Using Image Translation

Arunachalam Narayanaswamy[1(✉)], Subhashini Venugopalan[1], Dale R. Webster[2], Lily Peng[2], Greg S. Corrado[2], Paisan Ruamviboonsuk[3], Pinal Bavishi[2], Michael Brenner[4], Philip C. Nelson[1], and Avinash V. Varadarajan[2]

[1] Google Research, Mountain View, USA
arunachalam@google.com
[2] Google Health, Palo Alto, USA
[3] Rajavithi Hospital, Bangkok, Thailand
[4] Google Research, Cambridge, USA

Abstract. Model explanation techniques play a critical role in understanding the source of a model's performance and making its decisions transparent. Here we investigate if explanation techniques can also be used as a mechanism for scientific discovery. We make three contributions: first, we propose a framework to convert predictions from explanation techniques to a mechanism of discovery. Second, we show how generative models in combination with black-box predictors can be used to generate hypotheses (without human priors) that can be critically examined. Third, with these techniques we study classification models for retinal images predicting Diabetic Macular Edema (DME), where recent work [30] showed that a CNN trained on these images is likely learning novel features in the image. We demonstrate that the proposed framework is able to explain the underlying scientific mechanism, thus bridging the gap between the model's performance and human understanding.

1 Introduction

Visual recognition models are receiving increased attention in the medical domain [9,30], supplementing/complementing physicians and enabling screening at a larger scale, e.g. in drug studies [21]. While the application of deep neural net based models in the medical field has been growing, it is also important to develop techniques to make their decisions transparent. Further, these models provide a rich surface for scientific discovery. Recent works [18,22] show

A. Narayanaswamy and S. Venugopalan—Equal contribution.

Electronic supplementary material The online version of this chapter (https://doi.org/10.1007/978-3-030-59710-8_27) contains supplementary material, which is available to authorized users.

A. L. Martel et al. (Eds.): MICCAI 2020, LNCS 12261, pp. 273–283, 2020.
https://doi.org/10.1007/978-3-030-59710-8_27

that neural net models can make novel predictions previously unbeknownst to humans. However for such works, explanation techniques based on saliency are insufficient. Specifically, saliency maps only show spatial support i.e. "where" the model looks, but do not explain "what" is different (or "why").

There are a few critical differences between explanation methods for natural images and medical data. Firstly, while humans have a direct intuition for objects in natural images, only trained professionals typically read medical data with acceptable accuracy [13]. Further, such medical training and knowledge is accrued over time driven by advances in basic science. However, such knowledge is rarely injected when developing deep models for medical data. Many of these novel predictions are based on raw data and direct outcomes, not necessarily human labels. Such end-to-end models show improved performance while simultaneously reducing human involvement in knowledge distillation. Thus, there is a compelling need to develop explanation methods that can look at data without fully-known human interpretable annotations and generate hypotheses both to validate models and improve human understanding.

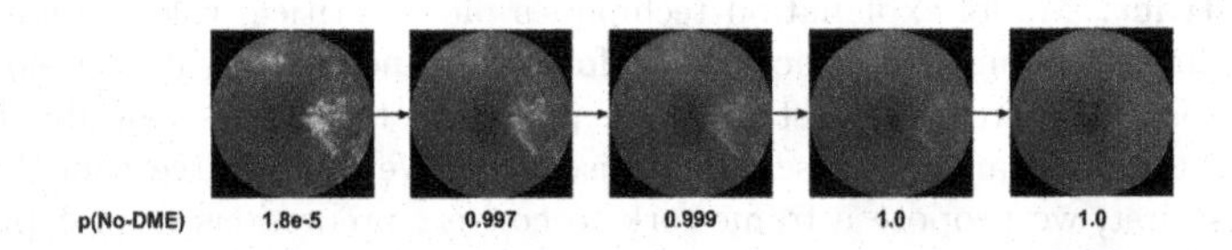

Fig. 1. An example of a transformation from left to right by successively applying our explanation technique based on unpaired image to image translation to modify the source image with diabetic macular edema (DME) (*leftmost column*) to no-DME (*rightmost column*). Probability of no-DME from an independent prediction model are presented below. Our method accurately shows "what" changes affect prediction.

In this work we propose to use image-to-image translation in combination with black-box prediction models to achieve this goal. Specifically, we first use existing techniques to identify salient regions relevant to the prediction task. We then develop image translation models that can accurately modify image regions of a source class to that of a target class to show what about the region influences prediction. Further, we are able to amplify the modifications to enhance human interpretability. Finally, based on the transformations observed, we identify a minimal set of hand-engineered features which when trained using a linear SVM achieves comparable performance to that of the CNN trained on the raw images. To demonstrate our approach we look at the Diabetic Macular Edema (DME) prediction models from [30]. In their work, the same CNN architecture shows drastically different performance when trained using labels from 2 different sources. We reproduce that and explain how. Our contributions are:

- A framework to convert predictions from black-box models to a mechanism of discovery.
- A demonstration of how image translation models in combination with black-box predictors can be used to generate hypotheses worth examining.

– A set of hand-engineered features, identified from the generated hypotheses, that account for the performance of a Diabetic Macular Edema classifier.

Related Work. The past few years have seen several methods for explaining classification models. While some focus on visualizing the concepts learned by the convolutional filters [1,17,19], a predominant number focus on generating saliency masks. These are more relevant to our work. Of the saliency methods, some are backpropagation based [6,12,25,28,29] and examine the model's gradients to produce a mask. Some others are pertubation based [5,6,20,23,27,32] treating the model as a black-box and observe its outputs while perturbing the inputs. In both cases the methods generate a heatmap to estimate *where* the prediction model is looking and reveal spatial support. While saliency maps are helpful to validate models, they don't provide the complete picture. In particular, they don't reveal *what* about these regions is different between two classes. Our method bridges this gap by generating counterfactuals akin to [2,4,11,15,24,26]. Specifically, we use image translations, to show sparse changes on the original image, to reveal subtle differences between classes. We also amplify these changes to enhance human interpretability while simultaneously producing realistic images that doctors can analyze (Figs. 1 and 6).

2 Approach

Given a deep prediction model that can distinguish between two image classes, we wish to discover features that are relevant for classification. One objective way to accomplish this is to identify a distilled set of features computed from the image that carry nearly the same predictive power as the model. In this work, instead of purely human driven hypotheses to discover features, our framework uses model explanation techniques as a tool for generating these hypotheses.

1. Input ablation to evaluate the importance of known regions. With human annotations of known landmarks/regions in medical images, it is possible to evaluate how relevant these are for prediction. Individually ablating each known landmark from the dataset, retraining from scratch, and evaluating the prediction model can help understand whether there are compact known salient regions in the data that influence the model.

2. Visual saliency to validate the model. Saliency techniques [25,28,29] allow for arbitrary spatial support. They can be used to validate the model based on known features or regions identified by the ablation experiments in the previous step. In natural images, with known object classes, the regions can be human interpretable but that may not always be the case for medical images.

3. Image-to-image translation to highlight "what" is different. Next, to show "what" is different between the 2 classes we propose unpaired image to image translation using Generative Adversarial Networks (GANs) [7]. Here, we use a slightly modified version of CycleGAN [33]. Let us refer to the two image classes as X and Y. The basic idea behind unpaired image translation is to learn

2 functions: $g : X \rightarrow Y$, and $f : Y \rightarrow X$, for translating images from one class to the other as determined by the GAN discriminator. CycleGAN [33] incorporates a cycle consistency loss in addition to the two GAN losses. This ensures that $f \circ g$ and $g \circ f$ ($\circ$ refers to composition) are close to identity functions i.e. $f(g(x)) \approx x$ and $g(f(y)) \approx y$. The total loss being optimized can be written as

$$L_{total} = L_{GAN}(G, D_y, X, Y) + L_{GAN}(F, D_x, X, Y) + \lambda L_{Cycle}$$
$$L_{Cycle} = \mathbb{E}_{x \sim X}(\|f(g(x) - x\|_1) + \mathbb{E}_{y \sim Y}(\|g(f(y) - y\|_1)$$

G, F refer to the generators, and D_x, D_y the discriminators within the GAN functions g and f respectively. L_{GAN} refers to the GAN loss function (we use a wasserstein loss [8]). Next, we modified the basic CycleGAN model architecture (above) to retain high frequency information during this translation:

- we have a 1×1 convolutional path from the input to the output, and
- we also model the functions as residual networks.

Such a construction allows the model to copy pixels from input to output if they aren't relevant to the classification. This allowed our model to quickly learn the identity function and then optimize for cycle consistency to highlight subtle differences between classes in the counterfactual.

To verify that the image transformation to the target class is successful we need a prediction model independent of the CycleGAN discriminator. It is necessary that the CycleGAN has no interaction with the prediction model to ensure that it cannot encode any class information directly into the translation functions (see [3]). We use a CNN, which we call M, trained on the same set of images (X and Y), completely independent of the CycleGAN model. In order to verify that our CycleGAN model succeeded, we evaluate area under the curve (AUC) of the translated and original images under model M.

4. Amplify differences from image transformation. Next we amplify the differences in the counterfactual by applying the translation function successively i.e $g(x), g(g(x)), \cdots g^4(x)$, (similarly for f). To evaluate this quantitatively we compute AUCs for all examples in $(x, y), (g(x), f(y)), \cdots (g^4(x), f^4(y))$ using the prediction model M. If the model M has an AUC of 0.85 on original images (x, y) with successive applications of f and g, we would expect this to drop to 0.15 if the image translation is able to perfectly map the two class of images. We also evaluate the images qualitatively (as in Figs. 1 and 6).

5. Identify and evaluate hand-engineered features. Recall that our goal is to discover and explain what about the data leads to the predictive power of the classification model. To this end, based on the visualizations generated from the image translation model, we designed specific hand-engineered features observed in the transformations. We then evaluate them using 2 simple classifiers, a linear SVM with $l1$ regularization, and an MLP with 1 hidden layer. We compare the performance of these classifiers on the hand-engineered features with that of the CNN model (M) on a held out evaluation set.

3 Diagnosing DME - 2 Sources of Labels

Diabetic Macular Edema (DME) [14], a leading cause of preventable blindness, is characterized by retinal thickening of the macula. The gold standard for DME diagnosis involves the evaluation of a 3D scan of the retina, based on Optical Coherence Tomography (OCT), by a trained expert. However, due to costs and availability constraints, screening for DME generally happens by examining a Color Fundus Photo (CFP), which is a 2D image of the retina. Human experts evaluating a CFP look for the presence of lipid deposits called hard exudates (HE) which is a proxy feature for retinal thickening [10]. This method of screening for DME through HE is known to have both poor precision and recall when compared to the gold standard diagnosis [16,31] (Fig. 2).

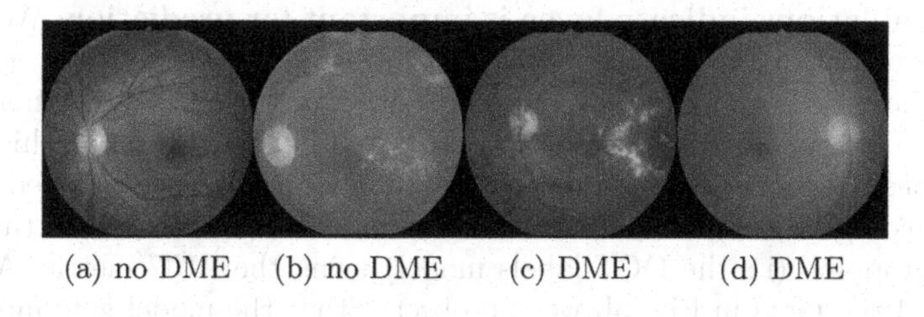

(a) no DME (b) no DME (c) DME (d) DME

Fig. 2. Sample color fundus photos (CFPs) with and without hard exudates (yellow lesions) and with and without diabetic macular edema (DME) as defined by the more accurate optical coherence tomography (OCT) measurement. (Color figure online)

Previous works [9,13] show that a CNN taking CFPs as input can be trained to predict the DME labels derived from human experts grading CFPs. We'll refer to such CNNs as HE-labels models. However, more recent work [30] shows that if we train the same CNN architectures using labels derived from human experts grading OCTs, the model (AUC:0.85) can significantly outperform the HE-labels model (AUC:0.74). We'll call these as OCT-labels models.

Key Question. we hope to answer is: What signal is the OCT-labels model capturing to produce such a remarkable increase in AUC, that has not been observed by human experts (and equivalently the HE-labels model)? Note that the only difference here is the source of the label (the architecture, and training fundus photos are the same in both cases).

DME Dataset. The DME dataset used in our experiments come from [30]. The dataset consisted of 7072 paired CFP and OCT images taken at the same time. The images were labeled by 2 medical doctors (details in [30]). Only the CFPs and the labels derived from the OCTs (and not the OCT images themselves) are used in training the CNN. We used the same datasplits as in [30] with 6039 images for training and validation, and 1033 images for test.

Annotations. For the ablation studies, we obtained expert annotations of two known "landmarks" in CFPs, namely optic disc and fovea. Optic disc is the bright ellpitical region where the optic nerve leaves the retina. Fovea is the (dark) area with the largest concentration of cones photoreceptors. Optic disc was labelled using an ellipse (four parameters) and fovea with a point (2 parameters).

4 Application

As a first step we reproduce results from [30] building a CNN model M based on inception-v3 (initializing weights from imagenet) and train on the DME dataset. This model achieves an AUC of 0.847. We also train a multi-task version of this model (which predicts other metadata labels) as described in [30] which gets an AUC of 0.89. Next we perform input ablations based on annotated landmarks.

1. Input ablations indicate fovea is important for prediction. We extract circular crops of different radii (0.25 to 5), measured relative to the optic disc diameter, around both landmarks (replacing other pixels with background color). Examples of cropped images are shown in Fig. 3 below each plot. This creates 16 "cropped" versions of the dataset. We train a model on each cropped version. We then compare the performance of these models against that of the model with no cropping (i.e the OCT-labels model) using the AUC metric. As noted in [30] and the plots in Fig. 3b we can observe that the model gets most of its information from the region surrounding the fovea (i.e the macula). This serves to reassure that the model is looking at the right region (macula).

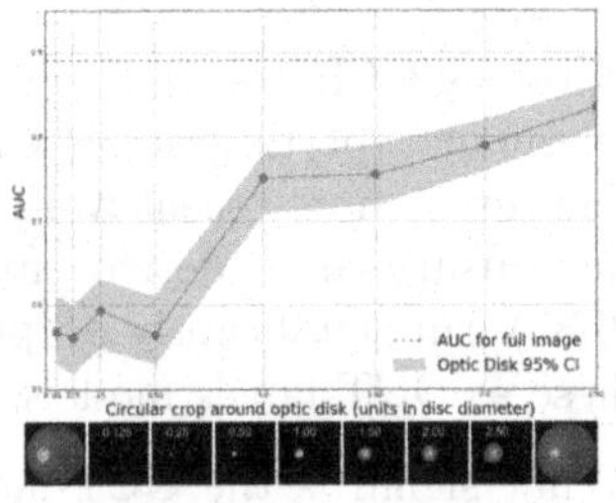

(a) Models trained on crops centered on the optic disc.

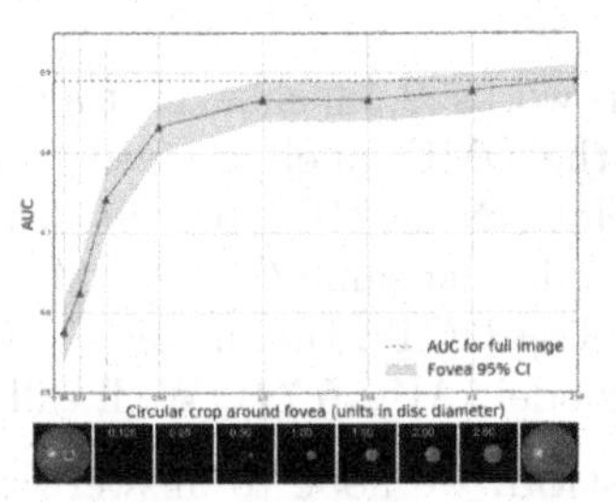

(b) Models trained on crops centered on the fovea.

Fig. 3. [Input ablations based on [30]] Performance (AUC) of models trained and evaluated on circular crops extracted by centering either on the (a) optic-disc, or the (b) fovea. Bottom panel of images show examples of cropped regions as the diameter of extraction is increased. The horizontal red dashed line is the performance of the model trained on images with no cropping (i.e the OCT-labels model). (Color figure online)

2. Saliency maps show where HE-labels and OCT-labels models differ. We applied GradCAM [25] on the HE-labels and OCT-labels models to visually see the differences between the two. Qualitative examples (as shown in Fig. 4)

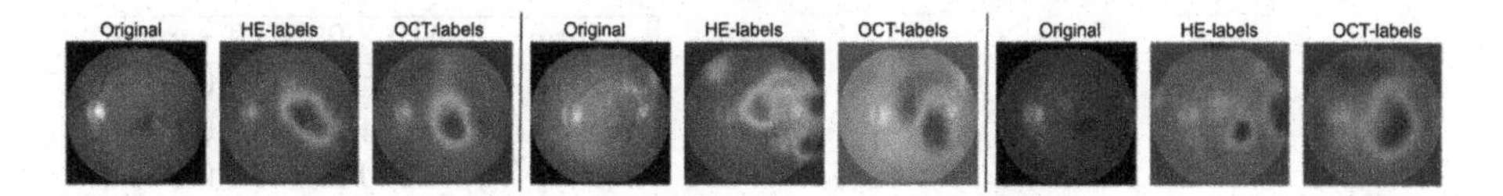

Fig. 4. Comparing visual saliency maps from applying GradCAM on the HE-labels and OCT-labels models on 3 sample images with DME from the tune set. Each example shows the original CFP (left), saliency from HE-labels model overlaid on the CFP (middle), and saliency from OCT-labels model overlaid on the CFP (right). OCT-labels model centers around the fovea (the dark region near the center of the image), and the HE-labels model focuses on hard-exudates (the yellow lesions in the image). [Best viewed electronically in high resolution] (Color figure online)

indicate that, the OCT-labels model focuses on the macula while the HE-labels model focuses on the hard-exudate (yellow lesions) locations. Recall that for the HE-labels model, the doctors use the presence of hard-exudates to determine testing for DME. The advantage with visual saliency methods are that they offer qualitative explanations without the need for landmarks. Our next step is to understand what the OCT-labels model picks up from the macula.

3. Modified CycleGANs learn to successfully translate between classes. We train a CycleGAN model as described in Sect. 2 on the training images with DME (X) and no-DME (Y). We evaluated AUC metric on the transformed images as measured by the independent prediction model M on a random subset of the evaluation set (200 images with near 50–50% split). Results in Fig. 5 (c) show that our model is able to successfully fool (to an extent) an independently trained supervised classifier on the same task. Specifically, Fig. 5 (a, b) and (c) show that with successive applications f is able to make M classify translated DME images as no-DME, and g is able to make M classify no-DME images as having DME.

Note that there is no explicit link between classifier M and the CycleGAN model. In general, we can imagine scenarios where the two may not agree, and hence the analysis in Fig. 5(c) is a critical component of the evaluation. The drop in classification model's AUC on repeated CycleGAN application (shown in Fig. 5) provides quantitative evidence that both models are learning similar high level features that distinguish the two classes in the dataset manifold.

4. Amplification highlights hard exudates and brightening in macula. With confidence in our CycleGAN transformations we next qualitatively analyze the changes introduced by the model in Fig. 6. With successive application of g and f two changes are observed:

- hard exudates are either added (for g) or removed (for f) consistently.
- fovea region is brightened (for g) or darkened (for f) consistently.

Hard Exudates. Figure 6 shows g (right images) adds hard exudates (yellow lesions) to images and f (left images) generally diminishes or removes hard

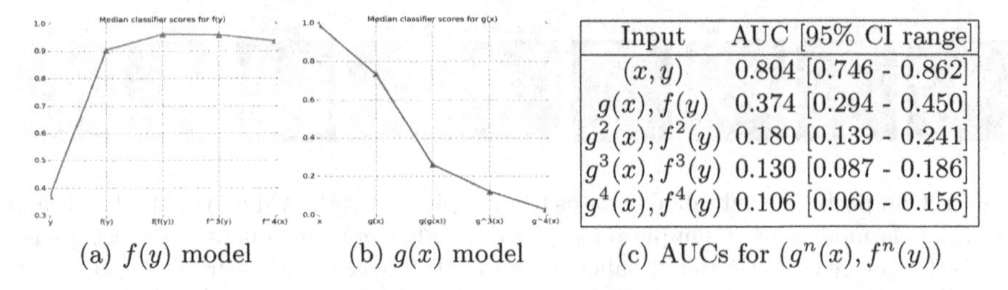

Input	AUC [95% CI range]
(x, y)	0.804 [0.746 - 0.862]
$g(x), f(y)$	0.374 [0.294 - 0.450]
$g^2(x), f^2(y)$	0.180 [0.139 - 0.241]
$g^3(x), f^3(y)$	0.130 [0.087 - 0.186]
$g^4(x), f^4(y)$	0.106 [0.060 - 0.156]

(a) $f(y)$ model (b) $g(x)$ model (c) AUCs for $(g^n(x), f^n(y))$

Fig. 5. CycleGAN validation using prediction model M on a random subset (200 images) of the tune set. Plots (a) and (b) show median prediction scores on successive applications of f (on DME images) and g (on no-DME images). Table (c) shows AUCs for all the transformed images $(g^n(x), f^n(y))$. These show that the CycleGAN is able to successfully fool prediction model M into thinking images are from the opposite class, and continues to improve with successive applications of g and f.

exudates completely. This is in line with our expectations since hard exudates are correlated (although not perfectly) with DME diagnosis.

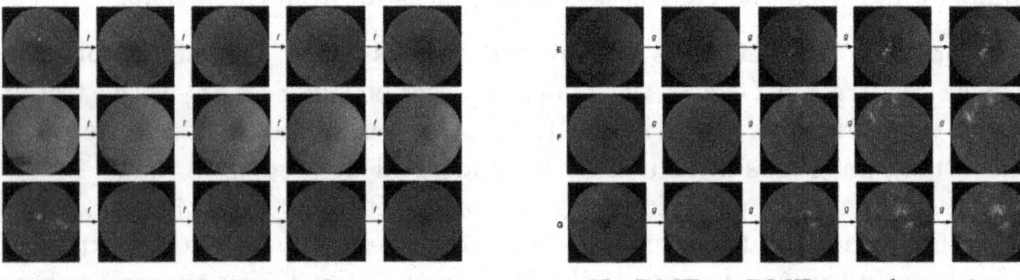

f: DME to No-DME transformations. g: No-DME to DME transformations.

Fig. 6. A selected sample of images showing successive application of f (left) and g (right). Notice that f not only removes hard exudates (yellow lesions) but also tends to darken the region around fovea. We believe this accentuates the foveal dip. Notice how g adds both the hard exudates and lightens the macula region. (Color figure online)

Fovea Brightening. The other is a very subtle difference only visible with successive application and with the images best viewed as a gif or video. For transformations in f (left images), in each subsequent image the fovea region gets darker than the surroundings. And in g the fovea region gets progressively lighter. Although this change is amplified (visible to the human eye) only with successive applications of functions f and g, a single application appears enough to convince classifier M (Fig. 5 (c)) that the classes have changed. This highlights one of the key features of our explanation method i.e we can amplify and enhance subtle changes to make them more human interpretable.

5. Converting CycleGAN hypotheses to hand-engineered features. Following the insights from amplification, we engineered a rather simple set of features to objectively evaluate whether the hypotheses given by the model is valid.

We took 10 concentric circles (5 mm apart) around the fovea, then computed the mean red, green and blue intensities within these discs to create our features. We train a linear SVM classifier and an MLP with 1 hidden layer on these features.

Results in Table 1 show that the presence(/absence) of hard exudates is complementary to our hand engineered features and boosts the overall AUC. This hand-engineered model used nothing but the average color intensities around the fovea. Our hand engineered features along with the presence(/absence) of hard exudates gets very close to explaining the totality of AUC of the CNN model M. This builds further confidence in our explanation method.

Table 1. Performance of SVM and MLP classifiers on features hypothesized by our image translation model on the tune set: either hand-engineered features alone, presence of hard-exudates alone, or combination of the two. AUC and standard deviation of 10 runs reported in percentage (higher is better).

Features	**SVM** (AUC)	**MLP** (AUC)
Hand-engineered features alone	72.4 ± 0.0	76.3 ± 0.3
Presence of hard exudates alone	74.1 ± 0.0	74.1 ± 0.0
Hand-engineered features + hard exudates' presence	$\mathbf{81.4 \pm 0.0}$	$\mathbf{82.2 \pm 0.2}$
M(raw pixels single task on cropped image)	CNN: 84.7	

5 Conclusions

In this paper, we propose a framework to convert novel predictions fueled by deep learning into direct insights for scientific discovery. We believe our image translation based explanation methods are a much more generic tool than ablation or image saliency based methods. Our method goes beyond spatial support to reveal the nature of change between classes which is evaluated objectively. We successfully applied this to explain the difference between 2 diabetic macular edema classifiers trained on different sources of labels. Our method is able to provide insights previously unknown to the scientific community.

References

1. Bau, D., Zhou, B., Khosla, A., Oliva, A., Torralba, A.: Network dissection: quantifying interpretability of deep visual representations. In: CVPR (2017)
2. Chang, C.H., Creager, E., Goldenberg, A., Duvenaud, D.: Explaining image classifiers by counterfactual generation. arXiv preprint arXiv:1807.08024 (2018)
3. Chu, C., Zhmoginov, A., Sandler, M.: Cyclegan, a master of steganography. arXiv preprint arXiv:1712.02950 (2017)
4. Dhurandhar, A., Chen, P.Y., Luss, R., Tu, C.C., Ting, P., Shanmugam, K., Das, P.: Explanations based on the missing: towards contrastive explanations with pertinent negatives. In: NeurIPS, pp. 592–603 (2018)

5. Fong, R., Patrick, M., Vedaldi, A.: Understanding deep networks via extremal perturbations and smooth masks. In: ICCV, pp. 2950–2958 (2019)
6. Fong, R.C., Vedaldi, A.: Interpretable explanations of black boxes by meaningful perturbation. In: ICCV, pp. 3429–3437 (2017)
7. Goodfellow, I., et al.: Generative adversarial nets. In: NeurIPS (2014)
8. Gulrajani, I., Ahmed, F., Arjovsky, M., Dumoulin, V., Courville, A.C.: Improved training of wasserstein GANs. In: NeurIPS, pp. 5767–5777 (2017)
9. Gulshan, V., et al.: Development and validation of a deep learning algorithm for detection of diabetic retinopathy in retinal fundus photographs. JAMA **316**(22), 2402–2410 (2016)
10. Harding, S., Broadbent, D., Neoh, C., White, M., Vora, J.: Sensitivity and specificity of photography and direct ophthalmoscopy in screening for sight threatening eye disease: the Liverpool diabetic eye study. BMJ **311**(7013), 1131–1135 (1995)
11. Joshi, S., Koyejo, O., Vijitbenjaronk, W., Kim, B., Ghosh, J.: Towards realistic individual recourse and actionable explanations in black-box decision making systems. arXiv preprint arXiv:1907.09615 (2019)
12. Kapishnikov, A., Bolukbasi, T., Viégas, F., Terry, M.: XRAI: better attributions through regions. In: ICCV, pp. 4948–4957 (2019)
13. Krause, J., et al.: Grader variability and the importance of reference standards for evaluating machine learning models for diabetic retinopathy. Ophthalmology **125**(8), 1264–1272 (2018)
14. Lee, R., Wong, T.Y., Sabanayagam, C.: Epidemiology of diabetic retinopathy, diabetic macular edema and related vision loss. Eye Vis. **2**(1), 1–25 (2015)
15. Liu, S., Kailkhura, B., Loveland, D., Han, Y.: Generative counterfactual introspection for explainable deep learning. arXiv preprint arXiv:1907.03077 (2019)
16. Mackenzie, S., et al.: SDOCT imaging to identify macular pathology in patients diagnosed with diabetic maculopathy by a digital photographic retinal screening programme. PLoS ONE **6**(5), e14811 (2011)
17. Mahendran, A., Vedaldi, A.: Understanding deep image representations by inverting them. In: CVPR, pp. 5188–5196 (2015)
18. Miller, A., Obermeyer, Z., Cunningham, J., Mullainathan, S.: Discriminative regularization for latent variable models with applications to electrocardiography. In: ICML. Proceedings of Machine Learning Research, PMLR (2019)
19. Mordvintsev, A., Olah, C., Tyka, M.: Deepdream-a code example for visualizing neural networks. Google Res. **2**(5) (2015)
20. Petsiuk, V., Das, A., Saenko, K.: Rise: randomized input sampling for explanation of black-box models (2018)
21. Pharmaceuticals, R.: Recursion Cellular Image Classification - Kaggle contest. www.kaggle.com/c/recursion-cellular-image-classification/data
22. Poplin, R., et al.: Prediction of cardiovascular risk factors from retinal fundus photographs via deep learning. Nat. Biomed. Eng. **2**(3), 158 (2018)
23. Ribeiro, M.T., Singh, S., Guestrin, C.: Why should i trust you?: explaining the predictions of any classifier. In: ACM SIGKDD (2016)
24. Samangouei, P., Saeedi, A., Nakagawa, L., Silberman, N.: Explaingan: model explanation via decision boundary crossing transformations. In: ECCV (2018)
25. Selvaraju, R.R., Cogswell, M., Das, A., Vedantam, R., Parikh, D., Batra, D.: Grad-cam: visual explanations from deep networks via gradient-based localization. In: ICCV, pp. 618–626 (2017)
26. Singla, S., Pollack, B., Chen, J., Batmanghelich, K.: Explanation by progressive exaggeration. arXiv preprint arXiv:1911.00483 (2019)

27. Smilkov, D., Thorat, N., Kim, B., Vigas, F., Wattenberg, M.: Smoothgrad: removing noise by adding noise (2017)
28. Springenberg, J.T., Dosovitskiy, A., Brox, T., Riedémiller, M.: Striving for simplicity: the all convolutional net (2014)
29. Sundararajan, M., Taly, A., Yan, Q.: Axiomatic attribution for deep networks (2017)
30. Varadarajan, A.V., et al.: Predicting optical coherence tomography-derived diabetic macular edema grades from fundus photographs using deep learning. Nat. Commun. **11**(1), 1–8 (2020)
31. Wang, Y.T., Tadarati, M., Wolfson, Y., Bressler, S.B., Bressler, N.M.: Comparison of prevalence of diabetic macular edema based on monocular fundus photography vs optical coherence tomography. JAMA Ophthalmol. **134**(2), 222–228 (2016)
32. Zeiler, M.D., Fergus, R.: Visualizing and understanding convolutional networks. In: Fleet, D., Pajdla, T., Schiele, B., Tuytelaars, T. (eds.) ECCV 2014. LNCS, vol. 8689, pp. 818–833. Springer, Cham (2014). https://doi.org/10.1007/978-3-319-10590-1_53
33. Zhu, J.Y., Park, T., Isola, P., Efros, A.A.: Unpaired image-to-image translation using cycle-consistent adversarial networks. In: ICCV, pp. 2223–2232 (2017)

Interpretable Deep Models for Cardiac Resynchronisation Therapy Response Prediction

Esther Puyol-Antón[1(✉)], Chen Chen[3], James R. Clough[1], Bram Ruijsink[1,2], Baldeep S. Sidhu[1,2], Justin Gould[1,2], Bradley Porter[1,2], Marc Elliott[1,2], Vishal Mehta[1,2], Daniel Rueckert[3], Christopher A. Rinaldi[1,2], and Andrew P. King[1]

[1] School of Biomedical Engineering and Imaging Sciences, King's College London, London, UK
esther.puyol_anton@kcl.ac.uk

[2] Guy's and St Thomas' Hospital, London, UK

[3] BioMedIA Group, Department of Computing, Imperial College London, London, UK

Abstract. Advances in deep learning (DL) have resulted in impressive accuracy in some medical image classification tasks, but often deep models lack interpretability. The ability of these models to explain their decisions is important for fostering clinical trust and facilitating clinical translation. Furthermore, for many problems in medicine there is a wealth of existing clinical knowledge to draw upon, which may be useful in generating explanations, but it is not obvious how this knowledge can be encoded into DL models - most models are learnt either from scratch or using transfer learning from a different domain. In this paper we address both of these issues. We propose a novel DL framework for image-based classification based on a variational autoencoder (VAE). The framework allows prediction of the output of interest from the latent space of the autoencoder, as well as visualisation (in the image domain) of the effects of crossing the decision boundary, thus enhancing the interpretability of the classifier. Our key contribution is that the VAE disentangles the latent space based on 'explanations' drawn from existing clinical knowledge. The framework can predict outputs as well as explanations for these outputs, and also raises the possibility of discovering new biomarkers that are separate (or disentangled) from the existing knowledge. We demonstrate our framework on the problem of predicting response of patients with cardiomyopathy to cardiac resynchronization therapy (CRT) from cine cardiac magnetic resonance images. The sensitivity and specificity of the proposed model on the task of CRT response prediction are 88.43% and 84.39% respectively, and we showcase the potential of our model in enhancing understanding of the factors contributing to CRT response.

Electronic supplementary material The online version of this chapter (https://doi.org/10.1007/978-3-030-59710-8_28) contains supplementary material, which is available to authorized users.

A. L. Martel et al. (Eds.): MICCAI 2020, LNCS 12261, pp. 284–293, 2020.
https://doi.org/10.1007/978-3-030-59710-8_28

Keywords: Interpretable ML · Cardiac resynchronization therapy · Cardiac MRI · Variational autoencoder

1 Introduction

Deep learning (DL) methods have achieved near-human accuracy levels in some classification tasks in the medical domain. However, most DL methods operate as 'black boxes', mapping a given input to a classification output, but offer little to no explanation as to how the output was decided upon. This has led to increasing research focus in recent years on techniques for interpretable DL [6]. In healthcare, interpretations (or explanations) are important in promoting clinicians' and patients' trust in automated decisions, as well as potentially for legal reasons [7].

Most existing work on interpretable DL has focused on "post-hoc" analysis of existing trained models [8]. Such approaches can be useful for visualising low-level features (e.g. parts of an image) that were important in producing the predicted output, but they do not offer a way to link these features with higher-level concepts that are well understood by clinicians. Furthermore, often there is a wealth of existing clinical knowledge that could be exploited in training a DL model to explain its decisions, but post-hoc analysis does not offer an easy way of linking this knowledge to the interpretation. An alternative, but less explored approach, which has the potential to address these limitations, is to include the need for an explanation into the training of the DL model. For example, Hind *et al.* [9] have proposed a generic framework for supervised machine learning that augments training data to include explanations elicited from domain experts. Similarly, Alvarez-Melis *et al.* [13] proposed a 'self-explaining neural network' in which interpretability is built in architecturally and more interpretable concepts are learnt during training. Both of these methods offer advantages in terms of explanations, but they do not offer an obvious way to link low level features with higher level explanatory concepts. Our work is based upon this type of approach but we seek to extend it to enable links to be made between low-level features and higher level concepts. We believe that such an interpretable approach is an essential characteristic to promote clinician trust in DL models. We propose a framework based on a variational autoencoder (VAE) to map images to low-dimensional latent vectors. The primary classification task is performed using these latent vectors. In addition, we incorporate clinical domain knowledge by using secondary classifier(s) in the latent space to encourage disentanglement of explanatory concepts within the learnt representation. By using a VAE we are able to decode and visualise these concepts to provide an explanatory basis for the primary model output. Simultaneously, we are able to link the high level concepts to features in image space. The disentanglement based on existing domain concepts also allows us to use the model to learn new undiscovered biomarkers.

Related Work on Interpretable DL: The idea of performing a classification task using fully connected layers from the latent space of a VAE has been proposed before. For example, Biffi *et al.* [2] adopted this approach to classify cardiac pathologies and used the decoder of the VAE to visualise the morphological features involved in the classification. Similarly, Clough *et al.* [5] proposed a similar architecture to detect the presence of coronary artery disease and also used concept activation vectors to quantify the importance of different explanatory concepts. Our work is methodologically distinct from [2,5] as we incorporate secondary classifier(s) as a way of incorporating existing clinical knowledge into the model as well as providing explanations of outputs.

Cardiac Resynchronisation Therapy: We illustrate our approach on the important clinical task of predicting response to cardiac resynchronization therapy (CRT). CRT is an established therapy for patients with medically refractory systolic heart failure and left ventricular dyssynchrony [1]. Current consensus guidelines [14] regarding selection for CRT focus on a limited set of patient characteristics including NYHA functional class, left ventricular ejection fraction (LVEF), QRS duration, type of bundle branch block, etiology of cardiomyopathy and atrial rhythm (sinus, atrial fibrillation). However, using these guidelines approximately 30% of patients do not respond to treatment [12,15]. The clinical research literature reveals a number of important insights into improving selection criteria. For example, it has been shown that strict left bundle branch block (LBBB) with type II contraction pattern is associated with increased response to CRT [10]. Another study demonstrated that the presence of septal flash (SF)[1] and apical rocking are also associated with improved response [11,18]. A limited number of papers have investigated the use of machine learning to predict response to CRT. Peressutti *et al.* [16] used supervised multiple kernel learning (MKL) to combine motion information derived from cardiac magnetic resonance (CMR) imaging and non-motion data to predict CRT response, achieving approximately 90% accuracy on a cohort of 34 patients. Cikes *et al.* [4] used unsupervised MKL to combine echocardiographic data and clinical parameters to phenogroup patients with HF with respect to both outcomes and response to CRT. To the best of our knowledge, no DL models have been proposed for predicting response to CRT.

Contributions: We present a new DL image-based classification framework based on a VAE that enhances the interpretability of the classifier for the application of CRT response prediction. The main novelty lies in the use of secondary classifier(s) that enable links to be made between low level image features and higher level concepts, which we believe is an important prerequisite for clinical translation of DL based cardiac diagnosis tools.

[1] An inward-outward motion of the septum in early systole (mainly during isovolumetric contraction).

2 Materials

A cohort of 73 patients fulfilling the conventional criteria for CRT was used in this study. The study was approved by the institutional ethics committee and all patients gave written informed consent. All patients underwent CMR imaging prior to CRT and 2D echocardiography imaging and clinical evaluation prior to CRT and at 6-month follow-up. CMR imaging was carried out on multiple scanners: Siemens Aera 1.5T, Siemens Biograph mMR 3T, Philips 1.5T Ingenia and P1.5T and 3T Achieva. The CMR multi-slice short-axis (SA) stack was used in this study, which had a slice thickness between 8 and 10 mm, an in-plane resolution between $0.94 \times 0.94\,\text{mm}^2$ to $1.5 \times 1.5\,\text{mm}^2$ and a temporal resolution of ~13–31 ms/frame. All patients were classified as responders or non-responders based on volumetric measures derived from 2D echocardiography acquired at the 6-month follow-up evaluation [14]. Patients were classified as responders if they had a reduction of ≥15% in left ventricular (LV) end-systolic volume after CRT, and non-responders otherwise. This information was used as the primary output label in training our proposed model. Based on a manual clinical assessment of the 2D echocardiography imaging acquired prior to CRT, those patients exhibiting SF were also identified by an experienced cardiologist. The SF information was used as an 'explanatory concept' in training our proposed model to generate meaningful explanations. Note that this information would not normally be available at inference time, as it results from a time-consuming expert inspection. From the cohort, there were 47/73 patients who were responders to CRT. 27 of the 47 responders had SF and 10 of the 26 non-responders had SF. The ratio of SF responders/non-responders was 27/37 (~73%), which is in line with the distribution reported by Parsai *et al.* [15]

3 Methods

Our CRT prediction model is illustrated in Fig. 1. The model consists of a segmentation network, followed by a VAE that incorporates multiple classifiers in the latent space. In the following sections we describe the different components of the pipeline.

Spatial-Temporal Normalisation: To correct for variation in acquisition protocols between vendors, all images were first resampled to an in-plane voxel size of 1.25×1.25 mm, and temporally resampled to $T = 25$ frames per cardiac cycle using piecewise linear warping of cardiac timings.

Automatic Segmentation Network: We used a U-net based architecture for automatic segmentation of the LV blood pool, LV myocardium and right ventricle (RV) blood pool from the SA slices in all frames through the cardiac cycle [3]. To take into account the inter-vendor differences in intensity distributions, the segmentation network was fine tuned with 300 images (multiple slices/time points from 20 CMR scans) from an independent clinical database.

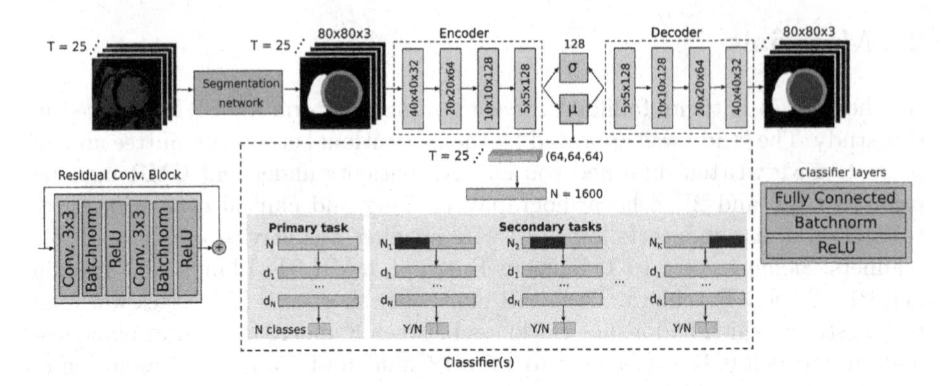

Fig. 1. Diagram showing the architecture of the joint VAE/classification model. The inputs to the VAE are CMR segmentations from $T = 25$ time points for the three top slices of the SA stack. The VAE consists of a series of residual convolutional blocks, with the image resolution and number of feature maps denoted in each block. For the multiple classification networks, the latent vectors belonging to the different time points are concatenated and used as inputs.

VAE Network: The VAE model is composed of an encoder that compresses the data into a latent space (of dimensionality 128), followed by a decoder network trained to reconstruct the original data from the latent representations. The encoder/decoder architecture has three channels corresponding to CMR segmentations from the top three slices.

Primary Task Classifier: Using the encoder network, the segmentations of the CMR images at all T time points from each subject were mapped into T latent vectors. The first fully connected layer of the classifier processes the latent vectors individually. The outputs of these layers are then concatenated into a single vector, which is used as input to three fully connected layers that predict CRT response for each subject.

Explanatory Concept Classifiers: The clinical domain knowledge is incorporate as secondary classifier(s). In the experiments presented in this paper, we illustrate this idea using a single explanatory concept (SF). However, in principle multiple explanatory concepts could be used, each of which would use a different (possibly overlapping) portion of the latent space. We always ensure that a part of the latent space is unused by any of the secondary classifiers, to ensure that this represents 'unknown' factors contributing to the primary output (i.e. CRT response). The secondary classifier(s) follow a similar structure to the primary task classifier.

We denote an input data sequence by $\mathbf{X} = [\mathbf{x}_1, \mathbf{x}_2, ...\mathbf{x}_T]$, and its corresponding latent mean and standard deviation vectors as $\mathbf{M} = [\mu_1, \mu_2, ...\mu_T]$ and $\mathbf{\Sigma} = [\sigma_1, \sigma_2, ...\sigma_T]$, where $(\mu_t, \sigma_t) = \text{Encoder}(\mathbf{x}_t)$. The decoded images are denoted by $\widetilde{\mathbf{X}} = [\tilde{\mathbf{x}}_1, \tilde{\mathbf{x}}_2, ...\tilde{\mathbf{x}}_T]$. For the primary classification task, the ground truth label is denoted by y and the predicted label by $\tilde{y} = \text{Classifier}(\mathbf{M})$, i.e. we use only the latent mean vector for classification. For the secondary tasks, we select a subset of the latent space $\mathbf{M_k} = [\mu_{l_k}, \mu_{l_k+1}, ...\mu_{l_k+N_k}]$, where $N_k < T$ is the size of the subset for secondary task k and l_k is the start of the subset for task k. This subset is used for classification of the explanatory concept(s) $\tilde{y_k} = \text{Classifier}(\mathbf{M_k})$. The joint loss function for the VAE and the primary and secondary classifiers can then be written as follows:

$$\mathcal{L}_{\text{total}} = \frac{1}{T}\sum_{t=1}^{t=T}[\mathcal{L}_{\text{re}}(\mathbf{x}_t, \tilde{\mathbf{x}}_t) + \beta\mathcal{L}_{\text{KL}}(\mu_t, \sigma_t)] + \gamma\mathcal{L}_{\text{cl}}(y, \tilde{y}) + \sum_{k=0}^{K}\alpha_k\mathcal{L}_{\text{cl}}(y_k, \tilde{y_k}) \quad (1)$$

where $\mathcal{L}_{\text{re}}$ is the cross-entropy between the input segmentations and the output predictions, $\mathcal{L}_{\text{cl}}$ is the binary cross entropy loss for the classification tasks (primary/secondary), $\mathcal{L}_{\text{KL}}$ is the Kullback-Leibler divergence between the latent variables and a unit Gaussian, and β, γ and α_k are constants that weight the components of the loss function. K is the number of secondary tasks and y_k are their corresponding ground truth labels, which are provided as clinical domain knowledge (see Sect. 2).

4 Experiments and Results

We applied the proposed pipeline for the primary classification task of predicting CRT response from pre-treatment CMR images. The explanatory concept used in the secondary classifier was SF. The inputs to the VAE were 80×80 segmentations of three basal slices of the SA stack, where each slice was treated as a channel of the network input. The data for each subject consist of $T = 25$ segmentations per slice, representing one full cardiac cycle.

Model Training: The model was trained in three stages. First, we trained only using the VAE loss (i.e. $\gamma = 0$ and $\alpha_k = 0$) using 10,000 subjects (including both healthy and cardiovascular diseases) from the UK Biobank database [17] for 500 epochs, and then fine tuned using CRT patient data for 300 epochs. Second, we trained both the VAE and the primary task classifier ($\alpha_k = 0$) for 500 epochs. Finally, we trained the VAE, the main and secondary classifiers together for 300 epochs. We used a grid search strategy to identify the optimal beta that was selected as a trade-off between disentanglement and reconstruction quality. For the secondary classifier, we used the first half of the latent space ($N_0 = 64$ and $l_0 = 0$). Data augmentation was used during all phases of the training (random rotation and translation). The model was trained on a NVIDIA GeForce GTX TITAN X using Adam optimiser with learning rate equal to 10^{-4} and batch size of 8. We set $\beta = 0.2$, $\gamma = 1$ and $\alpha_k = 0.9$.

Classification and Reconstruction Results: A 5-fold stratified cross-validation over the CRT patient data was employed to evaluate the proposed framework. For each fold, the decoder network was evaluated using the average Dice Score between the input segmentation and the predicted segmentation. The primary and secondary classifiers were evaluated using a receiver operating

Table 1. Comparison between baseline, primary task VAE classifier and primary/secondary task VAE classifier. McNemar's test was used to compute the p-values.

Methods	CRT					SF		
	BACC	SEN	SPE	Dice	p-value	BACC	SEN	SPE
Baseline	92.46	96.69	88.23	–	–	–	–	–
VAE + CRT	90.40	94.59	86.21	87.04	0.18	–	–	–
VAE + CRT + SF	86.41	88.43	84.39	85.85	0.06	80.24	73.42	87.06

characteristic curve (ROC) analysis, and based on this the balanced accuracy (BACC), sensitivity (SEN) and specificity (SPE) were computed for the optimal classifier selected using the Youden index. For comparison, we also evaluated a 'baseline' classifier (i.e. encoder + CRT classifier only) and also a version of our model featuring only the primary task classifier and no secondary task (i.e. VAE + primary classifier). Table 1 summarises the results for the proposed method and the comparative methods. To compare the performance of the different classification algorithms we used McNemar's test between the baseline method and the other methods.

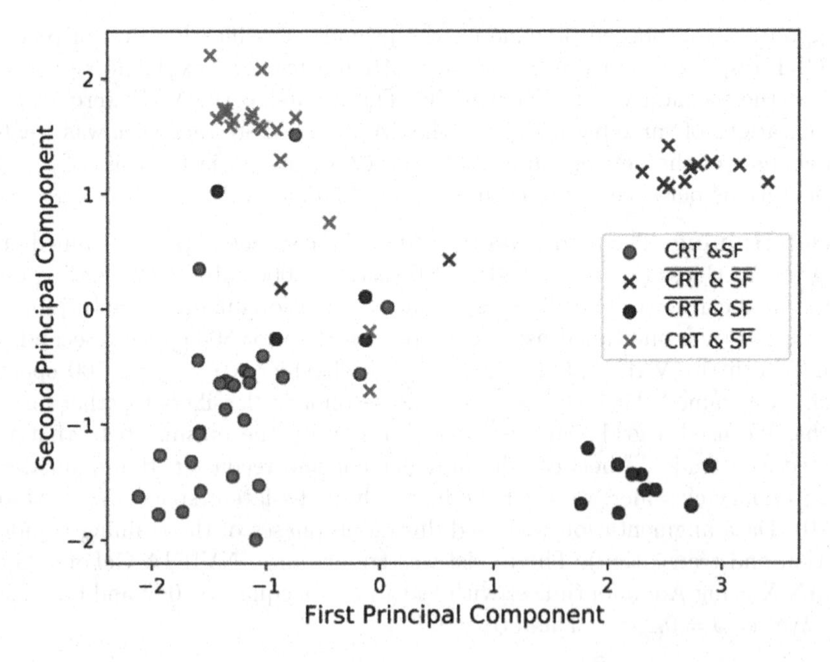

Fig. 2. PCA of the latent space vectors for the test cases, where red are CRT responders and blue non CRT responders, dots are subjects with SF and crosses subjects without SF

Disentanglement of the Latent Space: Figure 2 shows the first two PCA components of the latent space on the test database. It is visible that the primary and secondary classifier has enforced the disentanglement of the latent space between CRT responders and non responders.

Visualisation of the Explanatory Concepts: One of the strengths of our proposed model is that it enables visualisation in the image domain of the secondary classification task to investigate if the learned features correspond to the clinical domain knowledge. To illustrate this concept, we computed the mean point in the latent space from all subjects classified as having SF and reconstructed an image sequence using the VAE decoder. Figure 3 shows a two-dimensional M-Mode representation of a line crossing the LV and RV through the septum of the reconstructed images. The video of the full cardiac cycle is available in the supplementary material. An experienced cardiologist reviewed the video and the M-Mode representation and reported that it shows an irregular motion of the septal wall during contraction, most likely indicating SF.

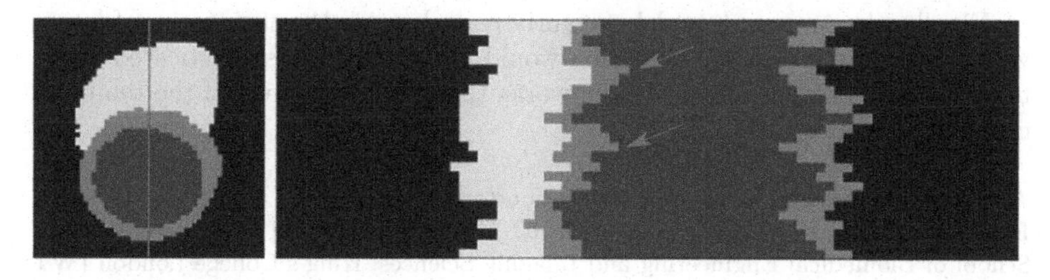

Fig. 3. Two dimensional M-Mode visualisation of the mid LV and RV over the full cardiac cycle. Red line indicates profile line selected to generate the two dimensional M-Mode image. Arrows indicate irregular motion of the septum during contraction, which likely corresponds to septal flash. (Color figure online)

5 Discussion

In this work we propose a model that not only performs classification, but also allows interpretation of features important in the classification. We achieve state-of-the-art performance for CRT response prediction, and this is the first time that DL has been used for this purpose. Our DL model is fully automated and requires no user interaction to generate the model input, unlike previous machine learning approaches for CRT response prediction [4,16]. Our key novelty is that we use additional secondary classifier(s) to encode existing clinical knowledge into the model to enable it to explain its output. Our results show that our explanatory classifier has similar performance to the baseline VAE, but offers the possibility to disentangle the latent space based on clinical knowledge and hence explain its decisions to clinicians.

To illustrate the ability of our method to incorporate clinical knowledge we showed in Fig. 3 that our model had learnt in a weakly supervised manner the concept of SF, which is associated with positive response to CRT treatment. In future work we aim to investigate extending the current framework to incorporate multiple explanatory concepts. Apart from allowing interpretation and thus improving trust, encoding of clinical concepts into the model offers a way to incorporate the extensive biophysical knowledge that is already known about a disease. It is likely that this will in the long run improve DL applicability. Furthermore, the proposed model has the potential to discover new explanatory factors related to CRT response by using the VAE decoder to visualise other portions of the latent space.

We used segmentations instead of intensity images to train our model as we achieved slightly higher performance in this way and the quality of reconstructed images and latent space interpolations was superior. However, segmentations do not offer the same level of detail as CMR intensity images and therefore this could cause a loss of information that could be important in predicting CRT response. In the future we will investigate the use of an adversarial loss to ensure high-quality intensity image reconstructions which can then be used to visualise both structural and textural features relevant to the classification.

Finally, the proposed model currently concatenates the time series of latent vectors. A further possible extension would be to use more sophisticated architectures such as recurrent neural networks that take advantage of the temporal correlations between frames.

Acknowledgements. This work was supported by the EPSRC (EP/R005516/1 and EP/P001009/1) and the Wellcome EPSRC Centre for Medical Engineering at the School of Biomedical Engineering and Imaging Sciences, King's College London (WT 203148/Z/16/Z). This research has been conducted using the UK Biobank Resource under Application Number 17806.

References

1. Abraham, W.T., et al.: Cardiac resynchronization in chronic heart failure. N. Engl. J. Med. **346**(24), 1845–1853 (2002)
2. Biffi, C., et al.: Explainable anatomical shape analysis through deep hierarchical generative models. IEEE Trans. Med. Imaging **39**(6), 2088–2099 (2020)
3. Chen, C., et al.: Improving the generalizability of convolutional neural network-based segmentation on CMR images. arXiv preprint arXiv:1907.01268 (2019)
4. Cikes, M., et al.: Machine learning-based phenogrouping in heart failure to identify responders to cardiac resynchronization therapy. Eur. J. Heart Fail. **21**(1), 74–85 (2019)
5. Clough, J.R., Oksuz, I., Puyol-Antón, E., Ruijsink, B., King, A.P., Schnabel, J.A.: Global and local interpretability for cardiac MRI classification. In: Shen, D., et al. (eds.) MICCAI 2019. LNCS, vol. 11767, pp. 656–664. Springer, Cham (2019). https://doi.org/10.1007/978-3-030-32251-9_72
6. Doshi-Velez, F., Kim, B.: Towards a rigorous science of interpretable machine learning. arXiv preprint arXiv:1702.08608 (2017)
7. Goodman, B., Flaxman, S.: European union regulations on algorithmic decision-making and a "right to explanation". AI Mag. **38**(3), 50–57 (2017)

8. Guidotti, R., Monreale, A., Ruggieri, S., Turini, F., Giannotti, F., Pedreschi, D.: A survey of methods for explaining black box models. ACM Comput. Surv. (CSUR) **51**(5), 1–42 (2018)
9. Hind, M., et al.: TED: teaching AI to explain its decisions. In: Proceedings of the 2019 AAAI/ACM Conference on AI, Ethics, and Society, pp. 123–129 (2019)
10. Jackson, T., et al.: A U-shaped type II contraction pattern in patients with strict left bundle branch block predicts super-response to cardiac resynchronization therapy. Heart Rhythm **11**(10), 1790–1797 (2014)
11. Marechaux, S., et al.: Role of echocardiography before cardiac resynchronization therapy: new advances and current developments. Echocardiography **33**(11), 1745–1752 (2016)
12. McAlister, F.A., et al.: Cardiac resynchronization therapy for patients with left ventricular systolic dysfunction: a systematic review. JAMA **297**(22), 2502–2514 (2007)
13. Melis, D.A., Jaakkola, T.: Towards robust interpretability with self-explaining neural networks. In: Advances in Neural Information Processing Systems, pp. 7775–7784 (2018)
14. Members, A.F., et al.: 2013 ESC guidelines on cardiac pacing and cardiac resynchronization therapy: the task force on cardiac pacing and resynchronization therapy of the European society of cardiology (ESC). Developed in collaboration with the European heart rhythm association (EHRA). Eur. Heart J. **34**(29), 2281–2329 (2013)
15. Parsai, C., et al.: Toward understanding response to cardiac resynchronization therapy: left ventricular dyssynchrony is only one of multiple mechanisms. Eur. Heart J. **30**(8), 940–949 (2009)
16. Peressutti, D., et al.: A framework for combining a motion atlas with non-motion information to learn clinically useful biomarkers: application to cardiac resynchronisation therapy response prediction. Med. Image Anal. **35**, 669–684 (2017)
17. Petersen, S.E., et al.: UK biobank's cardiovascular magnetic resonance protocol. J. Cardiovasc. Magn. Reson. **18**(1), 8 (2015)
18. Stankovic, I., et al.: Relationship of visually assessed apical rocking and septal flash to response and long-term survival following cardiac resynchronization therapy (PREDICT-CRT). Eur. Heart J.-Cardiovasc. Imaging **17**(3), 262–269 (2016)

Encoding Visual Attributes in Capsules for Explainable Medical Diagnoses

Rodney LaLonde[1(✉)], Drew Torigian[2], and Ulas Bagci[1]

[1] University of Central Florida, Orlando, USA
lalonde@Knights.ucf.edu
[2] University of Pennsylvania, Philadelphia, USA

Abstract. Convolutional neural network based systems have largely failed to be adopted in many high-risk application areas, including healthcare, military, security, transportation, finance, and legal, due to their highly uninterpretable "black-box" nature. Towards solving this deficiency, we teach a novel multi-task capsule network to improve the explainability of predictions by embodying the same high-level language used by human-experts. Our explainable capsule network, ***X-Caps***, encodes high-level visual object attributes within the vectors of its capsules, then forms predictions based solely on these human-interpretable features. To encode attributes, *X-Caps* utilizes a new routing sigmoid function to independently route information from child capsules to parents. Further, to provide radiologists with an estimate of model confidence, we train our network on a distribution of expert labels, modeling inter-observer agreement and punishing over/under confidence during training, supervised by human-experts' agreement. *X-Caps* simultaneously learns attribute and malignancy scores from a multi-center dataset of over 1000 CT scans of lung cancer screening patients. We demonstrate a simple 2D capsule network can outperform a state-of-the-art deep dense dual-path 3D CNN at capturing visually-interpretable high-level attributes and malignancy prediction, while providing malignancy prediction scores approaching that of non-explainable 3D CNNs. To the best of our knowledge, this is the first study to investigate capsule networks for making predictions based on radiologist-level interpretable attributes and its applications to medical image diagnosis. Code is publicly available at https://github.com/lalonderodney/X-Caps.

Keywords: Explainable AI · Lung cancer · Capsule networks

1 Introduction

In machine learning, predictive performance typically comes at the cost of *interpretability* [4,7,14,24]. While deep learning (DL) has impact many fields, there exist several high-risk domains which have yet to be comparably affected: military, security, transportation, finance, legal, and healthcare among others, often citing a lack of interpretability as the main concern [3,18,23]. As features become

A. L. Martel et al. (Eds.): MICCAI 2020, LNCS 12261, pp. 294–304, 2020.
https://doi.org/10.1007/978-3-030-59710-8_29

less *interpretable*, and the functions learned more complex, model predictions become more difficult to explain (Fig. 1b). While some works have begun to press towards this goal of explainable DL, the problem remains largely unsolved.

Interpretable vs. Explainable: There has been a recent push in the community to move away from the *post-hoc interpretations* of deep models and instead create explainable models from the outset [24,27]. Since the terms *interpretable* and *explainable* are often used interchangeably, we want to be explicit about our definitions for the purposes of this study. An *explainable* model is one which provides explanations for its predictions *at the human level* for a *specific task*. An *interpretable* model is one for which some conclusions can be drawn about the internals/predictions of the model; however, they are not explicitly provided by the model and are typically at a lower level. For example, in image classification, when a deep model predicts an image to be of a cat, saliency/gradient or other methods can attempt to *interpret* the model/prediction. However, the model is not *explaining* why the object in the image is a cat in the same way as a human. Humans classify objects based on a taxonomy of characteristics/attributes (*e.g.* cat equals four legs, paws, whiskers, fur, etc.). If our goal is to create *explainable* models, we should design models which explain their decisions using a similar set of "attributes" to humans.

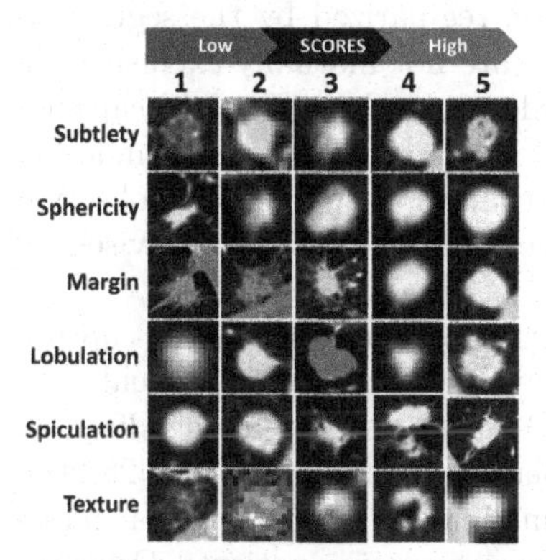

(a) Lung nodules with high-level visual attribute scores as determined by expert radiologists. Scores were given from 1 – 5 for six different visual attributes related to diagnosing lung cancer.

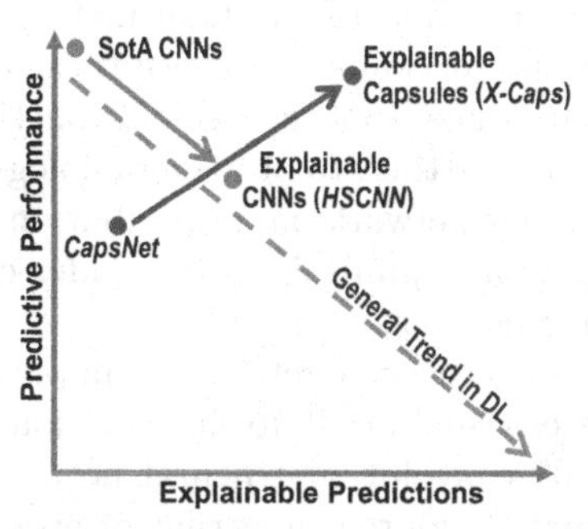

(b) A symbolic plot showing the general trade-off between explainability and predictive performance in deep learning (DL) [4,7,14,24]. Our proposed *X-Caps* rebuts the trend of decreasing performance from state-of-the-art (SotA) as explainability increases and shows it is possible to create more explainable models *and* increase predictive performance with capsule networks.

Fig. 1. Encoding visual attributes (a) for explainable predictions (b).

Why Capsule Networks? Capsule networks differ from convolutional neural networks (CNNs) by replacing the scalar feature maps with vectorized representations, responsible for encoding information (*e.g.* pose, scale, color) about each feature. These vectors are then used in a dynamic routing algorithm which seeks to maximize the agreement between lower-level predictions for the instantiation parameters (i.e. capsule vectors) of higher-level features. In their introductory work, a capsule network (*CapsNet*) was shown to produce promising results on the MNIST data set; but more importantly, was able to encode high-level visually-interpretable features of digits (*e.g.* stroke thickness, skew, localized-parts) within the dimensions of its capsule vectors [25].

Lung Cancer Diagnosis with a Multi-task Capsule Network: In diagnosing the malignancy of lung nodules, similar to describing why an image of a cat is catlike, radiologists explain their predictions through the language of high-level visual attributes (i.e., radiographical interpretations): subtlety (sub), sphericity (sph), margin (mar), lobulation (lob), spiculation (spi), and texture (tex), shown in Fig. 1a, which are known to be predictive (with inherent uncertainty) of malignancy [8]. To create a DL model with this same level of radiographical interpretation, we propose a novel multi-task capsule architecture, called ***X-Caps***, for learning visually-interpretable feature representations within capsule vectors, then predicting malignancy based solely on these interpretable features. By supervising different capsules to embed specific visually-interpretable features, multiple visual attributes are learned simultaneously, with their weights being updated by both the radiologists visual interpretation scores as well as their contribution to the final malignancy score, regularized by the segmentation reconstruction error. Since these attributes are not mutually-exclusive, we introduce a new routing sigmoid function to independently route child capsules to parents. Further, to provide radiologists with an estimate of model confidence, we train our network on a distribution of expert labels, modeling inter-observer agreement and punishing over/under confidence during training, supervised by human-experts' agreement.

We show even a relatively simple 2D capsule network can better capture high-level visual attribute information than the state-of-the-art deep dual-path dense 3D convolutional neural network (CNN) while also improving diagnostic accuracy, approaching that of even some black-box methods (*e.g.,* [28,29]). *X-Caps* simultaneously learns attribute and malignancy scores from a multi-center dataset of over 1000 CT scans of lung cancer screening patients. **Overall, the contributions of this study are summarized as:**

1. The first study to directly encode high-level visual attributes within the vectors of a capsule network to perform explainable image-based diagnosis *at the radiologist-level.*
2. Create a novel modification to the dynamic routing algorithm to independently route information from child capsules to parents when parent capsules are not mutually-exclusive.
3. Provide a meaningful confidence metric with our predictions at test by learning directly from expert label distributions to punish network over/under confidence. Visual attribute predictions are verified at test via the reconstruction branch of the network.

4. Demonstrate a simple 2D capsule network (*X-Caps*) trained from scratch outperforming a state-of-the-art deep pre-trained dense dual-path 3D CNN at capturing visually-interpretable high-level attributes and malignancy prediction, while providing malignancy prediction scores approaching that of non-explainable 3D CNNs.

2 Related Work

The majority of work in explainable deep learning has focused around *post hoc* deconstruction of already trained models (*i.e.* interpretation). These approaches typically rely on human-experts to examine their results and attempt to discover meaningful patterns. Zeiler and Fergus [31] attached a deconvolutional network to network layers to map activations back to pixel space for visualizing individual filters and activation maps, while also running an occlusion-based study of which parts of the input contribute most to the final predictions. *Grad-CAM* [26] is a popular approach which highlights the relative positive activation map of convolutional layers with respect to network outputs. *InfoGAN* [5] separates noise from the "latent code", maximizing the mutual information between the latent representations and the image inputs, encoding concepts such as rotation, width, and digit type for MNIST. In a similar way, capsule networks encode visually-interpretable concepts such as stroke thickness, skew, rotation, and others [25].

A number of recent studies have proposed using *CapsNet* for a variety of medical imaging classification tasks [1,11–13,19,22,30]. However, these methods nearly all follow the exact *CapsNet* architecture, or propose minor modifications which present nearly identical predictive performance [16,20]; hence, it is sufficient to compare only with *CapsNet* in reference to these works.

In the area of lung nodule malignancy, many DL-based approaches have been proposed [28,29], with further methods being developed with complicated post-processing techniques [9], curriculum learning methods [21], or gradient-boosting machines [32]. However, adding such techniques is beyond the scope of this study and would lead to an unwieldy enumeration of ablation studies necessary to understand the contributions between our proposed capsule architecture and such techniques. For a fair comparison in this study, we compare our method directly against *CapsNet* and explainable CNN approaches. *HSCNN* [27] creates one of the first explainable methods, by designing a dense 3D CNN which first predicts visual attribute scores then predicts malignancy from those features. This decreased the overall performance as compared to other 3D networks [29] but provided some explanations for the final malignancy predictions.

3 Capsules for Encoding Visual Attributes

Our approach, referred to as *explainable capsules*, or *X-Caps*, was designed to remain as similar as possible to our control network, *CapsNet*, while allowing us to have more control over the visually-interpretable features learned. *CapsNet* already showed great promise when trained on the MNIST data set for its ability

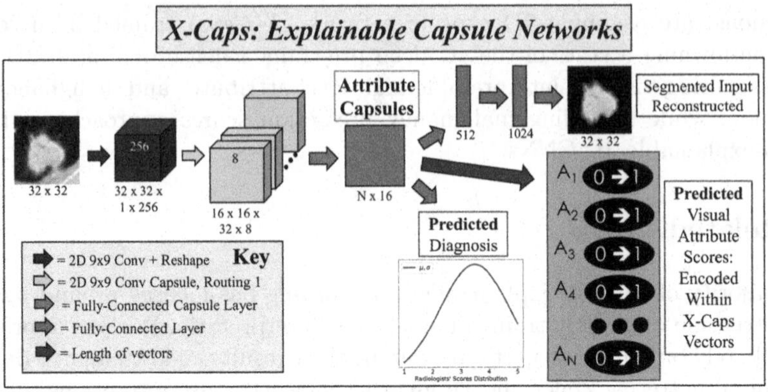

Fig. 2. *X-Caps*: Explainable Capsule Networks. The proposed network (1) predicts N high-level visual attributes of the nodule, (2) segments the nodule and reconstruct the input image, and (3) diagnoses the nodule on a scale of 1 to 5 based on the visually-interpretable high-level features encoded in the X-Caps capsule vectors. The malignancy diagnosis branch is attempting to model the distribution of radiologists' scores in both mean and variance.

to model high-level visually-interpretable features. With this study, we examine the ability of capsules to model *specific* visual attributes within their vectors, rather that simply hoping these are learned successfully in the more challenging lung nodule data. As shown in Fig. 2, *X-Caps* shares a similar overall structure as *CapsNet*, with the major differences being the addition of the supervised labels for each of the *X-Caps* vectors, the fully-connected layer for malignancy prediction, the reconstruction regularization also performing segmentation, and the modifications to the dynamic routing algorithm.

The first layer of our proposed network is a 2D convolutional layer which extracts low-level features. Next, we form our primary capsules of 32 capsule types with $8D$ vector capsules. Following this, we form our attribute capsules using a fully-connected capsule layer whose output is N $16D$ capsule types, one for each of the visual-attributes we want to predict. Unlike *CapsNet* where each of the parent capsules were dependant on one another (e.g. if the prediction is the digit 5 it cannot also be a 3), our parent capsules are not mutually-exclusive of each other (i.e. a nodule can score high or low in each of the attribute categories). For this reason, we needed to modify the dynamic routing algorithm presented in *CapsNet* to accommodate this significant difference. The key change is the "routing softmax" employed by *CapsNet* forces the contributions of each child to send their information to parents in a manner which sums to one, which in practice effectively makes them "choose" a parent to send their information to. However, when computing prediction vectors for independent parents, we want a child to be able to contribute to all parent capsules for attributes which are present in the given input. With that motivation, the specific algorithm, which we call "routing sigmoid", is computed as

$$r_{i,j} = \frac{\exp(b_{i,j})}{\exp(b_{i,j}) + 1}, \tag{1}$$

where $r_{i,j}$ are the routing coefficients determined by the dynamic routing algorithm for child capsule i to parent capsule j and the initial logits, $b_{i,j}$ are the prior probabilities that the prediction vector for capsule i should be routed to parent capsule j. Note the prior probabilities are initially set to 1 rather than 0 as in *CapsNet*, otherwise no routing could take place. The rest of the dynamic routing procedure follows the same as in [25].

Predicting Malignancy from Visually-Interpretable Capsule Vectors: In order to predict malignancy scores, we attach a fully-connected layer to our attribute capsules with output size equal to the range of scores. We wish to emphasize here, our final malignancy prediction is coming solely from the vectors whose magnitudes represent *visually-interpretable* feature scores. Every malignancy prediction score has a set of weights connected to the high-level attribute capsule vectors, and the activation from each tells us the exact contribution of the given visual attribute to the final malignancy prediction for that nodule. Unlike previous studies which look at the importance of these attributes on a global level, our method looks at the importance of each visual attribute in relation to a specific nodule being diagnosed. To verify the correctness of our attribute modeling, we reconstruct the nodules while varying the dimensions of the capsule vectors to ensure the desired visual attributes are being modeled. At test, these reconstructions give confidence that the network is properly capturing the attributes, and thus the scores can be trusted. Confidence in the malignancy prediction score, in addition to coming solely from these trusted attributes, is provided via an uncertainty modeling approach.

Previous works in lung nodule classification follow the same strategy of averaging radiologists' scores for visual attributes and malignancy, and then either attempt to regress this average or performing binary classification of the average as below or above 3. To better model the uncertainty inherently present in the labels due to inter-observer variation, we propose a different approach: we attempt to predict the *distribution* of radiologists' scores. Specifically, for a given nodule where we have at minimum three radiologists' score values for each attribute and for malignancy prediction, we compute the mean and variance of those values and fit a Gaussian function to them, which is in turn used as the ground-truth for our classification vector. Nodules with strong inter-observer agreement produce a sharp peak, in which case wrong or unsure (*i.e.*, low confidence score) predictions are severely punished. Likewise, for low inter-observer agreement nodules, we expect our network to output a more spread distribution and it will be punished for strongly predicting a single class label. This proposed approach allows us to model the uncertainty present in radiologists' labels in a way that no previous study has and provide a meaningful confidence metric at test time to radiologists.

Loss and Regularization: As in *CapsNet*, we also perform reconstruction of the input as a form of regularization. However, we extend the idea of regularization to perform a pseudo-segmentation, similar in nature to the reconstruction used by [15,17]. Whereas in true segmentation, the goal is to output a binary mask of pixels which belong to the nodule region, in our formulation we attempt to reconstruct only the pixels which belong to the nodule region, while the rest are mapped to zero. More specifically, we formulate this loss as

$$\mathcal{L}_r = \frac{\gamma}{H \times W} \sum_x^W \sum_y^H \|R^{x,y} - O_r^{x,y}\|, \text{ with } R^{x,y} = I^{x,y} \times S^{x,y} \mid S^{x,y} \in \{0,1\}, \quad (2)$$

where $\mathcal{L}_r$ is the supervised loss for the reconstruction regularization, γ is a weighting coefficient for the reconstruction loss, $R^{x,y}$ is the reconstruction target pixel, $S^{x,y}$ is the ground-truth segmentation mask value, and $O_r^{x,y}$ is the output of the reconstruction network, at pixel location (x, y), respectively, and H and W are the height and width, respectively, of the input image. This adds another task to our multi-task approach and an additional supervisory signal which can help our network distinguish visual characteristics from background noise. The malignancy prediction score, as well as each of the visual attribute scores also provide a supervisory signal in the form of

$$\mathcal{L}_a = \sum_n^N \alpha^n \|A^n - O_a^n\|, \; \mathcal{L}_m = \beta \sum_{x \in X} \varepsilon(\boldsymbol{O}_m) \log \left(\frac{\varepsilon(\boldsymbol{O}_m)}{\frac{1}{\sqrt{2\pi\sigma^2}} \exp\left(\frac{-(x-\mu)^2}{2\sigma^2}\right)} \right), \quad (3)$$

where $\mathcal{L}_a$ is the combined loss for the visual attributes, A^n is the average of the attribute scores given by at minimum three radiologists for attribute n, N is the total number of attributes, α^n is the weighting coefficient placed on the nth attribute, O_a^n is the network prediction for the score of the nth attribute, $\mathcal{L}_m$ is a KL divergence loss for the malignancy score, β is the weighting coefficient for the malignancy score, μ and σ are the mean and variance of radiologists' scores, and $\varepsilon = \exp(O_m^i)/\sum_{j=1}^N \exp(O_m^j)$ is the softmax over the network malignancy prediction vector $\boldsymbol{O}_m = \{O_m^1, ..., O_m^N\}$. In this way, the overall loss for *X-Caps* is simply $\mathcal{L} = \mathcal{L}_m + \mathcal{L}_a + \mathcal{L}_r$. For simplicity, the values of each α^n and β are set to 1, and γ is set to $0.005 \times 32 \times 32 = 0.512$.

4 Experiments, Results, Limitations, and Ablations

We performed our experiments on the LIDC-IDRI data set [2]. Five-fold stratified cross-validation was performed, with 10% of each training set used for validation and early stopping. *X-Caps* was trained with a batch size of 16 using Adam with an initial learning rate of 0.02 reduced by a factor of 0.1 after validation loss plateau. Consistent with the literature, nodules of mean radiologists' score 3 were removed (leaving 646 benign and 503 malignant nodules) and predictions were considered correct if within ±1 of the radiologists' classification [9,10]. The results summarized in Table 1 illustrate the prediction of visual attributes with the proposed *X-Caps* in comparison with an adapted version of *CapsNet*, a deep dense dual-path 3D explainable CNN (*HSCNN* [27]), and two state-of-the-art

Table 1. Prediction accuracy of visual attributes with capsule networks. Dashes (-) represent values which the given method could not produce.

	Attribute prediction accuracy %						Malignancy
	sub	sph	mar	lob	spi	tex	
Non-explainable methods							
3D Multi-Scale + RF [28]	-	-	-	-	-	-	86.84
3D Multi-Crop [29]	-	-	-	-	-	-	87.14
CapsNet [25]	-	-	-	-	-	-	77.04
Explainable methods							
3D Dual-Path *HSCNN* [27]	71.9	55.2	72.5	-	-	83.4	84.20
Proposed *X-Caps*	**90.39**	**85.44**	**84.14**	**70.69**	**75.23**	**93.10**	**86.39**

non-explainable methods which do not have extra post-processing or learning strategies. Compared methods results are from the original reported works.

Our results show that *X-Caps* outperforms the state-of-the-art explainable method (*HSCNN*) at attribute modeling (the main goal of both studies), while also producing higher malignancy prediction scores, approaching state-of-the-art non-explainable methods performance. Further, we wish to emphasize the significance of *X-Caps* providing increased predictive performance *and* explainability over *CapsNet*. This goes against the assumed trend in DL, illustrated with a symbolic plot in Fig. 1b, that explainability comes at the cost of predictive performance, a trend we observe with *HSCNN* being outperformed by less powerful (*i.e.* not dense or dual-path) but non-explainable 3D CNNs [28,29]. While *X-Caps* slightly under-performs the best non-explainable models, it is reasonable to suspect that future research into more powerful 3D capsule networks would allow explainable capsules to surpass these methods; we hope this study will promote such future works. As two limitations of our work, we did not tune the weight balancing terms between the different tasks and further investigation could lead to superior performance. Also, we found capsule networks can be somewhat fragile; often random initializations failed to converge to good performance. However, this may be due to the small/shallow network size and its relation to the Lottery Ticket Hypothesis [6].

Ablation Studies: To analyze the impact of each component of our proposed approach, we performed ablation studies for: (1) learning the distribution of radiologists' scores rather than attempting to regress the mean value of these scores, (2) removing the reconstruction regularization from the network, and (3) performing our proposed "routing sigmoid" over the original "routing softmax" proposed in [25]. The malignancy prediction accuracy for each of these ablations is (1) 83.09%, (2) 80.30%, and (3) 80.69%, respectively, as compared to the proposed model's accuracy of 86.39%. This shows retaining the agreement/disagreement information among radiologists proved useful, the reconstruction played a role in improving the network performance, and our proposed modifications to the dynamic routing algorithm were necessary for passing information from children to parents when the parent capsule types are independent.

5 Discussions and Concluding Remarks

Available studies for explaining DL models, typically focus on *post hoc* interpretations of trained networks, rather than attempting to build-in explainability. This is the first study for directly learning an interpretable feature space by encoding high-level visual attributes within the vectors of a capsule network to perform explainable image-based diagnosis. We approximate visually-interpretable attributes through individual capsule types, then predict malignancy scores directly based only on these high-level attribute capsule vectors, in order to provide malignancy predictions with explanations *at the human-level*, in the same language used by radiologists. Our proposed multi-task explainable capsule network, *X-Caps*, successfully approximated visual attribute scores better than the previous state-of-the-art explainable diagnosis system, while also achieving higher diagnostic accuracy. We believe our approach should be applicable to any image-based classification task where high-level attribute information is available to provide explanations about the final prediction.

Acknowledgments. This project is partially supported by the NIH funding: R01-CA246704 and R01-CA240639.

References

1. Afshar, P., Mohammadi, A., Plataniotis, K.N.: Brain tumor type classification via capsule networks. In: 2018 25th IEEE International Conference on Image Processing (ICIP), pp. 3129–3133. IEEE (2018)
2. Armato III, S., et al.: The Lung Image Database Consortium (LIDC) and Image Database Resource Initiative (IDRI): a completed reference database of lung nodules on CT scans. Med. Phys. **38**(2), 915–931 (2011)
3. Bloomberg, J.: Don't Trust Artificial Intelligence? Time To Open The AI 'Black Box', (11162018), Forbes Magazine. http://www.forbes.com/sites/jasonbloomberg/2018/09/16/dont-trust-artificial-intelligence-time-to-open-the-ai-black-box/#6ceaf3793b4a
4. Bologna, G.: A model for single and multiple knowledge based networks. Artif. Intell. Med. **28**(2), 141–163 (2003)
5. Chen, X., Duan, Y., Houthooft, R., Schulman, J., Sutskever, I., Abbeel, P.: Infogan: interpretable representation learning by information maximizing generative adversarial nets. In: Advances in Neural Information Processing Systems, pp. 2172–2180 (2016)
6. Frankle, J., Carbin, M.: The lottery ticket hypothesis: finding sparse, trainable neural networks. arXiv preprint arXiv:1803.03635 (2018)
7. Gilpin, L.H., Bau, D., Yuan, B.Z., Bajwa, A., Specter, M., Kagal, L.: Explaining explanations: an overview of interpretability of machine learning. In: 2018 IEEE 5th International Conference on Data Science and Advanced Analytics (DSAA), pp. 80–89. IEEE (2018)

8. Hancock, M., Magnan, J.: Lung nodule malignancy classification using only radiologist-quantified image features as inputs to statistical learning algorithms. J. Med. Imaging **3**(4), 044504 (2016)
9. Hussein, S., Cao, K., Song, Q., Bagci, U.: Risk stratification of lung nodules using 3D CNN-based multi-task learning. In: Niethammer, M., et al. (eds.) IPMI 2017. LNCS, vol. 10265, pp. 249–260. Springer, Cham (2017). https://doi.org/10.1007/978-3-319-59050-9_20
10. Hussein, S., Gillies, R., Cao, K., Song, Q., Bagci, U.: Tumornet: lung nodule characterization using multi-view convolutional neural network with gaussian process. In: 14th International Symposium on Biomedical Imaging (ISBI), pp. 1007–1010. IEEE (2017)
11. Iesmantas, T., Alzbutas, R.: Convolutional capsule network for classification of breast cancer histology images. In: Campilho, A., Karray, F., ter Haar Romeny, B. (eds.) ICIAR 2018. LNCS, vol. 10882, pp. 853–860. Springer, Cham (2018). https://doi.org/10.1007/978-3-319-93000-8_97
12. Jiménez-Sánchez, A., Albarqouni, S., Mateus, D.: Capsule networks against medical imaging data challenges. In: Stoyanov, D., et al. (eds.) LABELS/CVII/STENT -2018. LNCS, vol. 11043, pp. 150–160. Springer, Cham (2018). https://doi.org/10.1007/978-3-030-01364-6_17
13. Kandel, P., LaLonde, R., Ciofoaia, V., Wallace, M.B., Bagci, U.: Su1741 colorectal polyp diagnosis with contemporary artificial intelligence. Gastrointest. Endosc. **89**(6), AB403 (2019)
14. Kuhn, M., Johnson, K.: Applied Predictive Modeling, vol. 26. Springer, New York (2013). https://doi.org/10.1007/978-1-4614-6849-3
15. LaLonde, R., Bagci, U.: Capsules for object segmentation. arXiv preprint arXiv:1804.04241 (2018)
16. LaLonde, R., Kandel, P., Spampinato, C., Wallace, M.B., Bagci, U.: Diagnosing colorectal polyps in the wild with capsule networks. In: 17th International Symposium on Biomedical Imaging (ISBI). IEEE (2020)
17. LaLonde, R., Xu, Z., Jain, S., Bagci, U.: Capsules for biomedical image segmentation. arXiv preprint arXiv:2004.04736 (2020)
18. Lehnis, M.: Can We Trust AI If We Don't Know How It Works? (15062018), BBC News. http://www.bbc.com/news/business-44466213
19. Mobiny, A., Lu, H., Nguyen, H.V., Roysam, B., Varadarajan, N.: Automated classification of apoptosis in phase contrast microscopy using capsule network. IEEE Trans. Med. Imaging **39**(1), 1–10 (2019)
20. Mobiny, A., Van Nguyen, H.: Fast CapsNet for lung cancer screening. In: Frangi, A.F., Schnabel, J.A., Davatzikos, C., Alberola-López, C., Fichtinger, G. (eds.) MICCAI 2018. LNCS, vol. 11071, pp. 741–749. Springer, Cham (2018). https://doi.org/10.1007/978-3-030-00934-2_82
21. Nibali, A., He, Z., Wollersheim, D.: Pulmonary nodule classification with deep residual networks. Int. J. Comput. Assist. Radiol. Surgery 1–10 (2017). https://doi.org/10.1007/s11548-017-1605-6
22. Pal, A., Chaturvedi, A., Garain, U., Chandra, A., Chatterjee, R., Senapati, S.: CapsDeMM: capsule network for detection of munro's microabscess in skin biopsy images. In: Frangi, A.F., Schnabel, J.A., Davatzikos, C., Alberola-López, C., Fichtinger, G. (eds.) MICCAI 2018. LNCS, vol. 11071, pp. 389–397. Springer, Cham (2018). https://doi.org/10.1007/978-3-030-00934-2_44
23. Polonski, V.: People Don't Trust AI-Here's How We Can Change That, (10012018), Scientific American. http://www.scientificamerican.com/article/people-dont-trust-ai-heres-how-we-can-change-that/

24. Rudin, C.: Stop explaining black box machine learning models for high stakes decisions and use interpretable models instead. Nat. Mach. Intell. **1**(5), 206–215 (2019)
25. Sabour, S., Frosst, N., Hinton, G.E.: Dynamic routing between capsules. In: Advances in Neural Information Processing Systems, pp. 3856–3866 (2017)
26. Selvaraju, R.R., Cogswell, M., Das, A., Vedantam, R., Parikh, D., Batra, D.: Grad-cam: visual explanations from deep networks via gradient-based localization. In: Proceedings of the IEEE International Conference on Computer Vision, pp. 618–626 (2017)
27. Shen, S., Han, S.X., Aberle, D.R., Bui, A.A., Hsu, W.: An interpretable deep hierarchical semantic convolutional neural network for lung nodule malignancy classification. Expert Syst. Appl. **128**, 84–95 (2019)
28. Shen, W., Zhou, M., Yang, F., Yang, C., Tian, J.: Multi-scale convolutional neural networks for lung nodule classification. In: Ourselin, S., Alexander, D.C., Westin, C.-F., Cardoso, M.J. (eds.) IPMI 2015. LNCS, vol. 9123, pp. 588–599. Springer, Cham (2015). https://doi.org/10.1007/978-3-319-19992-4_46
29. Shen, W., et al.: Multi-crop convolutional neural networks for lung nodule malignancy suspiciousness classification. Pattern Recogn. **61**, 663–673 (2017)
30. Shen, Y., Gao, M.: Dynamic routing on deep neural network for thoracic disease classification and sensitive area localization. In: Shi, Y., Suk, H.-I., Liu, M. (eds.) MLMI 2018. LNCS, vol. 11046, pp. 389–397. Springer, Cham (2018). https://doi.org/10.1007/978-3-030-00919-9_45
31. Zeiler, M.D., Fergus, R.: Visualizing and understanding convolutional networks. In: Fleet, D., Pajdla, T., Schiele, B., Tuytelaars, T. (eds.) ECCV 2014. LNCS, vol. 8689, pp. 818–833. Springer, Cham (2014). https://doi.org/10.1007/978-3-319-10590-1_53
32. Zhu, W., Liu, C., Fan, W., Xie, X.: Deeplung: deep 3D dual path nets for automated pulmonary nodule detection and classification. In: 2018 IEEE Winter Conference on Applications of Computer Vision (WACV), pp. 673–681. IEEE (2018)

Interpretability-Guided Content-Based Medical Image Retrieval

Wilson Silva[1,2(✉)], Alexander Poellinger[3], Jaime S. Cardoso[1,2], and Mauricio Reyes[4,5]

[1] INESC TEC, Porto, Portugal
wilson.j.silva@inesctec.pt
[2] Faculty of Engineering, University of Porto, Porto, Portugal
[3] Department of Diagnostic, Interventional and Pediatric Radiology, Inselspital, Bern University Hospital, Bern, Switzerland
[4] Insel Data Science Center, Inselspital, Bern University Hospital, Bern, Switzerland
[5] ARTORG Center for Biomedical Research, University of Bern, Bern, Switzerland

Abstract. When encountering a dubious diagnostic case, radiologists typically search in public or internal databases for similar cases that would help them in their decision-making process. This search represents a massive burden to their workflow, as it considerably reduces their time to diagnose new cases. It is, therefore, of utter importance to replace this manual intensive search with an automatic content-based image retrieval system. However, general content-based image retrieval systems are often not helpful in the context of medical imaging since they do not consider the fact that relevant information in medical images is typically spatially constricted. In this work, we explore the use of interpretability methods to localize relevant regions of images, leading to more focused feature representations, and, therefore, to improved medical image retrieval. As a proof-of-concept, experiments were conducted using a publicly available Chest X-ray dataset, with results showing that the proposed interpretability-guided image retrieval translates better the similarity measure of an experienced radiologist than state-of-the-art image retrieval methods. Furthermore, it also improves the class-consistency of top retrieved results, and enhances the interpretability of the whole system, by accompanying the retrieval with visual explanations.

Keywords: Medical image retrieval · Interpretability · Chest X-ray

1 Introduction

Accessibility to medical imaging technologies has considerably increased over the last decade, leading to an increase in the number of images that need to

Electronic supplementary material The online version of this chapter (https://doi.org/10.1007/978-3-030-59710-8_30) contains supplementary material, which is available to authorized users.

A. L. Martel et al. (Eds.): MICCAI 2020, LNCS 12261, pp. 305–314, 2020.
https://doi.org/10.1007/978-3-030-59710-8_30

be analyzed by radiologists in their daily workflow [11]. As the ratio of diagnostic demand to the number of radiologists is increasing, the effective available time per diagnostic has been decreasing and became a critical issue when diagnostic needs to be supported by confirmatory evidence of a potential suspected diagnosis. Currently, when in doubt for a suspected condition, radiologists turn to public or internal image databases where similar disease-matching images can be searched and compared against. Such a task is time-consuming and ineffective since it requires several iterations to find the right matching image supporting a final diagnosis. Therefore, it is of great value to develop disease-targeted content-based image retrieval (CBIR) systems that automatically present disease-matching similar images to the one being analyzed. CBIR systems mainly consist of two tasks: feature representation, and feature indexing and search [10]. For feature representation, one seeks to find a low-dimensional description of the image that is suitable for characterizing it well enough. In contrast, in feature indexing and search, the objective is more related to the efficiency of the retrieval process.

The focus of this work is on the feature representation task. To date, feature representation is mainly performed in one of three different ways: based on statistical measures, hand-crafted features, or through learned features. As pointed out by Li *et al.* [10], one successful approach to do feature representation is the use of a pre-trained Convolutional Neural Network (CNN), with a following fine-tuning phase using the medical dataset related to the task, as done in several state-of-the-art works [7,14,18]. Li *et al.* also mentioned that there are other possibilities, such as training from scratch in the medical dataset, or combining extracted deep features with hand-crafted features. Furthermore, in the absence of a sizeable labelled dataset, unsupervised approaches have also been proposed to perform feature extraction [3]. In terms of computing similarity among feature representations, it was referred by Ghorbani *et al.* [6] and demonstrated by Zhang *et al.* [20] that the Euclidean distance (L2 distance) measured in the activation space of final layers is an effective perceptual similarity metric.

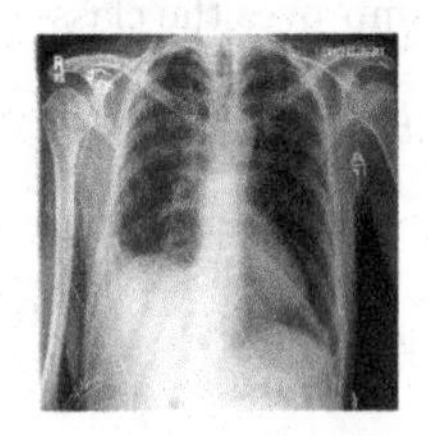

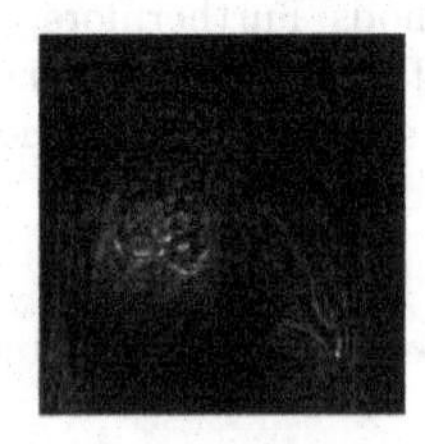

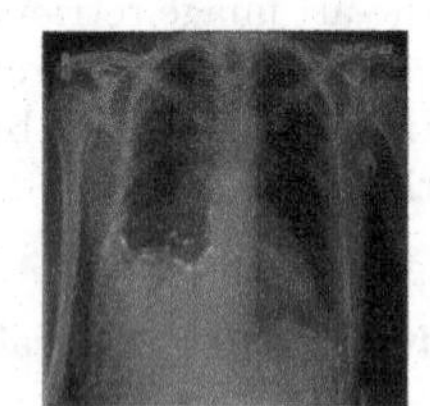

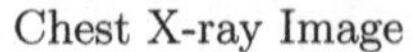

Chest X-ray Image   Saliency Map   Image with Saliency

Fig. 1. Chest X-ray image and corresponding disease-related saliency map. In the saliency map, brighter colors mean higher relevance. (Color figure online)

The state-of-the-art approaches fail to give particular attention to the regions that are determinant for the medical condition, performing an overall image to image comparison, equally weighting anatomical and pathophysiological related image information. It is then of great importance to find a way to direct the focus of the image retrieval methods to the clinically relevant regions of a medical image, ideally, in an unsupervised manner. In that sense, interpretability methods [15], namely those which produce visual explanations in the form of saliency maps [2,12,13,16] appear as a suitable solution to find these relevant regions without supervision. In Fig. 1, we illustrate the motivation behind our central idea, by presenting an example of a Chest X-ray image and its corresponding saliency map, which points out to the disease-related image regions.

In this work, we explore the use of interpretability saliency maps as an attention mechanism to focus the feature representations to image regions that characterize the class to which they belong. As a proof-of-concept, experiments were conducted for the pleural effusion condition in Chest X-Ray images. The evaluation of the retrieval quality of the proposed method is based on its ranking capabilities and class-consistency.

2 Materials and Methods

2.1 Data

For the experiments, we used the publicly available CheXpert dataset [9], which consists of 224,316 chest radiographs from 65,240 patients collected from the Stanford Hospital. Each case was labelled for the presence of 14 different observations, with training set labels being automatically generated from the associated radiology reports, while both validation (200 chest radiographs) and test (500 chest radiographs) sets were labelled by board-certified radiologists. Currently, the test set is not publicly available since a competition is running[1]. Thus, the validation set was used for the evaluation of the proposed approach. From the validation set, we create two different sets: a test set with the cases to be analyzed, and a catalogue, with well-curated cases to be retrieved. For the sake of this work, we focused on the Pleural Effusion condition. Multiple reasons justify this decision, namely, most of the images of the validation set having been acquired in the anteroposterior position, and the data being highly imbalanced for certain medical conditions. Our work was also supported by an experienced board-certified radiologist, who provided us a ranking ground-truth for the catalogue cases, and the localization of the condition.

2.2 Method

State-of-the-art image retrieval methods analyze the image as a whole, producing a feature representation that characterizes the image in its entirety. Our proposed method aims to refine this feature computation process, enforcing the focus to relevant regions, and consequently improving medical image retrieval.

[1] https://stanfordmlgroup.github.io/competitions/chexpert/.

As the focus mechanism used here is based on interpretability saliency maps, we named our approach as Interpretability-guided CBIR (IG-CBIR). The method[2] is presented in Fig. 2 and described in the following paragraphs.

Training: The training process can be divided into two different steps. In **Step 1**, we train a CNN model to classify images into Pleural Effusion or Non Pleural Effusion. The CNN model used was the well-known DenseNet-121 [8], which was initialized with the pre-trained weights from ImageNet [4], and afterwards was fine-tuned (all weights) with the CheXpert training set. To accommodate for the use of the pre-trained network, grayscale images were replicated (by concatenation), so that they became three-channel RGB-like images. Furthermore, image resolution (224×224) and pre-processing were the same as for the ImageNet pre-training process. The model was trained for 10 epochs, using the Adadelta optimizer [19], a batch size of 32, and the binary cross-entropy loss. The number of epochs was optimized by splitting training set into train and validation. In **Step 2**, the goal is to enforce the focus of the network in clinically relevant regions. To do so, we generate saliency maps, using one of the standard interpretability methods provided by the iNNvestigate toolbox [1] (Deep Taylor [12] was the one used in this work). Afterwards, these training saliency maps are used to fine-tune the previously trained CNN. This fine-tuning stage follows the same procedures as before, with the only difference being that the inputs now are the saliency maps, instead of the original images. In short, this training phase results

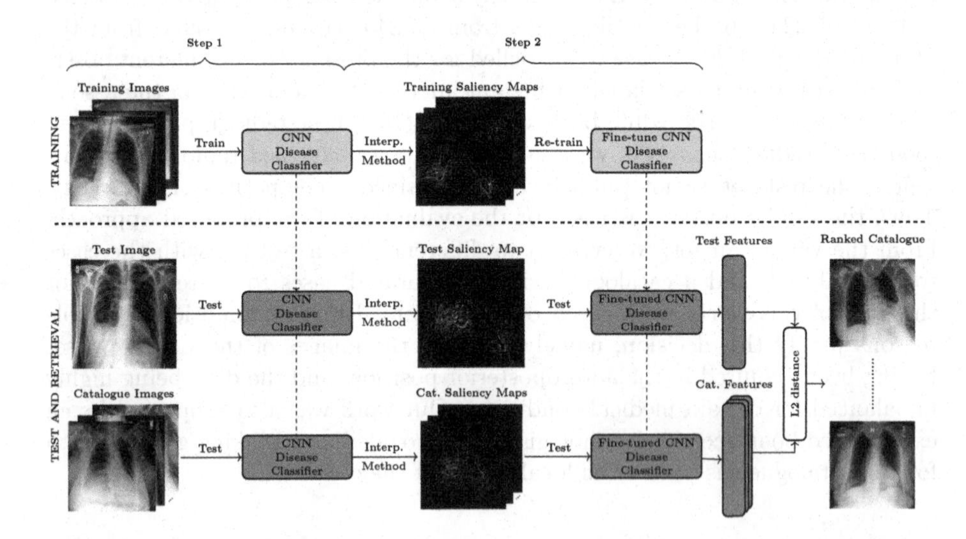

Fig. 2. Overview of the proposed approach. Blocks in light gray (■) mean CNNs are being trained (i.e., weights are being updated), whereas blocks in dark gray (■) represent trained CNNs (i.e., weights are fixed). In the saliency maps, brighter colors mean higher relevance. Blue circles indicate ranking positions. (Color figure online)

[2] Code available at https://github.com/wjsilva19/ig_cbir.

in two trained models: a first model to make predictions and generate saliency maps; and, a fine-tuned version of it to compute deep feature representations.

Test and Retrieval: The next step consists of using test and catalogue images as inputs to the first trained model (Fig. 2, lower part). With this, label predictions and saliency maps (for test and catalogue images) become available. Afterwards, these saliency maps are the input to the second trained model. During this stage, the features computed in the previous to last layer of the model are saved. The final step consists of calculating the Euclidean distance (L2) between the feature representations obtained for test and catalogue images, and rank the catalogue in terms of similarity to the test image (from most to least similar).

2.3 Evaluation

Baselines: We considered two types of baselines: a statistically-based, and a CNN-based. For the statistically-based baseline, we considered a well-known statistical measure of similarity, the structural similarity index (SSIM) [17]. The SSIM was computed directly between test and catalogue images, using its default values. Since high values of SSIM mean high similarity, the top retrieved images with this method are the ones with the highest similarity index. The second type of baseline is, like the proposed approach, based on deep learned features. As detailed in Li *et al.* [10], a current successful technique to learn feature representations of medical images is to use a CNN, pre-trained with natural images (e.g., ImageNet [4]), and fine-tune it in the application dataset (e.g., CheXpert). Afterwards, one can use the features computed in the last layers of the network to measure similarity. In practical terms, this CNN-based baseline consists of using the CNN disease classifier (from step 1 of Fig. 2) and saving the feature representations computed from the previous to last layer, which means that this baseline also works as ablation to assess the value of Step 2. Finally, the ranking of the catalogue images is performed in the same way as for the proposed approach, by computing and sorting the Euclidean distances between the input image and each catalogue image.

Assessing the Quality of the Retrieval: We considered two types of metrics, one to measure the quality of the ranking, and a second one to evaluate the class-consistency of the top retrieved images.

To measure the quality of the ranking, we used a standard metric in learning to rank tasks [5], the normalized Discounted Cumulative Gain (nDCG) - Eq. (1). The nDCG is the normalized version of the Discounted Cumulative Gain (DCG) - Eq. (2), being it normalized by the ideal/maximum possible value of DCG (IDCG). In both Eq. (1) and Eq. (2) the subscript p represents the number of retrieved images considered. Relevance values (rel_i) were assigned from 1 to 5.5, being 1 the least similar image according to the radiologist, and 5.5 the most similar one (i.e., the relevance of two contiguous positions differs by 0.5). This

$$\text{nDCG}_p = \frac{DCG_p}{IDCG_p} \quad (1) \qquad\qquad \text{DCG}_p = \sum_{i=1}^{p} \frac{2^{rel_i} - 1}{\log_2(i+1)} \quad (2)$$

was done with the goal of giving more importance to the first positions of the catalogue.

As images with high similarity ideally belong to the same class, we also measure the class-consistency of each method. For that, we considered a traditional retrieval evaluation measure, namely, precision - Eq. (3). In this context, relevant images are the ones that belong to the class of the test image.

$$\text{precision} = \frac{|\{\text{relevant images}\} \cap \{\text{retrieved images}\}|}{|\{\text{retrieved images}\}|} \quad (3)$$

3 Results

We performed five initial experiments, corresponding to five different sets of images, that resulted from the use of different seeds when doing the split of the validation data into test and catalogue (keeping the proportion of the classes). Due to time limitations of the radiologist, we considered catalogues of 10 images of size. In Fig. 3, we present the results in terms of nDCG for the top-4, top-7, and top-10 retrieved images[3].

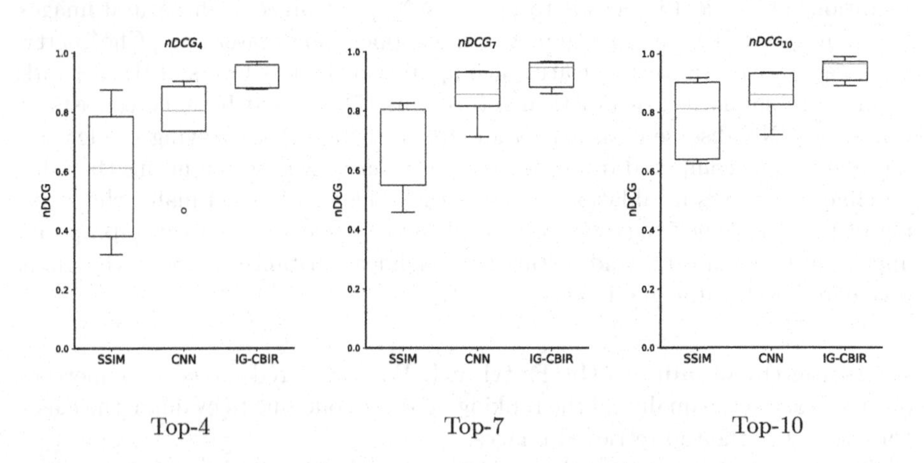

Fig. 3. Box-and-whisker plots regarding the nDCG results for Top-4, Top-7, and Top-10 retrieved images, respectively. SSIM is the statistically-based baseline, CNN is the CNN-based baseline, and IG-CBIR is the proposed interpretability-based approach.

To evaluate the class-consistency, we only needed dataset images and labels, and no expert-based ranking, hence we considered larger catalogues, computing

[3] Detailed results are provided in Table 1 of the Supplementary Material.

precision results for three different settings: top-4 images retrieved when catalogue has 10 images; top-7 images retrieved when catalogue has 20 images; and, top-11 images retrieved when catalogue has 30 images. Class-consistency results

Table 1. Precision results. Top-X means X retrieved images (X: 4, 7, 11). Cat-Y means catalogue of size Y (Y: 10, 20, 30). X is also the number of relevant images in catalogue Y. Results are presented as average [min, max].

Method	Top-4 (Cat-10)	Top-7 (Cat-20)	Top-11 (Cat-30)
SSIM-based	0.55 [0.25, 0.75]	0.40 [0.29, 0.57]	0.42 [0.27, 0.55]
CNN-based	0.85 [0.50, 1.00]	0.60 [0.29, 0.86]	0.69 [0.55, 0.82]
IG-CBIR (Proposed)	**0.95** [0.75, 1.00]	**0.77** [0.71, 0.86]	**0.80** [0.73, 0.82]

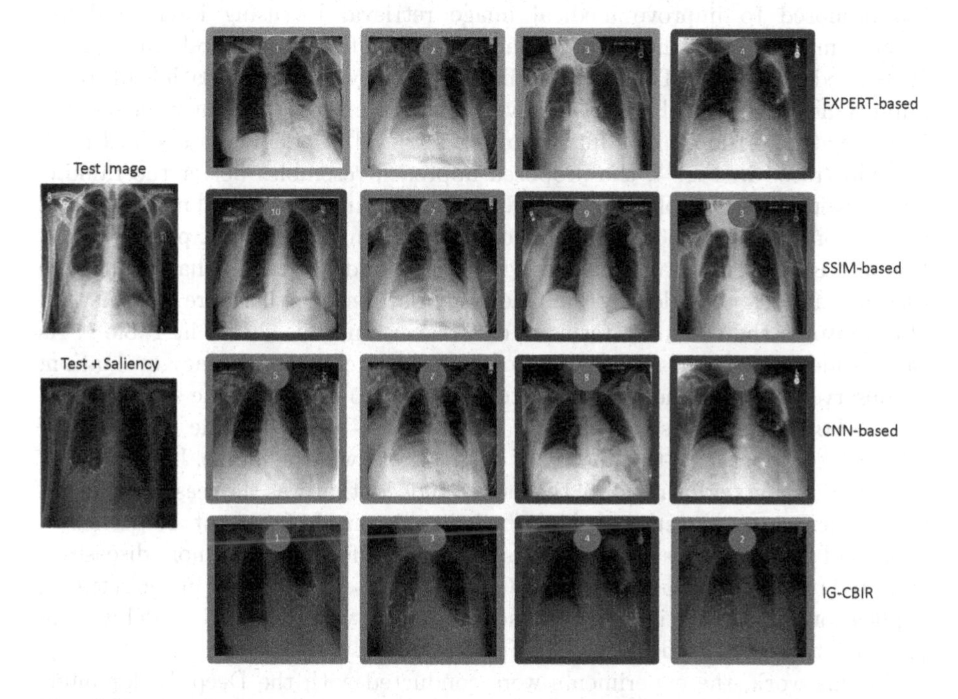

Fig. 4. Retrieved catalogue images for one example test image. From left to right: test image (and test image with saliency map superimposed) and most similar images sorted according to each method. From top to bottom: ground-truth defined by the radiologist, SSIM baseline results, CNN baseline results, and IG-CBIR results. IG-CBIR results are presented with image and saliency superimposed. Green boxes mean agreement with test image label whereas red boxes mean disagreement. Numbers on top of the images represent ranking position in the ground-truth based on expert rating. (Color figure online)

are presented in Table 1[4] in terms of average, minimum, and maximum values obtained by each method.

In Fig. 4, we present an example of a test image and the top-4 retrieved images given by each of the methods for that same test image. We show the test image and the test image with the corresponding saliency map superimposed (Fig. 4, left). It is important to point out that this saliency map is in agreement with the report of the radiologist: "Bilateral pleural effusion, stronger on the right side but also present, to a lesser extent, in the left side". On the right side, one can see the top-4 of the most similar catalogue images to the test one according to the expert and to each of the considered methods.

4 Discussion and Conclusion

We proposed to improve medical image retrieval by using interpretability saliency maps to focus the image retrieval system in the clinically relevant regions of the medical images. The proposed interpretability-based approach leads to an improvement in medical image retrieval, with the most significant improvement being related to the ranking quality of the retrieval. As demonstrated in Fig. 3, and illustrated in Fig. 4, the proposed approach resembles better the ranking order given by the radiologist than state-of-the-art image retrieval methods. It is also important to mention that the method's training process is expert-agnostic. We only use label information during training, and the labels that we use are the ones already provided by the CheXpert dataset. Furthermore, the method also improves the results in terms of class-consistency, as shown in Table 1. As we considered different sizes of catalogues for the class-consistency evaluation, we observed that the method seems to be robust to the catalogue size.

For both our proposed approach and the CNN-based baseline (or even any CNN-based approach), the quality of the retrieval will, of course, be limited by the classification performance of the model. Indeed, that was the reason for us not to consider conditions such as Atelectasis in this proof-of-concept. Nonetheless, as sizes of the databases grow, the classification performance for more diseases is expected to be in a suitable range for CNNs to be used in medical image retrieval applications. Classification performances obtained with our CNN model were in line with those reported by the CheXpert team [9].

In this work, the experiments were conducted with the Deep Taylor interpretability method. It can be noted that results may change according to the interpretability method to be used, since they produce very different saliency maps. Nonetheless, we are confident that the use of other interpretability saliency maps will also help in the refinement of the retrieval. As future work, we intend to explore different interpretability methods to generate the saliency maps, and also different ways of combining them to perform the fine-tuning stage, with the goal of improving the robustness of the method.

In conclusion, we have introduced a novel approach based on interpretability saliency maps to refine the quality of medical image retrieval achieved by deep

[4] Detailed results are provided in Table 2 of the Supplementary Material.

CNNs. As shown, this approach leads to a better similarity measure between medical images of the same condition, and, therefore, to a better image retrieval than that obtained using state-of-the-art approaches. Moreover, it also enhances the interpretability of the computer aided-diagnosis system, as it accompanies the retrieval with visual explanations.

Acknowledgements. This work was partially supported by the ERDF - European Regional Development Fund through the Operational Programme for Competitiveness and Internationalisation - COMPETE 2020 Programme and by National Funds through the Portuguese funding agency, FCT - Fundação para a Ciência e Tecnologia within project POCI-01-0145-FEDER-028857, and also by Fundação para a Ciência e Tecnologia within PhD grant number SFRH/BD/139468/2018.

References

1. Alber, M., et al.: Innvestigate neural networks. J. Mach. Learn. Res. **20**(93), 1–8 (2019)
2. Bach, S., Binder, A., Montavon, G., Klauschen, F., Müller, K.R., Samek, W.: On pixel-wise explanations for non-linear classifier decisions by layer-wise relevance propagation. PLoS ONE **10**(7), e0130140 (2015)
3. Bengio, Y., Courville, A.C., Vincent, P.: Unsupervised feature learning and deep learning: a review and new perspectives. CoRR, abs/1206.5538, vol. 1, 2012 (2012)
4. Deng, J., Dong, W., Socher, R., Li, L.J., Li, K., Fei-Fei, L.: ImageNet: a large-scale hierarchical image database. In: CVPR 2009 (2009)
5. Fernandes, K., Cardoso, J.S.: Hypothesis transfer learning based on structural model similarity. Neural Comput. Appl. **31**(8), 3417–3430 (2017). https://doi.org/10.1007/s00521-017-3281-4
6. Ghorbani, A., Wexler, J., Zou, J.Y., Kim, B.: Towards automatic concept-based explanations. In: Advances in Neural Information Processing Systems, pp. 9273–9282 (2019)
7. Hofmanninger, J., Langs, G.: Mapping visual features to semantic profiles for retrieval in medical imaging. In: Proceedings of the IEEE Conference on Computer Vision and Pattern Recognition, pp. 457–465 (2015)
8. Huang, G., Liu, Z., Van Der Maaten, L., Weinberger, K.Q.: Densely connected convolutional networks. In: Proceedings of the IEEE Conference on Computer Vision and Pattern Recognition, pp. 4700–4708 (2017)
9. Irvin, J., et al.: Chexpert: a large chest radiograph dataset with uncertainty labels and expert comparison. arXiv preprint arXiv:1901.07031 (2019)
10. Li, Z., Zhang, X., Müller, H., Zhang, S.: Large-scale retrieval for medical image analytics: a comprehensive review. Med. Image Anal. **43**, 66–84 (2018)
11. McDonald, R.J., et al.: The effects of changes in utilization and technological advancements of cross-sectional imaging on radiologist workload. Acad. Radiol. **22**(9), 1191–1198 (2015)
12. Montavon, G., Lapuschkin, S., Binder, A., Samek, W., Müller, K.R.: Explaining nonlinear classification decisions with deep Taylor decomposition. Pattern Recogn. **65**, 211–222 (2017)
13. Selvaraju, R.R., Cogswell, M., Das, A., Vedantam, R., Parikh, D., Batra, D.: Grad-cam: visual explanations from deep networks via gradient-based localization. In: Proceedings of the IEEE International Conference on Computer Vision, pp. 618–626 (2017)

14. Shin, H.C., et al.: Deep convolutional neural networks for computer-aided detection: CNN architectures, dataset characteristics and transfer learning. IEEE Trans. Med. Imaging **35**(5), 1285–1298 (2016)
15. Silva, W., Fernandes, K., Cardoso, J.S.: How to produce complementary explanations using an ensemble model. In: 2019 International Joint Conference on Neural Networks (IJCNN), pp. 1–8. IEEE (2019)
16. Springenberg, J.T., Dosovitskiy, A., Brox, T., Riedmiller, M.: Striving for simplicity: the all convolutional net. arXiv preprint arXiv:1412.6806 (2014)
17. Wang, Z., Bovik, A.C., Sheikh, H.R., Simoncelli, E.P.: Image quality assessment: from error visibility to structural similarity. IEEE Trans. Image Process. **13**(4), 600–612 (2004)
18. Wolterink, J.M., Leiner, T., Viergever, M.A., Išgum, I.: Automatic coronary calcium scoring in cardiac CT angiography using convolutional neural networks. In: Navab, N., Hornegger, J., Wells, W.M., Frangi, A.F. (eds.) MICCAI 2015. LNCS, vol. 9349, pp. 589–596. Springer, Cham (2015). https://doi.org/10.1007/978-3-319-24553-9_72
19. Zeiler, M.D.: Adadelta: an adaptive learning rate method. arXiv preprint arXiv:1212.5701 (2012)
20. Zhang, R., Isola, P., Efros, A.A., Shechtman, E., Wang, O.: The unreasonable effectiveness of deep features as a perceptual metric. In: Proceedings of the IEEE Conference on Computer Vision and Pattern Recognition, pp. 586–595 (2018)

Domain Aware Medical Image Classifier Interpretation by Counterfactual Impact Analysis

Dimitrios Lenis(✉), David Major, Maria Wimmer, Astrid Berg, Gert Sluiter, and Katja Bühler

VRVis Zentrum für Virtual Reality und Visualisierung Forschungs-GmbH, Vienna, Austria
lenis@vrvis.at

Abstract. The success of machine learning methods for computer vision tasks has driven a surge in computer assisted prediction for medicine and biology. Based on a data-driven relationship between input image and pathological classification, these predictors deliver unprecedented accuracy. Yet, the numerous approaches trying to explain the causality of this learned relationship have fallen short: time constraints, coarse, diffuse and at times misleading results, caused by the employment of heuristic techniques like Gaussian noise and blurring, have hindered their clinical adoption.

In this work, we discuss and overcome these obstacles by introducing a neural-network based attribution method, applicable to any trained predictor. Our solution identifies salient regions of an input image in a single forward-pass by measuring the effect of local image-perturbations on a predictor's score. We replace heuristic techniques with a strong neighborhood conditioned inpainting approach, avoiding anatomically implausible, hence adversarial artifacts. We evaluate on public mammography data and compare against existing state-of-the-art methods. Furthermore, we exemplify the approach's generalizability by demonstrating results on chest X-rays. Our solution shows, both quantitatively and qualitatively, a significant reduction of localization ambiguity and clearer conveying results, without sacrificing time efficiency.

Keywords: Explainable AI · XAI · Classifier decision visualization · Image inpainting

1 Introduction

The last decade's success of machine learning methods for computer-vision tasks has driven a surge in computer assisted prediction for medicine and biology.

VRVis is funded by BMVIT, BMDW, Styria, SFG and Vienna Business Agency in the scope of COMET - Competence Centers for Excellent Technologies (854174) which is managed by FFG. Thanks go to our project partner AGFA HealthCare for providing valuable input.

A. L. Martel et al. (Eds.): MICCAI 2020, LNCS 12261, pp. 315–325, 2020.
https://doi.org/10.1007/978-3-030-59710-8_31

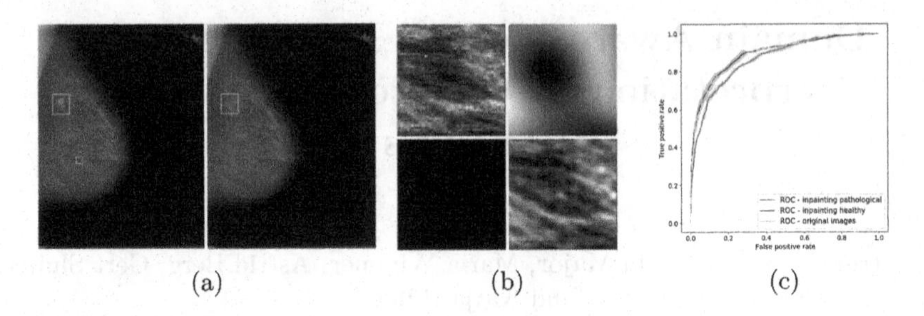

Fig. 1. Overview of marginalization: (a) original with annotated mass (red box) before and after marginalization by our method; (b) local comparisons with popular methods (clockwise): original, blurring [9], inpainting (ours), and averaging [28]; (c) ROC curves of the mammography classifier (green curve) vs. healthy pixel inpainting only in healthy/pathological (blue/red curves) structures. (Color figure online)

This has posed a conundrum. Current predictors, predominantly artificial neural networks (ANNs), learn a data-driven relationship between input image and pathological classification, whose validity, i.e. accuracy and specificity, we can quantitatively test. In contrast, this learned relationship's causality typically remains elusive [1,17,18]. A plethora of approaches have been proposed that aim to fill this gap by explaining causality through identifying and attributing salient image-regions responsible for a predictor's outcome [7–9,24,25,27].

Lacking a canonical mapping between an ANN's prediction and its domain, this form of reasoning is predominantly based on *local explanations* (LE), i.e. explicit attribution-maps characterizing image-prediction tuples [9,17]. Typically, these maps are loosely defined as regions with *maximal influence* towards the predictor, implying that any texture change within the attributed area will significantly change the prediction. Besides technical insight, these LE can provide a key benefit for clinical applications: by relating the ANN's algorithmic outcome to the user's a-priori understanding of pathology-causality, they can strengthen confidence in the predictor, thereby increasing its clinical acceptance. To achieve this goal, additional restrictions and clarifications are crucial. Qualitatively, such maps need to be *informative* for its users, i.e. narrow down regions of medical interest, hence coincide with medical knowledge and expectations [20]. Furthermore, the regions' characteristic, i.e. the meaning of *maximal influence*, must be clearly conveyed. Quantitatively, such LE need to be *faithful* to the underpinning predictor, i.e. dependent on architecture, parametrization, and preconditions [2].

The dominant class of methods follow a *direct approach*. Utilizing an ANN's assumed analytic nature and its layered architecture, they typically employ a modified backpropagation approach to backtrack the ANN's activation to the input image [25,29]. While efficiently applicable, the resulting maps lack a clear a-priori interpretation, are potentially incomplete, coarse, and may deliver misleading information [2,8,9,30]. Thereby they are potentially neither *informative* nor *faithful*, thus pose an inherent risk in medical environments.

In contrast, *reference based* LE approaches directly manipulate the input image and analyze the resulting prediction's differences [9]. They aim to assess an image-region's influence on prediction by counterfactual reasoning: how would the prediction score vary, if the region's image-information would be missing, i.e. its contribution marginalized? The prevailing heuristic approaches, e.g. Gaussian noise and blurring or replacement by a predefined colour [8,9,28], have been advanced to local neighborhood [30] and stronger conditional generative models [7,27]. Reference based LEs have the advantage of an a-priori clear and intuitively conveyable meaning of their result, hence address *informativeness* for end-users. However, their applicability for medical imaging hinges on the utilized marginalization technique, i.e. the mapping between potentially pathological tissue representations and their healthy equivalent. Resulting *prediction-neutral* regions need to depict healthy tissue per definition. Contradictory, the presented approaches introduce noise and thereby possibly pathological indications or anatomically implausible tissue (cf. Fig. 1). Hence, they violate the needed *faithfulness* [9].

While dedicated generative adversarial networks (GANs) for medical images deliver significantly improved results, applications are hindered by possible resolutions and limited control over the globally acting models [3–6]. In [21], the locally acting, but globally conditioned, per-pixel reconstruction of partial convolution inpainting (PCI) [19] is favoured over GANs, thereby enforcing anatomically sound, image specific replacements. While overcoming out-of-domain issues, this gradient descent based optimization method works iteratively, hence cannot be used in time restrictive environments.

Contribution: We introduce a *resource efficient* reference based *faithful* and *informative* attribution method for real time pathology classifier interpretation. Utilizing a specialized ANN and exploiting PCI's local per-pixel reconstruction, conditioned on a global healthy tissue representation, we are able to enforce anatomically sound, image specific marginalization, without sacrificing computational efficiency. We formulate the ANN's objective function as a quantitative prediction problem under strict area constraints, thereby clarifying the resulting attribution map's a-priori meaning. We evaluate the approach on public mammography data and compare against two existing state-of-the-art methods. Furthermore, we exemplify the method's generalizability by demonstrating results on a second unrelated task, namely chest X-ray data. Our solution shows, both quantitatively and qualitatively, a significant reduction of localization ambiguity and clearer conveying results without sacrificing time efficiency.

2 Methods

Given a pathology classifier's prediction for an input image, we want to estimate its cause by attributing the specific pixel-regions that substantially influenced the predictor's outcome. Informally, we search for the image-area that, if changed, results in a *sufficiently healthy* image able to *fool the classifier*. The resulting attribution-map needs to be *informative* for the user and *faithful* to its underpinning classifier. While we can quantitatively test for the latter, the former is an ill-posed problem. We therefore formalize as follows:

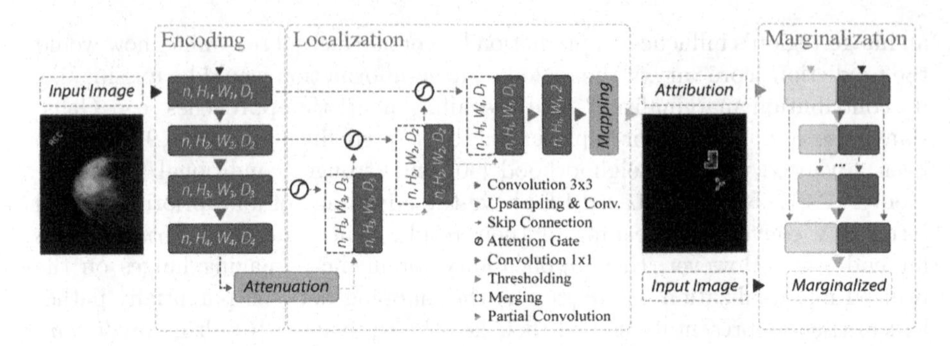

Fig. 2. Attribution framework: The input image is encoded using the classifier's features (left) and attenuated to enclose pathological regions (middle). During training, counterfactual images are produced by the marginalization-net (right), fed by thresholded attribution (pink blocks) and input image (blue blocks). (Color figure online)

Let I denote an image of a domain $\mathcal{I}$ with pixels on a discrete grid $m_1 \times m_2$, c a fixed pathology-class, and f a classifier capable of estimating $p(c|I)$, the probability of c for I. Also, let M denote the attribution-map for image I and class c, hence $M \in M^{m_1 \times m_2}(\{0,1\})$. Furthermore, assume a function $\pi(M)$ proficient in marginalizing all pixel regions attributed by M in I such that the result of the operation is still within the domain of f. Hence, $\pi(M)$ yields a new image similar to I, but where we know all regions attributed by M to be healthy per definition. Therefore, assuming I depicts a pathological case and M attributes only pathology pixel representations, $\pi(M)$ is a healthy counterfactual image to I. In any case $p(c|\pi(M))$ is well defined. Using this notation, we can formalize what an *informative* map $\hat{M}$ means, hence give it an a-priori, testable semantic meaning. We define it as

$$\hat{M} := \underset{M \in \hat{\mathcal{M}}}{\operatorname{argmin}}\, d(M) \quad \text{where} \quad \hat{\mathcal{M}} := \{p(c|\pi(M)) \leq \theta, d(M) \leq \delta, M \in \mathcal{S}\},$$

where θ is the classification-threshold, d a metric measuring the attributed area, δ a constant limiting the attributed area, and $\mathcal{S}$ the set of compact and connected masks. Any map of $M^{m_1 \times m_2}(\{0,1\})$ can be (differentiably) mapped into $\mathcal{S}$ by taking the smoothed maximum of a convolution with a Gaussian kernel [9,15]. In this form, $\hat{M}$ is clearly defined, and can be intuitively understood by end-users.

Solving for $\hat{M}$ requires choosing (i) an appropriate measure d (e.g. the map area in pixels), (ii) an appropriate size-limit δ (e.g. n times average mass-size for mammography), and (iii) a fitting marginalization technique $\pi(\cdot)$. In the following we describe how we solve for $\hat{M}$ through an ANN, and overcome the out-of-domain obstacles by partial convolution [19] for marginalization.

2.1 Architecture

Iteratively finding solutions for $\hat{M}$ is typically time-consuming [9,21]. Therefore, we develop a dedicated ANN, capable of finding the desired attribution in a single forward pass. To this end, the network learns on multiple resolutions, to combine relevant classifier-extracted features (cf. Fig. 2). Inspired by [8], we build on a U-Net architecture, where the down-sampling, encoding branch consists of the trained classifier without its classification layers. These features, $x_{i,j,l}$, are subsequentially passed through a feature-filter, performing $x_{i,j,l} \cdot \sigma((W_m \rho(W_l^\intercal x_{i,j,l} + b_l) + b_m))$ where ρ is an element-wise nonlinearity (namely a rectified linear unit), σ a normalization function (sigmoid function) and $W_.$ resp. $b_.$ linear transformation parameters. This is similar to additive attention, which, compared to multiplicative attention, has shown better performance on high dimensional input-features [23]. The upsampling branch consists of four consecutive blocks of: upsampling by a factor of two, followed by convolution and merging with attention-gate weighted features from the classifier of the corresponding resolution scale. After final upsampling back to input-resolution, we apply 1×1 conv. of depth two, resulting in two channels $c_{1,2}$. The final attribution-map $\hat{M}$ is derived through thresholding $\frac{|c_1|}{|c_1|+|c_2|}$. Intuitively, the network attenuates the classifier's final features, generating an initial localization. This coarse map is subsequently refined by additional weighting and information from higher resolution features (cf. Fig. 2). We train the network, by minimizing

$$\mathcal{L}(M) = \phi(M) + \psi(M) + \lambda \cdot \mathcal{R}(M), \text{ s.t. } d(M) \leq \delta$$

where $\phi(M) := -1 \cdot \log(p(c|\pi(M)))$, $\psi(M) := \log(\text{odds}(I)) - \log(\text{odds}(\pi(M)))$, and $\text{odds}(I) = \frac{p(c|I)}{1-p(c|I)}$, hence weigh the probability of the marginalized image, enforcing $p(c|\pi(M)) \leq \theta$. We introduced an additional regularization-term: a weighted version of total variation [22], which experimentally greatly improved convergence. All terms where normalized through a generalized logistic function. The inequality constraint was enforced by the method proposed in [14]. Note that after mapping into $\mathcal{S}$, any solution to $\mathcal{L}$ will also estimate $\hat{M}$, thereby yielding our desired attribution-map. The parametrization is task/classifier-dependent and will be described in the following sections.

2.2 Marginalization

As we need to derive $p(c|\pi(M))$, our goal is to marginalize arbitrary image regions marked by our network during its training process. Therefore, we aim for an image inpainting method to replace pathological tissue by healthy appearance. The result should resemble valid global anatomical appearance with high quality local texture. To address the these criteria we apply the U-Net like architecture with partial convolution blocks of [19] which gets an image and a hole mask as input (cf. Fig. 2). Partial convolution considers only unmasked inputs in a current sliding window to compute its output. Where it succeeded, hole mask positions are eliminated. This mechanism helps conditioning on local texture. The loss function ($\mathcal{L}_{PCI}$) balances local per-pixel reconstruction quality of masked/unmasked regions ($\mathcal{L}_{hole}/\mathcal{L}_{valid}$), against globally sound anatomical appearance ($\mathcal{L}_{perc}, \mathcal{L}_{style}$). An additional total variation term ($\mathcal{L}_{tv}$) ensures a smooth transition between hole and present image regions in the final result.

This yields $\mathcal{L}_{PCI} = \mathcal{L}_{valid} + 6 \cdot \mathcal{L}_{hole} + 0.05 \cdot \mathcal{L}_{perc} + 120 \cdot \mathcal{L}_{style} + 0.1 \cdot \mathcal{L}_{tv}$ where parametrization follows [19]. The architecture's contraction path consists of 8 partial convolution blocks with a stride of 2. The kernels of depth 64, 128, 256, 512, ..., 512 have sizes 7, 5, 5, 3, ..., 3. The expansion path, a mirrored version of the contraction path, contains upsampling layers with a factor of 2, kernel size of 3 at every layer, and a final filterdepth of 3. Each block contains batch normalization (BN) and ReLU/LeakyReLU (alpha = 0.2) activations in the contraction/expansion paths which are connected by skip connections. Zero padding of the input was applied to control resolution shrinkage and keep aspect ratio.

3 Experimental Setup

Datasets: We evaluated our framework on two different datasets, on mammography scans and on chest X-ray images. For mammography, we complemented the 1565 annotated, pathological CBIS-DDSM scans containing masses [16] with 2778 healthy DDSM images [10] and downsampled them to 576 × 448 pixels. Data was split into 1231/2000 mass/healthy samples for training, and into 334/778 scans for testing. There was no patient-wise overlap between the training/test data. We demonstrate generalization on a private collection of healthy and tuberculotic (TBC) frontal chest X-ray images, at a downsampled resolution of 256 × 256. We split healthy images into sets of 1700/135 for training respectively validation set, and TBC cases into 700/70. The test set contains 52 healthy and 52 TBC samples. No pixel-wise GT information was provided for this data.

Classifiers: The backbone of our mammography attribution network is a MobileNet [11] classifier for distinguishing between healthy samples and scans with masses. The network was trained using the Adam optimizer with batchsize of 4 and learning rate of 1e-5 for 250 epochs with early stopping. The network was pretrained with 50k 224 × 224 pixel patches from the training data for the same task. The TBC attribution utilized a DenseNet-121 [12] classifier for the binary classification task of healthy or TBC cases. It was trained using the SGD momentum optimizer with a batchsize of 32 and learning rate of 1e-5 for 2000 epochs. This network was pretrained on the CheXpert dataset [13].

Marginalization: The chest X-ray images have one magnitude smaller resolution than the mammography scans, thus we removed the bottom-most blocks from the contraction and expansion paths. Both inpainter networks were trained on healthy training samples with a batch size of 1 for mammography and 5 for chest X-ray. Training was done in two phases, the first phase with BN after each partial convolution layer and the second with BN only in the expansion path. The network for the mass classification task was trained with learning rates of 1e-5/1e-6 and for the TBC classification task of 2e-4/1e-5 for the two phases. For each image irregular masks were generated which mimic possible configurations during the attribution network training [19].

Attribution: We used the last four resolution-scales of each classifier, and in all cases the features immediately after the activation function, following the convolution. The weights of the pre-trained ANNs were kept fixed during the complete process. Filterdepths of the upsampling convolution blocks correspond to the equivalent down-sampling filters, filter-size is fixed to 1×1. Upsampling itself is done via neighborhood upsampling. We used standard gradient descent, and a cyclic learning rate [26], varying between 1e-6 and 1e-4, and trained for up to 5000 epochs with early stopping. We thresholded the masks at 0.55, and used a Gaussian RBF with σ = 5e-2, and a smoothing parameter of 30. All trainable weights where random-normal initialized.

4 Results and Conclusion

Marginalization: To evaluate the inpainter network we assessed how much the classification score of an image changes, when pathological tissue is replaced.

Thus, we computed ROC curves using the classifier on all test samples (i) without any inpainting as reference, and for comparison, randomly sampled inpainting (ii) only in healthy respective (iii) pathological scans over 10 runs (Fig. 1). The clear distance between the ROC curves of the mammography image classifiers without any inpainting, yielding an AUC of 0.89, and with inpainting in pathological regions, resulting in an AUC of 0.86, shows that the classifier is sensitive to changes around pathological regions of the image. Moreover, it is visible that the ROC curves of inpainting in healthy tissues with an AUC of 0.89 follow closely the unaffected classifier's ROC curve (Fig. 1). The AUC scores for the TBC classifier without and with inpainting in healthy tissue are 0.89 and 0.88 which proves the above mentioned observations. Pathological tissue inpainting was ommitted in this case due to the lack of pixel-wise annotations.

Attribution: We compared our attribution network against the gradient explanation *saliency map* [25] (SAL), and the network/gradient-derived *GradCAM* [24] visualizations. We limited our comparisons to these direct approaches, as they are widely used within medical imaging [13], and inherently valid [2]. Popular *reference based* approaches either utilize blurring, noise or some other heuristic [8,9,30], or were not available [7], therefore could not be considered. Quantitatively, we relate (i) the result-maps $\hat{M}$ to both organ, and ground truth (GT) annotations, and (ii) to each other. Particularly for (i) we studied the Hausdorff distances H between GT and $\hat{M}$ indicating location proximity. Lower values demonstrate better localization in respect to the pathology. Further, we performed a weak localization experiment [8,9]: per image, we derived bounding boxes (BB) for each connected component of GT and $\hat{M}$ attributions. A GT BB counts as found, if any $\hat{M}$ BB has an $IOU \leq 0.125$. We chose this threshold, as a proficient classifier presumably focuses on the masses' boundaries and neighborhoods, thereby limiting possible BB-overlap. We report average localization L. For (ii) we derived the area ratio A between $\hat{M}$ and organ-mask (breast-area) or whole image (chest X-ray). Again, lower values indicate a smaller thereby clearer map. Due to missing GT we could only derive (ii) for TBC. All measurements were performed on binary masks, hence GradCAM and SAL had to be thresholded. We chose the $50, 75, 90$ percentiles, i.e. compared 50, 25, 10%

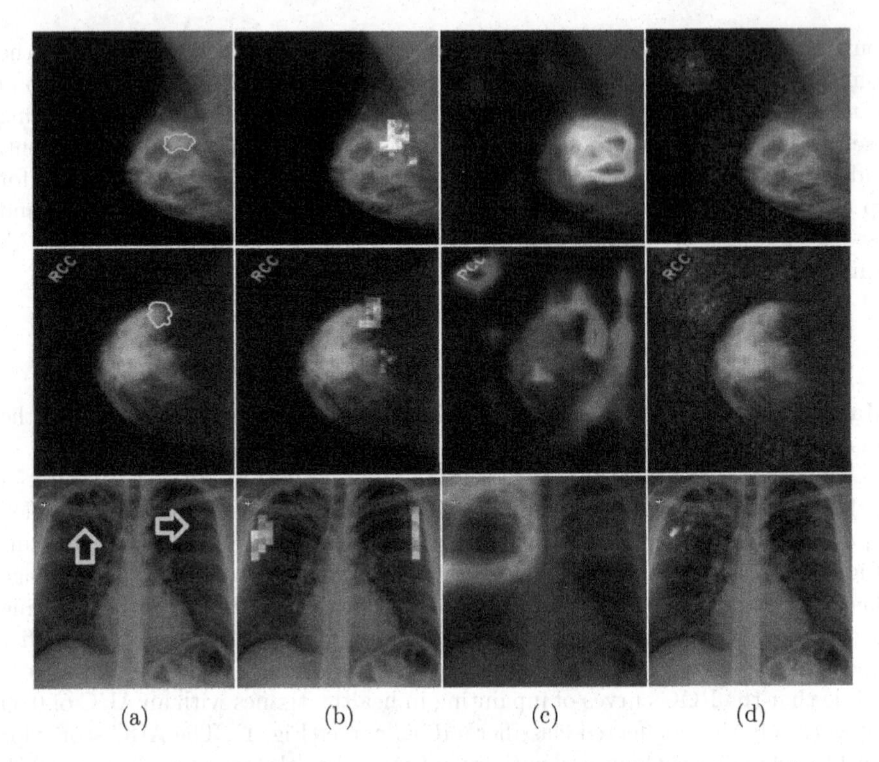

Fig. 3. Result attribution heatmaps for mammography and chest X-ray: (a) original image overlayed with annotation contours (and arrows for missing GT), (b) our attribution framework. (c) GradCAM [24] (d) Saliency [25].

of the map-points. Where multiple pathologies, or mapping results occurred we used the median for a robust estimation per image. Statistically significant difference between all resulting findings was formalized using Wilcoxon signed-rank tests, for $\alpha < 0.05$. Additionally we followed [2], and tested our network with randomised parametrization (labels have no effect in our case).

As seen in Table 1, our framework achieves significantly lower H, than either GradCAM or SAL at all threshold levels. Moreover, we report significantly better weak localization (L) which underlines the higher accuracy of our approach. Qualitatively our attribution-maps are tighter focused (c.f. Fig. 3(b)) and enclose the masses. The former is also expressed by the lower overlap values A. All p-values where significantly below 1e-2, hardening our results. Randomization of the ANN's weights yields pure noise maps, hence we pass [2]'s checks.

Timing: We estimated the time needed for a single attribution map, one forward pass, by averaging over ten times repeated map derivations for all images of the resp. test sets. These were compared with the analogous timings of GRAD and SAL. Additionally, as a reference for iterative methods, we compared with [21] that, using same marginalization technique, yields equivalent maps.

Table 1. Top: Hausdorff distances H and weak localization results L, relating maps $\hat{M}$ to GT; Bottom: relating maps $\hat{M}$ to the organ resp. image-size

P	H_{ours}	H_{grad}	H_{sal}	L_{ours}	L_{grad}	L_{sal}
50	**188.12** $\pm$ 68.3	296.29 $\pm$ 54.4	240.83 $\pm$ 36.2	**0.45**	0.06	0.27
75	**188.12** $\pm$ 68.3	274.86 $\pm$ 40.0	257.85 $\pm$ 38.6	**0.45**	0.23	0.30
90	**188.12** $\pm$ 68.3	243.80 $\pm$ 59.6	259.57 $\pm$ 43.7	**0.45**	0.28	0.25
P	A_{ours}^{mammo}	A_{grad}^{mammo}	A_{sal}^{mammo}	A_{ours}^{tbc}	A_{grad}^{tbc}	A_{sal}^{tbc}
50	**0.07** $\pm$ 0.04	1.10 $\pm$ 0.10	1.10 $\pm$ .14	**0.06** $\pm$ 0.0	0.50 $\pm$ 0.0	0.50 $\pm$ 0.0
75	**0.07** $\pm$ 0.04	0.55 $\pm$ 0.21	0.55 $\pm$ 0.2	**0.06** $\pm$ 0.0	0.25 $\pm$ 0.0	0.25 $\pm$ 0.0
90	**0.07** $\pm$ 0.04	0.22 $\pm$ 0.40	0.22 $\pm$ 0.43	**0.06** $\pm$ 0.0	0.10 $\pm$ 0.0	0.10 $\pm$ 0.0

Our model is capable of deriving 75 mammography maps per second (mps) utilizing a GPU (NVIDIA Titan RTX). This compares favourably to both GRAD and SAL, 50 resp. 31 mps, and significantly outperforms the iterative method (27 seconds per map). Considering the smaller X-ray images, these throughputs increase up to a factor of three, sufficient even for real time environments.

Conclusion: In this work, we proposed a novel neural network based attribution method for real time interpretation of pathology classifiers. Our reference based approach enforces domain aware marginalization, without sacrificing computational efficiency. Overcoming these common obstacles, our approach can provide further confidence, and thereby increase critical user acceptance. We compared our method with state-of-the-art techniques on two different tasks, and show favorable results throughout. This underlines the suitability of our approach as an interpretation tool in radiology workflows.

References

1. Adadi, A., Berrada, M.: Peeking inside the black-box: a survey on explainable artificial intelligence (XAI). IEEE Access **6**, 52138–52160 (2018)
2. Adebayo, J., Gilmer, J., Muelly, M., Goodfellow, I., Hardt, M., Kim, B.: Sanity checks for saliency maps. In: Proceedings of NIPS, pp. 9505–9515 (2018)
3. Andermatt, S., Horváth, A., Pezold, S., Cattin, P.: Pathology segmentation using distributional differences to images of healthy origin. In: Crimi, A., et al. (eds.) BrainLes 2018. LNCS, vol. 11383, pp. 228–238. Springer, Cham (2019). https://doi.org/10.1007/978-3-030-11723-8_23
4. Baumgartner, C., Koch, L., Tezcan, K., Ang, J., Konukoglu, E.: Visual feature attribution using Wasserstein GANs. In: Proceedings of CVPR, pp. 8309–8319 (2017)
5. Becker, A., et al.: Injecting and removing suspicious features in breast imaging with CycleGAN: a pilot study of automated adversarial attacks using neural networks on small images. Eur. J. Radiol. **120**, 108649 (2019)
6. Bermudez, C., Plassard, A., Davis, L., Newton, A., Resnick, S., Landman, B.: Learning implicit brain MRI manifolds with deep learning. In: SPIE Medical Imaging, vol. 10574, pp. 408–414 (2018)

7. Chang, C.H., Creager, E., Goldenberg, A., Duvenaud, D.: Explaining image classifiers by counterfactual generation. In: Proceedings of ICLR (2019)
8. Dabkowski, P., Gal, Y.: Real time image saliency for black box classifiers. In: Proceedings of NIPS, pp. 6967–6976 (2017)
9. Fong, R., Patrick, M., Vedaldi, A.: Understanding deep networks via extremal perturbations and smooth masks. In: Proceedings of ICCV (2019)
10. Heath, M., Bowyer, K., Kopans, D., Moore, R., Kegelmeyer, W.: The digital database for screening mammography. In: Proceedings of IWDM, pp. 212–218 (2000)
11. Howard, A., et al.: MobileNets: efficient convolutional neural networks for mobile vision applications. arXiv preprint arXiv:1704.04861 (2017)
12. Huang, G., Liu, Z., van der Maaten, L., Weinberger, K.Q.: Densely connected convolutional networks. In: Proceedings of CVPR (2017)
13. Irvin, J., et al.: CheXpert: a large chest radiograph dataset with uncertainty labels and expert comparison. AAAI (2019)
14. Kervadec, H., Dolz, J., Tang, M., Granger, E., Boykov, Y., Ayed, I.: Constrained-CNN losses for weakly supervised segmentation. Med. Image Anal. **54**, 88–99 (2019)
15. Lange, M., Zühlke, D., Holz, O., Villmann, T.: Applications of lp-norms and their smooth approximations for gradient based learning vector quantization. In: Proceedings of ESANN (2014)
16. Lee, R., Gimenez, F., Hoogi, A., Rubin, D.: Curated breast imaging subset of DDSM. The Cancer Imaging Archive, vol. 8 (2016)
17. Lipton, Z.: The mythos of model interpretability. ACM Queue **16**(3), 30:31–30:57 (2018)
18. Litjens, G., et al.: A survey on deep learning in medical image analysis. Med. Image Anal. **42**, 60–88 (2017)
19. Liu, G., Reda, F., Shih, K., Wang, T.C., Tao, A., Catanzaro, B.: Image inpainting for irregular holes using partial convolutions. In: Proceedings of ECCV, pp. 85–100 (2018)
20. Lombrozo, T.: The structure and function of explanations. Trends Cogn. Sci. **10**(10), 464–70 (2006)
21. Major, D., Lenis, D., Wimmer, M., Sluiter, G., Berg, A., Bühler, K.: Interpreting medical image classifiers by optimization based counterfactual impact analysis. In: Proceedings of ISBI (2020)
22. Peng, J., Kervadec, H., Dolz, J., Ayed, I.B., Pedersoli, M., Desrosiers, C.: Discretely-constrained deep network for weakly supervised segmentation (2019)
23. Schlemper, J., et al.: Attention gated networks: learning to leverage salient regions in medical images. Med. Image Anal. **53**, 197–207 (2019)
24. Selvaraju, R.R., Cogswell, M., Das, A., Vedantam, R., Parikh, D., Batra, D.: Grad-CAM: visual explanations from deep networks via gradient-based localization. In: Proceedings of ICCV (2017)
25. Simonyan, K., Vedaldi, A., Zisserman, A.: Deep inside convolutional networks: visualising image classification models and saliency maps. arXiv preprint arXiv:1312.6034 (2013)
26. Smith, L.N.: No more pesky learning rate guessing games. CoRR (2015)
27. Uzunova, H., Ehrhardt, J., Kepp, T., Handels, H.: Interpretable explanations of black box classifiers applied on medical images by meaningful perturbations using variational autoencoders. In: SPIE Medical Imaging, vol. 10949, pp. 264–271 (2019)

28. Zeiler, M.D., Fergus, R.: Visualizing and understanding convolutional networks. In: Fleet, D., Pajdla, T., Schiele, B., Tuytelaars, T. (eds.) ECCV 2014. LNCS, vol. 8689, pp. 818–833. Springer, Cham (2014). https://doi.org/10.1007/978-3-319-10590-1_53
29. Zhou, B., Khosla, A., Lapedriza, A., Oliva, A., Torralba, A.: Learning deep features for discriminative localization. In: Proceedings of CVPR, pp. 2921–2929 (2016)
30. Zintgraf, L., Cohen, T., Adel, T., Welling, M.: Visualizing deep neural network decisions: Prediction difference analysis. In: Proceedings of ICLR (2017)

Towards Emergent Language Symbolic Semantic Segmentation and Model Interpretability

Alberto Santamaria-Pang(✉), James Kubricht, Aritra Chowdhury, Chitresh Bhushan, and Peter Tu

Artificial Intelligence, GE Research, Niskayuna, NY 12309, USA
santamar@ge.com

Abstract. Recent advances in methods focused on the grounding problem have resulted in techniques that can be used to construct a symbolic language associated with a specific domain. Inspired by how humans communicate complex ideas through language, we developed a generalized Symbolic Semantic (S2) framework for interpretable segmentation. Unlike adversarial models (e.g., GANs), we explicitly model cooperation between two agents, a Sender and a Receiver, that must cooperate to achieve a common goal. The Sender receives information from a high layer of a segmentation network and generates a symbolic sentence derived from a categorical distribution. The Receiver obtains the symbolic sentences and cogenerates the segmentation mask. In order for the model to converge, the Sender and Receiver must learn to communicate using a private language. We apply our architecture to segment tumors in the TCGA dataset. A UNet-like architecture is used to generate input to the Sender network which produces a symbolic sentence, and a Receiver network cogenerates the segmentation mask based on the sentence. Our Sementation framework achieved similar or better performance compared with state-of-the-art segmentation methods. In addition, our results suggest direct interpretation of the symbolic sentences to discriminate between normal and tumor tissue, tumor morphology, and other image characteristics.

Keywords: Emergent language · Symbolic semantic segmentation · Interpretability · Explainability

1 Introduction

Current limitations in state-of-the-art Machine Learning (ML) and Artificial Intelligence (AI) include lack of interpretability and explainability; i.e., classical black-box approaches utilizing deep neural networks cannot provide evidence on how models behave. In medical applications, interpretability and explainability is a paramount requirement if we intend to rely on clinical diagnoses derived from automated systems.

Inspired by the symbol grounding problem [1], the current work investigates synergies between deep learning Semantic Segmentation and Emergent Language (EL) models.

A. L. Martel et al. (Eds.): MICCAI 2020, LNCS 12261, pp. 326–334, 2020.
https://doi.org/10.1007/978-3-030-59710-8_32

We further utilize general properties of EL architectures to facilitate model interpretability by demonstrating how black-box semantic segmentation can be extended to provide Symbolic Semantic (S^2) outputs. Corresponding sentences – drawn from a categorical distribution – are formed by integrating symbolic components into a conventional UNet-like architecture. We term this the Symbolic UNet (SUNet) framework.

Following description and analysis of the proposed framework, we explore the utility of symbolic segmentation masks towards direct data interpretability in clinical applications. Specifically, we utilize The Cancer Imaging Archive (TCGA) dataset to determine whether SUNet sentences correspond with meaningful semantics in neural imagery.

2 Literature Review

Interpretability and explainability of artificial intelligent (AI) systems is an important criterion for faster and wider adoption, especially in clinical applications [2]. Recently, several approaches for interpretable machine learning have been developed, with heavy emphasis on classification problems [3, 4]. The majority of the existing approaches focus on explanation of the representation learned by the deep learning system through saliency maps, class activation maps, occlusion study, etc. [5–9]. Recent work from Natekar *et al.* [6] implemented a network dissection approach on a segmentation network to locate internal functional regions that identify human-understandable concepts like core and enhancing tumor regions. In a similar vein, Couteaux *et al.* [7] used activation maximization through gradient ascent, similar to DeepDream, to identify features in input images that the network is most sensitive to for segmentation of liver CT images. While these approaches do provide some interpretability of how networks represent data, the identified features or internal network components (layers, individual units, etc.) are not particularly interpretable by human experts [10, 11].

In this work we present an approach, inspired by the symbol grounding problem [1], that explicitly generates Symbolic Semantic (S2) and segmentation outputs. Our framework is inspired by Havrylov *et al.* [12], where multi-agent cooperation showed emergence of artificial language in natural images. In this approach, the sequence of symbols is modelled using paired Long Short-Term Memory (LSTM) networks. Cogswell *et al.* [13] also introduced compositional generality in emergent languages among multiple agents, similar to Larazidou *et al.* [14] who presented a series of studies investigating properties of language protocols generated by the agents exposed to symbolic and image data. In this work, we extend emergent language models to provide fully interpretable segmentation. To the best of our knowledge, no prior work has attempted to automatically express segmentation in contextually meaningful (symbolic) sentences. The two main innovations of this works are: i) emergent language extension to any segmentation architecture and ii) interpretation of symbolic expressions derived from segmentation tasks.

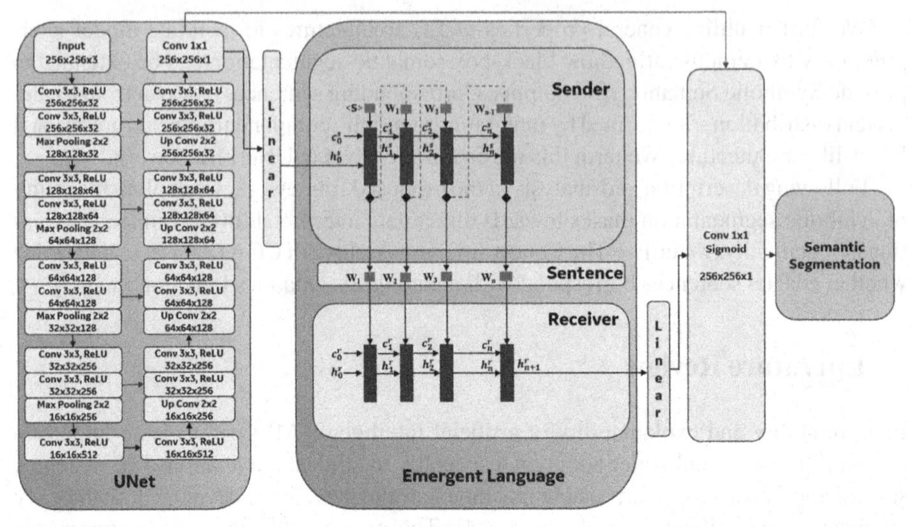

Fig. 1. Schematic of the proposed SUNet architecture composed of a UNet and Emergent Language (EL) network. Both networks are encouraged to cooperate in order to generate an appropriate segmentation mask. Once cooperation is achieved, semantic meaning of EL sentences are generated. These sentences are notably interpretable in the context of semantic segmentation output.

3 Methods

We present our S^2-Segmentation framework for simultaneous segmentation and emergent language generation. In general, we assume the following:

1. There is a segmentation network that provides a segmentation output $\boldsymbol{x}$.
2. There is a vocabulary $\mathrm{V} = \{w_1, w_2, \ldots, w_{N_V}\}$. A sentence S_{N_w} of length N_w is a sequence of words $\{w_1, w_2, \ldots, w_{N_w}\}$.
3. A Sender agent or network which receives the segmentation output $\boldsymbol{x}$ and generates a sentence S_{N_w} of length N_w, where $S_{N_w} = Sender(\boldsymbol{x})$.
4. A Receiver agent or network, which obtains the symbolic sentence S_{N_w} and generates an output $\boldsymbol{x}' = Receiver(S_{N_w})$.
5. The final segmentation is co-generated from: $\boldsymbol{x}$ and $\boldsymbol{x}'$.

To demonstrate applicability, we use a UNet network as implemented in [15], we omit the last sigmoid function (Fig. 1, left side) to generate an output $\boldsymbol{x}$. We include an Emergent Language (EL) network to generate a second output $\boldsymbol{x}'$. Segmentation is obtained by concatenating $\boldsymbol{x}$ and $\boldsymbol{x}'$ and applying a sigmoid function (Fig. 1, right side). We train from end-to-end the segmentation and EL networks and when both converge, we conclude that an interpretable symbolic language has emerged. Next, we outline how to co-train both networks while reusing the original segmentation loss function.

3.1 Sender and Receiver Network

We implemented a variant of the Sender and Receiver networks reported in Havrylov *et al.* [12] using stacked LSTM models; see Hochreiter [16].

Sender Network. Input is: i) a tensor $\boldsymbol{x}$ from the last convolutional layer output from the UNet, ii) a token <S> representing the start of the message. First, a linear transformation **Linear**($\boldsymbol{x}$) is applied and passed to the stacked LSTM network. The initial cell state $\boldsymbol{c}_0^s$ is set to zero. Unlike a conventional LSTM, the implemented LSTM samples a single symbol from a categorical distribution $\boldsymbol{w} \sim \mathbf{Cat}(\boldsymbol{p}_v^n)$, where $\boldsymbol{p}_v^n$ represents the class probabilities with respect to the symbols in the vocabulary V at iteration $\boldsymbol{n}$. Given that we are sampling from a categorical distribution, it is evident that this operation is not differentiable and therefore we cannot estimate the gradient during back propagation. To estimate a gradient during training, the Gumbel-Softmax trick [17] for discrete variables is implemented. Then, at each iteration $\boldsymbol{n}$, we estimate a single symbol or word w_i as follows:

$$w_i = G_\tau\left(p_i^n\right) = \frac{\left(\exp\left(\log\left(p_i^n\right) + g_i\right)/\tau\right)}{\sum_{j=1}^{V}\left(\exp\left(\log\left(p_j^n\right) + g_j\right)\right)/\tau)} \tag{1}$$

where τ is the temperature parameter that regulates the Gumbel-Soft-max operator G_τ (diamond icons in Fig. 1) and g_i is Gumbel(0, 1). The output of the Sender is the last hidden state $\boldsymbol{h}_{n+1}^s$ which encodes the sentence as a sequence of words w_i as: $\boldsymbol{h}_{n+1}^s = \mathbf{LSTM}(w_i, \boldsymbol{h}_n^s, \boldsymbol{c}_n^s)$. During inference, we do not apply the Gumbel-Soft-max operator [17], making $\boldsymbol{h}_{n+1}^s$ fully deterministic. Then, the generated sentence is represented as $S_{N_w} = Sender(\boldsymbol{x})$.

Receiver Network. Unlike the Sender, the Receiver is implemented as a standard LSTM model. Input to the Receiver are the Sender's hidden states (which encode the sentence S_{N_w}). The initial hidden and cell states $\boldsymbol{h}_0^r$ and $\boldsymbol{c}_0^r$ respectively are set to zero. We then apply a linear transform **Linear**($\boldsymbol{h}_{n+1}^r$) to the Receiver's last hidden state. The output of the receiver is $\boldsymbol{x}' = Receiver(S_{N_w})$. During this process, the Sender and Receiver are encouraged to establish a communication protocol and if successful we conclude that a new emergent language has been produced. In order to generate a deterministic output, during inference we encode the categorical variable as a one-hot vector.

Semantic Symbolic Segmentation: To generate the final segmentation, we concatenate $\boldsymbol{x}$ and $\boldsymbol{x}'$ as: **Concat**($\boldsymbol{x}, \boldsymbol{x}'$). We then apply a Convolution operator followed by batch normalization to produce a tensor of the same dimensions as $\boldsymbol{x}$. The final segmentation output is obtained by applying a Sigmoid function. It is worth mentioning that this method is generally applicable to any segmentation network with the original loss function. Mathematically, the Emergent Language model can be interpreted as a regularization network, which forces the original segmentation network to have an internal feedback and control mechanism. In our experiments we noticed an average of six percent increase in performance.

4 Results

Imaging data was obtained from The Cancer Imaging Archive (TCGA) dataset. We analyzed 110 subjects as described in Buda *et al.* [15]. We integrated our Emergent Language module as described in the previous section. We associate each symbol to an integer number, and a segmentation mask is generated after applying a single connected component step, as reported in [15]. Table 1 presents a performance comparison between the UNet model from [15] and our SUNet model. For the SUNet, different sentence length N_W and vocabulary size N_V was selected. The size of hidden dimension tensor and cell state tensor was set to 1024 and 2048 respectively (Sender and Receiver). We found best performance with a sentence length of 10 and vocabulary of 10K symbols. In Fig. 2, columns, w0 to w10, show a visualization of emergent symbols (per 2D section) and the last four columns correspond to area and eccentricity, tumor detection and ground truth. We observe that different patterns of symbols emerge if the 2D image corresponds to normal or tumor tissue. For example, for subject CS 4941 (left), the regular expressions associated with normal tissue are i) **657, 653, ***; ii) **657, 8927, 3330, ***; iii) **657, 3785, ***; whereas the expression associated with tumor presence are i) **8584, ***; ii) **1168, ***; iii) **3912, ***. Unlike the last symbols in the sentence, we notice that the initial symbols appear to best characterize tissue type. This is expected given that the LSTM model is hierarchical.

Table 1. Comparative performance from [15] and our proposed SUNet framework.

Prediction	Buda *et al.* [15]	$N_W = 4$ $N_V = 100$	$N_W = 10$ $N_V = 10\,\text{k}$	$N_W = 10$ $N_V = 50\,\text{k}$	$N_W = 20$ $N_V = 10\,\text{k}$	$N_W = 20$ $N_V = 1\,\text{k}$
Best validation (mean DSC)	81.4	89.1	**90.0**	89.3	81.5	78.8

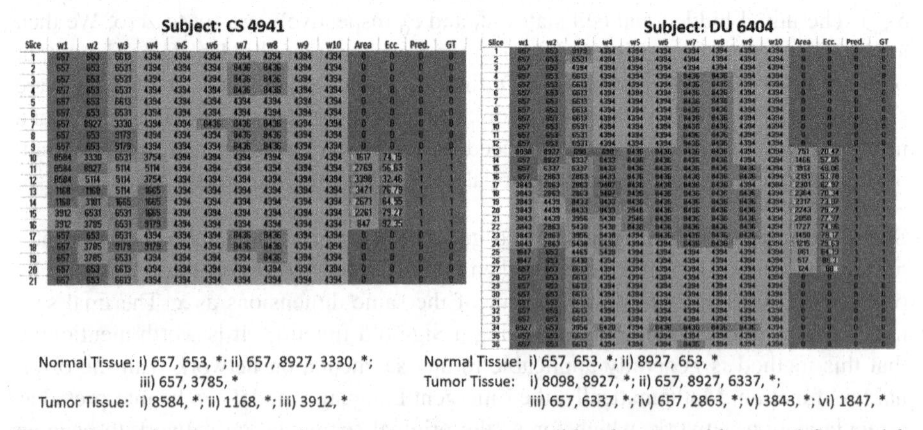

Fig. 2. Example of emergent language sentence across slides for two subjects and for model corresponding to $N_W = 10$ and $N_V = 10\text{k}$. Columns w1 through w10 show distribution of symbols. The last four columns correspond to area, eccentricity, tumor detection and ground truth.

4.1 Regression Performance

Results in the previous sections indicate that symbolic expressions can be used to predict segmentation masks on a fluid-attenuated inversion recovery (FLAIR) brain images, and those symbols are informative with respect to presence or absence of tumor. Further analyses were conducted to determine whether individual symbols in emergent language expressions can predict tumor presence (Tumor), morphology (Area, Eccentricity), location (Laterality, Location) and patient genome (miRNA). Specifically, we examined which expression symbol (i.e., first, second, etc.) is best at predicting each outcome for all candidate models. Results from this analysis are reported in Table 2.

Table 2. Linear and logistic regression results. The top row indicates model parameters, where N_W is the sentence length, and N_V is the vocabulary size. The following rows indicate prediction type, and columns indicate which model was most predictive of the outcome, and which specific symbol S^* performed best. Squared Pearson correlation coefficients are provided for continuous variables (Area, Eccentricity), and McFadden's pseudo-R^2 is reported for categorical outcomes (Tumor, Laterality, Location, Histology Type, Histology Grade and miRNA).

Prediction	$N_W = 4$ $N_V = 100$		$N_W = 10$ $N_V = 10$ k		$N_W = 10$ $N_V = 50$ k		$N_W = 20$ $N_V = 10$ k		$N_W = 20$ $N_V = 1$ k	
	S^*	R^2	S^*	R^2	S^*	R^2	S^*	R^2	S^*	R^2
Tumor	S_4	0.27	S_1	0.48	S_1	0.29	S_4	0.59	S_1	**0.67**
Area	S_3	0.27	S_2	0.43	S_1	0.33	$\mathbf{S_1}$	**0.44**	S_2	0.39
Eccentricity	S_4	0.06	S_2	0.11	S_1	0.10	S_1	0.15	$\mathbf{S_5}$	**0.19**
Laterality	S_3	0.22	S_1	0.37	S_4	0.22	$\mathbf{S_1}$	**0.41**	S_2	0.33
Location	S_4	0.11	$\mathbf{S_1}$	**0.23**	S_3	0.12	S_4	0.19	S_2	0.16
Histology Type	S_3	0.06	$\mathbf{S_2}$	**0.12**	S_4	0.06	S_1	0.10	S_2	0.08
Histology Grade	S_1	0.03	S_1	**0.09**	S_1	0.06	S_2	0.09	S_2	0.08
miRNA	S_3	0.07	$\mathbf{S_1}$	**0.13**	S_1	0.06	S_1	0.13	S_1	0.09

The statistical model used to predict each outcome varied. For binary data (Tumor), a binary logistic regression model was used. For outcomes with multiple classes (Laterality, Location, miRNA), a multinomial logistic regression was performed. Finally, linear regression was performed on continuous data (Area, Eccentricity). Both multinomial logistic and linear regression were performed on data where a tumor was present. We report squared Pearson correlation coefficient R^2 values for continuous outcomes and McFadden's pseudo-R^2 for categorical outcomes. Results indicate high correlations between expression symbols and tumor presence, area and laterality. Both tumor eccentricity and location were moderately correlated with emergent language symbols, although histology type, histology grade and patient genome (miRNA) achieved lesser correlations. When only four symbols were used ($V = 4$), expressions were least predictive of each outcome, as indicated by correspondingly lower correlations.

It is worth noting, however, that symbols occurring later in each expression (S_3 and S_4) were most informative with respect to each outcome, whereas the opposite trend was observed in the remaining models. With the exception of tumor eccentricity, all outcomes were best explained by either the first or second symbol in emergent language expressions (S_1 or S_2). This suggests that while symbols occurring later in the expression are important for capturing and refining semantics, the weight of explainability appears to fall on earlier elements.

Interpretability and Explainability. Figure 3 shows a visualization of symbol interpretability and quality of segmentation in the analyzed 2D images. We show monochromatic images after mapping from three color channels (FLAIR images) to grayscale. Green and red contours represent ground truth and predicted tumor contours, respectively. Predictions correspond to our SUNet model with 10 K symbols and sentence length 10. Post-processing was applied as reported in [15].

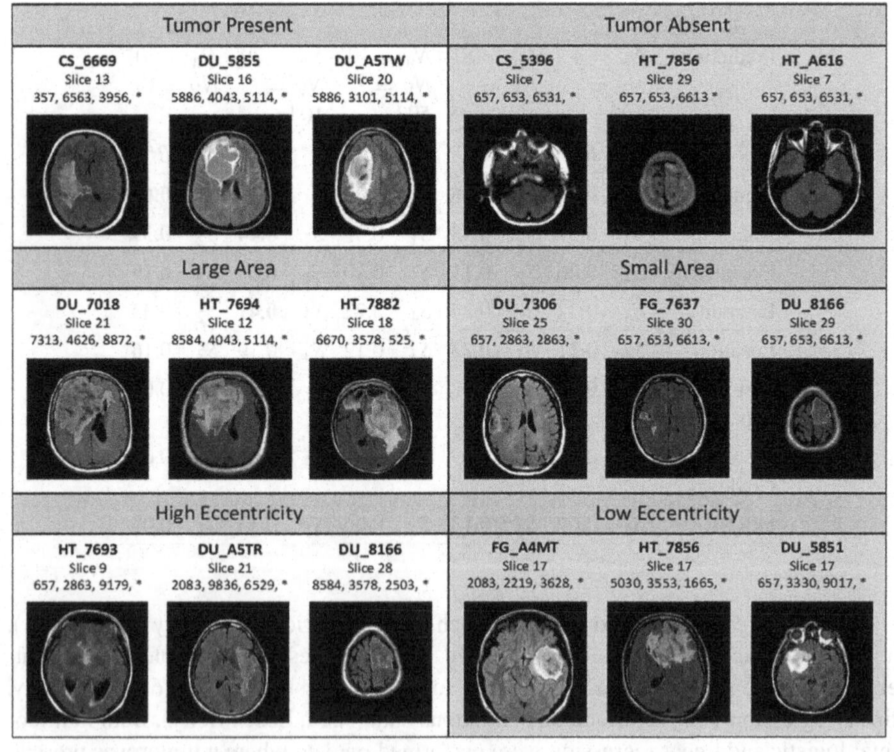

Fig. 3. Examples of brain images associated with (top) tumor presence, (middle) tumor size or area and (bottom) tumor eccentricity. Subject identifiers are reported in bold text, followed by the associated image slice. The first two emergent language symbols (S_1, S_2) are reported, followed by * indicating "all remaining symbols". (Color figure online)

From top to bottom we show images for: i) tumor present or absent, ii) tumor area and iii) tumor eccentricity across several representative subjects that correspond to the first two symbols within the generated ten symbol sentence. The top rows show tissue characterization for tumor or no tumor. The top left shows the segmented tumors along with the interpretable symbolic regular expressions: **357, 6563, 3956, ***; **5886, 4043, 5114, *** and **5886, 3101, 5114, ***, where the * indicates any remaining symbol in the sentence. We also notice that while we can generate a symbolic sentence, the quality of the segmentation highly matches the ground truth. This suggests that the Receiver network was able to: i) interpret correctly the sentence generated by the Sender network and ii) co-generate correctly the segmentation mask from the full sentence. The top-right row suggests a trend from the regular expressions: **657, 653, 6131, ***; **657, 653, 6613, *** and **657, 653, 6531, *** for bottom (slice 7), top (slice 29) and bottom (slice 7) areas of the brain respectively. It should be noted that R^2 value for the predictive model was 0.48 (high confidence). The middle rows correspond to area. On the left and right show examples of large and small tumor. For example, symbols **7313, 4626, 8872, ***; **8584, 4043, 5114, *** and **6670, 3578, 525, *** may be associated with large tumors, whereas **657, 2863, 2863, ***; **657, 653, 6613, *** and **657, 653, 6613, *** may be indicative of small tumors. Here the R^2 value was 0.43 (high confidence). The bottom row shows an example of prediction with medium-low confidence ($R^2 = 0.11$) for small tumors with high eccentricity and large tumors with high eccentricity.

5 Conclusions and Future Work

We have presented a generalized framework to enable Symbolic Semantic (S^2) Segmentation in medical imagery. Unlike standard semantic segmentation, our proposed S^2-Segmentation uses an Emergent Language model via Sender and Receiver agents to produce symbolic sentences. Such symbolic sentences are used by the Receiver agent to co-generate the final segmentation mask. The proposed framework can be applied to any segmentation network allowing direct interpretation of predictions. While integer symbols are not particularly intuitive, future work will build on recent success translating symbols to natural language through neural machine translation architectures. To the best of our knowledge, this is the first work to demonstrate feasibility of Emergent Language models towards semantic segmentation interpretation and explanation.

We implemented a Symbolic UNet (SUNet) architecture for tumor segmentation and interpretation using the TCGA dataset. Our segmentation results show high accuracy, close to results from human experts. Statistical analysis suggests feasibility in associating symbolic sentences with clinically relevant information, such as tissue type (tumor vs normal), object morphology (area, eccentricity), object localization (tumor laterality and location), tumor histology and genomics data. For future work we plan to extend validation to larger datasets and investigate different network architectures. Similarly, we plan carry on a more detailed analysis on the possibility of using symbolic expressions to drive interpretability in complex bioinformatics tasks towards personalized diagnostics and precision medicine linked with pathology data.

References

1. Harnad, S.: Minds, machines and turing: the indistinguishability of indistinguishables. J. Logic Lang. Inf. **9**(4), 425–445 (2000)
2. Kelly, C.J., Karthikesalingam, A., Suleyman, M., et al.: Key challenges for delivering clinical impact with artificial intelligence. BMC Med. **17**, 195 (2019)
3. Doshi-Velez, F., Kim, B.: Towards a rigorous science of interpretable machine learning. arXiv preprint arXiv:1702.08608 (2017)
4. Adadi, A., Berrada, M.: Peeking inside the black-box: a survey on explainable artificial intelligence (XAI). IEEE Access **6**, 52138–52160 (2018)
5. Gilpin, L.H., Bau, D., Yuan, B.Z., Bajwa, A., Specter, M., Kagal, L.: Explaining explanations: an overview of interpretability of machine learning. In: IEEE 5th International Conference on Data Science and Advanced Analytics (DSAA), Italy, pp. 80–89 (2018)
6. Natekar, P., Kori, A., Krishnamurthi, G.: Demystifying brain tumor segmentation networks: interpretability and uncertainty analysis. Front. Comput. Neurosci. **14**, 6 (2020)
7. Couteaux, V., Nempont, O., Pizaine, G., Bloch, I.: Towards interpretability of segmentation networks by analyzing DeepDreams. In: Suzuki, K., et al. (eds.) ML-CDS/IMIMIC -2019. LNCS, vol. 11797, pp. 56–63. Springer, Cham (2019). https://doi.org/10.1007/978-3-030-33850-3_7
8. Ribeiro, M.T., Singh, S., Guestrin, C.: "Why should I trust you?" Explaining the predictions of any classifier. In: Proceedings of the 22nd ACM SIGKDD International Conference on Knowledge Discovery and Data Mining, pp. 1135–1144 (2016)
9. Simonyan, K., et al.: Deep Inside Convolutional Networks: Visualising Image Classification Models and Saliency Maps. arXiv:1312.6034 (2013)
10. Yeh, C.K., Hsieh, C.Y., Suggala, A.S., Inouye, D., Ravikumar, P.: How Sensitive are Sensitivity-Based Explanations? arXiv preprint arXiv:1901.09392 (2019)
11. Kindermans, P.-J., et al.: The (un)reliability of saliency methods. In: Samek, W., Montavon, G., Vedaldi, A., Hansen, L.K., Müller, K.-R. (eds.) Explainable AI: Interpreting, Explaining and Visualizing Deep Learning. LNCS (LNAI), vol. 11700, pp. 267–280. Springer, Cham (2019). https://doi.org/10.1007/978-3-030-28954-6_14
12. Havrylov, S., Tritov, I.: Emergence of language with multi-agent games: learning to communicate with sequences of symbols. In: Advances in Neural Information Processing Systems, vol. 30, pp. 2149–2159 (2017)
13. Cogswell, M., Lu, J., Lee, S., Parikh, D., Batra, D.: Emergence of Compositional Language with Deep Generational Transmission. arXiv preprint arXiv:1904.09067, April 2019
14. Lazaridou, A., Herman, K.M., Tuyls, K., Clark, S.: Emergence of linguistic communication from referential games with symbolic and pixel input. arXiv preprint arXiv:1804.03984, April 2018
15. Buda, M., Saha, A., Mazurowski, M.A.: Association of genomic subtypes of lower-grade gliomas with shape features automatically extracted by a deep learning algorithm. Comput. Biol. Med. **109**, 218–225 (2019)
16. Hochreiter, S., Schmidhuber, J.: LSTM can solve hard long time lag problems. In: Advances in Neural Information Processing Systems 9. MIT Press, Cambridge (1997)
17. Jang, E., Gu, S., Poole, B.: Categorical reparameterization with Gumbel-Softmax. arXiv preprint arXiv:1611.01144 (2016)

Meta Corrupted Pixels Mining for Medical Image Segmentation

Jixin Wang[1], Sanping Zhou[1], Chaowei Fang[2], Le Wang[1], and Jinjun Wang[1](✉)

[1] Institute of Artificial Intelligence and Robotics, Xi'an Jiaotong University, Xi'an, Shaanxi, People's Republic of China
jinjun@mail.xjtu.edu.cn

[2] School of Artificial Intelligence, Xidian University, Xi'an, Shaanxi, People's Republic of China

Abstract. Deep neural networks have achieved satisfactory performance in piles of medical image analysis tasks. However the training of deep neural network requires a large amount of samples with high-quality annotations. In medical image segmentation, it is very laborious and expensive to acquire precise pixel-level annotations. Aiming at training deep segmentation models on datasets with probably corrupted annotations, we propose a novel Meta Corrupted Pixels Mining (MCPM) method based on a simple meta mask network. Our method is targeted at automatically estimate a weighting map to evaluate the importance of every pixel in the learning of segmentation network. The meta mask network which regards the loss value map of the predicted segmentation results as input, is capable of identifying out corrupted layers and allocating small weights to them. An alternative algorithm is adopted to train the segmentation network and the meta mask network, simultaneously. Extensive experimental results on LIDC-IDRI and LiTS datasets show that our method outperforms state-of-the-art approaches which are devised for coping with corrupted annotations.

Keywords: Meta Corrupted Pixels Mining · Deep neural network · Medical image segmentation

1 Introduction

Recent years have witnessed the blooming of Deep Neural Networks (DNNs) in medical image analysis, including image segmentation, image registration, image reconstruction [14], and etc. Due to the powerful representation capability of DNN, significant progress has been achieved in medical image analysis. However, training a DNN usually requires a large number of high-quality labeled samples, which is hard to acquire in various applications. For example, it is very expensive to generate a precise segment of input image, because the pathological tissue needs to be marked by professional radiologists [26,27]. As a result, a question was naturally raised: How can we train a powerful segmentation network only using a small number of high-quality labeled samples?

A. L. Martel et al. (Eds.): MICCAI 2020, LNCS 12261, pp. 335–345, 2020.
https://doi.org/10.1007/978-3-030-59710-8_33

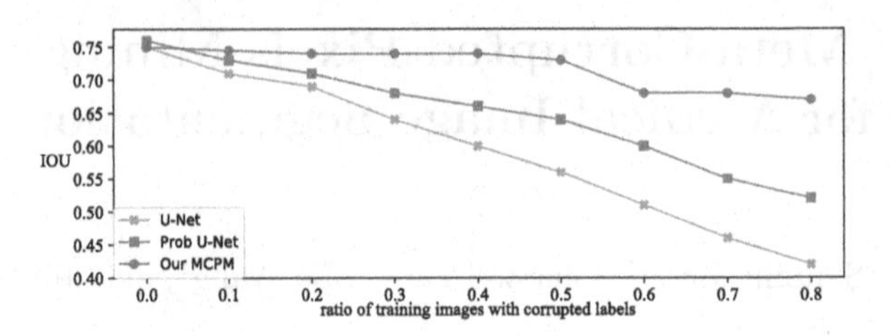

Fig. 1. Corrupted labels effect network's performance

To address this situation, researchers have paid much attention to train DNNs in a semi-supervised manner. For example, Yang et al. [23] presented an active learning method for 2D biomedical image segmentation, which can improve segmentation accuracy through suggesting the most effective rather than all samples for labeling. In [25], Zhao et al. applied a modified Mask R-CNN to volumetric data for instance segmentation, and they used bounding boxes for all instances and voxel-wise labels for a small proportion of instances. Nie et al. [17] proposed an attention based semi-supervised deep networks, which adopted the adversarial learning strategy to deal with the insufficient data problem in training complex networks. In practice, the success of these semi-supervised methods depends on mining a kind of knowledge which can be used to find out more accurate labels in the training process. However, most of the existing methods use a fixed prior knowledge to guide the pseudo label estimation. Therefore, they are very unstable when dealing with training samples with complex noise distributions. As shown in Fig. 1, the segmentation network's results are seriously affected when corrupted labels are taken as supervisory signals. This phenomenon reveals that mining corrupted labels is a critical issue in semi-supervised image segmentation.

In this paper, we design a novel Meta Corrupted Pixels Mining (MCPM) method for medical image segmentation, which can alleviate the impacts of corrupted labels in the training process. To achieve this goal, we design a simple meta mask network to protect the training of the segmentation network from the influence of pixels with incorrect labels. Specifically, the meta mask network absorbs in the loss value map of the segmentation prediction as input, and estimate a weight map indicating the importance of every pixel in the training of the segmentation network. Once the meta mask network is learned, small weights are allocated to pixels with corrupted labels. Therefore influences from these pixels are weakened when updating the segmentation network. In the training process, we update the segmentation network and meta mask network in an alternate manner, which can learn a powerful segmentation network from images with corrupted labels. The main contributions of this work can be highlighted as follows:

- We design a novel meta learning framework to mine pixels with corrupted labels during the process of training a segmentation network.

- Based on the fully convolutional structure, we build up a meta mask network which can automatically estimate pixel-wise importance factors for mitigating the influence of corrupted labels.
- Extensive experiments on both LIDC-IDRI and LiTS datasets indicate that our method achieves the state-of-the-art performance in medical image segmentation with incorrect labels.

2 Related Works

Because our method takes U-Net [19] as segmentation network and applies the meta learning regime [1] to mine pixels of corrupted labels, we briefly review a few existing works in terms of U-Net and meta learning in the following paragraphs.

Methods Based on U-Net. This type of methods aim to design a powerful network structure, which can obtain accurate segmentation results at the output layer. In [19], Ronneberger et al. proposed a well-known U-shaped structure for 2D medical image segmentation, in which the low-level and high-level feature are recursively concatenated together from top to down, to improve segmentation results. Inspired by this idea, a number of variants have been introduced in the past few years. For example, Milletari et al. [15] extended the U-shaped structure into 3D version and built an objective function and adopted Dice coefficient maximisation to supervise the training process. In [13], Kohl et al. proposed a generative segmentation model based on a combination of a U-Net, in which a conditional variational autoencoder is designed to produce an unlimited number of plausible hypotheses. Because its superior performance in medical image segmentation, we simply choose U-Net as our segmentation network. Then, we concentrate on designing a meta learning regime which can help learn a robust segmentation network from training samples with corrupted labels.

Methods Based on Meta Learning. This kind of methods aim to learn a kind of knowledge which can be used to guide the training of the network for solving the target problem [1], which has a wide application in the few-shot learning community. For example, a number of methods, such as FWL [8]. MentorNet [11] used the concept of meta learning to learn an adaptive weighting function to make the training process more robust to noisy images. However, the meta learners used in these methods have complex forms and require complicated inputs, which are very hard to be implemented in the training process. To overcome this problem, Ren et al. [18] proposed a novel meta learning algorithm which can learn an implicit function to assign weights to training samples based on their gradient directions. In [20], Shu et al. designed a meta weight network to lean an explicit function which can impose small weights to noisy samples, therefore the noisy samples will not severely affect the training process. The difference between our proposed model and the meta weight network is that, we design a simple meta mask network to learn a knowledge which can mine the pixels of corrupted labels, so as to learn a powerful segmentation network from low-quality labeled images.

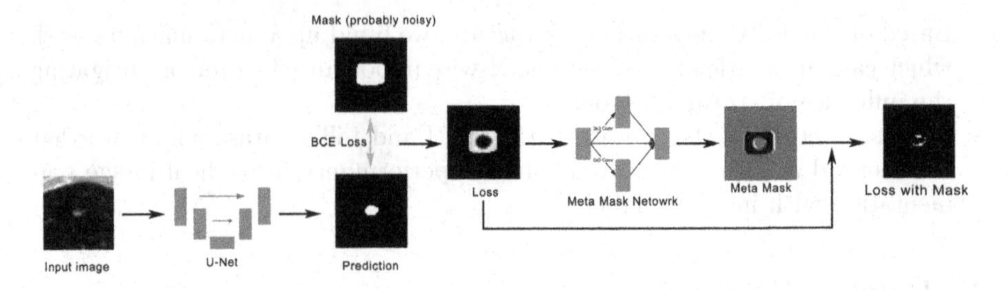

Fig. 2. The architecture and workflow of one loop

3 Meta Corrupted Pixels Mining Algorithm

We propose a novel MCPM method which can learn a powerful segmentation network from images with corrupted labels. Given a small set of images with clean labels and a large set of images with corrupted labels, our method is capable of identifying out the pixels with corrupted labels, and excluding them during the optimization procedure. As shown in Fig. 2, our network architecture is constituted by two modules: (1) a U-Net based module for segmentation; and (2) a meta mask network for mining pixels with corrupted labels. In the following paragraphs, we will introduce our method in detail.

3.1 Objective Functions

Let $\mathcal{S} = \{(\mathbf{X}^i, \mathbf{Y}^i)\}_{i=1}^N$ represent training images with probably noisy segmentation annotations, in which the width and height of training images are denoted by w and h respectively, and N indicates the number of training samples. Besides, $\mathbf{Y}^i \in \{0,1\}^{h \times w \times c}$ denote the corrupted labels, where c is the number of classes to be segmented out. First of all, we set up a segmentation network based on U-Net [19], which yields a pixel-level prediction $\mathbf{P}^i$ from input image $\mathbf{X}^i$. We define $\mathbf{P}^i = \mathcal{F}(\mathbf{X}^i; \mathbf{W})$ where $\mathbf{W}$ represents parameters of the segmentation network. To learn $\mathbf{W}$, an objective function is usually adopted to calculate pixel-wise loss values as function $\mathrm{L}_{xy}^i = \mathrm{loss}(\mathrm{P}_{xy}^i, \mathrm{Y}_{xy}^i)$, where $x \in [1, h]$ and $y \in [1, w]$ indicate the pixel coordinates. Here, the cross entropy loss function is used as the objective function.

As mentioned above, there might exist errors in segmentation annotations. These errors will severely hamper the optimization procedure, for example, providing incorrect gradient directions in the training process. A straightforward approach to cope with this issue is ignoring these pixels with incorrect labels through reweighting loss values. Inspired from [21], we design our meta mask network in a fully convolutional structure, which can learn an accurate mask map $\mathbf{R}^i$ for the input loss value map $\mathbf{L}^i$. We denote $\mathbf{R}^i = \mathcal{G}(\mathbf{L}^i; \mathbf{\Theta})$, where R_{xy}^i indicates the reweighting factor of the pixel at (x, y), and $\mathbf{\Theta}$ represents parameters of our meta mask network. Given a fixed $\mathbf{\Theta}$, the optimized solution to $\mathbf{W}$

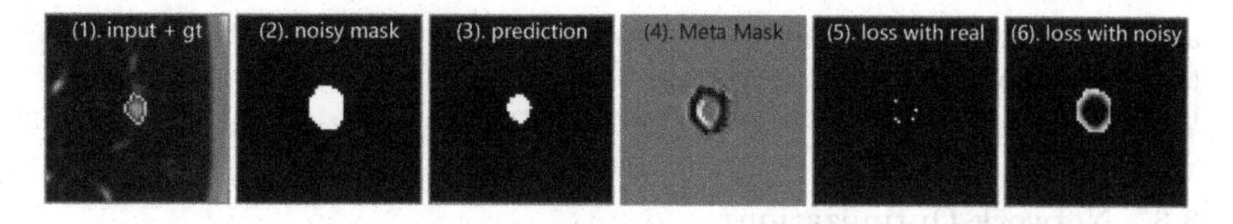

Fig. 3. Illustration of how meta mask network works in the training process

can be found through minimizing the following objective function:

$$\mathbf{W}^{\star}(\mathbf{\Theta}) = \arg\min_{\mathbf{W}} \frac{1}{Nhw} \sum_{i=1}^{N} \sum_{x=1}^{h} \sum_{y=1}^{w} \mathrm{R}_{xy}^{i} \mathrm{L}_{xy}^{i}. \tag{1}$$

To learn the parameters of our meta mask network, we introduce an additional meta dataset $\hat{\mathcal{S}} = \{(\hat{\mathbf{X}}^j, \hat{\mathbf{Y}}^j)\}_{j=1}^{M}$ which contains images with high-quality annotations. In particular, given an input image $\hat{\mathbf{X}}^j$ and optimized parameters $\mathbf{W}^{\star}(\mathbf{\Theta})$, the segmentation network will produce a pixel-wise prediction map $\hat{\mathbf{P}}^j = \mathcal{F}(\hat{\mathbf{X}}^j, \mathbf{W}^{\star}(\mathbf{\Theta}))$ at the output layer. Again, we can obtain a loss value map $\hat{\mathbf{L}}^j$ through comparing $\hat{\mathbf{P}}^j$ against $\hat{\mathbf{Y}}^j$ according to the cross entropy loss function. With the optimized $\mathbf{W}$, the optimized solution to $\mathbf{\Theta}$ can be acquired through minimizing the following objective function:

$$\mathbf{\Theta}^{*}(\mathbf{W}) = \arg\min_{\mathbf{\Theta}} \frac{1}{Mhw} \sum_{j=1}^{M} \sum_{x=1}^{h} \sum_{y=1}^{w} \hat{\mathrm{L}}_{xy}^{j}. \tag{2}$$

In the training process, we update $\mathbf{W}$ and $\mathbf{\Theta}$ in an alternation manner. As a result, the $\mathbf{\Theta}$ can cope with the varying $\mathbf{W}$, which is beneficial to effectively mine more corrupted pixels from the predictions of the segmentation network.

3.2 Meta Mask Network

We take a fully convolutional structure to design our meta mask network, which can explore more local information to locate the pixels with corrupted labels. The particularities of the network are two aspects: (1) It has two convolutional layers with kernels in size of 3×3 and 5×5, which can extract multi-scale context information from $\mathbf{L}^i$. (2) The resulting outputs and input are further fused through another 1×1 convolutional layer, giving rise to the final mask map $\mathbf{R}^i$. This simple structure can be trained under the guidance of a few high-quality labeled samples, which will in turn help train a powerful segmentation network by using a large number of low-quality labeled samples.

In Fig. 3, we visualize how our meta mask network alleviates the side effect of corrupted labels in the training process, in which: (1) shows the input image and ground truth annotations; (2) indicates the corrupted labels; (3) represents the predicted result obtained by the segmentation network; (4) denotes the mined pixels of corrupted labels. As we can observe in (5) and (6), the loss between (1) and (3) is very small, while the loss between (2) and (3) is large. This indicates that our meta network can help train a powerful segmentation network with a large number of images accompanied with corrupted labels.

3.3 Network Optimization

We employ the iterative optimization algorithm to train our model. It is implemented with a single loop and mainly contains the following steps.

- At first, $\mathbf{W}$ and $\mathbf{\Theta}$ are randomly initialized as $\mathbf{W}^0$ and $\mathbf{\Theta}^0$.
- For the t-th iteration, the parameters of the segmentation network are temporally renovated as in Eq. (3), via one step of gradient descent in minimizing the objective function (1),

$$\mathbf{W}'^{(t)}(\mathbf{\Theta}) = \mathbf{W}^{(t)} - \alpha \frac{1}{Nhw} \sum_{i=1}^{N} \sum_{x=1}^{h} \sum_{y=1}^{w} \mathrm{R}_{xy}^{i(t)} \frac{\partial \mathrm{L}_{xy}^{i}}{\partial \mathbf{W}} \bigg|_{\mathbf{W}^{(t)}}, \tag{3}$$

  where α is the learning rate. $\mathrm{R}_{xy}^{i(t)}$ is computed through feeding the loss value map into the meta mask network with parameters $\mathbf{\Theta}^{(t)}$.
- Then $\mathbf{\Theta}$ can be updated via optimizing the objective function (2),

$$\mathbf{\Theta}^{(t+1)} = \mathbf{\Theta}^{(t)} - \beta \frac{1}{Mhw} \sum_{j=1}^{M} \sum_{x=1}^{h} \sum_{y=1}^{w} \frac{\partial \hat{\mathrm{L}}_{xy}^{j}}{\partial \mathbf{W}'(\Theta)} \bigg|_{\mathbf{W}'^{(t)}} \frac{\partial \mathbf{W}'(\Theta)}{\partial \mathbf{\Theta}} \bigg|_{\mathbf{\Theta}^{(t)}}, \tag{4}$$

  where β is the learning rate.
- Finally, $\mathbf{W}$ is updated through minimizing objective function (1),

$$\mathbf{W}^{(t+1)} = \mathbf{W}^{(t)} - \alpha \frac{1}{Nhw} \sum_{i=1}^{N} \sum_{x=1}^{h} \sum_{y=1}^{w} \mathrm{R}_{xy}^{i(t+1)} \frac{\partial \mathrm{L}_{xy}^{i}}{\partial \mathbf{W}} \bigg|_{\mathbf{W}^{(t)}}. \tag{5}$$

  Here $\mathrm{R}_{xy}^{i(t+1)}$ is computed through feeding the loss value map into the meta mask network with updated parameters $\mathbf{\Theta}^{(t+1)}$.

3.4 Discussion

Under the guidance of a small meta set with clean annotations, the meta mask network is learned in a gradient descent by gradient descent manner as shown in (3) and (4). The update of parameters in the meta mask network is dependent on the gradients of losses calculated on pixels from both meta and training images. After putting (3) into (4), it can be easily observed that the ascending direction of the weight coefficient of every pixel relies on the inner product (it can also be interpreted as a similarity metric) between the gradient of the pixel (formulated as $\frac{\partial \mathrm{L}_{xy}^{i}}{\partial \mathbf{W}}|_{\mathbf{W}^{(t)}}$) and the average gradient of pixels of meta images (formulated as $\frac{1}{Mhw} \sum_{j=1}^{M} \sum_{x=1}^{h} \sum_{y=1}^{w} \frac{\partial \hat{\mathrm{L}}_{xy}^{j}}{\partial \mathbf{W}'(\Theta)}|_{\mathbf{W}'^{(t)}}$). A positive inner product pushes the parameters of the meta mask network towards a direction which can give rise to a larger weighting coefficient for the corresponding pixel; a negative inner product pushes the network towards the opposite direction. This is the reason why our method can effectively identify corrupted pixels.

4 Experiments

4.1 Datasets and Metrics

Two datasets are exploited to validate the superiority of our method in medical image segmentation with noisy annotations, including LIDC-IDRI (Lung Image Database Consortium and Image Database Resource Initiative) [2,3,7] and LiTS (Liver Tumor Segmentation) [9]. 64×64 patches covering lesions are cropped out as training or testing samples.

(1) The LIDC-IDRI dataset contains 1018 lung CT scans from 1010 patients with lesion masks annotated by four experts. 3591 patches are cropped out. They are split into a training set of 1906 images and a testing set of 1385 images. The remain 300 images are used as the meta set.
(2) The LiTS dataset includes 130 abdomen CT scans accompanied with annotations of liver tumors. 2214 samples are sampled from this dataset. 1471, 300 and 443 images are used for training, meta weight learning, and testing respectively.

Three metrics, including IOU (also referred as the Jaccard Index), Dice coefficient and Hausdorff distance, are employed for quantitatively measuring performances of segmentation algorithms.

Synthesizing Noisy Annotations. In practice, it is difficult to localize the boundary of the target region during the annotating procedure. Considering this phenomenon, we synthesize noisy annotations through creating masks which loosely encompasses target lesions. We use 2 operators to simulate corrupted annotations. 1) The dilation morphology operator is employed to extend the foreground region by several pixels (randomly drawn from $[0, 6]$). 2) The toolkit of deformation provided in ElasticDeform [6,19,22], which includes more complicate operations such as rotation, translation, deformation and morphology dilation, is used to contaminate ground-truths of training images. In our experiment, only a part of samples are contaminated with the above strategies. We denote the percent of images which are selected out to generate noisy labels as r.

4.2 Implementation Details

Adam and SGD is used to optimize to network parameters on LIDC-IDRI and LiTS, respectively. The learning rates α and β are initialized as 10^{-4} and 10^{-3} respectively, and decayed by 0.1 in 20th epoch and 40th epoch. The batch size is set as 128. All models are trained with 120 epochs.

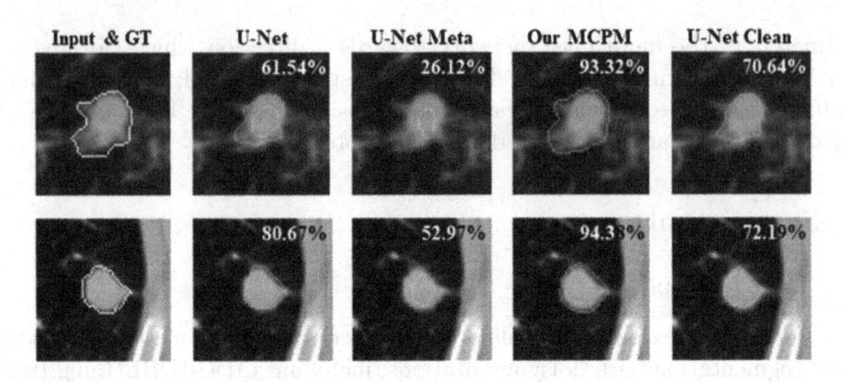

Fig. 4. Visualization of segmentation results. Green and red contours indicate the ground-truths and segmentation results, respectively. The Dice value is shown at the top-right corner, and our method produces much better results than other methods. (Color figure online)

4.3 Experimental Results

Comparison with Existing Methods. Without specification, r is set to 40% in experiments of this section which means annotations of 40% training images are contaminated. We compare our method against 7 existing segmentation models which are proposed to deal with ambiguous, low-quality or insufficient annotations on the LIDC-IDRI dataset: **Prob U-Net** [13], **Phi-Seg** [5], **UA-MT** [24] modified for 2D segmentation, **Curriculum** [12], **Few-Shot GAN** [16], **Quality Control** [4], U^2 **Net** [10], and **MWNet** [20] which is integrated with U-Net. All above models and the baseline U-Net are trained with mixed images of the training set and the meta set. We also implement another variant of U-Net which is trained merely using images from the meta set (indicated by 'U-Net Meta'). We also report the result of U-Net trained using images with clean labels (indicated by 'U-Net Clean'). As shown in Table 1. Our method performs significantly better than other methods. For example, the Dice value of our method surpasses that of the second best method MWNet by 3.4%. Additionally, our method outperforms baseline U-Net models by a large margin. It even achieves promising performance which is comparable to the result of 'U-Net Clean'. This indicates that the impact of incorrect annotations fabricated as in Sect. 4.1 is almost eliminated. Visualization examples are shown in Fig. 4.

Results with Various r-s. In this section, we vary the percent of noisy images r from 0 to 0.8. The segmentation results of four methods on LIDC-IDRI and LiTS are presented in Table 2. On the LIDC-IDRI dataset, our method performs better than other methods when noises are introduced into the training set. On the LiTS dataset, our method exceeds other methods consistently under all cases.

Table 1. Results of segmentation models on LIDC-IDRI.

Noisy	Dilation			ElasticDeform		
Model name	mIOU	Dice	Hausdorff	mIOU	Dice	Hausdorff
U-Net	62.53	75.56	1.9910	65.01	76.17	1.9169
U-Net Meta	60.91	72.21	2.0047	60.91	72.21	2.0047
Prob U-Net	66.42	78.39	1.8817	68.43	79.50	1.8757
Phi-Seg	67.01	79.06	1.8658	68.55	81.76	1.8429
UA-MT	68.18	80.98	1.8574	68.84	82.47	1.8523
Curriculum	67.78	79.54	1.8977	68.18	81.30	1.8691
Few-Shot GAN	67.74	78.11	1.9137	67.93	77.83	1.9223
Quality Control	65.00	76.50	1.9501	68.07	77.68	1.9370
U^2 Net	65.92	76.01	1.9666	67.20	77.05	1.9541
MWNet	71.56	81.17	1.7762	71.89	81.04	1.7680
Our MCPM	**74.69**	**84.64**	**1.7198**	**75.79**	**84.99**	**1.7053**
U-Net Clean	75.73	83.91	1.7051	75.73	83.91	1.7051

Table 2. Results (mIOU) of segmentation models using various r-s.

Dataset name	LIDC-IDRI					LiTS				
r	0.8	0.6	0.4	0.2	0	0.8	0.6	0.4	0.2	0
U-Net	42.64	51.23	62.53	69.88	75.73	37.18	43.55	46.41	51.20	61.07
Prob U-Net	52.13	60.81	66.42	71.03	**76.37**	40.16	45.90	49.22	53.97	60.60
MWNet	61.28	67.33	71.56	72.07	74.40	43.14	44.97	51.96	58.65	59.18
Our MCPM	**67.60**	**68.97**	**74.69**	**74.87**	75.26	**45.09**	**48.76**	**55.17**	**62.04**	**62.68**

5 Conclusion

We proposed a novel Meta Corrupted Pixels Mining method to alleviate the side effect of corrupted label in medical image segmentation. Given a small number of high-quality labeled images, the deduced learning regime make our meta mask network able to locate the pixels having corrupted labels, which can be used to help train a powerful segmentation network from a large number of low-quality labeled images. Extensive experiments on two datasets, LIDC-IDRI and LiTS, show that the proposed method can achieve the state-of-the-art performance in medical image segmentation.

Acknowledgments. This work is jointly supported by the National Key Research and Development Program of China under Grant No. 2017YFA0700800, the National Natural Science Foundation of China Grant No. 61629301, 61976171, and the Key Research and Development Program of Shaanxi Province of China under Grant No. 2020GXLH-Y-008.

References

1. Andrychowicz, M., et al.: Learning to learn by gradient descent by gradient descent. In: NeurIPS, pp. 3981–3989 (2016)
2. Armato III, S.G., et al.: Data from LIDC-IDRI. The cancer imaging archive, vol. 9, no. 7 (2015). https://doi.org/10.7937/K9/TCIA.2015.LO9QL9SX
3. Armato III, S.G., et al.: The lung image database consortium (LIDC) and image database resource initiative (IDRI): a completed reference database of lung nodules on CT scans. Med. Phys. **38**(2), 915–931 (2011)
4. Audelan, B., Delingette, H.: Unsupervised quality control of image segmentation based on Bayesian learning. In: Shen, D., et al. (eds.) MICCAI 2019. LNCS, vol. 11765, pp. 21–29. Springer, Cham (2019). https://doi.org/10.1007/978-3-030-32245-8_3
5. Baumgartner, C.F., et al.: PHiSeg: capturing uncertainty in medical image segmentation. arXiv preprint arXiv:1906.04045 (2019)
6. Çiçek, Ö., Abdulkadir, A., Lienkamp, S.S., Brox, T., Ronneberger, O.: 3D U-net: learning dense volumetric segmentation from sparse annotation. In: Ourselin, S., Joskowicz, L., Sabuncu, M.R., Unal, G., Wells, W. (eds.) MICCAI 2016. LNCS, vol. 9901, pp. 424–432. Springer, Cham (2016). https://doi.org/10.1007/978-3-319-46723-8_49
7. Clark, K., et al.: The cancer imaging archive (TCIA): maintaining and operating a public information repository. J. Digit. Imaging **26**(6), 1045–1057 (2013)
8. Dehghani, M., Mehrjou, A., Gouws, S., Kamps, J., Schölkopf, B.: Fidelity-weighted learning. arXiv preprint arXiv:1711.02799 (2017)
9. Han, X.: Automatic liver lesion segmentation using a deep convolutional neural network method. arXiv preprint arXiv:1704.07239 (2017)
10. Huang, C., Han, H., Yao, Q., Zhu, S., Zhou, S.K.: 3D U^2-net: a 3D universal U-net for multi-domain medical image segmentation. In: Shen, D., et al. (eds.) MICCAI 2019. LNCS, vol. 11765, pp. 291–299. Springer, Cham (2019). https://doi.org/10.1007/978-3-030-32245-8_33
11. Jiang, L., Zhou, Z., Leung, T., Li, L.J., Fei-Fei, L.: Mentornet: learning data-driven curriculum for very deep neural networks on corrupted labels. In: ICML, pp. 2304–2313 (2018)
12. Kervadec, H., Dolz, J., Granger, É., Ben Ayed, I.: Curriculum semi-supervised segmentation. In: Shen, D., et al. (eds.) MICCAI 2019. LNCS, vol. 11765, pp. 568–576. Springer, Cham (2019). https://doi.org/10.1007/978-3-030-32245-8_63
13. Kohl, S., et al.: A probabilistic u-net for segmentation of ambiguous images. In: NeurIPS, pp. 6965–6975 (2018)
14. Liu, H., Xu, J., Wu, Y., Guo, Q., Ibragimov, B., Xing, L.: Learning deconvolutional deep neural network for high resolution medical image reconstruction. Inf. Sci. **468**, 142–154 (2018)
15. Milletari, F., Navab, N., Ahmadi, S.A.: V-net: fully convolutional neural networks for volumetric medical image segmentation. In: 2016 Fourth International Conference on 3D Vision (3DV), pp. 565–571. IEEE (2016)
16. Mondal, A.K., Dolz, J., Desrosiers, C.: Few-shot 3D multi-modal medical image segmentation using generative adversarial learning. arXiv preprint arXiv:1810.12241 (2018)

17. Nie, D., Gao, Y., Wang, L., Shen, D.: ASDNet: attention based semi-supervised deep networks for medical image segmentation. In: Frangi, A.F., Schnabel, J.A., Davatzikos, C., Alberola-López, C., Fichtinger, G. (eds.) MICCAI 2018. LNCS, vol. 11073, pp. 370–378. Springer, Cham (2018). https://doi.org/10.1007/978-3-030-00937-3_43
18. Ren, M., Zeng, W., Yang, B., Urtasun, R.: Learning to reweight examples for robust deep learning. arXiv preprint arXiv:1803.09050 (2018)
19. Ronneberger, O., Fischer, P., Brox, T.: U-net: convolutional networks for biomedical image segmentation. In: Navab, N., Hornegger, J., Wells, W.M., Frangi, A.F. (eds.) MICCAI 2015. LNCS, vol. 9351, pp. 234–241. Springer, Cham (2015). https://doi.org/10.1007/978-3-319-24574-4_28
20. Shu, J., et al.: Meta-weight-net: learning an explicit mapping for sample weighting. arXiv preprint arXiv:1902.07379 (2019)
21. Szegedy, C., Ioffe, S., Vanhoucke, V., Alemi, A.A.: Inception-v4, inception-resnet and the impact of residual connections on learning. In: AAAI (2017)
22. van Tulder, G.: Package elsticdeform. https://github.com/gvtulder/elasticdeform
23. Yang, L., Zhang, Y., Chen, J., Zhang, S., Chen, D.Z.: Suggestive annotation: a deep active learning framework for biomedical image segmentation. In: Descoteaux, M., Maier-Hein, L., Franz, A., Jannin, P., Collins, D.L., Duchesne, S. (eds.) MICCAI 2017. LNCS, vol. 10435, pp. 399–407. Springer, Cham (2017). https://doi.org/10.1007/978-3-319-66179-7_46
24. Yu, L., Wang, S., Li, X., Fu, C.-W., Heng, P.-A.: Uncertainty-aware self-ensembling model for semi-supervised 3D left atrium segmentation. In: Shen, D., et al. (eds.) MICCAI 2019. LNCS, vol. 11765, pp. 605–613. Springer, Cham (2019). https://doi.org/10.1007/978-3-030-32245-8_67
25. Zhao, Z., Yang, L., Zheng, H., Guldner, I.H., Zhang, S., Chen, D.Z.: Deep learning based instance segmentation in 3D biomedical images using weak annotation. In: Frangi, A.F., Schnabel, J.A., Davatzikos, C., Alberola-López, C., Fichtinger, G. (eds.) MICCAI 2018. LNCS, vol. 11073, pp. 352–360. Springer, Cham (2018). https://doi.org/10.1007/978-3-030-00937-3_41
26. Zhou, S., Wang, J., Zhang, M., Cai, Q., Gong, Y.: Correntropy-based level set method for medical image segmentation and bias correction. Neurocomputing **234**, 216–229 (2017)
27. Zhou, S., Wang, J., Zhang, S., Liang, Y., Gong, Y.: Active contour model based on local and global intensity information for medical image segmentation. Neurocomputing **186**, 107–118 (2016)

UXNet: Searching Multi-level Feature Aggregation for 3D Medical Image Segmentation

Yuanfeng Ji[1,2], Ruimao Zhang[2], Zhen Li[3], Jiamin Ren[2], Shaoting Zhang[2], and Ping Luo[1(✉)]

[1] The University of Hong Kong, Pokfulam, Hong Kong
luoping@sensetime.com
[2] SenseTime Research, Shatin, Hong Kong
[3] Shenzhen Research Institute of Big Data, The Chinese University of Hong Kong, Shenzhen, Guangdong, China

Abstract. Aggregating multi-level feature representation plays a critical role in achieving robust volumetric medical image segmentation, which is important for the auxiliary diagnosis and treatment. Unlike the recent neural architecture search (NAS) methods that typically searched the optimal operators in each network layer, but missed a good strategy to search for feature aggregations, this paper proposes a novel NAS method for 3D medical image segmentation, named UXNet, which searches both the scale-wise feature aggregation strategies as well as the block-wise operators in the encoder-decoder network. UXNet has several appealing benefits. (1) It significantly improves flexibility of the classical UNet architecture, which only aggregates feature representations of encoder and decoder in equivalent resolution. (2) A continuous relaxation of UXNet is carefully designed, enabling its searching scheme performed in an efficient differentiable manner. (3) Extensive experiments demonstrate the effectiveness of UXNet compared with recent NAS methods for medical image segmentation. The architecture discovered by UXNet outperforms existing state-of-the-art models in terms of Dice on several public 3D medical image segmentation benchmarks, especially for the boundary locations and tiny tissues. The searching computational complexity of UXNet is cheap, enabling to search a network with best performance less than 1.5 days on two TitanXP GPUs.

1 Introduction

Volumetric medical image segmentation, which provides the detailed pixel-wise categorization of organ regions, is critical to a series of medical analysis, e.g., lung tumour detection [3,10], gland disease classification [9,19]. Recently, a family of deep models, including fully convolutional networks (FCNs) [14] and 3D

Electronic supplementary material The online version of this chapter (https://doi.org/10.1007/978-3-030-59710-8_34) contains supplementary material, which is available to authorized users.

A. L. Martel et al. (Eds.): MICCAI 2020, LNCS 12261, pp. 346–356, 2020.
https://doi.org/10.1007/978-3-030-59710-8_34

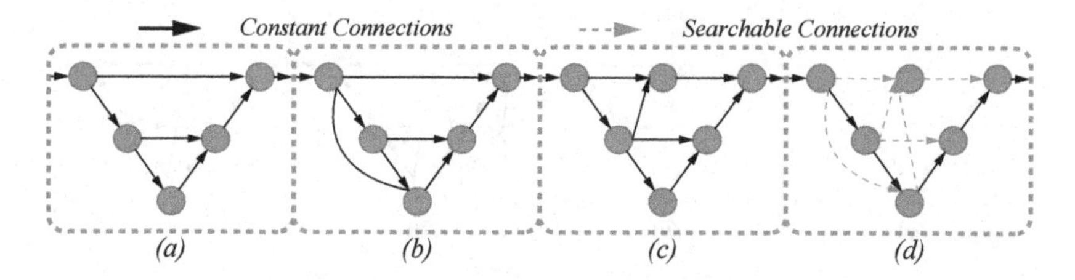

Fig. 1. Comparisons of UXNet with previous methods. UXNet's feature aggregation search space in (d) is more general than existing manual designs for medical image segmentation such as (a) UNet (Olaf et al. [15]), (b) Res/Dense-Unet (Li et al. [10], Yu et al. [19]), and (c) Deep Layer Aggregation (Zhou et al. [22]).

convolutions [17], has been proposed to improve the segmentation accuracy, by extracting the powerful feature representation of organ regions. However, a lot of the progress of deep models come from the design of neural network architectures, which heavily relies on expert domain knowledge.

Inspired by the AutoML [2,13], there has been significant interest in automatically searching the neural network architecture (NAS) through the given searching space. The goal of NAS is to discover better neural network architectures with the higher performance, the fewer parameters, and even lower computation cost [2,13]. For medical image segmentation, Weng et al. [18], Kim et al. [8], Bae et al. [1], Zhu et al. [23] explore to search the building blocks to construct UNet [15] structure in a gradient-based manner or reinforce learning methods. Yu et al. [20] further develop a more effective search strategy to allitterative the huge memory-cost problems caused by the 3D task. In the field of computer vision, Liu et al. [12] also search resolution sampling strategy, which is a operator configuration. Despite the successes of these methods, the searching schemes mainly focus on *searching the effective operators in different layers.* However, the huge variations in abnormalities' size, shape, location in 3D medical images usually require information from multi-level feature representations for the robust and dense prediction (*i.e.* multi-level feature aggregations).

In literature, many previous studies have demonstrated that aggregation of multi-level features could address the issues of the huge variations in abnormalities for more accurate segmentation [6,10,11,15,22]. Intuitively, merging the high-level and low-level features extracted from different layers helps to enrich semantic representation and capture detailed information. For example, the latest studies [6,7,10,21] designed to propagate coarse semantic context information back to the shallow layers through top-down and lateral skip-connections, where different layers have various size of 3D receptive field capturing the multi-scale context. More recently, Zhou et al. [22] exploit more effectively connected architectures via deeper aggregation strategy, which iteratively and hierarchically merges features across adjacent layers, yielding better performance. However, all of the existing aggregation strategies under the preset are *designed manually.*

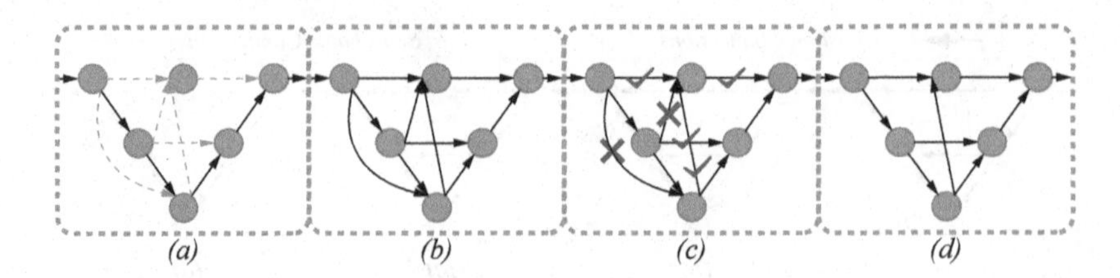

Fig. 2. An illustration of our multi-scale search architecture, including (a) all candidate searchable connections (grey dashed arrows) are initially regarded as equal, (b) joint optimization of the architecture weights (*i.e.* operators and feature aggregations) and the network weights by solving a bi-level optimization, (c) pruning the learned network according to some preset rules, and (d) the final searching result.

Specifically, they perform feature fusion among adjacent or all layers based on a fixed pattern, which may miss useful or involve useless information.

To address the above drawbacks, we advocate the idea that searching block-wise operators as well as scale-wise aggregations strategies are equally important for medical segmentation task, and investigate a novel searching method, UXNet, for more general aggregation search space as shown in Fig. 1. Concretely, during the searching process, UXNet allows each layer of the network to select optimum operation (*e.g.* traditional convolution or dilated convolution) with a *proper receptive field* to generate better feature representations. As shown in Fig. 2, based on the extracted multi-level feature representations, a searching strategy for multi-level feature aggregation is further conducted to discover a more efficient fusion method (i.e. which levels of the feature representations are selected to be aggregated in a specific node) for precise segmentation. Besides, the block-wise operators and scale-wise aggregations can be jointly searched in a differentiable manner through a continuous relaxation. Thus, the overall optimization process of NAS can be automatically driven by the segmentation accuracy, surpassing the methods of using a pre-fixed set of 3D receptive fields to construct the multi-scale context modelling.

The **main contributions** of this paper are three folds. (1) We present a novel architecture searching setting: searching for the optimal multi-level feature aggregation strategy to fuse the feature maps in UNet-like architecture for 3D medical image analysis. (2) A novel UXNet searching scheme is proposed by leveraging block-wise operation searching, as well as scale-wise aggregation searching in a uniform framework. (3) Extensive experiments demonstrate that UXNet outperforms existing state-of-the-art results on most challenging semantic segmentation benchmarks on the 3D Medical Segmentation Decathlon (MSD) challenge [16]. UXNet's computational complexity is cheap, such that the best-performing network can be searched less than 1.5 days on two TitanXP GPUs.

2 Methods

We illustrate the proposed UNXet in Fig. 3. The network has encoder and decoder architectures, which is the same as classical UNet, along with the *Searchable Building Block* (SBB) and *Multi-Scale Searchable Aggregation* (MSSA) architecture in-between. The former is applied to search the optimum operations in each layer, while the later is used to determine whether or not aggregate the feature maps from various levels in each node.

In practice, we input the volumetric image into the encoder network, producing convolution feature maps (i.e., $N_{0,0}, N_{0,1}, N_{0,2}, N_{0,3}, N_{0,4}$) at different levels. For each levels, input feature maps are fed into a SBB, which enables a flexible combination of various convolution and pooling operators (i.e. the yellow ellipse in Fig. 3), to do the transformation.

MSSA aggregates multi-scale information for assisting the segmentation of organ regions having various sizes. As illustrated in Fig. 3, MSSA includes several stages. At each stage, the feature maps from all of the levels are firstly regarded as the candidates for aggregation to generate feature maps of the next stage (e.g., at the 0th stage, $N_{0,0}, N_{0,1}, N_{0,2}, N_{0,3}, N_{0,4}$ is connected to $N_{1,1}$ at the 1st stage). It is worthy note that the encoder network also involves candidate dense connections for feature aggregation. Compared with the existing approaches that adjust the weight of the connection, MSSA further enables/disables the connection based on its importance to the final recondition task. It facilitates a more straightforward way to guide the search of a connection between feature maps at different stages. Besides, thanks to SBBs that preserve useful information, MSSA can simplify the search process by eliminating the unnecessary lowest-resolution feature maps at each stage (e.g., $N_{1,4}$ and $N_{2,3}$), while yielding better segmentation result.

2.1 Searchable Building Block

As illustrated in Fig. 3(a), SBB takes input as a conventional feature map. The block has two layers of convolution/pooling operations. In each layer, we search for an appropriate operation from all options (e.g. traditional convolution or dilated convolution), using the appropriate receptive field to capture useful image content at different levels.

In each SBB, there are three types of layers: (1) normal layer (i.e., $3 \times 3 \times 3$ conv, $3 \times 3 \times 1$ conv, $5 \times 5 \times 5$ conv, pseudo-3d conv, $2 * 3 \times 3 \times 3$ conv, $3 \times 3 \times 3$ conv with rate 2 or $5 \times 5 \times 5$ conv with rate 2; (2) reduction layer (i.e., max pooling, average pooling or $3 \times 3 \times 3$ conv with stride 2; (3) expansion layer (i.e., transpose conv or trilinear interpolation). We define the SBB in encoder to be the combination of normal and reduction layers for yielding higher level feature maps. The SBB in the decoder has normal and expansion layers to recover spatial resolution of feature maps.

Specifically, in each layer of SBB, we select from all candidate operations. We formulate this selection as a search process: $y = \sum_{i=1}^{n} \alpha_i O_i(x)$, where α_i is a learnable weight. Given a layer, we denote O as a set of candidate operations.

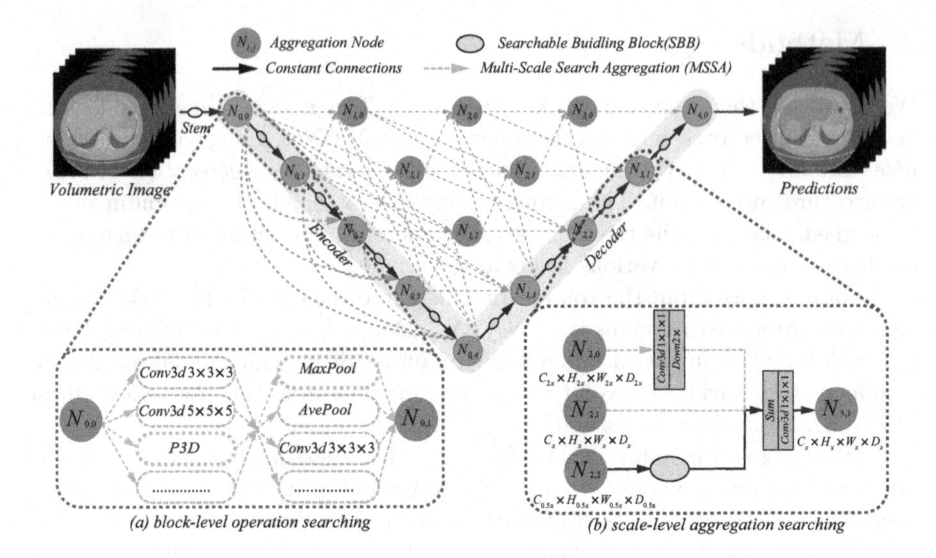

Fig. 3. The overall searching space of UXNet, our goal is to implement scale-wise aggregation searching over multi-stage to learn effective feature aggregation for precise segmentation in (b). Besides, the block-wise operation searching is also conducted to search optimum operations for each layer in (a). Blue circle and yellow oval represent the feature aggregation node and the network building block respectively. (Color figure online)

x is the input feature map to each candidate. Here, the feature map y plays as a weighted summation of outputs of all candidates. During the network training, we optimize learnable weights by using the continuous relaxation [2], where the operation having the maximum weight is selected.

2.2 Multi-scale Search Aggregation

We propose MSSA to learn to connect/disconnect the information propagation pathways. As illustrated in Fig. 3, we search aggregations over multiple stages. At each stage, all levels of feature maps have searchable connections (see the feature aggregation module in Fig. 3(b)) with feature maps at the next stage. To compute the feature map $N_{i,j}$ at the ith stage of the jth level, we use searchable connections to combine the multi-scale information as:

$$N_{i,j} = \begin{cases} \sum_{k=0}^{j-1}(\sigma(\beta_{i,k\rightarrow i,j})T(N_{i,k})) & i = 0 \\ \sum_{k=0}^{L-1}(\sigma(\beta_{i-1,k\rightarrow i,j})T(N_{i-1,k})) & i > 0 \end{cases} \quad (1)$$

where the weight β is learnable weight of each searchable connection, we use σ function to map it to $[0, 1]$, which is used to measure the importance of the propagated feature, and more larger score means the corresponding connection is more important. The $i = 0$ indicates the aggregations are occurred in the encoder

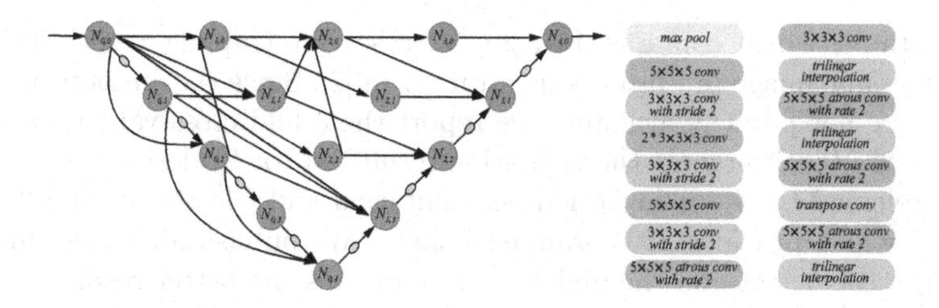

Fig. 4. The network discovered by our UXNet on the brain dataset of MSD challenge.

network, and L represents the number of candidate features in last stage. The T means a series of transformation to align feature maps with different level, we use a sum operator to aggregate all input features. Figure 1 shows that our multi-scale search space is able to cover existing human-designed networks for medical segmentation.

Furthermore, we minimize the discretization gap with a simple regulation to improve the sparsity, pushing the connection weight to the extreme (i.e., 0 or 1). That is, we aim to either enable or disable a specific pathway. The regulation is formulated as:

$$\mathcal{J} = -(\sigma(\beta) - 0.5)^2 \tag{2}$$

After searching the connections for aggregating multi-scale information, we prune the searchable pathway to satisfy the rules as follows (as illustrated in Fig. 1). (a) A threshold τ is pre-defined, when $\sigma(\beta) \geq \tau$, we keep the corresponding connections. (b) Once the aggregation node has no connection with features in the next stage, we delete the relevant connections.

2.3 Network Training

All of the learnable weights (i.e., α in SBB and β in MSSA) can be optimized efficiently using gradient descent. To further avoid overfitting the training data, we follow the bi-level optimization policy in [2,13]. That is, we divide the training data into two sets *trainA* and *trainB*, which are used to train the set of operation weights w (e.g., convolutional layers) and the set of connection weights α, β, respectively. We solve the objective $\mathcal{L}_{trainA}(w, \alpha, \beta)$ for w, by fixing the connection weights. Then we fix the operation weights and solve the objective $\mathcal{L}_{trainB}(w, \alpha, \beta)$ and $\mathcal{J}_{trainB}(\beta)$ for the connection weights. Here, $\mathcal{L}_{trainA}(w, \alpha, \beta)$ represents the cross-entropy loss. $\mathcal{L}_{trainB}(w, \alpha, \beta)$ and $\mathcal{J}_{trainB}(\beta)$ denotes the L2-regulation (see Eq. 2).

3 Experiments

3.1 Datasets and Settings

Following previous NAS-based methods, we evaluate UXNet on three subset of 3D Medical Segmentation Decathlon (MSD) challenge [16] (i.e. the brain, heart, and prostate), which contains 484, 20, 32 cases for training respectively. We adopt the same image pre-processing strategy in [7]. Since the annotation of test datasets are not publicly available, we report the 5-fold cross-validation results as in [1,7,8]. We also report the validation results on the 2D lesion segmentation dataset released by the Skin Lesion Segmentation and Classification 2018 challenge [5], which provides 2594 training images. We publish all results in terms of the dice coefficient and the higher score indicates the better result.

When learning network weights w in Sect. 2.3, we use Adam optimizer with an initial learning rate of 0.0003, and the betas range from 0.9 to 0.99. The initial values of α and β are set as 1 and 0. They are optimized using Adam optimizer with a learning rate of 0.003 and weight. When we finish searching and pruning the network, we retrain the derived network from scratch. The computation of UXNet is cheap by training 1.5 days on two TitanXP GPUs for brain task, which is cheaper than RONASMIS [1] that trained 3.1 days on one RTX 2080Ti GPU and SCNAS [8] that trained one day on four V100 GPUs. Please refer to appendix for more training details of each dataset.

3.2 Ablation Studies

We first conduct the ablation studies on the brain dataset of MSD challenge by removing the critical component of UXNet, i,e., SBB and MSSA. In such case, the model degrades to the original UNnet and achieves the score of 72.5%. It lags far behind our full model in terms of segmentation performance. As shown in Table 1, by adding the SBBs, the performance is promoted to 73.43%. It demonstrates that the SBB enables each layer of the network to produce rich presentation via optimum operation. Note that we adopt the UNet-Style connections to fuse the features in encoder and decoder network in this case. When the MSSA is enabled, the discovered optimal feature aggregation strategy could further improve the score to 74.57%.

We further evaluate the influence of hyper-parameter τ in Sect. 2.2. The higher value indicates the more sparse aggregations. Table 1 shows the results of using different τ to prune the over-connections. we find that using a too low or high threshold to clip the connections will decrease the model's performance. It indirectly illustrates that involving useless or missing useful information will lead to the poor performance.

3.3 Comparison with State-of-the-Arts

Medical Segmentation Decathlon. The MSD challenge contains ten tasks totally, which provide the multimodal images with a vast variant of shapes and locations. In Table 2, we compare our UXNet with state-of-the-art methods in terms of cross-validation results on the brain, heart prostate dataset as in [1,7,8]. We divide the compared methods into two groups. In this first group, the methods adopt human-design architecture for segmentation. Compared to this kind of methods, our UXNet achieves significant improvement by searching a optimal architecture automatically. Some visualization comparisons are also

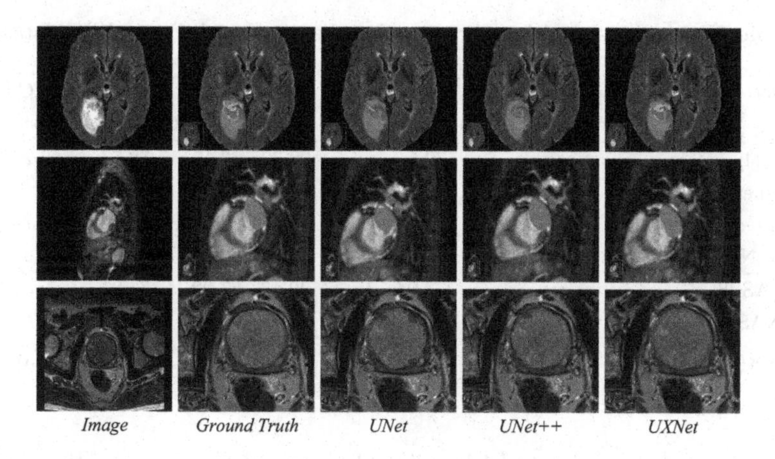

Fig. 5. The segmentation results of the UNet [15], UNet++ [22] and our UXNet on some challenging cases. Our method achieves more accurate segmentation results.

Table 1. Ablations studies of different setting on the brain dataset of MSD.

SBB	MSSA	τ	Dice (%)
×	×	-	72.50
✓	×	-	73.43
×	✓	0.75	73.65
✓	✓	0.60	73.90
✓	✓	0.75	**74.57**
✓	✓	0.90	74.35

available in Fig. 5. In this second group, the methods also apply the NAS technique to search architecture. They mainly focus on designing various searching strategies by using reinforcement learning or differentiable manner but not the task-specific searching space design. Thus, these kinds of methods did not consider the feature aggregation strategies into search space, limited the flexibility of the models to recognize the organs with narrow areas compared with proposed UXNet. For a fair comparison, we report the result of UXNet by using a single model with basic data augmentation (i.e., sliding window, flipping). The searched architecture still outperforms these state-of-the-art NAS-based methods. The Fig. 4 visualizes the best architecture found on the brain dataset. We observed that the atrous convolution with the large kennel is heavily used, suggesting the importance of learning with the large receptive field. It can also be seen that the low-level feature is more used compared to other methods, which indicate the importance of spatial details for medical segmentation.

Skin Lesion Segmentation and Classification. We also evaluate our UXNet on the 2D medical image segmentation. In Table 2, the results of our method still

Table 2. Comparison with different approaches on the MSD and ISIC dataset

Model	Auto search		MSD [16]				ISIC [16]
	Block	Aggregation	Brain Tumor	Heart	Prostate	Average	Lesion
Unet [4]	×	×	72.5	90.70	73.13	78.77	86.2
NNUnet [7]	×	×	74.00	92.70	74.54	80.41	-
Unet++ [22]	×	×	72.66	91.56	72.95	79.05	86.7
U-ResNet [8]	×	×	71.61	89.60	63.77	75.00	88.0
SCNAS [8]	✓	×	72.04	90.47	67.92	76.81	-
RONASMIS [1]	✓	×	74.14	92.72	75.71	80.85	-
UXNet($\tau = 0.75$)	✓	✓	**74.57**	**93.50**	**76.36**	**81.48**	**89.6**

outperforms other conventional baselines by a significant margin. Such results demonstrate the versatility of proposed UXNet on various segmentation tasks.

4 Conclusion

In this paper, we propose a general framework of neural architecture search for 3D medical image segmentation, termed UXNet, which searches scale-wise feature aggregation strategies as well as block-wise operators in the encoder-decoder path. The careful designed searching space achieves the robust segmentation results. In addition, the discovered segmentation architecture reveals the properties of information propagation for the specific dataset. The further will explore UXNet under the real-word hardware constraints, such as memory, speed, and power consumption. In addition, the task-oriented searching space design will also be a potential research direction.

Acknowledgments. This work was supported in part by the Key Area R&D Program of Guangdong Province with grant No. 2018B030338001, by the National Key R&D Program of China with grant No. 2018YFB1800800, by Natural Science Foundation of China with grant NSFC-61629101, by Guangdong Zhujiang Project No. 2017ZT07X152, ßby Shenzhen Key Lab Fund No. ZDSYS201707251409055, by Open Research Fund from Shenzhen Research Institute of Big Data No. 2019ORF01005, by NSFC-Youth 61902335, and by the General Research Fund No. 27208720, by STCSM (19511121400).

References

1. Bae, W., Lee, S., Lee, Y., Park, B., Chung, M., Jung, K.-H.: Resource optimized neural architecture search for 3D medical image segmentation. In: Shen, D., et al. (eds.) MICCAI 2019. LNCS, vol. 11765, pp. 228–236. Springer, Cham (2019). https://doi.org/10.1007/978-3-030-32245-8_26
2. Cai, H., Zhu, L., Han, S.: ProxylessNAS: direct neural architecture search on target task and hardware. arXiv preprint arXiv:1812.00332 (2018)

3. Chlebus, G., Schenk, A., Moltz, J.H., van Ginneken, B., Hahn, H.K., Meine, H.: Automatic liver tumor segmentation in CT with fully convolutional neural networks and object-based postprocessing. Sci. Rep. **8**(1), 15497 (2018)
4. Çiçek, Ö., Abdulkadir, A., Lienkamp, S.S., Brox, T., Ronneberger, O.: 3D U-net: learning dense volumetric segmentation from sparse annotation. In: Ourselin, S., Joskowicz, L., Sabuncu, M.R., Unal, G., Wells, W. (eds.) MICCAI 2016. LNCS, vol. 9901, pp. 424–432. Springer, Cham (2016). https://doi.org/10.1007/978-3-319-46723-8_49
5. Hardie, R.C., Ali, R., De Silva, M.S., Kebede, T.M.: Skin lesion segmentation and classification for isic 2018 using traditional classifiers with hand-crafted features. arXiv preprint arXiv:1807.07001 (2018)
6. Ibtehaz, N., Rahman, M.S.: MultiResUNet: rethinking the U-net architecture for multimodal biomedical image segmentation. Neural Netw. **121**, 74–87 (2020)
7. Isensee, F., et al.: NNU-net: self-adapting framework for u-net-based medical image segmentation. arXiv preprint arXiv:1809.10486 (2018)
8. Kim, S., et al.: Scalable neural architecture search for 3D medical image segmentation. In: Shen, D., et al. (eds.) MICCAI 2019. LNCS, vol. 11766, pp. 220–228. Springer, Cham (2019). https://doi.org/10.1007/978-3-030-32248-9_25
9. Kirschner, M., Jung, F., Wesarg, S.: Automatic prostate segmentation in mrimages with a probabilistic active shape model. MICCAI Grand Challenge: Prostate MR Image Segmentation (2012)
10. Li, X., Chen, H., Qi, X., Dou, Q., Fu, C.W., Heng, P.A.: H-DenseUNet: hybrid densely connected UNet for liver and tumor segmentation from CT volumes. IEEE Trans. Med. Imaging **37**(12), 2663–2674 (2018)
11. Lin, D., Zhang, R., Ji, Y., Li, P., Huang, H.: SCN: switchable context network for semantic segmentation of RGB-D images. IEEE Trans. Cybern. (2018)
12. Liu, C., et al.: Auto-deeplab: hierarchical neural architecture search for semantic image segmentation. In: Proceedings of the IEEE Conference on Computer Vision and Pattern Recognition, pp. 82–92 (2019)
13. Liu, H., Simonyan, K., Yang, Y.: Darts: differentiable architecture search. arXiv preprint arXiv:1806.09055 (2018)
14. Long, J., Shelhamer, E., Darrell, T.: Fully convolutional networks for semantic segmentation. In: Proceedings of the IEEE Conference on Computer Vision and Pattern Recognition, pp. 3431–3440 (2015)
15. Ronneberger, O., Fischer, P., Brox, T.: U-net: convolutional networks for biomedical image segmentation. In: Navab, N., Hornegger, J., Wells, W.M., Frangi, A.F. (eds.) MICCAI 2015. LNCS, vol. 9351, pp. 234–241. Springer, Cham (2015). https://doi.org/10.1007/978-3-319-24574-4_28
16. Simpson, A.L., et al.: A large annotated medical image dataset for the development and evaluation of segmentation algorithms. arXiv preprint arXiv:1902.09063 (2019)
17. Tran, D., Bourdev, L., Fergus, R., Torresani, L., Paluri, M.: Learning spatiotemporal features with 3D convolutional networks. In: Proceedings of the IEEE International Conference on Computer Vision, pp. 4489–4497 (2015)
18. Weng, Y., Zhou, T., Li, Y., Qiu, X.: NAS-Unet: neural architecture search for medical image segmentation. IEEE Access **7**, 44247–44257 (2019)
19. Yu, L., Yang, X., Chen, H., Qin, J., Heng, P.A.: Volumetric convnets with mixed residual connections for automated prostate segmentation from 3D MR images. In: Thirty-First AAAI Conference on Artificial Intelligence (2017)
20. Yu, Q., et al.: C2FNAS: coarse-to-fine neural architecture search for 3D medical image segmentation. arXiv preprint arXiv:1912.09628 (2019)

21. Zhang, R., Yang, W., Peng, Z., Wei, P., Wang, X., Lin, L.: Progressively diffused networks for semantic visual parsing. Pattern Recogn. **90**, 78–86 (2019)
22. Zhou, Z., Rahman Siddiquee, M.M., Tajbakhsh, N., Liang, J.: UNet++: a nested U-net architecture for medical image segmentation. In: Stoyanov, D., et al. (eds.) DLMIA/ML-CDS - 2018. LNCS, vol. 11045, pp. 3–11. Springer, Cham (2018). https://doi.org/10.1007/978-3-030-00889-5_1
23. Zhu, Z., Liu, C., Yang, D., Yuille, A., Xu, D.: V-nas: neural architecture search for volumetric medical image segmentation. In: 2019 International Conference on 3D Vision (3DV), pp. 240–248. IEEE (2019)

Difficulty-Aware Meta-learning for Rare Disease Diagnosis

Xiaomeng Li[1,2](✉), Lequan Yu[1,2], Yueming Jin[1], Chi-Wing Fu[1], Lei Xing[2], and Pheng-Ann Heng[1,3]

[1] Department of Computer Science and Engineering, The Chinese University of Hong Kong, Shatin, Hong Kong

[2] Department of Radiation Oncology, Stanford University, Stanford, USA
xmengli999@gmail.com

[3] Shenzhen Key Laboratory of Virtual Reality and Human Interaction Technology, Shenzhen Institutes of Advanced Technology, Chinese Academy of Sciences, Shenzhen, China

Abstract. Rare diseases have extremely low-data regimes, unlike common diseases with large amount of available labeled data. Hence, to train a neural network to classify rare diseases with a few per-class data samples is very challenging, and so far, catches very little attention. In this paper, we present a difficulty-aware meta-learning method to address rare disease classifications and demonstrate its capability to classify dermoscopy images. Our key approach is to first train and construct a meta-learning model from data of common diseases, then adapt the model to perform rare disease classification. To achieve this, we develop the difficulty-aware meta-learning method that dynamically monitors the importance of learning tasks during the meta-optimization stage. To evaluate our method, we use the recent ISIC 2018 skin lesion classification dataset, and show that with only five samples per class, our model can quickly adapt to classify unseen classes by a high AUC of 83.3%. Also, we evaluated several rare disease classification results in the public Dermofit Image Library to demonstrate the potential of our method for real clinical practice.

1 Introduction

Deep learning methods have become the de facto standard for many medical imaging analysis tasks, *e.g.*, anatomical structure segmentation and computer-aided disease diagnosis. One reason of the success is due to a large amount of labeled data to support the network training. Yet, there are about 7,000 known rare diseases [4] that typically catch little attention and the data is difficult to obtain. These conditions collectively affect about 400 million people worldwide [5] and were generally neglected by the medical imaging community. Taking the retinal diseases as an example, the glaucoma, diabetic retinopathy, age-related macular degeneration, and retinopathy of prematurity are relatively common in the clinical practice. Whereas, other retinal diseases, such as fundus pulverulentus and fundus albipunctatus are rare [13]. Generally, it is difficult to collect data for these rare diseases and obtain annotations from experienced

A. L. Martel et al. (Eds.): MICCAI 2020, LNCS 12261, pp. 357–366, 2020.
https://doi.org/10.1007/978-3-030-59710-8_35

physicians. This phenomenon raises the following question: *given the extremely low-data regime of rare diseases, can we transfer the inherent knowledge learned from the common diseases to support the automated rare disease diagnosis?*

A vanilla solution is transfer learning, *i.e.*, fune-tuning, a widely-used and effective approach, where the model is usually transferred from one dataset to another smaller one. However, fine-tuning a model on an extremely low-data regime will severely overfit to the few given data, *i.e.*, *less than five samples per class*; see the training curves shown in Fig. 3 in the experiments. Another possible solution to alleviate overfitting problem is to use a pretrained network to extract features, and then utilize another classifiers, *e.g.*, support vector machine (SVM) and k-nearest neighbors (KNN), to perform the classification. However, a series of experiments in Sect. 3 show that these methods have limited capacity in the classification of rare diseases. Recently, Zhao *et al.* [15] and Mondal *et al.* [12] tackle the low-data regime related issues, *i.e.*, one-shot or few-shot segmentation problems, in the medical image domain. However, both works relies on the large amount of unlabeled images, which are not appropriate to handle rare disease diagnosis. Maicas *et al.* [10] present a meta-learning method to learn a good initialization on a series of tasks, which can be used to pre-train medical image analysis models. In this regard, developing effective techniques for rare disease diagnosis in low-data regime is of vital importance.

Meta-learning techniques, or learning to learn, is the science of systematically observing how an algorithm performs on a wide range of learning tasks, and then learning from this experience, *i.e.*, meta-data, to learn new tasks much faster. In general, meta-learning updates a network by equally treating (averaging) the gradient directions of different randomly-sampled tasks. So, the meta-learning process often stops at a stage, at which "easy tasks" are well-learned and "difficulty tasks" are still being misclassified. This hinders the meta-learning and affects the results on rare disease classification, in which the rare disease samples are unseen in the training. This observation motivates us to consider a more effective meta-learning optimization. To this end, we propose a novel difficulty-aware meta-learning (DAML) method. Our method first train a meta-classifier on a series of related tasks (*e.g.*, common disease classification), instead of an individual single task, such that the transferable internal representations with the gradient-based learning rule can make rapid progress on the new tasks (*e.g.*, rare disease classification). More importantly, we discover that the contribution of each task sample to the meta-objective is various. To better optimize the meta-classifier, a dynamic modulating function over the learning tasks is formulated, where the function automatically down-weights the well-learned tasks and rapidly focuses on the hard tasks. Our method is evaluated on the ISIC 2018 Skin Lesion Dataset [1,9], where the data are annotated with seven lesion categories. Only training on the four skin lesion classes, our method achieves a promising result on classifying other three unseen classes, with an AUC of 83.3% under five samples setting. We also validate our method on several rare disease classification tasks using the public Dermofit Image Library[1] and achieve a high

[1] https://licensing.eri.ed.ac.uk/i/software/dermofit-image-library.html.

AUC of 82.67% under the five samples setting, demonstrating the potential of our method for real clinical practice.

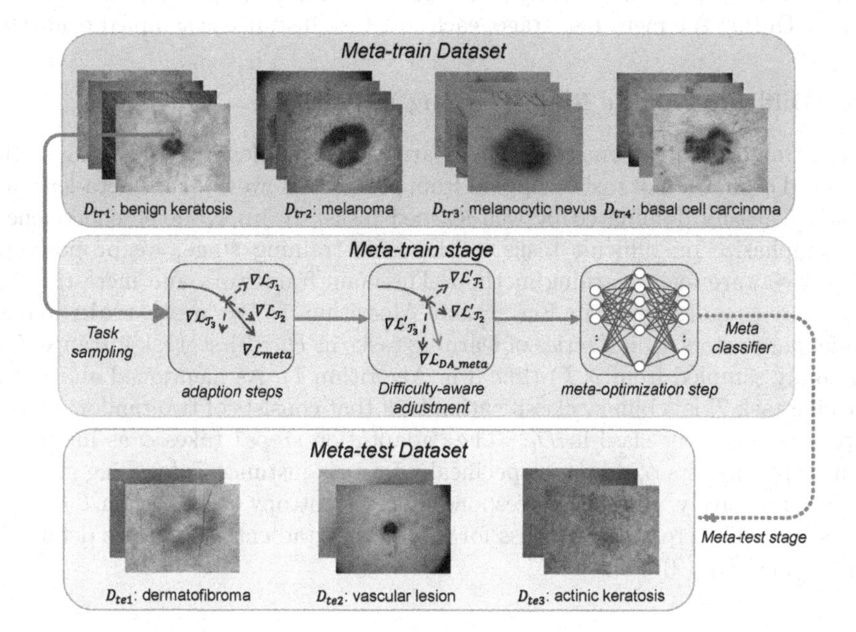

Fig. 1. The pipeline of our proposed difficulty-aware meta-learning (DAML) system. The meta-classifier (neural network) is explicitly trained on the meta-train dataset, such that given new tasks with only a few samples, the meta-classifier can rapidly adapt to the new tasks with a high accuracy. Our novel difficulty-aware meta-optimization scheme can dynamically down-weight the contribution of easy tasks and focus more to learn from hard tasks.

2 Method

We aim to train a neural network using meta-train data (common diseases), such that given new tasks associated with few data samples (meta-test data for rare diseases), we can quickly adapt the network model via a few steps of gradient descent to handle the new tasks; see Fig. 1 for the pipeline.

2.1 Problem Setting

We employ the ISIC 2018 Skin Lesion Analysis Towards Melanoma Detection Dataset [7,11], which has a total of 10,015 skin lesion images from seven skin diseases, including melanocytic nevus (6705), melanoma (1113), benign keratosis (1099), basal cell carcinoma (514), actinic keratosis (327), vascular lesion (142) and dermatofibroma (115). We simulate the problem by utilizing the four classes with largest amount of cases as common diseases (*i.e.*, meta-train dataset D_{tr}) and the left three classes as the rare diseases (*i.e.*, meta-test dataset D_{te}).

Task instance $\mathcal{T}_i$ is randomly sampled from distribution over tasks $p(\mathcal{T})$ and $D_{tr}, D_{te} \in p(\mathcal{T})$. During meta-train stage, learning task $\mathcal{T}_i$ are binary classification tasks and each task consists of two random classes with k samples per class in D_{tr}. During the meta-test stage, each test task instance is sampled from D_{te}.

2.2 Difficulty-Aware Meta-learning Framework

Current model-agnostic meta-earning learns the meta-classifier according to the averaged evaluation of tasks sampled from $p(\mathcal{T})$ [2]. However, this meta-learning process is easily dominated by well-learned tasks. To improve the effectiveness and emphasize on difficult tasks in the meta-training stage, we propose the difficulty-aware meta-learning method. The main framework and meta-training procedure are described in Fig. 1 and Algorithm 1. We **meta-train** a base model parameters ϕ on a series of learning tasks in D_{tr}. First, task instance $\mathcal{T}_i$ is randomly sampled from $p(\mathcal{T})$ (line 3 in Algorithm 1). As mentioned above, the learning task $\mathcal{T}_i$ is a binary classification task that consists of two random classes with k samples per class in D_{tr}. The "adaptation steps" takes ϕ as input and returns parameters ϕ_i' adapted specifically for task instance $\mathcal{T}_i$ by using gradient descent iteratively, with the corresponding cross entropy loss function $\mathcal{L}_{\mathcal{T}_i}$ for $\mathcal{T}_i$ (lines 5–8). The cross-entropy loss for $\mathcal{T}_i$ and the gradient descent are defined in Eq. (1) and Eq. (2).

$$\mathcal{L}_{\mathcal{T}_i}(f_{\phi_i}) = - \sum_{x_j, y_j \sim \mathcal{T}_i} y_j \log(f_{\phi_i}(x_j)) + (1 - y_j) \log(1 - f_{\phi_i}(x_j)), \tag{1}$$

$$\phi_i' \leftarrow \phi_i - \gamma \nabla_{\phi} \mathcal{L}_{\mathcal{T}_i}(f_{\phi_i}), \tag{2}$$

where γ is the inner loop adaptation learning rate and ϕ_i' is the adapted model parameters for task $\mathcal{T}_i$.

After the "adaptation steps", the model parameters are trained by optimizing for the performance respect to tasks adapted in the "adaptation steps". Concretely, a dynamically scaled cross-entropy loss **over the learning tasks** is formulated, where it automatically down-weights the easy tasks and focuses on hard tasks; as shown in Fig. 2. Formally, the difficulty-aware meta optimization function is defined as

$$\mathcal{L}_{DA\_meta} = \sum_{\mathcal{T}_i \sim p(\mathcal{T})} -\mathcal{L}_{\mathcal{T}_i}{}^{\eta} \log(\max(\epsilon, 1 - \mathcal{L}_{\mathcal{T}_i})), \tag{3}$$

$$\phi \leftarrow \phi - \alpha \nabla_{\phi} \sum_{\mathcal{T}_i \sim p(\mathcal{T})} \mathcal{L}_{\mathcal{T}_{DA\_meta}}, \tag{4}$$

where $\mathcal{L}_{\mathcal{T}_i}$ is the original cross entropy loss for task $\mathcal{T}_i$, η is a scaling factor and ϵ is a smallest positive integer satisfying $\max(\epsilon, 1 - \mathcal{L}_{\mathcal{T}_i}) > 0$. Then the meta-classifier is updated by performing Eq. (4). The whole meta-training procedure can be found in Algorithm 1.

Algorithm 1. Meta Learning Algorithm

Require: D_{tr}: Meta-train dataset
Require: α, γ: meta learning rate, inner-loop adaptation learning rate
1: Randomly initialize network weight ϕ
2: **while** not converged **do**
3: Sample batch of tasks from D_{tr}
4: **for** task $\mathcal{T}_i$ in batch **do**
5: **for** number of adaptation steps **do**
6: Evaluate $\mathcal{L}_{\mathcal{T}_i}(f_{\phi_i})$ with respect to k samples using Eq. (1).
7: Compute gradient descent for $\mathcal{T}_i$ using Eq. (2).
8: **end for**
9: Evaluate $\mathcal{L}_{DA\_meta}(f_{\phi_i})$ using Eq. (3).
10: **end for**
11: Update network weight ϕ using Eq. (4).
12: **end while**
13: **return** ϕ;

Intuitively, our difficulty-aware meta-loss dynamically reduces the contribution from easy tasks and focuses on the hard tasks, which in turn increases the importance of optimizing the misclassified tasks. For example, with $\eta = 3$, a task with cross-entropy loss 0.9 would have higher loss in the meta-optimization stage, while a task with lower loss would be given less importance. We analyze the effect of different values for hyperparameter η in the following experiments.

2.3 Meta-training Details

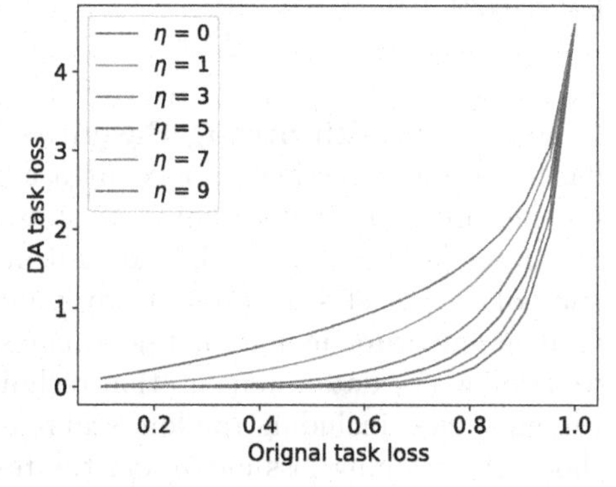

Fig. 2. Visualization of Eq. (3). The original task loss could be infinite. For clear visualization, we show the original task loss within [0, 1].

We employed the 4 conv blocks as the backbone architecture [14], which has 4 modules with a 3×3 conv and 64 filters, followed by a BN, a ReLU, and a 2×2 max-pooling. We used Adam optimizer with a meta-learning rate of 0.001 and divide by 10 for every 150 epochs. We totally trained 3000 iterations and the adopted the difficulty-aware optimization at around 1500 iterations. The batch size is 4, consisting of 4 tasks sampled from meta-train dataset. Each task consists of randomly k samples from 2 classes. We query 15 images from each of two classes to adapt parameters for $\mathcal{T}_i$, as the same protocol in [2,8,14]. During meta-test stage, the inference is performed by randomly sampling k samples from 2 classes from meta-test dataset, *i.e.*, D_{te}. The final report results is the averaged AUC over 30 runs.

3 Experiments and Results

We conduct experiments on ISIC 2018 skin lesion classification dataset[2]. We first compare our method with some strong baselines, *i.e.*, fine-tuning, feature extraction+classifiers, to validate the effectiveness of our method for classification in the extremely low-data regime. Then, we compare with other related methods to show the effectiveness of our novel difficulty-aware meta-learning framework. Next, we analyze the improvement of our proposed difficulty-aware meta-optimization loss, the effect of different scaling factors, as well as the importance of network architecture and data augmentation.

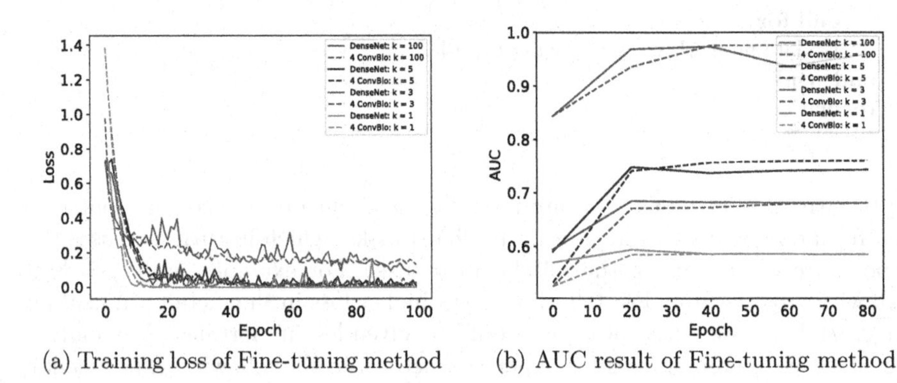

(a) Training loss of Fine-tuning method (b) AUC result of Fine-tuning method

Fig. 3. The training curve (a) and test AUC (b) when performing fine-tuning on meta-test dataset.

Comparison with Strong Baselines. We first show the results of the standard fine-tuning method on rare disease classification for comparison. We report the performance of fine-tuning with two different network architectures: the DenseNet [3] and 4 Conv Blocks [14] in Table 1, where the former one is the state-of-the-art classification network and the latter one is the most commonly used architecture in few-shot learning setting in computer vision. Note that we employ the data augmentation technique on both the pre-training and fine-tuning stages, including random scaling, rotation, and mirror flipping. Figure 3 shows the training loss curve and the test AUC performance when performing fine-tuning on few given samples. Each plot is the average over 30 runs. It is observed that when k is large, the fine-tuning could perform well and achieves the AUC at around 97%. However, with a smaller k, the model converges rapidly but the AUC performance reduces devastatingly. For example, the 4 Conv Blocks only achieves 58.49% AUC with $k = 1$. We also employ the pre-trained DenseNet as feature extractor and utilize another classifiers, *e.g.*, SVM and KNN, to coduct rare disease classification. As shown in Table 1, the performance of these

[2] https://challenge2018.isic-archive.com/task3/.

Table 1. The AUC performance of different methods on skin lesion dataset. Each result is averaged over 30 runs.

	Backbone	Sample #	AUC	Sample #	AUC	Sample #	AUC
ConvFeature + KNN	DenseNet	1	50.00%	3	56.07%	5	62.69%
ConvFeature + SVM	DenseNet	1	61.46%	3	61.68%	5	67.44%
Finetune + Aug	DenseNet	1	58.57%	3	68.05%	5	73.65%
Finetune + Aug	4 Conv Blocks	1	58.49%	3	68.13%	5	75.90%
Relation Net [14]	4 Conv Blocks	1	59.97%	3	62.87%	5	72.40%
MAML [2]	4 Conv Blocks	1	63.77%	3	77.98%	5	81.20%
Task sampling [10]	4 Conv Blocks	1	64.21%	3	78.40%	5	82.05%
DAML (ours)	4 Conv Blocks	1	**67.33%**	3	**79.60%**	5	**83.30%**

methods is inferior, indicating that these methods have limited capacity in tackling the classification task with just a few samples from unseen class. In other aspect, our method achieves 83.30% AUC with $k = 5$ and 79.60% AUC with $k = 3$, demonstrating the effectiveness of our method for disease classification in the extremely low-data regime.

Comparison with Other Methods. We also report the performance of the widely used few shot learning approaches, Relation Net [14], MAML [2] and Task sample [10] for the rare disease classification task in Table 1. Our method surpasses these three approaches, especially the Relation Net. The reason may be that our meta-train dataset only has four classes, limiting the representation capability of the feature extraction in the metric-based approach. It is worth noting that the meta-learning based approaches excels all the baseline methods and metric-based method, demonstrating the promising results of meta-learning approach for rare disease classification. Overall, our method further improves the original MAML method, which demonstrates the effectiveness of our difficulty-aware meta optimization procedure. Maicas *et al.* [10] proposes a meta-learning method that addresses the task sampling issue. From Table 1, we can see our method achieves better results than task sampling method [10] under all sample settings.

Ablation Study of Our Method. We provide detailed analysis on the effects of our difficulty-aware meta-optimization loss, network architectures and data augmentation under 5 samples setting, as shown in Table 2. First, we analyze the effects of η in our difficulty-aware loss. We found that our method can obviously improve the overall performance of the meta-learning and the performance is best when $\eta = 5$. We then explore the importance of network architecture. "conv blocks + 64f + Aug" refers to architectures consisting of 4 conv blocks with 64 feature maps with heavy data augmentation. More complicated networks, *i.e.*, residual blocks, would severely lead to the overfitting problem; as the comparison shown in Table 2. Moreover, the heavy data augmentation, *i.e.*, random rotation, flipping, scaling with crop, can only slightly improve the AUC results from 79.3% to 81.2%.

Table 2. Ablation study results under 5 samples setting.

Experiments setting	AUC
ResBlocks + 32f + no Aug	75.8%
ConvBlocks + 64f + no Aug	79.3%
ConvBlocks + 64f + Aug	81.2%
ConvBlocks + 64f + Aug, $\eta = 1$	81.7%
ConvBlocks + 64f + Aug, $\eta = 3$	82.4%
ConvBlocks + 64f + Aug, $\eta = 5$	**83.3%**
ConvBlocks + 64f + Aug, $\eta = 7$	83.1%

Table 3. Results of our method and other method for rare disease classification on the Public Dermofit Image Library.

	Backbone	Sample	AUC	Sample	AUC	Sample	AUC
MAML [2]	4 Conv Blocks	1	63.00%	3	74.03%	5	80.70%
Task sampling [10]	4 Conv Blocks	1	63.20%	3	76.10%	5	81.80%
DAML (ours)	4 Conv Blocks	1	**63.33%**	3	**77.15%**	5	**82.76%**

Validation on Real Clinical Data. We validated our method for rare diseases classification in the real clinical data from public Dermofit Image Library[3]. The disease classes we employed are *squamous cell carcinoma*, *haemangioma*, and *pyogenic granuloma*. As shown in Table 3, our method outperforms other related meta-learning methods. Our method achieves an average AUC of 82.67% under five samples setting, demonstrating the potential usage of our method for real clinical applications.

Discussions on Other Applications. Our method has the potential to be applied to other related applications such as federated learning. Medical data usually has privacy regulations, hence, it is often infeasible to collect and share patient data in a centralized data lake. This issue poses challenges for training deep convolutional networks, which often require large numbers of diverse training examples. Federated learning provides a solution, which allows collaborative and decentralized training of neural networks without sharing the patient data. For example, Li *et al.* [6] implemented and evaluated practical federated learning systems for brain tumor segmentation. Our method is also feasible to be tested for federated learning since our method only accesses training data (common dataset) during the training stage and can be fast adapted according to the test data (private dataset) during the inference time. Exploring the applications of our method in federated learning would be a future work of this paper.

[3] https://licensing.eri.ed.ac.uk/i/software/dermofit-image-library.html.

4 Conclusion

In this paper, we propose a novel difficulty-aware meta-learning method to tackle the extremely low-data regime problem, *i.e.*, rare disease classification. Our difficulty-aware meta learning approach optimizes the meta-optimization stage by down-weighting the well classified tasks and emphasizing on hard tasks. Extensive experiments demonstrated the superiority our method. Our results excels other strong baselines as well as other related methods. The clinical rare skin disease cases from Dermofit Image Library also validated our method for piratical usage.

Acknowledgements. The work described in the paper was supported in parts by the following grants from Key-Area Research and Development Program of Guangdong Province, China (2020B010165004), Hong Kong Innovation and Technology Fund (Project No. ITS/311/18FP & ITS/426/17FP) and National Natural Science Foundation of China (Project No. U1813204).

References

1. Codella, N.C., et al.: Skin lesion analysis toward melanoma detection challenge. In: IEEE International Symposium on Biomedical Imaging, pp. 168–172. IEEE (2018)
2. Finn, C., Abbeel, P., Levine, S.: Model-agnostic meta-learning for fast adaptation of deep networks. In: ICML, pp. 1126–1135. JMLR.org (2017)
3. Huang, G., Liu, Z., Van Der Maaten, L., Weinberger, K.Q.: Densely connected convolutional networks. In: Proceedings of the IEEE Conference on Computer Vision and Pattern Recognition, pp. 4700–4708 (2017)
4. Jia, J., Wang, R., An, Z., Guo, Y., Ni, X., Shi, T.: Rdad: a machine learning system to support phenotype-based rare disease diagnosis. Front. Genet. **9**, 587 (2018)
5. Khoury, M., Valdez, R.: Rare diseases, genomics and public health: an expanding intersection. Genomics and Health Impact Blog (2016)
6. Li, W., et al.: Privacy-preserving federated brain tumour segmentation. In: Suk, H.-I., Liu, M., Yan, P., Lian, C. (eds.) MLMI 2019. LNCS, vol. 11861, pp. 133–141. Springer, Cham (2019). https://doi.org/10.1007/978-3-030-32692-0_16
7. Li, X., Yu, L., Chen, H., Fu, C.W., Xing, L., Heng, P.A.: Transformation-consistent self-ensembling model for semisupervised medical image segmentation. IEEE Trans. Neural Netw. Learn. Syst. (2020)
8. Li, X., Yu, L., Fu, C.W., Fang, M., Heng, P.A.: Revisiting metric learning forfew-shot image classification. Neurocomputing **408**, 49–58 (2020)
9. Li, X., Yu, L., Fu, C.-W., Heng, P.-A.: Deeply supervised rotation equivariant network for lesion segmentation in dermoscopy images. In: Stoyanov, D., et al. (eds.) CARE/CLIP/OR 2.0/ISIC -2018. LNCS, vol. 11041, pp. 235–243. Springer, Cham (2018). https://doi.org/10.1007/978-3-030-01201-4_25
10. Maicas, G., Bradley, A.P., Nascimento, J.C., Reid, I., Carneiro, G.: Training medical image analysis systems like radiologists. In: Frangi, A.F., Schnabel, J.A., Davatzikos, C., Alberola-López, C., Fichtinger, G. (eds.) MICCAI 2018. LNCS, vol. 11070, pp. 546–554. Springer, Cham (2018). https://doi.org/10.1007/978-3-030-00928-1_62

11. Milton, M.A.A.: Automated skin lesion classification using ensemble of deep neural networks in isic 2018: Skin lesion analysis towards melanoma detection challenge. arXiv preprint arXiv:1901.10802 (2019)
12. Mondal, A.K., Dolz, J., Desrosiers, C.: Few-shot 3d multi-modal medical image segmentation using generative adversarial learning. arXiv preprint arXiv:1810.12241 (2018)
13. Skorczyk-Werner, A., et al.: Fundus albipunctatus. J. Appl. Genet. **56**(3), 317–327 (2015)
14. Sung, F., Yang, Y., Zhang, L., Xiang, T., Torr, P.H., Hospedales, T.M.: Learning to compare: relation network for few-shot learning. In: Proceedings of the IEEE Conference on Computer Vision and Pattern Recognition, pp. 1199-1208 (2018)
15. Zhao, A., Balakrishnan, G., Durand, F., Guttag, J.V., Dalca, A.V.: Data augmentation using learned transformations for one-shot medical image segmentation. In: Proceedings of the IEEE Conference on Computer Vision and Pattern Recognition, pp. 8543–8553 (2019)

Few Is Enough: Task-Augmented Active Meta-learning for Brain Cell Classification

Pengyu Yuan[1(✉)], Aryan Mobiny[1], Jahandar Jahanipour[1,2], Xiaoyang Li[1], Pietro Antonio Cicalese[1], Badrinath Roysam[1], Vishal M. Patel[3], Maric Dragan[2], and Hien Van Nguyen[1]

[1] University of Houston, Houston, TX, USA
pyuan2@uh.edu
[2] National Institutes of Health, Bethesda, MD, USA
[3] Johns Hopkins University, Baltimore, MD, USA

Abstract. Deep Neural Networks (or DNNs) must constantly cope with distribution changes in the input data when the task of interest or the data collection protocol changes. Retraining a network from scratch to combat this issue poses a significant cost. Meta-learning aims to deliver an adaptive model that is sensitive to these underlying distribution changes, but requires many tasks during the meta-training process. In this paper, we propose a tAsk-auGmented actIve meta-LEarning (AGILE) method to efficiently adapt DNNs to new tasks by using a small number of training examples. AGILE combines a meta-learning algorithm with a novel task augmentation technique which we use to generate an initial adaptive model. It then uses Bayesian dropout uncertainty estimates to actively select the most difficult samples when updating the model to a new task. This allows AGILE to learn with fewer tasks and a few informative samples, achieving high performance with a limited dataset. We perform our experiments using the brain cell classification task and compare the results to a plain meta-learning model trained from scratch. We show that the proposed task-augmented meta-learning framework can learn to classify new cell types after a single gradient step with a limited number of training samples. We show that active learning with Bayesian uncertainty can further improve the performance when the number of training samples is extremely small. Using only 1% of the training data and a single update step, we achieved 90% accuracy on the new cell type classification task, a 50% points improvement over a state-of-the-art meta-learning algorithm.

Keywords: Meta learning · Active · Brain cell

1 Introduction

The ability of Deep Neural Networks (or DNNs) to generalize to a given target concept is dependent on the amount of training data used to generate the

A. L. Martel et al. (Eds.): MICCAI 2020, LNCS 12261, pp. 367–377, 2020.
https://doi.org/10.1007/978-3-030-59710-8_36

model. This is a significant limitation, as many real-world classification tasks depend on a limited number of training samples for accurate classification. Various research groups have developed techniques that utilize different underlying principles to address this issue. These techniques include data augmentation [30] and generative models [12,17,38], which try to directly increase the number of training samples. Active learning is another technique which aims to select the most valuable samples to include in the training set [28,29]. Finally, transfer learning allows the model to adapt to new and unseen tasks by using its own previous knowledge, often reducing the amount of data needed for generalization [4,13,25,25].

Medical data is characterized by significant distribution shifts and small samples sets that negatively affect the quality of the generated model. In the cell classification task [37], there are several contributing factors for poor generalization; different biomarkers, unique cell morphologies, variations in stain intensity and image quality all contribute to the variability of the data. Each of these factors could be considered a unique parameter with which to create unique classification tasks. Traditional transfer learning methods pre-trained on various source tasks may not perform well; poor model initialization parameters coupled with unadjusted hyperparameters may cause the model to fall into a bad local minima. Mainstream transfer learning methods also require a time-consuming model retraining process [23]. In recent years, a more advanced model architecture called meta-learning has been developed to address these adaptability issues [34]. Meta-learning approaches try to generate a more robust model that can learn to quickly adapt to new tasks with minimal labeled samples. Although meta-learning does not require many labeled samples for each task, it requires many different tasks to effectively learn how to adapt; this may be a problem when the number of tasks is limited. This technique also selects training samples randomly for each new task, which may negatively impact the strength of the model if it is trained on easier samples.

In this paper, we utilize various strategies to create an adaptive framework called tAsk-auGmented actIve meta-LEarning (or AGILE) which allows the classifier to achieve high performance with few training samples and gradient updates for each new task. The experiments we perform on the brain cell type classification task show that the AGILE classifier can quickly adapt to new cell types by utilizing very few labeled training samples.

2 Related Works

Brain Cell Type Classification. The brain is a highly complex organ made up of myriad different cell types, each with their own unique properties [7]. Gene expression experiments have highlighted cell type composition based on the expression value of markers for five major cell types: neurons, astrocytes, oligodendrocytes, microglia, and endothelial cells [6,22]. Besides that, there are around 50–250 neuronal sub-cell types purported exist [2]. Different cell types may express different biomarkers or unique combinations of biomarkers, with some of then shared with other cell types. Correctly identifying the cell type using these biomarkers is essential for many medical researches such as schizophrenia [31] and brain cell type specific gene expression [22]. It is easy to train a deep neural network to identify one cell type but it is not easy to scale the network for hundreds of new cell types. Our meta-learning framework aims to provide an adaptive model that can rapidly adjust itself to new classification tasks.

Meta-learning. Meta-learning aims to study how meta-learning algorithms can acquire fast adaptation capability from a collection of tasks [3,5,26,34]. Meta-learning often consists of a meta-learner and a learner that learn at two levels of different time scales [14,27]. Santoro et al. Koch et al. [18], Vinyals et al. [35] and Snell et al. [33] proposed to learn a robust kernel function of feature embeddings to illustrate the similarities between different samples. Another popular approach is to directly optimize the meta-learner through the gradient descent [1,9,15,24]. Model agnostic meta-learning (MAML) [9] is proven to be one of the state-of-the-art approaches in the meta-learning field. These methods demonstrated human-level accuracy on many classification tasks. However most of them require a lot of tasks to train the meta-learner and it can not actively select training samples which might not be the optimal case in practice.

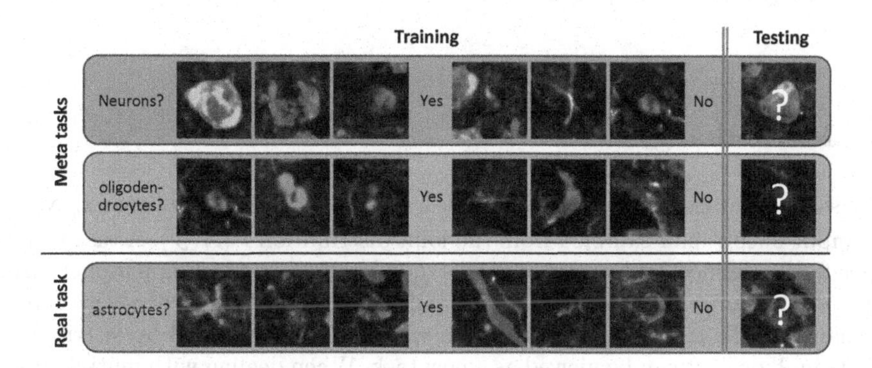

Fig. 1. Multi-task brain cell classification. Each task is a binary classification problem for a specific brain cell type. Cell to be classified is located in the center of the image. The model needs to be adapted to the unseen real task with few training samples.

Active Learning. Active learning has been used to interactively and efficiently query information to achieve optimal performance for the task of interest. These methods select training samples based on information theory [20], ensemble approaches [21], and uncertainty measurements [16,19]. However, these methods may not be effective for deep networks. Gal et al. [10,11], Sener et al. [28] and Fang et al. [8] did a lot of studies in finding heuristics of annotating new samples for deep networks. However, few of them have been applied to domain adaptation or meta-learning problems. Woodward et al. [36] added an active part in one-shot learning but did not utilize the best meta-learning structure. In this paper, we combine the advantages of both meta-learning and active-learning to get a fast adaptive model which can use the fewest data to achieve a good performance.

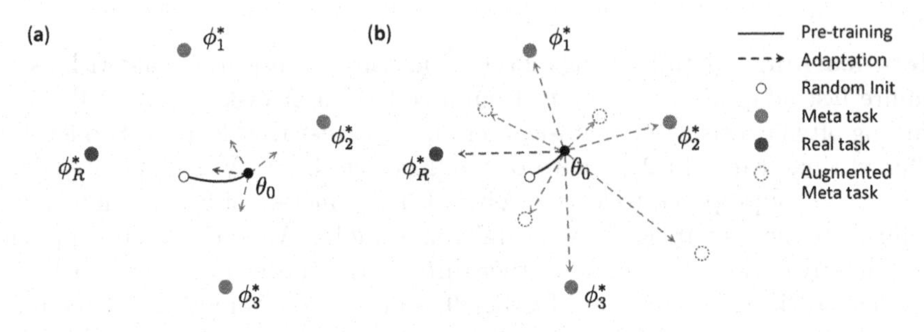

Fig. 2. Comparison of (a) transfer learning and (b) task-augmented meta-learning. (a) After pre-training on very few number of meta tasks, the model is pulled closer to the meta tasks and may be far from the real task. (b) By creating pseudo meta tasks to pre-train the meta-learning model, it gains the ability of adapting to the new tasks with only one gradient step updating.

3 Methodology

Consider a dataset consisting Q samples: $\mathcal{D} = \{(\mathbf{x}_q, \mathbf{y}_q)\}_{q=1}^{Q}$, where $(\mathbf{x}_q, \mathbf{y}_q)$ is an input-output pair sampled from the joint distribution $P(\mathcal{X}, \mathcal{Y})$. A task can be specifically defined by learning a model $f_\phi(\mathbf{x}) : \mathcal{X} \to \mathcal{Y}$ which is parameterized by ϕ to maximize the conditional probability $P_\phi(\mathcal{Y}|\mathcal{X})$. Thus whenever there is a change in the conditional distribution which is mainly caused by distribution shifts in $\mathcal{X}$ or $\mathcal{Y}$, it can be viewed as a new task. When dealing with mutiple tasks drawn from $P(\mathcal{T})$, each task $\mathcal{T}_i$ is associated with a unique dataset $\mathcal{D}_i$. We split these tasks into two parts: meta tasks $\mathcal{T}_{\text{meta}}$ and real tasks $\mathcal{T}_{\text{real}}$. Real tasks mean the model performance on these tasks is what we really care about and Meta tasks are what we used to pre-train the model. If there is no model adaptation, then the meta tasks are not needed. For each task $\mathcal{T}_i$, we have a train/test split $\mathcal{D}_i^{\text{train}}/\mathcal{D}_i^{\text{test}} \subset \mathcal{D}_i$. The goal for task $\mathcal{T}_i$ is to learn a set of parameters ϕ_i from the $\mathcal{D}_i^{\text{train}}$ to get the minimal loss on the test data, i.e. $\mathcal{L}(\phi_i, D_i^{test})$.

Our proposed framework AGILE has two phases, a task-augmented meta-learning method which learns to generate a strong model initialization which is sensitive to data distribution changes by using $\mathcal{T}_{\text{meta}}$, and an active learning process which selects the most informative samples using Bayesian dropout uncertainties when apply the adaptive model on $\mathcal{T}_{\text{real}}$.

3.1 Task-Augmented Meta-learning

The phase I of our AGILE framework is the task-augmented meta-learning module. For the meta-learning setting, we have a learner which operates at fast time-scale and parameterized by $\phi \in \Phi$, and a meta-learner at a slower time-scale parameterized by $\theta \in \Theta$. The goal of meta-learning is to learn meta-parameters θ that can produce good task-specific parameters ϕ for all M tasks after the fast adaptation:

$$\theta^* = \underset{\theta \in \Theta}{\operatorname{argmin}} \frac{1}{M} \sum_{i=1}^{M} \mathcal{L}\left(\mathcal{A}dapt(\theta, \mathcal{D}_i^{\text{train}}), \mathcal{D}_i^{\text{test}}\right) \quad (1)$$

where $\mathcal{A}dapt()$ function is an adaptation step completed by the learner. We employ model agnostic meta-learning (MAML)[9] which initializes the model at θ then updates it using training data for each task $\mathcal{D}_i^{\text{train}}$ as follows:

$$\phi_i \equiv \mathcal{A}dapt(\theta, \mathcal{D}_i^{\text{train}}) = \theta \overset{Z}{-} \alpha \nabla_\theta \mathcal{L}(\theta, \{(\mathbf{x}_k, \mathbf{y}_k)\}_{k=1}^{K}), \quad (\mathbf{x}, \mathbf{y}) \sim \mathcal{D}_i^{\text{train}} \quad (2)$$

where $\mathcal{L}$ is the loss function, α the learning rate for the learner, and $\overset{Z}{-}$ a short-hand notation for running a Z-step gradient descent which is relatively fast. A fixed number of K class-balanced samples $\{(\mathbf{x}_k, \mathbf{y}_k)\}_{k=1}^{K}$ randomly sampled from $\mathcal{D}_i^{\text{train}}$ are used to update the learner for every iteration.

Task Transformation. Meta-learner is trained on the meta tasks. Our experiments show that when the number of meta tasks is not big enough, it can not obtain the general fast adaptation ability on $\mathcal{T}_{\text{real}} \sim P(\mathcal{T})$. Similar as data augmentation, we applied a set of task-augmentation functions $\{\mathcal{G}_l\}_{l=1}^{L} : \mathcal{T} \rightarrow \mathcal{T}'$ on existing tasks to create new tasks. Because the task is specifically determined by the conditional probability $P(\mathcal{Y}|\mathcal{X})$, either change $\mathcal{X}$ or $\mathcal{Y}$ will lead to a new task. The task-augmentation functions we applied are (1) Flipping the label (2) Shuffling the order of input channels (3) Rotating the images. Considering the input image with c biomarkers $\mathbf{x} \in \mathbb{R}^{w \times h \times c}$ and the binary label $y \in \mathbb{R}$, flipping labels is achieved by:

$$y' = z(1-y) + (1-z)y, \quad \text{where } z \sim \text{Bernoulli}(p_f) \quad (3)$$

where p_f is the probability of flipping the label. Shuffling the input channels is selected with a probability of p_s. By Constructing c different one-by-one kernels $\{\mathbf{s}_{ij}\}_{i=1}^{c} \in \mathbb{R}^{1 \times 1 \times c}$ where $\mathbf{s}_{ij} = 1$ only if the jth bio-marker is placed at i^{th} channel after the random shuffling, the shuffled images are obtained by the convolution:

$$\mathbf{x}' = \mathbf{x} * \mathbf{s}_{ij}, \quad i, j = 1, 2, 3 \ldots c \quad (4)$$

The third task augmentation method we used is to rotate the images for $90°, 180°$ or $270°$ with the probability of p_r. In the experiment, we set $p_f = p_s = p_r$. Theoretically any data augmentation techniques can be applied here to slightly change the distribution of $\mathcal{X}$. By increasing the diversity of the meta tasks, the meta-learner can be well trained to extract useful features after adapting to training data in any tasks. The comparison between task augmented meta-learning and plain transfer learning is illustrated in Fig. 2.

3.2 Active Learning with Bayesian Uncertainty

The phase II of AGILE is to apply the pre-trained model with active learning. Previous meta-learning methods consider training and testing the meta-learning algorithm in the exactly same manner, which might not be practical. They fixed the number of training samples for each task. But for the real task $\{\mathcal{T}_{\text{real}}\}_{j=1}^{N}$, this number may vary, especially when users want to annotate some new samples to improve the performance of the model. When the number of training samples is extremely small, how to unsupervisedly select the most informative samples to annotate is an important question. First, we use a random number of samples to update the learner during the meta-learning training process, so that the model is forced to learn with a different training size.

$$\phi_i = \theta \overset{Z}{-} \alpha \nabla_\theta \mathcal{L}(\theta, \{(\mathbf{x}_k, \mathbf{y}_k)\}_{k=1}^{\widetilde{K}})), \quad (\mathbf{x}, \mathbf{y}) \sim \mathcal{D}_i^{\text{train}} \tag{5}$$

where $\widetilde{K}$ is varied between 1 sample from each class to the maximum number of samples K allowed by the label budget. Second, we applied Bayesian dropout [10,32] to get the uncertainties of the predictions over all unlabeled samples, which represents the current belief of the learner on the predictions:

$$H\left(\mathbf{y}|\mathbf{x}, \mathcal{D}_j^{\text{train}}\right) = -\sum_{\mathbf{y} \in \mathcal{Y}} p_{\text{MC}}\left(\mathbf{y}|\mathbf{x}\right) \log p_{\text{MC}}\left(\mathbf{y}|\mathbf{x}\right), \quad \text{where } (\mathbf{x}, \mathbf{y}) \in \mathcal{D}_j^{\text{test}} \tag{6}$$

$P_{\text{MC}}\left(\mathbf{y}|\mathbf{x}\right)$ is the approximation of Bayesian inference with Monte Carlo integration $\frac{1}{T}\sum_{t=1}^{T} p_{\phi_j}\left(\mathbf{y}|\mathbf{x}\right)$, where $p_{\phi_j}\left(\mathbf{y}|\mathbf{x}\right)$ is the conditional probability of predicted class for input $\mathbf{x}$ in task j and T is the number of Monte-Carlo experiments. The higher entropy H is, the higher uncertainty is in this prediction. For this Bayesian deep learning network, dropout is turned on both for the training and test time on $\mathcal{T}_{\text{real}}$ with a drop rate of 0.1. Based on the uncertainties, the model will only select hard samples for training which significantly improves the performance.

4 Experiments and Results

For this brain cell type classification problem, we collected 4000 cells from 5 major cell types imaged from rat brain tissue sections: neurons, astrocytes, oligodendrocytes, microglia, and endothelial cells. There are 800 cells for each cell type. 7 biomarkers are used as the feature channels: DAPI, Histones, NeuN, S100, Olig 2, Iba1 and RECA1. DAPI and Histones are used to indicate the location of the cells while others are biomarkers for classification of specific cell types. Each cell is located in the center of a patch with a size of 100×100 pixels. As shown in Fig. 1, for each cell type, we have a binary classification task,

800 cells from this cell type and 800 from others. The model needs to correctly identify whether the given cell is this cell type. And it has no prior knowledge about the usage of each biomarker. Out of 1600 cells, 60% of them are used as potential training samples and the rest are used for the test. Binary classification for neurons, oligodendrocytes and microglia are considered as the meta tasks. And the real tasks are to identify the remaining two cell types.

Table 1. Methods configuration comparison which differ mainly in the data they use and the training framework. Meta-learning methods are supposed to perform well with few training samples and little training time. (# means the number of)

Methods	Use data			in Real-train		# Meta tasks
	Meta-train	Meta-test	Real-train	# samples	# gradient updates	
Vanilla_limit	–	–	✓	16 (1%)	100	0
Vanilla_full	–	–	✓	960 (60%)	100	0
Transfer	✓	–	✓	16 (1%)	100	3
MAML	✓	✓	✓	16 (1%)	1	3
AGILE(phase I)	✓	✓	✓	16 (1%)	1	Many
AGILE(phase II)	✓	✓	✓	16 (1%)	1	Many
AGILE(phase II)	✓	✓	✓	160 (10%)	1	Many

Table 2. Quantitative results of different methods. Original method use all available training data (60%) and act as the upper bound while task-augmented MAML method get the highest accuracy using very few training data (1%).

Methods (Size %)	Precision	Recall	F1-score	Accuracy($\pm$ Std)	CI_{95}
Vanilla_limit (1%)	0.642	0.622	0.632	0.637($\pm$0.062)	0.632–0.642
Vanilla_full (60%)	0.937	**0.965**	**0.951**	**0.950**($\pm$0.021)	0.948–0.952
Transfer (1%)	0.447	0.433	0.440	0.449($\pm$0.085)	0.449–0.456
MAML (1%)	0.408	0.402	0.405	0.409($\pm$0.030)	0.406–0.412
AGILE(phase I) (1%)	0.791	0.790	0.791	0.791($\pm$0.054)	0.786–0.796
AGILE(phase II) (1%)	0.883	0.926	0.904	0.902($\pm$0.048)	0.898– 0.906
AGILE(phase II) (10%)	**0.950**	0.951	**0.951**	**0.950**($\pm$0.044)	0.946–0.954

We used three other methods for comparison. The first is the Vanilla method which is to train a network from scratch without any adaptation. We further split it into Vanilla_limit and Vanilla_full methods which use limited and all available training data respectively. They are acting like the lower bound and the upper bound of the performances. The second is the transfer learning model, which means the model is pre-trained on 3 meta tasks and then fine-tuned on 2 real takes. Both of these are trained with 100 gradient updates to ensure the model convergence. The third method is the plain MAML model [9] with limited data and only 1 gradient update for the purpose of fast adaptation with few training data. We chose MAML as our meta-learning baseline because it achieved highly competitive performances compared to other meta-learning methods [9]. All configuration differences are shown in Table 1. Note that all methods share the same neural network structure which has 4 convolution blocks. Each block consists of a convolutional layer with 32 filters of size 3 * 3, a batch normalization layer, and a max-pooling layer. Relu activation is used after each convolutional layer. After the last convolution block, a dense layer is added to project the 32 feature maps to only two classes. And all methods have the same learning rate $\alpha = 0.01$ for the classifier. The learning rate for meta-learner is 0.001. Adam optimizer is used for all optimization. Training iterations are set as 12000 for all methods. 12 replaced tasks are used in one iteration during the training on meta-tasks. The only differences between these methods are the data and the training/adaptation framework they use.

The quantitative results on the real tasks are summarized in Table 2. The upper bound we can get by using all training data is 95% classification accuracy while the lower bound is 63.7%. Plain transfer learning method and MAML methods are even worse than the lower bound because they were stuck in the local minimum after pre-trained on only 3 meta tasks. Our proposed task-augmented meta-learning is able to quickly adapt with few training samples and gradient updates. With only 16 training samples, the AGILE method can reach 90% accuracy. With 160 training samples, AGILE reaches the upper bound. The 95% confidence interval are presented in the Table 2.

The adaptation processes are shown in Fig. 3 (**a**). The solid line is the mean value and the shaded area shows the variance. Without any prior knowledge, the original model starts with the classification accuracy around 50%. Transferring an existing model has the lowest starting point because images labeled as "1" in one task should be labeled as "0" in other tasks. The original MAML method also does not help for this brain cell type binary classification problem. Our results show that MAML is unable to adapt effectively due to the small number of training tasks. In contrast, AGILE model has a sharp increase in accuracy after 1 gradient update. Note that although AGILE model is trained for maximal performance after one gradient step, it continues to improve with additional gradient steps.

Figure 3 (**b**) illustrate the relationship between the training size and the performances of AGILE method and the baseline methods. AGILE performs extremely well when the training size is small. Since AGILE selects samples based on their uncertainties, the newly added training samples might not be class-balanced. This is more practical because difficult classes require more examples for training. Our strategy contrasts against current meta-learning setting where all the training set is class-balanced by default. AGILE method reachs validation accuracy of 95% with 160 samples. Meantime, the vanilla model with 100 gradient updates get the accuracy around 84% and the vanilla model with only 1 update basically cannot learn anything.

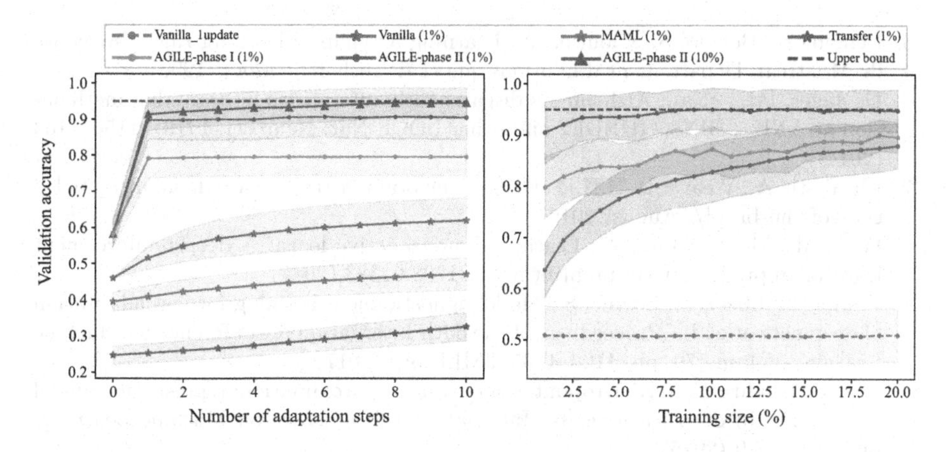

Fig. 3. (a) Comparison of different methods when adapting to new real task with only 1% labeled training data, AGILE method shows its fast adapting ability. Upper bound is obtained by training on all 60% training data with Vanilla method. (b) The impact of the training size on the AGILE method and the vanilla method.

5 Conclusion

In this paper, we proposed a fast adaptation framework AGILE combining data augmentation, meta learning and active learning to deliver a model which is sensitive to the data distribution/task changes and able to adjust itself with few training samples and few updating steps. The results show that only 10% of training data and 1 gradient update are enough to get the best performance on identifying unseen brain cell type. AGILE can be used in many diagnose systems or detection algorithms which have to deal with various input data.

Acknowledgments. This work was supported by the National Science Foundation (NSF-IIS 1910973), and the Intramural Research Program of the National Institute of Neurological Disorders and Stroke, National Institutes of Health (1R01NS109118-01A1).

References

1. Andrychowicz, M., et al.: Learning to learn by gradient descent by gradient descent. In: Advances in Neural Information Processing Systems, pp. 3981–3989 (2016)
2. Ascoli, G.A., Donohue, D.E., Halavi, M.: Neuromorpho org: a central resource for neuronal morphologies. J. Neurosci. **27**(35), 9247–9251 (2007)
3. Bengio, S., Bengio, Y., Cloutier, J., Gecsei, J.: On the optimization of a synaptic learning rule. In: Preprints Conference Optimality in Artificial and Biological Neural Networks, vol. 2. University of Texas (1992)
4. Bengio, Y.: Deep learning of representations for unsupervised and transfer learning. In: Proceedings of ICML Workshop on Unsupervised and Transfer Learning, pp. 17–36 (2012)

5. Bengio, Y., Bengio, S., Cloutier, J.: Learning a synaptic learning rule. Université de Montréal, Département d'informatique et de recherche *ldots* (1990)
6. De Jager, P.L., et al.: Alzheimer's disease: early alterations in brain dna methylation at ANK1, BIN1, RHBDF2 and other LOCI. Nat. Neurosci. **17**(9), 1156–1163 (2014)
7. Elizabeth, A., Weaver II, H.H.D.: Cells of the brain. https://www.dana.org/article/cells-of-the-brain/, August 2019
8. Fang, M., Li, Y., Cohn, T.: Learning how to active learn: A deep reinforcement learning approach. arXiv preprint arXiv:1708.02383 (2017)
9. Finn, C., Abbeel, P., Levine, S.: Model-agnostic meta-learning for fast adaptation of deep networks. In: Proceedings of the 34th International Conference on Machine Learning-Volume 70, pp. 1126–1135. JMLR.org (2017)
10. Gal, Y., Ghahramani, Z.: Dropout as a bayesian approximation: representing model uncertainty in deep learning. In: International Conference on Machine Learning, pp. 1050–1059 (2016)
11. Gal, Y., Islam, R., Ghahramani, Z.: Deep bayesian active learning with image data. In: Proceedings of the 34th International Conference on Machine Learning-Volume 70, pp. 1183–1192. JMLR.org (2017)
12. Goodfellow, I., et al.: Generative adversarial nets. In: Advances in Neural Information Processing Systems, pp. 2672–2680 (2014)
13. Gopalan, R., Li, R., Chellappa, R.: Domain adaptation for object recognition: an unsupervised approach. In: 2011 International Conference on Computer Vision, pp. 999–1006. IEEE (2011)
14. Hochreiter, S., Schmidhuber, J.: Long short-term memory. Neural Comput. **9**(8), 1735–1780 (1997)
15. Hochreiter, S., Younger, A.S., Conwell, P.R.: Learning to learn using gradient descent. In: Dorffner, G., Bischof, H., Hornik, K. (eds.) ICANN 2001. LNCS, vol. 2130, pp. 87–94. Springer, Heidelberg (2001). https://doi.org/10.1007/3-540-44668-0_13
16. Joshi, A.J., Porikli, F., Papanikolopoulos, N.: Multi-class active learning for image classification. In: 2009 IEEE Conference on Computer Vision and Pattern Recognition, pp. 2372–2379. IEEE (2009)
17. Kingma, D.P., Mohamed, S., Rezende, D.J., Welling, M.: Semi-supervised learning with deep generative models. In: Advances in Neural Information Processing Systems, pp. 3581–3589 (2014)
18. Koch, G., Zemel, R., Salakhutdinov, R.: Siamese neural networks for one-shot image recognition. In: ICML deep learning workshop, vol. 2. Lille (2015)
19. Li, X., Guo, Y.: Adaptive active learning for image classification. In: Proceedings of the IEEE Conference on Computer Vision and Pattern Recognition, pp. 859–866 (2013)
20. MacKay, D.J.: Information-based objective functions for active data selection. Neural Comput. **4**(4), 590–604 (1992)
21. McCallumzy, A.K., Nigamy, K.: Employing EM and pool-based active learning for text classification. In: Proceedings International Conference on Machine Learning (ICML), pp. 359–367. Citeseer (1998)
22. McKenzie, A.T., et al.: Brain cell type specific gene expression and co-expression network architectures. Sci. Rep. **8**(1), 1–19 (2018)
23. Pan, S.J., Yang, Q.: A survey on transfer learning. IEEE Trans. Knowl. Data Eng. **22**(10), 1345–1359 (2009)
24. Ravi, S., Larochelle, H.: Optimization as a model for few-shot learning (2016)

25. Saenko, K., Kulis, B., Fritz, M., Darrell, T.: Adapting visual category models to new domains. In: Daniilidis, K., Maragos, P., Paragios, N. (eds.) ECCV 2010. LNCS, vol. 6314, pp. 213–226. Springer, Heidelberg (2010). https://doi.org/10.1007/978-3-642-15561-1_16
26. Santoro, A., Bartunov, S., Botvinick, M., Wierstra, D., Lillicrap, T.: One-shot learning with memory-augmented neural networks. arXiv preprint arXiv:1605.06065 (2016)
27. Schweighofer, N., Doya, K.: Meta-learning in reinforcement learning. Neural Networks **16**(1), 5–9 (2003)
28. Sener, O., Savarese, S.: Active learning for convolutional neural networks: a core-set approach. arXiv preprint arXiv:1708.00489 (2017)
29. Settles, B.: Active learning literature survey. University of Wisconsin-Madison Department of Computer Sciences, Technical report (2009)
30. Shorten, C., Khoshgoftaar, T.M.: A survey on image data augmentation for deep learning. J. Big Data **6**(1), 60 (2019)
31. Skene, N.G., et al.: Genetic identification of brain cell types underlying schizophrenia. Nat. Genet. **50**(6), 825–833 (2018)
32. Smith, L., Gal, Y.: Understanding measures of uncertainty for adversarial example detection. arXiv preprint arXiv:1803.08533 (2018)
33. Snell, J., Swersky, K., Zemel, R.: Prototypical networks for few-shot learning. In: Advances in Neural Information Processing Systems, pp. 4077–4087 (2017)
34. Vilalta, R., Drissi, Y.: A perspective view and survey of meta-learning. Artif. Intell. Rev. **18**(2), 77–95 (2002)
35. Vinyals, O., Blundell, C., Lillicrap, T., Wierstra, D., et al.: Matching networks for one shot learning. In: Advances in Neural Information Processing Systems, pp. 3630–3638 (2016)
36. Woodward, M., Finn, C.: Active one-shot learning. arXiv preprint arXiv:1702.06559 (2017)
37. Zeng, H., Sanes, J.R.: Neuronal cell-type classification: challenges, opportunities and the path forward. Nat. Rev. Neurosci. **18**(9), 530 (2017)
38. Zhu, X., Goldberg, A.B.: Introduction to semi-supervised learning. Synth. Lect. Artif. Intell. Mach. Learn. **3**(1), 1–130 (2009)

Automatic Data Augmentation for 3D Medical Image Segmentation

Ju Xu[1], Mengzhang Li[1,2], and Zhanxing Zhu[1,3](✉)

[1] Center for Data Science, Peking University, Beijing, China
{xuju,mcmong,zhanxing.zhu}@pku.edu.cn
[2] Canon Medical Systems, Beijing, China
[3] School of Mathematical Sciences, Peking University, Beijing, China

Abstract. Data augmentation is an effective and universal technique for improving generalization performance of deep neural networks. It could enrich diversity of training samples that is essential in medical image segmentation tasks because 1) the scale of medical image dataset is typically smaller, which may increase the risk of overfitting; 2) the shape and modality of different objects such as organs or tumors are unique, thus requiring customized data augmentation policy. However, most data augmentation implementations are hand-crafted and suboptimal in medical image processing. To fully exploit the potential of data augmentation, we propose an efficient algorithm to automatically search for the optimal augmentation strategies. We formulate the coupled optimization w.r.t. network weights and augmentation parameters into a differentiable form by means of stochastic relaxation. This formulation allows us to apply alternative gradient-based methods to solve it, i.e. stochastic natural gradient method with adaptive step-size. To the best of our knowledge, it is the first time that differentiable automatic data augmentation is employed in medical image segmentation tasks. Our numerical experiments demonstrate that the proposed approach significantly outperforms existing build-in data augmentation of state-of-the-art models.

Keywords: Medical image segmentation · Data augmentation · AutoML

1 Introduction

In the past few years, deep neural network has achieved incredible progress in medical image segmentation tasks and promoted booming development of computer assisted intervention. This has benefitted research and clinical treatment of disease diagnosis, treatment design and prognosis evaluation [13,14]. Given the training data, researchers proposed various 2D/3D medical image segmentation models for supervised or semi-supervised tasks [6,8]. However, the performance of deep learning models heavily relies on large scale well-labeled data.

J. Xu and M. Li—Equal contributions.

A. L. Martel et al. (Eds.): MICCAI 2020, LNCS 12261, pp. 378–387, 2020.
https://doi.org/10.1007/978-3-030-59710-8_37

Currently, data augmentation is a widely used and effective technique to increase the amount and diversity of available data, and thus improving models' generalization performance. In the domain of natural image processing, typical data augmentation strategies include *manually* cropping, rotating or adding random noise to the original images. Besides thess ad-hoc approaches, generative models [7] and unsupervised learning models [15] are also employed for generating extra data. Unfortunately, those augmentation techniques might not be optimal for a specific task, and thus the customized data augmentation strategy is required. Recently, researchers proposed to search the augmentation policy by reinforcement learning [5] or density matching [10], inspired by previously works of automatic machine learning (AutoML) on neural architecture search (NAS [11,12,18]).

For medical image segmentation tasks, data augmentation techniques are also used in UNet and its variants nnUNet [8], R2U-Net [2], etc. However, these methods are simple and hand-made, and the improvement of segmentation accuracy is limited. In [16], the authors proposed to utilize reinforcement learning to search for augmentation strategies. However, it costs 768 GPU hours and it only search the probability of each augmentation strategy in [16]. Moreover, the difference between natural and medical images such as spatial contextual correlation, smaller scale of dataset and unique pattern of specified organs or tumor makes the augmentation strategies adopted in natural images difficult be transferred to medical domains.

In this paper, we propose an *automatic data augmentation* framework (ASNG) through searching the optimal augmentation policy, particularly for 3D medical image segmentation tasks. It's the first automatic data augmentation work in whole semantic segmentation filed. The contributions of our paper are as follows:

- It's the first time that we formulate the auto-augmentation problem into a bi-level optimization problem and apply an approximate algorithm to solve it
- The designed search space in medical image field is novel. Different from previous methods which searched for a fixed magnitude of operations, we search for an interval of magnitude
- Different from previous method which searched for a fixed augmentation strategy, the searched augmentation strategy of our method is dynamically changing during the training. Besides, we don't need to retrain the target network after the searching process
- Experiments demonstrate that our ASNG can indeed achieve the SOTA of the performance

2 Method

In our method, we formulate the problem of finding the optimal augmentation policy as a discrete search problem. Our method consists of two components:

the designed of search space and search algorithm. The search algorithm samples a data augmentation policy S from the search space consisting of proposed operations, and then decides the magnitude of the operation and the probability of applying this operation. The framework of our method can be seen in Fig. 1. We will elaborate the two components in the following.

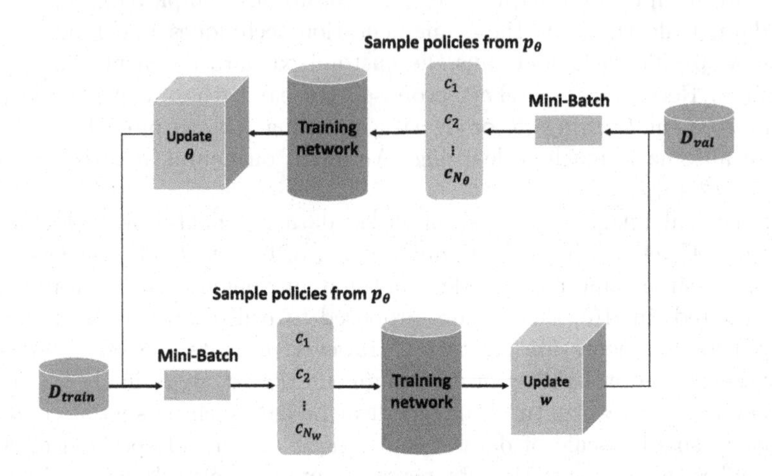

Fig. 1. The framework of our proposed method. D_{train}, D_{val} represent training dataset, validation dataset, respectively. p_θ is the distribution of c.

2.1 Search Space of Data Augmentation for 3D Medical Images

Since it is the first work for applying AutoAugment strategies in medical image area, we have to design the search space for our ASNG algorithm. In our search space, a policy consists of seven image operations to be applied in a sequential manner. Each image operation is associated with two hyperparameters: 1) the probability of applying this operation, and 2) the interval of magnitude for the image operation.

The seven image operations we used in our experiments are from batchgenerators, a popular Python image library[1], including Scale, RoateX/Y/Z, Alpha (magnitude of the elastic deformation), Sigma (scale of the elastic deformation), Gamma (same as gamma correction in photos or computer monitors). In order to increase the diversity of augmentation policies, we do not fix a specific magnitude for an operation like previous works [5], but set an interval of magnitude for an operation. Therefore we should decide the left boundary of interval (LB) and the right one (RB). To decrease the search complexity, we discretize the range of magnitude into 11 values with uniform spacing so that we can use a discrete search algorithm to find them. Besides the magnitude of transformation operation, we also search for the probability of conducting these transformations,

[1] https://github.com/MIC-DKFZ/batchgenerators.

i.e. the probability of applying scale transformation, rotation, gamma transformation, and elastic deformation, denoted as p_{scale}, p_{rot}, p_{gamma}, p_{eldef}, respectively. Similarly, we also discretize the probability of applying that operation into 11 values with uniform spacing. Table 1 summarizes the range of magnitudes and possibilities for the seven operations. Figure 2 shows one example of augmented image and label based on our method, in which the image transformations are from the defined search space.

We can easily observe that naively searching one augmentation strategy becomes a search problem with 11^{11} possibilities. The search space is so huge that an efficient algorithm is required, as proposed in the following.

Table 1. The range of parameters in strategies we will search.

Operation	LB	RB	Probability	Range
Scale	[0.5, 1.0]	[1.0, 1.5]	p_{scale}	[0, 1]
RotationX	$[\frac{-\pi}{6}, 0]$	$[0, \frac{-\pi}{6}]$	p_{rot}	[0, 1]
RotationY	$[\frac{-\pi}{6}, 0]$	$[0, \frac{-\pi}{6}]$	p_{rot}	[0, 1]
RotationZ	$[\frac{-\pi}{6}, 0]$	$[0, \frac{-\pi}{6}]$	p_{rot}	[0, 1]
Alpha	[0, 450]	[450, 900]	p_{eldef}	[0, 1]
Sigma	[0, 7]	[7, 14]	p_{eldef}	[0, 1]
Gamma	[0.5, 1]	[1, 1.5]	p_{gamma}	[0, 1]

2.2 Stochastic Relaxation Optimization of Policy Sampling

We denote $f(w, c)$ as the objective function, where $w \in \mathcal{W}$ are network parameters and $c \in \mathcal{C}$ are data augmentation strategies. f_{train} and f_{val} are the training and the validation loss, respectively. Both losses are determined not only by the augmentation policy c, but also the weight w. The goal for augmentation strategy search is to find c^* that minimizes the validation loss $f_{val}(w^*, c^*)$, where the weights are obtained by minimizing the training loss $w^* = \text{argmin}_w f_{train}(w, c^*)$. Thus augmentation strategy search is a bi-level optimization problem, we can write the problem as follows:

$$\min_c f_{val}(w^*(c), c) \tag{1}$$

$$s.t. \quad w^*(c) = \operatorname*{argmin}_w f_{train}(w, c) \tag{2}$$

Solving the above problem is not easy, since we cannot obtain the gradient w.r.t. c, thus it is hard to optimize c via gradient descent. Though simple grid search or reinforcement learning proposed in [5] can be utilized to search for c, the computational cost is extremely high if we evaluate the performance of every c. To this end, we propose to solve this optimization problem efficiently first by stochastic relaxation and then applying natural gradient descent [3], as described in the following.

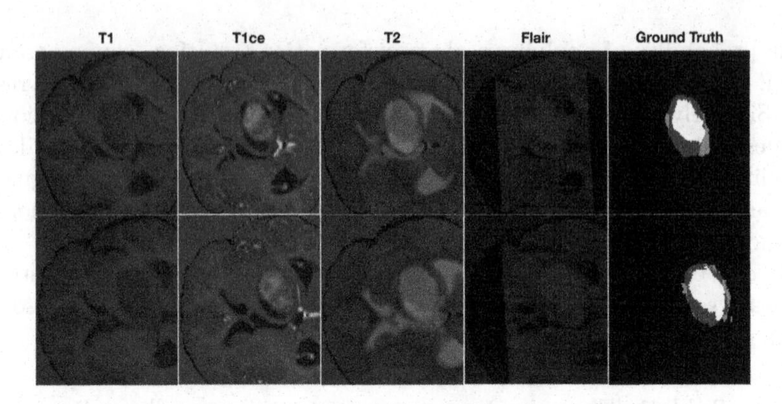

Fig. 2. Visualization of the proposed data augmentation on certain 2D section of Task01 BrainTumour dataset. **Top:** original image and label, **Bottom:** the image and label generated by the searched data augmentation policy, in which the operations are from Table 1.

Stochastic Relaxation. We turn the original optimization problem into an optimization of differentiable objective J by stochastic relaxation [1]. The basic idea of stochastic relaxation is: instead of directly optimizing w.r.t c, we consider a distribution $p_\theta(c)$ over c parametrized by θ, and minimize the expected value of the validation loss f_{val} w.r.t θ, i.e.,

$$\min_{\theta} J(\theta) = \int_{c \in C} f_{val}(w^*(c), c) p_\theta(c) dc \tag{3}$$

$$s.t. \quad w^*(c) = \operatorname*{argmin}_{w} f_{train}(w, c) \tag{4}$$

The stochastic relaxation makes J differentiable w.r.t both w and θ. Therefore we can update w and θ by gradient descent.

However, the gradient $\nabla_w J(w, \theta)$ is not tractable because we can not evaluate the mean performance of $c \in \mathcal{C}$ in a closed-form way. Here we estimate the gradient by Monte-Carlo (MC) using $\nabla_w J(w^t, c_i)$ with i.i.d. samples $c_i \sim p_{\theta^t}(c)$, $i = 1, \ldots, N_w$, namely:

$$G_w(w^t, \theta^t) = \frac{1}{N_w} \sum_{i=1}^{N_w} \nabla_w f_{train}(w^t, c_i) \tag{5}$$

Now we can approximate $\nabla_w J(w, \theta)$ with the stochastic gradient $G_w(w^t, \theta^t)$, w^t can be updated as

$$w^{t+1} = w^t - \epsilon_w G_w(w^t, \theta^t), \tag{6}$$

where ϵ_w is the learning rate for network parameters. Due to that the distance between two probability distribution is not Euclidean, updating θ directly by gradient descent like w is not appropriate. We then resort to natural gradient

(NG [3]) designed for parametric probability distributions,

$$\theta^{t+1} = \theta^t - \epsilon_\theta F(\theta_t)^{-1} \nabla_\theta J(w, \theta), \tag{7}$$

where $F(\theta_t)$ is the Fisher matrix, ϵ_θ is the learning rate. The probability distribution we utilized for $c \in \mathcal{C}$ is multinomial distribution. How to calculate the Fisher matrix can be seen in [1]. Similar with [1], we utilize adaptive step-size ϵ_θ to make the learning process faster. Monte-Carlo is also adopted to approximate $\nabla_\theta J(w, \theta)$, and then

$$\theta^{t+1} = \theta^t - \epsilon_\theta F(\theta_t)^{-1} \frac{1}{N_\theta} \sum_{j=1}^{N_\theta} \nabla_\theta f_{val}(w^{t+1}, c_j) \ln p_\theta(c_j) \tag{8}$$

We summarize the procedure of our proposed approach in Algorithm 1.

Algorithm 1. ASNG

1: **Input:** w^0, θ^0, ϵ_w. ϵ_θ, N_w, N_θ
2: **Input:** Training dataset D_{train}, validation dataset D_{val}, test dataset D_{test}.
3: **for** i=1 to epoch **do**
4:   **for** t=1 to T **do**
5:     Generate N_w policys according to p_{θ_t}
6:     Augment training data from D_{train} with N_w policys, respectively;
7:     Obtain the loss $f_{train}(w_t, c_i)$ $(i = 1, \ldots, N_w)$ on D_{train};
8:     Update w_t according to Equation 6, then obtain w_{t+1};
9:     Generate N_θ policys according to p_{θ_t};
10:     **for** j=1 to N_θ **do**
11:       Augment training data according to policy c_j;
12:       Update w_t to obtain $\hat{w_t}$;
13:       Obtain the validation loss $f_{val}(\hat{w_t})^j$ on D_{val};
14:       Restore the network parameters, $\hat{w_t} = w_t$;
15:     **end for**
16:     Utilize validation loss $f_{val}(\hat{w_t})^j$, policys c_j $(j = 1, \ldots, N_\theta)$ to update θ_t according to equation 8;
17:   **end for**
18: **end for**
19: Test the network on D_{test};
20: **return** final networks.

3 Implementation and Experiments

3.1 Datasets and Implementation Details

Datasets: We conduct the proposed method on three 3D segmentation tasks used in the medical segmentation decathlon challenge (MSD[2]): (1) Task01 Brain

[2] http://medicaldecathlon.com/.

Tumour (484 labeled images, 3 classes), (2) Task02 Heart (20 labeled images, 1 class) and (3) Task05 Prostate (32 labeled images, 2 classes). Each dataset is collected for a specified task, their various input sizes, voxel spacings and foreground targets are suitable for demonstrating our algorithm's generalization. We evaluate the performance with 5-fold cross validation as [4,9] since the ground truth labels for test dataset are not publicly available.

Compared Methods include 3D U-ResNet [17], SCNAS [9], nnUNet without data augmentation (nnUNet NoDA) and 3D nnUNet [8] with default data augmentation strategy[3]. 3D U-ResNet utilizes residual blocks and attention gates; and SCNAS is the latest neural architecture search model for 3D medical image segmentation, which applies a scalable gradient-based optimization to find the optimal model architecture. The method proposed in SCNAS can't utilized for differentiable autoaugmentation strategies search. In SCNAS, a mixed operation is created by adding all these operations in search space based on the importance of each operation. However, we can't add the transformation results of each augmentation strategy. There we don't apply the proposed method of SCNAS to our augmentation strategies search tasks. Note that the code of nnUnet has already implemented random augmentation. In nnUnet, the magnitude of operation is sampled from a predefined interval in every training epoch. AutoAugment [5] costs 5000 GPU hours to search for a policy. FastAutoAugment [10] needs to spilt the training dataset as K folds. However, dataset in medical image area is quite small. Training model in a small dataset will overfit. Therefore, we don't compare our method with AutoAugment and FastAutoAugment. The prediction result is inferenced using a sliding window with half the patch size ensuring 50% overlapping, i.e., each voxel is inferenced at least two times at test.

Implementation Details. We preprocess the data with same pipeline used in 3D nnUNet. We unify the identical voxel spacing values by proper interpolation due to different spacing values of each case, i.e. resampling them to $0.7\,\text{mm} \times 0.7\,\text{mm} \times 0.7\,\text{mm}$ firstly. We apply Z-score normalization of voxel value for each input channel separately; and grip the input patch whose size is set as $128 \times 128 \times 128$, and its foreground ratio is set larger than $1/3$ ensuring UNet variants could learn features of foreground.

Following the default data augmentation policy in nnUNet, we utilize scaling, rotation, elastic and gamma transformation both in training and test. The parameters of the operation and probability of conducting that operation are both within search space of our ASNG algorithm.

The code is implemented using PyTorch 1.0.0. The ADAM optimizer is utilized for training where the learning rate and weight decay are initialized as 3×10^{-4} and 3×10^{-5}, respectively, where it is reduced by 80% if the training loss is not reduced over 30 epochs. Besides ASNG, the training process of other benchmarks would last for 500 epochs if the learning rate is larger than 10^{-7}. Following [9] and [8], the loss function for 3D U-ResNet and SCNAS is Jaccard distance, for nnUNet and ASNG is sum of minus Dice similarity and Cross Entropy. Considering the training time, ASNG is trained for 50, 200 and 200 epochs on Brain Tumour, Heart and Prostate

[3] https://github.com/MIC-DKFZ/nnUNet/.

respectively, with batch size of 2. ASNG is evaluated more than 5 times in each public dataset. It takes about 10 d on one NVIDIA TITAN RTX GPU, compared with that one integrated nnUNet training procedure takes about 3 d. The sampling times T of ASNG is 2 because of the limited memory of GPUthough larger T could produce better numerical results.

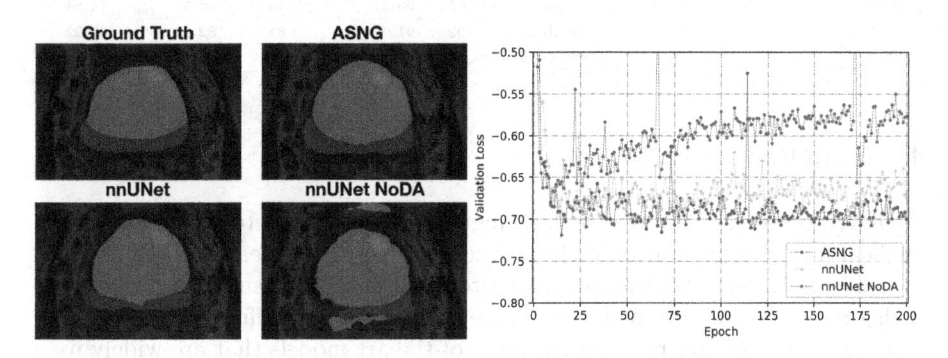

Fig. 3. Results on Task05 Prostate of selected architectures. **Left:** Example of inference, green mask represents peripheral zone and red mask represents transitional zone. **Right:** The trend of loss on validation set.

3.2 Experimental Results

Our result is shown in Table 2. Because of unavailable labels of test set and restricted online submission times, those Auto ML models on 3D Medical Image Segmentation tasks are all evaluated on validation set. In this paper we still follows this metric for fair comparison. ASNG outperforms other architectures especially 3D nnUNet, which is the best medical image segmentation framework with default data augmentation. It should be noted that since Heart and Prostate only have 20 and 32 labeled images, in [9] the obtained architecture of SCNAS based on the first fold of 484 labeled Brain Tumour images is transferred to Heart and Prostate tasks to avoid overfitting. Remarkably, our method, applied only on the basic network architecture, could still achieve best prediction accuracy. This clearly demonstrates the necessity and effectiveness of data augmentation policy search in 3D medical image segmentation.

Figure 3 shows the example of segmentations results and validation loss w.r.t. number of epochs in the Prostate task. We can observe that our method ASNG can produce better prediction and achieve more stable improvement during training than other compared methods.

In this paper [16], the proposed method utilizes reinforcement learning to search for augmentation strategies, which costs 768 GPU hours while our method costs less than 100. And their result (Dice 0.92) on task 02 is worse than ours (0.933). Besides, the first paper only searched the probability of each augmentation strategy while our method not only search the probability but also the magnitude.

Table 2. Average Dice similarity coefficients (%) for Brain tumor, Heart, and Prostate 3D segmentation tasks of MSD.

Label	Brain tumour				Heart	Prostate		
	Edema	Non-Enhancing	Enhancing	Average	Left atrium	Peripheral	Transitional	Average
U-ResNet	79.10	58.38	77.37	71.61	91.48	48.37	79.17	63.77
nnUNet NoDA	81.27	60.92	77.90	73.36	92.85	58.61	83.61	71.11
nnUNet	81.68	61.29	77.97	73.65	93.21	63.14	86.53	74.84
SCNAS	80.41	59.85	78.50	72.92	91.91	53.81	82.02	67.92
ASNG	**81.99**	**62.06**	**78.58**	**74.21**	**93.36**	**67.65**	**87.04**	**77.35**

4 Conclusion

We have proposed an automatic data augmentation strategy to accommodate 3D medical image segmentation tasks. By configuring proper search space followed by gradient-based optimization, the customized data augmentation strategy for each task could be obtained. The numerical results for different segmentation tasks show it could outperform the state-of-the-art models that are widely used in this area. Furthermore, the proposed approach shows that, compared with searching network architectures, searching for optimal data augmentation policy is also important. As for future work, designing better search space and accelerating the search process can be considered.

Acknowledgement. This project is supported by National Natural Science Foundation of China (No.61806009 and 61932001), PKU-Baidu Funding 2019BD005 and Beijing Academy of Artificial Intelligence (BAAI).

References

1. Akimoto, Y., Shirakawa, S., Yoshinari, N., Uchida, K., Saito, S., Nishida, K.: Adaptive stochastic natural gradient method for one-shot neural architecture search. arXiv preprint arXiv:1905.08537 (2019)
2. Alom, M.Z., Hasan, M., Yakopcic, C., Taha, T.M., Asari, V.K.: Recurrent residual convolutional neural network based on u-net (r2u-net) for medical image segmentation. arXiv preprint arXiv:1802.06955 (2018)
3. Amari, S.I.: Natural gradient works efficiently in learning. Neural Comput. **10**(2), 251–276 (1998)
4. Bae, W., Lee, S., Lee, Y., Park, B., Chung, M., Jung, K.-H.: Resource optimized neural architecture search for 3D medical image segmentation. In: Shen, D., et al. (eds.) MICCAI 2019. LNCS, vol. 11765, pp. 228–236. Springer, Cham (2019). https://doi.org/10.1007/978-3-030-32245-8_26
5. Cubuk, E.D., Zoph, B., Mane, D., Vasudevan, V., Le, Q.V.: Autoaugment: learning augmentation strategies from data. In: Proceedings of the IEEE Conference on Computer Vision and Pattern Recognition. pp. 113–123 (2019)
6. Ganaye, P.A., Sdika, M., Triggs, B., Benoit-Cattin, H.: Removing segmentation inconsistencies with semi-supervised non-adjacency constraint. Med. Image Anal. **58**, 101551 (2019)

7. Huang, S.W., Lin, C.T., Chen, S.P., Wu, Y.Y., Hsu, P.H., Lai, S.H.: Auggan: cross domain adaptation with gan-based data augmentation. In: Proceedings of the European Conference on Computer Vision (ECCV), pp. 718–731 (2018)
8. Isensee, F., Petersen, J., Kohl, S.A., Jäger, P.F., Maier-Hein, K.H.: NNU-net: Breaking the spell on successful medical image segmentation. arXiv preprint arXiv:1904.08128 (2019)
9. Kim, S., et al.: Scalable neural architecture search for 3D medical image segmentation. In: Shen, D., et al. (eds.) MICCAI 2019. LNCS, vol. 11766, pp. 220–228. Springer, Cham (2019). https://doi.org/10.1007/978-3-030-32248-9_25
10. Lim, S., Kim, I., Kim, T., Kim, C., Kim, S.: Fast autoaugment. In: Advances in Neural Information Processing Systems, pp. 6662–6672 (2019)
11. Liu, H., Simonyan, K., Yang, Y.: Darts: differentiable architecture search. arXiv preprint arXiv:1806.09055 (2018)
12. Pham, H., Guan, M.Y., Zoph, B., Le, Q.V., Dean, J.: Efficient neural architecture search via parameter sharing. arXiv preprint arXiv:1802.03268 (2018)
13. Ronneberger, O., Fischer, P., Brox, T.: U-Net: convolutional networks for biomedical image segmentation. In: Navab, N., Hornegger, J., Wells, W.M., Frangi, A.F. (eds.) MICCAI 2015. LNCS, vol. 9351, pp. 234–241. Springer, Cham (2015). https://doi.org/10.1007/978-3-319-24574-4_28
14. Tajbakhsh, N., et al.: Convolutional neural networks for medical image analysis: full training or fine tuning? IEEE Trans. Med. Imaging **35**(5), 1299–1312 (2016)
15. Xie, Q., Dai, Z., Hovy, E., Luong, M.T., Le, Q.V.: Unsupervised data augmentation. arXiv preprint arXiv:1904.12848 (2019)
16. Yang, D., Roth, H., Xu, Z., Milletari, F., Zhang, L., Xu, D.: Searching learning strategy with reinforcement learning for 3D medical image segmentation. In: Shen, D., et al. (eds.) MICCAI 2019. LNCS, vol. 11765, pp. 3–11. Springer, Cham (2019). https://doi.org/10.1007/978-3-030-32245-8_1
17. Yu, L., Yang, X., Chen, H., Qin, J., Heng, P.A.: Volumetric convnets with mixed residual connections for automated prostate segmentation from 3D MR images. In: Thirty-first AAAI conference on artificial intelligence (2017)
18. Zoph, B., Vasudevan, V., Shlens, J., Le, Q.V.: Learning transferable architectures for scalable image recognition. In: Proceedings of the IEEE Conference on Computer Vision and Pattern Recognition, pp. 8697–8710 (2018)

MS-NAS: Multi-scale Neural Architecture Search for Medical Image Segmentation

Xingang Yan[1], Weiwen Jiang[2], Yiyu Shi[2], and Cheng Zhuo[1](✉)

[1] ZheJiang University, Hangzhou, China
{21832147,czhuo}@zju.edu.cn
[2] University of Notre Dame, Notre Dame, USA
{wjiang2,yshi4}@nd.edu

Abstract. The recent breakthroughs of Neural Architecture Search (NAS) have motivated various applications in medical image segmentation. However, most existing work either simply rely on hyper-parameter tuning or stick to a fixed network backbone, thereby limiting the underlying search space to identify more efficient architecture. This paper presents a *Multi-Scale NAS* (MS-NAS) framework that is featured with multi-scale search space from network backbone to cell operation, and multi-scale fusion capability to fuse features with different sizes. To mitigate the computational overhead due to the larger search space, a partial channel connection scheme and a two-step decoding method are utilized to reduce computational overhead while maintaining optimization quality. Experimental results show that on various datasets for segmentation, MS-NAS outperforms the state-of-the-art methods and achieves 0.6–5.4% mIOU and 0.4–3.5% DSC improvements, while the computational resource consumption is reduced by 18.0–24.9%.

1 Introduction

Accurate segmentation of medical images is a crucial step in computer-aided diagnosis, surgical planning and navigation [1]. The recent breakthroughs in deep learning, such as UNet [11], have steadily improved segmentation efficiency, which not only defeats human visual systems, but also exceeds the conventional algorithms in both speed and accuracy [2,11,12]. However, in general, designers have to spend significant efforts through manual trial-and-error deciding network architecture, hyper-parameters, pre- and post-processing procedures [14]. Thus, it is highly desired to have an efficient network design procedure when segmenting for different modalities, subjects, and resolutions [13].

The recently proposed automated machine learning (AutoML) is well aligned with such demands to *automatically* design the neural network architecture instead of relying on human experiences and repeated manual tuning. More importantly, many works in *Neural Architecture Search* (NAS) have already identified more efficient neural network architectures in general computer vision tasks [6,14]. Such success has motivated various NAS applications in medical

A. L. Martel et al. (Eds.): MICCAI 2020, LNCS 12261, pp. 388–397, 2020.
https://doi.org/10.1007/978-3-030-59710-8_38

image segmentation [7–9,13]. However, most of them either simply apply Darts-like framework [4] with hyper-parameter tuning, or sticks to a fixed network backbone (e.g., UNet), thereby restricting the underlying optimization space to identify more efficient architectures for different modalities, such as CT, MRI, and PET [7–9].

In addition, as medical images are featured with inhomogeneous intensity, similar floorplan, and low semantic information, it is then a natural idea to utilize both high-level semantics and low-level features as a combined effort (i.e., multi-scale fusion) to suffice segmentation efficiency. The effectiveness of multi-scale fusion has already been partially proved by UNet [11], which fuses the features of the same sizes from encoders/decoders. Thus, most UNet-based NAS work implicitly embeds such fusion capability. However, the implicit embedding also restricts the fusion only to the features of the same size, thereby functioning as enforced operation and preventing further optimization. Intuitively speaking, *real* multi-scale fusion should fuse features of different sizes to provide more informative content for segmentation efficiency. To address the aforementioned concerns, this paper proposes a **Multi-Scale NAS** (MS-NAS) framework to design neural network for medical image segmentation, which is featured with:

- **Multi-scale search space:** The framework employs a *larger* search space at different scales, from network backbone, artificial module and cell, to operation, which can identify more optimal architecture for different tasks.
- **Multi-scale fusion:** The framework also explores the *real* multi-scale fusion operation to improve segmentation efficiency by concatenating features at different scales within each artificial module.

The proposed MS-NAS framework is an end-to-end solution to automatically determine the network backbone, cell type, operation parameters, and fusion scales. To facilitate such search, three types of cells, *expanding cells*, *contracting cells*, and *non-scaling cells*, are defined in the next section to compose the learned network architecture. With the optimized cell types, fusion scales, and operation connections, the framework can identify among various backbones, such as UNet, ResUNet, FCN, etc., the most effective architecture meeting the varying demands from modality to modality. Thus, our proposal is different to the prior NAS work for medical image segmentation, which often sticks to one network backbone and only fuses the features of the same scale [7–9]. Apparently, the proposed MS-NAS can be resource consuming when optimizing in such a large search space. We here employ a partial channel connections scheme [10] and a two-step decoding method to speed up the search procedure. As a result, the proposed MS-NAS framework is capable to optimize on various high-resolution segmentation datasets in larger search space with reduced computational cost.

Experimental results show that the proposed MS-NAS outperforms the prior NAS work [6,9] with 0.6–5.4% mIOU and 0.4–3.5% DSC improvement on average for various datasets while it achieves 18.0–24.9% computational cost reduction. It is noted that the framework can trade off between accuracy and complexity, thereby enabling desired flexibility to build networks for different tasks while maintaining the state-of-the-art performance.

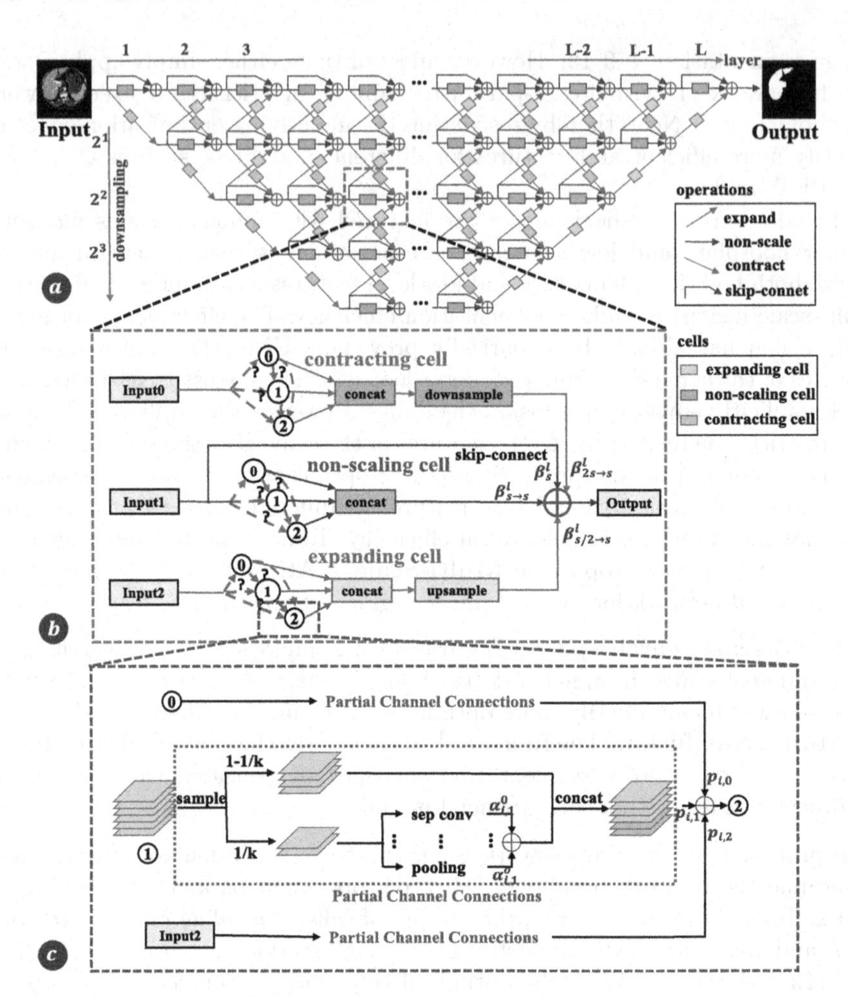

Fig. 1. Overview of the proposed architecture search space for medical image segmentation: (a) Search space for MS-NAS; (b) One artificial module containing three cells; and (c) Example illustration of partial channel connections.

2 Method

In this section, we first present the proposed multi-scale architecture search space for medical image segmentation. Then, we discuss optimization and decoding schemes to obtain the discrete architecture. Figure 1(a) provides an overview of the search space for MS-NAS as well as its components, which can be represented by a directed acyclic graph (DAG), with V vertice and E edges. Before we go into algorithm details, we would like to define the notations and components of the network from top to bottom for better discussions in the following sections.

For the searched network, a *sub-network* is defined as a path from input to segmentation output. Within the sub-network, the basic component is *artificial module* (as shown in Fig. 1(b)), which may contain three types of *cells*: (1) *Expanding cell* expands and up-samples the scale of feature map; (2) *Contracting cell* contracts and down-samples the scale of feature map; and (3) *Non-scaling cell* keeps the scale of feature map constant. With *module* and *cell* defined, we then define *operations* that connect different cells within the modules and from module to module. In addition to the commonly used operations, such as *pooling*, we can add three *operations*, expanding, contracting, and non-scaling, corresponding to the three cells above. A *skip-connect* operation within the cell is also employed to transfer the shallow features to the deep semantics.

2.1 Multi-scale Architecture Search Space

The search space of MS-NAS covers different scales, from network, module, cell, to operation. *Cell level* search is conducted at *local* to find the desired cell structures and connections from the cell search space. Each cell search space can be represented by a DAG consisting of N blocks, with each block representing the mapping from an input tensor X_{in} to an output tensor X_{out}. For the i_{th} block in a cell, we define a tuple (I_i, o_i) for such mapping: $I_i \in \mathcal{I}$, where $\mathcal{I}$ denotes a set of X_{in} and output tensors $X_1, \ldots X_{i-1}$ from the 1_{st} to the $i-1_{th}$ blocks; and $o_i \in \mathcal{O}$ is the operation applied to I_i, where $\mathcal{O}$ denotes a set of operators modified to facilitate search, including depth-wise-separable convolution, dilated convolution with a rate of 2, average pooling, skip connection, etc. An operator example of 3×3 *depth-wise-separable convolution* is shown in Fig. 2, with slightly changed operation order and additional ResNet-based skip connection. To reduce the memory cost during searching, a *partial channel connections* scheme [10] is embedded in the framework. In particular, the channels of the input tensor for a block are partitioned to two parts according to a hyper-parameter k. The output tensor of the i_{th} block then can be calculated as:

$$X_i = \sum_{X_j \in \mathcal{I}_i} \frac{\exp\{p_{i,j}\}}{\sum_{j'<i} \exp\{p_{i,j'}\}} \cdot f_{i,j}^{\mathrm{PC}}\left(\mathrm{X}_j; \mathrm{K}_{i,j}\right) \tag{1}$$

where (i, j) denotes the edge connecting blocks i and j, which is parameterized by a scalar $p_{i,j}$. $f_{i,j}^{\mathrm{PC}}\left(\mathrm{X}_j; \mathrm{K}_{i,j}\right)$ is an auxiliary function for partial channel connection:

$$f_{i,j}^{\mathrm{PC}}\left(\mathrm{X}_j; \mathrm{K}_{i,j}\right) = \sum_{o \in \mathcal{O}} \frac{\exp\left\{\alpha_{i,j}^{o}\right\}}{\sum_{o' \in \mathcal{O}} \exp\left\{\alpha_{i,j}^{o'}\right\}} \cdot o\left(\mathrm{K}_{i,j} * \mathrm{X}_j\right) + \left(1 - \mathrm{K}_{i,j}\right) * \mathrm{X}_j \tag{2}$$

where $K_{i,j}$ is a channel sampling mask. As shown in Fig. 1(c), $1/k$ portion of channels go through the operations selected from $\mathcal{O}$ while the rest remain unchanged. $\alpha_{i,j}^{o}$ parameterizes the operator in partial channel connection to control the contributions from different operators. This scheme helps reduce memory consumption during search while still maintaining a good convergence rate [10]. Finally, to weight the contributions from different edges when computing X_i, we use $\frac{\exp\{p_{i,j}\}}{\sum_{j'<i} \exp\{p_{i,j'}\}}$ for edge normalization. The output X_{out} of the cell is then computed as the concatenation of the tensors $X_1, X_2, \ldots X_N$ for the N blocks.

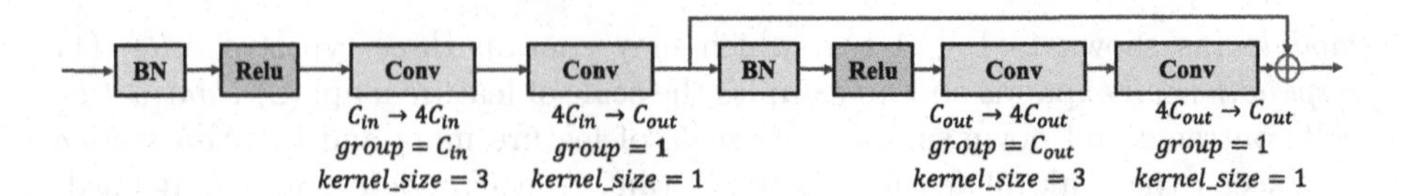

Fig. 2. An operator example of 3×3 *depth-wise-separable convolution*, where C_{in} and C_{out} denote the number of channels for input and output tensors of a block.

Network level search is conducted to find the desired network backbone within the entire network search space to determine the network connection of sub-networks. As shown in Fig. 1(a), the search space structure is gradually shrunk from top to bottom. Each path from input to output is unique and goes through different modules and different operations, resulting in different feature map scale changes. The search is then to find one or multiple sub-networks as well as their connections for the given hyper-parameters.

To facilitate the search procedure, we use continuous relaxation for network search [4]. With $Ep(*)$, $Ns(*)$, $Ct(*)$ defined as the operators for expanding, non-scaling and contracting, the connection between the cells at layer l can be parameterized by a scalar $\beta^l_{s_1 \to s_2}$ for the three operators, where s_1, s_2 indicate the sampling scale[1]. A *skip-connect* operations is parameterized by a scalar β^l_s without an explicit operator. Then the output feature map for the l_{th} layer is:

$$\begin{aligned} X^l_s = &Softmax(\beta^l_{\frac{s}{2} \to s}) Ep(X^{l-1}_{\frac{s}{2}}; \alpha_{ep}, p_{ep}) \\ &+ Softmax(\beta^l_{s \to s}) Ns(X^{l-1}_s; \alpha_{ns}, p_{ns}) \\ &+ Softmax(\beta^l_{2s \to s}) Ct(X^{l-1}_{2s}; \alpha_{ct}, p_{ct}) \\ &+ Softmax(\beta^l_s) X^{l-1}_s \end{aligned} \tag{3}$$

where $s \in \{2^0, 2^1, 2^2, 2^3, 2^4\}$ is the scaling ratio, and $Softmax(\beta) = \frac{\exp\{\beta\}}{\sum_{\beta \in B} \exp\{\beta\}}$. Note that zero operation is also accounted in our framework, which is simply disconnection between the blocks.

2.2 Optimization and Decoding

With the defined search space and relaxed parameters, we formulate the architecture search as an continuous optimization problem, similar to [4,6]. This can be effectively solved using stochastic gradient decent (SGD) method to obtain an approximate solution by optimizing the parameter sets of α, β and p as discussed in the previous subsection [6]. Note that even with a larger search space than prior NAS work, the embedding of partial channel connection helps significantly reduce memory usage and time to make the framework feasible for desired high-resolution image segmentation tasks. This will be demonstrated in the experimental results in Sect. 3.

[1] Without loss of generality, we use a factor of 2 for up- and down-sampling.

After architecture search, we still need to derive the final architecture from the relaxed variables. Due to the unique multi-scale nature of our framework, we here propose a two-step decoding approach for cell and network structure determination. In the first step, at the cell level, the normalized coefficients $\frac{\exp\{p_{i,j}\}}{\sum_{j'<i}\exp\{p_{i,j'}\}}$ and $\frac{\exp\{\alpha^{o}_{i,j}\}}{\sum_{o'\in\mathcal{O}}\exp\{\alpha^{o'}_{i,j}\}}$ are multiplied as the weight for each edge (i, j). Then the cell structure is identified by taking the operation associated to the edge with the highest weight. Simply extending this strategy to network structure is sub-optimal as the network has much larger size than a cell. On the other hand, conventional discrete optimization to decode the discrete architecture, such as dynamic programming, is infeasible due to a too high complexity of at least $O(V^2)$. Thus, here we propose a method to convert the network structure decoding to the top N_l longest path search, where N_l is the number of paths.

In the second step, at the network level, the selection of top N_l longest paths is based on the accumulated weights of all the edges in one path from input to output. As in Eq. (3), we use parameters β and $softmax$ function to formulate edge weights for the network. Kindly note that after the optimization, the sum of weights on the edges entering into a cell is always 1, which reflects the probability of strength or optimality. Inspired from this, the optimality or performance of a sub-network (corresponding to one path from input to output) can be partially measured by the sum of edge weights along the path. Therefore, to identify N_l top sub-networks is equivalent to find N_l longest paths in the DAG graph. This can be effectively solved by *Dijkstra* algorithm with a complexity of $O(E\log V)$. Kindly note that the larger N_l indicates more sub-networks to be included in the architecture, which is helpful in improving segmentation performance. As far as we know, this is the first work to incorporate MULTIPLE paths founded in a SuperNet (containing all searched paths) to make a better tradeoff between accuracy and hardware efficiency. Thus, the accuracy can approach that of SuperNet, while achieving high hardware efficiency. Experimental results in Sect. 3 will show that a small N_l can achieve higher performance than UNet, with less Flops.

The proposed two-step decoding method not only has the ability to identify a high-quality network architecture, but also provides freedom for designers to make trade-offs between segmentation performance and hardware usage by adjusting hyper-parameter N_l.

3 Experiments

3.1 Dataset and Experiment Setup

We employ three datasets from Grand Challenges to evaluate the proposed MC-NAS, including: (1) Sliver07 [16] dataset (liver CT scans, 8318 images, 20 training cases); (2) Promise12 [17] dataset (prostate MRI scans, 1377 images, 20 training cases); and (3) Chaos [18] dataset (liver, kidneys and spleen MRI scans, 1270 images, 20 training cases). The Chaos dataset also contains CT scans for livers, which is used for MS-NAS to search a network architecture. The searched architecture is then adapted to each datasets with limited training cases as reported through transfer learning. The performance of the proposed MS-NAS is compared with several state-of-the-art NAS frameworks, including AutoDeeplab, which is considered as one of the best NAS frameworks, NAS-UNet, one of the first few NAS frameworks using gradient optimization to search architecture for medical image segmentation, and a conventional UNet implementation [6,9,11]. For comparison purpose, all these methods are separately trained with evaluation conducted by 5-fold cross-validation for the same metrics, while the architecture identified by MS-NAS is transferred from the searched architecture on Chaos (CT) dataset with a very small subset for tuning.

In the proposed MS-NAS implementation, the number of network layers is 10, and the number of blocks in a cell is 3, yielding a search space of 3.89×10^9 paths, 4.22×10^8 cells, and 1.64×10^{18} possible network architectures. For contracting-cell, *maxpooling* with $stride = 2$ is used for $\frac{s}{2} \rightarrow s$ connection, while for expanding-cell, bilinear up-sampling is used for $2s \rightarrow s$ connection. For the partial channel connections, k is set to 4. N_l is varied from 3 to 5 to identify different architectures as tradeoff between accuracy and complexity, which are denoted as MS-NAS(N_l) accordingly.

The ground-truth and the input images are resized to 256×256. A total of 40 epochs of architecture search optimization are conducted, with the first 20 epochs optimizing cell parameters and the last 20 epochs for network architecture parameters. A SGD optimizer is employed with a momentum of 0.9, learning rate from 0.025 to 0.001, and a weight decay of 0.0003. The search procedure takes about 2 d to complete on a GTX 1080Ti GPU. Figure 3 plots the optimized cell and network structure with $N_l = 5$ paths, as indicated by the dashed arrows.

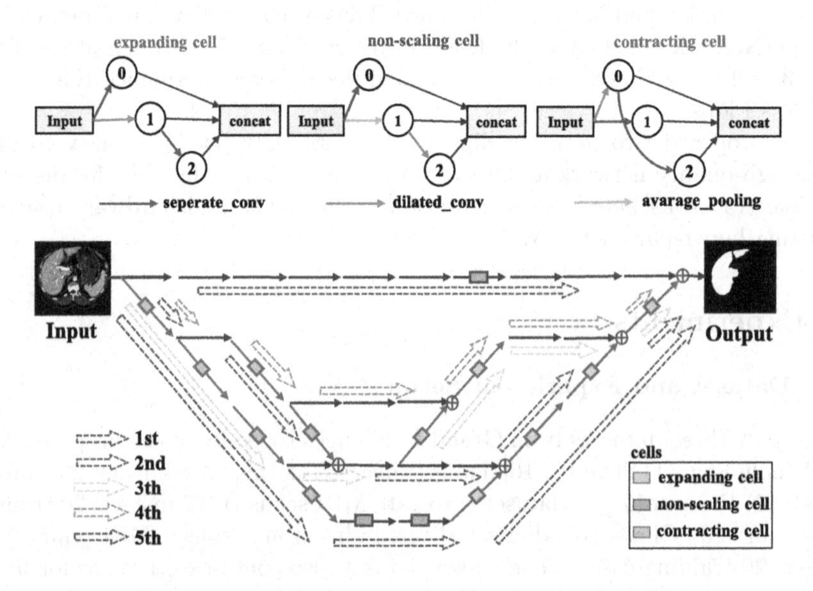

Fig. 3. Illustration of the optimized cell and network architecture.

3.2 Experimental Results

For quantitative evaluation, Table 1 presents the comparison among AutoDeeplab, UNet, NAS-UNet, and the proposed MS-NAS with N_l varied from 3 to 5 (denoted as MS-NAS(3) to MS-NAS(5)) on the metrics of mean Intersection over Union (mIOU) and Dice Similarity Coefficient (DSC) for the datasets of Silver07 and Promise12. It is clear that even with $N_l = 3$, MS-NAS significantly outperforms all the other frameworks. When $N_l = 5$, MS-NAS can achieve 1.2–4.9% improvement in average mIOU and 0.8–3.5% improvement in average DSC. It is worth noting that our proposal consumes the least computational resources, with 14.0–20.3% saving on parameter size and 18.0–24.9% saving in computational Flops. This simply indicates MS-NAS is capable to find a more optimal architecture with reduced overhead on a larger search space. Moreover, as shown in Table 2, the proposed method consistently achieves the best performance on the Chaos (MRI) dataset with 0.6–5.4% mIOU and 0.4–3.1% DSC improvement for MRI scans of all the organs. The hyper-parameter N_l trades off between accuracy and complexity to meet different demands from different tasks while maintaining the state-of-the-art performance. Finally, an example input image and the corresponding segmentation results from the Chaos (MRI) dataset are presented in Fig. 4, which qualitatively show consistently better segmentation performance by MS-NAS compared with other methods.

Table 1. Comparison of average mIOU, average DSC, model size, and computational flops for the datasets of Silver07 and Promise12.

Model	Sliver07		Promise12		Params (M)	Flops (G)
	mIOU(%)	DSC(%)	mIOU(%)	DSC(%)		
UNet	95.3±0.5	97.6±0.6	65.4±0.8	79.1±0.9	13.39	31.01
NAS-UNet	96.0±0.4	98.0±0.4	65.9±0.7	79.4±0.8	12.45	28.43
AutoDeepLab	95.2±0.6	97.5±0.6	64.2±0.9	78.2±0.9	14.45	33.06
MS-NAS(3)	96.7±0.4	98.3±0.4	70.1±0.6	82.4±0.6	10.52	21.33
MS-NAS(4)	97.1±0.4	98.4±0.4	70.8±0.6	82.9±0.7	10.52	21.33
MS-NAS(5)	**97.2±0.3**	**98.8±0.4**	**70.8±0.6**	**82.9±0.6**	11.51	23.31

Table 2. Comparison of average mIOU and average DSC for the Chaos (MRI) dataset.

Model	Liver		Right Kidney		Left Kidney		Spleen	
	mIOU(%)	DSC(%)	mIOU(%)	DSC(%)	mIOU(%)	DSC(%)	mIOU(%)	DSC(%)
UNet	88.1±0.4	93.6±0.4	76.8±0.8	86.0±0.9	73.3±0.7	84.6±0.8	79.8±0.5	88.7±0.6
NAS-UNet	88.3±0.4	93.7±0.4	77.6±0.8	87.5±0.8	74.0±0.7	85.4±0.7	80.2±0.5	89.3±0.5
AutoDeeplab	87.9±0.6	93.5±0.6	75.6±0.9	85.1±1.0	73.1±0.8	84.2±0.9	78.2±0.6	87.3±0.6
MS-NAS(3)	88.7±0.4	94.0±0.5	78.6±0.7	88.0±0.7	78.2±0.7	87.7±0.8	81.8±0.5	89.9±0.5
MS-NAS(4)	88.9±0.4	94.1±0.4	79.1±0.7	88.3±0.7	78.5±0.7	87.9±0.7	**83.0±0.5**	**90.7±0.5**
MS-NAS(5)	**88.9±0.4**	**94.1±0.4**	**79.3±0.6**	**88.4±0.7**	**79.4±0.7**	**88.5±0.7**	82.9±0.5	90.0±0.5

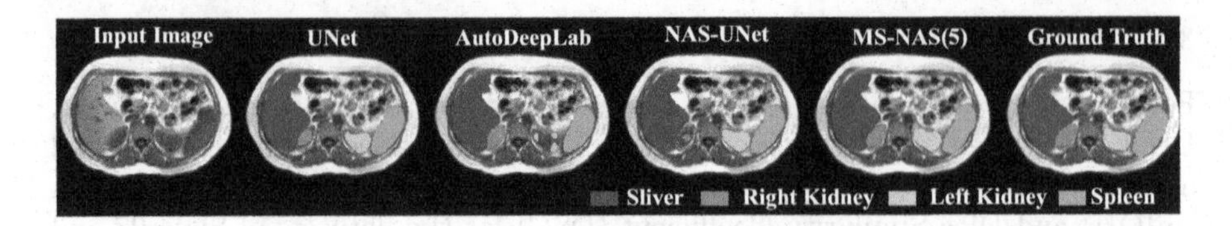

Fig. 4. Qualitative comparison of segmentation results for different methods.

4 Conclusion

In this paper, a multi-scale neural network architecture search framework is proposed and evaluated for medical image segmentation. In the proposed framework, multi-scale search space and multi-scale fusion of different tensor sizes are employed to achieve a larger search space and higher segmentation efficiency. To address the computational overhead caused by the larger search space, a partial channel connection scheme and a two-step decoding method are utilized to ensure high quality with reduced computational cost. Experimental results show that on various datasets with different modalities, MS-NAS can achieve consistently better performance than several state-of-the-art NAS frameworks with the least computational resource consumption.

Acknowledgement. This work was supported by National Key Research and Development Program Program of China [No.2018YFE0126300] and Key Area Research and Development Program of Guangdong Province [No.2018B030338001].

References

1. Lai, M.: Deep Learning for Medical Image Segmentation. arXiv preprint https://arxiv.org/pdf/1505.02000.pdf (2015)
2. Drozdzal, M., Vorontsov, E., Chartrand, G., Kadoury, S., Pal, C.: The importance of skip connections inbiomedical image segmentation. arXiv preprint https://arxiv.org/pdf/1608.04117.pdf (2016)
3. Pham, H., Guan, M.Y., Zoph, B., Le, Q.V., Dean, J.: Efficient neural architecture search via parameter sharing. arXiv preprint arXiv:1802.03268 (2018)
4. Liu, H., Simonyan, K., Yang, Y.: Darts: differentiable architecture search. arXiv preprint arXiv:1806.09055 (2018)
5. Zoph, B., Vasudevan, V., Shlens, J., Le Q.V.: Learning transferable architectures for scalable image recognition. In: CVPR (2018)
6. Liu, C., et al.: Auto-deeplab: Hierarchical neural architecture search for semantic image segmentation. arXiv preprint arXiv:1901.02985 (2019)
7. Dong, N., Xu, M., Liang, X., Jiang, Y., Dai, W., Xing, E.: Neural architecture search for adversarial medical image segmentation. In: MICCAI (2019)
8. Kim, S., et al.: Scalable neural architecture search for 3D medical image segmentation. In: MICCAI (2019)
9. Weng, Y., Zhou, T., Li, Y., Qiu, X.: NAS-Unet: neural architecture search for medical image segmentation. IEEE Access **7**, 44247–44257 (2019)

10. Xu, Y., et al.: PC-DARTS: partial channel connections for memory-efficient differentiable architecture search. arXiv preprint https://arxiv.org/abs/1907.05737v1 (2019)
11. Ronneberger, O., Fischer, P., Brox, T.: U-Net: convolutional networks for biomedical image segmentation. In: Navab, N., Hornegger, J., Wells, W.M., Frangi, A.F. (eds.) MICCAI 2015. LNCS, vol. 9351, pp. 234–241. Springer, Cham (2015). https://doi.org/10.1007/978-3-319-24574-4_28
12. Ibtehaz1and, N., Rahman, M.: MultiResUNet : rethinking the U-net architecture for multimodal biomedical image segmentation. arXiv preprint https://arxiv.org/abs/1902.04049 (2019)
13. Mortazi, A., Bagci, U.: Automatically designing CNN architectures for medical image segmentation. arXiv preprint https://arxiv.org/abs/1807.07663 (2018)
14. He, X., Zhao, K., Chu, X.: AutoML: a survey of the state-of-the-art. arXiv preprint https://arxiv.org/abs/1908.00709 (2019)
15. Zoph, B., Vasudevan, V., Shlens, J., Le, Q.V.: Learning transferable architectures for scalable image recognition. In: CVPR (2018)
16. Heimann, T., et al.: Comparison and evaluation of methods for liver segmentation from CT datasets. IEEE Trans. Med. **28**(8), 1251–1265 (2009)
17. Litjens, G., et al.: Evaluation of prostate segmentation algorithms for MRI: the promise12 challenge. Med. Image Anal. **18**(2), 359–373 (2014)
18. Kavur, A.E., Selver, M.A., Dicleış, O., Bar, M., Gezer, N.S.: CHAOS - Combined (CT-MR) healthy abdominal organ segmentation challenge data, April 2019. https://doi.org/10.5281/zenodo.3362844

Comparing to Learn: Surpassing ImageNet Pretraining on Radiographs by Comparing Image Representations

Hong-Yu Zhou, Shuang Yu, Cheng Bian(✉), Yifan Hu, Kai Ma, and Yefeng Zheng

Jarvis Lab, Tencent, Shenzhen, China
{hongyuzhou,shirlyyu,tronbian,ivanyfhu,kylekma,yefengzheng}@tencent.com

Abstract. In deep learning era, pretrained models play an important role in medical image analysis, in which ImageNet pretraining has been widely adopted as the best way. However, it is undeniable that there exists an obvious domain gap between natural images and medical images. To bridge this gap, we propose a new pretraining method which learns from 700k radiographs given no manual annotations. We call our method as *Comparing to Learn* (C2L) because it learns robust features by comparing different image representations. To verify the effectiveness of C2L, we conduct comprehensive ablation studies and evaluate it on different tasks and datasets. The experimental results on radiographs show that C2L can outperform ImageNet pretraining and previous state-of-the-art approaches significantly. Code and models are available at https://github.com/funnyzhou/C2L_MICCAI2020.

Keywords: Pretrained models · Self-supervised learning · Radiograph

1 Introduction

ImageNet [2] pretraining has been proved to be an effective way to perform 2D transfer learning for medical image analysis. Lots of experiments have shown that, compared with learning from scratch, pretrained models not only help to achieve higher accuracy but also speed up the model convergence. These benefits can be attributed to two factors: (a) effective learning algorithms designed for deep neural networks and (b) generalized feature representations learned from a great quantity of natural images. However, there exists an obvious domain gap between natural images and medical images, which raises a question whether we can have a pretrained model directly from medical images and how to approach this.

H.-Y. Zhou and S. Yu—Contributed equally.

Electronic supplementary material The online version of this chapter (https://doi.org/10.1007/978-3-030-59710-8_39) contains supplementary material, which is available to authorized users.

A. L. Martel et al. (Eds.): MICCAI 2020, LNCS 12261, pp. 398–407, 2020.
https://doi.org/10.1007/978-3-030-59710-8_39

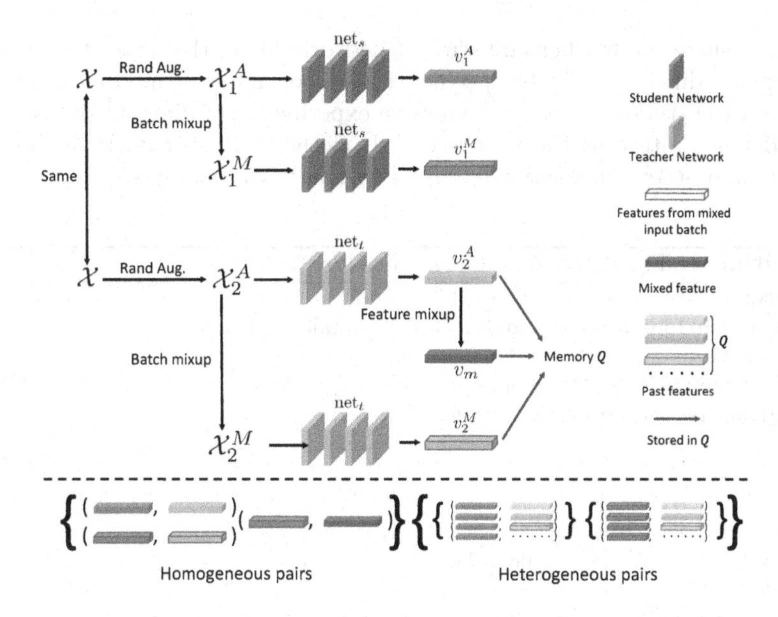

Fig. 1. Overview of proposed Comparing to Learn (C2L) framework. For clarify, we also demonstrate our definition of homogeneous and heterogeneous image pairs. Note that light color volumes in Q denote past features for v_2^A, v_m and v_2^M, respectively.

As is well known, we need domain experts' diagnosis to produce reliable medical annotations, which definitely help improve our model performance. On the other side, it is often difficult to access to a great number of doctors' conclusions considering limited medical resources as well as protection of patient privacy. So how to develop algorithms to learn from a vast amount of data without annotations has drawn more attention in the medical imaging community. Zhou *et al.* [13] proposed a self-supervised pretraining method *Model Genesis* which utilized medical images without manual labeling. On the chest X-ray classification task, Model Genesis is able to achieve comparable performance with ImageNet pretraining but still cannot beat it.

In this paper, we present a novel self-supervised pretraining approach focusing on providing pretrained 2D deep models for radiograph related tasks from massive unannotated data. We name our method as Comparing to Learn (C2L) because the goal is to learn general image representations by comparing different image features as the supervision. Different from Model Genesis [13] that resorts to an image restoration pretext task, the supervision signal of the proposed C2L comes from the self-defined representation similarity. Similar ideas have been adopted in [1,4,9], where most of them take advantage of the transitive invariance of images to produce self-supervised signal. On the contrary, in this paper, we mainly focus on feature level contrast and propose to construct homogeneous and heterogeneous data pairs by mixing image and feature batches. Moreover, a momentum-based teacher-student architecture is proposed for the contrastive learning, where the teacher and student networks share the same structure but are updated differently. To be specific, the teacher model is updated using both itself and the student network. Extensive experiments of different datasets and tasks demonstrate that the proposed C2L method can surpass ImageNet pre-training and other competitive baselines by non-trivial margins.

Algorithm 1. The detailed training procedure of C2L.

Require:
The original input image batch $\mathcal{X}$ which contains Z images;
1: $\mathcal{X}_1^A = \text{Augment}(\mathcal{X})$; $\mathcal{X}_2^A = \text{Augment}(\mathcal{X})$ // Generate two augmented batches
2: $\mathcal{X}_1^M = \text{mixup}(\mathcal{X}_1^A)$; $\mathcal{X}_2^M = \text{mixup}(\mathcal{X}_2^A)$ // Apply mixup within each batch
3: Initialize two networks net_s and net_t
4: // Q is a memory queue storing past features. Its length is N and D is the feature dimension
5: $\text{Q} = \text{Rand}(N, D)$
6: **for** $i = 1$ **to** K **do** // K is the number of training iterations
7: $\text{v}_1^A = \text{net}_s(\mathcal{X}_1^A)$; $\text{v}_1^M = \text{net}_s(\mathcal{X}_1^M)$ // we use v to stand for output feature vectors
8: $\text{v}_2^A = \text{net}_t(\mathcal{X}_2^A)$; $\text{v}_2^M = \text{net}_t(\mathcal{X}_2^M)$; $v_m = \text{mixup}(\text{v}_2^A)$
9: $\text{loss}_A = 0; \text{loss}_M = 0$
10: **for** $j = 1$ **to** Z **do**
11: // gt stands for the one-hot ground truth vector {1,0,0,...,0} whose size is N+1
12: $\text{loss}_A \mathrel{+}= \text{CE}(\{(v_{1,j}^A)^T v_{2,j}^A, (v_{1,j}^A)^T Q\}, gt)$
13: $\text{loss}_M \mathrel{+}= \text{CE}(\{(v_{1,j}^M)^T v_{2,j}^M, (v_{1,j}^M)^T Q\}, gt) + \text{CE}(\{(v_{1,j}^M)^T v_{m,j}, (v_{1,j}^M)^T Q\}, gt)$
14: **end for**
15: $\text{loss} = \text{loss}_A + \text{loss}_M$
16: Backward (loss) // Update net_s
17: $\text{net}_t = \text{Momentum}(\text{net}_t, \text{net}_s)$ // Update net_t
18: $\text{Q.insert}(v_2^A, v_2^M, v_m)$ // Update Q using current feature vectors.
19: **end for**

2 Proposed Method

In this section, we introduce the proposed Comparing to Learn (C2L) method in details. The overall workflow is provided in Fig. 1.

Batch Mixup and Feature Mixup. As shown in Fig. 1 and Algorithm 1, for each input image batch, we first use random augmentation (e.g., random cropping, rotation, and cutout [3]) to generate two augmented batches. Different from traditional image-level mixup [11], a batch-wise mixup operation is proposed to apply to each augmented batch. Suppose each batch $\mathcal{X}_i^A$ contains Z images where $i = \{1, 2\}$, we randomly shuffle $\mathcal{X}_i^A$ to construct its paired batch $\mathcal{X}_i^M$ which can be expressed as:

$$\mathcal{X}_i^M = \lambda * \mathcal{X}_i^A + (1 - \lambda) * \text{shuffle}(\mathcal{X}_i^A), \tag{1}$$

where $\lambda \sim$ Beta(1.0, 1.0) and Beta stands for the beta distribution. In practice, we found that *using the same mixing factor* λ and shuffling method for both batches ($i = \{1, 2\}$) actually helps improve the model performance. As for feature mixup in Fig. 1, we apply the same mixing strategy to the feature representations.

Teacher Network. An intuitive idea is to use the same model for both student and teacher networks. However, we found that such strategy does not work in practice which may lead to gradient explosion. Meanwhile, constructing teacher network using momentum update has been widely adopted as a way to produce stable predictions [4,8]. In our case, using momentum helps stabilize the training process and reduce the difficulty of network optimization. As shown in line 17 from Algorithm 1, the momentum function can be formalized as:

$$\text{net}_t = \theta * \text{net}_t + (1 - \theta) * \text{net}_s, \tag{2}$$

where we use an exponential factor θ to control the degree of momentum. We can see that the teacher model net_t is updated using both itself and the student network net_s. In practice, we pass $\mathcal{X}_1^A$ and $\mathcal{X}_1^M$ to the student network while $\mathcal{X}_2^A$ and $\mathcal{X}_2^M$ are passed to the teacher network. In Algorithm 1, we use $v_1^{\{A,M\}}$ to represent feature vectors from the student model and $v_2^{\{A,M\}}$ are used to represent the outputs of the teacher model.

Homogeneous and Heterogeneous Pairs. For constructing homogeneous pairs, we assume that applying data augmentation (including mixup operation) only slightly change the distribution of training data. Based on this, each homogeneous pair should contain the results after the same set of operations which includes random augmentation, batch mixup or feature mixup as shown in Fig. 1. For heterogeneous pairs, we simply contrast the current features with *all* preceding features stored in the memory queue.

Feature Comparison, Memory Q and Loss Function. As we have mentioned above, the goal of C2L is to minimize the distance between homogeneous representation pairs such as (v_1^A, v_2^A) and (v_1^M, v_2^M). Meanwhile, it is also necessary to maximize the difference between heterogeneous representations, in which we contrast current features with past features which are collected from past training iterations. To store these past representations, we employ a memory queue Q proposed in [4]. The reason why we use a large Q is that we hope to contrast current features with a great number of preceding features because more comparisons usually lead to better representations (as shown in Table 2). So the pairs of current features and past features can be formalized as (v_1^A, Q) and (v_1^M, Q), as shown in Fig. 1. For simplicity, we use v_i^Q to denote a specific feature vector in Q where $i = \{1, 2, ..., N\}$ and N is the length of queue. To be specific, we import a subscript j to index a feature vector given a batch of features. Then, we can convert this distance measurement problem to a naive classification problem. For image j in $\mathcal{X}_1^A$, $\{(v_{1,j}^A)^T v_{2,j}^A, (v_{1,j}^A)^T Q\}$ in line 12 can be further expressed as:

$$\left\{(v_{1,j}^A)^T v_{2,j}^A,\ (v_{1,j}^A)^T v_1^Q,\ (v_{1,j}^A)^T v_2^Q,\ ...,\ (v_{1,j}^A)^T v_N^Q\right\},\tag{3}$$

whose length is $N+1$. A similar case also exists in $\mathcal{X}_1^M$. For more comparison, we apply feature mixup to v_2^A. In Algorithm 1, we use v_m to represent the output of feature mixup. Similarly, we also compare v_1^M with v_m and Q which leads to another set of $N+1$ predictions:

$$\left\{(v_{1,j}^M)^T v_{m,j},\ (v_{1,j}^M)^T v_1^Q,\ (v_{1,j}^M)^T v_2^Q,\ ...,\ (v_{1,j}^M)^T v_N^Q\right\}.\tag{4}$$

It is worth noting that *the first item in each set should be larger than other items because the first item is an inner product of homogeneous representation.* Thus, we can apply a cross entropy loss (CE) to the above sets of predictions where a one hot vector {1,0,0,...,0} is treated as the ground truth for each set.

After we update the network parameters, we also update Q by inserting v_2^A, v_2^M, and v_m, respectively. Since Q is a queue and has a fixed size, the previous feature vectors are automatically removed. After we complete the training stage, only net$_s$ is extracted to become a pretrained model.

3 Datasets

3.1 Pretraining

ImageNet pretraining contains about one million labeled images which helps deep models learn general representations. In this paper, we make use of ChestX-ray14 [10], MIMIC-CXR [6], CheXpert [5] and MURA [7] as unlabeled data for network pretraining. Note that we only use ChestX-ray14 in ablation studies in order to choose appropriate hyperparameters. After that, we merge four datasets and discard their labels to perform unsupervised pretraining which has approximate 700k unlabeled radiographs.

ChestX-ray14. The training set contains 86k images while the validation set has 25k X-rays. For the ablation study, 70k images from the training set are used for self-supervised pretraining and the rest 16k images are used for fine-tuning to show the results of pretraining. Moreover, after we determine the appropriate hyperparameters, we merge the whole training set into the other three datasets. Overall, C2L uses about 700k unlabeled radiographs for model pretraining.

CheXpert. The training set has 220k images while the official validation set contains 234 images. Similar to ChestX-ray14, we only use the training set without labels for self-supervised pretraining.

MIMIC-CXR. The MIMIC-CXR dataset is a large publicly available dataset of chest radiographs in the JPEG format with structured labels derived from free-text radiology reports. The dataset contains 377,110 JPEG format images. In practice, we treat the whole dataset as an unlabeled database.

MURA. MURA is a dataset of bone X-rays. The training set contains 36k X-rays and the validation set contains 3k images. The whole dataset is used for C2L pretraining.

3.2 Fine-Tuning

We fine-tune our pretrained models on ChestX-ray14, CheXpert and Kaggle Pneumonia Detection and report their experimental results.

ChestX-ray14 and CheXpert. For ChestX-ray14, we use all labeled X-rays in the training set (86k) to fine-tune models pretrained with C2L and report experimental results on the validation set. The same setting also applies to CheXpert where we use the whole labeled training set (220k) for fine-tuning.

Kaggle Pneumonia Detection. This dataset is designed for diagnosing pneumonia automatically and accurately. We split the training set in Stage 1 into a local training set (80%) and a validation set (20%). The evaluation metric is mean average precision.

4 Implementation Details

For pretraining, we employ C2L to pretrain ResNet-18 and DenseNet-121. The default batch size is 256 and the size of each input image is 224×224. For input augmentation, we apply random crop, rotation (10°), grayscale and horizontal flip to each input batch. Moreover, we also add cutout to augmented images in order to increase the diversity of transformation for learning better representations. We use L2 normalization for each feature vector. The momentum factor θ is set to 0.999 and the length of queue Q is $2^{15} = 32768$. We use SGD as the default optimizer where the initial learning rate is 0.03 and its weight decay is 0.0001. We train each model for 240 epochs and the learning rate is divided by

Table 1. Ablation study of proposed batch mixup, feature mixup and mixed consistency loss loss_M. We use single **Mix.** to denote the traditional mixup method [11]. **Bat.** and **Feat.** are abbreviations for Batch and Feature, respectively. We report the performance of fourteen categories in ChestX-ray14. The best results are in bold while the second best are underlined.

	ImageNet	Mix	Bat. Mix	Bat. Mix. + loss_M	Bat. + Feat. Mix. + loss_M
Average	74.4	74.7	75.3	75.6	**76.3**
Atelectasis	80.0	<u>81.4</u>	80.1	80.9	**81.9**
Cardiomegaly	65.3	<u>68.2</u>	**68.4**	**68.4**	67.9
Effusion	74.9	74.9	74.3	<u>75.3</u>	**75.7**
Infiltration	68.4	67.1	66.9	<u>68.6</u>	**68.9**
Mass	79.4	79.6	80.1	<u>80.2</u>	**80.7**
Nodule	82.2	79.4	79.2	<u>80.3</u>	**82.9**
Pneumonia	72.1	73.2	73.3	<u>73.7</u>	**74.6**
Pneumothorax	77.7	80.7	80.6	<u>81.7</u>	**82.4**
Consolidation	69.6	70.9	**71.6**	<u>71.5</u>	71.2
Edema	76.4	73.6	<u>76.1</u>	74.9	**77.0**
Emphysema	64.9	66.2	<u>68.2</u>	<u>68.2</u>	**68.6**
Fibrosis	69.9	71.5	71.1	<u>72.0</u>	**72.1**
Pleural Thickening	79.5	81.3	<u>81.9</u>	**82.5**	81.7
Hernia	82.1	77.4	**82.9**	79.8	<u>82.8</u>

10 at 120, 160 and 200 epochs, respectively. When fine-tuning pretrained models on ChestX-ray14 and CheXpert, the input image size is set to 224×224 and we train both ResNet-18 and DenseNet-18 for 50 epochs. As for Kaggle Pneumonia Detection, we employ RetinaNet which uses ResNet-18 as backbone. The default image size is 512×512 and the batch size is 4.

5 Ablation Study

In this part, we conduct experiments on ChestX-ray14. As we have mentioned above, we use 70k unlabeled images for C2L pretraining and then use the rest labeled training set for fine-tuning. We report averaged AUROC performance and results of eight class are also provided (results of all fourteen classes are provided in the supplementary material). Note that to save space, the ablation study of θ (cf. Eq. 2) is put in the supplementary material.

We first report the ablation results of proposed mixup approaches in Table 1. We can see that the proposed batch mixup can already outperform the original mixup method [12] by 0.6 point on average performance. This is because using the same λ for shared batches may help maintain the consistency between batches. After adding mixed consistency loss, we can improve the batch mixup method by about 0.3 point. Since the goal of C2L is to learn powerful feature representations, we further apply mixup to generated features. Somewhat surprisingly, we can find that the propose feature mixup can be well integrated with loss_M and surpass batch mixup by approximate 1 point. In summary, the proposed mixup strategies can outperform the original mixup method by 1.6 points.

Another important component of C2L is the augmentation strategies. An appropriate augmentation method should reasonably increase the diversity of augmented batches. Such characteristic may help pretrained models to learn representations which are discriminative enough to distinguish different radiographs. In Table 2, we investigate the effects of widely adopted augmentation

Table 2. Influence of augmentation strategies and length of Q. **Mix.** represents the proposed mixup method which includes batch and feature mixup with mixed consistency loss. Note that we employ RandCrop (random crop) for all experiments.

RandCrop	Rotation	Jigsaw	Dropout	cutout	cutout + Mix.	Length of Q				Average
						2^{11}	2^{13}	2^{15}	2^{17}	
✓						✓				74.1
✓							✓			74.5
✓									✓	74.4
✓								✓		74.9
✓	✓							✓		75.2
✓	✓	✓						✓		75.0
✓	✓		✓					✓		74.9
✓	✓			✓				✓		75.4
✓	✓				✓			✓		**76.3**

strategies. It is normal to find that random rotation and cutout can enhance the performance by 0.3 point while adding jigsaw and dropout may degrade performance. Moreover, we find that simply increasing the length of Q can be harmful. We argue the reason is that a longer queue may contain more useless features and thus reduce the attention of other useful representations.

6 Fine-Tuning C2L Pretrained Models

In this section, we compare C2L pretrained models with Model Genesis, ImageNet pretraining and MoCo [4]. For pretraining datasets, we merge the training sets of ChestX-ray14 and CheXpert with radiographs in MIMIC-CXR and MURA to generate an unlabeled database containing approximate 700k images. For network architectures, we deploy ResNet-18 and DenseNet-121, both of which are widely used networks (Table 4).

Table 3. Results on ChestX-ray14. We use **MG** to represent Model Genesis [13] while **MoCo** is for Momentum Contrast [4].

	ResNet-18				DenseNet-121			
	MG	ImageNet	MoCo	C2L	MG	ImageNet	MoCo	C2L
Average	80.9	81.5	81.4	**83.5**	82.4	82.9	83.0	**84.4**
Atelectasis	79.2	80.1	79.8	**82.1**	80.7	81.2	81.7	**82.7**
Cardiomegaly	85.9	87.7	87.5	**89.7**	88.3	88.5	89.2	**90.5**
Effusion	85.7	86.2	87.0	**88.2**	87.0	86.7	86.6	**87.9**
Infiltration	67.8	68.9	68.5	**70.9**	68.9	69.6	70.2	**70.9**
Mass	81.9	82.5	83.0	**84.5**	83.6	84.4	84.0	**86.3**
Nodule	75.4	75.2	75.5	**77.2**	77.0	78.1	77.8	**79.8**
Pneumonia	74.0	74.3	74.5	**76.3**	74.4	75.1	75.7	**76.3**
Pneumothorax	85.1	85.8	85.1	**87.8**	87.0	86.8	86.5	**88.4**
Consolidation	78.3	78.6	77.9	**80.6**	80.0	79.3	79.8	**80.7**
Edema	86.9	87.4	87.2	**89.4**	87.7	88.2	88.6	**89.4**
Emphysema	89.7	89.8	90.0	**91.8**	91.0	91.6	90.7	**93.0**
Fibrosis	80.8	81.8	80.5	**83.8**	82.6	83.0	82.3	**85.1**
Pleural Thickening	76.1	76.2	76.4	**78.2**	76.8	77.2	77.5	**78.3**
Hernia	86.4	86.8	86.3	**88.8**	88.9	92.1	91.6	**92.2**

ChestX-ray14. We report fine-tuned AUROC results on the validation set. Besides ImageNet pretraining, we also perform experiments using Model Genesis (MG) [13] and recently proposed MoCo [4]. In Table 3, the proposed C2L method surpasses other approaches by a significant margin. In fact, although

MG and MoCo are able to achieve comparable results with ImageNet pretraining, they cannot surpass ImageNet pretrained model significantly. However, C2L pretrained model outperforms ImageNet pretraining by 2 points on ResNet-18. On DenseNet-121, C2L achieves 84.4% averaged AUROC which is 1.5 points higher than ImageNet pretraining.

Table 4. Results on CheXpert. We report the performance on six classes.

Method	Model	Average	Atelectasis	Cardiomegaly	Consolidation	Edema	Pleural Effusion
MG	ResNet-18	86.7	79.8	80.0	91.5	91.3	90.9
ImageNet	ResNet-18	87.0	80.3	79.6	91.9	91.7	91.5
MoCo	ResNet-18	87.1	80.3	79.4	92.5	92.0	91.1
C2L	ResNet-18	**88.2**	**81.1**	**81.4**	**93.0**	**92.9**	**92.6**
MG	DenseNet-121	87.5	80.6	81.0	92.7	91.9	91.1
ImageNet	DenseNet-121	87.9	81.5	81.9	92.4	92.1	91.7
MoCo	DenseNet-121	87.4	81.5	80.8	92.0	91.4	92.0
C2L	DenseNet-121	**89.3**	**83.3**	**83.0**	**93.6**	**92.7**	**93.8**

CheXpert. Similar to ChestX-ray14, we fine-tune the pretrained models using the training set. We can see that ImageNet pretraining performs better than MG while MoCo achieves comparable results. In contrast, C2L generates better pretrained models. On ResNet-18, C2L outperforms ImageNet pretraining by 1.2 points. When it comes to DenseNet-121, the performance gap becomes 1.4 point.

Table 5. Results of mean average precision (mAP) under different thresholds of predicted scores on Kaggle Pneumonia dataset. For each threshold, we only report predictions whose confidence scores are higher than this threshold.

	0.2	0.3	0.4	0.5	0.6	0.7	0.8
MG	12.5	16.0	18.5	20.7	21.4	21.1	20.4
ImageNet	13.5	17.1	19.9	21.4	22.2	22.1	21.1
MoCo	13.0	17.4	19.9	21.3	22.4	21.9	21.2
C2L	**14.8**	**18.4**	**21.3**	**22.4**	**23.9**	**23.1**	**22.5**

Kaggle Pneumonia Detection. We use ResNet-18 as the backbone of RetinaNet. In Table 5, it is obvious that C2L outperforms ImageNet pretraining on all thresholds significantly, especially on large thresholds. As for MoCo and MG, MoCo is marginally better than ImageNet pretraining while MG performs slightly worse.

7 Conclusion

We proposed a self-supervised pretraining method C2L (Comparing to Learn) to learn medical representations from unlabeled data. Our approach makes use of the relation between images as supervision signal and thus requires no extra manual labeling.

Acknowledgment. This work was funded by the Key Area Research and Development Program of Guangdong Province, China (No. 2018B010111001), National Key Research and Development Project (2018YFC2000702) and Science and Technology Program of Shenzhen, China (No. ZDSYS201802021814180).

References

1. Caron, M., Bojanowski, P., Joulin, A., Douze, M.: Deep clustering for unsupervised learning of visual features. In: Proceedings of the European Conference on Computer Vision, pp. 132–149 (2018)
2. Deng, J., Dong, W., Socher, R., Li, L.J., Li, K., Fei-Fei, L.: Imagenet: a large-scale hierarchical image database. In: 2009 IEEE Conference on Computer Vision and Pattern Recognition, pp. 248–255. IEEE (2009)
3. DeVries, T., Taylor, G.W.: Improved regularization of convolutional neural networks with cutout. arXiv preprint arXiv:1708.04552 (2017)
4. He, K., Fan, H., Wu, Y., Xie, S., Girshick, R.: Momentum contrast for unsupervised visual representation learning. arXiv preprint arXiv:1911.05722 (2019)
5. Irvin, J., et al.: CheXpert: a large chest radiograph dataset with uncertainty labels and expert comparison. In: Proceedings of the AAAI Conference on Artificial Intelligence, vol. 33, pp. 590–597 (2019)
6. Johnson, A.E., et al.: MIMIC-CXR: A large publicly available database of labeled chest radiographs. arXiv preprint arXiv:1901.07042 (2019)
7. Rajpurkar, P., et al.: MURA: Large dataset for abnormality detection in musculoskeletal radiographs. arXiv preprint arXiv:1712.06957 (2017)
8. Tarvainen, A., Valpola, H.: Mean teachers are better role models: Weight-averaged consistency targets improve semi-supervised deep learning results. In: Advances in Neural Information Processing Systems, pp. 1195–1204 (2017)
9. Wang, X., He, K., Gupta, A.: Transitive invariance for self-supervised visual representation learning. In: Proceedings of the IEEE International Conference on Computer Vision, pp. 1329–1338 (2017)
10. Wang, X., Peng, Y., Lu, L., Lu, Z., Bagheri, M., Summers, R.M.: ChestX-ray8: hospital-scale chest x-ray database and benchmarks on weakly-supervised classification and localization of common thorax diseases. In: Proceedings of the IEEE Conference on Computer Vision and Pattern Recognition, pp. 2097–2106 (2017)
11. Zhang, H., Cisse, M., Dauphin, Y.N., Lopez-Paz, D.: Mixup: Beyond empirical risk minimization. arXiv preprint arXiv:1710.09412 (2017)
12. Zhang, R., Isola, P., Efros, A.A.: Colorful image colorization. In: Leibe, B., Matas, J., Sebe, N., Welling, M. (eds.) ECCV 2016. LNCS, vol. 9907, pp. 649–666. Springer, Cham (2016). https://doi.org/10.1007/978-3-319-46487-9_40
13. Zhou, Z., et al.: Models genesis: generic autodidactic models for 3D medical image analysis. In: Shen, D., et al. (eds.) MICCAI 2019. LNCS, vol. 11767, pp. 384–393. Springer, Cham (2019). https://doi.org/10.1007/978-3-030-32251-9_42

Dual-Task Self-supervision for Cross-modality Domain Adaptation

Yingying Xue, Shixiang Feng, Ya Zhang(✉), Xiaoyun Zhang, and Yanfeng Wang

Cooperative Medianet Innovation Center,
Shanghai Jiao Tong University, Shanghai, China
{xueyingying,fengshixiang,ya_zhang,xiaoyun.zhang,wangyanfeng}@sjtu.edu.cn

Abstract. Data annotation is always an expensive and time-consuming issue for deep learning based medical image analysis. To ease the need of annotations, domain adaptation is recently introduced to generalize neural networks from a labeled source domain to unlabeled target domain without much performance degradation. In this paper, we propose a novel target domain self-supervision for domain adaptation by constructing an edge generation auxiliary task to assist primary segmentation task so as to extract better target representation and improve target segmentation performance. Besides, in order to leverage detailed information contained in low-level features, we propose a hierarchical low-level adversarial learning mechanism to encourage low-level features domain uninformative in a hierarchical way, so that the segmentation performance can benefit from low-level features without being affected by domain shift. Following these two proposed approach, we develop a cross-modality domain adaptation framework which employs the dual-task collaboration for target domain self-supervision, and encourages low-level detailed features domain uninformative for better alignment. Our proposed framework achieves state-of-the-art results on public cross-modality segmentation datasets.

Keywords: Domain adaptation · Dual-task collaboration · Hierarchical adversarial learning

1 Introduction

For medical image analysis based on deep learning, a great challenge remains that deep learning models require high-quality and large quantities of annotated images. This problem results in expensive data collection and repeatedly annotation workload. Furthermore, annotating different image modalities such as CT and MR of the same organ makes the issue more pronounced. Consequently, an annotation-efficient deep learning method, namely unsupervised domain adaptation (UDA), is introduced to address cross-modality medical image analysis.

Unsupervised domain adaptation generalizes the learning model trained on annotated source domain to another unlabeled target domain without any target label supervision. For semantic segmentation task, many existing UDA methods

A. L. Martel et al. (Eds.): MICCAI 2020, LNCS 12261, pp. 408–417, 2020.
https://doi.org/10.1007/978-3-030-59710-8_40

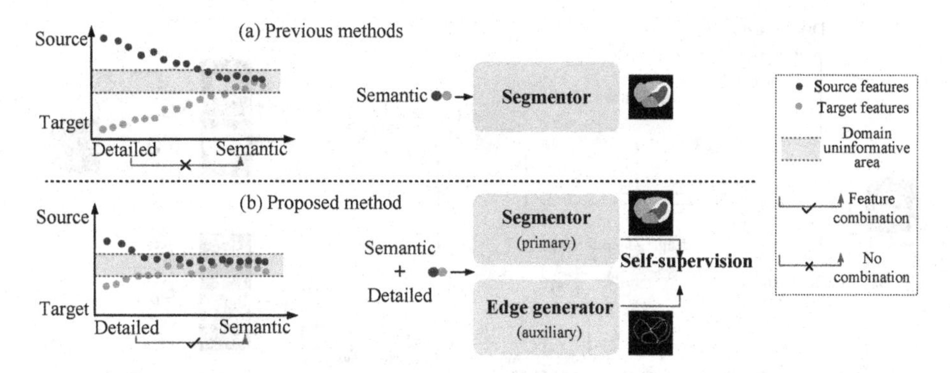

Fig. 1. Comparison between our proposed method and previous methods. (a) Previous methods where the detailed low-level domain informative features are not utilized for single way segmentation. (b) Proposed method encourages low-level detailed features to be domain uninformative for following segmentation, and employs self-supervision on target domain by constructing an edge generation task as auxiliary task.

[1–3,8] borrow the idea of image-to-image translation from CycleGAN [15] and multi-modal image translation network [6], so that the aligned images of two domains can be learned together under source domain supervision. Another main stream of UDA strategies employs adversarial learning to align source and target domain, where a common way is to follow [11] to utilize adversarial learning at the segmentation output [2,13], or at the segmentation entropy map [12,14], or in the VAE-based latent space [9].

These UDA methods have two drawbacks. Firstly, simple adversarial learning between source and target domain is not enough to completely align two domains, especially when unpaired source and target modalities vary much in medical images. Under unsupervised conditions, the edge region of target domain segmentation mask may be very inaccurate and has a high probability to over segmented or under segmented. Therefore, we propose a novel self-supervision on target domain to directly improve target domain performance. Specifically, we propose an auxiliary task that generates edges to assist primary segmentation task to improve prediction accuracy around contour. These two tasks are collaborated through a designed edge consistency function and their partially shared parameters, where two tasks share a common feature extractor and partial layers in decoder.

Secondly, existing methods employ segmentation task on aligned semantic features, without considering rich detailed information in low-level features, because domain information contained in low-level features can harm the adaptation performance. But detailed features can benefit medical image segmentation which are proved by the great success of U-net [10], and should also be considered. Therefore, to leverage detailed information in low-level features, while simultaneously reduce the adaptation degradation results from the skip-connection in U-net, we propose a hierarchical low-level adversarial learning mechanism to encourage low-level detailed features domain uninformative in a hierarchical way according to the content of domain.

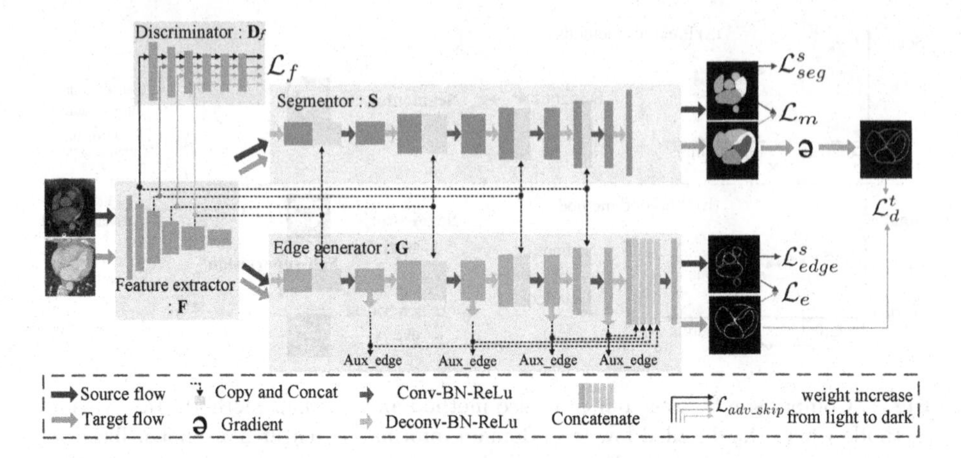

Fig. 2. Our proposed framework. The feature extractor F generates domain invariant features, and the hierarchical discriminator D_f differentiates input features accordingly. Segmentor S and edge generator G take features from corresponding layers of F to generate segmentation masks and edges. Two discriminators D_m and D_e (omitted in this figure) are employed at the output of S and G for adversarial learning.

In general, the comparison between our Dual-task and Hierarchical learning Network (DualHierNet) and previous UDA methods is shown in Fig. 1.

2 Methodology

Given N_s pixel-level labeled source domain data $\{X^s, Y^s\} = \{(x^s_i, y^s_i)\}_{i=1}^{N_s}$ and N_t unlabeled target domain data $X^t = \{x^t_i\}_{i=1}^{N_t}$, unsupervised domain adaptation aims to use these data to learn a source to target adaptation network to correctly segment target images without any target domain supervision.

The architecture of proposed DualHierNet is shown in Fig. 2. Target domain self-supervision is achieved through the edge consistency between partially shared primary segmentation task S and auxiliary edge generation task G. Also, the low-level features extracted by feature extractor F are encouraged to be domain invariant through the adversarial learning with discriminator D_f in a hierarchical way according to the domain content. Lastly, two discriminators D_m and D_e are employed on output semantic space to align generated segmentation masks and edges respectively.

2.1 Dual-Task Collaboration for Target Domain Self-supervision

Under unsupervised conditions, the edge region of target domain segmentation mask may be inaccurate and has a high probability to over or under segmented. We therefore propose a novel target self-supervision by constructing an auxiliary task to generate edges, and making it collaborate with primary segmentation task to obtain a more accurate target segmentation mask at the edge region.

Specifically, feature extractor F generates domain invariant features from input images of source and target domain, where this part will be illustrated in later subsection. Domain invariant features are input to both segmentor S and edge generator G. The edge generator G has a similar network structure to segmentor S, with low-level features of F copied and concatenated to corresponding high-level features. Besides, edge generator employs deep-supervision for the purpose of providing auxiliary supervision to improve edge generation quality by outputting upsampled features in G as auxiliary edges shown in Fig. 2. These auxiliary edges are fused together to obtain the final generated edge p_e.

For source domain supervision, we use a combination of weighted cross-entropy loss and Dice loss: $\mathcal{L}^s(p^s, y^s) = \mathcal{L}^s_{wCE} + \mathcal{L}^s_{Dice}$, where p^s and y^s are segmentation mask and ground truth. We employ multi-class and two-class segmentation for segmentor S and edge generator G:

$$\begin{aligned} \mathcal{L}^s_{seg} &= \mathcal{L}^s(p^s_m, y^s_m), \\ \mathcal{L}^s_{edge} &= \mathcal{L}^s(p^s_e, y^s_e) + \sum_{Ae} \mathcal{L}^s(p^s_{Ae}, y^s_e), \end{aligned} \tag{1}$$

where $\mathcal{L}^s_{seg}$ and $\mathcal{L}^s_{edge}$ are objective functions of S and G respectively. p^s_m and y^s_m is the segmentation mask and ground truth of source domain, and p^s_e and y^s_e is the generated edge and ground truth edge. Noted that y^s_e is obtained by calculating the first derivative of y^s_m. p^s_{Ae} are auxiliary edges shown in Fig. 2.

For target domain self-supervision, we encourage the segmentation mask p^t_m and generated edge p^t_e to keep consistency at the edges and we propose a dual-task consistency loss $\mathcal{L}^t_d$ on target domain. An operation ∂ calculates the first derivative of soft segmentation mask p^t_m on two spatial axes i, j to obtain a soft edge, which should possess structural consistency with generated edge p^t_e. The consistency loss and the soft edge calculation formula are:

$$\begin{aligned} \mathcal{L}^t_d &= \left\| p^t_e - \partial(p^t_m) \right\|^2_2, \\ \partial(p^t_m) &= \frac{1}{2}(\left|\frac{\partial p^t_m}{\partial i}\right| + \left|\frac{\partial p^t_m}{\partial j}\right|) \approx \frac{1}{2}(\sum_c \left|p^t_{m,i+1} - p^t_{m,i}\right| + \sum_c \left|p^t_{m,j+1} - p^t_{m,j}\right|), \end{aligned} \tag{2}$$

where the summation symbol is applied to channel dimension c. The soft edge $\partial(p^t_m)$ has a probability between [0,1].

2.2 Hierarchical Adversarial Learning for Better Alignment

Hierarchical Adversarial Learning. We follow the success of U-net in medical image segmentation [10] to combine low-level detailed features with high-level semantic features. However, low-level features are domain informative, and severe domain gap in detailed features can harm adaptation performance when combined with domain uninformative semantic features. We thus develop a hierarchical adversarial skip connection mechanism to make low-level detailed features domain invariant when concatenating them to semantic features simultaneously.

Specifically, feature extractor F maps input images to feature space, and we propose a hierarchical discriminator D_f to differentiate input domains accordingly. Features of each layer in F: l_1, l_2, l_3, l_4 and l_5, are gradually decreasing in domain information and increasing in semantic information. l_5 is directly input to following segmentor S and edge generator G, while l_1, l_2, l_3, l_4 are input to different layers of discriminator D_f in a hierarchical way according to their distinct resolutions for domain alignment. The objective function of layer $l_{k,k=1,2,\ldots,K}$ is formulated as follows where F and D_f play a min-max game:

$$\begin{aligned}\mathcal{L}_f &= \sum_{k=1}^{K} \gamma_k \mathcal{L}_{f,k}, \\ \mathcal{L}_{f,k} &= \mathbb{E}_{l_k^s \in F(X^s)}[\log(D_f(l_k^s))] + \mathbb{E}_{l_k^t \in F(X^t)}[\log(1 - D_f(l_k^t))],\end{aligned} \tag{3}$$

where γ_k increases as k decreases, indicating that lower layer features contained more domain information is assigned with larger weights for attention.

Output Alignment. Finally, two discriminators D_m and D_e are employed in output space to align segmentation mask p_m and generate edge p_e with adversarial learning. $\mathcal{L}_m$ and $\mathcal{L}_e$ are the adversarial objective as follows respectively:

$$\begin{aligned}\mathcal{L}_m &= \mathbb{E}_{x^s \sim X^s}[\log(D_m(p_m^s))] + \mathbb{E}_{x^t \sim X^t}[\log(1 - D_m(p_m^t))], \\ \mathcal{L}_e &= \mathbb{E}_{x^s \sim X^s}[\log(D_e(p_e^s))] + \mathbb{E}_{x^t \sim X^t}[\log(1 - D_e(p_e^t))].\end{aligned} \tag{4}$$

Therefore, with trade-off parameters $\lambda_0, \lambda_1, \lambda_2, \lambda_3$, the total objective function of the model is formulated as:

$$\begin{aligned}\min_{F,S,G} \max_{D_f,D_m,D_e} \; & \mathcal{L}_{seg}^s + \mathcal{L}_{edge}^s + \lambda_0 \mathcal{L}_d^t \\ & + \lambda_1 \mathcal{L}_f + \lambda_2 \mathcal{L}_m + \lambda_3 \mathcal{L}_e.\end{aligned} \tag{5}$$

3 Experiments and Results

Dataset and Implementation Details. The proposed framework is evaluated on the *Multimodality Whole Heart Segmentation Challenge* MMWHS2017 dataset [17] which consists of unpaired 20 CT and 20 MR volumes with pixel-level annotation of seven heart structures: left ventricle blood cavity (LV), right ventricle blood cavity (RV), left atrium blood cavity (LA), right atrium blood cavity (RA), myocardium of the left ventricle (Myo), ascending aorta (AA) and pulmonary artery (PA). We follow Pnp-AdaNet and SIFA [2,4] to use randomly selected sixteen MR volumes as source and sixteen CT as target for training. The remaining four CT volumes are for testing. Each volume is split into transverse view slices as inputs since doctors observe transverse view to diagnose cardiac diseases, and is augmented with flipping, rotation and scaling, and normalized to zero mean and unit variance and resized to 256 × 256. The volume metrics Dice score and Average Surface Distance (ASD) are employed for evaluation.

For fair comparison, 5-fold cross validation is employed. All annotations of CT are only used for evaluation without being presented during training.

We also validate our proposed method on another multi-modality cardiac dataset: MS-CMRSeg 2019 [16] which consists of 45 patients and each patient has cardiac images of three MR modalities: bSSFP, T2 and LGE. For fair comparison, we re-implement methods [1,13] under the same experiment setup with us, and follow [13] to combine labeled bSSFP and T2 as source and unlabeled LGE as target, where the target LGE is divided by competition, and use transverse view slices with the same preprocessing and augmentation as above.

The detailed dual-task architecture is shown in Fig. 2. Discriminators follow [7] to have 6 convolutional layers where the first 3 use instance normalization. Adam optimizers are utilized with a learning rate of 1.0×10^{-3} for segmentation and edge generation, but with a decay rate of 0.9 every 2 epochs for segmentation and no decay for edge generation, since we empirically found edge generation task converges slower than segmentation. The model is trained for 100 epochs with a batch size equals 4. Hyper-parameters λ_0 is 10, λ_1 is 1.0, while λ_2, λ_3 grow linearly from 0.0 to 1.0 as epoch increases to 40, and remain 1.0 subsequently.

Quantitative and Qualitative Analysis. For MMWHS2017 dataset, we validate our methods on seven structures and show the results in Table 1, and also follow [2,4] to validate on four left-side structures in Table 2. We compare with several state-of-the-art UDA methods including CyCADA [5], Pnp-AdaNet [4], BEAL [14], Cascaded U-net [1] and SIFA [2]. We re-implemented all above methods under the same experiment setup with five-fold cross validation shown in the mean $\pm$ std manner, and no post-processing is employed.

In Table 1, we first obtain the unadapted results by directly testing a source domain trained U-net on target domain, and a Dice score of 30.43% reflects severe domain shift between different modalities. A supervised target domain upper bound of 84.95% is also obtained through a supervised U-net. Our proposed method outperforms several UDA methods by a great margin and achieves superior performance of 73.68% in average Dice and 7.3 in average ASD. Note that our approach significantly improves the accuracy of LA by achieving a performance gain up to 9.4% in Dice, and even the most difficult structure to segment: Myo, is also improved to 64.03%. For four class segmentation of LV, LA, Myo and AA shown in Table 2, we achieve an average Dice of 76.98% and average ASD of 4.6, with great margin compared with other methods. Results on MS-CMRSeg shown in Table 3 prove the generalization ability of our method on cross-MR modalities by achieving an average Dice of 84.85%.

Visual results are shown in Fig. 3. Our DualHierNet has a smoother 3D heart with clearer contours, and better segmentation masks inside cardiac structures. For generated edges in lower part of Fig. 3, figures inside the red box are good examples that generated edge p_E^t and $\partial(p_M^t)$ are well constrained to be similar. Figures inside the blue box are poor examples, where the blue arrows point to boundary area that are distinct in p_E^t and $\partial(p_M^t)$. This usually results from incoherent annotation between two adjacent slices.

Table 1. Comparison results on *MMWHS2017* for 7 cardiac structures.

Methods	↑ Dice [%]								↓ ASD [voxel]							
	LV	RV	LA	RA	Myo	AA	PA	Average	LV	RV	LA	RA	Myo	AA	PA	Average
Upper Bound	88.02 ±4.34	86.01 ±3.91	85.42 ±5.28	87.84 ±6.01	81.27 ±4.24	82.22 ±6.12	83.91 ±7.71	84.95 ±5.72	5.4 ±3.2	4.7 ±2.9	4.2 ±2.4	4.6 ±3.6	5.4 ±3.7	2.3 ±1.8	4.8 ±2.6	4.5 ±3.7
W/o Adaptation	15.87 ±3.72	15.21 ±4.54	52.14 ±13.74	45.65 ±6.20	6.54 ±3.98	43.82 ±13.24	33.81 ±5.77	30.43 ±9.26	52.4 ±16.4	56.8 ±17.1	31.9 ±15.0	24.4 ±14.3	36.0 ±29.6	42.1 ±20.7	31.2 ±22.8	39.26 ±21.1
Pnp-AdaNet [4]	50.49 ±10.46	51.43 ±6.72	70.18 ±5.41	72.24 ±4.18	31.18 ±10.09	64.92 ±8.44	57.49 ±9.23	56.85 ±8.20	13.6 ±2.5	8.5 ±3.7	15.4 ±6.8	12.8 ±5.1	9.4 ±2.8	9.7 ±4.9	8.9 ±5.1	11.2 ±5.2
CyCADA [5]	55.49 ±9.75	54.16 ±13.47	66.90 ±10.05	68.70 ±5.42	50.03 ±8.80	68.42 ±9.67	65.77 ±6.96	61.35 ±9.58	10.6 ±3.4	8.1 ±4.2	9.3 ±2.3	9.4 ±5.5	8.2 ±4.7	8.6 ±4.6	7.4 ±3.8	8.8 ±4.2
BEAL [14]	65.23 ±7.85	58.01 ±9.22	64.00 ±8.44	62.04 ±9.91	56.10 ±6.80	72.05 ±9.80	61.95 ±9.67	62.77 ±8.21	10.2 ±3.8	7.9 ±5.1	9.2 ±2.1	10.1 ±4.8	8.4 ±3.6	8.2 ±4.3	7.0 ±3.9	8.7 ±4.3
Cascaded U-net [1]	72.35 ±8.33	60.14 ±9.24	67.75 ±9.18	69.88 ±8.74	60.74 ±5.65	73.25 ±7.42	62.45 ±8.77	66.65 ±9.33	9.6 ±4.1	7.7 ±5.6	8.8 ±3.0	9.7 ±3.6	8.2 ±3.0	8.2 ±3.1	6.7 ±4.2	8.4 ±4.0
SIFA [2]	79.56 ±8.17	**70.79** ±11.24	69.77 ±9.55	72.02 ±7.00	61.48 ±6.57	75.59 ±9.45	63.82 ±8.45	70.43 ±8.73	8.4 ±4.1	7.2 ±3.2	8.1 ±4.6	8.6 ±4.9	8.0 ±2.8	6.6 ±3.1	6.8 ±3.2	7.7 ±3.9
DualHierNet (Ours)	**81.58** ±9.70	70.23 ±8.45	**79.17** ±9.19	**73.30** ±8.29	**64.03** ±7.97	**80.25** ±7.24	**67.22** ±7.16	**73.68** ±8.93	**8.2** ±3.6	**7.2** ±3.0	**7.0** ±2.8	**8.4** ±2.9	**7.7** ±2.6	**6.4** ±3.2	**6.3** ±2.7	**7.3** ±3.4

Table 2. Comparison results on *MMWHS2017* for 4 cardiac structures.

Methods	↑ Dice [%]					↓ ASD [voxel]				
	LV	LA	Myo	AA	Average	LV	LA	Myo	AA	Average
Upper Bound	89.54 ±5.27	87.72 ±1.58	86.06 ±1.97	85.21 ±6.67	87.13 ±5.24	3.3 ±3.0	4.3 ±2.6	4.9 ±1.2	1.6 ±1.1	3.5 ±3.3
W/o Adaptation	13.42 ±2.31	46.05 ±20.32	12.25 ±10.92	20.39 ±10.96	23.03 ±18.47	47.1 ±17.9	22.8 ±19.9	24.5 ±12.1	42.7 ±12.6	34.3 ±19.8
Pnp-AdaNet [4]	52.32 ±21.00	75.75 ±4.35	28.73 ±13.31	73.86 ±7.48	57.67 ±13.85	9.2 ±3.9	13.6 ±3.6	8.8 ±4.3	11.5 ±2.9	10.8 ±4.8
CycADA [5]	61.90 ±10.78	68.95 ±5.26	50.83 ±7.06	74.08 ±7.30	63.94 ±8.59	11.4 ±7.0	6.3 ±2.3	14.7 ±3.8	10.8 ±3.2	10.8 ±5.1
BEAL [14]	68.49 ±11.23	62.77 ±9.47	57.93 ±7.59	75.47 ±9.67	66.17 ±10.25	9.8 ±5.0	7.0 ±3.4	8.9 ±4.9	7.7 ±5.2	8.4 ±4.9
Cascaded U-net [1]	70.05 ±9.60	65.28 ±10.16	60.66 ±9.74	77.36 ±6.28	68.34 ±9.31	9.2 ±4.2	6.8 ±4.9	8.2 ±4.9	7.6 ±3.4	8.0 ±4.8
SIFA [2]	77.39 ±11.04	73.35 ±12.20	60.68 ±5.02	84.55 ±4.05	73.99 ±8.24	4.2 ±2.5	5.5 ±3.8	5.9 ±3.7	**4.4** ±2.1	5.0 ±3.4
DualHierNet (Ours)	**83.42** ±7.46	**74.61** ±10.07	**65.19** ±6.33	**84.70** ±6.41	**76.98** ±7.84	**3.6** ±1.7	**5.3** ±2.0	**4.8** ±2.2	4.5 ±2.8	**4.6** ±2.3

Table 3. Comparison results on *MS-CMRSeg*.

Methods	↑ Dice [%]				↓ ASD [voxel]			
	LV	RV	Myo	Average	LV	RV	Myo	Average
Cascaded U-net [1]	82.05	80.90	81.47	81.47	1.8	1.9	1.8	1.8
U-net+GFRM [13]	85.42	83.64	79.62	82.89	1.6	1.7	1.8	1.7
DualHierNet (Ours)	**88.91**	**84.01**	**81.62**	**84.85**	**1.5**	**1.7**	**1.8**	**1.7**

Ablation Study. Firstly, we conduct an ablation experiment to evaluate the effectiveness of each component: (i) U-net with output adversarial learning (Base), (ii) Base equipped with dual-task collaboration (Base+Dual), (iii) Base with hierarchical adversarial learning (Base+Hier), and (iv) ours (Base+Dual+Hier). In Table 4, the performance is improved to 68.50% and

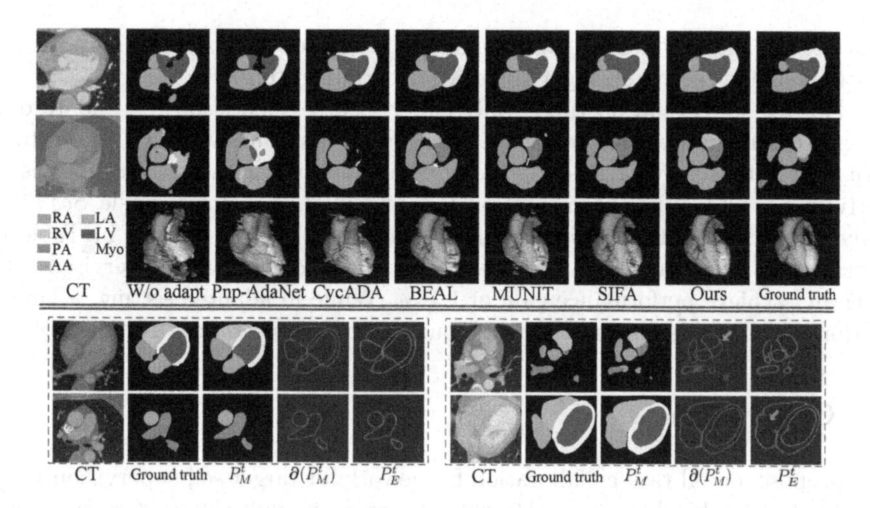

Fig. 3. Visual results of comparison and generated edges.

Table 4. Effects of each component.

Methods	LV	LA	Myo	AA	Avg Dice [%]
W/o Adaptation	13.42	46.05	12.25	20.39	23.03
Base	67.20	54.46	53.27	69.39	61.08
Base+Dual	76.82	60.62	57.48	79.09	68.50
Base+Hier	78.11	63.74	60.42	81.28	70.89
Base+Dual+Hier	**83.42**	**74.61**	**65.19**	**84.70**	**76.98**

Table 5. Effect of hierarchical weights.

$\gamma_1 : \gamma_2 : \gamma_3 : \gamma_4$	LV	RV	Myo	AA	Avg Dice [%]
1 : 2 : 3 : 4	78.10	67.05	58.14	79.32	70.65
1 : 1 : 1 : 1	79.24	69.95	60.33	80.90	72.61
4 : 3 : 2 : 1	**83.42**	**74.61**	**65.19**	**84.70**	**76.98**

Table 6. Dual-task self-supervision extended on supervised setting.

Target domain	Model	LV	RV	Myo	AA	Avg Dice [%]
Supervised	Seg+Seg	90.02	87.62	86.98	89.40	88.51
	Seg+Edge	**90.17**	**89.06**	**88.11**	**91.24**	**89.65**
Adapted	Seg+Seg	79.64	68.11	60.28	79.69	71.93
	Seg+Edge	**83.42**	**74.61**	**65.19**	**84.70**	**76.98**

70.89% equipped with our proposed dual-task self-supervision and hierarchical strategy respectively. The further improvement to 76.98% in our DualHierNet confirms the effect of using dual-task as self-supervision and hierarchically aligning low-level features.

Secondly, we experiment on choice of hierarchical weights γ_k shown in Table 5. When we assign larger weights to higher layers, only an average Dice of 70.65% is achieved. A Dice of 72.61% is achieved if each layer shares a same weight. When we enlarge the weights of shallow layers which contain more domain information, we can get a Dice of 76.98%. This further justify that low-level domain informative features should receive stronger adversarial learning attention.

Thirdly, we extend on target-only supervised segmentation to validate our proposed self-supervision. We replace *Seg+Edge* structure with two segmentors *Seg+Seg* so that they have nearly the same number of parameters. In supervised setting, *Seg+Edge* uses segmentation loss and dual_consist loss while *Seg+Seg* only uses segmentation loss to train the networks. Results shown in Table 6 reveals that auxiliary edge task assists segmentation even on supervised setting and achieve performance gain of 1.14%. While in adapted setting, a larger performance gain is obtained through our proposed dual-task self-supervision.

4 Conclusion

We propose a dual-task collaboration framework for target self-supervision with low-level hierarchical adversarial learning for cross-modality image segmentation. We develop a novel self-supervision by constructing an auxiliary task to generate edges to assist segmentation task, and we also design a hierarchical adversarial mechanism according to the content of domain. Our framework outperforms several adaptation methods on cross-modality datasets and the proposed dual-task architecture even achieves promising performance in supervised setting.

Acknowledgement. This work is supported by SHEITC (No. 2018-RGZN-02046), 111 plan (No. BP0719010), and STCSM (No. 18DZ2270700).

References

1. Chen, C., et al.: Unsupervised multi-modal style transfer for cardiac MR segmentation. In: Pop, M., et al. (eds.) STACOM 2019. LNCS, vol. 12009, pp. 209–219. Springer, Cham (2020). https://doi.org/10.1007/978-3-030-39074-7_22
2. Chen, C., Dou, Q., Chen, H., Qin, J., Heng, P.A.: Synergistic image and feature adaptation: towards cross-modality domain adaptation for medical image segmentation. In: Proceedings of the AAAI Conference on Artificial Intelligence, vol. 33, pp. 865–872 (2019)
3. Chen, Y.C., Lin, Y.Y., Yang, M.H., Huang, J.B.: Crdoco: pixel-level domain transfer with cross-domain consistency. In: Proceedings of the IEEE Conference on Computer Vision and Pattern Recognition, pp. 1791–1800 (2019)
4. Dou, Q., et al.: PNP-adanet: plug-and-play adversarial domain adaptation network at unpaired cross-modality cardiac segmentation. IEEE Access (2019)
5. Hoffman, J., et al.: Cycada: Cycle-consistent adversarial domain adaptation. In: International Conference on Machine Learning, pp. 1989–1998 (2018)
6. Huang, X., Liu, M.Y., Belongie, S., Kautz, J.: Multimodal unsupervised image-to-image translation. In: Proceedings of the European Conference on Computer Vision (ECCV), pp. 172–189 (2018)
7. Isola, P., Zhu, J.Y., Zhou, T., Efros, A.A.: Image-to-image translation with conditional adversarial networks. In: Proceedings of the IEEE Conference on Computer Vision and Pattern Recognition, pp. 1125–1134 (2017)

8. Jiang, J., et al.: Cross-modality (CT-MRI) prior augmented deep learning for robust lung tumor segmentation from small MR datasets. Med. Phys. **46**(10), 4392–4404 (2019)
9. Ouyang, C., Kamnitsas, K., Biffi, C., Duan, J., Rueckert, D.: Data efficient unsupervised domain adaptation for cross-modality image segmentation. In: Shen, D., et al. (eds.) MICCAI 2019. LNCS, vol. 11765, pp. 669–677. Springer, Cham (2019). https://doi.org/10.1007/978-3-030-32245-8_74
10. Ronneberger, O., Fischer, P., Brox, T.: U-net: convolutional networks for biomedical image segmentation. In: Navab, N., Hornegger, J., Wells, W.M., Frangi, A.F. (eds.) MICCAI 2015. LNCS, vol. 9351, pp. 234–241. Springer, Cham (2015). https://doi.org/10.1007/978-3-319-24574-4_28
11. Tsai, Y.H., Hung, W.C., Schulter, S., Sohn, K., Yang, M.H., Chandraker, M.: Learning to adapt structured output space for semantic segmentation. In: Proceedings of the IEEE Conference on Computer Vision and Pattern Recognition, pp. 7472–7481 (2018)
12. Vu, T.H., Jain, H., Bucher, M., Cord, M., Pérez, P.: Advent: adversarial entropy minimization for domain adaptation in semantic segmentation. In: Proceedings of the IEEE Conference on Computer Vision and Pattern Recognition, pp. 2517–2526 (2019)
13. Wang, J., Huang, H., Chen, C., Ma, W., Huang, Y., Ding, X.: Multi-sequence cardiac MR segmentation with adversarial domain adaptation network. In: Pop, M., et al. (eds.) STACOM 2019. LNCS, vol. 12009, pp. 254–262. Springer, Cham (2020). https://doi.org/10.1007/978-3-030-39074-7_27
14. Wang, S., Yu, L., Li, K., Yang, X., Fu, C.-W., Heng, P.-A.: Boundary and entropy-driven adversarial learning for fundus image segmentation. In: Shen, D., et al. (eds.) MICCAI 2019. LNCS, vol. 11764, pp. 102–110. Springer, Cham (2019). https://doi.org/10.1007/978-3-030-32239-7_12
15. Zhu, J.Y., Park, T., Isola, P., Efros, A.A.: Unpaired image-to-image translation using cycle-consistent adversarial networks. In: Proceedings of the IEEE International Conference on Computer Vision, pp. 2223–2232 (2017)
16. Zhuang, X.: Multivariate mixture model for myocardial segmentation combining multi-source images. IEEE Trans. Pattern Anal. Mach. Intell. **41**(12), 2933–2946 (2018)
17. Zhuang, X., Shen, J.: Multi-scale patch and multi-modality atlases for whole heart segmentation of MRI. Med. Image Anal. **31**, 77–87 (2016)

Dual-Teacher: Integrating Intra-domain and Inter-domain Teachers for Annotation-Efficient Cardiac Segmentation

Kang Li[1], Shujun Wang[1], Lequan Yu[2(✉)], and Pheng-Ann Heng[1,3]

[1] Department of Computer Science and Engineering, The Chinese University of Hong Kong, Shatin, Hong Kong
{kli,sjwang,pheng}@cse.cuhk.edu.hk

[2] Department of Radiation Oncology, Stanford University, Stanford, USA
ylqzd2011@gmail.com

[3] Guangdong Provincial Key Laboratory of Computer Vision and Virtual Reality Technology, Shenzhen Institutes of Advanced Technology, Chinese Academy of Sciences, Shenzhen, China

Abstract. Medical image annotations are prohibitively time-consuming and expensive to obtain. To alleviate annotation scarcity, many approaches have been developed to efficiently utilize extra information, *e.g.*, semi-supervised learning further exploring plentiful unlabeled data, domain adaptation including multi-modality learning and unsupervised domain adaptation resorting to the prior knowledge from additional modality. In this paper, we aim to investigate the feasibility of simultaneously leveraging abundant unlabeled data and well-established cross-modality data for annotation-efficient medical image segmentation. To this end, we propose a novel semi-supervised domain adaptation approach, namely Dual-Teacher, where the student model not only learns from labeled target data (*e.g.*, CT), but also explores unlabeled target data and labeled source data (*e.g.*, MR) by two teacher models. Specifically, the student model learns the knowledge of unlabeled target data from intra-domain teacher by encouraging prediction consistency, as well as the shape priors embedded in labeled source data from inter-domain teacher via knowledge distillation. Consequently, the student model can effectively exploit the information from all three data resources and comprehensively integrate them to achieve improved performance. We conduct extensive experiments on MM-WHS 2017 dataset and demonstrate that our approach is able to concurrently utilize unlabeled data and cross-modality data with superior performance, outperforming semi-supervised learning and domain adaptation methods with a large margin.

Keywords: Semi-supervised domain adaptation · Cross-modality segmentation · Cardiac segmentation

A. L. Martel et al. (Eds.): MICCAI 2020, LNCS 12261, pp. 418–427, 2020.
https://doi.org/10.1007/978-3-030-59710-8_41

1 Introduction

Deep convolutional neural networks (CNNs) have made great progress in various medical image segmentation applications [15,19]. The success is partially relied on massive datasets with abundant annotations. However, collecting and labeling such large-scaled dataset is prohibitively time-consuming and expensive, especially in medical area, since it requires diagnostic expertise and meticulous work [13]. Plenty of efforts have been devoted to alleviate annotation scarcity by utilizing extra supervision. Among them, semi-supervised learning and domain adaptation are two widely studied learning approaches and increasingly gain people's interests.

Semi-supervised learning (SSL) aims to leverage unlabeled data to reduce the usage of manual annotations [11,12,20]. For example, Lee *et al.* [12] proposed to generate the pseudo labels of unlabeled data by a pretrained model, and utilize them to further finetune the training model for performance improvements. Recently, self-ensembling methods [11,20] have achieved state-of-the-art performance in many semi-supervised learning benchmarks. Laine *et al.* [11] proposed the temporal ensembling method to encourage the consensus between the exponential moving average (EMA) predictions and current predictions for unlabeled data. Tarvainen *et al.* [20] proposed the mean-teacher framework to force prediction consistency between current training model and the corresponding EMA model. Although semi-supervised learning has made great progress on utilizing the unlabeled data within the same domain, it leaves rich cross-modality data unexploited. Considering that multi-modality data is widely available in medical imaging field, recent works have studied on domain adaptation (DA) to leverage the shape priors of another modality for enhanced segmentation performance [5,8,9,18] Among them, multi-modality learning (MML) exploits the labeled data from a related modality (*i.e.*, source domain) to facilitate the segmentation on the modality of interest (*i.e.*, target domain) [3,10,21,22]. Valindria *et al.* [21] proposed a dual-stream approach to integrate the prior knowledge from unpaired multi-modality data for improved multi-organ segmentation, and suggested X-shape achieving the leading performance among all architectures. Since multi-modality learning requires annotations on two modality data, unsupervised domain adaptation (UDA) extends it with a broader application potential [2,4,16]. In UDA setting, source domain annotations are still required, while none target domain annotation is needed. Contemporary unsupervised domain adaptation methods attempt to extract domain-invariant representations, where Dou *et al.* [4] investigated in feature space and Chen *et al.* [2] explored both feature-level and image-level in a synergistic manner.

All approaches mentioned above have exhibited their feasibility in medical area. However, semi-supervised learning simply concentrates on leveraging the unlabeled data affiliated to the same domain as labeled ones, ignoring the rich prior knowledge (*e.g.*, shape priors) cross modalities. While domain adaptation can utilize cross-modality prior knowledge, it still has considerable space for improvement. These motivate us to explore the feasibility of integrating the merits of both semi-supervised learning and domain adaptation by *concurrently leveraging all available data resources*, including limited labeled target data, abundant unlabeled target data and well-established labeled source data, to enhance the segmentation performance on target domain.

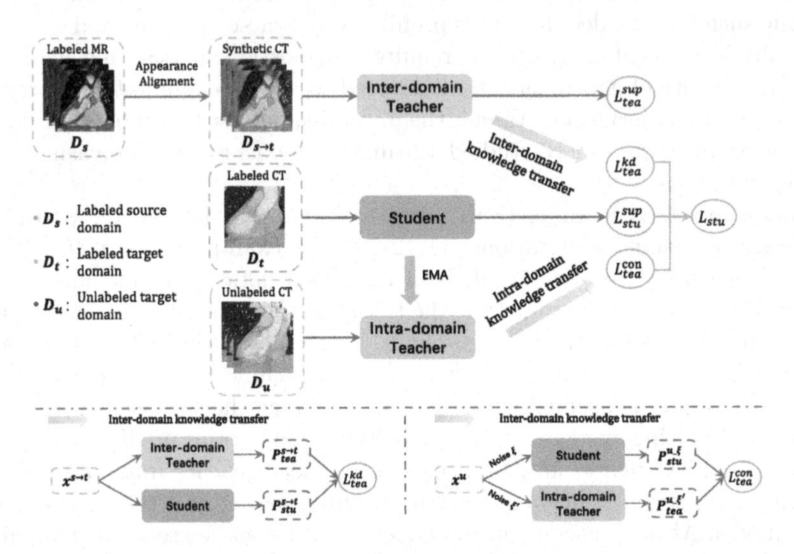

Fig. 1. Overview of our framework. The student model learns from $\mathcal{D}_t$ by the L_{stu}^{sup} loss, and concurrently acquires the knowledge of $\mathcal{D}_s$ from inter-domain teacher by knowledge distillation loss L_{stu}^{kd}, as well as the knowledge of $\mathcal{D}_u$ from intra-domain teacher by the consistency loss L_{stu}^{con}. In this way, the student model would integrate and leverage knowledge of $\mathcal{D}_s$, $\mathcal{D}_t$ and $\mathcal{D}_u$ simultaneously, leading to better generalization on target domain. In the inference phase, only the student model is used to predict.

In this paper, we propose a novel semi-supervised domain adaptation framework, namely Dual-Teacher, to simultaneously leverage abundant unlabeled data and widely-available cross-modality data to mitigate the need for tedious medical annotations. We implement it with the teacher-student framework [14] and adopt two teacher models in the network training, where one teacher guides the student model with intra-domain knowledge embedded in unlabeled target domain (*e.g.*, CT), while another teacher instructs the student model with inter-domain knowledge beneath labeled source domain (*e.g.*, MR). To be specific, our Dual-Teacher framework consists of three components: (1) intra-domain teacher, which employs the self-ensembling model of the student network to leverage unlabeled target data and transfers the acquired knowledge to student model by forcing prediction consistency; (2) inter-domain teacher, which adopts an image translation model, *i.e.*, CycleGAN [24], to narrow the appearance gap cross modalities and transfers the prior knowledge in the source domain to student model via knowledge distillation; and (3) student model, which not only directly learns from limited labeled target data, but also grasps auxiliary intra-domain and inter-domain knowledge transferred from two teachers. Our whole framework is trained in an end-to-end manner to seamlessly integrate the knowledge of all data resources into the student model. We extensively evaluated our approach on MM-WHS 2017 dataset [25], and achieved superior performance compared to semi-supervised learning methods and domain adaptation methods.

2 Methodology

In our problem setting, we are given a set of source images and their annotations in source domain (*e.g.*, labeled MR data) as $\mathcal{D}_s = \{(\mathbf{x}_i^s, y_i^s)\}_{i=1}^{m_s}$. In addition, we are also given a limited number of annotated target domain samples (*e.g.*, labeled CT data) as $\mathcal{D}_t = \{(\mathbf{x}_i^t, y_i^t)\}_{i=1}^{m_t}$, and abundant unlabeled target domain data (*e.g.*, unlabeled CT data) as $\mathcal{D}_u = \{(\mathbf{x}_i^u)\}_{i=1}^{m_u}$. Normally, we assume m_t is far less than m_u. Our goal is to exploit $\mathcal{D}_s$, $\mathcal{D}_t$ and $\mathcal{D}_u$ to enhance the performance in target domain (*e.g.*, CT). Figure 1 overviews our proposed Dual-Teacher framework, which consists of an inter-domain teacher model, an intra-domain teacher model, and a student model. The inter-domain teacher model and intra-domain teacher model explore the knowledge beneath $\mathcal{D}_s$ and $\mathcal{D}_u$, respectively, and simultaneously transfer the knowledge to the student model for comprehensive integration and thorough exploitation.

2.1 Inter-domain Teacher

Despite the consistent shape priors shared between source domain (*e.g.*, MR) and target domain (*e.g.*, CT), they are distinct in many aspects like appearance and image distribution [7,17]. Considering that, we attempt to reduce the appearance discrepancy first by using an appearance alignment module. Various image translation models can be adopted. Here we use CycleGAN [24] to translate source samples x^s to synthetic target-style samples $x^{s\to t}$ for synthetic target set $\mathcal{D}_{s\to t}$. After appearance alignment, we input synthetic samples $x^{s\to t}$ into the inter-domain teacher, which is implemented as a segmentation network. With the supervision of corresponding labels y^s, the inter-domain teacher is able to learn the prior knowledge in source domain by $\mathcal{L}_{tea}^{seg}$ following

$$\mathcal{L}_{tea}^{seg} = \mathcal{L}_{\text{ce}}\left(y^s, p_{tea}^{s\to t}\right) + \mathcal{L}_{\text{dice}}\left(y^s, p_{tea}^{s\to t}\right), \tag{1}$$

where $\mathcal{L}_{\text{ce}}$ and $\mathcal{L}_{\text{dice}}$ denote cross-entropy loss and dice loss, respectively, and $p_{tea}^{s\to t}$ represents the inter-domain teacher predictions taking $x^{s\to t}$ as inputs. To transfer the acquired knowledge from inter-domain teacher to the student, we further feed the same synthetic samples $x^{s\to t}$ into both inter-domain teacher model and student model. Since the inter-domain teacher has acquired reliable source domain knowledge from its annotations, we encourage the student model to produce similar outputs as inter-domain teacher model via knowledge distillation loss $\mathcal{L}_{tea}^{kd}$. Following previous works [1,6], we formulate $\mathcal{L}_{tea}^{kd}$ as

$$\mathcal{L}_{\text{tea}}^{kd} = \mathcal{L}_{\text{ce}}\left(p_{tea}^{s\to t}, p_{stu}^{s\to t}\right), \tag{2}$$

where $p_{tea}^{s\to t}$ and $p_{stu}^{s\to t}$ represent the predictions of inter-domain teacher model and student model, respectively.

2.2 Intra-domain Teacher

As $\mathcal{D}_u$ has no expert-annotated labels to directly guide network learning, recent works [20] propose to temporally ensemble the models in different training steps for reliable predictions. Inspired by them, we design the intra-domain teacher model following the same network architecture as student model and its weights θ' are updated as the exponential moving average (EMA) of the student model weights θ in different training steps. Specifically, at training step t, the weights of intra-domain teacher model θ'_t are updated as

$$\theta'_t = \alpha\theta'_{t-1} + (1-\alpha)\theta_t, \tag{3}$$

where α is the EMA decay rate to control updating rate. To transfer the knowledge from intra-domain teacher to the student, we add different noise ξ and ξ' to the same unlabeled sample and feed them into intra-domain teacher model and student model, respectively. Given small perturbation operations, *e.g.*, Gaussian noise, the outputs between the student model and the corresponding EMA model (*i.e.*, the intra-domain teacher model) should be the same. Therefore, we encourage them to generate consistent predictions via consistency loss $\mathcal{L}_{\text{tea}}^{con}$ as

$$\mathcal{L}_{tea}^{con} = \mathcal{L}_{\text{mse}}\left(f\left(x^u;\theta,\xi\right), f\left(x^u;\theta',\xi'\right)\right), \tag{4}$$

where $\mathcal{L}_{\text{mse}}$ denotes the mean squared error loss. $f\left(x^u;\theta,\xi\right)$ and $f\left(x^u;\theta_t,\xi'\right)$ represent the outputs of the student model (with weight θ and noise ξ) and intra-domain teacher model (with weight θ' and noise ξ'), respectively.

2.3 Student Model and Overall Training Strategies

For the student model, it explicitly learns from $\mathcal{D}_t$ with the supervision of its labels via the segmentation loss $\mathcal{L}_{stu}^{seg}$. Meanwhile, it also concurrently acquires the knowledge of $\mathcal{D}_s$ and $\mathcal{D}_u$ from two teacher models and comprehensively integrates them as a united cohort. In particular, the student model attains inter-domain knowledge by knowledge distillation loss $\mathcal{L}_{tea}^{kd}$ as Eq. (2), and intra-domain knowledge by prediction consistency loss $\mathcal{L}_{tea}^{con}$ as Eq. (4). Overall, the training objective for the student model is formulated as

$$\begin{aligned} \mathcal{L}_{stu}^{seg} &= \mathcal{L}_{\text{ce}}\left(y^t, p_{stu}^t\right) + \mathcal{L}_{\text{dice}}\left(y^t, p_{stu}^t\right) \\ \mathcal{L}_{stu} &= \mathcal{L}_{stu}^{seg} + \lambda_{kd}\mathcal{L}_{tea}^{kd} + \lambda_{con}\mathcal{L}_{tea}^{con}, \end{aligned} \tag{5}$$

where λ_{kd} and λ_{con} are hyperparameters to balance the weight of $\mathcal{L}_{tea}^{kd}$ and $\mathcal{L}_{tea}^{con}$.

Our whole framework is updated in an end-to-end manner. We first optimize the inter-domain teacher model, then update the intra-domain teacher model with the EMA parameters of the student network, and optimize the student model in the last. In this way, no pre-training stage would be required and the student model updates its parameters synchronously along with teacher models in an online manner.

3 Experiments

Dataset and Pre-processing. We evaluated our method in Multi-modality Whole Heart Segmentation (MM-WHS) 2017 dataset [25], which provided 20 annotated MR and 20 annotated CT volumes. We employed CT as target domain and MR as source domain, and randomly split 20 CT volumes into four folds to perform four-fold cross validation. In each fold, we validated on five CT volumes, and took 20 MR volumes as $\mathcal{D}_s$, five randomly chosen CT volumes as $\mathcal{D}_t$ and the remaining 10 CT volumes as $\mathcal{D}_u$ to train our framework. For pre-processing, we resampled all data with unit spacing and cropped them into 256 × 256 centering at the heart region, following previous work [2]. To avoid overfitting, we applied on-the-fly data augmentation with random affine transformations and random rotation. We evaluated our method with dice coefficient on all seven heart substructures, including the left ventricle blood cavity (LV), the right ventricle blood cavity (RV), the left atrium blood cavity (LA), the right atrium blood cavity (RA), the myocardium of the left ventricle (MYO), the ascending aeorta (AA), and the pulmonary artery (PA) [25].

Implementation Details. In our framework, the student model and two teacher models were implemented with the same network backbone, U-Net [19]. We empirically set λ_{kd} as 0.1 for inter-domain teacher. For intra-domain teacher, we closely followed the experiment configurations in previous work [23], where the EMA decay rate α was set to 0.99 and the hyperparameter λ_{con} was dynamically changed over time with the function $\lambda_{con}(t) = 0.1 * e^{\left(-5(1-t/t_{\max})^2\right)}$, where t and t_{max} denote the current and the last training epoch respectively and t_{max} is set to 50. To optimize the appearance alignment module, we followed the setting in [24] and used Adam optimizer with learning rate 0.0001 to optimize the student model and two teacher models until the network converge.

Table 1. Comparison with other methods. The dice of all heart substructures and the average of them are reported here.

Method		Avg	Dice of heart substructures						
			MYO	LA	LV	RA	RV	AA	PA
Supervised-only ($\mathcal{D}_t$)		0.7273	0.7113	0.7346	0.8086	0.7099	0.6524	0.8707	0.6037
UDA ($\mathcal{D}_s, \mathcal{D}_u$)	Dou *et al.*[4]	0.6635	0.5664	0.7655	0.7654	0.6230	0.6600	0.7138	0.5505
	Chen *et al.*[2]	0.7138	0.6573	0.8290	0.8306	0.7804	0.7082	0.7089	0.4827
MML ($\mathcal{D}_s, \mathcal{D}_t$)	Finetune	0.7313	0.7533	0.8081	0.7825	0.6412	0.5928	0.8466	0.6943
	Joint training	0.7875	0.7816	0.8312	0.8469	0.7699	0.7008	0.8802	0.7019
	X-shape [21]	0.7643	0.7317	0.8361	0.8432	0.7259	0.7453	0.8968	0.5709
SSL ($\mathcal{D}_u, \mathcal{D}_t$)	MT [20]	0.8165	0.7764	0.8712	0.8748	0.7930	0.7051	0.9274	0.7677
SSDA ($\mathcal{D}_s, \mathcal{D}_u, \mathcal{D}_t$)	**Ours**	**0.8604**	**0.8143**	**0.8784**	**0.9054**	**0.8449**	**0.8342**	**0.9412**	**0.8043**

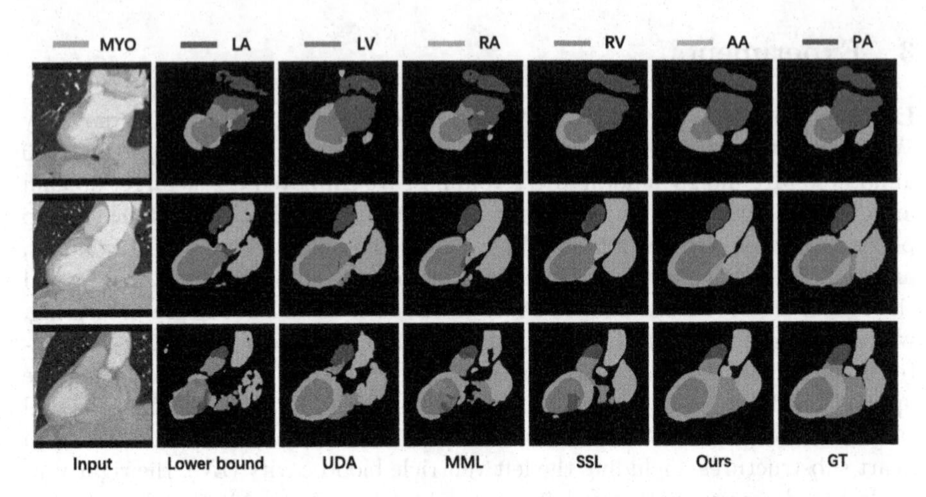

Fig. 2. Visual comparisons with other methods. Due to page limit, we only present the methods with best mean dice in MML and UDA (*i.e.*, Joint-training and Chen *et al.* [2]). As observed, our predictions are more similar to the ground truth than others.

Comparison with Other Methods. To demonstrate the effectiveness of our proposed semi-supervised domain adaptation method (SSDA) for leveraging unlabeled data and cross-modality data, we compare with both semi-supervised learning methods and domain adaptation methods. We first compare with the model trained with only limited labeled CT data $\mathcal{D}_t$ (referred as Supervised-only), and take mean-teacher (MT) method [20] in semi-supervised learning (SSL) for comparison. For domain adaptation methods, besides straightforward methods like finetune and joint training, we also compare with X-shape model [21] in multi-modality learning (MML). Meanwhile, we consider two unsupervised domain adaptation methods (UDA) for comparisons, *i.e.*, Dou *et al.* [4] and Chen *et al.* [2], which achieve the state-of-the-art performance in cardiac segmentation.

As presented in Table 1, the supervised-only model achieves 72.73% in mean dice by taking only limited labeled target data $\mathcal{D}_t$ in network training. When two types of data resources are available, UDA methods achieve comparable mean dice to the supervised-only model by utilizing $\mathcal{D}_s$ and $\mathcal{D}_u$. Compared with supervised-only method, MML-based Joint training and SSL-based MT [20] methods further improve the segmentation performance with 6.02% and 8.92% in mean dice, respectively, demonstrating the effectiveness of leveraging cross-modality data $\mathcal{D}_s$ and unlabeled data $\mathcal{D}_u$ for improving segmentation performance. By simultaneously exploiting all of data resources, our Dual-Teacher outperforms the unsupervised domain adaptation, multi-modality learning and semi-supervised learning methods by a large margin, *i.e.*, 14.66%, 7.29% and 4.39% increase in mean dice respectively, validating the feasibility of our proposed semi-supervised domain adaptation approach.

We also present visual comparisons in Fig. 2. Due to page limit, we only present the predictions of the methods with best mean dice in MML (*i.e.*, Joint-training) and UDA (*i.e.*, Chen *et al.* [2]). It is observed that our method better identifies heart substructures with clean and accurate boundary, and produces less false positive predictions and more similar results to the ground truth compared with other methods.

Table 2. Analysis of our method. We report the mean dice of all cardiac substructures.

Methods		Mean Dice
No-Teacher	Baseline	0.7330
	GAN-baseline	0.7510
One-Teacher	W/o inter-domain teacher	0.8477
	W/o intra-domain teacher	0.7984
Dual-teacher (Ours)		0.8604

Analysis of Our Method. We further compare with other methods, which also utilize all three types of data in SSDA, and analyze the key components of our method in Table 2. For $\mathcal{D}_s$, $\mathcal{D}_t$, $\mathcal{D}_u$ in SSDA, one straightforward method is to train $\mathcal{D}_s$ and $\mathcal{D}_t$ jointly, and deploy Pseudo-label method [12] to utilize $\mathcal{D}_u$, which is considered as our Baseline. A more effective version of baseline (referred as GAN-baseline) is using appearance alignment module (*e.g.*, CycleGAN [24]) on $\mathcal{D}_s$ to minimize appearance difference, and then following the previous routine by joint training synthetic target data $\mathcal{D}_{s \to t}$ along with $\mathcal{D}_t$ and applying Pseudo-label method [12] for $\mathcal{D}_u$. For the Baseline and GAN-baseline, no teacher-student scheme is applied. Moreover, we conduct other experiments: (i) without inter-domain teacher, where we substitute it as a joint-training network attached with appearance alignment module to tackle $\mathcal{D}_s$ and $\mathcal{D}_t$, and (ii) without intra-domain teacher, where we replace it with Pseudo-label method [12] to handle $\mathcal{D}_u$.

The results are shown in Table 2. Without any knowledge transfer from teacher models, neither the knowledge in $\mathcal{D}_s$ or that in $\mathcal{D}_u$ would be well-exploited. Since GAN-baseline adopts special treatments to narrow appearance gap, it performs better than the baseline model, but it still has large room for improvement compared to our method. Without the intra-domain teacher, the pseudo label bias will gradually accumulated and deteriorate the segmentation performance with 6.20% lower than our Dual-Teacher framework in mean dice. Without the inter-domain teacher, the performance is 1.27% lower than our method in mean dice, indicating that the prior knowledge of $\mathcal{D}_s$ are not effectively utilized. These comparison results show that each teacher model plays a crucial role in our framework and further improvements could be achieved when combining them together.

4 Conclusion

We present a novel annotation-efficient semi-supervised domain adaptation framework for multi-modality cardiac segmentation. Our method integrates the inter-domain teacher model to leverage cross-modality priors from source domain, and the intra-domain teacher model to exploit the knowledge embedded in unlabeled target data. Both teacher models transfer the learnt knowledge into the student model, thereby seamlessly combining the merits of semi-supervised learning and domain adaptation. We extensively evaluated our method in MM-WHS 2017 dataset. Our method can simultaneously utilize cross-modality data and unlabeled data, and outperforms state-of-the-art semi-supervised and domain adaptation methods.

Acknowledgments. The work described in this paper was supported by Key-Area Research and Development Program of Guangdong Province, China under Project No. 2020B010165004, Hong Kong Innovation and Technology Fund under Project No. ITS/426/17FP and ITS/311/18FP and National Natural Science Foundation of China under Project No. U1813204.

References

1. Anil, R., Pereyra, G., Passos, A., Ormandi, R., Dahl, G.E., Hinton, G.E.: Large scale distributed neural network training through online distillation. arXiv preprint arXiv:1804.03235 (2018)
2. Chen, C., Dou, Q., Chen, H., Qin, J., Heng, P.A.: Synergistic image and feature adaptation: towards cross-modality domain adaptation for medical image segmentation. In: Proceedings of the AAAI Conference on Artificial Intelligence, vol. 33, pp. 865–872 (2019)
3. Dou, Q., Liu, Q., Heng, P.A., Glocker, B.: Unpaired multi-modal segmentation via knowledge distillation. In: IEEE Transactions on Medical Imaging (2020)
4. Dou, Q., Ouyang, C., Chen, C., Chen, H., Heng, P.A.: Unsupervised cross-modality domain adaptation of convnets for biomedical image segmentations with adversarial loss. In: Proceedings of the 27th International Joint Conference on Artificial Intelligence, pp. 691–697 (2018)
5. Ghafoorian, M., et al.: Transfer learning for domain adaptation in MRI: application in brain lesion segmentation. In: Descoteaux, M., Maier-Hein, L., Franz, A., Jannin, P., Collins, D.L., Duchesne, S. (eds.) MICCAI 2017. LNCS, vol. 10435, pp. 516–524. Springer, Cham (2017). https://doi.org/10.1007/978-3-319-66179-7_59
6. Hinton, G., Vinyals, O., Dean, J.: Distilling the knowledge in a neural network. arXiv preprint arXiv:1503.02531 (2015)
7. Hoffman, J., et al.: Cycada: Cycle-consistent adversarial domain adaptation. In: International Conference on Machine Learning, pp. 1989–1998 (2018)
8. Huo, Y., et al.: Synseg-net: synthetic segmentation without target modality ground truth. IEEE Trans. Med. Imaging **38**(4), 1016–1025 (2018)
9. Jiang, J., et al.: Tumor-aware, adversarial domain adaptation from CT to MRI for lung cancer segmentation. In: Frangi, A.F., Schnabel, J.A., Davatzikos, C., Alberola-López, C., Fichtinger, G. (eds.) MICCAI 2018. LNCS, vol. 11071, pp. 777–785. Springer, Cham (2018). https://doi.org/10.1007/978-3-030-00934-2_86

10. Jue, J., Jason, H., Neelam, T., Andreas, R., Sean, B.L., Joseph, D.O., Harini, V.: Integrating cross-modality hallucinated MRI with CT to aid mediastinal lung tumor segmentation. In: Shen, D., et al. (eds.) MICCAI 2019. LNCS, vol. 11769, pp. 221–229. Springer, Cham (2019). https://doi.org/10.1007/978-3-030-32226-7_25
11. Laine, S., Aila, T.: Temporal ensembling for semi-supervised learning. arXiv preprint arXiv:1610.02242 (2016)
12. Lee, D.H.: Pseudo-label: The simple and efficient semi-supervised learning method for deep neural networks. In: Workshop on Challenges in Representation Learning, ICML. vol. 3, p. 2 (2013)
13. Litjens, G., et al.: A survey on deep learning in medical image analysis. Med. Image Anal. **42**, 60–88 (2017)
14. Liu, F., Deng, C., Bi, F., Yang, Y.: Dual teaching: a practical semi-supervised wrapper method. arXiv preprint arXiv:1611.03981 (2016)
15. Milletari, F., Navab, N., Ahmadi, S.A.: V-net: fully convolutional neural networks for volumetric medical image segmentation. In: 2016 Fourth International Conference on 3D Vision (3DV), pp. 565–571. IEEE (2016)
16. Orbes-Arteainst, M., et al.: Knowledge distillation for semi-supervised domain adaptation. In: Zhou, L., et al. (eds.) OR 2.0/MLCN -2019. LNCS, vol. 11796, pp. 68–76. Springer, Cham (2019). https://doi.org/10.1007/978-3-030-32695-1_8
17. Pan, S.J., Yang, Q.: A survey on transfer learning. IEEE Trans. Knowl. Data Eng. **22**(10), 1345–1359 (2009)
18. Perone, C.S., Ballester, P., Barros, R.C., Cohen-Adad, J.: Unsupervised domain adaptation for medical imaging segmentation with self-ensembling. NeuroImage **194**, 1–11 (2019)
19. Ronneberger, O., Fischer, P., Brox, T.: U-net: convolutional networks for biomedical image segmentation. In: Navab, N., Hornegger, J., Wells, W.M., Frangi, A.F. (eds.) MICCAI 2015. LNCS, vol. 9351, pp. 234–241. Springer, Cham (2015). https://doi.org/10.1007/978-3-319-24574-4_28
20. Tarvainen, A., Valpola, H.: Mean teachers are better role models: weight-averaged consistency targets improve semi-supervised deep learning results. In: Advances in Neural Information Processing Systems, pp. 1195–1204 (2017)
21. Valindria, V.V., et al.: Multi-modal learning from unpaired images: Application to multi-organ segmentation in CT and MRI. In: 2018 IEEE Winter Conference on Applications of Computer Vision (WACV), pp. 547–556. IEEE (2018)
22. Van Tulder, G., de Bruijne, M.: Learning cross-modality representations from multi-modal images. IEEE Trans. Med. Imaging **38**(2), 638–648 (2018)
23. Yu, L., Wang, S., Li, X., Fu, C.-W., Heng, P.-A.: Uncertainty-aware self-ensembling model for semi-supervised 3D left atrium segmentation. In: Shen, D., et al. (eds.) MICCAI 2019. LNCS, vol. 11765, pp. 605–613. Springer, Cham (2019). https://doi.org/10.1007/978-3-030-32245-8_67
24. Zhu, J.Y., Park, T., Isola, P., Efros, A.A.: Unpaired image-to-image translation using cycle-consistent adversarial networks. In: Proceedings of the IEEE International Conference on Computer Vision, pp. 2223–2232 (2017)
25. Zhuang, X., et al.: Evaluation of algorithms for multi-modality whole heart segmentation: an open-access grand challenge. Med. Image Anal. **58**, 101537 (2019)

Test-Time Unsupervised Domain Adaptation

Thomas Varsavsky[1,2](✉), Mauricio Orbes-Arteaga[1,3], Carole H. Sudre[1,4], Mark S. Graham[1], Parashkev Nachev[5], and M. Jorge Cardoso[1]

[1] School of Biomedical Engineering and Imaging Sciences, KCL, London, UK
[2] Department of Medical Physics and Biomedical Engineering, UCL, London, UK
ucabtmv@ucl.ac.uk
[3] Biomediq A/S, Copenhagen, Denmark
[4] Dementia Research Centre, Institute of Neurology, UCL, London, UK
[5] High Dimensional Neurology Group, Institute of Neurology, UCL, London, UK

Abstract. Convolutional neural networks trained on publicly available medical imaging datasets (source domain) rarely generalise to different scanners or acquisition protocols (target domain). This motivates the active field of domain adaptation. While some approaches to the problem require labelled data from the target domain, others adopt an unsupervised approach to domain adaptation (UDA). Evaluating UDA methods consists of measuring the model's ability to generalise to unseen data in the target domain. In this work, we argue that this is not as useful as adapting to the test set directly. We therefore propose an evaluation framework where we perform test-time UDA on each subject separately. We show that models adapted to a specific target subject from the target domain outperform a domain adaptation method which has seen more data of the target domain but not this specific target subject. This result supports the thesis that unsupervised domain adaptation should be used at test-time, even if only using a single target-domain subject.

Keywords: Domain adaptation · One-shot · Brain MRI

1 Introduction

Recent years have seen huge progress in performance in brain MRI segmentation, classification and synthesis largely thanks to the application of convolutional neural networks to these problems. The organisation of challenges such as BRATS [12] and the MICCAI 2017 White Matter Hyperintensity Challenge [10] have allowed the community to benchmark their segmentation algorithms on research data. In these cases, training data is usually preprocessed following a consistent protocol with techniques such as skull stripping, bias field correction, histogram normalisation and co-registration. Efforts are often put in place to ensure a certain degree of standardisation across the centres providing data, in terms of scanners parameters such as field strength, manufacturer, echo time,

A. L. Martel et al. (Eds.): MICCAI 2020, LNCS 12261, pp. 428–436, 2020.
https://doi.org/10.1007/978-3-030-59710-8_42

relaxation time and contrast agent. In addition, individuals generally have similar pre-clinical conditions and pathological presentations. When applied to data from clinical practice that presents much more heterogeneous acquisition conditions, the performance of algorithms trained on challenge data degrades. Performance can improve if algorithms are fine-tuned on labelled data in the target domain, but these can be expensive to acquire and rely on relative homogeneity of acquisition parameters in the target domain. If no labels are available then unsupervised domain adaptation may be used, which has seen growing interest in recent years e.g. [9,14].

Domain is not always a clear binary label. Scans of a particular MR modality (e.g. T1-weighted) may come from the same scanner in the same hospital but may use different acquisition parameters. Variability can be so large that each image can almost be considered its own domain.

When evaluating domain adaptation methods for segmentation, there is often a training set, a validation set and a test set for both source and target domains. Methods are judged on their ability to generalise from seen data in the source domain to unseen data in the target domain. In this work we argue for a different evaluation criterion, namely how well a model performs on the data it adapts to. We call this *"test-time unsupervised domain adaptation"*. When this test-time adaptation is performed on an individual subject we call it *"one-shot unsupervised domain adaptation"*. We present a domain adaptation method which leverages a combination of adversarial learning and consistency under augmentation to work in this one-shot case. We apply this methodology on multiple sclerosis lesion segmentation but it is designed to be applicable to other tasks in medical imaging.

Related Work: Our work considers the use of existing unsupervised domain adaptation methods when only a single *unlabelled* sample from the target domain is available. In this work we use the same data, pre-processing and segmentation task as in [18], where the authors tackle one-shot supervised domain adaptation, adapting to a target domain using a single *labelled* subject.

It is worth mentioning the framework proposed by Zhao et al. [19] and highlighting the difference to this work. The authors consider the variability between single-modality brain MRIs to be quantifiable by an additive intensity transform and a spatial transform to a brain atlas. They use this technique to create an entire labelled dataset from a single brain with an associated anatomical parcellation (hence the term "one-shot"). While the intensity transform tackles the variation in acquisition parameters, the spatial transform covers variations in anatomy. Although this and follow-up work produce realistic training data in the context of brain parcellation, such scheme cannot be trivially extended to application to pathologies in which the variability in presentation, location and extent is far greater. This is especially true in lesion segmentation, where a lesion prior cannot be produced from a non-linear deformations of an atlas.

Neural style-transfer methods were recently applied for unsupervised domain adaptation of cardiac MRI in [11]. The style of the target domain is matched to that of a single subject in the source domain by simultaneously minimising a style loss $l_{style}(\hat{y}, y)$ and a content loss $l_{content}(\hat{y}, x)$ where $\hat{y}$ is the generated style-transferred image, x is the image from the target domain and y is the image from the source domain. This method relies on finding an image in the source domain which most closely resembles the target image based on a Wasserstein distance metric. This method is similar to ours in that adaptation is performed on each individual test subject as its own optimisation problem.

Recent advances in self-supervised learning have led to large improvements in semi-supervised learning. Methods such as [2] use self-supervised tasks such as solving jigsaw puzzles to perform domain adaptation. Promoting invariance in networks outputs under data augmentation is another self-supervised task which was shown to work well for domain adaptation in [4] and which we refer to as Mean Teacher. It was adapted for use in medical image segmentation in [14]. In [13] the authors showed improvements over Mean Teacher using a simpler paired consistency method. They used paired data as a form of "ground-truth augmentation". When paired data is not available, which is most common in practice, small adjustments to this method can lead to substantial improvements. The method of [13] was chosen to demonstrate the value of test-time UDA, as it reported better results than domain adversarial learning and Mean Teacher on a related task. However, note that our domain adaptation methodology is not bound to a particular method.

2 Domain Adversarial Learning and Paired Consistency

We adapt the method for domain adaptation described in [13] which consists of domain adversarial learning and consistency training. In domain adversarial training we seek to find a feature representation $\phi_\theta(x)$ which contains as little information about d - the domain of x - as possible and the most information about the label y. We do so by including a domain discriminator $D_\gamma(x)$ which predicts a domain $\hat{d}$ and is trained by minimising the binary cross-entropy between this prediction and the ground-truth domain d, $\mathcal{L}_{adv} = l_{bce}(D_\gamma(\phi_\theta(x)), d)$. We use the gradient reversal layer from [5] to guarantee that the network weights θ change in the direction which minimises the supervised loss $\mathcal{L}_{sup}$ and maximises the adversarial loss $\mathcal{L}_{adv}$ where $\mathcal{L}_{sup} = l(\mathcal{M}(x), y_s)$ (we use the dice loss for l).

Consistency training is a simple semi-supervised learning method which works by enforcing invariance to data augmentation. A model $\mathcal{M}$ is trained to produce a prediction $\hat{y}_s$ on some source data x_s which has an associated label y_s using a regular supervised loss $\mathcal{L}_{sup}$. An image from the target domain x_T is passed to the same model $\mathcal{M}$ to obtain $\hat{y}_T$. The same image is passed through the model after augmentation $g(x_T)$ (details about the choice of g in Sect. 3) to produce $\hat{y}_T^{aug}$. The paired consistency loss $\mathcal{L}_{pc}$ aims at minimising the difference between $\hat{y}_T$ and $\hat{y}_T^{aug}$. Following the guidance from [14] and [13], the soft dice is used as $\mathcal{L}_{pc}$, defined as $\mathcal{L}_{pc}(\hat{y}, \hat{y}^{aug}) = \sum_{i=1}^{N} \hat{y}_i \hat{y}_i^{aug} / (\sum_{i=1}^{N} \hat{y}_i + \sum_{i=1}^{N} \hat{y}_i^{aug})$. By enforcing predictions to be invariant to some noise of perturbation δ,

$y(x) = y(x + \delta)$, we encourage the decision boundary of our classifier to fall in regions of low density.

The right hand side of Fig. 1 (right) depicts the benefits of domain adversarial learning. In frame a) we see a source and target domain represented by green and red ovals respectively. They contain representations of foreground and background pixels shown as grey crosses and red dots. Frame b) shows what happens when domain adversarial learning is used. The domains become indistinguishable which makes the ovals overlap. However, when the decision boundary is drawn to separate the two classes it is done only by looking at the source domain. In frame c) we introduce paired consistency. The unlabelled points are near the labelled ones, they will be assigned the label of their nearest cluster which allows the boundary to be redrawn in an area of low density. We include some t-SNE plots of our learned features in Figure 3 of the Supplementary Material which clearly show the positive effect of domain adaptation to the separability of lesion and background across both domains.

The method proposed in [13] achieved consistency training using what they denote as "ground-truth augmentation". This means two registered scans of the same patient using different acquisition parameters. In this work, we avoid this requirement by providing stronger augmentation and dropping the third output of their domain discriminator which sought to find a feature space which contained no information about whether an image was source, target or target augmented. Note that this minor change significantly reduces the data requirements of the model.

Implementation Details. We use a simple 2D U-Net with five levels as the backbone of our model. Each encoding block has two 2D convolutions with kernel sizes of 3×3, a stride of 1, and padding of 1 (except the first which has a padding of 2 and kernel size 5). The blocks have gradually increasing number of filters: 64, 96, 128, 256, 512. We use instance norm and leaky ReLU after each convolution in each block as in [7]. We use max pooling between each encoder block and bilinear upsampling between each decoder block and the standard concatenation of feature vectors from the same depth.

For the domain discriminator we use a small VGG-style convolutional neural network with four convolutions of kernel size 3 and stride of 2 each followed by a batch norm operation and three fully connected layers of size 28800, 256 and 128 respectively with 0.5 dropout in between. We follow the suggestion from [9] to feed in a concatenated vector of multi-depth features as input to the discriminator. Specifically, we take the activations from each depth of the decoder (excluding the center of the U-Net) and use bilinear interpolation to make them the same shape as the penultimate depth in the spatial dimension. We then concatenate on the channels dimension. All code is written in PyTorch and will be made available at the time of publication.

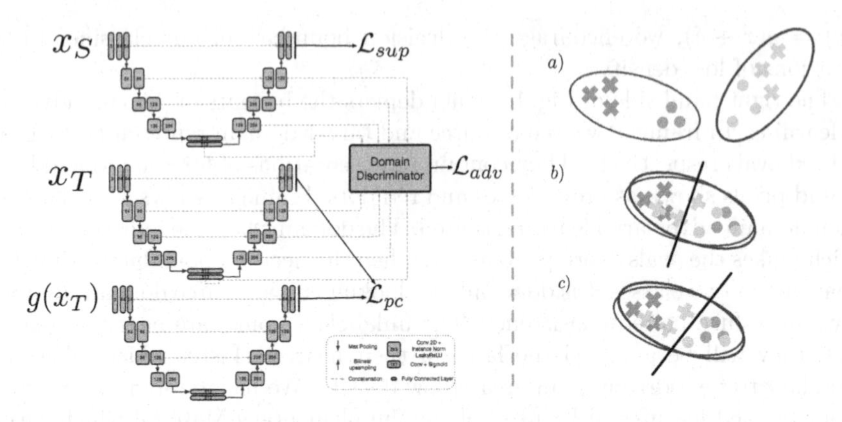

Fig. 1. Left: Our domain adaptation method uses a paired consistency loss $\mathcal{L}_{pc}$ which encourages predictions from the target image x_T to be invariant to some augmentation g. The backbone is a single 2D U-Net (parameters are shared) with features from each depth being interpolated bilinearly, concatenated and fed to a domain discriminator which uses an adversarial loss $\mathcal{L}_{adv}$ to maximise domain confusion. Right: In a) we depict representations of pixels in some feature space, the green circle is source and red target with crosses and circles depicting foreground and background. b) shows what happens when we introduce an adversarial loss, the feature spaces are shifted such that they are indistinguishable from the source domain but the decision boundary is drawn with only souce data. In c) we show the effect of the PC loss in moving the decision boundary to an area of low-density (Color figure online)

3 Experiments

In the proposed test-time UDA, an unusual approach to train/val/test splits is taken. In fact, part of the data for which we train the paired consistency component of our model $\mathcal{M}$ is the one on which the labelling quality is tested. Please note that the labels of the test set are never used during training. In order to prevent data leakage, all hyperparameters tuning strategies and model selection steps were performed on a completely separate dataset (results not shown). Each UDA run was trained for exactly 15,000 iterations using a batch size of 20 with the exception of the supervised baseline which had a validation subject to allow for model selection. We used the Adam optimiser with a learning rate of 1×10^{-3} with no learning rate policy. A separate Adam optimiser with learning rate 1×10^{-4} was used for the discriminator. In order to further validate our model we submit results to the online validation server for the ISBI 2015 challenge. We provide results for the first timepoint of each of the test subjects in the supplementary material.

Augmentation. In [14] the authors used random affine transforms (rotating, scaling, shearing and translating) as well as random elastic deformations where an affine grid is warped and applied to the image. Their method does augmentation on the output of a neural network but this output does not need to be differentiated. We use all these augmentations but exclude elastic deformation, as it is difficult to implement in a differentiable manner (a requirement of the proposed method). Following the recommendations in [13] we use augmentation which is realistic, valid and smooth. To this end, we also add bias field augmentation [6] and k-space augmentation [15] as extra transformations, as they have been shown to produce realistic variations in MRIs.

Data. Domain adaptation is here applied to multiple sclerosis lesion segmentation as an exemplary task. We use as source domain data from two separate MICCAI challenges on multiple sclerosis lesion segmentation MS2008 [17] and MS2016 [3]. Data from ISBI2015 [1] is used as target domain. The FLAIR sequences from each of these datasets are skull-stripped (using HD-BET [8]), bias-field corrected using the N4 algorithm and registered to MNI space as in [18].

3.1 Results

We present results from five different methods. First, there is a lower bound provided by using a model trained on the source domain and applied to data from the target domain, which we refer to as *no adaptation*. The highest expected performance is provided by training a model on the target domain images and labels, fine-tuned from a model trained on the source domain, which we refer to as *supervised*. When we use paired consistency and adversarial learning to domain adapt to a single subject on the target domain, this is denoted as *One-shot UDA*. We compare this against a model which sees this and two more subjects from the target domain, and refer to it as *Test-time UDA*. A comparison was also made against a traditional approach to domain adaptation where the model trains on target domain data which excludes the test subject; we refer to this variant as *Classic UDA*. In Table 1 of the supplementary material we show results for each of these methods evaluated on a variety of metrics. These were chosen to match those in [1]. The LFPR is the lesion false positive rate and LTPR is the lesion true positive rate which are implemented as in [17]. We follow the recommendations of the MICCAI Grand Challenges, specifically the method described in [16], to provide a single rank score comparing all methods. Note that this ranking method provides a single summary metric that incorporates a per-metric non-parametric statistical significance model (Fig. 2).

Table 1. Results on metrics described in [1]. The metrics are ranked using the scheme from [16] to provide a rank score. The proposed test-time methods are labelled (ours).

Method	Rank	Dice	Hausdorff	LFPR	LTPR	PPV	Sensitivity	Vol Diff
Supervised	1.71	0.67 ± 0.1	37. ± 8.	0.52 ± 0.2	0.61 ± 0.2	0.67 ± 0.2	0.73 ± 0.2	0.44 ± 0.2
Test-time UDA (ours)	2.43	0.61 ± 0.2	48. ± 5.	0.54 ± 0.2	0.57 ± 0.2	0.54 ± 0.2	0.76 ± 0.09	0.72 ± 1.0
One-shot UDA (ours)	2.71	0.60 ± 0.2	47. ± 11	0.52 ± 0.2	0.51 ± 0.2	0.54 ± 0.2	0.76 ± 0.09	0.92 ± 2.0
Classic UDA	3.86	0.56 ± 0.1	47. ± 14	0.55 ± 0.2	0.58 ± 0.2	0.49 ± 0.2	0.73 ± 0.2	0.78 ± 0.5
No adaptation	4.29	0.57 ± 0.1	55. ± 7.	0.68 ± 0.08	0.55 ± 0.2	0.49 ± 0.2	0.76 ± 0.08	0.76 ± 0.7

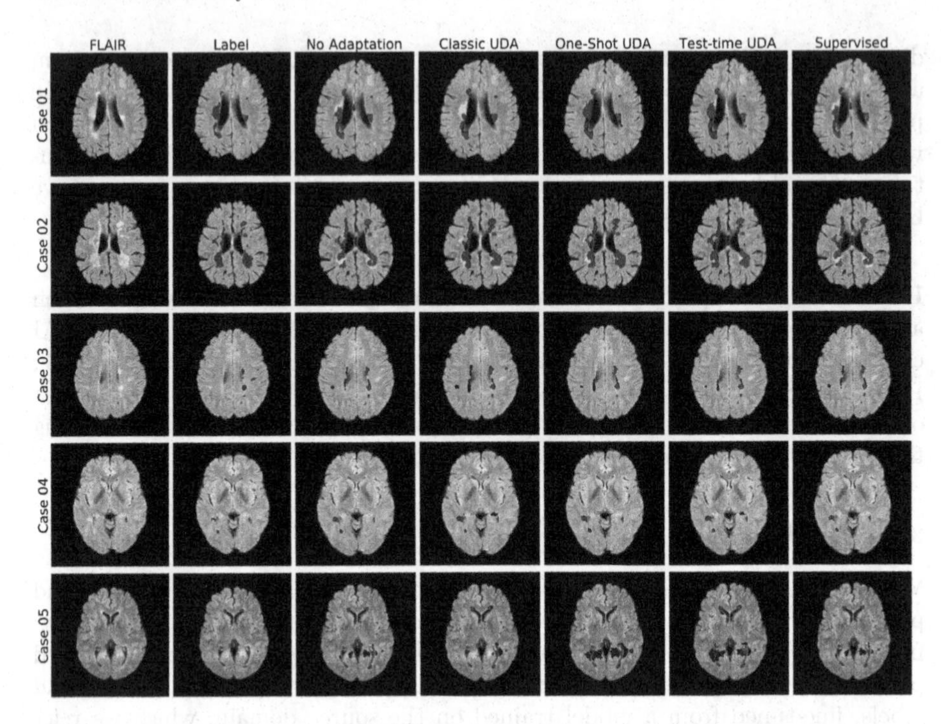

Fig. 2. Some qualitative results comparing no adaptation, classic unsupervised domain adaptation, one-shot unsupervised domain adaptation, test-time unsupervised domain adaptation, and the hypothetical gold-standard using supervised learning. Red denotes the ground-truth annotation, true positives are shown in green, false negatives are in yellow and false positives are in blue. (Color figure online)

4 Discussion

The results in Table 1 show a clear ordering with Supervised as the best performing method, as expected, followed by test-time UDA, one-shot UDA, classic UDA and finally no adaptation. These results reveal that learning enough information about a domain shift, i.e. Classic UDA, is not enough to get the best performance on each test subject in the target domain. By domain-adapting to each test subject, we are adapting to the subjects individual anatomical and pathological presentation. It is also worth mentioning that our One-shot unsupervised domain adaptation achieved a dice of 0.60 on the ISBI training set which is comparable to the 0.58 reported on the ISBI holdout set in [18] despite not using a single label from ISBI. Results in Table 2 show the performance of Test-time UDA against a Supervised baseline, Classic UDA and One-shot UDA. Classic UDA outperformed One-shot, but test-time UDA was best of all. Future work will include experiments on brain tumour segmentation and compare additional UDA methods in the Classic, One-shot and Test-time settings.

Table 2. Results from the ISBI 2015 holdout set hosted at https://smart-stats-tools.org/lesion-challenge. We ran our three UDA methods on the first timepoint of each of the 14 test subjects. Note that one of the limitations of this form of validation is the low inter-rater disagreement reported in Carass et al. The same ranking scheme was used as in the training set, however the symmetric distance was used instead of the Hausdorff. The Classic UDA outperformed One-shot but test-time UDA was best of all.

Method	Rank	Dice	LFPR	LTPR	PPV	TPR	Volume difference
Valverde et al. (Supervised)	1.50	0.60 ± 0.2	0.22 ± 0.2	0.41 ± 0.2	0.73 ± 0.2	0.54 ± 0.2	5829 ± 7900
Test-time UDA (ours)	4.25	0.51 ± 0.2	0.53 ± 0.2	0.25 ± 0.2	0.59 ± 0.2	0.51 ± 0.2	6947 ± 8800
Classic UDA	4.42	0.49 ± 0.2	0.54 ± 0.2	0.28 ± 0.2	0.55 ± 0.2	0.48 ± 0.2	5784 ± 7500
One-shot UDA (ours)	4.50	0.48 ± 0.2	0.52 ± 0.3	0.28 ± 0.1	0.52 ± 0.3	0.51 ± 0.2	7009 ± 7700

5 Conclusion

Existing approaches to unsupervised domain adaptation in medical image segmentation adapt to subjects in a target domain. The performance of these algorithms is then measured based on how well they generalise to unseen subjects in this target domain. When looking through scans in a hospital PACS system there is a large amount of heterogeneity in acquisition parameters. As an example, at our local hospital (anonymous), we found more than 1400 different brain MRI sequences being used. We can thus think of each of these scans as its own domain, which motivates what we call "test-time unsupervised domain adaptation". Note that this is not an algorithmic modification, but simply a training and testing framework, where a domain adaptation algorithm is trained and evaluated on the same target data. We perform experiments using a modern domain adaptation technique which combines the benefits of domain adversarial learning and consistency regularisation. Our experiments on multiple sclerosis lesions suggest that using domain adaptation on a single subject can be more effective than classic domain adaptation on more subjects.

References

1. Carass, A., et al.: Longitudinal multiple sclerosis lesion segmentation: resource and challenge. NeuroImage **148**, 77–102 (2017)
2. Carlucci, F.M., et al.: Domain generalization by solving jigsaw puzzles. In: Proceedings of the IEEE Conference on Computer Vision and Pattern Recognition, pp. 2229–2238 (2019)
3. Commowick, O., et al.: Objective evaluation of multiple sclerosis lesion segmentation using a data management and processing infrastructure. Sci. Rep. **8**(1), 1–17 (2018)
4. French, G., Mackiewicz, M., Fisher, M.: Self-ensembling for visual domain adaptation. arXiv preprint arXiv:1706.05208 (2017)
5. Ganin, Y., et al.: Domain-adversarial training of neural networks. J. Mach. Learn. Res. **17**(1), 2030–2090 (2016)

6. Gibson, E., et al.: NiftyNet: a deep-learning platform for medical imaging. arXiv preprint arXiv:1709.03485 (2017)
7. Isensee, F., et al.: nnu-net: Self-adapting framework for u-net-based medical image segmentation. arXiv preprint arXiv:1809.10486 (2018)
8. Isensee, F., et al.: Automated brain extraction of multisequence MRI using artificial neural networks. Hum. Brain Mapp. **40**(17), 4952–4964 (2019)
9. Kamnitsa, K., et al.: Unsupervised domain adaptation in brain lesion segmentation with adversarial networks. In: Niethammer, M., et al. (eds.) IPMI 2017. LNCS, vol. 10265, pp. 597–609. Springer, Cham (2017). https://doi.org/10.1007/978-3-319-59050-9_47
10. Kuijf, H.J., et al.: Standardized assessment of automatic segmentation of white matter hyperintensities; results of the WMH segmentation challenge. IEEE Trans. Med. Imaging **38**(11), 2556–2568 (2019)
11. Ma, C., Ji, Z., Gao, M.: Neural style transfer improves 3D cardiovascular MR image segmentation on inconsistent data. In: Shen, D., et al. (eds.) MICCAI 2019. LNCS, vol. 11765, pp. 128–136. Springer, Cham (2019). https://doi.org/10.1007/978-3-030-32245-8_15
12. Menze, B.H., et al.: The multimodal brain tumor image segmentation benchmark (brats). IEEE Trans. Med. Imaging **34**(10), 1993–2024 (2014)
13. Orbes-Arteaga, M., et al.: Multi-domain adaptation in brain MRI through paired consistency and adversarial learning. In: Wang, Q., et al. (eds.) DART/MIL3ID -2019. LNCS, vol. 11795, pp. 54–62. Springer, Cham (2019). https://doi.org/10.1007/978-3-030-33391-1_7
14. Perone, C.S., et al.: Unsupervised domain adaptation for medical imaging segmentation with self-ensembling. NeuroImage **194**, 1–11 (2019)
15. Shaw, R., et al.: MRI k-space motion artefact augmentation: model robustness and task-specific uncertainty. In: MIDL, pp. 427–436 (2019)
16. Simpson, A.L., et al.: A large annotated medical image dataset for the development and evaluation of segmentation algorithms. arXiv preprint arXiv:1902.09063 (2019)
17. Styner, M., Lee, J., Chin, B., Chin, M., Commowick, O., Tran, H.: 3D segmentation in the clinic: a grand challenge ii: Ms lesion segmentation (2008)
18. Valverde, S., et al.: One-shot domain adaptation in multiple sclerosis lesion segmentation using convolutional neural networks. NeuroImage Clin. **21**, 101638 (2019)
19. Zhao, A., et al.: Data augmentation using learned transformations for one-shot medical image segmentation. In: Proceedings of the IEEE Conference on Computer Vision and Pattern Recognition, pp. 8543–8553 (2019)

Self Domain Adapted Network

Yufan He[1(✉)], Aaron Carass[1], Lianrui Zuo[1,3], Blake E. Dewey[1], and Jerry L. Prince[1,2]

[1] Department of Electrical and Computer Engineeringz, The Johns Hopkins University, Baltimore, MD 21218, USA
yhe35@jhu.edu
[2] Department of Computer Science, The Johns Hopkins University, Baltimore, MD 21218, USA
[3] Laboratory of Behavioral Neuroscience, National Institute on Aging, National Institute of Health, Baltimore, MD 20892, USA

Abstract. Domain shift is a major problem for deploying deep networks in clinical practice. Network performance drops significantly with (target) images obtained differently than its (source) training data. Due to a lack of target label data, most work has focused on unsupervised domain adaptation (UDA). Current UDA methods need both source and target data to train models which perform image translation (harmonization) or learn domain-invariant features. However, training a model for each target domain is time consuming and computationally expensive, even infeasible when target domain data are scarce or source data are unavailable due to data privacy. In this paper, we propose a novel self domain adapted network (SDA-Net) that can rapidly adapt itself to a single test subject at the testing stage, without using extra data or training a UDA model. The SDA-Net consists of three parts: adaptors, task model, and auto-encoders. The latter two are pre-trained offline on labeled source images. The task model performs tasks like synthesis, segmentation, or classification, which may suffer from the domain shift problem. At the testing stage, the adaptors are trained to transform the input test image and features to reduce the domain shift as measured by the auto-encoders, and thus perform domain adaptation. We validated our method on retinal layer segmentation from different OCT scanners and T1 to T2 synthesis with T1 from different MRI scanners and with different imaging parameters. Results show that our SDA-Net, with a single test subject and a short amount of time for self adaptation at the testing stage, can achieve significant improvements.

Keywords: Unsupervised domain adaptation · Self supervised learning · Segmentation · Synthesis

1 Introduction

The success of deep networks relies on the assumption that test data (target) and training data (source) are generated from the same distribution. In real scenarios such as multi-center longitudinal studies—even with a pre-defined imaging

A. L. Martel et al. (Eds.): MICCAI 2020, LNCS 12261, pp. 437–446, 2020.
https://doi.org/10.1007/978-3-030-59710-8_43

protocol—subjects will occasionally be scanned by different scanners with different imaging parameters. The model that is trained on the source data will incur a significant performance drop on those scans, which can vary from the source domain in different ways (the domain shift problem), and each target domain may only contain a few subjects.

Unsupervised domain adaptation (UDA), which reduces the domain shift without any target labels, is often used for solving domain shift problems. UDA can be categorized into two types. The first type is data harmonization in the pixel domain, which translates the target image to be similar to the source image. This includes methods like histogram matching, style transfer [8], and Cycle-Gan [26]. Ma et al. [18] used style transfer [8] to reduce the effect of domain shift in MR heart image segmentation. Seeböck et al. [21] used a Cycle-Gan to improve OCT lesion segmentation. The second type of UDA learns domain-invariant features. In particular, the network is re-trained to produce domain-invariant features from the source and target data such that those features are similar, as measured by metrics like maximum mean discrepancy [17], or are indistinguishable by domain classifiers [7] or discriminators in adversarial training [6,23,25]. A combination of these two approaches is proposed in [14]. However, those methods (except histogram matching and style transfer) 1) require retraining a UDA model for each target domain which is time consuming and computationally expensive; 2) require a fair amount of target data from each domain to train the model, which may not be feasible in clinical practice; or 3) require source data which may not be available for people deploying a pre-trained model due to data privacy. Some work [2,19] has addressed problem (2) but no method has addressed all three problems.

Can we design a model that can be rapidly adapted to a single test subject during inference without using extra data? If so, we can directly deploy the trained model on images from various target domains without accessing the source data or retraining a UDA model. In this paper, we propose a new deep model for this purpose and name it the self domain adapted network (SDA-Net). Adapting a pre-trained classifier to test images was first proposed in [15] for multi-face detection, and Assaf et al. [22] trained a super-resolution model on a single test image. However, those methods cannot be easily modified for our task. The SDA-Net consists of three parts: 1) a task network (T) which performs our task (synthesis, segmentation, or classification); 2) a set of auto-encoders (AEs), which are used as alignment measurements; and 3) a set of adaptors which perform domain adaptation on each test subject during inference. The core idea is to align the target and source domain in the pixel, network feature, and network output [23] spaces. We have two major differences with the previous methods: 1) only one target test subject is used for training the adaptors at testing stage, while T and AEs with high training cost are frozen; and 2) the alignment is measured by the AEs. AEs have been used for anomaly detection under a core assumption: abnormal inputs will have larger reconstruction error than normal inputs [9]. We extend this and define the source domain as normal and the target domain as abnormal and use the reconstruction error as an alignment measurement. The adaptors perform domain adaptation by minimizing the AEs' reconstruction error on the target data.

Implementation of this general framework faces several obstacles. Firstly, AEs have a strong generalization ability such that abnormal inputs with low reconstruction error can be far away from the source features[9]. Secondly, features from different classes can collapse to one by the adaptors, as illustrated in Fig. 2. Thirdly, deep networks can hallucinate features [4], which is a severe problem for medical data. We avoid the first problem by focusing on tasks with relatively minor domain shifts like scanner differences; thus, we can assume that initial target features are close to the source. With this assumption and with specially designed adaptors and training loss, we can successfully address the other problems.

2 Method

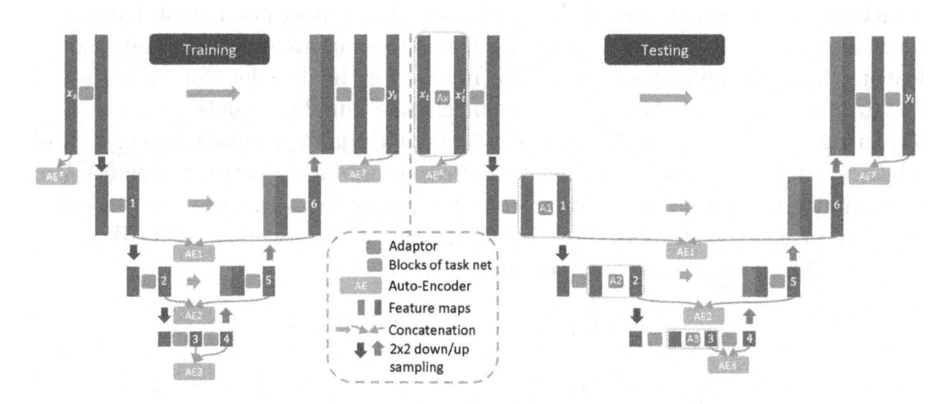

Fig. 1. During source domain training with (x_s, y_s) (left figure), the adaptors are not used. We first train the task network then freeze the weights and train AEs. During inference (right figure), the adaptors transform input target domain image x_t and intermediate features to minimize AE's reconstruction loss.

Task Network. We consider two tasks in this paper: retinal OCT layer segmentation and T1 to T2 MRI image synthesis. We do not focus on designing the best task network for these tasks, however, since this is not the main focus of our paper. Instead, we simply use a residual U-Net [12], a variation of the widely used U-Net [20], as our task network for both tasks (with the only difference being the output channels and output activation). The network has three 2×2 max-pooling and 64 channels for all intermediate features as shown in Fig. 1. We can replace the task network with any specific state-of-the-art structure.

Multi-level Auto-Encoders. A set of fully convolutional auto-encoders (AEs), $\{AE^x, \{AE^i\}_{i=1,2,3}, AE^y\}$ are used. AE^x and AE^y are trained to reconstruct the source data x_s and T's output y'_s, respectively, to encode information in

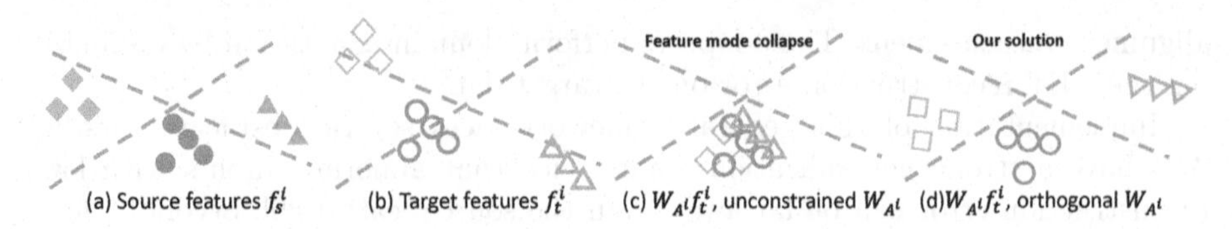

Fig. 2. (a) A trained task prediction model (dotted line) separates source features f_s^i into different classes (solid points with different colors) (b) Target features f_t^i (hollow points) have domain shifts. (c) f_t^i is transformed by a linear transformation W_{A^i} without constraints, which causes feature mode collapse, and is avoided by imposing orthogonality constraints on W_{A^i} in (d).

the highest resolution level. As shown in Fig. 1, we use three AEs $\{AE^i\}_{i=1,2,3}$ to encode T's intermediate features $\{f_s^i\}_{i=1,\cdots,6}$ (all 64 channels) at the lower three resolution levels. $\{AE^i\}$'s input is the concatenation of the task network feature f_s^i and f_s^{7-i} with the same spatial resolution at level i. The AEs' network structure is a modification of T, where two 2×2 max-pooling and instance normalization are used, while the long skip connections are removed. The encoder feature channel numbers are 64, 32, 16 (inverse for the decoder) for $\{AE^i\}$'s and 32, 16, 8 for AE^x and AE^y (AE^y's input channel number equals y_s''s).
Training. The task network T is first trained with pairs of source images x_s and labels y_s under standard training procedure by updating network weights W_T to minimize prediction error (we use cross entropy for segmentation and mean squared error for synthesis). Then we freeze the task network weights W_T, and train the AEs to minimize the reconstruction error $\mathcal{L}_{AE}$.

$$
\begin{aligned}
y_s', \{f_s^i\}_{i=1,\cdots,6} &= T(x_s; W_T) \\
\mathcal{L}_{AE} &= \sum_{i=1}^{3} |AE^i(\{f_s^i, f_s^{7-i}\}) - \{f_s^i, f_s^{7-i}\}|_2^2 \\
&\quad + |AE^x(x_s) - x_s|_2^2 + |AE^y(y_s') - y_s'|_2^2
\end{aligned}
$$

Domain Adaptor. The adaptors consist of an image adaptor A^x which transforms the input target image x_t in the pixel-domain and three feature adaptors $\{A^i\}_{i=1,2,3}$ which transform the intermediate features $\{f_t^i\}_{i=1,2,3}$ from T in the feature domain (A^x also influences f_t^i by transforming x_t). To make the adaptors trainable by a single subject and to prevent hallucination, we limit the transformation ability of the adaptors. The image adaptor A^x is a pure histogram manipulator with three convolutional layers, where each layer has a 1×1 convolution followed by leaky ReLU and instance normalization. The output channels of each layer are 64, 64, 1. Each feature adaptor A^i is a 1×1 convolution with 64 input and output channels and the weight W_{A^i} is a 64×64 linear transformation matrix. For a 64-channel feature map, each pixel has a feature vector of length 64. Consider two 1D feature vectors f_a, f_b and their transformation $W_{A^i} f_a$, $W_{A^i} f_b$, we prevent the feature mode collapse (illustrated in Fig. 2) by keeping the L_2 distance between them $(f_a - f_b)^T W_{A^i}^T W_{A^i} (f_a - f_b) = (f_a - f_b)^T (f_a - f_b)$, which requires orthogonality of W_{A^i} such that $W_{A^i}^T W_{A^i} = I$. We impose orthogonality on W_{A^i} by using the Spectral Restricted Isometry Property Regularization [1], which minimizes the spectral norm of $W_{A^i}^T W_{A^i} - I$. We define it as the orthogonal loss $\mathcal{L}_{orth}$ in Eq. 1, and the implementation details are in [1]. We train the adaptors with $\mathcal{L}_A = \mathcal{L}_{AE} + \lambda_{orth}\mathcal{L}_{orth}$. The overall algorithm in the testing stage is described in Algorithm 1.

$$\mathcal{L}_{orth} = \sum_{i=1}^{3} \sigma(W_{A^i}^T W_{A^i} - I) = \sum_{i=1}^{3} \sup_{z \in R^n, z \neq 0} \left| \frac{||W_{A^i} z||^2}{||z||^2} - 1 \right| \tag{1}$$

Algorithm 1: Self domain adapted network

Input: Single test subject scan x_t

1 Load pre-trained task network $T(\cdot; W_T)$ and $\{AE^i\}_{i=1,2,3}, AE^x, AE^y$;

2 Configure learning rate η and loss weight λ_{orth} ;

3 Initialize A^x with Kaiming normal [10] and $\{A^i\}_{i=1,2,3}$ with identity;

4 **while** iter < 5 *and* $0.95 * \mathcal{L}_A^{iter-1} > \mathcal{L}_A^{iter}$ **do**

5 Obtain adapted image, adapted features and prediction: $x_t', \{f_t^i\}_{i=1,\cdots,6}, y_t'$ from $T(x_t, A^x, \{A^i\}_{i=1,2,3}; W_A, W_T)$;

6 Calculate AE reconstruction loss $\mathcal{L}_{AE} = \sum_{i=1}^{3} |AE^i(\{f_t^i, f_t^{7-i}\}) - \{f_t^i, f_t^{7-i}\}|_2^2 + |AE^x(x_t') - x_t'|_2^2 + |AE^y(y_t') - y_t'|_2^2$;

7 Calculate orthogonality loss $\mathcal{L}_{orth} = \sum_{i=1}^{3} \sigma(W_{A^i}^T W_{A^i} - I)$;

8 Update $A^x, \{A^i\}_{\{i=1,2,3\}}$'s weights $W_A = W_A - \eta \nabla_{W_A} \mathcal{L}_A$;

9 **end**

Output: Target prediction y_t' from $T(x_t, A^x, \{A^i\}_{i=1,2,3}; W_A, W_T)$

---

3 Experiments

We validated our SDA-Net on two tasks: retinal layer segmentation [11,12] and T1 to T2 synthesis [5,16]. The hyper-parameters for both tasks are the same (except $\lambda_{orth} = 1, 5$ for segmentation and synthesis respectively). The task network, AEs, and adaptors were trained with the Adam optimizer with a learning rate 0.001, batch size 2, and no augmentation. The task network training was stopped based on the source validation set and the AEs were trained for 20 epochs, both using the source training set. The adaptors test time training is in Algorithm 1.

Retinal Layer Segmentation in OCT. We used retinal images from two OCT scanners: Spectralis and Cirrus. Eight retinal layers were manually segmented and the images were pre-processed with retina flattening [12]. Spectralis 2D images [13] were used as the source dataset for training SDA-Net (588 train, 147 validation, 980 test) and Cirrus images were used as testing target dataset (6 subjects, each with 8 images). We used SDA-Net to segment each target subject independently. We compare to image harmonization methods without retraining the task network: 1) NA: No adaptation; 2) M&H: 3×3 median filter and histogram matching [21]; 3) St[1]: Style transfer using pre-trained vgg19 [8,18]; 4) Cyc[2]: Cycle-Gan [26]; The Cycle-Gan trained from a single Cirrus subject is not usable. Thus, we trained the Cycle-Gan with 588 Cirrus (48 test Cirrus images and an additional 540 Cirrus images) and the source training set. We used one image from source training set as the reference image for (2) and (3). We can further improve our results by simply changing the first 1×1 convolution of the image adaptor A^x to 3×3 for pixel-domain noise removal. We show this result as Ours-3×3. We also tested SDA-Net on the source testing subject (20 subjects each with 49 slices). The Dice scores of eight layers and qualitative results are shown in Table 1 and Fig. 3. The results show that our method improves the target domain results while not significantly affecting the source domain results.

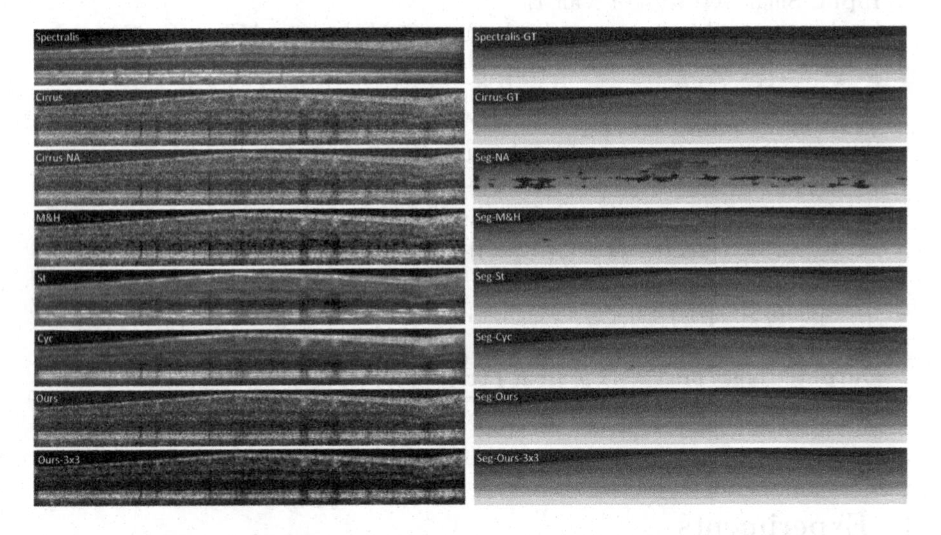

Fig. 3. Visualization of source, target and harmonized target images (Cirrus NA is the original Cirrus) using baselines and our methods (left). The ground-truth (GT) and segmentation (Seg) are shown on the right.

T1 to T2 Synthesis in MRI. In the synthesis experiments, we used four datasets which have paired T1-T2 scans. All the scans were N4 corrected [24], registered to MNI space, and white matter peak normalized. 21 axial slices were extracted from each subject (equally extracted from slice number 60 to 120, 3 mm slice distance). Source dataset (T1 MPRAGE from Philips Achieva 3T,T_R=3000 ms, T_E=6 ms) were used for training the SDA-Net (630 train, 84 validate, 315 test slices). The target test set comes from the IXI dataset (https://brain-development.org/ixi-dataset/), which has scans from three different clinical

[1] https://pytorch.org/tutorials/advanced/neural_style_tutorial.html.

[2] https://github.com/junyanz/pytorch-CycleGAN-and-pix2pix.

Table 1. Dice scores (Std. Dev.) for each method. The left part compares domain adaptation performance while the right part illustrates the performance of task model (NA) and our method on source testing data.

Layer	Target test results						Source test results	
	NA	M&H	St	Cyc	Ours	Ours-3 × 3	Ours	NA
RNFL	0.615(0.177)	0.688(0.143)	**0.724(0.109)**	0.709(0.126)	0.682(0.149)	0.698(0.132)	0.903(0.040)	0.906(0.039)
GCIP	0.742(0.092)	0.821(0.050)	**0.841(0.044)**	0.825(0.042)	0.818(0.055)	0.837(0.041)	0.923(0.032)	0.927(0.031)
INL	0.715(0.041)	0.753(0.033)	0.767(0.052)	0.769(0.032)	0.759(0.029)	**0.773(0.034)**	0.822(0.043)	0.829(0.042)
OPL	0.612(0.063)	0.632(0.057)	0.644(0.059)	0.670(0.050)	0.671(0.051)	**0.704(0.047)**	0.854(0.029)	0.856(0.028)
ONL	0.845(0.031)	0.859(0.022)	0.866(0.026)	0.878(0.026)	0.892(0.020)	**0.914(0.018)**	0.925(0.021)	0.927(0.023)
IS	0.803(0.035)	0.814(0.022)	0.835(0.024)	0.811(0.034)	0.830(0.018)	**0.838(0.019)**	0.822(0.033)	0.818(0.041)
OS	0.841(0.026)	0.846(0.028)	0.833(0.047)	0.849(0.026)	0.855(0.024)	**0.856(0.028)**	0.839(0.034)	0.836(0.039)
RPE	0.820(0.034)	0.828(0.032)	0.811(0.042)	**0.837(0.038)**	0.834(0.032)	0.825(0.035)	0.892(0.040)	0.890(0.040)
Overall	0.749(0.089)	0.780(0.076)	0.790(0.070)	0.794(0.067)	0.793(0.076)	**0.806(0.070)**	0.873(0.041)	0.874(0.042)

Table 2. MSE and SSIM (Std. Dev.) for synthesized T2 evaluated on four datasets (best result is in bold for each row). HH, GH, and IOP compares domain adaptation performance while Source illustrates the performance of task model (NA) and our method on source testing data.

		NA	Hist	St	Cyc	Ours
MSE	HH	0.223(0.034)	0.193(0.049)	0.292(0.264)	0.186(0.042)	**0.168(0.040)**
	GH	0.271(0.041)	0.237(0.077)	0.301(0.204)	0.240(0.061)	**0.233(0.055)**
	IOP	0.329(0.053)	0.286(0.064)	0.398(0.158)	0.297(0.072)	**0.279(0.046)**
	Source	0.092(0.047)	–	–	–	0.098(0.047)
SSIM	HH	0.605(0.051)	0.683(0.053)	0.575(0.170)	**0.700(0.045)**	0.693(0.039)
	GH	0.595(0.045)	**0.671(0.060)**	0.601(0.134)	0.656(0.041)	0.658(0.043)
	IOP	0.493(0.067)	0.594(0.067)	0.472(0.099)	**0.620(0.060)**	0.564(0.056)
	Source	0.774(0.050)	–	–	–	0.768(0.048)

centers (HH, GH, IOP). We used the first 30 subjects (630 slices) from HH (T1 SPGR from Philips Intera 3T,T_R=9.6 ms, T_E=4.6 ms), the first 30 from IOP (GE 1.5T, unknown parameters) and the first 30 from GH (T1 SPGR from Philips Gyroscan Intera 1.5T,T_R=9.8 ms, T_E=4.6 ms) for testing T1. As in the OCT segmentation task, we compare to baselines: 1) NA; 2) Hist: 2D histogram matching; 3) St; 4) Cyc: We train three Cycle-Gans for each clinical center, using all source training slices and test slices from each center separately. The reference image for (2) and (3) is the i-th slice of the first source training subject, where i is the slice number (1 to 21) of the input slice. We calculated the MSE and SSIM on the synthesized T2 from target (30 subjects for each target domain) and source test sets (15 subjects). The results are shown in Table 2 and Fig. 4.

4 Discussion and Conclusion

The SDA-Net adapts itself to a single subject for about 30s (testing time is 5s without adaptation on the same Nvidia GPU) for both tasks and shows

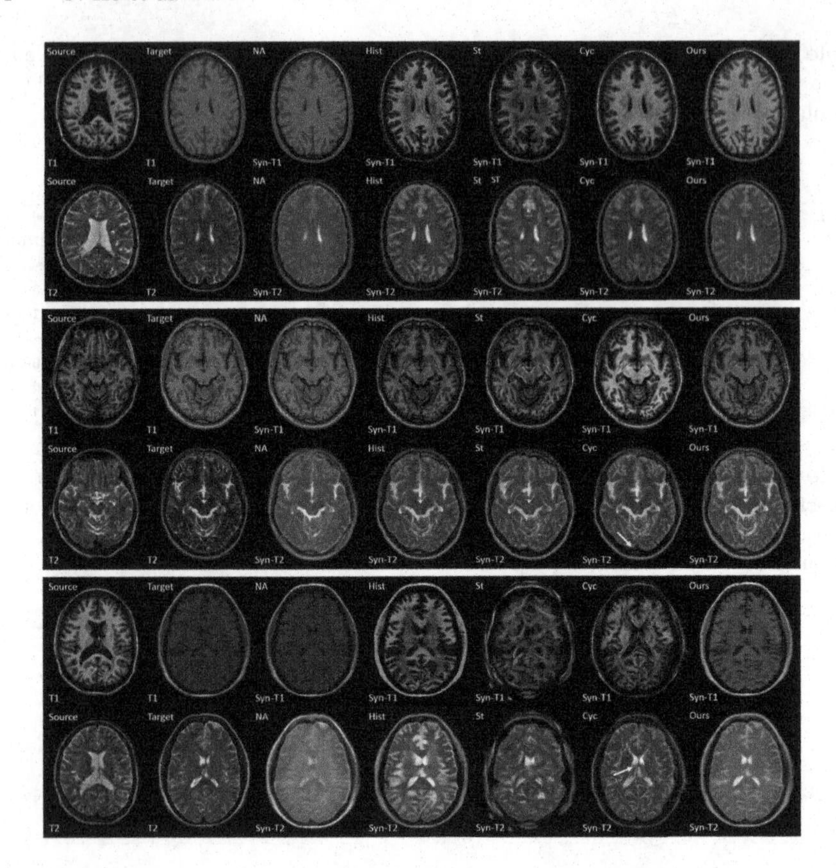

Fig. 4. Visualization of harmonized T1 and synthesized T2 from HH (top two rows), GH (middle two rows) and IOP (bottom two rows). The first (second) column are paired source (target) T1 and T2. The yellow arrows show a geometry shift and feature hallucination from Cycle-Gan. The blue arrows show artefacts from histogram matching. (Color figure online)

comparable results with Cycle-Gan, which requires extra target data and off-line training. As shown in Fig. 4, vanilla histogram matching can produce artefacts. Style-transfer can produce artistic results since the content and style are not completely disentangled [8]. The Cycle-Gan can cause geometry shift and hallucinate features [4]. A complicated pixel-domain transformation needs the model to extract high level features, which needs a fair amount of training data and labels. In order to train on a few images and avoid geometry shift and hallucination, we re-use those high level features extracted by the task network and transform them with A^i, while keeping the pixel-domain adaptor A^x as simple as possible: only 1×1 convolutions (Ours) or with a single 3×3 kernel (Ours-3×3). Despite the simplicity of A^x, SDA-Net achieves significant segmentation and synthesis improvements (Ours) as shown in Table 1 and Table 2. For segmentation, Ours-3×3 shows that a task specific adaptor can further improve the results. The major limitation of the work is that we only focus on problems with minor domain shift where we assume the features from task network can be re-used and not far from source features. Although we are not solving problems like using an MRI model for CT images [19], we argue that in real practice the minor domain shift from scanners and imaging parameters are the most common problems, and we propose a convenient and novel way to alleviate it. Future work will be improving the adaptation results by incorporating self-supervised methods [3] and improved auto-encoders [9].

Acknowledgments. This work is supported by NIH grants R01-EY024655 (PI: J.L. Prince), R01-NS082347 (PI: P.A. Calabresi) and in part by the Intramural research Program of the NIH, National Institute on Aging.

References

1. Bansal, N., Chen, X., Wang, Z.: Can we gain more from orthogonality regularizations in training deep networks? In: Advances in Neural Information Processing Systems, pp. 4261–4271 (2018)
2. Benaim, S., Wolf, L.: One-shot unsupervised cross domain translation. In: Advances in Neural Information Processing Systems, pp. 2104–2114 (2018)
3. Carlucci, F.M., D'Innocente, A., Bucci, S., Caputo, B., Tommasi, T.: Domain generalization by solving jigsaw puzzles. In: Proceedings of the IEEE Conference on Computer Vision and Pattern Recognition, pp. 2229–2238 (2019)
4. Cohen, J.P., Luck, M., Honari, S.: Distribution matching losses can hallucinate features in medical image translation. In: Frangi, A.F., Schnabel, J.A., Davatzikos, C., Alberola-López, C., Fichtinger, G. (eds.) MICCAI 2018. LNCS, vol. 11070, pp. 529–536. Springer, Cham (2018). https://doi.org/10.1007/978-3-030-00928-1_60
5. Dar, S.U., Yurt, M., Karacan, L., Erdem, A., Erdem, E., Çukur, T.: Image synthesis in multi-contrast MRI with conditional generative adversarial networks. IEEE Trans. Med. Imaging **38**(10), 2375–2388 (2019)
6. Dou, Q., Ouyang, C., Chen, C., Chen, H., Heng, P.A.: Unsupervised cross-modality domain adaptation of convnets for biomedical image segmentations with adversarial loss. In: Proceedings of the 27th International Joint Conference on Artificial Intelligence, pp. 691–697 (2018)
7. Ganin, Y., Lempitsky, V.: Unsupervised domain adaptation by backpropagation. In: Bach, F., Blei, D. (eds.) Proceedings of the 32nd International Conference on Machine Learning. Proceedings of Machine Learning Research, vol. 37, pp. 1180–1189. PMLR, 07–09 Jul 2015
8. Gatys, L.A., Ecker, A.S., Bethge, M.: Image style transfer using convolutional neural networks. In: Proceedings of the IEEE Conference on Computer Vision and Pattern Recognition, pp. 2414–2423 (2016)
9. Gong, D., et al.: Memorizing normality to detect anomaly: memory-augmented deep autoencoder for unsupervised anomaly detection. In: Proceedings of the IEEE International Conference on Computer Vision, pp. 1705–1714 (2019)
10. He, K., Zhang, X., Ren, S., Sun, J.: Delving deep into rectifiers: Surpassing human-level performance on imagenet classification. In: Proceedings of the IEEE International Conference on Computer Vision, pp. 1026–1034 (2015)

11. He, Y., et al.: Deep learning based topology guaranteed surface and MME segmentation of multiple sclerosis subjects from retinal OCT. Biomed. Opt. Express **10**(10), 5042–5058 (2019)
12. He, Y., et al.: Fully convolutional boundary regression for retina OCT segmentation. In: Shen, D., et al. (eds.) MICCAI 2019. LNCS, vol. 11764, pp. 120–128. Springer, Cham (2019). https://doi.org/10.1007/978-3-030-32239-7_14
13. He, Y., Carass, A., Solomon, S.D., Saidha, S., Calabresi, P.A., Prince, J.L.: Retinal layer parcellation of optical coherence tomography images: data resource for multiple sclerosis and healthy controls. Data Brief **22**, 601–604 (2018)
14. Hoffman, J., et al.: Cycada: cycle-consistent adversarial domain adaptation. In: International Conference on Machine Learning, pp. 1989–1998 (2018)
15. Jain, V., Learned-Miller, E.: Online domain adaptation of a pre-trained cascade of classifiers. In: CVPR 2011, pp. 577–584. IEEE (2011)
16. Jog, A., Carass, A., Roy, S., Pham, D.L., Prince, J.L.: MR image synthesis by contrast learning on neighborhood ensembles. Med. Image Anal. **24**(1), 63–76 (2015)
17. Long, M., Cao, Y., Wang, J., Jordan, M.I.: Learning transferable features with deep adaptation networks. arXiv preprint arXiv:1502.02791 (2015)
18. Ma, C., Ji, Z., Gao, M.: Neural style transfer improves 3D cardiovascular MR image segmentation on inconsistent data. In: Shen, D., et al. (eds.) MICCAI 2019. LNCS, vol. 11765, pp. 128–136. Springer, Cham (2019). https://doi.org/10.1007/978-3-030-32245-8_15
19. Ouyang, C., Kamnitsas, K., Biffi, C., Duan, J., Rueckert, D.: Data efficient unsupervised domain adaptation for cross-modality image segmentation. In: Shen, D., et al. (eds.) MICCAI 2019. LNCS, vol. 11765, pp. 669–677. Springer, Cham (2019). https://doi.org/10.1007/978-3-030-32245-8_74
20. Ronneberger, O., Fischer, P., Brox, T.: U-net: convolutional networks for biomedical image segmentation. In: Navab, N., Hornegger, J., Wells, W.M., Frangi, A.F. (eds.) MICCAI 2015. LNCS, vol. 9351, pp. 234–241. Springer, Cham (2015). https://doi.org/10.1007/978-3-319-24574-4_28
21. Seeböck, P., et al.: Using cyclegans for effectively reducing image variability across oct devices and improving retinal fluid segmentation. In: 2019 IEEE 16th International Symposium on Biomedical Imaging (ISBI 2019), pp. 605–609. IEEE (2019)
22. Shocher, A., Cohen, N., Irani, M.: "zero-shot" super-resolution using deep internal learning. In: Proceedings of the IEEE Conference on Computer Vision and Pattern Recognition, pp. 3118–3126 (2018)
23. Tsai, Y.H., Hung, W.C., Schulter, S., Sohn, K., Yang, M.H., Chandraker, M.: Learning to adapt structured output space for semantic segmentation. In: Proceedings of the IEEE Conference on Computer Vision and Pattern Recognition, pp. 7472–7481 (2018)
24. Tustison, N.J., et al.: N4itk: improved N3 bias correction. IEEE Trans. Med. Imaging **29**(6), 1310–1320 (2010)
25. Tzeng, E., Hoffman, J., Saenko, K., Darrell, T.: Adversarial discriminative domain adaptation. In: Proceedings of the IEEE Conference on Computer Vision and Pattern Recognition, pp. 7167–7176 (2017)
26. Zhu, J.Y., Park, T., Isola, P., Efros, A.A.: Unpaired image-to-image translation using cycle-consistent adversarial networks. In: Proceedings of the IEEE International Conference on Computer Vision, pp. 2223–2232 (2017)

Entropy Guided Unsupervised Domain Adaptation for Cross-Center Hip Cartilage Segmentation from MRI

Guodong Zeng[1], Florian Schmaranzer[2,3], Till D. Lerch[2,3], Adam Boschung[2,3], Guoyan Zheng[4(✉)], Jürgen Burger[1], Kate Gerber[1], Moritz Tannast[5], Klaus Siebenrock[2], Young-Jo Kim[6], Eduardo N. Novais[6], and Nicolas Gerber[1]

[1] sitem Center for Translational Medicine and Biomedical Entrepreneurship, University of Bern, Bern, Switzerland

[2] Department of Orthopaedic Surgery, Inselspital, University of Bern, Bern, Switzerland

[3] Department of Diagnostic, Interventional and Paediatric Radiology, Inselspital, University of Bern, Bern, Switzerland

[4] Institute of Medical Robotics, School of Biomedical Engineering, Shanghai Jiao Tong Unviersity, Shanghai, China
guoyan.zheng@sjtu.edu.cn

[5] Department of Orthopaedic Surgery and Traumatology, University Hospital of Fribourg, Fribourg, Switzerland

[6] Department of Orthopaedic Surgery, Boston Children's Hospital, Harvard Medical School, Boston, USA

Abstract. Hip cartilage damage is a major predictor of the clinical outcome of surgical correction for femoroacetabular impingement (FAI) and hip dysplasia. Automatic segmentation for hip cartilage is an essential prior step in assessing cartilage damage status. Deep Convolutional Neural Networks have shown great success in various automated medical image segmentations, but testing on domain-shifted datasets (e.g. images obtained from different centers) can lead to severe performance losses. Creating annotations for each center is particularly expensive. Unsupervised Domain Adaptation (UDA) addresses this challenge by transferring knowledge from a domain with labels (source domain) to a domain without labels (target domain). In this paper, we propose an entropy-guided domain adaptation method to address this challenge. Specifically, we first trained our model with supervised loss on the source domain, which enables low-entropy predictions on source-like images. Two discriminators were then used to minimize the gap between source and target domain with respect to the alignment of feature and entropy distribution: the feature map discriminator D_F and the entropy map discriminator D_E. D_F aligns the feature map of different domains, while D_E matches the target segmentation to low-entropy predictions like those from the source domain. The results of comprehensive experiments on cross-center MRI hip cartilage segmentation show the effectiveness of this method.

A. L. Martel et al. (Eds.): MICCAI 2020, LNCS 12261, pp. 447–456, 2020.
https://doi.org/10.1007/978-3-030-59710-8_44

Keywords: Unsupervised domain adaptation · Hip cartilage · Deep learning · Entropy · Segmentation · MRI

1 Introduction

The automatic and accurate segmentation of hip cartilage is an essential step in assessing cartilage thickness, surface area and volume size. Such information allows the assessment of the hip cartilage status, which is an important predictor of the clinical outcome after surgical correction for femoroacetabular impingement (FAI) and hip dysplasia [1]. Recently, the deep convolutional neural network has become a powerful tool for medical image segmentation [2] and has already been successfully applied for hip cartilage segmentation [3]. However, a well-trained model with cartilage data from one center (source domain) can lead to severe performance losses when tested on data from another center (target domain). This domain shift problem can be due to differences in the imaging device or the image acquisition process. An intuitive but ineffective way to address this problem is to create new annotations and train a new model for each center. A model that can be trained with annotations from only the source domain, which can be generalised to the target domain and without the need for additional annotations, would greatly reduce the adaptation workload and is thus highly desirable.

Unsupervised Domain Adaptation (UDA) aims to reduce the domain shift and improve performance on the target domain when only annotations from the source domain are available. Recent work can be primarily categorized as image adaptation [4,5], feature adaptation [6,7] and their mixture [8]. Image adaptation methods align the image appearance by a pixel-to-pixel transformation, typically realized by synthesizing source-like images from target images in an unsupervised way, e.g. CycleGAN [9]. The synthesized source-like images can then be tested directly on pre-trained source models [5]. Its weakness is the strong dependence on the image synthesis quality. Performance decreases sharply if the algorithm cannot synthesize source-like images with sufficient similarity. Feature adaptation methods learn domain invariant features by minimizing the feature distance between different domains. Recent studies use an adversarial training strategy to realize this [7]. In particular, a discriminator is used to classify the feature maps from different domains, while the segmentation network attempts to fool the discriminator. This type of method usually includes a separate path of feature encoding for each modality, and results in a complex training procedure [7]. Other work has investigated the combination of feature adaptation and image adaptation [8], which has led to very promising results in CT heart segmentation by domain adaptation from MRI. But the combination method deals with the unsupervised image translation and feature alignment in one network, which makes its architecture very complicated.

In this paper, we propose an entropy-guided unsupervised domain adaptation method and successfully demonstrate its application in the task of cross-center segmentation of the MRI hip cartilage. The idea of entropy regularization has been shown to be useful in semi-supervised learning [10,11]. To the best of our knowledge, we are the first to successfully apply the entropy-guided unsupervised domain adaptation method for medical image segmentation. Specifically, we first train our model with supervised loss on the labelled source data, leading to low-entropy predictions on the source-like images. Applying this pre-trained source model directly to the data from the target domain, would result in high entropy predictions and severe performance losses. To reduce this effect, we integrate two discriminators, namely the feature map discriminator D_F and the entropy map discriminator D_E, into our framework to reduce the gap and shift between domains. D_F aligns the feature maps of different domains, while D_E matches the target segmentation to the low-entropy prediction like those from the source domain. This allows the segmentation network to learn domain invariant feature maps and output similarly low-entropy predictions to transfer knowledge from the source domain to the target domain. It should be noted that our UDA framework does not involve transformation of the image appearance and that the different domains share all layers of the segmentation network instead of separating the paths for each domain.

The main contributions of this paper are summarized as follows: (1) We present the first domain adaptation framework for cross-center segmentation of hip cartilage, which can significantly reduce the workload for data annotation. (2) We propose to use a feature map discriminator and an entropy map discriminator to learn domain invariant features and reduce domain shift by enforcing the low-entropy output on the target domain like those from the source domain. (3) We validate the effectiveness of our method in the challenging task of cross-center hip MRI cartilage segmentation, and our approach overperforms the state-of-the-art methods.

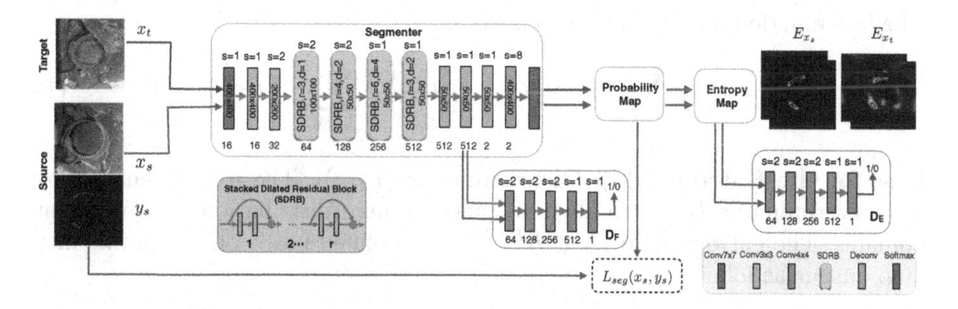

Fig. 1. Method overview: red arrows and blue arrows represent the target and source domain, respectively. The segmenter is trained with supervised loss on the source domain, and unsupervised adversarial loss between domains by two discriminators, i.e. the feature map discriminator D_F and the entropy map discriminator D_E. (Color figure online)

2 Methods

Overview of Our Method. Assume we have labelled source data $X^s = \{(x_i^s, y_i^s)\}_{i=1}^N$, and unlabelled target data $X^t = \{(x_j^t)\}_{j=1}^M$. To facilitate naming, x_s and y_s are directly used to represent the images and the ground-truth segmentation from the source domain, instead of x_i^s and y_i^s. Similarly, x_t denotes the target images. As shown in Fig. 1, our framework consists of a segmenter S and two discriminators, D_F and D_E. The segmenter is fully shared by the source and target domains. It takes an image x as input and outputs a probability map after the last softmax layer: $S(x) = P_x \subset R^{H \times W \times C}$, where H, W, C represent the height, the width of image and the class number of segmentation, respectively. The segmenter is optimized with a supervised loss $L_{seg}(x_s, y_s)$ on the source domain and unsupervised adversarial loss from both domains. Two discriminators, D_F and D_E, distinguish the feature map distribution and the entropy map distribution between different domains (label 1 for the source and label 0 for the target). The segmenter learns domain-invariant features and outputs entropy maps that cannot be distinguished by the discriminators.

Loss for the Feature Map Discriminator (L_{D_F}). The goal of the feature map discriminator D_F is to align the feature map distribution from different domains, which is typically achieved by adversarial training in recent [12]. Considering that the lower level feature maps mainly capture only simple features (such as edges and shape), while the deeper feature maps usually encode more abstract concepts, we choose to align the output of the second last convolution layer from the segmenter as shown in Fig. 1 and refer to it as $S_F(x)$. It is easy to extend this to other or more layers. D_F takes the feature map from both domains as input and classifies from which domain the feature map comes (label 1 for the source and label 0 for the target). Let L_D be the cross-entropy loss. The loss function for D_F is then defined as:

$$L_{D_F} = \frac{1}{|X_s|} \sum_{x_s \in X_s} L_D(S_F(x_s), 1) + \frac{1}{|X_t|} \sum_{x_t \in X_t} L_D(S_F(x_t), 0) \tag{1}$$

Loss for the Entropy Map Discriminator (L_{D_E}). The goal of the entropy map discriminator D_E is to align the entropy map distribution from different domains. The entropy map $E_x \in R^{H \times W \times C}$ is defined based on the probability map output as follows:

$$E_x^{(h,w,c)} = -p_x^{(h,w,c)} \cdot log(p_x^{(h,w,c)}) \tag{2}$$

where $p_x^{(h,w,c)}$ represents each voxel in the probability map $P_x \subset R^{H \times W \times C}$. The entropy value is minimal when the probability is at 0 and 1, and it is maximal when the probability is at 0.5.

Based on the entropy maps from both domains, E_{x_s} and E_{x_t}, shown in Fig. 1, we can observe that models trained on the source data typically produce low-entropy predictions on source-like images, but high entropy and noisy predictions on images of the target domain. Additionally, the entropy maps usually show high entropy along the cartilage boundary. If we can force the entropy maps from different domains to be similar, the segmentation from the target domain may look more similar to the segmentation from the source domain. To realise this, we construct another discriminator D_E to align the entropy distribution. D_E takes the entropy map of both domains as input and classifies from which domain the entropy map comes (label 1 for the source and label 0 for the target). Let L_D be the cross entropy loss. The loss function for D_E is defined as:

$$L_{D_E} = \frac{1}{|X_s|} \sum_{x_s \in X_s} L_D(E_{x_s}, 1) + \frac{1}{|X_t|} \sum_{x_t \in X_t} L_D(E_{x_t}, 0) \tag{3}$$

Loss for the Segmenter (L_S). The segmenter is optimized by supervised loss on the source domain and unsupervised adversarial loss between domains to fool the feature map discriminator and the entropy map discriminator. For the supervised loss on the source domain, we use a mixture of the cross entropy loss and the dice coefficient loss [13], which is defined as:

$$\begin{aligned} L_{seg(x_s,y_s)} = &-\sum_{h=1}^{H}\sum_{w=1}^{W}\sum_{c=1}^{C} y_s^{(h,w,c)} \cdot log(p_s^{(h,w,c)}) \\ &-\lambda_1 \sum_{h=1}^{H}\sum_{w=1}^{W}\sum_{c=1}^{C} \frac{2y_s^{(h,w,c)} \cdot \hat{y}_s^{(h,w,c)}}{y_s^{(h,w,c)} \cdot y_s^{(h,w,c)} + \hat{y}_s^{(h,w,c)} \cdot \hat{y}_s^{(h,w,c)}} \end{aligned} \tag{4}$$

where $y_s^{(h,w,c)}$ and $\hat{y}_s^{(h,w,c)}$ represent the ground truth and the predicted segmentation on the source domain in the form of one-hot vector with C-class, respectively, and $p_s^{(h,w,c)}$ represents the predicted probability map on the source domain from the segmenter. For the unsupervised adversarial loss, the segmenter attempts to fool the two discriminators. Finally, the loss for the segmenter L_S is defined as:

$$L_S = \frac{1}{|X_s|} \sum_{x_s \in X_s} L_{seg}(x_s, y_s) + \frac{1}{|X_t|} \sum_{x_t \in X_t} (\lambda_2 L_D(E_{x_t}, 1) + \lambda_3 L_D(S_F(x_t), 1)) \tag{5}$$

Network Architecture. As shown in Fig. 1, our segmenter takes advantage of the dilated residual blocks [14], which can extract representative features with a large receptive field. The segmenter consists of 3 convolution layers, 4 stacked dilated residual blocks (SDRB), again 3 convolution layers, a deconvolution layer and a final Softmax layer. Each convolution layer is immediately followed by a batch normalization (BN) and a ReLU layer. Each SDRB contains r repetitions of the basic dilated residual blocks, and all convolution layers inside use dilation factor of d. The basic dilated residual block contains two convolution layers with residual connection. We also show the channel number, stride, and output feature map size at the bottom, top, and center of each convolution layer or SDRB block, respectively. Note that "s = 2" applied in SDRB indicates that the stride is set to 2 only for the first convolution layer in the block. The discriminator networks, i.e. D_F and D_E, contain 5 convolution layers and all of them are followed by an instance normalization layer (IN) and a LeakyReLU layer (Leak = 0.2) except for the last output layer.

Implementation Details. The proposed method was implemented in Python using the Pytorch framework on a desktop with a 3.6 GHz Intel(R) i7 CPU and a GTX 1080 Ti graphics card with 11 GB GPU memory. We trained our network from scratch, and the parameters were updated by the stochastic gradient descent(SGD) algorithm (momentum = 0.9, weight decay = 0.005). The input image size was 400×400 and the batch size was 4. We trained the network for a total of 30 epochs. The initial learning rate was 1×10^{-3} and halved by every 5 epochs.

Table 1. Quantitative comparison results between our method and other state-of-the-art unsupervised domain adaptation methods for the task of cross-center hip cartilage segmentation. The best results are highlighted with bold font.

Methods	Adaptation	DICE (%)	HD (mm)	ASD (mm)
No Adaption	Train on Source Data Only	46.46 ± 31.73	51.10 ± 25.10	3.15 ± 4.37
CycleGAN [9]	Image Appearance Translation	8.86 ± 10.69	64.67 ± 26.87	21.48 ± 8.63
MCD [15]	Max Classifier Discrepancy	59.74 ± 27.79	32.43 ± 26.21	2.10 ± 3.03
ADDA [12]	Feature Alignment	67.25 ± 12.48	24.48 ± 9.14	1.23 ± 1.11
Ours	Feature Alignment + Entropy Alignment	**72.82 ± 3.43**	**14.98 ± 6.15**	**0.43 ± 0.16**
Target Model	Train on Target Data Directly	81.30 ± 4.58	10.48 ± 2.15	0.37 ± 0.12

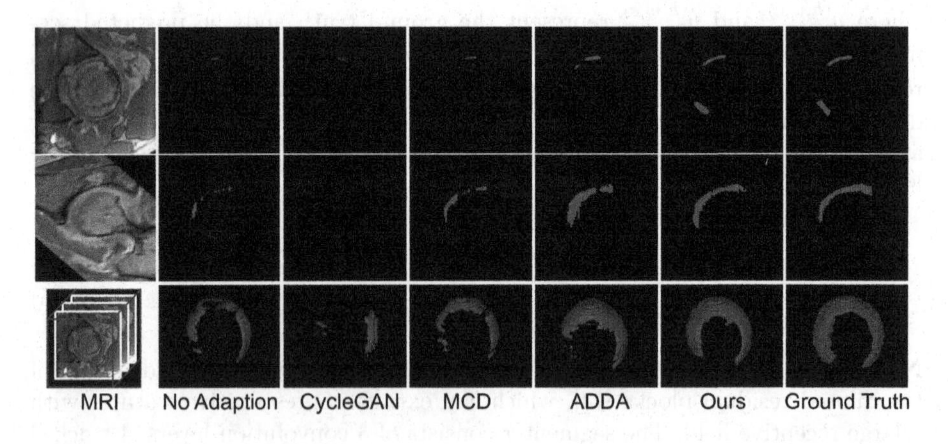

Fig. 2. Qualitative comparison of segmentation results by different methods. The first and second row show segmentation results in axial and coronal view respectively, while the last row shows comparison in 3D models.

3 Experiments and Results

The efficacy of the proposed method was evaluated in cross-centre hip cartilage segmentation from MRI. For all experiments described below, we used the Dice Coefficients (DICE), the Hausdorff Distance (HD) and the Average Surface Distance (ASD) as evaluation metrics.

3.1 Dataset and Preprocessing

The source dataset obtained from University Hospital of Bern contains 25 hip MR images, of which 20 cases were used for training and another 5 cases for validation. The target dataset from Boston Children's Hospital of Harvard Medical School contains 21 hip MR images, of which 14 cases were used for training and another 7 cases for testing. Slice by slice manual segmentation by experienced clinicians was used to create the ground truth segmentation from both centers. Note that the ground truth segmentation from the target domain was not used during the training phase. To deal with the different image size and the heterogeneous spacing between images from different centers, all MR images were resampled to a voxel spacing of $0.25 \times 0.25 \times 1\,\text{mm}^3$ and then cropped to $400 \times 400 \times 80$ pixels around the center of the femoral head. The femur head center was automatically detected by our previous introduced method [3], which reported a mean distance error of 2.78 mm, which was accurate enough to crop the regions containing the hip cartilage.

This could help to reduce the unnecessary calculation on unrelated tissues and focus on the cartilage regions. Each 3D MRI was normalized as zero mean and unit variance. Then, 2D slices were extracted from each 3D MRI in axial view. In total, we had $20 \times 80 = 1600$ slices from the source domain and $14 \times 80 = 1120$ slices from the target domain for training. The images were randomly flipped and augmented with random Gaussian noise $N(0, 0.05)$. For testing on the target domain, $7 \times 80 = 560$ slice images were used.

3.2 Experimental Results

Validation Study. We compared the proposed method with three state-of-the-art methods: CycleGAN [9], MCD[15], and ADDA [12]. CycleGAN code from the original paper was used for unsupervised image translation. The target images were then translated into source-like images and tested directly on the source model. For MCD, the results were also obtained by executing the official code provided by [15]. In order to save GPU memory, allowing for execution on a single GPU machine, we replaced the original fully connected discriminator in ADDA with a fully convolutional discriminator which uses the same discriminator architecture as shown in Fig. 1. The quantitative results are presented in Table 1. Our method outperformed CycleGAN, MCD and ADDA in all metrics, reporting an average DICE of 72.82%, an average HD of 14.98 mm and an average ASD of 0.43 mm. Interestingly, the results obtained by CycleGAN (8.86% for DICE) were unexpectedly far lower than those obtained from the source model (46.46% for DICE) without domain adaptation. This is because the synthesised image from the target domain is not sufficiently similar to the source image, which is illustrated in Fig. 3. A comparison of segmentation outputs from different methods is shown in Fig. 2. We can observe that the results from the source model without adaptation in the second column are very poor predictions, especially in the top regions of the cartilage 3D model, while the output from CycleGAN was close to zero. The results of MCD and ADDA produced better predictions, but they were still not in a meaningful structure or morphology. Our method provided good predictions with results more similar to the ground truth in the 3D view.

Ablation Study. To investigate the role of different discriminators in our method, we conducted a study to compare the performance of three variants of our method: a) No adaptation: trained on source data only; b) trained on source data with feature alignment between domains; c) trained on source data with feature alignment and entropy alignment between domains. The data and experiment setups used were the same as those described above. The quantitative results are displayed in Table 2. Compared with the baseline model without adaptation, the integration of the feature map alignment increased the average dice from 46.46% to 62.61% and the combination with the entropy map alignment further improved the average dice to 72.82%. These experimental results clearly demonstrate the effectiveness of each component of the proposed method.

Table 2. Quantitative comparison of three different variants of our method for cross-center hip cartilage segmentation. The best results are highlighted with bold font.

Methods	Lseg	L_{D_F}	L_{D_E}	DICE (%)	HD (mm)	ASD (mm)
No Adaptation	✓			46.46	51.10	3.15
+ Feature Alignment	✓	✓		62.61	23.95	1.13
+ Entropy Alignment	✓	✓	✓	**72.82**	**14.98**	**0.43**

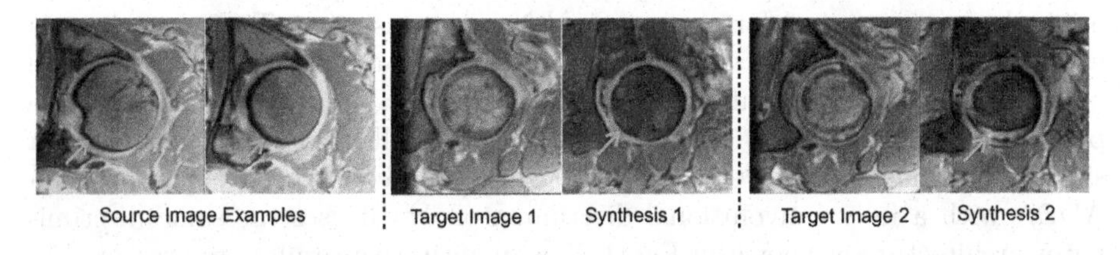

Fig. 3. Illustration of the poor image synthesis by CycleGAN. The source images usually show darker edges, but are brighter inside the femoral head, i.e. the round bone to which the red arrows point. In the images synthesized by CycleGAN from the target, however, the entire femoral head including the borders are dark. This inconsistent appearance will result in poor segmentation of the synthesized images when tested on a pre-trained model using source data. (Color figure online)

4 Discussion and Conclusion

In summary, we presented an entropy guided unsupervised domain adaptation method and successfully applied it to the task of cross-center hip MRI cartilage segmentation, without the need for additional time-consuming annotations on the target domain. Our method consists of a segmenter and two discriminators, the feature map discriminator and the entropy map discriminator. The feature map discriminator aligns the feature map distribution to learn domain invariant features, while the entropy map discriminator forces low-entropy predictions on the target domain like those on source domain. The segmenter is optimized by supervised loss on labelled source data and unsupervised adversarial loss to fool the discriminators between domains. Our results have shown that the present method achieves better results than the state-of-the-art unsupervised domain adaptation methods.

Acknowledgement. This study was partially supported by Swiss National Science Foundation (grant P1BEP3_181643), National Key Research and Development Program of China via Project 2019YFC0120603.

References

1. Haefeli, P.C., Albers, C.E., Steppacher, S.D., Tannast, M., Büchler, L.: What are the risk factors for revision surgery after hip arthroscopy for femoroacetabular impingement at 7-year followup? Clin. Orthop. Relat. Res.® **475**(4), 1169–1177 (2017)
2. Çiçek, Ö., Abdulkadir, A., Lienkamp, S.S., Brox, T., Ronneberger, O.: 3D U-net: learning dense volumetric segmentation from sparse annotation. In: Ourselin, S., Joskowicz, L., Sabuncu, M.R., Unal, G., Wells, W. (eds.) MICCAI 2016. LNCS, vol. 9901, pp. 424–432. Springer, Cham (2016). https://doi.org/10.1007/978-3-319-46723-8_49
3. Schmaranzer, F., et al.: Automatic MRI-based three-dimensional models of hip cartilage provide improved morphologic and biochemical analysis. Clin. Orthop. Relat. Res.® **477**(5), 1036–1052 (2019)
4. Chen, C., Dou, Q., Chen, H., Heng, P.-A.: Semantic-aware generative adversarial nets for unsupervised domain adaptation in chest X-Ray segmentation. In: Shi, Y., Suk, H.-I., Liu, M. (eds.) MLMI 2018. LNCS, vol. 11046, pp. 143–151. Springer, Cham (2018). https://doi.org/10.1007/978-3-030-00919-9_17
5. Chen, C., et al.: Unsupervised multi-modal style transfer for cardiac MR segmentation. arXiv preprint arXiv:1908.07344 (2019)
6. Kamnitsas, K., et al.: Unsupervised domain adaptation in brain lesion segmentation with adversarial networks. In: Niethammer, M., et al. (eds.) IPMI 2017. LNCS, vol. 10265, pp. 597–609. Springer, Cham (2017). https://doi.org/10.1007/978-3-319-59050-9_47
7. Dou, Q., Ouyang, C., Chen, C., Chen, H., Heng, P.A.: Unsupervised cross-modality domain adaptation of convnets for biomedical image segmentations with adversarial loss. In: Proceedings of the 27th International Joint Conference on Artificial Intelligence, pp. 691–697 (2018)

8. Chen, C., Dou, Q., Chen, H., Qin, J., Heng, P.: Unsupervised bidirectional cross-modality adaptation via deeply synergistic image and feature alignment for medical image segmentation. IEEE Trans. Med. Imag. (2020)
9. Zhu, J.Y., Park, T., Isola, P., Efros, A.A.: Unpaired image-to-image translation using cycle-consistent adversarial networks. In: Proceedings of the IEEE International Conference on Computer Vision, pp. 2223–2232 (2017)
10. Springenberg, J.T.: Unsupervised and semi-supervised learning with categorical generative adversarial networks. In: ICLR (2016)
11. Grandvalet, Y., Bengio, Y.: Semi-supervised learning by entropy minimization. In: Advances in Neural Information Processing Systems, pp. 529–536 (2005)
12. Tzeng, E., Hoffman, J., Saenko, K., Darrell, T.: Adversarial discriminative domain adaptation. In: Proceedings of the IEEE Conference on Computer Vision and Pattern Recognition, pp. 7167–7176 (2017)
13. Milletari, F., Navab, N., Ahmadi, S.A.: V-net: Fully convolutional neural networks for volumetric medical image segmentation. In: 2016 Fourth International Conference on 3D Vision (3DV), pp. 565–571. IEEE (2016)
14. Yu, F., Koltun, V., Funkhouser, T.: Dilated residual networks. In: Proceedings of the IEEE Conference on Computer Vision and Pattern Recognition, pp. 472–480 (2017)
15. Saito, K., Watanabe, K., Ushiku, Y., Harada, T.: Maximum classifier discrepancy for unsupervised domain adaptation. In: Proceedings of the IEEE Conference on Computer Vision and Pattern Recognition, pp. 3723–3732 (2018)

User-Guided Domain Adaptation for Rapid Annotation from User Interactions: A Study on Pathological Liver Segmentation

Ashwin Raju[1,2], Zhanghexuan Ji[1,3], Chi Tung Cheng[4], Jinzheng Cai[1], Junzhou Huang[2], Jing Xiao[5], Le Lu[1], ChienHung Liao[4], and Adam P. Harrison[1](✉)

[1] PAII Inc., Bethesda, MD, USA
adampharrison070@paii-labs.com
[2] The University of Texas at Arlington, Arlington, TX, USA
[3] University at Buffalo, Buffalo, NY, USA
[4] Chang Gung Memorial Hospital, Linkou, Taiwan, ROC
[5] PingAn Technology, Shenzhen, China

Abstract. Mask-based annotation of medical images, especially for 3D data, is a bottleneck in developing reliable machine learning models. Using minimal-labor (UIs) to guide the annotation is promising, but challenges remain on best harmonizing the mask prediction with the UIs. To address this, we propose the user-guided domain adaptation (UGDA) framework, which uses prediction-based adversarial domain adaptation (PADA) to model the combined distribution of UIs and mask predictions. The UIs are then used as anchors to guide and align the mask prediction. Importantly, UGDA can both learn from unlabelled data and also model the high-level semantic meaning behind different UIs. We test UGDA on annotating pathological livers using a clinically comprehensive dataset of 927 patient studies. Using only extreme-point UIs, we achieve a mean (worst-case) performance of 96.1% (94.9%), compared to 93.0% (87.0%) for deep extreme points (DEXTR). Furthermore, we also show UGDA can retain this state-of-the-art performance even when only seeing a fraction of available UIs, demonstrating an ability for robust and reliable UI-guided segmentation with extremely minimal labor demands.

Keywords: Liver segmentation · Interactive segmentation · User-guided domain adaptation

1 Introduction

Reliable computer-assisted segmentation of anatomical structures from medical images can allow for quantitative biomarkers for disease diagnosis, prognosis, and

Electronic supplementary material The online version of this chapter (https://doi.org/10.1007/978-3-030-59710-8_45) contains supplementary material, which is available to authorized users.

A. L. Martel et al. (Eds.): MICCAI 2020, LNCS 12261, pp. 457–467, 2020.
https://doi.org/10.1007/978-3-030-59710-8_45

progression. Given the extreme labor to fully annotate data, especially for 3D volumes, a considerable body of work focuses on weakly-supervised segmentation solutions [24]. Solutions that can leverage user interactions UIs, *e.g.*, extreme-points, scribbles, and boundary marks, are an important such category.

The main challenge is effectively leveraging UIs to constrain or guide the mask generation. Classic approaches, like the random walker (RW) algorithm [5], do so via propagating seed regions using intensity similarities. Later approaches add additional constraints, *e.g.*, based on presegmentations [6] or learned probabilities [7]. With the advent of deep-learning, harmonizing mask predictions with the UIs continues to be a challenge. Deep extreme points (DEXTR) [17], which requires the user to click on the extreme boundary points of an object, is a popular and effective approach. But DEXTR only adds the extreme point annotations as an additional channel when training the segmentor, meaning the predicted mask may not agree with the UIs. Later work uses expectation-maximization strategies that alternate between network training and then regularization via RW [22] or dense conditional random fields (CRFs) [2,20,27]. However, over and above their computational demands, the intensity-based RW or CRF regularization may not capture high-level semantics and any guidance on the mask predictions still remains highly indirect. DeepIGeoS [28] offers an alternative that uses a deep-CRF optimizer to propagate scribbles. However, DeepIGeos would only allow boundary annotations or extreme points to be treated as simple seed regions, neglecting their rich semantic meaning.

To address these issues, we propose user-guided domain adaptation (UGDA). Our new method uses prediction-based adversarial domain adaptation (PADA) [26] to guide mask predictions by the UIs. UGDA's advantage is that it is equipped to model the high-level meaning behind different types of UIs and how they should impact the ultimate mask prediction. Importantly, the UIs are used as anchors when adapting the mask. Another advantage is that, like PADA, UGDA can learn from and exploit completely unlabelled data, in addition to those accompanied by UIs. Without loss of generality, we focus on using DEXTR-style extreme point UIs because of their intuitiveness and effectiveness [19,22]. But other types, *e.g.*, boundary corrections, are equally possible in addition to, or instead of, extreme points. The only constraint is that we assume a fully-supervised dataset is available in order to model the interplay between mask and UIs. But such data can originate from sources other than the target dataset, *e.g.*, from public data. To the best of our knowledge, we are the first to use domain adaptation as a mechanism to drive UI-based segmentation.

We test UGDA on an extremely challenging pathological liver segmentation dataset collected directly from the picture archiving and communication system (PACS) of Chang Gung Memorial hospital (CGMH). This dataset comprises 927 patient studies, all with hepatocellular carcinoma (HCC), intrahepatic cholan- giocellular carcinoma (ICC), metastasized, or benign lesions and is *one of the most challenging datasets to date for pathological liver segmentation.* Liver volumetry analysis is a crucial pre-requisite prior to many hepatic procedures, such as liver transplantation [18,25] or resection [14,15]. Despite success with fully-automated solutions, modern deep solutions [11,12,29] are often trained only on public datasets, *e.g.*, LiTS [1], which only represents HCC and metastasized lesions and does not fully represent patient/co-morbidity distributions. Thus, they may not generalize to all datasets, such as ours. Using *only extreme-point UIs*, we achieve state-of-the-art mean (worst-case) Dice-Sørensen coefficient (DSC) scores of 96.1% (94.5%) on our dataset, compared to 93.0% (79.0%) and 93.1% (87.0%) for a strong a fully-supervised baseline and DEXTR [17], respectively. We also show that UGDA can improve over PADA [26] by 1.3% DSC and that UGDA can even perform robustly when only shown incomplete sets of UIs. Finally, we demonstrate predicted masks align extraordinarily well with UIs, allowing the user to interact with confidence and with minimal frustration.

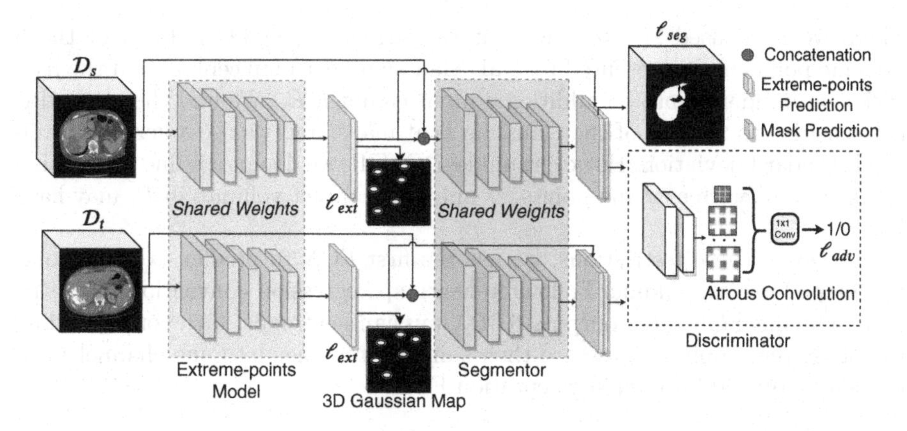

Fig. 1. Overview. UGDA chains together (1) an initial FCN that predicts the object's extreme points, and (2) a second FCN that accepts the first's predictions to predict a mask. For source data, $\mathcal{D}_s$, where the mask label is present, we compute fully-supervised loss on both the extreme-point and mask predictions. For target data, $\mathcal{D}_t$, we compute a fully-supervised loss when UIs are available. For all $\mathcal{D}_t$ volumes, whether UI-labelled or completely unlabelled, we use ADA to guide the mask predictions based on the extreme-point anchors.

2 Methods

We aim to produce reliable mask predictions on a target dataset or deployment scenario, given only minimal UIs. More formally, we assume we are given a dataset composed of both UI-labelled and completely unlabelled volumes, $\mathcal{D}_t = \{X_i, E_i\}_{i=1}^{N_w} \bigcup \{X_i\}_{i=1}^{N_u}$, with X_i and E_i denoting the images and extreme points UIs, respectively. In addition, we also assume a fully-supervised source dataset with masks is also available, $\mathcal{D}_s = \{\mathcal{X}_i, Y_i\}_{i=1}^{N_s}$. As long as the masks and extreme points describe the same anatomical structure, $\mathcal{D}_s$ may originate from entirely different sources, *e.g.*, public data. The goal is to use only the extreme point UIs to robustly annotate $\mathcal{D}_t$. Figure 1 outlines our workflow, which uses user-guided domain adaptation (UGDA) to efficiently and effectively exploit the extreme-point UIs.

2.1 Supervised Workflow

The backbone of UGDA is two 3D FCNs chained together, where the first FCN predicts extreme-points while the second predicts a full mask. Working backward, the second FCN acts very similarly to DEXTR [17], where the latter predicts a mask given an input image along with the extreme-point UIs:

$$\hat{Y} = s(X, E), \tag{1}$$

where we have used $s(.)$ to represent the segmentation FCN. Each of the 6 extreme points are represented by a 3D Gaussian heat map centered on the user clicks and rendered into an additional input channel, E_i. We find the method is not sensitive to the size of the Gaussian heat maps, and we use a kernel with 5-pixel standard deviation. However, unlike DEXTR, we do not assume all training volumes come with extreme-points, as only some target volumes in $\mathcal{D}_t$ may have UIs.

To loosen these restrictions, we use the first FCN to predict extreme-point heatmaps for each volume. Following heatmap regression conventions [30], the extreme-point FCN, $h(.)$, outputs 6 3D Gaussian heatmaps, each corresponding to one extreme point. These are then summed together into one channel prior to being inputted into our segmentation FCN:

$$\hat{Y} = s(X, \hat{E}), \tag{2}$$

$$\hat{E} = h(X), \tag{3}$$

where for convenience we have skipped showing the summation of 6 heatmaps into one channel. By relying on predictions, this allows our system to operate even with unlabelled data. Ground-truth extreme-points can be generated from any UIs performed on $\mathcal{D}_t$. Extreme-points for $\mathcal{D}_s$ can be deterministically generated from the full masks, simulating the UIs.

In terms of loss, if we use $\mathcal{D}_e$ to denote any input volume associated with extreme-point UIs, whether from $\mathcal{D}_s$ or $\mathcal{D}_t$, then a supervised loss can be formulated:

$$\mathcal{L}_{sup} = \mathcal{L}_{seg} + \mathcal{L}_{ext}, \tag{4}$$

$$\mathcal{L}_{ext} = \frac{1}{N_e} \sum_{X,E \in \mathcal{D}_e} \ell_{ext}\left(h(X), E\right), \tag{5}$$

$$\mathcal{L}_{seg} = \frac{1}{N_s} \sum_{X,Y \in \mathcal{D}_s} \ell_{seg}\left(s(X, h(X)), Y\right), \tag{6}$$

where N_e denotes the cardinality of $\mathcal{D}_e$. Following heat-map regression practices [30], we implement ℓ_{ext} using mean-squared error. For ℓ_{seg} we use a summation of cross entropy and DSC losses, which has experienced success in segmentation tasks [11]. While we focus on extreme points for this work, other types of UIs, such as boundary corrections, can be readily incorporated in this framework.

2.2 User-Guided Domain Adaptation

Similar to DEXTR [17], (6) indirectly guides mask predictions by using extreme-point heat maps as an additional input channel for the segmentor. Additionally, for UI-labelled volumes in $\mathcal{D}_t$, the supervised loss in (5) encourages the extreme-point predictions to actually match the UIs. However, the mask prediction may contradict the UIs because there is no penalty for disagreement between the two. Thus, an additional mechanism is needed to align the mask with UIs. We opt for an adversarial domain adaption approach to penalize discordant mask predictions. While image translation-based adversarial domain adaptation methods show excellent results [10,13], these are unsuited to our task because we are concerned with adapting the *prediction-space* to produce a mask well-aligned with the UIs. Thus, we use (PADA) [26].

More specifically, we use a discriminator, $d(.)$, to learn the distribution and interplay between masks and extreme-points. Treating samples from $\mathcal{D}_s$ as the "correct" distribution, the discriminator loss can be expressed as

$$\mathcal{L}_d = \frac{1}{N_s}\sum_{\mathcal{D}_s} \ell_{bce}(d(\{\hat{Y}, \hat{E}\}, \mathbf{1})) + \frac{1}{N_t}\sum_{\mathcal{D}_t} \ell_{bce}(d(\{\hat{Y}, \hat{E}\}, \mathbf{0})), \tag{7}$$

where ℓ_{bce} denotes the cross-entropy loss. Importantly, to model their combined distribution, the discriminator accepts both the UI and mask predictions. Following standard adversarial training, here gradients only flow through the discriminator. UGDA then attempts to fool the discriminator by predicting extreme-point/mask pairs for $\mathcal{D}_t$ that match $\mathcal{D}_s$'s distribution. More formally, an adversarial loss is set up for volumes in $\mathcal{D}_t$:

$$\mathcal{L}_{adv} = \frac{1}{N_t}\sum_{\mathcal{D}_t} \ell_{bce}(d(\{\hat{Y}, \hat{E}\}, \mathbf{1})). \tag{8}$$

Note that compared to (7), the "label" for $\mathcal{D}_t$ has been switched from $\mathbf{0}$ to $\mathbf{1}$. Like standard PADA setups, gradients do not flow through the discriminator weights in (8). Importantly, *gradients also do not flow through the extreme-point predictions when the UIs are present.* Consequently, when UIs are available, extreme-point predictions are only influenced by the supervised loss in (5) to match the UIs. While there is no strict guarantee, our results demonstrate almost perfect matching. Thus, the extreme-point predictions act as anchors, while the adversarial loss in (8) guides the mask predictions to properly align with the UIs. This alignment is more than simply making mask predictions agree with the UIs, as by modelling their interplay, PADA also guides mask regions far away from UIs. Finally, the use of PADA provides another important benefit, as completely unlabelled volumes in $\mathcal{D}_t$ can seamlessly contribute to the learning process in (7) and (8). In fact, UGDA can be seen as integrating domain adaptation learning processes [26] in addition to DEXTR-style guidance [17] from UIs. Thus, the overall training objective for UGDA is to minimize the following total loss:

$$\mathcal{L} = \mathcal{L}_{sup} + \lambda_{adv}\mathcal{L}_{adv}, \tag{9}$$

where we have purposely kept loss weighting to only the adversarial component to reduce hyper-parameter tuning.

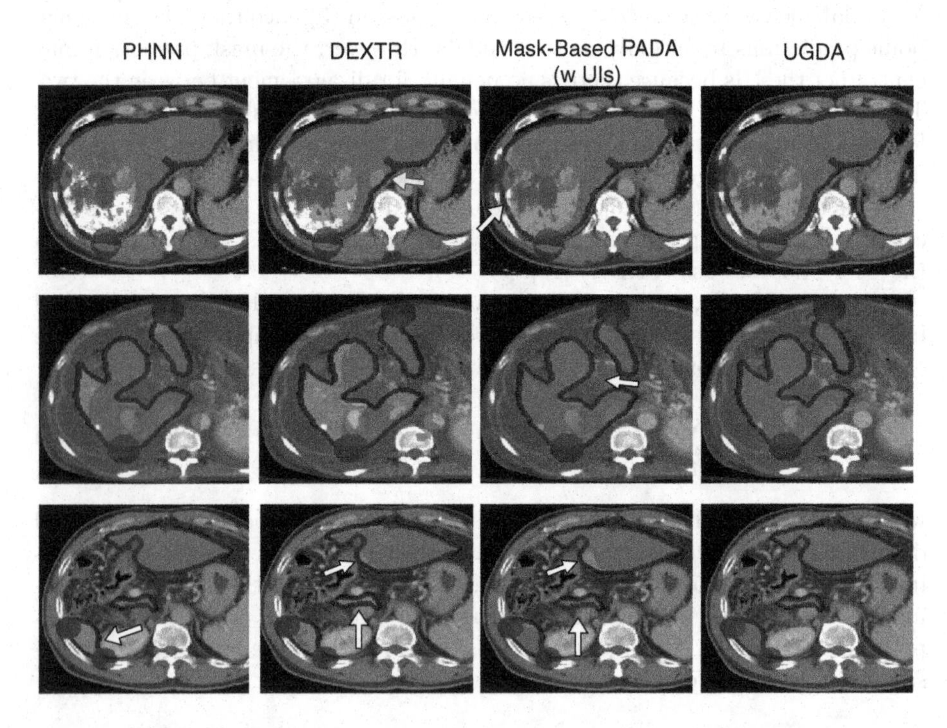

Fig. 2. Qualitative Results. Liver mask ground truth and predictions are rendered in green contour and blue mask, respectively, and Gaussian heatmaps centered on the extreme point UIs are shown in red. As can be seen, UGDA can much better align the masks with the UIs. Arrows highlight selected baseline prediction errors that UGDA corrects. (Best viewed in color) (Color figure online)

3 Experiments

Dataset Description. We test UGDA on segmenting pathological livers, using a target dataset of 927 venous-phase computed tomography (CT) studies from the PACS of CGMH. The only selection criterion was of patients with biospied or resecuted liver lesions, with CT scans taken within one month before the procedure. Patients directly reflect clinical distributions and represent ICC, HCC, benign or metastasized lesions, along with co-occurring maladies, such as liver fibrosis, splenomegaly, or embolized lesions. This serves as $\mathcal{D}_t$. From these, we selected 47 and 100 studies as validation and test sets, respectively, and delineated the patient livers. These 147 CTs are called *evaluation volumes*. We annotated the remainder using only extreme point UIs. For $\mathcal{D}_s$ we collected 235 fully-labelled venous-phase CT studies from public datasets [1,3,4,9], which, unlike $\mathcal{D}_t$, comprises both healthy and pathological livers and only represents HCC and metastasized tumors. Corresponding extreme-point "UIs" were generated from the full masks following [17]. For internal validation, we also split $\mathcal{D}_s$ into 70%, 20%, and 10% for training, testing and validation, respectively.

Implementation Details. For UGDA's two FCN architectures, we use a 3D version of the deeply-supervised progressive holistically nested network (PHNN) [8], which provides an efficient and decoder-free pipeline. As backbone, we use a 3D generalization of VGG-16 [23]. We first train a fully-supervised baseline on $\mathcal{D}_s$ using (4), then we finetune after convergence using (9). This dual-PHNN baseline is very strong on our public data, achieving a DSC score of 96.9% on the $\mathcal{D}_s$ test set. For discriminator, we use a 3D version of a popular architecture [16] using atrous convolution, which has proved a useful discriminator for liver masks [21]. We do not use the multi-level discriminator variant proposed by Tsai *et al.* [26], as its added complexity does not seem to result in noticeable improvements for liver-based PADA [21]. Specific details on the choice of hyper parameters are listed in the supplementary material.

Evaluation Protocols. We evaluate UGDA on how well it can annotate $\mathcal{D}_t$ using only extreme-point UIs. To do this, we include the evaluation volumes and their extreme point UIs within the training procedure, *but hide their masks*. For evaluation, we measure DSC scores and also the mask-extreme-point agreement (MXA). The latter measures the average distance between all six of a predicted *mask's* extreme-points vs. the ground-truth extreme points. This directly measures how well the model can produce a mask prediction that actually matches the extreme-point UIs.

Comparisons. We test against the user-interactive DEXTR [17] using the same PHNN backbone. DEXTR can only be trained on $\mathcal{D}_s$ because it requires fully-labelled training data, but during inference it sees $\mathcal{D}_t$'s extreme-point UIs. The DEXTR authors claim their approach can generalize well to new data. We also test against two mask-based PADA [26] variants. The first variant, called "mask-based PADA (no UIs)", matches published practices [26] and just uses a single network (PHNN in our case) to directly predict masks, using a discriminator to penalize ill-behaving mask predictions on unlabeled data. It does not incorporate UIs. The second variant, simply called "mask-based PADA (w UIs)", is almost identical to UGDA, meaning the mask and extreme-point predictions are still trained with the supervised losses of (6) and (5), respectively, but the discriminator in (7) and (8) only models mask predictions without seeing the extreme-point predictions. Comparing this variant to UGDA reveals the impact of using extreme-points as anchors to guide the domain adaptation of the masks.

Finally, we also compare against UGDA variants trained with only a fraction of the UIs available in $\mathcal{D}_t$, with the remainder being left unlabelled. This reveals how well UGDA can operate in scenarios when only a fraction of target volumes have UI-labels. When doing so, we remove the same percentage of UI-labels from both non-evaluation and evaluation volumes.

Results. Table 1 outlines the performance of all variants in annotating $\mathcal{D}_t$. As can be seen, compared to its performance on $\mathcal{D}_s$, the fully-supervised dual PHNN's performance drops from 96.9% to 93.0% due to the major differences between public liver datasets and our PACS-based clinical target dataset. This suggests that other strategies are required, *e.g.*, exploiting minimal-labor UIs.

Table 1. DSC and MXA mean and standard deviation scores. In parentheses are the fraction of UI-labelled $\mathcal{D}_t$ volumes used for training (or inference for DEXTR).

Model	% UIs	Mean DSC	Mean MXA (mm)
Dual PHNN [8]	n/a	93.0 ± 3.2	4.3 ± 1.2
DEXTR [17]	100%	93.1 ± 2.4	3.9 ± 1.2
Mask-based PADA (no UIs) [26]	0%	94.8 ± 1.8	3.4 ± 1.6
Mask-based PADA (w UIs) [26]	100%	95.5 ± 1.0	2.5 ± 1.0
UGDA	25%	95.8 ± 0.8	1.7 ± 0.8
UGDA	50%	96.0 ± 0.9	1.4 ± 0.9
UGDA	100%	$\mathbf{96.1 \pm 0.8}$	$\mathbf{1.1 \pm 0.9}$

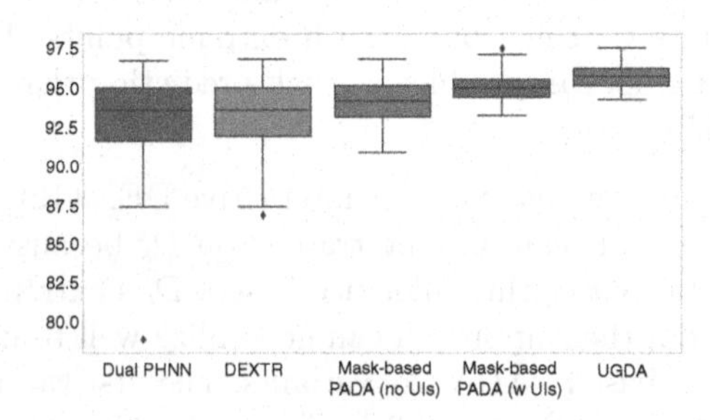

Fig. 3. Box and whisker plot of pathological liver segmentation DSC scores across the 100 $\mathcal{D}_t$ test set volumes.

As the table also demonstrates, by exploiting UIs, DEXTR can significantly boost the MXA, but its lackluster DSC scores suggest that the resulting masks, while aligning better with the extreme-points, still do not properly capture the liver extent. On the other hand, both mask-based PADA variants perform better, indicating that modelling mask distributions on top of DEXTR-style UI guidance can more robustly annotate $\mathcal{D}_t$. Finally, UGDA performs best, demonstrating that modelling the *interplay* between the UIs and mask predictions can boost performance even further. Importantly, UGDA's MXA is extremely good, which shows that the mask predictions well match the UIs. These mean scores are bolstered by Fig. 3's box-and-whisker plots, which demonstrate that UGDA provides important boosts in reliability, with an extremely robust worst-case performance of 94.9% DSC, compared to 93.2% for the mask-based PADA (w UIs) variant. Viewing the qualitative examples in Fig. 2 reinforces these quantitative improvements. In particular, UGDA is able to ensure that masks both agree with extreme-points and provide robust predictions away from the UIs.

Finally, as Table 1 also demonstrates, UGDA can perform almost as well when only a fraction of $\mathcal{D}_t$ is UI-labelled, outperforming both DEXTR and the mask-based PADA, both of which see all 100% of the UIs. These results indicate that UGDA can operate well even in scenarios with *extremely minimal* UI annotation. providing further evidence of its very high versatility.

4 Conclusion

We presented user-guided domain adaptation (UGDA), an effective approach to use UIs to rapidly annotate 3D medical volumes. We tested UGDA on arguably the most challenging and comprehensive pathological liver segmentation dataset to date, demonstrating that by only using extreme-point UIs we can achieve DSC scores of 96.1%, outperforming both DEXTR [17] and conventional mask-based PADA [26]. Future work includes testing on other anatomical structures, incorporating additional types of UIs, and adapting UGDA for real-time interaction.

References

1. Bilic, P., et al.: The liver tumor segmentation benchmark (LiTS). arXiv preprint arXiv:1901.04056 (2019)
2. Can, Y.B., Chaitanya, K., Mustafa, B., Koch, L.M., Konukoglu, E., Baumgartner, C.F.: Learning to segment medical images with scribble-supervision alone. In: Stoyanov, D., et al. (eds.) DLMIA/ML-CDS -2018. LNCS, vol. 11045, pp. 236–244. Springer, Cham (2018). https://doi.org/10.1007/978-3-030-00889-5_27
3. chaos: Chaos - combined (ct-mr) healthy abdominal organ segmentation (2019). https://chaos.grand-challenge.org/Combined_Healthy_Abdominal_Organ_Segmentation
4. Gibson, E., et al.: Multi-organ Abdominal CT Reference Standard Segmentations, February 2018. https://doi.org/10.5281/zenodo.1169361
5. Grady, L.: Random walks for image segmentation. IEEE Trans. Pattern Anal. Mach. Intell. **28**(11), 1768–1783 (2006)

6. Grady, L., Funka-Lea, G.: An energy minimization approach to the data driven editing of presegmented images/volumes. In: Larsen, R., Nielsen, M., Sporring, J. (eds.) MICCAI 2006. LNCS, vol. 4191, pp. 888–895. Springer, Heidelberg (2006). https://doi.org/10.1007/11866763_109
7. Harrison, A.P., Birkbeck, N., Sofka, M.: IntellEditS: intelligent learning-based editor of segmentations. In: Mori, K., Sakuma, I., Sato, Y., Barillot, C., Navab, N. (eds.) MICCAI 2013. LNCS, vol. 8151, pp. 235–242. Springer, Heidelberg (2013). https://doi.org/10.1007/978-3-642-40760-4_30
8. Harrison, A.P., Xu, Z., George, K., Lu, L., Summers, R.M., Mollura, D.J.: Progressive and multi-path holistically nested neural networks for pathological lung segmentation from CT images. In: Descoteaux, M., Maier-Hein, L., Franz, A., Jannin, P., Collins, D.L., Duchesne, S. (eds.) MICCAI 2017. LNCS, vol. 10435, pp. 621–629. Springer, Cham (2017). https://doi.org/10.1007/978-3-319-66179-7_71
9. Heimann, T., et al.: Comparison and evaluation of methods for liver segmentation from CT datasets. IEEE Trans. Med. Imaging **28**(8), 1251–1265 (2009)
10. Hoffman, J.,et al.: CyCADA: cycle-consistent adversarial domain adaptation. In: Dy, J., Krause, A. (eds.) Proceedings of the 35th International Conference on Machine Learning. Proceedings of Machine Learning Research, vol. 80, pp. 1989–1998. PMLR, Stockholmsmässan, Stockholm Sweden, 10–15 July 2018 (2018)
11. Isensee, F., et al.: nnU-Net: self-adapting framework for u-net-based medical image segmentation. arXiv preprint arXiv:1809.10486 (2018)
12. Li, X., Chen, H., Qi, X., Dou, Q., Fu, C.W., Heng, P.A.: H-DenseUNet: hybrid densely connected UNet for liver and tumor segmentation from CT volumes. IEEE Trans. Med. Imaging **37**(12), 2663–2674 (2018)
13. Li, Y., Yuan, L., Vasconcelos, N.: Bidirectional learning for domain adaptation of semantic segmentation. In: IEEE Conference on Computer Vision and Pattern Recognition, CVPR 2019, Long Beach, CA, USA, 16–20 June 2019, pp. 6936–6945. Computer Vision Foundation/IEEE (2019). https://doi.org/10.1109/CVPR.2019.00710
14. Lim, M., Tan, C., Cai, J., Zheng, J., Kow, A.: CT volumetry of the liver: where does it stand in clinical practice? Clin. Radiol. **69**(9), 887–895 (2014)
15. Lodewick, T.M., Arnoldussen, C.W., Lahaye, M.J., van Mierlo, K.M., Neumann, U.P., Beets-Tan, R.G., Dejong, C.H., van Dam, R.M.: Fast and accurate liver volumetry prior to hepatectomy. HPB **18**(9), 764–772 (2016)
16. Lv, F., Lian, Q., Yang, G., Lin, G., Pan, S.J., Duan, L.: Domain adaptive semantic segmentation through structure enhancement. In: Leal-Taixé, L., Roth, S. (eds.) ECCV 2018. LNCS, vol. 11130, pp. 172–179. Springer, Cham (2019). https://doi.org/10.1007/978-3-030-11012-3_13
17. Maninis, K., Caelles, S., Pont-Tuset, J., Van Gool, L.: Deep extreme cut: from extreme points to object segmentation. In: 2018 IEEE/CVF Conference on Computer Vision and Pattern Recognition, pp. 616–625, June 2018
18. Nakayama, Y., et al.: Automated hepatic volumetry for living related liver transplantation at multisection CT. Radiology **240**(3), 743–748 (2006)
19. Papadopoulos, D.P., Uijlings, J.R.R., Keller, F., Ferrari, V.: Extreme clicking for efficient object annotation. In: 2017 IEEE International Conference on Computer Vision (ICCV), pp. 4940–4949, October 2017
20. Rajchl, M., et al.: DeepCut: object segmentation from bounding box annotations using convolutional neural networks. IEEE Trans. Med. Imaging **36**(2), 674–683 (2017)

21. Raju, A., et al.: Co-heterogeneous and adaptive segmentation from multi-source and multi-phase CT imaging data: a study on pathological liver and lesion segmentation. In: ECCV 2020 (2020)
22. Roth, H., et al.: Weakly supervised segmentation from extreme points. In: Zhou, L., et al. (eds.) LABELS/HAL-MICCAI/CuRIOUS -2019. LNCS, vol. 11851, pp. 42–50. Springer, Cham (2019). https://doi.org/10.1007/978-3-030-33642-4_5
23. Simonyan, K., Zisserman, A.: Very deep convolutional networks for large-scale image recognition. In: International Conference on Learning Representations (2015)
24. Tajbakhsh, N., Jeyaseelan, L., Li, Q., Chiang, J., Wu, Z., Ding, X.: Embracing Imperfect Datasets: A Review of Deep Learning Solutions for Medical Image Segmentation (2019)
25. Taner, C.B., et al.: Donor safety and remnant liver volume in living donor liver transplantation. Liver Transplant. **14**(8), 1174–1179 (2008)
26. Tsai, Y.H., Hung, W.C., Schulter, S., Sohn, K., Yang, M.H., Chandraker, M.: Learning to adapt structured output space for semantic segmentation. In: Proceedings of the IEEE Conference on Computer Vision and Pattern Recognition, pp. 7472–7481 (2018)
27. Wang, G., et al.: Interactive medical image segmentation using deep learning with image-specific fine tuning. IEEE Trans. Med. Imaging **37**(7), 1562–1573 (2018)
28. Wang, G., et al.: DeepIGeoS: a deep interactive geodesic framework for medical image segmentation. IEEE Trans. Pattern Anal. Mach. Intell. **41**(7), 1559–1572 (2019)
29. Zhang, J., Xie, Y., Zhang, P., Chen, H., Xia, Y., Shen, C.: Light-weight hybrid convolutional network for liver tumor segmentation. In: Proceedings of the 28th International Joint Conference on Artificial Intelligence, Macao, China, pp. 10–16 (2019)
30. Zhou, X., Zhuo, J., Krähenbühl, P.: Bottom-up object detection by grouping extreme and center points. In: CVPR (2019)

SALAD: Self-supervised Aggregation Learning for Anomaly Detection on X-Rays

Behzad Bozorgtabar[1,2,3(✉)], Dwarikanath Mahapatra[4], Guillaume Vray[1], and Jean-Philippe Thiran[1,2,3]

[1] Signal Processing Laboratory 5 EPFL, Lausanne, Switzerland
{behzad.bozorgtabar,guillaume.vray,jean-philipe.thiran}@epfl.ch
[2] Department of Radiology, Lausanne University Hospital, Lausanne, Switzerland
[3] Center of Biomedical Imaging, Lausanne, Switzerland
[4] Inception Institute of Artificial Intelligence, Abu Dhabi, UAE

Abstract. Deep anomaly detection models using a supervised mode of learning usually work under a closed set assumption and suffer from overfitting to previously seen rare anomalies at training, which hinders their applicability in a real scenario. In addition, obtaining annotations for X-rays is very time consuming and requires extensive training of radiologists. Hence, training anomaly detection in a fully unsupervised or self-supervised fashion would be advantageous, allowing a significant reduction of time spent on the report by radiologists. In this paper, we present SALAD, an end-to-end deep self-supervised methodology for anomaly detection on X-Ray images. The proposed method is based on an optimization strategy in which a deep neural network is encouraged to represent prototypical local patterns of the normal data in the embedding space. During training, we record the prototypical patterns of normal training samples via a memory bank. Our anomaly score is then derived by measuring similarity to a weighted combination of normal prototypical patterns within a memory bank without using any anomalous patterns. We present extensive experiments on the challenging NIH Chest X-rays and MURA dataset, which indicate that our algorithm improves state-of-the-art methods by a wide margin.

Keywords: Anomaly detection · X-rays · Self-supervised learning · Deep similarity metric

1 Introduction

Currently, supervised based deep learning approaches are ubiquitous and achieve promising results for abnormality detection in X-ray images [21]. However, many

Electronic supplementary material The online version of this chapter (https://doi.org/10.1007/978-3-030-59710-8_46) contains supplementary material, which is available to authorized users.

A. L. Martel et al. (Eds.): MICCAI 2020, LNCS 12261, pp. 468–478, 2020.
https://doi.org/10.1007/978-3-030-59710-8_46

real-world datasets of radiographs often have long-tailed label distributions. On these datasets, deep neural networks have been found to perform poorly on rare classes of anomalies. This particularly has a pernicious effect on the deployed model if, at test time, we place more emphasis on minority classes of abnormal X-ray images. For example, in detecting rare lung opacities, e.g., such as pneumonia in chest X-rays (CXR), normal X-rays are much easier to acquire. Besides, examining radiographs and reporting work for the signs of abnormalities are very time consuming and require qualified radiologists.

Anomaly detection based methods [1,2,8,13] can be significantly useful in large-scale disease screening and spotting candidate regions for anomalies. Classical anomaly detection (AD) methods such as One-Class SVM (OC-SVM) [19], Local Outlier Factor [3], or Isolation Forest [12] often fail to be effective on high-dimensional data or be scaled to large datasets. To alleviate this concern, state-of-the-art methods like Deep SVDD [16] and Deep SAD [17] consider learning deep CNN features as an alternative to classical one-class anomaly detection. However, the former suffers from the well-known problem of mode collapse, while the latter tends to ignore the underlying structure of the images as the pre-trained weights from autoencoder are sensitive to biased low-level features. Reconstruction-based methods [8,23,25] use well-established convolutional autoencoders to compress and reconstruct single-class normal samples, but autoencoders can sometimes reconstruct abnormal samples well, yielding miss detection of anomalies at test time. New methods built upon generative adversarial networks (GANs) [5,18,20] have shown promising anomaly detection performance by using GANs' ability to learn a manifold of normal samples. However, generated samples by GANs do not always lie at the boundary of real data distribution, which is necessary to distinguish normal images from abnormal ones. Recently, self-supervised methods [4,7,10,24] have been proposed to use unlabeled data in a task-agnostic way for extracting generalizable features, where the dataset can be labeled by exploiting the relations between different input samples, rather than requiring external labels. For example, self-supervised deep methods [6,7] proposed to train a classifier for which a self-labeled multi-class dataset is created by applying a set of geometric transformations to the images. However, these methods are domain-specific and cannot generalize over other data types.

Contribution. In this paper, we propose SALAD, short for **S**elf-supervised **A**ggregation **L**earning for **A**nomaly **D**etection on X-rays, a new training scheme that derives an aggregation learning from measuring the similarity between the estimated features of normal samples, to improve clustering and form prototypical patterns. We present a principled formulation to bypass tedious annotations and remove potential bias introduced by training. We show our method's superiority to existing anomaly detection methods on X-ray datasets. We also highlight the limitations of the current state-of-the-art methods.

2 Method

The merit of our approach is self-supervised representation learning, where the feature representation of each X-ray image is pushed closer to its similar neighbors, forming well-clustered features (prototypical patterns) in the latent space. This intuition is illustrated in Fig. 1. To do so, we propose to minimize the *entropy* of each sample feature point's similarity distribution to other nearby samples. Learning feature similarity would require obtaining image embedding in the entire dataset. To avoid this, we use a memory bank [22] to record and use the features. In every iteration, the memory bank is updated with the mini-batch features. The clustering objective will help us identify abnormal samples if they have different characteristics compared to normal prototypical patterns.

The backbone of our anomaly detection model is based on deep auto-encoder, where the encoder $f_{\theta_{enc}} : \mathcal{X} \rightarrow \mathcal{Z}$ is a convolutional neural network that represents input images $\{x_i \in \mathcal{X}\}_{i=1}^{N}$ in an informative latent domain $\mathcal{Z}$. The encoded representation performs as a query to compare with the relevant items in the feature memory bank. The decoder $g_{\theta_{dec}} : \mathcal{Z} \rightarrow \mathcal{X}$ is an up-sampling convolutional neural network that reconstructs the samples given their latent representations.

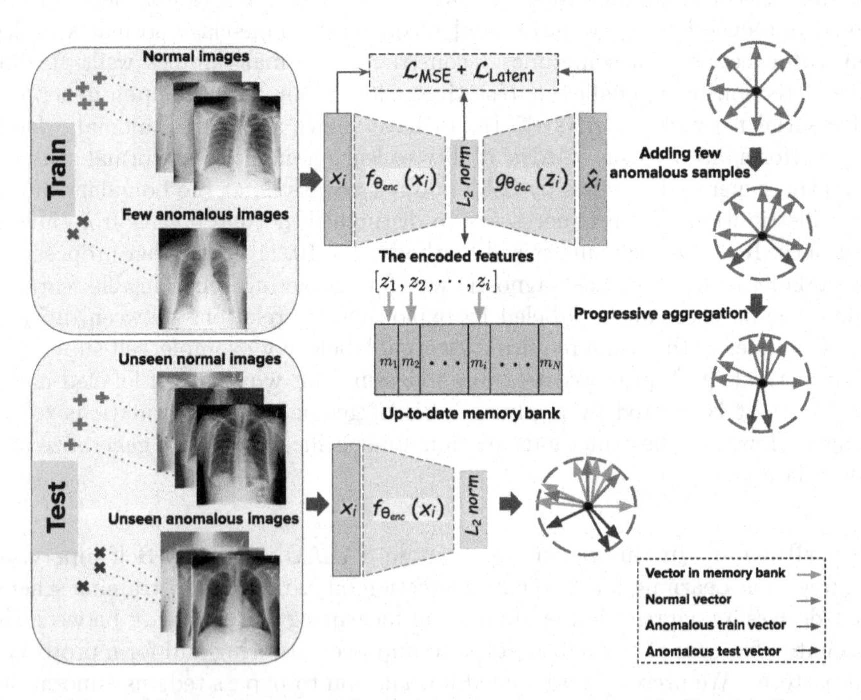

Fig. 1. The SALAD pipeline. The training process starts with forming the prototypical normal patterns (top). At test time, we measure the similarity between the test sample and normal patterns recorded in a memory bank (bottom).

Pre-training. For initialization, we establish an autoencoder pre-training routine using the image reconstruction loss (mean squared error), i.e. $\mathcal{L}_{\text{mse}} = \min_{\theta_{enc},\theta_{dec}} \|x - g_{\theta_{dec}} \circ f_{\theta_{enc}}(x)\|_2^2$. In addition, we impose a constraint on the lower-dimensional representation of the data in which features of the same X-ray image under random data augmentations are invariant, while the features of different images are scattered. To do so, we jointly optimize the training of network with reconstruction loss and a sample specific loss $\mathcal{L}_{\text{ss}}$ [22] to enforce a unique representation for each image:

$$\min_{\theta_{enc}} \mathcal{L}_{\text{ss}} = -\sum_{i \in \mathcal{B}_{\text{spl}}} \log\Big(\sum_{j \in AUG_i} p_{i,j}\Big) \quad \text{s.t.} \quad p_{i,j} = \frac{\exp(z_j^T z_i/\tau)}{\sum_{k=1}^{N} \exp(z_k^T z_i/\tau)} \tag{1}$$

where $\tau \in (0, 1]$ denotes a fixed temperature hyperparameter. $\mathcal{B}_{\text{spl}}$ denotes the set of samples in the mini-batch. z_i is the feature representation and AUG_i denotes the set of randomly augmented versions of the image x_i.

Training. The learned feature representation at the pre-training stage may not preserve the similarity of different images. Therefore, we add the aggregation loss $\mathcal{L}_{\text{agg}}$ (Eq. 2) to enforce consistency between samples lying in a neighborhood in latent space. We define aggregation loss as the (negative) log-likelihood that a specific sample will be identified as a member of the set of adjacent samples sharing the same prototypical pattern. This is achieved by the entropy measurement of the probability vector in Eq. 1. The more similar the samples are, the less relative entropy they have. We progressively increase entropy to consider larger prototypical neighborhood for the samples and form clusters (see Fig. 2). Finally, the proposed loss $\mathcal{L}_{\text{salad}}$ (Eq. 3) joins all training losses:

$$\min_{\theta_{enc}} \mathcal{L}_{\text{agg}} = -\sum_{i \in \mathcal{B}_{\text{ps}}} \log\Big(\sum_{j \in \mathcal{N}_k(z_i)} p_{i,j}\Big) \tag{2}$$

$$\min_{\theta_{enc},\theta_{dec}} \mathcal{L}_{\text{salad}} = \min_{\theta_{enc},\theta_{dec}} \mathcal{L}_{\text{mse}} + \lambda \min_{\theta_{enc}} \overbrace{(\mathcal{L}_{\text{ss}} + \mathcal{L}_{\text{agg}})}^{\mathcal{L}_{\text{latent}}} \tag{3}$$

where $\mathcal{N}_k(z_i)$ denotes the top-k neighbours determined by the lowest cosine distance with respect to the embedding vector z_i. λ is a hyperparameter to scale the losses ($\mathcal{L}_{\text{latent}}$) used in the latent space and $\mathcal{B}_{\text{ps}}$ denotes the set of prototypical samples in a mini-batch.

Memory Bank. Similarly to [22,24], we first initialize the memory bank with random unit vectors and then update its values m_i using a weighted moving average scheme $m_i \leftarrow (1-t)\, m_i + t z_i$ considering the up-to-date features z_i, where t is the fixed hyperparameter.

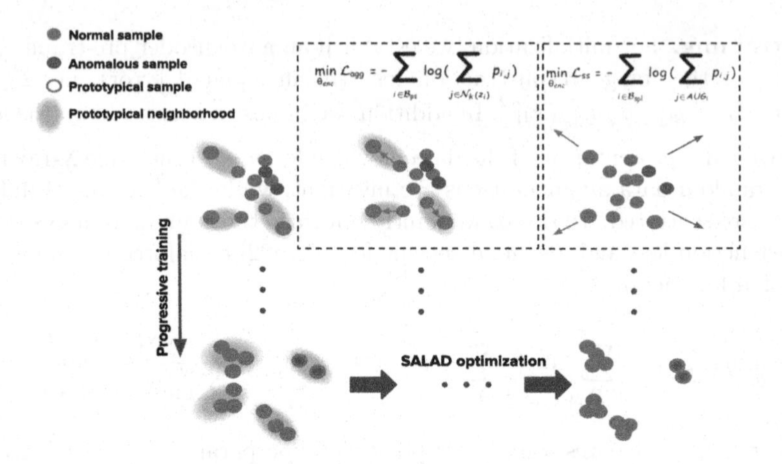

Fig. 2. An overview of a proposed progressive training strategy. We gradually increase sample neighborhoods to form prototypical patterns.

Inference. In the testing phase, an X-ray image is passed through the trained encoder and its representation is compared with the most relevant normal patterns in the memory bank for computing an anomaly score. Motivated by the weighted k-nearest neighbors (kNN), each vote w_i is obtained from the top K nearest feature vectors in the memory bank. An anomaly score $\mathcal{A}(\cdot)$ is calculated by:

$$\mathcal{A}(x_i) = \frac{1}{K}\sum_{k=1}^{K} w_{i,k} \quad \text{s.t.} \quad w_{i,k} = \frac{\arccos\left(d\left(z_i, m_k\right)\right)}{\sum_{j=1}^{N} \arccos\left(d\left(z_i, m_j\right)\right)} \tag{4}$$

where $d(\cdot,\cdot)$ denotes a cosine similarity, which computes similarity measurement between the test query feature $z_i = f_{\theta_{enc}}(x_i)$ and the elements stored in the memory bank $\{m_j\}_{j=1}^{N}$. $\mathcal{A}(x_i)$ is normalized to $[0, 1]$. Ideally, the anomaly scores of anomalous images should be significantly larger than the scores from normal images. We also discard anomalous trained patterns in a memory bank as they can lead to adverse effects if anomalous prototypical patterns are similar to learned abnormal patterns.

3 Experimental Results

Datasets and Repartition. We validated our proposed method for classification of normal versus abnormal X-ray scans using two challenging public X-ray datasets, i.e., the NIH clinical center chest X-ray dataset [21] and the MURA (musculoskeletal radiograph) dataset [14]. In NIH dataset [21], each radiographic image is assigned with diagnostic labels corresponding to 14 cardiothoracic or pulmonary diseases. We combine all CXRs with at least one of these 14 diseases into an aggregate abnormal class. For a fair comparison with [20], we followed the same train, validation, and test subsets as in [20] so there was no patient ID overlap among the subsets. The MURA dataset [14] contains upper limb X-rays images labeled whether they contain anomaly or not. This dataset is composed of seven classes of body parts: finger, hand, wrist, forearm, elbow, humerus, and shoulder. There are a total of 40'005 X-ray images from 11'967 unique patients. We present a preprocessing pipeline, including the X-ray image carrier detection and unsupervised body part segmentation, using hysteresis thresholding by producing a binary mask (see Fig. 3). The splitting of the MURA dataset has been done based on the patient ID and the body part. This implies that all images of a specific body part from a given patient will be present in the same set. The patient's body parts are grouped into normal, abnormal, and mixed (meaning there are both normal and abnormal X-rays for that body part of a patient). The train set is composed mainly of normal samples (50% of all the patient's body part and 95% of normal samples) with few abnormal samples (5%). The remaining normal, abnormal, and mixed samples are equally split between the validation and the test sets (see Fig. 4).

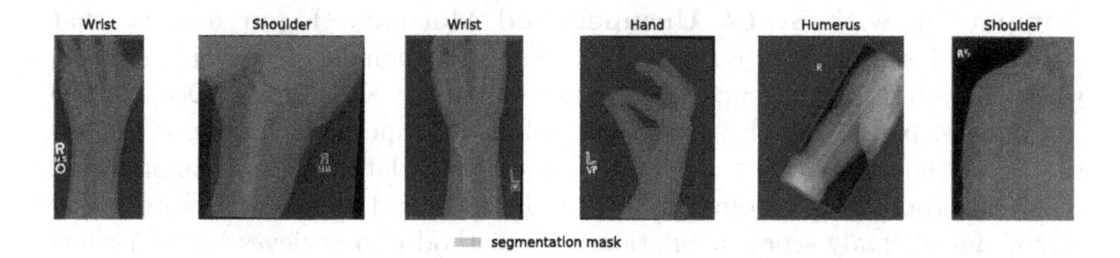

Fig. 3. Examples of segmentation results of the musculoskeletal X-rays.

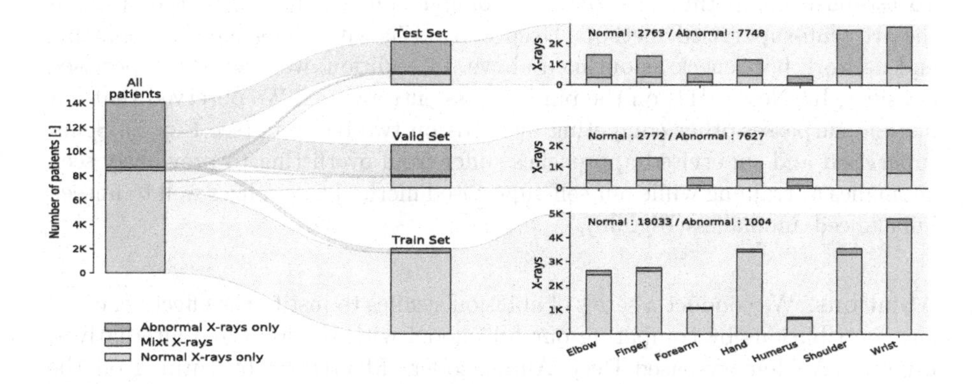

Fig. 4. A visual summary of the applied data split scheme on the MURA dataset.

Implementation Details and Evaluation Metrics. For the NIH dataset, we base our network architecture on the U-Net [15], consisting of a 6-layer convolutional encoder network and a 6-layer up-sampling convolutional decoder network without skip connections (both have batch normalization and leaky ReLU after each layer). The last encoder output features are projected to a 200-dimensional space, and L2 normalized. We use Adam optimizer, ($\beta_1 = 0.5, \beta_2 = 0.999$) and with a base learning rate of 0.0001. We pre-train the network for 50 epochs. Then, we train the network progressively with $\mathcal{L}_{\text{salad}}$ in 10 rounds with 50 epochs per round. The images were resized to 256×256 and we set $\tau = 0.1$, $\lambda = 0.25$, $t = 0.5$ and $K = 100$, respectively. The batch size was also set to 16. These optimum values are determined experimentally. We apply a slightly different experimental setup on the MURA dataset, and we replace the encoder and decoder with a ResNet-18 [11] and a mirrored ResNet-18, respectively. Besides, the musculoskeletal X-ray images are resized and padded so that their major axis is 512 pixels long while keeping the aspect ratio. We adopt area under the ROC curve (AUC) and Area Under Precision-Recall Curve (AUPRC) as our evaluation metrics.

Comparison with SOTA Unsupervised Methods. Figure 5 shows that our method significantly outperforms all recent anomaly detection methods, whether trained in an unsupervised mode, including OCGAN [20], Deep SVDD [16] and Deep Autoencoder (DAE) or with a self-supervised fashion (Geometric [7]). Although we use a few labeled anomalous data, the label information is not incorporated into our training, and we discard anomalous prototypical vectors for anomaly score calculation. Our method also achieves better performance compared to (GAN-GP) [9], where we replace GAN objective in [20] with gradient penalty (Fig. 6a).

Comparison to Methods that Use a Small Pool of Labeled Anomalies. To establish competing methods, we compare our method with the state-of-the-art semi-supervised method, Deep SAD [17], with the same data splitting and network bottleneck as outlined above. In addition, we train the supervised classifier, ResNet-18 [11] on the binary cross-entropy loss. We observed that our method surpasses other competing methods on two test sets (see Fig. 5). Semi-supervised and supervised approaches suffer from overfitting to previously seen anomalies at training while our self-supervised method generalizes well to unseen imbalanced anomalies (Fig. 6b).

Ablations. We conduct a series of ablation studies to justify the effectiveness of our contributions by comparing our full model with the following alternatives, using: 1. A Memory-based Deep Autoencoder (MemDAE) by turning off the proposed loss terms ($\mathcal{L}_{\text{agg}}$, $\mathcal{L}_{\text{ss}}$) and without using anomalous samples (Table 1 and Fig. 6a); 2. Our method without the loss term $\mathcal{L}_{\text{agg}}$; 3. Our method without the loss term $\mathcal{L}_{\text{mse}}$. We observed that our method trained with each of the

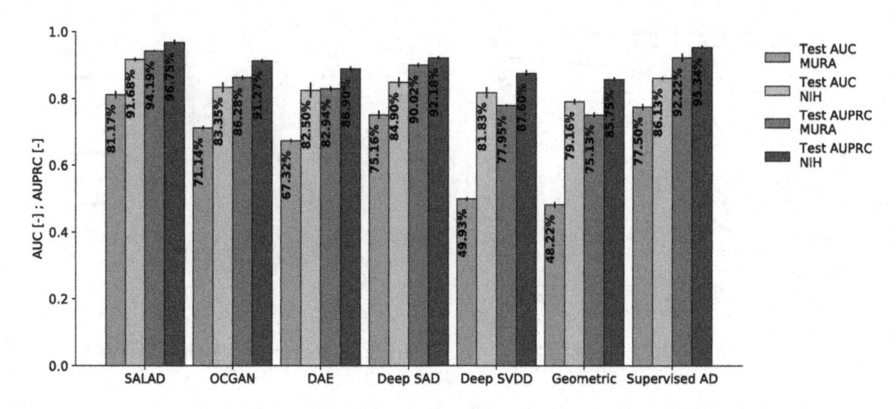

Fig. 5. AUC and AUPRC. Comparison of different anomaly detection methods. The bar height represents the mean (AUC or AUPRC) over four replicates of training, while the error bar is a 95% confidence interval computed as 1.96 std.

proposed loss terms, resulting in a notable performance gain over all the metrics, e.g., a gain of about 5.8% in AUC, compared to MemDAE on the MURA dataset (Table 1). Nevertheless, our baseline method (MemDAE) without anomalous samples, which is trained solely with MSE loss, outperforms all previous anomaly detection methods. We also conduct sensitivity analysis to investigate the effect of included labeled anomalies during training on final performance. To do so, we increased the ratio of known anomalous samples up to 15% and observed that our method is not very sensitive to an anomalous ratio (Table 1). This can be explained by the fact that SALAD does not require label information. Instead, it uses anomalous samples to have a better separation of prototypical patterns.

Table 1. Evaluation of the proposed approach and baselines. Ablation study for anomalous ratio (AR) and the loss terms used in training.

Method	AR (%)	NIH		MURA	
		AUC	AUPRC	AUC	AUPRC
MemDAE	0	0.8778 ± 0.0072	0.9198 ± 0.0024	0.7660 ± 0.0102	0.8823 ± 0.0035
SALAD w/o $\mathcal{L}_{mse}$	5	0.8693 ± 0.0095	0.9121 ± 0.0032	0.7765 ± 0.0094	0.8892 ± 0.0085
SALAD w/o $\mathcal{L}_{agg}$	5	0.8824 ± 0.0105	0.9332 ± 0.0062	0.7923 ± 0.0061	0.9128 ± 0.0026
SALAD (ours)	5	0.9091 ± 0.0069	0.9552 ± 0.0049	0.8117 ± 0.0028†	0.9419 ± 0.0075†
SALAD w/o $\mathcal{L}_{mse}$	10	0.8739 ± 0.0071	0.9287 ± 0.0024	0.7882 ± 0.0104	0.8926 ± 0.0031
SALAD w/o $\mathcal{L}_{agg}$	10	0.8913 ± 0.0056	0.9426 ± 0.0035	0.8012 ± 0.0042	0.9196 ± 0.0023
SALAD (ours)	10	0.9167 ± 0.0031†	0.9674 ± 0.0051†	0.8195 ± 0.0087	0.9502 ± 0.0085
SALAD w/o $\mathcal{L}_{mse}$	15	0.8774 ± 0.0095	0.9290 ± 0.0023	0.7904 ± 0.0025	0.8985 ± 0.0082
SALAD w/o $\mathcal{L}_{agg}$	15	0.8975 ± 0.0105	0.9494 ± 0.0055	0.8084 ± 0.0014	0.9214 ± 0.0103
SALAD (ours)	15	0.9189 ± 0.0076	0.9705 ± 0.0011	0.8245 ± 0.0017	0.9582 ± 0.0108

† Final models used for comparison against SOTA methods.

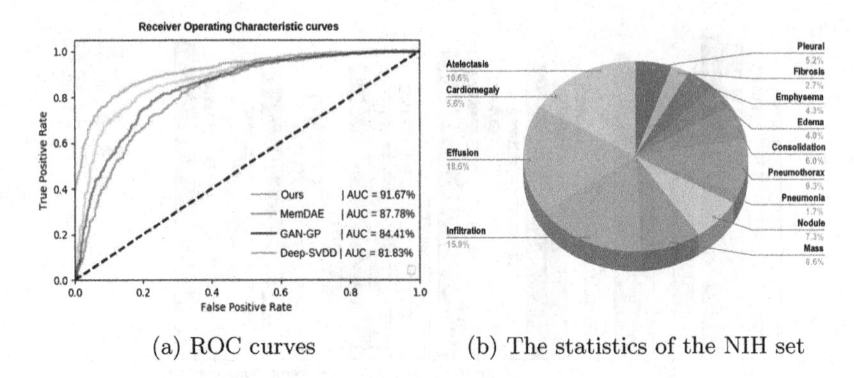

(a) ROC curves (b) The statistics of the NIH set

Fig. 6. (a) ROC curves comparison performances on the NIH dataset. (b) The NIH dataset statistics used in our experiments.

4 Conclusion and Future Work

In this work, we proposed SALAD, a self-supervised aggregation based learning framework for X-ray anomaly detection. This paper's novelty lies in jointly deep representation learning of X-ray images as well as aggregation criterion to distill out anomalous data. We use progressive training to enforce consistency between similar data samples in the embedding space to facilitate the formation of prototypical normal patterns. Hence, abnormal X-ray samples appear less likely to be represented by the normal learned patterns. SALAD achieves state-of-the-art anomaly detection results across all tested learning regimes, including unsupervised methods and those trained with small amounts of labeled data. As future work, we envision the broad application of our approach across different image modalities and beyond anomaly detection where the annotation is very costly, e.g., unsupervised domain adaptation.

References

1. Alaverdyan, Z., Jung, J., Bouet, R., Lartizien, C.: Regularized siamese neural network for unsupervised outlier detection on brain multiparametric magnetic resonance imaging: application to epilepsy lesion screening. Med. Image Anal. **60**, 101618 (2020)
2. Baur, C., Wiestler, B., Albarqouni, S., Navab, N.: Fusing unsupervised and supervised deep learning for white matter lesion segmentation. In: International Conference on Medical Imaging with Deep Learning, pp. 63–72 (2019)
3. Breunig, M.M., Kriegel, H.P., Ng, R.T., Sander, J.: Lof: identifying density-based local outliers. In: Proceedings of the 2000 ACM SIGMOD International Conference on Management of Data, pp. 93–104 (2000)
4. Chen, T., Kornblith, S., Norouzi, M., Hinton, G.: A simple framework for contrastive learning of visual representations. arXiv preprint arXiv:2002.05709 (2020)
5. Davletshina, D., et al.: Unsupervised anomaly detection for x-ray images. arXiv preprint arXiv:2001.10883 (2020)

6. Gidaris, S., Singh, P., Komodakis, N.: Unsupervised representation learning by predicting image rotations. arXiv preprint arXiv:1803.07728 (2018)
7. Golan, I., El-Yaniv, R.: Deep anomaly detection using geometric transformations. In: Advances in Neural Information Processing Systems, pp. 9758–9769 (2018)
8. Gong, D., Liu, L., Le, V., Saha, B., Mansour, M.R., Venkatesh, S., Hengel, A.V.D.: Memorizing normality to detect anomaly: Memory-augmented deep autoencoder for unsupervised anomaly detection. In: Proceedings of the IEEE International Conference on Computer Vision, pp. 1705–1714 (2019)
9. Gulrajani, I., Ahmed, F., Arjovsky, M., Dumoulin, V., Courville, A.C.: Improved training of wasserstein gans. In: Advances in Neural Information Processing Systems, pp. 5767–5777 (2017)
10. He, K., Fan, H., Wu, Y., Xie, S., Girshick, R.: Momentum contrast for unsupervised visual representation learning. In: Proceedings of the IEEE/CVF Conference on Computer Vision and Pattern Recognition, pp. 9729–9738 (2020)
11. He, K., Zhang, X., Ren, S., Sun, J.: Deep residual learning for image recognition. In: Proceedings of the IEEE Conference on Computer Vision and Pattern Recognition, pp. 770–778 (2016)
12. Liu, F.T., Ting, K.M., Zhou, Z.H.: Isolation forest. In: 2008 Eighth IEEE International Conference on Data Mining, pp. 413–422. IEEE (2008)
13. Norlander, E., Sopasakis, A.: Latent space conditioning for improved classification and anomaly detection. arXiv preprint arXiv:1911.10599 (2019)
14. Rajpurkar, P., et al.: Mura: Large dataset for abnormality detection in musculoskeletal radiographs. arXiv preprint arXiv:1712.06957 (2017)
15. Ronneberger, O., Fischer, P., Brox, T.: U-net: convolutional networks for biomedical image segmentation. In: Navab, N., Hornegger, J., Wells, W., Frangi, A. (eds.) MICCAI 2015. Lecture Notes in Computer Science, vol. 9351, pp. 234–241. Springer, Cham (2015). https://doi.org/10.1007/978-3-319-24574-4_28
16. Ruff, L., et al.: Deep one-class classification. In: International Conference on Machine Learning, pp. 4393–4402 (2018)
17. Ruff, L., et al.: Deep semi-supervised anomaly detection. In: International Conference on Learning Representations (2020). https://openreview.net/forum?id=HkgH0TEYwH
18. Schlegl, T., Seeböck, P., Waldstein, S.M., Schmidt-Erfurth, U., Langs, G.: Unsupervised anomaly detection with generative adversarial networks to guide marker discovery. In: Niethammer, M., et al. (eds.) IPMI 2017. LNCS, vol. 10265, pp. 146–157. Springer, Cham (2017). https://doi.org/10.1007/978-3-319-59050-9_12
19. Schölkopf, B., Platt, J.C., Shawe-Taylor, J., Smola, A.J., Williamson, R.C.: Estimating the support of a high-dimensional distribution. Neural Comput. **13**(7), 1443–1471 (2001)
20. Tang, Y.X., Tang, Y.B., Han, M., Xiao, J., Summers, R.M.: Abnormal chest x-ray identification with generative adversarial one-class classifier. In: 2019 IEEE 16th International Symposium on Biomedical Imaging (ISBI 2019), pp. 1358–1361. IEEE (2019)
21. Wang, X., Peng, Y., Lu, L., Lu, Z., Bagheri, M., Summers, R.M.: Chestx-ray8: hospital-scale chest x-ray database and benchmarks on weakly-supervised classification and localization of common thorax diseases. In: Proceedings of the IEEE Conference on Computer Vision and Pattern Recognition, pp. 2097–2106 (2017)
22. Wu, Z., Xiong, Y., Yu, S.X., Lin, D.: Unsupervised feature learning via non-parametric instance discrimination. In: Proceedings of the IEEE Conference on Computer Vision and Pattern Recognition, pp. 3733–3742 (2018)

23. Zhai, S., Cheng, Y., Lu, W., Zhang, Z.: Deep structured energy based models for anomaly detection. In: International Conference on Machine Learning, pp. 1100–1109 (2016)
24. Zhuang, C., Zhai, A.L., Yamins, D.: Local aggregation for unsupervised learning of visual embeddings. In: Proceedings of the IEEE International Conference on Computer Vision, pp. 6002–6012 (2019)
25. Zong, B., et al.: Deepautoencoding gaussian mixture model for unsupervised anomaly detection (2018)

Scribble-Based Domain Adaptation via Co-segmentation

Reuben Dorent[1(✉)], Samuel Joutard[1], Jonathan Shapey[1,2,3], Sotirios Bisdas[3], Neil Kitchen[3], Robert Bradford[3], Shakeel Saeed[4], Marc Modat[1], Sébastien Ourselin[1], and Tom Vercauteren[1]

[1] School of Biomedical Engineering and Imaging Sciences, King's College London, London, UK
reuben.dorent@kcl.ac.uk
[2] Wellcome/EPSRC Centre for Interventional and Surgical Sciences, University College London, London, UK
[3] National Hospital for Neurology and Neurosurgery, London, UK
[4] UCL Ear Institute, University College London, London, UK

Abstract. Although deep convolutional networks have reached state-of-the-art performance in many medical image segmentation tasks, they have typically demonstrated poor generalisation capability. To be able to generalise from one domain (e.g. one imaging modality) to another, domain adaptation has to be performed. While supervised methods may lead to good performance, they require to fully annotate additional data which may not be an option in practice. In contrast, unsupervised methods don't need additional annotations but are usually unstable and hard to train. In this work, we propose a novel weakly-supervised method. Instead of requiring detailed but time-consuming annotations, scribbles on the target domain are used to perform domain adaptation. This paper introduces a new formulation of domain adaptation based on structured learning and co-segmentation. Our method is easy to train, thanks to the introduction of a regularised loss. The framework is validated on Vestibular Schwannoma segmentation (T1 to T2 scans). Our proposed method outperforms unsupervised approaches and achieves comparable performance to a fully-supervised approach.

Keywords: Domain adaptation · Weak supervision · Regularised loss

1 Introduction

Deep Neural Networks (DNNs) are achieving state-of-the-art performance for many medical image segmentation tasks. However, deep networks still lack in their generalisation capability when confronted with new datasets.

Domain adaptation (DA) approaches have been developed to ensure that networks trained on a source domain can be successfully used on a target domain. A first supervised solution consists of annotating (a sufficient number of) new images from the target domain and fine-tune a network initially

A. L. Martel et al. (Eds.): MICCAI 2020, LNCS 12261, pp. 479–489, 2020.
https://doi.org/10.1007/978-3-030-59710-8_47

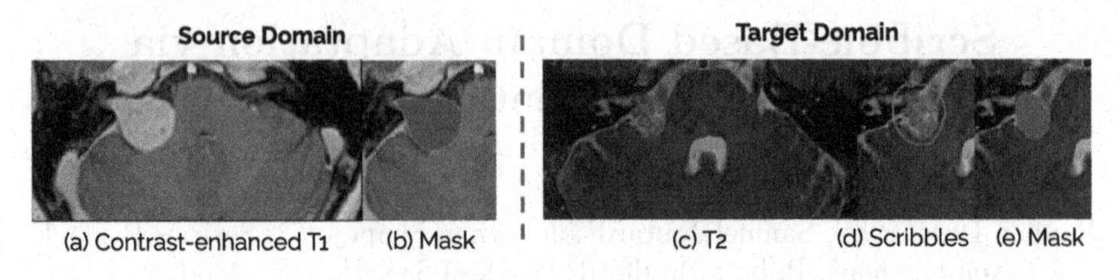

Fig. 1. Examples of Vestibular Schwanomma tumours. T1-c (a) and T2 (c) scans from the source and target domain are shown with their segmentation (b + e). Source masks (b) and target scribbles (e) are used at training stage.

trained on the source domain [8,15]. Although easy to implement, stable during training, and achieving satisfying performance, such supervised techniques may not be a practical option given the time and expertise required to manually segment additional medical images. For this reason, unsupervised methods, based on self-supervised learning and adversarial learning, have been proposed. Self-supervised techniques [18,19] typically use pretext tasks to learn task-agnostic feature representations that are adapted to the target domain. Example of self-supervision includes optimising for prediction consistency across different strongly augmented versions of the same target data [18]. Although these techniques have shown promising results, they have only been tested on relatively similar source and target domain. Alternatively or concurrently, adversarial learning has been used to ensure that the learned feature representations are similar across the two domains via a discriminator network [6,7,13,17,18,23]. Relying on a complex and unstable adversarial optimisation procedure based on many heuristics, successfully training these models is particularly challenging and time-consuming. Moreover, they are often limited to 2D models due to high memory requirements.

In parallel, efforts have been done to help clinicians segment medical images more efficiently. In particular, semi-automated segmentation has been shown to be a reliable option [25]. Based on efficient user interactions such as scribbles, DNN predictions are fine-tuned at an image-specific level [24]. Fine-tuning is performed for each new test image and is typically *forgotten* on purpose afterwards as the image-specific nature implies a poorer generalisation capability. Looking beyond single images to streamline the annotation task, weakly-supervised methods based on scribbles have been introduced. Networks trained using scribbles are used to perform inference on unseen and unlabelled data. A standard modelling approach is to rely on Conditional Random Fields (CRFs) with DNN outputs being used as unary weights [4,10,16]. The optimisation procedure typically alternates between proposing a one-hot crisp segmentation proposal extending from the scribbles (e.g. via a mean-field or graph-cut approach) and training the DNN with supervision provided by these proposals. A recent work [22] has shown that this two-step alternate optimisation can be efficiently approximated by a direct loss minimisation problem exploiting a regularised loss formulation.

In this work, we propose a novel weakly-supervised domain adaptation method. The contributions of this work are four-fold. First, we introduce a new formulation of domain adaptation as a co-segmentation problem. Secondly, we present a new structured learning approach to propagate information across domains. Thirdly, we show that alternating the proposal generation and network training can be approximated by directly minimising a regularised loss. Fourthly, we evaluate our framework on a challenging problem, unpaired cross-modality domain adaptation. Our method demonstrates the benefits of leveraging source data and obtained similar results compared to a fully-supervised approach.

2 Conditional Random Fields for Structured Predictions

In this section, we briefly present Conditional Random Fields (CRFs) for semantic segmentation and define some notations used in the remainder of this work. CRFs have been commonly used in image segmentation for their ability to produce structured predictions.

Let $\mathbf{Y}$ be the random variable representing the overall label assignment, i.e. the segmentation, of a random image $\mathbf{Y} \in \mathbb{R}^N$, where N is the number of voxels. For each voxel k, Y_k is an element of the set of C possible classes $\mathcal{L} = \{l_1, \ldots, l_C\}$. The general idea of CRFs is to model the pair $(\mathbf{X}, \mathbf{Y})$ as a graph where the nodes (i.e. voxels) are associated with voxel-wise labels and the edges are associated with the similarity between the voxels. Specifically, a CRF is characterised by a Gibbs distribution $P(\mathbf{Y} = \hat{\mathbf{y}}|\mathbf{X}) \propto \exp(-E_I(\hat{\mathbf{y}}|\mathbf{X}))$. Here $E_I(\hat{\mathbf{y}}|\mathbf{X})$ is the Gibbs energy and represents the cost associated to the label configuration $\hat{\mathbf{y}} \in \mathcal{L}^N$. Given an observed image $\mathbf{x}$, the optimal segmentation $\hat{\mathbf{y}}^*$ minimises the assignment cost. In the fully-connected pairwise CRF model, the problem is defined as:

$$\hat{\mathbf{y}}^* = \arg\min_{\hat{\mathbf{y}} \in \mathcal{L}^N} \left\{ E_I(\hat{\mathbf{y}}|\mathbf{x}) = \sum_{k \in [\![1;N]\!]} \psi_u(\hat{y_k}|\mathbf{x}) + \sum_{k,l \in [\![1;N]\!]} \psi_p(\hat{y_k}, \hat{y_l}|\mathbf{x}) \right\} \quad (1)$$

where $\psi_u(\hat{y_k}|\mathbf{x})$ and $\psi_p(\hat{y_k}, \hat{y_l}|\mathbf{x})$ are the unary and pairwise potentials.

Partial annotations, such as scribbles, provide known class values for a subset of voxels. Since each voxel depends on its neighbours, the sparse annotation information can be propagated within the image. Let $\mathbf{y} = (y_i)_{i \in \Omega_a} \in \mathcal{L}^{|\Omega_a|}$ be a partial annotation, where Ω_a is the set of annotated voxels (i.e. the scribbles). The optimisation problem then becomes a constrained one:

$$\begin{aligned} \hat{\mathbf{y}}^* = \arg\min_{\hat{\mathbf{y}} \in \mathcal{L}^N} & \left\{ \sum_{k \in [\![1;N]\!]} \psi_u(\hat{y_k}|\mathbf{x}) + \sum_{k,l \in [\![1;N]\!]} \psi_p(\hat{y_k}, \hat{y_l}|\mathbf{x}) \right\} \\ \text{subject to} : \quad & \forall k \in \Omega_a,\ \hat{y}_k = y_k \end{aligned} \quad (2)$$

The problem is typically solved by graph-cut [3] for submodular problems or mean-field inference [2,14] for the general case.

Recent works have proposed to combine the strengths of deep learning and structured learning via CRFs [4,10,16,24]. The common idea consists in defining

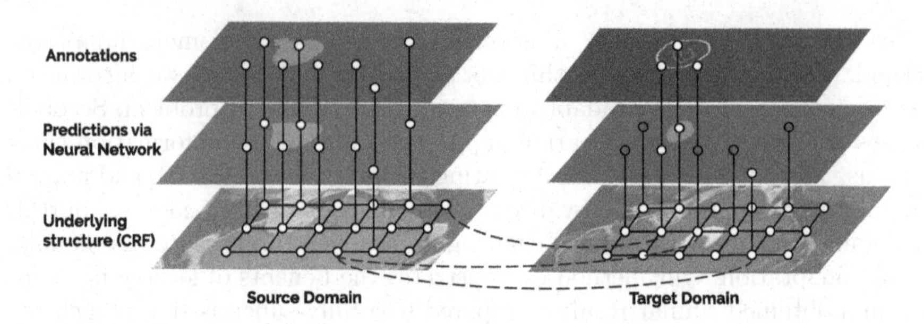

Fig. 2. Overview of the proposed graphical model. Each image voxel is a node. Annotations impose constraints on the predictions using a neural network f_θ. All the nodes are connected together within each image (image-specific CRF) and between images (domain adaptation; blue dashed lines). Only a few of these connections are shown. Although only two images are represented, all the images are connected to each other within and between domains. (Color figure online)

the unary potentials ψ_u with a neural network f_θ parameterised by the weights θ. Existing methods typically alternate between proposal generation, i.e. solving (2), and network parameters learning with supervision from these proposals. Recently this alternate optimisation has been replaced by a direct optimisation via a regularised loss [22], thereby reducing the optimisation complexity, the computational cost during training and at inference, and avoiding learning from synthetically generated labels. The formulation in [22] reads:

$$\arg\min_{\theta} \{ \sum_{k \in \Omega_a} H(y_k, p_k) + R(\mathbf{p}_\theta) \} \tag{3}$$

where $\mathbf{p}_\theta = f_\theta(\mathbf{x})$ is the softmax output of the network, H is the cross-entropy and R is a regularisation term that encourages spatial and image intensity consistency. In the next section, we provide more details about this regularisation term and we show how we adapt it for domain adaptation purposes.

3 The Scribble Domain Adaptation Model

Group Co-segmentation Formulation. In weakly-supervised domain adaptation, we are given a source domain $\mathcal{D}_s = \{(\mathbf{x}_i^s, \mathbf{y}_i^s)\}$ of n_s fully-labelled samples and a target domain $\mathcal{D}_t = \{(\mathbf{x}_i^t, \mathbf{y}_i^t)\}$ of n_t partially annotated samples. We denote $\Omega_{i,a}$ the set of annotated voxels for each image $\mathbf{x}_i^s$ ($\Omega_{i,a}$ representing the entire image) or $\mathbf{x}_i^t$ ($\Omega_{i,a}$ representing scribbles). Figure 1 shows an example of scribbles used in this work.

The overall objective is to predict accurate segmentation for the target data using a neural network f_θ. Since the annotations are partial on the target domain, we use a graphical model (a CRF) to include prior contextual information and perform structured predictions. This allows for propagating the partial annotation information within a particular image. Given data from the target domain, we aim to minimise each image-specific Gibbs energy E_I, as defined in (1). However, this basic formulation does not include the other important source of information we have access to: The fully-annotated data from the source domain.

Inspired by co-segmentation [9,11], we extend the image-specific CRF to a dataset-level CRF. Specifically, in addition to including typical image-specific pairwise potentials, each node (i.e. voxel) of each image is connected to every nodes of every other images, as shown in Fig. 2. The annotation information is then propagated between images, including from the fully-annotated images to the partially-annotated images. Consequently, knowledge is transferred from the source domain to the target domain, i.e. domain adaptation is performed. For this reason, we denote E_{DA} the proposed energy term associated to pairs of images. Our proposed optimisation problem can be defined as follows:

$$\begin{aligned} &\underset{\theta,(\hat{\mathbf{y}}_\mathbf{i})_i \in \mathcal{L}^{S\times N}}{\arg\min} \Big\{ \sum_{i\in[\![1;S]\!]} \big(E_I(\hat{\mathbf{y}}_\mathbf{i}|\mathbf{x}_\mathbf{i}) + \sum_{j\in[\![1;S]\!],j\neq i} E_{DA}(\hat{\mathbf{y}}_\mathbf{i},\hat{\mathbf{y}}_\mathbf{j}|\mathbf{x}_\mathbf{i},\mathbf{x}_\mathbf{j})\big)\Big\} \\ &\text{subject to}\ : \quad \forall i \in [\![1;S]\!],\ \forall k \in \Omega_a,\ \hat{y}_{i,k} = y_{i,k} \end{aligned} \tag{4}$$

where $S = n_s + n_t$ is the total number of scans, and indices i, j correspond to image index while k, l are voxels index. Note that the constraints impose that the proposals for the source training data are exactly their fully-annotated masks.

Image-Specific Gibbs Energy. We used a standard formulation of the unary and pairwise potentials for the image-specific energy E_I defined in (2). Similarly to [4,16,24], the DNN f_θ is used to compute the unary potentials:

$$\forall k \in [\![1;N]\!],\quad \psi_u(\hat{y}_{i,k}|\mathbf{x}_\mathbf{i}) = -\log P_\theta(\hat{y}_{i,k}|\mathbf{x}_\mathbf{i}) = H(\hat{y}_{i,k}, p_{i,k;\theta}) \tag{5}$$

where $\mathbf{p}_{\mathbf{i};\theta} = f_\theta(\mathbf{x}_\mathbf{i})$ is the probability given by softmax output of the DNN and H the cross-entropy. For the image-specific pairwise potentials, we follow the typical choice of using the Potts model and a bilateral filtering term:

$$\forall k,l \in [\![1;N]\!],\ \psi_p\left(\hat{y}_{i,k},\hat{y}_{i,l}|\mathbf{x}\right) = [\hat{y}_{i,k} \neq \hat{y}_{i,l}] \exp\Big(-\frac{(x_{i,k}-x_{i,l})^2}{2\sigma_\alpha^2} - \frac{d(x_{i,k},x_{i,l})^2}{2\sigma_\beta^2}\Big)$$

where $d(.,.)$ denotes the Euclidean distance between the pixel locations. By denoting W_i the affinity matrix of an image $\mathbf{x}_\mathbf{i}$ [22], E_i can be relaxed as:

$$\sum_{k,l\in[\![1;N]\!]} \psi_p\left(\hat{y}_{i,k},\hat{y}_{i,l}|\mathbf{x}_\mathbf{i}\right) = \hat{\mathbf{y}}_\mathbf{i}{}^T W_i(1-\hat{\mathbf{y}}_\mathbf{i}) \triangleq R_I(\hat{\mathbf{y}}_\mathbf{i}) \tag{6}$$

Domain Adaptation Gibbs Energy. The DA Gibbs energy E_{DA} only involves pairwise potential associated with voxels from different images. By minimising E_{DA}, we expect to assign similar labels to voxels with similar visual features representation across the datasets. Since the domains are shifted, the image intensity distributions are different between the two domains. Consequently the image intensity cannot be used as features. Instead, we propose to use the features extracted from the DNN. Specifically, we used the output of the penultimate convolution, i.e. just before the softmax regression. The domain adaptation cost is then defined as:

$$E_{DA}(\hat{\mathbf{y}}_\mathbf{i}, \hat{\mathbf{y}}_\mathbf{j}|\mathbf{x}_\mathbf{i}, \mathbf{x}_\mathbf{j}) = \sum_{k,l \in [\![1;N]\!]} [\hat{y}_{i,k} \neq \hat{y}_{j,l}] \exp\Big(- \frac{(g_{i,k} - g_{j,l})^2}{2\sigma_\gamma^2} \Big) \quad (7)$$

where $\mathbf{g}_i = g_\theta(\mathbf{x}_i)$. Note that the spatial position is not taken into account here. Again, the domain adaptation Gibbs energy can be relaxed as:

$$E_{DA}(\hat{\mathbf{y}}_\mathbf{i}, \hat{\mathbf{y}}_\mathbf{j}|\mathbf{x}_\mathbf{i}, \mathbf{x}_\mathbf{j}) = [\hat{\mathbf{y}}_\mathbf{i}, \hat{\mathbf{y}}_\mathbf{j}]^T W_{i-j}(1 - [\hat{\mathbf{y}}_\mathbf{i}, \hat{\mathbf{y}}_\mathbf{j}]) \triangleq R_{DA}(\hat{\mathbf{y}}_\mathbf{i}, \hat{\mathbf{y}}_\mathbf{j}; \theta) \quad (8)$$

4 Optimization via a Regularised Loss

In this section, we propose a method to optimise the parameters θ of the DNN. Similarly to [22], we show that the optimization problem can be approximated with a regularised loss. Let $\mathbf{p}^\mathbf{t}_{\mathbf{i};\theta} = f_\theta(\mathbf{x}^\mathbf{t}_\mathbf{i})$ and $\mathbf{p}^\mathbf{s}_{\mathbf{i};\theta} = f_\theta(\mathbf{x}^\mathbf{s}_\mathbf{i})$ be the outputs of the network for a target domain image $\mathbf{x}^\mathbf{t}_\mathbf{i}$ and a source domain image $\mathbf{x}^\mathbf{s}_\mathbf{i}$. We denote $H_{\Omega_a}(\mathbf{u}, \mathbf{v}) = \sum_{k \in \Omega_a} H(u_k, v_k)$. By combining (4), (6) and (8), the optimisation problem is defined as:

$$\begin{aligned} \arg\min_{\theta, (\hat{\mathbf{y}}_\mathbf{i})_{i \in [\![1;n]\!]}} \Big\{ &\sum_i \big(H_\Omega\left(\hat{\mathbf{y}}_\mathbf{i}^\mathbf{s}, \mathbf{p}^\mathbf{s}_{\mathbf{i};\theta}\right) + R_I(\hat{\mathbf{y}}_\mathbf{i}^\mathbf{s})\big) \\ &+ \sum_i \big(H_\Omega\left(\hat{\mathbf{y}}_\mathbf{i}^\mathbf{t}, \mathbf{p}^\mathbf{t}_{\mathbf{i};\theta}\right) + R_I(\hat{\mathbf{y}}_\mathbf{i}^\mathbf{t})\big) + \sum_{i,j} R_{DA}(\hat{\mathbf{y}}_\mathbf{i}, \hat{\mathbf{y}}_\mathbf{j}; \theta) \Big\} \quad (9) \\ \text{subject to}: \quad &\forall i \in [\![1;S]\!],\ \forall k \in \Omega_a,\ \hat{y}_{i,k} = y_{i,k} \end{aligned}$$

By adding a null negative entropy term $-\sum_i H\left(\hat{\mathbf{y}}_\mathbf{i}^\mathbf{t}, \hat{\mathbf{y}}_\mathbf{i}^\mathbf{t}\right) = 0$ and integrating the constraints directly in the formulation, (9) can be rewritten as:

$$\arg\min_\theta \Big\{ \sum_{i,j} u(\mathbf{p}_{\mathbf{i},\theta}) + \mathbb{1}_{\mathbf{x}_\mathbf{i} \in \mathcal{D}^t} \min_{\hat{\mathbf{y}}_\mathbf{i}^\mathbf{t}} \left\{ KL(\hat{\mathbf{y}}^\mathbf{t}_\mathbf{i}, \mathbf{p}_{\mathbf{i};\theta}) + R(\hat{\mathbf{y}}_\mathbf{i}^\mathbf{t}, \hat{\mathbf{y}}_\mathbf{j}^\mathbf{t}) \right\} \Big\} \quad (10)$$

where $u(\mathbf{p}_{\mathbf{i},\theta}) = H_{\Omega_{a,i}}(\mathbf{y}_\mathbf{i}, \mathbf{p}_{\mathbf{i},\theta})$, $R(\hat{\mathbf{y}}_\mathbf{i}^\mathbf{t}, \hat{\mathbf{y}}^\mathbf{t}_\mathbf{j}) = R_I(\hat{\mathbf{y}}_\mathbf{i}^\mathbf{t}) + R_{DA}(\hat{\mathbf{y}}_\mathbf{i}^\mathbf{t}, \hat{\mathbf{y}}_\mathbf{j}^\mathbf{t})$ and KL denotes the Kullback–Leibler divergence. Given that full annotations are provided for the source domain, the inner minimisation with respect to the proposals, $\hat{\mathbf{y}}^\mathbf{s}_\mathbf{i}$, only relates to the target data $(\mathbf{x}^t_i, \mathbf{y}^t_i)$.

The inner problem corresponds to minimising a divergence between the network output $\mathbf{p}^\mathbf{t}_{\mathbf{i};\theta}$ and the proposal $\hat{\mathbf{y}}_\mathbf{i}^\mathbf{t}$ together with a regularisation term. This discrepancy is null if the proposal is equal to the network output. We thus expect the optimal proposal to be close to the network output, i.e. $\hat{\mathbf{y}}_\mathbf{i}^{\mathbf{t}*} \approx \mathbf{p}^\mathbf{t}_{\mathbf{i};\theta}$. We assume that equality stands, which allows us to reformulate the problem as:

$$\arg\min_\theta \Big\{ \mathcal{L}(\theta) = \sum_{\substack{(\mathbf{x}_\mathbf{i}, \mathbf{y}_\mathbf{i}) \\ (\mathbf{x}_\mathbf{j}, \mathbf{y}_\mathbf{j})}} H_{\Omega_{a,i}}(\mathbf{y}^\mathbf{t}_\mathbf{i}, \mathbf{p}_{\mathbf{i};\theta}) + \mathbb{1}_{\mathbf{x}_\mathbf{i} \in \mathcal{D}^t}(R_I(\mathbf{p}_{\mathbf{i};\theta}) + R_{DA}(\mathbf{p}_{\mathbf{i};\theta}, \mathbf{p}_{\mathbf{j};\theta})) \Big\} \quad (11)$$

The parameters θ are directly optimised via a stochastic gradient descent. The high-dimensional filtering method proposed by [1] is used to reduce the quadratic complexity of the computation of R_I and R_{DA} to a linear one.

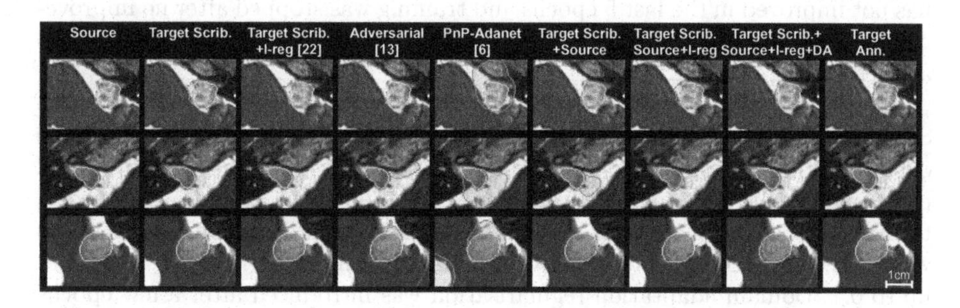

Fig. 3. Qualitative evaluation of different networks for Vestibular Schwannoma segmentation on T2 scans. Segmentation results (green curves) and the ground truth (yellow curves) are shown. (Color figure online)

5 Experiments

Experimental Setup. We conducted experiments on Vestibular Schwanomma (VS), a benign brain tumour arising from the vestibulocochlear nerve, the main nerve connecting the brain and inner ear. Current MR protocols include contrast-enhanced T1-weighted (T1-c) and high-resolution T2 scans. T1-c are generally currently used for segmenting the tumour as offering a better contrast, see Fig. 1. However, T2 imaging could be a reliable, safer and lower-cost alternative to T1-c [5,21].

In this work, we propose to segment VS images using T2 images only as input. The source domain data corresponds to 150 T1-c scans with the full set of annotations and the target domain training data corresponds to 30 T2 scans with scribble annotations only. Specifically, on average 1% of the T2 scans and 7% of the tumour has been annotated. 4 T2 scans and 20 T1 scans were used as validation set. For testing, 50 T2 scans (target domain) have been manually fully segmented. Images had an in-plane resolution of $0.4 \times 0.4 \times \text{mm}^2$, a slice thickness of $1.0 - 1.5$mm and were cropped manually with a bounding box of size $100 \times 50 \times 50\text{mm}^3$, covering the full axial brain length as shown in Fig. 1a, 1c.

Implementation Details. Our models were implemented in PyTorch using TorchIO [20][1]. A 2.5D U-Net was used for all our experiments, similar to [26]. A PyTorch GPU implementation of the high-dimensional filtering [12] was

[1] Code available at: https://github.com/KCL-BMEIS/ScribbleDA.

employed. We used the Adam optimizer with weight decay 10^{-5}. At each iteration, two images from the source domain and two images from the target domain are randomly selected and fed to the network. The initial learning rate 5.10^{-4} was reduced by a factor of 5 whenever the moving average of the validation loss has not improved in the last 5 epochs and training was stopped after no improvements in the last 10 epochs. Rotation, scaling and white noise augmentation were applied during training.

Concerning the regularisation terms, a typical value of α was chosen (15). Similar results were obtained for different values of β ($\{0.5, 0.05, 0.005\}$), the ones reported correspond to $\beta = 0.05$. In order to reduce the computational complexity, only two channels were used to compute the pairwise distance in the DA regularisation term. Specifically, at each training iteration, two channels were chosen randomly among the total number of channels (here 48). γ was set up to 0.1. Domain adaptation regularisation was introduced after a few epochs (70). Finally, we observed large improvements by using the Dice loss instead of the cross-entropy, thus we reported scores with the Dice loss.

Model Comparison. Firstly, we studied each component of our method independently. As a baseline, we trained a model on the target scribbles only (Target Scrib) and with the regularised loss [22] (Target Scrib+I-reg). Then the source data was used during training without (Target Scrib+I-reg+Source) and with (Target Scrib+I-reg+Source+DA) the cross-modality DA regularisation. Secondly, we compared our method with a fully-supervised approach trained using the same 30 T2 scans with the full set of annotations (Target Ann.). Thirdly, we compared our approach with two well-established unsupervised DA methods based on adversarial learning [13] and designed specifically for cross-modality DA [6]. Quantitative results are reported in Table 1 using the Dice and average symmetric surface distance (ASSD) between segmentation results and the ground truth. Examples of outputs are presented in Fig. 3.

Table 1. Quantitative evaluation of different networks for Vestibular Schwannoma segmentation. I-reg: The image-specific regularised loss proposed by [22]. DA: Our proposed Domain Adaptation regularised loss.

Method/Training	Test on source		Tets on target	
	Dice (%)	ASSD (mm^3)	Dice (%)	ASSD (mm^3)
Source	**93.7 (3.3)**	0.3 (0.5)	28.2 (33.0)	13.8 (10.8)
Target Scrib	46.9 (33.8)	10.9 (10.9)	77.6 (17.9)	2.1 (3.0)
Target Scrib+I-reg [22]	58.4 (29.5)	9.0 (8.5)	76.9 (18.8)	1.4 (2.0)
Source+Adversarial [13]	87.8 (8.9)	1.6 (1.5)	9.3 (18.9)	24.9 (16.2)
PnP-Adanet [6]	79.3 (15.2)	3.4 (3.1)	27.3 (21.1)	13.3 (4.2)
Target Scrib+Source	92.4 (4.6)	0.4 (0.4)	75.1 (18.6)	2.7 (4.5)
Target Scrib+Source+I-reg	93.2 (3.7)	0.3 (0.5)	76.7 (17.9)	1.6 (2.5)
Target Scrib+Source+I-reg+DA	93.3 (4.0)	**0.2 (0.2)**	**83.4 (10.4)**	**0.8 (0.8)**
Target Ann.	63.7 (33.9)	8.3 (11.2)	81.6 (13.1)	1.8 (2.8)

Results. Firstly, the ablation study shows that adding the cross DA regularisation brings significant improvements on the target domain compared to the other models trained using the target scribbles. Interestingly, including the source data during training only leads to improvements when the DA regularisation is employed. This shows the effectiveness of our DA method. Moreover, note that our technique didn't degrade the performance on the source domain. Secondly, our method obtained comparable performance to a fully-supervised model. Thus, scribble-based DA is a reliable option for performing supervised DA. Thirdly, both unsupervised methods failed on our problem. Since the inner brain and tumour appearance vary greatly between the contrast-enhanced T1 and T2 scans, our problem is too challenging for unsupervised approaches, highlighting the need for supervision.

6 Conclusion

This paper proposes a novel approach for weakly-supervised domain adaptation. Based on co-segmentation and structured learning, we introduced a new formulation for domain adaptation with scribbles. Our approach is mathematically grounded, easy to implement, new and relies on reasonable assumptions. We validated our method on challenging experiments: unpaired cross-modality brain lesion segmentation. Our model achieved comparable performance to a model trained on a fully-annotated data and outperformed existing unsupervised techniques. This work shows that scribbles is a reliable option for performing domain adaptation.

Acknowledgement. This work was supported by the Engineering and Physical Sciences Research Council (EPSRC) [NS/A000049/1] and Wellcome Trust [203148/Z/16/Z]. TV is supported by a Medtronic/Royal Academy of Engineering Research Chair [RCSRF1819\7\34].

References

1. Adams, A., Baek, J., Davis, M.A.: Fast high-dimensional filtering using the permutohedral lattice. Comput. Graph. Forum **29**, 753–762 (2010)
2. Baque, P., Bagautdinov, T., Fleuret, F., Fua, P.: Principled parallel mean-field inference for discrete random fields. In: The IEEE Conference on Computer Vision and Pattern Recognition (CVPR), June 2016
3. Boykov, Y., Funka-Lea, G.: Graph cuts and efficient ND image segmentation. Int. J. Comput. Vis. **70**(2), 109–131 (2006). https://doi.org/10.1007/s11263-006-7934-5
4. Can, Y.B., et al.: Learning to segment medical images with scribble-supervision alone. In: Stoyanov, D., et al. (eds.) DLMIA/ML-CDS -2018. LNCS, vol. 11045, pp. 236–244. Springer, Cham (2018). https://doi.org/10.1007/978-3-030-00889-5_27
5. Coelho, D.H., Tang, Y., Suddarth, B., Mamdani, M.: MRI surveillance of vestibular schwannomas without contrast enhancement: clinical and economic evaluation. Laryngoscope **128**(1), 202–209 (2018). https://doi.org/10.1002/lary.26589
6. Dou, Q., et al.: Pnp-adanet: plug-and-play adversarial domain adaptation network with a benchmark at cross-modality cardiac segmentation. ArXiv (2018)

7. Ganin, Y., et al.: Domain-adversarial training of neural networks. In: Csurka, G. (ed.) Domain Adaptation in Computer Vision Applications. ACVPR, pp. 189–209. Springer, Cham (2017). https://doi.org/10.1007/978-3-319-58347-1_10
8. Ghafoorian, M., et al.: Transfer learning for domain adaptation in MRI: application in brain lesion segmentation. In: Descoteaux, M., Maier-Hein, L., Franz, A., Jannin, P., Collins, D.L., Duchesne, S. (eds.) MICCAI 2017. LNCS, vol. 10435, pp. 516–524. Springer, Cham (2017). https://doi.org/10.1007/978-3-319-66179-7_59
9. Hochbaum, D.S., Singh, V.: An efficient algorithm for co-segmentation. In: 2009 IEEE 12th International Conference on Computer Vision, pp. 269–276, September 2009
10. Ji, Z., Shen, Y., Ma, C., Gao, M.: Scribble-based hierarchical weakly supervised learning for brain tumor segmentation. In: Shen, D., et al. (eds.) MICCAI 2019. LNCS, vol. 11766, pp. 175–183. Springer, Cham (2019). https://doi.org/10.1007/978-3-030-32248-9_20
11. Joulin, A., Bach, F., Ponce, J.: Discriminative clustering for image co-segmentation. In: 2010 IEEE Computer Society Conference on Computer Vision and Pattern Recognition, pp. 1943–1950, June 2010
12. Joutard, S., Dorent, R., Isaac, A., Ourselin, S., Vercauteren, T., Modat, M.: Permutohedral attention module for efficient non-local neural networks. In: Shen, D., et al. (eds.) MICCAI 2019. LNCS, vol. 11769, pp. 393–401. Springer, Cham (2019). https://doi.org/10.1007/978-3-030-32226-7_44
13. Kamnitsas, K., et al.: Unsupervised domain adaptation in brain lesion segmentation with adversarial networks. In: Niethammer, M., et al. (eds.) IPMI 2017. LNCS, pp. 597–609. Springer, Cham (2017). https://doi.org/10.1007/978-3-319-59050-9_47
14. Krähenbühl, P., Koltun, V.: Efficient inference in fully connected crfs with gaussian edge potentials. In: Advances in Neural Information Processing Systems, vol. 24, pp. 109–117. Curran Associates, Inc. (2011)
15. Kushibar, K., et al.: Supervised domain adaptation for automatic sub-cortical brain structure segmentation with minimal user interaction. Sci. Rep. **9**(1), 6742 (2019)
16. Lin, D., Dai, J., Jia, J., He, K., Sun, J.: Scribblesup: scribble-supervised convolutional networks for semantic segmentation. In: The IEEE Conference on Computer Vision and Pattern Recognition (CVPR), June 2016
17. Mahmood, F., Chen, R., Durr, N.J.: Unsupervised reverse domain adaptation for synthetic medical images via adversarial training. IEEE Trans. Med. Imaging **37**(12), 2572–2581 (2018)
18. Orbes-Arteaga, M., et al.: Multi-domain adaptation in brain MRI through paired consistency and adversarial learning. In: Wang, Q., et al. (eds.) DART/MIL3ID -2019. LNCS, vol. 11795, pp. 54–62. Springer, Cham (2019). https://doi.org/10.1007/978-3-030-33391-1_7
19. Perone, C.S., Ballester, P., Barros, R.C., Cohen-Adad, J.: Unsupervised domain adaptation for medical imaging segmentation with self-ensembling. NeuroImage **194**, 1–11 (2019)
20. Pérez-García, F., Sparks, R., Ourselin, S.: Torchio: a python library for efficient loading, preprocessing, augmentation and patch-based sampling of medical images in deep learning (2020)
21. Shapey, J., et al.: An artificial intelligence framework for automatic segmentation and volumetry of vestibular schwannomas from contrast-enhanced t1-weighted and high-resolution t2-weighted MRI. J. Neurosurg. JNS **1**, 1–9 (2019)

22. Tang, M., Perazzi, F., Djelouah, A., Ben Ayed, I., Schroers, C., Boykov, Y.: On regularized losses for weakly-supervised CNN segmentation. In: The European Conference on Computer Vision (ECCV), September 2018
23. Tzeng, E., Hoffman, J., Darrell, T., Saenko, K.: Adversarial discriminative domain adaptation. In: Computer Vision and Pattern Recognition (CVPR) (2017)
24. Wang, G., et al.: Interactive medical image segmentation using deep learning with image-specific fine tuning. IEEE Trans. Med. Imaging **37**(7), 1562–1573 (2018)
25. Wang, G., et al.: Deepigeos: a deep interactive geodesic framework for medical image segmentation. IEEE Trans. Pattern Anal. Mach. Intell. **41**(07), 1559–1572 (2019)
26. Wang, G., et al.: Automatic segmentation of vestibular schwannoma from t2-weighted MRI by deep spatial attention with hardness-weighted loss. In: Shen, D., et al. (eds.) MICCAI 2019. LNCS, vol. 11765, pp. 264–272. Springer, Cham (2019). https://doi.org/10.1007/978-3-030-32245-8_30

Source-Relaxed Domain Adaptation for Image Segmentation

Mathilde Bateson(✉), Hoel Kervadec, Jose Dolz, Hervé Lombaert, and Ismail Ben Ayed

ETS, Montréal, Canada
mathilde.bateson.1@ens.etsmtl.ca

Abstract. Domain adaptation (DA) has drawn high interests for its capacity to adapt a model trained on labeled source data to perform well on unlabeled or weakly labeled target data from a different domain. Most common DA techniques require the concurrent access to the input images of both the source and target domains. However, in practice, it is common that the source images are not available in the adaptation phase. This is a very frequent DA scenario in medical imaging, for instance, when the source and target images come from different clinical sites. We propose a novel formulation for adapting segmentation networks, which relaxes such a constraint. Our formulation is based on minimizing a label-free entropy loss defined over target-domain data, which we further guide with a domain-invariant prior on the segmentation regions. Many priors can be used, derived from anatomical information. Here, a class-ratio prior is learned via an auxiliary network and integrated in the form of a Kullback–Leibler (KL) divergence in our overall loss function. We show the effectiveness of our prior-aware entropy minimization in adapting spine segmentation across different MRI modalities. Our method yields comparable results to several state-of-the-art adaptation techniques, even though is has access to less information, the source images being absent in the adaptation phase. Our straight-forward adaptation strategy only uses one network, contrary to popular adversarial techniques, which cannot perform without the presence of the source images. Our framework can be readily used with various priors and segmentation problems.

Keywords: Image segmentation · Domain adaptation · Entropy minimization

1 Introduction

Semantic segmentation, or the pixel-wise annotation of an image, is a key first step in many clinical applications. Since the introduction of deep learning methods, automated methods for segmentation have outstandingly improved in many natural and medical imaging problems [14]. Nonetheless, the pixel-level ground-truth labelling necessary to train these networks is time-consuming, and deep-learning methods tend to under-perform when trained on a dataset with an

A. L. Martel et al. (Eds.): MICCAI 2020, LNCS 12261, pp. 490–499, 2020.
https://doi.org/10.1007/978-3-030-59710-8_48

underlying distribution different from the target images. To circumvent those impediments, methods learning robust networks with less supervision have been popularized in computer vision.

Domain Adaptation (DA) adresses the transferability of a model trained on an annotated source domain to another target domain with no or minimal annotations. The presence of a domain shift, such as those produced by different protocols, vendors, machines in medical imagining, often leads to a big performance drop (see Fig. 1). Adversarial strategies are currently the prevailing techniques to adapt networks to a different target domain, both in natural [4,9,10,23] and medical [2,6,7,17] images. These methods can either be generative, by transforming images from one domain to the other [25], or can minimize the discrepancy in the feature or output spaces learnt by the model [3,17,18].

One major limitation of adversarial techniques is that, by design, they require the concurrent access to both the source and target data during the adaptation phase. In medical imaging, this may not be always feasible when the source and target data come from different clinical sites, due to, for instance, privacy concerns or the loss or corruption of source data. Amongst alternative approaches to adversarial techniques, self training [26] and the closely-related entropy minimization [15,19,20] were investigated in computer vision. As confirmed by the low entropy prediction maps in Fig. 1, a model trained on an imaging modality tends to produce very confident predictions on within-sample examples, whereas uncertainty remains high on unseen modalities. As a result, enforcing high confidence in the target domain as well can close the performance gap. This is the underlying motivation for entropy minimization, which was first introduced in semi-supervised [5] and unsupervised [13] learning. To prevent the well-known collapse of entropy minimization to a trivial solution with a single class, the recent domain-adaptation methods in [19,20] further incorporate a criterion encouraging diversity in the prediction distributions. However, similarly to adversarial approaches, the entropy-based methods in [19,20] require access to the source data (both the images and labels) during the adaptation phase via a standard supervised cross-entropy loss. The latter discourages the trivial solution of minimizing the entropy alone on the unlabeled target images.

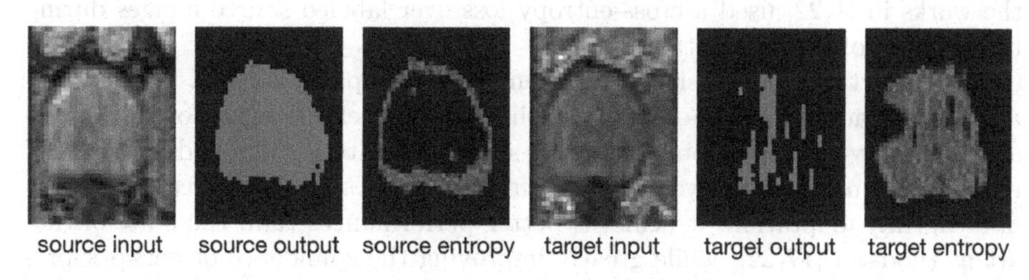

Fig. 1. Visualization of 2 aligned slice pairs in source (Water) and target modality (In-Phase): the domain shift in the target produces a drop in confidence and accuracy.

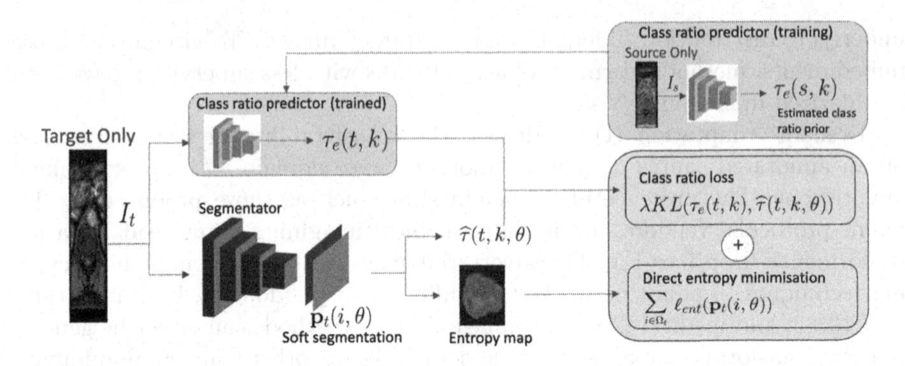

Fig. 2. Overview of our framework for Source-Relaxed Domain Adaptation: we leverage entropy minimization and a class-ratio prior to relax the need for a concurrent access to the source and target data.

We propose a domain-adaptation formulation tailored to a setting where the source data is unavailable (neither images, nor labeled masks) during the training of the adaptation phase. Instead, our method only requires the parameters of a model previously trained on the source data for initialisation. Our formulation is based on minimizing an label-free entropy loss defined over target-domain data, which we further guide with a domain-invariant prior on the segmentation regions. Many priors can be used, derived from anatomical information. Here, a class-ratio prior is learned via an auxiliary network and integrated in the form of a Kullback–Leibler (KL) divergence in our overall loss function. Unlike the recent entropy-based methods in [19,20], our overall loss function relaxes the need to access to the source images and labels during adaptation, as we do not use a source-based cross-entropy loss. Our class-ratio prior is related to several recent works in the context of semi- and weakly-supervised learning [11,24], which showed the potential of domain-knowledge priors for guiding deep networks when labeled data is scarce. Also, the recent works in [1,22] integrated priors on class-ratio/size in domain adaptation but, unlike our work, in the easier setting where one has access to source data (both the images and labels). In fact, the works in [1,22] used a cross-entropy loss over labeled source images during the training of the adaptation phase.

We report comprehensive experiments and comparisons with state-of-the-art domain-adaptation methods, which show the effectiveness of our prior-aware entropy minimization in adapting spine segmentation across different MRI modalities. Surprisingly, even though our method does not have access to source data during adaptation, it achieves better performances than the state-of-the-art methods in [17,22], while greatly improving the confidence of network predictions. Our framework can be readily used for adapting any segmentation problems. Our code is publicly and anonymously available[1]. To the best of our

[1] https://github.com/mathilde-b/SRDA.

knowledge, we are the first to investigate domain adaptation for segmentation without direct access to the source data during the adaptation phase (Fig. 2).

2 Method

We consider a set of S source images $I_s : \Omega_s \subset \mathbb{R}^d \to \mathbb{R}$, $d \in \{2, 3\}$, $s = 1, \ldots, S$. The ground-truth K-class segmentation of I_s can be written, for each pixel (or voxel) $i \in \Omega_s$, as a simplex vector $\mathbf{y}_s(i) = (y_s^1(i), \ldots, y_s^K(i)) \in \{0, 1\}^K$. For domain adaptation (DA) problems, the network is usually first trained on the source domain only, by minimizing a standard supervised loss with respect to network parameters θ:

$$\mathcal{L}_s(\theta, \Omega_s) = \frac{1}{|\Omega_s|} \sum_{s=1}^{S} \ell(\mathbf{y}_s(i), \mathbf{p}_s(i, \theta)) \tag{1}$$

where $\mathbf{p}_s(i, \theta) = (p_s^1(i, \theta), \ldots, p_s^K(i, \theta)) \in [0, 1]^K$ is the softmax output of the network at pixel/voxel i in image I_s, and ℓ is the standard cross-entropy loss: $\ell(\mathbf{y}_s(i), \mathbf{p}_s(i, \theta)) = -\sum_k y_s^k(i) \log p_s^k(i, \theta)$.

Given T images of the target domain, $I_t : \Omega_t \subset \mathbb{R}^{2,3} \to \mathbb{R}$, $t = 1, \ldots, T$, the first loss term in our adaptation phase encourages high confidence in the softmax predictions of the target, which we denote $\mathbf{p}_t(i, \theta) = (p_t^1(i, \theta), \ldots, p_t^K(i, \theta)) \in [0, 1]^K$. This is done by minimizing the entropy of each of these predictions:

$$\ell_{ent}(\mathbf{p}_t(i, \theta)) = -\sum_k p_t^k(i, \theta) \log p_t^k(i, \theta) \tag{2}$$

However, it is well-known from the semi-supervised and unsupervised learning literature [5,8,13] that minimizing this entropy loss alone may result into degenerate trivial solutions, biasing the prediction towards a single dominant class. To avoid such trivial solutions, the recent domain-adaptation works in [19,20] integrated a standard supervised cross-entropy loss over the source data, i.e., Eq. (1), during the training of the adaptation phase. However, this requires access to the source data (both the images and labels) during the adaptation phase. To relax this requirement, we embed domain-invariant prior knowledge to guide unsupervised entropy training during the adaptation phase, which takes the form of a class-ratio (i.e. region proportion) prior. We express the (unknown) true class-ratio prior for class k and image I_t as: $\tau_{GT}(t, k) = \frac{1}{|\Omega_t|} \sum_{i \in \Omega_t} y_t^k(i)$. As the ground-truth labels are unavailable in the target domain, this prior cannot be computed directly. Instead, we train an auxiliary network on source data to produce an estimation of class-ratio in the target[2], which we denote $\tau_e(t, k)$. Furthermore, the class-ratio can be approximated from the the segmentation network's output for target image I_t as follows: $\hat{\tau}(t, k, \theta) = \frac{1}{|\Omega_t|} \sum_{i \in \Omega_t} p_t^k(i, \theta)$

[2] Note that many other estimators could be used, e.g., using region statistics from the source domain or anatomical prior knowledge.

To match these two probabilities representing class-ratios, we integrate a KL divergence with the entropy in Eq. (2), minimizing the following overall loss during the training of the adaptation phase:

$$\min_{\theta} \sum_{t} \sum_{i \in \Omega_t} \ell_{ent}(\mathbf{p}_t(i,\theta)) + \lambda \mathrm{KL}(\tau_e(t,k), \widehat{\tau}(t,k,\theta)) \tag{3}$$

Clearly, minimizing our overall loss in Eq. (3) during adaptation does not use the source images and labels, unlike the recent entropy-based domain adaptation methods in [19,20].

3 Experiments

3.1 Cross-modality Adaption with Entropy Minimization

Dataset. We evaluated the proposed method on the publicly available MICCAI 2018 IVDM3Seg Challenge[3] dataset of 16 manually annotated 3D multi-modal magnetic resonance scans of the lower spine, in a study investigating intervertebral discs (IVD) degeneration. We first set the water modality (Wat) as the source and the in-phase (IP) modality as the target domain (Wat → IP), then reverted the adaptation direction (IP → Wat). 13 scans were used for training, and the remaining 3 scans were kept for validation. The slices were rotated in the transverse plane, and no other pre-processing was performed. The setting is binary segmentation (K=2), and the training is done with 2D slices.

Benchmark Methods. We compare our loss to the recent one adopted in [22]:

$$\mathcal{L}_s(\theta, \Omega_s) + \lambda \sum_{t} \sum_{i \in \Omega_T} KL(\tau_e(t,k), \widehat{\tau}_t(t,k,\theta))$$

Note that, in [22], the images from the source and target domain must be present concurrently in this framework, which we denote *AdaSource*. We also compared to the state-of-the art adversarial method in [17], denoted *Adversarial*. A model trained on the source only with Eq. (1), *NoAdaptation*, was used as a lower bound. A model trained with the cross-entropy loss on the target domain, referred to as *Oracle*, served as an upper bound.

Learning and Estimating the Class-Ratio Prior. We learned an estimation of the ground-truth class-ratio prior via an auxiliary regression network R, which is trained on the images I_s from the source domain S, where the ground-truth class-ratio $\tau_{GT}(s,k)$ is known. R is trained with the squared $\mathcal{L}_2$ loss: $\min_{\tilde{\theta}} \sum_{s=1 \dots S} \left(R(I_s, \tilde{\theta}) - \tau_{GT}(s,k) \right)^2$. The estimated class-ratio prior $\tau_e(t,k)$ of an image I_t in the target domain is obtained by inference. We added weak supervision in the form of image-level tag information by setting $\tau_e(t,k) = (1,0)$ for the target images that do not contain the region of interest. Note that we used

[3] https://ivdm3seg.weebly.com/.

exactly the same class-ratio priors and weak supervision for the method in [22], for a fair comparison. Note, also, that we adopted and improved significantly the performance of the adversarial method in [17] by using the same weak supervision information based on image-level tags, for a fair comparison[4].

Training and Implementation Details. For all methods, the segmentation network employed was ENet [16], trained with the Adam optimizer [12], a batch size of 12 for 100 epochs, and an initial learning rate of 1×10^{-3}. For all adaptation models, a model trained on the source data with Eq(1) for 100 epochs was used as initialization. The λ parameter in Eq(3) was set empirically to 1×10^{-2}. For *AdaSource*, the batches used were non-aligned random slices in each domain. To learn the class-ratio prior, a ResNeXt101 [21] regression network is used, optimized with SGD, a learning rate of 5×10^{-6}, and a momentum of 0.9.

Evaluation. The Dice similarity coefficient (DSC) and the Hausdorff distance (HD) were used as evaluation metrics in our experiments.

3.2 Discussion

Table 1 reports quantitative metrics. As expected, the model *NoAdaptation*, which doesn't use any adaptation strategy but instead is trained using Eq(1) on the source modality, can't perform well on a different target modality. The mean DSC reached is of 46.7% on IP, and 63.7% on Wat, and a very high standard deviation is observed in both case, showing a high subject variability. This is confirmed in Fig. 4, where it can be seen that the output segmentation is poor on 2 out of 3 subjects. Moreover, as can be observed in Fig. 3, where the evolution of the training in terms of validation DSC is shown, the DSC is very unstable on the target domain throughout learning. We also observe that the adaptation task isn't symmetrically difficult, as the performance drop is much bigger in one direction (Wat $\rightarrow$ IP). The performance of *Oracle*, the upper baseline is also lower in IP. As the visualisation in Fig. 1 show, the higher contrast in Wat images makes the segmentation task easier.

All models using adaptation techniques yield substantial improvement over the lower baseline. First, *Adversarial* achieves a mean DSC of 65.3% on IP and 77.3% on Wat. Nonetheless, the models without adversarial strategies yield better results: *AdaSource* achieves a mean DSC of 67.0% on IP and 78.3% on Wat. Interestingly, our model *AdaEnt* shows comparable performance, with a mean DSC of 67.0% on IP and 77.8% on Wat. These results show that having access to more information (i.e., source data) doesn't necessarily help for the adaptation task. For both models *AdaSource* and *AdaEnt*, the DSC comes close to the *Oracle*'s, the upper baseline, reaching respectively 82% and 82% of its performance respectively on IP, and 87% and 89% of its performance respectively on Wat. This demonstrates the efficiency of the using a class-ratio prior

[4] For the model in [17], pairs of source and target images were not used if neither had the region of interest as this confuses adversarial training, reducing its performance.

matching with a KL divergence. Moreover, in Fig. 3, we can observe that both these adaptation methods yield rapidly high validation DSC measures (first 20 epochs). This suggests that integrating such a KL divergence helps the learning process in domain adaptation. Finally, the HD values confirm the trend across the different models. Improvement over the lower baseline model (2.45 pixels on IP, 1.44 on Wat), is substantial for both *AdaSource* (1.34 pixels on IP, 1.14 pixels on Wat), as well as for *AdaEnt* (1.33 pixels on IP, 1.17 on Wat).

Table 1. Quantitative comparisons of performance on the *target* domain for the different models (mean ± std) show the efficiency of our source-relaxed formulation.

Wat (Source) → IP (Target)			IP (Source) → Wat (Target)		
Method	DSC (%)	HD (pix)	Method	DSC (%)	HD (pix)
No Adaptation	46.7 ± 10.8	2.45 ± 0.16	No Adaptation	63.7± 9.1	1.44 ± 0.2
Adversarial[17]	65.3 ± 5.5	1.67 ± 1.64	Adversarial[17]	77.3 ± 7.6	1.15 ± 0.2
AdaSource [22]	67.0 ± 7.2	1.34 ± 0.15	AdaSource [22]	**78.3 ± 3.5**	**1.14 ± 0.1**
AdaEnt (Ours)	**67.0 ± 6.1**	**1.33 ± 0.17**	AdaEnt (Ours)	77.8 ± 2.2	1.17 ± 0.1
Oracle	82.3 ± 1.2	1.09 ± 0.16	Oracle	89.0 ± 2.7	0.90 ± 0.1

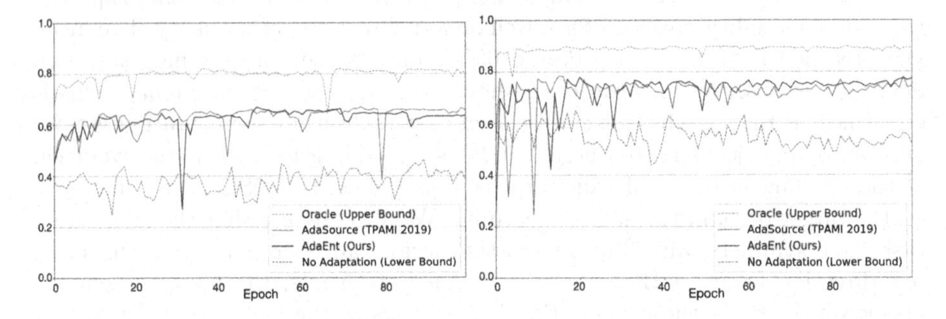

Fig. 3. Evolution of validation DSC over training for the different models. Comparison of the proposed model to the lower and upper bounds, and to the adaptation with access to source in [22] is shown. Water → In-Phase *(left)*, In-Phase → Water *(right)*

Qualitative segmentations and corresponding prediction entropy maps are depicted in Fig. 4, from the easiest to the hardest subject in the validation set. Without adaptation, a model trained on source data only can't recover the structure of the IVD on the target data, and is very uncertain, as revealed by the high activations in the prediction entropy maps. The output segmentation masks are noisy, with very irregular edges. As expected, the segmentation masks obtained using both adaptation formulations are much closer to the ground truth one, and have much more regular edges. Nonetheless, the entropy maps produced from *AdaSource* predictions still show high entropy activations inside and close to the IVD structures. On the contrary, those produced from *AdaEnt* look like edge detection results with high entropy activations only present along the IVD borders. Interestingly, it can be seen that even the *Oracle*'s segmentation predictions are more uncertain. This isn't surprising, as *AdaEnt* is the only model trained to directly minimize the entropy of the predictions. The visual results confirm *AdaEnt*'s remarkable ability to produce accurate predictions with high confidence.

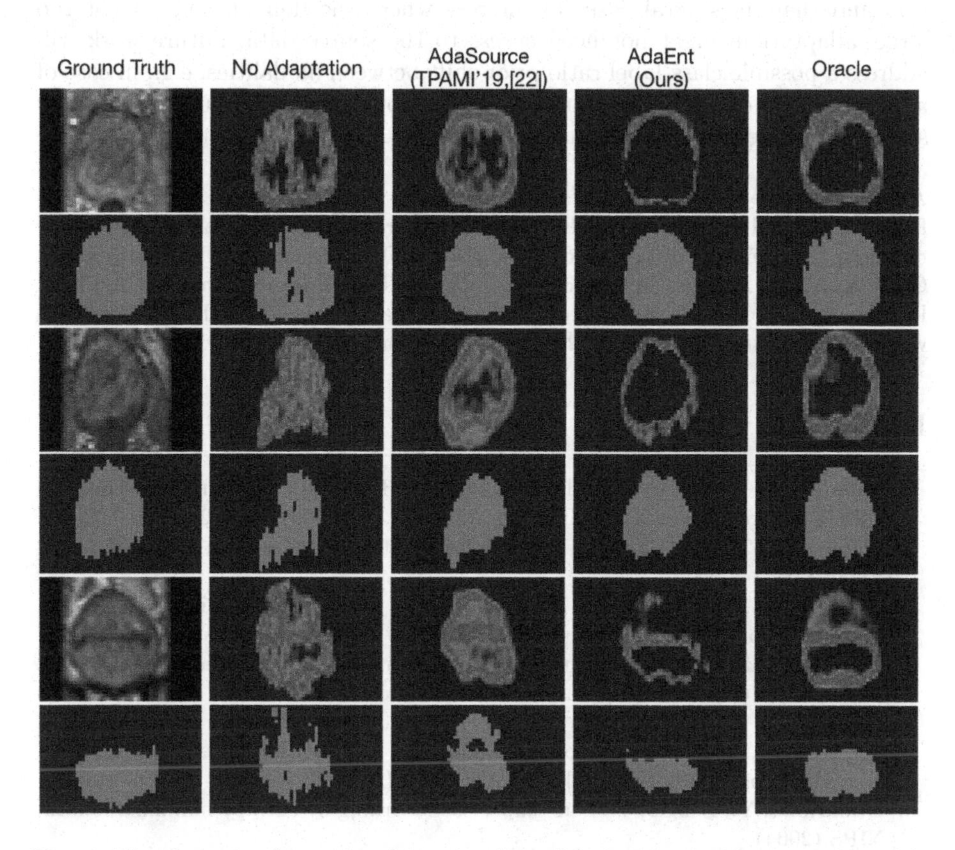

Fig. 4. Visual results for each subject in the validation set for the several models (Water → In-Phase). First column shows an input slice and the corresponding semantic segmentation ground-truth. The other columns show segmentation results (bottom) along with prediction entropy maps produced by the different models (top).

4 Conclusion

In this paper, we proposed a simple formulation for domain adaptation (DA), which removes the need for a concurrent access to the source and target data, in the context of semantic segmentation for multi-modal magnetic resonance images. Our approach substitutes the standard supervised loss in the source domain by a direct minimization of the entropy of predictions in the target domain. To prevent trivial solutions, we integrate the entropy loss with a class-ratio prior, which is built from an auxiliary network. Unlike the recent domain-adaptation techniques, our method tackles DA without resorting to source data during the adaptation phase. Interestingly, our formulation achieved better performances than related state-of-the-art methods with access to both source and target data. This shows the effectiveness of our prior-aware entropy minimization and that, in several cases of interest where the domain shift is not too large, adaptation might not need access to the source data. Future work will address a possible class-label ratio prior shift between modalities, e.g.. in case of a different field of view. Our proposed adaptation framework is usable with any segmentation network architecture.

Acknowledgment. This work is supported by the Natural Sciences and Engineering Research Council of Canada (NSERC), Discovery Grant program, by the The Fonds de recherche du Québec - Nature et technologies (FRQNT) grant, the Canada Research Chair on Shape Analysis in Medical Imaging, the ETS Research Chair on Artificial Intelligence in Medical Imaging, and NVIDIA with a donation of a GPU. The authors would like to thank the MICCAI 2018 IVDM3Seg organizers for providing the data.

References

1. Bateson, M., Kervadec, H., Dolz, J., Lombaert, H., Ayed, I.B.: Constrained domain adaptation for segmentation. In: Shen, D., et al. (eds.) MICCAI 2019. LNCS, vol. 11765, pp. 326–334. Springer, Cham (2019). https://doi.org/10.1007/978-3-030-32245-8_37
2. Chen, Y., Li, W., Van Gool, L.: Road: reality oriented adaptation for semantic segmentation of urban scenes. In: CVPR (2018)
3. Dou, Q., et al.: Pnp-adanet: plug-and-play adversarial domain adaptation network at unpaired cross-modality cardiac segmentation. IEEE Access **7**, 99065–99076 (2019)
4. Gholami, A., et al.: A novel domain adaptation framework for medical image segmentation. In: MICCAI Brainlesion Workshop (2018)
5. Grandvalet, Y., Bengio, Y.: Semi-supervised learning by entropy minimization. In: NIPS (2004)
6. Hoffman, J., et al.: Cycada: cycle-consistent adversarial domain adaptation. In: ICML (2018)
7. Hong, W., Wang, Z., Yang, M., Yuan, J.: Conditional generative adversarial network for structured domain adaptation. In: CVPR (2018)
8. Jabi, M., Pedersoli, M., Mitiche, A., Ben Ayed, I.: Deep clustering: On the link between discriminative models and k-means. IEEE TPAMI, 1 (2019)
9. Javanmardi, M., Tasdizen, T.: Domain adaptation for biomedical image segmentation using adversarial training. In: ISBI (2018)
10. Kamnitsas, K., et al.: Unsupervised domain adaptation in brain lesion segmentation with adversarial networks. In: Niethammer, M., et al. (eds.) IPMI 2017. LNCS, vol. 10265, pp. 597–609. Springer, Cham (2017). https://doi.org/10.1007/978-3-319-59050-9_47
11. Kervadec, H., Dolz, J., Tang, M., Granger, E., Boykov, Y., Ayed, I.B.: Constrained-CNN losses for weakly supervised segmentation. MedIA **54**, 88–99 (2019)

12. Kingma, D., Ba, J.: Adam: a method for stochastic optimization. In: ICLR (2014)
13. Krause, A., Perona, P., Gomes, R.G.: Discriminative clustering by regularized information maximization. In: NIPS (2010)
14. Litjens, G., et al.: A survey on deep learning in medical image analysis. MedIA **42**, 60–88 (2017)
15. Morerio, P., Cavazza, J., Murino, V.: Minimal-entropy correlation alignment for unsupervised deep domain adaptation. In: ICLR (2018)
16. Paszke, A., Chaurasia, A., Kim, S., Culurciello, E.: Enet: a deep neural network architecture for real-time semantic segmentation. arXiv 1606.02147 (2016)
17. Tsai, Y., Hung, W., Schulter, S., Sohn, K., Yang, M., Chandraker, M.: Learning to adapt structured output space for semantic segmentation. In: CVPR (2018)
18. Tzeng, E., et al.: Adversarial discriminative domain adaptation. In: CVPR (2017)
19. Vu, T.H., Jain, H., Bucher, M., Cord, M., Pérez, P.: Advent: adversarial entropy minimization for domain adaptation in semantic segmentation. In: CVPR (2019)
20. Wu, X., Zhang, S., Zhou, Q., Yang, Z., Zhao, C., Latecki, L.J.: Entropy minimization vs. diversity maximization for domain adaptation. arXiv 2002.01690 (2020)
21. Xie, S., Girshick, R., Dollar, P., Tu, Z., He, K.: Aggregated residual transformations for deep neural networks. In: CVPR (2017)
22. Zhang, Y., David, P., Foroosh, H., Gong, B.: A curriculum domain adaptation approach to the semantic segmentation of urban scenes. IEEE TPAMI **42**, 1823–1841 (2019)
23. Zhao, H., et al.: Supervised segmentation of un-annotated retinal fundus images by synthesis. IEEE TMI **38**, 46–56 (2019)
24. Zhou, Y., et al.: Prior-aware neural network for partially-supervised multi-organ segmentation. In: ICCV (2019)
25. Zhu, J.Y., Park, T., Isola, P., Efros, A.A.: Unpaired image-to-image translation using cycle-consistent adversarial networks. In: 2017 IEEE International Conference on Computer Vision (ICCV) (2017)
26. Zou, Y., Yu, Z., Kumar, B.V.K.V., Wang, J.: Unsupervised domain adaptation for semantic segmentation via class-balanced self-training. In: ECCV (2018)

Region-of-Interest Guided Supervoxel Inpainting for Self-supervision

Subhradeep Kayal[1(✉)], Shuai Chen[1], and Marleen de Bruijne[1,2]

[1] Department of Radiology & Nuclear Medicine, Erasmus MC, Biomedical Imaging Group Rotterdam, Rotterdam, The Netherlands
subhradeep.kayal@gmail.com

[2] Department of Computer Science, University of Copenhagen Machine Learning Section, Copenhagen, Denmark

Abstract. Self-supervised learning has proven to be invaluable in making best use of all of the available data in biomedical image segmentation. One particularly simple and effective mechanism to achieve self-supervision is inpainting, the task of predicting arbitrary missing areas based on the rest of an image. In this work, we focus on image inpainting as the self-supervised proxy task, and propose two novel structural changes to further enhance the performance. Our method can be regarded as an efficient addition to self-supervision, where we guide the process of generating images to inpaint by using supervoxel-based masking instead of random masking, and also by focusing on the area to be segmented in the primary task, which we term as the region-of-interest. We postulate that these additions force the network to learn semantics that are more attuned to the primary task, and test our hypotheses on two applications: brain tumour and white matter hyperintensities segmentation. We empirically show that our proposed approach consistently outperforms both supervised CNNs, without any self-supervision, and conventional inpainting-based self-supervision methods on both large and small training set sizes.

Keywords: Self-supervision · Inpainting · Deep learning · Segmentation · Brain tumor · White matter hyperintensities

1 Introduction and Motivation

Self-supervised learning points to methods in which neural networks are explicitly trained on large volumes of data, whose labels can be determined automatically and inexpensively, to reduce the need for manually labeled data. Many ways of performing self-supervision exist, amongst which a popular way is the *pre-train and fine-tune* paradigm where: (1) a convolutional neural network is pre-trained on a proxy task for which labels can be generated easily, and (2) it is then fine-tuned on the main task using labeled data. Utilizing a suitable and complex proxy task, self-supervision teaches the network robust and transferable visual

A. L. Martel et al. (Eds.): MICCAI 2020, LNCS 12261, pp. 500–509, 2020.
https://doi.org/10.1007/978-3-030-59710-8_49

features, which alleviates overfitting problems and aides its performance when fine-tuned on the main task [5].

In the medical imaging domain a variety of proxy tasks have been proposed, such as sorting 2D slices derived from 3D volumetric scans [14], predicting 3D distance between patches sampled from an organ [12], masking patches or volumes within the image and learning to predict them [3], and shuffling 3D blocks within an image and letting a network predict their original positions [16]. Recently, state-of-the-art results were achieved on several biomedical benchmark datasets by networks which were self-supervised using a sequence of individual proxy tasks [15].

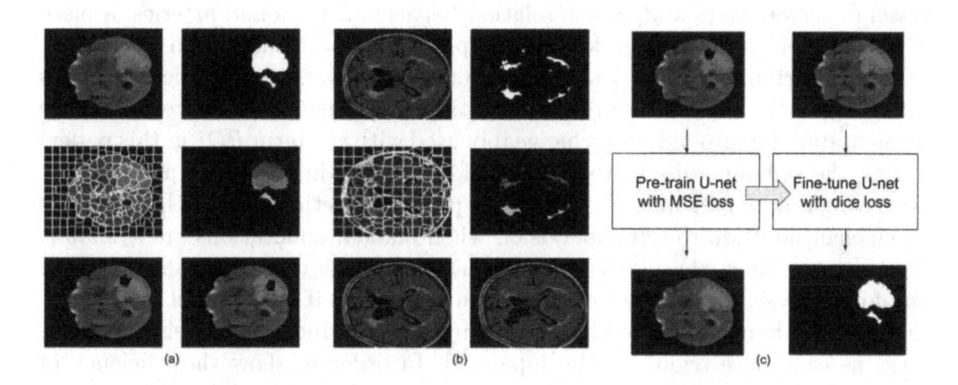

Fig. 1. Proposed ROI-guided inpainting. (a) Examples from BraTS 2018 dataset (left to right from top to bottom): original FLAIR image-slice, ground-truth segmentation map, FLAIR image-slice with superpixels overlaid, region-of-interest (ROI) influenced superpixels, examples of synthesized images to be inpainted. (b) Examples from White Matter Hyperintensities 2017 dataset. Notice that the ground-truth segmentations are much smaller in size. (c) first a U-net is pre-trained on the inpainting task with MSE loss, next it is fine-tuned on the main segmentation task with Dice loss.

Prior works in self-supervision literature have designed the proxy task largely uninfluenced by the downstream task in focus. However, since the features that the network learns are dependent on where it is focusing on during the self-supervision task, it might be beneficial to bias or *guide* the proxy task towards areas that are of interest to the main task. Specifically for image segmentation, these would be the foreground areas to be segmented in the main task, which we term as the *region-of-interests or ROIs*.

We experiment with the proxy task of inpainting [11], where the network must learn to fill-in artificially created gaps in images. In the context of biomedical imaging, a network that learns to inpaint healthy tissue will learn a different set of semantics than one which inpaints various kinds of tumours. Thus, if the main task is that of segmenting tumours, it can be hypothesized that having a network inpaint tumourous areas as a proxy task will likely teach it semantics attuned to segmenting tumours, and thereby be more beneficial for the main task than learning general semantics. In other words, by increasing the frequency of inpainting tumours, we can teach the network features which are more related to the tumour segmentation task.

Furthermore, in prior work the selection of regions to mask has largely been uninformed and random. We try to improve upon this situation by selecting regions which are homogeneous. Masking such regions could force the network to learn more about the anatomical meaning and relation to other structures of the masked tissue. For example, masking small regions in a lung CT scan would only require the model to correctly interpolate the structures (airways, vessels) around the masked region. In contrast, when a full airway or vessel branch is masked, inpainting requires understanding of the relation between branches in vessel or airway trees and/or the relation between airways and arteries, a piece of information that has been found to improve airway segmentation [8].

The contributions of this work are twofold. Firstly, this paper demonstrates that guiding the inpainting process with the main class(es) of interest (i.e., the segmentation foreground, interchangeably used with the term *ROI* in this paper) during the self-supervised pre-training of a network improves its performance over using random regions. Therefore, the proposed method can be thought of as an efficient addition to self-supervision when manual annotations are available. Secondly, we show that instead of inpainting regions of regular shapes in an uninformed way, further performance gain is possible if the masked regions are chosen to be homogeneous. This is done by constructing supervoxels and using these as candidate regions to be inpainted. In order to show the efficiency of these proposed changes, we conduct empirical analyses on two popularly used public datasets for biomedical image segmentation.

2 Methods

The proposed method (Fig. 1) utilizes supervoxelization to create candidate regions, followed by selecting only those supervoxels which have an overlap with (any of) the foreground class(es). The selected supervoxels are utilized in the inpainting process, where we use them as masks to suppress areas in an image to train a network to predict (or *inpaint*) them based on their surroundings. Since we control the parameters of this process, it can be used to create an arbitrarily large amount of synthetic training images for pre-training.

2.1 Region-of-Interest Guided Inpainting

Inpainting is an effective proxy task for self-supervision, which proceeds by training a network to reconstruct an image from a masked version of it. In this section, we explain our proposed masking approach, followed by the description of the network in Sect. 2.2.

Supervoxelization: While previous works primarily use random grids and cubes as candidate regions to inpaint, the first step in our proposed approach is to select regions based on some notion of homogeneity. One way of achieving this is to construct supervoxels, which may be defined as homogeneous groups of voxels that share some common characteristics. A particularly efficient algorithm to construct such supervoxels is *SLIC* or *simple linear iterative clustering* [1].

For 3D medical images, SLIC can cluster voxels based on their intensity values, corresponding to the various modalities, and spatial coordinates of the voxel within the image. SLIC has two main hyperparameters: one, *compactness*, controls the balance between emphasis on intensity values and spatial coordinates (larger values make square/cubic grids), and the other defines the maximum number of supervoxels. Examples in the second row of Fig. 1, subfigure (a) and (b).

In this work, we use SLIC with intensity values corresponding to the two modalities we used in our experiments, FLAIR and T1 (or contrast enhanced T1), in order to construct supervoxels. The exact parameter settings for supervoxelization are described later in Sect. 3.3.

ROI-Guided Masking for Inpainting Image Synthesis: Once the supervoxel labels have been created, the next step is to retain only those supervoxels which have an overlap with the region-of-interest. To achieve this, we first convert the segmentation map to a binary one by considering all foreground areas to be a class with a label value as 1 and the background as 0, since there may be multiple regions-of-interest in a multi-class segmentation setting. Then an elementwise *and* operation is performed between the resulting binary segmentation map and the generated supervoxel. For all the supervoxels that remain, training images for the inpainting task can be synthesized by masking an area corresponding to such a *ROI-guided supervoxel*, with the original unmasked image being the target for the network. Some examples of this are in the second row of Fig. 1, subfigure (a) and (b).

By constructing a training set for the inpainting task in this fashion, we are essentially increasing the frequency of inpainting regions which are important to the main task more than random chance. This is what, we posit, will bring about improvements in the performance of the network on the main task.

Formally, let $D_{train} = \{(I_i, S_i)\}_{i=1..n}$ be the training dataset containing n images with I_i being a 3D multi-modal training image and S_i being the segmentation ground-truth label, containing zero values representing background. If f is a supervoxelization algorithm (in our case, SLIC), then a ROI-guided supervoxelized image is given by $R_i = f(I_i) \odot S_i$, where $\odot$ signifies elementwise multiplication.

R_i contains supervoxel regions having non-zero labels corresponding to foregound supervoxels. Then, the synthetic dataset for inpainting, D_{inp} is constructed as:

$$D_{inp} = \left\{ \left\{ \left(I_i \odot r_{ij}^0, I_i \right) \right\}_{r_{ij} \in R_i, j=1..m_i} \right\}_{i=1..n} \tag{1}$$

where r_{ij} is a single supervoxel region in the set R_i, which contains a total of m_i supervoxels, and r_{ij}^0 is the corresponding inverted region-mask, containing 0 for voxels belonging to the region and 1 everywhere else. $I_i \odot r_{ij}$ is then the masked image input to the network and I_i is the expected output to reconstruct, the target for the inpainting task. Thus, the maximum cardinality of D_{inp} can be $n \times m_i$.

Examples D_{inp} are in the last row of Fig. 1, subfigure (a) and (b).

2.2 Training Strategy

Network: For all the experiments, a shallow 3D U-net [4] containing 3 resolution levels has been used, with a batch-normalization layer after every convolution layer. In our experiments we find that 3 layers provide sufficient capacity for both the inpainting and the segmentation task. Since we use two modalities for our experiments, the U-net has two input channels.

If we were to use an image reconstruction proxy task, a U-net would learn to copy over the original image because of its skip connections, and would not be useful in learning features. In our task of inpainting the network never sees the masked regions and, therefore, cannot memorize it, making the use of a U-net reasonable.

Pre-training: In order to pre-train the network, it is fitted to the D_{inp} dataset by minimizing the mean squared error (MSE) between the masked and the original images using the *Adam* [6] optimizer. We call this model *inpainter U-net.*

Fine-Tuning: The inpainter U-net is then fine-tuned on the (main) segmentation task using the original labeled training dataset, D_{train}, by optimizing the Dice segmentation overlap objective on the labeled images. If the data is multi-modal, the inpainter U-net will be trained to produce multi-channel outputs, in which case we would need to replace the last 3D convolutional layer to have a single-channel output for segmentation.

More details about the network parameters are provided in Sect. 3.3.

3 Experimental Settings

3.1 Data

For our experiments, we use two public datasets containing 3D MRI scans and corresponding manual segmentations.

BraTS 2018 [7]: 210 MRI scans from patients with high-grade glioma are randomly split three times into 150, 30 and 30 scans for training, validation and testing, respectively, using a 3-fold Monte-carlo cross-validation scheme. To be able to easily compare our method against baselines, we focus on segmenting the whole tumour and use two of the four modalities, *FLAIR* and *T1-gd*, which have been found to be the most effective at this task [13].

White Matter Hyperintensities (WMH) 2017 [10]: The total size of the dataset is 60 FLAIR and T1 scans, coming from 3 different sites, with corresponding manual segmentations of white matter hyperintensities. We employ a

3-fold Monte-carlo cross-validation scheme again, splitting the dataset into 40, 10 and 10 for training, validation and testing, respectively, and use both of the available modalities for our experiments.

To study the effect of training set sizes on the proposed approach, experiments were performed on the full training dataset as well as smaller fractions of it. For BraTS, we perform experiments on 25%, 50% and 100% of the training data, while for WMH, which is much smaller in size, we only perform an extra set of experiments with 50% of the data. To keep the comparisons fair, we use the same subset of the training data in the pre-training procedure as well. Note that, even though self-supervision by inpainting (with or without supervoxels) could be applied to unlabeled data as well, in our experiments we only use fully labeled training samples to facilitate comparison.

3.2 Baseline Methods

We term the technique proposed in this paper as *roi-supervoxel* to denote the use of the segmentation map and supervoxelization to guide the inpainting process used for pre-training. In order to validate its effectiveness, it is tested against the following baselines: *vanilla-unet*: a U-net without any pre-training; *restart-unet*: a U-net pre-trained on the main (segmentation) task and fine-tuned on the same task for an additional set of epochs; *noroi-grid*: the more traditional inpainting mechanism where random regular sized cuboids are masked; *roi-grid*: a similar process as *roi-supervoxel*, except for the use of regular cuboids overlapping with the segmentation map, instead of supervoxel regions, for masking; *noroi-supervoxel*: where random supervoxels are masked.

3.3 Settings

Inpainting Parameters: The inpainting process starts by creating the supervoxel regions using SLIC[1]. We fix these the compactness value at 0.15 and choose the maximum number of supervoxels to be 400, by visual inspection of the nature of the supervoxels that contain the tumour and the white matter hyperintensities for the two datasets. For example, between a setting where one supervoxel is part tumour and part background, versus another where one supervoxel fully represents tumour, we choose the latter case.

We then use either the supervoxels or simple cuboids (for the baseline methods) as areas to be masked, and the question arises of how many and how large areas to choose as masks to construct synthetic images for D_{inp}. Too small a volume, and it might be trivial for a network to inpaint it; too large, and it might not be a feasible task. For our experiments, we choose masks whose volume is at least 1500 voxels. For constructing cuboids, we randomly generate cuboids which are at least 12 units in each dimension (as 12^3 is more than 1500, but 11^3 is not).

[1] We use the implementation in https://scikit-image.org/docs/dev/api/skimage.segmentation.html?highlight=slic#skimage.segmentation.slic.

Finally, we ensure that the size of D_{inp} is roughly 10 times that of the D_{train}, by choosing masks which fit the volume criteria as they are generated, and producing at most 10 synthetic images on-the-fly for a single real input image.

Network Parameters: The input size to the 3D U-net is 160 × 216 × 32, such that each input image is centre-cropped to 160 × 216 ($X - Y$ axes) to tightly fit the brain region in the scan, while we use the overlapping tile strategy in the Z-axis as inspired by the original U-net. Each of the 3 resolution levels consists of two 3 × 3 × 3 convolution layers using zero-padding and *ReLU* activation, except for the last layer which is linear in the inpainter U-net and *sigmoid* in the fine-tuned U-net. The number of feature channels are 16, 32 and 64 at the varying resolution levels. The feature maps in the upsampling path are concatenated with earlier ones through skip-connections.

Optimization Parameters: The inpainter U-net is optimized on MSE while fine-tuning is performed using a Dice objective, both using *Adam*. The learning rate is 0.0001 and 0.001 for BraTS and WMH datasets, respectively. We used a batch-size of 4, as permitted by our GPU memory. For pre-training, we use 100 epochs while for fine-tuning we employ another 150, both without the possibility of early stopping, saving the best performing model based on the validation loss at every epoch.

To foster open-science, all of the code will be released[2].

4 Results and Discussion

The segmentation results are shown in Table 1. It can be observed that the proposed method (*roi-supervoxel*) outperforms the basic U-net (*vanilla-unet*) by a large margin, and traditional inpainting based pre-training (*noroi-grid*) by a small, but significant, margin.

The deductions from the empirical results can be summarized as follows:

Restarts Improve U-net Performance: It can be observed that for both datasets, the performance of the *restart-unet* is better than that of the *vanilla-unet*. This is in line with observations in literature [9], where warm restarts have aided networks to find a more stable local minimum. Based on this observation, we argue that for any proposed method involving pre-training models, the results should always be compared to such a *restarted* model.

Adding ROI Information to the Inpainting Proxy Task is Helpful: For both the datasets, the performance of the *roi-supervoxel* method exceeds that of all other baselines. Importantly, it exceeds the performance of the *restart-unet* and the *noroi-grid*, which is the traditional inpainting procedure, by 3.2% and 5% (relative) respectively for BraTS, and 4.9% and 2.9% for WMH, when all of the data is used. Also important to note is that the performance of methods which use the region-of-interest information to generate the masked areas is always better than those which do not.

[2] https://github.com/DeepK/inpainting.

Table 1. Dice-scores on BraTS 2018 and WMH 2017. The results represent the mean and standard deviation (in brackets) of the Dice coefficient averaged over the three folds. The top results are depicted in **bold**. *: indicates that a result is significantly worse than *roi-supervoxel* ($p < 0.05$) in the same column, p-values calculated by a two-sided t-test.

Fraction training data					
Method	BraTS			WMH	
	.25	**.50**	**1.0**	**.50**	**1.0**
vanilla-unet	0.257 (.05)*	0.585 (.03)*	0.784 (.02)	0.576 (.05)*	0.745 (0.02)*
restart-unet	0.302 (.05)*	0.607 (.03)*	0.793 (.02)	0.610 (.05)*	0.776 (.03)*
noroi-grid	0.311 (.06)*	0.611 (.04)*	0.780 (.03)*	0.632 (.04)*	0.791 (.03)
roi-grid	0.354 (.06)	0.620 (.04)	0.795 (.03)	0.653 (.04)	0.812 (.03)
noroi-supervoxel	0.340 (.06)	0.621 (.04)	0.791 (.02)	0.650 (.04)	0.797 (.03)
roi-supervoxel	**0.363 (.06)**	**0.646 (.04)**	**0.814 (.03)**	**0.671 (.04)**	**0.814 (.03)**

Inpainting is More Beneficial When the Size of the Training Set is Smaller: The difference in performance between the inpainting-assisted methods and *vanilla-unet* is larger when the size of the training dataset is smaller. For example, for BraTS, the difference between *vanilla-unet* and *roi-supervoxel* (our proposed approach) is as large as 41.2% (relative) when the size of the training dataset is 25% of the total. This trend is also observed between the methods with and without ROI information.

Supervoxels Help More When Areas to be Segmented are Larger Rather Than Finer: ROI-guided inpainting can be postulated to have a better chance of affecting the downstream performance when the ROI itself is larger. Taking into account that tumours in BraTS are, on-average, larger than the hyperintensities to be segmented in the WMH dataset, it can be observed that the performance difference between the inpainting methods using supervoxels (*roi-supervoxel* and *noroi-supervoxel*) versus the ones which do not (*roi-grid* and *noroi-grid*) is smaller in the case of WMH than for BraTS. For example, when using all of the training data, the difference in performance between *roi-supervoxel* and *roi-grid* is 3% (relative) for BraTS but only 0.25% for WMH. This could likely be alleviated by problem specific selection of parameters for SLIC, which we did not explore. This would ensure that the supervoxels are not too large as compared to the ROI, in which case the effect of ROI would not be significant.

These results show that our approach is promising. An important point to note is that a similar approach may be valuable in other forms of local self-supervision techniques like jigsaw puzzle solving [16], where the shuffling could be guided by the ROI and the tiles could be picked by ensuring homogeneity constraints.

Although efficient, this method does have some limitations: firstly, its efficiency depends on the parameters of the supervoxelization process and a poor choice of parameters could lead to limited performance gain; secondly, although sizeable synthetic datasets can be created in this process, the reliance on ROI means that we still need segmentation annotations. Perhaps one way of solving the second problem would be using co-training [2] to label all of the data and then employ our method using the entire corpus.

5 Conclusion

In summary, this work explores the use of supervoxels and foreground segmentation labels, termed the *region-of-interest (ROI)*, to guide the proxy task of inpainting for self-supervision. Together, these two simple changes have been found to add a significant boost in the performance of a convolutional neural network for segmentation (as much as a relative gain of 5% on the BraTS 2018 dataset), in comparison to traditional methods of inpainting-based self-supervision.

Acknowledgements. This research was partly funded by the Netherlands Organisation for Scientific Research (NWO), as well as by the China Scholarship Council (File No.201706170040).

References

1. Achanta, R., Shaji, A., Smith, K., Lucchi, A., Fua, P., Süsstrunk, S.: Slic superpixels compared to state-of-the-art superpixel methods. IEEE Trans. Pattern Anal. Mach. Intell. **34**(11), 2274–2282 (2012)
2. Blum, A., Mitchell, T.: Combining labeled and unlabeled data with co-training. In: Proceedings of the Eleventh Annual Conference on Computational Learning Theory, COLT 1998, pp. 92–100. Association for Computing Machinery, New York (1998)
3. Chen, L., Bentley, P., Mori, K., Misawa, K., Fujiwara, M., Rueckert, D.: Self-supervised learning for medical image analysis using image context restoration. Med. Image Anal. **58**, 101539 (2019)
4. Çiçek, Ö., Abdulkadir, A., Lienkamp, S.S., Brox, T., Ronneberger, O.: 3D U-Net: learning dense volumetric segmentation from sparse annotation. In: Ourselin, S., Joskowicz, L., Sabuncu, M.R., Unal, G., Wells, W. (eds.) MICCAI 2016. LNCS, vol. 9901, pp. 424–432. Springer, Cham (2016). https://doi.org/10.1007/978-3-319-46723-8_49
5. Jing, L., Tian, Y.: Self-supervised visual feature learning with deep neural networks: a survey. CoRR abs/1902.06162 (2019)
6. Kingma, D.P., Ba, J.: Adam: a method for stochastic optimization. In: Bengio, Y., LeCun, Y. (eds.) 3rd International Conference on Learning Representations, Conference Track Proceedings, ICLR 2015, San Diego, CA, USA, 7–9 May 2015 (2015)
7. Kuijf, H.J.: Standardized assessment of automatic segmentation of white matter hyperintensities and results of the WMH segmentation challenge. IEEE Trans. Med. Imaging **38**(11), 2556–2568 (2019)
8. Lo, P., Sporring, J., Ashraf, H., Pedersen, J.J., de Bruijne, M.: Vessel-guided airway tree segmentation: a voxel classification approach. Med. Image Anal. **14**(4), 527–538 (2010)

9. Loshchilov, I., Hutter, F.: SGDR: stochastic gradient descent with warm restarts. In: 5th International Conference on Learning Representations, Conference Track Proceedings, ICLR 2017, Toulon, France, 24–26 April 2017 (2017)
10. Menze, B.H.: The multimodal brain tumor image segmentation benchmark (BRATS). IEEE Trans. Med. Imaging **34**(10), 1993–2024 (2015)
11. Pathak, D., Krähenbühl, P., Donahue, J., Darrell, T., Efros, A.A.: Context encoders: feature learning by inpainting. In: IEEE Conference on Computer Vision and Pattern Recognition, CVPR 2016, Las Vegas, NV, USA, 27–30 June 2016, pp. 2536–2544 (2016)
12. Spitzer, H., Kiwitz, K., Amunts, K., Harmeling, S., Dickscheid, T.: Improving cytoarchitectonic segmentation of human brain areas with self-supervised siamese networks. In: Frangi, A.F., Schnabel, J.A., Davatzikos, C., Alberola-López, C., Fichtinger, G. (eds.) MICCAI 2018. LNCS, vol. 11072, pp. 663–671. Springer, Cham (2018). https://doi.org/10.1007/978-3-030-00931-1_76
13. van Tulder, G., de Bruijne, M.: Why does synthesized data improve multi-sequence classification? In: Navab, N., Hornegger, J., Wells, W.M., Frangi, A.F. (eds.) MICCAI 2015. LNCS, vol. 9349, pp. 531–538. Springer, Cham (2015). https://doi.org/10.1007/978-3-319-24553-9_65
14. Zhang, P., Wang, F., Zheng, Y.: Self supervised deep representation learning for fine-grained body part recognition. In: IEEE 14th International Symposium on Biomedical Imaging (ISBI 2017), pp. 578–582 (2017)
15. Zhou, Z., et al.: Models genesis: generic autodidactic models for 3D medical image analysis. In: Shen, D., et al. (eds.) MICCAI 2019. LNCS, vol. 11767, pp. 384–393. Springer, Cham (2019). https://doi.org/10.1007/978-3-030-32251-9_42
16. Zhuang, X., Li, Y., Hu, Y., Ma, K., Yang, Y., Zheng, Y.: Self-supervised feature learning for 3D medical images by playing a Rubik's cube. In: Shen, D., et al. (eds.) MICCAI 2019. LNCS, vol. 11767, pp. 420–428. Springer, Cham (2019). https://doi.org/10.1007/978-3-030-32251-9_46

Harnessing Uncertainty in Domain Adaptation for MRI Prostate Lesion Segmentation

Eleni Chiou[1,2](✉), Francesco Giganti[3,4], Shonit Punwani[5], Iasonas Kokkinos[2], and Eleftheria Panagiotaki[1,2]

[1] Centre for Medical Image Computing, UCL, London, UK
eleni.chiou.17@ucl.ac.uk
[2] Department of Computer Science, UCL, London, UK
[3] Department of Radiology, UCLH NHS Foundation Trust, London, UK
[4] Centre for Medical Imaging, Division of Medicine, UCL, London, UK
[5] Centre for Medical Imaging, UCL, London, UK

Abstract. The need for training data can impede the adoption of novel imaging modalities for learning-based medical image analysis. Domain adaptation methods partially mitigate this problem by translating training data from a related source domain to a novel target domain, but typically assume that a one-to-one translation is possible. Our work addresses the challenge of adapting to *a more informative target domain* where multiple target samples can emerge from a single source sample. In particular we consider translating from mp-MRI to VERDICT, a richer MRI modality involving an optimized acquisition protocol for cancer characterization. We explicitly account for the inherent uncertainty of this mapping and exploit it to generate multiple outputs conditioned on a single input. Our results show that this allows us to extract systematically better image representations for the target domain, when used in tandem with both simple, CycleGAN-based baselines, as well as more powerful approaches that integrate discriminative segmentation losses and/or residual adapters. When compared to its deterministic counterparts, our approach yields substantial improvements across a broad range of dataset sizes, increasingly strong baselines, and evaluation measures.

Keywords: Domain adaptation · Image synthesis · GANs · Segmentation · MRI

1 Introduction

Domain adaptation can be used to exploit training samples from an existing, densely-annotated domain within a novel, sparsely-annotated domain, by

Electronic supplementary material The online version of this chapter (https://doi.org/10.1007/978-3-030-59710-8_50) contains supplementary material, which is available to authorized users.

A. L. Martel et al. (Eds.): MICCAI 2020, LNCS 12261, pp. 510–520, 2020.
https://doi.org/10.1007/978-3-030-59710-8_50

bridging the differences between the two domains. This can facilitate the training of powerful convolutional neural networks (CNNs) for novel medical imaging modalities or acquisition protocols, effectively compensating for the limited amount of training data available to train CNNs in the new domain.

The assumption underlying most domain adaptation methods is that one can align the two domains either by extracting domain-invariant representations (features), or by establishing a 'translation' between the two domains at the signal level, where in any domain the 'resident' and the translated signals are statistically indistinguishable.

In particular for medical imaging, [23] and [13] rely on adversarial training to align the feature distributions between the source and the target domain for medical image classification and segmentation respectively. Pixel-level distribution alignment is performed by [2,11,27,28], who use CycleGAN [29] to map source domain images to the style of the target domain; they further combine the translation network with a task-specific loss to penalize semantic inconsistency between the source and the synthesized images. The synthesized images are used to train models for image segmentation in the target domain. Ouyang et al. [17] perform adversarial training to learn a shared, domain-invariant latent space which is exploited during segmentation. They show that their approach is effective in cases where target-domain data is scarce. Similarly, [26] embed the input images from both domains onto a domain-specific style space and a shared content space. Then, they use the content-only images to train a segmentation model that operates well in both domains. However, their approach does not necessarily preserve crucial semantic information in the content-only images.

These methods rely on the strong assumption that the two domains can be aligned - our work shows that accuracy gains can be obtained by acknowledging that this can often be only partially true, and mitigating the resulting challenges. As a natural image example, an image taken at night can have many day-time counterparts, revealed by light; similarly in medical imaging, a better imaging protocol can reveal structures that had previously passed unnoticed. In technical terms, the translation can be one-to-many, or, stated in probabilistic terms, multi-modal [10,14,30]. Using a one-to-one translation network in such a setting can harm performance, since the translation may predict the mean of the underlying multi-modal distribution, rather than provide diverse, realistic samples from it.

In our work we accommodate the inherent uncertainty in the cross-domain mapping and, as shown in Fig. 1, generate multiple outputs conditioned on a single input, thereby allowing for better generalization of the segmentation network in the target domain. As in recent studies [2,11,27,28], we use GANs [6] to align the source and target domains, but go beyond their one-to-one, deterministic mapping approaches. In addition, inspired by [2,8,11,27], we enforce semantic consistency between the real and synthesized images by exploiting source-domain lesion segmentation supervision to train target-domain networks operating on the synthesized images. This results in training networks that can generate diverse outputs while at the same time preserving critical structures -

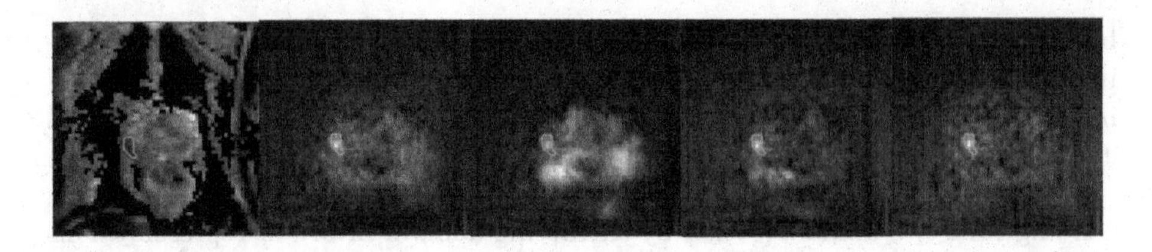

Fig. 1. One-to-many mapping from one mp-MRI image (left) to four VERDICT-MRI translations: our network can generate samples with both local and global structure variation, while at the same time preserving the critical structure corresponding to the prostate lesion, shown as a red circle. We note that the lesion area is annotated by a physician on the leftmost image, but is not used as input to the translation network - instead the translation network learns to preserve lesion structures thanks to the end-to-end discriminative training (details in text). (Color figure online)

such as the lesion area in Fig. 1. We further accommodate the statistical discrepancies between real and synthesized data by introducing residual adapters (RAs) [5,22] in the segmentation network. These capture domain-specific properties and allow the segmentation network to generalize better across the two domains.

We demonstrate the effectiveness of our approach in prostate lesion segmentation and an advanced diffusion weighted (DW)-MRI method called VERDICT-MRI (Vascular, Extracellular and Restricted Diffusion for Cytometry in Tumors). VERDICT-MRI is a non-invasive imaging technique for cancer microstructure characterisation [12,18,20]. The method has been recently in clinical trial to supplement standard multi-parametric (mp)-MRI for prostate cancer diagnosis. Compared to the naive DW-MRI from mp-MRI acquisitions, VERDICT-MRI has a richer acquisition protocol to probe the underlying microstructure and reveal changes in tissue features similar to histology. A recent study [12] has demonstrated that the intracellular volume fraction (FIC) maps obtained with VERDICT-MRI differentiate better benign and malignant lesions compared to the apparent diffusion coefficient (ADC) map obtained with the naive DW-MRI from mp-MRI acquisitions. However, the limited amount of available labeled training data does not allow the training of robust deep neural networks that could directly exploit the information in the raw VERDICT-MRI [4]. On the other hand, large scale clinical mp-MRI datasets exist [1,15]. As shown experimentally, our approach largely improves the generalization capabilities of a lesion segmentation model on VERDICT-MRI by exploiting labeled mp-MRI data.

2 Method

Our approach relies on a unified network for cross-modal image synthesis and segmentation, that is trained end-to-end with a combination of objective functions. As shown in Fig. 2, at the core of this network is an image-to-image translation network that maps images from the source ('S') to the target ('T') domain.

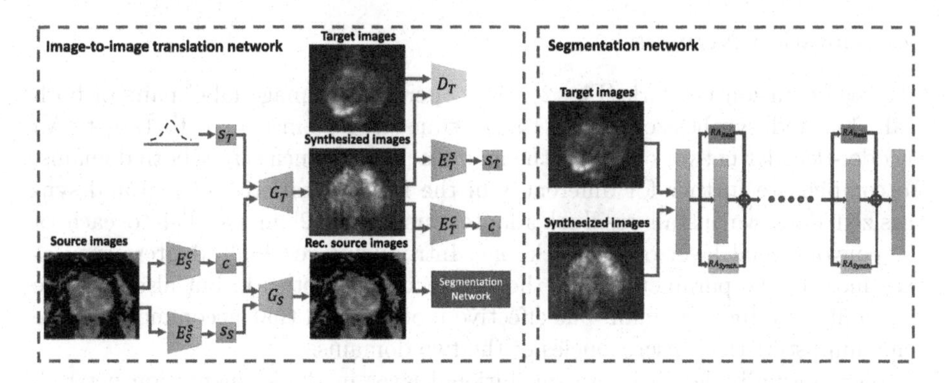

Fig. 2. Overview of our domain adaptation framework: we train a noise-driven domain translation network in tandem with a discriminatively supervised segmentation network in the target domain; GAN-type losses align the translated samples with the target distribution, while residual adapters allow the segmentation network to compensate for remaining discrepancies. Please see text for details.

The translation network is trained in tandem with a segmentation network that operates in the target domain, and is trained with both the synthesized and the few real annotated target-domain images. Beyond these standard components, our approach relies on three additional components: firstly, we sample a latent variable from a Gaussian distribution when translating to the target domain; this represents structures that cannot be accounted by a deterministic mapping, and can result in one-to-many translation when needed. Secondly, we introduce residual adapters (RAs) to a common backbone network for semantic segmentation, allowing the discriminative training to accommodate any remaining discrepancies between the real and synthesized target domain images. Finally, we use a dual translation network from the target to the source domain, allowing us to use cycle-consistency in domain adaptation [10,16,29]; the cycle constraint allows us to disentangle the deterministic, transferable part from the stochastic, non-transferable part, which is filled in by Gaussian sampling, as mentioned earlier.

2.1 Problem Formulation

Having provided a broad outline of our method, we now turn to a more detailed technical description. We consider the problem of domain adaptation in prostate lesion segmentation. We assume that the source domain, $\mathcal{X}_S$, contains N_S images, $x_S \in \mathcal{X}_S$, with associated segmentation masks, $y_S \in \mathcal{Y}_S$. Similarly, the sparsely labeled target domain, $\mathcal{X}_T$, consists of N_T images, $x_T \in \mathcal{X}_T$. A subset $\tilde{\mathcal{X}}_T$ of $\mathcal{X}_T$ comes with associated segmentation masks, $y_T \in \mathcal{Y}_T$. The proposed framework consists of two main components, i.e. an image-to-image translation network and a segmentation network described below.

Segmentation Network

The segmentation network (Fig. 2), Seg, operates on image-label pairs of both real, $\mathcal{X}_T$, and synthesized data, $\mathcal{X}_{S\rightarrow T}$, translated from source to target. An encoder-decoder network [3,24] is the main backbone which serves both domains. To compensate further for differences in the feature statistics of real and synthesized data we install residual adapter modules [22] in parallel to each of the convolutional layer of the backbone. Introducing residual adapters ensures that most of the parameters stay the same with the network, but also that the new unit introduces a small, but effective modification that accommodates the remaining statistical discrepancies of the two domains.

More formally, let ϕ_l be a convolutional layer in the segmentation network and $\mathbf{F}^l \in \mathbb{R}^{k\times k\times C_i\times C_o}$ be a set of filters for that layer, where $k \times k$ is the kernel size and C_i, C_o are the number of input and output feature channels respectively. Let also $\mathbf{Z}_i^l \in \mathbb{R}^{1\times 1\times C_i\times C_o}$ be a set of domain-specific residual adapter filters of domain i, where $i \in \{1, 2\}$, installed in parallel with the existing set of filters $\mathbf{F}_l$. Given an input tensor $\mathbf{x}_l \in \mathbb{R}^{H\times W\times C_i}$, the output $\mathbf{y}_l \in \mathbb{R}^{H\times W\times C_o}$ of layer l is given by

$$\mathbf{y}_l = \mathbf{F}^l * \mathbf{x} + \mathbf{Z}_i^l * \mathbf{x}. \tag{1}$$

We train the segmentation network by optimizing the following objective

$$\begin{aligned}\mathcal{L}_{Seg}(Seg, \tilde{\mathcal{X}_T}, \mathcal{Y}_T, \mathcal{X}_{S\rightarrow T}, \mathcal{Y}_S) = \\ \mathcal{L}_{DSC}(Seg, \tilde{\mathcal{X}_T}, \mathcal{Y}_T) + \mathcal{L}_{DSC}(Seg, \mathcal{X}_{S\rightarrow T}, \mathcal{Y}_S).\end{aligned} \tag{2}$$

The dice loss, $\mathcal{L}_{DSC}$, is given by

$$\mathcal{L}_{DSC}(Seg, \mathcal{X}, \mathcal{Y}) = -\frac{2\sum_{(\mathbf{x},\mathbf{y})\in(\mathcal{X},\mathcal{Y})}\sum_{k=1}^{K} Seg(\mathbf{x})_k \mathbf{y}_k}{\sum_{(\mathbf{x},\mathbf{y})\in(\mathcal{X},\mathcal{Y})}\sum_{k=1}^{K}(Seg(\mathbf{x})_k^2 + \mathbf{y}_k^2)}, \tag{3}$$

where K the number of voxels in the input images. We adopt this objective function since it is a differentiable approximation of a criterion that is well-adapted to our task.

Diverse Image-to-Image Translation Network

Recently, several studies [10,30] have pointed out that cross-domain mapping is inherently multi-modal and proposed approaches to produce multiple outputs conditioned on a single input. Here we use MUNIT [10] to illustrate the key idea. As it is illustrated in Fig. 2 the image-to-image translation network consists of content encoders E_S^c, E_T^c, style encoders E_S^s, E_T^s, generators G_S, G_T and domain discriminators D_S, D_T for both domains. The content encoders E_S^c, E_T^c map images from the two domains onto a domain-invariant content space $\mathcal{C}$ ($E_S^c : \mathcal{X}_S \rightarrow \mathcal{C}$, $E_T^c : \mathcal{X}_T \rightarrow \mathcal{C}$) and the style encoders E_S^s, E_T^s map the images onto domain-specific style spaces $\mathcal{S}_S$ ($E_S^s : \mathcal{X}_S \rightarrow \mathcal{S}_S$) and $\mathcal{S}_T$ ($E_T^s : \mathcal{X}_T \rightarrow \mathcal{S}_T$). The content code can be understood as the underlying anatomy which is the information that we want transfer during the translation while the style codes capture information related to the imaging modalities. Image-to-image translation is performed by combining the content code extracted from a given input and a random style code sampled from the target-style space. For instance, to translate an image $x_S \in \mathcal{X}_S$ to $\mathcal{X}_T$ we first extract its content code $c = E_S^c(x_S)$. The generator G_T uses the extracted content code c and a randomly drawn style code $s_T \in \mathcal{S}_T$ to produce the final output $x_{S\to T} = G_T(c, s_T)$. By sampling random style codes from the style spaces $\mathcal{S}_S$ and $\mathcal{S}_T$ the generators G_S and G_T are able to produce diverse outputs. We train the networks with a loss function that consists of domain adversarial, self-reconstruction, latent reconstruction, cycle-consistency and segmentation losses.

Domain Adversarial Loss. We utilize GANs to match the distribution between the synthesized and the real images of the two domains. The adversarial discriminators D_T, D_S aim at discriminating between real and synthesized images, while the generators G_T, G_S aim at generating realistic images that fool the discriminators. For G_T and D_T the GAN loss is defined as

$$\begin{aligned} &\mathcal{L}_{GAN}^T(E_S^c, G_T, D_T, \mathcal{S}_T, \mathcal{X}_S) = \\ &\quad \mathbb{E}_{x_S\sim\mathcal{X}_S, s_T\sim\mathcal{S}_T}[\log(1 - D_T(G_T(E_S^c(x_S), s_T)))] + \mathbb{E}_{x_T\sim\mathcal{X}_T}[\log(D_T(x_T))]. \end{aligned} \tag{4}$$

Self-reconstruction Loss. Given the encoded content and style codes of a source-domain image the generator G_S should be able to decode them back to the original one.

$$\mathcal{L}_{recon}^S(G_S, E_S^s, E_S^c, \mathcal{X}_S) = \mathbb{E}_{x_S\sim\mathcal{X}_S}[\|G_S(E_S^c(x_S), E_S^s(x_S)) - x_S\|_1]. \tag{5}$$

Latent Reconstruction Loss. To encourage the translated image to preserve the content of the source image, we require that a latent code c sampled from the latent distribution can be reconstructed after decoding and encoding.

$$\begin{aligned} &\mathcal{L}_{recon}^{c_S}(E_S^c, G_T, E_T^c, \mathcal{X}_S, \mathcal{S}_T) = \\ &\quad \mathbb{E}_{x_S\sim\mathcal{X}_S, s_T\sim\mathcal{S}_T}[\|E_T^c(G_T(E_S^c(x_s), s_T)) - E_S^c(x_s)\|_1]. \end{aligned} \tag{6}$$

Similarly, to align the style representation with a Gaussian prior distribution, we enforce the same constrain for the latent style code.

$$\begin{aligned} &\mathcal{L}_{recon}^{s_T}(E_S^c, G_T, E_T^s, \mathcal{X}_S, \mathcal{S}_T) = \\ &\quad \mathbb{E}_{x_S\sim\mathcal{X}_S, s_T\sim\mathcal{S}_T}[\|E_T^s(G_T(E_S^c(x_s), s_T)) - s_T)\|_1]. \end{aligned} \tag{7}$$

Cycle-Consistency Loss. To facilitate training we enforce cross-cycle consistency which implies that if we translate an image to the target domain and then translate it back to the source domain using the extracted source-domain style code, we should be able to obtain the original image.

$$\begin{aligned} &\mathcal{L}_{cyc}^S(E_S^c, E_S^s, G_T, E_T^c, G_S, \mathcal{X}_S, \mathcal{S}_T) = \\ &\quad \mathbb{E}_{x_S\sim\mathcal{X}_S, s_T\sim\mathcal{S}_T}[\|G_S(E_T^c(G_T(E_S^c(x_S), s_T)), E_S^s(x_S)) - x_S\|_1]. \end{aligned} \tag{8}$$

$\mathcal{L}_{GAN}^S$, $\mathcal{L}_{recon}^T$, $\mathcal{L}_{recon}^{c_T}$, $\mathcal{L}_{recon}^{s_S}$, $\mathcal{L}_{cyc}^T$ are defined in a similar way.

Segmentation loss. To enforce the generator to preserve the lesions, we enrich the network with segmentation supervision on the synthesized images. The segmentation loss on the synthesized images is given by

$$\mathcal{L}_{Seg}^{Synth}(Seg, G_T, E_S^c, \mathcal{X}_S, \mathcal{Y}_S, \mathcal{S}_T) = \mathcal{L}_{DSC}(Seg, G_T(E_S^c(\mathcal{X}_S), \mathcal{S}_T), \mathcal{Y}_S). \quad (9)$$

The full objective is given by

$$\begin{aligned} \min_{E_S^c, E_S^s, E_T^c, E_T^s, G_S, G_T} \max_{D_S, D_T} & \lambda_{GAN}(\mathcal{L}_{GAN}^S + \mathcal{L}_{GAN}^T) + \lambda_x(\mathcal{L}_{recon}^S + \mathcal{L}_{recon}^T) \\ & + \lambda_c(\mathcal{L}_{recon}^{c_S} + \mathcal{L}_{recon}^{c_T}) + \lambda_s(\mathcal{L}_{recon}^{s_S} + \mathcal{L}_{recon}^{s_T}) \\ & + \lambda_{cyc}(\mathcal{L}_{cyc}^S + \mathcal{L}_{cyc}^T) + \mathcal{L}_{Seg}^{Synth}, \end{aligned} \quad (10)$$

where λ_{GAN}, λ_x, λ_c, λ_s, λ_{cyc} are weights that control the importance of each term.

2.2 Implementation Details

We implement our model using Pytorch [21]. The content encoders consist of several convolutional layers and residual blocks followed by instance normalization [25]. The style encoders consist of convolutional layers followed by fully connected layers. The decoders include residual blocks followed by upsampling and convolutional layers. The residual blocks are followed by adaptive instance normalization (AdaIN) [9] layers to adjust the style of the output image. The affine parameters of AdaIN are generated by a multilayer perceptron from a given style code. The discriminators consist of several convolutional layers. The encoder of the segmentation network is a standard ResNet [7] consisting of several convolutional layers while the decoder consists of several upsampling and convolutional layers. For training we use Adam optimizer, a batch size of 32 and a learning rate of 0.0001. We make our code available at https://github.com/elchiou/DA.

2.3 Datasets

VERDICT-MRI: We use VERDICT-MRI data collected from 60 men with a suspicion of cancer. VERDICT-MRI images were acquired with pulsed-gradient spin-echo sequence (PGSE) using an optimized imaging protocol for VERDICT prostate characterization with 5 b-values (90, 500, 1500, 2000, 3000 s/mm^2) in 3 orthogonal directions [19]. Images with b = 0 s/mm^2 were also acquired before each b-value acquisition. The DW-MRI sequence was acquired with a voxel size of $1.25 \times 1.25 \times 5$ mm^3, 5 mm slice thickness, 14 slices, and field of view of 220×220 mm^2 and the images were reconstructed to a 176×176 matrix size. A dedicated radiologist, highly experienced in prostate mp-MRI, contoured the lesions on VERDICT-MRI using mp-MRI for guidance.

DW-MRI from mp-MRI Acquisition: We use DW-MRI data from the ProstateX challenge dataset [15] which consists of training mp-MRI data acquired from 204 patients. The DW-MRI data were acquired with a single-shot echo planar imaging sequence with a voxel size of $2 \times 2 \times 3.6$ mm^3, 3.6 mm slice thickness. Three b-values were acquired (50, 400, 800 s/mm^2), and subsequently, the ADC map and a b-value image at b = 1400 s/mm^2 were calculated by the scanner software. In this study, we use DW-MRI data from 80 patients. Since the ProstateX dataset provides only the position of the lesion, a dedicated radiologist manually annotated the lesions on the ADC map using as reference the provided position of the lesion.

Table 1. Average recall, precision, dice similarity coefficient (DSC), and average precision (AP) across 5 folds. The results are given in mean (±std) format.

Model	Recall	Precision	DSC	AP
VERDICT-MRI only	67.1 (± 14.2)	59.6 (± 11.5)	62.4 (± 13.4)	63.5 (± 13.1)
Finetuning	68.4 (± 12.4)	62.5 (± 13.5)	64.7 (± 11.2)	65.8 (± 14.7)
RAs	66.6 (± 11.6)	67.0 (± 8.8)	65.7 (± 10.2)	66.6 (± 12.6)
MUNIT	65.2 (± 10.2)	64.2 (± 13.7)	64.4 (± 11.3)	68.2 (± 12.0)
CycleGAN + $\mathcal{L}_{Seg}^{Synth}$	64.5 (± 10.4)	66.1 (± 10.1)	64.8 (± 8.7)	70.1 (± 9.8)
CycleGAN + $\mathcal{L}_{Seg}^{Synth}$ + RAs	60.9 (± 10.7)	**74.0** (± 11.8)	66.6 (± 13.6)	71.6 (± 11.3)
MUNIT + $\mathcal{L}_{Seg}^{Synth}$ (Ours)	**71.8** (± 7.8)	68.0 (± 6.8)	69.8 (± 7.9)	73.5 (± 8.1)
MUNIT + $\mathcal{L}_{Seg}^{Synth}$ + RAs (Ours)	69.2 (± 8.6)	71.2 (± 9.7)	**69.9** (± 9.0)	**75.4** (± 9.7)

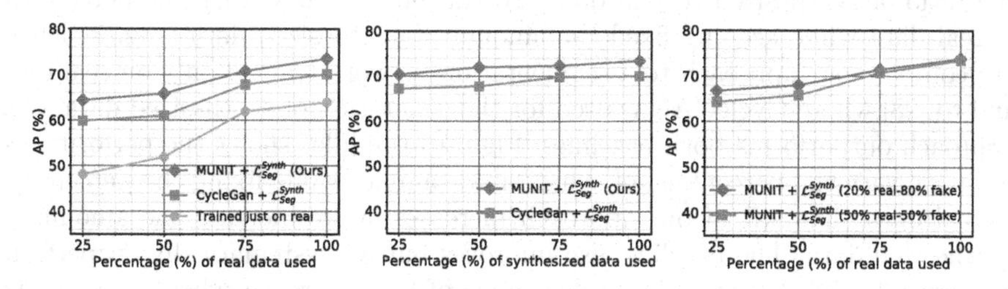

Fig. 3. Impact of the ratio of synthesized to real data on the performance. (Right) Average precision (AP) as a function of the percentage of real samples used given a constant number of synthesized ones. (Middle) AP as a function of the number of synthesized examples used given a constant number of real ones. (Left) AP as a function of the percentage of real data used given a constant number of synthesized ones. Here, the ratio of real to synthesized data in a mini-batch also varies during training.

3 Results

In this section we evaluate the performance of our approach and the impact of the ratio of synthesized to real data on the performance. In the supplementary material we provide qualitative results and quantitative results related to the effect of sampling random style codes on the performance.

Performance Evaluation. We first compare our approach to several baselines. i)VERDICT-MRI only: we train the segmentation network only on VERDICT-MRI. ii) Finetuning: we pre-train on mp-MRI and then perform finetuning using the VERDICT-MRI data. iii) RAs: we pre-train on mp-MRI, then we install RAs in parallel to each of the convolutional layers of the pre-trained network and update them using VERDICT-MRI. iv) MUNIT: we use MUNIT to map from source to target without segmentation supervision. v) CycleGAN + $\mathcal{L}_{Seg}^{Synth}$: we use CycleGAN and segmentation supervision to perform the translation, an approach similar to the one proposed in [28]. vi) CycleGAN + $\mathcal{L}_{Seg}^{Synth}$ + RAs: we use (v) for the translation and introduce RAs to the segmentation network. We evaluate the performance based on the average recall, precision, dice similarity coefficient (DSC), and average precision (AP). We report the results in Table 1. The proposed approach yields substantial improvements and outperforms all baselines including CycleGAN, which indicates that accommodating the uncertainty in the cross-domain mapping allows us to learn better representations for the target domain. Compared to the naive MUNIT without segmentation supervision, $\mathcal{L}_{Seg}^{Synth}$, our approach performs better since it successfully preserves the lesions during the translation. Finally, introducing RAs in the segmentation networks further improves the performance of both CyclgeGAN + $\mathcal{L}_{Seg}^{Synth}$ and MUNIT + $\mathcal{L}_{Seg}^{Synth}$.

Impact of the Ratio of Synthesized to Real Data on the Performance. Using synthesized data is motivated by the fact that annotating large datasets can be challenging in medical applications. We therefore evaluate the impact of the ratio of synthesized to real data. To this end, we first vary the percentage of real data while keeping fixed the amount of synthesized data (Fig. 3 (left)). We compare our approach to a segmentation network trained only on real data and to [28] where CycleGAN is used for the generation of synthesized data. Our approach outperforms both baselines. Figure 3 (middle) shows the performance when we vary the percentage of synthesized samples while fixing the percentage of real ones. The AP of our approach increases as we increase the amount of synthesized data. The baseline also improves but we systematically outperform it. Figure 3 (right) shows the performance of our approach when we vary the percentage of real data while fixing the percentage of synthesized. Here, we also vary the ratio of real to synthesized data in a mini-batch during training. Note that when the percentage of real data is small, a large ratio of synthesized to real data in the mini-batch delivers better results.

4 Conclusion

In this work we propose a domain adaptation approach for lesion segmentation. Our approach exploits the inherent uncertainty in the cross-domain mapping to generate multiple outputs conditioned on a single input allowing the extraction of richer representations for the task of interest in the target domain. We demonstrate the effectiveness of our approach in lesion segmentation on VERDICT-MRI, which is an advanced imaging modality for prostate cancer characterization. However, our approach is quite general can be applied in other application where the amount of labeled training data is limited.

Acknowledgments. This research is funded by EPSRC grand EP/N021967/1. We gratefully acknowledge the support of NVIDIA Corporation with the donation of the GPU used for this research.

References

1. Ahmed, H.U., Bosaily, A., et al.: Diagnostic accuracy of multi-parametric MRI and TRUS biopsy in prostate cancer (PROMIS): a paired validating confirmatory study. Lancet **389**, 815–822 (2017)
2. Cai, J., Zhang, Z., Cui, L., Zheng, Y., Yang, L.: Towards cross-modal organ translation and segmentation: a cycle and shape consistent generative adversarial network. MedIA **52**, 174–184 (2019)
3. Chen, L.C., Zhu, Y., Papandreou, G., Schroff, F., Adam, H.: Encoder-decoder with atrous separable convolution for semantic image segmentation. In: ECCV (2018)
4. Chiou, E., Giganti, F., Bonet-Carne, E., Punwani, S., Kokkinos, I., Panagiotaki, E.: Prostate cancer classification on VERDICT DW-MRI using convolutional neural networks. In: Shi, Y., Suk, H.-I., Liu, M. (eds.) MLMI 2018. LNCS, vol. 11046, pp. 319–327. Springer, Cham (2018). https://doi.org/10.1007/978-3-030-00919-9_37
5. Chiou, E., Giganti, F., Punwani, S., Kokkinos, I., Panagiotaki, E.: Domain adaptation for prostate lesion segmentation on VERDICT-MRI. In: ISMRM (2020)
6. Goodfellow, I., et al.: Generative adversarial nets. In: NIPS (2014)
7. He, K., Zhang, X., Ren, S., Sun, J.: Deep residual learning for image recognition. In: CVPR (2016)
8. Hoffman, J., et al.: CyCADA: cycle-consistent adversarial domain adaptation. In: ICML (2018)
9. Huang, X., Belongie, S.: Arbitrary style transfer in real-time with adaptive instance normalization. In: ICCV (2017)
10. Huang, X., Liu, M.Y., Belongie, S., Kautz, J.: Multimodal unsupervised image-to-image translation. In: ECCV (2018)
11. Jiang, J., et al.: Tumor-aware, adversarial domain adaptation from CT to MRI for lung cancer segmentation. In: Frangi, A.F., Schnabel, J.A., Davatzikos, C., Alberola-López, C., Fichtinger, G. (eds.) MICCAI 2018. LNCS, vol. 11071, pp. 777–785. Springer, Cham (2018). https://doi.org/10.1007/978-3-030-00934-2_86
12. Johnston, E.W., Bonet-Carne, E., et al.: VERDICT-MRI for prostate cancer: intracellular volume fraction versus apparent diffusion coefficient. Radiology **291**, 391–397 (2019)
13. Kamnitsas, K., et al.: Unsupervised domain adaptation in brain lesion segmentation with adversarial networks. In: Niethammer, M., et al. (eds.) IPMI 2017. LNCS, vol. 10265, pp. 597–609. Springer, Cham (2017). https://doi.org/10.1007/978-3-319-59050-9_47
14. Lee, H.Y., Tseng, H.Y., Huang, J.B., Singh, M., Yang, M.H.: Diverse image-to-image translation via disentangled representations. In: ECCV (2018)
15. Litjens, G., Debats, O., Barentsz, J., Karssemeijer, N., Huisman, H.: Computer-aided detection of prostate cancer in MRI. TMI **33**, 1083–1092 (2014)
16. Liu, M.Y., Breuel, T., Kautz, J.: Unsupervised image-to-image translation networks. In: NIPS (2017)
17. Ouyang, C., Kamnitsas, K., Biffi, C., Duan, J., Rueckert, D.: Data efficient unsupervised domain adaptation for cross-modality image segmentation. In: Shen, D., et al. (eds.) MICCAI 2019. LNCS, vol. 11765, pp. 669–677. Springer, Cham (2019). https://doi.org/10.1007/978-3-030-32245-8_74

18. Panagiotaki, E., Chan, R.W., Dikaios, N., et al.: Microstructural characterization of normal and malignant human prostate tissue with vascular, extracellular, and restricted diffusion for cytometry in tumours magnetic resonance imaging. Investigate Radiol. **50**, 218–227 (2015)
19. Panagiotaki, E., Ianus, A., Johnston, E., et al.: Optimised VERDICT MRI protocol for prostate cancer characterisation. In: ISMRM (2015)
20. Panagiotaki, E., Walker-Samuel, S., Siow, B., et al.: Noninvasive quantification of solid tumor microstructure using VERDICT MRI. Cancer Res. **74**, 1902–1912 (2014)
21. Paszke, A., Gross, S., Chintala, S., Chanan, G., et al.: Automatic differentiation in pytorch. In: Autodiff Workshop, NIPS (2017)
22. Rebuffi, S., Vedaldi, A., Bilen, H.: Efficient parametrization of multi-domain deep neural networks. In: CVPR (2018)
23. Ren, J., Hacihaliloglu, I., Singer, E.A., Foran, D.J., Qi, X.: Adversarial domain adaptation for classification of prostate histopathology whole-slide images. In: Frangi, A.F., Schnabel, J.A., Davatzikos, C., Alberola-López, C., Fichtinger, G. (eds.) MICCAI 2018. LNCS, vol. 11071, pp. 201–209. Springer, Cham (2018). https://doi.org/10.1007/978-3-030-00934-2_23
24. Ronneberger, O., Fischer, P., Brox, T.: U-Net: convolutional networks for biomedical image segmentation. In: Navab, N., Hornegger, J., Wells, W.M., Frangi, A.F. (eds.) MICCAI 2015. LNCS, vol. 9351, pp. 234–241. Springer, Cham (2015). https://doi.org/10.1007/978-3-319-24574-4_28
25. Ulyanov, D., Vedaldi, A., Lempitsky, V.S.: Improved texture networks: Maximizing quality and diversity in feed-forward stylization and texture synthesis. In: CVPR (2017)
26. Yang, J., Dvornek, N.C., Zhang, F., Chapiro, J., Lin, M.D., Duncan, J.S.: Unsupervised domain adaptation via disentangled representations: application to cross-modality liver segmentation. In: Shen, D., et al. (eds.) MICCAI 2019. LNCS, vol. 11765, pp. 255–263. Springer, Cham (2019). https://doi.org/10.1007/978-3-030-32245-8_29
27. Zhang, Y., Miao, S., Mansi, T., Liao, R.: Task driven generative modeling for unsupervised domain adaptation: application to X-ray image segmentation. In: Frangi, A.F., Schnabel, J.A., Davatzikos, C., Alberola-López, C., Fichtinger, G. (eds.) MICCAI 2018. LNCS, vol. 11071, pp. 599–607. Springer, Cham (2018). https://doi.org/10.1007/978-3-030-00934-2_67
28. Zhang, Z., Yang, L., Zheng, Y.: Translating and segmenting multimodal medical volumes with cycle and shape consistency generative adversarial network. In: CVPR (2018)
29. Zhu, J., Park, T., Isola, P., Efros, A.A.: Unpaired image-to-image translation using cycle-consistent adversarial networks. In: ICCV (2017)
30. Zhu, J.Y., et al.: Toward multimodal image-to-image translation. In: NIPS (2017)

Deep Semi-supervised Knowledge Distillation for Overlapping Cervical Cell Instance Segmentation

Yanning Zhou[1(✉)], Hao Chen[1], Huangjing Lin[1], and Pheng-Ann Heng[1,2]

[1] Department of Computer Science and Engineering,
The Chinese University of Hong Kong, Hong Kong SAR, China
{ynzhou,hchen,hjlin,pheng}cse.cuhk.edu.hk

[2] Guangdong Provincial Key Laboratory of Computer Vision and Virtual Reality Technology, Shenzhen Institutes of Advanced Technology,
Chinese Academy of Sciences, Shenzhen, China

Abstract. Deep learning methods show promising results for overlapping cervical cell instance segmentation. However, in order to train a model with good generalization ability, voluminous pixel-level annotations are demanded which is quite expensive and time-consuming for acquisition. In this paper, we propose to leverage both labeled and unlabeled data for instance segmentation with improved accuracy by knowledge distillation. We propose a novel Mask-guided Mean Teacher framework with Perturbation-sensitive Sample Mining (MMT-PSM), which consists of a teacher and a student network during training. Two networks are encouraged to be consistent both in feature and semantic level under small perturbations. The teacher's self-ensemble predictions from K-time augmented samples are used to construct the reliable pseudo-labels for optimizing the student. We design a novel strategy to estimate the sensitivity to perturbations for each proposal and select informative samples from massive cases to facilitate fast and effective semantic distillation. In addition, to eliminate the unavoidable noise from the background region, we propose to use the predicted segmentation mask as guidance to enforce the feature distillation in the foreground region. Experiments show that the proposed method improves the performance significantly compared with the supervised method learned from labeled data only, and outperforms state-of-the-art semi-supervised methods. Code: https://github.com/SIAAAAAA/MMT-PSM.

1 Introduction

Pap smear test is the recommended procedure for earlier cervical cancer screening worldwide [18]. By estimating the cell type and the cytological features, e.g., nuclei size, nuclear cytoplasmic ratio and multi-nucleation, it provides clear

Electronic supplementary material The online version of this chapter (https://doi.org/10.1007/978-3-030-59710-8_51) contains supplementary material, which is available to authorized users.

A. L. Martel et al. (Eds.): MICCAI 2020, LNCS 12261, pp. 521–531, 2020.
https://doi.org/10.1007/978-3-030-59710-8_51

guidance for clinical management and further treatment [22]. Automatic cervical cell segmentation can free doctors from time-consuming work and reduce the intra-/inter-observer variability [10,15,23,32]. Specifically, Deep Learning (DL) methods show promising results for cell nuclei segmentation [1,19,32]. However, optimizing the DL methods heavily relies on numerous data with expensively dense annotations by experts, which limits the model to acquire higher accuracy and better generalization ability. Since unlabeled data is easily accessible, how to leverage both limited labeled and large amounts of unlabeled data raises researchers' attention to improve the performance further for medical image analysis [29].

Several works have been done in medical image community for Semi-Supervised Learning (SSL) on classification and segmentation [2,3,8,16,17,24,25,30]. Bai *et al.* [2] proposed a self-training strategy by alternatively assigning labels to unlabeled data and optimizing the model parameters. Nie *et al.* [17] introduced an adversarial learning training strategy by selecting informative regions in unlabeled data to train the segmentation network. Shi *et al.* [21] created more reliable ensemble targets for feature and label predictions via the graph to encourage features mapped in the same cluster being more compact. Knowledge distillation [12], which was first used in model compression by encouraging the small model to mimic the behavior of a deeper model, has demonstrated excellent improvements mostly for classification setups [6,20,26] and shown the potential benefit for semi-supervised learning [26] and domain adaptation [9]. Chen *et al.* [7] extended it to the detection scenario with proposal-based method, and presented to learn a compact detector by distilling from both features and predictions. However, directly using entire feature maps will inevitably introduce the noise from the background. To eliminate the noise in background, Wang *et al.* [28] conducted feature distillation within the region close to objects based on prior knowledge. Other approaches [5,13] added consistent regularization either in region-based or relation-based. Although achieving promising progress, they do not consider the informative degree for each sample, which is one of the bottlenecks for further improving the performance. In medical imaging, researchers attempted to apply knowledge distillation to segmentation problems. Wang *et al.* [27] employed the teacher student network in 3D optical microscope images via knowledge distillation. Another approach [30] introduced uncertainty estimation into knowledge distillation for 3D left atrium segmentation. Instance segmentation, however, is a more challenging task that requires an additional detection step to distinguish the individual instances [11]. The potential of the knowledge distillation has not been well explored on it.

In this paper, we propose a novel deep semi-supervised knowledge distillation framework called Mask-guided Mean Teacher with Perturbation-sensitive Sample Mining (MMT-PSM) for overlapping cervical cell instance segmentation, which conducts both semantic and feature distillation. The proposed end-to-end trainable framework consists of a teacher model and a student model under the same backbone. Given a sample with different small perturbations, the proposed method encourages the predictions from two networks being consistent. The mean prediction of the K-time augmented samples from the teacher network are considered as the pseudo-label to supervise the student network. A perturbation-sensitive sample mining strategy is used to resolve the meaningless guidance from easy cases in unbalanced and massive data. Furthermore, we propose the mask-guided feature distillation which encourages the feature consistency only for the foreground region to alleviate the side effect in the noisy background. We perform comprehensive evaluation on cervical cell segmentation task. Results indicate that the proposed algorithm significantly improves the instance segmentation accuracy, consistently across different numbers of labeled data, and also outperforms other state-of-the-art semi-supervised methods (Fig. 1).

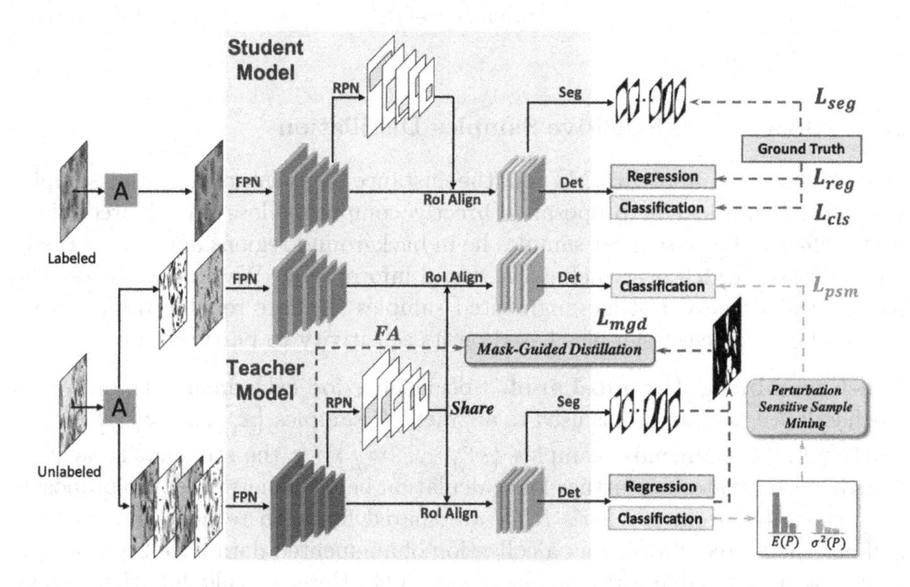

Fig. 1. Overview of the proposed framework. Annotated data is passed through the student to calculate standard supervised losses. Meanwhile, K-time data augmentation (A) is applied to the unlabeled data, which are fed through the teacher to generate soft pseudo-labels to optimize the student. The perturbation sensitivity is estimated on the K teacher's predictions to select samples for optimizing the student. The predicted masks are used as a guidance for feature distillation (i.e., Mask-Guided Distillation). FA denotes the Feature Adaptation layer.

2 Method

2.1 Mean Teacher Framework for Instance Segmentation

Formally, let $\mathcal{D}_L = \{(x_i, y_i)\}_{i=1}^{N}$ denote the labeled set and $\mathcal{D}_U = \{(x_i)\}_{i=N+1}^{N+M}$ denote the unlabeled set. The goal of semi-supervised learning is to improve the performance by leveraging the hidden information in $\mathcal{D}_U$. In this work, we adopt Mask R-CNN [11] as the instance segmentation model for both the student and the teacher, which consists of four modules: 1) A shared Feature Pyramid Network (FPN) extracts features as inputs for the other modules, 2) a Region Proposal Network (RPN) equipped with RoI Align layer to generate the object proposals, 3) a detection branch (Det) and 4) a segmentation branch (Seg) which take features and proposals as inputs, and then predict the detection scores, the spacial revision vectors and the segmentation results, respectively. We use Mean Teacher algorithm (MT) [26] as our basic framework, which consists of a teacher and a student model sharing the same architecture and encourages the predictions being consistent under small perturbations. Instead of optimizing the teacher by SGD, exponential moving average (EMA) weight in the student is used to form a better teacher model [26]: $\theta'_t = \alpha\theta'_{t-1} + (1-\alpha)\theta_{t-1}$, where θ'_{t-1} and θ_{t-1} are the teacher's and student's weights in $(t-1)$ step, and α controls the updating speed.

2.2 Perturbation-Sensitive Samples Distillation

One difficulty in applying MT on the instance segmentation is the sample-imbalanced problem in proposals. Directly computing loss on all predictions is not effective because most samples lie in background regions and can be easily distinguished, which overwhelms the useful information. We propose to use the mean predictions of K-time augmented samples as more reliable targets from the teacher and select samples based on its sensitivity to perturbations.

Self-Ensembling Pseudo-Label. Specifically, for each image x_i in $\mathcal{D}_U$, a stochastic Augmentor (A) is used to augment K samples $\{x^{\mathcal{T}}_{i,1}, \ldots, x^{\mathcal{T}}_{i,K}\}$ for the teacher and L augmented samples $\{x^{\mathcal{S}}_{i,1}, \ldots, x^{\mathcal{S}}_{i,L}\}$ for the student. To acquire the same candidates for further loss calculation between networks, the proposals $\mathcal{R}_{x^{\mathcal{T}}_i}$ generated from teacher's RPN are shared for both teacher and student. Self-ensemble predictions from a collection of augmented data have been considered as a more reliable target in classification [4]. Here, we calculate the average predictions across K augmented samples to generate the soft pseudo-label in teacher network:

$$\overline{P}_i = \frac{1}{K}\sum_{k=1}^{K} f_{cls}\left(x^{\mathcal{T}}_{i,k}, \mathcal{R}_{x^{\mathcal{T}}_i}; \theta'\right). \tag{1}$$

$f_{cls}(\cdot;\theta')$ denotes the classification sub-branch in teacher Det. A sharpen function $S(\overline{P}_i) = \overline{P}^t_i / \sum_{j=1}^{c} \overline{P}^t_j$ is further used to implicitly achieve entropy minimization [4], in which c denotes the number of categories. We set $t = 0.5$ in our study. See the supplementary material for augmentation details and ablation study of the temperature.

Perturbation-Sensitive Sample Mining. We hypothesize that perturbation-sensitive samples, which have larger prediction accuracy gaps between teacher and student, are more informative and beneficial for training. Firstly, the class with the maximum categorical probability in the self-ensembling prediction is assigned as its hard pseudo-label. Then we calculate the variance among K augmented samples as its degree of perturbation sensitivity:

$$Var(x_i^{\mathcal{T}}) = \frac{1}{K}\sum_{k=1}^{K}\left(f_{cls}\left(\hat{x}_{i,k}^{\mathcal{T}}, \mathcal{R}_{x_i^{\mathcal{T}}}; \theta'\right) - \overline{P}_i\right)^2. \quad (2)$$

All samples whose hard pseudo-labels are foreground classes remain. Meanwhile, background samples are sorted by descending according to the variances and kept the Top-s, where s is the number of foreground samples. The perturbation-sensitive sample mining loss $\mathcal{L}_{psm}$ is calculated on the selected samples as follow:

$$\mathcal{L}_{psm} = \frac{1}{ML}\sum_{i=N+1}^{N+M}\sum_{l=i}^{L} w\mathcal{L}_{ce}\left(f_{cls}\left(x_{i,l}^{\mathcal{S}}, \mathcal{R}_{x_i^{\mathcal{T}}}; \theta\right), S(\overline{P}_i)\right). \quad (3)$$

$f_{cls}(\cdot;\theta)$ denotes the classification sub-branch in student Det, $\mathcal{R}_{x_i^{\mathcal{T}}}$ and $\overline{P}_i$ denote the proposals and soft pseudo-labels for remained perturbation-sensitive samples, and $\mathcal{L}_{ce}$ is the cross-entropy loss. w is a class-balanced weight and is set empirically as 1.5 for the background and 1 for others.

2.3 Mask-Guided Feature Distillation

Study shows [20] that intermediate representations from the teacher can also improve the training process and final performance of the student in the classification task. However, directly minimize the difference in entire feature maps could harm the performance since it would introduce the noise in the background region. Therefore, we design to force the student only mimicking the teacher under the guidance of semantic segmentation results.

Firstly, an adaptation layer is added after each output stage of FPN, which is proved to be advantageous for feature distillation [20]. Here we use a 1×1 convolution as the adaptation layer and reduce the input feature dimension by half. Then the instance masks and bounding box's locations from the teacher are used to generate binary semantic masks. Let $Z_{tijc}^{\mathcal{S}}$ and $Z_{tijc}^{\mathcal{T}}$ denotes the student's and the teacher's feature value in the c-th channel at location (i,j) from the adaptation layer after FPN's t-th stage. We aim to encourage the consistency by minimizing feature distance through the mask-guided distillation loss:

$$\mathcal{L}_{mgd} = \frac{1}{CT}\sum_{t=1}^{T}\frac{\sum_{i=1}^{W}\sum_{j=1}^{H}\sum_{c=1}^{C} M_{tij} \circ \left\|Z_{tijc}^{\mathcal{T}} - Z_{tijc}^{\mathcal{S}}\right\|_2^2}{\sum_{i=1}^{W}\sum_{j=1}^{H} M_{tij}}. \quad (4)$$

Here M_{tij} denotes the corresponding semantic mask.

Total Loss for Optimization. The total loss can be defined as:

$$\mathcal{L}_{total} = \mathcal{L}_{sup} + \lambda(t)(\mathcal{L}_{psm} + \gamma\mathcal{L}_{mgd}), \quad (5)$$

where $\mathcal{L}_{sup} = \mathcal{L}_{cls} + \mathcal{L}_{reg} + \mathcal{L}_{seg}$ [11], γ is a balanced weight which we set to 5. $\lambda(t)$ is a piecewise weight function that guarantees the loss dominated by $\mathcal{L}_{sup}$ at beginning, gradually increases during training, and declines slowly at last.

2.4 Implementation Details

We use IR-Net [32] as our base model, which utilizes instance relations on Mask R-CNN [11]. The Augmentor (A) consists of both color and location transformations. Specifically, each sample is first randomly adjusted brightness, contrast and Hue, and then conducted random erasing [31]. After that, half of them are flipped. For the first 1000 iterations, only $\mathcal{D}_L$ is used. The teacher model is initiated by copying the parameters in the student at the 990th iter, which prevents the framework from degenerating by a poor teacher. We set $\alpha = \min(1 - /(t - 990), 0.99)$ to let the teacher have a larger update rate at the beginning when the student improves quickly. During training, each mini-batch includes both labeled and unlabeled images with a ratio of $1:1$. The sigmoid-shaped function $\lambda(t) = e^{-125(1-t/1250)^2}$ is used for $t \in [1000, 1250]$ and $\lambda(t) = e^{-12(1-(T-t)/250)^2}$ for $t \in [T - 250, T]$, where T is the total iterations. Pytorch is adopted to implement our framework. The learning rate is initiated to 1e-2 and decayed to 1e-3 and 1e-4 after 5000 and 7000 iterations. We adopt SGD algorithm to optimize the network, and one Titan XP GPU is used for training. The pseudo code of the proposed MMT-PSM can be found in the supplementary material.

3 Experiments and Results

Dataset and Evaluation Metrics. The liquid-based Pap test specimen was collected from 82 patients and imaged in $\times 40$ resolution with ~ 0.2529 μm per pixel. This is used as labeled dataset $\mathcal{D}_L$ with totally 4439 cytoplasm and 4789 nuclei annotations. Then the dataset is divided in patient-level with the ratio of $7:1:2$ for train, valid and test set. An overlapping ratio of 0.75 is used to crop images into 1000×1000 for the training set, while the valid, as well as the test set, are non-overlapping cropped. In sum, the number of images for train, valid and test is 961, 50 and 98, respectively. Apart from that, 4371 images from other patients with a resolution of 1000×1000 are randomly cropped from whole slide images as the unlabeled dataset $\mathcal{D}_U$.

We use Average Jaccard Index (AJI) [14] and mean Average Precision (mAP) [11] for quantitative evaluation. Results are calculated on cytoplasm (Cyto.), nuclei (Nuc.) and the average (Avg.). AJI is commonly used in cell nuclei segmentation task, which measures the ratio of the aggregated intersection and aggregated union for all the predictions and ground truths in the image. mAP is the mean of the average precision under different IOU thresholds, which is widely used in the general detection and instance segmentation tasks.

Evaluation on Different Dataset Settings. Firstly, we evaluate the impact of leveraging unlabeled images by our proposed model under different amounts of labeled samples. Our proposed method (MMT-PSM) is compared with the state-of-the-art fully supervised method, named IR-Net [32], which utilizes the instance relation for mask refinement and duplication removal based on the Mask R-CNN [11] structure. We evaluate the performance of the proposed MMT-PSM

Table 1. Evaluation of MMT-PSM with different numbers of labeled data.

% labeled	Method	AJI[%]			mAP[%]		
		Cyto.	Nuc.	Avg.	Cyto.	Nuc.	Avg.
10%	IR-Net	66.81	54.07	60.44	37.36	30.14	33.75
	MMT-PSM	**70.00**	**56.90**	**63.45**	**39.12**	**32.09**	**35.61**
20%	IR-Net	70.15	54.42	62.29	40.28	**31.75**	36.02
	MMT-PSM	**71.56**	**56.49**	**64.03**	**41.88**	31.34	**36.61**
40%	IR-Net	70.70	**57.35**	64.03	40.00	31.65	35.83
	MMT-PSM	**71.75**	57.23	**64.49**	**42.30**	**32.22**	**37.26**
80%	IR-Net	72.44	57.72	65.08	41.35	33.07	37.21
	MMT-PSM	**74.26**	**59.58**	**66.92**	**45.49**	**34.76**	**40.13**
100%	IR-Net	73.38	58.38	65.88	43.44	32.64	38.04
	MMT-PSM	**73.45**	**60.43**	**66.94**	**46.01**	**35.02**	**40.52**

with a varying number of labeled data from 96 to 961 and 4371 unlabeled data. The IR-Net is trained with the same labeled data only. As shown in Table 1, results from the proposed MMT-PSM achieve relatively consistent improvements on both metrics. It improves average AJI by 2.98%, 1.74%, 0.46%, 1.84%, and 1.06%, and also improves average mAP by 1.86%, 0.59%, 1.44%, 2.92%, and 2.48% for mAP compared with those only trained on the same number of labeled data, which demonstrates the effectiveness of the proposed SSL method.

Table 2. Quantitative comparisons with state-of-the-arts on the test set.

Method	AJI[%]			mAP[%]		
	Cyto.	Nuc.	Avg.	Cyto.	Nuc.	Avg.
IR-Net [32]	73.38	58.38	65.88	43.44	32.64	38.04
ODKD [7]	72.64	56.34	64.49	44.23	**35.50**	39.87
FFI [28]	**73.89**	59.30	66.60	44.60	34.72	39.66
MMT-PSM	73.45	**60.43**	**66.94**	**46.01**	35.02	**40.52**

Comparison with Other Semi-supervised Methods. We implement and adapt several state-of-the-arts methods for comparison: 1). Chen *et al.* [7] improved object detection by knowledge distillation (ODKD) with weighted cross-entropy loss for the imbalanced data problem. Meanwhile, feature imitation is conducted in all regions. 2). Wang *et al.* [28] proposed fine-grained feature imitation (FFI) which firstly estimated the object anchor locations and then let the student's features be closed to teacher's on the selected regions. Note that we used the same network backbone [32] on these methods with 961 labeled data and 4371 unlabeled data for fair comparison. As can be seen in

Table 3. Quantitative analysis of different components on the test set.

Method	AJI[%]			mAP[%]		
	Cyto.	Nuc.	Avg.	Cyto.	Nuc.	Avg.
IR-Net [32]	73.38	58.38	65.88	43.44	32.64	38.04
MMT-PSM (w/o $\mathcal{L}_{mgd}$)	**75.01**	59.23	**67.12**	45.58	33.16	39.37
MMT-PSM (w/o $\mathcal{L}_{psm}$)	74.38	59.46	66.92	44.39	33.75	39.07
MMT-PSM	73.45	**60.43**	66.94	**46.01**	**35.02**	**40.52**

Table 2, all the SSL methods outperforms the supervised method on most of the evaluation indicators. Compared with fully supervised methods, results from ODKD improves 1.85% mAP but decreases AJI. The reason is it penalizes the classification and feature discrepancy in all regions. Therefore it is inevitable to introduce the noise. FFI selects the proposals closed to objects for feature distillation, hence achieves better results. Furthermore, the proposed MMT-PSM achieves the best performance over the state-of-the-art SSL methods, illustrating that our method has the keen ability to distillate the information both in feature space and semantic predictions.

Ablation Study of the Proposed Method. We also conduct the ablation study for the impact of proposed components: 1). MMT-PSM (w/o $\mathcal{L}_{mgd}$) denotes the proposed method without the mask-guided feature distillation, and 2). MMT-PSM (w/o $\mathcal{L}_{psm}$) denotes the proposed method without the perturbation-sensitive sample mining for knowledge distillation. Results are shown in Table 3. Utilizing perturbation-sensitive samples measured in the teacher network as the pseudo-labels for optimizing the student improves 1.24% for average AJI and 1.33% for mAP. Meanwhile, forcing features from the teacher and the student being consistent in the foreground region also increases 1.04% average AJI and 1.03% mAP. Lastly, combining two components in our mean teacher framework achieves the competitive performance by 66.94% AJI and 40.52% mAP.

Qualitative Evaluation. We also visualize different methods' results from challenging cases including the heavily occlusion of cytoplasm and blurred regions. As can be seen in Fig. 2, each closed curve denotes an individual instance. Compared with other methods, our proposed MMT-PSM has the better ability to recognize the translucent cervical cells in low contrast areas.

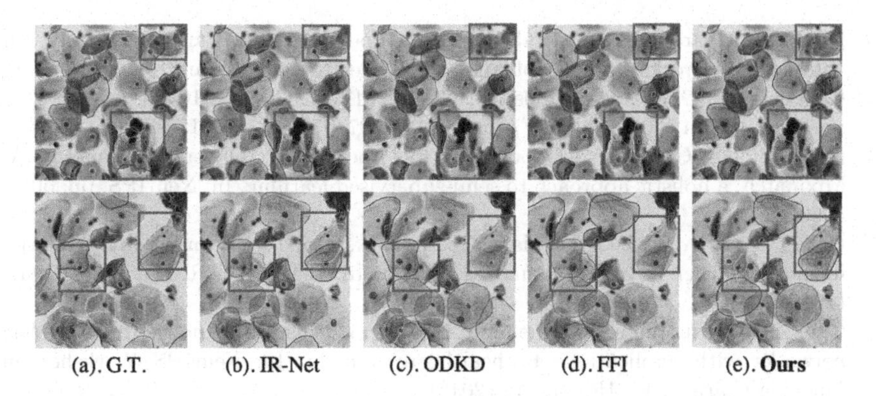

Fig. 2. Qualitative comparisons of semi-supervised cervical cell instance segmentation on the test set. Red rectangles highlight the main difference among different methods. (Color figure online)

4 Conclusion

In this paper, we propose a novel mask-guided mean teacher framework with perturbation-sensitive sample mining which conducts knowledge distillation for semi-supervised cervical cell instance segmentation. The proposed method encourages the network to output consistent feature maps and predictions under small perturbations. Only samples with high grade of perturbation sensitivity are selected for semantic distillation, which prevents the meaningless guidance from easy background cases. In addition, the segmentation mask is used as guidance for better feature distillation. Experiments demonstrate our proposed method effectively leverage the unlabeled data and outperforms other SSL methods. Our proposed MMT-PSM framework is general and can be easily adapted to other semi-supervised medical image instance segmentation tasks.

Acknowledgments. The work described in the paper was supported in parts by the following grants from Key-Area Research and Development Program of Guangdong Province, China (2020B010165004), Hong Kong Innovation and Technology Fund (Project No. ITS/041/16), National Natural Science Foundation of China (Project No. U1813204) and Shenzhen Science and Technology Program (JCYJ20170413162256793).

References

1. Al-Kofahi, Y., Zaltsman, A., Graves, R., Marshall, W., Rusu, M.: A deep learning-based algorithm for 2-d cell segmentation in microscopy images. BMC Bioinform. **19**(1), 1–11 (2018)
2. Bai, W., et al.: Semi-supervised learning for network-based cardiac MR image segmentation. In: Descoteaux, M., Maier-Hein, L., Franz, A., Jannin, P., Collins, D.L., Duchesne, S. (eds.) MICCAI 2017. LNCS, vol. 10434, pp. 253–260. Springer, Cham (2017). https://doi.org/10.1007/978-3-319-66185-8_29

3. Baur, C., Albarqouni, S., Navab, N.: Semi-supervised deep learning for fully convolutional networks. In: Descoteaux, M., Maier-Hein, L., Franz, A., Jannin, P., Collins, D.L., Duchesne, S. (eds.) MICCAI 2017. LNCS, vol. 10435, pp. 311–319. Springer, Cham (2017). https://doi.org/10.1007/978-3-319-66179-7_36
4. Berthelot, D., Carlini, N., Goodfellow, I., Papernot, N., Oliver, A., Raffel, C.A.: Mixmatch: a holistic approach to semi-supervised learning. In: NeurIPS, pp. 5050–5060 (2019)
5. Cai, Q., Pan, Y., Ngo, C.W., Tian, X., Duan, L., Yao, T.: Exploring object relation in mean teacher for cross-domain detection. In: IEEE CVPR, pp. 11457–11466 (2019)
6. Che, Z., Purushotham, S., Khemani, R., Liu, Y.: Distilling knowledge from deep networks with applications to healthcare domain. In: NeurIPS Workshop on Machine Learning for Healthcare (2015)
7. Chen, G., Choi, W., Yu, X., Han, T., Chandraker, M.: Learning efficient object detection models with knowledge distillation. In: NeurIPS, pp. 742–751 (2017)
8. Cui, W., et al.: Semi-supervised brain lesion segmentation with an adapted mean teacher model. In: Chung, A.C.S., Gee, J.C., Yushkevich, P.A., Bao, S. (eds.) IPMI 2019. LNCS, vol. 11492, pp. 554–565. Springer, Cham (2019). https://doi.org/10.1007/978-3-030-20351-1_43
9. Ge, Y., Chen, D., Li, H.: Mutual mean-teaching: Pseudo label refinery for unsupervised domain adaptation on person re-identification. In: International Conference on Learning Representations (2020)
10. GençTav, A., Aksoy, S., ÖNder, S.: Unsupervised segmentation and classification of cervical cell images. Pattern Recogn. **45**(12), 4151–4168 (2012)
11. He, K., Gkioxari, G., Dollár, P., Girshick, R.: Mask r-cnn. In: IEEE CVPR, pp. 2961–2969 (2017)
12. Hinton, G., Vinyals, O., Dean, J.: Distilling the knowledge in a neural network. arXiv preprint arXiv:1503.02531 (2015)
13. Jeong, J., Lee, S., Kim, J., Kwak, N.: Consistency-based semi-supervised learning for object detection. In: NeurIPS. pp. 10758–10767 (2019)
14. Kumar, N., Verma, R., Sharma, S., Bhargava, S., Vahadane, A., Sethi, A.: A dataset and a technique for generalized nuclear segmentation for computational pathology. IEEE Trans. Med. Imaging **36**(7), 1550–1560 (2017)
15. Lu, Z., Carneiro, G., Bradley, A.P.: An improved joint optimization of multiple level set functions for the segmentation of overlapping cervical cells. IEEE Trans. Image Proc. **24**(4), 1261–1272 (2015)
16. Meier, R., Bauer, S., Slotboom, J., Wiest, R., Reyes, M.: Patient-specific semi-supervised learning for postoperative brain tumor segmentation. In: Golland, P., Hata, N., Barillot, C., Hornegger, J., Howe, R. (eds.) MICCAI 2014. LNCS, vol. 8673, pp. 714–721. Springer, Cham (2014). https://doi.org/10.1007/978-3-319-10404-1_89
17. Nie, D., Gao, Y., Wang, L., Shen, D.: ASDNet: attention based semi-supervised deep networks for medical image segmentation. In: Frangi, A.F., Schnabel, J.A., Davatzikos, C., Alberola-López, C., Fichtinger, G. (eds.) MICCAI 2018. LNCS, vol. 11073, pp. 370–378. Springer, Cham (2018). https://doi.org/10.1007/978-3-030-00937-3_43
18. Papanicolaou, G.N.: A new procedure for staining vaginal smears. Science **95**(2469), 438–439 (1942)
19. Raza, S.E.A., et al.: Micro-Net: a unified model for segmentation of various objects in microscopy images. Med. Image Anal. **52**, 160–173 (2019)

20. Romero, A., Ballas, N., Kahou, S.E., Chassang, A., Gatta, C., Bengio, Y.: Fitnets: hints for thin deep nets. In: ICLR (2015)
21. Shi, X., Su, H., Xing, F., Liang, Y., Qu, G., Yang, L.: Graph temporal ensembling based semi-supervised convolutional neural network with noisy labels for histopathology image analysis. Medical Image Anal. **60**, 101624 (2020)
22. Solomon, D., et al.: The 2001 bethesda system: terminology for reporting results of cervical cytology. JAMA **287**(16), 2114–2119 (2002)
23. Song, Y., et al.: Accurate cervical cell segmentation from overlapping clumps in pap smear images. IEEE Trans. Med. Imaging **36**(1), 288–300 (2017)
24. Su, H., Shi, X., Cai, J., Yang, L.: Local and global consistency regularized mean teacher for semi-supervised nuclei classification. In: Shen, D., et al. (eds.) MICCAI 2019. LNCS, vol. 11764, pp. 559–567. Springer, Cham (2019). https://doi.org/10.1007/978-3-030-32239-7_62
25. Su, H., Yin, Z., Huh, S., Kanade, T., Zhu, J.: Interactive cell segmentation based on active and semi-supervised learning. IEEE Trans. Med. Imaging **35**(3), 762–777 (2015)
26. Tarvainen, A., Valpola, H.: Mean teachers are better role models: Weight-averaged consistency targets improve semi-supervised deep learning results. In: NeurIPS, pp. 1195–1204 (2017)
27. Wang, H., et al.: Segmenting neuronal structure in 3d optical microscope images via knowledge distillation with teacher-student network. In: ISBI, pp. 228–231 (2019)
28. Wang, T., Yuan, L., Zhang, X., Feng, J.: Distilling object detectors with fine-grained feature imitation. In: IEEE CVPR, pp. 4933–4942 (2019)
29. Wu, F., et al.: Towards a new generation of artificial intelligence in China. Nat. Mach. Intell. **2**, 312–316 (2020)
30. Yu, L., Wang, S., Li, X., Fu, C.-W., Heng, P.-A.: Uncertainty-aware self-ensembling model for semi-supervised 3D left atrium segmentation. In: Shen, D., et al. (eds.) MICCAI 2019. LNCS, vol. 11765, pp. 605–613. Springer, Cham (2019). https://doi.org/10.1007/978-3-030-32245-8_67
31. Zhong, Z., Zheng, L., Kang, G., Li, S., Yang, Y.: Random erasing data augmentation. In: AAAI, pp. 13001–13008 (2020)
32. Zhou, Y., Chen, H., Xu, J., Dou, Q., Heng, P.-A.: IRNet: instance relation network for overlapping cervical cell segmentation. In: Shen, D., et al. (eds.) MICCAI 2019. LNCS, vol. 11764, pp. 640–648. Springer, Cham (2019). https://doi.org/10.1007/978-3-030-32239-7_71

DMNet: Difference Minimization Network for Semi-supervised Segmentation in Medical Images

Kang Fang and Wu-Jun Li(✉)

National Key Laboratory for Novel Software Technology,
Department of Computer Science and Technology, Nanjing University,
National Institute of Healthcare Data Science at Nanjing University, Nanjing, China
fangk@lamda.nju.edu.cn, liwujun@nju.edu.cn

Abstract. Semantic segmentation is an important task in medical image analysis. In general, training models with high performance needs a large amount of labeled data. However, collecting labeled data is typically difficult, especially for medical images. Several semi-supervised methods have been proposed to use unlabeled data to facilitate learning. Most of these methods use a self-training framework, in which the model cannot be well trained if the pseudo masks predicted by the model itself are of low quality. Co-training is another widely used semi-supervised method in medical image segmentation. It uses two models and makes them learn from each other. All these methods are not end-to-end. In this paper, we propose a novel end-to-end approach, called difference minimization network (DMNet), for semi-supervised semantic segmentation. To use unlabeled data, DMNet adopts two decoder branches and minimizes the difference between soft masks generated by the two decoders. In this manner, each decoder can learn under the supervision of the other decoder, thus they can be improved at the same time. Also, to make the model generalize better, we force the model to generate low-entropy masks on unlabeled data so the decision boundary of model lies in low-density regions. Meanwhile, adversarial training strategy is adopted to learn a discriminator which can encourage the model to generate more accurate masks. Experiments on a kidney tumor dataset and a brain tumor dataset show that our method can outperform the baselines, including both supervised and semi-supervised ones, to achieve the best performance.

Keywords: Semantic segmentation · Semi-supervised learning

1 Introduction

Semantic segmentation is of great importance in medical image analysis, because it can help detect the location and size of anatomical structures and aid in making therapeutic schedule. With the development of deep learning, deep neural

This work is supported by the NSFC-NRF Joint Research Project (No. 61861146001).

A. L. Martel et al. (Eds.): MICCAI 2020, LNCS 12261, pp. 532–541, 2020.
https://doi.org/10.1007/978-3-030-59710-8_52

networks especially fully convolutional networks (FCN) [12] have shown promising performance in segmenting both natural images and medial images. The models in these methods have millions of parameters to be optimized, thus a large amount of labeled data with pixel-level annotations is typically needed for training such models to achieve promising performance. However, it is generally difficult to collect a large amount of labeled data in medical image analysis. One main reason is that annotating medical images needs expertise knowledge but few experts have time for annotation. Another reason is that it is time-consuming to annotate medical images.

Semi-supervised learning can utilize a large amount of unlabeled data to improve model performance. semiFCN [2] proposes a semi-supervised network-based approach for medical image segmentation. In semiFCN, a network is trained to predict pseudo masks. The predicted pseudo masks are then used to update the network in turn. ASDNet [14] trains a confidence network to select regions with high confidence in soft masks for updating the segmentation network. Zhou *et al.* [18] propose to jointly improve the performance of disease grading and lesion segmentation by semi-supervised learning with an attention mechanism. Souly *et al.* [17] use weakly labeled data and unlabeled data to train a generative adversarial network (GAN) [8], which can force real data to be close in feature space and thus cluster together. These methods all use a self-training framework, in which the model is updated using pseudo masks predicted by the model itself. If the pseudo masks predicted by the model itself have low quality, the model will be updated using data with noise. On the other hand, co-training [4] uses two models and each model is updated using unlabeled data with pseudo masks predicted by the other model and labeled data with ground truth. In this manner, each model in co-training is supervised by the other model. So the two models can be improved in turn. Several methods [9,15] explore co-training in deep learning. But they are not end-to-end methods.

In this paper, we propose a novel end-to-end approach, called difference minimization network (DMNet), for semi-supervised semantic segmentation in medical images. The contributions of our method can be listed as follows:

- DMNet is a semi-supervised segmentation model, which can be trained with a limited amount of labeled data and a large amount of unlabeled data.
- DMNet adopts the widely used encoder-decoder structure [1,7,16], but it has two decoder branches with a shared encoder. DMNet minimizes the difference between the soft masks predicted by the two decoders to utilize unlabeled data. Unlike co-training which is often not end-to-end, the two decoders in DMNet can be updated at the same time in an end-to-end way.
- DMNet uses the *sharpen* [3] operation to force the model to generate predictions with low entropy on unlabeled data, which can improve the model performance.
- DMNet adopts adversarial learning derived from GAN for further improvement.
- Experiments on a kidney tumor dataset and a brain tumor dataset show that our method can outperform other baselines to achieve the best performance.

2 Notation

We use $\boldsymbol{X} \in \mathcal{R}^{H \times W}$ to denote an image in the labeled training set, and $\boldsymbol{Y} \in \{0,1\}^{H \times W \times K}$ to denote the corresponding ground-truth label which is encoded into a one-hot format. Here, K is the number of classes, H and W are the height and width of the image respectively. DMNet has two segmentation branches, and we denote the class probability maps generated by the two segmentation branches as $\hat{\boldsymbol{Y}}^{(1)}, \hat{\boldsymbol{Y}}^{(2)} \in \mathcal{R}^{H \times W \times K}$. Furthermore, we denote an unlabeled image as $\boldsymbol{U} \in \mathcal{R}^{H \times W}$. We use $[1:N]$ to denote $[1, 2, \cdots, N]$.

3 Method

The framework of DMNet is shown in Fig. 1, which is composed of a segmentation network with two decoder branches, a *sharpen* operation for unlabeled data and a discriminator for both labeled and unlabeled data. Each component will be described detailedly in the following subsections.

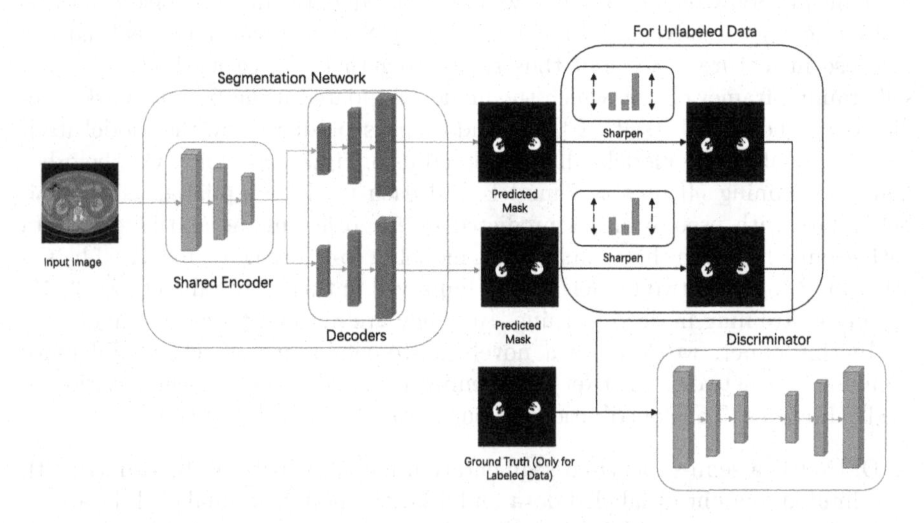

Fig. 1. The framework of DMNet

3.1 Segmentation Network

As shown in Fig. 1, the segmentation network in DMNet adopts the widely used encoder-decoder architecture, which is composed of a shared encoder and two different decoders. By sharing an encoder, our segmentation network has some advantages. First, it can save GPU memory compared to the architecture in which two decoders use separate encoders. Second, since the encoder is shared by two decoders, it can be updated by the information from two decoders. Therefore it can learn better features from the difference between soft masks generated by two decoders, which can lead to better performance. This will be verified by our experimental results in Sect. 4. The two decoders in DMNet use different architectures to introduce diversity. By adopting different architectures, the two decoders will not typically output exactly the same segmentation masks and they can learn from each other. By using labeled and unlabeled data in turn, DMNet can utilize unlabeled data adequately to improve segmentation performance. DMNet is a general framework, and any segmentation network with an encoder-decoder architecture, such as UNet [16], VNet [13], SegNet [1] and DeepLab v3+ [7], can be used in DMNet. In this paper, we adopt UNet [16] and DeepLab v3+ [7] for illustration. The shared encoder can extract latent representation with high-level semantic information of the input image. Then we use the ground truth to supervise the learning of segmentation network for labeled data while minimizing the difference between the masks generated by the two decoders to let them learn from each other for unlabeled data.

We use Dice loss [13] to train our segmentation network on labeled data, which is defined as follows:

$$L_{dice}(\hat{\boldsymbol{Y}}^{(1)}, \hat{\boldsymbol{Y}}^{(2)}, \boldsymbol{Y}; \theta_s) = \sum_{i=1}^{2} \left(1 - \frac{1}{K} \sum_{k=1}^{K} \frac{2 \sum_{h=1}^{H} \sum_{w=1}^{W} \boldsymbol{Y}_{h,w,k} \hat{\boldsymbol{Y}}_{h,w,k}^{(i)}}{\sum_{h=1}^{H} \sum_{w=1}^{W} (\boldsymbol{Y}_{h,w,k} + \hat{\boldsymbol{Y}}_{h,w,k}^{(i)})} \right),$$

where $\boldsymbol{Y}_{h,w,k} = 1$ when the pixel at position (h, w) belongs to class k, and other values in $\boldsymbol{Y}_{h,w,k}$ is set to be 0. $\hat{\boldsymbol{Y}}_{h,w,k}^{(i)}$ is the probability that the pixel at position (h, w) belongs to class k predicted by the segmentation branch i. θ_s is the parameter of the segmentation network.

The loss function used for unlabeled data is described in Sect. 3.3.

3.2 Sharpen Operation

Given an unlabeled data $\boldsymbol{U}$, our segmentation network can generate soft masks $\hat{\boldsymbol{Y}}^{(1)}$ and $\hat{\boldsymbol{Y}}^{(2)}$. To make the predictions of the segmentation networks have low entropy or high confidence, we adopt the *sharpen* operation [3] to reduce the entropy of predictions on unlabeled data, which is defined as follows:

$$Sharpen(\hat{\boldsymbol{Y}}_{h,w,c}^{(i)}, T) = \frac{(\hat{\boldsymbol{Y}}_{h,w,c}^{(i)})^{1/T}}{\sum_{i=1}^{K} (\hat{\boldsymbol{Y}}_{h,w,i}^{(i)})^{1/T}} \quad \forall h \in [1:H], w \in [1:W], T \in (0,1),$$

where $\hat{\boldsymbol{Y}}^{(i)}$ is the soft mask predicted by decoder branch i and temperature T is a hyperparameter.

3.3 Difference Minimization for Semi-supervised Segmentation

As described in Sect. 3.1, two decoders can generate two masks on unlabeled data. If the two masks vary from each other, it means the model is unsure about the predictions and thus the model cannot generalize well. Therefore, we minimize the difference between the two masks to make the two decoders generate consistent masks on the same unlabeled data. In other words, the two decoders can learn under the supervision of each other.

More specifically, given an unlabeled data $\boldsymbol{U}$, the two decoder branches can generate two probability masks $\hat{\boldsymbol{Y}}^{(1)}$ and $\hat{\boldsymbol{Y}}^{(2)}$ which are processed by the *sharpen* operation. Since dice loss can measure the similarity of two segmentation masks and the loss can be backpropogated through two terms, we extend dice loss to the unlabeled setting and get the corresponding loss L_{semi} as follows:

$$L_{semi}(\boldsymbol{U};\theta_s) = 1 - \frac{1}{K}\sum_{k=1}^{K}\frac{2\sum_{h=1}^{H}\sum_{w=1}^{W}\hat{\boldsymbol{Y}}_{h,w,k}^{(1)}\hat{\boldsymbol{Y}}_{h,w,k}^{(2)}}{\sum_{h=1}^{H}\sum_{w=1}^{W}(\hat{\boldsymbol{Y}}_{h,w,k}^{(1)}+\hat{\boldsymbol{Y}}_{h,w,k}^{(2)})}.$$

From the definition of L_{semi}, we can see that the two decoders can be updated by minimizing the difference between the masks they generate.

3.4 Discriminator

In DMNet, we also adopt adversarial learning to learn a discriminator. Unlike the original discriminator in GAN which discriminates whether an image is generated or is real, our discriminator adopts a fully convolutional network (FCN). The FCN discriminator is composed of three convolutional layers whose stride is 2 for downsampling and three corresponding upsampling layers. Each convolutional layer is followed by a ReLU layer. It can discriminate whether a region or some pixels are predicted or from ground truth.

Adversarial Loss for Discriminator. The objective function of discriminator can be written as follows:

$$\begin{aligned}L_{dis}(\hat{\boldsymbol{Y}}^{(1)},\hat{\boldsymbol{Y}}^{(2)},\boldsymbol{Y};\theta_d) = {} & L_{bce}(D(\hat{\boldsymbol{Y}}^{(1)}),\mathbf{0};\theta_d) + L_{bce}(D(\hat{\boldsymbol{Y}}^{(2)}),\mathbf{0};\theta_d)\\ & + L_{bce}(D(\boldsymbol{Y}),\mathbf{1};\theta_d),\end{aligned}$$

where θ_d is the parameter of the discriminator $D(\cdot)$. $\mathbf{1}$ and $\mathbf{0}$ are tensors filled with 1 or 0 respectively, with the same size as that of the outputs of $D(\cdot)$. The term $L_{bce}(D(\boldsymbol{Y}),\mathbf{1})$ in $L_{dis}(\hat{\boldsymbol{Y}}^{(1)},\hat{\boldsymbol{Y}}^{(2)},\boldsymbol{Y};\theta_d)$ is used only when the input data is labeled and is ignored when the input data is unlabeled data. L_{bce} is defined as follows:

$$L_{bce}(\boldsymbol{A},\boldsymbol{B};\theta) = -\sum_{h=1}^{H}\sum_{w=1}^{W}\boldsymbol{B}_{h,w}\log\boldsymbol{A}_{h,w} - \sum_{h=1}^{H}\sum_{w=1}^{W}[(1-\boldsymbol{B}_{h,w})\log(1-\boldsymbol{A}_{h,w})],$$

where θ is the parameter of $\boldsymbol{A}$.

Adversarial Loss for Segmentation Network. In the adversarial learning scheme, the segmentation network tries to fool the discriminator. Hence, there is an adversarial loss L_{adv} for segmentation network to learn consistent features:

$$L_{adv}(\boldsymbol{O};\theta_s) = L_{bce}(D(\hat{\boldsymbol{Y}}^{(1)}),\mathbf{1};\theta_s) + L_{bce}(D(\hat{\boldsymbol{Y}}^{(2)}),\mathbf{1};\theta_s),$$

where $\boldsymbol{O}$ denotes either a labeled image or an unlabeled image, $\hat{\boldsymbol{Y}}^{(1)}$ and $\hat{\boldsymbol{Y}}^{(2)}$ are the corresponding masks predicted by the two decoder branches in the segmentation network.

3.5 Total Loss

Based on the above results, the loss function for the segmentation network can be written as follows:

$$L_S = L_{dice} + \lambda_1 L_{adv} + \lambda_2 L_{semi},$$

where λ_1 and λ_2 are two balance parameters. By integrating the discriminator, the objective of DMNet can be written as follows:

$$L = \min_{\theta_s, \theta_d} \{L_S + L_{dis}\}.$$

4 Experiments

We adopt two real datasets to evaluate DMNet and other baselines, including supervised baselines and semi-supervised baselines.

4.1 Dataset and Evaluation Metric

We conduct our experiments on the KiTS19[1] dataset and BraTS18[2] dataset. KiTS19 dataset is a kidney tumor dataset. It contains 210 labeled 3D computed tomography (CT) images for training and validation, and 90 CT images whose annotation is not published for testing. In our experiments, we use the 210 CT images with annotation to verify the effectiveness of our DMNet.

BraTS18 dataset is a brain tumor dataset. It contains 385 labeled 3D MRI scans and each MRI scan has four modalities (T1, T1 contrast-enhanced, T2 and FLAIR). We use T1, T1 contrast-enhanced and T2 modality to form a three-channel input. This dataset divides the brain tumor into four categories: whole tumor, tumor core, enhancing tumor structures and cystic/necrotic components. In our experiments, we combine these four categories so there are two classes in our experiment: tumor and background.

For each patient in KiTS19 and BraTS18, we choose one slice with its ground-truth label as a labeled image, and choose two slices as unlabeled images by discarding their labels. We split all labeled data into three subsets for training, validation and testing according to the proportion of 7:1:2. The unlabeled data is used for training only. Training data, validation data and testing data have no patient-level overlap to make sure that our model has never seen slices from validation patient or testing patient during training.

Mean Intersection over Union (mIoU) [11] can measure the similarity of any two shapes and is widely used in semantic segmentation. We also adopt mIoU as the evaluation metric.

[1] https://kits19.grand-challenge.org/.

[2] https://www.med.upenn.edu/sbia/brats2018.html.

4.2 Implementation Detail

We use Pytorch[3] to implement DMNet on a workstation with an Intel (R) CPU E5-2620V4@2.1G of 8 cores, 128G RAM and an NVIDIA (R) GPU TITAN Xp. Our encoder network is ResNet101 [10] and we use it for all experiments. In the training phase, we resize the input image to 224 × 224 for KiTS19 and 240 × 240 for BraTS18, and randomly flip it horizontally with a probability of 0.5. In the inference phase, we use the average result of two segmentation branches as the final result. We train our model from scratch using Adam algorithm. The initial learning rate for segmentation network and discriminator is set to be 1e-4 and 1e-5, respectively. The weight decay is set to be 5e-5. We train our model for 150 epochs and decrease the learning rate according to a *poly* scheme [6]. In our experiment, β in *poly* is set to be 0.9. Without explicit statement, we set λ_1 and λ_2 to be 0.01 and 0.1 respectively and set temperature T to be 0.5.

4.3 Baselines

Several semi-supervised methods are adopted as baselines for comparison. More specifically, we compare DMNet to semiFCN [2] and SDNet [5]. semiFCN is a relatively early method in semi-supervised segmentation used for medical image analysis. SDNet is a state-of-the-art method in medical image segmentation. We carefully reimplement semiFCN and SDNet. We adopt ResNet101 as backbone for both methods for fair comparison.

We also design several supervised counterparts of DMNet to demonstrate the usefulness of unlabeled data and design some semi-supervised counterparts to demonstrate the usefulness of each component of DMNet. *Supervised DMNet without adv* denotes a supervised variant which adopts only labeled data for training without adversarial learning. *Supervised DMNet with adv* denotes a supervised variant which adopts only labeled data for training but the adversarial learning is adopted. Both variants do not minimize the difference between two decoder branches. *Separate DMNet* denotes a semi-supervised variant which adopts two separate encoders. That's to say, *Separate DMNet* is composed of two separate encoder-decoder networks. *DMNet_wo_adv_wo_sharpen* denotes a semi-supervised variant which does not adopt the adversarial training strategy and *sharpen* operation. *DMNet_wo_sharpen* denotes a semi-supervised variant which does not adopt the *sharpen* operation on unlabeled data but adopts adversarial learning.

4.4 Comparison with Baselines

We compare our DMNet to baselines, including semiFCN [2] and SDNet [5], on KiTS19 dataset and BraTS18 dataset. The results are shown in Table 1. From the results, we can see that our DMNet outperforms these methods and achieves the best results, when trained with different amount of labeled data. DMNet

[3] https://pytorch.org/.

has obvious advantage over other methods when the amount of labeled data is limited. When we use only 10% of the labeled data and all unlabeled data, DMNet can achieve 88.4% and 78.7% mIoU on KiTS19 and BraTS18, which outperforms semiFCN by 12.3% and 15.1%, and outperforms SDNet by 5.2% and 3.6%, respectively.

Table 1. mIoU on test set of KiTS19 and BraTS18 by different methods using 10%, 30%, 50% and 100% of the labeled data

	KiTS19				BraTS18			
	10%	30%	50%	100%	10%	30%	50%	100%
semiFCN [2]	78.7%	84.4%	86.7%	87.9%	68.4%	78.8%	77.9%	82.7%
SDNet [5]	84.0%	85.9%	89.0%	89.9%	76.0%	80.2%	80.8%	82.9%
DMNet	**88.4%**	**89.9%**	**90.2%**	**90.9%**	**78.7%**	**85.0%**	**85.4%**	**87.0%**

4.5 Ablation Study

We also perform ablation study on BraTS18 to show the effectiveness of each component used in DMNet.

Table 2 shows the results of *Supervised DMNet without adv* trained with 100% of the labeled data using different loss functions. From the results of Table 2, we can see that Dice loss can surpass the performance of cross entropy loss.

Table 2. Comparison between different loss functions

Loss function	mIoU
Cross entropy	81.1%
Dice loss	**84.5%**

Table 3 shows the results of DMNet and its variants introduced in Subsect. 4.3. From the results of *Separate DMNet*, we can see that our architecture design, in which the two decoders share an encoder, has better performance than the architecture in which two decoders use separate encoders. Therefore, it proves that the architecture of DMNet has advantages. More specifically, it can save GPU memory and achieve better performance. Comparing the results between *DMNet_wo_adv_wo_sharpen* and *DMNet_wo_sharpen*, and the results between *Supervised DMNet without adv* and *Supervised DMNet with adv*, we can see that adversarial learning strategy can improve the performance whether in supervised setting or semi-supervised setting. From the results of *DMNet_wo_sharpen* and *DMNet*, we can see that the *sharpen* operation can benefit the learning on unlabeled data. Comparing the results of *DMNet* to those of supervised variants, we can conclude that the proposed DMNet can utilize unlabeled data to improve the segmentation performance, especially when the amount of labeled data is limited. When only 10% of labeled data is available, DMNet can improve the mIoU from 67.0% to 78.7%. When all labeled data is available, in which case the amount of unlabeled data is almost the same as that of labeled data, DMNet can also improve the mIoU from 84.2% to 87.0%.

Table 3. Comparison between DMNet and its variants

Method	Amount of labeled data			
	10%	30%	50%	100%
Supervised DMNet without adv	59.3%	75.8%	79.4%	84.5%
Supervised DMNet with adv	67.0%	76.9%	79.8%	84.2%
Separate DMNet	76.1%	84.2%	84.4%	85.0%
DMNet_wo_adv_wo_sharpen	75.8%	82.0%	82.5%	86.8%
DMNet_wo_sharpen	76.9%	82.3%	83.9%	86.9%
DMNet	**78.7%**	**85.0%**	**85.4%**	**87.0%**

5 Conclusion

In this paper, we propose a novel semi-supervised method, called DMNet, for semantic segmentation in medical image analysis. DMNet can be trained with a limited amount of labeled data and a large amount of unlabeled data. Hence, DMNet can be used to solve the problem that it is typically difficult to collect a large amount of labeled data in medical image analysis. Experiments on a kidney tumor dataset and a brain tumor dataset show that DMNet can outperform other baselines, including both supervised ones and semi-supervised ones, to achieve the best performance.

References

1. Badrinarayanan, V., Kendall, A., Cipolla, R.: SegNet: a deep convolutional encoder-decoder architecture for image segmentation. IEEE Trans. Pattern Anal. Mach. Intell. **39**(12), 2481–2495 (2017)
2. Bai, W., et al.: Semi-supervised learning for network-based cardiac MR image segmentation. In: Descoteaux, M., Maier-Hein, L., Franz, A., Jannin, P., Collins, D.L., Duchesne, S. (eds.) MICCAI 2017. LNCS, vol. 10434, pp. 253–260. Springer, Cham (2017). https://doi.org/10.1007/978-3-319-66185-8_29
3. Berthelot, D., Carlini, N., Goodfellow, I.J., Papernot, N., Oliver, A., Raffel, C.: MixMatch: a holistic approach to semi-supervised learning. CoRR (2019)
4. Blum, A., Mitchell, T.M.: Combining labeled and unlabeled data with co-training. In: Proceedings of Annual Conference on Computational Learning Theory (COLT) (1998)

5. Chartsias, A., et al.: Factorised spatial representation learning: application in semi-supervised myocardial segmentation. In: Frangi, A.F., Schnabel, J.A., Davatzikos, C., Alberola-López, C., Fichtinger, G. (eds.) MICCAI 2018. LNCS, vol. 11071, pp. 490–498. Springer, Cham (2018). https://doi.org/10.1007/978-3-030-00934-2_55
6. Chen, L., Papandreou, G., Kokkinos, I., Murphy, K., Yuille, A.L.: Semantic image segmentation with deep convolutional nets and fully connected CRFs. In: Proceedings of International Conference on Learning Representations (ICLR) (2015)
7. Chen, L., Zhu, Y., Papandreou, G., Schroff, F., Adam, H.: Encoder-decoder with atrous separable convolution for semantic image segmentation. In: Proceedings of European Conference on Computer Vision (ECCV) (2018)
8. Goodfellow, I.J., et al.: Generative adversarial nets. In: Proceedings of Neural Information Processing Systems (NIPS) (2014)
9. Han, B., Yao, Q., Yu, X., Niu, G., Xu, M., Hu, W., Tsang, I.W., Sugiyama, M.: Co-teaching: robust training of deep neural networks with extremely noisy labels. In: Proceedings of Neural Information Processing Systems (NIPS) (2018)
10. He, K., Zhang, X., Ren, S., Sun, J.: Deep residual learning for image recognition. In: Proceedings of Computer Vision and Pattern Recognition (CVPR) (2016)
11. Jaccard, P.: Étude comparative de la distribution florale dans une portion des alpes et des jura. Bull. Soc. Vaudoise Sci. Nat. **37**, 547–579 (1901)
12. Long, J., Shelhamer, E., Darrell, T.: Fully convolutional networks for semantic segmentation. In: Proceeding of Computer Vision and Pattern Recognition (CVPR) (2015)
13. Milletari, F., Navab, N., Ahmadi, S.: V-net: fully convolutional neural networks for volumetric medical image segmentation. In: Proceeding of 3D Vision (3DV) (2016)
14. Nie, D., Gao, Y., Wang, L., Shen, D.: ASDNet: attention based semi-supervised deep networks for medical image segmentation. In: Frangi, A.F., Schnabel, J.A., Davatzikos, C., Alberola-López, C., Fichtinger, G. (eds.) MICCAI 2018. LNCS, vol. 11073, pp. 370–378. Springer, Cham (2018). https://doi.org/10.1007/978-3-030-00937-3_43
15. Qiao, S., Shen, W., Zhang, Z., Wang, B., Yuille, A.L.: Deep co-training for semi-supervised image recognition. In: Proceedings of European Conference on Computer Vision (ECCV) (2018)
16. Ronneberger, O., Fischer, P., Brox, T.: U-Net: convolutional networks for biomedical image segmentation. In: Navab, N., Hornegger, J., Wells, W.M., Frangi, A.F. (eds.) MICCAI 2015. LNCS, vol. 9351, pp. 234–241. Springer, Cham (2015). https://doi.org/10.1007/978-3-319-24574-4_28
17. Souly, N., Spampinato, C., Shah, M.: Semi and weakly supervised semantic segmentation using generative adversarial network. CoRR (2017)
18. Zhou, Y., et al.: Collaborative learning of semi-supervised segmentation and classification for medical images. In: Proceeding of Computer Vision and Pattern Recognition (CVPR) (2019)

Double-Uncertainty Weighted Method for Semi-supervised Learning

Yixin Wang[1,2,3], Yao Zhang[1,2,3], Jiang Tian[3], Cheng Zhong[3], Zhongchao Shi[3], Yang Zhang[4], and Zhiqiang He[1,2,4](✉)

[1] Institute of Computing Technology, Chinese Academy of Sciences, Beijing, China
[2] University of Chinese Academy of Sciences, Beijing, China
{wangyixin19,zhangyao215}@mails.ucas.ac.cn
[3] AI Lab, Lenovo Research, Beijing, China
[4] Lenovo Corporate Research and Development, Lenovo Ltd., Beijing, China
hezq@lenovo.com

Abstract. Though deep learning has achieved advanced performance recently, it remains a challenging task in the field of medical imaging, as obtaining reliable labeled training data is time-consuming and expensive. In this paper, we propose a double-uncertainty weighted method for semi-supervised segmentation based on the teacher-student model. The teacher model provides guidance for the student model by penalizing their inconsistent prediction on both labeled and unlabeled data. We train the teacher model using Bayesian deep learning to obtain double-uncertainty, i.e. segmentation uncertainty and feature uncertainty. It is the first to extend segmentation uncertainty estimation to feature uncertainty, which reveals the capability to capture information among channels. A learnable uncertainty consistency loss is designed for the unsupervised learning process in an interactive manner between prediction and uncertainty. With no ground-truth for supervision, it can still incentivize more accurate teacher's predictions and facilitate the model to reduce uncertain estimations. Furthermore, our proposed double-uncertainty serves as a weight on each inconsistency penalty to balance and harmonize supervised and unsupervised training processes. We validate the proposed feature uncertainty and loss function through qualitative and quantitative analyses. Experimental results show that our method outperforms the state-of-the-art uncertainty-based semi-supervised methods on two public medical datasets.

Keywords: Semi-supervised segmentation · Uncertainty · Teacher-student model

1 Introduction

There are great progresses in medical image segmentation using deep learning, such as U-Net [11] and V-Net [9], in the past few years. The accuracy and

Y. Zhang and Z. He—Equal contribution. This work was done when Yixin Wang and Yao Zhang were interns at AI Lab, Lenovo Research.

A. L. Martel et al. (Eds.): MICCAI 2020, LNCS 12261, pp. 542–551, 2020.
https://doi.org/10.1007/978-3-030-59710-8_53

robustness of these models heavily depend on the quantity and quality of the training data. However, it is well known that high quality annotated medical data is quite expensive due to the domain knowledge prerequisite. As a result, semi-supervised learning has been recently explored to leverage unlabeled data for medical image segmentation [1,8,10,15]. Teacher-student framework, a popular semi-supervised method, has been successfully applied to medical image segmentation tasks on some organs [2,12,14]. Concretely, the teacher model is a Temporal Ensemble [5] of the current model with perturbations, which yields more accurate targets [13]. The student model learns from the teacher model through penalizing inconsistent prediction between them, defined as consistency loss. However, with no ground-truth given for unlabeled training data, it is hard to judge whether the teacher model provides accurate prediction.

To alleviate this problem, uncertainty measures are considered to be the optimal strategy, due to their capability to detect when and where the model is likely to make false predictions or the input is out-of-distribution. Recent explorations on uncertainty estimations include Bayesian uncertainty estimation via test-time dropout [4] and network ensembling [6]. In the medical domain, Yu et al. [14] proposed UA-MT for left atrium segmentation, which exploits the uncertainty of the teacher model by filtering out unreliable predictions and providing only confident voxel-level predictions for student. Sedai et al. [12] adaptively weighted regions with uncertain soft labels to guide the student model in OCT images task. These methods rely on a manually set threshold to control the information flow from teacher to student. Thus, they are incapable of tackling incorrect prediction with low uncertainty and likely filtrate valuable guidance for student by mistake.

In this paper, we propose a double-uncertainty weighted method for semi-supervised learning, using a better captured uncertainty to make the teaching-learning process accurate and reliable. The teacher model is trained using Monte-Carlo dropout [4] as an approximation of Bayesian Neural Network (BNN) to obtain a double-uncertainty, which consists of segmentation uncertainty and feature one. The former is based on the model's prediction output, while the latter captures uncertainty information existing in each convolution kernel when they detect and extract features. Based on uncertainty estimations and model's predictions, a learnable uncertainty consistency loss is proposed to modify teacher's predictions in an interactive manner. Total loss is the weighted sum of supervised segmentation loss trained from labeled data and consistency loss on both labeled and unlabeled data. This balance is crucial [7], but little-noticed in existing semi-supervised approaches. We take the double-uncertainty of each prediction into consideration, namely assigning smaller loss weight to trustless results.

To the best of our knowledge, we are the first to explore and evaluate feature uncertainty, which reveals the internal mechanism of a model's decision making. Without ground-truth, our newly designed consistency loss still incentivizes teacher's prediction closer to target probabilities and reduces prediction uncertainties. Furthermore, our designed uncertainty weight effectively benefits from unlabeled data and avoids uncertain prediction disturbing supervised training.

We conduct exhaustive experiments on the datasets of 2018 Atrial Segmentation Challenge[1] for left atrium(LA) segmentation and MICCAI 2019 KiTS Challenge[2] for kidney segmentation. Our method outperforms the state-of-the-art uncertainty-based semi-supervised segmentation approaches.

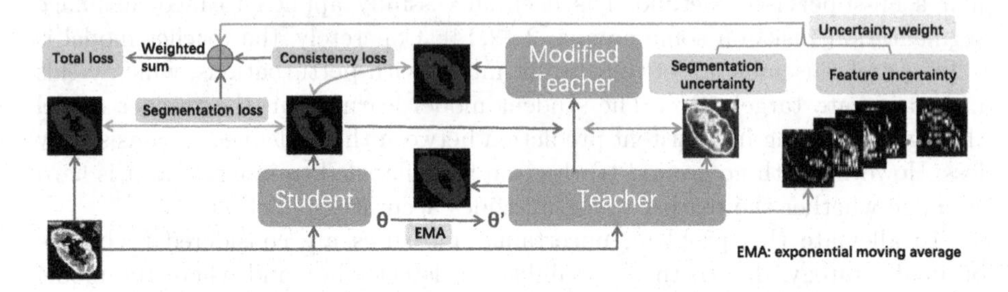

Fig. 1. Overview of our model. Teacher model utilizes EMA weights of the student model and generates predictions and double-uncertainty estimates. Uncertainty modifies the teacher in an interactive manner and serves as a weight. Student learns from the teacher by minimizing the weighted sum of the segmentation loss computed from labeled data and consistency loss on both labeled and unlabeled data.

2 Method

In this section, we first explore how feature uncertainty exists in channels and propose an approach to evaluating it. Then, we detail the method that exploits the feature and segmentation uncertainties into our semi-supervised model. Specifically, we propose a learnable uncertainty consistency loss to form more accurate and less uncertain predictions from the teacher. We further discuss how this double-uncertainty can be interpreted as consistency loss weight to benefit unsupervised training. The whole framework is shown in Fig. 1.

2.1 Feature Uncertainty

Uncertainty information can be excavated not only from prediction outputs, but also feature extracting process of convolution kernel. Each channel of a feature map can be considered as a feature detector focusing on 'what' and 'where' is an informative part. If a convolution kernel can extract meaningful features, then the activated regions in its corresponding channel should be basically consistent under multiple inference operations for the same input. Therefore, the uncertainty of each channel implies their learning capability separately. Furthermore, the uncertainty of a feature map can indicate the performance of a model on each

[1] http://atriaseg2018.cardiacatlas.org/data/.
[2] https://kits19.grand-challenge.org/data/.

given input. Following [4], we train our model with dropout and random noise at T inference times. For each input, we obtain T intermediate feature maps $F^t \in \mathbb{R}^{H \times W \times C}$ at the same layer, which can be considered as a set of $H \times W$ dimensional vectors of the corresponding channels $c \in \{1...C\}$. The uncertainty of channel c is defined by:

$$u_c = \sigma\Big(\sum_p \sum_q g\,(x_c^p, x_c^q)\Big) \quad p, q \in \{1 \ldots T\}, \tag{1}$$

$$U_c = mean\Big(\sum_{h,w} u_c^{h,w}\Big) \quad h \in \{1 \ldots H\}, w \in \{1 \ldots W\}, \tag{2}$$

where x_c^p and x_c^q are two $H \times W$ dimensional vectors of the same channel c at different inference time p and q. Function $g(x, y)$ computes an absolute value of difference between these two vectors. Through min-max normalization operation σ, each channel's uncertainty map $u_c \in \mathbb{R}^{H \times W}$ can be estimated, and uncertainty value U_c is obtained by Eq. 2. This relative uncertainty among channels represents their information capture ability. Higher uncertainty values imply flatter distributions over extracted channel's features. Accordingly, the feature uncertainty of the model at this layer can be obtained by:

$$U_f = \frac{1}{C} \sum_i^C (U_i - \min U_c)\,. \tag{3}$$

This equation calculates an overall estimation of the model's feature extraction ability.

2.2 Double-Uncertainty Weighted Model

Following [14], we adopt Mean Teacher [13] as our framework and baseline V-Net [9] as teacher and student networks. The weights of the student model at training step t are denoted by θ_t. The teacher model uses the exponential moving average(EMA) weights of the student model as $\theta'_t = \alpha\theta'_{t-1} + (1-\alpha)\theta_t$, where α is a hyper-parameter called EMA decay. It is utilized to update the weights of student model with gradient descent. This ensembling over training steps improves the quality of the predictions. Meanwhile, random noise on the input and dropout in training process of the teacher model serve as training regularization to improve results. The student aims to learn by minimizing the supervised segmentation loss on labeled data and consistency loss on both labeled and unlabeled data. This consistency enforces consistent prediction between the student and teacher model. Inspired by [3], the teacher model is also considered as an uncertainty estimation branch in parallel with itself as a prediction branch. We design a learnable uncertainty consistency loss to modify teacher's predictions and facilitate the model to produce less uncertain estimates. Meanwhile, uncertainty is utilized as a weight to balance the training between labeled data and unlabeled data, which significantly benefits the optimization process.

Learnable Uncertainty Consistency Loss. Segmentation uncertainty U_s is captured like [4] using the following equation:

$$u_i = \frac{1}{T}\sum_{t=1}^{T} \text{Softmax}(\rho_t) \qquad u_v = -\sum_{i=1}^{M} u_i \log u_i, \tag{4}$$

where ρ_t is the prediction logits of class i at t^{th} times and M is the number of classes which equals to two in our tasks. u_i is the average softmax probability of T times stochastic dropout sampling from teacher model and u_v is uncertainty metric on voxel v. The whole segmentation uncertainty U_s for the given input is obtained by the mean value of each voxel's uncertainty estimate. The prediction probabilities of the teacher model are adjusted as t_i' by interpolating between the original prediction t_i and the prediction of student model s_i voxel by voxel.

$$t_i' = (1 - u_v)t_i + u_v s_i, \tag{5}$$

$$\mathcal{L}_\text{c} = -\frac{1}{V}\sum_{v}^{V}(\sum_{i}^{M} \log\left(t_{v,i}'\right) s_{v,i} + \beta \log(1 - u_{v,i})), \tag{6}$$

where v represents the vth voxel and β is a hyper-parameter for the uncertainty log penalty, preventing the teacher model from producing high uncertainty estimates all the time. The designed consistency loss $\mathcal{L}_\text{c}$ not only represents whether the teacher's prediction is reliable enough for the student on voxel level, but also produces an interesting dynamic process to be optimized. When $u_{v,i} \to 1$, which means the prediction of teacher on voxel v is extremely untrustworthy, then $t_i' \to s_i$, narrows the distance between two models and mitigates teacher's guidance. Conversely, if $u_{v,i} \to 0$, $t_i' \to t_i$, then teacher's prediction remains unchanged and keeps its trustful guidance. The above interaction mechanism guarantees the consistency loss can be reduced by the teacher model when providing less uncertainty guidance.

Uncertainty Weight. The overall loss is a combination of supervised segmentation loss $\mathcal{L}_\text{s}$ and proposed consistency loss $\mathcal{L}_\text{c}$, calculated by:

$$\mathcal{L} = \mathcal{L}_\text{s} + \lambda\mathcal{L}_\text{c}. \tag{7}$$

Proper scheduling of consistency loss weight λ is very important [7]. Most of the existing choice of λ is a time-dependent Gaussian weighting function $\omega(t) = 0.1 * e^{\left(-5(1-S/L)^2\right)}$, where S and L represent the current training step and ramp-up length separately [2]. However, this trade-off ignores the proportion and characteristics of unlabeled data, which is incapable of making the most of the teacher model for guidance. Therefore, we take uncertainty estimates into consideration.

$$\lambda = \frac{\omega(t)}{U_f} \log\frac{1}{U_s} = -\frac{\omega(t)}{U_f}\log U_s, \tag{8}$$

where U_f and U_s represent feature uncertainty and segmentation uncertainty. It is noted that if batch size is bigger than one, U_f and U_s are the average of each input in a mini-batch. While $\omega(t)$ grasps overall share of consistency loss to ramp up from zero along a Gaussian curve, this double-uncertainty serves as weight priors to scale loss targeting on each prediction. Given an input with large uncertainty value U_f and U_s predicted by teacher model, λ drops in case of wrong guidance disturbing network training. To avoid poor local minima due to extremely small U_s, we use a *log* function to restrict its value. Double-uncertainty controls training loss together, leading to a convincing weighted result.

3 Experiment

The proposed method is evaluated on the datasets of 2018 Atrial Segmentation Challenge for left atrium (LA) segmentation and MICCAI 2019 KiTS Challenge for kidney segmentation. We compared our method with advanced supervised methods and state-of-the-art semi-supervised methods separately. We adopt Dice, Jaccard, the 95% Hausdorff Distance (95HD) and the Average Surface Distance (ASD) as our assessment metrics.

3.1 Implementation Details

The implementation is based on Pytorch using an NVIDIA Tesla V100 32 GB GPU. The model is trained using the SGD optimizer and a batch size of 4 with a gradually decaying learning rate of 0.01, which is divided by 10 after each 2500 training steps. Stochastic dropout with $p = 0.5$ is applied to layers of the encoder and decoder for $T = 16$ times. The supervised segmentation loss is a summation of Cross Entropy Loss and Dice Loss. The 4th upsampling layer with 32 channels is adopted for feature uncertainty estimation. EMA decay α is set as 0.99, referring to [13,14]. Hyperparameters β controls a log penalty and just a small value can prevent the teacher model from producing high uncertainty estimates. We test various values and 0.001 is adopted.

LA Segmentation. This dataset includes 100 3D gadolinium-enhanced magnetic resonance imaging scans (GE-MRIs) with segmentation masks. We select 80 samples as a training set, and the rest 20 data for testing. For better comparison, we use the same data preprocessing method as [14]. Table 1 shows under 16 and 8 labeled data, our method outperforms the SOTA semi-supervised method UA-MT [14] on the four index. Note that our method trained with only 16 labeled data performs close to baseline V-Net trained with all 80 labeled data.

Kidney Segmentation. This challenging dataset is a collection of contrast-enhanced 3D abdominal CT scans along with their segmentation ground-truth. We split the 210 given scans into 160 for training and 50 for testing. We extract 3D patches centering at the kidney region. To make the results more convincing, a comprehensive comparison with existing methods is conducted. Baseline

Table 1. Comparison with advanced supervised and semi-supervised methods.

Dataset	Labeled	Method	Dice	Jaccard	95HD	ASD
LA	80	Baseline V-Net [9]	0.9025	0.8240	8.29	1.91
	16	UA-MT [14]	0.8888	0.8021	7.32	2.26
		Ours	**0.8965**	**0.8135**	**7.04**	**2.03**
	8	UA-MT [14]	0.8425	0.7348	13.83	3.36
		Ours	**0.8591**	**0.7575**	**12.67**	**3.31**
Kidney	4	Baseline V-Net [9]	0.8173	0.7291	8.90	2.75
		MT-Sedai et al. [12]	0.8593	0.7248	9.26	3.06
		UA-MT [14]	0.8713	0.7866	11.74	3.56
		Ours	**0.8879**	**0.8169**	**8.04**	**2.34**

V-Net [9] is selected as a supervised-only method. For semi-supervised methods, besides UA-MT, we choose another latest uncertainty guided semi-supervised method [12], which achieves great success in the OCT image task. For a fair comparison, we re-implement this method with the same Mean Teacher architecture, which is referred to as MT-Sedai et al. The same V-Net is adopted as a Bayesian network to estimate uncertainty in all these methods. Table 1 shows our method achieves a high accuracy using only 4 labeled scans, which ranks top among all compared methods. From Dice comparison between the supervised-only method and semi-supervised methods under different labeled/unlabeled scans in Fig. 2(a), it is observed that our method improves performance significantly, especially when the amount of labeled data is small.

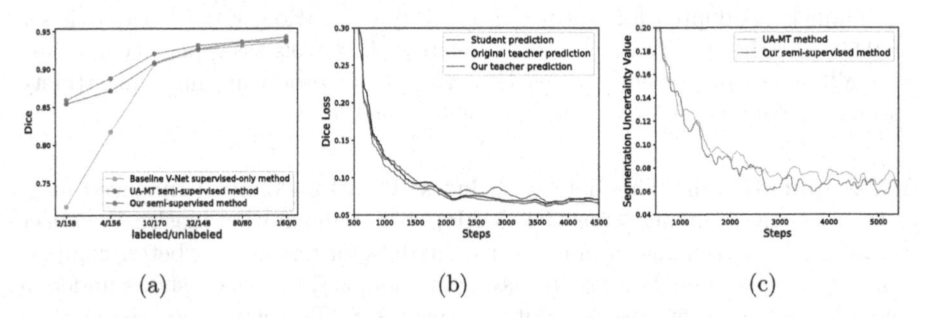

Fig. 2. (a) Results of kidney segmentation with different labeled/unlabeled data. (b) Smoothened Dice loss of training data for our modified teacher model, original teacher model and student model. (c) Mean value of segmentation uncertainty from teacher model after each training step.

3.2 Uncertainty Validation

Figure 3 shows an example of LA segmentation task under 16 labeled data. Experiments show the choice of layers to obtain feature uncertainty does not affect the segmentation results above. Thus, we choose 4th upsampling layer for better visualization of uncertainty maps. We randomly select three samples separately from the early, middle and late training stages and calculate their feature uncertainty. Figure 3(left) compares feature uncertainty maps of their first four channels (32 channels in total). Taking the early stage sample as an example, four channels' corresponding uncertainty values are [0.2453, 0.3327, 0.2782, 0.3206]. Obviously, the first channel has a relatively better ability to capture the target's features. From feature uncertainty map, uncertainty regions are mainly in the vicinity of the target, especially at its boundary and tissue joint. Conversely, the second channel is proved to be highly uncertain. Its uncertainty map indicates its corresponding convolution kernel is not sure about what and where to see. This provides a way to analyze channels' quality. Figure 3(right) shows all the uncertainty value of their 32 channels. It can be seen that 1) relative uncertainty among channels remains basically consistent. 2) as trained, feature uncertainty decreases (These three samples have diminishing feature uncertainty values of [0.3109, 0.2838, 0.2565], using Eq. 2).

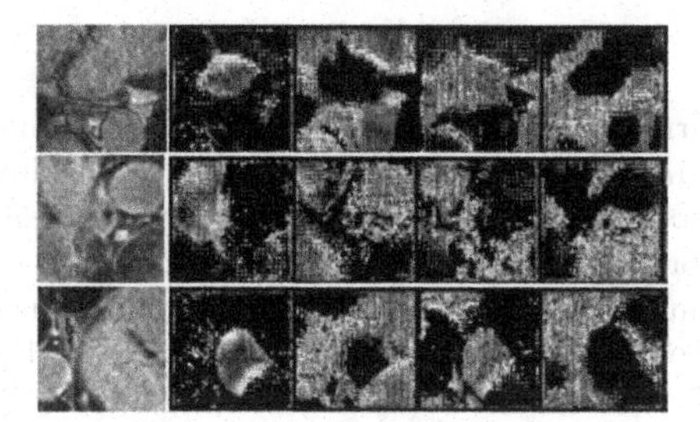

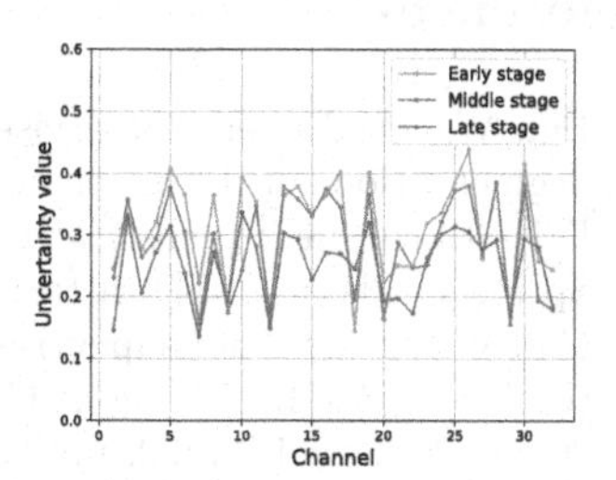

Fig. 3. Feature uncertainty. Left: three input images from early, middle, late stage and their first four channels' feature uncertainty maps (brighter regions represent more uncertainty). Right: feature uncertainty value of 32 channels.

Consistency loss defines the prediction inconsistency between student and teacher model. Thus, to investigate the ability of our designed consistency loss equals to verify our modification to the teacher model. By calculating Dice Loss of training data with the ground-truth label, Fig. 2(b) shows our modified teacher achieves better predictions. A more accurate teacher speeds up the feedback loop, leading to better test performance, as shown in Table 2. Additionally, the interaction between uncertainty estimates and model's predictions in our proposed loss function facilitates teachers to produce less uncertain results, which is proven in Fig. 2(c). We also validate the proposed uncertainty weight in Table 2. Particularly, with uncertainty weight, UA-MT [14] is improved as well.

Table 2. Contributions of proposed consistency loss and uncertainty weight.

Method	10 labeled	32 labeled
MSE loss [13]	0.8944	0.9215
UA-MT loss [14]	0.9081	0.9252
UA-MT loss [14] + Weight	0.9098	0.9346
Our Consistency loss	0.9182	0.9358
Our Consistency loss + Weight	**0.9206**	**0.9365**

4 Conclusion

In this paper, we make novel contributions to employ uncertainty estimates to the teacher-student model for semi-supervised learning. It is the first time that feature uncertainty is proposed and evaluated. Our learnable uncertainty consistency loss facilitates a more accurate teacher model with less uncertainty. Uncertainty is also utilized as a consistency weight to balance unsupervised training. Comprehensive experimental analysis on two medical datasets shows the significant improvements of our method using limited labeled images.

References

1. Bai, W., et al.: Semi-supervised Learning for Network-Based Cardiac MR Image Segmentation. In: Descoteaux, M., Maier-Hein, L., Franz, A., Jannin, P., Collins, D.L., Duchesne, S. (eds.) MICCAI 2017, Part II. LNCS, vol. 10434, pp. 253–260. Springer, Cham (2017). https://doi.org/10.1007/978-3-319-66185-8_29
2. Cui, W., et al.: Semi-supervised brain lesion segmentation with an adapted mean teacher model. In: Chung, A.C.S., Gee, J.C., Yushkevich, P.A., Bao, S. (eds.) IPMI 2019. LNCS, vol. 11492, pp. 554–565. Springer, Cham (2019). https://doi.org/10.1007/978-3-030-20351-1_43
3. DeVries, T., Taylor, G.W.: Learning confidence for out-of-distribution detection in neural networks. CoRR abs/1802.04865 (2018). http://arxiv.org/abs/1802.04865
4. Gal, Y., Ghahramani, Z.: Dropout as a bayesian approximation: representing model uncertainty in deep learning. In: Proceedings of the 33nd International Conference on Machine Learning, ICML 2016, New York City, NY, USA, 19–24 June 2016, pp. 1050–1059 (2016). http://proceedings.mlr.press/v48/gal16.html
5. Laine, S., Aila, T.: Temporal ensembling for semi-supervised learning. In: 5th International Conference on Learning Representations, ICLR 2017, Toulon, France, 24–26 April 2017, Conference Track Proceedings (2017). https://openreview.net/forum?id=BJ6oOfqge
6. Lakshminarayanan, B., Pritzel, A., Blundell, C.: Simple and scalable predictive uncertainty estimation using deep ensembles. In: Advances in Neural Information Processing Systems 30: Annual Conference on Neural Information Processing Systems 2017, 4–9 December 2017, Long Beach, CA, USA, pp. 6402–6413 (2017). http://papers.nips.cc/paper/7219-simple-and-scalable-predictive-uncertainty-estimation-using-deep-ensembles

7. Lee, D.H.: Pseudo-label: the simple and efficient semi-supervised learning method for deep neural networks. ICML 2013 Workshop: Challenges in Representation Learning (WREPL), July 2013
8. Li, X., Yu, L., Chen, H., Fu, C., Heng, P.: Semi-supervised skin lesion segmentation via transformation consistent self-ensembling model. In: British Machine Vision Conference 2018, BMVC 2018, 3–6 September 2018, p. 63. Northumbria University, Newcastle (2018). http://bmvc2018.org/contents/papers/0162.pdf
9. Milletari, F., Navab, N., Ahmadi, S.: V-net: Fully convolutional neural networks for volumetric medical image segmentation. In: Fourth International Conference on 3D Vision, 3DV 2016, Stanford, CA, USA, 25–28 October 2016, pp. 565–571 (2016). https://doi.org/10.1109/3DV.2016.79, https://doi.org/10.1109/3DV.2016.79
10. Mondal, A.K., Dolz, J., Desrosiers, C.: Few-shot 3d multi-modal medical image segmentation using generative adversarial learning. CoRR abs/1810.12241 (2018). http://arxiv.org/abs/1810.12241
11. Ronneberger, O., Fischer, P., Brox, T.: U-net: convolutional networks for biomedical image segmentation. In: Navab, N., Hornegger, J., Wells, W.M., Frangi, A.F. (eds.) MICCAI 2015, Part III. LNCS, vol. 9351, pp. 234–241. Springer, Cham (2015). https://doi.org/10.1007/978-3-319-24574-4_28
12. Sedai, S., et al.: Uncertainty guided semi-supervised segmentation of retinal layers in OCT images. In: Shen, D., et al. (eds.) MICCAI 2019, Part I. LNCS, vol. 11764, pp. 282–290. Springer, Cham (2019). https://doi.org/10.1007/978-3-030-32239-7_32
13. Tarvainen, A., Valpola, H.: Mean teachers are better role models: weight-averaged consistency targets improve semi-supervised deep learning results. In: Advances in Neural Information Processing Systems 30: Annual Conference on Neural Information Processing Systems 2017, 4–9 December 2017, Long Beach, CA, USA, pp. 1195–1204 (2017). http://papers.nips.cc/paper/6719-mean-teachers-are-better-role-models-weight-averaged-consistency-targets-improve-semi-supervised-deep-learning-results
14. Yu, L., Wang, S., Li, X., Fu, C.-W., Heng, P.-A.: Uncertainty-aware self-ensembling model for semi-supervised 3D left atrium segmentation. In: Shen, D., et al. (eds.) MICCAI 2019, Part II. LNCS, vol. 11765, pp. 605–613. Springer, Cham (2019). https://doi.org/10.1007/978-3-030-32245-8_67
15. Zhang, Y., Yang, L., Chen, J., Fredericksen, M., Hughes, D.P., Chen, D.Z.: Deep adversarial networks for biomedical image segmentation utilizing unannotated images. In: Descoteaux, M., Maier-Hein, L., Franz, A., Jannin, P., Collins, D.L., Duchesne, S. (eds.) MICCAI 2017, Part III. LNCS, vol. 10435, pp. 408–416. Springer, Cham (2017). https://doi.org/10.1007/978-3-319-66179-7_47

Shape-Aware Semi-supervised 3D Semantic Segmentation for Medical Images

Shuailin Li[1], Chuyu Zhang[1], and Xuming He[1,2](✉)

[1] ShanghaiTech University, Shanghai, China
{lishl,hexm}@shanghaitech.edu.cn, zcy_ai@whu.edu.cn
[2] Shanghai Engineering Research Center of Intelligent Vision and Imaging, Shanghai, China

Abstract. Semi-supervised learning has attracted much attention in medical image segmentation due to challenges in acquiring pixel-wise image annotations, which is a crucial step for building high-performance deep learning methods. Most existing semi-supervised segmentation approaches either tend to neglect geometric constraint in object segments, leading to incomplete object coverage, or impose strong shape prior that requires extra alignment. In this work, we propose a novel shape-aware semi-supervised segmentation strategy to leverage abundant unlabeled data and to enforce a geometric shape constraint on the segmentation output. To achieve this, we develop a multi-task deep network that jointly predicts semantic segmentation and signed distance map (SDM) of object surfaces. During training, we introduce an adversarial loss between the predicted SDMs of labeled and unlabeled data so that our network is able to capture shape-aware features more effectively. Experiments on the Atrial Segmentation Challenge dataset show that our method outperforms current state-of-the-art approaches with improved shape estimation, which validates its efficacy. Code is available at https://github.com/kleinzcy/SASSnet.

Keywords: Geometric constraints · Semantic segmentation · Semi-supevised learning

1 Introduction

Semantic object segmentation is a fundamental task in medical image analysis and has been widely used in automatic delineation of regions of interest in 3D medical images, such as cells, tissues or organs. Recently, tremendous progress has been made in medical semantic segmentation [15] thanks to modern deep convolutional networks, which achieve state-of-the-art performances in many

S. Li and C. Zhang—Both authors contributed equally to the work. This work was supported by Shanghai NSF Grant (No. 18ZR1425100).

A. L. Martel et al. (Eds.): MICCAI 2020, LNCS 12261, pp. 552–561, 2020.
https://doi.org/10.1007/978-3-030-59710-8_54

real-world tasks. However, training deep neural networks often requires a large amount of annotated data, which is particularly expensive in medical segmentation problems. In order to reduce labeling cost, a promising approach is to adopt a semi-supervised learning [1,2] framework that typically utilizes a small labeled dataset and many unlabeled images for effective model training.

Recent efforts in semi-supervised segmentation have been focused on incorporating unlabeled data into convolutional network training, which can be largely categorized into two groups. The first group of those methods mainly consider the generic setting of semi-supervised segmentation [3,7–9,11,16,18–20]. Most of them adopt adversarial learning or consistency loss as regularization in order to leverage unlabeled data for model learning. The adversarial learning methods [7,11,19,20] enforces the distributions of segmentation of unlabeled and labeled images to be close while the consistency loss approaches [3,8,9,16,18] utilize a teacher-student network design and require their outputs being consistent under random perturbation or transformation of input images. To cope with difficult regions, Nie et al. [11] utilize adversarial learning to select regions of unlabeled images with high confidence to train the segmentation network. Yu et al. [18] introduce an uncertainty map based on the mean-teacher framework [16] to guide student network learning. Despite their promising results, those methods lack explicit modeling of the geometric prior of semantic objects, often leading to poor object coverage and/or boundary prediction.

The second group of semi-supervised methods attempt to address the above drawback by incorporating a strong anatomical prior on the object of interest in their model learning [6,20]. For instance, Zheng et al. [20] introduce the Deep Atlas Prior (DAP) model that encodes a probabilistic shape prior in its loss design. He et al. [6] propose an auto-encoder to learn priori anatomical features on unlabeled dataset. However, such prior typically assumes properly aligned input images, which is difficult to achieve in practice for objects with large variation in pose or shape.

In this work, we propose a novel shape-aware semi-supervised segmentation strategy to address the aforementioned limitations. Our main idea is to incorporate a more flexible geometric representation in the network so that we are able to enforce a global shape constraint on the segmentation output, and meanwhile to handle objects with varying poses or shapes. Such a "shape-aware" representation enables us to capture the global shape of each object class more effectively. Moreover, by exploiting consistency of the geometric representations between labeled and unlabeled images, we aim to design a simple and yet effective semi-supervised learning strategy for deep segmentation networks.

To achieve this, we develop a multi-task deep network that jointly predicts semantic segmentation and signed distance map (SDM) [4,12,13,17] with a shared backbone network module. The SDM assigns each pixel a value indicating its signed distance to the nearest boundary of target object, which provides a shape-aware representation that encodes richer features of object shape and surface. To utilize the unlabeled data, we then introduce an adversarial loss between the predicted SDMs of labeled and unlabeled data for semi-supervised learning.

This allows the model to learn shape-aware features more effectively by enforcing similar distance map distributions on the entire dataset. In addition, the SDM naturally imposes more weights on the interior region of each semantic class, which can be viewed as a proxy of confidence measure. In essence, we introduce an implicit shape prior and its regularization based on an adversarial loss for semi-supervised volumetric segmentation.

We evaluate our approach on the Atrial Segmentation Challenge dataset with extensive comparisons to prior arts. The results demonstrate that our segmentation network outperforms the state-of-the-art methods and generates object segmentation with high-quality global shapes.

Our main contributions are three-folds: (1) We propose a novel shape-aware semi-supervised segmentation approach by enforcing geometric constraints on labeled and unlabeled data. (2) We develop a multi-task loss on segmentation and SDM predictions, and impose global consistency in object shapes through adversarial learning. (3) Our method achieves strong performance on the Atrial Segmentation Challenge dataset with only a small number of labeled data.

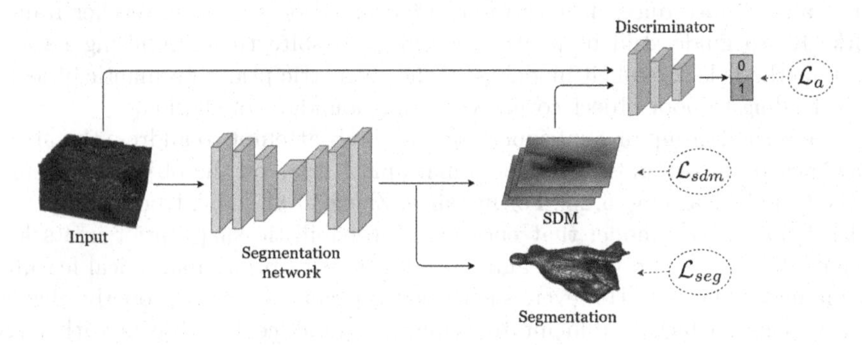

Fig. 1. Overview of our method. Our network takes as input a 3D volume, and predicts a 3D SDM and a segmentation map. Our learning loss consists of a multi-task supervised term and an adversarial loss on the SDM predictions.

2 Method

2.1 Overview

We aim to build a deep neural network for medical image segmentation in a semi-supervised setting in order to reduce annotation cost. Due to lack of annotated images, our key challenge is to regularize the network learning effectively from a set of unlabeled ones. In this paper, we tackle this problem by utilizing the regularity in geometric shapes of the target object class, which provides an effective constraint for both segment prediction and network learning.

Specifically, we propose to incorporate a shape-aware representation of object segments into the deep network prediction. In particular, we develop a multi-task segmentation network that takes a 3D image as input and jointly predicts a segmentation map and a SDM of object segmentation. Based on this SDM representation, we then design a semi-supervised learning loss for training the segmentation network. Our loss mainly consists of two components, one for the network predictions on the labeled set while the other enforcing consistency between the SDM predictions on the labeled and unlabeled set. To achieve effective consistency constraint, we adopt an adversarial loss that encourages the segmentation network to produce segment predictions with similar distributions on both datasets. Figure 1 illustrates the overall pipeline of our semi-supervised segmentation network. Below we will introduce the detailed model design in Sect. 2.2, followed by the learning loss and network training in Sect. 2.3.

2.2 Segmentation Network

In order to encode geometric shape of a target semantic class, we propose a multi-task segmentation network that jointly predicts a 3D object mask and its SDM for the input 3D volume. Our network has a V-Net [10] structure that consists of an encoder module and a decoder module with two output branches, one for the segmentation map and the other for the SDM. For notation clarity, we mainly focus on the single-class setting below[1].

Specifically, we employ a V-Net backbone as in [18], and then add a light-weighted SDM head in parallel with the original segmentation head. Our SDM head is composed by a 3D convolution block followed by the *tanh* activation. Given an input image $\mathbf{X} \in \mathbb{R}^{H \times W \times D}$, the segmentation head generates a confidence score map $\mathbf{M} \in [0,1]^{H \times W \times D}$ and the SDM head predicts a SDM $\mathbf{S} \in [-1,1]^{H \times W \times D}$ as follows:

$$\mathbf{M} = f_{\text{seg}}(\mathbf{X}; \theta), \qquad \mathbf{S} = f_{\text{sdm}}(\mathbf{X}; \theta) \tag{1}$$

where θ are the parameters of our segmentation network, and each element of $\mathbf{S}$ indicates the signed distance of a corresponding voxel to its closest surface point after normalization [17].

2.3 Shape-Aware Semi-supervised Learning

We now introduce our semi-supervised learning strategy for the segmentation network. While prior methods typically rely on the segmentation output $\mathbf{M}$, we instead utilize the shape-aware representation $\mathbf{S}$ to regularize the network training. To this end, we develop a multi-task loss consisting of a supervised loss $\mathcal{L}_s$ on the labeled set and an adversarial loss $\mathcal{L}_a$ on the entire set to enforce consistency of the model predictions.

[1] It is straightforward to generalize our formulation to the multi-class setting by treating each semantic class separately for SDMs.

Formally, we assume a standard semi-supervised learning setting, in which the training set contains N labeled data and M unlabeled data, where $N \ll M$. We denote the labeled set as $\mathcal{D}^l = \{\mathbf{X}_n, \mathbf{Y}_n, \mathbf{Z}_n\}_{n=1}^N$ and unlabeled set as $\mathcal{D}^u = \{\mathbf{X}_m\}_{m=N+1}^{N+M}$, where $\mathbf{X}_n \in \mathbb{R}^{H\times W\times D}$ are the input volumes, $\mathbf{Y}_n \in \{0,1\}^{H\times W\times D}$ are the segmentation annotations and $\mathbf{Z}_n \in \mathbb{R}^{H\times W\times D}$ are the groundtruth SDMs derived from $\mathbf{Y}_n$. Below we first describe the supervised loss on $\mathcal{D}^l$ followed by the adversarial loss that utilizes the unlabeled set $\mathcal{D}^u$.

Supervised Loss $\mathcal{L}_s$. On the labeled set, we employ a dice loss l_{dice} and a mean square loss l_{mse} for the segmentation and SDM output of the multi-task segmentation network, respectively:

$$\mathcal{L}_s(\theta) = \mathcal{L}_{seg} + \alpha \mathcal{L}_{sdm} \tag{2}$$

$$\mathcal{L}_{seg} = \frac{1}{N}\sum_{i=1}^{N} l_{dice}(f_{seg}(\mathbf{X}_i;\theta), \mathbf{Y}_i); \quad \mathcal{L}_{sdm} = \frac{1}{N}\sum_{i=1}^{N} l_{mse}(f_{sdm}(\mathbf{X}_i;\theta), \mathbf{Z}_i) \tag{3}$$

where $\mathcal{L}_{seg}$ denotes the segmentation loss and $\mathcal{L}_{sdm}$ is the SDM loss, and α is a weighting coefficient balancing two loss terms.

Adversarial Loss $\mathcal{L}_a$. To regularize the model learning with the unlabeled data, we introduce an adversarial loss that enforces the consistency of SDM predictions on the labeled and unlabeled set. To this end, we propose a discriminator network to tell apart the predicted SDMs from the labeled set, which should be high-quality due to the supervision, and the ones from the unlabeled set. Minimizing the adversarial loss induced by this discriminator enables us to learn effective shape-aware features that generalizes well to the unlabeled dataset.

Specifically, we adopt a similar discriminator network D as [14], which consists of 5 convolution layers followed by an MLP. The network takes a SDM and input volume as input, fuses them through convolution layers, and predicts its class probability of being labeled data. Given the discriminator D, we denote its parameter as ζ and define the adversarial loss as follows,

$$\mathcal{L}_a(\theta,\zeta) = \frac{1}{N}\sum_{n=1}^{N} \log D(\mathbf{X}_n, \mathbf{S}_n;\zeta) + \frac{1}{M}\sum_{m=N+1}^{N+M} \log\left(1 - D(\mathbf{X}_m, \mathbf{S}_m;\zeta)\right) \tag{4}$$

where $\mathbf{S}_n = f_{sdm}(\mathbf{X}_n;\theta)$ and $\mathbf{S}_m = f_{sdm}(\mathbf{X}_m;\theta)$ are the predicted SDMs.

Overall Training Pipeline. Our overall training objective $\mathcal{V}(\theta,\zeta)$ combines the supervised and the adversarial loss defined above and the learning task can be written as,

$$\min_{\theta}\max_{\zeta} \mathcal{V}(\theta,\zeta) = \mathcal{L}_s(\theta) + \beta\mathcal{L}_a(\theta,\zeta) \tag{5}$$

where β is a weight coefficient that balances two loss terms. We adopt a standard alternating procedure to train the entire network, which includes the following two subproblems.

Given a fixed discriminator $D(\cdot;\zeta)$, we minimize the overall loss w.r.t the segmentation network parameter θ. To speed up model learning, we simplify the loss in two steps: Firstly, we ignore the first loss term in Eq. (4) due to high-quality SDM predictions on the labeled set, i.e., $\mathbf{S}_n \approx \mathbf{Z}_n$, and additionally, we adopt a similar surrogate loss for the generator as in [5]. Hence the learning problem for the segmentation network can be written as,

$$\min_{\theta} \mathcal{L}_s(\theta) - \frac{\beta}{M} \sum_{m=N+1}^{N+M} \log(D(\mathbf{X}_m, f_{sdm}(\mathbf{X}_m;\theta);\zeta)) \tag{6}$$

On the other hand, given a fixed segmentation network, we simply minimize the binary cross entropy loss induced by Eq. (5) to train the discriminator, i.e., $\min_\zeta -\mathcal{V}(\theta,\zeta)$, or $\max_\zeta \mathcal{L}_a(\theta,\zeta)$. To stablize the overall training, we use an annealing strategy based on a time-dependent Gaussian warm-up function to slowly increase the loss weight β (See Sect. 3 for details).

Table 1. Quantitative comparisons of semi-supervised segmentation models on the LA dataset. All models use the V-Net as backbone network. Results on two different data partition settings show that our SASSNet outperforms the state-of-the-art results consistently.

Method	# Scans used		Metrics			
	Labeled	Unlabeled	Dice [%]	Jaccard [%]	ASD [voxel]	95HD [voxel]
V-Net	80	0	91.14	83.82	1.52	5.75
V-Net	16	0	86.03	76.06	3.51	14.26
DAP [20]	16	64	87.89	78.72	2.74	9.29
ASDNet [11]	16	64	87.90	78.85	**2.08**	9.24
TCSE [9]	16	64	88.15	79.20	2.44	9.57
UA-MT [18]	16	64	88.88	80.21	2.26	7.32
UA-MT (+NMS)	16	64	89.11	80.62	2.21	**7.30**
SASSNet (ours)	16	64	89.27	80.82	3.13	8.83
SASSNet (+NMS)	16	64	**89.54**	**81.24**	2.20	8.24
V-Net	8	0	79.99	68.12	5.48	21.11
DAP [20]	8	72	81.89	71.23	3.80	15.81
UA-MT [18]	8	72	84.25	73.48	3.36	13.84
UA-MT(+NMS)	8	72	84.57	73.96	2.90	12.51
SASSNet(ours)	8	72	86.81	76.92	3.94	12.54
SASSNet(+NMS)	8	72	**87.32**	**77.72**	**2.55**	**9.62**

3 Experiments and Results

We validate our method on the Left Atrium (LA) dataset from Atrial Segmentation Challenge[2] with detailed comparisons to prior arts. The dataset contains

[2] http://atriaseg2018.cardiacatlas.org/.

100 3D gadolinium-enhanced MR imaging scans (GE-MRIs) and LA segmentation masks, with an isotropic resolution of $0.625 \times 0.625\times\ 0.625\,\text{mm}^3$. Following [18], we split them into 80 scans for training and 20 scans for validation, and apply the same pre-processing methods.

Implementation Details and Metrics. The segmentation network is trained by a SGD optimizer for 6000 iterations, with an initial learning rate (lr) 0.01 decayed by 0.1 every 2500 iterations. The discriminator uses $4 \times 4 \times 4$ kernels with stride 2 in its convolutional layers and an Adam optimizer with a constant lr 0.0001. We use a batch size of 4 images and a single GPU with 12 Gb RAM for the model training. In all our experiments, we set α as 0.3 and β as a time-dependent Gaussian warming-up function $\lambda(t) = 0.001 * e^{-5(1-\frac{t}{t_{max}})^2}$ where t indicates number of iterations.

During testing, we take the segmentation map output $\mathbf{M}$ for evaluation. In addition, an non-maximum suppression (NMS) is applied as the post process in order to remove isolated extraneous regions. We use the standard evaluation metrics, including Dice coefficient (Dice), Jaccard Index (Jaccard), 95% Hausdorff Distance (95HD) and Average Symmetric Surface Distance (ASD).

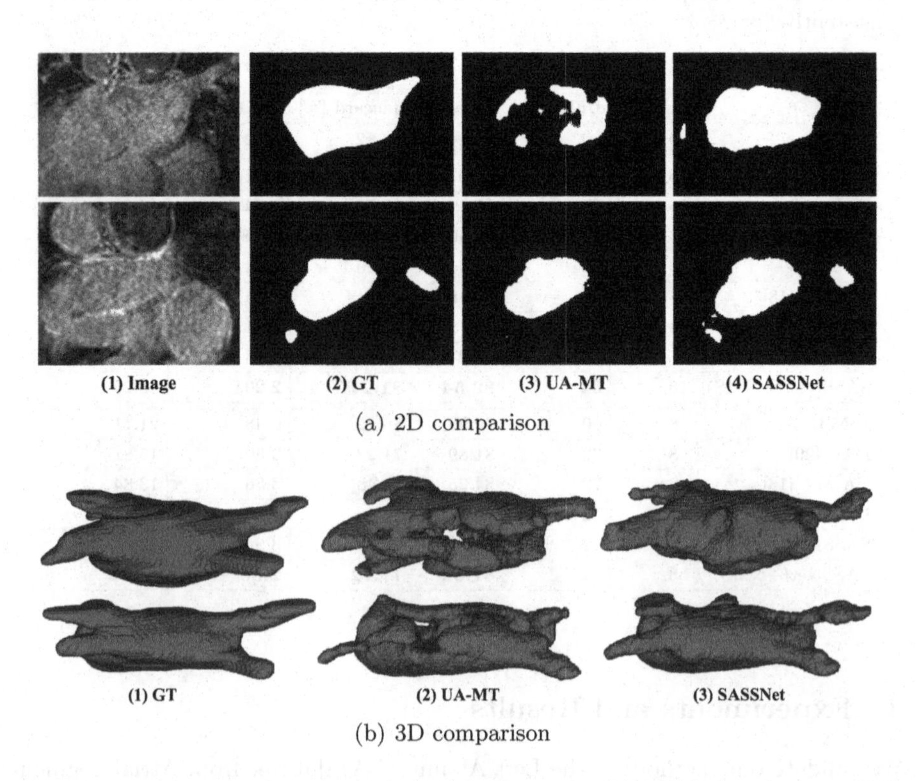

Fig. 2. 2D and 3D Visualization of the segmentations by UA-MT [18] and our method, where GT denotes groundtruth segmetnation.

Quantitative Evaluation and Comparison. We evaluate our method in two different settings with comparisons to several recent semi-supervised segmentation approaches, including DAP [20], ASDNet [11], TCSE [9] and UA-MT [18]. Table 1 presents a summary of the quantitative results, in which we first show the upper-bound performance achieved by a fully-supervised network, followed by two individual settings.

The first setting follows the work [18], which takes 20% of training data as labeled data (16 labeled), and the others as unlabeled data for semi-supervised training. We can see that this setting is relative easy as the model trained with 20% of data already achieves good performance (86.03% in Dice). Among the semi-supervised methods, the DAP performs worst, indicating the limitation of an atlas-based prior, while UA-MT achieves the top performance in the previous methods. Our method outperforms all the other semi-supervised networks in both Dice (89.54%) and Jaccard (81.24%), and achieves competitive results on other metrics. In particular, our SASSNet surpasses UA-MT in Dice without resorting to a complex multiple network architecture.

To validate the robustness of our method, we also consider a more challenging setting in which we only have 8 labeled images for training. The second half of Table 1 show the comparison results, where SASSNet outperforms UA-MT with a large margin (Dice: +2.56% without NMS and +3.07% with NMS). Without NMS, our SASSNet tends to generate more foreground regions, which leads to slightly worse performance on ASD and 95HD. However, it also produce better segmentation preserving the original object shape. By contrast, UA-MT often misses inner regions of target objects and generates irregular shapes. Figure 2 provides several qualitative results for visual comparison.

Table 2. Effectiveness of our proposed modules on the LA dataset. All the models use the same V-Net as the backbone, and we conduct an ablative study to show the contribution of each component module.

Method	# Scans used		Metrics				Cost
	Labeled	Unlabeled	Dice [%]	Jaccard [%]	ASD [voxel]	95HD [voxel]	Params [M]
V-Net	8	0	79.99	68.12	5.48	21.11	187.7
V-Net +SDM	8	0	81.12	69.75	6.93	25.58	187.9
V-Net +SDM +GAN	8	72	86.81	76.92	3.94	12.54	249.7
UA-MT [18]	8	72	84.25	73.48	3.36	13.84	375.5
V-Net +SDM +MT	8	72	84.97	74.14	6.12	22.20	375.8

Ablative Study. We conduct several detailed experimental studies to examine the effectiveness of our proposed SDM head and the adversarial loss (GAN). Table 2 shows the quantitative results of different model settings. The first row is a V-Net trained with only the labeled data, which is our base model. We first add a SDM head, denoted as V-Net+SDM, and as shown in the second row, such joint learning improves segmentation results by 1.1% in Dice. We then add the unlabeled data and our adversarial loss, denoted as V-Net+SDM+GAN, which significantly improves the performance (5.7% in Dice).

We also compare our semi-supervised learning strategy with two methods in the mean-teacher (MT) framework (last two rows). One is the original UA-MT and the other is our segmentation network with the MT consistency loss. Our SASSNet outperforms both methods with higher Dice and Jaccard scores, which indicates the advantage of our representation and loss design. Moreover, our network has a much simpler architecture than those two networks.

4 Conclusion

In this paper, we proposed a shape-aware semi-supervised segmentation approach for 3D medical scans. In contrast to previous methods, our method exploits the regularity in geometric shapes of the target object class for effective segment prediction and network learning. We developed a multi-task segmentation network that jointly predicts semantic segmentation and SDM of object surfaces, and a semi-supervised learning loss enforcing consistency between the predicted SDMs of labeled and unlabeled data. We validated our approach on the Atrial Segmentation Challenge dataset, which demonstrates that our segmentation network outperforms the state-of-the-art methods and generates object segmentation with high-quality global shapes.

References

1. Bai, W., et al.: Semi-supervised learning for network-based cardiac MR image segmentation. In: Descoteaux, M., Maier-Hein, L., Franz, A., Jannin, P., Collins, D.L., Duchesne, S. (eds.) MICCAI 2017. LNCS, vol. 10434, pp. 253–260. Springer, Cham (2017). https://doi.org/10.1007/978-3-319-66185-8_29
2. Baur, C., Albarqouni, S., Navab, N.: Semi-supervised deep learning for fully convolutional networks. In: Descoteaux, M., Maier-Hein, L., Franz, A., Jannin, P., Collins, D.L., Duchesne, S. (eds.) MICCAI 2017. LNCS, vol. 10435, pp. 311–319. Springer, Cham (2017). https://doi.org/10.1007/978-3-319-66179-7_36
3. Bortsova, G., Dubost, F., Hogeweg, L., Katramados, I., de Bruijne, M.: Semi-supervised medical image segmentation via learning consistency under transformations. In: Shen, D., et al. (eds.) MICCAI 2019. LNCS, vol. 11769, pp. 810–818. Springer, Cham (2019). https://doi.org/10.1007/978-3-030-32226-7_90
4. Dangi, S., Linte, C.A., Yaniv, Z.: A distance map regularized CNN for cardiac cine MR image segmentation. Med. Phys. **46**(12), 5637–5651 (2019)
5. Goodfellow, I., et al.: Generative adversarial nets. In: NIPS, pp. 2672–2680 (2014)
6. He, Y., et al.: DPA-DenseBiasNet: semi-supervised 3D fine renal artery segmentation with dense biased network and deep priori anatomy. In: Shen, D., et al. (eds.) MICCAI 2019. LNCS, vol. 11769, pp. 139–147. Springer, Cham (2019). https://doi.org/10.1007/978-3-030-32226-7_16
7. Hung, W.C., Tsai, Y.H., Liou, Y.T., Lin, Y.Y., Yang, M.H.: Adversarial learning for semi-supervised semantic segmentation. In: BMVC (2018)
8. Laine, S., Aila, T.: Temporal ensembling for semi-supervised learning. arXiv preprint arXiv:1610.02242 (2016)

9. Li, X., Yu, L., Chen, H., Fu, C.W., Heng, P.A.: Semi-supervised skin lesion segmentation via transformation consistent self-ensembling model. arXiv preprint arXiv:1808.03887 (2018)
10. Milletari, F., Navab, N., Ahmadi, S.A.: V-net: fully convolutional neural networks for volumetric medical image segmentation. In: 3DV, pp. 565–571. IEEE (2016)
11. Nie, D., Gao, Y., Wang, L., Shen, D.: ASDNet: attention based semi-supervised deep networks for medical image segmentation. In: Frangi, A.F., Schnabel, J.A., Davatzikos, C., Alberola-López, C., Fichtinger, G. (eds.) MICCAI 2018. LNCS, vol. 11073, pp. 370–378. Springer, Cham (2018). https://doi.org/10.1007/978-3-030-00937-3_43
12. Park, J.J., Florence, P., Straub, J., Newcombe, R., Lovegrove, S.: DeepSDF: learning continuous signed distance functions for shape representation. In: CVPR, pp. 165–174 (2019)
13. Perera, S., Barnes, N., He, X., Izadi, S., Kohli, P., Glocker, B.: Motion segmentation of truncated signed distance function based volumetric surfaces. In: WACV, pp. 1046–1053. IEEE (2015)
14. Radford, A., Metz, L., Chintala, S.: Unsupervised representation learning with deep convolutional generative adversarial networks. arXiv preprint arXiv:1511.06434 (2015)
15. Taghanaki, S.A., Abhishek, K., Cohen, J.P., Cohen-Adad, J., Hamarneh, G.: Deep semantic segmentation of natural and medical images: a review. arXiv preprint arXiv:1910.07655 (2019)
16. Tarvainen, A., Valpola, H.: Mean teachers are better role models: weight-averaged consistency targets improve semi-supervised deep learning results. In: NIPS, pp. 1195–1204 (2017)
17. Xue, Y., et al.: Shape-aware organ segmentation by predicting signed distance maps. arXiv preprint arXiv:1912.03849 (2019)
18. Yu, L., Wang, S., Li, X., Fu, C.-W., Heng, P.-A.: Uncertainty-aware self-ensembling model for semi-supervised 3D left atrium segmentation. In: Shen, D., et al. (eds.) MICCAI 2019. LNCS, vol. 11765, pp. 605–613. Springer, Cham (2019). https://doi.org/10.1007/978-3-030-32245-8_67
19. Zhang, Y., Yang, L., Chen, J., Fredericksen, M., Hughes, D.P., Chen, D.Z.: Deep adversarial networks for biomedical image segmentation utilizing unannotated images. In: Descoteaux, M., Maier-Hein, L., Franz, A., Jannin, P., Collins, D.L., Duchesne, S. (eds.) MICCAI 2017. LNCS, vol. 10435, pp. 408–416. Springer, Cham (2017). https://doi.org/10.1007/978-3-319-66179-7_47
20. Zheng, H., et al.: Semi-supervised segmentation of liver using adversarial learning with deep atlas prior. In: Shen, D., et al. (eds.) MICCAI 2019. LNCS, vol. 11769, pp. 148–156. Springer, Cham (2019). https://doi.org/10.1007/978-3-030-32226-7_17

Local and Global Structure-Aware Entropy Regularized Mean Teacher Model for 3D Left Atrium Segmentation

Wenlong Hang[1,2], Wei Feng[1], Shuang Liang[3](✉), Lequan Yu[4], Qiong Wang[5], Kup-Sze Choi[6], and Jing Qin[6]

[1] School of Computer Science and Technology, Nanjing Tech University, Nanjing, China
[2] State Key Laboratory for Novel Software Technology, Nanjing University, Nanjing, China
[3] Smart Health Big Data Analysis and Location Services Engineering Lab of Jiangsu Province, Nanjing University of Posts and Telecommunications, Nanjing, China
shuang.liang@njupt.edu.cn
[4] Department of Radiation Oncology, Stanford University, Stanford, CA 94305, USA
[5] CAS Key Laboratory of Human-Machine Intelligence-Synergy Systems, Shenzhen Institutes of Advanced Technology, Shenzhen, China
[6] School of Nursing, Hong Kong Polytechnic University, Hong Kong, China

Abstract. Emerging self-ensembling methods have achieved promising semi-supervised segmentation performances on medical images through forcing consistent predictions of unannotated data under different perturbations. However, the consistency only penalizes on independent pixel-level predictions, making structure-level information of predictions not exploited in the learning procedure. In view of this, we propose a novel structure-aware entropy regularized mean teacher model to address the above limitation. Specifically, we firstly introduce the entropy minimization principle to the student network, thereby adjusting itself to produce high-confident predictions of unannotated images. Based on this, we design a local structural consistency loss to encourage the consistency of inter-voxel similarities within the same local region of predictions from teacher and student networks. To further capture local structural dependencies, we enforce the global structural consistency by matching the weighted self-information maps between two networks. In this way, our model can minimize the prediction uncertainty of unannotated images, and more importantly that it can capture local and global structural information and their complementarity. We evaluate the proposed method on a publicly available 3D left atrium MR image dataset. Experimental results demonstrate that our method achieves outstanding segmentation performances than the state-of-the-art approaches in scenes with limited annotated images.

Keywords: Self-ensembling · Entropy minimization · Structural consistency · Segmentation.

W. Hang and W. Feng—are co-first authors.

A. L. Martel et al. (Eds.): MICCAI 2020, LNCS 12261, pp. 562–571, 2020.
https://doi.org/10.1007/978-3-030-59710-8_55

1 Introduction

Accurate segmentation of left atrium (LA) from 3D magnetic resonance (MR) images is essential for obtaining its morphological information, which provides support for the diagnosis and treatment of various cardiovascular diseases [5]. Deep learning has achieved promising performance on LA segmentation [1], but its performance depends on the availability of abundant annotated images. For 3D medical images, it is difficult to obtain abundant annotated data because manually annotating data slice by slice is costly and time consuming.

Deep semi-supervised learning methods have been proposed by utilizing limited annotated data together with abundant unannotated data to achieve better generalization for medical image segmentation [2,6,11,12,16]. Inspired by the success of self-ensembling method, Li *et al.* [6] embedded the transformation consistency into Π-model (TCSE) to boost the generalization capability of network. The mean teacher (MT) model [12] was extended to an adapted MT model with soft Dice consistency loss (MT-Dice) [2] for brain lesion segmentation. Subsequently, Yu *et al.* [16] proposed an uncertainty-aware MT model (UA-MT) to generate more reliable predictions by encouraging low uncertainty of the teacher network. However, these methods focus only on the pixel-level consistency of prediction, ignoring the underlying structural information [13]. More recently, Liu *et al.* [7,8] proposed to learn structural relationship between pixels by measuring their spatial distance. Based on this, Kim *et al.* [4] exploited the inter-pixel relationship of prediction to optimize MT model for semantic segmentation. Other approaches [15] leveraged the entropy map of prediction to obtain object border information for pixel-wise image segmentation. Despite the success in some applications, current self-ensembling methods fail to fully exploit the complementarity of rich spatial and geometric structural information in prediction.

Herein, we propose a novel deep semi-supervised learning method named *local and global structure-aware entropy regularized mean teacher* (LG-ER-MT) for 3D LA segmentation. In general, our method inherits the robustness of MT model in the same way that encourages the segmentation consistency under different perturbations. Concretely, since the entropy minimization can adjust network to generate high-confident predictions [9], we design an *entropy regularized mean teacher* (ER-MT) model to penalize voxel-level prediction uncertainty of unannotated data. To further exploit structure-level information of 3D MR images, we calculate inter-voxel similarities within small volumes stochastically sampled from prediction to obtain local spatial structural information, while utilizing weighted self-information (*i.e.*, the disentanglement of Shannon Entropy [14]) to acquire global geometric structure information. We simultaneously introduce local and global structural information to MT model to capture their complementarity. By encouraging structural consistencies between teacher and student networks, we further improves the generalization capacity of network. Experiments on MICCAI 2018 Atrial Segmentation Challenge dataset demonstrate that our method can achieve state-of-the-art performances.

2 Methodology

In this section, we first introduce the proposed entropy regularized mean teacher (ER-MT) model. Then, we elaborate on both local and global structural consistencies. Figure 1 illustrates the pipeline of our local and global structure-aware entropy regularized mean teacher (LG-ER-MT) model for 3D LA segmentation. The student network leverages the entropy minimization principle to adjust itself to make more precise segmentation of unannotated data. The local and global structural consistencies jointly enhance the generalization capability of the framework.

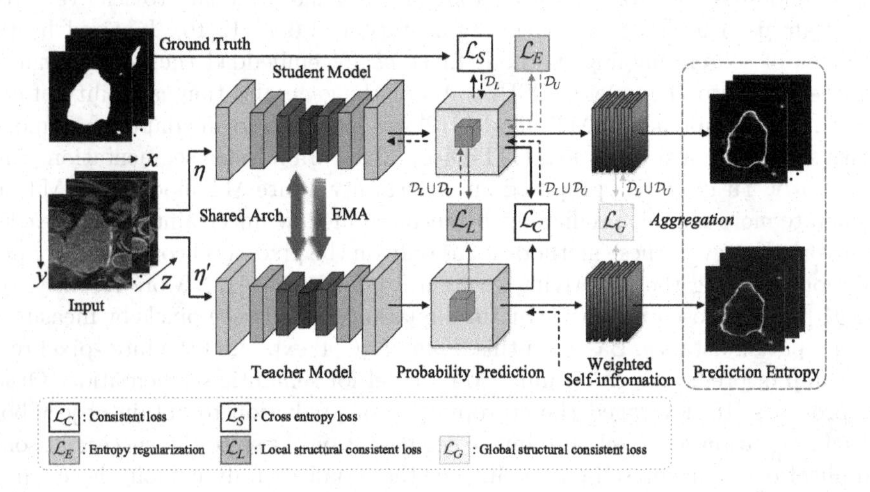

Fig. 1. Overview of our LG-ER-MT framework for semi-supervised segmentation. It jointly encourages local and global structural consistancy on both annotated data $\mathcal{D}_L$ and unannotated data $\mathcal{D}_U$, as well as generates high-confident prediction of $\mathcal{D}_U$. The prediction entropy maps equal to the voxel-wise aggregation of weighted self-information.

2.1 Entropy Regularized Mean Teacher (ER-MT)

Supposing that the training set $\mathcal{D}$ consists of L annotated data and U unannotated data, $\mathcal{D}_L = \{(\mathbf{x}_i, \boldsymbol{y}_i)\}_{i=1}^{L}$ and $\mathcal{D}_U = \{\mathbf{x}_i\}_{i=L+1}^{L+U}$ represent annotated data set and unannotated data set, where $\mathbf{x}_i$ is a $H \times W \times D$ dimensional input volume and $\boldsymbol{y}_i \in \{0,1\}^{H \times W \times D}$ is the corresponding ground truth. Let $\mathbf{p}_{i,v}^{s}$ and $\mathbf{p}_{i,v}^{t}$ represent prediction vectors of the v-th voxel of the i-th input volume from student and teacher networks, respectively. We denote $p_{i,v,c}^{s}$ as the predicted probability score of the class c, $c \in \{0,1\}$.

For image segmentation tasks, low-density regions are generally distributed in the pixels of an object border [3]. Considering that the entropy minimization principle can adjust the network passing through the low-density regions to make high-confident predictions in most situations [9], we introduce this principle to the student network for the first time in order to achieve more accurate segmentation of unannotated images. The entropy loss $\mathcal{L}_E$ is formulated as follows:

$$\mathcal{L}_E = -\frac{1}{U \cdot \log |c|} \sum_{i \in \mathcal{R}} \sum_{v \in \mathcal{V}} \sum_{c \in \{0,1\}} p^s_{i,v,c} \log p^s_{i,v,c} \tag{1}$$

where $\mathcal{R} = \{L+1, L+2, \ldots, L+U\}$ and $\mathcal{V} = \{1, 2, \ldots, H \times W \times D\}$.

By incorporating the entropy minimization principle with MT model, we design an entropy regularized mean teacher (ER-MT) model. The optimization problem of ER-MT can be represented as:

$$\mathcal{L}_{ER-MT} = \mathcal{L}_S + \lambda_c \cdot \mathcal{L}_C + \lambda_e \cdot \mathcal{L}_E \tag{2}$$

where $\mathcal{L}_S$ is the cross-entropy loss used to evaluate the segmentation performance of annotated images. $\mathcal{L}_C$ denotes the conventional consistency loss that is defined as the expected distance between the predictions of teacher and student networks. We adopt mean squared error (MSE) to calculate $\mathcal{L}_C$ on both annotated and unannotated data. λ_c and λ_e are the regularization parameters for balancing the trade-off with other terms. $\mathcal{L}_E$ enables the student network more precisely segment unannotated images, which ultimately improves the generalization capability of MT model.

2.2 Local and Global Structure-Aware Entropy Regularized Mean Teacher Model (LG-ER-MT)

In the field of medical image, MT model and its variants treat 3D segmentation as a classification task performed voxel by voxel, thus they employ the original consistency loss generally used in semi-supervised classification. However, image segmentation differs from general classification problems in that network prediction has structure-level characteristics. If ignored, it is difficult for MT models to achieve higher performance. To this end, we exploit local and global structural consistencies to jointly learn spatial and geometric structural information to promote the generalization capability of MT model for 3D LA segmentation.

Local Structural Consistency Loss. To obtain spatial structural information, one possible way is to calculate the similarity (distance) of each pairwise voxels in prediction. However, it is unrealistic to apply it in 3D MR image segmentation due to its high computational cost $\mathcal{O}\left((H \times W \times D)^2\right)$. Accordingly, we propose to stochastically sample few local volumes from prediction in each mini-batch and use them to compute the inter-voxel similarity. Figure 2 illustrates the schematic of stochastic sampling and similarity computing.

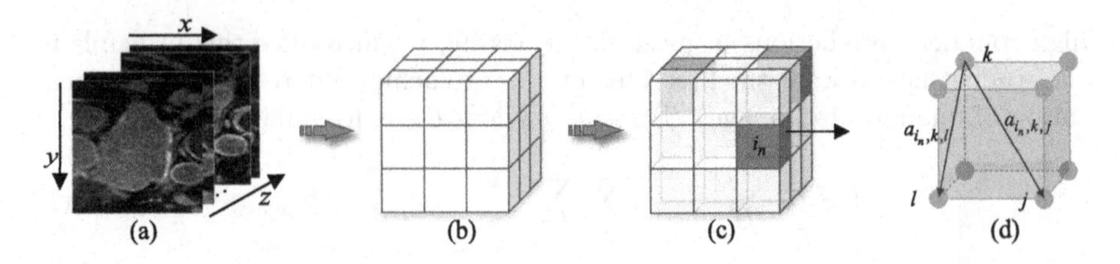

Fig. 2. Schematic of stochastic sampling. (a) 3D MR image. (b) dividing network prediction into sub-volumes. (c) stochastically sampled volumes (indicated in blue). (d) inter-voxel similarity, *e.g.*, $a_{i_n,k,l}$. (Color figure online)

After sampling N local volumes, we calculate the local structural consistency loss, which can be formulated as:

$$\mathcal{L}_L = \frac{1}{L+U} \sum_{i \in \mathcal{R}} \sum_{n \in \mathcal{N}} \frac{1}{(H_n \times W_n \times D_n)^2} \sum_{k,l \in \mathcal{V}} \left\| a^s_{i_n,k,l} - a^t_{i_n,k,l} \right\|^2 \\ \text{and} \quad a_{i_n,k,l} = \left(\mathbf{p}_{i_n,k}\right)^T \mathbf{p}_{i_n,l} / \left(\|\mathbf{p}_{i_n,k}\|_2 \|\mathbf{p}_{i_n,l}\|_2\right). \tag{3}$$

Here, $\mathcal{R} = \{1, 2, \ldots, L+U\}$, $\mathcal{N} = \{1, 2, \ldots, N\}$, $\mathcal{V} = \{1, 2, \ldots, H_n \times W_n \times D_n\}$. All the local volumes are with the same dimension $H_n \times W_n \times D_n$. For the i-th prediction from student network s and teacher network t , $\mathbf{p}_{i_n,k}$ represents the prediction vector of the k-th voxel located in the n-th local volume, and $a_{i_n,k,l}$ denotes the cosine similarity (distance) between the k-th voxel and l-th voxel in the same local volume. $\mathcal{L}_L$ is designed to encourage inter-voxel similarities of local volumes between teacher and student networks to be consistent. By this way, notably, the computational cost can be driven down to $\mathcal{O}\left(N \cdot (H_n \times W_n \times D_n)^2\right)$, where $(H_n, W_n, D_n) \ll (H, W, D)$ and N is often small (*e.g.*, 4).

Global Structural Consistency Loss. Recall that the local structural consistency merely focuses on the spatial structural relationship within small volumes of prediction. The dependencies of local structural information have been previously ignored but not trivial. Since the prediction entropy of an image can reflect the results similar to the object border detection, it is reasonable to encourage the prediction entropy between teacher and student networks to be consistent, which is conductive to match the global geometric structural information. Given that the prediction entropy equals to the voxel-wise linear aggregation of weighted self-information [14], we intuitively use the latter (in higher dimensional space) to exploit the global geometric structural information.

Formally, we denote $\mathbf{I}^s_{i,v}$ as the weighted self-information of the v-th voxel of the i-th input from the student network, which equals to $-\mathbf{p}^s_{i,v} \circ \log \mathbf{p}^s_{i,v}$. The notation $\circ$ is Hadamard product and log is the logarithmic operation on each element. We formulate the global structural consistency loss $\mathcal{L}_G$ as minimizing

MSE of the weighted self-information between teacher and student networks.

$$\mathcal{L}_G = \frac{1}{L+U} \sum_{i \in \mathcal{R}} \frac{1}{H \times W \times D} \sum_{v \in \mathcal{V}} \left\| \mathbf{I}_{i,v}^s - \mathbf{I}_{i,v}^t \right\|^2 \tag{4}$$

where $\mathcal{R} = \{1, 2, \ldots, L+U\}$ and $\mathcal{V} = \{1, 2, \ldots, H \times W \times D\}$. $\mathcal{L}_G$ encourages the global geometric structure between the two networks to be consistent, which can further improve the generalization capability of our model.

LG-ER-MT Framework. Based on the above discussions, we integrate Eq. (2), Eq. (3) and Eq. (4) in a unified local and global structure-aware entropy regularized MT (LG-ER-MT) framework as:

$$\mathcal{L}_{LG-ER-MT} = \mathcal{L}_{ER-MT} + \lambda_l \cdot \mathcal{L}_L + \lambda_g \cdot \mathcal{L}_G \tag{5}$$

where λ_l and λ_g are the tradeoff parameters for local and global structural consistency loss, respectively. With the proposed LG-ER-MT framework, we can obtain more accurate segmentation results through suppressing voxel-level prediction uncertainty of unannotated data as well as capturing local and global structural information and their complementarity.

3 Experiments and Results

Dataset and Pre-processing. The proposed LG-ER-MT method was extensively evaluated on the Atrial Segmentation Challenge dataset[1]. It provides 100 3D gadolinium-enhanced MR imaging scans with an isotropic resolution of 0.625 mm × 0.625 mm × 0.625 mm and their corresponding ground truth. In the experiment, 80 scans were used for training and the remaining 20 scans for testing. All the scans were cropped centering at the heart region and normalized as zero mean and unit variance.

Implementation Details. The framework[2] was implemented using PyTorch and trained on two RTX 2080Ti GPUs. According to [16], we randomly cropped 112 × 112 × 80 sub-volumes and used the standard data augmentation techniques. V-Net [10] was used as our network backbone for both teacher and student networks. After *L-Stage* 5 layer and *R-Stage* 1 layer in V-Net, we added two dropout layers with the dropout rate of 0.5. Following [12], the parameter value of exponential moving average (EMA) was set to 0.99. The network parameters were trained for a total of 6000 iterations and updated by the SGD optimizer. The initial learning rate was set to 1*e*-2 and then divided it by 10 every 2500 iterations. We randomly choose the unannotated scans as many as the annotated scans in each epoch, and set the batch size to 4, containing 2 annotated scans and 2 unannotated scans. In our preliminary experiments, we found

[1] https://atriaseg2018.cardiacatlas.org/.

[2] Model is available in https://github.com/3DMRIs/LG-ER-MT.

that the results are not sensitive to the hyper-parameters, so we empirically set the hyper-parameters. The time-dependent Gaussian warming up function $\lambda(t) = \exp\left(-5(1 - t/t_{\max})^2\right)$ was used to ramp up hyper-parameters λ_c, λ_e,λ_l and λ_g from 0 to 0.1, 0.01, 0.01 and 0.01, respectively. Here, t denotes the current step and $t_{\max}$ is the maximum training step. In particular, it is critical to determine the size of stochastic sampled volumes when computing the local structural consistency loss. In each mini-batch, we stochastically sample 4 local volumes with a size of $16 \times 16 \times 16$ to balance the efficiency of the local structural information and the computational complexity.

Evaluation of 3D LA Segmentation. All methods were evaluated on four metrics, i.e., Dice, Jaccard, average surface distance (ASD), and 95% Hausdorff Distance (95HD). The 20% of training scans (*i.e.*, 16) were used as annotated data and the remaining training scans (*i.e.*, 64) as unannotated data. Table 1 lists the segmentation results of comparison methods on 20 test scans. We compared our LG-ER-MT with the baseline method V-Net and the original MT model [12], as well as several state-of-the-art semi-supervised segmentation methods, including adversarial learning based semi-supervised method (ASDNet) [11], TCSE [6], MT-Dice [2], and UA-MT [16]. These methods all used V-Net as the network backbone.

Table 1. Comparison results of different segmentation methods.

Method	# Scans used		Metrics			
	Annotated	Unannotated	Dice [%]	Jaccard [%]	ASD [voxel]	95HD [voxel]
V-Net	80	0	91.14	83.82	1.52	5.75
	16	0	86.03	76.06	3.51	14.26
ASDNet [11]	16	64	87.90	78.85	2.08	9.24
TCSE [6]	16	64	88.15	79.20	2.44	9.57
MT [12]	16	64	88.12	79.03	2.65	10.92
MT-Dice [2]	16	64	88.32	79.37	2.76	10.50
UA-MT [16]	16	64	88.88	80.21	2.26	7.32
LG-ER-MT	16	64	**89.62**	**81.31**	**2.06**	**7.16**

Firstly, we performed a quantitative evaluation of our methods. The first two rows in Table 1 show the segmentation performances of supervised V-Net using 80 and 16 annotated data, where the former results can be considered as the upper-bound performance. Our LG-ER-MT outperforms ASDNet and TCSE, both of which have proven to be effective in semi-supervised segmentation. Besides, our method achieves better segmentation results than MT and MT-Dice in almost all cases. Compared to the state-of-the-art method UA-MT, LG-ER-MT improves by 0.74% Dice and 1.1% Jaccard, while reducing the metrics ASD and 95HD. It can be seen that our LG-ER-MT is approaching the results of the supervised V-Net using 80 annotated data.

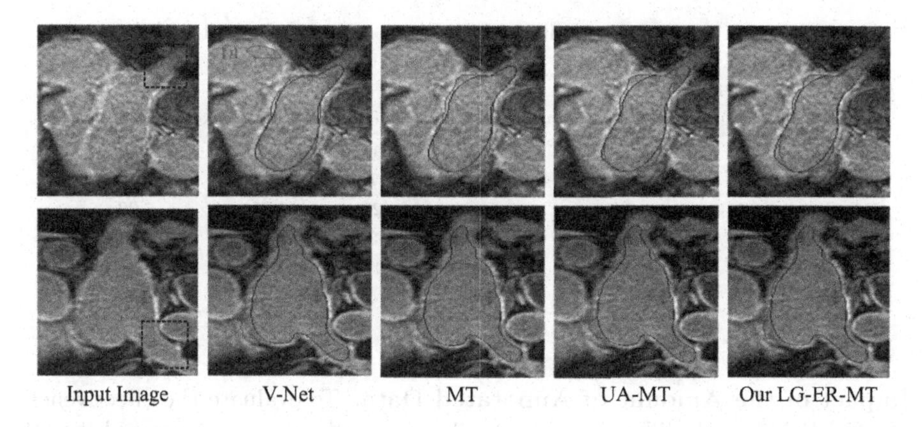

Fig. 3. Visualization of segmentation results. Green box labels PVs. Red and blue colors show the predictions and ground truths, respectively. (Color figure online)

We next qualitatively evaluated our method. Figure 3 gives the segmentation examples of the supervised method V-Net (using 16 annotated data) and three semi-supervised methods MT, UA-MT, and our LG-ER-MT. Note that the pulmonary veins (PVs), as indicated by the green box, are difficult to precisely recognize due to the limited MR resolution and ambiguous borders. Our LG-ER-MT method produces segmentation results closer to the ground truth with more accurate borders and shapes.

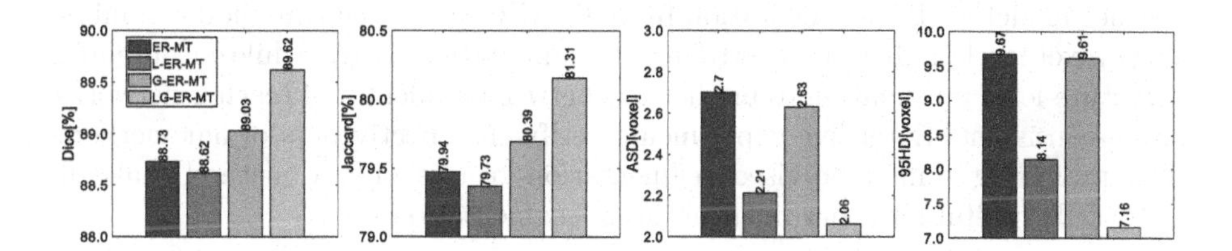

Fig. 4. Ablation study on different components.

To verify the efficacy of entropy loss as well as local and global structural consistency losses, we performed the ablation study as follows: 1) ER-MT; 2) ER-MT with the local structural consistency loss (L-ER-MT), 3) ER-MT with the global structural consistency loss (G-ER-MT), and 4) LG-ER-MT. Figure 4 demonstrates that three key components are able to bring performance improvement for semi-supervised 3D LA segmentation. Compare to ER-MT, local structural consistency loss can improve ASD and 95HD, while Global structural consistency loss can bring performance improvement on all metrics. Notably, the better performance of LG-ER-MT reveals the role of structure-level information complementarity.

Table 2. Comparison results with different amount of annotated data.

Method	# Scans used		Metrics			
	Annotated	Unannotated	Dice [%]	Jaccard [%]	ASD [voxel]	95HD [voxel]
UA-MT	8	72	84.25	73.48	**3.36**	13.84
LG-ER-MT	8	72	**85.54**	**75.12**	3.77	**13.29**
UA-MT	16	64	88.88	80.21	2.26	7.32
LG-ER-MT	16	64	**89.62**	**81.31**	**2.06**	**7.16**
UA-MT	24	56	90.16	82.18	2.73	8.9
LG-ER-MT	24	56	**90.33**	**82.42**	**2.06**	**6.92**

Impact of the Amount of Annotated Data. To evaluate the effectiveness of LG-ER-MT with different amounts of annotated data, we compared it with UA-MT, as shown in Table 2. Concretely, we investigated two other cases, where 10% training scans (*i.e.*, 8) and 30% training scans (*i.e.*, 24) were used as annotated data, and the remaining training scans as unannotated data, respectively. Compared with UA-MT, LG-ER-MT improves by 1.29% Dice and 1.64% Jaccard using 8 annotated data. The metrics ASD and 95HD are decreased by 0.67 and 1.98, when using 24 annotated data. These promising results further validate the effectiveness of our method for semi-supervised 3D LA segmentation.

4 Conclusion

We propose a novel local and global structure-aware entropy regularized mean teacher model for LA segmentation from 3D MR images. Our method can minimize voxel-level prediction uncertainty of unannotated images, while encouraging structure-level information of predictions between student and teacher networks to be consistent. Extensive experiments verify the effectiveness of our method. The promising semi-supervised segmentation results of our method make it potentially useful for other medical image segmentation tasks.

Acknowledgments. This work is supported in part by the Key-Area Research and Development Program of Guangdong Province, China (2020B010165004), the National Natural Science Foundation of China (61802177), the Open Project of State Key Laboratory for Novel Software Technology at Nanjing University (KFKT2019B10), the CAS Key Laboratory of Human-Machine Intelligence-Synergy Systems (2014DP173025), and the Hong Kong Research Grants Council under General Research Fund scheme (15205919).

References

1. Bian, C., et al.: Pyramid network with online hard example mining for accurate left atrium segmentation. In: Pop, M., et al. (eds.) STACOM 2018. LNCS, vol. 11395, pp. 237–245. Springer, Cham (2019). https://doi.org/10.1007/978-3-030-12029-0_26

2. Cui, W., et al.: Semi-supervised brain lesion segmentation with an adapted mean teacher model. In: Chung, A.C.S., Gee, J.C., Yushkevich, P.A., Bao, S. (eds.) IPMI 2019. LNCS, vol. 11492, pp. 554–565. Springer, Cham (2019). https://doi.org/10.1007/978-3-030-20351-1_43
3. French, G., Aila, T., Laine, S., Mackiewicz, M., Finlayson, G.: Consistency regularization and cutmix for semi-supervised semantic segmentation. arXiv preprint arXiv:1906.01916 (2019)
4. Kim, J., Jang, J., Park, H.: Structured consistency loss for semi-supervised semantic segmentation. arXiv preprint arXiv:2001.04647 (2020)
5. Lang, R.M., et al.: Recommendations for cardiac chamber quantification by echocardiography in adults: an update from the American society of echocardiography and the European association of cardiovascular imaging. Eur. Heart J.-Card. Img. **16**(3), 233–271 (2015)
6. Li, X., Yu, L., Chen, H., Fu, C.W., Heng, P.A.: Transformation consistent self-ensembling model for semi-supervised medical image segmentation. arXiv preprint arXiv:1903.00348 (2019)
7. Liu, Y., Chen, K., Liu, C., Qin, Z., Luo, Z., Wang, J.: Structured knowledge distillation for semantic segmentation. In: CVPR, pp. 2604–2613 (2019)
8. Liu, Y., Shu, C., Wang, J., Shen, C.: Structured knowledge distillation for dense prediction. arXiv preprint arXiv:1903.04197 (2019)
9. Long, M., Cao, Y., Cao, Z., Wang, J., Jordan, M.I.: Transferable representation learning with deep adaptation networks. IEEE TPAMI **41**(12), 3071–3085 (2018)
10. Milletari, F., Navab, N., Ahmadi, S.A.: V-net: fully convolutional neural networks for volumetric medical image segmentation. In: 2016 Fourth International Conference on 3D Vision (3DV), pp. 565–571. IEEE (2016)
11. Nie, D., Gao, Y., Wang, L., Shen, D.: ASDNet: attention based semi-supervised deep networks for medical image segmentation. In: Frangi, A.F., Schnabel, J.A., Davatzikos, C., Alberola-López, C., Fichtinger, G. (eds.) MICCAI 2018. LNCS, vol. 11073, pp. 370–378. Springer, Cham (2018). https://doi.org/10.1007/978-3-030-00937-3_43
12. Tarvainen, A., Valpola, H.: Mean teachers are better role models: weight-averaged consistency targets improve semi-supervised deep learning results. In: NIPS, pp. 1195–1204 (2017)
13. Tsai, Y.H., Hung, W.C., Schulter, S., Sohn, K., Yang, M.H., Chandraker, M.: Learning to adapt structured output space for semantic segmentation. In: CVPR, pp. 7472–7481 (2018)
14. Vu, T.H., Jain, H., Bucher, M., Cord, M., Pérez, P.: DADA: depth-aware domain adaptation in semantic segmentation. In: ICCV, pp. 7364–7373 (2019)
15. Wang, S., Yu, L., Li, K., Yang, X., Fu, C.-W., Heng, P.-A.: Boundary and entropy-driven adversarial learning for fundus image segmentation. In: Shen, D., et al. (eds.) MICCAI 2019. LNCS, vol. 11764, pp. 102–110. Springer, Cham (2019). https://doi.org/10.1007/978-3-030-32239-7_12
16. Yu, L., Wang, S., Li, X., Fu, C.-W., Heng, P.-A.: Uncertainty-aware self-ensembling model for semi-supervised 3D left atrium segmentation. In: Shen, D., et al. (eds.) MICCAI 2019. LNCS, vol. 11765, pp. 605–613. Springer, Cham (2019). https://doi.org/10.1007/978-3-030-32245-8_67

Improving Dense Pixelwise Prediction of Epithelial Density Using Unsupervised Data Augmentation for Consistency Regularization

Minh Nguyen Nhat To[1], Sandeep Sankineni[2], Sheng Xu[3], Baris Turkbey[2], Peter A. Pinto[4], Vanessa Moreno[5], Maria Merino[5], Bradford J. Wood[3], and Jin Tae Kwak[1](✉)

[1] School of Electrical Engineering, Korea University, Seoul 02841, Korea
jkwak@korea.ac.kr

[2] Molecular Imaging Program, National Cancer Institute, National Institutes of Health, Bethesda, MD 20892, USA

[3] Center for Interventional Oncology, National Cancer Institute, National Institutes of Health, Bethesda, MD 20892, USA

[4] Urologic Oncology Branch, National Cancer Institute, National Institutes of Health, Bethesda, MD 20892, USA

[5] Laboratory of Pathology, National Cancer Institute, National Institutes of Health, Bethesda, MD 20892, USA

Abstract. Although the amount of medical data keeps increasing, data annotations are scarce and often very difficult to obtain. It is even harder for the case that involves multi-modal imaging or data such as radiology-pathology correlation. In this regard, semi-supervised learning has the potential to leverage unlabeled data for improved medical image analysis. Herein, we propose a semi-supervised learning framework for the dense pixelwise prediction in multi-parametric magnetic resonance imaging (mpMRI). The proposed method predicts the epithelial density in mpMRI per-voxel basis. The ground truth annotations are only obtainable from the corresponding pathology images, which are often unavailable for mpMRI. Introducing unsupervised data augmentation and supervised training signal annealing strategies during training, the proposed method utilizes both labeled and unlabeled mpMRI in an efficient and effective manner. The experimental results demonstrate that the proposed framework is effective in improving the stability and accuracy of the density prediction. The proposed framework achieves the mean absolute error of 6.493, compared to 7.353 by the supervised learning counterpart, outperforming other competing methods. The results suggest that the semi-supervised learning framework could aid in resolving the scarcity of medical data and annotations, in particular for radiology-pathology correlation.

Keywords: Semi-supervised learning · Unsupervised data augmentation · Consistency regularization · Epithelial density · Radiology-pathology correlation

A. L. Martel et al. (Eds.): MICCAI 2020, LNCS 12261, pp. 572–581, 2020.
https://doi.org/10.1007/978-3-030-59710-8_56

1 Introduction

Training effective machine learning and deep learning models on a small amount of dataset has been a long-standing challenge [1, 2]. The larger the dataset, the greater the power of such models is [3]. Although large-scale datasets become more available [4], it is often costly and time-consuming to obtain annotations, i.e., ground truths, that are essential for model training and testing purposes. This is much severe for medical imaging where substantial domain knowledge is a prerequisite for the task. Therefore, an alternative manner to avoid the extensive data annotation and to improve the efficiency and effectiveness of the model training holds great potential for improving medical image analysis.

Semi-supervised learning (SSL) is a promising learning approach to leverage both (scarce) labeled and (plenty of) unlabeled data [5]. SSL has been successfully applied to several medical imaging applications. For instance, [6] proposed an alternative way to update model weights using both labeled and unlabeled data and achieved superior cardiac segmentation in MR imaging; [7] used prior knowledge to enforce consistency between similar labeled and unlabeled inputs in latent space for improving multiple Sclerosis lesion segmentation; a student-teacher algorithm was applied to retinal segmentation in [8] where a teacher model generates soft pixelwise labels on unlabeled data that are used for training a student model.

Most of SSL algorithms have been developed for a single-source imaging. In medical imaging, multi-modal imaging is prevalent, and it is even harder to obtain high quality labeled datasets. For example, radiology-pathology correlation is an important subject of study to integrate the radiologic and pathologic findings for optimal patient care [9]. Previously, the relationship between prostate multi-parametric magnetic resonance imaging (mpMRI) and histopathology has been investigated [10, 11], suggesting that histopathological analysis in prostate mpMRI could aid a quantitative assessment of tissue microstructures. In such cases, the annotations in both imaging modalities are required but are often scarce in both or either of imaging modalities.

In this study, we propose an SSL approach of deep neural networks for improved histopathology analysis in prostate mpMRI. The approach aims to predict the epithelial density in mpMRI per-voxel basis, which are only available from the corresponding pathology images. A deep neural network is trained based upon a limited amount of the labeled mpMRI slices and pathology images and a larger amount of the unlabeled mpMRI slices. Introducing unsupervised data augmentation (UDA) [12], the approach utilizes the unlabeled mpMRI slices in an efficient and effective fashion. The introduction of a new training strategy, called Supervised Training Signal Annealing (STSA), further facilitates efficient training of the deep neural network by controlling the amount of the labeled data used during training. We also formulate the objective functions in a way that the contribution of the labeled and unlabeled data is adaptively regularized during training. Finally, the proposed method achieves the robust and accurate dense pixelwise prediction of the epithelial density in prostate mpMRI.

2 Methodology

The pipeline of the proposed approach is illustrated in Fig. 1. It utilizes a deep residual neural network (DRNN) to predict epithelial density in prostate mpMRI. The DRNN is trained using the labeled mpMRI slices and pathology images in a supervised manner as well as using the unlabeled mpMRI slices in an unsupervised fashion by introducing UDA for the dense pixelwise prediction (PixUDA).

2.1 PixUDA:Unsupervised Data Augmentation for Dense Pixelwise Prediction

The core concept of PixUDA is to utilize unsupervised data augmentation (UDA) to improve the robustness and accuracy of the dense pixelwise prediction in deep learning models. PixUDA attempts to minimize the disagreement between the pixelwise predictions and their ground truths from the labeled dataset, meanwhile maximizing the consistency between the predictions of different transformations of the same data from the unlabeled dataset.

Formally, let $\mathcal{S} = \{(x_i, y_i) : i = 1, \ldots, B\}$ be a set of B labeled dataset and $\mathcal{U} = \{\widehat{x}_i : i = 1, \ldots, \gamma B\}$ be a set of γB unlabeled dataset, where x_i and $\widehat{x}_i$ are labeled and unlabeled inputs, respectively, y_i is the ground truth label of x_i, and γ is a hyperparameter denoting the size ratio between $\mathcal{S}$ and $\mathcal{U}$. Given an input x_i, a deep learning model $\mathcal{F}_\theta$ produces an output $\mathcal{F}_\theta(x_i)$ in the same dimensional space with y_i, where θ is a set of parameters trained on both labeled and unlabeled data. Let $\tau \sim \mathcal{T}$ and $\tau' \sim \mathcal{T}$ be a distinct set of arithmetic augmentations drawn from a family of arithmetic functions $\mathcal{T} = \{t_i : i = 1, \ldots, n\}$ where n is the number of arithmetic functions. We assume that, for any augmentations τ and τ', $\tau(x_i)$, $\tau'(x_i)$ and x_i would share the same ground truth label even though the ground truth label is not explicitly available.

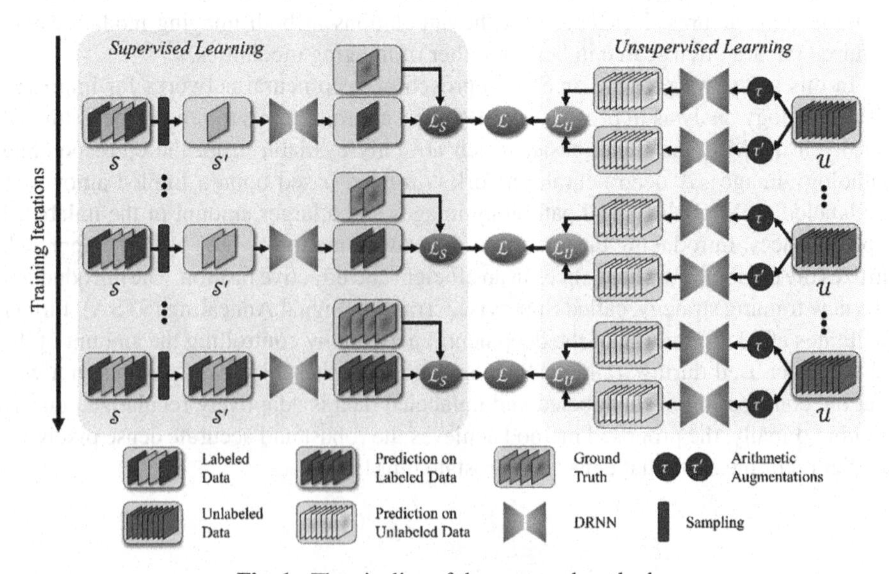

Fig. 1. The pipeline of the proposed method.

Objective Functions. The overall objective function for PixUDA includes a supervised loss $\mathcal{L}_S$ and an unsupervised loss $\mathcal{L}_U$ as follows:

$$\mathcal{L} = \mathcal{L}_S + \mathcal{L}_U. \tag{1}$$

where $\mathcal{L}_S$ is the loss between the prediction map $\mathcal{F}_\theta(\boldsymbol{x})$ and ground truth density map $\boldsymbol{y}$ and $\mathcal{L}_U$ measures the discrepancy between the predictions on two different arithmetic augmentations of the same unlabeled data – $\mathcal{F}_\theta(\tau(\hat{x}))$ and $\mathcal{F}_\theta(\tau'(\hat{x}))$.

We define the two loss functions based upon the Gaussian log-likelihood function [13]. Given an input $\boldsymbol{x}$, let us assume that the likelihood of the output can be written as a Gaussian with mean $\mathcal{F}_\theta(\boldsymbol{x})$ and a noise $\boldsymbol{\sigma}$ as:

$$p(y|\mathcal{F}_\theta(\boldsymbol{x})) = \mathcal{N}\left(\mathcal{F}_\theta(\boldsymbol{x}), \boldsymbol{\sigma}^2\right). \tag{2}$$

Then, the log-likelihood can be written as:

$$\log p(y|\mathcal{F}_\theta(\boldsymbol{x})) \propto -\frac{1}{2\boldsymbol{\sigma}^2}\|y - \mathcal{F}_\theta(\boldsymbol{x})\|^2 - \log \boldsymbol{\sigma}. \tag{3}$$

Finally, the negative log-likelihood forms the loss function $\mathcal{L}_S$ as:

$$\mathcal{L}_S = \frac{1}{\boldsymbol{\sigma}_S^2 B}\sum_{i=1}^{B}\frac{1}{\boldsymbol{K}_i}\sum_{\boldsymbol{K}_i}\left(\mathcal{F}_\theta(\boldsymbol{x}_i) - \boldsymbol{y}_i\right)^2 + \log \boldsymbol{\sigma}_S \tag{4}$$

where $\boldsymbol{K}_i$ is a set of voxels in $\boldsymbol{y}_i$ under consideration. Similarly, the loss function $\mathcal{L}_U$ can be given as:

$$\mathcal{L}_U = \frac{1}{\sigma_U^2 \gamma B}\sum_{i=1}^{\gamma B}\frac{1}{K_i}\sum_{K_i}\left(\mathcal{F}_\theta(\tau(\hat{x}_i)) - \mathcal{F}_\theta(\tau'(\hat{x}_i))\right)^2 + \log \boldsymbol{\sigma}_U. \tag{5}$$

Both $\boldsymbol{\sigma}_S$ and $\boldsymbol{\sigma}_U$ are noise scalars. In minimizing the overall objection function, these serve as the relative weights of the two loss functions. The larger the noise is, the smaller the weight or the contribution of the corresponding loss is. Both loss functions are regularized by the second term in each function, i.e., $\log \boldsymbol{\sigma}_S$ and $\log \boldsymbol{\sigma}_U$.

Augmentations. The PixUDA utilizes augmentations for both labeled and unlabeled datasets. The augmentations can be divided into two types: geometric and arithmetic augmentations. The geometric augmentations are operations that perform spatial deformations such as cropping and rotation. The arithmetic augmentations numerically transform the pixel intensity in an input image. As for the geometric augmentations, we utilize a random scaling, translation, rotation, shearing, perspective transform, and horizontal flipping. The arithmetic augmentations consist of a Gaussian blur and contrast normalization. For both supervised and unsupervised training, we use an identical set of geometric and arithmetic augmentations. During training, the geometric augmentations are first applied to the inputs, including MRI slices and ground truth density maps (for the supervised learning). Arithmetic augmentations are then applied to the MRI slices only. It is noteworthy to emphasize that the inputs to both τ and τ' in (5) are transformed by the same set of geometric augmentations.

2.2 Deep Residual Neural Network Architecture

For the epithelial density estimation in mpMRI, we specifically design a deep residual neural network (DRNN) that follows the encoder-decoder architecture and adopts densely connected blocks and residual connections for improving efficacy. The encoder comprises of five convolutional blocks. The first block contains a 7×7 convolutional layer and a max-pooling layer of stride 2. The following four blocks utilize densely connected layers of a stacked 1×1 and 3×3 convolutions. Three of the four blocks are followed by a transition layer, including a 1×1 convolution and a max-pooling layer of stride 2. The highest number of feature maps that the five blocks produce are 8, 32, 64, 128, and 256, respectively. The decoder includes four decoding blocks and a prediction block. Each block contains a 2×2 transposed convolution layer and a residual layout of $1 \times 1, 3 \times 3$, and 1×1 convolutions. The four decoding blocks output 512, 256, 128, and 32 feature maps, respectively. The prediction block generates the prediction map, in which the value ranges from -1 to 1, using a 1×1 convolution layer and a Tanh activation layer.

2.3 STSA: Supervised Training Signal Annealing

Overfitting is a critical and widespread issue in training supervised models for medical imaging, in particular due to the small size of the labeled dataset. This issue might persist in SSL and hinder the effective usage of the unlabeled datasets. To address this issue, we propose a simple yet effective training strategy, called Supervised Training Signal Annealing (STSA), to restrict the over-training on the labeled dataset in the early training stage. Formally, let $B_t = B \times \frac{t}{\alpha T} + 1$ be the number of the labeled samples within S that is used for the loss computation where T is the total number of the iterations, t is the current iteration, and α is a coefficient that controls the rate of the labeled samples. At an iteration t, B_t samples are randomly sampled from the B labeled dataset for training. The number of the labeled data revealed to the model is linearly increased across the training procedure, which reduces the model's liability on the labeled dataset at the early training stage while encouraging the model to learn more from the unlabeled dataset.

We note that our method is different from the others that gradually increase the batch size [14], as the batch size remains the same across the training in STSA where only the percentage of the labeled data changes. STSA also differs from approaches that increase the percentage of the whole dataset. In contrast, the size of the whole labeled dataset remains the same in the proposed method. The key idea is to control how frequently the labeled training samples can be seen by the model. STSA is similar to the training scheme proposed in [12] where the training signal of the labeled dataset is only backpropagated if and only if the model's confidence in the prediction reaches a prescheduled threshold. However, their ablation studies showed inconsistent and unreliable results on this technique. Setting the threshold value and scheduling threshold is rather ambiguous.

3 Experiments and Results

3.1 Datasets

This study includes 242 patients who gave informed consent and had biopsy-proven adenocarcinoma of the prostate between Aug. 2011 and Sep. 2014. 42 of them, who underwent mpMRI, T2-weighted MRI (T2W), and apparent diffusion coefficient (ADC), and robot-assisted radical prostatectomy, were included in the labeled dataset, while the rest constitutes the unlabeled dataset. The labeled dataset from 42 patients is divided into the train set (25 patients from Aug. 2011 to Nov. 2013) and the test set (17 patients from Dec. 2013 to Sep. 2014). For the labeled dataset, epithelial density maps are generated from the digital pathology analysis of the hematoxylin and eosin (H&E) stained whole-mount tissue specimens obtained from the robot-assisted radical prostatectomy. The resultant density maps are further registered with the corresponding mpMRI per voxelwise, then serve as ground truths. Specifically, the patient-specific model (PSM) is created from the mpMRI for each patient using 3-D computer-aided design software and printer. Following the prostatectomy, whole-mount tissue specimens are cut in the mold as matching the location of the mpMRI slices. The corresponding mpMRI slices and pathology images are registered using the outer contours and internal fiducial structures within the prostate via thin-plate splines image registration. The details of the entire procedure are available in [11]. Regions of interest (ROIs) on mpMRI and pathology images are identified and confirmed by experienced pathologists and radiologists. 312 ROIs within 125 slices are identified and used for the supervised training. In sum, the training set consists of 71 labeled and 1855 unlabeled mpMRI slices. The test set includes 54 labeled mpMRI slices.

3.2 Implementation Details

For each pair of T2W and ADC, a voxel intensity is normalized as: $x = \frac{x-Q_1}{Q_3-Q_1}$ where x is a voxel intensity in an MR slice, Q_1 and Q_3 is the value at 25th and 75th quantile, respectively, of the intensity distribution within the prostate. Each epithelial density map is linearly scaled to [0, 1] using the minimum and maximum density of the whole training set. All images are cropped to the spatial size of 256 × 256 voxels, which is sufficiently large to include the whole prostate. Adam optimizer is employed with parameters $\beta_1 = 0.9$, $\beta_1 = 0.999$, and $\varepsilon = 1.0e^{-8}$. The decayed cycle learning rate schedule is used with the initial smallest and largest learning rate assigned to $1.0e^{-4}$ and $1.0e^{-3}$, respectively, and the cycle length decay and cycle magnitude decay factors are set to 0.9 and 0.95, respectively. The size of the labeled and unlabeled dataset in each iteration is 8 and 16 ($\gamma = 2$), respectively. Every model is trained for 1800 iterations. For STSA, α is set to 0.9. The model at the last iteration is used to evaluate the test set.

3.3 Results and Discussions

We report the performance of the proposed method in Table 1. The proposed method achieved the mean absolute error (MAE) of 6.493 (lower is better) and structural similarity index (SSIM) [15] of 0.841 (higher is better). Without PixUDA, the performance of the model was substantially decreased (MAE of 7.353 and 0.811 SSIM), indicating the utility of the unlabeled dataset. Moreover, we examined the effectiveness of PixUDA on a variety of network architectures. For the encoder-decoder framework, U-Net [16] and U-Net-pp [17] were employed with a different number of parameters. As for the backbone network, VGG, ResNet, and DenseNet were utilized. In a head-to-head comparison, the proposed method outperformed all the other competing networks as shown in Table 1. The evaluation of the comparative experiments further emphasizes the importance of PixUDA. Regardless of the network architecture, the networks with PixUDA always resulted in a superior performance for both MAE and SSIM. All the supervised approaches, i.e., without PixUDA, obtained $\geq$7.244 MAE and $\leq$0.817 SSIM. However, the networks equipped with PixUDA attained $\leq$7.160 MAE and $\geq$0.828 SSIM. These results demonstrate the effectiveness and generalizability of PixUDA in improving the model performance. Furthermore, we qualitatively evaluated the epithelial density prediction by the proposed method (Fig. 2). The results were compared to the ground truth density maps and the predictions by the DRNN without PixUDA, i.e., supervised learning. The density predictions were comparable to the ground truth maps, in particular for the low and intermediate density regions, and superior to the predictions by the supervised method.

Table 1. Results of epithelial density prediction.

Network	Backbone	#Parameters (Millions)	PixUDA	Metrics	
				MAE	SSIM
DRNN(proposed)		3.385	–	7.353	0.811
			✓	**6.493**	**0.841**
U-Net	VGG	3.345	–	7.423	0.807
			✓	**7.113**	**0.828**
U-Net-pp	VGG	3.414	–	7.460	0.807
			✓	**6.909**	**0.829**
U-Net	VGG	7.763	–	7.411	0.811
			✓	**6.759**	**0.829**
U-Net-pp	VGG	8.733	–	7.509	0.810
			✓	**6.870**	**0.829**
U-Net	ResNet	2.765	–	7.244	0.817
			✓	**7.160**	**0.834**
U-Net-pp	ResNet	3.649	–	7.426	0.810
			✓	**7.003**	**0.836**
U-Net	DenseNet	2.640	–	7.569	0.807
			✓	**7.011**	**0.829**
U-Net-pp	DenseNet	2.837	–	7.244	0.813
			✓	**6.850**	**0.832**

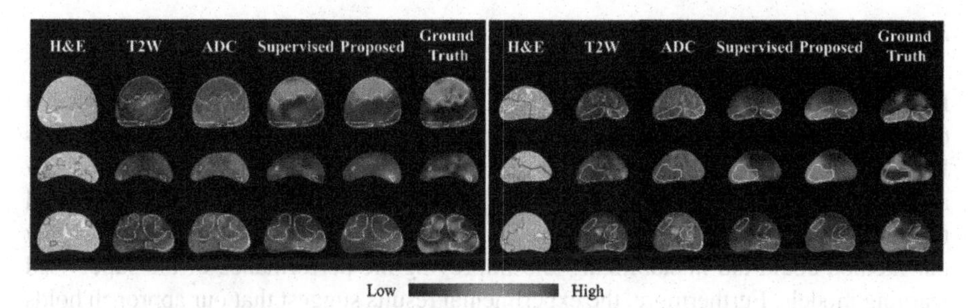

Fig. 2. Qualitative assessment of epithelial density prediction. The columns "Supervised" and "Proposed" are the predictions by the DRNN without and with PixUDA, respectively.

Ablation Study. To better understand the effect of the number of the unlabeled dataset and the availability of PixUDA and STSA, we conducted extensive ablation experiments using the proposed DRNN and the two best variants of U-Net and U-Net-pp (with 7.763 and 8.733 million parameters, respectively). The results are presented in Fig. 3. Utilizing the unlabeled dataset, i.e., PixUDA, we obtained a superior performance of the proposed DRNN (Fig. 3a); the increase in the number of the unlabeled dataset progressively improved the prediction performance with and without STSA. The models trained with STSA outperformed the ones without STSA. Similar trends were found for both U-Net variants (Fig. 3b). PixUDA and STSA independently contribute to the improvement in the performance. With a sufficient number of the unlabeled dataset, the models, equipped with both PixUDA and STSA, performed the best.

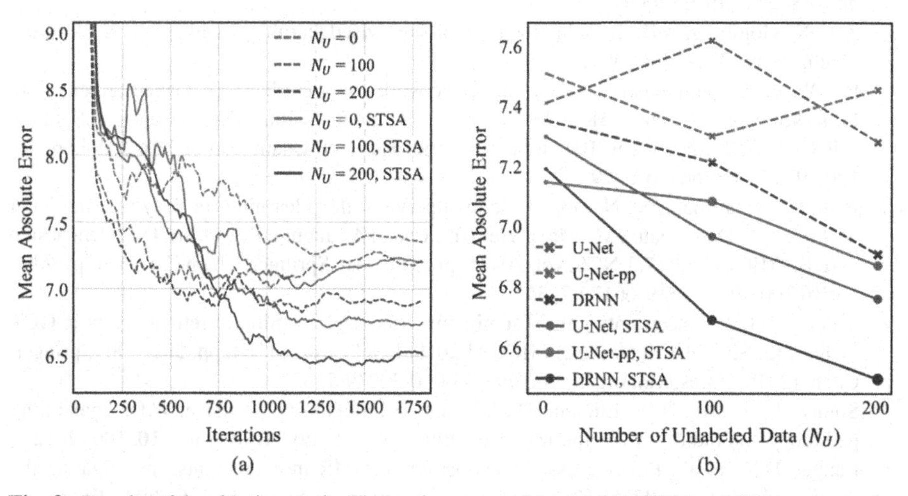

Fig. 3. Results of the ablation study. The performance of the (a) (b) DRNN and (b) two competing methods on the test set with and without PixUDA and STSA. N_U is the number of unlabeled dataset. For $N_U = 100$ and 200, γ is set to 1 and 2, respectively.

4 Conclusions

Herein, we address the issue of data scarcity in the dense pixelwise prediction of epithelial density by introducing an approach of SSL, called PixUDA. Our approach employs UDA for consistency regularization to utilize the large unlabeled dataset efficiently and effectively. We also propose a training scheme, called STSA, which is essential to fully exploit PixUDA. The results in the comparative experiments demonstrate that our method could aid in stabilizing and improving the performance of the supervised learning models. Furthermore, the experimental results suggest that our approach holds great potential to be a general approach for dense pixelwise prediction tasks.

Funding. This work was supported by the National Research Foundation of Korea (NRF) grant funded by the Korea government (MSIP) (No. 2016R1C1B2012433).

References

1. Jain, A.K., Chandrasekaran, B.: 39 Dimensionality and sample size considerations in pattern recognition practice. Handb. Stat. **2**, 835–855 (1982)
2. Cawley, G.C., Talbot, N.L.: On over-fitting in model selection and subsequent selection bias in performance evaluation. J. Mach. Learn. Res. **11**, 2079–2107 (2010)
3. Raudys, S.J., Jain, A.K.: Small sample size effects in statistical pattern recognition: recommendations for practitioners. IEEE Trans. Pattern Anal. Mach. Intell. **13**, 252–264 (1991)
4. Deng, J., Dong, W., Socher, R., Li, L.-J., Li, K., Fei-Fei, L.: Imagenet: a large-scale hierarchical image database. In: 2009 IEEE Conference on Computer Vision and Pattern Recognition, pp. 248–255. IEEE (2009)
5. Zhu, X., Goldberg, A.B.: Introduction to semi-supervised learning. Synth. Lect. Artif. Intell. Mach. Learn. **3**, 1–130 (2009)
6. Bai, W., et al.: Semi-supervised learning for network-based cardiac MR image segmentation. In: Descoteaux, M., Maier-Hein, L., Franz, A., Jannin, P., Collins, D.L., Duchesne, S. (eds.) MICCAI 2017. LNCS, vol. 10434, pp. 253–260. Springer, Cham (2017). https://doi.org/10.1007/978-3-319-66185-8_29
7. Baur, C., Albarqouni, S., Navab, N.: Semi-supervised deep learning for fully convolutional networks. In: Descoteaux, M., Maier-Hein, L., Franz, A., Jannin, P., Collins, D.L., Duchesne, S. (eds.) MICCAI 2017. LNCS, vol. 10435, pp. 311–319. Springer, Cham (2017). https://doi.org/10.1007/978-3-319-66179-7_36
8. Sedai, S., et al.: Uncertainty guided semi-supervised segmentation of retinal layers in OCT images. In: Shen, D., et al. (eds.) MICCAI 2019. LNCS, vol. 11764, pp. 282–290. Springer, Cham (2019). https://doi.org/10.1007/978-3-030-32239-7_32
9. Sorace, J., Aberle, D.R., Elimam, D., Lawvere, S., Tawfik, O., Wallace, W.D.: Integrating pathology and radiology disciplines: an emerging opportunity? BMC Med. **10**, 100 (2012)
10. Langer, D.L., et al.: Prostate tissue composition and MR measurements: investigating the relationships between ADC, T2, K trans, ve, and corresponding histologic features. Radiology **255**, 485–494 (2010)
11. Kwak, J.T., et al.: Prostate cancer: a correlative study of multiparametric MR imaging and digital histopathology. Radiology **285**, 147–156 (2017)
12. Xie, Q., Dai, Z., Hovy, E., Luong, M.-T., Le, Q.V.: Unsupervised data augmentation. arXiv preprint (2019)

13. BenTaieb, A., Hamarneh, G.: Uncertainty driven multi-loss fully convolutional networks for histopathology. In: Cardoso, M.J., et al. (eds.) LABELS/CVII/STENT -2017. LNCS, vol. 10552, pp. 155–163. Springer, Cham (2017). https://doi.org/10.1007/978-3-319-67534-3_17
14. Smith, S.L., Kindermans, P.-J., Le, Q.V.: Don't decay the learning rate, increase the batch size. In: International Conference on Learning Representations (2018)
15. Wang, Z., Bovik, A.C., Sheikh, H.R., Simoncelli, E.P.: Image quality assessment: from error visibility to structural similarity. IEEE Trans. Image Process. **13**, 600–612 (2004)
16. Ronneberger, O., Fischer, P., Brox, T.: U-net: convolutional networks for biomedical image segmentation. In: Navab, N., Hornegger, J., Wells, W.M., Frangi, A.F. (eds.) MICCAI 2015. LNCS, vol. 9351, pp. 234–241. Springer, Cham (2015). https://doi.org/10.1007/978-3-319-24574-4_28
17. Zhou, Z., Rahman Siddiquee, M.M., Tajbakhsh, N., Liang, J.: UNet++: a nested U-net architecture for medical image segmentation. In: Stoyanov, D., et al. (eds.) DLMIA/ML-CDS -2018. LNCS, vol. 11045, pp. 3–11. Springer, Cham (2018). https://doi.org/10.1007/978-3-030-00889-5_1

Knowledge-Guided Pretext Learning for Utero-Placental Interface Detection

Huan Qi[1(✉)], Sally Collins[2], and J. Alison Noble[1]

[1] Institute of Biomedical Engineering (IBME), University of Oxford, Oxford, UK
qh.scdapm@gmail.com

[2] Nuffield Department of Women's and Reproductive Health, University of Oxford, Oxford, UK

Abstract. Modern machine learning systems, such as convolutional neural networks rely on a rich collection of training data to learn discriminative representations. In many medical imaging applications, unfortunately, collecting a large set of well-annotated data is prohibitively expensive. To overcome data shortage and facilitate representation learning, we develop Knowledge-guided Pretext Learning (KPL) that learns anatomy-related image representations in a pretext task under the guidance of knowledge from the downstream target task. In the context of utero-placental interface detection in placental ultrasound, we find that KPL substantially improves the quality of the learned representations without consuming data from external sources such as ImageNet. It outperforms the widely adopted supervised pre-training and self-supervised learning approaches across model capacities and dataset scales. Our results suggest that pretext learning is a promising direction for representation learning in medical image analysis, especially in the small data regime.

1 Introduction

Recent years have seen breakthroughs in the field of medical image analysis driven by deep convolutional neural networks (CNNs). Despite advances in architectural refinement, incorporating more representative training data remains the most plausible way to improve model generalizability [1]. Recent successful medical deep-learning systems (e.g. [2,3]) rely on a large amount of annotated data to perform effective representation learning. For many medical applications, however, it is not applicable to build datasets of similar orders of magnitudes as those large-scale studies did [4].

Placenta accreta spectrum disorders (PASD) are adverse obstetric conditions of particular rarity. Ultrasonography-based prenatal assessment of PASD requires examination of structural and vascular abnormalities near the utero-placental interface (UPI). Manually localizing UPI can be challenging and time-consuming even for obstetric specialists due to the UPI's variable shape and length compounded by low contrast, as shown in Fig. 1. These issues render learning-based detection of UPI clinically useful in assisting PASD assessment

A. L. Martel et al. (Eds.): MICCAI 2020, LNCS 12261, pp. 582–593, 2020.
https://doi.org/10.1007/978-3-030-59710-8_57

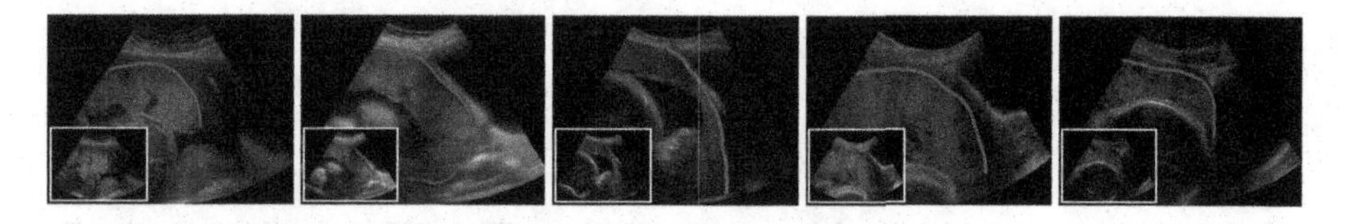

Fig. 1. Placental ultrasound samples. UPIs are annotated as red curves, featured by the indistinct appearance with variable shape and length.

[5]. Unfortunately, the low incidence rate makes collecting large-scale annotated data prohibitively expensive in practice.

To facilitate feature learning under data shortage, we develop Knowledge-guided Pretext Learning (KPL) that learns anatomy-related image representations without using data from external sources such as ImageNet. KPL works with a commonly used pretext task known as *solving jigsaw puzzles* [6,7]. Particularly, KPL tackles this task only in the region of interest, which is guided by supervision signals from the target task. We find that KPL substantially improves the quality of the learned representations for the downstream UPI detection task, outperforming widely adopted supervised pre-training and self-supervised learning approaches across model capacities and dataset scales. Our results demonstrate the potential of pretext learning in medical image analysis, especially in the small data regime.

Related Work: Transfer learning (TL) is a widely adopted technique in medical image analysis, whereby CNN architectures and their weights are first pre-trained on external sources (X_e, Y_e) (e.g. ImageNet) and then fine-tuned on downstream medical imaging data (X_t, Y_t) (Fig. 2b). Compared to a random initialization approach (Fig. 2a), TL is reported to benefit from feature reuse [1,8]. However, recent studies have challenged some common beliefs about TL. For instance, He *et al.* showed that TL indeed helped achieve faster convergence, but did not necessarily yield better performance compared to random initialization [9]. Raghu *et al.* demonstrated that randomly initialized small models could perform comparably to large ImageNet pre-trained models on two medical imaging tasks [8]. Their findings imply that good performance on external data is not a reliable indicator of success on downstream medical imaging tasks. These results raise questions about the applicability of current TL approaches.

Alternatively, self-supervised learning (SSL) aims to obtain semantically relevant image representations without any supervision. Instead, it devises pretext tasks by applying a pre-defined transformation Φ to input images X_t and training a model to predict properties of the transformation from the transformed images $\Phi(X_t)$ [11], for which ground truth is freely available (Fig. 2c). Such pretext task includes, but are not limited to, solving patch-wise jigsaw puzzles [6,7], predicting image rotation [12,13], determining relative position [14] and absolute position [15]. However, since routine placental ultrasound scans are sector-shaped, directly applying a generic SSL transformation can be problematic because an SSL model can easily over-fit to sector boundaries without

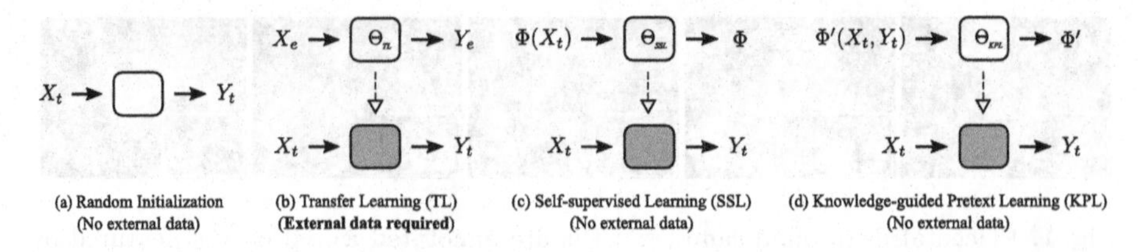

Fig. 2. Four popular feature-related training setups in medical imaging and computer vision. (a) depicts a random initialization approach by initializing the network with zero-mean Gaussian weights [10]; (b) illustrates the widely-adopted transfer learning (TL) approach, which requires well-annotated external data; (c) denotes a generic self-supervised learning (SSL) approach, where features are learnt using freely available supervisions derived from certain types of transformation Φ; (d) shows the proposed knowledge-guided pretext learning approach, where the transformation Φ' is guided by knowledge from the downstream task. Neither SSL or KPL requires any external data.

actually learning useful semantics. To overcome this issue, we propose to solve pretext tasks under the guidance of Y_t (i.e. knowledge from downstream tasks, Fig. 2-d). The resulting transformation $\Phi'(X_t, Y_t)$ only performs feature learning in the region of interest determined by Y_t, reducing the risk of over-fitting to sector boundaries.

2 Methods

`Jigsaw` Pretext Learning: Noroozi *et al.* proposed to learn image representation in a self-supervised manner by solving jigsaw puzzles induced from the given image [6]. As shown in Fig. 3a, it works by first cropping an input image X_t into a grid of 3×3 non-overlapping patches. After shuffling these patches according to a permutation Φ randomly selected from a set of pre-defined permutations, a CNN Θ_{SSL} is then trained to *solve this puzzle* by predicting Φ given these unordered patches $\Phi(X_t)$. Unfortunately, sector boundaries are more discriminative than the actual visual contents in this puzzle-solving scenario, which likely renders the learnt representation irrelevant for downstream tasks.

To this end, we propose to solve jigsaw puzzles by repurposing the downstream supervision signal Y_t to guide representation learning. We refer to this method as knowledge-guided pretext learning (KPL). Figure 3b illustrates the process of solving jigsaw puzzles under KPL. We first perform a spatial clustering that groups UPI pixels into N_k segments, as encoded in colours. A sampling step follows by cropping N_k square patches from X_t, whose centres are randomly drawn from pixels on N_k UPI segments respectively in a specific order. For simplicity, we take the order from top left to bottom right along UPI. Similar to `Jigsaw`-under-SSL, a random shuffle Φ' is applied to these patches and a CNN Θ_{KPL} is then trained to predict this permutation. As illustrated in Fig. 4a, Θ_{KPL} is an N_k-way Siamese CNN with shared parameters that map each patch into a 128-d vector. The resulting N_k vectors are concatenated to predict Φ' as

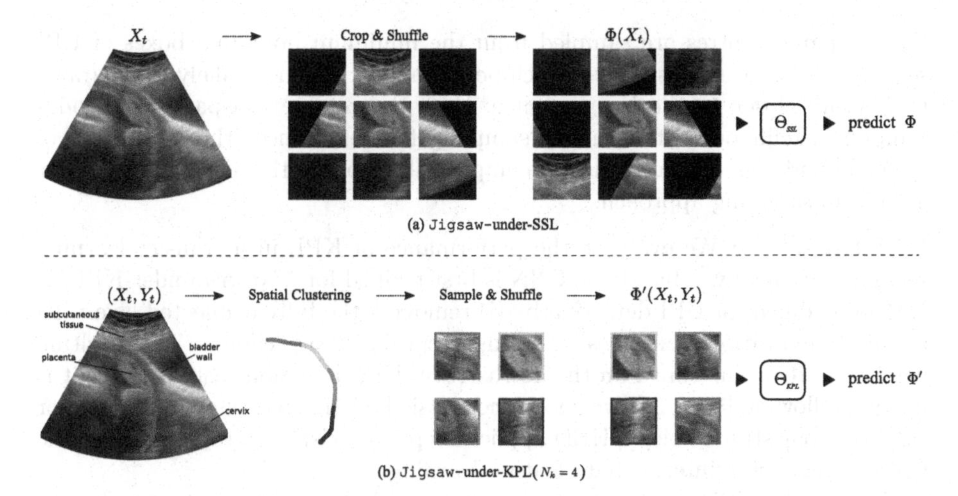

Fig. 3. (a)Jigsaw-under-SSL solves a generic puzzle by shuffling a grid of 3×3 non-overlapping patches without considering the visual contents. (b) Jigsaw-under-KPL solves a specific puzzle guided by Y_t, which reduces the occurrence of sector boundaries.

a classification task [16]. The total number of permutations $|P|$ is at most $N_k!$ (e.g. a maximum of 24 permutations for $N_k = 4$). As illustrated in Fig. 1 and Fig. 3b, the UPI neighbourhood carries rich semantics about placenta anatomy. We expect that the sampled patches also contain key information relevant to the downstream UPI detection task.

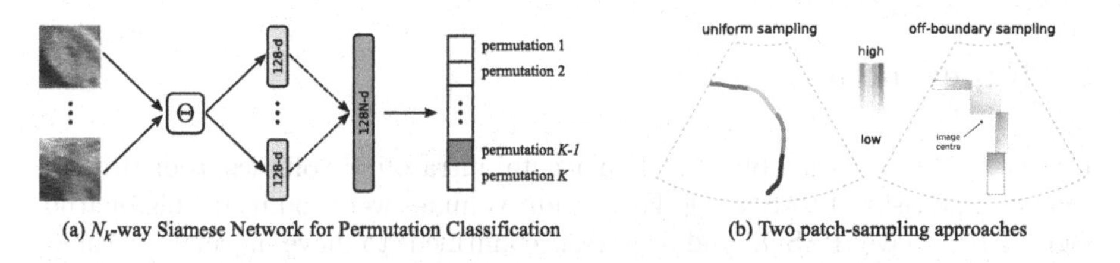

Fig. 4. (a) presents a N_k-way Siamese architecture to predict permutation given a set of N patches. Each patch is encoded by a 128-d vector. (b) illustrates two patch-sampling strategies. The proposed off-boundary sampling assigns higher weights to positions closer to the image centre, thus further reduces the occurrence of sector boundaries.

Off-boundary Sampling: As opposed to SSL, the proposed Jigsaw-under-KPL conducts feature learning only in the region of interest through the guidance of the downstream task, thus alleviates the risk of over-fitting to sector boundaries. However, patches drawn from top-left and bottom-right UPI segments may still contain sector boundaries when sampled uniformly. To further reduce this risk, we propose a heuristic off-boundary sampling approach. As illustrated in

Fig. 4b, patch centres are sampled from the minimum bounding boxes of UPI segments. Moreover, positions in each bounding box are more likely to be drawn if they are closer to the image centre. Here the colour transparency encodes sampling likelihood. Off-boundary sampling further reduces the occurrence of sector boundaries. Our experiments empirically support this design choice over a uniform sampling approach.

UPI Detection: We evaluate the performance of KPL in a transfer learning setting. Specifically, a backbone CNN is first trained for `Jigsaw`-under-KPL. It is then modified for UPI detection by (i) removing the fully-connected layer; (ii) adding computational modules according to specific design choices. The resulting network is then fine-tuned on the downstream UPI detection task. No weight is frozen. Following [5,17,18], we adopt the classic holistically-nested edge detector (HED) in our study. Briefly, HED applies deep supervision on top of a backbone CNN to regularize intermediate feature learning.

Concretely, UPI detection is formulated as a binary segmentation task with $Y_t \in \{0,1\}^{H \times W}$ being the reference UPI mask, with UPI pixels taking the value 1. Given an input image X_t, a UPI detection network predicts the probability $Pr(\hat{Y}_t^p = 1|X_t)$ for each pixel position p in X_t, where $\hat{Y}_t$ is the network output given X_t. Following [5], we adopt a weighted cross-entropy loss:

$$\mathcal{L}(\hat{Y}_t, Y_t) = \begin{cases} -\frac{\omega}{|Y_t|}\sum_p \log Pr(\hat{Y}_t^p = 1|X_t) & \text{if } Y_t^p = 1 \\ -\frac{1}{|Y_t|}\sum_p \log(1 - Pr(\hat{Y}_t^p = 1|X_t)) & \text{if } Y_t^p = 0 \end{cases} \tag{1}$$

where $\omega = |Y_t^-|/|Y_t^+|$ is a class-balancing weight that equalizes the expected weight update for both classes, with $|Y_t^+|$ nd $|Y_t^-|$ denoting the amount of UPI and non-UPI pixels in Y_t [19].

3 Experiments

Dataset: We had available 101 3D placental ultrasound volumes from 101 subjects at high risks of PASD [20]. Forty-eight volumes were confirmed histopathologically to have PASD, and the rest confirmed to have normal placentation. Static transabdominal 3D ultrasound volumes of the placental bed were obtained. Data usage was approved by the local research ethics committee. The median gestation was 30 1/7 weeks. All volumes were sliced along the sagittal plane. The resulting 11,166 2D image planes had the UPI manually annotated by X (a computer scientist) under the guidance of Y (an obstetric specialist). The median spatial resolution was 392×553 px, with a spatial sampling rate of 0.33 mm/pixel. A subject-level random split was performed to have 7,340 images from 67 subjects for training, 1,954 images from 17 subjects for validation and 1,872 images from 17 subjects for testing.

Model Setup: To investigate the effects of representation learning on downstream transfer task performance, we evaluate on the following setups: (i) training from random initialization; (ii) transfer learning from ImageNet pre-trained weights; (iii) transfer learning from CheXpert pre-trained weights [21]; (iv) transfer learning from `Jigsaw`-under-SSL; (v) transfer learning from `Jigsaw`-under-KPL. We further test on three HED instances with different backbones to investigate the effects of model capacity on downstream task performance. These backbones include: (i) Tiny-VGG[1]; (ii) VGG-13 [22]; (iii) VGG-19 [22], whose number of parameters before and after removing the fully-connected layers are listed in Table 2.

Table 1. Comparison of the ODS scores on the test set for UPI detection under five setups. Three HED instances with different backbones were listed. *Rand. Init.* denotes training from random initialization. TL-($\cdot$) denotes transfer learning from the corresponding external database. Performance of `Jigsaw`-under-SSL with $|P| = 90$ and `Jigsaw`-under-KPL with $N_k = 5$ and $|P| = 120$ are reported.

	Rand. Init.	TL-ImageNet	TL-CheXpert	SSL ($\lvert P\rvert = 90$)	KPL ($\lvert P\rvert = 120$)
Tiny-VGG	0.562	0.583	0.569	0.574	0.578
VGG-13	0.566	0.576	0.560	0.562	0.571
VGG-19	0.582	0.595	0.583	0.596	**0.605**

Table 2. Configurations of three backbone CNNs, including depth, number of parameters and ImageNet Top-1 and Top-5 accuracies on the validation set.

	Depth	# Params with/without FC layers	Top-1 Acc.	Top-5 Acc.
Tiny-VGG	13	70.0 M/1.1 M	63.6%	85.1%
VGG-13	13	126.9 M/9.0 M	71.6%	90.4%
VGG-19	19	137.0 M/19.1 M	74.2%	91.8%

Table 3. AUC scores of three backbone CNNs on the validation set of CheXpert using the *U-Ones* approach in [21]. Column 2-6 report the binary classification AUC scores for five chest-related clinical observations.

AUC	Cardiomegaly	Edema	Consolidation	Atelectasis	Pleural effusion
Tiny-VGG	0.812	0.897	0.901	0.848	0.915
VGG-13	0.820	0.875	0.911	0.858	0.919
VGG-19	0.827	0.857	0.903	0.880	0.902

[1] Tiny-VGG builts on VGG-13 [22], with 16-32-64-128-256 channels in five blocks. See Appendix.

Implementation: Basic geometry and contrast jittering were applied for data augmentation (see Appendix). For SSL and KPL, an input image was resized to 390×390 px and patches with size 120×120 px were drawn. We used the Adam optimizer with a mini-batch size of 16. Models were trained for 16 epochs, and the weights were stored for subsequent fine-tuning on the downstream task. For UPI detection task, an input image was resized to 360×360 px. We used a mini-batch size of 6 and optimized models with Adam for 20 epochs on NVIDIA Tesla P100 GPUs. The learning rate was set to 9e-4 and weight-decay was set to 1.1e-5. We evaluate UPI detection using a standard edge detection metric: the best F-measure on the dataset for a fixed prediction threshold (optimal dataset scale, or ODS [23]). All models were implemented with PyTorch.

Effects of Feature Learning: Table 1 compares the performance of UPI detection in terms of ODS score, under five setups. For TL-ImageNet and TL-CheXpert, backbone CNNs were first pre-trained on the corresponding external datasets before fine-tuned on UPI data. Table 2 and Table 3 display the pre-training results in terms of Top-1 and Top-5 accuracies for ImageNet and class-wise AUC scores for CheXpert. For SSL and KPL, an important hyper-parameter is the permutation number $|P|$, which is determined using the validation set. According to results in Fig. 5a-b, we set $|P| = 90$ for `Jigsaw`-under-SSL and $|P| = 120$ ($N_k = 5$) for `Jigsaw`-under-KPL. Note that $|P|$ can be as high as 9! for SSL, yet we find that a very large $|P|$ does not help with the downstream task. On the other hand, increasing $|P|$ up to 120 helps with UPI detection under KPL, which indicates that KPL is able to learn good representations that benefit the downstream task.

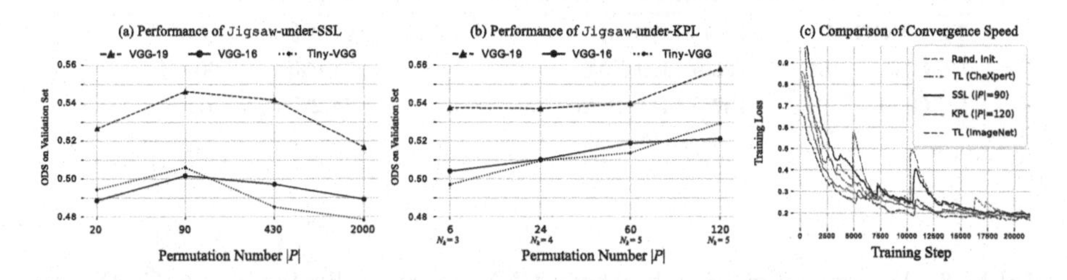

Fig. 5. (a) and (b) display performance of transfer learning from `Jigsaw`-under-SSL and `Jigsaw`-under-KPL respectively, by varying the permutation number $|P|$. ODS scores on the validation set are reported. (c) presents the training losses of HED (VGG-19) instances under different setups. Note that ImageNet pre-trained weights and KPL achieve the best convergence speeds in terms of training steps.

According to Table 1, UPI detection generally benefits from transfer learning, as opposed to the conclusion drawn by [8] for image classification. This result implies that localization tasks may be more feature-sensitive than classification tasks. The proposed KPL outperforms the *Rand. Init.*, TL-CheXpert and SSL approaches in terms of ODS and performs comparably to TL-ImageNet. The best performance (ODS = 0.605) was achieved by KPL under VGG-19. Interestingly, TL-ImageNet achieves a better performance than TL-CheXpert. One may expect the opposite since CheXpert is a medical imaging database after all. In terms of convergence speed, KPL and TL-ImageNet achieve the best performance compared to the rest, as illustrated in Fig. 5c. This result further indicates that KPL can produce good representations that generalize well for the downstream task.

Effects of Backbone Capacity: According to Table 1, HED instances powered by VGG-19 generally outperform those powered by Tiny-VGG and VGG-13. Tiny-VGG and VGG-13 achieve similar performances across different setups. Note that Tiny-VGG has the same depth as VGG-13 but with fewer channels in each block. This result indicates that increasing model depth can be more beneficial than increasing channel numbers for UPI detection.

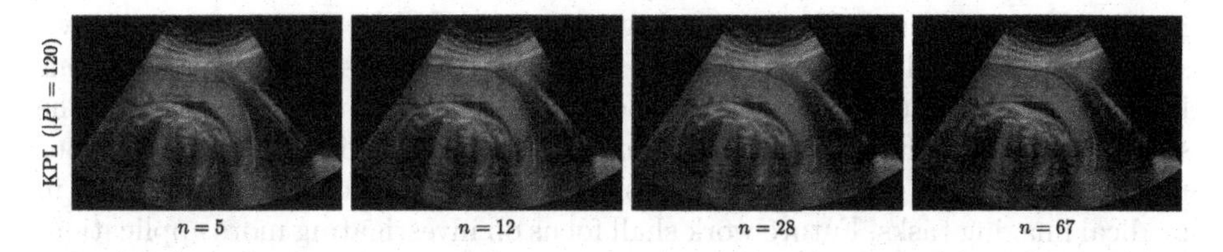

Fig. 6. KPL-powered UPI detection samples with an increase of training subjects ($n \in \{5, 12, 28, 67\}$). Red curves represent reference UPI masks and blue curves represent model predictions.

Table 4. Comparison of the ODS scores in the small data regime. HED models are powered by VGG-19 backbones.

# Subject (# plane)	Rand. Init.	TL-ImageNet	SSL ($\|P\| = 90$)	KPL ($\|P\| = 120$)
5 (640)	0.442	**0.470**	0.417	0.445
12 (1,350)	0.467	0.481	0.452	**0.521**
28 (3,025)	0.495	0.535	0.546	**0.567**
67 (7,340)	0.582	0.595	0.596	**0.605**

As shown in Table 2 and Table 3, three backbone CNNs achieve significantly different classification accuracies on ImageNet but perform comparably to each other on CheXpert. This result further validates that performance on an external database is not predictive of performance on the target task [8]. It is worth noting that the proposed KPL method does not exploit any external database and performs consistently well with different backbones.

Performance in the small data regime: We further evaluate the effect of feature learning in the small data regime by training HEDs with VGG-19 backbones on data from $n \in \{5, 12, 28\}$ subjects respectively. Test results are displayed in Table 4. Figure 6 displays some exemplar detection results. Compared to *Rand. Init.*, TL (ImageNet) and SSL, the proposed KPL method can substantially boost detection performance under data shortage. For instance, with only 12 subjects, KPL can learn rich representations that transfer very well to UPI detection (ODS=0.521), even outperforming *Rand. Init.* trained on 28 subjects.

4 Conclusion

In this paper, we propose knowledge-guided pretext learning (KPL) for utero-placental interface detection (UPI) in placental ultrasound. KPL only attends the region of interest for representation learning, as guided by downstream supervision. Experiments demonstrate that UPI detection benefits from representations learnt from KPL in terms of detection performance and convergence speed. In the small data regime, KPL outperforms popular feature learning strategies without using any external data. Our results show the potential of pretext learning in medical imaging tasks. Future work shall focus on investigating more application-specific pretext learning tasks.

Acknowledgements. Huan Qi is supported by a China Scholarship Council doctoral research fund (grant No. 201608060317). The NIH Eunice Kennedy Shriver National Institute of Child Health and Human Development Human Placenta Project UO1 HD 087209, EPSRC grant EP/M013774/1, and ERC-ADG-2015 694581 are also acknowledged.

Appendix

See Tables 5, 6, and 7.

Table 5. Network configurations of three backbone CNNs: Tiny-VGG, VGG-13 and VGG-19. Each 'conv3-x' module contains a 3×3 convolution layer with x channels, followed by a batch-norm layer and a ReLU non-linearity. 'FC-X' means that the network outputs X class scores, where X is determined by specific applications: $X = 1000$ for IMAGENET and $X = 5$ for CHEXPERT.

VGG-Net Configuration		
Tiny-VGG	VGG-13	VGG-19
conv3-16 conv3-16	conv3-64 conv3-64	conv3-64 conv3-64
Maxpool		
conv3-32 conv3-32	conv3-128 conv3-128	conv3-128 conv3-128
Maxpool		
conv3-64 conv3-64	conv3-256 conv3-256	conv3-256 conv3-256 conv3-256 conv3-256
Maxpool		
conv3-128 conv3-128	conv3-512 conv3-512	conv3-512 conv3-512 conv3-512 conv3-512
Maxpool		
conv3-256 conv3-256	conv3-512 conv3-512	conv3-512 conv3-512 conv3-512 conv3-512
Maxpool		
FC-4096		
FC-4096		
FC-X		
Softmax		

Table 6. Hyper-parameter settings for ImageNet and CheXpert image classification pre-training.

Hyper-parameter	ImageNet	CheXpert
Learning rate	0.1	1e-4
Weight decay	1e-4	0
SGD momentum	0.9	0.9
Mini-batch size	64	16
Total epochs	90	10
Learning rate scheduler	Decay by 10 every 30 epochs	N.A

Table 7. Basic geometry and contrast jittering were applied as the routine data augmentation steps for SSL, KPL and subsequent UPI detection. For SSL and KPL, data augmentation were applied to image patches. For UPI detection, data augmentation were applied to image planes.

Geometry jittering (randomly apply one or two approaches below)
1. Crop and pad on each side by up to 10% of the image height/width
2. Scale each side by $weight \in [0.85, 1.15]$
Contrast jittering (randomly select one of the approaches below)
1. Gamma Contrast Correction with $gamma \in (0.5, 2.0)$
2. Sigmoid Contrast Correction with $gain \in (3, 10)$ and $cutoff \in (0.4, 0.6)$
3. Log Contrast Correction with $gain \in (0.6, 1.4)$
4. Linear Contrast Correction with $alpha \in (0.4, 1.6)$

References

1. Shin, H.-C., et al.: Deep convolutional neural networks for computer-aided detection: CNN architectures, dataset characteristics and transfer learning. IEEE Trans. Med. Imaging **35**(5), 1285–1298 (2016)
2. Gulshan, V., et al.: Development and validation of a deep learning algorithm for detection of diabetic retinopathy in retinal fundus photographs. Jama **316**(22), 2402–2410 (2016)
3. Esteva, A., et al.: Dermatologist-level classification of skin cancer with deep neural networks. Nature **542**(7639), 115–118 (2017)
4. Lee, H., et al.: An explainable deep-learning algorithm for the detection of acute intracranial haemorrhage from small datasets. Nat. Biomed. Eng. **3**(3), 173 (2019)
5. Qi, H., et al.: UPI-Net: semantic contour detection in placental ultrasound. In: ICCV-VRMI (2019)
6. Noroozi, M., Favaro, P.: Unsupervised learning of visual representations by solving jigsaw puzzles. In: Leibe, B., Matas, J., Sebe, N., Welling, M. (eds.) ECCV 2016. LNCS, vol. 9910, pp. 69–84. Springer, Cham (2016). https://doi.org/10.1007/978-3-319-46466-4_5
7. Taleb, A., et al.: Multimodal self-supervised learning for medical image analysis. arXiv preprint arXiv:1912.05396 (2019)

8. Raghu, M., et al.: Transfusion: understanding transfer learning for medical imaging. In: NeurIPS (2019)
9. He, K., et al.: Rethinking imagenet pre-training. In: ICCV (2019)
10. Szegedy, C., et al.: Going deeper with convolutions. In: CVPR (2015)
11. Misra, I., van der Maaten, L.: Self-supervised learning of pretext-invariant representations. arXiv preprint arXiv:1912.01991 (2019)
12. Gidaris, S., et al.: Unsupervised representation learning by predicting image rotations. In: ICLR (2018)
13. Tajbakhsh, N., et al.: Surrogate supervision for medical image analysis: effective deep learning from limited quantities of labeled data. In: ISBI (2019)
14. Doersch, C., et al.: Unsupervised visual representation learning by context prediction. In: ICCV (2015)
15. Bai, W., et al.: Self-supervised learning for cardiac MR image segmentation by anatomical position prediction. In: Shen, D., et al. (eds.) MICCAI 2019. LNCS, vol. 11765, pp. 541–549. Springer, Cham (2019). https://doi.org/10.1007/978-3-030-32245-8_60
16. Goyal, P., et al.: Scaling and benchmarking self-supervised visual representation learning. In: ICCV (2019)
17. Xie, S., Tu, Z.: Holistically-nested edge detection. In: ICCV (2015)
18. Yu, Z., et al.: Casenet: deep category-aware semantic edge detection. In: CVPR (2017)
19. Lawrence, S., Burns, I., Back, A., Tsoi, A.C., Giles, C.L.: Neural network classification and prior class probabilities. In: Montavon, G., Orr, G.B., Müller, K.-R. (eds.) Neural Networks: Tricks of the Trade. LNCS, vol. 7700, pp. 295–309. Springer, Heidelberg (2012). https://doi.org/10.1007/978-3-642-35289-8_19
20. Collins, S.L., et al.: Influence of power doppler gain setting on virtual organ computer-aided analysis indices in vivo: can use of the individual sub-noise gain level optimize information? Ultrasound Obstetr. Gynecol. **40**(1), 75–80 (2012)
21. Irvin, J., et al.: CheXpert: a large chest radiograph dataset with uncertainty labels and expert comparison. In: AAAI (2019)
22. Simonyan, K., Zisserman, A.: Very deep convolutional networks for large-scale image recognition. arXiv preprint arXiv:1409.1556 (2014)
23. Arbelaez, P., et al.: Contour detection and hierarchical image segmentation. IEEE Trans. Pattern Anal. Mach. Intell. **33**(5), 898–916 (2010)

Self-supervised Depth Estimation to Regularise Semantic Segmentation in Knee Arthroscopy

Fengbei Liu[1(✉)], Yaqub Jonmohamadi[2], Gabriel Maicas[1], Ajay K. Pandey[2], and Gustavo Carneiro[1]

[1] Australian Institute for Machine Learning, School of Computer Science, University of Adelaide, Adelaide, Australia
{fengbei.liu,gabriel.maicas,gustavo.carneiro}@adelaide.edu.au

[2] School of Electrical Engineering and Robotics, Science and Engineering Faculty, Queensland University of Technology, Brisbane, Australia
{y.jonmo,a2.pandey}@qut.edu.au

Abstract. Intra-operative automatic semantic segmentation of knee joint structures can assist surgeons during knee arthroscopy in terms of situational awareness. However, due to poor imaging conditions (e.g., low texture, overexposure, etc.), automatic semantic segmentation is a challenging scenario, which justifies the scarce literature on this topic. In this paper, we propose a novel self-supervised monocular depth estimation to regularise the training of the semantic segmentation in knee arthroscopy. To further regularise the depth estimation, we propose the use of clean training images captured by the stereo arthroscope of routine objects (presenting none of the poor imaging conditions and with rich texture information) to pre-train the model. We fine-tune such model to produce both the semantic segmentation and self-supervised monocular depth using stereo arthroscopic images taken from inside the knee. Using a data set containing 3868 arthroscopic images captured during cadaveric knee arthroscopy with semantic segmentation annotations, 2000 stereo image pairs of cadaveric knee arthroscopy, and 2150 stereo image pairs of routine objects, we show that our semantic segmentation regularised by self-supervised depth estimation produces a more accurate segmentation than a state-of-the-art semantic segmentation approach modeled exclusively with semantic segmentation annotation.

Keywords: Semantic segmentation · Self-supervised depth estimation · Monocular depth estimation · Multi-task learning · Arthroscopy · Knee

1 Introduction

Knee arthroscopy is a minimally invasive surgery (MIS) conducted via small incisions that reduce surgical trauma and post-operation recovery time [18]. Despite

A. L. Martel et al. (Eds.): MICCAI 2020, LNCS 12261, pp. 594–603, 2020.
https://doi.org/10.1007/978-3-030-59710-8_58

these advantages, arthroscopy has some drawbacks, namely: limited access and loss of direct eye contact with the surgical scene, limited field of view (FoV) of the arthroscope, tissues too close to the camera (e.g., 10 mm away) being only partially visible in the camera FoV, diminished hand-eye coordination, and prolonged learning curves and training periods [19]. In this scenario, surgeons can only confidently identify the femur due to its distinctive shape, while other structures, such as meniscus, tibia, and anterior cruciate ligament (ACL), remain challenging to be recognised. This limitation increases surgical operation time and may lead to unintentional tissue damage due to un-tracked camera movements. The automatic segmentation of these tissues has the potential to help surgeons by providing contextual awareness of the surgical scene, reducing surgery time, and decreasing the learning curve [14].

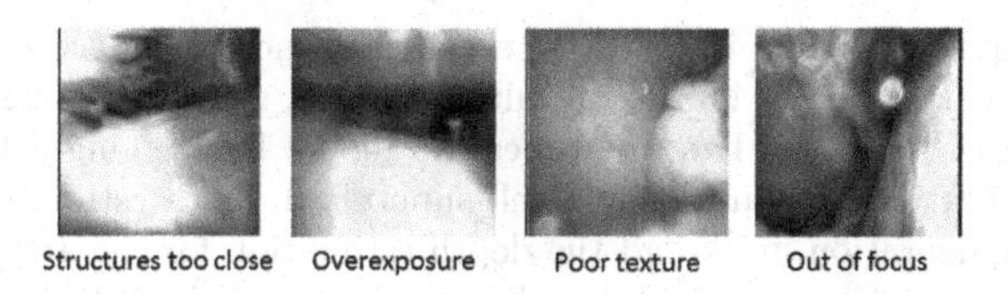

Fig. 1. Challenging imaging conditions from knee arthroscopy.

Deep learning (DL) semantic segmentation has been intensively studied by the computer vision community [1,2,10,11,16]. For arthroscopy, we are aware of just one method that produces automatic semantic segmentation of knee structures [8]. These semantic segmentation approaches tend to be prone to overfitting, depending on the data set available for the training process. As a consequence, there is an increasing interest in the development of regularisation methods, such as the ones based on multi-task learning (MTL) [17]. For instance, fusing semantic segmentation and depth estimation has been shown to be an effective approach [4], but it requires the manual annotation for the training of the segmentation and depth tasks. Considering that obtaining the depth ground truth for knee arthroscopy is challenging, self-supervised techniques such as [5,6] are highly favourable as they do not require ground truth depth. A similar approach has been successfully explored in robotic surgery [21], but not for knee arthroscopy. Moreover, self-supervised depth estimation techniques have been recently combined with semantic segmentation for training regularisation in non-medical imaging approaches [3,15]. Nevertheless, these approaches rely on data sets that contain stereo images captured from street or indoor scenes, where visual objects are far from the camera, contain rich texture, and images have few recording issues, such as overexposure and focus problems. On the other hand, knee arthroscopy images generally suffer from under or overexposure and focus problems, where visual objects are too close to the camera and contain poor texture, as shown in Fig. 1.

In this paper, we present an MTL approach for jointly estimating semantic segmentation and depth, where our aim is to use self-supervised depth

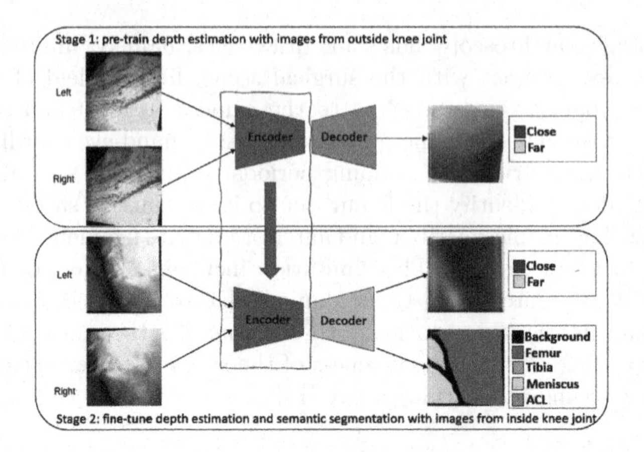

Fig. 2. The proposed method is first pre-trained with the self-supervised depth estimation using stereo arthroscopic images of routine objects, where the images contain none of the issues of Fig. 1. Stage two fine-tunes the model by training a fully supervised semantic segmentation regularised by a self-supervised depth estimation. The output contains the segmentation mask and the depth estimation for the arthroscopic image. A detailed structure of encoder and decoder are shown in Fig. 4. (Color figure online)

estimation from stereo images to regularise the semantic segmentation training from knee arthroscopy. Contrary to [15] that uses outdoor scenes, we tackle the segmentation of challenging arthroscopy images (Fig. 1). To this end, we pre-train our model on images of routine objects that do not show any of the issues displayed in Fig. 1. Then, we fine-tune our model with an MTL loss formed by the fully supervised semantic segmentation and the self-supervised depth estimation, as shown in Fig. 2. Using a data set containing 3868 arthroscopic images (with semantic segmentation annotations), 2000 stereo pairs captured during five cadaveric experiments and 2150 stereo image pairs of routne objects, we demonstrate that our method achieves higher accuracy in semantic segmentation (for the visual classes Femur, Meniscus, Tibia, and ACL) than state-of-the-art pure semantic segmentation methods.[1]

2 Proposed Method

2.1 Data Sets

We use three data sets: 1) the pre-training depth estimation data set $\mathcal{D}^{pre} = \{(\mathbf{I}^l, \mathbf{I}^r)_{k,n}\}_{k=1,n=1}^{|\mathcal{D}^{pre}|,N_k^{pre}}$, where l and r represent the left and right images of a stereo pair, k indexes the out-of-the-knee scene, and N_k^{pre} denotes the number of frames in the kth scene; 2) the fine-tuning depth estimation and semantic segmentation data sets, respectively denoted by $\mathcal{D}^{dep} = \{(\mathbf{I}^l, \mathbf{I}^r)_{k,n}\}_{k=1,n=1}^{|\mathcal{D}^{dep}|,N_k^{dep}}$

[1] The code will be available at https://github.com/ThomasLiu1021/geo-sem.

and $\mathcal{D}^{seg} = \{(\mathbf{I}, \mathbf{y})_{k,n}\}_{k=1,n=1}^{|\mathcal{D}^{seg}|,N_k^{seg}}$, where k indexes a human knee, and N_k^{pre} and N_k^{seg} denote the number of frames in the kth knee. In these data sets, colour images are denoted by $\mathbf{I} : \Omega \rightarrow \mathbb{R}^3$, where Ω represents the image lattice, and the semantic annotation is represented by $\mathbf{y} : \Omega \rightarrow \mathcal{Y}$, with $\mathcal{Y} = \{\text{Background, Femur, Tibia, Meniscus, ACL}\}$.

2.2 Data Set Acquisition

The arthroscopy images were acquired with a monocular Stryker endoscope (4.0 mm diameter) and a custom built stereo arthroscope using two muC103A cameras and a white LED for illumination (see Fig. 3). The Stryker endoscope has resolution 1280 × 720 with FoV of 30°, and the custom built camera has resolution 384 × 384 and FoV of 87.5°. Stryker images were cropped to have resolution 720 × 720 and then down-sampled to 384 × 384. Two clinicians performed the semantic segmentation annotations for classes femur, ACL, tibia and meniscus of 3868 images taken from four cadavers (where for one of the cadavers we used images from both knees) – see annotation details in Table 1.

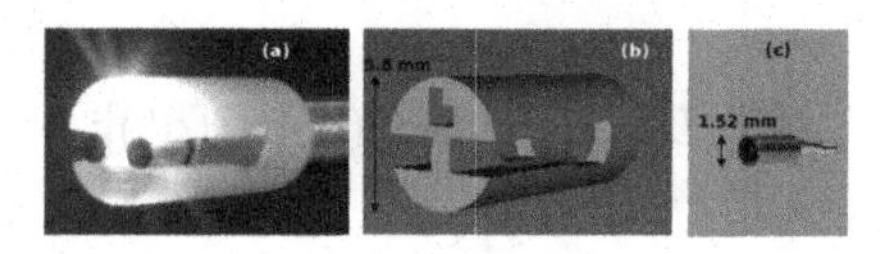

Fig. 3. The custom built camera. The camera tip is shown in (a), the 3D design is displayed in (b), and the muC103A camera is in (c).

Table 1. Percentage of training images per cadaver containing each of the structures [8].

Structure Cadaver knee	Femur	ACL	Tibia	Meniscus	Number of images
1	40%	0%	7%	0%	99
2	32%	20%	5%	9%	1043
3	30%	14%	8%	10%	1768
4-left	47%	3%	4%	6%	459
4-right	33%	8%	9%	12%	489
Total	33%	13%	7%	9%	3868

We also collected 2000 stereo pairs captured during these five cadaveric experiments. The data set with images acquired of routine objects contains 2050 stereo images pairs used for pre-training the depth estimator and 100 stereo image pairs to validate the depth output performance (see an example of this type of image in Fig. 2). To fine tune the depth estimation method, we grab video frames from original arthroscopy stereo camera video by every two seconds.

Note that there is no disparity ground truth available for any of the data sets above, so we cannot estimate the performance of the depth estimator.

2.3 Model for Semantic Segmentation and Self-supervised Depth Estimation

The goal of our proposed network is to simultaneously estimate semantic segmentation and depth estimation from a single image. Motivated by [8], the model backbone is the U-net++ [23], which predicts semantic segmentation and depth at four different levels, where the features are shared between these two tasks, as shown in Fig. 2.

In the model depicted in Fig. 4, each module $F_{i,j}(\mathbf{x}; \theta_{i,j})$ consists of blocks of convolutional layers (the input is represented by $\mathbf{x}$ and weights are represented by $\theta_{i,j}$), where the index i denotes the down-sampling layer and j represents the convolution layer of the dense block along the same skip connections (horizontally in the model). These modules are defined by

$$\mathbf{x}_{i,j} = \begin{cases} F_{i,j}(\mathbf{x}_{i-1,j}; \theta_{i,j}) & \text{if } j = 0 \\ F_{i,j}([[\mathbf{x}_{i,l}]_{l=0}^{j-1}, U(\mathbf{x}_{i+1,j-1}), U(\mathbf{d}_i^l), U(\mathbf{d}_i^r)]; \theta_{i,j}) & \text{if } j > 0 \end{cases}, \quad (1)$$

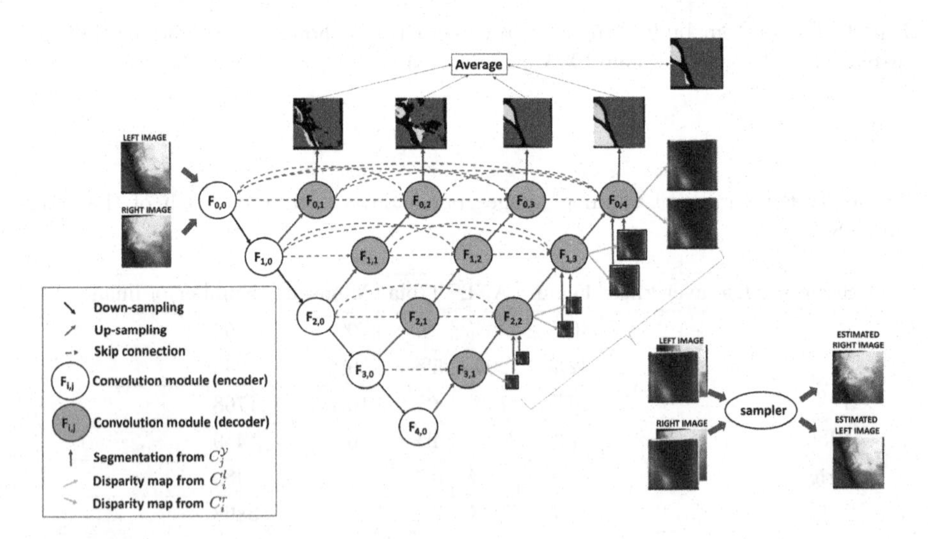

Fig. 4. We extend Unet++ [23] to produce multi-level semantic segmentation (on top), and multi-resolution disparity (inverse depth) estimations between the left and right images from the stereo pair (on the right hand side). During training, all levels and resolutions of semantic segmentation and depth estimation are used, and for testing, we only output the result from the finest segmentation level and depth resolution.

where $U(.)$ denotes an up-sampling layer (using bilinear interpolation), [.] represents a concatenation layer, and $\mathbf{d}_i^{\{l,r\}}$ is the disparity map that is defined only when $(i,j) \in \{(2,2),(1,3),(0,4)\}$ (otherwise it is empty), as described below in Eq. 3. The input image $\mathbf{I}$ enters the model at $F_{0,0}(\mathbf{I},\theta_{0,0})$. Each encoder convolution module (white nodes in Fig. 2) consists of a 3×3 filter followed by max pooling, and each decoder convolution module (green nodes in Fig. 2) comprises bi-linear upsampling with scale factor 2, followed by two layers of 3×3 filters, batch normalization and ReLU. The semantic segmentation output consists of

$$\tilde{\mathbf{y}}_j = C_j^{\mathcal{Y}}(\mathbf{x}_{0,j};\theta_j^{\mathcal{Y}}), \tag{2}$$

where $j \in \{1,2,3,4\}$, and $C_j^{\mathcal{Y}}(.)$ is a convolutional layer parameterised by $\theta_j^{\mathcal{Y}}$ that outputs the estimation of the semantic segmentation $\tilde{\mathbf{y}}_j : \Omega \to \mathcal{Y}$ for the jth convolutional layer. In particular, $C_j^{\mathcal{Y}}(.)$ is formed by a 1×1 convolution filter followed by pixel-wise softmax activation. The left and right disparity maps are obtained from

$$\mathbf{d}_i^{\{l,r\}} = C_i^{\{l,r\}}(\mathbf{x}_{i,j};\theta_i^{\{l,r\}}) \tag{3}$$

where $(i,j) \in \{(3,1),(2,2),(1,3),(0,4)\}$, and $C_i^{\{l,r\}}(.)$ is a convolutional layer parameterised by $\theta_i^{\{l,r\}}$ that outputs the estimation of the left and right disparity maps $\mathbf{d}_i^{\{l,r\}} : \Omega_i \to \mathbb{R}$ for the resolution at the ith down-sampling layer with Ω_i representing the image lattice at the same layer. The nodes $C_i^{\{l,r\}}(\mathbf{x}_{i,j};\theta_i^{\{l,r\}})$ consist of a 3×3 convolution filter with sigmoid activation to estimate the disparity result.

The **training for the supervised semantic segmentation** for a particular image $\mathbf{I}^l$ with annotation $\mathbf{y}$ and the average semantic segmentation results from the intermediate layers $\bar{\mathbf{y}} = \sum_{j=1}^{4} \tilde{\mathbf{y}}_j$ from (2) is based on the minimisation of the following loss function [12]:

$$\ell_{se}(\mathbf{y},\bar{\mathbf{y}}_j) = \alpha_{ce}\ell_{ce}(\mathbf{y},\bar{\mathbf{y}}) + (1 - \ell_{Dice}(\mathbf{y},\bar{\mathbf{y}})), \tag{4}$$

where $\ell_{ce}(\mathbf{y},\bar{\mathbf{y}})$ is the pixel-wise cross entropy loss computed between the annotation $\mathbf{y}$ and the average of the estimated semantic segmentation $\bar{\mathbf{y}}$, $\ell_{Dice}(\mathbf{y},\bar{\mathbf{y}})$ denotes the Dice loss [12], with α_{ce} being set to 0.5. The **inference for the supervised semantic segmentation** is based solely on the segmentation result from the last layer $\tilde{\mathbf{y}}_4$ from (2).

The **self-supervised depth estimation training** [6] uses rectified stereo pair images $\mathbf{I}^{\{l,r\}}$ to predict the disparity maps $\{\mathbf{d}_i^{\{l,r\}}\}_{i=0}^{3}$ to match the left-to-right and right-to-left images. The loss to be minimised is defined as

$$\begin{aligned}\ell_d(\mathbf{I}^l,\mathbf{I}^r) = \sum_{i=0}^{3} \Big[\alpha_{ap}\Big(\sum_{m\in\{l,r\}} \ell_{ap}^m(\mathbf{I}^l,\mathbf{I}^r,\mathbf{d}_i^m)\Big) + \alpha_{lr}\Big(\sum_{m\in\{l,r\}} \ell_{lr}^m(\mathbf{I}^l,\mathbf{I}^r,\mathbf{d}_i^m)\Big) \\ + \alpha_{ds}\Big(\sum_{m\in\{l,r\}} \ell_{ds}^m(\mathbf{I}^l,\mathbf{I}^r,\mathbf{d}_i^m)\Big)\Big],\end{aligned} \tag{5}$$

where

$$\ell^l_{ap}(\mathbf{I}^l,\mathbf{I}^r,\mathbf{d}^l_i) = \frac{1}{|\Omega_i|}\Big[\sum_{\omega\in\Omega_i}\Big(\gamma\Big(\frac{1-SSIM(\mathbf{I}^l(\omega),\tilde{\mathbf{I}}^l(\omega))}{2}\Big)+(1-\gamma)|\mathbf{I}^l(\omega)-\tilde{\mathbf{I}}^l(\omega)|\Big)\Big], \tag{6}$$

where $\ell^r_{ap}(\mathbf{I}^l,\mathbf{I}^r,\mathbf{d}^r_i)$ is similarly defined, $SSIM(.)$ represents the structural similarity index [20], $|\Omega_i|$ denotes the size of the image lattice at the ith resolution, $\tilde{\mathbf{I}}^l$ is the reconstructed left image using the right image re-sampled from the disparity map $\mathbf{d}^l_i$. Also in (5), we have

$$\ell^l_{lr}(\mathbf{I}^l,\mathbf{I}^r,\mathbf{d}^l_i) = \sum_{\omega\in\Omega_i}\Big|\mathbf{d}^l_i(\omega)-\mathbf{d}^l_i(\omega+\mathbf{d}^r_i(\omega))\Big|, \tag{7}$$

and similarly for $\ell^r_{lr}(\mathbf{I}^l,\mathbf{I}^r,\mathbf{d}^r_i)$ – this loss minimises the ℓ_1-norm between the left disparity map $\mathbf{d}^l_i$ and the transformed right-to-left disparity map. The last loss term in (5) is defined by

$$\ell^l_{ds}(\mathbf{I}^l,\mathbf{I}^r,\mathbf{d}^l_i) = \frac{1}{|\Omega_i|}\sum_{\omega\in\Omega_i}|\partial_x\mathbf{d}^l_i(\omega)|\times e^{-\|\partial_x\mathbf{I}^l(\omega)\|}+|\partial_y\mathbf{d}^l_i(\omega)|\times e^{-\|\partial_y\mathbf{I}^l(\omega)\|}, \tag{8}$$

and similarly for $\ell^r_{ds}(\mathbf{I}^l,\mathbf{I}^r,\mathbf{d}^r_i)$ – this loss penalises large disparity changes in smooth regions of the image, and when there are large image changes, there can be large transitions in the disparity maps. The **inference for the depth estimation** relies on the result for the finer scale $\mathbf{d}^{\{l,r\}}_0$.

Model pre-training is done with the data set $\mathcal{D}^{pre}$ by minimising the depth estimation loss (5), where we learn the model parameters $\{\theta_{i,j}\}_{i,j\in\{0,1,2,3,4\}}$ in (1) and disparity module parameters $\{\theta^{\{l,r\}}_i\}_{i\in\{0,1,2,3\}}$ in (3). After pre-training, we add the layers $\{C^{\mathcal{Y}}_j\}_{j\in\{1,2,3,4\}}$ and perform an **end-to-end training of all model parameters** with $\mathcal{D}$ by summing the losses in (4) and (5).

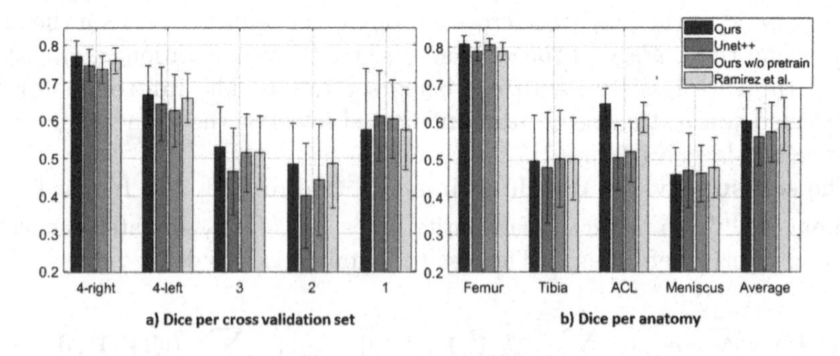

Fig. 5. Dice results over each test set (left) and each anatomy (right), and the final average over all sets and anatomies (rightmost) for all methods tested in this paper.

3 Experiments and Results

We implement our model in Pytorch [13]. The encoder for the model consists of the ResNet50 [7]. **Pre-training** takes 200 epochs with batch size 32, where initial learning rate is 10^{-4} and halved at 80 and 120 epochs, and we use Adam [9] optimizer. Data augmentation includes random horizontal and vertical flipping, random gamma from [0.8,1.2], brightness [0.5,2.0], and colour shifts [0.8,1.2] by sampling from uniform distributions. For **fine-tuning** of segmentation and depth using arthroscopic images, we use the pre-trained encoder and re-initialise the decoder. The training takes 120 epochs with batch size 12. We use polynomial learning decay [22] with $\gamma = 0.9$ and weight decay 10^{-5}. The data augmentation for segmentation includes horizontal and vertical flipping, random brightness contrast change and non-rigid transformation, including elastic transformation (the elastic transformation was particularly important to avoid over-fitting the training set) and depth data augmentation is the same as pre-training stage. For the inference time, the network takes 50 ms to process a single test image and output the segmentation mask and depth.

We assess the performance of our method using the Dice coefficient computed on the testing set in a leave one out cross validation experiment (i.e., we train with 4 knees and test with the remaining one from Table 1). In Fig. 5 we show the mean and standard deviation of the Dice results over each test set and each anatomy, and the final average over all sets and anatomies. We compare our newly proposed method (labelled as Ours) against the pure semantic segmentation model Unet++ [8,23], our method without the pre-training stage (labelled as Ours w/o pretrain), and the joint semantic segmentation and depth estimation method designed for computer vision applications by Ramirez et al. [15]. The results indicate that our method (mean Dice of 0.603 ± 0.159) is significantly better than Unet++ (mean Dice of 0.560 ± 0.152), with a Wilcoxon signed rank test showing a p-value < 0.05, indicating that the use of depth indeed improves the segmentation result from a pure segmentation method [8,23]. In fact, our method produces significant gains in the segmentation of ACL (arguably the most challenging anatomy in the experiment). Our method that uses pre-training is better than the one without pre-training (mean Dice of 0.573 ± 0.157), but not

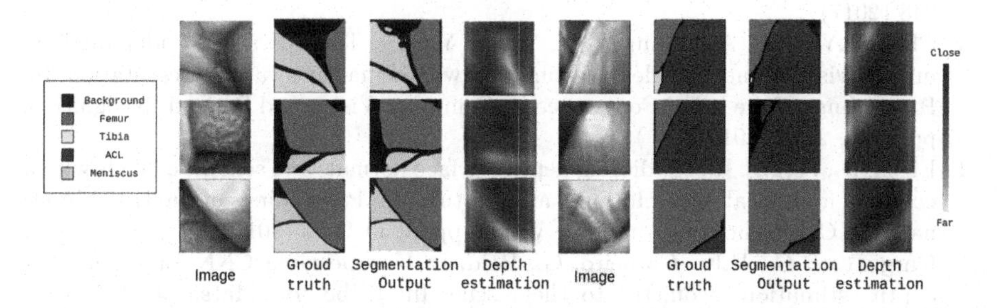

Fig. 6. Examples of results, including original arthroscopy image, segmentation ground truth, proposed method segmentation prediction and unsupervised depth estimation.

significantly so given that p-value is > 0.05. An interesting point is that even though our method without pre-training is better than the pure segmentation approach, it still cannot produce accurate segmentation for ACL. Finally, compared to the method by Ramirez et al. [15] (mean Dice of 0.595 ± 0.141) ours is slightly better, indicating that both methods are competitive. Figure 6 shows a few segmentation and depth estimation results. Note that we cannot validate depth estimation because we do not have ground truth available for it.

4 Conclusion

In this paper, we proposed a method to improve semantic segmentation using self-supervised depth estimation for arthroscopic images. Our network architecture is end-to-end trainable and does not require depth annotation. We also showed that the use of arthroscopic images of normal objects to pre-train the model can mitigate the challenging image conditions presented by this problem. By using geometry information, the model provides a slight improvement in terms of semantic segmentation accuracy. Future work will focus on improving the segmentation accuracy for the non-femural anatomies.

Acknowledgements. We acknowledge several technical discussions that influenced this paper with Ravi Garg and Adrian Johnston. This work was supported by the Australia India Strategic Research Fund (Project AISRF53820) and in part by the Australian Research Council through under Grant DP180103232. The cadaver studies is covered by the Queensland University of Technology Ethics Approval under project1400000856.

References

1. Badrinarayanan, V., Kendall, A., Cipolla, R.: SegNet: a deep convolutional encoder-decoder architecture for image segmentation. IEEE Trans. Pattern Anal. Mach. Intell. **39**(12), 2481–2495 (2017)
2. Chen, L.C., Papandreou, G., Kokkinos, I., Murphy, K., Yuille, A.L.: DeepLab: semantic image segmentation with deep convolutional nets, atrous convolution, and fully connected CRFs. IEEE Trans. Pattern Anal. Mach. Intell. **40**(4), 834–848 (2017)
3. Chen, P.Y., Liu, A.H., Liu, Y.C., Wang, Y.C.F.: Towards scene understanding: unsupervised monocular depth estimation with semantic-aware representation. In: Proceedings of the IEEE Conference on Computer Vision and Pattern Recognition. pp. 2624–2632 (2019)
4. Eigen, D., Fergus, R.: Predicting depth, surface normals and semantic labels with a common multi-scale convolutional architecture. In: Proceedings of the IEEE International Conference on Computer Vision, pp. 2650–2658 (2015)
5. Garg, R., B.G., V.K., Carneiro, G., Reid, I.: Unsupervised CNN for single view depth estimation: geometry to the rescue. In: Leibe, B., Matas, J., Sebe, N., Welling, M. (eds.) ECCV 2016. LNCS, vol. 9912, pp. 740–756. Springer, Cham (2016). https://doi.org/10.1007/978-3-319-46484-8_45

6. Godard, C., Mac Aodha, O., Brostow, G.J.: Unsupervised monocular depth estimation with left-right consistency. In: Proceedings of the IEEE Conference on Computer Vision and Pattern Recognition, pp. 270–279 (2017)
7. He, K., Zhang, X., Ren, S., Sun, J.: Deep residual learning for image recognition. In: Proceedings of the IEEE Conference on Computer Vision and Pattern Recognition, pp. 770–778 (2016)
8. Jonmohamadi, Y., et al.: Automatic segmentation of multiple structures in knee arthroscopy using deep learning. IEEE Access **8**, 51853–51861 (2020)
9. Kingma, D.P., Ba, J.: Adam: a method for stochastic optimization. arXiv preprint arXiv:1412.6980 (2014)
10. Lin, G., Milan, A., Shen, C., Reid, I.: Refinenet: multi-path refinement networks for high-resolution semantic segmentation. In: Proceedings of the IEEE Conference on Computer Vision and Pattern Recognition, pp. 1925–1934 (2017)
11. Long, J., Shelhamer, E., Darrell, T.: Fully convolutional networks for semantic segmentation. In: Proceedings of the IEEE Conference on Computer Vision and Pattern Recognition, pp. 3431–3440 (2015)
12. Milletari, F., Navab, N., Ahmadi, S.A.: V-net: fully convolutional neural networks for volumetric medical image segmentation. In: 2016 Fourth International Conference on 3D Vision (3DV), pp. 565–571. IEEE (2016)
13. Paszke, A., et al.: Automatic differentiation in PyTorch (2017)
14. Price, A., Erturan, G., Akhtar, K., Judge, A., Alvand, A., Rees, J.: Evidence-based surgical training in orthopaedics: how many arthroscopies of the knee are needed to achieve consultant level performance? The Bone Joint J. **97**(10), 1309–1315 (2015)
15. Zama Ramirez, P., Poggi, M., Tosi, F., Mattoccia, S., Di Stefano, L.: Geometry meets semantics for semi-supervised monocular depth estimation. In: Jawahar, C.V., Li, H., Mori, G., Schindler, K. (eds.) ACCV 2018. LNCS, vol. 11363, pp. 298–313. Springer, Cham (2019). https://doi.org/10.1007/978-3-030-20893-6_19
16. Ronneberger, O., Fischer, P., Brox, T.: U-Net: convolutional networks for biomedical image segmentation. In: Navab, N., Hornegger, J., Wells, W.M., Frangi, A.F. (eds.) MICCAI 2015. LNCS, vol. 9351, pp. 234–241. Springer, Cham (2015). https://doi.org/10.1007/978-3-319-24574-4_28
17. Ruder, S.: An overview of multi-task learning in deep neural networks. arXiv preprint arXiv:1706.05098 (2017)
18. Siemieniuk, R.A., et al.: Arthroscopic surgery for degenerative knee arthritis and meniscal tears: a clinical practice guideline. BMJ **357**, j1982 (2017)
19. Smith, R., Day, A., Rockall, T., Ballard, K., Bailey, M., Jourdan, I.: Advanced stereoscopic projection technology significantly improves novice performance of minimally invasive surgical skills. Surg. Endosc. **26**(6), 1522–1527 (2012)
20. Wang, Z., Bovik, A.C., Sheikh, H.R., Simoncelli, E.P.: Image quality assessment: from error visibility to structural similarity. IEEE Trans. Image Process. **13**(4), 600–612 (2004)
21. Ye, M., Johns, E., Handa, A., Zhang, L., Pratt, P., Yang, G.Z.: Self-supervised siamese learning on stereo image pairs for depth estimation in robotic surgery. arXiv preprint arXiv:1705.08260 (2017)
22. Zhao, H., Shi, J., Qi, X., Wang, X., Jia, J.: Pyramid scene parsing network. In: Proceedings of the IEEE Conference on Computer Vision and Pattern Recognition, pp. 2881–2890 (2017)
23. Zhou, Z., Rahman Siddiquee, M.M., Tajbakhsh, N., Liang, J.: UNet++: a nested U-Net architecture for medical image segmentation. In: Stoyanov, D., et al. (eds.) DLMIA/ML-CDS -2018. LNCS, vol. 11045, pp. 3–11. Springer, Cham (2018). https://doi.org/10.1007/978-3-030-00889-5_1

Semi-supervised Medical Image Classification with Global Latent Mixing

Prashnna Kumar Gyawali(✉), Sandesh Ghimire, Pradeep Bajracharya, Zhiyuan Li, and Linwei Wang

Rochester Institute of Technology, Rochester, USA
pkg2182@rit.edu

Abstract. Computer-aided diagnosis via deep learning relies on large-scale annotated data sets, which can be costly when involving expert knowledge. Semi-supervised learning (SSL) mitigates this challenge by leveraging unlabeled data. One effective SSL approach is to regularize the local smoothness of neural functions via perturbations around single data points. In this work, we argue that regularizing the global smoothness of neural functions by filling the void in between data points can further improve SSL. We present a novel SSL approach that trains the neural network on linear mixing of labeled and unlabeled data, at both the input and latent space in order to regularize different portions of the network. We evaluated the presented model on two distinct medical image data sets for semi-supervised classification of thoracic disease and skin lesion, demonstrating its improved performance over SSL with local perturbations and SSL with global mixing but at the input space only. Our code is available at https://github.com/Prasanna1991/LatentMixing.

Keywords: Semi-supervised learning · Mixup · Chest x-ray · Skin images.

1 Introduction

Medical image analysis via deep learning has achieved strong performance when supervised with a large labeled data set. Collecting such data sets is however costly in the medical domain since it involves expert knowledge. Semi-supervised learning (SSL) mitigates this challenge by leveraging unlabeled data.

An important goal in SSL is to avoid over-fitting the network function to small labeled data. A common inductive bias to guide this is the assumption of *smoothness* or *consistency* of the network function, *i.e.*, nearby points and points of the same manifold should have the same label predictions. For instance, self-ensembling [6] penalizes inconsistent predictions of unlabeled data under local perturbations, and virtual adversarial training [8] maintains consistency by forcing predictions of different adversarially-perturbed inputs to be the same.

By considering perturbations around single data points, these approaches regularize only the *local* smoothness of the network function in the vicinity of available data points: no constraint is imposed on the global behavior of the network function in between data points [7]. To better exploit the structure of

A. L. Martel et al. (Eds.): MICCAI 2020, LNCS 12261, pp. 604–613, 2020.
https://doi.org/10.1007/978-3-030-59710-8_59

unlabeled data, we consider a strategy of *mixup* which was recently proposed to train a deep network on a linear mixing of pairs of input data and their corresponding labels [12]. By filling the void between input samples, this strategy regularizes the *global* smoothness of the function and was shown to improve the generalization of state-of-the-art neural architectures in both supervised [12] and semi-supervised learning [1]. This mixup strategy was recently extended to the latent space, showing improvement over mixing in the input space only, in a supervised setting.

We argue that the mixup strategy – training a network on linear mixing of input data and their labels – can be interpreted as regularizing the network to approximate a linear interpolation function in between data points. The gain of performance brought by mixing in the latent space, therefore, is partly owing to relaxing this linearity constraint to a portion of the network between the selected latent space and the output space. We also hypothesize that, since high-level representations in deep-networks encode important information for discriminative tasks, mixing at the latent space may provide novel training signals for SSL.

Therefore, we propose to extend this regularization, *i.e.*, regularizing different portions of the network between the latent space and output space, for SSL and demonstrate its first application in medical image classification. In this approach, we perform linear mixing of pairs of labeled and unlabeled data – both in the input and latent space – along with their corresponding labels: for the latter, the label is *guessed* and continuously updated from an average of predictions of augmented samples for each unlabeled data point. We evaluate the presented SSL model on two distinct medical image classification tasks: multi-label classification of thoracic disease using Chexpert lung X-ray images [5], and skin disease classification using Skin Lesion images [3,10]. We compare the performance of the presented method with both a supervised baseline, and several SSL methods including mixup at the input space [12], standard self-ensembling in the input space [6], and recently-introduced self-ensembling at the latent space [4]. We further provide ablation studies and analyze the effect of function smoothing achieved by the presented method.

2 Related Work

SSL in Medical Image Analysis: Many recent semi-supervised works in medical image analysis have focused on explicitly regularizing the local smoothness of the neural function [2,4,9]. For instance, in [2], a siamese architecture for both labeled and unlabeled data points was proposed to encourage consistent segmentation under a given class of transformations. In [9], ensemble diversity was enforced with the use of adversarial samples to improve semi-supervised semantic image segmentation. In [4], the disentangled stochastic latent space was learned to improve self-ensembling for semi-supervised classification of chest X-ray images. In these works, each data point was subjected to local perturbations, *e.g.*, elastic deformations [2], virtual adversarial direction [9], or sampling from latent posterior distributions [4], for *local* smoothness regularization.

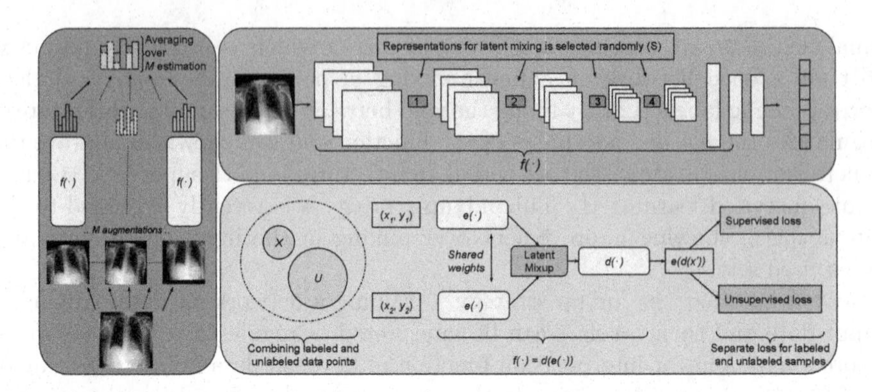

Fig. 1. Schematic diagram of the presented SSL method. During training, we continuously guess labels for the unlabeled data points (left) and then perform SSL via mixing at the input and latent space (bottom right). On the top right, we demonstrate the layers in the deep network where latent representations can be mixed.

In [7], the idea of promoting *global* smoothness in SSL was explored by constructing a teacher graph network. Similar approaches exploiting the global smoothness of neural functions, however, has not been studied in medical images.

Regularization with the Mixup Strategy: The mixup strategy was first presented in [12] to improve generalization of supervised models by mixing the data pairs at the input space. It was recently extended in a semi-supervised setting where the mixing is considered for both labeled and unlabeled data points [1]. In the meantime, a similar idea was also extended to the mixing of hidden representations [11], demonstrating improvement over mixing at the input space, although only in supervised learning.

To our knowledge, this is the first semi-supervised classification network that employs the mixup strategy at the latent space, and the first time this type of approaches is applied to semi-supervised medical image classification.

3 Methodology

We consider a set of labeled training examples $\mathcal{X}$ with the corresponding labels $\mathcal{Y}$, and a set of unlabeled training samples $\mathcal{U}$. We aim to learn parameters θ for the mapping function $f : \mathcal{X} \rightarrow \mathcal{Y}$, approximated via a deep neural network. Along the course of the training, we first *guess* and continuously update the labels for unlabeled data points (Sect. 3.1). We then perform linear mixing between labeled and unlabeled data points, both in the input and latent space, along with their corresponding actual or guessed labels (Sect. 3.2). Finally, the SSL model is trained on the mixed data sets using different losses depending on whether the mixed data point is closer to labeled or unlabeled data (Sect. 3.3). Figure 1 summarizes the key components of this semi-supervised learning process.

3.1 Guessing Labels

We guess the labels for unlabeled data by augmenting M separate copies of data batch u_b, and computing the average of the model's prediction as:

$$q_b = \frac{1}{M} \sum_{m=1}^{M} f(u_{b,m}; \theta) \tag{1}$$

The label guessing in this manner implicitly works as consistency regularization as the input transformations are assumed to leave class semantics unaffected. The guessed labels are continuously changed as the neural function $f(\mathbf{x})$ is updated over the course of the training.

3.2 Input and Latent Mixup

Since the mapping function $f(\mathbf{x})$ is approximated by deep neural network, we can decompose this function as $f(\mathbf{x}) = d_l(e_l(\mathbf{x}))$, where e_l represents the part of the neural network that encodes the input data to some latent representation at layer l, and d_l denotes the part of neural network that decodes such latent representation to the output $f(\mathbf{x})$. Inspired by [11], we determine a set of eligible layers $\mathcal{S}$ in the neural network from which we randomly select a layer l and apply mixup in that layer (schematics in top-right; Fig. 1). For each batch, we combine and shuffle labeled and unlabeled data points to obtain a pair of random mini-batches ($\mathbf{x}_1$, $\mathbf{y}_1$) and ($\mathbf{x}_2$, $\mathbf{y}_2$). We pass these pairs to e_l to obtain latent pairs ($e_l(\mathbf{x}_1)$, $\mathbf{y}_1$) and ($e_l(\mathbf{x}_2)$, $\mathbf{y}_2$), and then perform mixup at this latent layer to produce the mixed minibatch as ($e_l(\mathbf{x})'$, $\mathbf{y}'$) as:

$$\begin{aligned} \lambda &\sim \text{Beta}(\alpha, \alpha) \\ \lambda' &= \max(\lambda, 1 - \lambda) \\ e_l(\mathbf{x})' &= \lambda' \cdot e_l(\mathbf{x}_1) + (1 - \lambda') \cdot e_l(\mathbf{x}_2) \\ \mathbf{y}' &= \lambda' \cdot \mathbf{y}_1 + (1 - \lambda') \cdot \mathbf{y}_1 \end{aligned} \tag{2}$$

where α is the positive shape parameter of the Beta distribution, treated as hyperparameter in this work. Because the mixing could occur between labeled and unlabeled data, it is important to ensure that the mixed data fairly represent the distribution of both labeled and unlabeled data. Furthermore, as will be described in Sect. 3.3, different losses will be used to reflect a different treatment of the actual and guess labels due to their difference in reliability. It is thus also important to know whether each mixed data point is closer to labeled or unlabeled data. To do so, we use λ' instead of λ in Eqs. (2) to ensure that $e_l(\mathbf{x})'$ is always closer to $e_l(\mathbf{x}_1)$ than to $e_l(\mathbf{x}_2)$, allowing us to rely on the knowledge of $\mathbf{x}_1$ to determine which loss to apply on the mixed data point.

Depending upon $\mathcal{S}$, we achieve different mixup strategies. For example, when $\mathcal{S} = \{0\}$, we only mix at the input space. When $\mathcal{S} = \{0, 1\}$, we mix at the input space and latent layer 1. When $\mathcal{S} = \{1\}$, we mix only at the latent layer 1.

3.3 Supervised and Unsupervised Loss

To treat the actual and guessed labels differently because the latter are less reliable, we use different losses for data points that are closer to labeled versus unlabeled data. For data points in a batch $\mathcal{B}$ that are closer to labeled data, the loss term $\mathcal{L}_{\mathcal{X}}$ is the cross-entropy loss:

$$\mathcal{L}_{\mathcal{X}} = \sum_{(\mathcal{B} \cap \mathcal{X})} \sum_{l \sim \mathcal{S}} \ell(d_l(e_l(\mathbf{x})'), \mathbf{y}') \tag{3}$$

For data points in $\mathcal{B}$ that are closer to unlabeled data, the loss function is defined as a L_2 loss because it is considered to be less sensitive to incorrect predictions:

$$\mathcal{L}_{\mathcal{U}} = \sum_{(\mathcal{B} \cap \mathcal{U})} \sum_{l \sim \mathcal{S}} \| d_l(e_l(\mathbf{x})') - \mathbf{y}' \|_2^2 \tag{4}$$

After obtaining *mixed* latent representation, the network is optimized by minimizing the sum of these two losses:

$$\mathcal{L} = \mathcal{L}_{\mathcal{X}} + \lambda_{\mathcal{U}} \cdot \mathcal{L}_{\mathcal{U}} \tag{5}$$

where $\lambda_{\mathcal{U}}$ is the weight term for the unsupervised loss.

4 Experiments

We first test the effectiveness of the presented SSL approach on two distinct benchmark data sets for medical image classifications, in comparison to a supervised baseline and alternative SSL models. We then analyze the effect of mixing at different latent layers, and perform ablation studies to assess the impact of different hyperparameters and the depth of latent mixing on the presented method. Finally, we discuss the effect of function smoothing achieved by the presented SSL strategy.

4.1 Data Sets

We evaluate the presented model on two open-sourced large-scale medical dataset: Chexpert [5] and ISIC 2018 Skin Lesion Analysis [3,10].

Chexpert X-Ray Image Classification: Chexpert comprises of 224316 chest radiograph images from more than 60000 patients with labels for 14 different pathology categories. For pre-processing, we removed all uncertain and lateral-view samples from the data set, and re-sized the images to 128×128 in dimension. To ensure a fair comparison, we used the publicly available data splits for the labeled training set (ranging from 100 to 500 samples), unlabeled set, validation set, and test set [4]. For data augmentation, we rotated an image in the range of (−10°, 10°) and shifted (horizontal and vertical) it in the range of (0, 0.1) fraction of the image.

Table 1. Mean AUROC of 14 categories in the Chexpert data and seven categories in the skin data. The reported values are the average of five random seeds runs.

Model	Chexpert (k)					Skin (k)		
	100	200	300	400	500	350	600	1200
Supervised baseline	0.5576	0.6166	0.6208	0.6343	0.6353	0.7707	0.7991	0.8538
Input mixup	0.6491	0.6627	0.6731	0.6779	0.6823	0.8504	0.8609	0.9040
Latent mixup	**0.6523**	0.6632	**0.6747**	0.6795	0.6836	0.8536	0.8736	0.9036
Input+Latent mixup	0.6512	**0.6641**	0.6739	**0.6796**	**0.6847**	**0.8666**	**0.8768**	**0.9073**

Table 2. Mean AUROC for classification for 14 categories in the Chexpert data. The average of five randomly-seeded runs is reported by the presented method, whereas the best result is reported for the other method based on [4].

Model	Chexpert (k)				
	100	200	300	400	500
Image-space self-ensembling (noise)	0.6012	0.6277	0.6444	0.6550	0.6626
Image-space self-ensembling (augmentation)	0.6089	0.6301	0.6423	0.6530	0.6617
Latent-space self-ensembling	0.6200	0.6386	0.6484	0.6637	0.6697
Input + Latent mixup *(ours)*	**0.6512**	**0.6641**	**0.6739**	**0.6796**	**0.6847**

Skin Image Classification: ISIC 2018 skin data set comprises of 10015 dermoscopic images with labels for seven different disease categories. Three sets of labeled training data (350, 600, and 1200) were created considering class balance. The same data re-sizing and data augmentation strategies as applied to X-ray images were applied here.

4.2 Implementation Details

In our experiments, we use the AlexNet-inspired network from [4] to match their model implementation and training procedure closely. The network consists of five convolution blocks, followed by three fully-connected layers. All the models were trained up to 256 epochs with a learning rate of 1e-4 and decayed by a factor of 10 at the 50th and 125th epochs. For label guessing, we used $M = 2$ copies of unlabeled data. For Chexpert, unless mentioned otherwise, we used a set of eligible layers $\mathcal{S} = \{0, 2, 4\}$, mixing parameter $\alpha = 1.0$ for input mixup and $\alpha = 2.0$ for latent mixup, and $\lambda_{\mathcal{U}} = 75$ for the weight on unsupervised loss. For the skin data set, we used a set of eligible layers $\mathcal{S} = \{0, 1\}$, mixing parameter $\alpha = 1.0$ for both input and latent mixup, and $\lambda_{\mathcal{U}} = 50$ for the weight on unsupervised loss. We used the separately held out validation set to determine the best model along the course of the training, and report the results on the test set.

4.3 Results

Comparison Studies: In both data sets, we first evaluate the SSL performance of the presented model in comparison with two baselines: a fully-supervised baseline where we train the network with a supervised cross-entropy loss without mixing, and input mixup where SSL is performed with mixing at the input space only. The results are presented in Table 1. For the presented approach, we present two versions: mixing only at the latent space (latent mixup SSL), and combining both input and latent mixing (input + latent mixup SSL). As shown, mixing in the latent space in general improved the SSL performance over the baseline methods. Among the alternatives involving latent mixup, combined input and latent mixing yielded the best performance in three out of five cases in the Chexpert data set, and in all cases in the skin dataset.

Table 3. Effect of hyperparameters (left) and the latent depth for mixing (right).

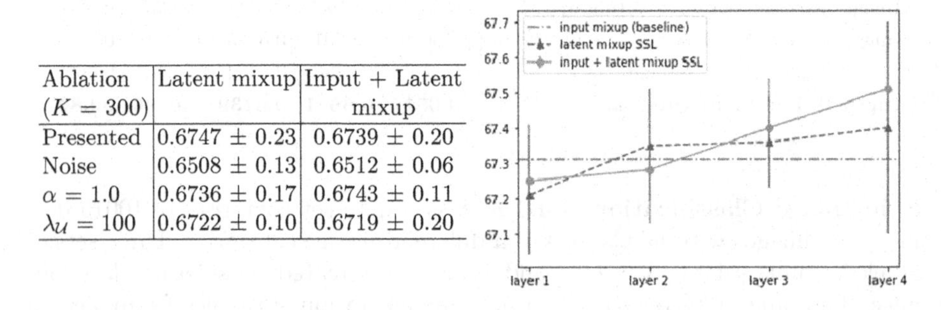

Ablation ($K = 300$)	Latent mixup	Input + Latent mixup
Presented	0.6747 ± 0.23	0.6739 ± 0.20
Noise	0.6508 ± 0.13	0.6512 ± 0.06
$\alpha = 1.0$	0.6736 ± 0.17	0.6743 ± 0.11
$\lambda_{\mathcal{U}} = 100$	0.6722 ± 0.10	0.6719 ± 0.20

Using the Chexpert data set, we further compared the presented model with existing SSL methods that focused on regularizing *local* smoothness of the network function via perturbations around single data points: self-ensembling at the input space [6] using Gaussian noise perturbations (with std = 0.15, image-space self-ensembling (noise)) or augmentation with random translation and rotation (image-space self-ensembling (augmentation)), and ensembling at the disentangling latent space (latent-space self-ensembling) [4]. The results, as presented in 2, showed a clear improvement of the presented method, supporting the advantage of regularizing the global in addition to local smoothness of neural functions.

Ablation Studies: We study the effect of different hyperparameters and elements in the presented SSL method, using a labeled dataset of size 300. The results are shown in Table 3 (left). While each had certain effect on the model performance, the most notable difference came from the data augmentation strategy used in the presented SSL method: replacing the presented data augmentation with image-level noises notably reduced the model performance, although still at a level higher than the ensembling baselines presented in Table 2.

Fig. 2. Decision boundary of SSL learning on two-moon toy data, where yellow dots represent the labeled data and the rest are unlabeled data.

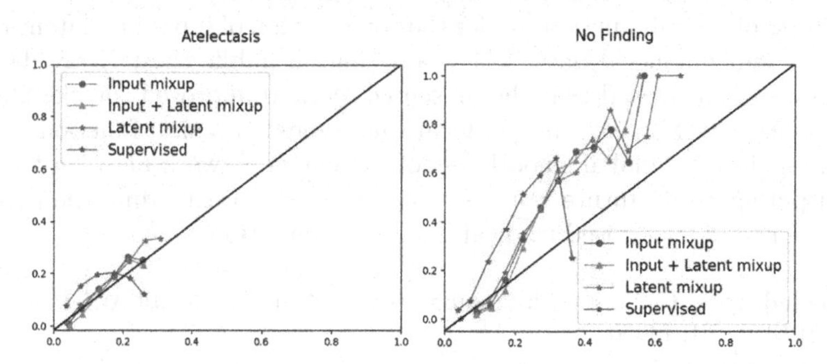

Fig. 3. Reliability diagram of the networks on classifying two class labels from X-ray images, trained with $K = 300$ labeled data. Perfect calibration is indicated by the diagonal line representing identity function.

In Table 3 (right), we show how the model performance was affected by the depth of latent space at which the mixing was performed, in comparison to a fixed baseline (green dashed) of mixing at the input space only. As shown, mixing at the deeper layers of the network appeared to be more beneficial in general. This implies that it may be more appropriate to apply the linearity constraint, considering its limited function capacity, to the later portion of a deep neural network. It may also suggest that, since higher-level representations are more task-related, mixing in such space could help in generalization.

4.4 The Effect of Function Smoothing

Finally, we explore the effect of function smoothing brought by the presented SSL method. Starting with a two-moon toy data set, we observed in Fig. 2 that mixing in the latent space increases the smoothness of the decision boundary in comparison to mixing at the input space only, an observation similar to [11] for supervised learning. In addition, it also provided a broader range of uncertainty (broader region of low confidence) compared to mixing in input space only.

While it is not feasible to visualizing the decision boundary for the deep neural network in the presented medical image classification, we instead investigated the effect of a more smoothed confidence measure as observed in the toy data. To do so, we consider the calibration of the model via the reliability diagram. Figure 3 shows examples of the network in classifying two class labels: as shown, in general, the mixup strategy improves the calibration of the network compared to a supervised baseline, while mixing at the latent space tends to further marginally improve the calibration compared to mixing at the input space alone.

5 Conclusion

We presented a novel semi-supervised learning method that regularizes the global smoothness of neural functions under the combination of input and latent mixing of labeled and unlabeled data. The evaluation on public chest X-ray data and skin disease data showed that the presented method improved the classification performance over SSL focusing on local smoothness of neural functions, as well as SSL regularizing global smoothness of the entire network between the input and output space. In future work, we are interested in extending the presented method for semi-supervised medical image segmentation.

Acknowledgement. This work is supported by NSF CAREER ACI-1350374 and NIH NHLBI R15HL140500.

References

1. Berthelot, D., Carlini, N., Goodfellow, I., Papernot, N., Oliver, A., Raffel, C.A.: Mixmatch: a holistic approach to semi-supervised learning. In: Advances in Neural Information Processing Systems, pp. 5050–5060 (2019)
2. Bortsova, G., Dubost, F., Hogeweg, L., Katramados, I., de Bruijne, M.: Semi-supervised medical image segmentation via learning consistency under transformations. In: Shen, D., et al. (eds.) MICCAI 2019. LNCS, vol. 11769, pp. 810–818. Springer, Cham (2019). https://doi.org/10.1007/978-3-030-32226-7_90
3. Codella, N., et al.: Skin lesion analysis toward melanoma detection 2018: a challenge hosted by the international skin imaging collaboration (isic). arXiv preprint arXiv:1902.03368 (2019)
4. Gyawali, P.K., Li, Z., Ghimire, S., Wang, L.: Semi-supervised learning by disentangling and self-ensembling over stochastic latent space. In: Shen, D., et al. (eds.) MICCAI 2019. LNCS, vol. 11769, pp. 766–774. Springer, Cham (2019). https://doi.org/10.1007/978-3-030-32226-7_85
5. Irvin, J., et al.: Chexpert: a large chest radiograph dataset with uncertainty labels and expert comparison. In: AAAI (2019)
6. Laine, S., Aila, T.: Temporal ensembling for semi-supervised learning. In: ICLR (2017)
7. Luo, Y., Zhu, J., Li, M., Ren, Y., Zhang, B.: Smooth neighbors on teacher graphs for semi-supervised learning. In: Proceedings of the IEEE Conference on Computer Vision and Pattern Recognition, pp. 8896–8905 (2018)
8. Miyato, T., Maeda, S.I., Koyama, M., Ishii, S.: Virtual adversarial training: a regularization method for supervised and semi-supervised learning. IEEE Trans. Pattern Anal. Mach. Intell. **41**(8), 1979–1993 (2018)

9. Peng, J., Estrada, G., Pedersoli, M., Desrosiers, C.: Deep co-training for semi-supervised image segmentation. Pattern Recogn. **107**, 107269 (2020)
10. Tschandl, P., Rosendahl, C., Kittler, H.: The ham10000 dataset, a large collection of multi-source dermatoscopic images of common pigmented skin lesions. Sci. Data **5**, 180161 (2018)
11. Verma, V., et al.: Manifold mixup: better representations by interpolating hidden states. In: International Conference on Machine Learning, pp. 6438–6447 (2019)
12. Zhang, H., Cisse, M., Dauphin, Y.N., Lopez-Paz, D.: Mixup: beyond empirical risk minimization. arXiv preprint arXiv:1710.09412 (2017)

Self-Loop Uncertainty: A Novel Pseudo-Label for Semi-supervised Medical Image Segmentation

Yuexiang Li(✉), Jiawei Chen, Xinpeng Xie, Kai Ma, and Yefeng Zheng

Tencent Jarvis Lab, Shenzhen, China
vicyxli@tencent.com

Abstract. Witnessing the success of deep learning neural networks in natural image processing, an increasing number of studies have been proposed to develop deep-learning-based frameworks for medical image segmentation. However, since the pixel-wise annotation of medical images is laborious and expensive, the amount of annotated data is usually deficient to well-train a neural network. In this paper, we propose a semi-supervised approach to train neural networks with limited labeled data and a large quantity of unlabeled images for medical image segmentation. A novel pseudo-label (namely self-loop uncertainty), generated by recurrently optimizing the neural network with a self-supervised task, is adopted as the ground-truth for the unlabeled images to augment the training set and boost the segmentation accuracy. The proposed self-loop uncertainty can be seen as an approximation of the uncertainty estimation yielded by ensembling multiple models with a significant reduction of inference time. Experimental results on two publicly available datasets demonstrate the effectiveness of our semi-supervised approach.

Keywords: Semi-supervised learning · Pseudo-label · Jigsaw puzzles

1 Introduction

Deep neural networks often require large quantity of labeled images to achieve satisfactory performance. However, since annotating medical images requires experienced physicians to spend hours or days to investigate, which is laborious and expensive, the labeled medical images are often very deficient, especially for the tasks requiring pixel-wise annotations (e.g., segmentation). To tackle this problem, many researches [1,2,17,19] have been proposed to improve the segmentation performance of deep neural networks through exploiting the information from unlabeled data. Using pseudo-labels of unlabeled data (generated automatically by a segmentation algorithm via uncertainty estimation) is one of the potential solutions, which has been extensively studied. The most popular approaches are: 1) softmax probability map [1], 2) Monte Carlo (MC) dropout [17,19], and 3) uncertainty estimation via network ensemble [8]. Specifically,

A. L. Martel et al. (Eds.): MICCAI 2020, LNCS 12261, pp. 614–623, 2020.
https://doi.org/10.1007/978-3-030-59710-8_60

Bai et al. [1] proposed a semi-supervised approach for the cardiac magnetic resonance volume segmentation. The proposed approach first used a limited number of labeled data to train the neural network and then utilized the softmax probability maps predicted by the neural network as the pseudo-label for the unlabeled volumes to augment the training set. In a more recent study, Sedai et al. [17] proposed an uncertainty guided semi-supervised learning framework for the segmentation of retinal layers in optical coherence tomography images. The pseudo-label for semi-supervised learning was generated using the Monte Carlo (MC) dropout [4], which can be viewed as an approximation of Bayesian uncertainty. Uncertainty estimation via model ensemble [8] is another form of approximation of Bayesian uncertainty, which separately trained K networks and combined the softmax probability map of each network k by averaging as the ensemble uncertainty (i.e., $\frac{1}{K}\sum_{k=1}^{K} p_k$, where p is the probability map).

Due to the variety of existing uncertainty estimation methods, Jungo et al. [7] conducted experiments to evaluate the reliability and limitation of existing approaches and concluded several observations. Two of them cause our interests: 1) the widely-used MC-dropout-based approaches are heavily dependent on the influence of dropout on the segmentation performance; 2) the computational-expensive ensemble method yields the most reliable results and is typically a good choice if the resources allow. To this end, an efficient way to yield the reliable ensemble uncertainty is worthwhile to investigate.

In this paper, we propose a novel pseudo-label, namely self-loop uncertainty, for the semi-supervised medical image segmentation. The proposed self-loop uncertainty is generated by recurrently optimizing the encoder of a fully convolutional network (FCN) with a self-supervised sub-task (e.g., Jigsaw puzzles). The benefits of integrating self-supervised learning into our framework can be summarized in two folds: 1) the self-supervised learning sub-task encourages the neural network to deeply mine the information from raw data and benefits the image segmentation task; 2) the same network at different stages during the self-supervised sub-task optimization can be seen as different models, which leads our self-loop uncertainty to an approximation of ensemble uncertainty with much lower computational cost. We evaluate the proposed semi-supervised learning approach on two medical image segmentation tasks—nuclei segmentation and skin lesion segmentation. Experimental results show that our self-loop uncertainty can significantly improve the segmentation accuracy of the neural network, which outperforms the currently widely-used pseudo-label (e.g., softmax probability map and MC dropout).

2 Method

The proposed semi-supervised segmentation framework is illustrated in Fig. 1. The training set for our semi-supervised framework consists of labeled data D_L and unlabeled data D_U. The proposed semi-supervised framework involves three losses (i.e., $\mathcal{L}_{SEG}$, $\mathcal{L}_{UG}$, and $\mathcal{L}_{SS}$) to supervise the network training with D_L and D_U, respectively. The colored arrows in Fig. 1 represent the information flows of

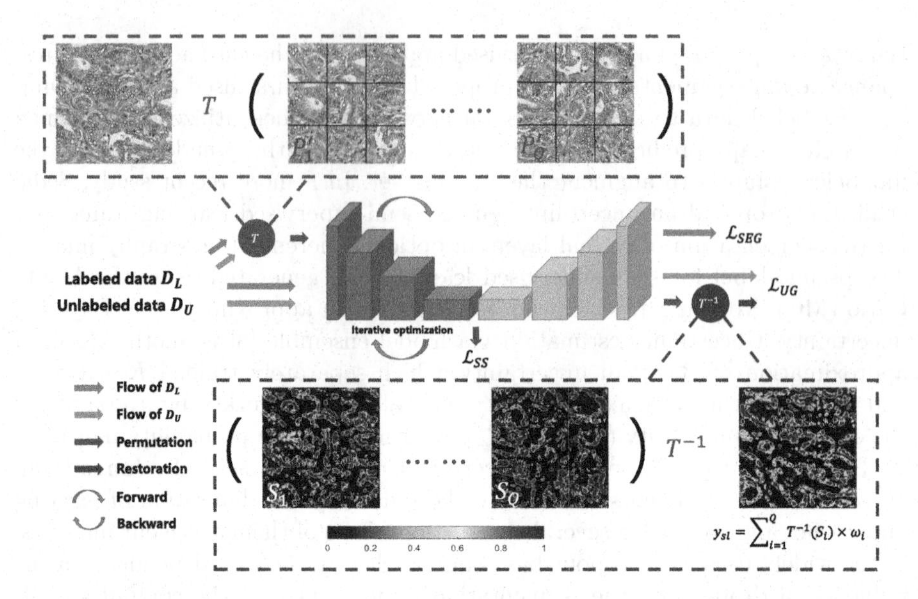

Fig. 1. The pipeline of our semi-supervised segmentation framework. The proposed framework recurrently optimizes the encoder part of FCN by addressing the self-supervised learning task (i.e., supervised by $\mathcal{L}_{SS}$) to generate the pseudo-label for the unlabeled data. There are two losses, i.e., segmentation loss $\mathcal{L}_{SEG}$ and uncertainty-guided loss $\mathcal{L}_{UG}$, adopted in our framework to supervise the segmentation of labeled and unlabeled data. Our framework generates Q permutations ($P_1^{'}$, ... $P_Q^{'}$) for an image (either labeled or unlabeled) and yields corresponding Q segmentation predictions (S_1, ... S_Q) for the estimation of self-loop uncertainty y_{sl} (as illustrated in Algorithm 1). (Color figure online)

D_U (orange) and D_L (cyan). For a batch containing images from D_L and D_U, we calculate the supervised segmentation loss $\mathcal{L}_{SEG}$ (i.e., binary cross-entropy loss in our experiment) for labeled data with pixel-wise annotation to ensure the FCN has the segmentation capacity, self-supervised loss $\mathcal{L}_{SS}$ for both D_L and D_U to exploit rich information from raw data and generate the self-loop uncertainty, and uncertainty-guided loss $\mathcal{L}_{UG}$ for the unlabeled images to boost the segmentation performance of FCN with unlabeled data.

2.1 Self-supervised Sub-task

As aforementioned, the self-supervised loss $\mathcal{L}_{SS}$ aims to exploit rich information contained in raw data and generate the self-loop uncertainty. Various pretext tasks, such as rotation prediction [5] and colorization [9], can be adopted to achieve this goal. In this study, we use Jigsaw puzzles [14] consisting of translation and rotation transformations as the self-supervised sub-task to recurrently optimize the encoder of an FCN and yield the self-loop uncertainty.

Similar to the standard Jigsaw puzzles, we partition the image into several tiles, e.g., nine tiles for 3 × 3 Jigsaw puzzles. To formulate the Jigsaw puzzles

Algorithm 1. Generation of self-loop uncertainty.

1: **Input:**
2: Network weights: θ_e of the encoder and θ_d of the decoder.
3: Unlabeled data: $x \in D_U$.
4: **Function:**
5: $f(x;\theta)$ neural network forward function.
6: *update*(.) backpropagation to update the neural network weights.
7: $T(.)$ permuted transformation of Jigsaw puzzles.
8: $T^{-1}(.)$ inverse-permuted transformation.
9: $\mathcal{L}_{SS}(p,g)$ calculation of the self-supervised loss with prediction p and self-supervised signal g.
10: **Procedure**†**:**
11: Q permutations are randomly selected from $\mathbb{P}'$: $\{P_1', ..., P_Q' \in \mathbb{P}'\}$.
12: **for** $i \in \{1, ..., Q\}$ **do**
13: $p_i \leftarrow f(T_{P_i'}(x); \theta_e)$; $S_i \leftarrow f(T_{P_i'}(x); \{\theta_e, \theta_d\})$;
14: $l_i \leftarrow \mathcal{L}_{SS}(p_i, g_i)$; $\theta_e^i \leftarrow update(l_i)$;
15: $\theta_e \leftarrow \theta_e^i$.
16: **end for**
17: $y_{sl} = \sum_{i=1}^{Q} T_{P_i'}^{-1}(S_i) \times norm(\omega_i)$, where $norm(.) = \frac{\omega_i}{\sum_{i=1}^{Q} \omega_i}$ and $\omega_i = 1 - \frac{l_i}{\sum_{i=1}^{Q} l_i}$.
† S is the segmentation prediction of FCN.
† l is the calculated self-supervised loss.
18: **Output:** self-loop uncertainty y_{sl} of input x.

sub-task, we permute the tiles using the approach proposed by [14]—a small subset $\mathbb{P}'$ of the large permutation pool, i.e., $\mathbb{P} = (P_1, P_2, ..., P_{9!})$ is formed by selecting the K permutations with the largest Hamming distance between each other. In each training iteration, the input image is repeatedly disarranged (Q times in total where $Q \ll K$, $Q = 10$ and $K = 100$ in our experiments) by one randomly selected permutation from $\mathbb{P}'$. Meanwhile, the encoder of FCN is recurrently updated to identify the selected permutation from the K options for each disarranged image, which can be seen a classification task with K categories; therefore, we employ the cross-entropy loss as $\mathcal{L}_{SS}$ to supervise the sub-task.

The Jigsaw puzzles transformation adopted in our approach has two differences, compared to the one in [14]. First, to increase the diversity of permutation, each of the tiles is randomly rotated by an angle $a \in \{0°, 90°, 180°, 270°\}$ besides the translation transformation. Second, to integrate the Jigsaw puzzles task into the end-to-end semi-supervised framework, the input of self-supervised sub-task is required to have the same size as that of the target segmentation task. Hence, instead of using the shared-weight neural network for each tile, the permuted tiles are first assembled to an image of the same size of the original image (i.e., $\{P_1', ..., P_Q'\}$ shown in Fig. 1) and then fed as input to the neural network for the permutation classification.

2.2 Estimation of Self-loop Uncertainty for Unlabeled Data

The generation procedure of our self-loop uncertainty is presented in Algorithm 1. The self-supervised sub-task is able to recurrently optimize the neural network in an iteration, as the self-supervised signal can be self-driven without manual annotation. The different stages (i.e., $\{\theta_e^i, \theta_d\}, i \in \{1, ..., Q\}$) of self-supervised optimization are seen as different models, which enable the proposed self-loop uncertainty to approximate the ensemble uncertainty with a single neural network. The permutated images go through the FCN and yield a set of segmentation predictions $S_i, i \in \{1, ..., Q\}$. Since the calculated self-supervised loss (l) can explicitly represent the difficulty of puzzled image for neural network to restore, we formulate l as the confidence of corresponding segmentation result S (via $norm(.)$ and ω defined in Algorithm 1) to revise its contribution to the final pseudo-label. Our self-loop uncertainty thereby is the weighted average of the segmentation predictions produced by different stages of self-supervised optimization.

Uncertainty-Guided Loss. The set of segmentation predictions $\{S_1, ..., S_Q\}$ is presented in Fig. 1, where the red color represents the high score of foreground. The weight-averaged self-loop uncertainty y_{sl} can be used as the guidance to maintain the reliable predictions (i.e., high score) as target for the neural network to learn from unlabeled data. To achieve this goal, we adopt the mean squared error (MSE) loss as the uncertainty-guided loss $\mathcal{L}_{UG}$ for the network optimization with unlabeled data and pseudo-labels y_{sl}, which can be defined as:

$$\mathcal{L}_{UG}(S_x, y_{sl}) = \frac{\sum_{H \times W} \mathbb{I}(y_{sl} > th) \|S_x - y_{sl}\|^2}{\sum_{H \times W} \mathbb{I}(y_{sl} > th)} \tag{1}$$

where $\mathbb{I}$ is the indicator function; H and W are the image height and width, respectively; S_x is the segmentation prediction of input image x; and th is the threshold to select the high score target.

2.3 Objective Function

Assuming a batch contains N labeled data ($\{(x_j, y_j)\}_{j=1}^N$) and M unlabeled data $\{x_j\}_{j=N+1}^{N+M}$, where $x_j \in \mathbb{R}^{H \times W \times C}$ is the input image (H, W, and C are the height, width, and channel of the image, respectively) and $y_j \in \{0, 1\}^{H \times W}, j = 1, 2, \ldots, N$ is the ground-truth annotation, the objective function $\mathcal{L}$ for this batch can be formulated as:

$$\mathcal{L} = \sum_{j=1}^{N} \mathcal{L}_{SEG}(x_j, y_j) + \sum_{j=N+1}^{N+M} \mathcal{L}_{UG}(x_j, y_{sl}) + \sum_{j=1}^{N+M} \sum_{i=1}^{Q} \mathcal{L}_{SS}(T_{P_i'}(x_j), g_i). \tag{2}$$

During network optimization, for the unlabeled data, we first fixed the decoder of FCN and recurrently update the encoder with $\mathcal{L}_{SS}$ to generate y_{sl}.

Then, the weight of the whole FCN is optimized by $\mathcal{L}_{UG}$. In other words, an unsynchronized optimization of the encoder and decoder happens when using the unlabeled data. For the labeled data, on the other hand, the network is optimized with the $\mathcal{L}_{SEG}$ and $\mathcal{L}_{SS}$ simultaneously.

3 Experiments

MoNuSeg Dataset [13]. The dataset consists of diverse H&E stained tissue images captured from seven different organs (e.g., breast, liver, kidney, prostate, bladder, colon and stomach), which were collected from 18 institutes. The dataset has a public training set and a public test set for network training and evaluation, respectively. The training set contains 30 histopathological images with hand-annotated nuclei, while the test set consists of 14 images. The size of the histopathological images is 1000×1000 pixels.

ISIC Dataset [3]. The ISIC dataset is widely-used to assess the segmentation accuracy of skin lesion areas of automatic segmentation algorithms. The dataset contains 2,594 dermoscopic images. The skin lesion area of each image has been manually annotated by the data provider. The image size varies from around 1000×1000 pixels to 4000×3000 pixels. We resize all the images to a uniform size of 512×512 pixels for network training and validation. The dataset is randomly separated to training and test sets according to the ratio of 75:25.

Evaluation Criterion. The F1 score, i.e., the unweighted average classification accuracy of the foreground and background tissues, which is widely-used in the area of nuclei [11,15,20] and skin lesion [10,12,18] segmentation, is adopted as the metric to evaluate the segmentation performance.

Baselines. Three popular uncertainty approaches—softmax probability map [1], Monte Carlo (MC) dropout [17,19], and uncertainty estimation via ensembling networks [8]—are involved as baselines in this study. Similar to [17], we set the dropout rate to 0.2 and forward the image through the neural network for ten times to generate MC dropout uncertainty. The ensemble uncertainty is generated by ensembling ten models trained with different network initializations. Consistent with the baselines, we generate ten permutations for an image to iteratively optimize the neural network and accordingly yield the self-loop uncertainty. The widely used ResUNet-18 [6,16] is used as the backbone for uncertainty estimation. For fair comparison, all the baselines are trained according to the same protocol.

3.1 Evaluation of Pseudo-Label Quality

Compared to skin lesion segmentation, which contains a single target in each image, the annotation of nucleus is more difficult and laborious. Hence, we mainly use the MoNuSeg dataset to evaluate the quality of pseudo-label yielded by different approaches in this section.[1]

To quantitatively validate the accuracy of different pseudo-labels, we calculate the F1 score between the pseudo-labels and ground-truth and present the results in Table 1. The pseudo-labels are generated with different amounts (i.e., 20% and 50%) of labeled data D_L and the remaining training set is used as unlabeled data D_U. As shown in Table 1, our self-loop uncertainty outperforms all the baselines under different amounts of labeled data, which are +2.27% and +2.88% higher than the runner-up (i.e., MC Dropout) with 20% and 50% labeled data, respectively. The pesudo-labels yielded by uncertainty via ensembling models achieve lower accuracy among the baselines. The underlying reason may be that the MoNuSeg training set only contains 30 histopathological images, which make the amount of labeled data (i.e., 20% and 50%) insufficient to well train the neural network. Therefore, the ensembling of multiple unsatisfactory models cannot improve the accuracy of uncertainty estimation.

Table 1. F1 score (%) between ground-truth and the pseudo-labels generated by different uncertainty approaches with different amounts of labeled data. The superscript of SL is the number of permutations Q generated for self-supervised learning. (MC D.—MC Dropout, SL—Self-loop)

Amount of D_L	Softmax	MC D.	Ensemble	SL^3	SL^6	SL^{10}
20%	67.48	72.42	67.46	73.90	74.68	**75.24**
50%	69.53	73.58	70.01	76.51	76.77	**76.85**

Ablation Study. We conduct an ablation study to investigate the relationship between the number of permutation Q and the quality of pseudo-label. Q is set to 3, 6, 10, respectively, for the generation of self-loop uncertainty. As shown in Table 1, the self-loop uncertainty generated with a larger Q achieves the higher F1 score. However, the improvement of F1 score provided by increasing Q from 6 to 10 becomes marginal (e.g., +0.08% using 50% labeled data), which illustrates that Q may not be the larger the better for practical applications, when taking the computational cost into account.

3.2 Segmentation Performance Evaluation

To validate the effectiveness of pseudo-labels, we evaluate the performance of different semi-supervised frameworks on the test sets of MoNuSeg and ISIC. The semi-supervised approaches are trained with different portions (i.e., 20% and 50%) of labeled data. The evaluation results are listed in Table 2. The performance of fully-supervised approach with 100% labeled data is also assessed as the

[1] For visual comparison between pseudo-labels, please refer to *arxiv version*.

upper bound for the semi-supervised approaches. To validate the effectiveness of self-supervised sub-task, the self-loop uncertainty without $\mathcal{L}_{SS}$ is also involved for comparison. We pass the ten permutated images through the FCN without self-supervised optimization and yield the uncertainty by averaging the segmentation predictions. Due to lack of extra information exploited by self-supervised sub-task, the improvements yielded without $\mathcal{L}_{SS}$ significantly decrease.

Nuclei Segmentation. As shown in Table 2, the performance of fully-supervised approach significantly drops from 79.30% to 75.87% and 71.51%, respectively, with the reductions (i.e., −50% and −80%) of manual annotations. The application of pseudo-labels provides a consistent improvement to the segmentation accuracy. Among them, the proposed self-loop uncertainty yields the largest improvements, especially under the condition with 20% annotated data, i.e., +5.6% higher than the fully-supervised approach. Furthermore, we notice that our semi-supervised framework trained with 50% labeled data achieves comparable F1 score (79.10%) to that of 100% fully-supervised approach (79.30%), which demonstrates the potential of our approach for reducing the workload of manual annotations.

Table 2. F1 score (%) yielded by different semi-supervised approaches on the two publicly available datasets.

	MoNuSeg			ISIC		
	20%	50%	100%	20%	50%	100%
Fully-supervised	71.51	75.87	**79.30**	81.49	84.86	**86.58**
Softmax [1]	73.65	76.18	-	82.81	85.11	-
MC Dropout [17]	75.31	77.98	-	83.68	85.74	-
Ensemble [8]	73.33	76.87	-	83.27	86.06	-
Self-loop w/o $\mathcal{L}_{SS}$	74.70	77.78	-	82.70	85.22	-
Self-loop (Ours)	77.11	79.10	-	84.92	86.17	-

Skin Lesion Segmentation. Similar trends of improvement are observed on the ISIC test set. Due to the extra information provided by the unlabeled data, the semi-supervised approaches outperform the fully-supervised one with limited annotated data (20% and 50%). The framework adopted our self-loop uncertainty as pseudo-labels achieves the highest F1 scores, i.e., 84.92% and 86.17% with 20% and 50% labeled data, respectively, and the latter is comparable to that of fully-supervied approach with 100% annotations (i.e., 86.17%). As ISIC has much more training data, compared to MoNuSeg, the ensemble-uncertainty-based framework achieves a comparable F1 score of 86.06% with 50% labeled data. However, it is worthwhile to mention that the generation of ensemble uncertainty requires 10 times of inferences during the test phase, as well as the MC dropout. Conversely, the proposed self-loop uncertainty can be generated with a single inference, which significantly saves the computational cost.

4 Conclusion

In this paper, we proposed a semi-supervised approach to train neural networks with limited labeled data and a large quantity of unlabeled images for medical image segmentation. A novel pseudo-label (namely self-loop uncertainty), generated by recurrently optimizing the neural network with a self-supervised task, is adopted as the ground-truth for the unlabeled images to augment the training set and boost the segmentation accuracy.

Acknowledgement. This work is supported by the Natural Science Foundation of China (No. 61702339), the Key Area Research and Development Program of Guangdong Province, China (No. 2018B010111001), National Key Research and Development Project (2018YFC2000702) and Science and Technology Program of Shenzhen, China (No. ZDSYS201802021814180).

References

1. Bai, W., et al.: Semi-supervised learning for network-based cardiac MR image segmentation. In: Descoteaux, M., Maier-Hein, L., Franz, A., Jannin, P., Collins, D.L., Duchesne, S. (eds.) MICCAI 2017. LNCS, vol. 10434, pp. 253–260. Springer, Cham (2017). https://doi.org/10.1007/978-3-319-66185-8_29
2. Bortsova, G., Dubost, F., Hogeweg, L., Katramados, I., de Bruijne, M.: Semi-supervised medical image segmentation via learning consistency under transformations. In: Shen, D., et al. (eds.) MICCAI 2019. LNCS, vol. 11769, pp. 810–818. Springer, Cham (2019). https://doi.org/10.1007/978-3-030-32226-7_90
3. Codella, N., et al.: Skin lesion analysis toward melanoma detection 2018: a challenge hosted by the International Skin Imaging Collaboration (ISIC). arXiv preprint arXiv:1902.03368 (2019)
4. Gal, Y., Ghahramani, Z.: Dropout as a Bayesian approximation: representing model uncertainty in deep learning. In: International Conference on Machine Learning, pp. 1050–1059 (2016)
5. Gidaris, S., Singh, P., Komodakis, N.: Unsupervised representation learning by predicting image rotations. In: International Conference on Learning Representations (2018)
6. He, K., Zhang, X., Ren, S., Sun, J.: Deep residual learning for image recognition. In: IEEE Conference on Computer Vision and Pattern Recognition, pp. 770–778 (2016)
7. Jungo, A., Reyes, M.: Assessing reliability and challenges of uncertainty estimations for medical image segmentation. In: Shen, D., et al. (eds.) MICCAI 2019. LNCS, vol. 11765, pp. 48–56. Springer, Cham (2019). https://doi.org/10.1007/978-3-030-32245-8_6
8. Lakshminarayanan, B., Pritzel, A., Blundell, C.: Simple and scalable predictive uncertainty estimation using deep ensembles. In: Annual Conference on Neural Information Processing Systems, pp. 6402–6413 (2017)

9. Larsson, G., Maire, M., Shakhnarovich, G.: Colorization as a proxy task for visual understanding. In: IEEE Conference on Computer Vision and Pattern Recognition, pp. 840–849 (2017)
10. Li, Y., Shen, L.: Skin lesion analysis towards melanoma detection using deep learning network. Sensors **18**(2), 556 (2018)
11. Luna, M., Kwon, M., Park, S.H.: Precise separation of adjacent nuclei using a Siamese neural network. In: Shen, D., et al. (eds.) MICCAI 2019. LNCS, vol. 11764, pp. 577–585. Springer, Cham (2019). https://doi.org/10.1007/978-3-030-32239-7_64
12. Nasr-Esfahani, E., et al.: Dense pooling layers in fully convolutional network for skin lesion segmentation. Comput. Med. Imaging Graph. **78**, 101658 (2019)
13. Naylor, P., Lae, M., Reyal, F., Walter, T.: Segmentation of nuclei in histopathology images by deep regression of the distance map. IEEE Trans. Med. Imaging **38**(2), 448–459 (2018)
14. Noroozi, M., Favaro, P.: Unsupervised learning of visual representations by solving jigsaw puzzles. In: Leibe, B., Matas, J., Sebe, N., Welling, M. (eds.) ECCV 2016. LNCS, vol. 9910, pp. 69–84. Springer, Cham (2016). https://doi.org/10.1007/978-3-319-46466-4_5
15. Oda, H., et al.: BESNet: boundary-enhanced segmentation of cells in histopathological images. In: Frangi, A.F., Schnabel, J.A., Davatzikos, C., Alberola-López, C., Fichtinger, G. (eds.) MICCAI 2018. LNCS, vol. 11071, pp. 228–236. Springer, Cham (2018). https://doi.org/10.1007/978-3-030-00934-2_26
16. Ronneberger, O., Fischer, P., Brox, T.: U-Net: convolutional networks for biomedical image segmentation. In: Navab, N., Hornegger, J., Wells, W.M., Frangi, A.F. (eds.) MICCAI 2015. LNCS, vol. 9351, pp. 234–241. Springer, Cham (2015). https://doi.org/10.1007/978-3-319-24574-4_28
17. Sedai, S., et al.: Uncertainty guided semi-supervised segmentation of retinal layers in OCT images. In: Shen, D., et al. (eds.) MICCAI 2019. LNCS, vol. 11764, pp. 282–290. Springer, Cham (2019). https://doi.org/10.1007/978-3-030-32239-7_32
18. Tang, Y., Yang, F., Yuan, S., Zhan, C.: A multi-stage framework with context information fusion structure for skin lesion segmentation. In: International Symposium on Biomedical Imaging, pp. 1407–1410 (2019)
19. Yu, L., Wang, S., Li, X., Fu, C.-W., Heng, P.-A.: Uncertainty-aware self-ensembling model for semi-supervised 3D left atrium segmentation. In: Shen, D., et al. (eds.) MICCAI 2019. LNCS, vol. 11765, pp. 605–613. Springer, Cham (2019). https://doi.org/10.1007/978-3-030-32245-8_67
20. Zhou, Y., Onder, O.F., Dou, Q., Tsougenis, E., Chen, H., Heng, P.-A.: CIA-Net: robust nuclei instance segmentation with contour-aware information aggregation. In: Chung, A.C.S., Gee, J.C., Yushkevich, P.A., Bao, S. (eds.) IPMI 2019. LNCS, vol. 11492, pp. 682–693. Springer, Cham (2019). https://doi.org/10.1007/978-3-030-20351-1_53

Semi-supervised Classification of Diagnostic Radiographs with NoTeacher: A Teacher that is Not Mean

Balagopal Unnikrishnan[1], Cuong Manh Nguyen[1], Shafa Balaram[1,2], Chuan Sheng Foo[1], and Pavitra Krishnaswamy[1](✉)

[1] Institute for Infocomm Research, A*STAR, Singapore
{balagopalu,pavitrak}@i2r.a-star.edu.sg
[2] National University of Singapore, Singapore

Abstract. Deep learning approaches offer strong performance for radiology image classification, but are bottlenecked by the need for large labeled training datasets. Semi-supervised learning (SSL) methods that can leverage small labeled datasets alongside larger unlabeled datasets offer potential for reducing labeling cost. However, few studies have demonstrated gains of SSL for real-world radiology image classification. Here, we adapt three leading SSL methods (Mean Teacher, Virtual Adversarial Training, Pseudo-labeling) for radiograph classification, and characterize their performance on two public X-Ray and CT classification benchmarks. We observe that Mean Teacher can achieve good performance gains in the low labeled data regime, but is sensitive to hyperparameters and susceptible to confirmation bias. To address these issues, we introduce a novel SSL method named NoTeacher. This method incorporates a probabilistic graphical model to maximize mutual agreement between student networks, thereby eliminating the need for a teacher network. We show that NoTeacher outperforms contemporary SSL baselines by enforcing better consistency regularization, and achieves over 90% of the fully supervised AUROC with less than 5% labeling budget.

Keywords: Semi-supervised deep learning · Classification · Multi-label · X-rays · CT · Mean teacher

1 Introduction

Deep learning approaches offer state-of-the-art performance for a range of image classification applications in radiology. Recent successes include chest radiograph diagnosis, brain tumor prognostication, fracture detection, and breast cancer screening [7,11,14,21]. However, these efforts typically require large labeled

B. Unnikrishnan and C. M. Nguyen—Equal Contribution.

Electronic supplementary material The online version of this chapter (https://doi.org/10.1007/978-3-030-59710-8_61) contains supplementary material, which is available to authorized users.

A. L. Martel et al. (Eds.): MICCAI 2020, LNCS 12261, pp. 624–634, 2020.
https://doi.org/10.1007/978-3-030-59710-8_61

training datasets assembled through resource-intensive labeling by specialized domain experts. Further, as discordance between expert raters can result in noisy labels, even more resource-intensive consensus rating across multiple blinded raters is often required [6].

One way to reduce this labeling burden is to use semi-supervised learning (SSL) approaches. These approaches leverage large numbers of unlabeled images alongside smaller numbers of labeled images for model development. The best performing approaches are typically *consistency-based*; these encourage a classifier's predictions to be consistent with a target on the unlabeled data. For instance, the popular Mean Teacher (MT) method [26] enforces consistency between predictions from two networks, termed the student and teacher, where the teacher is a time-averaged version of the student and is used for inference at test time. MT and other SSL approaches have been applied to computer vision classification benchmarks and to radiology image segmentation [3,10,29]. By contrast, few studies have demonstrated gains of these SSL methods for radiology image classification.

Here, we focus on semi-supervised learning for the under-studied radiograph (X-Ray and CT) classification task that typically involves detection of multiple abnormality types in the same image. We adapt three leading semi-supervised deep learning methods (Pseudo-labeling [9], Virtual Adversarial Training [15], and Mean Teacher [26]) to this multi-label setting and characterize performance using a realistic semi-supervised annotation and evaluation process. We observe that MT offers good performance gains in the low labeled data regime, but note its sensitivity to hyperparameters and vulnerability for confirmation bias. These MT limitations are in part due to reliance on a time-averaged student as a consistency target (teacher) that essentially leads the model to enforce self-consistency. To address these issues, we introduce the NoTeacher method that instead enforces consistency between two independent networks. Our method is derived by marginalizing out a latent consistency variable in a probabilistic graphical model. Results on the NIH-14 Chest X-Ray and RSNA Brain Hemorrhage CT datasets [4,22] show that NoTeacher outperforms competing methods on both datasets with minimal hyperparameter tuning, and achieves over 90% of the fully supervised AUROC with less than 5% labeling budget.

Related Work: We briefly introduce the methods characterized in this work. Pseudo-labeling (PSU) [9] is a classic SSL algorithm that is simple and commonly used. It is an iterative algorithm where one trains a model on labeled data, uses this trained model to infer pseudo-labels on the unlabeled data, then includes these pseudo-labels to enlarge the training set in the next iteration. Virtual Adversarial Training (VAT) [15] regularizes the output posterior distribution to be isotropically smooth around each input image. Mean Teacher (MT) [26] enforces consistency between two networks: a student model and a teacher model whose parameters are the exponential moving average (EMA) of the student's.

Aside from these methods that we characterized, other SSL methods have been proposed. For example, MixMatch [2] is a consistency-based SSL method which emphasizes data augmentation. However, the Mix-Up augmentation technique cannot be directly applied to medical images. Further, GAN-based methods [23] learn the underlying distribution from unlabeled data, but do not easily scale to high resolution images typical in radiology [12,13] and patch-wise adaptations [8] neglect the necessary global context. Deep Co-training [20] combines consistency and adversarial training in a multi-view framework, but the official implementation is not released and the method is not easily adapted to our multi-label setting. Finally, most related is GraphXNet [1], a graph-based label propagation method for X-Ray classification, but it does not support the multi-label setting and was not compared to other SSL methods.

2 Methods

Adaptations for Multi-label Classification: Radiology images may be associated with more than one label, where each label presents a binary classification task. Therefore, in our experiments, we used multi-task neural networks, trained using the sum of binary cross-entropy losses, one per label. The adaptation of VAT for our multi-label setting is described in the Supplement.

Mean Teacher Background: To provide context for our NoTeacher method, we briefly review Mean Teacher [26]. We consider a semi-supervised single-label classification task with image input x and binary output $y \in \{0,1\}$. Mean Teacher (MT) employs two networks with identical architecture: a student model F_S and a teacher model F_T. A schematic illustration of the MT model is provided in Fig. 1 (a). Given a batch $\mathbf{x}$ of training data, MT applies random augmentations η_S, η_T to generate inputs $\mathbf{x}_S$ and $\mathbf{x}_T$ for the corresponding model. During the feed-forward pass, MT computes a total loss combining the usual supervised cross-entropy (CE) classification loss and a consistency loss (mean-squared error MSE on the posteriors):

$$\mathfrak{L}_{\mathrm{MT}} = \mathrm{CE}\left(\mathbf{y}, \mathbf{f}_S^{\mathrm{L}}\right) + \lambda_{\mathrm{cons}} \mathrm{MSE}\left(\mathbf{f}_S, \mathbf{f}_T\right), \tag{1}$$

where $\mathbf{f}_S, \mathbf{f}_T$ are posterior outputs from the student and teacher networks, $\mathbf{f}_S^{\mathrm{L}}$ is the student's posterior output on the labeled data, and λ_{cons} is a consistency weight hyperparameter. The student model is updated directly by backpropagation using gradients of the loss $\mathfrak{L}_{\mathrm{MT}}$. Meanwhile, the teacher model updates its parameters by computing an EMA over the student's parameters. MT improves over supervised learning when the teacher generates better expected targets, or pseudo labels, to train its student. Recent papers have adapted MT for medical imaging tasks such as MR segmentation [19,29] and nuclei classification [24].

However, because the teacher model is essentially a temporal ensemble of the student model in the parameter space, MT has two potential drawbacks. First, enforcing consistency of the student model with its historical self may lead to confirmation bias or unwanted propagation of label noise [5]. Second, the teacher model is sensitive to the choice of the EMA hyperparameter, causing performance degradation when this hyperparameter is set outside a narrow range, as seen in realistic evaluation regimes [18]. Moreover, since MT does not establish a systematic method to compute the consistency weight λ_{cons}, the process to tune this hyperparameter for varied datasets is unclear. Several variants of MT therefore try to enforce consistency using other regularization schemes [19,24].

NoTeacher Overview: To address the above challenges, we introduce a new method. Figure 1(b) illustrates the overall framework of NoTeacher (NoT), where we have made two major changes: (a) we removed the EMA update so that the networks are completely detached and (b) we trained the model with a novel loss function $\mathfrak{L}_{\mathrm{NoT}}$ based on a probabilistic graphical model to enhance consistency. Since the two networks are treated equally, we index them numerically as F_1 and F_2.

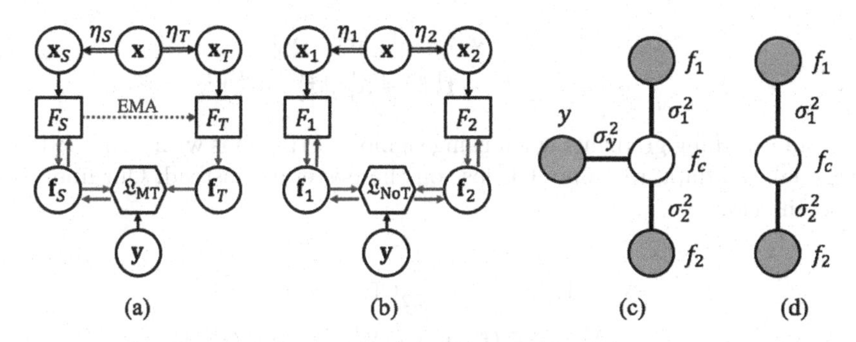

Fig. 1. Training procedure of (a) MT and (b) NoT for one iteration of semi-supervised learning. Double-line arrows denote random data augmentations, purple dotted arrow represents EMA update, blue and red arrows are forward and backward passes, respectively. NoT graphical model on a single (c) labeled and (d) unlabeled image.

NoTeacher Graphical Model and Loss: For each image x, the networks F_1, F_2 take as inputs two different images x_1, x_2 generated by applying random augmentations η_1, η_2, respectively. Since x_1, x_2 are generated from the same image, the network outputs $f_1 = F_1(x_1)$ and $f_2 = F_2(x_2)$ should be similar. Furthermore, if the image is labeled, then those outputs should also match the label. Because label y is binary, the network outputs can be interpreted as *posteriors* of the label, e.g., $f_1 = \Pr(y = 1|x_1)$. Inspired by previous works in semi-supervised regression [16] and kernel learning [30], we consider y, f_1, f_2 as random variables and design an undirected graphical model to impose probabilistic constraints on them. Figure 1 (c) and (d) show the NoT graphical models for labeled and unlabeled images respectively. The observed variables y, f_1, f_2 are represented by separate nodes, each is connected only to a latent variable called the *consensus function* $f_c \in [0, 1]$. As its name implies, f_c enforces the mutual agreement of the posteriors on both labeled and unlabeled data. When the label is available, f_c acts as an information relay between the posteriors and y. For analytical tractability, the differences $f_1 - f_c$ and $f_2 - f_c$ are assumed to follow Gaussian distributions $\mathcal{N}(0, \sigma_1^2)$ and $\mathcal{N}(0, \sigma_2^2)$ respectively. To account for labeling noise, we also assume the difference $y - f_c$ follows a Gaussian distribution $\mathcal{N}(0, \sigma_y^2)$.

Given a training batch of n_L labeled and n_U unlabeled images, the likelihood can be expressed as follows

$$\begin{aligned} p\left(\mathbf{y}|\mathbf{x}\right) \propto \exp\left(-\lambda_{y,1}^{\text{L}}\|\mathbf{f}_1^{\text{L}}-\mathbf{y}\|^2-\lambda_{y,2}^{\text{L}}\|\mathbf{f}_2^{\text{L}}-\mathbf{y}\|^2\right)\cdot \\ \exp\left(-\lambda_{1,2}^{\text{L}}\|\mathbf{f}_1^{\text{L}}-\mathbf{f}_2^{\text{L}}\|^2-\lambda_{1,2}^{\text{U}}\|\mathbf{f}_1^{\text{U}}-\mathbf{f}_2^{\text{U}}\|^2\right), \end{aligned} \tag{2}$$

where $\mathbf{f}_\bullet^\text{L}, \mathbf{f}_\bullet^\text{U}$ are vectors containing the posteriors on labeled and unlabeled data, respectively, $\mathbf{y}$ is the vector of labels, and $\lambda_{y,1}^\text{L}, \lambda_{y,2}^\text{L}, \lambda_{1,2}^\text{L}, \lambda_{1,2}^\text{U}$ are derived as detailed in Supplement. Maximizing the likelihood in (2) yields the following loss function:

$$\begin{aligned} \mathfrak{L}_{sq} =& \lambda_{y,1}^{\text{L}}\|\mathbf{f}_1^{\text{L}}-\mathbf{y}\|^2+\lambda_{y,2}^{\text{L}}\|\mathbf{f}_2^{\text{L}}-\mathbf{y}\|^2 \\ &+\lambda_{1,2}^{\text{L}}\|\mathbf{f}_1^{\text{L}}-\mathbf{f}_2^{\text{L}}\|^2+\lambda_{1,2}^{\text{U}}\|\mathbf{f}_1^{\text{U}}-\mathbf{f}_2^{\text{U}}\|^2. \end{aligned} \tag{3}$$

To avoid vanishing gradients when using sigmoid activations with a squared loss, on the labeled data, we apply CE loss on the posteriors instead. Our final NoT loss is therefore

$$\begin{aligned} \mathfrak{L}_{\text{NoT}} =& \lambda_{y,1}^{\text{L}}\text{CE}\left(\mathbf{y},\mathbf{f}_1^{\text{L}}\right)+\lambda_{y,2}^{\text{L}}\text{CE}\left(\mathbf{y},\mathbf{f}_2^{\text{L}}\right) \\ &+\lambda_{1,2}^{\text{L}}\text{MSE}\left(\mathbf{f}_1^{\text{L}},\mathbf{f}_2^{\text{L}}\right)+\lambda_{1,2}^{\text{U}}\frac{n_\text{U}}{n_\text{L}}\text{MSE}\left(\mathbf{f}_1^{\text{U}},\mathbf{f}_2^{\text{U}}\right), \end{aligned} \tag{4}$$

where we set $n_\text{L} = n_\text{U}$ to cancel out the fraction in the last term. These changes also enable a fair comparison with MT in our experiments. The first two terms of $\mathfrak{L}_{\text{NoT}}$ represent the supervised losses while the last two terms enforce mutual agreement between the classifiers, thus enhancing consistency of the predictions.

3 Experiment Setup

Datasets: We now describe the two datasets used in our experiments, and provide statistical breakdowns in the Supplement.

NIH-14 Chest X-Ray [28]*:* The first dataset we used comprises 112,120 frontal chest X-Ray images, where images are annotated for presence of one or more of 14 pathologies (14 binary labels). 53.9% of images are normal (negative for all 14 labels) and 46.1% abnormal; 40.1% of abnormal images are positive for more than one pathology. We used publicly available training (70%), validation (10%) and testing (20%) splits with no patient overlaps [31].

RSNA Brain CT [4]*:* The second dataset we used is a collection of 19,530 CT brain exams from the Stage 1 training dataset of the RSNA 2019 Challenge, where images are annotated for presence of one or more of 5 types of intracranial hemorrhage (5 binary labels). We focus on slice-level classification, and consider only one study per patient. 85.8% of images are normal (negative for all 5 labels) while the remaining 14.2% are positive for bleeds; 30.1% of the abnormal images are positive for multiple bleeds. We derived random training (60%), validation (20%) and testing (20%) splits with no patient overlaps. We obtained pre-processed images from [17]. The pre-processing steps used are in accordance with leading solutions in the RSNA Challenge [25] and include: (a) converting the raw pixel values to Hounsfield Units using the slope and rescale intercept from the original DICOM files, (b) windowing to restrict all pixel values to the width around the center as reported in the individual DICOM files, (c) normalizing to range [0, 255], and (d) resizing to 256 × 256.

Design of Realistic Semi-supervised Experiments: We model a realistic annotation process where clinical raters extract a finite pool of N images for model development, and proceed to label them systematically. First, we consider a labeling budget of $L = L_T + L_V$ images, where L_T denotes the labeled subset of the training data, and L_V denotes the labeled validation set. A practical labeling budget implies that $L_T \gg L_V$ with L_V very small. Second, for a given L, we need to randomly sample a subset for labeling from the overall pool of N images. Implicitly, this means the subset of L images cannot be chosen based on their labels i.e., it is infeasible to fit any stratified class distribution requirements. For lower budgets, this constraint requires repeatedly sampling and labeling images until representative numbers are obtained for each class. Third, increases in L require maintaining the existing labeled images, and progressively adding on images and labeling them (implicitly decreasing unlabeled set size).

For our experiments, we had to align the above process to publicly available data collections that were already split into training, validation and test sets. Hence, for a given labeling budget L, we randomly sampled L images proportionately from the training and validation splits. Then, for the unlabeled set, we only considered the remaining portion of the training split to ensure that the unlabeled validation split does not inform training. All our experiments assume a fixed-size held out test set and do not count the test set definition as part of labeling budget for model development. The above practical requirements preclude the conveniences of large validation sets, stratified labeling, balanced class distributions, fixed unlabeled budgets that are often encountered in standard SSL benchmarking papers [18], and introduce additional challenges.

Implementation Details: We describe the supervised training setup and hyperparameter tuning procedures. SSL validation details are in Supplement.

Supervised Training Backbone: We use the same backbone network architecture for supervised (SUP) and semi-supervised methods to ensure fair comparisons: DenseNet121 for NIH-14 Chest X-Ray as per [22] and DenseNet169 for RSNA Brain CT (used by the leading RSNA Challenge solution [25]). In each case, we initialized the network with weights pretrained on ImageNet. For train-time augmentations, we employed random horizontal flipping, resizing, and center cropping. We normalized all images based on ImageNet statistics, and used the Adam optimizer (learning rate 10^{-4}, $\beta = [0.9, 0.999]$, $\epsilon = 10^{-8}$, and weight decay 10^{-5}).

Hyperparameter Tuning: We tuned hyperparameters in accordance with literature norms for all semi-supervised methods to ensure fair comparisons. In particular, we tuned ϵ for VAT, EMA decay and consistency weight for MT, and considered variations required for different labeling budgets. For NoT, even though there are four weights, only the ratio between the weights is of consequence – hence only the ratio hyperparameter requires tuning. As the two networks have similar architecture, we selected $\sigma_1^2 = \sigma_2^2$ and varied the labeling noise σ_y^2. The tuning process and final parameter choices are provided in Supplement.

Parameter Averaging: In order to enable fair comparison with MT, we keep an EMA copy for the supervised baseline and VAT; and report the best result from either the trained model or the EMA copy. This way, performance gains reported are not just because of averaging but also due to the consistency mechanism.

4 Results

We systematically evaluate performance of each SSL algorithm as a function of varying labeling budget L. We start L at 500 images of the development dataset or the number required to have at least 1 positive image per label (whichever is higher). For each labeling budget, we also evaluate performance of a comparable supervised baseline (SUP) trained purely on the labeled images. All experiments maintain the same held-out test set for rigorous comparison. As both classification objectives are multi-label tasks, we compute the per-class AUROC and report average across all classes.

Performance vs. Labeling Budget: Figures 2 and 3 show results on the NIH-14 X-Ray and RSNA Brain CT datasets respectively. Detailed performance numbers are provided in Supplement.

For low labeling budgets, (a) the semi-supervised methods offer strong gains over the supervised baseline, and (b) our NoT method outperforms the other semi-supervised methods. To surpass 90% of the fully supervised AUROC, NoT requires less than 5% labeling budget for NIH-14 X-Ray dataset and less than 2.5% labeling budget for RSNA CT dataset. First, For NIH-14, with 5% labeling budget, NoT gains 6.4% over the corresponding supervised baseline and over 3.1% vs. other SSL methods. NoT also outperforms the comparable GraphXNet result at 5% labeling budget [1]. For RSNA CT, with 2.5% labeling budget, NoT gains 3.7% over the supervised baseline (SUP) and over 2.6% vs. other SSL methods. Second, for higher labeling budgets, performance of all semi-supervised methods converges, suggesting saturation of gain from unlabeled data. Across all methods, NoT can achieve over 90% of the fully supervised AUROC with less than 5% labeling budget. Third, at 5% labeling budget for NIH-14 and 2.5% labeling budget for RSNA CT, we compute the AUROC gain of each SSL method over the corresponding supervised baseline (SUP). We plot these gains for each class in a heatmap format, where conditions are ordered by number of images in the labeled set. For rarer conditions with lower prevalence in the labeled subsets, NoT gains much more than other SSL methods.

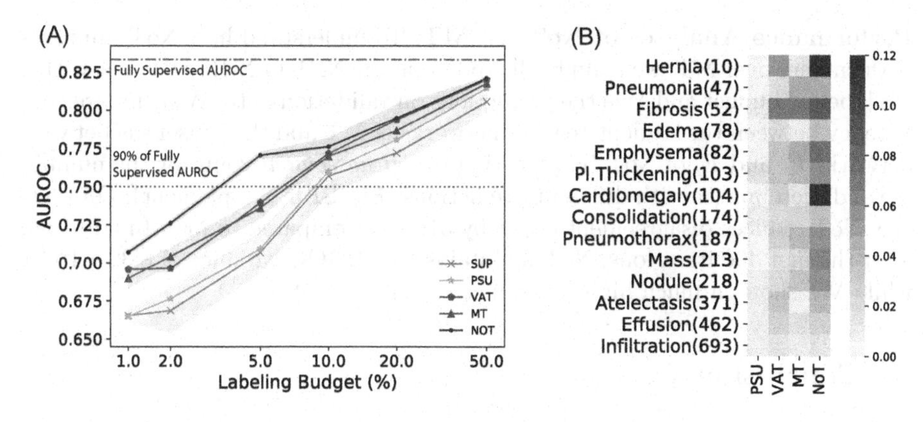

Fig. 2. Performance Results for the NIH-14 Chest X-Ray Dataset.(A) Average AUROC vs. Labeling Budget. AUROC evaluated with 1177, 1569, 3923, 7846, 15693, 39234 labeled images (with L_T : L_V set to 70:10). Fully supervised baseline (dash-dotted line) is based on 78468 images. (B) Class-wise SSL vs. SUP AUROC gains for 5% labeling budget. Hernia (10) indicates 10 images with hernia in labeled set.

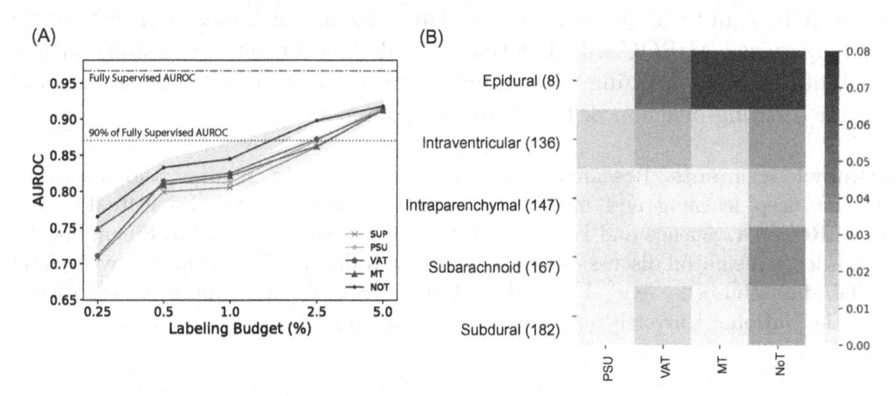

Fig. 3. Performance Results for the RSNA Brain CT Dataset. (A) Average AUROC vs. Labeling Budget. AUROC evaluated with 749, 1777, 3495, 6744, and 17242 images. Labeling budget set at scan level and selected scans have all slices labeled. L_T : L_V set to 60:20. Fully supervised baseline (dash-dotted line) is based on 352839 images. (B) Class-wise SSL vs. SUP AUROC gains for 2.5% labeling budget. Epidural (8) indicates 8 images with epidural bleed in labeled set.

Connections to Co-training and Label Propagation: We posit that these gains arise from the multi-view formulation of NoT. While VAT, PSU and even MT are essentially single-network models (the MT teacher network is learned passively via EMA), NoT is a multi-view learning technique which benefits from having multiple views of the data. Being a co-training method, NoT also has connections with label propagation [27].

Performance Analyses of NoT vs. MT: To understand how NoT improves performance over MT, we train both models on the NIH-14 X-Ray dataset with a 5% labeling budget and save the predictions on validation data. We compare consistency between the student-teacher networks of MT and the consensus between networks F_1 and F_2 of NoT by reporting the *disagreement* count or the number of validation images with different predictions (Fig. S1 in Supplement). On average, NoT reduces disagreement count by 51.87 % compared to MT. In addition, after the first 400 iterations, NoT maintains an AUROC variance of 8.39×10^{-5}, while MT shows a much higher variance of 3.42×10^{-4}.

5 Conclusion

We adapted and characterized three leading semi-supervised methods for multi-label radiograph classification using a realistic annotation and evaluation process. To further improve the best of these methods, MT, we introduce the NoTeacher method (NoT) to better enforce consistency and reduce confirmation bias. We demonstrate that NoT provides strong performance gains on two public X-Ray and CT classification benchmarks, and achieves over 90% of the fully supervised AUROC with less than 5% labeling budget. Our results suggest feasibility for deep learning with minimal supervision on radiology images and provide a strong benchmark for future developments.

Acknowledgements. Research efforts were supported by funding and infrastructure for deep learning and medical imaging research from the Institute for Infocomm Research, Science and Engineering Research Council, A*STAR, Singapore. We acknowledge insightful discussions with Jayashree Kalpathy-Cramer and Praveer Singh at the Massachusetts General Hospital, Boston USA. We also thank Ashraf Kassim from the National University of Singapore for his support.

References

1. Aviles-Rivero, A.I., Papadakis, N., et al.: GraphX^{NET} - chest X-ray classification under extreme minimal supervision. In: International Conference on Medical Image Computing and Computer-Assisted Intervention, pp. 504–512 (2019)
2. Berthelot, D., et al.: Mixmatch: a holistic approach to semi-supervised learning. In: Advances in Neural Information Processing Systems, pp. 5050–5060 (2019)
3. Feng, Z., et al.: Semi-supervised learning for pelvic MR image segmentation based on multi-task residual fully convolutional networks. In: IEEE International Symposium on Biomedical Imaging, pp. 885–888 (2018)
4. Flanders, A.E., et al.: Construction of a machine learning dataset through collaboration: the RSNA 2019 brain CT hemorrhage challenge. Radiol. Artif. Intell. **2**(3), e190211 (2020)
5. Ke, Z., et al.: Dual student: breaking the limits of the teacher in semi-supervised learning. In: IEEE International Conference on Computer Vision, pp. 6728–6736 (2019)

6. Langlotz, C.P., et al.: A roadmap for foundational research on artificial intelligence in medical imaging: from the 2018 NIH/RSNA/ACR/The Academy Workshop. Radiology **291**(3), 781–791 (2019)
7. Lao, J., et al.: A deep learning-based radiomics model for prediction of survival in glioblastoma multiforme. Sci. Rep. **7**(1), 1–8 (2017)
8. Lecouat, B., et al.: Semi-supervised deep learning for abnormality classification in retinal images. In: NeurIPS Machine Learning for Health Workshop (2018)
9. Lee, D.H.: Pseudo-label: The simple and efficient semi-supervised learning method for deep neural networks. In: ICML Workshop on Challenges in Representation Learning, vol. 3 (2013)
10. Li, X., et al.: Transformation-consistent self-ensembling model for semisupervised medical image segmentation. In: IEEE Transactions on Neural Networks and Learning Systems (2020)
11. Lindsey, R., et al.: Deep neural network improves fracture detection by clinicians. Proc. Natl. Acad. Sci. **115**(45), 11591–11596 (2018)
12. Madani, A., et al.: Semi-supervised learning with generative adversarial networks for chest X-ray classification with ability of data domain adaptation. In: IEEE International Symposium on Biomedical Imaging, pp. 1038–1042 (2018)
13. Madani, A., et al.: Deep echocardiography: data-efficient supervised and semi-supervised deep learning towards automated diagnosis of cardiac disease. Nat. Partner J. Digit. Med. **1**(1), 1–11 (2018)
14. McKinney, S.M., et al.: International evaluation of an AI system for breast cancer screening. Nature **577**(7788), 89–94 (2020)
15. Miyato, T., et al.: Virtual adversarial training: a regularization method for supervised and semi-supervised learning. IEEE Trans. Pattern Anal. Mach. Intell. **41**(8), 1979–1993 (2018)
16. Nguyen, C.M., et al.: Partial Bayesian co-training for virtual metrology. IEEE Trans. Ind. Inform. **16**(5), 2937–2945 (2019)
17. Oh, R.: RSNA Train/Test png (256 × 256). Kaggle dataset. https://kaggle.com/richul/rsna_png_128_128
18. Oliver, A., Odena, A., Raffel, C.A., et al.: Realistic evaluation of deep semi-supervised learning algorithms. In: Advances in Neural Information Processing Systems, pp. 3235–3246 (2018)
19. Perone, C.S., Cohen-Adad, J.: Deep semi-supervised segmentation with weight-averaged consistency targets. In: Stoyanov, D., et al. (eds.) DLMIA/ML-CDS - 2018. LNCS, vol. 11045, pp. 12–19. Springer, Cham (2018). https://doi.org/10.1007/978-3-030-00889-5_2
20. Qiao, S., Shen, W., Zhang, Z., Wang, B., Yuille, A.: Deep co-training for semi-supervised image recognition. In: Ferrari, V., Hebert, M., Sminchisescu, C., Weiss, Y. (eds.) ECCV 2018. LNCS, vol. 11219, pp. 142–159. Springer, Cham (2018). https://doi.org/10.1007/978-3-030-01267-0_9
21. Rajpurkar, P., Irvin, J., et al.: Deep learning for chest radiograph diagnosis: a retrospective comparison of the CheXNeXt algorithm to practicing radiologists. Pub. Libr. Sci. Med. **15**(11), e1002686 (2018)
22. Rajpurkar, P., Irvin, J., et al.: CheXNet: radiologist-level pneumonia detection on chest X-rays with deep learning. arXiv preprint arXiv:1711.05225 (2017)
23. Salimans, T., et al.: Improved techniques for training GANs. In: Advances in Neural Information Processing Systems, pp. 2234–2242 (2016)

24. Su, H., Shi, X., Cai, J., Yang, L.: Local and global consistency regularized mean teacher for semi-supervised nuclei classification. In: Shen, D., et al. (eds.) MICCAI 2019. LNCS, vol. 11764, pp. 559–567. Springer, Cham (2019). https://doi.org/10.1007/978-3-030-32239-7_62
25. Tao, S.: RSNA intracranial hemorrhage detection. GitHub repository (2019). https://github.com/SeuTao/RSNA2019_Intracranial-Hemorrhage-Detection
26. Tarvainen, A., Valpola, H.: Mean teachers are better role models: weight-averaged consistency targets improve semi-supervised deep learning results. In: Advances in Neural Information Processing Systems, pp. 1195–1204 (2017)
27. Wang, W. and Zhou, Z.H.: A new analysis of co-training. In: International Conference on Machine Learning, pp. 1135–1142 (2010)
28. Wang, X., et al.: Hospital-scale chest X-ray database and benchmarks on weakly-supervised classification and localization of common thorax diseases. In: IEEE Conference on Computer Vision and Pattern Recognition, pp. 3462–3471 (2017)
29. Yu, L., Wang, S., Li, X., Fu, C.-W., Heng, P.-A.: Uncertainty-aware self-ensembling model for semi-supervised 3D left atrium segmentation. In: Shen, D., et al. (eds.) MICCAI 2019. LNCS, vol. 11765, pp. 605–613. Springer, Cham (2019). https://doi.org/10.1007/978-3-030-32245-8_67
30. Yu, S., et al.: Bayesian co-training. J. Mach. Learn. Res. **12**, 2649–2680 (2011)
31. Zech, J.: Reproduce-chexnet. GitHub repository (2018). https://github.com/jrzech/reproduce-chexnet

Predicting Potential Propensity of Adolescents to Drugs via New Semi-supervised Deep Ordinal Regression Model

Alireza Ganjdanesh[1], Kamran Ghasedi[1], Liang Zhan[1], Weidong Cai[2], and Heng Huang[1,3](✉)

[1] Electrical and Computer Engineering, University of Pittsburgh, Pittsburgh, PA 15261, USA
{alireza.ganjdanesh,liang.zhan,heng.huang}@pitt.edu, kamran.ghasedi@gmail.com

[2] School of Computer Science, University of Sydney, Sydney, NSW 2006, Australia
tom.cai@sydney.edu.au

[3] JD Finance America Corporation, Mountain View, USA

Abstract. Addiction to drugs between young people is one of the most severe problems in the real world, and it imposes a huge financial and emotional burden on their families and societies. Therefore, predicting potential inclination to drugs at earlier ages can prevent lots of detriments. In this paper, we propose a new semi-supervised deep ordinal regression model to predict the possible propensity of adolescents to marijuana using the diffusion MRI-derived mean diffusivity (MD) from 148 Regions of Interest (ROIs). The traditional deep ordinal regression models cannot be directly applied to our biomedical problem which only has a small number of labeled data, not enough to train the deep learning models. Thus, we design a semi-supervised learning mechanism for deep ordinal regression, such that both labeled and unlabeled data can be used to enhance the model training. In our experiments, we use the ABCD dataset, which contains MRI images of the adolescents under study and their answers in the Likert scale to a questionnaire containing questions about Marijuana. Experimental results on the ABCD dataset validate the superior performance of our new method. Our study provides an inexpensive way to predict the drug tendency using brain MRI data.

Keywords: Adolescent · Marijuana · Deep learning · Ordinal regression · Semi-supervised learning · Diffusion MRI · Mean diffusivity

1 Introduction

Predicting a potential tendency of adolescents to drugs in the future enables us to take effective preventive actions against the risk of their addiction to drugs.

This work was partially supported by U.S. NSF IIS 1836945, IIS 1836938, IIS 1845666, IIS 1852606, IIS 1838627, IIS 1837956.

A. L. Martel et al. (Eds.): MICCAI 2020, LNCS 12261, pp. 635–645, 2020.
https://doi.org/10.1007/978-3-030-59710-8_62

One of the approaches to do so is to study brain condition and its possible correlation with different behavioral patterns. In this regard, a study called Adolescents Brain Cognitive Development (ABCD)[1] is in progress, which is the largest long-term study of brain development in the United States. This study aims to monitor the brain condition of children from the age of 9–10 to primary stages of adulthood using diffusion and functional Magnetic Resonance Imaging (dMRI and fMRI respectively), and by doing so analyzing the factors that impact different aspects of the young people's life such as the potential inclination to drugs. ABCD dataset is the fruit of this project. It contains dMRI and fMRI images of the cases under study. Also, these cases have answered some questionnaires on the Likert Scale, and their answers can reflect their viewpoint regarding drugs. Therefore, developing a method that can predict answers of a new case to the questionnaires based on their MRI features can be a solution to the goal of drug tendency prediction.

In ABCD data, the answers to the questions (labels) are on the Likert Scale where answers 1, 2, 3, 4, and 5 correspond to Strongly Disagree, Disagree, Neither Agree nor Disagree, Agree, and Strongly Agree respectively. This is neither a traditional regression problem nor a multi-class classification one because the typical regression and classification tasks don't consider the order relations between response variables. But in this study, the answers' values have order meanings. For example, if the correct answer is 5, predicting 4 should have a lower negative impact than predicting 1. Thus, the ordinal regression models should be used for such an answer prediction.

In recent research, the deep ordinal regression models have achieved much better results in various applications than traditional ordinal regression methods. However, these deep ordinal regression models cannot be applied to our study. Because the existing deep learning methods require large amount labeled data to train satisfied models but the ABCD dataset does not contain the answers for all the cases whose MRI data are available. This is not surprising, and at the most circumstances in biomedical applications, unlabeled data are abundant and labeled data are rare because in biomedical research providing labels is expensive or difficult as it need human expert supervision.

To address the above challenging problems, in this paper, we focus on designing new semi-supervised deep learning model that addresses the ordinal regression task, and at the same time, reduces the need for massive labeled training data. Our new approach shows superior performance on predicting the possible propensity of adolescents to marijuana using the diffusion MRI-derived mean diffusivity (MD) from 148 Regions of Interest (ROIs) of ABCD data.

2 Related Work

Ordinal Regression refers to the supervised machine learning problems in which the labels are categorical, and concurrently, the categories have meaningful order

[1] https://www.addictionresearch.nih.gov/abcd-study.

between them. In the literature, there are several types of methods proposed for ordinal regression problems.

In the first category, the methods try to address the problem from the regression perspective. They learn some mapping function that maps the samples to real numbers and suggest a way to find some decision boundaries to determine the rank of a given sample based on the interval that its mapped value lies in [15]. In the second category, the related methods try to reformulate the ordinal regression problem so that it enables us to leverage the power of prominent classification methods. They split the problem into a sequence of binary classification sub-problems and determine the class of the input based on the aggregation of the answers of the binary classifiers [3,9,10].

However, all of the mentioned methods are based on handcrafted features. In recent years, deep learning has obtained the state of the art results on different tasks in classification such as object recognition [7]. The main reason for its success is its superior ability to learn how to extract useful features for classification. As a result, modern approaches to the ordinal regression problem have focused on designing their methods based on deep learning. Niu *et al.* [13] proposed the first solution to ordinal regression using deep learning in the context of age estimation. They converted the problem with R possible ranks to R − 1 binary classification problems such that the i-th problem determines whether the rank of a sample is bigger than i or not. They have used a Convolutional Neural Network (CNN) as the feature extractor for their network. Liu *et al.* [11] suggested the second deep learning based method for ordinal regression based on the idea of the first category by mapping the samples to real numbers, but they did not use these real numbers to determine the rank of the samples. Rather, they proposed a loss function on these numbers to show the network the natural order between the ranks. Similar to [13], Liu *et al.* used CNN for feature extraction, and they called their method as CNNPOR.

Semi-supervised learning aims to reduce the need for labeled data for training models, especially deep neural networks. The main importance of semi-supervised techniques is in the areas that labeled data is scarce and there is ample unlabeled data. Numerous methods have been proposed in this regard, and the general idea of almost all of them is to add a term to the loss function that is calculated using the unlabeled data set that ultimately benefits generalization of the trained network [1,4,8,14,16,17]. Sajjadi *et al.* [14], Laine and Aila [8], and Tarvainen and Valpola [16] addressed consistency regularization for semi-supervised training. Berthelot *et al.* [1] applied sharpening to enforce the predictions to have lower entropy and impose entropy regularization [4]. Verma *et al.* [17] and Berthelot *et al.* [1] used MixUp idea to make the network more robust. The problem is that these methods both are originally intended for image data or their performance is validated on image data. In addition, they have been used to improve the classification task. Our method addresses the semi-supervised training in the ordinal regression task.

3 Proposed Method

3.1 Motivations and Model Design

In this section, we propose our method for solving an ordinal regression problem with the semi-supervised mechanism. Because our problem is semi-supervised learning, we first explain how to use the labeled set, *i.e.* supervised learning, and then we will focus on incorporating the unlabeled data to enhance the performance of our model.

Our method should use the labeled data of the training set to not only provide the neural network model the corresponding rank of each input but also suggest a trick to teach the natural order between the possible ranks to it. We address the former by framing it as a classification task and do so for the latter by introducing a mapping and imposing an order between mapped values of inputs from different ranks.

To leverage the unlabeled part of the training set, we add new terms to the loss function of the supervised training so that it encourages the network to make more consistent predictions, predict more confident scores, and have convex behavior with the inputs and their corresponding labels. We show a quantitative description of our approach in the following subsections.

3.2 Problem Formulation

Let us consider an ordinal regression problem with rank R and a set of labeled samples $L = \{(x_i, y_i)|x_i \in \mathcal{X}, y_i \in \mathcal{Y}\}$ where $\mathcal{Y} = \{1, 2, ..., R\}$ along with a set of unlabeled ones $U = \{(x_i)|x_i \in \mathcal{X}\}$. We denote the subset of samples with rank k as $\mathcal{X}_k$. We show the development of our loss function step by step, and then we will provide our proposed architecture.

3.3 Loss Functions

Cross Entropy Loss. We convert the labels in the labeled part of the training set into the one-hot format (a vector with the length R and its k-th element being one and other elements being zero if the rank of a sample is k) and define a part of the loss function as the cross-entropy between the targets and softmax of the outputs of the network to minimize the KL divergence between the target distribution and $P_{model}(y|x;\theta)$ where θ is the vector of the model parameters.

$$\mathcal{L}_{CE} = \frac{1}{|L|} \sum_{(x_i,y_i)\in L} H(y_i, P_{model}(y|x_i;\theta)). \tag{1}$$

Ordinal Loss. To guide the network to learn the natural order between the ranks, we map the samples to real numbers and define a loss over the mapped values of the samples from different ranks. Considering the activations of the penultimate layer of the deep neural network for an input x as $f(x)$, we define a linear mapping $\mathcal{M}(f(x))$ from f to real numbers. Then, we enforce the network to generate larger values for the samples of the class k than the values for the ones of the class $k-1$. To do so, given a batch of samples $X_k \subseteq \mathcal{X}_k$ and $X_{k-1} \subseteq \mathcal{X}_{k-1}$ we add the following term to the loss function:

$$\mathcal{L}_{Ordinal} = \sum_{k=2}^{R} \sum_{\substack{x_k \in X_k \\ x_{k-1} \in X_{k-1}}} ReLU(1 - \mathcal{M}(f(x_k)) + \mathcal{M}(f(x_{k-1}))) \tag{2}$$

where $ReLU(x)$ is Rectified Linear Unit [12] with output equal to x given $x > 0$ and zero otherwise. The advantage of this loss function is that it only needs to consider pairwise comparison between adjacent ranks because if $\mathcal{M}(f(x_{k-1})) < \mathcal{M}(f(x_k))$ and $\mathcal{M}(f(x_k)) < \mathcal{M}(f(x_{k+1}))$, then $\mathcal{M}(f(x_{k-1})) < \mathcal{M}(f(x_{k+1}))$ $(x_i \in \mathcal{X}_i)$, i.e, adjacent ranks comparison implies farther ranks comparison.

3.4 Semi-supervised Learning

Consistency Regularization. Consistency regularization aims to make the prediction of the network for a sample and its augmented versions, varieties of the sample that have the same conceptual meaning in the problem context, as close as possible. For example, in image classification, the classifier should output the same distribution for an image that is a rotated version of the original one because rotation does not change the class of a sample. To apply consistency regularization, we produce a guessed label for each unlabeled sample in two steps. At first, we generate several augmented samples from the unlabeled sample by adding Gaussian noise with a small variance to it. After that, we enter the original sample as well as augmented ones to the network and determine the average of the output distributions of the network as the guessed label for the original sample.

One of the techniques for training consistent classifier in a semi-supervised training is to motivate the network to show convex behavior in its predictions, i.e make a similar prediction for a linear combination of two unlabeled samples to the same linear combination of its predictions for them. [17] To implement this idea, we generate new samples by mixing samples in the dataset. Given two samples $(x_1, p_1), (x_2, p_2)$, we produce (x_3, p_3) as following:

$$\beta \sim Beta(\alpha, \alpha) \tag{3}$$

$$\beta := max(1-\beta, \beta) \tag{4}$$

$$x_3 = \beta * x_1 + (1-\beta) * x_2 \tag{5}$$

$$p_3 = \beta * p_1 + (1-\beta) * p_2 \tag{6}$$

Entropy Minimization. Entropy minimization idea is originated in the information theory context where the uncertainty of a distribution is measured with its entropy. As a result, minimizing the entropy of the network output distribution is equivalent to enforcing the network to make more confident predictions, and we use sharpening to do so. If we denote the network output distribution prediction with vector p, the sharpened vector q gets calculated as following:

$$q_i = \frac{p_i^{\frac{1}{T}}}{\sum_{j=1}^{R} p_j^{\frac{1}{T}}} \tag{7}$$

where R is the length of p (number of the possible ranks in the ordinal regression problem), and T is the distribution temperature.

Based on the above ideas and to leverage their advantages, we use the Algorithm 1 to prepare inputs for our loss function for the semi-supervised training that we will introduce in the next subsection.

Algorithm 1 Mixing Up Labeled and Unlabeled Set

Input: Input: A set of labeled samples $\mathcal{L} = \{(x_{l_i}, y_i)\}$, a set of unlabeled samples $\mathcal{U} = \{(x_{u_j})\}$ for $1 \leq i, j \leq N$, α, T, Gaussian noise standard deviation σ, number of augmentations M, Deep Network net

for $m = 0$ to M **do**
  if m is 0 **then**
    $y_{pred_{j,m}} = net(x_{u_{j,m}})$
  else
    $noise \sim \text{Gaussian}(0, \sigma)$
    $x_{u_{j,m}} = x_{u_j} + noise$
    $y_{pred_{j,m}} = net(x_{u_{j,m}})$
  end if
end for
for $j = 1$ to N **do**
  $y_{pred_j} = Average(y_{pred_{j,0}}, \ldots, y_{pred_{j,M}})$
  $y_{guess_j} = Sharpen(y_{pred_j}, T)$
end for
$\mathcal{U}_1 = \{(x_{u_j}, y_{guess_j}) \mid 1 \leq j \leq N\}$
$\mathcal{C} = Shuffle(Concatenation(\mathcal{L}, \mathcal{U}_1))$
$\mathcal{L} = \{(MixUp(\mathcal{L}, \mathcal{C}[1 : N])\}$
$\mathcal{U} = \{(MixUp(\mathcal{U}_1, \mathcal{C}[N + 1 : 2N])\}$
Return $\mathcal{L}, \mathcal{U}$

Semi-supervised Training. Now we introduce the loss functions that we use to perform the semi-supervised training.

$$\mathcal{L}_l = \frac{1}{|\hat{\mathcal{L}}|} \sum_{(x_i, y_i) \in \hat{\mathcal{L}}} H(y_i, P_{model}(y|x_i; \theta)) \tag{8}$$

$$\mathcal{L}_u = \frac{1}{R|\hat{\mathcal{U}}|} \sum_{(x_j, y_j) \in \hat{\mathcal{U}}} \|y_j - P_{model}(y|x_j; \theta)\|_2^2 \tag{9}$$

In these equations, $\mathcal{L}_l$ has the notion of consistency regularization, and $\mathcal{L}_u$ aims to push the network to show convex behavior.

3.5 Proposed Loss Function and Model Architecture

We train our model based on the following loss function:

$$\mathcal{L} = \mathcal{L}_{CE} + c_1 * \mathcal{L}_{Ordinal} + c_2 * \mathcal{L}_l + c_3 * \mathcal{L}_u \tag{10}$$

where c_1, c_2, and c_3 are hyperparameters. This loss function combines all of the motivations for ordinal regression and semi-supervised learning mentioned in the above sections. As we discussed, our experiments were on non-image data. Therefore, we used Multi-Layer Perceptron as the DNN feature extractor. Because our method solves the ordinal regression problem in a semi-supervised setting, we name it ORSS. Figure 1 shows the proposed procedure for training the network. We have shown the calculation path for each loss function with distinct colors. For example, the path used to compute the cross-entropy loss is the yellow one in Fig. 1.a.

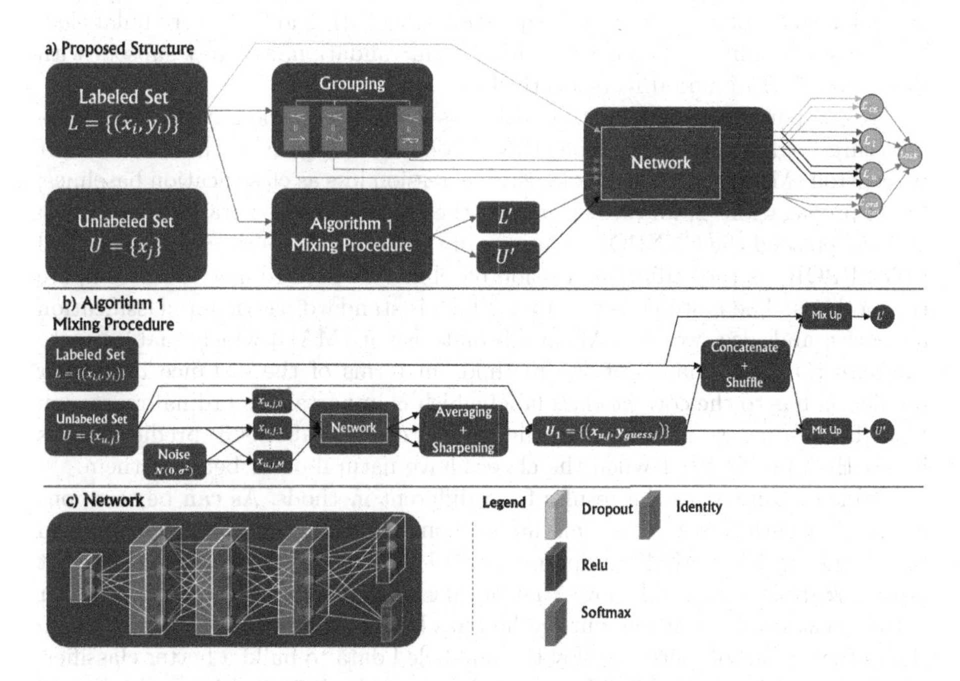

Fig. 1. The architecture proposed for ORSS problem. **a)** The path for calculation of each part of the loss function is specified with a separate color. **b)** Block diagram of the Algorithm 1. **c)** The network used for prediction. (Color figure online)

4 Experiments

We evaluated our method on 10731 subjects (mean age = 118.96 ± 7.5 months) from the ABCD cohort. We extracted mean diffusivity (MD) from 148 Regions of Interest (ROIs) for each subject. Then, we put these measurements in a vector with length 148 and concatenated with three extra measures to the vector: the mean MD for left hemisphere, the mean MD for right hemisphere, and the mean MD for the whole brain. In aggregate, we obtained a MD vector with the length of 151 (=148+3) for each person, from the dMRI images.

ABCD dataset contains questionnaires containing questions that ask the opinion of people regarding statements about drugs on the Likert Scale. For example, one of the questions is 'Marijuana helps a person relax and feel better.' If one of the adolescents answers completely agree to this question, it may suggest that they may have an inclination to drugs in the future.

In our experiments, our goal was to predict the answers to the question mentioned above using brain MRI features. As the answers are on the Likert scale, our task is Ordinal Regression, and because we did not have answers for some people whom we had dMRI features of them, we tried semi-supervised training to enhance the performance of our model. In summary, we had dMRI features for 10731 people (a vector of length 151 for each person). Among them, we had answers of 3663 ones to the question (labeled), and 7068 were unlabeled. We randomly split the labeled part into train, validation, and test subsets with the ratio 0.7, 0.15, and 0.15 respectively.

We compared our model with multi-class logistic regression, K-nearest neighbor, thresholded ridge regression, thresholded lasso regression, and multi-layer perceptron (MLP) with softmax logistic regression loss as classification baselines; Label propagation [2] and MixMatch as the semi-supervised training baselines; and we replaced the CNNPOR structure of the Liu *et al.* approach [11], specified as MLPPOR, as the ordinal regression baseline. We used to metrics to compare the methods. The first one is accuracy which is standard metric for classification networks, and also, we used Mean Absolute Error (MAE) which enables us to compare the performance of the methods in terms of the distance that their prediction has to the correct class label which is important in ordinal regression tasks because as we mentioned earlier, if the correct label is 5, predicting 4 is better than predicting 1 when the classes have natural order between them.

Table 1 summarizes the results from different methods. As can be seen, our method outperforms all other methods when comparing by both accuracy and MAE. Having lower MAE compared to MLPPOR which is the state of the art ordinal regression method shows that our method can effectively learn the order between classes. In addition, our method has better accuracy performance which shows that it can properly employ the unlabeled data to build a better classifier. For deep models, we used 5 different random seeds for initialization and reported the average of the results as the performance of the model.

We observed that using Dropout [5] ($p = 0.5$) in the input layer improves the performance of the network, but applying Dropout for other layers had negative impacts on the performance. In addition, hyperparameter setting ($K = 5$, $alpha = 0.5$, $T = 0.5$, $c_1 = 2$, $c_2 = 1$, and $c_3 = 1$) yielded the best result when our metric was accuracy, and ($K = 5$, $alpha = 0.5$, $T = 0.5$, $c_1 = 2$, $c_2 = 1$, and $c_3 = 1$) was the best one when the metric was MAE. We employed Adam optimizer [6] with parameters learning rate $= 0.0001$, $\beta_1 = 0.9$, $\beta_2 = 0.99$, and weight decay equal to 0.0001.

Table 1. Results on the ABCD dataset.

Methods	Accuracy	MAE
Multi-class logistic regression	34.7%	1.2309
K-nearest neighbour	33.3%	1.2818
Thresholded ridge regression	31.9%	1.1655
Thresholded lasso regression	31.6%	1.1582
Label propagation (semi-supervised baseline)	33.4%	1.3255
MLP with softmax logistic regression loss	35.7%	1.2432
MLPPOR [11]	36.5%	1.1380
MixMatch [1]	36.4%	1.1571
ORSS (Ours)	**38.6%**	**1.0810**

4.1 Ablation Study

In order to further analyze the importance of each term in the loss function, we removed each of them one at a time and examined the performance of the method. We did not perform all hyperparameter search again and used the best setting that we obtained for the 'MAE' metric above. At each time, we change one of the 'c_i's ($i = 1, 2, 3$) to zero while keeping all other hyperparameters unchanged. Again, we reported the results by averaging the performance of 5 different initializations. The results are shown in Table 2.

Table 2. Results on the ABCD dataset.

Ablation	Accuracy	MAE
$c_1 = 0$	36.5%	1.1869
$c_2 = 0$	36.2%	1.1570
$c_3 = 0$	37.1%	1.1680

5 Conclusion

In this paper, we proposed a new framework for semi-supervised training of an ordinal regression problem. We developed the idea behind each part of the proposed network and loss function extensively and showed that our method outperforms modern methods of ordinal regression and semi-supervised learning on the ABCD dataset. In future, we will investigate more advanced brain MRI features and conduct extensively experiments on more adolescent behavior correlations as well as evaluate the gender effect on this problem.

Acknowledgment. Data used in the preparation of this article were obtained from the Adolescent Brain Cognitive Development (ABCD) Study(https://abcdstudy.org), held in the NIMH Data Archive (NDA). This is a multisite, longitudinal study designed to recruit more than 10,000 children age 9–10 and follow them over 10 years into early adulthood. The ABCD Study is supported by the National Institutes of Health and additional federal partners under aware numbers U01DA041022, U01DA041028, U01DA041048, U01DA041089, U01DA041106, U01DA041117, U01DA041120, U01DA041134, U01DA041148, U01DA041156, U01DA041174, U01DA041123, U01DA041147. A full list of supporters is available at here(https://abdstudy.org/nih--collaborators). ABCD consortium investigators designed and implemented the study and/or provided data but did not necessarily participate in the analysis or writing of this report. This manuscript reflects the views of the authors and may not reflect the opinions or views of the NIH or ABCD consortium investigators. The ABCD data repository grows and changes over time. The ABCD data used in this report came from the NIMH Data Archive Digital Object Identifier (DOI)(https://doi.org/10.15154/1503885). (NDA Release 2.0). DOI can be found at here(https://nda.nih.gov/study.html?tab=cohort&id=693). (NDA Release 2.0)

References

1. Berthelot, D., Carlini, N., Goodfellow, I., Papernot, N., Oliver, A., Raffel, C.A.: Mixmatch: a holistic approach to semi-supervised learning. In: Advances in Neural Information Processing Systems, pp. 5050–5060 (2019)
2. Chapelle, O., Schlkopf, B., Zien, A.: Semi-Supervised Learning, 1st edn. The MIT Press, Cambridge (2010)
3. Frank, E., Hall, M.: A simple approach to ordinal classification. In: De Raedt, L., Flach, P. (eds.) ECML 2001. LNCS (LNAI), vol. 2167, pp. 145–156. Springer, Heidelberg (2001). https://doi.org/10.1007/3-540-44795-4_13
4. Grandvalet, Y., Bengio, Y.: Semi-supervised learning by entropy minimization. In: Advances in Neural Information Processing Systems, pp. 529–536 (2005)
5. Hinton, G.E., Srivastava, N., Krizhevsky, A., Sutskever, I., Salakhutdinov, R.R.: Improving neural networks by preventing co-adaptation of feature detectors. arXiv preprint arXiv:1207.0580 (2012)
6. Kingma, D.P., Ba, J.: Adam: a method for stochastic optimization. arXiv preprint arXiv:1412.6980 (2014)
7. Krizhevsky, A., Sutskever, I., Hinton, G.E.: ImageNet classification with deep convolutional neural networks. In: Advances in Neural Information Processing Systems, pp. 1097–1105 (2012)
8. Laine, S., Aila, T.: Temporal ensembling for semi-supervised learning. arXiv preprint arXiv:1610.02242 (2016)
9. Li, L., Lin, H.T.: Ordinal regression by extended binary classification. In: Advances in Neural Information Processing Systems, pp. 865–872 (2007)
10. Lin, H.T., Li, L.: Reduction from cost-sensitive ordinal ranking to weighted binary classification. Neural Comput. **24**(5), 1329–1367 (2012)
11. Liu, Y., Wai Kin Kong, A., Keong Goh, C.: A constrained deep neural network for ordinal regression. In: Proceedings of the IEEE Conference on Computer Vision and Pattern Recognition, pp. 831–839 (2018)
12. Nair, V., Hinton, G.E.: Rectified linear units improve restricted Boltzmann machines. In: Proceedings of the 27th International Conference on Machine Learning (ICML-10), pp. 807–814 (2010)

13. Niu, Z., Zhou, M., Wang, L., Gao, X., Hua, G.: Ordinal regression with multiple output CNN for age estimation. In: Proceedings of the IEEE Conference on Computer Vision and Pattern Recognition, pp. 4920–4928 (2016)
14. Sajjadi, M., Javanmardi, M., Tasdizen, T.: Regularization with stochastic transformations and perturbations for deep semi-supervised learning. In: Advances in Neural Information Processing Systems, pp. 1163–1171 (2016)
15. Shashua, A., Levin, A.: Taxonomy of large margin principle algorithms for ordinal regression problems. Adv. Neural Inf. Process. Syst. **15**, 937–944 (2002)
16. Tarvainen, A., Valpola, H.: Mean teachers are better role models: weight-averaged consistency targets improve semi-supervised deep learning results. In: Advances in Neural Information Processing Systems, pp. 1195–1204 (2017)
17. Verma, V., Lamb, A., Kannala, J., Bengio, Y., Lopez-Paz, D.: Interpolation consistency training for semi-supervised learning. arXiv preprint arXiv:1903.03825 (2019)

Deep Q-Network-Driven Catheter Segmentation in 3D US by Hybrid Constrained Semi-supervised Learning and Dual-UNet

Hongxu Yang[1(✉)], Caifeng Shan[2], Alexander F. Kolen[3], and Peter H. N. de With[1]

[1] Eindhoven University of Technology, Eindhoven, The Netherlands
h.yang@tue.nl
[2] Shandong University of Science and Technology, Qingdao, China
[3] Philips Research, Eindhoven, The Netherlands

Abstract. Catheter segmentation in 3D ultrasound is important for computer-assisted cardiac intervention. However, a large amount of labeled images are required to train a successful deep convolutional neural network (CNN) to segment the catheter, which is expensive and time-consuming. In this paper, we propose a novel catheter segmentation approach, which requests fewer annotations than the supervised learning method, but nevertheless achieves better performance. Our scheme considers a deep Q learning as the pre-localization step, which avoids voxel-level annotation and which can efficiently localize the target catheter. With the detected catheter, patch-based Dual-UNet is applied to segment the catheter in 3D volumetric data. To train the Dual-UNet with limited labeled images and leverage information of unlabeled images, we propose a novel semi-supervised scheme, which exploits unlabeled images based on hybrid constraints from predictions. Experiments show the proposed scheme achieves a higher performance than state-of-the-art semi-supervised methods, while it demonstrates that our method is able to learn from large-scale unlabeled images.

Keywords: Catheter segmentation · Deep reinforcement learning · Dual-UNet · Semi-supervised learning · Hybrid constraint

1 Introduction

Catheter segmentation in 3D ultrasound (US) images is of great importance in computer-assisted intervention, such as RF-ablation or cardiac TAVI procedures [13–15]. Catheter segmentation in 3D US is usually treated as a voxel-wise classification task, which assigns a semantic category to every 3D voxels (regions) by data-driven methods, such as using Deep Learning (DL) [8]. Typically, DL methods encounter challenges like limited training samples, expensive annotations and imbalanced classes. These factors degrade the DL performances for catheter segmentation in 3D US in different aspects [12]. To overcome the class-imbalance

A. L. Martel et al. (Eds.): MICCAI 2020, LNCS 12261, pp. 646–655, 2020.
https://doi.org/10.1007/978-3-030-59710-8_63

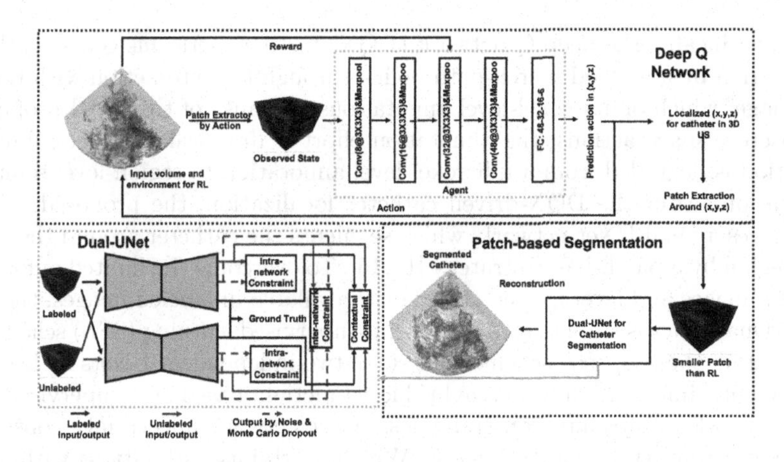

Fig. 1. Schematic view of the proposed framework.

issue, a patch-based strategy was introduced to address the strongly imbalanced class distributions [13,15]. However, this approach requires an accurate annotation for the training datasets, which is expensive for 3D US images and laborious for clinical experts. Moreover, a patch-based strategy requires the Convolutional Neural Network (CNN) to exhaustively segment the patches on the whole image, which is computationally expensive for real-time clinical usage. In the past years, several methods [2,7,9,10,16,17] have been proposed to also utilize the unlabeled images and thereby improve the segmentation performance for medical imaging. Adversarial learning has been studied to make use of unlabeled images, by either enforcing the network to perform segmentation on unlabeled images similar to the labeled images [17], or selecting the most reliable segmentations on unlabeled images to train the segmentation network [9]. Li *et al.* [7] proposed to extend the Π-model for semi-supervised learning, which achieved promising results in skin lesion segmentation. A multi-task-based framework was proposed to exploit unlabeled images by attention-based image reconstruction [2]. An uncertainty-aware self-ensembling model was proposed [10,16] to make use of certainty estimations for the segmentations of unlabeled images, which enhances the segmentation performance with limited annotations. Although uncertainty-aware methods [10,16] achieved better performances than state-of-the-art methods (SOTAs), they are all based on the mean-teacher approach with exponential moving averaging on parameter updating, which encounters a parameter-correlation between teacher and student models [4]. As a consequence, the reliability of used voxels may not be stable enough for SSL.

To address above challenges on annotation effort and segmentation strategies for patch-based CNN, we propose a deep Q learning-driven [3] catheter segmentation framework in 3D US by a semi-supervised approach that leverages reinforcement learning as a pre-localization step for catheter segmentation, which makes use of limited labeled and abundant unlabeled images to train a patch-based CNN. The proposed two-stage deep learning scheme is shown in Fig. 1.

More specifically, the deep Q network (DQN) method performs coarse catheter localization by a learned search-path policy through the interaction with the 3D US image, which omits voxel-level annotations. Because of the DQN properties for discrete space actions, the annotation effort is drastically reduced, but the detection is degraded from a full voxel-level annotation to region-level semantic annotation. After the DQN-driven catheter localization, the proposed scheme introduces a Dual-UNet network, which segments the catheter around the localized region by a patch-based strategy [13,15]. To overcome the limited amount of labeled images and leverage the abundantly available unlabeled images, the proposed Dual-UNet is trained by a novel semi-supervised learning (SSL) scheme. In more detail, the proposed scheme consists of two independent UNets by learning the semantic information from a hybrid loss function, based on a supervised loss, intra-network uncertainty constraint loss, inter-network uncertainty constraint loss, and a contextual constraint loss. With uncertainty estimations within two UNets, the most reliable voxel-level predictions of unlabeled images are used as the semi-supervised signal to enhance the segmentation performance. To compensate the voxel-level constraints from the uncertainty estimation and enhance the usage of semantic information, we also introduce a contextual constraint, which enforces the network to learn the contextual information from unlabeled images. From our experiments, the proposed method was extensively compared to the SOTAs on an *ex-vivo* dataset for cardiac catheterization. The results demonstrate that our method performs better than supervised learning method from same back-bone, while outperforming the SSL SOTAs.

Exploiting the proposed scheme for catheter segmentation, this paper presents the following contributions. We propose a novel catheter segmentation scheme based on the DNQ-driven Dual-UNet, which is trained by the proposed SSL approach. The proposed SSL training strategy is based on hybrid constraints and employs an unlabeled signal to improve the discriminating capacity of the CNN. The method leverages fewer annotations than a supervised learning approach, thereby reducing the training challenges. Moreover, the proposed scheme achieves a higher efficiency because of the use of a coarse-to-fine strategy.

2 Methods

The proposed segmentation method is shown in Fig. 1. First, catheter center is detected by the DQN, which provides an estimation for the catheter location. Second, with the obtained catheter center, Dual-UNet is applied on local patches around the estimated location, which is trained by proposed SSL scheme.

2.1 Deep Q Learning as a Pre-localization Step for Segmentation

The task of extracting a 3D US region containing the catheter can be modeled under the reinforcement learning framework. The agent with its current observation state s interacts with the environment $\mathcal{E}$, by performing successive actions $a \in \mathcal{A}$ to maximize the expected reward r. In this paper, we define the observation state in Cartesian coordinates as a 3D observation patch with size of d^3 voxels w.r.t. the patch center point (x, y, z). To interact with $\mathcal{E}$, the action space $\mathcal{A}$ has six elements, which are defined as $\{\pm a_x, \pm a_y, \pm a_z\}_{scale}$ w.r.t. three different axes with a resolution of step *scale*. With the observed state s, the agent makes a decision of an action from $\mathcal{A}$, to iteratively update the location of the 3D patch. After each action, a reward RL system is characterized as $r = sign(D(Pt_g, Pt_{t-1})/scale - D(Pt_g, Pt_t)/scale)$, where D denotes the Euclidean distance between two points, Pt_g is the ground-truth point, Pt_t is the current state, Pt_{t-1} is the previous state and *scale* represents the step scale in $\mathcal{E}$ [1,3]. As a result, reward $r \in \{-1, 0, +1\}$ indicates whether the agent invokes a patch to move or leave it to the target point. With the obtained reward, the optimized action-selection policy can be characterized by learning a state-action value function $Q(s, a)$, which can be approximated by the Deep Q Network (DQN). To train the DQN, the loss function is defined as:

$$\mathcal{L}_{\text{DQN}} = E_{s,r,a,\hat{a}\sim M}[(r + \gamma \cdot \max_{\hat{a}} Q(\hat{a}, \hat{s}; \widetilde{\omega}) - Q(a, s; \omega))^2], \tag{1}$$

where the future reward discount parameter γ is set to be 0.9, $\hat{a}$ and $\hat{s}$ are action and observed state in the next step, respectively. Parameter M is the experience replay to de-correlate the random samples. Parameters ω and $\widetilde{\omega}$ are trainable parameters of the Q networks for current and target network, respectively. The architecture of the adapted Q network is depicted in Fig. 1, where four recent patches are used as the input. Search is terminated after local oscillation.

2.2 Semi-supervised Catheter Segmentation by Dual-UNet

With coarse localization of the catheter in a 3D US volume, the catheter is then segmented by the proposed patch-based Dual-UNet, which is trained by a hybrid constrained framework. The proposed Dual-UNet structure is motivated by the mean-teacher architecture, which learns the network parameters by updating a student network from a teacher network [10,16]. Intuitively, this method introduces two networks whose parameters are highly correlated due to the strategy of performing an exponential moving average on the updating process. As a result, the obtained knowledge is biased and may not be discriminative enough [4]. Alternatively, we propose to use two independent networks, to learn the discriminating information by knowledge interaction through uncertainty constraints. Moreover, a contextual-level prediction constraint is introduced to compensate the information usage from voxel-level uncertainty estimations.

With the localized catheter from the DNQ, patches around it are extracted for semantic segmentation (or training). We studied the task of catheter segmentation in 3D US volumetric data, where training patches contain N labeled patches and M unlabeled patches. In this paper, labeled patches are denoted by $\{(x_i, y_i)\}_{i=1}^N$ and unlabeled patches are denoted as $\{x_j\}_{j=1}^M$, where $x \in R^{V^3}$ is the 3D input patch and $y \in \{0, 1\}^{V^3}$ is the corresponding ground truth ($V = 48$

as suggested by [15]). The task of semi-supervised learning is to minimize the following loss function with inputs of patches and ground truth (if applicable):

$$\begin{aligned}
&\min_{\omega_1,\omega_2} \Big(\sum_{i=1}^{N}(L_{sup}(f_1(x_i;\omega_1),y_i) + L_{sup}(f_2(x_i;\omega_2),y_i))\\
&+\alpha\sum_{j=1}^{M}(L_{intra}(f_1(x_j;\omega_1),f_1(x_j;\omega_1,\xi_1)) + L_{intra}(f_2(x_j;\omega_2),f_2(x_j;\omega_2,\xi_2))\\
&+L_{inter}(f_1(x_j;\omega_1),f_2(x_j;\omega_2),f_1(x_j;\omega_1,\xi_1),f_2(x_j;\omega_2,\xi_2)))\\
&-\beta\sum_{i,j=1,1}^{N,M} L_c(f_1(x_i;\omega_1),f_2(x_i;\omega_2),f_1(x_j;\omega_1),f_2(x_j;\omega_2))\Big),
\end{aligned} \tag{2}$$

where ω_1, ω_2 are trainable parameters, L_{sup} denotes supervised loss (e.g. cross-entropy loss + Dice loss), L_{intra} is the uncertainty constraint to measure the consistency between outputs from each individual network $f_q(\cdot)$ under different perturbations, where $q \in \{1,2\}$. Here, ξ involves the dropout layers and Gaussian random noise of the input. Function L_{inter} is an uncertainty constraint to measure the consistency between the two networks $f_1(\cdot)$ and $f_2(\cdot)$, which makes use of information from two independent networks to enhance the performance. Function L_c is the cross-entropy loss w.r.t. the prediction with a label or without label for Adversarial learning [17]. Coefficient α and β were empirically selected as 0.1 and 0.002 to balance the losses.

Uncertainty Intra-network Constraint L_{intra}: Without annotation for an input patch, the prediction from networks to guide the unsupervised learning may be unreliable. To generate a reliable prediction from history and guide the network to gradually learn from the more reliable prediction, we design an uncertainty constraint for each individual network. Given an input patch, T predictions are generated by a forward pass, based on a Monte Carlo Dropout and input with Gaussian noise [5]. Therefore, the estimated probability map is obtained by the average of T times prediction for an input patch, i.e. $\hat{P}_q$ for network $f_q(\cdot)$. Based on the above probability maps, the uncertainty is obtained as $\hat{U}_q = -\hat{P}_q log(\hat{P}_q)$ and the loss constraint for network $f_q(\cdot)$ is formulated by:

$$L_{intra} = \frac{\sum(\mathcal{I}(\hat{U}_q < \tau_1) \odot ||f_q(x;\omega_q) - \hat{P}_q||)}{\sum\mathcal{I}(\hat{U}_q < \tau_1)}, \tag{3}$$

where $\mathcal{I}$ is a binary indicator function, $\tau_1 = -0.5ln(0.5)$ is a threshold to measure the uncertainty, which selects the most reliable voxels by binary voxel-level multiplication $\odot$. Parameter $f_q(x;\omega_q)$ is the prediction from network $f_q(\cdot)$ with $q \in \{1,2\}$. By following this approach, the proposed strategy is approximately equal to the mean-teacher method with the history step as unity.

Uncertainty Inter-network Constraint L_{inter}: Besides the above uncertainty constraint for each network, we also propose an uncertainty constraint to measure the consistency between two individual networks with the purpose to constrain the knowledge and avoid bias. The proposed uncertainty constraint on two networks can enhance the overall stability of the model by comparing the predictions between two networks. From the proposed networks in Fig. 1 and the above definitions, different predictions can be obtained, expressed as P_q and $\hat{P}_q$ for normal prediction and Bayesian prediction. Moreover, corresponding binary predictions are obtained as C_q and $\hat{C}_q$, respectively, which are thresholded by 0.5. This leads to a stable prediction, which is formally written as

$$\mathcal{S}_q = \mathcal{I}(C_q \odot \hat{C}_q) \odot (\mathcal{I}(U_q < \tau_2) \oplus \mathcal{I}(\hat{U}_q < \tau_2)), \tag{4}$$

where $\tau_2 = -0.7ln(0.7)$ is a threshold to select the more reliable voxels for $f_q(\cdot)$. Furthermore, we also define the voxel-level probability distance $D_q = ||f_q(x;\omega_q) - \hat{P}_q||$, which indicates the network-wise consistency. Symbol $\oplus$ stands for a voxel-based logical OR. With the above, the uncertainty inter-network constraint L^1_{inter} (where $L_{inter} = L^1_{inter} + L^2_{inter}$) for $f_1(\cdot)$ is formulated by:

$$L^1_{inter} = \frac{\sum(((\mathcal{S}_1 \odot \mathcal{S}_2 \odot \mathcal{I}(D_1 > D_2)) \oplus (\overline{\mathcal{S}_1 \odot \mathcal{S}_2} \odot \mathcal{S}_2)) \odot ||P_1 - P_2||)}{\sum((\mathcal{S}_1 \odot \mathcal{S}_2 \odot \mathcal{I}(D_1 > D_2)) \oplus (\overline{\mathcal{S}_1 \odot \mathcal{S}_2} \odot \mathcal{S}_2))}, \tag{5}$$

where $||\cdot||$ is the probability distance at the voxel level and $\overline{(\cdot)}$ is binary NOT operation. This uncertainty constraint enables the unsupervised signal communication between two individual networks for $f_1(\cdot)$ and vice versa for $f_2(\cdot)$.

Contextual Constraint L_c: The above constraints only consider voxel-level consistency of paired predictions, while ignoring the differences between labeled and unlabeled predictions at the contextual level. To enhance the predictions similarity at contextual level, we also introduce a contextual constraint, which is based on the implementation of adversarial learning. The labeled and unlabeled predictions are classified by an adversarial classifier, which is learned by the loss L_c, i.e. BCE, based on whether it has annotation or not. Meanwhile, L_c is maximized by Eq. (2) to enhance the contextual similarity between predictions.

3 Experimental Results

Materials and Implementation Details: We collected 88 3D US volumes from 8 porcine hearts. During the data collection, the tissues were placed in water tanks with an RF-ablation catheter (diameter ranging from 2.3 mm to 3.3 mm) inside the chambers. The obtained images were re-sampled to have a volume size of $160 \times 160 \times 160$ voxels with padding applied, which leads to a voxel size ranging from 0.3^3–0.9^3 mm. All the volumes were manually annotated at voxel level. Moreover, the catheter centers were also marked as the target location for DQN. To validate proposed method, 60 volumes were randomly selected as training set and 28 volumes were used as testing images. To train the DQN, 60 volumes with target location were used to learn the path-search policy. To train the Dual-UNet, 10%, 20% and 30% of 60 images were selected as the labeled images, while the remainder were the unlabeled images for SSL.

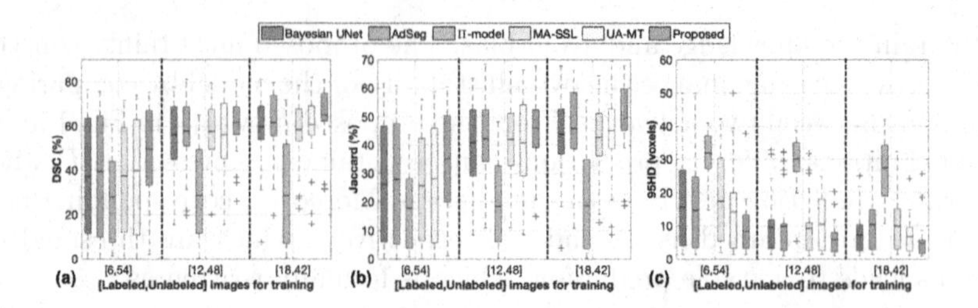

Fig. 2. Comparison to SOTAs by boxplot: (a) Dice Score, (b) Jaccard Score, (c) 95HD. Higher DSC/Jaccard and lower 95HD show better results for segmentation.

Table 1. Segmentation performance by ablation studies in Dice Score (DSC), Jaccard Score (Jaccard) and 95% Hausdorff Distance (95HD), which are shown in mean±std.

Method	*# Images Used*		*Metrics*		
	Labeled	Unlabeled	DSC (%)↑	Jaccard (%)↑	95HD (voxels)↓
DU-L_{intra}	6	54	42.4±25.1	29.8±19.2	13.9±11.2
DU-$L_{intra+inter}$	6	54	47.9±23.5	34.2±19.4	11.7±8.3
Proposed	6	54	**49.4±20.7**	**35.1±17.3**	**8.5±7.8**
DU-L_{intra}	12	48	58.8±17.0	43.5±15.9	8.2±9.4
DU-$L_{intra+inter}$	12	48	59.6±13.2	43.7±13.1	7.8±8.5
Proposed	12	48	**61.8±13.4**	**45.9±13.3**	**5.7±4.6**
DU-L_{intra}	18	42	62.6±14.0	46.8±13.4	6.2±6.6
DU-$L_{intra+inter}$	18	42	63.8±13.3	48.1±13.3	5.2±5.9
Proposed	18	42	**65.0±13.5**	**49.4±13.3**	**4.7±5.1**
UNet[15]	60	0	63.0±14.4	47.3±13.2	5.5±7.1

We implemented our framework in *TensorFlow*, using a standard PC with a TITAN 1080Ti GPU. We trained the DQN with Adam [6] optimizer (learning rate 1e–4) for 40 epochs. Replay memory was 1e5. Parameters of the target network were updated for every 2,500 steps. When considering the efficiency and accuracy of the DQN, we defined the input state space to 55^3 voxels for resized images with size 128^3, which promises the observations can contain sufficient catheter contextual information. To construct the Dual-UNet as the segmentation network, we employ Compact-UNet [15] as our backbone, together with a joint cross-entropy and Dice loss. To adapt the UNet as a Bayesian network and generate uncertainty prediction, dropout layers with rate 0.5 were inserted prior to the convolutional layers. Gaussian random noise was also considered during uncertainty estimation. For the uncertainty estimation, $T = 8$ was used to balance the efficiency and quality of the estimation. As for L_c, the classifier was constructed based on the contextual encoder from [15] with fully connected layers 128-32-1. Training was terminated after 10,000 steps with mini-batch size of

4 using the Adam optimizer, which includes 2 labeled and 2 unlabeled patches. The ramp-up weight was applied on τ_1 and τ_2 to control the training process.

Results Analysis: As for the experiments of using DQN as pre-selection, we compared localization accuracy for different volume sizes as environment: 128^3 and 160^3 voxels, since this would use different amounts of contextual information within the fixed observation space. The metric is the Euclidean distance between the detected catheter center point and ground-truth center point by voxels in the resolution of 160^3 voxels. Statistical performance of the 128^3 case is 4.3 ± 2.6 voxels, while it becomes 4.7 ± 4.8 voxels for the 160^3 voxels case (measured by mean $\pm$ std.). From the result, 128^3 voxels' environment provides a better detection accuracy without failure because of the increased contextual information that is observed by the agent for a fixed input size.

With the detected catheter center point, patches with size of 48^3 are extracted around the point for semantic segmentation (i.e. 2^3 patches). Segmentation results are obtained based on DQN detection. To evaluate the performance, we consider the Dice Score (DSC), Jaccard Score and 95% Hausdorff Distance (95HD) as the evaluation metrics [11]. We have implemented several state-of-the-art (SOTA) SSL methods for comparison, which include Bayesian UNet [5], Π-model [7] Adversarial-based segmentation (AdSeg) [17], multi-task attention-based SSL (MA-SSL) [2] and uncertainty-aware-based mean-teacher (UA-MT) [16] (all of them are based on Compact-UNet [15]). Results are shown in Fig. 2, which shows the proposed method outperforms SOTA SSL approaches by at least 4% DSC on average. As the number of labeled images increases, the segmentation performances are improved w.r.t available supervised information except for the Π-model, which is due to the unreliable information of the predicted unlabeled images. Because of the unreliable information in challenging 3D US images, the Π-model obtains a lower performance than a simple Bayesian UNet. Compared to MA-SSL and UA-MT, our method achieves a better performance, since it exploits inter/intra-network uncertainty information and boosts the information usage of unlabeled images. To perform ablation studies, the loss components in Eq. (2) are gradually introduced based on the Dual-UNet (denoted as DU), which are shown in Table 1. The ablation studies show the proposed constraint component can make use of unlabeled information at different levels, and therefore improve the performances. As can be observed, using 18 labeled images and 42 unlabeled images, the proposed scheme achieves a higher performance than the supervised learning method. From the experiments, the proposed two-stage scheme achieves around 1.2 s per volume (0.7 + 0.5 s). As a comparison, exhaustive patch-based segmentation spends >10 s per volume [15], while a voxel-of-interest-based CNN method costs around 10 s [14].

4 Conclusion

In this paper, we have proposed a catheter segmentation scheme using a DQN-driven semi-supervised learning for US-guided cardiac intervention therapy. In the proposed method, we design a DQN to coarsely localize the catheter in the

3D US, which is then segmented by an SSL-trained Dual-UNet. With extensive comparison, the proposed method outperforms the SOTAs, while it achieves a higher performance than the supervised learning approach with fewer annotations. Future work will investigate a more complex model with a better supervised loss but to obtain a higher efficiency.

References

1. Alansary, A., et al.: Evaluating reinforcement learning agents for anatomical landmark detection. Med. Image Anal. **53**, 156–164 (2019)
2. Chen, S., Bortsova, G., García-Uceda Juárez, A., van Tulder, G., de Bruijne, M.: Multi-task attention-based semi-supervised learning for medical image segmentation. In: Shen, D., et al. (eds.) MICCAI 2019. LNCS, vol. 11766, pp. 457–465. Springer, Cham (2019). https://doi.org/10.1007/978-3-030-32248-9_51
3. Ghesu, F.C., et al.: Multi-scale deep reinforcement learning for real-time 3D-landmark detection in CT scans. IEEE Trans. Pattern Anal. Mach. Intell. **41**(1), 176–189 (2017)
4. Ke, Z., Wang, D., Yan, Q., Ren, J., Lau, R.W.: Dual student: breaking the limits of the teacher in semi-supervised learning. In: Proceedings of the IEEE CVPR, pp. 6728–6736 (2019)
5. Kendall, A., Gal, Y.: What uncertainties do we need in Bayesian deep learning for computer vision? In: NeurIPS, pp. 5574–5584 (2017)
6. Kingma, D.P., Ba, J.: Adam: a method for stochastic optimization. arXiv preprint arXiv:1412.6980 (2014)
7. Li, X., Yu, L., Chen, H., Fu, C.W., Heng, P.A.: Semi-supervised skin lesion segmentation via transformation consistent self-ensembling model. arXiv preprint arXiv:1808.03887 (2018)
8. Litjens, G., et al.: A survey on deep learning in medical image analysis. Med. Image Anal. **42**, 60–88 (2017)
9. Nie, D., Gao, Y., Wang, L., Shen, D.: ASDNet: attention based semi-supervised deep networks for medical image segmentation. In: Frangi, A.F., Schnabel, J.A., Davatzikos, C., Alberola-López, C., Fichtinger, G. (eds.) MICCAI 2018. LNCS, vol. 11073, pp. 370–378. Springer, Cham (2018). https://doi.org/10.1007/978-3-030-00937-3_43
10. Sedai, S., et al.: Uncertainty guided semi-supervised segmentation of retinal layers in OCT images. In: Shen, D., et al. (eds.) MICCAI 2019. LNCS, vol. 11764, pp. 282–290. Springer, Cham (2019). https://doi.org/10.1007/978-3-030-32239-7_32
11. Taha, A.A., Hanbury, A.: Metrics for evaluating 3D medical image segmentation: analysis, selection, and tool. BMC Med. Imaging **15**(1), 29 (2015)
12. Yang, H., Shan, C., Kolen, A.F., de With, P.H.N.: Catheter detection in 3D ultrasound using triplanar-based convolutional neural networks. In: 2018 25th IEEE ICIP, pp. 371–375. IEEE (2018)
13. Yang, H., Shan, C., Kolen, A.F., de With, P.H.N.: Automated catheter localization in volumetric ultrasound using 3D patch-wise U-Net with focal loss. In: 2019 IEEE ICIP, pp. 1346–1350. IEEE (2019)
14. Yang, H., Shan, C., Kolen, A.F., de With, P.H.N.: Catheter localization in 3D ultrasound using voxel-of-interest-based convnets for cardiac intervention. Int. J. Comput. Assisted Radiol. Surg. **14**(6), 1069–1077 (2019)

15. Yang, H., Shan, C., Tan, T., Kolen, A.F., de With, P.H.N.: Transferring from *ex-vivo* to *in-vivo*: Instrument localization in 3D cardiac ultrasound using pyramid-UNet with hybrid loss. In: Shen, D., et al. (eds.) MICCAI 2019. LNCS, vol. 11768, pp. 263–271. Springer, Cham (2019). https://doi.org/10.1007/978-3-030-32254-0_30
16. Yu, L., Wang, S., Li, X., Fu, C.-W., Heng, P.-A.: Uncertainty-aware self-ensembling model for semi-supervised 3D left atrium segmentation. In: Shen, D., et al. (eds.) MICCAI 2019. LNCS, vol. 11765, pp. 605–613. Springer, Cham (2019). https://doi.org/10.1007/978-3-030-32245-8_67
17. Zhang, Y., Yang, L., Chen, J., Fredericksen, M., Hughes, D.P., Chen, D.Z.: Deep adversarial networks for biomedical image segmentation utilizing unannotated images. In: Descoteaux, M., Maier-Hein, L., Franz, A., Jannin, P., Collins, D.L., Duchesne, S. (eds.) MICCAI 2017. LNCS, vol. 10435, pp. 408–416. Springer, Cham (2017). https://doi.org/10.1007/978-3-319-66179-7_47

Domain Adaptive Relational Reasoning for 3D Multi-organ Segmentation

Shuhao Fu[1](✉), Yongyi Lu[1], Yan Wang[1], Yuyin Zhou[1], Wei Shen[1], Elliot Fishman[2], and Alan Yuille[1]

[1] Johns Hopkins University, Baltimore, USA
fushuhao6@gmail.com, yylu1989@gmail.com, wyanny.9@gmail.com, zhouyuyiner@gmail.com, shenwei1231@gmail.com, alan.l.yuille@gmail.com

[2] Johns Hopkins University School of Medicine, Baltimore, USA
efishman@jhmi.edu

Abstract. In this paper, we present a novel unsupervised domain adaptation (UDA) method, named Domain Adaptive Relational Reasoning (DARR), to generalize 3D multi-organ segmentation models to medical data collected from different scanners and/or protocols (domains). Our method is inspired by the fact that the spatial relationship between internal structures in medical images is relatively fixed, *e.g.*, a spleen is always located at the tail of a pancreas, which serves as a latent variable to transfer the knowledge shared across multiple domains. We formulate the spatial relationship by solving a jigsaw puzzle task, *i.e.*, recovering a CT scan from its shuffled patches, and jointly train it with the organ segmentation task. To guarantee the transferability of the learned spatial relationship to multiple domains, we additionally introduce two schemes: 1) Employing a super-resolution network also jointly trained with the segmentation model to standardize medical images from different domain to a certain spatial resolution; 2) Adapting the spatial relationship for a test image by test-time jigsaw puzzle training. Experimental results show that our method improves the performance by 29.60% DSC on target datasets on average without using any data from the target domain during training.

Keywords: Unsupervised domain adaptation · Relational reasoning · Multi-organ segmentation

1 Introduction

Multi-organ segmentation in medical images, *e.g.*, CT scans, is a crucially important step for many clinical applications such as computer-aided diagnosis of abdominal disease. With the surge of deep convolutional neural networks (CNN),

---

Electronic supplementary material The online version of this chapter (https://doi.org/10.1007/978-3-030-59710-8_64) contains supplementary material, which is available to authorized users.

A. L. Martel et al. (Eds.): MICCAI 2020, LNCS 12261, pp. 656–666, 2020.
https://doi.org/10.1007/978-3-030-59710-8_64

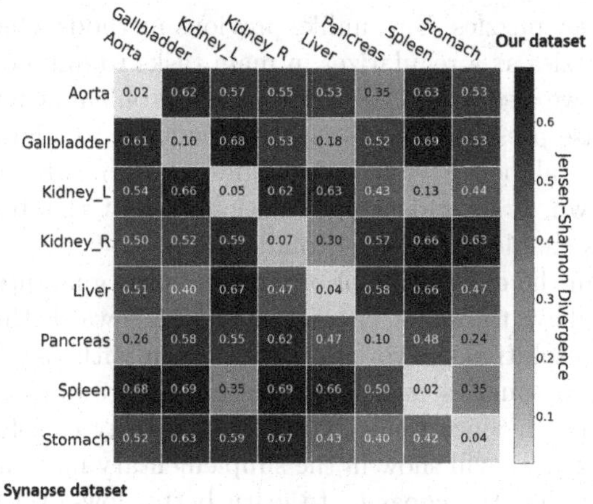

Synapse dataset \ Our dataset	Aorta	Gallbladder	Kidney_L	Kidney_R	Liver	Pancreas	Spleen	Stomach
Aorta	0.02	0.62	0.57	0.55	0.53	0.35	0.63	0.53
Gallbladder	0.61	0.10	0.68	0.53	0.18	0.52	0.69	0.53
Kidney_L	0.54	0.66	0.05	0.62	0.63	0.43	0.13	0.44
Kidney_R	0.50	0.52	0.59	0.07	0.30	0.57	0.66	0.63
Liver	0.51	0.40	0.67	0.47	0.04	0.58	0.66	0.47
Pancreas	0.26	0.58	0.55	0.62	0.47	0.10	0.48	0.24
Spleen	0.68	0.69	0.35	0.69	0.66	0.50	0.02	0.35
Stomach	0.52	0.63	0.59	0.67	0.43	0.40	0.42	0.04

Jensen-Shannon Divergence (scale 0.1–0.6)

Fig. 1. We split each case into $3 \times 3 \times 3$ equally large patches and count the number of occurrences of an organ voxel appearing in each patch for Synapse and our dataset. We then calculate the Jensen–Shannon divergence of the two datasets. Smaller value means the row entry and the column entry are closer.

intensive studies of automatic segmentation methods have been proposed. But more evidence pointed out the problem of performance degradation when transferring domains [6] *e.g.*, testing and training data come from different CT scanners or suffer from a high deviation of scanning protocols between clinical sites. For example, training a well-known V-Net [20] on our in-house dataset and directly testing it on a public MSD spleen dataset [25] yields 43.12% performance drop in terms of DSC. The reason is that their reconstruction and acquisition parameters are different, *e.g.*, pitch/table speeds are 0.55–0.65/25.0–32.1 for the in-house dataset, and 0.984–1.375/39.37–27.50 for MSD spleen dataset. In the context of large-scale applications, generalization capability to deal with scans acquired with different scanners or protocols (*i.e.*, different domains) as compared to the training data is desirable for machine learning models when deploying to real-world conditions.

In this paper, we focus on unsupervised domain adaptation (UDA) for deviating acquisition scanners/protocols in 3D abdominal multi-organ segmentation on CT scans. We propose a domain adaptive relational reasoning (DARR) by fully leveraging the organ location information. More concretely, the relative locations of organs remain stable in medical images [29]. As an example shown in Fig. 1, we calculate the Jensen–Shannon divergence matrix of the location probability distribution of the 8 organs between Synapse dataset and our dataset. The co-occurrence of the same organ appearing in the same location is high. Such relational configuration is deemed as weak cues for segmentation task, which is easier to learn, and thus better in transfer [28]. We aim at learning the spatial relationship of organs via recovering a CT scan from its shuffled patches, *a.k.a*, solving jigsaw puzzles. But, unlike previous methods which simply treated solving jigsaw puzzles as a regularizer in main tasks to mitigate the spatial correlation issue [4], we also solve the jigsaw puzzle problem at test-time, based on one single test case presented. This can help us learn to adapt to a new target domain since the unlabeled test case provides us a hint about the distribution where it was drawn. It is worthwhile mentioning that this test-time relational reasoning process enables one model to adapt all.

To better learn the correlation of organs, we must guarantee that data from different domains have the same spatial resolution. Towards this end, we further propose a super-resolution network to jointly train with the segmentation network and the jigsaw puzzles, which can obtain high-resolution output from its low-resolution version. Since there exists a multiplicity of solutions for a given low-resolution voxel, we will show in the supplementary material that our super-resolution network has the capacity to learn better low-level features, *i.e.*, the deviation of voxels' Hounsfield Units within an organ is reduced, and that of inter-organ is enlarged.

Our proposed DARR performs test-time relative position training, which enjoys the following benefits: (1) establishing a naturally existed common constraint in medical images, so that it can easily adapt to unknown domains; (2) mapping data from different domain sites to the same spatial resolution and encouraging a more robust low-level feature for segmenting organs and learning organ relation; (3) free of re-training the network on source domain when adapting to new domains; and (4) outperforming baseline methods by a large margin, *e.g.*, with even over 29% improvement in terms of mean DSC when adapting our model to multiple target datasets.

2 Related Work

(Unsupervised) Domain adaptation (UDA) has recently gained considerable interests in computer vision primarily for classification [3], detection [13,33] and semantic segmentation [2,24,34]. A key principle of unsupervised domain adaptation is to learn domain invariant features by minimizing cross-domain differences either in feature-level or image-level [26,27]. Inspired by the success of CycleGAN [32] in unpaired image-to-image translation, many recent image adaptation methods are built upon modified CycleGAN frameworks to mitigate the impact of domain gap [1,5,14,19,21]. CyCADA [14] poses unsupervised domain adaptation as style transfer with adversarial learning to close the gap in appearance between the source and target domains. Similar adversarial learning techniques are applied in cross-modality medical data [6,10,12,17,18]. SIFA [6] is among the latest GAN-based methods dedicated to adapt MR/CT cardiac and multi-organ segmentation networks, which conducts both image-level and feature-level adaptations with a shared encoder structure.

More recently, there have been multiple self-training/pseudo-label based methods for unsupervised domain adaptation [3,15,31,34,35]. [31] proposes a semi-supervised 3d abdominal multi-organ segmentation by first training a

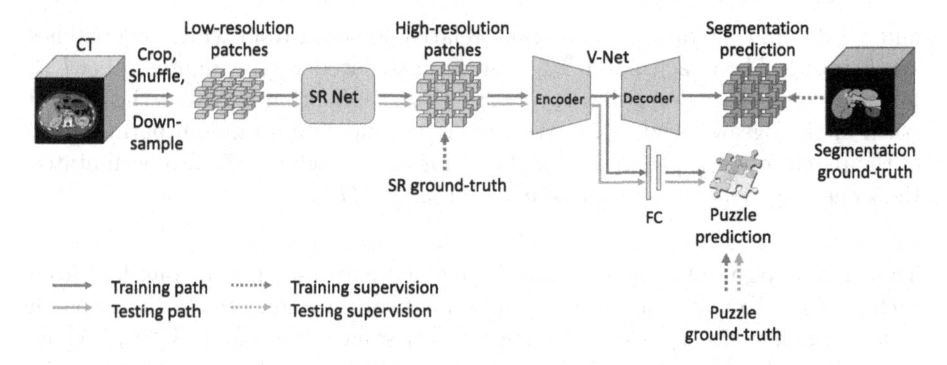

Fig. 2. An overview of our model. Our framework consists of three components, a super-resolution network that upsamples low-resolution images to high resolution, a standard V-Net that performs the segmentation task, and a puzzle module to learn the spatial relations among patches.

teacher model in source dataset in a fully-supervised manner and compute the pseudo-labels on the target dataset. Then a student model is trained on the union of both datasets. However, domain shift is not delicately addressed in this method, thus it hampers its usage on domain adaptation tasks. Another important class for unsupervised domain adaptation is based on self-supervised learning [4,7]. The key challenge for self-supervised learning is identifying a suitable self supervision task. Patch relative positions [9], local context [22], color [30], jigsaw puzzles [28] and even recognizing scans of the same patient [16] have been used in self-supervised learning. In this paper, we aim at learning the spatial relationship of organs via recovering a CT scan from its shuffled patches.

3 Method

Problem Definition. Our goal is to develop a framework that enables machine learning trained on one source domain to adapt to multiple target domains during testing. An overview of our architecture is shown in Fig. 2. Our framework consists of three components, a super-resolution network that upsamples low-resolution images to high resolution, a standard V-Net [20] that performs the segmentation task and a puzzle module to learn the spatial relations among patches. We adopt the generator network from [23] with the subpixel upsampling method as our super-resolution module and we will show the details of the puzzle module in the following section.

We first define some notations in the paper. We parametrize the super-resolution network as θ_{sr}, the encoder part of V-Net as θ_{en} - which is shared by the puzzle module, the decoder part of V-Net as θ_{de} and the puzzle module as θ_p. Suppose we are partitioning an image $\mathbf{I}$ into $W \times H \times L$ patches where each patch can be denoted as $\mathbf{i}_{xyz}$. The index $x \in \{1, ..., W\}$, $y \in \{1, ..., H\}$

and $z \in \{1, ..., L\}$ indicate the original relative location from which the patches are cropped. Then each patch $\mathbf{i}_{xyz}$ can be associated with a unique label l_{xyz} following the row-major policy $l_{xyz} = x + Wy + WHz$ that serves as the ground truth in the jigsaw puzzle task. We use $\{\mathbf{i}_a\}$ to indicate a random permutation of the patch set $\{\mathbf{i}_{xyz}\}$, and the label l_{xyz} for each patch $\mathbf{i}_{xyz}$ is also permutated the same way, denoted as l_a, where $a \in \{1, ..., WHL\}$.

Training Stage. Our network can be trained end-to-end, with one loss from each module. To train the super-resolution network, we squeeze the image patch $\mathbf{i}_a$ to a smaller size $\mathbf{i}'_a$ and minimize a mean square loss $L_{sr}(\mathbf{i}'_a, \mathbf{i}_a; \theta_{sr})$ which makes the output patch as close as possible to the original patch. The segmentation network produces a cross-entropy loss $L_{seg}(\mathbf{i}_a, \mathbf{g}_a; \theta_{sr}, \theta_{en}, \theta_{de})$, where $\mathbf{g}_a$ is the ground truth segmentation mask. The third loss, $L_p(\mathbf{i}_a, l_a; \theta_{sr}, \theta_{en}, \theta_p)$, is given by the puzzle task that classifies the correct location of the patches. Note that the former two losses L_{seg} and L_{sr} are only trained on the training dataset, while the puzzle loss L_p can be utilized on both training and testing set because it does not require any manually labeled data.

Overall, we can obtain the optimal model through

$$\underset{\theta_{sr}\ \theta_{en}\ \theta_{de}\ \theta_p}{\arg\min} \quad L_{seg}(\mathbf{i}_a, \mathbf{g}_a; \theta_{sr}, \theta_{en}, \theta_{de}) + \lambda_{sr} L_{sr}(\mathbf{i}'_a, \mathbf{i}_a; \theta_{sr}) + \lambda_p L_p(\mathbf{i}_a, l_a; \theta_{sr}, \theta_{en}, \theta_p), \tag{1}$$

where λ_{sr} and λ_p are loss weights.

Adaptive Testing. During testing, our goal is to adapt our feature extractor to the target domain, or a target image, through optimizing the self-supervised learning task. By minimizing the puzzle loss L_p on the testing data for a few iterations, the feature extractor is able to reason about the spatial relations among organs, and thus improving the performance on the unseen target domain.

Jigsaw Puzzle Solver. Medical images share a strong spatial relationship that organs are organized in specific locations inside the body with similar relative scales. With this prior knowledge, it is natural to investigate a self-supervised learning task that solves for the relative locations given arbitrarily cropped 3D patches. We select a Jigsaw Puzzle Solver in our case, as it has been proven to be helpful in initializing 3D segmentation models [28]. During training, the permuted set of patches $\{\mathbf{i}'_a\}$ are passed through the super-resolution network and the shared feature extractor (the encoder part) of V-Net to generate corresponding features, denoted as $\{\mathbf{f}_a\}$. Following the previous work [28], all features are then flattened into 1D vectors and concatenated together according to the permuted order, forming a long vector. After two fully-connected layers, the puzzle module outputs a vector in size $(WHL)^2$, which can be reshaped into a matrix of size $WHL \times WHL$. We apply a softmax function on each row so that each row a of the matrix indicates the probability of patch $\{\mathbf{i}'_a\}$ belonging to the WHL locations. We use negative log-likelihood loss as the puzzle loss in our model.

4 Experiments

Datasets. We train the proposed DARR model on our high-resolution multi-organ dataset with 90 cases and adapt it to five different public medical datasets, including 1) multi-organ dataset: Synapse dataset[1] (30 cases); and 2) 4 single-organ datasets [25]: Spleen (41 cases), Liver (131 cases), Pancreas (282 cases) and NIH Pancreas dataset[2] (82 cases). **For the Synapse dataset**, we evaluate on 8 abdominal organs, including Aorta, Gallbladder, Kidney (L), Kidney (R), Liver, Pancreas, Spleen and Stomach, which are also annotated in our multi-organ dataset. **For all other datasets**, we directly evaluate on the target organ. Each dataset is randomly split into 80% of training data and 20% of testing data. Note that unlike other domain generalization methods which use data from the target domain for training, DARR only sees target domain data during testing. We use Dice-Sørensen coefficient (DSC) as the evaluation metric. For each target dataset, we report the average DSC of all cases.

Implementation Details. We set puzzle-related hyperparameters $W = H = L = 3$ in all experiments, which leads to a puzzle composed of 27 patches. The loss-related weights are set as $\lambda_{sr} = 30$ and $\lambda_p = 0.1$, which are consistent across all experiments.

We use 3D V-Net as our backbone architecture, which is initialized with a standard V-Net pre-trained on our in-house dataset. The puzzle module shares the same encoder with the segmentation branch with an additional classification head. We use two fully-connected layers to generate puzzle prediction. The whole network is then finetuned with Adam solver for another 40000 iterations with the batch size of 1 and a learning rate of 0.0003. Each patch has size $64 \times 64 \times 64$, and is squeezed into size $64 \times 64 \times 16$ before feeding into DARR.

For each target dataset, we further train a supervised V-Net model with their ground-truth labels and test directly on the same target dataset. These results serve as our upper bound performance and can be used to calculate the performance degradation for source-to-target adaptation.

During testing, the DARR is first finetuned with a puzzle module only from each target image for 30 iterations with a learning rate of 1e–5 and SGD solver. Then we fix the network parameters and output the segmentation results via a forward pass through the segmentation branch. After predicting one target image, the model is rolled back to the original model, and the above test-time jigsaw puzzle training is repeated for the next target image. No further post-processing strategies are applied.

Results and Discussions. We compare our DARR with state-of-the-art methods, *i.e.*, GAN-based methods [6], self-learning-based methods [31], and meta-learning-based methods [11]. The performance comparison on different datasets

[1] https://www.synapse.org/#!Synapse:syn3193805/wiki/217789.

[2] https://wiki.cancerimagingarchive.net/display/Public/Pancreas-CT.

Table 1. Domain generalization results (DSC %) on all five target datasets.

	Synapse	MSD liver	MSD pancreas	MSD spleen	NIH pancreas	Average
Lower Bound	52.42	88.10	24.42	43.31	10.55	43.76
SIFA [6]	62.33	91.58	24.68	89.37	18.76	57.35
VNET-Puzzle	60.99	89.28	36.22	60.25	31.32	55.61
VNET-SR	61.27	88.83	47.09	80.82	35.71	62.74
DARR	69.77	92.33	56.12	88.61	59.96	73.36
Upper Bound	68.81	91.55	72.56	86.43	86.37	81.14

is shown in Table 1. To measure the performance gain after adaptation, we also provide results trained on our in-house dataset and tested directly on the target datasets without DARR (denoted as "Lower Bound" in Table 1). We observe that our method improves Lower Bound results by 29.60% on average and outperforms all other methods by a large margin. It is worth noting that our method even outperforms Upper Bound results on Synapse dataset, MSD Liver dataset, and MSD Spleen dataset, without using any target domain data in training. This result indicates that our method, which captures spatial relations among organs, is able to bridge the domain gap between multi-site data.

Comparison with Self-learning. Following [31], we first train a teacher model on our multi-organ dataset in a fully-supervised manner and compute the pseudo-labels on the Synapse dataset. Then a student model is trained on the union of both datasets. By evaluating the Synapse dataset, we find that the student model yields a lower segmentation performance than that of the teacher model. This indicates that simply using self-learning may not effectively distill information from data of a different source site.

Comparison with Meta-learning Model-Agnostic Learning Methods. The MASF [11] splits the source domain into multiple non-overlapping training and testing sets and trains a meta-learning model-agnostic model viewing the smaller set as different tasks. It also utilizes delicately designed losses to align intra-class features and separate inter-class features. Nevertheless, MASF does not transfer well from the source domain to the target domains. It is only able to transfer large organs like the liver and stomach while performs poorly in detecting the other small organs. This further confirms that the domain gaps among datasets are substantial, especially in multi-organ segmentation and cannot be easily solved by Meta-learning methods.

Comparison with GAN-Based Methods. The SIFA [6] is dedicated to adapt MR/CT cardiac and multi-organ segmentation networks. It conducts both image-level and feature-level adaptations based on a modified CycleGAN [32]. We use the generated target domain images and their corresponding ground

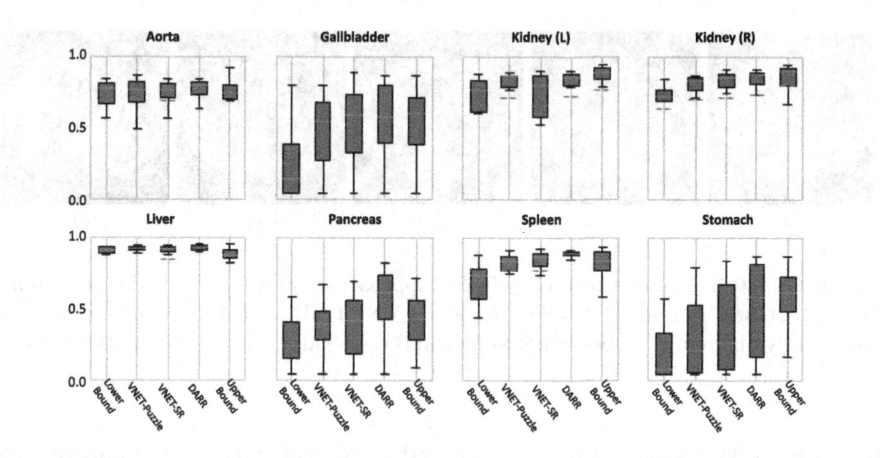

Fig. 3. Ablation study on key components of DARR with Synapse dataset.

truth in the source domain to train a target segmentation network. Here we apply DeepLab-v2 [8] for training the segmentation network after image adaptation of SIFA. From Table 1 we can see that our VNET-SR already outperforms SIFA and achieves inspiring results with average Dice increased to 62.74%. Our full DARR recovers the performance degradation further by an average of 29.6% compared with lower bound, and outperforms SIFA by a significant margin (only 13.59% by SIFA), which shows the superior performance of DARR.

Ablation Study. In this section, we evaluate how each component contributes to our model. We compare different variants of our method (using V-Net as the backbone model): 1) **VNET-Puzzle**, which integrates an additional puzzle module to adaptively learn the spatial relations among image patches; 2) **VNET-SR**, which employs a super-resolution module before the segmentation network; and 3) our proposed **DARR** with both the puzzle module and the super-resolution module applied. As can be seen from Table 1, compared with Lower Bound (which simply uses bilinear upsampling strategies to overcome the resolution divergence among datasets), VNET-SR consistently achieves performance gains on all 5 different target datasets. Especially, for more challenging datasets, the performance improvement can be significant and substantial, *e.g.*, 8.86% on the Synapse dataset, 22.68% on the MSD Pancreas dataset and 25.16% on the NIH Pancreas dataset. This finding indicates the efficacy of our super-resolution module for handling the resolution differences among multi-site data. In addition, VNET-Puzzle also consistently outperforms Lower Bound by a large margin, *e.g.*, 60.99% vs. 52.42% for the Synapse dataset, 36.22% vs. 24.42% for the MSD Pancreas dataset, and 10.55% vs. 31.32% for the NIH Pancreas dataset. Equipped with both the puzzle module and the super-resolution module, our DARR can even lead to additional performance gains compared with VNET-Puzzle and VNET-SR. For instance, we observe an improvement of 17.36% on the Synapse dataset, 31.70% on thee MSD Pancreas dataset, and over 40% on

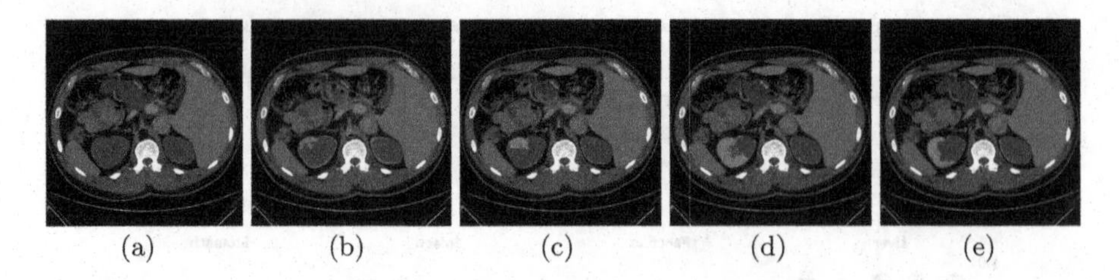

Fig. 4. Qualitative comparison of different approaches. From left to right: a) Ground Truth, b) Lower Bound, c) VNET-Puzzle, d) VNET-SR, e) DARR. Our method can successfully distinguish between left kidney and right kidney after learning their spatial relations.

both the MSD Spleen dataset and the NIH Pancreas dataset. We also provide component comparison results in box plots (see Fig. 3) for the Synapse dataset, which suggests a general statistical improvement among all tested organs. To further demonstrate the efficacy of the proposed DARR, a qualitative comparison is illustrated in Fig. 4, where the spatial location of both kidneys is successfully identified by DARR.

5 Conclusions

We proposed an unsupervised domain adaptation method to generalize 3D multi-organ segmentation models to medical images collected from different scanners and/or protocols (domains). This method, named Domain Adaptive Relational Reasoning, is inspired by the fact that the spatial relationship between internal structures in medical images are relatively fixed. We formulated the spatial relationship by solving jigsaw puzzles and utilized two schemes, *i.e.*, spatial resolution standardisation and test-time jigsaw puzzle training, to guarantee its transferability to multiple domains. Experimental results on five public datasets demonstrate the superiority of our method.

Acknowledgement. We especially Chen Wei for her valuable discussions and ideas. This work was supported by the Lustgarten Foundation for Pancreatic Cancer Research.

References

1. Almahairi, A., Rajeswar, S., Sordoni, A., Bachman, P., Courville, A.: Augmented cyclegan: learning many-to-many mappings from unpaired data. In: Proceedings of ICML, June 2018
2. Bolte, J.A., et al.: Unsupervised domain adaptation to improve image segmentation quality both in the source and target domain. In: Proceedings of CVPR Workshops (2019)
3. Busto, P.P., Iqbal, A., Gall, J.: Open set domain adaptation for image and action recognition. IEEE Trans. Pattern Anal. Mach. Intell. (2018)

4. Carlucci, F.M., D'Innocente, A., Bucci, S., Caputo, B., Tommasi, T.: Domain generalization by solving jigsaw puzzles. In: Proceedings of CVPR (2019)
5. Chang, H., Lu, J., Yu, F., Finkelstein, A.: Pairedcyclegan: asymmetric style transfer for applying and removing makeup. In: Proceedings of the IEEE Conference on Computer Vision and Pattern Recognition, pp. 40–48 (2018)
6. Chen, C., Dou, Q., Chen, H., Qin, J., Heng, P.A.: Unsupervised bidirectional cross-modality adaptation via deeply synergistic image and feature alignment for medical image segmentation. IEEE Trans. Med. Imaging (2020)
7. Chen, L., Bentley, P., Mori, K., Misawa, K., Fujiwara, M., Rueckert, D.: Self-supervised learning for medical image analysis using image context restoration. Med. Image Anal. **58**, 101539 (2019)
8. Chen, L.C., Papandreou, G., Kokkinos, I., Murphy, K., Yuille, A.L.: DeepLab: Semantic image segmentation with deep convolutional nets, atrous convolution, and fully connected CRFs. IEEE Trans. Pattern Anal. Mach. Intell. **40**(4), 834–848 (2017)
9. Doersch, C., Gupta, A., Efros, A.A.: Unsupervised visual representation learning by context prediction. In: Proceedings of the IEEE International Conference on Computer Vision, pp. 1422–1430 (2015)
10. Dou, Q., et al.: PnP-AdaNet: plug-and-play adversarial domain adaptation network at unpaired cross-modality cardiac segmentation. IEEE Access **7**, 99065–99076 (2019)
11. Dou, Q., Castro, D.C., Kamnitsas, K., Glocker, B.: Domain generalization via model-agnostic learning of semantic features. In: Advances in Neural Information Processing Systems (NeurIPS) (2019)
12. Dou, Q., Ouyang, C., Chen, C., Chen, H., Heng, P.A.: Unsupervised cross-modality domain adaptation of convnets for biomedical image segmentations with adversarial loss. In: Proceedings of the 27th International Joint Conference on Artificial Intelligence (IJCAI), pp. 691–697 (2018)
13. Ganin, Y., Lempitsky, V.S.: Unsupervised domain adaptation by backpropagation. In: Bach, F.R., Blei, D.M. (eds.) Proceedings of ICML (2015)
14. Hoffman, J., et al.: Cycada: cycle-consistent adversarial domain adaptation. arXiv preprint arXiv:1711.03213 (2017)
15. Inoue, N., Furuta, R., Yamasaki, T., Aizawa, K.: Cross-domain weakly-supervised object detection through progressive domain adaptation. In: Proceedings of CVPR (2018)
16. Jamaludin, A., Kadir, T., Zisserman, A.: Self-supervised learning for spinal MRIs. In: Cardoso, M.J., et al. (eds.) DLMIA/ML-CDS -2017. LNCS, vol. 10553, pp. 294–302. Springer, Cham (2017). https://doi.org/10.1007/978-3-319-67558-9_34
17. Joyce, T., Chartsias, A., Tsaftaris, S.A.: Deep Multi-class Segmentation Without Ground-truth Labels (2018)
18. Kamnitsas, K., et al.: Unsupervised domain adaptation in brain lesion segmentation with adversarial networks. In: Niethammer, M., et al. (eds.) IPMI 2017. LNCS, vol. 10265, pp. 597–609. Springer, Cham (2017). https://doi.org/10.1007/978-3-319-59050-9_47
19. Liu, M.Y., Breuel, T., Kautz, J.: Unsupervised image-to-image translation networks. In: Advances in Neural Information Processing Systems, pp. 700–708 (2017)
20. Milletari, F., Navab, N., Ahmadi, S.: V-net: fully convolutional neural networks for volumetric medical image segmentation. In: Proceedings of 3DV (2016)
21. Murez, Z., Kolouri, S., Kriegman, D., Ramamoorthi, R., Kim, K.: Image to image translation for domain adaptation. In: Proceedings of CVPR (2018)

22. Pathak, D., Krahenbuhl, P., Donahue, J., Darrell, T., Efros, A.A.: Context encoders: feature learning by inpainting. In: Proceedings of the IEEE Conference on Computer Vision and Pattern Recognition, pp. 2536–2544 (2016)
23. Sánchez, I., Vilaplana, V.: Brain MRI super-resolution using 3D generative adversarial networks. CoRR abs/1812.11440 (2018)
24. Sankaranarayanan, S., Balaji, Y., Jain, A., Nam Lim, S., Chellappa, R.: Learning from synthetic data: addressing domain shift for semantic segmentation. In: Proceedings of CVPR (2018)
25. Simpson, A.L., et al.: A large annotated medical image dataset for the development and evaluation of segmentation algorithms. CoRR abs/1902.09063 (2019)
26. Sun, B., Saenko, K.: Deep CORAL: correlation alignment for deep domain adaptation. In: Hua, G., Jégou, H. (eds.) ECCV 2016. LNCS, vol. 9915, pp. 443–450. Springer, Cham (2016). https://doi.org/10.1007/978-3-319-49409-8_35
27. Tzeng, E., Hoffman, J., Zhang, N., Saenko, K., Darrell, T.: Deep domain confusion: maximizing for domain invariance. arXiv preprint arXiv:1412.3474 (2014)
28. Wei, C., et al.: Iterative reorganization with weak spatial constraints: solving arbitrary jigsaw puzzles for unsupervised representation learning. In: Proceedings of CVPR (2019)
29. Yao, J., Summers, R.M.: Statistical location model for abdominal organ localization. In: Yang, G.-Z., Hawkes, D., Rueckert, D., Noble, A., Taylor, C. (eds.) MICCAI 2009. LNCS, vol. 5762, pp. 9–17. Springer, Heidelberg (2009). https://doi.org/10.1007/978-3-642-04271-3_2
30. Zhang, R., Isola, P., Efros, A.A.: Split-brain autoencoders: unsupervised learning by cross-channel prediction. In: Proceedings of the IEEE Conference on Computer Vision and Pattern Recognition, pp. 1058–1067 (2017)
31. Zhou, Y., et al.: Semi-supervised 3D abdominal multi-organ segmentation via deep multi-planar co-training. In: Proceedings of WACV (2019)
32. Zhu, J.Y., Park, T., Isola, P., Efros, A.A.: Unpaired image-to-image translation using cycle-consistent adversarial networks. In: Proceedings of ICCV (2017)
33. Zhu, X., Pang, J., Yang, C., Shi, J., Lin, D.: Adapting object detectors via selective cross-domain alignment. In: Proceedings of CVPR (2019)
34. Zou, Y., Yu, Z., Vijaya Kumar, B.V.K., Wang, J.: Unsupervised domain adaptation for semantic segmentation via class-balanced self-training. In: Ferrari, V., Hebert, M., Sminchisescu, C., Weiss, Y. (eds.) ECCV 2018. LNCS, vol. 11207, pp. 297–313. Springer, Cham (2018). https://doi.org/10.1007/978-3-030-01219-9_18
35. Zou, Y., Yu, Z., Liu, X., Kumar, B.V., Wang, J.: Confidence regularized self-training. In: Proceedings of ICCV (2019)

Realistic Adversarial Data Augmentation for MR Image Segmentation

Chen Chen[1(✉)], Chen Qin[1,2], Huaqi Qiu[1], Cheng Ouyang[1], Shuo Wang[4], Liang Chen[1,5], Giacomo Tarroni[1,3], Wenjia Bai[4,5], and Daniel Rueckert[1]

[1] BioMedIA Group, Department of Computing, Imperial College London, London, UK
chen.chen15@imperial.ac.uk
[2] Institute for Digital Communications, University of Edinburgh, Edinburgh, UK
[3] CitAI Research Centre, Department of Computer Science, City, University of London, London, UK
[4] Data Science Institute, Imperial College London, London, UK
[5] Department of Brain Sciences, Imperial College London, London, UK

Abstract. Neural network-based approaches can achieve high accuracy in various medical image segmentation tasks. However, they generally require large labelled datasets for supervised learning. Acquiring and manually labelling a large medical dataset is expensive and sometimes impractical due to data sharing and privacy issues. In this work, we propose an adversarial data augmentation method for training neural networks for medical image segmentation. Instead of generating pixel-wise adversarial attacks, our model generates plausible and realistic signal corruptions, which models the intensity inhomogeneities caused by a common type of artefacts in MR imaging: bias field. The proposed method does not rely on generative networks, and can be used as a plug-in module for general segmentation networks in both supervised and semi-supervised learning. Using cardiac MR imaging we show that such an approach can improve the generalization ability and robustness of models as well as provide significant improvements in low-data scenarios.

Keywords: Image segmentation · Adversarial data augmentation · MR

1 Introduction

Segmentation of medical images is an important task for diagnosis, treatment planning and clinical research [1]. Recent years have witnessed the fast development of deep learning for medical imaging with neural networks being applied to a variety of medical image segmentation tasks [2,3]. Deep learning-based approaches in general require a large-scale labelled dataset for training, in order to achieve good model generalization ability and robustness on unseen test

Electronic supplementary material The online version of this chapter (https://doi.org/10.1007/978-3-030-59710-8_65) contains supplementary material, which is available to authorized users.

A. L. Martel et al. (Eds.): MICCAI 2020, LNCS 12261, pp. 667–677, 2020.
https://doi.org/10.1007/978-3-030-59710-8_65

cases. However, acquiring and manually labelling such large medical datasets is extremely challenging, due to the difficulties that lie in data collection and sharing, as well as to the high labelling costs [4].

To address the aforementioned problems, one of the commonly adopted strategies is data augmentation, which aims to increase the diversity of the available training data without collecting and manually labelling new data. Conventional data augmentation methods mainly focus on applying simple *random* transformations to labelled images. These random transformations include intensity transformations (e.g. pixel-wise noise, image brightness and contrast adjustment) and geometric transformations (e.g. affine, elastic transformations). Recently, there is a growing interest in developing generative network-based methods for data augmentation [5–8], which have been found effective for one-shot brain segmentation [5] and low-shot cardiac segmentation [7]. Unlike conventional data augmentation, which generates new examples in an uninformative fashion and does not account for complex variations in data, this generative network-based method is data-driven, learning optimal image transformations from the underlying data distribution in the real world [7]. However, in practice, training generative networks is not trivial due to their sensitivity to hyper-parameters tuning [9] and it can suffer from the mode collapse problem.

In this work, we introduce an effective adversarial data augmentation method for medical imaging without resorting to generative networks. Specifically, we introduce a realistic intensity transformation function to amplify intensity non-uniformity in images, simulating potential image artefacts that may occur in clinical MR imaging (i.e. bias field). Our work is motivated by the observations that MR images often suffer from low-frequency intensity corruptions caused by inhomogeneities in the magnetic field. This artefact cannot be easily eliminated [10,11] and can be regarded as a physical attack to neural networks, which have been reported to be sensitive to intensity perturbations [12,13]. To efficiently improve the model generalizability and robustness, we apply adversarial training to directly search for optimal intensity transformations that benefit model training. By continuously generating these realistic, 'hard' examples, we prevent the network from over-fitting and, more importantly, encourage the network to defend itself from intensity perturbations by learning robust semantic features for the segmentation task.

Our main contributions can be summarised as follows: (1) We introduce a realistic adversarial intensity transformation model for data augmentation in MRI, which simulates intensity inhomogeneities which are common artefacts in MR imaging. The proposed data augmentation is complementary to conventional data augmentation methods. (2) We present a simple yet effective framework based on adversarial training to learn adversarial transformations and to regularize the network for segmentation robustness, which can be used as a plug-in module in general segmentation networks, see Sect. 2.2. More importantly, unlike conventional adversarial example construction [14,15], generating adversarial bias fields does not require manual labels, which makes it applicable for both supervised and semi-supervised learning. (3) We demonstrate the efficacy of the proposed method on a public cardiac MR segmentation dataset in challenging low-data settings. In this scenario, the proposed method greatly outperforms competitive baseline methods, see Sect. 3.2.

Related Work. Recent studies have shown that adversarial data augmentation, which generates adversarial data samples during training, is effective to improve model generalization and robustness[15,16]. Most existing works are based on designing attacks with pixel-wise noise, i.e. by adding gradient-based adversarial noise [14,17–20]. More recently, there have been studies showing that neural networks can also be fragile to other, more natural form of transformations that can occur in images, such as affine transformations [21–23], illumination changes [23], and small deformations [13,24]. In medical imaging, designing and constructing realistic adversarial perturbations, which can be used for improving medical image segmentation networks, has not been explored in depth.

2 Adversarial Data Augmentation with Robust Optimization

In this work, we aim at generating realistic adversarial examples to improve model generalization ability and robustness, given a limited number of training examples. To achieve the goal, we first introduce a physics-based intensity transformation model that can simulate intensity inhomogeneities in MR images. We then propose an adversarial training method, which finds effective adversarial transformation parameters to augment training data, and then regularizes the network with a distance loss function which penalizes network's sensitivity to such adversarial perturbations. Since our method is based on virtual adversarial training (VAT) [19], we will first briefly review VAT before introducing our method.

2.1 Virtual Adversarial Training

VAT is a regularization method based on adversarial data augmentation, which can prevent the model from over-fitting and improve the generalization performance and robustness [19]. Given an input image $\mathbf{I} \in \mathbb{R}^{H\times W\times C}$ (H,W,C denote image height, width, and number of channels, respectively) and a classification network $f_{cls}(\cdot;\theta)$, VAT first finds a small adversarial noise $\mathbf{r}^{\text{adv}} \in \mathbb{R}^{H\times W\times C}$ to construct its adversarial example $\mathbf{I}^{adv} = \mathbf{I}+\mathbf{r}^{\text{adv}}$ (as shown in Fig. 1A), with the goal of maximising the Kullback–Leibler (KL) divergence $\mathcal{D}_{\text{KL}}$ between an original probabilistic prediction $f_{cls}(\mathbf{I};\theta)$ and its perturbed prediction $f_{cls}(\mathbf{I}+\mathbf{r}^{\text{adv}};\theta)$. The adversarial example is then used to regularize the network for robust feature learning.

The adversarial noise can be generated by taking the gradient of $\mathcal{D}_{\text{KL}}$ with respect to a random noise vector: $\mathbf{r}^{\text{adv}} = \epsilon \cdot \frac{r'}{\|r'\|_2}$,$r' = \nabla_{\mathbf{r}}\mathcal{D}_{\text{KL}}[f(\mathbf{I};\theta) \parallel f(\mathbf{I}+\mathbf{r};\theta)]$. Here ϵ is a hyper-parameter that controls the strength of perturbation. After finding adversarial examples, one can utilize them for robust learning, which penalizes the network's sensitivity to local perturbations. This is achieved by adding $\mathcal{D}_{\text{KL}}$ to its main objective function.

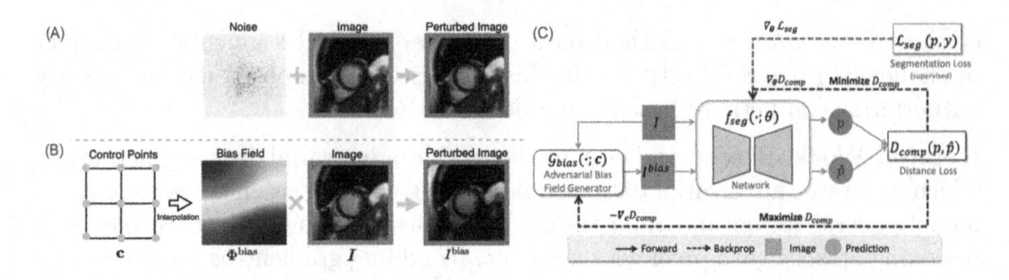

Fig. 1. (A) Adversarial example construction with additive gradient-based noise in VAT [19]; (B) Adversarial example construction with a multiplicative control point-based bias field (proposed); (C) Adversarial training with bias field perturbation.

2.2 Adversarial Training by Modelling Intensity Inhomogeneities

In this work, we extend the VAT approach by introducing a new type of adversarial attack, namely intensity inhomogeneities (bias field) that often occur in MR imaging. In MR imaging, a bias field is a low frequency field that smoothly varies across images, introducing intensity non-uniformity across the anatomy being imaged. The model for the intensity non-uniformity can be defined as follows [10,25]: $\mathbf{I}^{\text{bias}} = \mathcal{G}_{\text{bias}}(\mathbf{I}; \mathbf{c}) = \mathbf{I} \times \Phi^{\text{bias}}(\mathbf{c})$. Here, the intensity of the image $\mathbf{I}$ is perturbed with a multiplication with the bias field $\Phi^{\text{bias}} \in \mathbb{R}^{H \times W}$. As the bias field is typically composed of low frequencies and thus slowly varying across the image, it can be modelled using a set of uniformly distributed k by k points $\mathbf{c} = \{\mathbf{c}_{(i)}\}_{1 \ldots k \times k}$ [10], see Fig. 1B. A smooth bias field at the finest resolution is obtained by interpolating scattered control points with a third-order B-spline smoothing [26].

While one can repeatedly sample random bias fields for data augmentation, this might be computationally inefficient as it may generate images which are of no added value for model optimization. We therefore would like to construct adversarial examples (perturbed by bias field as described above) targeting the weakness of the network in an intelligent way. This allows the use of the generated adversarial examples to improve the model performance and robustness, which can be achieved via the following min-max game:

$$\begin{aligned} &\min_{\theta} \max_{\mathbf{c}} \quad \mathcal{D}_{\text{comp}}[f_{seg}(\mathbf{I}; \theta), f_{seg}(\mathcal{G}_{\text{bias}}(\mathbf{I}; \mathbf{c}); \theta)] \\ &\text{subject to} \quad \forall (x, y) \in \mathbb{R}^2, \ \Phi^{\text{bias}}_{(x,y)} > 0; \ |\Phi^{\text{bias}} - \mathbf{1}|_{\infty} \leq \alpha, 0 < \alpha < 1. \end{aligned} \tag{1}$$

As shown in Fig. 1C, given a segmentation network $f_{seg}(\cdot; \theta)$ and an input image $\mathbf{I}$, we first find optimal values for control points $\mathbf{c}$ in the search space to construct an adversarial bias field, so that it **maximizes** the distance measured by $\mathcal{D}_{\text{comp}}$ between the original prediction and the prediction after perturbation: $\mathbf{p} = f_{seg}(\mathbf{I}; \theta), \hat{\mathbf{p}} = f_{seg}(\mathcal{G}_{\text{bias}}(\mathbf{I}; \mathbf{c}); \theta)$, with θ fixed. We then optimize the parameters θ in the network to **minimize** the distance between the original prediction and the prediction after the generated adversarial bias attack $f_{seg}(\mathcal{G}_{\text{bias}}(\mathbf{I}; \mathbf{c}^{adv}); \theta)$.

Finding Adversarial Bias Fields. To find the optimal values for the control points $\mathbf{c}$ for adversarial example construction, we use the gradient descent algorithm and search the values of control points in its log space for numerical stability [10,25], which allows to produce positive bias fields. Specifically, similar to the projected gradient decent (PGD) attack construction in [15], we first randomly initialize the values of control points and then apply a projected gradient ascent algorithm to iteratively update $\mathbf{c}$ with n steps: $\mathbf{c} \leftarrow \Pi(\mathbf{c} + \xi \cdot \mathbf{c}'/\|\mathbf{c}'\|_2)$ where $\mathbf{c}' = \nabla_{\mathbf{c}} \mathcal{D}_{\text{comp}}[f_{seg}(\mathbf{I};\theta), f_{seg}(\mathcal{G}_{\text{bias}}(\mathbf{I};\mathbf{c});\theta)]$. Π denotes the projection function which projects $\mathbf{c}$ onto the feasible set, and ξ is the step size. For neural networks, gradients $\mathbf{c}'$ can be efficiently computed with back-propagation. Φ^{bias} is updated by first interpolating the coarse-grid control points (log values at the current iteration) to its finest grid using B-spline convolution, and then taking the exponential function for value recovering. Finally, the generated bias field is rescaled to meet the magnitude constraint in Eq. 1.

Composite Distance Function $\mathcal{D}_{\text{comp}}$. Here, we propose a composite distance function $\mathcal{D}_{\text{comp}}$ to enhance its discrimination ability between the original prediction $\mathbf{p}$ (short for $f_{seg}(\mathbf{I};\theta)$) and the prediction after perturbation $\hat{\mathbf{p}}$, for *semantic segmentation* tasks. This composite loss consists of (1) the original $\mathcal{D}_{\text{KL}}$ used in VAT, which measures the difference between distributions and (2) a contour-based loss function $\mathcal{D}_{\text{contour}}$ [27] which is specifically designed to capture mismatch between object boundaries: $\mathcal{D}_{\text{comp}}(\mathbf{p},\hat{\mathbf{p}}) = \mathcal{D}_{\text{KL}}[\mathbf{p} \,\|\, \hat{\mathbf{p}}] + w\mathcal{D}_{\text{contour}}(\mathbf{p},\hat{\mathbf{p}})$; $\mathcal{D}_{\text{contour}}(\mathbf{p},\hat{\mathbf{p}}) = \sum_{m \in M} \sum_{S_{x,y}} \|S(\mathbf{p}^m) - S(\hat{\mathbf{p}}^m)\|_2$. M denotes foreground channels, $S_{x,y}$ denote two Sobel filters in x- and y-direction for edge extraction and w controls the relative importance of both terms.

Optimizing Segmentation Network. After constructing the adversarial examples, one can compute $\mathcal{D}_{\text{comp}}$ and apply it to regularizing the network, encouraging the network to be less sensitive to adversarial perturbations, and thus produce consistent predictions. Since this algorithm uses probabilistic predictions (produced by the network) rather than manual labels for adversary construction, it can be applied to both labelled (l) and unlabelled data (u) for supervised and semi-supervised learning [19]. The loss functions for the two scenarios are defined as: $\mathcal{L}_{\text{SU}} = \mathcal{L}_{\text{seg}}(\mathbf{p}^{(l)}, \mathbf{y}_{gt}^{(l)}) + \lambda_l \mathcal{D}_{\text{comp}}(\mathbf{p}^{(l)}, \hat{\mathbf{p}}^{(l)})$; $\mathcal{L}_{\text{SE}} = \mathcal{L}_{\text{SU}} + \lambda_u \mathcal{D}_{\text{comp}}(\mathbf{p}^{(u)}, \hat{\mathbf{p}}^{(u)})$. $\mathcal{L}_{\text{seg}}$ denotes a general task-related segmentation loss function for supervised learning (e.g. cross-entropy loss) and $\mathbf{y}_{gt}^{(l)}$ denotes ground truth.

3 Experiments

To test the efficacy of the proposed method, we applied it to training a segmentation network for the left ventricular myocardium from MR images in low-data settings. We compared the results with several competitive baseline methods.

3.1 Dataset and Experiment Settings

ACDC Dataset. Experiments were performed on a public benchmark dataset for cardiac MR image segmentation: The Automated Cardiac Diagnosis Challenge (ACDC) dataset [28][1]. This dataset was collected from 100 subjects which were evenly classified into 5 groups: 1 normal group (NOR) and 4 pathological groups with cardiac abnormalities: dilated cardiomyopathy(DCM); hypertrophic cardiomyopathy (HCM); myocardial infarction with altered left ventricular ejection fraction (MINF); abnormal right ventricle (ARV). The left ventricular myocardium in end-diastolic and end-systolic frames were manually labelled.

Image Pre-processing. We used the same image preprocessing as in [7]. In addition, all images were centrally cropped into 128×128, given that the heart is generally located in the center of the image. This saves computational costs.

Random Data Augmentation (Rand Aug). We applied a strong random data augmentation method to our training data as a basic setting. Random affine transformation (i.e. scaling, rotation, translation), random horizontal and vertical flipping, random global intensity transformation (brightness and contrast) [7] and elastic transformation were applied.

Training Details. For ease of comparison, same as [7], we adopted the commonly-used 2D U-net as our segmentation network, which takes 2D image slices as input. The Adam optimizer with a batch size of 20 was used to update network parameters. For the proposed method, we first trained the network with the default data augmentation (Rand Aug) for 10,000 iterations (learning rate=$1e^{-3}$), and then finetuned the network by adding the proposed adversarial training using a smaller learning rate ($1e^{-5}$) for 2,000 iterations. The common standard cross-entropy loss function was used as $\mathcal{L}_{\text{seg}}$. For bias field construction, we adopted the B-spline convolution kernel (order=3) with 4×4 control points. The kernel was provided by AirLab library [29]. We empirically set: $\alpha = 0.3$, $w = 0.5$, $\lambda_l = 1$ and $\lambda_u = 0.1$. Besides, we found that in our experiments, one step searching in the inner loop produced sufficient improvement. Thus, we set $n = 1, \xi = 1$ to save computational cost. All the experiments were performed on an Nvidia® GeForce® 2080 Ti with Pytorch.

3.2 Experiments and Results

Experiment 1: Low-shot Learning. In this experiment, the proposed method was evaluated in both *supervised* learning and *semi-supervised* learning scenarios, where only 1 or 3 labelled subjects are available. Specifically, we used the same data splitting setting as in [7]. The ACDC dataset was split into 4 subsets: a labelled set (where N_l images were sampled from for training), unlabelled training set (N = 25), validation set (N = 2), test set (N = 20). N denotes the number of subjects. Details of the low-data setting can be found in [7]. For one-shot learning (N_l = 1) and three-shot learning (N_l = 3) in both supervised

[1] https://www.creatis.insa-lyon.fr/Challenge/acdc/databases.html.

Table 1. Comparison of the proposed method (Adv Bias) to other data augmentation methods.

Setting	Method	# labelled subjects	
		1	3
Supervised	No Aug	0.293	0.544
	Rand Aug	0.560	0.796
	+Mixup[30]	0.575	0.801
	+VAT[19]	0.570	0.811
	+Adv Bias	**0.650**	**0.826**
Semi-supervised	+VAT[19]	0.625	0.826
	+Adv Bias	0.692	**0.830**
	cGANs[7]	**0.710**	0.823

Table 2. Segmentation performance of the proposed method and baseline methods across five populations. All were trained with NOR cases only.

Population	Rand Aug	+Mixup	+VAT	**Adv Bias** (Proposed)
NOR	0.911	0.901	0.909	**0.912**
DCM	0.831	0.803	0.843	**0.871**
HCM	0.871	0.881	**0.891**	0.890
MINF	0.805	0.789	0.824	**0.847**
ARV	0.843	0.844	0.843	**0.853**
Average	0.841	0.833	0.853	**0.868**

and semi-supervised settings, we trained the network for five times, each with a different labelled set.

We compared the proposed method (**Adv Bias**) with several competitive data augmentation methods including **VAT** [19], an effective data mixing-based method (**Mixup**) [30] for supervised learning and the state-of-the-art semi-supervised generative model-based method(**cGANs**) [7]. For VAT and Mixup, we used the set of hyperparameters that achieved the best performance on the validation set and applied the same training procedure. For cGANs, we report the results of one-shot and three-shot learning in their original paper for reference, which were tested on the same test set. Table 1 compares the segmentation accuracy obtained by different data augmentation methods. Each reported value is the average Dice score of 20 test cases. In the supervised learning setting (no access to unlabelled images), when only one or three labelled subject was available, the proposed method clearly outperformed all baseline methods. For semi-supervised learning, the proposed methods outperformed VAT, especially when only one labelled subject is available (0.686 vs 0.625). The proposed method achieves competitive results compared to the semi-supervised GAN-based method (cGANs) as well. Of note, cGANs adopts two additional GANs to sample geometric transformations and intensity transformations from unlabelled images. This is why it was only compared in the semi-supervised learning setting here. On the contrary, our approach is applicable to both low-shot supervised learning and semi-supervised learning. In addition, cGANs contains more parameters than our method and thus it might be less computationally efficient.

Experiment 2: Learning From Limited Population. In this experiment, we trained the network using only normal healthy subjects (NOR) and evaluated its performance on pathological cases. 20 healthy subjects were split into 14/2/4 subjects for training, validation and test. This setting simulates a practical data scarcity problem, where pathological cases are rarer, compared to healthy data. As shown in Table 2, while the conventional method (Rand Aug) achieved excellent performance on the test healthy subjects (NOR), its performance dropped on pathological cases. Interestingly, applying Mixup did not help

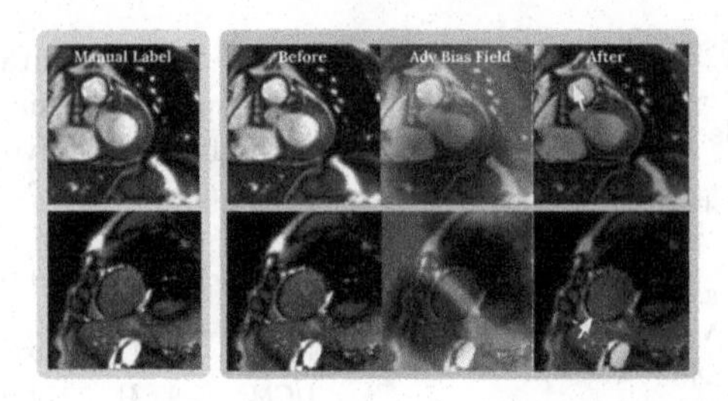

Fig. 2. Visualization of generated adversarial examples and failed network predictions. Before/After: network prediction before/after bias field attack (Adv Bias Field).

to solve this population shift problem, but rather slightly reduced the average performance compared to the baseline, from 0.841 to 0.833. This might be due to the fact that Mixup generates unrealistic images through its linear combination of paired images, which may modify semantic features and affect representation learning for *precise* segmentation. By contrast, our method outperformed both Mixup and VAT, yielding substantial and consistent improvements across five different populations. Notably, we attained evident improvement on the most challenging MINF images (0.805 vs 0.847), where the shape of the myocardium is clearly irregular. As shown in Fig. 2, the proposed method does not only generate adversarial examples during training, but also increases the variety of image styles while preserving the shape information. Augmenting images with various styles can encourage the network to learn high-level shape-based representation instead of texture-based representation, leading to improved network robustness on unseen classes, as discussed in [31]. By contrast, VAT only introduces imperceptible noise, failing to model realistic image appearance variations.

Ablation Study. To get a better understanding of the effectiveness of adversarial bias field, we compared it to data augmentation using random bias field, using experiment setting 2. Results clearly showed that training with adversarial bias field improved the model generalization ability, increasing the Dice score from 0.852 to 0.868, (see supple. Table 1). On the other hand, applying $\mathcal{D}_{\text{comp}}$ to regularize the network improved the average Dice score from 0.859 to 0.868, compared to the one trained with only $\mathcal{D}_{\text{KL}}$ (see supple. Table 2). Unlike random-based approach, constructing adversarial attacks considers both the posterior probability information estimated by the model and semantic information from images. In experiments, we found these attacks focused on attacking challenging images on which the network was uncertain, e.g.. object boundary is not clear or there is another similar structure presented, see Fig. 2. In the same spirit of online hard example mining, utilizing these borderline examples during training helps the network to improve its generalization and robustness ability. Please find more details in the supplementary material.

4 Discussion and Conclusion

In this work, we presented a realistic adversarial data augmentation method to improve the generalization and robustness for neural network-based medical image segmentation. We demonstrated that by modelling bias field and introducing adversarial learning, the proposed method is able to promote the learning of robust semantic features for cardiac image segmentation. It can also alleviate the data scarcity problem, as demonstrated in the low-data setting and cross-population experiments. The proposed method does not rely on generative networks but instead employs a small set of explainable and controllable parameters to augment data with image appearance variations which are realistic for MR. It can be easily extended for multi-class segmentation and used in general segmentation networks for improving model generalization and robustness.

Acknowledgements. This work was supported by the SmartHeart EPSRC Programme Grant(EP/P001009/1) and the EPSRC Programme Grant (EP/R005982/1).

References

1. Smistad, E., Falch, T.L., Bozorgi, M., Elster, A.C., Lindseth, F.: Medical image segmentation on GPUs-a comprehensive review. Med. Image Anal. **20**(1), 1–18 (2015)
2. Shen, D., Guorong, W., Suk, H.-I.: Deep learning in medical image analysis. Ann. Rev. Biomed. Eng. **19**, 221–248 (2017)
3. Litjens, G., et al.: A survey on deep learning in medical image analysis. Med. Image Anal. **42**, 60–88 (2017)
4. Tajbakhsh, N., Jeyaseelan, L., Li, Q., Chiang, J., Wu, Z., Ding, X.: Embracing imperfect datasets: a review of deep learning solutions for medical image segmentation. Arxiv, August 2019
5. Zhao, A., Balakrishnan, G., Durand, F., Guttag, J.V., Dalca, A.V.: Data augmentation using learned transformations for one-shot medical image segmentation. In: CVPR, pp. 8543–8553 (2019)
6. Liu, J., Shen, C., Liu, T., Aguilera, N., Tam, J.: Active appearance model induced generative adversarial network for controlled data augmentation. In: Shen, D., et al. (eds.) MICCAI 2019. LNCS, vol. 11764, pp. 201–208. Springer, Cham (2019). https://doi.org/10.1007/978-3-030-32239-7_23
7. Chaitanya, K., Karani, N., Baumgartner, C.F., Becker, A., Donati, O., Konukoglu, E.: Semi-supervised and task-driven data augmentation. In: Chung, A.C.S., Gee, J.C., Yushkevich, P.A., Bao, S. (eds.) IPMI 2019. LNCS, vol. 11492, pp. 29–41. Springer, Cham (2019). https://doi.org/10.1007/978-3-030-20351-1_3
8. Xing, Y., et al.: Adversarial pulmonary pathology translation for pairwise chest x-ray data augmentation. In: Shen, D., et al. (eds.) MICCAI 2019. LNCS, vol. 11769, pp. 757–765. Springer, Cham (2019). https://doi.org/10.1007/978-3-030-32226-7_84
9. Lei, N., et al.: A geometric understanding of deep learning. Engineering **6**, 361–374 (2020)
10. Tustison, N.J., et al.: N4ITK: improved N3 bias correction. TMI **29**(6), 1310–1320 (2010)

11. Khalili, N., et al.: Automatic brain tissue segmentation in fetal MRI using convolutional neural networks. Magn. Reson. Imaging **64**, 7–89 (2019)
12. Paschali, M., Conjeti, S., Navarro, F., Navab, N.: Generalizability vs. robustness: adversarial examples for medical imaging. In: MICCAI (2018)
13. Chen, L., Bentley, P., Mori, K., Misawa, K., Fujiwara, M., Rueckert, D.: Intelligent image synthesis to attack a segmentation CNN using adversarial learning. In: Burgos, N., Gooya, A., Svoboda, D. (eds.) SASHIMI 2019. LNCS, vol. 11827, pp. 90–99. Springer, Cham (2019). https://doi.org/10.1007/978-3-030-32778-1_10
14. Goodfellow, I.J., Shlens, J., Szegedy, C.: Explaining and harnessing adversarial examples. In: ICLR (2015)
15. Madry, A., Makelov, A., Schmidt, L., Tsipras, D., Vladu, A.: Towards deep learning models resistant to adversarial attacks. In: ICLR, June 2017
16. Volpi, R., Namkoong, H., Sener, O., Duchi, J.C., Murino, V., Savarese, S.: Generalizing to unseen domains via adversarial data augmentation. In: NeurIPS, pp. 5339–5349 (2018)
17. Carlini, N., Wagner, D.A.: Towards evaluating the robustness of neural networks. In: 2017 IEEE Symposium on Security and Privacy, SP 2017, San Jose, CA, USA, 22–26 May 2017, pp. 39–57 (2017)
18. Tramèr, F., Boneh, D.: Adversarial training and robustness for multiple perturbations. In: NIPS, April 2019
19. Miyato, T., Maeda, S.-I., Koyama, M., Ishii, S.: Virtual adversarial training: a regularization method for supervised and Semi-Supervised learning. In: TPAMI (2018)
20. Paschali, M., Conjeti, S., Navarro, F., Navab, N.: Generalizability vs. robustness: adversarial examples for medical imaging. In: MICCAI, March 2018
21. Kanbak, C., Moosavi-Dezfooli, S.-M., Frossard, P.: Geometric robustness of deep networks: analysis and improvement. In: CVPR, pp. 4441–4449 (2018)
22. Engstrom, L., Tran, B., Tsipras, D., Schmidt, L., Madry, A.: Exploring the landscape of spatial robustness. In: Chaudhuri, K., Salakhutdinov, R. (eds.) ICML, Proceedings of Machine Learning Research, vol. 97, pp. 1802–1811, Long Beach, California (2019). PMLR
23. Zeng, X., et al.: Adversarial attacks beyond the image space. In: CVPR, pp. 4302–4311 (2019)
24. Alaifari, R., Alberti, G.S., Gauksson, T.: Adef: an iterative algorithm to construct adversarial deformations. In: 7th International Conference on Learning Representations, ICLR 2019, New Orleans, LA, USA, 6–9 May 2019 (2019)
25. Sled, J.G., Zijdenbos, A.P., Evans, A.C.: A nonparametric method for automatic correction of intensity nonuniformity in MRI data. IEEE Trans. Med. Imaging **17**(1), 87–97 (1998)
26. Gallier, J., Gallier, J.H.: Curves and Surfaces in Geometric Modeling: Theory and Algorithms. Morgan Kaufmann, San Francisco (2000)
27. Chen, C., et al.: Unsupervised multi-modal style transfer for cardiac MR segmentation. In: Pop, M., et al. (eds.) STACOM 2019. LNCS, vol. 12009, pp. 209–219. Springer, Cham (2020). https://doi.org/10.1007/978-3-030-39074-7_22
28. Bernard, O., Lalande, A., Jodoin, P.-M.: Deep learning techniques for automatic MRI cardiac multi-structures segmentation and diagnosis: is the problem solved? TMI **0062**(11), 2514–2525 (2018)
29. Sandkühler, R., Jud, C., Andermatt, S., Cattin, P.C.: AirLab: autograd image registration laboratory. Arxiv (2018)

30. Zhang, H., Cisse, M., Dauphin, Y.N., Lopez-Paz, D.: Mixup: beyond empirical risk minimization. In: ICLR (2018)
31. Geirhos, R., Rubisch, P., Michaelis, C., Bethge, M., Wichmann, F.A., Brendel, W.: Imagenet-trained CNNs are biased towards texture; increasing shape bias improves accuracy and robustness. In: ICLR (2019)

Learning to Segment Anatomical Structures Accurately from One Exemplar

Yuhang Lu[1,2](✉), Weijian Li[1,3], Kang Zheng[1], Yirui Wang[1], Adam P. Harrison[1], Chihung Lin[4], Song Wang[2], Jing Xiao[5], Le Lu[1], Chang-Fu Kuo[4], and Shun Miao[1]

[1] PAII Inc., Bethesda, MD, USA
yuhang@email.sc.edu
[2] University of South Carolina, Columbia, SC, USA
[3] University of Rochester, Rochester, NY, USA
[4] Chang Gung Memorial Hospital, Linkou, Taiwan, R.O.C.
[5] Ping An Technology, Shenzhen, China

Abstract. Accurate segmentation of critical anatomical structures is at the core of medical image analysis. The main bottleneck lies in gathering the requisite expert-labeled image annotations in a scalable manner. Methods that permit to produce accurate anatomical structure segmentation without using a large amount of fully annotated training images are highly desirable. In this work, we propose a novel contribution of *Contour Transformer Network* (CTN), a one-shot anatomy segmentor including a naturally built-in human-in-the-loop mechanism. Segmentation is formulated by learning a contour evolution behavior process based on graph convolutional networks (GCN). Training of our CTN model requires only one labeled image exemplar and leverages additional unlabeled data through newly introduced loss functions that measure the global shape and appearance consistency of contours. We demonstrate that our one-shot learning method significantly outperforms non-learning-based methods and performs competitively to the state-of-the-art fully supervised deep learning approaches. With minimal human-in-the-loop editing feedback, the segmentation performance can be further improved and tailored towards the observer desired outcomes. This can facilitate the clinician designed imaging-based biomarker assessments (to support personalized quantitative clinical diagnosis) and outperforms fully supervised baselines.

Keywords: Contour Transformer Network · One-shot segmentation · Graph convolutional network

---

Electronic supplementary material The online version of this chapter (https://doi.org/10.1007/978-3-030-59710-8_66) contains supplementary material, which is available to authorized users.

A. L. Martel et al. (Eds.): MICCAI 2020, LNCS 12261, pp. 678–688, 2020.
https://doi.org/10.1007/978-3-030-59710-8_66

1 Introduction

Obtaining manual image segmentation labels or masks has often been an obstacle in scaling up medical image segmentation applications. Without abundant pixel-level fully annotated image data, the state-of-the-art CNN-based segmentation methods cannot achieve their best performances [5,7,10,21,22,27,29]. However, annotating segmentation masks for medical images is very time-consuming and requires specialized expertise on human anatomy and its variations [28]. How to train an accurate segmentation model with less labeled data demands prompt solutions. In this paper, we tackle the problem on high-resolution anatomical structure X-ray images and propose *Contour Transformer Network*, allowing to learn from only one labeled exemplar image.

Several one-shot or few-shot segmentation methods have been proposed for natural images [8,12,19,24,33] by extracting information from a few support images to guide the segmentation of query images in testing. Nevertheless, the training process still relies on large-scale annotated datasets such as PASCAL VOC [9] and MS-COCO [16]. This condition renders them inapplicable directly to the medical imaging domain, because such equivalent datasets do not exist yet. Our problem is defined under a very different setting that only one pixel-level annotated training image instance is available. Other problem settings could be found in [20,34] where they attempt to alleviate the label shortage problem via data augmentation. In contrast, our method is to train the segmentation model with only one labeled exemplar and a set of unlabeled images.

The main challenge of one-shot segmentation is the lack of ground truth image mask or contour labels. Regular training strategies of comparing predictions with ground truth labels are no longer applicable. We adopt a new training scheme in CTN. Because of the inherent regularized nature of anatomical structures, the same anatomy in different (X-ray) images may share some common features or properties, such as the anatomical structure's *shape*, *appearance* and *gradients* along the structural object boundary. Although different images are not directly comparable, we can compare their common features only and use the exemplar segmentation to guide other unlabeled images partially, thus making CTN trainable in a one-shot setting.

To leverage these shared anatomical properties, we represent the image segmentation problem as learning a contour evolution behavior. Thus three differentiable contour-based loss functions are proposed to describe the common features. For each unlabeled image, CTN takes the exemplar contour as an initialization, then gradually evolves it to minimize the weighted loss. Furthermore, we offer a naturally built-in human-in-the-loop mechanism to allow CTN to learn from extra partial labels. If any part in the predicted contour is inaccurate, users can correct them by drawing line segments, then CTN will format these corrections as partial contours and incorporate them back into the training via an additional Chamfer loss. In this way, we can improve and refine the segmentation performance with minimum annotation costs.

In summary, our contributions are three folds. (1) We propose a CNN-based image segmentation framework that could be trained with only one labeled image. (2) We describe the contour perceptual loss and the contour bending loss as two new optimization loss functions, to measure the similarity of two contours in terms of the appearance or shape cues, respectively. (3) We demonstrate that CTN achieves the state-of-the-art one-shot segmentation results; performs competitively when compared to fully supervised alternatives; and can outperform them with minimal human-in-the-loop feedback, on three datasets.

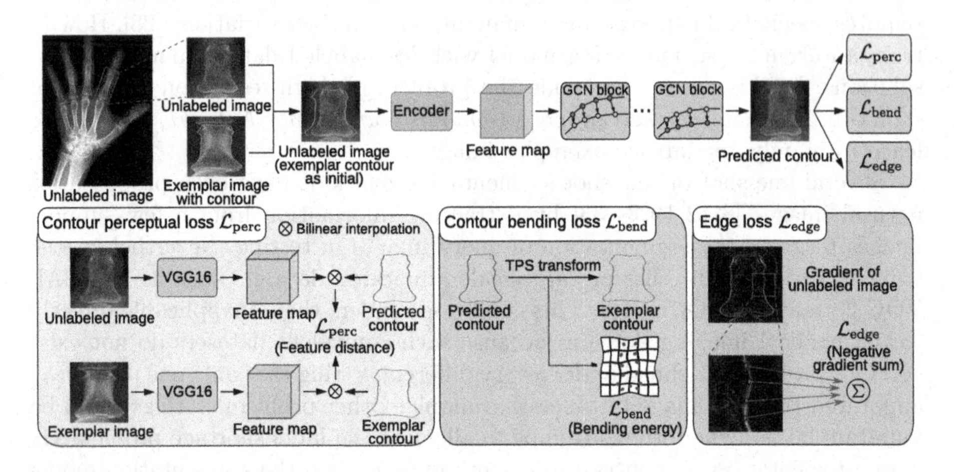

Fig. 1. Contour Transformer Network. CTN is trained to fit a contour to the object boundary by learning from one labeled exemplar. It takes the exemplar and an unlabeled image as input, and predict a contour that has similar contour features with the exemplar. Three losses are proposed to make this network "one-shot" trainable.

2 Methods

The problem of anatomical structure segmentation on images can be decomposed into two steps: ROI (Region of Interest) cropping; and ROI segmentation. ROI detection has been well-studied in past literature [4,6,15,29,31,32], so we focus on achieving very high segmentation accuracy by taking the detected/cropped ROI (with noise and errors) as input.

Assuming that a set of images $\mathbf{I}$ contains the same type of anatomical structure and only one of them is labeled, called the *exemplar*. Our goal is to learn a segmentation model for this structure from $\mathbf{I}$. As mentioned above, we frame image segmentation as a process of contour evolution. Each contour is represented by N uniformly spaced vertices. Denote the exemplar image and its contour by I_E and C_E, respectively. For any unlabeled image $I \in \mathbf{I}$, its contour is $C = \{\mathbf{p}_1, \mathbf{p}_2, \ldots, \mathbf{p}_N\}$. The exemplar contour is placed at the center of I as the initial location of C; next CTN is employed to estimate the point-wise offsets from the initial to the correct location, which is formulated by

$F_{\boldsymbol{\theta}}(I, I_E, C_E) = \{\Delta\mathbf{p}_1, \Delta\mathbf{p}_2, \ldots, \Delta\mathbf{p}_N\}$ where $F_{\boldsymbol{\theta}}$ denotes the CTN model with weights $\boldsymbol{\theta}$.

Inspired by [17], we use a *CNN-GCN network architecture* to model the contour evolution in CTN. From Fig. 1, CTN consists of two parts of an image encoding block and cascaded contour evolution blocks. It takes an unlabeled image I, an exemplar image I_E and its ground truth contour C_E as input, and predicts the contour C of I. (1) We first place C_E at the center of I as the initial location of C, then the encoder outputs a feature map encoding the local image appearance of I. ResNet-50 [11] is used as the backbone of CNN encoder. (2) The cascaded GCN blocks are then employed to evolve the contour C step by step. The GCN takes the contour graph with vertex features as input. Each vertex in the contour is connected to four neighboring vertices, two on each side. These vertex features are extracted from the feature map of I at the vertex locations via interpolation. Each GCN block takes the output contour of the previous block and updates it by predicting the point-wise coordinate offsets. We use five GCN blocks with the same multi-layer GCN architecture, although weights are not shared. The output of the $5th$ block is the predicted contour of CTN. (3) Three one-shot trainable losses are utilized to optimize CTN, as the contour perceptual loss L_{perc}, the contour bending loss L_{bend} and the edge loss L_{edge}. The total loss of CTN in the one-shot setting is written as: $L = \lambda_1 L_{perc} + \lambda_2 L_{bend} + \lambda_3 L_{edge}$ where $\lambda_1, \lambda_2, \lambda_3$ are the weighting factors of the three losses. We describe the three employed losses in detail as follows.

Contour Perceptual Loss. We propose a new contour perceptual loss to measure the *appearance dissimilarity* between the visual patterns of the exemplar contour C_E and the predicted contour C, on the exemplar image I_E or the target image I, respectively. Partially motivated by the original perceptual loss [14] developed for image super-resolution, modeling the image perceptual similarities in the feature space of VGG-Net [26], we measure the contour perceptual similarities in the graph feature space. In particular, graph features are extracted from the ImageNet pre-trained VGG-16 feature maps of the two images along the two contours, and their L1 distance is calculated as the contour perceptual loss: $L_{perc} = \sum_{i=1,\ldots,N} \|P(\mathbf{p}_i) - P_E(\mathbf{p}_i')\|_1$ where $\mathbf{p}_i \in C$, $\mathbf{p}_i' \in C_E$, and P and P_E denote the VGG-16 features of I and I_E, respectively. The VGG-16 baseline network weights are trained on ImageNet dataset [23].

The contour perceptual loss is used to guide the evolution of the contour in CTN, by having several advantages. (1) Since VGG-16 network features can capture the image pattern of a neighboring area with spatial contexts (*i.e.*, network receptive field), the contour perceptual loss enjoys a relatively large capturing range (*i.e.*, the convex region around the minimum), making the CTN training optimization easier. (2) The backbone VGG-16 model is trained on ImageNet [23] for classification tasks, so that its learned features are less sensitive to noises and illumination variations, which also benefits the training of CTN.

Contour Bending Loss. If we operate under the assumption that an exemplar contour is broadly informative to other data samples, then it should be beneficial to use the exemplar shape to ground any predictions on other samples. To this end, we propose a novel contour bending loss to measure the *shape dissimilarity* between contours. The loss is calculated as the bending energy of thin-plate spline (TPS) warping [1] that maps C_E to C. It is worth noting that TPS warping achieves the minimum bending energy among all warpings that map C_E to C. Since the bending energy measures the magnitude of the second order derivatives of the warping function, the contour bending loss penalizes more on the local and acute shape changes, often associated with mis-segmentations. Given C and C_E, the TPS bending energy can be calculated as follows. Define $\mathbf{p}_i = (x_i, y_i)$, $\mathbf{p}'_i = (x'_i, y'_i)$, and $\mathbf{K} = \left(\|\mathbf{p}'_i - \mathbf{p}'_j\|_2^2 \cdot log\|\mathbf{p}'_i - \mathbf{p}'_j\|_2\right)$, $\mathbf{P} = (\mathbf{1}, \mathbf{x}', \mathbf{y}')$, $\mathbf{L} = \begin{bmatrix} \mathbf{K} & \mathbf{P} \\ \mathbf{P}^T & \mathbf{0} \end{bmatrix}$ where $\mathbf{x}' = \{x'_1, x'_2, \ldots, x'_N\}^T$, $\mathbf{y}' = \{y'_1, y'_2, \ldots, y'_N\}^T$. The TPS bending energy is written as $L_{bend} = \max\left[\frac{1}{8\pi}(\mathbf{x}^T\mathbf{H}\mathbf{x} + \mathbf{y}^T\mathbf{H}\mathbf{y}), 0\right]$ where $\mathbf{x} = \{x_1, x_2, \ldots, x_N\}^T$, $\mathbf{y} = \{y_1, y_2, \ldots, y_N\}^T$, and $\mathbf{H}$ is the $N \times N$ upper left submatrix of $\mathbf{L}^{-1}$ [30].

Edge Loss. Although the contour perceptual and bending losses can achieve robust segmentation, they are inherently insensitive to (very) small segmentation fluctuations, such as minimal deviations from the correct boundary by a few pixels. Therefore, in order to obtain desirably high segmentation accuracy and adequately facilitate the downstream workflows like rheumatoid arthritis quantification [13], we employ an edge loss that measures the image gradient magnitude along the computed contour and attracts the contour toward edges in the image naturally. The edge loss is written as: $L_{edge} = -\frac{1}{N}\sum_{\mathbf{p}\in C}\|\nabla I(\mathbf{p})\|_2$ where ∇ is the gradient operator.

2.1 Human-in-the-Loop

More labels are always helpful to enhance the model's generalization ability and robustness, if available. Benefiting from the contour-based setting, CTN offers a natural way to incorporate additional user labels with a human-in-the-loop mechanism. Assuming that we have a CTN model trained with one exemplar image, we intend to finetune it with more segmentation annotations. We run this model on a set of unlabeled images first, and select any number of images with inaccurate predictions as new instances. Instead of drawing the whole contour from scratch on these new images, the annotator only needs to redraw some partial contours, to correct the previously undesirable predictions. The point-wise training of CTN makes it feasible to learn from these partial corrections.

A **partial contour matching loss** is proposed to utilize the partial ground truth contours during the CTN training. Denote $\hat{\mathbf{C}}$ as a set of partial contours in image I, each element of which is an individual contour segment. For each contour segment $\hat{C}_i \in \hat{\mathbf{C}}$, we build the point correspondence between $\hat{C}_i$ and C. For each $\hat{C}_i$, we find two points in the predicted contour C, closest to the start and end points of $\hat{C}_i$, then each predicted point between the two points are assigned to the closest corrected point. Denote the corresponding predicted contour segment by C_i ($C_i \in C$). We define the distance between C and $\hat{C}_i$ as the Chamfer distance from C_i to $\hat{C}_i$: $D(\hat{C}_i, C) = \sum_{\mathbf{p}\in C_i}\min_{\hat{\mathbf{p}}\in\hat{C}_i}\|\mathbf{p} - \hat{\mathbf{p}}\|_2$ and the partial matching loss of C is defined as $L_{pcm} = \frac{1}{N}\sum_{\hat{C}_i\in\hat{\mathbf{C}}} D(\hat{C}_i, C)$. In the human-in-the-loop scenario, we combine all losses to train the CTN, and rewrite the loss function as $\hat{L} = \lambda_1 L_{perc} + \lambda_2 L_{bend} + \lambda_3 L_{edge} + \lambda_4 L_{pcm}$, which allows CTN to be trained with fully labeled, partially labeled and unlabeled images simultaneously and seamlessly. Whenever new labels are available, we can use $\hat{L}$ to finetune the one-shot CTN model.

3 Experimental Results

Datasets. We evaluate our method on three X-ray image datasets of knee, lung and phalanx, respectively. The *knee* dataset contains 212 knee X-ray images from the Osteoarthritis Initiative (OAI) database[1], 100 for training and 112 for testing. The *lung* dataset is the public JSRT [25] of 247 posterior-anterior chest radiographs, 124 for training and 123 for testing. The *phalanx* dataset comes from hand X-ray images of patients with rheumatoid arthritis. 202 ROIs of proximal phalanx are extracted automatically from hand joint detection [13], randomly split into 100 training and 102 testing images.

Table 1. Performances of CTN and seven existing methods on three datasets.

Method		Knee		Lung		Phalanx	
		IoU(%)	HD(px)	IoU(%)	HD(px)	IoU(%)	HD(px)
Non- learning	MorphACWE [2,18]	65.89	54.07	76.09	55.35	74.33	69.13
	MorphGAC [3,18]	87.42	15.78	70.79	45.67	82.15	24.73
One-shot	CANet [33]	29.22	175.86	56.90	73.46	60.90	67.13
	Brainstorm [34]	90.17	29.07	77.13	43.28	80.05	30.30
	CTN (Ours)	**97.32**	**6.01**	**94.75**	**12.16**	**96.96**	**8.19**
Fully supervised	UNet [21]	96.60	7.14	95.38	12.48	95.76	10.10
	DeepLab [5]	97.18	5.41	96.18	10.81	97.63	6.52
	HRNet [29]	96.99	5.18	95.99	10.44	97.47	7.03

We evaluate the accuracy of segmentation masks by Intersection-over-Union (IoU) metric and the corresponding object contour distance by the Hausdorff distance (HD). For comparative methods not explicitly outputting the anatomy contours, we extract the external contour of the largest segmented image region. The hyperparameters are $N = 1000$, $\lambda_1 = 1$, $\lambda_2 = 0.25$, $\lambda_3 = 0.1$, $\lambda_4 = 1$. All networks are trained using Adam optimizer with a learning rate of 1×10^{-4}, a weight decay of 1×10^{-4} and a batch size of 12 for 500 epochs. The one-shot training and human-in-the-loop finetuning settings are the same.

The proposed CTN is compared with seven previous methods. The quantitative results are reported in Table 1; qualitative results are given in Fig. 2.

[1] https://nda.nih.gov/oai/.

Comparison with Non-Learning-Based Methods. We first compare with two non-learning based methods of MorphACWE [2,18] and MorphGAC [3,18]. Both approaches are based on active contour models (ACMs), which evolves an initial contour to the object by minimizing an energy function. The initial contour is the same as ours. Quantitative results in Table 1 show that our method significantly outperforms them. Specifically, in average CTN achieves 16.22% higher IoU and 19.94 pixels less in HD than MorphGAC, the better of the two. Visualization results in Fig. 2 show that these two approaches cannot localize anatomical structures accurately, especially when the boundary of such structures are not clearly contrasted, such as in the lung image. Both methods are based on ACMs and predict contours by minimizing some hand-crafted energy functions for a *single* image. In contrast, CTN learns from an exemplar contour to guide the contour transformation for all images in the entire training set.

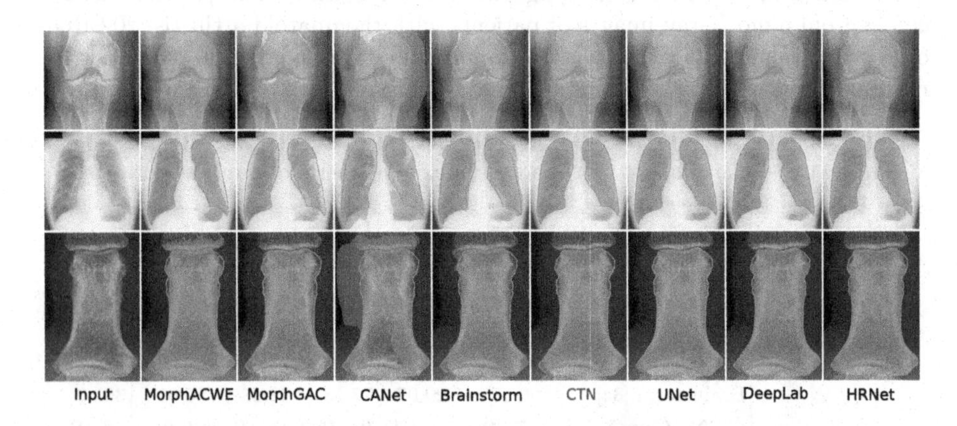

Fig. 2. Segmentation results of three example images using eight methods. From top to bottom, these images are from the knee, lung and phalanx testing sets, respectively. The ground truth boundaries are drawn in **green** line for the ease of comparison. (Color figure online)

Comparison With Other One-Shot Methods. Next, we compare with two state-of-the-art one-shot segmentation methods: CANet [33] and Brainstorm [34]. All one-shot approaches (including ours) use the same exemplar image. For each training set, we compare the distance of each image to all other images in the VGG network feature space and the exemplar is selected to be the image with the smallest distance. CANet is proposed to perform one-shot segmentation on unseen objects in testing, but it requires a fully annotated dataset to train. Such that, we use the specific model [33] trained on PASCAL VOC 2012 dataset for comparison. From Table 1, CANet achieves only 49.01% IoU on average. We speculate that the poor performance is caused by the domain gap between natural images and medical images. Brainstorm [34] learns an image augmentation model and is one-shot trainable. We follow its default procedure to train the segmentation models on three datasets. It yields reasonable results with averaged 82.45% IoU and 34.22 HD, but still dramatically lower than ours.

Comparison with Fully Supervised Methods. Last, we evaluate and compare the performance of three fully supervised methods: UNet [21], DeepLab v3+ [5] and HRNet W18 [29]. All of them are trained for 500 epochs using all training image annotations in three datasets. On average, CTN performs comparably or better than UNet, and falls behind DeepLab (the best of the three), by 0.66% in IoU and 1.21 pixels in HD, respectively. It demonstrates that CTN while using only one training image can usually compete head-to-head with the state-of-the-art fully supervised baselines [5,21,29]. Note that the heatmap-based segmentation methods predict the per-pixel labels, which could cause the loss of integrity of object boundaries, *e.g.*, some small "islands" in the lung masks of Fig. 2. On the other hand, CTN naturally retains the integrity of the object segmentation, as an important aspect in assessing visual segmentation quality.

Incorporating Simulated Human Corrections. To evaluate the effectiveness of the human-in-the-loop mechanism, we empirically simulate different degrees of human-computer interactions. For each dataset, we first train a CTN model with the default exemplar and run inference on the training set. We sort all training images by their HD segmentation errors from high to low. Three subsets are formed by selecting the top 10%, 25% or 100% training images; and fine-tune the initial one-shot CTN model using these training subsets augmented by the ground truth contours, respectively. This protocol results in four CTN models. From Fig. 3, we observe that CTN consistently improves with more human corrections. Specifically, when using 25% such corrected samples, CTN starts to outperform DeepLab using all training images (IoUs of 97.17% vs 97.0%, and HDs of 7.01 vs 7.58). With all samples, CTN reaches 97.33% on IoU and 6.5 on HD. These results indicate that the human-in-the-loop mechanism can potentially help CTN achieves better performance than fully supervised methods with considerably less annotation efforts.

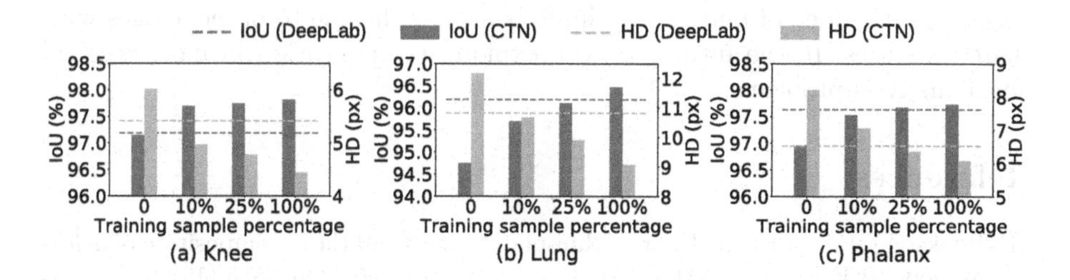

Fig. 3. Using 0, 10%, 25% and 100% human corrections to finetune the one-shot CTN model, respectively ("0" means no finetuning).

Ablation Study. We conduct an ablation experiment to validate the effectiveness of the three proposed losses. The results are shown in Table 2 where the performance of our method is indeed impaired if any loss is removed, with mean

IoU reductions of 4.32%, 4.12%, and 1.72% for L_{perc}, L_{bend}, and L_{edge}, respectively. This validates the necessity of all three losses. An exception is the knee dataset when L_{bend} is removed. Knee X-ray images share similar appearance features along the contour so that they can be segmented robustly with just the contour perceptual loss and edge loss. Thus, adding contour bending loss leads to statistically insignificant decreases (*i.e.*, IoUs of 97.32% vs 97.50%, HDs of 6.01 vs 5.87) in this particular scenario. However, such a regularization effect by the contour bending loss is generally desired to alleviate the worst-case scenarios and proves useful in the other two datasets.

Table 2. Ablation study. Remove one loss each time and re-train the model.

L_{perc}	L_{bend}	L_{edge}	Knee		Lung		Phalanx	
			IoU(%)	HD(px)	IoU(%)	HD(px)	IoU(%)	HD(px)
	✓	✓	94.62	8.28	87.45	26.51	94.01	15.81
✓		✓	97.50	5.87	84.93	36.74	94.24	26.13
✓	✓		94.43	11.90	92.99	16.22	96.45	9.84
✓	✓	✓	97.32	6.01	94.75	12.17	96.96	8.19

4 Conclusion

In this paper, we propose a novel one-shot segmentation method, Contour Transformer Network, which takes one labeled exemplar and a set of unlabeled images to train a segmentation model for anatomical structures in medical images. The key idea that enables one-shot training is to guide the segmentation of unlabeled images by utilizing their shared features with the exemplar image but not ground truth masks. Experiments on three datasets demonstrate that CTN performs competitively to state-of-the-art fully supervised approaches and outperforms them with minimal human corrections. Although CTN is for anatomical structures, the idea of one-shot training is also applicable to other images with shared features. In the future, we will explore its application in more medical image analysis problems.

References

1. Bookstein, F.L.: Principal warps: thin-plate splines and the decomposition of deformations. IEEE Trans. Pattern Anal. Mach. Intell. **11**(6), 567–585 (1989)
2. Caselles, V., Kimmel, R., Sapiro, G.: Geodesic active contours. Int. J. Comput. Vis. **22**(1), 61–79 (1997)
3. Chan, T.F., Vese, L.A.: Active contours without edges. IEEE Trans. Image Process. **10**(2), 266–277 (2001)
4. Chen, H., et al.: Anatomy-aware Siamese network: exploiting semantic asymmetry for accurate pelvic fracture detection in x-ray images. arXiv preprint arXiv:2007.01464 (2020)

5. Chen, L.-C., Zhu, Y., Papandreou, G., Schroff, F., Adam, H.: Encoder-decoder with atrous separable convolution for semantic image segmentation. In: Ferrari, V., Hebert, M., Sminchisescu, C., Weiss, Y. (eds.) ECCV 2018. LNCS, vol. 11211, pp. 833–851. Springer, Cham (2018). https://doi.org/10.1007/978-3-030-01234-2_49
6. Chen, R., Ma, Y., Chen, N., Lee, D., Wang, W.: Cephalometric landmark detection by attentive feature pyramid fusion and regression-voting. In: Shen, D., et al. (eds.) MICCAI 2019. LNCS, vol. 11766, pp. 873–881. Springer, Cham (2019). https://doi.org/10.1007/978-3-030-32248-9_97
7. Dolz, J., Gopinath, K., Yuan, J., Lombaert, H., Desrosiers, C., Ayed, I.B.: Hyperdense-net: a hyper-densely connected CNN for multi-modal image segmentation. IEEE Trans. Med. Imaging **38**(5), 1116–1126 (2018)
8. Dong, N., Xing, E.: Few-shot semantic segmentation with prototype learning. In: Proceedings of the British Machine Vision Conference, vol. 1, p. 6 (2018)
9. Everingham, M., Van Gool, L., Williams, C.K.I., Winn, J., Zisserman, A.: The pascal visual object classes (VOC) challenge. Int. J. Comput. Vis. **88**(2), 303–338 (2010)
10. Harrison, A.P., Xu, Z., George, K., Lu, L., Summers, R.M., Mollura, D.J.: Progressive and multi-path holistically nested neural networks for pathological lung segmentation from CT images. In: Descoteaux, M., Maier-Hein, L., Franz, A., Jannin, P., Collins, D.L., Duchesne, S. (eds.) MICCAI 2017. LNCS, vol. 10435, pp. 621–629. Springer, Cham (2017). https://doi.org/10.1007/978-3-319-66179-7_71
11. He, K., Zhang, X., Ren, S., Sun, J.: Deep residual learning for image recognition. In: Proceedings of the IEEE Conference on Computer Vision and Pattern Recognition, pp. 770–778 (2016)
12. Hu, R., Dollár, P., He, K., Darrell, T., Girshick, R.: Learning to segment every thing. In: Proceedings of the IEEE Conference on Computer Vision and Pattern Recognition, pp. 4233–4241 (2018)
13. Huo, Y., Vincken, K.L., van der Heijde, D., De Hair, M.J., Lafeber, F.P., Viergever, M.A.: Automatic quantification of radiographic finger joint space width of patients with early rheumatoid arthritis. IEEE Trans. Biomed. Eng. **63**(10), 2177–2186 (2015)
14. Johnson, J., Alahi, A., Fei-Fei, L.: Perceptual losses for real-time style transfer and super-resolution. In: Leibe, B., Matas, J., Sebe, N., Welling, M. (eds.) ECCV 2016. LNCS, vol. 9906, pp. 694–711. Springer, Cham (2016). https://doi.org/10.1007/978-3-319-46475-6_43
15. Li, W., et al.: Structured landmark detection via topology-adapting deep graph learning. arXiv preprint arXiv:2004.08190 (2020)
16. Lin, T., Maire, M., et al.: Microsoft COCO: common objects in context. arXiv preprint arXiv:1405.0312 (2014)
17. Ling, H., Gao, J., Kar, A., Chen, W., Fidler, S.: Fast interactive object annotation with curve-GCN. In: Proceedings of the IEEE Conference on Computer Vision and Pattern Recognition, pp. 5257–5266 (2019)
18. Marquez-Neila, P., Baumela, L., Alvarez, L.: A morphological approach to curvature-based evolution of curves and surfaces. IEEE Trans. Pattern Anal. Mach. Intell. **36**(1), 2–17 (2013)
19. Michaelis, C., Ustyuzhaninov, I., Bethge, M., Ecker, A.S.: One-shot instance segmentation. arXiv preprint arXiv:1811.11507 (2018)
20. Oliveira, A., Pereira, S., Silva, C.A.: Augmenting data when training a CNN for retinal vessel segmentation: How to warp? In: IEEE 5th Portuguese Meeting on Bioengineering, pp. 1–4 (2017)

21. Ronneberger, O., Fischer, P., Brox, T.: U-Net: convolutional networks for biomedical image segmentation. In: Navab, N., Hornegger, J., Wells, W.M., Frangi, A.F. (eds.) MICCAI 2015. LNCS, vol. 9351, pp. 234–241. Springer, Cham (2015). https://doi.org/10.1007/978-3-319-24574-4_28
22. Roth, H.R., et al.: Spatial aggregation of holistically-nested convolutional neural networks for automated pancreas localization and segmentation. Med. Image Anal. **45**, 94–107 (2018)
23. Russakovsky, O., et al.: ImageNet large scale visual recognition challenge. Int. J. Comput. Vis. **115**(3), 211–252 (2015)
24. Shaban, A., Bansal, S., Liu, Z., Essa, I., Boots, B.: One-shot learning for semantic segmentation. arXiv preprint arXiv:1709.03410 (2017)
25. Shiraishi, J., et al.: Development of a digital image database for chest radiographs with and without a lung nodule: receiver operating characteristic analysis of radiologists' detection of pulmonary nodules. Am. J. Roentgenol. **174**(1), 71–74 (2000)
26. Simonyan, K., Zisserman, A.: Very deep convolutional networks for large-scale image recognition. arXiv preprint arXiv:1409.1556 (2014)
27. Sinha, A., Dolz, J.: Multi-scale guided attention for medical image segmentation. arXiv preprint arXiv:1906.02849 (2019)
28. Tajbakhsh, N., Jeyaseelan, L., Li, Q., Chiang, J., Wu, Z., Ding, X.: Embracing Imperfect Datasets: A Review of Deep Learning Solutions for Medical Image Segmentation (2019)
29. Wang, J., et al.: Deep high-resolution representation learning for visual recognition. arXiv preprint arXiv:1908.07919 (2019)
30. Wang, S., Munsell, B., Richardson, T.: Correspondence establishment in statistical shape modeling: Optimization and evaluation. In: Statistical Shape and Deformation Analysis, pp. 67–87. Elsevier (2017)
31. Wang, Y., et al.: Weakly supervised universal fracture detection in pelvic x-rays. In: Shen, D., et al. (eds.) MICCAI 2019. LNCS, vol. 11769, pp. 459–467. Springer, Cham (2019). https://doi.org/10.1007/978-3-030-32226-7_51
32. Wu, W., Qian, C., Yang, S., Wang, Q., Cai, Y., Zhou, Q.: Look at boundary: a boundary-aware face alignment algorithm. In: Proceedings of the IEEE Conference on Computer Vision and Pattern Recognition, pp. 2129–2138 (2018)
33. Zhang, C., Lin, G., Liu, F., Yao, R., Shen, C.: CANet: class-agnostic segmentation networks with iterative refinement and attentive few-shot learning. In: Proceedings of the IEEE Conference on Computer Vision and Pattern Recognition, pp. 5217–5226 (2019)
34. Zhao, A., Balakrishnan, G., Durand, F., Guttag, J.V., Dalca, A.V.: Data augmentation using learned transformations for one-shot medical image segmentation. In: Proceedings of the IEEE Conference on Computer Vision and Pattern Recognition, pp. 8543–8553 (2019)

Uncertainty Estimates as Data Selection Criteria to Boost Omni-Supervised Learning

Lorenzo Venturini[1(✉)], Aris T. Papageorghiou[2], J. Alison Noble[1], and Ana I. L. Namburete[1]

[1] Institute of Biomedical Engineering, Department of Engineering Science, University of Oxford, Oxford, UK
lorenzo.venturini@new.ox.ac.uk

[2] Nuffield Department of Women's and Reproductive Health, University of Oxford, Oxford, UK

Abstract. For many medical applications, large quantities of imaging data are routinely obtained but it can be difficult and time-consuming to obtain high-quality labels for that data. We propose a novel uncertainty-based method to improve the performance of segmentation networks when limited manual labels are available in a large dataset. We estimate segmentation uncertainty on unlabeled data using test-time augmentation and test-time dropout. We then use uncertainty metrics to select unlabeled samples for further training in a semi-supervised learning framework. Compared to random data selection, our method gives a significant boost in Dice coefficient for semi-supervised volume segmentation on the EADC-ADNI/HARP MRI dataset and the large-scale INTERGROWTH-21st ultrasound dataset. Our results show a greater performance boost on the ultrasound dataset, suggesting that our method is most useful with data of lower or more variable quality.

Keywords: Uncertainty · Omni-supervised learning · Boosting

1 Introduction

A major challenge in supervised medical image segmentation is the scarcity of accurately labeled data [13]. While imaging is routinely used in clinical settings, it is much harder to obtain accurate segmentation labels for imaging data. Producing accurate manual labels, especially for 3D segmentation tasks, is time-consuming and often not part of a standard clinical protocol.

Machine learning methods such as convolutional neural networks (CNNs) typically rely on large amounts of labeled data to achieve optimal performance, but generating reliable manual labels for medical datasets requires significant time investment from trained clinicians, which is often not available.

Several approaches have been proposed to reduce the time investment required to generate labels, such as weak supervision with bounding boxes [11]

A. L. Martel et al. (Eds.): MICCAI 2020, LNCS 12261, pp. 689–698, 2020.
https://doi.org/10.1007/978-3-030-59710-8_67

or annotated landmarks [1]. There have also been attempts to use additional unlabeled data to improve a network's performance. Active learning methods [19] can prompt a human labeler to only provide additional manual labels on examples that would lead to the greatest performance improvements. Adversarial networks have also been proposed to improve performance with limited training labels [20].

Another method that has been proposed to exploit the presence of unlabeled data to improve a neural network's performance is omni-supervised learning [16]. In general, training a neural network using its own predictions does not improve its performance [11], but combining the predictions of multiple independent learners leads to a prediction superior to any of them [5,7]. Similarly, aggregating predictions generated from different transformations (such as reflections and rotations) of the data can also lead to improvements [12,16]. In other words, "a set of weak learners can create a strong learner" [17]. These two principles, of *model diversity* and *data diversity*, drive the concept of omni-supervised learning.

Omni-supervised learning aggregates network-generated segmentations of the unlabeled dataset using both types of diversity and uses them additional labels for further training. Huang et al. [8] used this to improve localisation in 3D neurosonography images. In that work, a subset of the predictions generated on unlabeled data are used for the next iteration of training. Selecting an appropriate subset of predictions, however, has been ignored in the literature. Previous work using CNNs for omni-supervised learning has either selected a subset of the unlabeled dataset at random [8] or used weak heuristics [16], such as ensuring that the number of labeled pixels is similar to the average in the training data. In this work, we explore different selection criteria and evaluate their performance.

1.1 Uncertainty Estimation

CNNs for binary segmentation typically use sigmoid activation functions at their output, so each voxel is given a "soft" classification ranging from 0 (confident background) to 1 (confident foreground). The network thus expresses some uncertainty in its segmentation, but CNNs also often confidently make incorrect predictions [4], especially when they are presented with noisy or ambiguous data, or when the data presented to them differs in appearance from the training data.

A distinction can be made between the two sources of error: (1) error caused by image noise, ambiguity, and inconsistent labeling in the training set, and (2) error caused by the classifier being shown data that appears different to that in the training set. The first type, known as *aleatoric uncertainty* [10], is due to ambiguity and noise in the data itself. Since this is a measure of genuine uncertainty in the data, this type of uncertainty is unlikely to reduce with increased training data. On the other hand, the second source of error, known as *epistemic uncertainty* [10], is due to the model parameters and overfitting to data not represented in the training set. The amount of epistemic uncertainty can be used to give a measure of how different any given example is from the training data.

Gal and Ghahramani [6] have proposed a method to estimate the uncertainty present in the model, using *dropout uncertainty*. Their method applies dropout at test time and measures differences in output to give a measure of the model's uncertainty. Wang et al. [18] proposed using test-time augmentation to improve estimates of aleatoric uncertainty, using a similar method: measuring differences in output across augmented versions of the same data. The principle behind these differing approaches comes from different sources of uncertainty. Epistemic uncertainty, introduced by the model's parameters, is estimated by making changes to the model. Aleatoric uncertainty is introduced by the data, so it can be estimated by varying the data.

Omni-supervised learning requires data diversity and model diversity, which are also methods to measure aleatoric and epistemic uncertainty, respectively. In this paper we show that these two measures can form the selection criteria of data for omni-supervised models, without adding complexity.

We propose a scheme to incorporate these metrics of uncertainty in an omni-supervised framework to select data to use for subsequent training iterations. We harness data and model diversity in omni-supervised learning to obtain uncertainty estimates at little additional computational cost. We then experiment with data selection for further training based on these metrics, and compare the segmentation performance of different data-selection methods with random selection. Our novel data-selection scheme uses information from the unlabeled data to improve segmentation performance with no additional supervision.

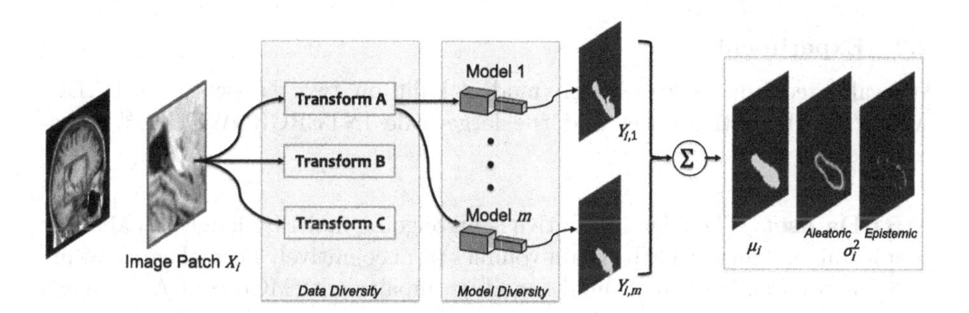

Fig. 1. Data diversity consisting of transformations and model diversity using m models are used to generate independent predictions of the same sample X_i, which can be aggregated to generate a stronger prediction.

2 Methods

2.1 Uncertainty Estimation

To measure epistemic uncertainty on 3D data we implemented a minor variation on a conventional 3D U-Net [3] adding dropout to obtain uncertainty estimates. After each max-pool layer or before each upsampling layer, some of the previous layer's weights are dropped out ($p = 0.5$).

Using test-time dropout, we generated N_p independent predictions for each unlabeled volume. The uncertainty in the segmentation of each voxel is then given by the variance in the voxel's predicted label across different segmentations[1]. A global measure of the epistemic uncertainty in a volume can be obtained simply by summing the uncertainty of all pixels

$$\text{uncertainty} = \sum_i \sigma_n^2 (x_i)$$

where $\sigma_n^2 (x_i)$ is the variance of the softmax score of each pixel x_i over n samples.

While epistemic uncertainty arises from the model, and can be measured by varying the model at test time, aleatoric uncertainty arises from the data, and can be measured by varying the data at test time. The augmentation methods used are described in Sect. 2.2.

While higher uncertainty across both metrics, in general, appears to point to a lower-quality segmentation, it is also important to increase the diversity of the training dataset and reduce the differences between the training data and the unlabeled data. We hypothesise, therefore, that selecting volumes segmented with a lower level of epistemic uncertainty will not lead to an improvement in segmentation performance in the next round of training.

The same is not true of *aleatoric* uncertainty, which is linked to the data, rather than the model: higher aleatoric uncertainty is likely to link to a lower signal-to-background ratio, or more challenging data. Therefore, we expect that selecting volumes with lower aleatoric uncertainty will lead to better results.

2.2 Experiments

We validated our hypothesis on models built on two datasets, the EADC-ADNI/HARP MRI dataset and the large-scale INTERGROWTH-21st ultrasound dataset.

MRI Dataset: The ADNI initiative is an ongoing multisite longitudinal study that acquires structural MRI brain volumes from cognitively normal aging adults (CN), as well as those with mild cognitive impairment (MCI) and Alzheimer's disease (AD) [9]. 63 sites participate in acquisition, using different scanners and acquisition methods. The EADC/HARP dataset consists of a subset of 135 volumes selected from the ADNI dataset, with expert 3D manual segmentations of the hippocampus. We also used 680 additional unlabeled volumes from the ADNI dataset, each from different subjects, removing any volumes already in the HARP dataset. All volumes were brain-extracted, linearly registered to MNI152 image space [2] and normalized by dividing each pixel by the 99th percentile value. $64 \times 64 \times 64$ patches were extracted around the hippocampus in each hemisphere, leaving 270 labeled patches and 1360 unlabeled patches. Of the 135 labeled volumes, 80 for training, 20 for validation, and 35 for testing. Five-fold cross-validation was used.

[1] As most voxels distant from anatomical boundaries are consistently segmented, the uncertainty (or segmentation variance) of most voxels is 0.

Ultrasound Dataset: We used data from the INTERGROWTH-21st study [15], a longitudinal study which includes 3D ultrasound volumes acquired from optimally healthy pregnant women. For this study, we used 948 neurosonography volumes in the range of 14–25 gestational weeks. The volumes were registered to a common coordinate system using similarity transforms [14]. The cerebellum was manually segmented by the same annotator on 146 volumes: 86 for training, 20 for validation, and 40 for testing. Five-fold cross-validation was used. The remaining volumes were kept unlabeled.

We trained a model for each dataset using the labeled volumes in the training set, and used the model to generate segmentations. The aleatoric and epistemic uncertainties of each segmentation were estimated using test-time augmentation and test-time dropout.

An aggregated segmentation was produced for each volume by taking the median predicted value for each pixel. N_{unl} volumes were then selected for the second stage of omni-supervised training: these volumes (with corresponding annotations) are added to the training set and the model is retrained using the expanded training set. We experimented with different selection criteria for the automatically segmented volumes:

1. Randomly select the volumes, as a control (similar to current methods [8])
2. Select the volumes with the lowest epistemic uncertainty
3. Select the volumes with the lowest aleatoric uncertainty.
4. Select the volumes with the lowest sum of epistemic and aleatoric uncertainty.

We explored the effect of different amounts of initial labeled data. More training data would be expected to reduce epistemic uncertainty, as the training set more closely resembles the distribution of possible images and the network's overall performance improves, but should not affect aleatoric uncertainty estimates.

Implementation Details: The augmentation used consists of simple similarity transforms. Since the images are aligned to a common reference space (in both datasets), the only axis they can be reflected across to maintain alignment is the midsagittal plane (MSP), the median plane that runs vertically between the brain's hemispheres. Small random **translations** (up to ± 5 voxels along each axis), **rotations** (up to $10°$ around each axis), and **scaling** (up to $\pm 10\%$ linear zoom) were also used. Nearest-neighbour interpolation was used for computational efficiency. The volumes were **reflected** with 50% probability, while the degree of all other augmentations was sampled from a uniform distribution. Once a prediction had been obtained on an augmented volume, the inverse transform T_x^{-1} was applied to return to a common reference space. Aleatoric uncertainty was estimated using the variance of the resulting predictions, after test-time augmentation. The variance was measured across N_p predictions per volume. This augmentation scheme was used for all experiments.

We selected $N_{unl} = 200$ for the MRI dataset and $N_{unl} = 150$ volumes for the ultrasound dataset, to be similar in size to the labeled training dataset in line with previous work on omni-supervised learning [8]. We also selected $N_p = 40$, to allow an accurate estimate of variance without using excessive computational resources.

3 Results

3.1 Uncertainty Quantification

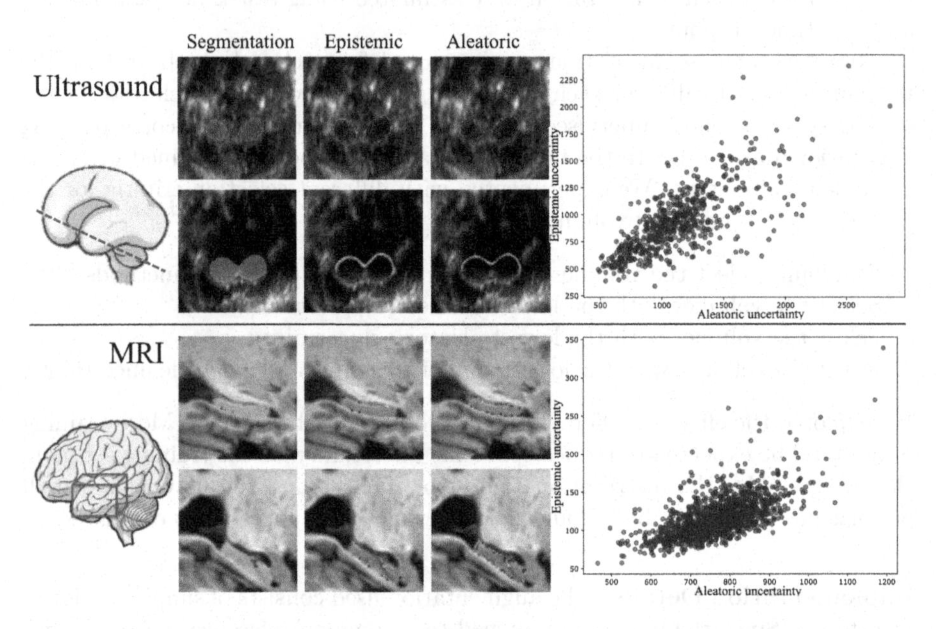

Fig. 2. Visualisation of aleatoric and epistemic uncertainty estimates in two example images per dataset, and scatterplots (with transparency for overlapping points) showing uncertainty across both datasets.

Figure 2 shows global estimates of aleatoric and epistemic uncertainty of every volume in the unlabeled dataset. Due to less well-defined edges in ultrasound and higher variability in image quality, the ultrasound dataset shows higher uncertainty in both metrics as seen by the higher values on the axes. The bottom ultrasound example shows a strong shadow obscuring part of the cerebellum, and uncertainty is accordingly higher in that area.

The plots in Fig. 2 show a correlation between the estimates of aleatoric and epistemic uncertainty ($r = 0.72$ in the ultrasound dataset, and $r = 0.69$ in the MRI dataset). We believe this is partially driven by boundary effects: classification is more difficult at boundary voxels, so structures with a higher surface area are likely to show higher uncertainty in both cases.

3.2 Selection Methods

Table 1. The segmentation performance of retraining a 3D CNN using different selection methods for the additional labeled data. The "IOV" row indicates the intra-observer variability of manual segmentations obtained by the same individual.

	Ultrasound		MRI	
Volume selection	Dice coeff.	Hausdorff (mm)	Dice coeff.	Hausdorff (mm)
Fully supervised	0.673 ± 0.046	12.25 ± 3.57	0.848 ± 0.015	3.97 ± 0.25
Random	0.700 ± 0.023	11.44 ± 1.75	0.849 ± 0.015	3.73 ± 0.18
Epistemic	0.689 ± 0.040	10.01 ± 1.48	0.847 ± 0.015	4.15 ± 0.32
Aleatoric	$\mathbf{0.727 \pm 0.014}$	$\mathbf{8.54 \pm 1.60}$	$\mathbf{0.851 \pm 0.015}$	$\mathbf{3.69 \pm 0.25}$
Aleat. + epist.	0.697 ± 0.015	11.10 ± 1.51	0.848 ± 0.015	4.20 ± 0.35
IOV	0.764 ± 0.060	3.96 ± 0.99	N/A	N/A

Table 1 shows the performance of each data selection method. In ultrasound, random volume selection leads to an improved Dice coefficient ($p < 0.02$ in a 1-tailed t-test) in the validation set over a fully-supervised network, but does not lead to a significant improvement in Hausdorff distance. Selecting the volumes with the lowest epistemic uncertainty leads to no significant improvement ($p = 0.08$) in Dice coefficient or Hausdorff distance. Selecting volumes based on the lowest aleatoric uncertainty, on the other hand, led to significant improvements across both metrics ($p < 0.01$). Selecting volumes based on the average of aleatoric and epistemic uncertainty seems to yield accuracy roughly in the middle of the two methods ($p < 0.03$). This does not represent an improvement over aleatoric selection only.

In the MRI dataset, effects are smaller: the effect size is not significant ($p = 0.1$) for random volume selection, but remains significant ($p < 0.01$) for aleatoric selection. The same pattern as with the ultrasound dataset emerges, with larger performance improvement with aleatoric selection over random selection.

One possible explanation for the small effect size is that the labeled data alone is sufficient to generalize to the test set, and the network is already near the maximum achievable performance on this dataset. Figure 3a shows that reducing the number of labeled training examples to 10 or 20 (from the original 180) seems to have only a limited impact on segmentation performance.

4 Discussion

We observe an increase in Dice coefficient using omni-supervised learning over a fully-supervised implementation. Selecting the volumes with the lowest epistemic uncertainty, measured by test-time dropout, seems to lead to worse performance than random selection, especially in Hausdorff distance. The volumes with the lowest epistemic uncertainty are those that most closely resemble the training

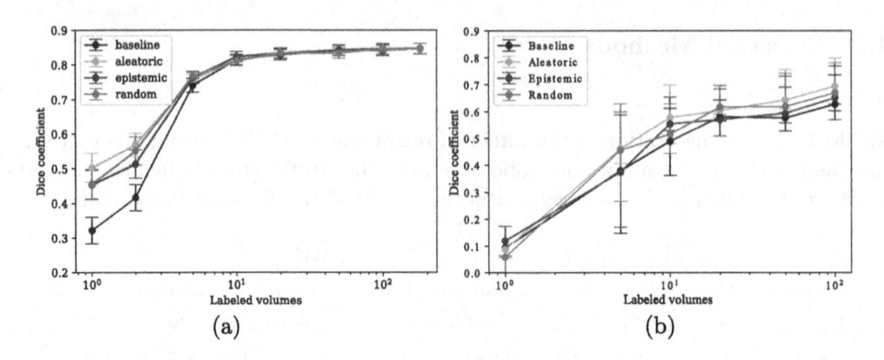

Fig. 3. Change in Dice coefficient with a changing number of labeled training examples for the (a) MRI dataset, and (b) the ultrasound dataset.

data, and we believe that adding these volumes to the training set does not improve the generalisation of the network. Instead, it introduces a bias in the training data towards the segmentations it has generated in the previous round. Further evidence of this is seen in the results selecting samples with the average of aleatoric and epistemic uncertainty: it is lower than with aleatoric uncertainty alone, suggesting that including epistemic uncertainty in this selection does not add value.

On the other hand, using aleatoric uncertainty leads to a significant improvement in performance over a random selection, especially on the ultrasound dataset. Aleatoric uncertainty is used as an estimate of the amount of uncertainty inherent in the data itself. In principle, it is unrelated to how different the data appears from training and only a measure of the quality of the data itself. Lower aleatoric uncertainty, therefore, can be thought of as correlating to clearer data and possibly better segmentation performance, without compromising the generalisability of the segmentation method.

The performance improvement is significantly larger in the ultrasound dataset. Direct comparisons are difficult since it's a different structure, but we believe that this is due to the lower signal-to-background ratio of ultrasound imaging, and the presence of artifacts such as shadows, in some volumes. The anatomical boundaries of the cerebellum in the volumes we examined were also ambiguous, which is itself a source of aleatoric uncertainty. We believe that the greater variation in image quality in ultrasound imaging accounts for the increased performance gain from using aleatoric uncertainty-based selection. This implies that a scheme similar to the one produced here may lead to a larger performance boost in datasets with data of lower, or more variable, quality.

Results on MRI nonetheless show a small, but significant increase in Dice coefficient. We believe this is partly because the fully-supervised network is already close to the highest performance a CNN-based method could achieve: Fig. 3 shows only a modest performance reduction from a 10-fold reduction in training examples. With fewer training examples, the performance boost is largon two medical datasets with the same segmentation CNNer.

5 Conclusion

We demonstrate a novel uncertainty-based data selection scheme for omni-supervised learning, demonstrating its effectiveness for different segmentation tasks with different imaging modalities. We found a significant improvement for both segmentation problems which increases with fewer training labels, and found a larger performance boost in ultrasound. Our results suggest that our method is of most use in datasets with less labeled data, and with images of lower or more variable quality.

Acknowledgment. We would like to thank Nicola Dinsdale for her help with data preparation and analysis of the MRI dataset. This work is supported by funding from the Engineering and Physical Sciences Research Council (EPSRC) and Medical Research Council (MRC) [grant number EP/L016052/1]. A. T. Papageorghiou is supported by the National Institute for Health Research (NIHR) Oxford Biomedical Research Centre. A. Namburete is grateful for support from the UK Royal Academy of Engineering under the Engineering for Development Research Fellowships scheme. J. A. Noble acknowledges the National Institutes of Health (NIH) through the National Institute on Alcohol Abuse and Alcoholism (NIAAA) (U01 AA014809-14), We thank the INTERGROWTH-21st Consortium for permission to use 3D ultrasound volumes of the fetal brain.

References

1. Bearman, A., Russakovsky, O., Ferrari, V., Fei-Fei, L.: What's the point: semantic segmentation with point supervision. In: Leibe, B., Matas, J., Sebe, N., Welling, M. (eds.) ECCV 2016. LNCS, vol. 9911, pp. 549–565. Springer, Cham (2016). https://doi.org/10.1007/978-3-319-46478-7_34
2. Bocchetta, M., et al.: Harmonized benchmark labels of the hippocampus on magnetic resonance: The EADC-ADNI project. Alzheimer's and Dementia (2015). https://doi.org/10.1016/j.jalz.2013.12.019
3. Çiçek, Ö., Abdulkadir, A., Lienkamp, S.S., Brox, T., Ronneberger, O.: 3D U-Net: learning dense volumetric segmentation from sparse annotation. In: Ourselin, S., Joskowicz, L., Sabuncu, M.R., Unal, G., Wells, W. (eds.) MICCAI 2016. LNCS, vol. 9901, pp. 424–432. Springer, Cham (2016). https://doi.org/10.1007/978-3-319-46723-8_49
4. Denker, J.S., LeCun, Y.: Transforming neural-net output levels to probability distributions. In: Advances in Neural Information Processing Systems, vol. 3 (1991)
5. Dietterich, T.G.: Ensemble methods in machine learning. In: Kittler, J., Roli, F. (eds.) MCS 2000. LNCS, vol. 1857, pp. 1–15. Springer, Heidelberg (2000). https://doi.org/10.1007/3-540-45014-9_1
6. Gal, Y., Ghahramani, Z.: Dropout as a Bayesian approximation: representing model uncertainty in deep learning, June 2015. http://arxiv.org/abs/1506.02142
7. Hinton, G., Vinyals, O., Dean, J.: Distilling the Knowledge in a Neural Network, March 2015. http://arxiv.org/abs/1503.02531
8. Huang, R., Noble, J.A., Namburete, A.I.L.: Omni-supervised learning: scaling up to large unlabelled medical datasets. In: Frangi, A.F., Schnabel, J.A., Davatzikos, C., Alberola-López, C., Fichtinger, G. (eds.) MICCAI 2018. LNCS, vol. 11070, pp. 572–580. Springer, Cham (2018). https://doi.org/10.1007/978-3-030-00928-1_65

9. Jack, C.R., et al.: The Alzheimer's Disease Neuroimaging Initiative (ADNI): MRI methods (2008). https://doi.org/10.1002/jmri.21049
10. Kendall, A., Gal, Y.: What Uncertainties Do We Need in Bayesian Deep Learning for Computer Vision? (2017). http://papers.nips.cc/paper/7141-what-uncertainties-do-we-need
11. Khoreva, A., Benenson, R., Hosang, J., Hein, M., Schiele, B.: Simple does it: weakly supervised instance and semantic segmentation. In: Proceedings - 30th IEEE Conference on Computer Vision and Pattern Recognition, CVPR 2017 (2017). https://doi.org/10.1109/CVPR.2017.181
12. Krizhevsky, A., Sutskever, I., Hinton, G.E.: ImageNet classification with deep convolutional neural networks. In: Advances in Neural Information Processing Systems (2012)
13. Makropoulos, A., Counsell, S.J., Rueckert, D.: A review on automatic fetal and neonatal brain MRI segmentation. NeuroImage (2017). https://doi.org/10.1016/j.neuroimage.2017.06.074
14. Namburete, A.I., Xie, W., Yaqub, M., Zisserman, A., Noble, J.A.: Fully-automated alignment of 3D fetal brain ultrasound to a canonical reference space using multi-task learning. Med. Image Anal. **46**, 1–14 (2018). https://doi.org/10.1016/J.MEDIA.2018.02.006. https://www.sciencedirect.com/science/article/pii/S1361841518300306http://www.ncbi.nlm.nih.gov/pubmed/29499436
15. Papageorghiou, A.T., et al.: International standards for fetal growth based on serial ultrasound measurements: the Fetal Growth Longitudinal Study of the INTERGROWTH-21st Project. Lancet **384**(9946), 869–879 (2014). https://doi.org/10.1016/S0140-6736(14)61490-2. http://linkinghub.elsevier.com/retrieve/pii/S0140673614614902
16. Radosavovic, I., Dollar, P., Girshick, R., Gkioxari, G., He, K.: Data distillation: towards omni-supervised learning. In: Proceedings of the IEEE Computer Society Conference on Computer Vision and Pattern Recognition (2018). https://doi.org/10.1109/CVPR.2018.00433
17. Schapire, R.E.: The strength of weak learnability. Mach. Learn. (1990). https://doi.org/10.1023/A:1022648800760
18. Wang, G., Li, W., Aertsen, M., Deprest, J., Ourselin, S., Vercauteren, T.: Aleatoric uncertainty estimation with test-time augmentation for medical image segmentation with convolutional neural networks. Neurocomputing **338**, 34–45 (2019). https://doi.org/10.1016/J.NEUCOM.2019.01.103. https://www.sciencedirect.com/science/article/pii/S0925231219301961
19. Yang, L., Zhang, Y., Chen, J., Zhang, S., Chen, D.Z.: Suggestive annotation: a deep active learning framework for biomedical image segmentation. In: Descoteaux, M., Maier-Hein, L., Franz, A., Jannin, P., Collins, D.L., Duchesne, S. (eds.) MICCAI 2017. LNCS, vol. 10435, pp. 399–407. Springer, Cham (2017). https://doi.org/10.1007/978-3-319-66179-7_46
20. Zhang, Y., Yang, L., Chen, J., Fredericksen, M., Hughes, D.P., Chen, D.Z.: Deep Adversarial Networks for Biomedical Image Segmentation Utilizing Unannotated Images. In: Descoteaux, M., Maier-Hein, L., Franz, A., Jannin, P., Collins, D.L., Duchesne, S. (eds.) MICCAI 2017. LNCS, vol. 10435, pp. 408–416. Springer, Cham (2017). https://doi.org/10.1007/978-3-319-66179-7_47

Extreme Consistency: Overcoming Annotation Scarcity and Domain Shifts

Gaurav Fotedar(✉), Nima Tajbakhsh(✉), Shilpa Ananth, and Xiaowei Ding

VoxelCloud Inc., Los Angeles, USA
{gauravfo,ntajbakhsh,shilpa,xding}@voxelcloud.io

Abstract. Supervised learning has proved effective for medical image analysis. However, it can utilize only the small labeled portion of data; it fails to leverage the large amounts of unlabeled data that is often available in medical image datasets. Supervised models are further handicapped by domain shifts, when the labeled dataset fails to cover different protocols or ethnicities. In this paper, we introduce *extreme consistency*, which overcomes the above limitations, by maximally leveraging unlabeled data from the same or a different domain in a teacher-student semi-supervised paradigm. Extreme consistency is the process of sending an extreme transformation of a given image to the student network and then constraining its prediction to be consistent with the teacher network's prediction for the original image. The extreme nature of our consistency loss distinguishes our method from related works that yield suboptimal performance by exercising only mild prediction consistency. Our method is 1) auto-didactic, as it requires no extra expert annotations; 2) versatile, as it handles both domain shift and limited annotation problems; 3) generic, as it is readily applicable to classification, segmentation, and detection tasks; and 4) simple to implement, as it requires no adversarial training. We evaluate our method for the tasks of lesion and retinal vessel segmentation in skin and fundus images. Our experiments demonstrate a significant performance gain over both modern supervised networks and recent semi-supervised models. This performance is attributed to the strong regularization enforced by extreme consistency, which enables the student network to learn how to handle extreme variants of both labeled and unlabeled images. This enhances the network's ability to tackle the inevitable same- and cross-domain data variability during inference.

Keywords: Limited annotation · Domain shift · Semi-supervised learning

---

Electronic supplementary material The online version of this chapter (https://doi.org/10.1007/978-3-030-59710-8_68) contains supplementary material, which is available to authorized users.

A. L. Martel et al. (Eds.): MICCAI 2020, LNCS 12261, pp. 699–709, 2020.
https://doi.org/10.1007/978-3-030-59710-8_68

1 Introduction

Deep learning has proved effective in automating the diagnosis and quantification of various diseases and conditions. However, the vast majority of such models have limited clinical value, because they are trained and evaluated on labeled datasets that lack adequate annotation or show poor data diversity. The former problem is mainly caused by limited annotation budgets, while the latter, which causes the domain shift problem, arises when a dataset has poor coverage of ethnicities, imaging protocols, and scanner devices. In practice, securing a large, diverse labeled dataset is hardly feasible if not impossible; therefore, it is critical to devise learning strategies that increase model robustness against such dataset limitations through the effective utilization of unlabeled data.

To mitigate both limited annotation and domain shift problems, we propose a semi-supervised learning (SSL) method, called *Extreme Consistency*. The main idea behind extreme consistency is to regularize the model by forcing predictions for the original data and its extreme transformations to be consistent. This idea is fundamentally different from data augmentation (see Table 2(c)), because the latter can be applied only to labeled data and, further, lacks explicit consistency constraints. We realize our method through a teacher-student paradigm, where we feed an extreme variant of a given image to the student network and constrain its prediction to be consistent with the teacher's prediction for the original image. Importantly, extreme consistency relies on a diverse set of intensity-based, geometric, and image mixing transformations, which are blended based on a prior probability. If the combined transformation consists of only intensity manipulations, consistency loss reduces to an invariance constraint; that is, the predictions for the original and transformed image must be identical. However, inclusion of geometric and mixing augmentation will elevate this to an equivariance constraint; i.e., predictions must exhibit the same geometric transformations applied to the original images. Owing to the stochastic nature of diverse transformations, extreme consistency effectively spawns numerous invariance and equivariance constraints, providing strong regularization and thus higher-performing models. Critically, we exclusively apply diverse transformations to the inputs of the student network; thus, the teacher will receive "easy" examples that have only mild augmentation, which facilitates a healthy reference prediction for the consistency loss. Our experimental results on three public datasets demonstrate that the exclusive and diverse transformations make critical contributions to the success of our method, enabling a new level of performance that surpasses recent semi-supervised models [6,8,16,19], as well as supervised models based on U-Net and UNet++ [27].

Extreme consistency offers the following advantages: First, our method is auto-didactic, because extreme consistency uses the teacher's predictions on unlabeled data, requiring no additional expert annotations. Second, our method is versatile, handling both domain shift and limited annotation problems. This is because extreme consistency exposes the model to extreme variants of labeled and unlabeled data, which enhances the capability to cope with future same- and cross-domain data disparity. Third, our method is generic because extreme consistency can be applied to classification, segmentation, and detection tasks by subjecting logits, masks, and saliency maps to the extreme consistency loss, respectively. Note that saliency or center maps are used in recent object detectors such as CenterNet [9]. Fourth, our method is simple to implement, requiring neither adversarial training nor any changes in the architecture of the base model.

Our contributions include 1) a novel semi-supervised approach based on extreme consistency, 2) a cross-dataset image mixing transformation to expand the unlabeled data for stronger consistency regularization, 3) a comprehensive ablation study on the individual components of our method, 4) extensive evaluation using three public datasets for the problems of limited annotations and domain shifts, and 5) demonstrated gains over several semi-supervised baselines as well as supervised baselines trained only with labeled data.

2 Related Work

In this section, we briefly review the semi-supervised learning and unsupervised domain adaptation methods that are most relevant to our work, and refer the readers to [22] for a comprehensive review of these methodologies.

Early SSL methods were based on adversarial zero-sum games, which, at their equilibrium, generate similar feature embedding [1,20] or similar predictions [17,26] for labeled and unlabeled data. Recent SSL methods have, however, strived for simplicity and practicality. One of the popular recent themes is to train the model so it generates consistent results for a given image undergoing certain transformations. The consistency loss in [6,24] encourages invariance against only noise perturbations, while the consistency loss in [2,16] encourages equivariance against geometric transformations. Finally, the consistency loss in [19] is similar to [2,16], but benefits from a teacher-student paradigm. Our method is also based on consistency loss; however, it is not limited to certain geometric transformations (flip and rotation in [16]) or certain types of additive or multiplicative noise perturbations (Gaussian noise in [6,24]). This is because specific transformations may only be effective for a limited range of applications. For instance, consistency under perturbation may not be as effective in ultrasound images or optical coherence tomography scans that have inherent speckle noise or low signal-to-noise ratio. A concurrent work similar to ours is the arXiv submission [21], which considers only image classification and tackles the limited annotation problem in the context of natural images. By contrast, our paper tackles image segmentation in medical images, benefits from image mixing augmentation, and targets both domain shift and scarce annotation problems.

The domain adaptation methods fall into two major categories. The first category consists of methods that convert labeled source images such that they take the appearance and style of the target domain images [5,13], whereas the methods in the second category convert the target domain images into the source domain images [4,11]. Despite the differences, methodologies in both categories heavily rely on different variants of adversarial networks. By contrast, our approach is easy to train, requiring no sophisticated adversarial training.

3 Method

Our semi-supervised learning method follows the teacher-student paradigm. Specifically, our architecture consists of a student network N_s, whose weights are updated through backpropagation, and a teacher network N_t whose weights simply track the weights of the student network through an exponential moving average filter. During training, the student network has two learning objectives: 1) minimizing the target supervised loss for the labeled data, 2) minimizing an unsupervised consistency loss by generating predictions that are consistent with the reference predictions generated by the teacher network. The novel aspect of our method is the notion of extreme consistency, which maximally utilizes unlabeled data during training.

Extreme Consistency: Extreme consistency is the process of feeding an extreme variant of a given image to the student network and constraining its prediction to be consistent with the teacher's prediction for the original image. The extreme nature of our consistency loss distinguishes our method from similar works (e.g., [6,16,24]) that exercise prediction consistency under a limited set of transformations. We take the prediction consistency to extremes through a diverse set of transformations, which effectively spawns numerous invariance and equivariance constraints on the student network. Invariance constraints are imposed when the two networks receive variants of the same input that differ only in intensity values with no change in the spatial layout; otherwise, the applied transformation leads to an equivariance constraint; that is, the resulting predictions from the two networks must reflect the same geometric differences as do the original and transformed images. We hypothesize that extreme consistency provides stronger regularization, thereby enabling a higher-performing student network.

Diverse Transformations: We generate diverse variants of an input image through an extreme image transformation module T_e. For this purpose, we propose an image mixing augmentation, which, when combined with other intensity manipulations (e.g., posterization, equalization, etc.) and geometric transformations (rotation, scale, and translation), can effectively generate a diverse set of out-of-distribution images. As discussed in [7], these extremely distorted images can potentially improve the performance of an SSL method. We briefly explain our image mixing methods and refer the reader to our experiments section for the detailed description of the T_e module.

Cross-Dataset Image Mixing: Our image mixing technique aims to expand the unlabeled dataset by cutting a random region from a labeled image and pasting it into an unlabeled image. The prediction for a mixed image is compared for consistency against the prediction for the labeled image only in the pasted region, and against the prediction result for the unlabeled image in all other pixels. Our image mixing method is inspired by [10,25], but differs from [25], which is data augmentation for image classification, and from [10], which mixes only same-domain images.

Exclusive Transformations: Applying diverse transformations to the inputs of both student and teacher networks, however, either leads to model divergence during training or results in only a marginal improvement over a supervised baseline. We have experimentally determined that it is critical to use the diverse transformations exclusively for the inputs to the student network and apply only mild transformations to the inputs of the teacher network. Intuitively, this design makes sense because the success of the consistency loss, especially for unlabeled data, relies on a healthy reference prediction, which in our case is generated by the teacher network. It is therefore necessary to feed "easy" examples to the teacher while sending the "hard" examples to the student. Omitting this critical design choice may have shifted previous works towards consistency based on just a narrow set of transformations, leading to sub-optimal improvements.

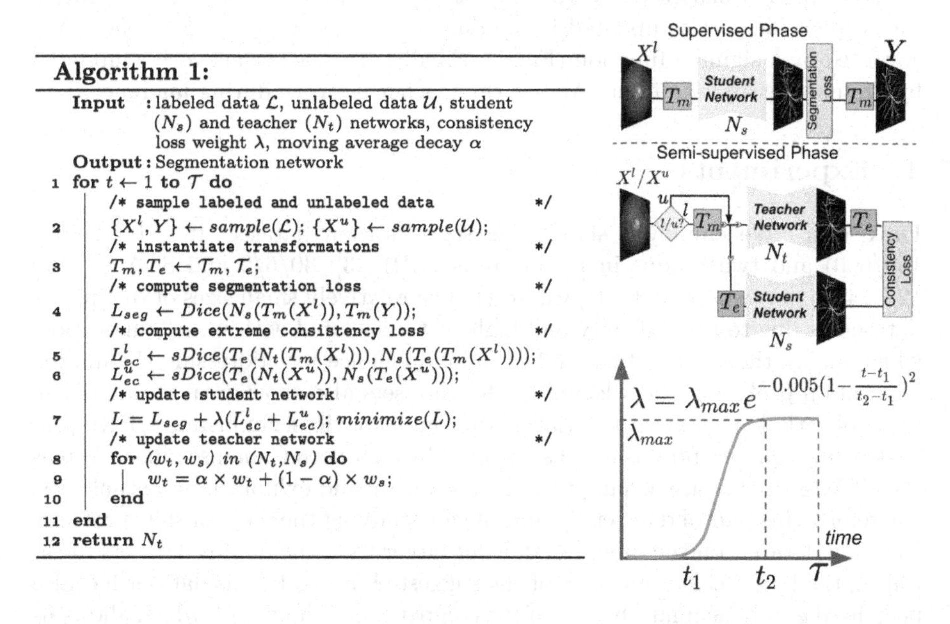

Fig. 1. (left) Algorithm 1 shows the pseudocode of our semi-supervised learning method based on extreme consistency. $sDice$ denotes soft dice, T_m denotes mild transformation, and T_e denotes extreme transformation. (top right) Schematic of the supervised phase (Lines 2–4) and semi-supervised phase (Lines 5–6). (bottom right) Sigmoid ramp-up function used for consistency loss weight (Line 7).

Pseudocode: Figure 1 presents the pseudocode of our SSL method. In each iteration, a batch of labeled $\{X^l, Y\}$ and unlabeled images $\{X^u\}$ are sampled. During the supervised phase, the labeled batch is first sent to T_m for mild augmentation, and then to the student network to compute the segmentation loss $L_{seg} = Dice(N_s(T_m(X^l)), T_m(Y))$. For unsupervised learning, we compute the extreme consistency losses L^l_{ec} and L^u_{ec} for both the labeled and unlabeled data. For the former, this is realized by first applying extreme transformations T_e to the batch of images before feeding it to the student network $N_s(T_e(T_m(X^l)))$. To generate the reference predictions for the consistency loss, the original labeled batch is sent to the teacher network and the equivalent extreme transformations $T_e(N_t(T_m(X^l)))$ are applied to its predictions. The extreme consistency loss L^l_{ec} is then computed by comparing the two outputs through a soft Dice function (both outputs are probabilistic). The extreme consistency loss for the unlabeled batch L^u_{ec} is computed similarly, with the difference being that no mild transformation T_m is applied to the unlabeled data. This choice is made because the student and teacher networks do not have the ground truth for unlabeled images during training; hence, applying mild transformations may degrade the quality of reference predictions for the unlabeled data. Once the three losses are computed, the student network is updated by minimizing $L_{seg} + \lambda(L^l_{ec} + L^{ul}_{ec})$, where λ is governed by a sigmoid function (Fig. 1). Finally, the teacher network is updated by the moving average filter. We use the teacher network during inference.

4 Experiments

Datasets: We employ a skin lesion segmentation dataset (ISIC'17: 2,000/150/600) and two fundus image datasets (HRF [3]: 30/6/9 and STARE [12]: 12/4/4) in our experiments (tr/val/test). The relatively small sizes of the fundus datasets is due to the difficulty and high cost associated with vessel annotation, which makes them a practical and meaningful choice for addressing the limited annotation problem. The selected fundus datasets also show great variability in terms of pathologies as well as variations in scanner devices. For instance, HRF consists of high-quality fundus images acquired by a modern Canon scanner whereas the STARE images are acquired by older scanners and exhibit more pathologies. Therefore, this pair of datasets is suitable for studying the domain shift problem. The skin dataset, on the other hand, is far larger than the fundus datasets, facilitating the large-scale evaluation of the suggested method. This dataset has also been used as a benchmark by one of the competing methods [16], which allows us to draw more direct comparisons with it.

Experiment Protocols: For the limited annotation problem, we use all three datasets by dividing their training sets into labeled and unlabeled subsets, resulting in $\mathcal{L}/\mathcal{U}$ splits of 50/1,950 for ISIC, 3/27 for HRF, and 2/10 for STARE. For the domain shift problem, we use HRF as the source and STARE as the target dataset; that is, $\mathcal{L}$ and $\mathcal{U}$ consist of the training splits of HRF and STARE, respectively. For realistic evaluation, we choose the best segmentation model based on the validation split of the *source* dataset, and then evaluate the best model on the test split of the *target* dataset. Similar data splits and evaluation strategies have been adopted in previous works.

Competing Baselines: For the limited annotation problem, we compare our method with two SSL methods, [6] and [16], which use consistency loss based on noise perturbations and affine transformations, respectively. The primary baselines for the domain shift problem are unsupervised domain adaptation methods suggested in [8] and [19]. The secondary baselines are the above SSL methods ([6] and [16]), which can be used for domain adaptation as well.

Implementation Details: For a fair comparison, as suggested in [18], we have implemented our method as well as all competing baselines in the same codebase. Thus, all methods benefit from the same data augmentation and further operate on the same segmentation network (exception being [19] requiring GroupNorm [23]). We have also optimized the weight function of the consistency loss for each method and dataset, by evaluating different configurations on the validation split. See our supplementary material for the best hyper-parameters.

Mild and Extreme Augmentations: Mild augmentation consists of a sequential application of four transformations: color jittering, channel shuffling, flipping and affine transformations, which each can be selected with a probability of 50%. Extreme augmentation further distorts mildly augmented images by applying at least 3 intensity and 1 geometric transformations followed by image mixing augmentation with a probability of 50% For the detailed list of augmentations and their parameters refer to the supplementary material.

Segmentation Architecture: We use a U-Net with 4 blocks in the encoder and decoder. Each block consists of 2 Conv-Bn-Relu followed by a 2×2 max-pooling layer in the encoder and an up-sampling layer based on bi-linear interpolation in the decoder. To avoid overfitting, we place a dropout layer with 50% drop rate right after each encoder block. We use the Dice loss during training. Being optimized on the validation set, this U-Net performs at par with the sophisticated, heavy-weight UNet++ [27] (Table 1). This observation is in line with similar findings in [14]. We thus use the U-Net for the remaining experiments.

Results for Limited Annotation: Table 1(a) summarizes the results. Each model is trained and evaluated in 10 trials. According to our statistical analyses, our method has significantly outperformed the best-performing SSL method for HRF and ISIC'17. Notably, our re-implementation of [16] achieves a $\sim$2-point increase in Dice over the lower bound U-Net for the ISIC'17 dataset, which is consistent with the original publication [16]. Our method has further outperformed the second-best method in STARE in both the mean and standard deviation.

Table 1. 10-trial comparison between our method and the competing baselines for (a) limited annotation and (b) the domain shift problem. In both tables, the goal is to bridge the performance gap between the lower bound (LB) and upper bound (UB) performance levels. To mimic limited annotation, we have divided the training split D_{tr} of each of the datasets into a small labeled subset $\mathcal{L}$ and a large subset $\mathcal{U}$ whose annotations are removed. As described in the text, the sizes of D_{tr}, $\mathcal{L}$, $\mathcal{U}$ differ from one dataset to another. For domain adaptation, we use HRF as the source and STARE as the target dataset. We use the training split of STARE after removing the labels (unsupervised), denoted as $\not{D}^S_{tr}$. Statistical significance between ours and the second best method is based on a two tailed T-test.

(a) Limited annotation

Method	Tr set	Dice: $\mu(\sigma)$ STARE	HRF	ISIC'17
LB (U-Net)	$\mathcal{L}$	0.806(0.003)	0.807(0.012)	0.775(0.005)
LB (UNet++)	$\mathcal{L}$	0.797(0.007)	0.812(0.003)	0.781(0.004)
Li et al. [16]	$\mathcal{L}$+$\mathcal{U}$	0.809(0.006)	0.811(0.006)	**0.797(0.005)**
Cui et al. [6]	$\mathcal{L}$+$\mathcal{U}$	**0.814(0.013)**	**0.813(0.005)**	0.774(0.016)
Proposed	$\mathcal{L}$+$\mathcal{U}$	**0.824(0.007)**	**0.828(0.002)**	**0.806(0.005)**
UB (U-Net)	D_{tr}	0.836(0.006)	0.838(0.001)	0.840(0.001)
UB (UNet++)	D_{tr}	0.827(0.005)	0.839(0.003)	0.832(0.002)
p-value		0.079	p <0.001	p <0.05

(b) Domain shift: **HRF →STARE**

Method	Tr set	Val set	Test Set	Dice: $\mu(\sigma)$
LB (U-Net)	D^H_{tr}	D^H_{val}	D^S_{test}	0.678(0.010)
LB (UNet++)	D^H_{tr}	D^H_{val}	D^S_{test}	0.655(0.014)
Li et al. [16]	$D^H_{tr}+\not{D}^S_{tr}$	D^H_{val}	D^S_{test}	0.712(0.005)
Cui et al. [6]	$D^H_{tr}+\not{D}^S_{tr}$	D^H_{val}	D^S_{test}	0.704(0.012)
Dong et al. [8]	$D^H_{tr}+\not{D}^S_{tr}$	D^H_{val}	D^S_{test}	0.668(0.017)
Perone et al. [19]	$D^H_{tr}+\not{D}^S_{tr}$	D^H_{val}	D^S_{test}	**0.723(0.012)**
Proposed	$D^H_{tr}+\not{D}^S_{tr}$	D^H_{val}	D^S_{test}	**0.736(0.004)**
UB (U-Net)	D^S_{tr}	D^S_{val}	D^S_{test}	0.833(0.005)
UB (UNet++)	D^S_{tr}	D^S_{val}	D^S_{test}	0.827(0.005)
p-value				p <0.05

Results for Domain Shift: The results are summarized in Table 1(b). Each model is trained and evaluated in 10 trials. Our method achieves a ~6-point increase in Dice over the lower bound performance (U-Net), on average ~3-point increase over the SSL methods, and a 1-point increase over the best performing domain adaptation method. All improvements are statistically significant. These results demonstrate the ability of our model to handle domain shifts.

Ablation Study: Table 2 summarizes the results of our ablation study conducted on the HRF dataset. **First**, we studied the necessity of diverse transformations by using one transformation type at a time, observing significant drops in Dice ($p < 0.05$) compared to using diverse transformations. We further compared diverse transformation with and without our image mixing method (the last two rows in Table 2(a)), observing a substantial decrease in average Dice without image mixing. It is also clear that image mixing significantly reduces the standard deviation of Dice, which translates to far more stable training sessions and reproducible models. **Second**, we removed the exclusion principle, applying the diverse transformations to the inputs of both student and teacher networks. This change led to a significant drop in Dice (3%), degrading the performance to that of the lower bound supervised model, motivating the need for the exclusive transformations. **Third**, we excluded unlabeled data, computing segmentation and consistency losses using only labeled data. This configuration led to 1.4% drop in Dice. **Fourth**, we studied the effect of replacing the teacher with the student network, which is known as the Π model [15], observing a 0.6% drop in

Dice. **Fifth**, we trained the lower bound U-Net with extreme data augmentation (T_e), observing no significant gains, which rules out the possibility that the gains by our SSL method are due to extensive data augmentation. These experiments underline the importance of each individual component of our method.

Table 2. 10-trial ablation study for HRF. We have split the results into three tables: (a) the necessity of diverse transformations and our image mixing method (last two rows); (b) impact of exclusive transformations (+Exclusive), use of unlabeled data (+$\mathcal{U}$), and the necessity of a mean teacher network (+Teacher); (c) whether a supervised model with extreme augmentation is as effective as our method.

(a) Diverse transformations

+Inten.	+Geo.	+Mixing	Dice: $\mu(\sigma)$	ΔDice(%)
✓	✗	✗	0.817(0.002)	-1.3%
✗	✓	✗	0.814(0.020)	-1.6%
✗	✗	✓	0.817(0.008)	-1.3%
✓	✓	✗	0.818(0.019)	-1.3%
✓	✓	✓	**0.828(0.002)**	0.0%

(b) Exclusive transformations + others

+Exclusive	+$\mathcal{U}$	+Teacher	Dice: $\mu(\sigma)$	ΔDice(%)
✗	✓	✓	0.803(0.005)	-3.0%
✓	✗	✓	0.816(0.002)	-1.4%
✓	✓	✗	0.823(0.004)	-0.6%
✓	✓	✓	**0.828(0.002)**	0.0%

(c) Supervised model with extreme augmentation

model	T_m	T_e	Dice: $\mu(\sigma)$
Sup. U-Net	✓		0.807(0.012)
Sup. U-Net	✓	✓	0.801(0.007)
Ours: SSL U-Net	✓	✓	0.828(0.002)

5 Conclusion

We presented Extreme Consistency as a simple yet effective semi-supervised learning framework. Although our method is intuitive and simple to implement, it demonstrates a substantial performance gain over several baselines across multiple datasets in handling both limited annotation and domain shift problems. Through extensive ablation studies, we further demonstrated that each component of our method contributes to its success. Our future work will explore applications of extreme consistency beyond image segmentation and the currently used datasets.

References

1. Baur, C., Albarqouni, S., Navab, N.: Semi-supervised deep learning for fully convolutional networks. In: Descoteaux, M., Maier-Hein, L., Franz, A., Jannin, P., Collins, D.L., Duchesne, S. (eds.) MICCAI 2017. LNCS, vol. 10435, pp. 311–319. Springer, Cham (2017). https://doi.org/10.1007/978-3-319-66179-7_36
2. Bortsova, G., Dubost, F., Hogeweg, L., Katramados, I., de Bruijne, M.: Semi-supervised medical image segmentation via learning consistency under transformations. In: Shen, D., et al. (eds.) MICCAI 2019. LNCS, vol. 11769, pp. 810–818. Springer, Cham (2019). https://doi.org/10.1007/978-3-030-32226-7_90
3. Budai, A., Bock, R., Maier, A., Hornegger, J., Michelson, G.: Robust vessel segmentation in fundus images. Int. J. Biomed. Imaging **2013** (2013)

4. Chen, C., Dou, Q., Chen, H., Heng, P.-A.: Semantic-aware generative adversarial nets for unsupervised domain adaptation in chest X-ray segmentation. In: Shi, Y., Suk, H.-I., Liu, M. (eds.) MLMI 2018. LNCS, vol. 11046, pp. 143–151. Springer, Cham (2018). https://doi.org/10.1007/978-3-030-00919-9_17
5. Chen, C., Dou, Q., Chen, H., Qin, J., Heng, P.A.: Synergistic image and feature adaptation: towards cross-modality domain adaptation for medical image segmentation. In: Proceedings of the AAAI Conference on Artificial Intelligence, vol. 33, pp. 865–872 (2019)
6. Cui, W., et al.: Semi-supervised brain lesion segmentation with an adapted mean teacher model. In: Chung, A.C.S., Gee, J.C., Yushkevich, P.A., Bao, S. (eds.) IPMI 2019. LNCS, vol. 11492, pp. 554–565. Springer, Cham (2019). https://doi.org/10.1007/978-3-030-20351-1_43
7. Dai, Z., Yang, Z., Yang, F., Cohen, W.W., Salakhutdinov, R.R.: Good semi-supervised learning that requires a bad GAN. In: Advances in neural information processing systems, pp. 6510–6520 (2017)
8. Dong, N., Kampffmeyer, M., Liang, X., Wang, Z., Dai, W., Xing, E.: Unsupervised domain adaptation for automatic estimation of cardiothoracic ratio. In: Frangi, A.F., Schnabel, J.A., Davatzikos, C., Alberola-López, C., Fichtinger, G. (eds.) MICCAI 2018. LNCS, vol. 11071, pp. 544–552. Springer, Cham (2018). https://doi.org/10.1007/978-3-030-00934-2_61
9. Duan, K., Bai, S., Xie, L., Qi, H., Huang, Q., Tian, Q.: Centernet: object detection with keypoint triplets. arXiv preprint arXiv:1904.08189 (2019)
10. French, G., Aila, T., Laine, S., Mackiewicz, M., Finlayson, G.: Consistency regularization and Cutmix for semi-supervised semantic segmentation. arXiv preprint arXiv:1906.01916 (2019)
11. Giger, M.L.: Whole brain segmentation and labeling from CT using synthetic MR images. J. Am. Coll. Radiol. **15**(3), 512–520 (2018). https://doi.org/10.1016/j.jacr.2017.12.028
12. Hoover, A., Kouznetsova, V., Goldbaum, M.: Locating blood vessels in retinal images by piecewise threshold probing of a matched filter response. IEEE Trans. Med. Imaging **19**(3), 203–210 (2000)
13. Huo, Y., et al.: Synseg-net: synthetic segmentation without target modality ground truth. IEEE Trans. Med. Imaging **38**(4), 1016–1025 (2018)
14. Isensee, F., Kickingereder, P., Wick, W., Bendszus, M., Maier-Hein, K.H.: No New-Net. In: Crimi, A., Bakas, S., Kuijf, H., Keyvan, F., Reyes, M., van Walsum, T. (eds.) BrainLes 2018. LNCS, vol. 11384, pp. 234–244. Springer, Cham (2019). https://doi.org/10.1007/978-3-030-11726-9_21
15. Laine, S., Aila, T.: Temporal ensembling for semi-supervised learning. arXiv preprint arXiv:1610.02242 (2016)
16. Li, X., Yu, L., Chen, H., Fu, C.W., Heng, P.A.: Transformation consistent self-ensembling model for semi-supervised medical image segmentation. arXiv preprint arXiv:1903.00348 (2019)
17. Mondal, A.K., Dolz, J., Desrosiers, C.: Few-shot 3D multi-modal medical image segmentation using generative adversarial learning. arXiv preprint arXiv:1810.12241 (2018)
18. Oliver, A., Odena, A., Raffel, C.A., Cubuk, E.D., Goodfellow, I.: Realistic evaluation of deep semi-supervised learning algorithms. In: Advances in Neural Information Processing Systems, pp. 3235–3246 (2018)
19. Perone, C.S., Ballester, P., Barros, R.C., Cohen-Adad, J.: Unsupervised domain adaptation for medical imaging segmentation with self-ensembling. NeuroImage **194**, 1–11 (2019)

20. Sedai, S., Mahapatra, D., Hewavitharanage, S., Maetschke, S., Garnavi, R.: Semi-supervised segmentation of optic cup in retinal fundus images using variational autoencoder. In: Descoteaux, M., Maier-Hein, L., Franz, A., Jannin, P., Collins, D.L., Duchesne, S. (eds.) MICCAI 2017. LNCS, vol. 10434, pp. 75–82. Springer, Cham (2017). https://doi.org/10.1007/978-3-319-66185-8_9
21. Sohn, K., et al.: Fixmatch: simplifying semi-supervised learning with consistency and confidence. arXiv preprint arXiv:2001.07685 (2020)
22. Tajbakhsh, N., Jeyaseelan, L., Li, Q., Chiang, J., Wu, Z., Ding, X.: Embracing imperfect datasets: a review of deep learning solutions for medical image segmentation. arXiv preprint arXiv:1908.10454 (2019)
23. Wu, Y., He, K.: Group normalization. In: Proceedings of the European Conference on Computer Vision (ECCV), pp. 3–19 (2018)
24. Yu, L., Wang, S., Li, X., Fu, C.-W., Heng, P.-A.: Uncertainty-aware self-ensembling model for semi-supervised 3D left atrium segmentation. In: Shen, D., et al. (eds.) MICCAI 2019. LNCS, vol. 11765, pp. 605–613. Springer, Cham (2019). https://doi.org/10.1007/978-3-030-32245-8_67
25. Yun, S., Han, D., Oh, S.J., Chun, S., Choe, J., Yoo, Y.: Cutmix: regularization strategy to train strong classifiers with localizable features. In: Proceedings of the IEEE International Conference on Computer Vision, pp. 6023–6032 (2019)
26. Zhang, Y., Yang, L., Chen, J., Fredericksen, M., Hughes, D.P., Chen, D.Z.: Deep adversarial networks for biomedical image segmentation utilizing unannotated images. In: Descoteaux, M., Maier-Hein, L., Franz, A., Jannin, P., Collins, D.L., Duchesne, S. (eds.) MICCAI 2017. LNCS, vol. 10435, pp. 408–416. Springer, Cham (2017). https://doi.org/10.1007/978-3-319-66179-7_47
27. Zhou, Z., Rahman Siddiquee, M.M., Tajbakhsh, N., Liang, J.: UNet++: a nested U-Net architecture for medical image segmentation. In: Stoyanov, D., et al. (eds.) DLMIA/ML-CDS -2018. LNCS, vol. 11045, pp. 3–11. Springer, Cham (2018). https://doi.org/10.1007/978-3-030-00889-5_1

Spatio-Temporal Consistency and Negative Label Transfer for 3D Freehand US Segmentation

Vanessa Gonzalez Duque[1(✉)], Dawood Al Chanti[1], Marion Crouzier[2], Antoine Nordez[2], Lilian Lacourpaille[2], and Diana Mateus[1]

[1] École Centrale de Nantes, Laboratoire des Sciences du Numérique de Nantes LS2N, UMR CNRS, 6004 Nantes, France
vanessa.gonzalezduque@ls2n.fr

[2] Université de Nantes, Laboratoire "Movement - Interactions - Performance", MIP, EA 4334, 44000 Nantes, France

Abstract. The manual segmentation of multiple organs in 3D ultrasound (US) sequences and volumes towards their quantitative analysis is very expensive and time-consuming. Fully supervised segmentation methods still require the collection of large volumes of annotated data while unlabeled images are abundant. In this work, we propose a novel semi-automatic deep learning approach modeled as a weak-label learning problem: given a few 2-D annotations for selected slices, the goal is to propagate the masks to the entire sequence. To this end, we make use of both positive and negative constraints induced by incomplete labels to penalize the segmentation loss function. Our model is composed of one encoder and two decoders to model the segmentation and an auxiliary reconstruction task. Moreover, we consider the spatio-temporal information by deploying a Convolutional Long Short Term Memory module. Our findings suggest that the reconstruction decoder and the Spatio-temporal information lead to a better geometrical estimation of the mask shape. We apply the model to the task of low-limb muscle segmentation in a dataset of 44 patients and 6160 images.

Keywords: 3-D ultrasound · Weakly supervised learning · Guided back-propagation · Convolutional LSTM · Fully convolutional neural networks.

This work has been supported in part by the European Regional Development. Fund, the Pays de la Loire region on the Connect Talent scheme (MILCOM Project) and Nantes Métropole (Convention 2017–10470).

Electronic supplementary material The online version of this chapter (https://doi.org/10.1007/978-3-030-59710-8_69) contains supplementary material, which is available to authorized users.

A. L. Martel et al. (Eds.): MICCAI 2020, LNCS 12261, pp. 710–720, 2020.
https://doi.org/10.1007/978-3-030-59710-8_69

1 Introduction

Duchenne Muscular Dystrophy (DMD) is a degenerative muscular disorder in which muscle fibers are replaced with fat. Treatment follow-up is commonly done through imaging of the lower limb muscles under Magnetic Resonance (MR) [16,20]. Due to the early onset of the disease, patients are often children, for whom MRI is impractical. 3-D US-imaging is rapidly evolving [8] offering an inexpensive and portable alternative, yet needs further clinical validation. An identified imaging bio-marker for DMD evolution is muscle volume [19]. Quantifying such information often requires the segmentation of the 3-D images. Towards the validation of 3-D ultrasound as a viable alternative for DMD follow-up, we propose an automatic segmentation algorithm for 3-D freehand ultrasound images.

Ultrasound image segmentation is strongly influenced by the quality of data. There are characteristic US properties [15], which make the segmentation task complicated such as attenuation, speckle, shadows, or missing boundaries due to the orientation dependence of the acquisition. These challenges exist not only for automatic segmentation algorithms but also for the clinical experts who annotate the images. In particular, the manual and automatic segmentation of muscles, in general, is recognized as a challenging task [3] given the high variability of shapes and relative positions between muscles and among individuals. Besides these anatomical variabilities, contrast and texture differences between individual muscles are hardly discernible. Thereby, for 3-D US or ultrafast acquisitions, it is impractical to label every image.

It is essential to develop advanced methods that handle the above challenges to make a more objective and accurate image assessment. Herein, we propose a deep learning-based segmentation approach that relies on the spatial coherence of an image sequence (or contiguous slices of a 3D volume) to better exploit incompletely annotated data. We design a network architecture based on the encoder-decoder topology, built using depthwise separable convolutions. Moreover, we rely on Spatio-temporal information to partially compensate for the missing sequential annotations and recover better boundaries.

Similar to [18], for every muscle to segment, we bring all the negative evidence available from other muscles' masks to constrain the prediction area. Moreover, we further propagate the information across the sequence by means of a Convolutional Long Short Term Memory (CLSTM)[25] placed in the bottleneck of an encoder-decoder architecture. The CLSTM captures the possible short and long range muscle deformations while preventing to propagate incorrect or noisy information via the gated mechanism learning. To improve the network convergence and to help over-fitting prevention, separable depth-wise convolutions [26] are favorable as they have fewer parameters. Finally, we enforce the encoding path to learn better US image representations via an auxiliary reconstruction decoder trained in an unsupervised manner. This auxiliary task allows preserving boundaries and the spatial structure, which in turn improves the pixel-level predictions.

In this study, we focus on the segmentation of the lower limb Gastrocnemius Medialis (GM) muscle. Our method is evaluated over a total of 44 participants and 6160 images to demonstrate our model performance for muscle segmentation. We performed an ablation study to evaluate the effectiveness of each novelty. We show that the proposed method produces high-quality segmentation results, with an average dice similarity coefficient of up to 94.5% under fully supervised training and up to 70.8% using only 50% of the annotations.

2 Related Work

The well-known U-net architecture [21] has been successfully extended in [2] to fuse features between the encoding and the decoding path in a non-linear way using LSTMs. Such connections have the advantage of enhancing feature propagation and encourage feature reuse. However, both [21] and [2] are limited by their inability to incorporate temporal information, that can facilitate the segmentation task with sequential or volumetric data. To exploit the dynamics, we integrated a CLSTM within the U-Net, similar to [1,4,23]. CLSTM can perceive the entire spatio-temporal context and provide more discriminative features. Our integration of CLSTM is done at the bottleneck of the encoding path, wherein two CLSTMs for the spatio-temporal encoding path are deployed and followed by another two CLSTMs for the temporal decoding path. Instead of using the CLSTM in a fully supervised way as in [1,4,23], we propose to exploit it for feature propagation when dealing with incomplete masks. Moreover, we integrate prior information, e.g the masks of other easier to segment organs provided by clinical experts, to guide the network back-propagation. In this way, an interactive semi-automatic segmentation can be leveraged.

Weak segmentation methods exist that reduce the cost of full pixel-wise image annotations. Kolesnikov *et al.* [12] proposed to expand segmentation seeds from weak localization cues to objects that match image boundaries. The segmentation results are, however, dependent on the combination of different loss terms. Dai *et al.* [6] investigated the use of bounding boxes obtained from region proposal networks [14] as a weaker form of annotations to train a Fully Convolutional Network (FCN) for segmentation. Despite the competitive performance demonstrated for object recognition, this method requires a huge amount of bounding boxes, up to 123 k. Medical data cannot afford such amount of annotation. Simple yet effective approaches [17,18] address the weak supervision by compensating the missing annotations for a specific organ with the incorporation of background labels. As missing annotations are usually considered as background pixel classes, prior knowledge of other known classes is leveraged to restrict the area of prediction for the missing class. In the absence of expert annotations for the muscle of interest on a given image, we also rely on available labels from easier-to-segment muscles to build a true-negative mask that constraints the segmentation loss. However, different to [17,18] we also consider the propagation of spatio-temporal features with the CLSTM to improve the segmentation of 3D or sequential ultrasound data.

To improve learning under the incomplete and weak annotations, we rely on an auxiliary unsupervised reconstruction task. Such multi-task learning [10,13]

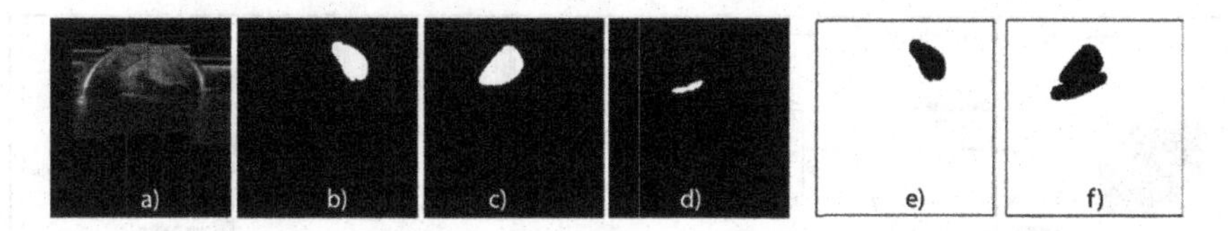

Fig. 1. a) 2-D US image; its manual segmentation masks (b):GM, c):GL, d):SOL); e) The background of annotated GM. f) Generated true negative background.

has been shown to improve the performance of semantic segmentation, among other tasks. Using the same encoding path, we explore an auxiliary and parallel unsupervised reconstruction decoder that pushes the network to recover the sequential US slices. Our experiments show that the reconstruction task helps to efficiently preserve the geometrical and appearance structure of the segmented mask, yielding better shape estimation. To handle multi-task learning, we propose a principled way of solving multiple loss functions to simultaneously learn multiple objectives instead of a naive weighted sum combination [7]. The novelty lies in updating the network parameters twice for each iteration Whenever the segmentation network parameters are stuck in a saddle point, the reconstruction task's optimizer has the opportunity to re-update the gradient to a better position. This is similar to a model first trained for reconstruction and then fine-tuned for segmentation, except training and fine-tuning happen concurrently.

3 Method

On the long term our aim is to assist volumetric and other quantitative measurements of muscles during the follow-up of DMD patients, under freehand ultrasound acquisitions. Towards this goal, we address here the problem of segmenting for healthy controls one of the three lower-limb muscles: the Gastrocnemius Medialis (GM) and Lateralis (GL), and the Soleus (SOL). Given the difficulties of manually annotating such difficult sequences, we present a FCN model and a training strategy that rely on incomplete 2-D annotations, where only some of the slices are annotated and not necessarily with all the muscle masks (see Fig. 1). To train a deep learning model under these constraints, we devise a training strategy capable of handling and propagating partial annotations while exploiting all the available information. In particular, the location of other muscles masks is advantageous for constraining the extent of the foreground prediction.

We propose a spatio-temporal multi-task approach performing two important and complementary tasks: 1) segmentation and 2) image reconstruction. The segmentation relies on a spatio-temporal U-Net with a CLSTM in the bottleneck ensuring the propagation of information across slices. Furthermore, two competing masks are considered: the foreground mask containing the muscle of interest, and the background mask filled with negative evidence from other anno-

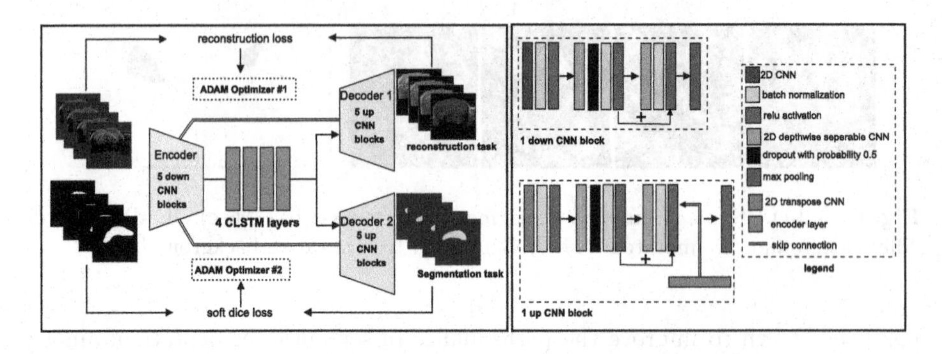

Fig. 2. Schematic representation of our network architecture.

tated organs. The purpose of the auxiliary reconstruction task is to compress and store the important spatio-temporal information into a compact representation.

Model Architecture. The core of the FCN model is a combination of two decoders sharing the same encoding path (see Fig. 2). The encoding path extracts compact low-resolution features with convolutional blocks. The feature maps from the last encoder layer are fed to a CLSTM module to capture the spatio-temporal transition within inter-slices and help to compensate missing annotations. The output of CLSTM is then passed to two decoders, the first focusing on reconstructing the original image, while the second on the segmentation task. The last layer of the reconstruction decoder is mapped into one channel while the segmentation decoder is mapped onto C maps, where C represents the number of classes. Afterward, the output feature maps C are passed to a pixel-wise softmax layer to generate the probabilities of the predicted masks.

Architecture Details: Our model is composed of an encoder with five residual depthwise-separable convolutional blocks and max-pooling operations, gradually projecting the gray-scale channel into 16, 32, 64, 128 and 256 feature maps at each layer respectively. The CLSTM at the bottleneck is composed of 4 stacked cells of 256 feature maps. The decoder mirrors this structure.

Loss Functions. The reconstruction objective function is the average Mean Square Error (st-MSE) between an input sequence $\mathbf{X} = \{\mathbf{x}_1, \ldots, \mathbf{x}_T\}$ of T frames, and the corresponding output reconstructions $\hat{\mathbf{X}} = \{\hat{\mathbf{x}}_1, \ldots, \hat{\mathbf{x}}_T\}$, which is achieved by the reconstruction network $f_\theta(.)$ parameterized by θ. Formally:

$$\text{st-MSE}(\mathbf{X}, \hat{\mathbf{X}}, \theta) = \sum_{t=1}^{T} (\mathbf{x}_t - f_\theta(\mathbf{x}_t)) \tag{1}$$

where $\mathbf{x}_t$ is the t-th 2-D slice, and $f_\theta(\mathbf{x}_t)$ denotes the reconstruction output.

The segmentation loss is the Soft Dice Coefficient (SDC) adapted to suit sequential data, we refer to as (st-SDC). Consider an input sequence $\mathbf{X}$ of T slices, where some slices are provided with expert annotations $\mathbf{y}_i^a$ and other are provided with prior negatives $\mathbf{y}_i^n$ generated from other muscles' masks. We introduce a binary variables δ_i to denote which masks to consider for each frame, such that for each image $\mathbf{x}_i$ we have $(\mathbf{y}_i^a, \mathbf{y}_i^n, \delta_i)$. The masks estimated by the segmentation network $g_\omega(.)$ parameterized by ω are $\hat{\mathbf{Y}} = \{g_\omega(\mathbf{x}_1), \ldots, g_\omega(\mathbf{x}_T)\}$, where $g_\omega(\mathbf{x}_t)$ is the output of the segmentation branch. Then the segmentation loss is formalized and conditioned as follow:

$$\text{dice}(\mathbf{y}_t, g_\omega(\mathbf{x}_t)) = \left(\frac{2\sum_{\text{pixels}} \mathbf{y}_t\, g_\omega(\mathbf{x}_t)}{\sum \mathbf{y}_t^2 + \sum g_\omega(\mathbf{x}_t)^2}\right) \tag{2}$$

$$\text{SDC}(\mathbf{y}_t, \hat{\mathbf{y}}_t, \omega) = \begin{cases} 1 - \text{dice}(\mathbf{y}_t^a, g_\omega(\mathbf{x}_t)), & \text{if } \delta_t = 1 \\ \text{dice}(\mathbf{y}_t^n, g_\omega(\mathbf{x}_t)) - 1, & \text{if } \delta_t = 0 \end{cases} \tag{3}$$

$$\text{st-SDC}(\mathbf{Y}, \hat{\mathbf{Y}}, \omega) = \frac{1}{T}\sum_{t=1}^{T} (\text{SDC}(\mathbf{y}_t, \hat{\mathbf{y}}_t, \omega)) \tag{4}$$

The loss is conditioned upon the type of available information at time t eq. (3).

Multi-task Learning. Multi-task learning is typically treated as a weighted sum of two criteria. However, as different tasks may conflict, finding suitable weighting hyperparameters is complex. Instead, we follow a multi-objective approach, similar to [22]. To this end, we optimize each objective function separately using two ADAM optimizers [11]. Since both networks $f_\theta(.)$ and $g_\omega(.)$ share the same encoder path, we alternatively update the networks parameters θ and ω. With the proposed multi-task training, the encoder learns to extract a compact representation that not only serves the segmentation task, but also the reconstruction. Conditioning the encoder to preserve the Spatio-temporal information required to reconstruct a sequence of images, favors a better geometrical and textural representations in the bottleneck. The improved representation leads then to better segmentation masks, especially in the boundaries.

4 Experimental Validation

Data Acquisition. A total of 59 acquisitions taken from 44 healthy volunteers aged between 18 and 45 years old are recorded [5]. Every acquisition consists of a sequence of 2-D B-mode US images (Fig. 3-a) acquired with a Supersonix Ultrasound machine and a 40 mm linear VERMON probe. Images are recorded every 5 mm in low speed mode. The probe is followed with an optical tracking system and then the 3-D volume is reconstructed (Fig. 3-c) using stradwin software [9]. The B-mode images are of size 227×544 with a pixel spacing of 0.176 mm/pixel. Volumes are recovered to fill a grid of $1372 \times 632 \times 2270$. The manual annotation of the GM, GL and SOL muscles is performed over 300 ± 96 muscles in B-mode US images (Fig. 3-b). In some cases, it was possible to segment GM and GL only. Hence, a second acquisition was acquired to segment the SOL independently. After volume reconstruction (Fig. 3-c), 3-D annotations (Fig. 3-d) are obtained through a surface fitting algorithm [24]. There was a difference of only 3% in the volumes computed from two expert's annotations, validating

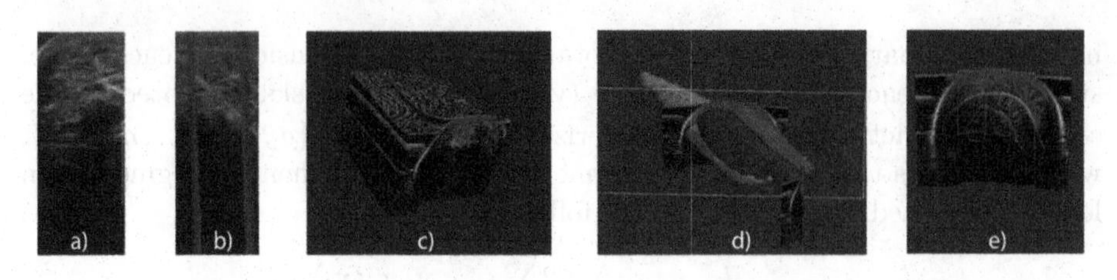

Fig. 3. a) B-mode image. b) manual annotation over B-mode. c) 3D volume. d) 3D segmented volumes. e) cross-section with GM,GL and SOL annotations

the repeatability of the annotation process. The masks used for training were extracted as transverse-sections of the surface models (Fig. 3-e) computed from the more expert examiner annotations.

Experimental Setup. For each participant, we extract low-resolution cross-sections (300×400 pixels) from the reconstructed volumes and select a sub-volume with 140 images. The data split consists of 29 train sequences ($29 \times 140 = 4060$ images) coming from participants in which the ground-truth mask of the GM, GL and SOL was provided over a single slice. For validation and test purposes, we used the data of other 5 and 10 (700 and 1400 images) participants with two sequences (coming from different acquisitions). During training we consider sequences of $T = 30$ successive 2D US slices sampled randomly from the 140-slice volumes. Batches consider 16 such sequences at a time.

In our experimental setting, we consider different ratios of annotations to train our model. The 100% annotation setting corresponds to the 140 images available for a sub-volume along with their ground truth for the muscle of interest (e.g. GM). However, having 30% of annotations means that only 42 masks (e.g. GM) out of 140 images are given to the network. Instead of ignoring the 98 images with un-annotated GM masks, we generated a true negative mask from the prior information in the SOL and GL annotations. This mimics the situation where two muscles (here SOL and GL) have been previously segmented but the current muscle of interest is GM.

In our experimental validation, we perform first an ablation study. We then present the results for different amounts of annotation ratios. Finally, we compare our model performance with the upper bound setting of full supervision with different % of annotation, but without prior information from other muscles. We use the validation set for hyper-parameter tuning, and report the performance on the test set. We use Dice (DSC), mean Intersection over Union (mIoU), and Hausdorff distance error (HDE) to quantify the accuracy of the predicted segmentation map, and present also qualitative results.

4.1 Ablation Study

To validate the contribution of the novel model components with respect to a plain U-Net, we perform a comparative study with full supervision (100% of

Table 1. Comparison of the baseline (U-Net) for segmenting SOL muscle verses our model and its different variants.

Ablation studies	DSC (%)	mIoU (%)
U-Net	78	76
U-Net-S	82	79
U-Net-S-R	87	85
U-Net-S-R-CLSTM + weighted average loss (ours)	**89**	**87**
U-Net-S-R-CLSTM + multi-objective loss (ours)	**91**	**89**

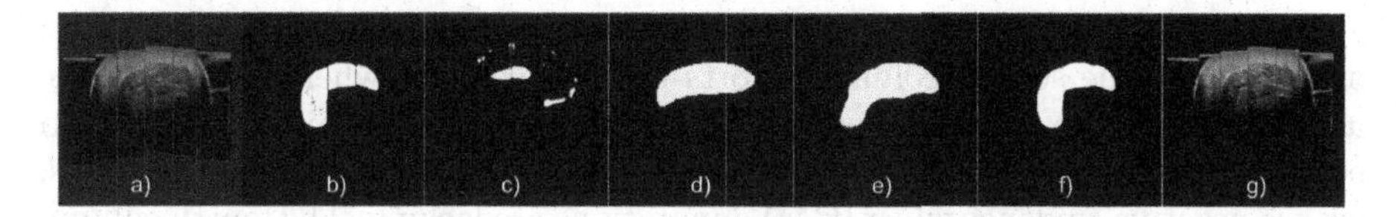

Fig. 4. Input: a) 2D US input slice at t_1, b) annotated SOL mask. **Prediction:** c) U-Net, d) U-Net with separable depthwise convolution, e) U-Net with separable depthwise convolution and reconstruction decoder, f) *Our Model.* g) Reconstructed slice.

the SOL masks for training are used). SOL was chosen being the hardest of the three muscles to segment. First, we replace the fully convolutional operators with separable depthwise convolutions, we refer to this model as *U-Net-S*. Then, we integrate the second decoder for reconstruction in the *U-Net-S-R* model. Finally, we evaluate our full model including the *CLSTM* module on top of the *U-Net-S-R* and we refer to as *U-Net-S-R-CLSTM*.

Table 1 demonstrates the effectiveness of using separable depth-wise convolutions when deploying a U-Net architecture over a moderate size dataset. Moreover, our finding suggests that the additional reconstruction decoder improves both the mIoU and the DSC measures. The CLSTM dramatically improves the performance as it exploits the full Spatio-temporal structure of the input data. Finally, Fig. 4 shows the visual differences of the predicted mask from different models. The geometrical structure is better preserved with our network due to the reconstruction decoder.

4.2 Amount of Annotations and Negative Priors

To cope with the difficulty of manual annotations, we evaluate the model's capacity to learn from fewer annotations and to exploit the available true negative knowledge from other muscle masks. DSC, mIoU, and HDE evaluation metrics for the GM muscle with prior knowledge are evaluated. In general, our model obtains reasonable DSC scores with fewer annotations.

To analyse the contribution of negative priors, we compare the above results to our model trained only with the ground truth GM masks, without the prior knowledge from other muscles. The results are presented in Table 2.

Table 2. Performance of the proposed model under different percentage of annotations using supervised and weakly-supervised settings.

Annotation percentage	Without negative prior	With negative prior			Performance gain w.r.t DSC score
%	DSC	DSC	mIoU	HDE	Ratio over 100
100	—	94.5	91	2.42	—
90	**91.4**	90.1	88	2.85	−1.44
70	**88.8**	83.4	76	4.20	−6.47
50	66.2	**70.8**	61	4.87	+6.49
30	43.1	**50.6**	42	6.50	+14.82

We evaluate the performance under different % of the relevant muscle (GM) annotations. The second column reports the performance when only using the available % of the relevant muscle (GM) annotations for training. The third column reports the results if, in addition to the available % of GM annotations, we use negative evidence when (GM) annotations are lacking. The fourth column presents the gain performance of the contribution with different % of negative evidence while lacking of positive evidence.

The contribution of prior negative knowledge is relevant when the % of un-annotated data is less than 50%, in which the gain performance increase gradually from 6.49% to 14.82%. Otherwise, the gain performance is −1.44% and −6.47% when 90% and 70% of annotations are available. When a sufficient amount of annotations are available to train a fully supervised model, the negative prior tends to degrade the generalization performance as it introduces bias to the system, as the network weight updates start focusing on the textural and geometrical patterns of the background instead of those of a certain muscle. On the other hand, the negative prior is valuable only when a small amount of labeled data is available, as it allows updating the loss with incomplete yet useful information.

A second way to transfer knowledge from one muscle to another is fine-tuning. We train our model over 100% annotations of GL muscle. We then fine-tune the model with 50% of the GM annotations. We obtain 68.8% of DSC, which is comparable but less interesting than the 70.8% DSC obtained with our model in the experiments (Table 2 with 50% annotations).

5 Discussion and Conclusions

In this paper, we proposed a deep learning approach to segment muscles in 3D freehand ultrasound data. Our model benefits from the spatio-temporal structure of the data at the feature level, as well from an auxiliary reconstruction task. For the latter, we also present a multi-objective training strategy that avoids the need of weighting competing losses. We explore different means to transfer prior knowledge from complementary masks and study the behavior of the different components under fewer annotations. Experimental results show that with a sufficient amount of supervision, the Spatio-temporal consistency enforced through the CLSTM, as well as the addition of a parallel reconstruction decoder are effective tools to improve the segmentation results. The use of complementary masks (negative priors) is the most useful when the amount of the annotated ground truth is relatively small (up to 1000 images in our case). Future work aims at improving the system to handle patient data. Duchenne US images from patients are harder because muscle is replaced by fat. We also contemplate validating the usability when using the proposed method as mask initialization to reduce the time of expert's segmentation. The proposed methodology may also be useful for the segmentation other anatomies requiring volume measurements and for other image analysis tasks dealing with sequential image data.

References

1. Arbelle, A., Raviv, T.R.: Microscopy cell segmentation via convolutional LSTM networks. In: 2019 IEEE 16th International Symposium on Biomedical Imaging (ISBI 2019), pp. 1008–1012. IEEE (2019)
2. Azad, R., Asadi-Aghbolaghi, M., Fathy, M., Escalera, S.: Bi-directional ConvLSTM U-net with Densley connected convolutions. In: Proceedings of the IEEE International Conference on Computer Vision Workshops (2019)
3. Cerrolaza, J.J., et al.: Deep learning with ultrasound physics for fetal skull segmentation. In: 2018 IEEE 15th International Symposium on Biomedical Imaging (ISBI 2018), pp. 564–567. IEEE (2018)
4. Chen, J., Yang, L., Zhang, Y., Alber, M., Chen, D.Z.: Combining fully convolutional and recurrent neural networks for 3D biomedical image segmentation. In: Advances in Neural Information Processing Systems, pp. 3036–3044 (2016)
5. Crouzier, M., Lacourpaille, L., Nordez, A., Tucker, K., Hug, F.: Neuromechanical coupling within the human triceps surae and its consequence on individual force-sharing strategies. J. Exp. Biol. **221**, 21 (2018)
6. Dai, J., He, K., Sun, J.: Boxsup: exploiting bounding boxes to supervise convolutional networks for semantic segmentation. In: Proceedings of the IEEE International Conference on Computer Vision, pp. 1635–1643 (2015)
7. Eigen, D., Fergus, R.: Predicting depth, surface normals and semantic labels with a common multi-scale convolutional architecture. In: Proceedings of the IEEE International Conference on Computer Vision, pp. 2650–2658 (2015)
8. Fenster, A., Downey, D.B., Cardinal, H.N.: Three-dimensional ultrasound imaging. Phys. Med. Biol. **46**(5), R67 (2001)
9. Gee, A., Prager, R., Treece, G., Cash, C., Berman, L.: Processing and visualizing three-dimensional ultrasound data. Br. J. Radiol. **suppl–77**(2), S186–S193 (2004)
10. Kendall, A., Gal, Y., Cipolla, R.: Multi-task learning using uncertainty to weigh losses for scene geometry and semantics. In: Proceedings of the IEEE Conference on Computer Vision and Pattern Recognition, pp. 7482–7491 (2018)
11. Kingma, D.P., Ba, J.: Adam: a method for stochastic optimization. In: International Conference on Learning Representations (2014)
12. Kolesnikov, A., Lampert, C.H.: Seed, expand and constrain: three principles for weakly-supervised image segmentation. In: Leibe, B., Matas, J., Sebe, N., Welling, M. (eds.) ECCV 2016. LNCS, vol. 9908, pp. 695–711. Springer, Cham (2016). https://doi.org/10.1007/978-3-319-46493-0_42
13. Kuga, R., Kanezaki, A., Samejima, M., Sugano, Y., Matsushita, Y.: Multi-task learning using multi-modal encoder-decoder networks with shared skip connections. In: Proceedings of the IEEE International Conference on Computer Vision Workshops, pp. 403–411 (2017)

14. Li, B., Yan, J., Wu, W., Zhu, Z., Hu, X.: High performance visual tracking with Siamese region proposal network. In: Proceedings of the IEEE Conference on Computer Vision and Pattern Recognition, pp. 8971–8980 (2018)
15. Liu, S., et al.: Deep learning in medical ultrasound analysis: a review. Engineering **5**(2), 261–275 (2019)
16. Loram, I.D., Maganaris, C.N., Lakie, M.: Use of ultrasound to make noninvasive in vivo measurement of continuous changes in human muscle contractile length. J. Appl. Physiol. **100**(4), 1311–1323 (2006)
17. Lu, Z., Fu, Z., Xiang, T., Han, P., Wang, L., Gao, X.: Learning from weak and noisy labels for semantic segmentation. IEEE Trans. Pattern Anal. Mach. Intell. **39**(3), 486–500 (2016)
18. Petit, O., Thome, N., Charnoz, A., Hostettler, A., Soler, L.: Handling missing annotations for semantic segmentation with deep ConvNets. In: Stoyanov, D., et al. (eds.) DLMIA/ML-CDS -2018. LNCS, vol. 11045, pp. 20–28. Springer, Cham (2018). https://doi.org/10.1007/978-3-030-00889-5_3
19. Pichiecchio, A., et al.: Muscle ultrasound elastography and MRI in preschool children with duchenne muscular dystrophy. Neuromuscul. Disord. **28**(6), 476–483 (2018)
20. Pillen, S., Arts, I.M., Zwarts, M.J.: Muscle ultrasound in neuromuscular disorders. Muscle Nerve: Official J. Am. Assoc. Electrodiagnostic Med. **37**(6), 679–693 (2008)
21. Ronneberger, O., Fischer, P., Brox, T.: U-Net: convolutional networks for biomedical image segmentation. In: Navab, N., Hornegger, J., Wells, W.M., Frangi, A.F. (eds.) MICCAI 2015. LNCS, vol. 9351, pp. 234–241. Springer, Cham (2015). https://doi.org/10.1007/978-3-319-24574-4_28
22. Sener, O., Koltun, V.: Multi-task learning as multi-objective optimization. In: Advances in Neural Information Processing Systems, pp. 527–538 (2018)
23. Stollenga, M.F., Byeon, W., Liwicki, M., Schmidhuber, J.: Parallel multi-dimensional LSTM, with application to fast biomedical volumetric image segmentation. In: Advances in Neural Information Processing Systems, pp. 2998–3006 (2015)
24. Treece, G.M., Prager, R.W., Gee, A.H., Berman, L.: Surface interpolation from sparse cross sections using region correspondence. IEEE Trans. Med. Imaging **19**(11), 1106–1114 (2000)
25. Xingjian, S., Chen, Z., Wang, H., Yeung, D.Y., Wong, W.K., Woo, W.c.: Convolutional LSTM network: a machine learning approach for precipitation nowcasting. In: Advances in Neural Information Processing Systems, pp. 802–810 (2015)
26. Zhang, X., Zhou, X., Lin, M., Sun, J.: Shufflenet: an extremely efficient convolutional neural network for mobile devices. In: Proceedings of the IEEE Conference on Computer Vision and Pattern Recognition, pp. 6848–6856 (2018)

Characterizing Label Errors: Confident Learning for Noisy-Labeled Image Segmentation

Minqing Zhang[1], Jiantao Gao[1,3], Zhen Lyu[2], Weibing Zhao[1], Qin Wang[1], Weizhen Ding[1], Sheng Wang[4], Zhen Li[1(✉)], and Shuguang Cui[1]

[1] Shenzhen Research Institute of Big Data, The Chinese University of Hong Kong, Shenzhen, Guangdong, China
lizhen@cuhk.edu.cn
[2] Warshel Institute for Computational Biology, The Chinese University of Hong Kong, Shenzhen, Guangdong, China
[3] School of Mechatronic Engineering and Automation, Shanghai University, Shanghai, China
[4] Tencent AI Lab, Bellevue, USA

Abstract. Convolutional neural networks (CNNs) have achieved remarkable performance in image processing for its mighty capability to fit huge amount of data. However, if the training data are corrupted by noisy labels, the resulting performance might be deteriorated. In the domain of medical image analysis, this dilemma becomes extremely severe. This is because the medical image annotation always requires medical expertise and clinical experience, which would inevitably introduce subjectivity. In this paper, we design a novel algorithm based on the teacher-student architecture for noisy-labeled medical image segmentation. Creatively, We introduce confident learning (CL) method to identify the corrupted labels and endow CNN an anti-interference ability to the noises. Specifically, the CL technique is introduced to the teacher model to characterize the suspected wrong-labeled pixels. Since the noise identification maps are a little away from sufficient precision, the spatial label smoothing regularization technique is utilized to generate soft-corrected masks for training the student model. Since our method identifies and revises the noisy labels of the training data in a pixel-level rather than simply assigns lower weights to the noisy masks, it outperforms the state-of-the-art method in the noisy-labeled image segmentation task on the JSRT dataset, especially when the training data are severely corrupted by noises.

1 Introduction

Convolutional neural network (CNN)-based methods currently dominate various computer vision tasks, such as face recognition [9] and object detection [10], attesting their capabilities to tackle challenging problems. In the domain of medical image analysis, CNN-based methods also achieved great successes

A. L. Martel et al. (Eds.): MICCAI 2020, LNCS 12261, pp. 721–730, 2020.
https://doi.org/10.1007/978-3-030-59710-8_70

in numerous clinical practice, including lung nodule detection [6] and pediatric bone age assessment [15]. These applications brought an intelligent and efficient diagnosis process to reality and the success is usually based on well-performed models.

To obtain a well-trained CNN model, clear-labeled data are crucial. If the labels in the training set are corrupted by noises, the resulting performance might become moderate. To avoid these potential noises, a third party is usually consulted to provide a review for the preliminary annotation and uncover the disputed-labeled data. This review was found to be tedious and inefficient. Still, noise-free labels could not be guaranteed. Besides, unlike the natural image labeling where noises mainly come from random errors, noises of medical image labeling have broad sources. These sources can be separated into two parts. First, high-resolution medical images often require pixel-level manual annotations for segmentation tasks. This onerous labeling process would unavoidably bring in random noises such as missed, wrong or inaccurate annotations. Second, to accelerate the labeling process, medical experts usually cooperate. This cooperation would induce individual subjectivity based on different clinical experience and personal opinions. This subjectivity would also introduce noises.

CNN models are struggling with the noises, meanwhile it is usually time-consuming to obtain the noise-free labels. Researchers have proposed various methods to alleviate the negative effects brought by the noises so that the model performance will not be severely deteriorated even with their presence. Ren et al. [11] uncovered the noisy labels by the gradient and assigned lower weights to the samples with these noisy labels. Goldberger et al. [3] designed an adaptation layer to model the process where the latent true labels were corrupted into the noisy ones. Jiang et al. introduced MentorNet [7] to discover the 'correct-ish' samples and concentrate more on them. For the medical image applications, Xue et al. [14] designed an online uncertainty sample mining method and a re-weighting strategy to eliminate the disturbance from the noisy labels. These researches provide practical solutions to the noisy label problem. However, most studies focus on the classification task because it is the most basic problem in natural image processing domain. However, the medical image domain also requires segmentation tasks to provide the radiologists with pixel-level analysis results. Yet, this task is not comprehensively studied considering the existing difficulties. In the literature, Zhu et al. [16] proposed an end-to-end trainable architecture to automatically evaluate the label quality and, in an image-level, assign lower weights to the noisy labels. This approach did improve the performance of the model trained with noisy labels. Nevertheless, the corrupted areas of each label mask remain unknown. Besides, the image-level re-weighting strategy targets the whole image, while the noises might only corrupt the pixels in a local region. Consequently, the ability of noise identification in this method might be limited.

In this paper, a novel method is proposed to endow the model interpretability and anti-interference ability to the noises existing in the medical image segmentation tasks. Specifically, our algorithm is based on the teacher-student architecture [5]. First, we creatively applied confident learning (CL) technique [8] to the teacher model to characterize the label noises from the training data. As the pixel-level noise identification maps could only roughly distinguish the corrupted area, the spatial label smoothing regularization (SLSR) technique [1] is utilized as a soft label correction method to train the student model.

To summarize, our research has a dual contribution. First, CL is applied to the segmentation task. The noises identified by CL can help visualize the specific area corrupted by the noisy labels. Second, a soft pixel-level label correction module based on SLSR is designed to train the student model. The algorithm was evaluated on the public JSRT dataset [13]. The result outperforms the *state-of-the-art* method in the noisy-labeled image segmentation tasks. Such anti-interference ability becomes more obvious as the noises increase.

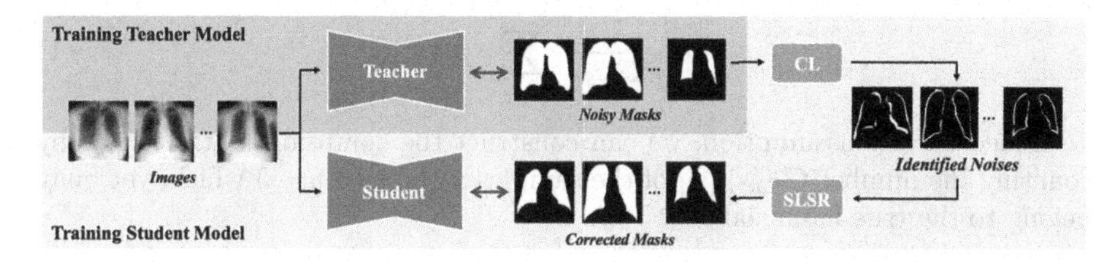

Fig. 1. The pipeline of the teacher-student architecture method. CL module identifies the label noises based on the teacher model. SLSR revises the noises. The student model trained with the soft-corrected masks provides the segmentation results.

2 Method

Figure 1 illustrates the pipeline of our method. The method consists of three parts: a teacher-student architecture containing two segmentation models, CL and SLSR. First, a teacher model is trained with the noisy-labeled data. Then, the CL module will identify the label noises at a pixel-level from the training set. Afterwards, SLSR will correct the noisy labels smoothly. Finally, a student model can be further supervised by the soft-corrected masks.

2.1 Segmentation Network

We adopt a U-Net [12] network with residual blocks [4] as the segmentation network for our framework. The modified U-Net consists of a top-down contracting path and a bottom-up expansive path. The contracting path follows the repeated application of downsampling blocks and residual blocks, while the expansive path consists of repeated upsampling blocks and residual blocks. The downsampling block is a 2×2 convolution with stride 2 and doubles the output channels, and the upsampling block is a 2×2 transposed convolution with stride 2 and reduces the output channels by half. The residual blocks consist of multiple 3×3 convolutions with identity shortcut connections [4]. Each convolution layer in the network is followed by batch normalization and rectified linear unit.

2.2 The Confident Learning Module

Based on the assumption of Angluin [2], CL [8] can identify the label errors in the datasets and improve the training with noisy labels by estimating the joint distribution between the noisy (observed) labels $\tilde{y}$ and the true (latent) labels y^*. Remarkably, no hyper-parameters and few extra computations are required.

Specifically, given a training set $\mathbf{X} = (\mathbf{x}, \tilde{y})^n$ containing n samples $\mathbf{x}$ with noisy label $\tilde{y}$, the predicted probabilities $\hat{p}$ of m classes can be obtained through the teacher model. Assuming that a sample $\mathbf{x}$ labeled $\tilde{y}= i$ has large enough predicted probabilities $\hat{p}_j(\mathbf{x}) \geq t_j$, it can be suspected that the annotation is wrong and $\mathbf{x}$ might belong to the true latent label $y^* = j$. Here, we select the threshold t_j as the average predicted probabilities $\hat{p}_j(\mathbf{x})$ of all samples labeled $\tilde{y}= j$:

$$t_j := \frac{1}{|\mathbf{X}_{\tilde{y}=j}|} \sum_{\mathbf{x} \in \mathbf{X}_{\tilde{y}=j}} \hat{p}_j(\mathbf{x}). \tag{1}$$

Based on this assumption, we can construct the confusion matrix $\mathbf{C}_{\tilde{y},y^*}$ by counting the number $\mathbf{C}_{\tilde{y},y^*}[i][j]$ of the samples $\mathbf{x}$ (labeled $\tilde{y}= i$) which, yet, may belong to the true latent label $y^* = j$:

$$\begin{aligned} \mathbf{C}_{\tilde{y},y^*}[i][j] &:= \left|\hat{\mathbf{X}}_{\tilde{y}=i,y^*=j}\right| \quad , where \\ \hat{\mathbf{X}}_{\tilde{y}=i,y^*=j} &:= \{\mathbf{x} \in \mathbf{X}_{\tilde{y}=i} : \hat{p}_j(\mathbf{x}) \geq t_j,\ j = \underset{k \in M: \hat{p}_k(\mathbf{x}) \geq t_k}{\arg\min} \ \hat{p}_k(\mathbf{x})\}. \end{aligned} \tag{2}$$

The required confusion matrix $\mathbf{C}_{\tilde{y},y^*}$ is then normalized, and the joint distribution $\mathbf{Q}_{\tilde{y},y^*}$ between the noisy labels and the true labels could be obtained:

$$\mathbf{Q}_{\tilde{y},y^*}[i][j] = \frac{\frac{\mathbf{C}_{\tilde{y},y^*}[i][j]}{\sum_{b=1}^{m} \mathbf{C}_{\tilde{y},y^*}[i][b]} \cdot |\mathbf{X}_{\tilde{y}=i}|}{\sum_{a,b=1}^{m} (\frac{\mathbf{C}_{\tilde{y},y^*}[a][b]}{\sum_{b=1}^{m} \mathbf{C}_{\tilde{y},y^*}[a][b]} \cdot |\mathbf{X}_{\tilde{y}=a}|)}. \tag{3}$$

Finally following the confusion matrix $\mathbf{C}_{\tilde{y},y^*}$ or the joint distribution $\mathbf{Q}_{\tilde{y},y^*}$, the wrong-labeled sample set $\tilde{\mathbf{X}}$ can be discovered by the following four options:

Option 1: Confusion Matrix $\mathbf{C}_{\tilde{y},y^*}$. The wrong-labeled samples are selected from the off-diagonals of confusion matrix $\mathbf{C}_{\tilde{y},y^*}$.

Option 2: Joint Distribution $\mathbf{Q}_{\tilde{y},y^*}$. For each class $i \in 1, 2, \ldots m$, the $n \cdot \sum_{j=1, j\neq i}^{m} (\mathbf{Q}_{\tilde{y},y^*}[i][j])$ samples with the lowest self-confidence $\hat{p}_i(\mathbf{x} \in \mathbf{X}_{\tilde{y}=i})$ are selected as the wrong-labeled samples.

Option 3: $\mathbf{C}_{\tilde{y},y^*} \bigcap \mathbf{Q}_{\tilde{y},y^*}$. The element-wise set intersection of the option 1 and 2.

Option 4: $\mathbf{C}_{\tilde{y},y^*} \bigcup \mathbf{Q}_{\tilde{y},y^*}$. The element-wise set union of the option 1 and 2.

2.3 Spatial Label Smoothing Regularization

After identifying the wrong-labeled sample set $\tilde{\mathbf{X}}$ from the training set, traditional methods based on the CL technique clean the data by pruning. However, the samples of the medical images often undergo the onerous labeling process. Therefore, in our binary segmentation task, rather than directly removing the noisy-labeled data, the SLSR technique is introduced to correct the noisy labels $\tilde{y}$. Specifically, based on the wrong-labeled pixels $\tilde{\mathbf{X}}$ identified by the CL module, SLSR revises the corresponding noisy labels $\tilde{y}$. The corrected labels $\dot{y}$ of noisy samples are defined as:

$$\dot{y}(\mathbf{x}) = \tilde{y}(\mathbf{x}) + \mathbb{1}(\mathbf{x} \in \tilde{\mathbf{X}}) \cdot (-1)^{\tilde{y}} \cdot \epsilon, \tag{4}$$

where $\epsilon \in [0, 1]$ is the hyper-parameter. If $\epsilon = 0$, there will be no modification to the noisy labels, while $\epsilon = 1$ indicates that the noisy labels $\tilde{y}$ will be revised by the output of the CL module directly. Since CL may have uncertainties, ϵ is chosen as 0.8 empirically in our experiments. Based on the training data with the soft-corrected labels, the segmentation network can be further trained by the cross-entropy loss:

$$\mathbf{L} = \sum_{\mathbf{x} \in \mathbf{X}} \log(\hat{p}(\mathbf{x})) \cdot \dot{y}(\mathbf{x}) = \sum_{\mathbf{x} \in \mathbf{X}} \log(\hat{p}(\mathbf{x})) \cdot (\tilde{y}(\mathbf{x}) + \mathbb{1}(\mathbf{x} \in \tilde{\mathbf{X}}) \cdot (-1)^{\tilde{y}} \cdot \epsilon). \tag{5}$$

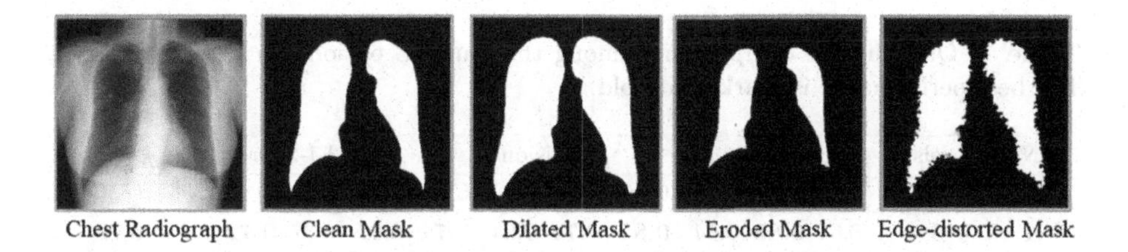

Fig. 2. An example radiograph with clean mask and three types of corrupted masks.

3 Experiments and Results

3.1 Dataset and Noisy Labels Synthesis

Our method was evaluated on the JSRT [13] dataset containing 247 chest radiographs. Each radiograph has a $2,048 \times 2,048$ resolution and was annotated with three anatomical structures: clavicle, heart, and lung. The data were resized into 256×256 pixels and randomly split into training (197) and validation (50) set.

To better simulate the label noises brought by the manual annotating process, a variety of digital image processing techniques, including dilating, eroding, and edge-distorting, were applied to corrupt the original (clean) masks, which is shown in Fig. 2. To comprehensively investigate our algorithm, we generate several noisy-labeled training sets. For each training set, the proportion and extent of the synthesized noisy-labeled data were various, with their emergence being controlled by two variables, α and β. Specifically, given a noisy-labeled training set with α and β, the data were corrupted by α proportion, with each mask being dilated, eroded, or edge-distorted by $\beta \pm 3$ pixels. It is noted that edge-distorted was implemented by randomly dilating or eroding the pixels on the boundary with a circle ($radius = \lceil \beta/5 \pm 2 \rceil$ pixel). To generate training sets with several noisy levels, we set α as 0.3, 0.5, and 0.7 respectively, representing three different proportions of noisy-labeled data. Simultaneously, two sets of β, termed A and B (A: $\beta_{clavicle} = 5, \beta_{heart} = 10, \beta_{lung} = 15$; B: $\beta_{clavicle} = 10, \beta_{heart} = 15, \beta_{lung} = 20$), were used to distinguish the size difference for each anatomical structure in the following experiments.

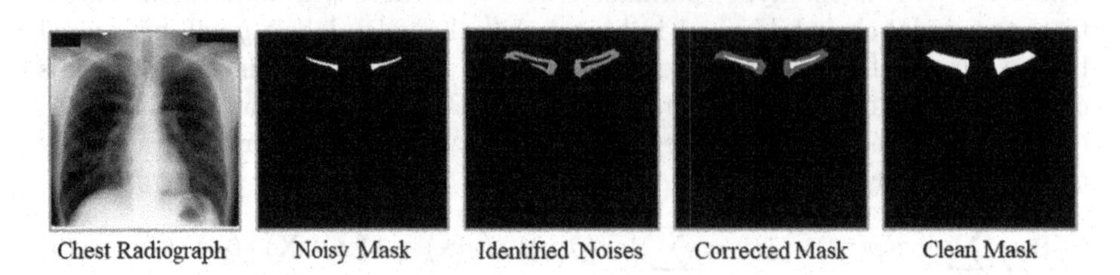

Fig. 3. The noises in the clavicle mask identified by the CL module.

Table 1. Quantitative comparisons among the four CL options on the training set. The best performance is marked in bold.

Noise levels		Recall				Precision				F1-Score			
		C	Q	C∩Q	C∪Q	C	Q	C∩Q	C∪Q	C	Q	C∩Q	C∪Q
$\alpha = 0.3$	$\beta = A$	0.64	**0.83**	0.64	**0.83**	**0.71**	0.64	**0.71**	0.62	0.67	**0.72**	0.67	0.71
	$\beta = B$	0.69	**0.77**	0.68	**0.77**	0.76	0.74	**0.77**	0.73	0.72	**0.75**	0.72	**0.75**
$\alpha = 0.5$	$\beta = A$	0.55	0.70	0.52	**0.73**	0.69	**0.74**	**0.74**	0.70	0.60	**0.72**	0.59	0.71
	$\beta = B$	0.53	**0.64**	0.53	**0.64**	0.67	**0.72**	0.67	0.71	0.59	**0.68**	0.59	0.67
$\alpha = 0.7$	$\beta = A$	0.49	**0.66**	0.49	**0.66**	0.67	**0.70**	0.67	**0.70**	0.56	**0.68**	0.56	0.67
	$\beta = B$	0.52	**0.57**	0.52	**0.57**	0.68	**0.69**	0.68	0.66	0.59	**0.62**	0.59	0.61

3.2 Experimental Settings

The experiments were implemented in PyTorch. Horizontal flipping was applied for data augmentation. All models were trained for 1,000 epochs with Adam optimizer (learning rate = 0.001 and decayed by 0.95 every 50 epochs). The same data split and training settings were applied to all the following experiments.

3.3 Identifying Error from Noisy Labels

To comprehensively investigate the CL module in our algorithm, we test all four CL implementing options on the datasets corrupted by different noise levels. Each option was evaluated by ***Recall***, ***Precision***, and ***F1-Score*** to estimate their abilities to identify the noise. The quantitative comparisons of these options are shown in Table 1. It can be observed that compared with $\mathbf{C}_{\tilde{y},y^*}$, $\mathbf{Q}_{\tilde{y},y^*}$ achieved higher resulting ***Recall*** and ***Precision*** over almost all noise levels. This means that the option with normalized joint distribution, $\mathbf{Q}_{\tilde{y},y^*}$ can characterize the noises more thoroughly with fewer false positives. By comparison, the noise identification result of $\mathbf{C} \cap \mathbf{Q}$ excluded some true positive noises from $\mathbf{Q}_{\tilde{y},y^*}$, and $\mathbf{C} \cup \mathbf{Q}$ introduced some false positive noises from $\mathbf{C}_{\tilde{y},y^*}$. Therefore, in Sect. 3.4, we chose $\mathbf{Q}_{\tilde{y},y^*}$ to implement the CL module in the experiments for its superior ***F1-Score*** under all noise settings. Figure 3 exhibits the efficacy of noise identification from the CL module.

Table 2. Quantitative comparisons of our methods and the state-of-the-art method pick-and-learn [16] under all noise settings. The best performance is marked in bold.

Noise levels		Methods	Clavicle	Heart	Lung	Average
No noise		Baseline(CE)	0.910	0.939	0.977	0.942
		Baseline + CL	0.920	0.945	0.978	0.948
		Baseline + CL + SLSR	**0.935**	**0.949**	**0.979**	**0.954**
		Pick-and-Learn [16]	**0.935**	0.945	0.978	0.953
$\alpha = 0.3$	$\beta = A$	Baseline(CE)	0.875	0.908	0.948	0.910
		Baseline + CL	0.880	0.925	0.957	0.921
		Baseline + CL + SLSR	0.893	**0.938**	**0.972**	**0.934**
		Pick-and-Learn [16]	**0.894**	0.931	**0.972**	0.932
	$\beta = B$	Baseline(CE)	0.857	0.892	0.938	0.896
		Baseline + CL	0.860	0.921	0.948	0.910
		Baseline + CL + SLSR	**0.890**	**0.932**	**0.968**	**0.930**
		Pick-and-Learn [16]	0.876	0.925	0.967	0.923
$\alpha = 0.5$	$\beta = A$	Baseline(CE)	0.806	0.891	0.898	0.865
		Baseline + CL	0.825	0.908	0.928	0.887
		Baseline + CL + SLSR	**0.856**	**0.927**	**0.966**	**0.916**
		Pick-and-Learn [16]	0.844	0.925	0.965	0.911
	$\beta = B$	Baseline(CE)	0.718	0.861	0.868	0.816
		Baseline + CL	0.730	0.878	0.903	0.837
		Baseline + CL + SLSR	**0.786**	**0.913**	**0.958**	**0.886**
		Pick-and-Learn [16]	0.764	0.911	0.956	0.877
$\alpha = 0.7$	$\beta = A$	Baseline(CE)	0.762	0.826	0.806	0.798
		Baseline + CL	0.774	0.849	0.895	0.839
		Baseline + CL + SLSR	**0.812**	**0.903**	**0.957**	**0.891**
		Pick-and-Learn [16]	0.769	0.896	0.948	0.871
	$\beta = B$	Baseline(CE)	0.614	0.785	0.718	0.706
		Baseline + CL	0.641	0.805	0.831	0.759
		Baseline + CL + SLSR	**0.745**	**0.885**	**0.948**	**0.859**
		Pick-and-Learn [16]	0.630	0.874	0.940	0.815

3.4 Robust Training with Noisy-Labeled Data

In this section, we conducted experiments to perform ablation analysis of our method and compared it with others. All methods were trained on the training set with different noise levels and tested on the clean validation set. In the experiments, ***Dice*** coefficient is introduced to quantify the matching score between predicted and clean masks. Table 2 illustrates the experimental results of the baseline U-Net model without teacher-student architecture trained with the cross-entropy loss, our method (with or without CL module and SLSR), and the leading **Pick-and-Learn** method in noisy-label alleviation. Although the baseline method achieves high ***Dice*** in all three segmentation tasks without noise, as the noisy levels increase, the segmentation performance decreases sharply, especially for a small anatomical structure like clavicle. By contrast, with the aid of the CL module and SLSR, our method can retain high ***Dice*** even with the interference of strong noises. Surprisingly, our method also outperforms the baseline method on the"original" dataset. To explain this, we visualized the label noise identification maps generated by the CL module. As shown in Fig. 4, it can be observed that the "original" dataset is not completely "clean" due to human errors. In our method, these errors can be corrected and thus elevating the ***Dice***. On the whole, our method can help fix the label errors existing in the dataset and thus significantly reduce the labeling requirements. Finally, our method was also compared with the leading **Pick-and-Learn** method in noisy-label alleviation. As our method could ultimately correct the noisy labels rather than simply reducing their weights, it can achieve higher ***Dice*** compared with **Pick-and-Learn**. The qualitative comparisons of the validation set of each method with intensive noises ($\alpha = 0.7, \beta = B$) are shown in Fig. 5.

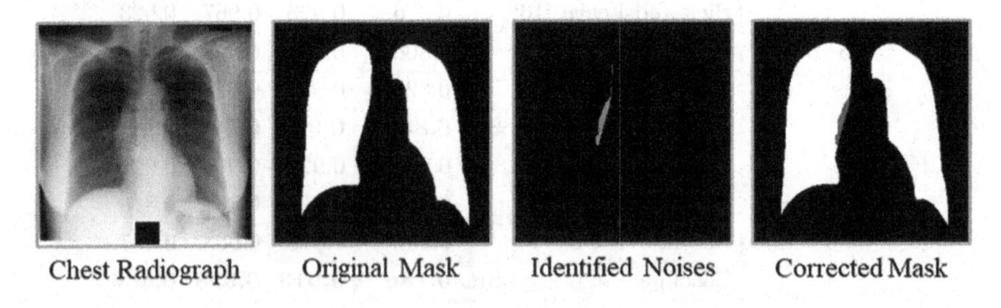

Fig. 4. An example of label error existing in the JSRT dataset. The error is identified by the CL module.

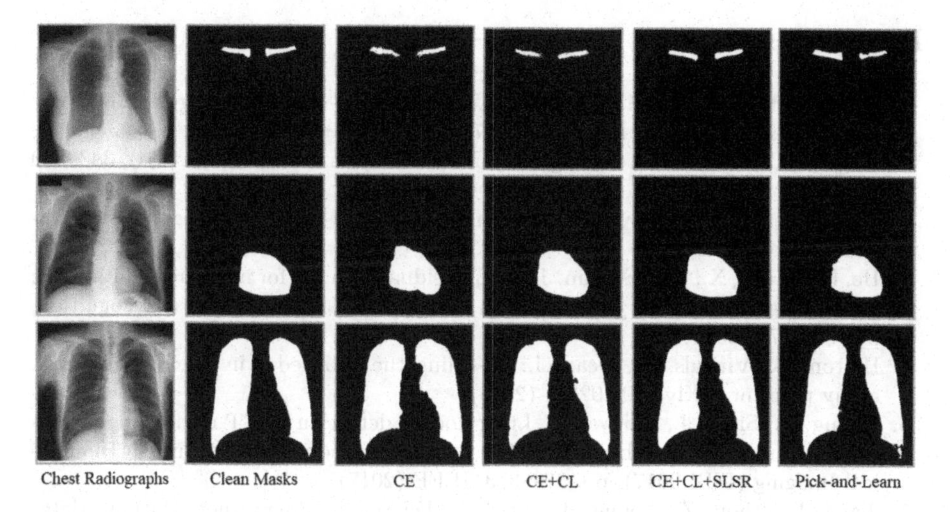

Fig. 5. Qualitative comparisons on the validation set with noise level, $\alpha = 0.7, \beta = B$.

4 Conclusion

In this paper, we propose a two-stage method to address the noisy label issue in the medical image segmentation. The method consists of a teacher-student architecture, a confident learning (CL) module, and a spatial label smoothing regularization (SLSR) technique. This is the first time that CL is involved in the segmentation tasks. This statistic-based technique can identify the label errors in the training set by estimating the joint distribution between the noisy (observed) labels and the true (latent) labels. After recognition of the noises, SLSR will correct the noisy labels smoothly instead of directly pruning these significant data away. The efficacy and necessity of the CL module and SLSR were supported by extensive experiments. Based on the experimental results, our model keeps positive segmentation performance against the increase of the noise levels. The method identifies and corrects the noisy labels in a pixel level instead of directly deleting it. Therefore it outperforms the *state-of-the-art* method in segmentation tasks and could be employed to correct the datasets with noisy labels, especially when treating the intensive noises.

Acknowledgement. The work was supported in part by the Key Area R&D Program of Guangdong Province with grant No. 2018B030338001, by the National Key R&D Program of China with grant No. 2018YFB1800800, by Natural Science Foundation of China with grant NSFC-61629101, by Guangdong Zhujiang Project No. 2017ZT07X152, by Shenzhen Key Lab Fund No. ZDSYS201707251409055, by NSFC-Youth 61902335, by Guangdong Province Basic and Applied Basic Research Fund Project Regional Joint Fund-Key Project 2019B1515120039 and CCF-Tencent Open Fund.

References

1. Ainam, J.P., Qin, K., Liu, G., Luo, G.: Sparse label smoothing regularization for person re-identification. IEEE Access **7**, 27899–27910 (2019)
2. Angluin, D., Laird, P.: Learning from noisy examples. Mach. Learn. **2**(4), 343–370 (1988)
3. Goldberger, J., Ben-Reuven, E.: Training deep neural-networks using a noise adaptation layer (2016)
4. He, K., Zhang, X., Ren, S., Sun, J.: Deep residual learning for image recognition. In: Proceedings of the IEEE Conference on Computer Vision and Pattern Recognition, pp. 770–778 (2016)
5. Hinton, G., Vinyals, O., Dean, J.: Distilling the knowledge in a neural network. arXiv preprint arXiv:1503.02531 (2015)
6. Huang, X., Shan, J., Vaidya, V.: Lung nodule detection in CT using 3D convolutional neural networks. In: 2017 IEEE 14th International Symposium on Biomedical Imaging (ISBI 2017), pp. 379–383. IEEE (2017)
7. Jiang, L., Zhou, Z., Leung, T., Li, L.J., Fei-Fei, L.: Mentornet: learning data-driven curriculum for very deep neural networks on corrupted labels. arXiv preprint arXiv:1712.05055 (2017)
8. Northcutt, C.G., Jiang, L., Chuang, I.L.: Confident learning: estimating uncertainty in dataset labels. arXiv preprint arXiv:1911.00068 (2019)
9. Parkhi, O.M., Vedaldi, A., Zisserman, A.: Deep face recognition (2015)
10. Redmon, J., Divvala, S., Girshick, R., Farhadi, A.: You only look once: unified, real-time object detection. In: Proceedings of the IEEE Conference on Computer Vision and Pattern Recognition, pp. 779–788 (2016)
11. Ren, M., Zeng, W., Yang, B., Urtasun, R.: Learning to reweight examples for robust deep learning. arXiv preprint arXiv:1803.09050 (2018)
12. Ronneberger, O., Fischer, P., Brox, T.: U-net: convolutional networks for biomedical image segmentation. In: Navab, N., Hornegger, J., Wells, W.M., Frangi, A.F. (eds.) MICCAI 2015. LNCS, vol. 9351, pp. 234–241. Springer, Cham (2015). https://doi.org/10.1007/978-3-319-24574-4_28
13. Shiraishi, J., et al.: Development of a digital image database for chest radiographs with and without a lung nodule: receiver operating characteristic analysis of radiologists' detection of pulmonary nodules. Am. J. Roentgenol. **174**(1), 71–74 (2000)
14. Xue, C., Dou, Q., Shi, X., Chen, H., Heng, P.A.: Robust learning at noisy labeled medical images: applied to skin lesion classification. In: 2019 IEEE 16th International Symposium on Biomedical Imaging (ISBI 2019), pp. 1280–1283. IEEE (2019)
15. Zhang, M., Wu, D., Liu, Q., Li, Q., Zhan, Y., Zhou, X.S.: Multi-Task convolutional neural network for joint bone age assessment and ossification center detection from hand radiograph. In: Suk, H.-I., Liu, M., Yan, P., Lian, C. (eds.) MLMI 2019. LNCS, vol. 11861, pp. 681–689. Springer, Cham (2019). https://doi.org/10.1007/978-3-030-32692-0_78
16. Zhu, H., Shi, J., Wu, J.: Pick-and-learn: automatic quality evaluation for noisy-labeled image segmentation. In: Shen, D., et al. (eds.) MICCAI 2019. LNCS, vol. 11769, pp. 576–584. Springer, Cham (2019). https://doi.org/10.1007/978-3-030-32226-7_64

Leveraging Undiagnosed Data for Glaucoma Classification with Teacher-Student Learning

Junde Wu[1,2], Shuang Yu[1(✉)], Wenting Chen[1], Kai Ma[1], Rao Fu[2], Hanruo Liu[3], Xiaoguang Di[2], and Yefeng Zheng[1]

[1] Tencent Healthcare, Shenzhen, Tencent, China
shirlyyu@tencent.com

[2] Control and Simulation Center, Harbin Institute of Technology, Harbin, China

[3] Beijing Tongren Hospital, Capital Medical University, Beijing, China

Abstract. Recently, deep learning has been adopted to the glaucoma classification task with performance comparable to that of human experts. However, a well trained deep learning model demands a large quantity of properly labeled data, which is relatively expensive since the accurate labeling of glaucoma requires years of specialist training. In order to alleviate this problem, we propose a glaucoma classification framework which takes advantage of not only the properly labeled images, but also undiagnosed images without glaucoma labels. To be more specific, the proposed framework is adapted from the teacher-student-learning paradigm. The teacher model encodes the wrapped information of undiagnosed images to a latent feature space, meanwhile the student model learns from the teacher through knowledge transfer to improve the glaucoma classification. For the model training procedure, we propose a novel training strategy that simulates the real-world teaching practice named as "Learning To Teach with Knowledge Transfer (L2T-KT)", and establish a"Quiz Pool" as the teacher's optimization target. Experiments show that the proposed framework is able to utilize the undiagnosed data effectively to improve the glaucoma prediction performance.

Keywords: Glaucoma classification · Teacher-Student Learning · Unlabeled data

1 Introduction

Glaucoma is the leading cause of irreversible vision loss that primarily damages the optic cup/disc and surrounding optic nerve [17]. Recently, deep learning methods have achieved rapid advancement and been widely adopted to the automatic glaucoma classification using fundus images [5,11–13]. However, a substantially large amount of properly labeled data is generally required for training deep learning models, which might not be easily accessed as the accurate grading of glaucoma, especially at the early stage, requires years of expertise for glaucoma specialists.

A. L. Martel et al. (Eds.): MICCAI 2020, LNCS 12261, pp. 731–740, 2020.
https://doi.org/10.1007/978-3-030-59710-8_71

Beyond the publicly available glaucoma classification datasets, on the other hand, there have been several high-quality publicly available datasets for cup/disc segmentation, but without image-level glaucoma labels [2,4,14]. The Cup-to-Disc Ratio (CDR) parameter, which can be easily computed from the cup/disc masks, is one of the most important clinical parameters for the diagnosis of glaucoma. Generally, patients with a CDR value higher than 0.6 are considered as glaucoma suspects and a higher CDR value indicates a higher probability of having glaucoma [9]. This inspires us to take advantage of the images with only cup/disc segmentation masks to improve the glaucoma classification performance.

In order to properly utilize the undiagnosed images, we propose to transfer the knowledge of pixel-wise cup/disc labels to the learner model via a teacher-student learning paradigm, which has been a popular and effective way to incorporate any prior information. However, most existing teacher-student learning methods learn a compact student from a stronger but more complex teacher, for the purpose of knowledge distillation [3,15,18]. Some other methods [7,20] learn a teacher to improve the student's training speed or performance, but they assume that the ground-truth labels of all the training samples are available, which is not true under our scenario. To the best of our knowledge, there is still a research gap of how to utilize the teacher model to learn from undiagnosed data to improve the student model's performance on glaucoma classification.

In this paper, we aim to address this research gap by proposing a novel training strategy, named "Learning To Teach with Knowledge Transfer (L2T-KT)" with a reserved quiz pool, imitating the real-world teaching practice. In L2T-KT, the teacher learns to encode the undiagnosed fundus images and the corresponding cup/disc masks to a latent feature space with the ultimate goal of improving the student's performance on the quiz pool. Meanwhile, the student is updated by the supervision of the teacher through knowledge transfer. Three major contributions are made with this paper. Firstly, we propose to adapt the teacher-student learning paradigm to the glaucoma screening task and verify the feasibility to utilize undiagnosed images to improve the glaucoma classification performance. Secondly, we propose a novel training strategy of L2T-KT and quiz pool to update the teacher model with undiagnosed images, which enables the teacher to extract potential important features from the undiagnosed images and further improve the performance of the student model via knowledge transfer. Finally, the proposed method can be easily extended to learn from totally unlabeled images or transductive learning to improve the model performance.

2 Methodology

Consider all the collected fundus images as dataset $\mathcal{D}$ that can be divided into the primary training set $\mathcal{D}_d$ with glaucoma classification labels, and auxiliary training set $\mathcal{D}_m$ with cup/disc masks but without glaucoma labels. The primary training set is denoted as: $(x_1^d, y_1), (x_2^d, y_2)...(x_n^d, y_n) \in \mathcal{D}_d$, where x_i^d denotes the fundus image, $y_i \in \{0,1\}$ denotes the glaucoma label for the input image x_i^d. Meanwhile, the auxiliary training data is denoted as $(x_1^m, m_1), (x_2^m, m_2)...(x_c^m, m_c) \in \mathcal{D}_m$, where x_i^m is the fundus image, m_i denotes the optic cup/disc mask. Furthermore, the primary training dataset is further divided into textbook pool $\mathcal{D}_{TP}$

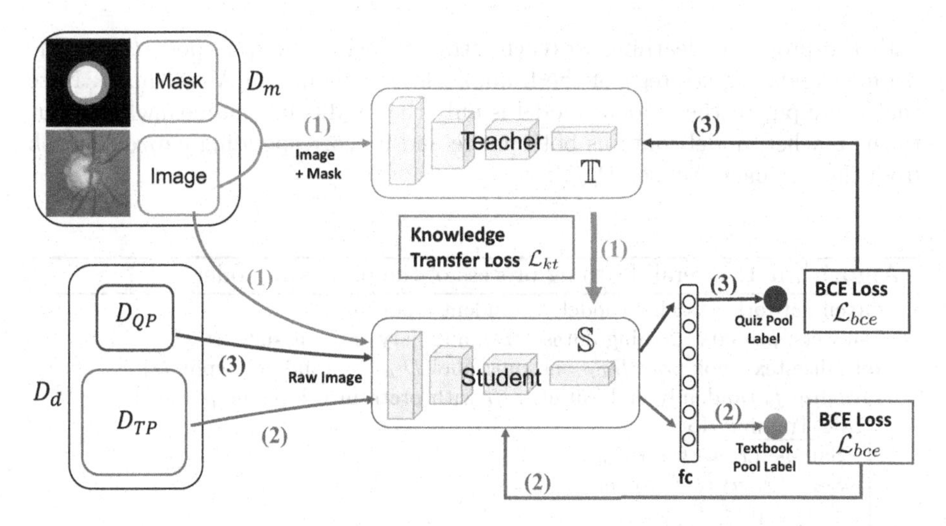

Fig. 1. Framework and data flow of the proposed L2T-KT framework. Stage (1), train the student model with knowledge transfer loss; Stage (2), update the student model with textbook pool data using binary cross entropy loss; Stage (3), train the teacher model with quiz pool data using binary cross entropy loss.

(for training the student model) and quiz pool $\mathcal{D}_{QP}$ (for updating the teacher model). Provided with these datasets, the target of this research is to construct a framework that can learn a mapping function $f(x_i^d) = y_i$ using both primary dataset $\mathcal{D}_d$ and auxiliary dataset $\mathcal{D}_m$, and thus is expected to outperform the mapping function $g(x_i^d) = y_i$ that learns only using the primary training dataset. Since the teacher-student learning paradigm has been widely used for knowledge distillation and proved effective for extracting latent information, we construct a deep learning model based on teacher-student learning for the glaucoma screening task.

As shown in Fig. 1, the overall framework contains two networks: the teacher model f_t and the student model f_s. The teacher model is a convolutional neural network which encodes the fundus images together with the corresponding cup/disc masks into a latent feature space. And the student model shares the same feature extraction backbone as that of the teacher. In this paper, the state-of-the-art classification network EfficientNet (B4) is adopted as the feature extraction backbone [16]. Different from the teacher model, the student model contains a fully connected layer (fc) to make predictions for glaucoma.

The proposed framework is optimized via an iterative three-stage training strategy. In the first stage, the student model is supervised by the teacher model using data from the auxiliary dataset $\mathcal{D}_m$ with knowledge transfer loss, as marked by the cyan color data flow in Fig. 1. In the second stage, the student model is further optimized by the ground-truth glaucoma labels from the textbook pool D_{TP} with binary cross entropy (BCE) loss, as marked by the green data flow. In the last stage, as marked by the red data flow, the teacher model is updated with the proposed 'learning to teach' strategy using the quiz pool D_{QP}. The detailed updating strategy of the framework is explained in Algorithm 1. Note that the input to the student model is only the fundus image, meanwhile input to the teacher model contains both image and its corresponding cup/disc mask from the auxiliary dataset $\mathcal{D}_m$.

Algorithm 1: Overall learning process of the proposed model

Given networks: teacher model f_t, student model f_s;
Datasets: primary training dataset D_d, auxiliary training dataset D_m ;
Initialize textbook pool D_{TP} and quiz pool D_{QP} by randomly split D_d;
Initialize f_s randomly and initialize f_t with pretrained baseline parameters;
while Training **do**
- Sample (x, m) from $\mathcal{D}_m$;
- Send (x, m) to f_t to get $\mathbb{T}$;
- Send x to f_s to get $\mathbb{S}$;
- Update f_s with knowledge transfer loss $\mathcal{L}_{kt}$ by Eqn. 3;
- Sample (x, y) from D_{TP};
- Update f_s with BCE loss by Eqn. 4;
- Sample (x, y) from D_{QP};
- Update f_t through L2T-KT by Eqn. 5 and Eqn.6;
- Update D_{QP} by Eqn. 1, Eqn. 2;

end

2.1 Quiz Pool

In the real-world teaching practice, students have access to the textbook content, but have no access to the answers of the quiz problems, which are used by the teacher to evaluate the student's performance and update the teaching strategy based on the evaluation scores. Inspired by this scenario, we propose to split the primary data set $\mathcal{D}_d$ into two subsets, the textbook pool $\mathcal{D}_{TP}$ and the quiz pool $\mathcal{D}_{QP}$. The student model learns with the ground-truth glaucoma labels from the textbook pool $\mathcal{D}_{TP}$; meanwhile, the student's performance is evaluated on the quiz pool $\mathcal{D}_{QP}$. The evaluation score is used to update the teacher model.

In this paper, $\mathcal{D}_{TP}$ and $\mathcal{D}_{QP}$ are split with two different approaches. The first method is the static quiz pool, where $\mathcal{D}_{QP}$ is randomly selected from the auxiliary set $\mathcal{D}_d$ and kept the same during the training procedure. In this paper, 20% of samples are randomly selected from $\mathcal{D}_d$ as the quiz pool $\mathcal{D}_{QP}$.

In the second method, we propose to update the quiz pool dynamically during the training process, i.e., the dynamic quiz pool. Practically, teachers often reserve the important or difficult contents for the quiz problems. Similar to this idea, a dynamic quiz pool is established, which focuses on the difficult cases and positive cases, since missing the positive cases is at higher risk for glaucoma screening. The pool is dynamically updated depending on the samples' difficulty reported by comparing the student's predictions with the glaucoma labels. The difficulty of individual samples reported by the student can be obtained with:

$$\psi^s = (1 - y)y' + y(1 - y'), \tag{1}$$

where $y \in \{0, 1\}$ denotes the ground-truth glaucoma label and y' denotes the student's prediction. Then, the probability of a sample from D_d being selected into the dynamic quiz pool is calculated by:

$$P_d = (\alpha(1 - y) + y) \cdot (\psi^s)^\gamma \cdot \left(1 + e^{-\sigma[(1-\bar{\psi}^q)-\mu]}\right)^{-1}, \tag{2}$$

where α denotes the relative importance of negative samples compared to that of positive samples, and $(\psi^s)^\gamma$ encourages the pool to focus on the difficult samples. In this paper, α and γ are empirically set to 0.7 and 2, respectively. The last term is a shifted sigmoid function that controls the average difficulty $\bar{\psi}^q$ of the quiz pool within a reasonable range. It encourages the quiz pool to retain the test content if it is challenging to the student, while changing the pool if it is too easy. In this work, σ and μ are set to 16 and 0.5, respectively. In addition, when a new sample is added to the quiz pool, the easiest sample will be dropped, so as to keep a constant size of the quiz pool. In the implementation, the quiz pool is updated every epoch.

2.2 Student Update Through Knowledge Transfer

The student model is trained on both textbook pool $\mathcal{D}_{TP}$ of the primary dataset with glaucoma label and the auxiliary dataset $\mathcal{D}_m$ without glaucoma label.

In the first stage, the student model is trained with the auxiliary set $\mathcal{D}_m$ and supervised by the teacher model with knowledge transfer. More specifically, the color fundus images are first concatenated with the corresponding cup/disc masks and then fed to the teacher model, which will encode the input to the latent feature maps $\mathbb{T}$. Meanwhile, the same set of color fundus images (without masks) will be sent to the student model as well to get the latent feature maps $\mathbb{S}$. Then the knowledge transfer (KT) loss between $\mathbb{T}$ and $\mathbb{S}$ can be computed by learning the domain-invariant latent representations with Centered Kernel Alignment (CKA) [10], as below:

$$\mathcal{L}_{kt} = \text{CKA}\left(\mathbb{T}\,\mathbb{T}^T, \mathbb{S}\,\mathbb{S}^T\right) = \frac{\|\mathbb{S}^T\,\mathbb{T}\|_F^2}{\|\mathbb{T}^T\,\mathbb{T}\|_F\,\|\mathbb{S}^T\,\mathbb{S}\|_F}, \tag{3}$$

where $\|\cdot\|_F$ denotes the Frobenius norm.

In the second stage, the student model is further trained with data from the textbook pool $\mathcal{D}_{TP}$, which contains fundus images and the ground-truth glaucoma labels. At this stage, the student model is directly supervised by the ground-truth glaucoma labels with binary cross entropy (BCE) loss:

$$\mathcal{L}_{bce} = -\left(y \log y' + (1 - y) \log(1 - y')\right), \tag{4}$$

where y is the ground-truth glaucoma label, and y' is the student's prediction.

2.3 Teacher Update Trough L2T-KT

Following the real-world teaching practice where teachers often update their teaching strategies based on students' feedback, we propose to update the teacher model parameters based on the student's performance on the constructed quiz pool D_{QP}. Formally speaking, consider a teacher network with parameters θ_t as f_{θ_t} and a student network with parameters θ_s as f_{θ_s}. The response of f_{θ_t} to a concatenated fundus image and mask $(x, m) \in \mathcal{D}_m$ is $f_{\theta_t}(x, m)$. The response of f_{θ_s} to the raw fundus image $x \in (x, m)$ is $f_{\theta_s}(x)$. The first step of L2T-KT training strategy is computing the update of the student parameters θ_s with the knowledge transfer loss between $f_{\theta_t}(x, m)$ and $f_{\theta_s}(x)$, which can be expressed in the gradient descent format as:

$$\hat{\theta_s} = \theta_s - \lambda_s \frac{\partial \mathcal{L}_{kt}[f_{\theta_t}(x, m), f_{\theta_s}(x)]}{\partial \theta_s}, \tag{5}$$

where λ_s denotes the learning rate of the student model and is set as 3×10^{-4}. After the knowledge transfer and student parameter update, we denote the refreshed student as $f_{\hat{\theta_s}}$.

The teacher's goal is learning to teach the student to achieve better performance on the quiz. In other words, the optimization target of the teacher is the refreshed student $f_{\hat{\theta_s}}$ to perform better than f_{θ_s} on the same $\mathcal{D}_{QP}$. Let $f_{\hat{\theta_s}}(q)$ denotes the prediction of $f_{\hat{\theta_s}}$ over a random sample $q \in \mathcal{D}_{QP}$ and y_q denotes the ground-truth label of q, the teacher can be optimized by minimizing the BCE loss between $f_{\hat{\theta_s}}(q)$ and y_q with gradient descent. It is theoretically feasible because as shown in Eqn. 5, teacher parameter θ_t is a variable of the updated student's parameter $\hat{\theta_s}$. Therefore, the partial derivative of $f_{\hat{\theta_s}}(q)$ w.r.t θ_t can be computed, and θ_t can be updated via:

$$\hat{\theta_t} = \theta_t - \lambda_t \frac{\partial \mathcal{L}_{bce}[f_{\hat{\theta_s}}(q), y_q]}{\partial \theta_t}, \tag{6}$$

where λ_t is the learning rate of the teacher model and set as 3×10^{-4}. We compute the partial derivative of $f_{\hat{\theta_s}}(q)$ w.r.t the teacher parameter θ_t, rather than its own parameter $\hat{\theta_s}$ as commonly used. That is because we aim at making the teacher to learn how to teach a better student, but not making the student to learn by itself. Note that the refreshed student is only temporarily used in L2T-KT, which will not change the parameters of the original student.

3 Experiments

Datasets. The data utilized in this work mainly originates from two sources: the primary dataset $\mathcal{D}_d$ with glaucoma labels from Beijing Tongren Hospital with approval obtained from the institutional review board, and the auxiliary dataset $\mathcal{D}_m$ with cup/disc segmentation masks from publicly available dataset RIGA [2]. The primary dataset contains in total of 3,830 fundus images graded by certified

Table 1. Performance comparison (%) of leveraging undiagnosed data under different settings of the quiz pool.

	Static	Dynamic	Acc	Sen	Spec	AUC
Baseline*			90.69	90.10	93.78	95.77
Proposed†	√		93.02	90.70	**94.53**	98.16
Proposed‡		√	**93.29**	**96.03**	91.42	**98.29**

*: Baseline method using purely labeled data;
†: Proposed method: labeled data + undiagnosed data + static quiz pool;
‡: Proposed method: labeled data + undiagnosed data+dynamic quiz pool;

glaucoma specialists, including 1,586 glaucoma and 2,244 non-glaucoma images. We randomly selected 60% images as the training set, 15% as the validation set and the rest 25% as the test set to evaluate the model performance. The RIGA dataset contains 650 fundus images with pixel-wise cup/disc masks labeled by experts, but image-level glaucoma labels are not provided [2].

3.1 Ablation Studies

Ablation studies have been conducted to evaluate the effectiveness of the proposed framework under different setups of the quiz pool, including the static quiz pool and dynamic quiz pool. The comparison baseline method utilizes the same backbone as that of the teacher/student model, i.e., EfficientNet-B4, and trained with the glaucoma/non-glaucoma labels of the primary training set. Four metrics are adopted to evaluate the model performance, including accuracy (Acc), sensitivity (Sen), specificity (Spec) and area under the receiver operating characteristic curve (AUC).

Table 1 shows the quantitative comparisons of the baseline method and the proposed framework trained on undiagnosed auxiliary dataset under different settings of the quiz pool. Compared with the baseline method using purely labeled data, training using both labeled and undiagnosed data with the proposed L2T-KT framework with a static quiz pool increases the AUC with 2.39% and accuracy with 2.33%. In addition, by changing the static quiz pool to a dynamic quiz pool, the model performance is further improved, with an obvious improvement on the model sensitivity, since the dynamic quiz pool favors the positive cases. Clinically, for the glaucoma screening task, a higher sensitivity measure is much more important than specificity, so as not to miss the potential glaucoma patients.

3.2 Auxiliary Data Setting

We have also evaluated the model performance under different settings of the auxiliary dataset. The evaluation is conducted on both the private dataset and

Table 2. Performance comparison (%) with other methods.

	LAG				Private			
	Acc	Sen	Spec	AUC	Acc	Sen	Spec	AUC
Li et al. [12] (supervised)	95.3	95.4	**95.2**	97.5	-	-	-	-
Fu et al. [8] (supervised)	93.88	96.79	92.29	98.27	91.94	91.38	92.30	96.70
Pinto et al. [6] (semi)	92.75	92.30	93.16	97.11	91.85	92.51	89.79	96.12
Pinto et al. [6] (trans)	93.77	93.26	94.68	97.95	92.03	92.75	**92.60**	97.45
Ghamdi et al. [1] (semi)	94.11	97.43	92.29	98.16	91.76	93.19	90.83	96.88
Ghamdi et al. [1] (trans)	95.01	97.75	93.52	98.73	92.57	94.10	91.57	97.39
Auxiliary-GT* (proposed)	95.81	98.40	94.22	99.49	93.29	96.03	91.42	98.29
Auxiliary-Psd† (proposed)	95.47	**98.72**	93.70	99.32	92.84	95.01	91.42	97.84
Transductive‡ (proposed)	**96.04**	**98.72**	94.75	**99.51**	**93.64**	**96.37**	91.82	**98.41**

*: RIGA dataset with ground-truth cup/disc mask is used as undiagnosed auxiliary set;

†: RIGA dataset with pseudo cup/disc mask is used as totally unlabeled auxiliary set;

‡: Test set with pseudo cup/disc mask is used as totally unlabeled auxiliary set;

a publicly available glaucoma classification dataset LAG [12], which contains 1,711 glaucoma images and 3,143 non-glaucoma images. Apart from the undiagnosed auxiliary set with ground-truth cup/disc masks (RIGA), we have also alternatively trained on the totally unlabeled auxiliary set, by producing pseudo masks of RIGA images using a state-of-the-art cup/disc segmentation algorithm [19]. As Table 2 lists, by using an auxiliary set with pseudo cup/disc masks, the model performance degenerates slightly compared with the standard undiagnosed auxiliary set using ground-truth cup/disc masks, with the AUC value drop of 0.17% and 0.45% for LAG and private set, respectively. However, compared with the baseline model in Table 1, the proposed framework using pseudo labels still surpasses that of the baseline model with a remarkable margin, with an AUC improvement of 2.07% for the private set.

The performance of the proposed method can be further improved when it is authorized to get access to the raw images of the test dataset, i.e., in the transductive learning scenario. When taking the raw fundus images of the test set and their pseudo cup/disc masks as the auxiliary data (denoted as 'Transductive' in Table 2), the proposed method achieves the best performance, with an AUC score of 99.51% and 98.41% for the LAG and private set, respectively. This indicates the expandability and effectiveness of the proposed L2T-KT framework.

3.3 Comparing with State-of-the-Art

The proposed framework has been compared with other state-of-the-art methods on the glaucoma classification task, including two fully-supervised methods [8,12] and two semi-supervised methods [1,6]. The semi-supervised methods take RIGA or the test set as the unlabeled datasets, which are denoted as 'semi' and 'trans', respectively. As listed in Table 2, the proposed method achieves the best performance on both the private and LAG datasets, especially for the sensitivity metric, indicating the effectiveness of the proposed method in exploiting extra undiagnosed data. In addition, different from the existing methods of designing complex architectures or using model ensembling, the proposed method adopts a plug-and-play training strategy without any change on the backbone network. At the prediction stage, only the student network is used, ensuring fast computational speed during inference.

4 Conclusion

Many publicly available datasets contain images with cup/disc masks but without image-level glaucoma labels. In order to fully exploit those undiagnosed images for the glaucoma screening task, we proposed a novel training strategy Learning to Teach with Knowledge Transfer (L2T-KT), which enabled the model to learn from those undiagnosed images through teacher-student paradigm. Detailed experiments revealed that the proposed method could not only improve the glaucoma screening performance through learning from the data with cup/disc masks, but also could be easily extended to learn from the completely unlabeled data with pseudo labels and improved the test set performance via transductive learning. Future works will continue to explore the potential of the proposed method and optimize the time efficiency at the training stage.

Acknowledgment. This work was funded by the Key Area Research and Development Program of Guangdong Province, China (No. 2018B010111001), National Key Research and Development Project (No. 2018YFC2000702) and Science and Technology Program of Shenzhen, China (No. ZDSYS201802021814180).

References

1. Al Ghamdi, M., Li, M., Abdel-Mottaleb, M., Shousha, M.A.: Semi-supervised transfer learning for convolutional neural networks for glaucoma detection. In: IEEE International Conference on Acoustics, Speech and Signal Processing, pp. 3812–3816 (2019)
2. Almazroa, A., et al.: Retinal fundus images for glaucoma analysis: The RIGA dataset. In: SPIE Conference on Medical Imaging (2018)
3. Ba, J., Caruana, R.: Do deep nets really need to be deep? In: Advances in Neural Information Processing Systems, pp. 2654–2662 (2014)
4. Carmona, E.J., Rincón, M., García-Feijoó, J., Martínez-de-la Casa, J.M.: Identification of the optic nerve head with genetic algorithms. Artif. Intell. Med. **43**(3), 243–259 (2008)
5. Chen, X., Xu, Y., Wong, D.W.K., Wong, T.Y., Liu, J.: Glaucoma detection based on deep convolutional neural network. In: 37th Annual International Conference of the IEEE Engineering in Medicine and Biology Society, pp. 715–718 (2015)
6. Diaz-Pinto, A., Colomer, A., Naranjo, V., Morales, S., Xu, Y., Frangi, A.F.: Retinal image synthesis and semi-supervised learning for glaucoma assessment. IEEE Trans. Med. Imaging **38**(9), 2211–2218 (2019)

7. Fan, Y., Tian, F., Qin, T., Li, X., Liu, T.: Learning to teach. In: 6th International Conference on Learning Representations (2018)
8. Fu, H., Cheng, J., Xu, Y., Zhang, C., Wong, D.W.K., Liu, J., Cao, X.: Disc-aware ensemble network for glaucoma screening from fundus image. IEEE Trans. Med. Imaging **37**(11), 2493–2501 (2018)
9. Garway-Heath, D.F., Ruben, S.T., Viswanathan, A., Hitchings, R.A.: Vertical cup/disc ratio in relation to optic disc size: its value in the assessment of the glaucoma suspect. Br. J. Ophthalmol. **82**(10), 1118–1124 (1998)
10. Kornblith, S., Norouzi, M., Lee, H., Hinton, G.E.: Similarity of neural network representations revisited. In: 36th International Conference on Machine Learning (2019)
11. Li, A., Cheng, J., Wong, D.W.K., Liu, J.: Integrating holistic and local deep features for glaucoma classification. In: 38th Annual International Conference of the IEEE Engineering in Medicine and Biology Society, pp. 1328–1331. IEEE (2016)
12. Li, L., Xu, M., Wang, X., Jiang, L., Liu, H.: Attention based glaucoma detection: a large-scale database and CNN model. In: IEEE Conference on Computer Vision and Pattern Recognition, pp. 10571–10580 (2019)
13. Li, Z., He, Y., Keel, S., Meng, W., Chang, R.T., He, M.: Efficacy of a deep learning system for detecting glaucomatous optic neuropathy based on color fundus photographs. Ophthalmology **125**(8), 1199–1206 (2018)
14. Lowell, J., Hunter, A., Steel, D., Basu, A., Ryder, R., Fletcher, E., Kennedy, L.: Optic nerve head segmentation. IEEE Trans. Med. Imaging **23**(2), 256–264 (2004)
15. Rusu, A.A., et al.: Policy distillation. In: 4th International Conference on Learning Representations (2016)
16. Tan, M., Le, Q.V.: EfficientNet: rethinking model scaling for convolutional neural networks. In: 36th International Conference on Machine Learning (2019)
17. Tham, Y.C., Li, X., Wong, T.Y., Quigley, H.A., Aung, T., Cheng, C.Y.: Global prevalence of glaucoma and projections of glaucoma burden through 2040: a systematic review and meta-analysis. Ophthalmology **121**(11), 2081–2090 (2014)
18. Urban, G., et al.: Do deep convolutional nets really need to be deep and convolutional? In: 5th International Conference on Learning Representations (2017)
19. Wang, S., Yu, L., Yang, X., Fu, C.W., Heng, P.A.: Patch-based output space adversarial learning for joint optic disc and cup segmentation. IEEE Trans. Med. Imaging **38**(11), 2485–2495 (2019)
20. Wu, L., Tian, F., Xia, Y., Fan, Y., Qin, T., Lai, J., Liu, T.: Learning to teach with dynamic loss functions. In: Advances in Neural Information Processing Systems, pp. 6466–6477 (2018)

Difficulty-Aware Glaucoma Classification with Multi-rater Consensus Modeling

Shuang Yu[1(✉)], Hong-Yu Zhou[1], Kai Ma[1], Cheng Bian[1], Chunyan Chu[1], Hanruo Liu[2], and Yefeng Zheng[1]

[1] Tencent Healthcare, Tencent, Shenzhen, China
{shirlyyu,hongyuzhou,kylekma,tronbian,yefengzheng}@tencent.com
[2] Beijing Tongren Hospital, Capital Medical University, Beijing, China

Abstract. Medical images are generally labeled by multiple experts before the final ground-truth labels are determined. Consensus or disagreement among experts regarding individual images reflects the gradeability and difficulty levels of the image. However, when being used for model training, only the final ground-truth label is utilized, while the critical information contained in the raw multi-rater gradings regarding the image being an easy/hard case is discarded. In this paper, we aim to take advantage of the raw multi-rater gradings to improve the deep learning model performance for the glaucoma classification task. Specifically, a multi-branch model structure is proposed to predict the most sensitive, most specifical and a balanced fused result for the input images. In order to encourage the sensitivity branch and specificity branch to generate consistent results for consensus labels and opposite results for disagreement labels, a consensus loss is proposed to constrain the output of the two branches. Meanwhile, the consistency/inconsistency between the prediction results of the two branches implies the image being an easy/hard case, which is further utilized to encourage the balanced fusion branch to concentrate more on the hard cases. Compared with models trained only with the final ground-truth labels, the proposed method using multi-rater consensus information has achieved superior performance, and it is also able to estimate the difficulty levels of individual input images when making the prediction.

Keywords: Multi rater · Retinal imaging · Glaucoma classification · Uncertainty estimation

1 Introduction

Glaucoma is the leading cause of irreversible vision loss world widely and is projected to affect around 111 million people by year 2040 [16]. Therefore, the screening and treatment of glaucoma in the early stage play an important role to prevent vision loss. Recent years, there has been an increasing trend in the automatic classification of glaucoma with deep learning methods, including [8,9,12].

S. Yu and H.-Y. Zhou—Contributed equally.

A. L. Martel et al. (Eds.): MICCAI 2020, LNCS 12261, pp. 741–750, 2020.
https://doi.org/10.1007/978-3-030-59710-8_72

However, the reference standard and guideline for glaucoma diagnosis are often not well defined and may vary from one center to another, which might result in disagreement among graders and negatively affect the grading procedure [4,12]. It is reported that the sensitivity of individual graders for glaucoma ranged from 29.2% to 73.9%, with specificity ranged from 75.8% to 92.6% [12]. Therefore, the inter-rater variability problem constitutes a major impact for the grading procedure of glaucoma.

Glaucoma images, in fact medical images in general, are usually labeled by multiple experts independently, so as to avoid the subjective bias or potential labeling noise of each rater resulted by different levels of expertise, negligence of subtle symptoms, quality of images, etc. [13]. The final ground-truth label then can be obtained by fusing individual labels using majority vote, average or other fusion strategies [13]. However, at model training stage, only the final ground-truth label is utilized to train the model and those intermediate labels generated by individual raters are neglected, which contain important information regarding the gradeability or difficulty levels of the images.

Recently, there have been emerging research works paying attention to the multi-rater labels and inter-rater variability. Alain *et al.* [7] studied the effect of common label fusion techniques on the uncertainty estimation of segmentation tasks. It was observed that the models trained with fused 'ground truth' label tended to under-estimate the uncertainty, meanwhile uncertainty generated by models trained with individual labels was able to reflect the underlying expert disagreement [7]. Similar influence of the fused final label was observed by Jensen *et al.* [6] for the skin disease classification task as well, which reported that the classification model trained with fused final label would be over-confident, while the model trained with the label sampling method using inter-rater variability was better calibrated. To better utilize the individual ratings, Guan *et al.* [2] proposed to predict the labels of each rater individually and then learn the respective weight to make the final prediction. In another similar work, Sudre *et al.* [15] proposed to model the individual raters' performance together with their consensus status, which achieved better performance compared with training using the fused final label.

Although those recent studies achieved better performance, the critical information contained in the raw multi-rater gradings regarding the image being an easy/hard case is usually neglected or discarded during the training procedure. It is observed that images with consensus labels generally tend to be easy cases while disagreement gradings tend to be hard or highly uncertain ones, as the labeling consensus among individual graders is highly correlated to the grade-ability and difficulty levels of the images being graded [11,13]. Therefore, we believe that the model performance can be further boosted by utilizing the multi-rater agreement/disagreement information.

This research aims to fill the performance gap by leveraging the multi-rater consensus information for the glaucoma classification task. Instead of predicting the labels from individual raters, we propose to use a multi-branch structure to generate three predictions under different sensitivity settings, one with the

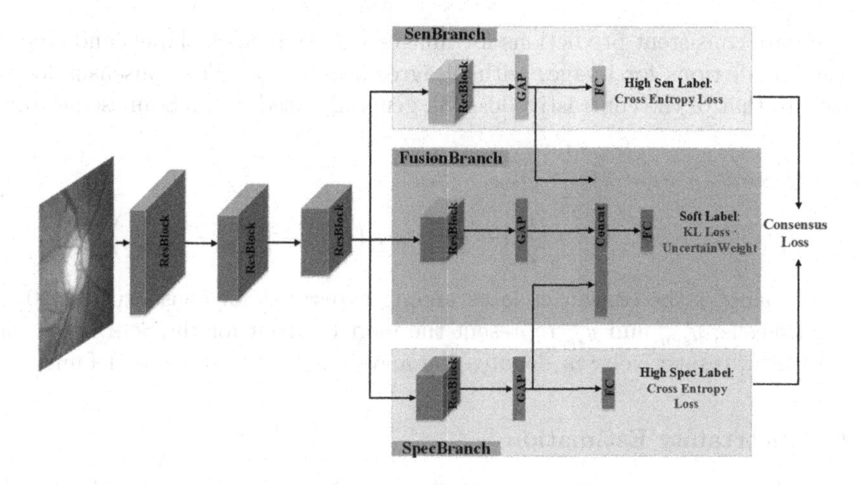

Fig. 1. Framework of the proposed system.

best sensitivity, one with the best specificity, and one in-between, respectively. It also fulfills the clinical requirement of different sensitivity levels for various application scenarios. In addition, a consensus loss is proposed to encourage the sensitivity branch and specificity branch to generate consistent predictions for images with consensus labels and contradictory predictions for images with disagreement labels. Moreover, cosine similarity between the predictions of the two branches contains important uncertainty information and serves as an indicator of the difficulty level for the input image, which is further utilized to encourage the model to focus more on the hard cases and improve the model performance.

2 Method

Figure 1 shows the framework of the proposed system, which consists of three branches, corresponding to three different levels of sensitivity and specificity settings, including the sensitivity branch (SenBranch), the specificity branch (SpecBrach) and the balanced fusion branch (FusionBranch), respectively. The three branches share the same weights for the first three ResNet blocks (ResBlock), with ResNet18 [5] being adopted as the backbone. And then, each branch contains a ResNet block, global average pooling (GAP) and fully connected layers (FC). The extracted GAP features from the SenBranch and SpecBranch are concatenated together with that from the FusionBranch and then fed to the FC layer for the final glaucoma prediction.

2.1 Consensus Loss

In order to take advantage of the agreement/disagreement among individual raters, a consensus loss is proposed to encourage the SenBranch and SpecBranch to generate consistent predictions for images with agreement labels and contradictory predictions for images with disagreement labels. The consensus loss is similar to that of the contrastive loss [3], generally used in the Siamese network:

$$\begin{aligned} \mathcal{L}_{con}(y'_{sen}, y'_{sp}, a) =& \frac{1}{2}a \left\| y'_{sen} - y'_{sp} \right\|_2^2 \\ &+ \frac{1}{2}(1-a)\left\{max\left(0, m - \left\| y'_{sen} - y'_{sp} \right\|_2\right)\right\}^2, \end{aligned} \tag{1}$$

where a denotes the consensus label among experts, 1 for consensus and 0 for non-consensus; y'_{sen} and y'_{sp} represent the model output for the SenBranch and SpecBranch, respectively; m denotes the margin and is set to 1 by default.

2.2 Uncertainty Estimation

The prediction consensus between the SenBranch and SpecBranch indicates the difficulty level, i.e., uncertainty, of the images. In this paper, cosine similarity is adopted to measure the distance between the predictions of the two branches. Then, the uncertainty of the model prediction can be estimated with:

$$u = 0.5(1 - Similarity) = 0.5\left(1 - \frac{y'_{sen} \cdot y'_{sp}}{\left\| y'_{sen} \right\| \left\| y'_{sp} \right\|}\right). \tag{2}$$

The obtained uncertainty is further utilized to adjust the relative weight of individual samples for the training of the FusionBranch, so as to encourage the model to concentrate more on the difficult samples.

2.3 Loss Function for Multi-branch

For the training of the proposed model, each branch is optimized individually with the same batch of images, but with different corresponding labels and loss functions. The SenBranch and SpecBranch are trained with the most sensitive and most specifical labels for individual images, the labels of which are determined with a random sampling procedure by assigning different probabilities for individual ratings in the labeling pool. For the SenBranch, the glaucoma labels are set with a higher probability than that of the non-glaucoma labels by repeating the glaucoma labels twice in the label pool; vice versa for the SpecBranch. Both branches are optimized with cross entropy loss and consensus loss, as in:

$$\mathcal{L}_s(y'_s, y_s, a) = -\sum_{i=1}^{2} y_{si} log\left(y'_{si}\right) + \alpha \mathcal{L}_{con}(y'_{sen}, y'_{sp}, a), \tag{3}$$

where y_s and y'_s denote the glaucoma label and model output for the SenBranch or SpecBranch; α denotes the relative weight of the consensus loss, empirically set as 0.5 in this research.

For the FusionBranch, instead of training on the final ground-truth, it is trained with soft labels generated from individual rater's grading weighted by their respective accuracy: $y = \frac{\sum_{i=1}^{m} w_i r_i}{\sum_{i=1}^{m} w_i}$, where m is the total number of raters for the image; r_i denotes the raw labels by individual raters; w_i is the weight of the corresponding rater, which is determined by the rating accuracy against the ground-truth label. Furthermore, the soft labels used for the FusionBranch training are clipped to the range of (0.01, 0.99), so as to avoid the potential problems of hard labels.

The Kullback–Leibler divergence loss (KL loss) is then adopted to optimize the model parameters of the FusionBranch. In addition, in order to encourage the model to emphasize the difficult and highly uncertain samples, the estimated uncertainty value is utilized to adjust the relative weight of individual samples, as in:

$$\mathcal{L}_f(y, y') = KL(y||y') = \frac{\sum_{i=1}^{n} \sum_{j=1}^{2} (1 + u_i) y_{ij} \left(log y_{ij} - log y'_{ij} \right)}{\sum_{i=1}^{n} (1 + u_i)}, \quad (4)$$

where y and y' denote the soft label and model output of the FusionBranch, respectively; n is the total number of samples in a training batch; and u_i is the uncertainty weight for the corresponding sample obtained with Eq. 2.

3 Experimental Results

A total of 6,318 color fundus images with acceptable image quality were collected from Beijing Tongren Hospital, with approval obtained from the institutional review board of Tongren Hospital. The images were labeled following the adjudication process, with two certified ophthalmologists and one senior glaucoma specialist involved in the grading procedure. Each image was independently labeled by two certified ophthalmologists in the first stage. If consensus label was reached, then the grading process was completed. Otherwise, the image would be passed to the senior glaucoma specialist, who had access to the individual ratings of the first stage and graded the image based on his or her expertise. After the adjudication grading procedure, 2,171 images are graded with consensus glaucoma label, 2,315 images with consensus non-glaucoma label, 781 images with non-consensus glaucoma label and 1,051 images with non-consensus non-glaucoma label. At the model training stage, 60% of the images are randomly selected as the training set, 15% as the validation set and the rest 25% are reserved for test purpose.

Apart from the private dataset, two publicly available datasets, REFUGE (test set) [10] and DRISHTI [14], are also adopted. The REFUGE test set contains 40 glaucoma images and 360 non-glaucoma images, and the performance of two individual experts who are not part of the ground-truth labeling group is also reported [10]. Meanwhile, the DRISHTI dataset contains 70 glaucoma images and 31 non-glaucoma images [14]. Each image is independently graded by five experts with majority vote being adopted as the ground truth. Note that both datasets are used for direct model inference without any further training or fine-tuning, so as to verify the generalization capability of the proposed model.

Table 1. The ablation study results of multi-rater consensus model (%).

Methods			Acc	Sen	Spec	F1	AUC
Expert 1			86.71	88.11	85.50	86.03	-
Expert 2			83.78	84.76	82.94	82.92	-
Baseline			89.47	88.12	90.64	88.60	95.83
MultiBr	-	-	90.53	89.60	91.35	89.79	96.93
MultiBr	ConLoss	-	91.35	88.66	93.68	90.50	97.43
MultiBr	-	Uncerty	91.70	90.28	92.94	91.02	97.52
MultiBr	ConLoss	Uncerty	**92.54**	**90.96**	**93.92**	**91.89**	**97.94**

As pathologies of glaucoma concentrate on the optic disc and surrounding regions, the three-disc-diameter region around the disc center is cropped as the region-of-interest (ROI) and resized to the dimension of 256×256 pixels before being fed to the network. All experiments are performed on an NVIDIA Tesla P40 GPU with 24 GB of memory. The Adam optimizer is adopted to optimize the model with a batch size of 32 and maximum training epochs of 50. The initial learning rate is set as 2×10^{-4} and halved every 15 epochs. Data augmentation strategies, including random cropping, rotation, horizontal flipping and color jitting, are utilized during the training procedure, so as to increase the diversity of the training data.

3.1 Ablation Studies

Comprehensive ablation studies have been performed to evaluate the effectiveness of different modules proposed in this research, including the multi-branch structure (MultiBr), consensus loss (ConLoss) and uncertainty loss (Uncerty). The comparison baseline model shares the same backbone as the proposed model, i.e., ResNet18, and it is trained with the final ground-truth label. Five metrics are adopted to evaluate the model performance, including accuracy (Acc), sensitivity (Sen), specificity (Spec), F1 score (F1) and area under curve (AUC).

Detailed results of the ablation studies are listed in Table 1. We have also evaluated the performance of the two graders in the first stage. On the test set, the two experts achieve an F1 score of 86% and 82.9%, respectively, indicating the apparent challenges of glaucoma labeling even for certified ophthalmologists. In contrast, the baseline model trained with ground-truth surpasses the performance of the stage one raters, with an F1 score of 88.6%. By introducing the multi-branch model structure and taking advantage of raw gradings, the F1 score is improved by 1.19% over the baseline, with an AUC score of 96.93%. The effectiveness of consensus loss and uncertainty loss is also verified, yielding an F1 improvement of 0.71% and 1.23%, respectively. At last, the proposed multi-branch model combining consensus loss and uncertainty loss achieves the best performance with an F1 score of 91.89% and AUC value of 97.94%.

Table 2. Comparison with other multi-rater research method (%).

Methods	Consensus Data			Non-Consensus Data			All Data		
-	Sen	Spec	AUC	Sen	Spec	AUC	Sen	Spec	AUC
Expert 1	100	100	-	55.06	52.99	-	88.11	85.50	-
Expert 2	100	100	-	42.38	44.72	-	84.76	82.94	-
Baseline	94.68	97.11	99.42	69.90	76.32	80.10	88.12	90.64	95.83
RandLabel [6]	94.86	98.47	99.58	53.43	69.77	64.82	83.58	89.73	95.13
IndiRaters [2]	95.78	97.79	99.63	71.94	78.57	82.71	89.47	91.81	96.81
Proposed	**96.33**	**98.64**	**99.76**	**76.02**	**83.46**	**87.49**	**90.96**	**93.92**	**97.94**
SenBr	*99.63*	85.06	99.65	*98.47*	34.59	86.43	*99.32*	69.36	97.49
SpecBr	88.26	*100*	99.68	44.90	*93.23*	85.45	76.79	*97.89*	97.50

3.2 Comparison with Existing Methods

We have also compared the proposed method with other research works that utilize the multi-rater labels, including the random sampling method (RandLabel) used by Jensen *et al.* [6] and individual rater modeling (IndiRaters) proposed by Guan *et al.* [2]. As listed in Table 2, the comparison is individually performed on the consensus data, non-consensus data and all data combined together. The proposed method achieves the optimal performance across all the three scenarios, especially for the non-consensus data, with a dramatical performance gain of 4.1% for sensitivity and 4.9% for specificity over the current best methods for the non-consensus data. Furthermore, all the comparison methods achieve a superior performance on the consensus data, with both sensitivity and specificity close to or above 95%. However, for the non-consensus data, there is a dramatical drop of model performance, even for the proposed model. This implies that the images with consensus label tend to be easy or typical cases, either typical normal cases or typical glaucoma images in the advanced stage. In contrast, for the images that grading experts hold different opinions, i.e., the non-consensus data, there is a high probability that the images are hard cases or non-typical cases, difficult for both human graders and deep learning models. Moreover, we have also evaluated the performance of the proposed SenBranch and SpecBranch in Table 2. Compared with the FusionBranch, the SenBranch achieves a sensitivity of 99.32%, and the SpecBranch achieves a specificity of 97.89%, reflecting the effectiveness of the SenBranch/SpecBranch in fulfilling the designated purpose.

3.3 Model Performance on Public Datasets

In order to verify the generalization capability of the proposed model, we have also tested the model performance on two publicly available datasets, REFUGE and DRISHTI. Note that no further training or fine-tuning is performed on the two datasets. As listed in Table 3, the model achieves the state-of-the-art performance on the DRISHTI dataset, with an AUC improvement of 9.2% over the current best performance [1]. Comparing to the individual expert's grading,

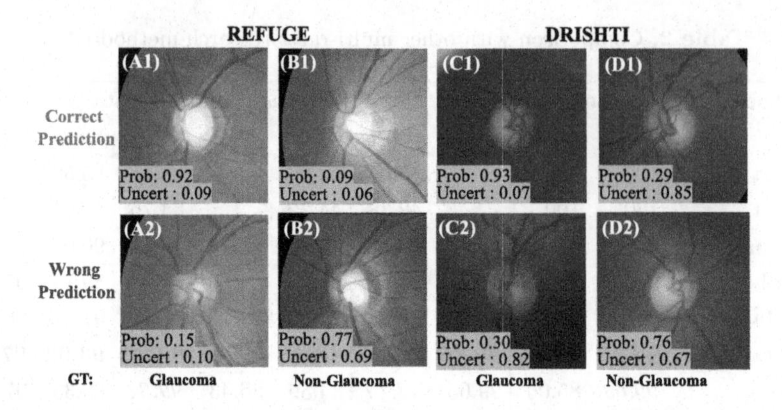

Fig. 2. Representative results on the REFUGE and DRISHTI datasets. Top row: correct predictions; bottom row: wrong predictions. (A1-2), (B1-2): images from the REFUGE test set; (C1-2), (D1-2): images from the DRISHTI dataset.

the sensitivity and F1 score of the proposed model exceed two out of the five grading experts. As for the REFUGE test dataset (Table 4), the model achieves an AUC value of 96.83%, which is better than the 3rd place solution on the challenge leaderboard. Concerning that the model is not trained or fine-tuned on the REFUGE data, the result is satisfactory. Especially, when comparing the model performance with two experts, the SenBranch achieves higher sensitivity and specificity than both of the experts.

Table 3. Performance comparison of glaucoma classification on DRISHTI (%).

Methods	Acc	Sen	Spec	F1	AUC
Expert 1	91.09	95.71	80.65	93.71	-
Expert 2	87.13	81.43	100	89.76	-
Expert 3	91.09	92.86	87.10	93.53	-
Expert 4	86.14	92.86	70.97	90.28	-
Expert 5	85.15	80.00	96.77	88.19	-
Sivaswamy *et al.* (2015) [14]	-	81.0	72.0	-	79.0
Diaz-Pinto *et al.* (2019) [1]	75.25	74.19	71.43	-	80.41
Proposed	**86.14**	**91.43**	**74.19**	**90.14**	**89.63**
SenBr	78.22	*98.57*	32.26	86.25	87.70
SpecBr	74.26	68.57	*87.10*	78.69	84.75

Figure 2 demonstrates several representative results for the correct and wrong predictions on the REFUGE and DRISHTI datasets. In the same time of predicting for the glaucoma probability, the proposed model is also able to estimate the difficulty level, i.e., uncertainty, via the outputs of SenBranch and SpecBranch. Especially, we have also checked the raw gradings of individual experts for the

Table 4. Performance comparison of glaucoma classification on REFUGE (%).

Methods	Acc	Sen	Spec	AUC
Expert 1 [10]	90.50	85.00	91.11	-
Expert 2 [10]	90.75	85.00	91.39	-
1st Place [10]	-	97.52	85.00	**98.85**
2nd Place [10]	-	**97.60**	85.00	98.17
3rd Place [10]	-	95.00	85.00	96.44
FusionBr	**98.0**	82.50	**99.72**	96.83
SenBr	92.0	*92.50*	91.94	96.44
SpecBr	96.75	67.50	*100*	93.47

listed wrong predictions. For images A2 and B2 from REFUGE, the ground-truth labels are glaucoma and non-glaucoma, respectively. However, the two independent ophthalmologists unanimously graded the two images as non-glaucoma and glaucoma, same as the model prediction. For images C2 and D2 from DRISHTI, the uncertainty values estimated by the model are high. When referring to the raw gradings by five experts, 4/5 of the experts labeled C2 as glaucoma and 1/5 as non-glaucoma; 3/5 experts labeled D2 as non-glaucoma and 2/5 as glaucoma, indicating that the glaucoma grading is challenging for experts as well.

4 Conclusion

In this paper, we proposed to leverage the multi-rater consensus information contained in the raw expert gradings to enhance the model performance. Ablation studies have validated the effectiveness of the proposed method. It has achieved the state-of-the-art classification performance on the publicly available DRISHTI dataset and satisfactory performance on the REFUGE test set with direct model inference. The proposed model has achieved comparable or better performance than the experts. Future works will continue to explore the potential influence of multi-rater consensus on other deep learning related tasks.

Acknowledgment. This work was funded by the Key Area Research and Development Program of Guangdong Province, China (No. 2018B010111001), National Key Research and Development Project (No. 2018YFC2000702) and Science and Technology Program of Shenzhen, China (No. ZDSYS201802021814180).

References

1. Diaz-Pinto, A., Morales, S., Naranjo, V., Köhler, T., Mossi, J.M., Navea, A.: CNNs for automatic glaucoma assessment using fundus images: an extensive validation. Biomed. Eng. Online **18**(1), 29 (2019)
2. Guan, M.Y., Gulshan, V., Dai, A.M., Hinton, G.E.: Who said what: modeling individual labelers improves classification. In: Thirty-Second AAAI Conference on Artificial Intelligence (2018)

3. Hadsell, R., Chopra, S., LeCun, Y.: Dimensionality reduction by learning an invariant mapping. In: Proceedings of the IEEE Conference on Computer Vision and Pattern Recognition, vol. 2, pp. 1735–1742. IEEE (2006)
4. Hammel, N., et al.: A study of feature-based consensus formation for glaucoma risk assessment. Investigative Ophthalmol. Vis. Sci. **60**(9), 164–164 (2019)
5. He, K., Zhang, X., Ren, S., Sun, J.: Deep residual learning for image recognition. In: Proceedings of the IEEE Conference on Computer Vision and Pattern Recognition, pp. 770–778 (2016)
6. Jensen, M.H., Jørgensen, D.R., Jalaboi, R., Hansen, M.E., Olsen, M.A.: Improving uncertainty estimation in convolutional neural networks using inter-rater agreement. In: Shen, D., et al. (eds.) MICCAI 2019. LNCS, vol. 11767, pp. 540–548. Springer, Cham (2019). https://doi.org/10.1007/978-3-030-32251-9_59
7. Jungo, A., et al.: On the effect of inter-observer variability for a reliable estimation of uncertainty of medical image segmentation. In: Frangi, A.F., Schnabel, J.A., Davatzikos, C., Alberola-López, C., Fichtinger, G. (eds.) MICCAI 2018. LNCS, vol. 11070, pp. 682–690. Springer, Cham (2018). https://doi.org/10.1007/978-3-030-00928-1_77
8. Li, Z., He, Y., Keel, S., Meng, W., Chang, R.T., He, M.: Efficacy of a deep learning system for detecting glaucomatous optic neuropathy based on color fundus photographs. Ophthalmology **125**(8), 1199–1206 (2018)
9. Liu, H., et al.: Development and validation of a deep learning system to detect glaucomatous optic neuropathy using fundus photographs. JAMA Ophthalmol. **137**(12), 1353–1360 (2019)
10. Orlando, J.I., et al.: REFUGE challenge: a unified framework for evaluating automated methods for glaucoma assessment from fundus photographs. Med. Image Anal. **59**, 101570 (2020)
11. Paletz, S.B., Chan, J., Schunn, C.D.: Uncovering uncertainty through disagreement. Appl. Cogn. Psychol. **30**(3), 387–400 (2016)
12. Phene, S., et al.: Deep learning and glaucoma specialists: the relative importance of optic disc features to predict glaucoma referral in fundus photographs. Ophthalmology **126**(12), 1627–1639 (2019)
13. Schaekermann, M., Beaton, G., Habib, M., Lim, A., Larson, K., Law, E.: Understanding expert disagreement in medical data analysis through structured adjudication. In: Proceedings of the ACM on Human-Computer Interaction. (CSCW), vol. 3, pp. 1–23 (2019)
14. Sivaswamy, J., et al.: A comprehensive retinal image dataset for the assessment of glaucoma from the optic nerve head analysis. JSM Biomed. Imaging Data Papers **2**(1), 1004 (2015)
15. Sudre, C.H., et al.: Let's agree to disagree: learning highly debatable multirater labelling. In: Shen, D., et al. (eds.) MICCAI 2019. LNCS, vol. 11767, pp. 665–673. Springer, Cham (2019). https://doi.org/10.1007/978-3-030-32251-9_73
16. Tham, Y.C., Li, X., Wong, T.Y., Quigley, H.A., Aung, T., Cheng, C.Y.: Global prevalence of glaucoma and projections of glaucoma burden through 2040: a systematic review and meta-analysis. Ophthalmology **121**(11), 2081–2090 (2014)

Intra-operative Forecasting of Growth Modulation Spine Surgery Outcomes with Spatio-Temporal Dynamic Networks

William Mandel[1], Stefan Parent[2], and Samuel Kadoury[1,2](✉)

[1] MedICAL, Polytechnique Montreal, Montreal, QC, Canada
samuel.kadoury@polymtl.ca
[2] Sainte-Justine Hospital Research Center, Montreal, QC, Canada

Abstract. Vertebral Body Growth Modulation (VBGM) allows to treat mild to severe spinal deformations by tethering vertebral bodies together, helping to preserve lower back flexibility. Forecasting the outcome of VBGM from skeletally immature patients remains elusive with several factors involved in corrective vertebral tethering, but could help orthopaedic surgeons plan and tailor VBGM procedures prior to surgery. We introduce a novel intra-operative framework forecasting the outcomes during VBGM surgery in scoliosis patients. The method is based on spatial-temporal corrective networks, which learns the similarity in segmental corrections between patients and integrates a long-term shifting mechanism designed to cope with timing differences in onset to surgery dates, between patients in the training set. The model captures dynamic geometric dependencies in scoliosis patients, as well as ensuring long-term dependancy with temporal dynamics in curve evolution. The loss function of the network introduces a regularization term based on learned group-average piecewise-geodesic path to ensure the generated corrective transformations are coherent with regards to the observed evolution of spine corrections at follow-up exams. The network was trained on 695 3D spine models and tested on 72 patients using a set of pre-operative spine reconstructions as inputs. The spatio-temporal network predicted outputs with errors of 2.1 ± 0.9 mm in 3D anatomical landmarks, and yielding geometries similar to ground-truth reconstructions.

Keywords: Idiopathic scoliosis · Image-guided spine surgery · Vertebral tethering · Spatio-temporal networks · Outcome prediction

1 Introduction

Progressive forms of adolescent idiopathic scoliosis (AIS) can either be treated conservatively using a brace or in some cases can necessitate surgical correction and fusion to prevent further progression of the pathology [13]. Unfortunately, there is no proven method available to identify affected children or adolescents at risk of AIS which may require treatment due to progression. Consequently,

Supported by the Canada Research Chairs and NSERC Discovery Grants.

A. L. Martel et al. (Eds.): MICCAI 2020, LNCS 12261, pp. 751–760, 2020.
https://doi.org/10.1007/978-3-030-59710-8_73

the application of treatment is often delayed until a significant deformity is detected, resulting in less than optimal treatment [2]. Surgery can correct the deformity, but with the limitation of having to fuse the spine over segments of vertebrae, this significantly reduces spinal mobility and increases the risk of pain and osteoarthritis in later years [4].

With the advent of new vertebral body growth modulation (VBGM) techniques, patients can receive early management of their curves to prevent further progression and reduce the need for surgical spine fusion [14,15]. VBGM relates to the instrumentation of segments with vertebral implants which links together vertebral bodies with cables in polypropylene. Once attached, the fusion-less approach induces external forces on the spinal column's outer curve, modulating the pressure spread applied on growth plates in the spine. It has shown promise for skeletally immature patients [6] and allows to retain spine flexibility [16].

A prediction model identifying patients responding to VBGM could have an impact for patients at risk of progression. There is thus a clinical need for a predictive model guiding therapeutic approaches during bone maturation stages. Recent studies have shown the promise of using either geometric-based 3D parameters and a linear regression model [12], while spatio-temporal models were proposed in order to predict the progression of the disease [9] or fusion-based surgical outcomes [11]. However, these focused on single geometric models, with no temporal integration of corrective changes. Previous works in medical imaging using recurrent networks focused on extracting flows for motion in image sequences [1] or exploiting serial time data to classify surgical gestures [7]. On the other hand, separating anatomical changes from dynamical flows in long-term data with varying time gaps following surgery has not been explored.

We present a predictive network forecasting patient outcomes during intra-operative VBGM procedures, combining the 3D spines at diagnosis, pre-surgery, and in the OR in the decubitus position prior to instrumentation to predict the correction after 1 and 2 years (Fig. 1). The approach trains the spatio-temporal corrective network (STCN), which is regularized with a piecewise-geodesic trajectory described in latent space. The path is trained end-to-end from a dataset of follow-up reconstructions of the vertebral column obtained during AIS assessment following VBGM surgery. The contributions of this paper are: 1) the introduction of a dynamic network for isolating the correction transforms associated to modulation forces during bone growth, 2) learning spatial relationships between spine segments during correction and 3) accommodating for varying onset and surgical timing with long-term shifting.

2 Methods

2.1 Spine Modeling

The inputs to the STCN are 3D reconstructions of the spine, modeled by $\mathbf{S} = \{\mathbf{s}_1, \ldots, \mathbf{s}_m\}$, describing a vector of m segments (vertebrae). For each level $\mathbf{s}_i$, mesh-based geometries are obtained with vertex coordinates possessing direct

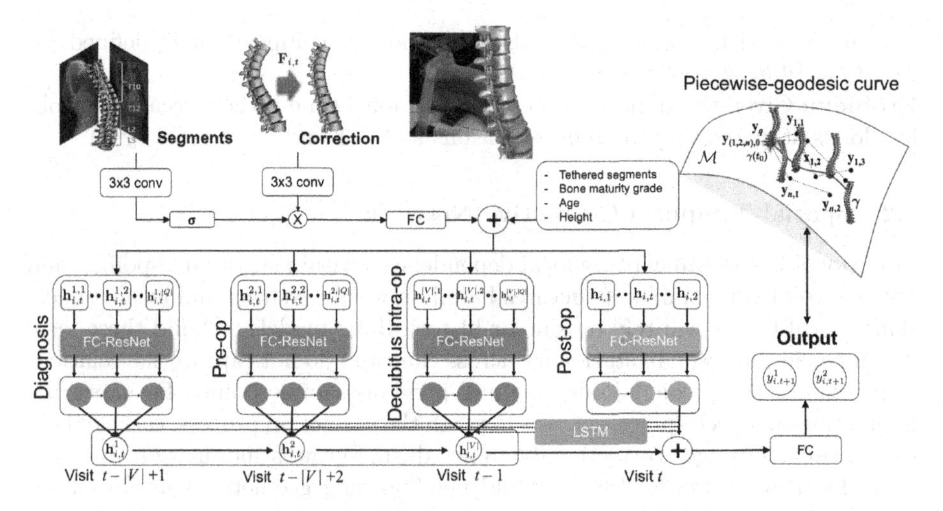

Fig. 1. The proposed STCN architecture for prediction of spine surgery outcomes. (Top) The spine correction gating process which allows to extract the changes directly related to spine segments representations by modulating the propagation of geometric data, using fully-connected layers (FC) and convolutional layers (Conv) to extract information. (Left) In order to accommodate for varying timings between disease onset and surgical date, an LSTM attention module allows to shift long-term dependencies. For each patient visit, an FC-ResNet module allows to model the changes between time visits. (Middle) The short-term dependency is captured with a final module. (Right) A united prediction of the geometric spine shape at 1 and 2 yrs is generated simultaneously from the multi-task framework. The prediction is regularized with a pre-trained trajectory model, capturing the natural evolution of spinal curvature in AIS, treated with VBGM.

links between each object. Every shape $\mathbf{s}_i$ is also marked with a series of anatomical points to determine a vertebral coordinate system, with $A = [T_1, T_2, \ldots, T_m]$, and $T_i = \{\theta, \tau\}$ defines a 6-dof transformation between levels. Therefore, the global spine appearance is described as an array of the previous rigid transforms linked to every vertebra, taking into account the entire set of registrations:

$$\mathbf{y}_i = [T_1, \mathbf{s}_1; T_1 \circ T_2, \mathbf{s}_2, \ldots, T_1 \circ T_2 \circ \ldots \circ T_m, \mathbf{s}_m] \tag{1}$$

with $\mathbf{y}_i$ as the feature array that models the position and orientation of the vertebral constellation, based on recursive compositions. Corrections to the spine vector $\mathbf{y}_i$ can be defined by an array inter-vertebral transforms applied to each level. A global representation of the deformation is obtained by extending $\mathbf{y}_i$ to an absolute vector describing the geometric spine. In this scenario, rigid transforms are defined from the lower vertebra's coordinate system, corresponding to the main direction of the vertebral object, described with respect to the centroid of the mesh model. The 6-dof registration combines both the displacement t and orientation θ, such that $T = \{\theta, \tau\}$ of $\mathbf{s}_i$ can be determined as

$y = \theta x + \tau$ with $x, y, \tau \in \Re^3$. The operation of composition is defined as: $T_1 \circ T_2 = \{\theta_1\theta_2, \theta_1\tau_2 + \tau_1\}$.

Problem: Given the spine $\mathbf{y}_i$ representations until time t, the forecasting problem looks at generating the representations at $t = 1$ and $t = 2$ years ($\mathbf{y}_{i,t}$).

2.2 Spatial-Temporal Corrective Network

Traditional CNNs can capture local dependencies within single time-points, and are able to identify similar structures based on weight sharing and local connections. Local CNNs will rely on similar historical 3D models to build these local dependencies but remain stationary across time and do not capture the relationship between the predicted outcome and previous sample points. By capturing interactions between spine segments through the correction process, this can help to determine stronger corrective dynamics during growth modulation. Inspired by traffic flow networks [18], we build neighbouring geometric transformation matrices to maintain the dependency within the correction vectors.

In this work, spine changes are exploited directly to drive the correction information propagation between segments. Hence, a correction gating mechanism is proposed to capture in a hierarchical fashion, the dependency in the geometrical changes in spine segments. We treat a specific spine segment i denoted as $\mathbf{Y}_{i,t} \in \mathbb{R}^{R \times R \times 2}$, with time point t and R indicating the neighbouring segmental size. We extract the associated correction transformations from previous l time points (i.e., time interval $t - l + 1$ to t). The measured transformation matrices are subsequently stacked and defined as $\mathbf{F}_{i,t} \in \mathbb{R}^{R \times R \times 2l}$, and $2l$ is the number of transformation matrices for every patient visit. Due to the fact combined transformation matrices integrate previous dynamic interactions for spine segment i, a CNN is used to capture the geometric correction interactions between segments, denoted as $\mathbf{F}_{i,t}$, with $\mathbf{F}_{i,t}^{(0)}$ defined as the input. Hence, for all layers k in the network, the correction matrices are given as:

$$\mathbf{F}_{i,t}^{(k)} = \text{ReLU}(\mathbf{W}_f^{(k)} * \mathbf{Y}_{i,t}^{(k-1)} + \mathbf{b}_f^{(k)}) \tag{2}$$

where the learned parameters are $\mathbf{W}_f^{(k)}$ and $\mathbf{b}_f^{(k)}$. For every layer, similarity between corrections in spine segments is captured by limiting the geometric information $\mathbf{Y}_{i,t}^{(k)}$ going through the network gates, by modulating the inputs:

$$\mathbf{Y}_{i,t}^{(k)} = \text{ReLU}(\mathbf{W}^k * \mathbf{Y}_{i,t}^{k-1} + \mathbf{b}^k \otimes \sigma(\mathbf{F}_t^{i,k-1})) \tag{3}$$

with $\otimes$ performing the product between each element of the tensors. Following the convolutional layers, both a flattening and a FC layer are used to obtain the representation $\mathbf{y}_{i,t}$. The output representation of segment i is denoted as $\mathbf{h}_{i,t}$.

In the AIS population, the time difference between visits can range in months, requiring long-term information to be propagated. To handle variations in time differences between AIS onset and surgery, we implement an attention mechanism which, for each predicted time, considers long-term information to improve prediction capabilities. Here, $q \in Q$ are time intervals (defined in months) used to handle the variations in examination and VBGM surgery time. Furthermore, a FC-ResNet is used to retrain the geometric features for each serial visit $v \in V$:

$$\mathbf{h}_{i,t}^{v,q} = \text{FC-ResNet}([\mathbf{y}_{i,t}^{v,q}; \mathbf{e}_{i,t}^{v,q}], \mathbf{h}_{i,t}^{v,q-1}) \tag{4}$$

with $\mathbf{h}_{i,t}^{v,q}$ representing the spine region i at time q from a previous visit v for the predicted time t; $\mathbf{e}_{i,t}^{v,q}$ are the features related to the additional external parameters used as inputs to the model such as the number of tethered segments, age, height and bone maturity status, incorporated with $\mathbf{y}_{i,t}^{v,q}$. Therefore, the term $\mathbf{h}_{i,t}^{v,q}$ includes long-term and short-term geometric temporal changes.

In order to handle the time shift between onset and surgery time, we implement a mechanism to focus on variations in previous visits and obtain a representation which is weighted from each previous visit. Specifically, the representation $\mathbf{h}_{i,t}^{v}$ for each previous visit is found by the sum of all visits within the time frame v close to the input case and denoted as:

$$\mathbf{h}_{i,t}^{v} = \sum_{q \in Q} \alpha_{i,t}^{v,q} \mathbf{h}_{i,t}^{v,q} \tag{5}$$

which averages each previous representation with a weighted term $\alpha_{i,t}^{v,q}$, assigning the importance of the time gap q in a visit $v \in V$. Specifically, the definition of the weight parameters are as:

$$\alpha_{i,t}^{v,q} = \frac{\exp(\Phi(\mathbf{h}_{i,t}^{v,q}, \mathbf{h}_{i,t}))}{\sum_{v \in V} \exp(\Phi(\mathbf{h}_{i,t}^{v,q}, \mathbf{h}_{i,t}))} \tag{6}$$

which compares the learned spatio-temporal representation using the content-based score term $\Phi(\mathbf{n}, \mathbf{m}) = \mathbf{v}^T \tanh(\mathbf{W_H n} + \mathbf{W_x m} + \mathbf{b_x})$, with $\mathbf{W_H}$, $\mathbf{W_x}$, $\mathbf{v}$ and $\mathbf{b_x}$ as the learned parameters. The term Φ measures the similarity in correction flows between pairs of spines. As a final step, we use a LSTM taking as input the result of each independent module. This allows to keep the temporal information present by exploiting the time interval representations:

$$\widehat{\mathbf{h}}_{i,t}^{v} = \text{LSTM}(\mathbf{h}_{i,t}^{v}, \widehat{\mathbf{h}}_{i,t}^{v-1}). \tag{7}$$

Here, the representation of the long-term correction similarity is captured by the final time $\widehat{\mathbf{h}}_{i,t}^{V}$, which combines the predominant features of past visits.

2.3 Training with a Piecewise-Geodesic Correction Trajectory

To predict the shape at future time, the long-term ($\widehat{\mathbf{h}}_{i,t}^{V}$) and short-term ($\mathbf{h}_{i,t}$) representations are combined to a single $\mathbf{h}_{i,t}^{c}$, which is subsequently sent to a fully connected layer to generate the output values for each spine segment i, denoted as $y_{i,t+1}^{1}$ and $y_{i,t+1}^{2}$ (1 and 2 yrs). The output predictions are given as:

$$[y_{i,t+1}^{1}, y_{i,t+1}^{2}] = \tanh(\mathbf{W}_{fa} \mathbf{h}_{i,t}^{c} + \mathbf{b}_{fa}) \tag{8}$$

with $\mathbf{W}_{fa}$ and $\mathbf{b}_{fa}$ as the learned parameters. To ensure the generated models are regular with latent space representations, we constrain the generation of predicted models $y_{i,t}$ to map closer to a continuous piecewise-geodesic trajectory denoted as $\gamma(t)$. This maps the data in latent space to model the spatiotemporal evolution in correction with VBGM procedures [11]. The curve embedding is generated from neighbors showing similar features in the training set from pre-operative models within latent space, yielding neighborhoods $\mathcal{N}(\mathbf{y}_{i,t})$ with data-points across the correction continuum. Hence, a Riemannian space from samples $\mathbf{y}_{i,t}$ is created, with i a patient model acquired at time t ($t = 0$ as the diagnosis). Assuming the domain manifold is well represented and the piecewise-geodesic curve spanning the time spectrum, a regression of the labelled samples can be obtained in $\mathbb{R}^D$, generating regular paths in the spatio-temporal domain $\mathbb{R}^d$.

A discretized regression technique is used [3] for implementation purposes. The geodesic path is obtained from the neighborhoods of samples obtained along γ, by optimizing a loss function which minimizes the geodesic distance between samples and the regressed curve, as well as the L_2 metric of the first and second derivatives of γ. These last two terms captures the velocity and acceleration of the regressed path, which ensures a continuous curve with smooth transitions during changes of direction. This allows to navigate through the path, capturing the mean rigid transformations linked to correction, which are obtained from n patient training cases at follow-up. Hence, the loss function integrating the network output combined with the piecewise-geodesic curve is defined as:

$$\mathcal{L} = \sum_{i=1}^{n} \beta(y^1_{i,t+1} - \widehat{y}^1_{i,t+1})^2 + (1-\beta)(y^2_{i,t+1} - \widehat{y}^2_{i,t+1})^2 + \lambda((y^1_{i,t+1} - \gamma(1)) + (y^2_{i,t+1} - \gamma(2))) \tag{9}$$

with β balancing the outcome reconstructions and λ weighting the regularization term. The network allows to generate complete shapes composed of a series linked vertebrae, both at one or two years, based on actual corrected models $(\widehat{y}^1_{i,t+1}, \widehat{y}^2_{i,t+1})$. Each vertebral model is annotated with anatomical points, representing the precise morphological changes during the tethering process.

3 Experiments

The spatio-temporal correction network was trained end-to-end using a dataset of 695 3D spine models obtained from pre-, intra-op and post-op X-ray images. The cohort included 139 patients showing thoracic and thoraco/lumbar spine deformities. Patients were imaged at five time points: 2 prior to surgery, 1 during VBGM and 2 at follow-up. A mono-centric IRB-approved prospective trial enrolled AIS cases with a curve angulation in the 30–60° range. Timings between initial onset and surgery ranged between 1 and 11 months. All of the vertebral levels had landmark points placed on the endplates and pedicle extremities by a radiologist. The two 3D reconstructions used prior to surgery were acquired in a standing position, using the EOS imaging system (Paris, France). During VBGM

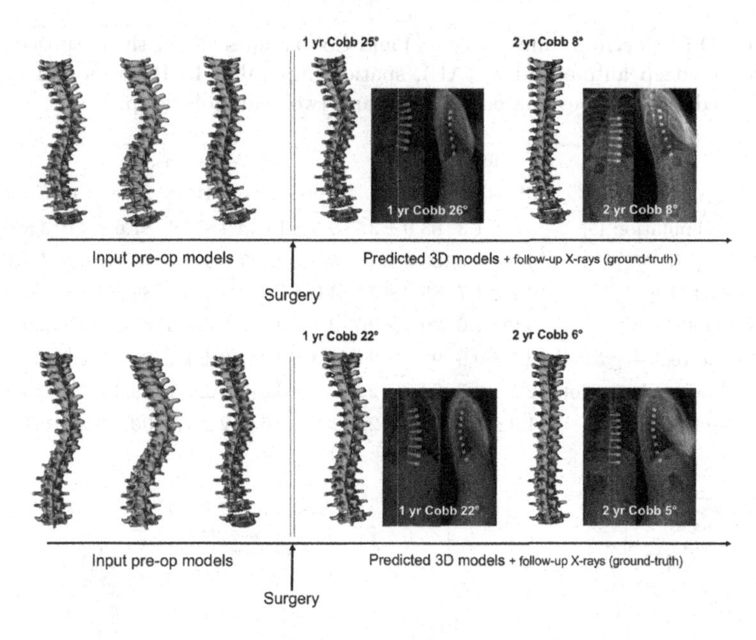

Fig. 2. Two sample results of predicted 3D spine models, at 1 and 2 years following surgery. The input 3D models are shown on left side, while the forecasted 3D models with the actual follow-up X-rays are shown on the right, comparing Cobb angulations.

surgery, once the patient is positioned on the operating table in a decubitus position, a set of C-arm images (0 and 90°) were acquired before tethering, used to generate the third 3D spine model. All geometric spine were reconstructed in 3D automatically using an unsupervised method [8]. Batch size was set at 32, with a learning rate of 0.001, a dropout rate of 0.5, $\beta = 0.5$, $\lambda = 0.3$ and using the Adam optimizer. The timing parameters were set at $|Q| = 1$ and $|V| = 5$.

The accuracy of the predictive network was assessed in 72 unseen patients undergoing VBGM surgery (age 13±4, mean Cobb angle 48° ±11°), with follow-up at $t = 1, 2$ years. For the forecasted spine shapes, the 3D RMS error in anatomical points, the Dice score from segmented vertebrae and differences in main angulation were computed. Table 1 shows the results, including from a deep auto-encoder [17], a spatio-temporal manifold framework [10] and a ST-ResNet LSTM [19]. Results from biomechanical simulations are also shown. Figure 2 presents examples in generated 3D models for 13 y.o. and 12 y.o. patients at 1 and 2 years (with thoracic deformations), which are more frequent in the AIS population. Spine models forecasted at follow-up demonstrate that the STCN framework allows to capture not only the morphology of vertebrae segments, but also generates consistent and coherent models between time points.

Table 1. 3D RMS errors (mm), Dice (%) and Cobb angles (°) for the proposed STCN, compared to deep auto-encoders (AE), spatio-temporal (ST) ResNets and manifold models. Predictions are evaluated with one and two years follow-up.

	1-year visit			2-year visit		
	3D RMS	Dice	Cobb	3D RMS	Dice	Cobb
Biomec. simulation [5]	3.8 ± 1.3	83.0 ± 3.7	3.5 ± 1.0	4.4 ± 2.4	81.6 ± 4.0	3.8 ± 0.9
Deep AE [17]	4.9 ± 3.5	80.1 ± 4.3	5.7 ± 2.7	7.0 ± 4.7	72.1 ± 6.0	7.4 ± 4.1
ST-ResNet [19]	3.4 ± 1.7	85.2 ± 3.6	4.0 ± 2.4	4.2 ± 3.8	82.5 ± 3.9	4.9 ± 3.2
ST-Manifold [11]	3.0 ± 1.0	90.4 ± 2.7	2.1 ± 0.8	3.2 ± 1.5	87.6 ± 3.3	2.9 ± 1.0
STCN (no reg. + gating)	2.5 ± 0.8	91.4 ± 2.6	2.0 ± 0.6	2.9 ± 1.1	88.2 ± 2.9	2.7 ± 0.7
STCN (no regularization)	2.2 ± 0.7	92.2 ± 2.5	1.9 ± 0.6	2.5 ± 1.1	89.3 ± 3.0	2.5 ± 0.5
Proposed STCN	1.9 ± 0.8	93.0 ± 2.5	1.7 ± 0.5	2.2 ± 1.1	90.1 ± 2.9	2.1 ± 0.5

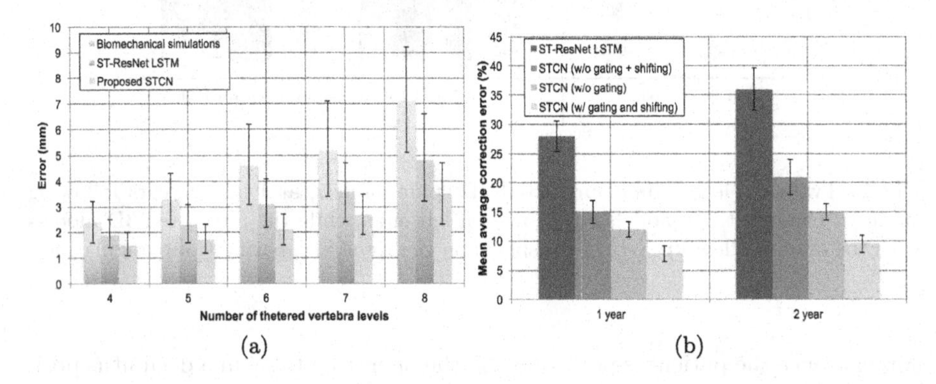

Fig. 3. (a) Effect of different tethering levels, comparing results with a ST-ResNet approach. (b) Evaluation of gating and time shifting mechanisms at 1 and 2 years.

To assess the reliability of the trained network with regards to increasing tethered vertebral levels and changes in predicted outcomes, we computed the error of the 1 and 2 year predictions with increasing levels of implanted levels (for both lumbar and thoracic regions). Figure 3 presents the improvement of the STCN network with regards to increasing tethered vertebrae, as well as with the gating and shifting mechanisms, which help to handle the higher variability in the correction effects, compared to state-of-the-art networks.

4 Conclusion

In this paper, we proposed a spatio-temporal corrective network (STCN) for spine surgery using vertebral body tethering, capturing the dynamic changes between 3 pre-operative time points (diagnosis, pre-op and intra-op decubitus) to predict the correction one and two years following surgery. Anatomical coherence is ensured with a piecewise-geodesic growth trajectory model used in the training of the network, yielding results with statistically insignificant differences to ground-truth follow-up 3D models. The LSTM-based network models spatio-temporal evolution of correction changes, which can be helpful for planning of VBGM and guiding surgery. Future work will consist of training the model with multi-centric data and real-time intra-operative evaluation.

References

1. Bai, W., et al.: Recurrent neural networks for aortic image sequence segmentation with sparse annotations. In: Frangi, A.F., Schnabel, J.A., Davatzikos, C., Alberola-López, C., Fichtinger, G. (eds.) MICCAI 2018. LNCS, vol. 11073, pp. 586–594. Springer, Cham (2018). https://doi.org/10.1007/978-3-030-00937-3_67
2. Beauséjour, M., Roy-Beaudry, M., Goulet, L., Labelle, H.: Patient characteristics at the initial visit to a scoliosis clinic: a cross-sectional study in a community without school screening. Spine **32**(12), 1349–1354 (2007)
3. Boumal, N., Absil, P.A.: A discrete regression method on manifolds and its application to data on SO (n). IFAC Proc. Volumes **44**(1), 2284–2289 (2011)
4. Cheng, J.C., et al.: Adolescent idiopathic scoliosis. Nat. Rev. Dis. Primers **1**(1), 1–21 (2015)
5. Cobetto, N., Parent, S., Aubin, C.E.: 3D correction over 2 years with anterior vertebral body growth modulation: a finite element analysis of screw positioning, cable tensioning and postop functional activities. Clinical Biome. **51**, 26–33 (2018)
6. Crawford III, C.H., Lenke, L.G.: Growth modulation by means of anterior tethering resulting in progressive correction of juvenile idiopathic scoliosis: a case report. JBJS **92**(1), 202–209 (2010)
7. DiPietro, R., et al.: Recognizing surgical activities with recurrent neural networks. In: Ourselin, S., Joskowicz, L., Sabuncu, M.R., Unal, G., Wells, W. (eds.) MICCAI 2016. LNCS, vol. 9900, pp. 551–558. Springer, Cham (2016). https://doi.org/10.1007/978-3-319-46720-7_64
8. Humbert, L., de Guise, J., Aubert, B., Godbout, B., Skalli, W.: 3D reconstruction of the spine from biplanar X-rays using parametric models based on transversal and longitudinal inferences. Med. Eng. Phy. **31**(6), 681–87 (2009)
9. Kadoury, S., Mandel, W., Roy-Beaudry, Nault, M.L., Parent S: 3-D morphology prediction of progressive spinal deformities from probabilistic modeling of discriminant manifolds. IEEE Trans. Med. Imaging **36**(5), 1194–1204 (2017)
10. Mandel, W., Turcot, O., Knez, D., Parent, S., Kadoury, S.: Spatiotemporal manifold prediction model for anterior vertebral body growth modulation surgery in idiopathic scoliosis. In: Frangi, A.F., Schnabel, J.A., Davatzikos, C., Alberola-López, C., Fichtinger, G. (eds.) MICCAI 2018. LNCS, vol. 11073, pp. 206–213. Springer, Cham (2018). https://doi.org/10.1007/978-3-030-00937-3_24
11. Mandel, W., Turcot, O., Knez, D., Parent, S., Kadoury, S.: Prediction outcomes for anterior vertebral body growth modulation surgery from discriminant spatiotemporal manifolds. Int. J. Comput. Assist. Radiol. Surg. **14**(9), 1565–1575 (2019). https://doi.org/10.1007/s11548-019-02041-w
12. Nault, M.L., Mac-Thiong, J.M., Roy-Beaudry, M., Turgeon, I., Parent, S.: Three-dimensional spinal morphology can differentiate between progressive and nonprogressive patients with adolescent idiopathic scoliosis at the initial presentation: a prospective study. Spine **39**(10), E601 (2014)
13. Parent, S., Newton, P., Wenger, D.: Adolescent idiopathic scoliosis: etiology, anatomy, natural history, and bracing. Instr. Course Lect. **54**, 529–536 (2005)

14. Samdani, A.F., Ames, R.J., Kimball, J.S., Pahys, J.M., Grewal, H., Pelletier, G.J., Betz, R.R.: Anterior vertebral body tethering for idiopathic scoliosis: two-year results. Spine **39**(20), 1688–1693 (2014)
15. Samdani, A.F., et al.: Anterior vertebral body tethering for immature adolescent idiopathic scoliosis: one-year results on the first 32 patients. Eur. Spine J. **24**(7), 1533–1539 (2014). https://doi.org/10.1007/s00586-014-3706-z
16. Skaggs, D.L., Akbarnia, B.A., Flynn, J.M., Myung, K., Sponseller, P., Vitale, M.: A classification of growth friendly spine implants. J. Pediatr. Orthop. **34**(3), 260–274 (2014)
17. Thong, W., Parent, S., Wu, J., Aubin, C.-E., Labelle, H., Kadoury, S.: Three-dimensional morphology study of surgical adolescent idiopathic scoliosis patient from encoded geometric models. Eur. Spine J. **25**(10), 3104–3113 (2016). https://doi.org/10.1007/s00586-016-4426-3
18. Yao, H., Tang, X., Wei, H., Zheng, G., Li, Z.: Revisiting spatial-temporal similarity: A deep learning framework for traffic prediction. In: Proceedings of the AAAI Conference on Artificial Intelligence, vol. 33, pp. 5668–5675 (2019)
19. Zhang, J., Zheng, Y., Qi, D.: Deep spatio-temporal residual networks for citywide crowd flows prediction. In: Thirty-First AAAI Conference on Artificial Intelligence (2017)

Self-supervision on Unlabelled OR Data for Multi-person 2D/3D Human Pose Estimation

Vinkle Srivastav[1(✉)], Afshin Gangi[1,2], and Nicolas Padoy[1]

[1] ICube, University of Strasbourg, CNRS, IHU Strasbourg, Strasbourg, France
{srivastav,padoy}@unistra.fr

[2] Radiology Department, University Hospital of Strasbourg, Strasbourg, France

Abstract. 2D/3D human pose estimation is needed to develop novel intelligent tools for the operating room that can analyze and support the clinical activities. The lack of annotated data and the complexity of state-of-the-art pose estimation approaches limit, however, the deployment of such techniques inside the OR. In this work, we propose to use knowledge distillation in a teacher/student framework to harness the knowledge present in a large-scale non-annotated dataset and in an accurate but complex multi-stage teacher network to train a lightweight network for joint 2D/3D pose estimation. The teacher network also exploits the unlabeled data to generate both hard and soft labels useful in improving the student predictions. The easily deployable network trained using this effective self-supervision strategy performs on par with the teacher network on *MVOR+*, an extension of the public MVOR dataset where all persons have been fully annotated, thus providing a viable solution for real-time 2D/3D human pose estimation in the OR.

Keywords: Human pose estimation · Knowledge distillation · Data distillation · Operating room · Low-resolution images

1 Introduction

The current surge in deep learning research has made its way into the operating room (OR) with the goal to develop novel intelligent context-aware assistance systems [13,19]. Human pose estimation (HPE), a key requirement to build such systems, is a computer vision task aiming at localizing human body parts, also called keypoints, either in 2D or 3D. Owing to the current advancements in deep learning, HPE has made remarkable progress and works reliably on *in the wild* images. These advances can be attributed to the emergence of large scale supervised datasets, such as COCO [12] and Human3.6 [9], and to the modern deep learning architectures. These architectures employ multi-stage deep neural networks to achieve state-of-the-art performance. For example, top approaches for the task of 2D pose estimation on the COCO dataset use a two-stage approach, in which the first stage

Electronic supplementary material The online version of this chapter (https://doi.org/10.1007/978-3-030-59710-8_74) contains supplementary material, which is available to authorized users.

A. L. Martel et al. (Eds.): MICCAI 2020, LNCS 12261, pp. 761–771, 2020.
https://doi.org/10.1007/978-3-030-59710-8_74

determines the person's bounding boxes and the second stage estimates the keypoints for each bounding box. This multi-stage design using high-capacity deep neural networks in both stages helps to achieve better accuracy. The run-time performance of such a design is however considerably low. More practical solutions such as Mask-RCNN [7] or OpenPose [3] use low-capacity and single-stage networks for the same task to achieve better run-time performance, but the accuracy of these system is low compared to the multi-stage systems. Therefore, one key challenge for the deployment of HPE network inside the OR is not only to give accurate predictions, but also to be fast using a light-weight and single-stage end-to-end design, as needed for real-time applications.

Another key challenge is the decrease in the accuracy of state-of-the art approaches when applied to a different target domain, such as the OR [17]. This decrease can be compensated by finetuning the approach on manual annotations from the target domain, as done by [1,6,10,11] for the OR. Obtaining manually annotated data is however time-consuming and expensive. Specifically, in the case of the operating room, utilizing crowd sourcing technique such as Amazon Mechanical turk would be impractical due to obvious privacy concerns. Therefore, techniques that can harness knowledge from non-annotated datasets captured in the target environment are extremely useful.

In this paper, our work lies at the intersection of knowledge distillation [8,22] and data distillation [15] and exploits these techniques to solve the task of multi-person 2D/3D HPE. We use knowledge distillation to transfer knowledge from an accurate, larger, and multi-stage teacher network to a practical, smaller, and single-stage student network. The idea of knowledge distillation has been adapted to different problems. In [8], authors use the probability output vector of a teacher network as soft-labels to train a student network for multi-class classification. The student network learns jointly from the soft-labels generated by the teacher output and from the hard-labels given by the ground truth. Similarly, in [22], authors use this approach for single-person 2D HPE by jointly training the student network from the soft output heatmaps of a teacher network and the hard ground truth heatmaps. Both the student and the teacher network work on the fully supervised dataset, and the soft output of the teacher network serves as additional useful labels along with the hard labels obtained from the supervised ground truth.

In this work, we aim at applying knowledge distillation when only *non-annotated* data is available in the target domain. Instead of using supervised annotations, we propose to use data distillation to generate labels automatically from the non-annotated dataset. We run a complex teacher network which ensembles output predictions on geometrically transformed input images. Unlike the standard use of data distillation [15], which only exploits hard predictions obtained by removing the low confidence keypoints, we also use the soft predictions from the confidence value for each keypoint.

As student network, we use a low capacity single stage network based on Mask-RCNN. The architecture of the student network is inspired from [5] and further extended to effectively use the *hard-set* and *soft-set* for joint 2D/3D

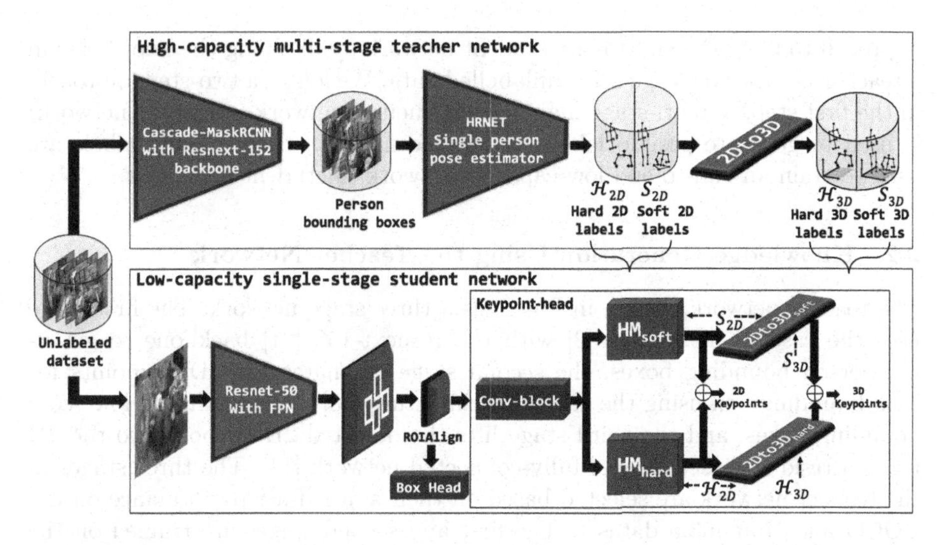

Fig. 1. Proposed self-supervised methodology for joint 2D/3D keypoint estimation using the teacher/student paradigm. The teacher network is a three-stage network which uses the unlabelled dataset to extract person bounding boxes, estimate 2D keypoints, and regress 2D keypoints to 3D. It generates soft and hard pseudo labels to be used by the student network. The student network is a single-stage network and effectively utilizes the soft and hard pseudo labels to jointly estimate the 2D and 3D keypoints.

multi-person HPE in the OR. By utilizing our approach, the student network reaches an accuracy on par with the teacher network.

Another specific issue in the OR is to preserve the privacy of patients and clinicians while performing computer vision tasks. Human pose estimation on low-resolution images has been suggested to improve the privacy [16], like for activity recognition [4]. We therefore also extend our approach to deliver accurate poses on low-resolution images. This turns out to be very effective on the OR images compared to the teacher network, even with a 12× downsampling factor.

2 Methodology

2.1 Problem Overview

Given a monocular RGB input image $\mathcal{I}$ of size $W \times H$, our task is to detect the 2D and 3D body keypoints for multiple persons using a single efficient end-to-end network. The 2D keypoints $\mathcal{P}_{2D} \in \mathbb{R}^{m \times n \times 2}$ are in image coordinates, and the 3D keypoints $\mathcal{P}_{3D} \in \mathbb{R}^{m \times n \times 3}$ are in the root-relative coordinates (where the root-joint is set to be the origin and all other joints are measured w.r.t root-joint). Here, m is the number of persons, and $n = 17$ is the number of joints for each pose. We also consider low-resolution images for the same task, which have very small input sizes. To tackle the problem, we utilize a teacher/student approach to train the end-to-end student network by distilling the knowledge in a teacher network on large-scale unlabelled data. We follow a two-step approach: in the first step, a multi-stage high-capacity neural network (a teacher network) is used to generate pseudo labels; in the second step, these pseudo labels are used to train an end-to-end low-capacity network (a student network).

2.2 Knowledge Generation Using the Teacher Network

The teacher network, shown in Fig. 1, is a three-stage network: The first stage uses the cascade-mask-rcnn [2] with the resnext-152 [21] backbone to generate person bounding boxes, the second stage estimates the 2D keypoints for each bounding box using the HRNet architecture [18] after discarding low-score bounding boxes, and the third stage lifts the detected 2D keypoints to the 3D using a residual-based 2-layer fully-connected network [14]. The three stages in the teacher network are selected based on their state-of-art performance on the COCO and Human3.6 dataset. The first and second stages are trained on the COCO dataset [12] and the third stage is trained on the Human3.6 dataset [9]. Multi-level scaling and flipping transformations are applied in the first and the second stage to obtain good quality person bounding boxes and 2D keypoints. However, errors can still be present in the keypoints and are encoded in the keypoint confidence scores. Therefore, we construct two sets of pseudo-labels: the soft-set $\mathcal{S}$ and the hard-set $\mathcal{H}$. The soft-set $\mathcal{S} = \{\mathcal{S}_{2D}, \mathcal{S}_{3D}\}$ consists of soft 2D keypoints and soft 3D keypoints. Soft 2D keypoints $\mathcal{S}_{2D} \in \mathbb{R}^{m \times n \times 3}$ are obtained by storing the confidence value for each keypoint along with their coordinates. The last dimension in $\mathbb{R}^{m \times n \times 3}$ represents the channel for the confidence value. $\mathcal{S}_{2D}$ is sent to the third stage to obtain the soft 3D keypoints $\mathcal{S}_{3D}$. Similarly, the hard-set $\mathcal{H} = \{\mathcal{H}_{2D}, \mathcal{H}_{3D}\}$ consists of hard 2D keypoints and hard 3D keypoints. $\mathcal{H}_{2D}$ is obtained by only keeping the high confidence 2D keypoints and discarding the low confidence keypoints. $\mathcal{H}_{3D}$ is obtained by passing the $\mathcal{H}_{2D}$ to the lifting network. We show in the experiments that these two sets provide useful learning signals when used to train the student. In the next section, we show how we exploit these two sets of pseudo labels for effectively training the student network.

2.3 Knowledge Distillation in the Student Network

The student network presented in Fig. 1 is an end-to-end network based on Mask-RCNN that jointly predicts the 2D and 3D poses. We replace the mask head of the Mask-RCNN network with a keypoint-head for joint 2D and 3D pose estimation. The keypoint-head accepts the fixed size proposals from the ROIAlign layer and passes them through 8 conv-block layers to generate the features. These features are upsampled using a deconv and bi-linear upsampling layer into two branches to generate 17 channel heatmaps corresponding to each body joint. The first branch upsamples the features to generate the heatmaps HM_{soft}, and the second branch upsamples them to generate the heatmaps HM_{hard}. The HM_{soft}

and HM_{hard} heatmaps are connected to their respective lifting networks i.e $2Dto3D_{soft}$ and $2Dto3D_{hard}$ to lift the incoming 2D keypoints to 3D.

Training: Training of the network follows the same framework as Mask-RCNN along with the additional losses coming from the keypoint-head. In the keypoint-head, we compute 2D and 3D losses L_{2D} and L_{3D} to estimate the 2D and 3D keypoints. L_{2D} consists of soft and hard 2D keypoint losses. The soft 2D keypoint loss L_{2Dsoft} is obtained by first multiplying HM_{soft} with the corresponding confidence values from the last channel of $\mathcal{S}_{2D}$ and then computing its cross-entropy loss with $\mathcal{S}_{2D}$. The hard 2D keypoint loss L_{2Dhard} is obtained by calculating the cross-entropy loss between HM_{hard} and $\mathcal{H}_{2D}$. Similarly, the 3D loss L_{3D} consists of soft and hard 3D keypoint losses. Soft 3D keypoint loss L_{3Dsoft} is obtained by taking the smooth L1 loss between $\mathcal{S}_{3D}$ and the output of $2Dto3D_{soft}$ using the input $\mathcal{S}_{2D}$, and hard 3D keypoint loss L_{3Dhard} is obtained by taking the smooth L1 loss between $\mathcal{S}_{3D}$ and the output of $2Dto3D_{hard}$ using the input $\mathcal{H}_{2D}$. All four losses are added together to obtain the loss for the keypoint-head L_{kpt}. The overall loss is the sum of L_{kpt} with the standard Faster-RCNN loss, i.e. the bounding box classification and regression loss, and the region proposal loss.

Inference: During inference, the 2D keypoints are computed by taking the arg-max over each channel from the mean output of HM_{soft} and HM_{hard}, and the 3D keypoints are computed by calculating the 2D keypoints from HM_{soft} and HM_{hard} using arg-max, passing the 2D keypoints to the respective 2Dto3D lifting network, and averaging the 3D output.

3 Experiments and Results

3.1 Training and Testing Dataset

We use the public MVOR dataset [17] as a test set. We also use an unlabeled dataset of 80k images to generate pseudo labels and train our networks. The unlabeled dataset consists of images captured on different days than MVOR to ensure that there is no overlap with the test set. We split the dataset into 77k train and 3k validation images. Note that the public MVOR dataset [17] is not fully annotated: not all persons are annotated and 2D keypoints were only annotated for 10 upper body parts. To use this dataset in the standardized COCO evaluation framework for bounding boxes and 2D keypoints, we extend the dataset with additional annotations. Before the extension, MVOR consists of 4699 person bounding boxes, 2926 2D upper body poses with 10 keypoints, and 1061 3D upper body poses. The extended MVOR dataset, called *MVOR+*, consists of 5091 person bounding boxes and 5091 body poses with 17 keypoints in the COCO format. The original size of all the images is 640×480. We also conduct experiments with downsampled images using the scaling factors $8\times$, $10\times$, and $12\times$, yielding images of size 80×64, 64×48, and 53×40.

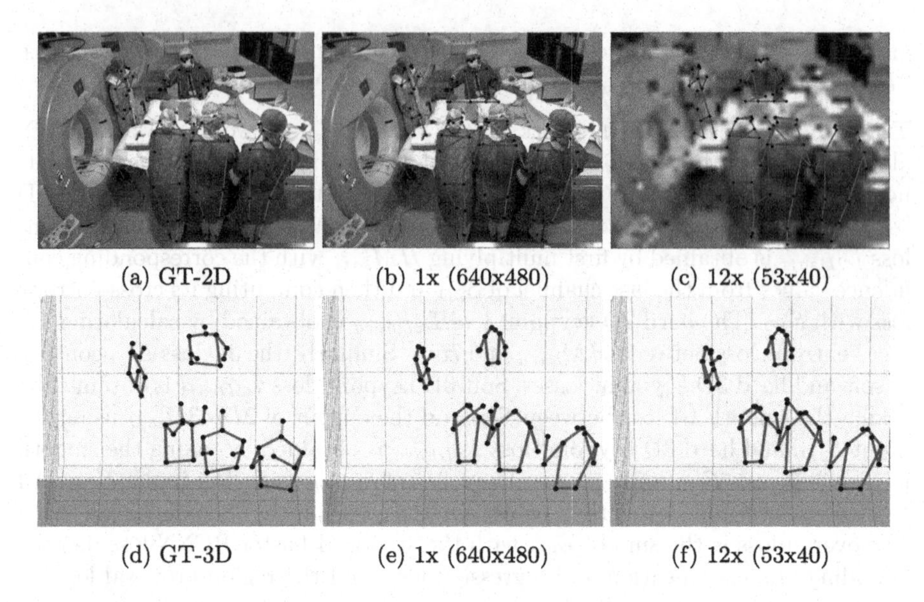

(a) GT-2D (b) 1x (640x480) (c) 12x (53x40)

(d) GT-3D (e) 1x (640x480) (f) 12x (53x40)

Fig. 2. Qualitative results for 2D and 3D keypoints estimation from the student network (ORPose_all) at original and downsampled image sizes. GT-2D and GT-3D are the visualization results from 2D and 3D ground truth keypoints respectively. Since we are not predicting the scale of the 3D pose in the camera frame, we use the depth of the root node from the ground truth as scale to generate this visualization.

3.2 Evaluation Metrics

We use the Average Precision (AP) metric from COCO to evaluate bounding box detection and 2D keypoint estimation. We use the 3D mean per joint position error (MPJPE) in millimeters to evaluate the 3D keypoints. The 2D keypoint AP is computed using 17 keypoints in COCO format whereas the MPJPE metric is computed using the 8 keypoint from upper-body pose (shoulder, elbow, hand, hip). High AP is desired for the bounding box detection and 2D keypoint estimation, while a low MPJPE score is desired for 3D keypoint estimation. For 3D evaluation, the 3D ground-truth keypoints are expressed in the camera coordinate frame and each joint in the 3D pose is subtracted from the pelvis root-joint (taken as the mean of left and right hips) to obtain the root-relative pose.

3.3 Experiments

The student network is trained differently for two sets of experiments, yielding the networks ORPose_fixed_sx ($\mathbf{s}$ = 1, 8, 10, 12) and ORPose_all. ORPose_fixed_sx is trained using either images of the original size ($\mathbf{s} = 1$) or low-resolution images at a fixed scaling factor ($\mathbf{s}$ = 8, 10, 12). When feeding the networks, low-resolution images are first upsampled to match the original input size.

Evaluation of networks ORPose_fixed_sx is done on the same scale they are trained on. In the second experiment, ORPose_all is trained using original and downsampled images with a random downsampling factor. This is similar to the scaling data augmentation technique, but we consider here a very low-resolution scenario where the input image is downsampled up to 12×. We choose the random scale such that for 30% of the training time there is no downsampling, for 35% the downsampling scale is randomly chosen between 2 and 8, and for the remaining 35% of training time downsampling scale is randomly chosen between 8 and 12. The intention to train the network using this strategy is to obtain a single model that can work on high-resolution images and should also perform considerably better on the low-resolution images. The base learning rate for ORPose_fixed_1x is set to 1e−3 for 5k total number of iterations with a step decay of 0.1 after 2k, 3k, and 4k iterations; the base learning rate for ORPose_fixed_sx ($\mathbf{s} = 8, 10, 12$) is set to 1e−2 for 10k total number of iterations with a step decay of 0.1 after 7k, 8k, and 9k iterations; the base learning rate for ORPose_all is set to 1e−1 for 20k total number of iterations with step decay of 0.1 after 14k, 16k, and 18k iterations. The downsampling and upsampling operation is performed using bilinear interpolation. We use a detectron2 framework [20] to run all the experiments on two V100 NVidia GPUs using the distributed data parallelism framework of PyTorch. We use a batch size of 32 and the stochastic gradient solver as the optimizer for all the experiments.

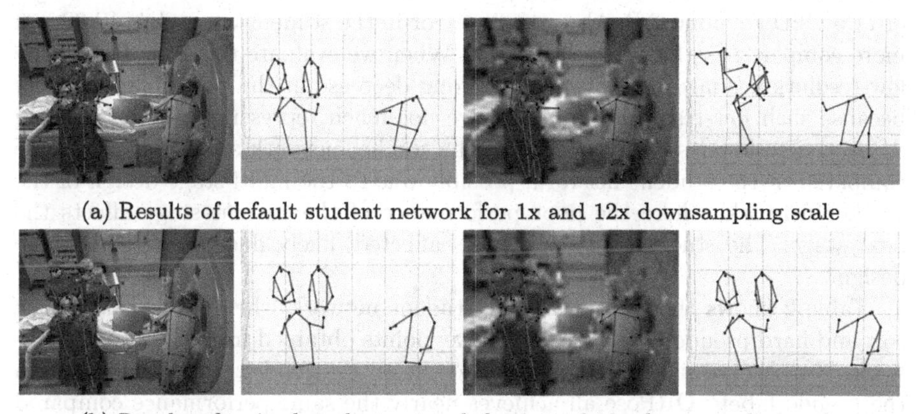

(a) Results of default student network for 1x and 12x downsampling scale

(b) Results of trained student network for 1x and 12x downsampling scale

Fig. 3. Comparative qualitative results for the default and the trained student networks. (a) The default student network uses the pre-trained COCO and Human3.6 weights. (b) The trained student network exploits the soft and hard pseudo labels obtained from the teacher network. The left side shows the 2D/3D visualization results at 1× scale, and the right side shows the 2D/3D visualization results at 12× scale.

Table 1. Baseline results on *MVOR+* for teacher and student networks when no training is performed on OR data. Higher AP and lower MPJPE are better. Student and teacher networks are evaluated at original and low-resolution sizes. The aim is to train the student to reach the same performance as the teacher at high resolution (1×).

Network	#Params	GFLOPs	Scale	Box-AP	Keypoint2D-AP	Keypoint3D-MPJPE
Teacher	250.1M	1048.8	1×	**53.81**	**57.78**	**134.88**
			8×	39.03	29.28	170.25
			10×	31.90	18.60	203.89
			12×	24.38	8.89	260.83
Student	67.9M	215.0	1×	**52.77**	**46.17**	**147.17**
			8×	40.21	27.02	168.10
			10×	34.12	20.06	181.69
			12×	29.19	14.42	194.35

3.4 Results

Table 1 shows the results for the teacher and the student networks on *MVOR+*, along with the network parameter complexity, before training on OR data. These networks are initialized from the COCO and Human3.6 pre-trained network weights. We evaluated both on the original image size (1×) and downsampled images of scale 8×, 10×, and 12×. As shown in the Table 1, there exists a margin of 11.6% 2D keypoint AP. Also, the 3D error in the student network is 12.30 mm more compared to the teacher network. When we evaluate these models on the low-resolution images, we observe a strong decrease in the performance, likely because such low-resolution images were not much represented in the training dataset. The low-resolution results of the teacher network are somewhat worse compared to the student network, possibly due to the multi-stage design of the teacher network, where the poor performance of the current stage affects the next stage. The student network is less affected, likely due to its single-stage design.

Table 2 shows the results for our student network when trained using the soft and hard pseudo labels for 2D/3D keypoints obtained from the teacher network. We observe improved performance in all the models when trained with the pseudo labels. ORPose_all achieves nearly the same performance compared to the models trained for specific scale low-resolution images. Performance of ORPose_all on the high-resolution images nearly reaches the teacher network and on the low-resolution images this network performs much better. This is illustrated in the qualitative results shown in Fig. 2 and Fig. 3. Additional qualitative results for other model variants are available in the supplementary material.

Table 2. Results of our student network evaluated at original size and low resolution images. ORPose_fixed_sx (s = 1, 8, 10, 12) are trained and evaluated at fixed scale. ORPose_all is a single model trained on random size low resolution and high resolution images, and evaluated on original size images and fixed scale downsampled images.

Student network	Scale	Box-AP	Keypoint2D-AP	Keypoint3D-MPJPE
ORPose_fixed_1×	1×	**50.87**	55.20	134.23
ORPose_fixed_8×	8×	49.50	53.50	137.40
ORPose_fixed_10×	10×	49.01	51.98	137.71
ORPose_fixed_12×	12×	48.23	49.88	138.83
ORPose_all	1×	50.59	**55.80**	**134.13**
	8×	49.57	53.31	136.45
	10×	49.25	52.12	136.95
	12×	47.54	49.51	138.35

Table 3. Ablation study on the student network, by comparing to a single branch trained using hard, soft and hard+soft labels. We achieve the best result when using our proposed two-branch design for both 2D and 3D keypoint estimation.

Student Network	Scale	Box-AP	Keypoint2D-AP	Keypoint3D-MPJPE
Single-branch(hard)	1×	50.61	54.73	145.77
Single-branch(soft)	1×	**51.04**	54.70	134.20
Single-branch(hard+soft)	1×	50.95	55.11	152.28
Double-branch(hard+soft)	1×	50.59	**55.80**	**134.13**

3.5 Ablation Study

To evaluate the effect of soft-labels on the student network, we keep only one branch for 2D/3D keypoint estimation i.e only one heatmap layer and one 2Dto3D network. We train this single branch keypoint-head with only hard labels, only soft labels, and both hard and soft labels. To train for both the hard and soft labels, the 2D losses are computed using the same heatmap layer and 3D losses are computed using the same 2Dto3D network. As shown in Table 3, we observe that training with the hard labels hurt the 3D keypoint estimation, and training using only the soft labels achieves good overall results. 2D keypoint estimation is however inferior compared to our two-branch design trained for soft and hard losses.

4 Conclusion

In this work, we tackle joint 2D/3D pose estimation from monocular RGB images and propose a self-supervised approach to train an end-to-end and easily deployable model for the OR. We use data distillation to exploit non-annotated data and knowledge distillation to benefit from the high-quality predictions of a multi-stage high capacity pose estimation model. Our approach does not require any ground truth poses from the OR and evaluation on the *MVOR+* dataset suggests its effectiveness. We further demonstrate that the proposed network can yield accurate results on low-resolution images, as needed to ensure privacy, even using a downsampling rate of 12×.

Acknowledgements. This work was supported by French state funds managed by the ANR within the Investissements d'Avenir program under reference ANR-16-CE33-0009 (DeepSurg). The authors would also like to thank the members of the Interventional Radiology Department at University Hospital of Strasbourg for their help in generating the dataset.

References

1. Belagiannis, V., et al.: Parsing human skeletons in an operating room. Mach. Vis. Appl. **27**(7), 1035–1046 (2016). https://doi.org/10.1007/s00138-016-0792-4
2. Cai, Z., Vasconcelos, N.: Cascade R-CNN: high quality object detection and instance segmentation. IEEE Trans. Pattern Anal. Mach. Intell. (2019). https://doi.org/10.1109/TPAMI.2019.2956516
3. Cao, Z., Simon, T., Wei, S.E., Sheikh, Y.: Realtime multi-person 2D pose estimation using part affinity fields. In: Proceedings of the IEEE conference on computer vision and pattern recognition. pp. 7291–7299 (2017)
4. Chou, E., et al.: Privacy-preserving action recognition for smart hospitals using low-resolution depth images. NeurIPS Workshop on Machine Learning for Health (ML4H) (2018)
5. Dabral, R., Gundavarapu, N.B., Mitra, R., Sharma, A., Ramakrishnan, G., Jain, A.: Multi-person 3D human pose estimation from monocular images. In: 2019 International Conference on 3D Vision (3DV). pp. 405–414. IEEE (2019)
6. Hansen, L., Siebert, M., Diesel, J., Heinrich, M.P.: Fusing information from multiple 2D depth cameras for 3D human pose estimation in the operating room. Int. J. Comput. Assist. Radiol. Surg. **14**(11), 1871–1879 (2019)
7. He, K., Gkioxari, G., Dollár, P., Girshick, R.: Mask R-CNN. In: Proceedings of the IEEE International Conference on Computer Vision. pp. 2961–2969 (2017)
8. Hinton, G., Vinyals, O., Dean, J.: Distilling the knowledge in a neural network. arXiv preprint arXiv:1503.02531 (2015)
9. Ionescu, C., Papava, D., Olaru, V., Sminchisescu, C.: Human3. 6m: Large scale datasets and predictive methods for 3D human sensing in natural environments. IEEE Trans. Pattern Anal. Mach. Intell. **36**(7), 1325–1339 (2013)
10. Kadkhodamohammadi, A., Gangi, A., de Mathelin, M., Padoy, N.: Articulated clinician detection using 3D pictorial structures on RGB-D data. Med. Image Anal. **35**, 215–224 (2017)
11. Kadkhodamohammadi, A., Gangi, A., de Mathelin, M., Padoy, N.: A multi-view RGB-D approach for human pose estimation in operating rooms. In: IEEE Winter Conference on Applications of Computer Vision (WACV). pp. 363–372. IEEE (2017)
12. Lin, T.Y., et al.: Microsoft coco: common objects in context. In: Fleet, D., Pajdla, T., Schiele, B., Tuytelaars, T. (eds.) ECCV 2014. LNCS, vol. 8693, pp. 740–755. Springer, Cham (2014). https://doi.org/10.1007/978-3-319-10602-1_48

13. Main-Hein, L., et al.: Surgical data science: enabling next-generation surgery. Nature Biomed. Eng. **1**, 691–696 (2017)
14. Martinez, J., Hossain, R., Romero, J., Little, J.J.: A simple yet effective baseline for 3D human pose estimation. In: Proceedings of the IEEE International Conference on Computer Vision. pp. 2640–2649 (2017)
15. Radosavovic, I., Dollár, P., Girshick, R., Gkioxari, G., He, K.: Data distillation: towards omni-supervised learning. In: Proceedings of the IEEE conference on computer vision and pattern recognition. pp. 4119–4128 (2018)
16. Srivastav, V., Gangi, A., Padoy, N.: Human pose estimation on privacy-preserving low-resolution depth images. In: Shen, D. (ed.) MICCAI 2019. LNCS, vol. 11768, pp. 583–591. Springer, Cham (2019). https://doi.org/10.1007/978-3-030-32254-0_65
17. Srivastav, V., Issenhuth, T., Abdolrahim, K., de Mathelin, M., Gangi, A., Padoy, N.: Mvor: A multi-view RGB-D operating room dataset for 2D and 3D human pose estimation. In: MICCAI-LABELS workshop (2018)
18. Sun, K., Xiao, B., Liu, D., Wang, J.: Deep high-resolution representation learning for human pose estimation. In: Proceedings of the IEEE Conference on Computer Vision and Pattern Recognition (2019)
19. Vercauteren, T., Unberath, M., Padoy, N., Navab, N.: Cai4cai: the rise of contextual artificial intelligence in computer-assisted interventions. Proc. IEEE **108**(1), 198–214 (2019)
20. Wu, Y., Kirillov, A., Massa, F., Lo, W.Y., Girshick, R.: Detectron2. https://github.com/facebookresearch/detectron2 (2019)
21. Xie, S., Girshick, R., Dollár, P., Tu, Z., He, K.: Aggregated residual transformations for deep neural networks. In: Proceedings of the IEEE Conference on Computer Vision and Pattern Recognition. pp. 1492–1500 (2017)
22. Zhang, F., Zhu, X., Ye, M.: Fast human pose estimation. In: Proceedings of the IEEE Conference on Computer Vision and Pattern Recognition. pp. 3517–3526 (2019)

Knowledge Distillation from Multi-modal to Mono-modal Segmentation Networks

Minhao Hu[1,2], Matthis Maillard[2(✉)], Ya Zhang[1(✉)], Tommaso Ciceri[2], Giammarco La Barbera[2], Isabelle Bloch[2], and Pietro Gori[2]

[1] CMIC, Shanghai Jiao Tong University, Shanghai, China
ya_zhang@sjtu.edu.cn
[2] LTCI, Télécom Paris, Institut Polytechnique de Paris, Paris, France
matthis.maillard@telecom-paris.fr

Abstract. The joint use of multiple imaging modalities for medical image segmentation has been widely studied in recent years. The fusion of information from different modalities has demonstrated to improve the segmentation accuracy, with respect to mono-modal segmentations, in several applications. However, acquiring multiple modalities is usually not possible in a clinical setting due to a limited number of physicians and scanners, and to limit costs and scan time. Most of the time, only one modality is acquired. In this paper, we propose KD-Net, a framework to transfer knowledge from a trained multi-modal network (teacher) to a mono-modal one (student). The proposed method is an adaptation of the generalized distillation framework where the student network is trained on a subset (1 modality) of the teacher's inputs (n modalities). We illustrate the effectiveness of the proposed framework in brain tumor segmentation with the BraTS 2018 dataset. Using different architectures, we show that the student network effectively learns from the teacher and always outperforms the baseline mono-modal network in terms of segmentation accuracy.

1 Introduction

Using multiple modalities to automatically segment medical images has become a common practice in several applications, such as brain tumor segmentation [11] or ischemic stroke lesion segmentation [10]. Since different image modalities can accentuate and better describe different tissues, their fusion can improve the segmentation accuracy. Although multi-modal models usually give the best results, it is often difficult to obtain multiple modalities in a clinical setting due to a limited number of physicians and scanners, and to limit costs and scan time. In many cases, especially for patients with pathologies or for emergency, only one modality is acquired.

M. Hu and M. Maillard—Contributed equally to this paper.

Electronic supplementary material The online version of this chapter (https://doi.org/10.1007/978-3-030-59710-8_75) contains supplementary material, which is available to authorized users.

A. L. Martel et al. (Eds.): MICCAI 2020, LNCS 12261, pp. 772–781, 2020.
https://doi.org/10.1007/978-3-030-59710-8_75

Two main strategies have been proposed in the literature to deal with problems where multiple modalities are available at training time but some or most of them are missing at inference time. The first one is to train a generative model to synthesize the missing modalities and then perform multi-modal segmentation. In [13], the authors have shown that using a synthesized modality helps improving the accuracy of classification of brain tumors. Ben Cohen et al. [1] generated PET images from CT scans to reduce the number of false positives in the detection of malignant lesions in livers. Generating a synthesized modality has also been shown to improve the quality of the segmentation of white matter hypointensities [12]. The main drawback of this strategy is that it is computationally cumbersome, especially when many modalities are missing. In fact, one needs to train one generative network per missing modality in addition to a multi-modal segmentation network.

The second strategy consists in learning a modality-invariant feature space that encodes the multi-modal information during training, and that allows for all possible combinations of modalities during inference. Within this second strategy, Havaei et al. proposed HeMIS [4], a model that, for each modality, trains a different feature extractor. The first two moments of the feature maps are then computed and concatenated in the latent space from which a decoder is trained to predict the segmentation map. Dorent et al. [3], inspired by HeMIS, proposed U-HVED where they introduced skip-connections by considering intermediate layers, before each down-sampling step, as a feature map. This network outperformed HeMIS on BraTS 2018 dataset. In [2], instead of fusing the layers by computing mean and variance, the authors learned a mapping function from the multiple feature maps to the latent space. They claimed that computing the moments to fuse the maps is not satisfactory since it makes each modality contribute equally to the final result which is inconsistent with the fact that each modality highlights different zones. They obtained better results than HeMIS on BraTS 2015 dataset. This second strategy has good results only when one or two modalities are missing, however, when only one modality is available, it has worse results than a model trained on this specific modality. This kind of methods is therefore not suitable for a clinical setting where only one modality is usually acquired, such as pre-operative neurosurgery or radiotherapy.

In this paper, in contrast to the previously presented methods, we propose a framework to *transfer knowledge* from a multi-modal network to a mono-modal one. The proposed method is based on *generalized knowledge distillation* [9] which is a combination of distillation [5] and privileged information [14]. *Distillation* has originally been designed for classification problems to make a small network (Student) learn from an ensemble of networks or from a large network (Teacher). It has been applied to image segmentation in [8,15] where the *same input modalities* have been used for the Teacher network and the Student network. In [15], the Student learns from the Teacher only thanks to a loss term between their outputs. In [8], the authors also constrained the intermediate layers of the Student to be similar to the ones of the Teacher. With a different perspective, the framework of *privileged information* was designed to boost the performance of a Student model by learning from both the training data and a Teacher model with privileged and additional information. In generalized knowledge distillation, one uses distillation to extract useful knowledge from the privileged information of the Teacher [9]. In our case, Teacher and Student have the same architecture (i.e. same number of parameters) but the Teacher can learn from multiple input modalities (additional information) whereas the Student from only one. The proposed framework is based on two encoder-decoder networks, which have demonstrated to work well in image segmentation [7], one for the Student and one for the Teacher. Importantly, the proposed framework is generic since it works for different architectures of the encoder-decoder networks. Each encoder summarizes its input space to a latent representation that captures important information for the segmentation. Since the Teacher and the Student process different inputs but aim at extracting the same information, we make the assumption that their first layers should be different, whereas the last layers and especially the latent representations (i.e. bottleneck) should be similar. By forcing the latent space of the Student to resemble the one of the Teacher, we make the hypothesis that the Student should learn from the additional information of the Teacher. To the best of our knowledge, this is the first time that the generalized knowledge distillation strategy is adapted to guide the learning of a mono-modal student network using a multi-modal teacher network. We show the effectiveness of the proposed method using the BraTS 2018 dataset [11] for brain tumor segmentation.

The paper is organized as follows. First, we present the proposed framework, called KD-Net and illustrated in Fig. 1, and how the Student learns from the Teacher and the reference segmentation. Then, we present the implementation details and the results on the BraTS 2018 dataset [11].

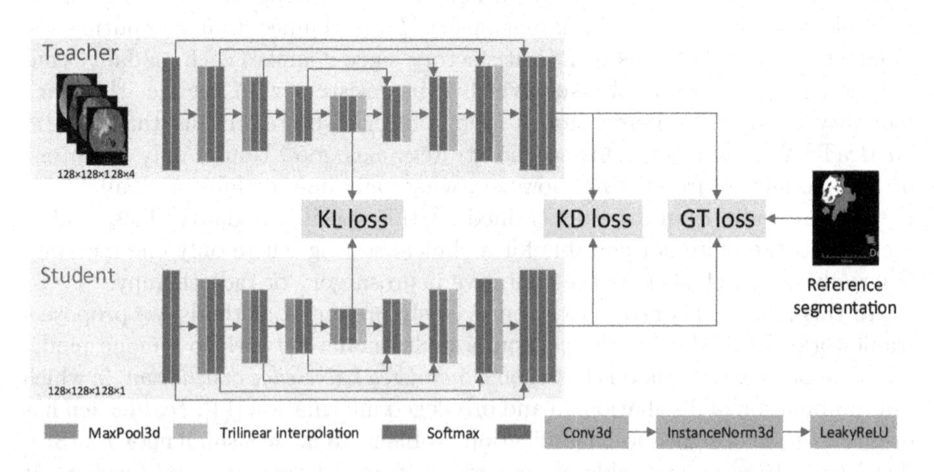

Fig. 1. Illustration of the proposed framework. Both Teacher and Student have the same architecture adapted from nnUNet [7]. First, the Teacher is trained using only the reference segmentation (GT loss). Then, the student network is trained using all proposed losses: KL loss, KD loss and GT loss.

2 KD-Net

The goal of the proposed framework is to train a mono-modal segmentation network (Student) by leveraging the knowledge from a well-trained multi-modal segmentation network (Teacher). Except for the number of input channels, both networks have the same encoder-decoder architecture with skip connections. The multi-modal input $\mathrm{x}^{\mathrm{i}} = \{x_n^i, n = 1...N\}$ is the concatenation of the N modalities for the ith sample of the dataset. Let E_t and D_t (resp. E_s and D_s) denote the encoder and decoder parts of the Teacher (resp. Student). The Teacher network $f_t(\mathrm{x}^{\mathrm{i}}) = D_t \circ E_t(\mathrm{x}^{\mathrm{i}})$ receives as input multiple modalities whereas the student network $f_s(x_k^i) = D_s \circ E_s(x_k^i)$ only one modality x_k^i, k being a fixed integer between 1 and N.

We first train the Teacher, using only the reference segmentation as target. Then, we train the Student using three different losses: the knowledge distillation term, the dissimilarity between the latent spaces, and the reference segmentation loss. Note that the weights of the Teacher are frozen during the training of the Student and the error of the Student is not back-propagated to the Teacher.

The first two terms allow the Student to learn from the Teacher by using the soft prediction of the latter as target and by forcing the encoded information (i.e. bottleneck) of the Student to be similar to the one of the Teacher. The last term makes the predicted segmentation of the Student similar to the reference segmentation.

2.1 Generalized Knowledge Distillation

Following the strategy of generalized knowledge distillation [9], we transfer useful knowledge from the additional information of the Teacher to the Student using the soft label targets of the Teacher. These are computed as follows:

$$s_i = \sigma(f_t(\mathrm{x}^{\mathrm{i}})/T) \tag{1}$$

where σ is the softmax function and T, the temperature parameter, is a strictly positive value. The parameter T controls the softness of the target, and the higher it is, the softer the target. The idea of using soft targets is to uncover relations between classes that would be harder to detect with hard labels. The effectiveness of using a temperature parameter to soften the labels was demonstrated in [5].

The knowledge distillation loss is defined as:

$$L_{KD} = \sum_i [(1 - Dice(s_i, \sigma(f_s(x_k^i)))) + BCE(s_i^*, \sigma(f_s(x_k^i))] \tag{2}$$

where $Dice$ is the Dice score, BCE the binary cross-entropy measure and s_i^* the binary prediction of the teacher. We need to binarize s_i since the soft labels cannot be used in the binary cross-entropy. The dice score ($Dice$) measures the similarity of the shape of two ensembles. Hence, it globally measures how the Teacher and Student's segmentation maps are close to each other. By contrast, the binary cross-entropy (BCE) is computed for each pixel independently and therefore it is a local measure. We use the combination of these two terms to globally and locally measure the distance between the Student prediction and the Teacher soft labels.

2.2 Latent Space

We speculate that Teacher and Student, having different inputs, should also encode differently the information in the first layers, the ones related to low-level image properties, such as color, texture and edges. By contrast, the deepest layers closer to the bottleneck, and related to higher level properties, should be more similar. Furthermore, we make the assumption that an encoder-decoder network encodes the information to correctly segment the input images in its latent space. Based on that, we propose to force the Student to learn from the additional information of the Teacher encoded in its bottleneck (and partially in the deepest layers) by making their latent representations as close as possible. To this end, we apply the Kullback-Leibler (KL) divergence as a loss function between the teacher and student's bottlenecks:

$$L_{KL}(p,q) = \sum_i \sum_j q_i(j) \log\left(\frac{q_i(j)}{p_i(j)}\right) \tag{3}$$

where p_i (resp. q_i) are the flattened and normalized vector of the bottleneck $E_s(x_k^i)$ (resp $E_t(\mathrm{x}^i)$). Note that this function is not symmetric and we put the vectors in that order because we want the distribution of the Student's bottleneck to be similar to the one of the Teacher.

2.3 Objective Function

We add a third term to the objective function to make the predicted segmentation as close as possible to the reference segmentation. It is the sum of the Dice loss (*Dice*) and the binary cross-entropy (*BCE*) for the same reasons as in Sect. 2.1. We call it L_{GT}:

$$L_{GT} = \sum_i \left[(1 - Dice(y_i, \sigma(f_s(x_k^i)))) + BCE(y_i, \sigma(f_s(x_k^i))\right]. \tag{4}$$

where y_i denotes the reference segmentation of the ith sample in the dataset.

The complete objective function is then:

$$L = \lambda L_{KD} + (1-\lambda)L_{GT} + \alpha L_{KL} \tag{5}$$

with $\lambda \in [0,1]$ and $\alpha \in \mathbb{R}^+$. The imitation parameter λ balances the influence of the reference segmentation with the one of the Teacher's soft labels. The greater the λ value, the greater the influence of the Teacher's soft labels. The α parameter is instead needed to balance the magnitude of the KL loss with respect to the other two losses.

3 Results and Discussion

3.1 Dataset

We evaluate the performance of the proposed framework on a publicly available dataset from the BraTS 2018 Challenge [11]. It contains MR scans from 285 patients with four modalities: T1, T2, T1 contrasted-enhanced (T1ce) and Flair. The goal of the challenge is to segment three sub-regions of brain tumors: whole tumor (WT), tumor core (TC) and enhancing tumor (ET). We apply a central crop of size $128 \times 128 \times 128$ and a random flip along each axis for data augmentation. For each modality, only non-zero voxels have been normalized by subtracting the mean and dividing by standard deviation. Due to memory and time constraint, we subsample the images to the size $64 \times 64 \times 64$.

3.2 Implementation Details

We adopt the encoder-decoder architecture described in Fig. 1. Empirically, we found that the best parameters for the objective function are $\lambda = 0.75$, $T = 5$ and $\alpha = 10$. We used Adam optimizer for 500 epochs with a learning rate equal to 0.0001 that is multiplied by 0.2 when the validation loss has not decreased for 50 epochs. We run a three fold cross validation on the 285 training cases of BraTS 2018. The training of the baseline, the Teacher or the Student takes approximately 12 h on a NVIDIA P100 GPU.

3.3 Results

In our experiments, the Teacher uses all four modalities (T1, T2, T1ce and Flair concatenated) and the Student uses only T1ce. We choose T1ce for the Student since this is the standard modality used in pre-operative neurosurgery or radiotherapy.

Model Comparison: To demonstrate the effectiveness of the proposed framework, we first compare it to a baseline model. Its architecture is the same as the encoder-decoder network in Fig. 1 and it is trained using only the T1ce modality as input. We also compare it to two other models, U-HVED and HeMIS, using only T1ce as input. Results were directly taken from [3]. The results are visible in Table 1. Our method outperforms U-HVED and HeMIS in the segmentation of all three tumor components. KD-Net also seems to obtain better results than the method proposed in [2] (again when using only T1ce as input). The authors show results on the BraTS 2015 dataset and therefore they are not directly comparable to KD-Net. Furthermore, we could not find online their code. Nevertheless, the results of HeMIS [4] on BraTS 2015 (in [2]) and on BraTS 2018 (in [3]) suggest that the observations of BraTS 2018 seem to be more difficult to segment. Since the method proposed in [2] has worst results than ours on a dataset that seems easier to segment, this should also be the case for the BraTS 2018 dataset. However, this should be confirmed.

Table 1. Comparison of 3 models using the dice score on the tumor regions. Results of U-HVED and HeMIS are taken from the article [3], where the standard deviations were not provided.

Model	ET	TC	WT
Baseline (nnUnet [7])	68.1 ± 1.27	80.28 ± 2.44	$\mathbf{77.06 \pm 1.47}$
Teacher (4 modalities)	69.47 ± 1.86	80.77 ± 1.18	88.48 ± 0.79
U-HVED	65.5	66.7	62.4
HeMIS	60.8	58.5	58.5
Ours	$\mathbf{71.67 \pm 1.22}$	$\mathbf{81.45 \pm 1.25}$	76.98 ± 1.54

Ablation Study: To evaluate the contribution of each loss term, we did an ablation study by removing each term from the objective function defined in Eq. 5. Table 2 shows the results using either 0 or 4 skip-connections both in the Student and Teacher networks. We observe that both the KL and KD loss improves the results with respect to the baseline model, especially for the enhanced tumor and tumor core. This also demonstrates that the proposed framework is generic and it works with different encoder-decoder architectures. More results can be found in the supplementary material.

Table 2. Ablation study of the loss terms. We compare the results of the model trained with 3 different objective functions: the baseline using only the GT loss, KD-Net trained with only the KL term and KD-Net with the complete objective function. We also tested it with 0 or 4 skip-connections for both the Student and the Teacher.

Skip connections	Model	Loss	ET	TC	WT
4	Baseline	GT	68.1 ± 1.27	80.28 ± 2.44	77.06 ± 1.47
4	Teacher	GT	69.47 ± 1.86	80.77 ± 1.18	88.48 ± 0.79
4	KD-Net	GT+KL	70.00 ± 1.51	80.85 ± 1.82	$\mathbf{77.08 \pm 1.29}$
4	KD-Net	GT+KD	69.22 ± 1.19	80.54 ± 1.66	76.83 ± 1.36
4	KD-Net	GT+KL+KD	$\mathbf{71.67 \pm 1.22}$	$\mathbf{81.45 \pm 1.25}$	76.98 ± 1.54
0	Baseline	GT	42.95 ± 3.42	69.44 ± 1.37	69.41 ± 1.52
0	Teacher	GT	42.59 ± 2.54	69.79 ± 1.63	75.93 ± 0.33
0	KD-Net	GT+KL	$\mathbf{47.59 \pm 0.98}$	$\mathbf{70.96 \pm 1.73}$	71.41 ± 1.2
0	KD-Net	GT+KD	44.8 ± 1.1	70.12 ± 2.42	70.19 ± 1.4
0	KD-Net	GT+KL+KD	46.23 ± 2.91	70.73 ± 2.47	$\mathbf{71.93 \pm 1.26}$

Qualitative Results: In Fig. 2, we show some qualitative results of the proposed framework and compare them with the ones obtained using the baseline method. We can see that the proposed framework allows the Student to discard some outliers and predict segmentation labels of higher quality. In the experiments, the student uses as input only T1ce, which clearly highlights the enhancing tumor. Remarkably, it seems that the Student learns more in this region (see

Fig. 2 and Table 1). The knowledge distilled from the Teacher seems to help the Student learn more where it is supposed to be "stronger". More qualitative results can be found in the supplementary material.

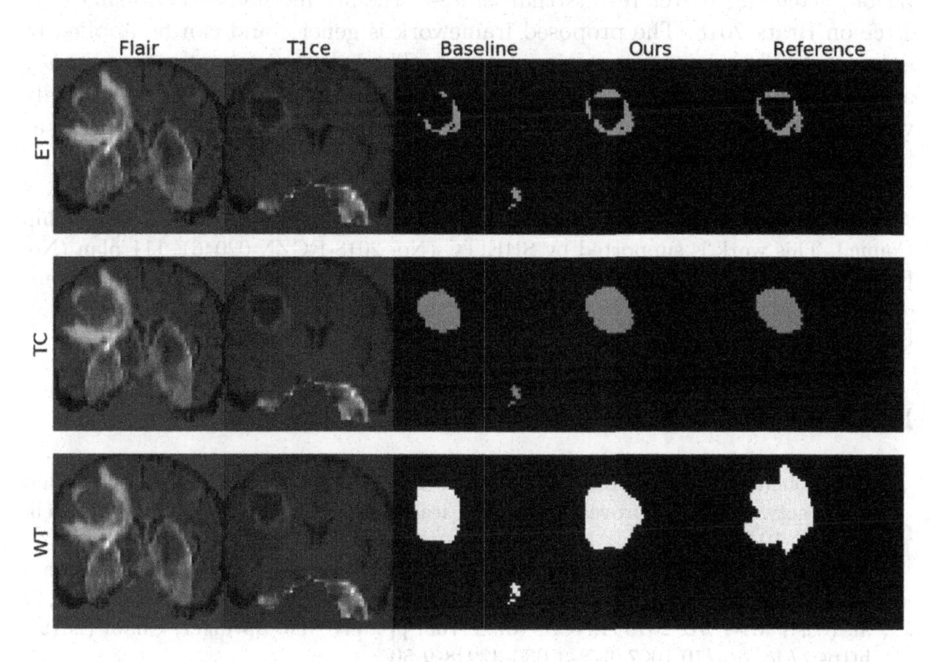

Fig. 2. Qualitative results obtained using the baseline and the proposed framework (Student). We show the slice of a subject with the corresponding 3 segmentation labels.

Observations: It is important to remark that we also tried to expand the Student network by first synthesizing another modality, such as the Flair, from the T1ce and then using it, together with the T1ce, for segmenting the tumor labels. Results were actually worse than the baseline and the computational time quite prohibitive. We also tried sharing the weights between the Teacher and the Student in the deepest layers of the networks to help transferring the knowledge. The intuition behind it was that since the bottlenecks should be the same, the information in the deepest layers should be handled identically. The results were almost identical, but slightly worse, to the ones obtained with the proposed framework presented in Fig. 1. In this paper, we used the nnUNet [7] as network for the Student and Teacher, but theoretically any other encoder-decoder architecture, such as the one in [6], could be used.

4 Conclusions

We present a novel framework to transfer knowledge from a multi-modal segmentation network to a mono-modal one. To this end, we propose to use a twofold approach. We employ the strategy of generalized knowledge distillation and, in addition, we also constrain the latent representation of the Student to be similar to the one of the Teacher. We validate our method in brain tumor segmentation, achieving better results than state-of-the-art methods when using only T1ce on Brats 2018. The proposed framework is generic and can be applied to any encoder-decoder segmentation network. The gain in segmentation accuracy and robustness to errors produced by the proposed framework makes it highly valuable for real-world clinical scenarios where only one modality is available at test time.

Acknowledgment. M. Hu is grateful for financial support from China Scholarship Council. This work is supported by SHEITC (No. 2018-RGZN-02046), 111 plan (No. BP0719010), and STCSM (No. 18DZ2270700). M. Maillard was supported by a grant of IMT, Fondation Mines-Télécom and Institut Carnot TSN, through the "Futur & Ruptures" program.

References

1. Ben-Cohen, A., et al.: Cross-modality synthesis from CT to PET using FCN and GAN networks for improved automated lesion detection. Eng. Appl. Artif. Intell. **78**, 186–194 (2018)
2. Chen, C., Dou, Q., Jin, Y., Chen, H., Qin, J., Heng, P.A.: Robust multimodal brain tumor segmentation via feature disentanglement and gated fusion. In: Chen, D., et al. (eds.) MICCAI 2019. LNCS, vol. 11766, pp. 447–456. Springer, Cham (2019). https://doi.org/10.1007/978-3-030-32248-9_50
3. Dorent, R., Joutard, S., Modat, M., Ourselin, S., Vercauteren, T.: Hetero-modal variational encoder-decoder for joint modality completion and segmentation. In: Shen, D., et al. (eds.) MICCAI 2019. LNCS, vol. 11765, pp. 74–82. Springer, Cham (2019). https://doi.org/10.1007/978-3-030-32245-8_9
4. Havaei, M., Guizard, N., Chapados, N., Bengio, Y.: HeMIS: Hetero-Modal Image Segmentation. In: Ourselin, S., Joskowicz, L., Sabuncu, M.R., Unal, G., Wells, W. (eds.) MICCAI 2016. LNCS, vol. 9901, pp. 469–477. Springer, Cham (2016). https://doi.org/10.1007/978-3-319-46723-8_54
5. Hinton, G., Vinyals, O., Dean, J.: Distilling the Knowledge in a Neural Network. Deep Learning and Representation Learning Workshop: NIPS (2015)
6. Ibtehaz, N., Rahman, M.S.: MultiResUNet: rethinking the U-Net architecture for multimodal biomedical image segmentation. Neural Networks **121**, 74–87 (2020)
7. Isensee, F., Kickingereder, P., Wick, W., Bendszus, M., Maier-Hein, K.H.: No new-net. In: Crimi, A., et al. (eds.) BrainLes 2018. LNCS, vol. 11384, pp. 234–244. Springer, Cham (2019). https://doi.org/10.1007/978-3-030-11726-9_21
8. Liu, Y., Chen, K., Liu, C., Qin, Z., Luo, Z., Wang, J.: Structured knowledge distillation for semantic segmentation. In: CVPR. pp. 2604–2613 (2019)
9. Lopez-Paz, D., Bottou, L., Schlkopf, B., Vapnik, V.: Unifying distillation and privileged information. In: ICLR (2016)
10. Maier, O., et al.: ISLES 2015 - a public evaluation benchmark for ischemic stroke lesion segmentation from multispectral MRI. Med. Image Anal. **35**, 250–269 (2017)
11. Menze, B.H., et al.: The multimodal brain tumor image segmentation benchmark (BRATS). IEEE Trans. Med. Imag. **34**(10), 1993–2024 (2015)

12. Orbes-Arteaga, M., et al.: Simultaneous synthesis of FLAIR and segmentation of white matter hypointensities from T1 MRIs. In: MIDL (2018)
13. van Tulder, G., de Bruijne, M.: Why does synthesized data improve multi-sequence classification? In: Navab, N., Hornegger, J., Wells, W.M., Frangi, A.F. (eds.) MICCAI 2015. LNCS, vol. 9349, pp. 531–538. Springer, Cham (2015). https://doi.org/10.1007/978-3-319-24553-9_65
14. Vapnik, V., Izmailov, R.: Learning using privileged information: similarity control and knowledge transfer. J. Mach. Learn. Res. **16**(61), 2023–2049 (2015)
15. Xie, J., Shuai, B., Hu, J.F., Lin, J., Zheng, W.S.: Improving fast segmentation with teacher-student learning. In: British Machine Vision Conference (BMVC) (2018)

Heterogeneity Measurement of Cardiac Tissues Leveraging Uncertainty Information from Image Segmentation

Ziyi Huang[1], Yu Gan[2], Theresa Lye[1], Haofeng Zhang[3], Andrew Laine[4], Elsa D. Angelini[4,5], and Christine Hendon[1](✉)

[1] Department of Electrical Engineering, Columbia University, New York, NY, USA
[2] Department of Electrical and Computer Engineering, The University of Alabama, Tuscaloosa, AL, USA
[3] Department of Industrial Engineering and Operations Research, Columbia University, New York, NY, USA
[4] Department of Biomedical Engineering, Columbia University, New York, NY, USA
[5] NIHR Imperial Biomedical Research Centre, ITMAT Data Science Group, Imperial College London, London, UK

Abstract. Identifying arrhythmia substrates and quantifying their heterogeneity has great potential to provide critical guidance for radio frequency ablation. However, quantitative analysis of heterogeneity on cardiac optical coherence tomography (OCT) images is lacking. In this paper, we conduct the first study on quantifying cardiac tissue heterogeneity from human OCT images. Our proposed method applies a dropout-based Monte Carlo sampling technique to measure the model uncertainty. The heterogeneity information is extracted by decoupling the intra/inter-tissue heterogeneity and tissue boundary uncertainty from the uncertainty measurement. We empirically demonstrate that our model can highlight the subtle features from OCT images, and the heterogeneity information extracted is positively correlated with the tissue heterogeneity information from corresponding histology images.

Keywords: Optical coherence tomography · Deep learning · Heterogeneity.

1 Introduction

Arrhythmia is a major type of cardiovascular disease that afflicts millions of patients in the United States [17]. A standard intervention to treat arrhythmia is radio-frequency ablation (RFA), an intra-cardiac procedure that directs a catheter and delivers heat to areas where irregular rhythms are observed. Current clinical guidance of RFA is based on low-resolution imaging modalities with limited tissue composition information [26]. For many patients, an additional procedure is required to achieve a chronic successful termination of the arrhythmia. Thus, precise intervention is challenging. Knowledge of patients' heart structure

A. L. Martel et al. (Eds.): MICCAI 2020, LNCS 12261, pp. 782–791, 2020.
https://doi.org/10.1007/978-3-030-59710-8_76

could help to optimize the intervention strategy by identifying arrhythmia substrates and avoiding critical structures. In addition, previous research has suggested that heterogeneity within the myocardium, such as fibrosis and adipose, are substrates that are potential mechanisms for the generation and maintenance of arrhythmias [1,11,28]. Thus, it is important to identify arrhythmia substrates and quantify their heterogeneity.

Optical coherence tomography (OCT) [14] is a depth-resolved imaging modality that can provide micron-level-resolution images in real time. It can detect micro-structures within cardiac tissues [4,8–10,21] and has great potential for precise guidance of RFA. Some forward-viewing OCT catheters have been designed to image cardiac tissues while the catheter is in contact with the cardiac surface [5,30]. As an interferometry imaging system which highlights the differences of reflective index of adjacent tissues, OCT can depict boundaries among tissue layers, opening a great possibility to segment various tissue compositions using either conventional machine learning or deep learning methods. However, quantitative analysis of heterogeneity within cardiac OCT images is still lacking.

Cardiac tissue heterogeneity refers to intermittent and spatially-varying structural distributions within the myocardium [11], or adipose-infiltrated fibrotic myocardium [28]. Histology heterogeneity assessment is used as a reference for cardiac fibrosis analysis, according to the conventional approach considering histology imaging as the gold standard [20]. OCT imaging provides an alternative non-invasive imaging solution, but quantification of heterogeneity in cardiac OCT images remains particularly challenging. First, unlike histology images, the cellular walls of cardiac muscle cells are not visible in OCT images. Second, there are subtle differences in pixel intensity and texture contrasts between normal myocardium and fibrotic myocardium. Third, OCT images are in grayscale, which have less tissue information than colored histology images. Other imaging modalities have a too low spatial resolution to identify cardiac fibrosis heterogeneity. These modalities rely upon indirect measures to assess cardiac tissue, such as mechanical deformation, strain, and wall thickness [2], which are only proxies to quantify cardiac fibrosis heterogeneity.

In this paper, we propose a novel deep learning framework for cardiac tissue heterogeneity measurement. We evaluate our method by comparing the heterogeneity information from OCT images with its corresponding histology images heterogeneity. In summary, this paper has following contributions:

1) This is the first paper, to the best of our knowledge, that attempts to extract the tissue heterogeneity information from OCT images via deep learning-based uncertainty measurement.
2) We empirically demonstrate that our model can highlight subtle features from OCT images, offering a way to extract the tissue heterogeneity information in a non-invasive way for real-time imaging and processing.

2 Methodology

Our goal is to measure the heterogeneity of human cardiac tissue on OCT images. As illustrated in Fig. 1, our proposed model consists of three major modules: 1) robust learning on region-based labels, 2) uncertainty measurement, and 3) heterogeneity information extraction. In the robust learning module, we use focal loss [19] to robustly learn the representative features from the region-based labels. Then, a novel dropout-based Monte Carlo sampling technique is applied for uncertainty measurement to get the uncertainty map of the prediction. Finally, the tissue heterogeneity information is extracted by decoupling the intra/inter-tissue heterogeneity and tissue boundary uncertainty from the uncertainty maps.

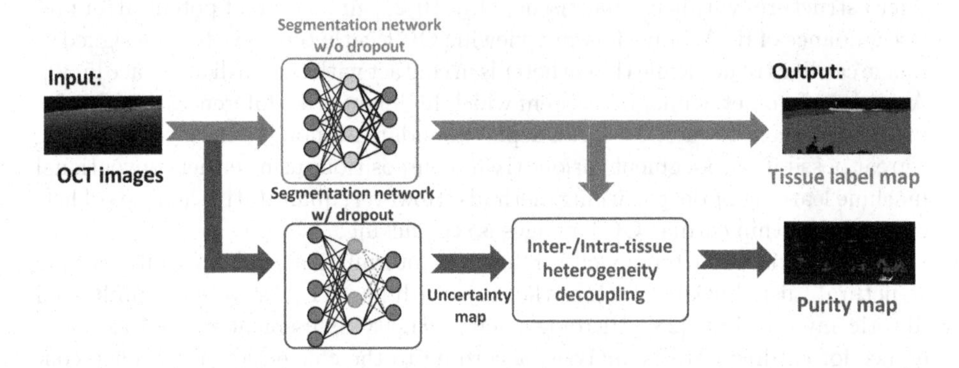

Fig. 1. The framework of our proposed algorithm in testing. For each image, an uncertainty map is generated by the comparison of predictions from the network with dropout layers on or off. Based on the uncertainty map, a separate purity map is calculated to highlight the tissue heterogeneity information.

2.1 Segmentation Framework

We use the ReLayNet [24] architecture as the base of our network. ReLayNet is an end-to-end image segmentation framework that achieves state-of-the-art performance. Inspired by [16], we add dropout layers in the three inner encoder blocks in the training phase. The introduction of dropout addresses the overfitting issue for small datasets and also lays a foundation for uncertainty quantification in the next section. The network is jointly optimized by the following loss function:

$$L = w \cdot DL + FL \tag{1}$$

where w is a trade-off parameter. FL is the focal loss [19] and DL is the Dice loss defined as:

$$DL = 1 - \frac{1}{K} \sum_{k=1}^{K} \frac{2 \sum_{x \in \Omega} (p_k(x) q_k(x))}{\sum_{x \in \Omega} (p_k(x))^2 + \sum_{x \in \Omega} (q_k(x))^2} \tag{2}$$

where $p_k(x)$ is the predicted probability of class K at the pixel position $x \in \Omega$ with $\Omega \subset \mathbb{Z}^2$, $q_k(x)$ is the one-hot ground-truth label, and K is the number of classes. Based on Dice coefficient, Dice loss evaluates pixel-wise agreement between the prediction and the ground truth.

2.2 Uncertainty Measurement

We use dropout [27] based Monte Carlo sampling technique [7,13,26] for uncertainty measurement. For each segment, a pixel-wise uncertainty map is generated to present the posterior distribution for the tissue prediction map. During the training process, the dropout layers are opened to robustly learn the discriminative features. At the test time, as shown in Fig. 1, they are turned off to provide a baseline prediction *pred*. Then, served as proximal inference, these layers are turned on again N_{it} times to get the Monte Carlo sampling maps $pred^i_{MC}$, $i \in \{1, 2, ..., N_{it}\}$. For each pixel position $x \in \Omega$, the uncertainty map UM is generated as:

$$UM(x) = \frac{1}{N_{it}} \sum_{i=1,...,N_{it}} (\mathbf{1}_{\{pred^i_{MC}(x) \neq pred(x)\}}) \tag{3}$$

where $\mathbf{1}_A$ is the indicator function of A. The value in the uncertainty map encodes the empirical confidence level of the network on its prediction. More specifically, a high uncertainty value on a pixel indicates a highly unreliable prediction, which corresponds to a low confidence.

2.3 Heterogeneity Measurement

In cardiac OCT images, the uncertainty of prediction comes from the intra/inter-tissue heterogeneity and the boundary effect. The boundary effect is caused by the subtle boundaries between two different tissue types. To remove this unwanted boundary effect, we generate a purity map by multiplying the uncertainty map with a boundary weight mask. This boundary weight mask is generated from the tissue prediction map via assigning smaller weights on the boundary. Similar to the weighting map in [24], lower weights are assigned to pixels at the boundary and higher weights are assigned to pixels in the central region of a tissue. Therefore, this boundary mask highlights the tissue boundary regions, decoupling the tissue heterogeneity and the boundary effect from the uncertainty map. Hence, the purity map is more informative to evaluate the tissue heterogeneity features. The whole process to generate a purity map is presented in Fig. 1.

Our tissue heterogeneity information is measured by homogeneity values [3]. Following previous research [3,6,15,29], the homogeneity values of both purity maps and tissue images (histology images) are calculated from gray-level co-occurrence matrix (GLCM). In particular, we average four directions ($0°, 45°, 90°$ and $135°$) to make the descriptor rotation invariant [15]. Then, we extract the homogeneity from the GLCM descriptors by using the following equation [3,6]:

$$Homogeneity = \sum_{i,j} \frac{G(i,j)}{1 + |i - j|} \tag{4}$$

where $G(i, j)$ represents the (i, j) value of the GLCM. The homogeneity value can be used to measure the closeness of the distribution of pixel intensities in the GLCM [18], representing the amount of local variations in an image [3]. We compare the homogeneity values of purity maps with those of histology images. A high similarity value, which is positive correlation, between values from these two groups indicates that our purity maps can extract the tissue heterogeneity information.

3 Experimental Results

3.1 Dataset and Experimental Setup

We acquired an *in-vitro* cohort of 185 images taken from 15 human atria and ventricles from the Thorlabs OCT system. The samples were acquired through an National Disease Research Interchange (NDRI) approved protocol from Columbia University [8]. Upon imaging, sections of samples were stained with Masson's Trichrome. We used white-light images acquired simultaneously with the OCT images to guarantee that histological slides and OCT volumes are from the same regions. This, along with inking, was used to match OCT with histology images. Each OCT image is of size 512 $\times$ 600 or 512 $\times$ 800 pixels with a field of view of 2.51 mm $\times$ 4 mm. Based on histology and the guidance from a pathologist, two investigators labeled the OCT images into the following tissue types: endocardium, myocardium, artifacts, adipose & fibrosis, and other tissue types. These investigators were blind to our algorithm results.

In order to fully utilize our dataset, we perform 6-fold cross-validation in evaluation. The images are randomized by donors. We have two pre-processing steps: first, we crop each OCT image into a depth of 360 pixels ($\sim$1.76 mm); next, we partition the images into a set of overlapping patches with a size of 360 $\times$ 64 pixels for data augmentation.

3.2 Quantitative Results

Table 1. Comparison of segmentation results

Method	Accuracy (AC)	Dice coefficient (DC)
Proposed method	0.726 $\pm$ 0.08	0.605 $\pm$ 0.04
RelayNet architecture	0.698 $\pm$ 0.10	0.596 $\pm$ 0.03

Learning from Region-Based Labels: We empirically set the $epoch$ = 500, $learning\ rate$ = 0.001, and w = 0.5 (Eq. 1). Then, we use both Dice coefficient (DC) and accuracy (AC) to evaluate the segmentation performance of our proposed model. Table. 1 compares the experimental results averaged over the 6-fold cross-validation sets with the baseline algorithm, RelayNet [24], based on a similar network structure but without dropout and focal loss. However, these evaluation metrics cannot fully reflect the segmentation performance. The manual labels are region-based, which is not accurate enough for pixel-wise evaluation. Among six validation sets, the best accuracy is above 0.8. In addition, compared with the baseline algorithm, the use of focal loss and dropout layers improves both DC and AC. These results demonstrate that our network has the ability to robustly learn representative features from region-based labels.

Visual results demonstrate the learning ability of our model on region-based labels. Figure 2 illustrates the predicted tissue maps, on two myocardium examples with some adipose and fibrosis regions. Our algorithm pinpoints isolated fibrosis in Fig. 2(c) and the large adipose region in Fig. 2(g). Overall, the prediction results are consistent with both OCT images and histology findings.

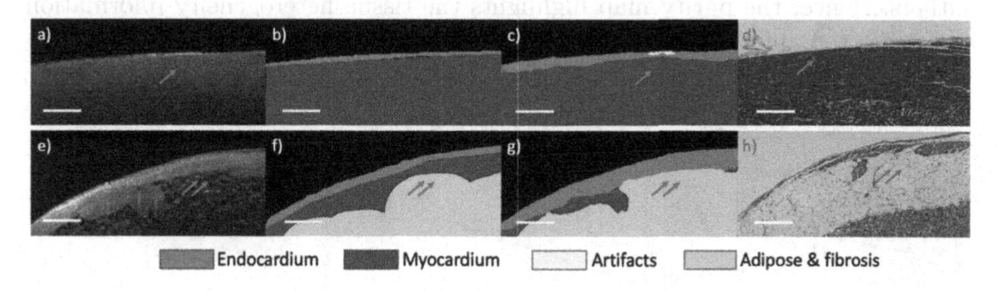

Fig. 2. Segmentation results from the proposed method on two myocardium regions. (a, e) original OCT images; (b, f) manual labels; (c, g) prediction results; (d, h) corresponding histology images. Arrows highlight locations of excellent agreement between our tissue prediction results and histology images. Scale bar: 500 µm

Uncertainty Measurement: We set N_{it} to 10, which is the smallest number leading to a stable uncertainty map. In Fig. 3, we compare an uncertainty map with its histology to show the effectiveness of our uncertainty measurement. In concordance with the results in [16], we observe that the uncertainty map has a high value at the tissue boundaries and subtle features that are hard to visualize. Therefore, the homogeneous regions appear low uncertainty to the model. These results visually demonstrate that our proposed algorithm can provide accurate localization of model uncertainty and this uncertainty is related to the tissue heterogeneity.

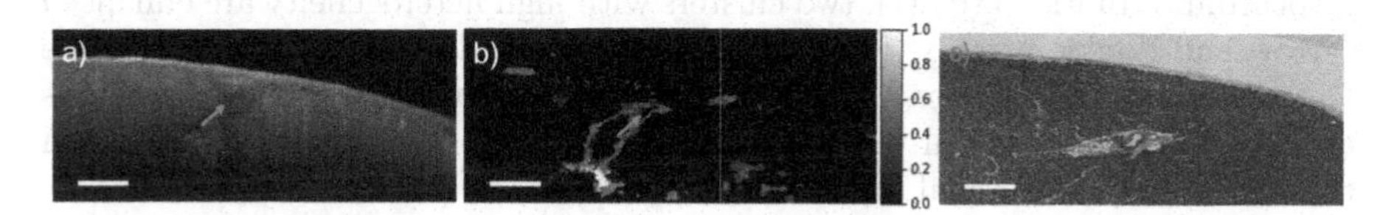

Fig. 3. An uncertainty map from a human myocardium region. (a) the original OCT image; (b) the uncertainty map; (c) the corresponding histology image. The uncertainty map highlights heterogeneous regions caused by fibrosis and adipose. Scale bar: 500 µm.

Heterogeneity Measurement: The positive correlation between purity map heterogeneity and the tissue heterogeneity can be observed in the scatter diagram. Figure 4 shows the normalized heterogeneity results on a cross-validation set with 95% confidence interval. The homogeneity from the purity map has the strongest correlation relationship with tissue homogeneity. The correlation coefficient obtained on these testing images is 0.7117, which achieves the highest value among other methods (OCT images: 0.3501, Uncertainty maps: 0.5959). From the interpreting table in [22,25], the homogeneity values obtained from the purity maps have a high positive correlation with the homogeneity values obtained from the histology images. Therefore, tissue boundary removal is essential to highlight purity map information within tissue regions. In addition, from the results of 95% confidence interval, using a purity map for heterogeneity measurement has less outliers. Hence, the purity map highlights the tissue heterogeneity information and it has the ability for tissue heterogeneity measurement.

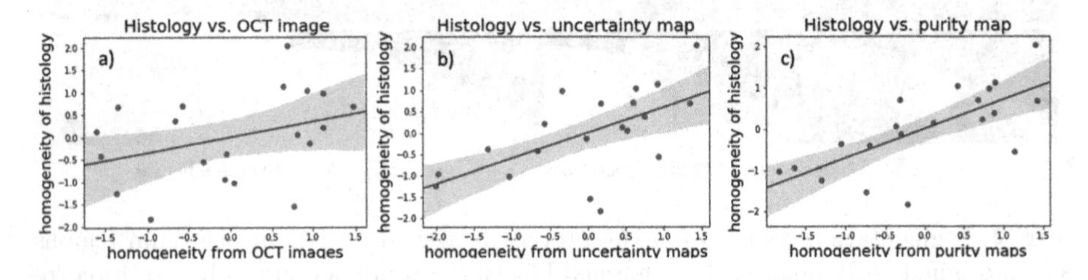

Fig. 4. Scatter diagram for normalized homogeneity values. (a) homogeneity values of OCT images vs. histology images; (b) homogeneity values of uncertainty maps vs. histology images; (c) homogeneity values of purity maps vs. histology images. Compared with the OCT images and uncertainty maps, homogeneity obtained from purity maps has the highest correlation with tissue homogeneity.

Our algorithm shows good performance on quantifying both inter-tissue heterogeneity and intra-tissue heterogeneity. Figure 5 shows two purity maps from different human hearts, demonstrating two scenarios of intra/inter-tissue heterogeneity. A high value in the purity map indicates a high heterogeneity and a low value indicates a high homogeneity. In Fig. 5 (a)-(c), the purity map has low values (high homogeneity) all through the homogeneous cardiac regions. Meanwhile, it highlights heart fibers that are not well aligned due to the dilation of the myocardium. In Fig. 5 (d)-(f), two clusters with high heterogeneity are enhanced (arrow and double arrow). The purity map successfully highlights these subtle features in the OCT images. This intra/inter-tissue heterogeneity information is entirely determined by our purity maps, which cannot be observed directly from the labels since all manual labels are region-based.

Cardiac heterogeneity measurement from OCT images is very challenging and complicated. Figure 6 shows two representative outlier cases from Fig. 4 (c). In Fig. 6 (a)-(c), the adipose regions in the purity map are more heterogeneous than the histology images. It is caused by the fixation and dehydration in the

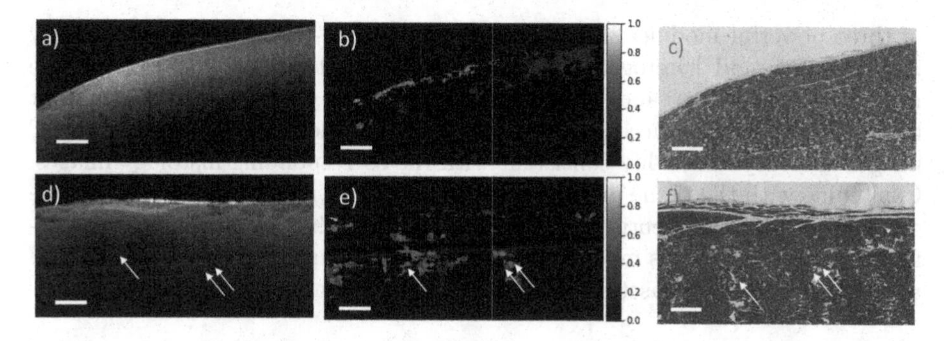

Fig. 5. Purity maps from human myocardium regions. (a, d) original OCT images; (b, e) purity maps; (c, f) corresponding histology images. The purity maps highlight the heterogeneity within the human myocardium regions. Scale bar: 500 μm.

histology process. During the histology process, scattering changes and tissue architectural distortion are inevitable due to the shrinkage of epithelial, muscle, and connective tissue layers [12]. Structural information may get lost after the shrinkage, especially in highly heterogeneous regions. Therefore, a lower value of heterogeneity is observed, leading to a low correlation value. In addition, stain variation in histology could also result in a lower correlation value. In Fig. 6, the loose and dense collagen regions in (a)-(c) are purple and blue, while in (d)-(f), they are pink and pale with isolated light blue strands. This is due to the concentration of the stains and the timings of staining [23]. This issue may be mitigated by improving the histology protocol with better control on the amount of stain used in processing.

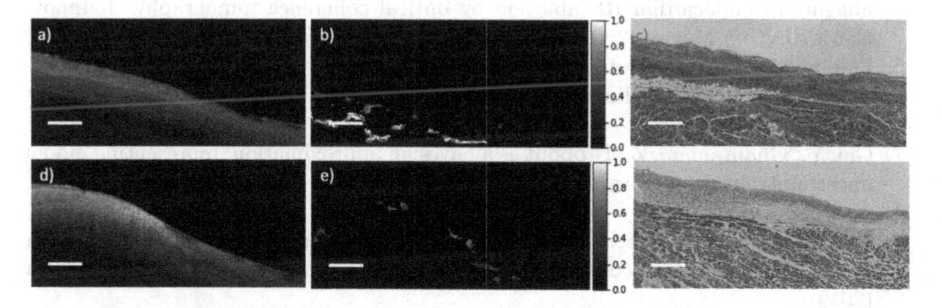

Fig. 6. Examples of two outliers from Fig. 4 (c). (a, d) original OCT images; (b, e) purity maps; (c, f) corresponding histology images. Scale bar: 500 μm.

4 Conclusion

In this paper, we propose the first deep learning framework for cardiac tissue heterogeneity measurement on OCT images. Our proposed algorithm consists of three powerful modules: robust learning on region-based labels, uncertainty measurement, and heterogeneity information extraction. We show that these modules are necessary and benefit from each other. Our results indicate that the purity maps can successfully highlight the subtle features from the OCT images and they are highly consistent with the corresponding histology images. Combining with the tissue prediction map, the purity map could be used for the guidance of RFA intervention by identifying regions that have diffusive adipose-fibrosis or heterogeneous myofiber alignment. In the future, we will evaluate our method on a larger dataset and further improve our model in 3D.

Acknowledgements. The study was funded in part by the National Institute of Health (4DP2HL127776-02 and 1R01HL149369-01, CPH), the National Science Foundation Career Award (1454365, CPH).

References

1. Aslanidi, O.V., Boyett, M.R., Dobrzynski, H., Li, J., Zhang, H.: Mechanisms of transition from normal to reentrant electrical activity in a model of rabbit atrial tissue: interaction of tissue heterogeneity and anisotropy. Biophysical J. **96**(3), 798–817 (2009)
2. Baues, M., et al.: Fibrosis imaging: current concepts and future directions. Adv. Drug Deliv. Rev. **121**, 9–26 (2017)
3. Buch, K., et al.: Using texture analysis to determine human papillomavirus status of oropharyngeal squamous cell carcinomas on CT. AJNR Am. J. Neuroradiol. **36**(7), 1343–1348 (2015)
4. Cua, M., et al.: Morphological phenotyping of mouse hearts using optical coherence tomography. J. Biomed. Opt. **19**(11), 116007 (2014)
5. Fleming, C.P., Rosenthal, N., Rollins, A.M., Arruda, M.: First in vivo real-time imaging of endocardial RF ablation by optical coherence tomography. J. Innov. Card. Rhythm Manag. **2**, 199–201 (2011)
6. Fujima, N., et al.: The utility of MRI histogram and texture analysis for the prediction of histological diagnosis in head and neck malignancies. Cancer Imaging **19**(1), 5 (2019)
7. Gal, Y., Ghahramani, Z.: Dropout as a bayesian approximation: representing model uncertainty in deep learning. In: International Conference on Machine Learning. pp. 1050–1059 (2016)
8. Gan, Y., Lye, T.H., Marboe, C.C., Hendon, C.P.: Characterization of the human myocardium by optical coherence tomography. J. Biophotonics **12**(12), e201900094 (2019)
9. Gan, Y., Tsay, D., Amir, S.B., Marboe, C.C., Hendon, C.P.: Automated classification of optical coherence tomography images of human atrial tissue. J. Biomed. Opt. **21**(10), 101407 (2016)
10. Goergen, C.J., et al.: Optical coherence tractography using intrinsic contrast. Opt. Lett. **37**(18), 3882–3884 (2012)
11. Haissaguerre, M., et al.: Intermittent drivers anchoring to structural heterogeneities as a major pathophysiological mechanism of human persistent atrial fibrillation. J. Physiol. **594**(9), 2387–2398 (2016)

12. Hsiung, P.L., Nambiar, P.R., Fujimoto, J.G.: Effect of tissue preservation on imaging using ultrahigh resolution optical coherence tomography. J. Biomed. Opt. **10**(6), 064033 (2005)
13. Hu, S., et al.: Supervised uncertainty quantification for segmentation with multiple annotations. In: Shen, D., et al. (eds.) MICCAI 2019. LNCS, vol. 11765, pp. 137–145. Springer, Cham (2019). https://doi.org/10.1007/978-3-030-32245-8_16
14. Braunmühl, T.: Optical coherence tomography. Der Hautarzt **66**(7), 499–503 (2015). https://doi.org/10.1007/s00105-015-3607-z
15. Kather, J.N., et al.: Multi-class texture analysis in colorectal cancer histology. Scientific Reports **6**, 27988 (2016)
16. Kendall, A., Badrinarayanan, V., Cipolla, R.: Bayesian segnet: Model uncertainty in deep convolutional encoder-decoder architectures for scene understanding. arXiv preprint arXiv:1511.02680 (2015)
17. Khurshid, S., et al.: Frequency of cardiac rhythm abnormalities in a half million adults. Circ. Arrhythm Electrophysiol. **11**(7), e006273 (2018)
18. Laplante, P.: Encyclopedia of Image Processing. CRC Press, United States (2018)
19. Lin, T.Y., Goyal, P., Girshick, R., He, K., Dollár, P.: Focal loss for dense object detection. In: IEEE International Conference on Computer Vision. pp. 2980–2988 (2017)
20. López, B., et al.: Circulating biomarkers of myocardial fibrosis: the need for a reappraisal. J. Am. Coll. Cardiol. **65**(22), 2449–2456 (2015)
21. Lye, T.H., Iyer, V., Marboe, C.C., Hendon, C.P.: Mapping the human pulmonary venoatrial junction with optical coherence tomography. Biomed. Opt. Express **10**(2), 434–448 (2019)
22. Mukaka, M.: A guide to appropriate use of correlation coefficient in medical research. Malawi Med. J. **24**(3), 69–71 (2012)
23. Rotimi, O., Cairns, A., Gray, S., Moayyedi, P., Dixon, M.: Histological identification of helicobacter pylori: comparison of staining methods. J. Clin. Pathol. **53**(10), 756–759 (2000)
24. Roy, A.G., et al.: Relaynet: retinal layer and fluid segmentation of macular optical coherence tomography using fully convolutional networks. Biomed. Opt. Express **8**(8), 3627–3642 (2017)
25. Schober, P., Boer, C., Schwarte, L.A.: Correlation coefficients: appropriate use and interpretation. Anesthesia & Analgesia **126**(5), 1763–1768 (2018)
26. Sedai, S., Antony, B., Mahapatra, D., Garnavi, R.: Joint segmentation and uncertainty visualization of retinal layers in optical coherence tomography images using bayesian deep learning. In: Stoyanov, D., et al. (eds.) OMIA/COMPAY -2018. LNCS, vol. 11039, pp. 219–227. Springer, Cham (2018). https://doi.org/10.1007/978-3-030-00949-6_26
27. Srivastava, N., Hinton, G., Krizhevsky, A., Sutskever, I., Salakhutdinov, R.: Dropout: a simple way to prevent neural networks from overfitting. J. Mach. Learn. Res. **15**(1), 1929–1958 (2014)
28. Tereshchenko, L.G., et al.: Infiltrated atrial fat characterizes underlying atrial fibrillation substrate in patients at risk as defined by the aric atrial fibrillation risk score. Int. J. Cardiol. **172**(1), 196–201 (2014)
29. Wei, L., Gan, Q., Ji, T.: Cervical cancer histology image identification method based on texture and lesion area features. Comput. Assist. Surg. **22**(sup1), 186–199 (2017)
30. Zhao, X., et al.: Integrated RFA/PSOCT catheter for real-time guidance of cardiac radio-frequency ablation. Biomed. Opt. Express **9**(12), 6400–6411 (2018)

Efficient Shapley Explanation for Features Importance Estimation Under Uncertainty

Xiaoxiao Li[1(✉)], Yuan Zhou[3], Nicha C. Dvornek[1,3], Yufeng Gu[5], Pamela Ventola[4], and James S. Duncan[1,2,3]

[1] Biomedical Engineering, Yale University, New Haven, CT, USA
xiaoxiao.li@aya.yale.edu
[2] Electrical Engineering, Yale University, New Haven, CT, USA
[3] Radiology and Biomedical Imaging, Yale School of Medicine, New Haven, CT, USA
[4] Child Study Center, Yale School of Medicine, New Haven, CT, USA
[5] College of Information Science and Electronic Engineering, Zhejiang University, Hangzhou, China

Abstract. Complex deep learning models have shown their impressive power in analyzing high-dimensional medical image data. To increase the trust of applying deep learning models in medical field, it is essential to understand why a particular prediction was reached. Data feature importance estimation is an important approach to understand both the model and the underlying properties of data. Shapley value explanation (SHAP) is a technique to fairly evaluate input feature importance of a given model. However, the existing SHAP-based explanation works have limitations such as 1) computational complexity, which hinders their applications on high-dimensional medical image data; 2) being sensitive to noise, which can lead to serious errors. Therefore, we propose an uncertainty estimation method for the feature importance results calculated by SHAP. Then we theoretically justify the methods under a Shapley value framework. Finally we evaluate our methods on MNIST and a public neuroimaging dataset. We show the potential of our method to discover disease related biomarkers from neuroimaging data.

1 Introduction

Deep learning models, trained on extremely large data sets, have even exceeded human performance in many tasks. Even in the medical field, there are impressive results. Due to the difficulty of interpreting complex deep learning models, some simple ones, such as linear regression and random forest, are often preferred

This work was supported by NIH Grant [R01NS035193, R01MH100028].
Our code is publicly available at: https://github.com/xxlya/DistDeepSHAP/.

Electronic supplementary material The online version of this chapter (https://doi.org/10.1007/978-3-030-59710-8_77) contains supplementary material, which is available to authorized users.

A. L. Martel et al. (Eds.): MICCAI 2020, LNCS 12261, pp. 792–801, 2020.
https://doi.org/10.1007/978-3-030-59710-8_77

in clinical usage. One common property of linear regression and random forest is their interpretability on feature importance. Without sacrificing the benefits of using deep learning models to improve task performance, many efforts have been made to estimate feature importance scores for deep learning models. There are three main approaches for feature importance estimation: 1) gradient-based methods, such as Simple Gradient (SG) [1], Integrated Gradient (IG) [2], LRP [3], and DeepLIFT [4]; 2) sensitivity based methods, such as LIME [5] and SHAP [6]; 3) methods that mimic deep learning models using tree- or rule-based models [7]. The Shapley value is a means of fairly portioning the collective profit attained by a coalition of players, based on the relative contributions of the players in a game [8]. In this work, we focus on SHAP-based explanation [6].

Although reliability is necessary for model explanations to be trustworthy, relatively few studies have focused on quantifying the uncertainty and robustness of explanation methods. For example, it has been shown that multiple importance estimation methods incorrectly attribute the scores when a constant vector shift is applied to the input [9]. The attributions provided by interpretation methods may themselves contain significant uncertainty [10] and imperceptibly small perturbations of the input can significantly alter the explanations [11].

In order to apply sampling based uncertainty estimation, we modified the original formulation of DeepSHAP [6]. Different from DeepSHAP that back-propagates the prediction difference between input and the a point estimate of references, we consider the reference values as a distribution and show that the Shapley value can be estimated by bootstrap sampling. Therefore, uncertainty of feature importance scores could be measured. Our experiments quantify the performance of different uncertainty estimation methods and their impact on uncertainty-related error reduction. Our key contributions are summarized as follows: 1) Propose a Shapley value estimation framework with uncertainty estimation; 2) Evaluate uncertainty estimation using both human interpretation and quantitative methods; and 3) Apply our method to a subgroup biomarker detection problem in neuroimaging.

2 Preliminaries

2.1 Shapley Value for Feature Importance Estimation

Consider a cooperative game with N players aiming at maximizing a payoff, and let $S \subset \mathcal{N} = \{1, ..., N\}$ be a subset consisting of $|S|$ players. Suppose the prediction function $f : \mathbb{R}^N \to \mathbb{R}$ is the model to be explained, which outputs an importance score for a specific class. Denote $v_x : 2^N \to \mathbb{R}$ (which maps a subset of $\mathcal{N}$ to a real number) as the importance score evaluation function given a subset of features of input $x = [x_1, \ldots, x_N]^T \in \mathbb{R}^N$. We have $f(x) = v_x(\mathcal{N})$. The prediction power for the ith feature is the weighted sum of all possible marginal contributions:

$$\phi_i^x = \frac{1}{N} \sum_{S \subseteq \mathcal{N} \setminus \{i\}} \binom{N-1}{|S|}^{-1} (v_x(S \cup \{i\}) - v_x(S)). \quad (1)$$

While Shapley values give a more accurate interpretation of the importance of each player in a coalition, their calculation is expensive. When the number of features (i.e., players in the game) is a large N, the computational complexity will be 2^N, which is especially expensive. Therefore, computing accurate Shapley values is a challenging task.

In classification scenario, f is the output score for a specific class and $v_x(S)$ is computed by feeding x into f with $\{x_i : i \notin S\}$ replaced by some reference values, which come from samples of some other reference classes.

2.2 Propagating Shapley Values

DeepSHAP [6], which is built on DeepLIFT [4], is a fast approximation to Shapley value by recursively passing DeepLIFT's multipliers. With uninformative reference $r \in \mathbb{R}^N$ (i.e. background of images) and prediction model f and $y = f(x)$, we define $\Delta y = f(x) - f(r)$, $\Delta x_i = x_i - r_i$. DeepLIFT has $\sum_{i=1}^{N} m_{x_i f} \Delta x_i = \Delta y$, where $m_{x_i f}$ is the multiplier and the importance score of x_i is $m_{x_i f} \Delta x_i$. Suppose $f = g \circ h$, and $h : \mathbb{R}^N \to \mathbb{R}^M : x \mapsto [h_1(x), \ldots, h_M(x)]^T$ is a hidden layer with neurons $h_1, \ldots, h_M$, where $h_j : \mathbb{R}^N \to \mathbb{R}$ (corresponding outputs are $z_1, \ldots, z_M$ and $z_j = h_j(x)$), by the chain rule, the multiplier $m_{x_i f}$ is calculated as:

$$m_{x_i f} = \sum_j m_{x_i h_j} m_{z_j g}. \quad (2)$$

Therefore, the output prediction can be decomposed and backpropagated to each feature dimension. DeepSHAP approximation replaces reference input r in DeepLIFT with $E[x]$ and equates ϕ_i to $m_{x_i f} \Delta x_i$. By chain rule and linear approximation, DeepSHAP calculates Shapley value as:

$$\begin{aligned} m_{x_i f} &= \sum_{j_1} m_{x_i h_{j_1}^{(1)}} \sum_{j_2} m_{z_{j_1}^{(1)} h_{j_2}^{(2)}} \cdots \sum_{j_L} m_{z_{j_{L-1}}^{(L-1)} h_{j_L}^{(L)}} \quad \text{(Chain rule)}, \\ \phi_i^x &= m_{x_i f}(x_i - E[x_i]), \end{aligned} \quad (3)$$

where $h_{j_l}^{(l)}$ and $z_{j_l}^{(l)}$ are the j_lth hidden neuron function and output at the lth layer, for $l \in \{1, \cdots, L\}$ and $\sum_{j_L} h_{j_L}^{(L)}(z_{j_{L-1}}^{(L-1)}) = y = f(x)$. The *Rescale* and *RevealCancel* mechanisms [4] in DeepLIFT can make sure that the attributes are correctly propagated to the input. Since the Shapley values for the simple network components can be efficiently solved analytically if they are either linear, max pooling, or an activation function with just one input, this composition rule enables a fast approximation to the original Shapley value for the whole model.

3 Proposed Approach

3.1 DeepSHAP with Reference Distribution

In DeepSHAP, the "missing" features are set to the sample mean of the dataset. Hence, the estimated Shapley value may change when the given dataset changes.

When the reference distribution (empirical distribution of the dataset) is given, we show that a more accurate approach is to obtain Shapley values for samples from the reference distribution and name it as distribution-based DeepSHAP (DistDeepSHAP). The improvement over the original DeepSHAP can be easily seen, because we can rewrite the Shapley value as an average over contributions from all the reference samples.

Next, we introduce an alternative formulation of the Shapley value. Let $\pi(\mathcal{N})$ be the set of all ordered permutations of $\mathcal{N}$. Let $Pre^i(O)$ be the set of players which are predecessors of player i in the order $O \in \pi(\mathcal{N})$, we have

$$\phi_i^x = \frac{1}{N!} \sum_{O \in \pi(\mathcal{N})} (v_x(Pre^i(O) \cup \{i\}) - v_x(Pre^i(O))). \tag{4}$$

Given a single reference $\hat{x}$ sampled from data distribution $\mathcal{D}$ and using $f(x)$ to replace v_x, we define the single reference SHAP as

$$\phi_i^{x|\hat{x}} = \frac{1}{N!} \sum_{O \in \pi(\mathcal{N})} \{f(\tau(x, \hat{x}, Pre^i(O) \cup \{i\})) - f(\tau(x, \hat{x}, Pre^i(O)))\}, \tag{5}$$

where

$$\tau(x, \hat{x}, P) = (v_1, v_2, \ldots, v_N), \quad v_j = \begin{cases} x_j, \ j \in P \\ \hat{x}_j, \ j \notin P \end{cases}. \tag{6}$$

Since the reference $\hat{x}$ is a random variable, $\phi_i^{x|\hat{x}}$ is an induced random variable by $\hat{x}$. We estimate Eq. (4) by:

$$\phi_i^x = \mathbb{E}_{\hat{x} \in \mathcal{D}} [\phi_i^{x|\hat{x}}], \tag{7}$$

It is obvious that Eq. (5) is a typical Shapley value format as Eq. (4). In practice, we can borrow the efficient approximation to Shapley value based on DeepSHAP (Eq. (3), [6]). Hence, a single reference SHAP value (Eq. (5)) can be efficiently computed as:

$$\phi_i^{x|\hat{x}} = m_{x_i f}(x_i - \hat{x}_i). \tag{8}$$

3.2 Uncertainty of Shapley Values

To estimate the uncertainty of Shapley values, we collect samples of $\phi_i^{x|\hat{x}}$ based on randomly drawn $\hat{x}$ and use the percentiles to measure its uncertainty, u_i, which is associated with each individual Shapley value ϕ_i^x produced by Eq. (8). In particular, we calculate confidence intervals $CI_{i,\gamma} = [c_{i,\frac{\alpha}{2}}, c_{i,1-\frac{\alpha}{2}}]$ with lower bounds $c_{i,\frac{\alpha}{2}}$ and upper bounds $c_{i,1-\frac{\alpha}{2}}$ at confidence level $\gamma = 1 - \alpha$ for each assigned Shapley value. The detailed methods are presented in Algorithm 1. When the training dataset is not available, we can use the testing dataset to approximate the reference distribution.

Algorithm 1. Estimating Shapley value ϕ_i^x and its uncertainty u_i^x for the ith feature given input x

Input: x, a given instance; R, number of repeats; y, prediction model output with respect to x; $\hat{X}$, a set of $\hat{x}$ (i.e. training data); and Φ_i^x, a list to store $\phi_i^{x|\hat{x}}$ calculated by different samples; $1-\alpha$, confidence level.

1: $\Phi_i^x \leftarrow$ empty list
2: **for** $r = 1$ to R **do**
3: choose a random instance $\hat{x}$ from $\hat{X}$
4: $\phi_i^{x|\hat{x}} \leftarrow m_{x_i y}(x_i - \hat{x}_i)$ ▷ $m_{x_i y}$ is calculated by Eq. (3)
5: Add $\phi_i^{x|\hat{x}}$ to list Φ_i^x
6: **end for**
7: $\phi_i^x \leftarrow \text{MEAN}(\Phi_i^x)$
8: $u_i^x \leftarrow c_{i,1-\frac{\alpha}{2}} - c_{i,\frac{\alpha}{2}}$
9: **Output:** ϕ_i^x, u_i^x

3.3 Evaluation of Uncertainty

In general settings, it is difficult to evaluate uncertainty estimates for feature importance estimation methods, since we typically do not have per-feature ground-truth to evaluate against. To quantitatively and qualitatively assess the accuracy of the uncertainty estimates provided by DistDeepSHAP, we propose a calibration-based method. The intuition is that the uncertainty estimated on a single input x should reflect the distribution of important scores of all the instances within the class of x. Specifically, we randomly sample m instances within the same class of x and estimate the feature importance scores of those instances. Intuitively, calibration means that given the $1-\alpha$ confidence interval of a feature, the feature importance scores should occur within this interval with probability $1-\alpha$.

4 Experiments and Results

4.1 Validation on MNIST Dataset

In order to show the feasibility of our proposed method, we test the explanation results on the MNIST dataset [12], which is intuitive for human judgment. We flattened the MNIST images to vectors. Denote dropout layer with ratio 0.5 as 'Drop', Relu non-linear activation as 'Relu' and fully-connected layer as 'FC'. Then, we trained a Multilayer Perceptron (MLP) classifier (Input(784) → Drop(Relu(FC(512))) → Drop(Relu(FC(512))) → FC(10)) achieving 97.12% accuracy. Given the pre-trained classifier, we compare the feature importance estimation results using DistDeepSHAP with the alternative methods (including GuideBackProp [13], Integrated Gradient [2], DeepLIFT [4] and DeepSHAP [6]) and the corresponding uncertainty estimation results using our proposed DistDeepSHAP in Fig. 1. We set $\alpha = 0.1$ and repeat sampling times $R = 100$[1].

[1] The investigation on the repeat sampling times is left in the Appendix (see supplementary material).

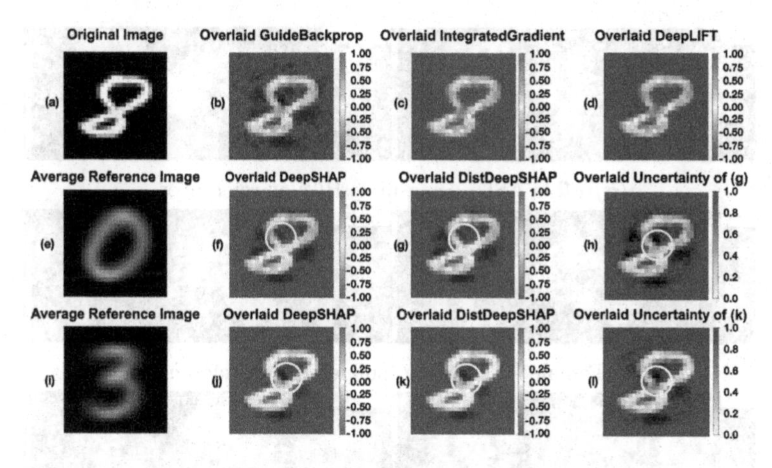

Fig. 1. Explanation results on one digit '8' shown in (a). The color-bar next to each subfigure indicates the scale. (b) (c) (d) show the overlaid pixel importance score to identify this '8' using GuideBackProp [13], Integrated Gradient [2] and DeepLIFT [4] seperately. For SHAP-based methods, we can choose another class as the reference. With digits '0's (e) as reference, (f) shows the overlaid Shapley values using DeepSHAP and (g) shows the mean of SHAP scores estimated by our proposed DistDeepSHAP, (h) provides the uncertainty of the Shapley values computed by DistDeepSHAP. With digits '3's (i) as reference, (j) shows the overlaid Shapley values using DeepSHAP and (k) shows the mean of SHAP scores estimated by our proposed DistDeepSHAP, (l) provides the uncertainty of the Shapley values computed by DistDeepSHAP. All values are normalized to range [0, 1].

DeepLIFT uses value 0 as reference. Different from DeepLIFT, DeepSHAP and DistDeepSHAP, GuideBackProp and Integrated Gradient do not need a reference. As we expected, our proposed DistDeepSHAP achieved similar feature estimation results to DeepSHAP. However, DistDeepSHAP can provide additional uncertainty estimations of importance score as shown in Fig. 1(h) and Fig. 1(l). For human interpretation, "central cross" (circled in Fig. 1) is an important pattern to identify digit 8. However, SHAP-based methods marked the central part as negative evidence, which disagrees with human interpretation, and also was different from the results using GuideBackProp, Integrated Gradient and DeepLIFT. We noticed that Fig. 1(h) and Fig. 1(l) assigned the central part high uncertainty values. Therefore, we could use this additional uncertainty information to help judge whether the interpretation result is reliable.

4.2 Application on Autism Resting-State fMRI

Data and Preprocessing. The study was carried out using resting-state fMRI (rs-fMRI) data from the Autism Brain Imaging Data Exchange dataset (ABIDE I preprocessed, [14]). We downloaded Regions of Interests (ROIs) fMRI series of the top four largest sites (UM1, NYU, USM, UCLA1) from the preprocessed ABIDE dataset with Configurable Pipeline for the Analysis of Connectomes

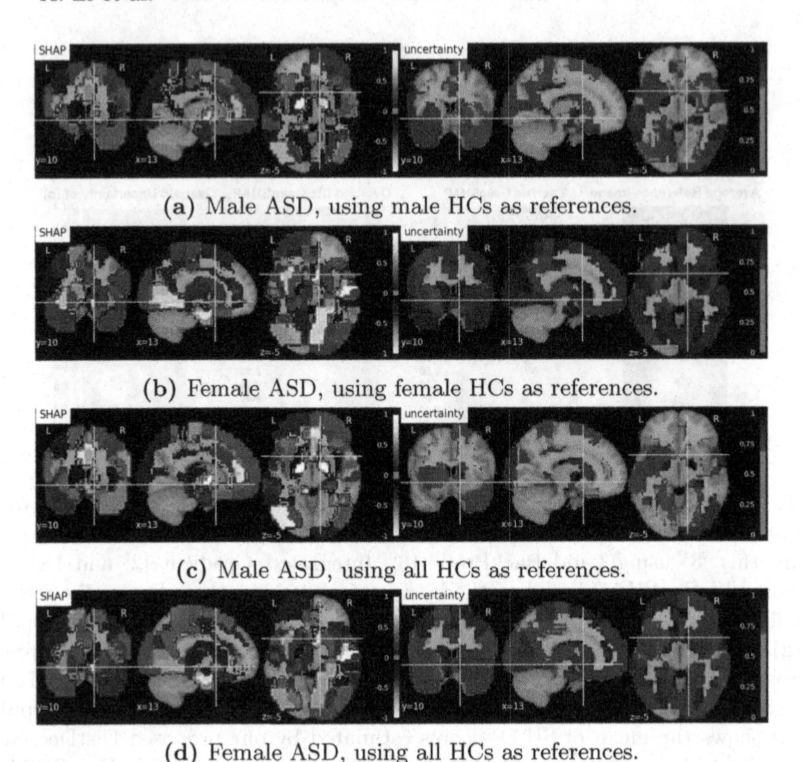

(a) Male ASD, using male HCs as references.

(b) Female ASD, using female HCs as references.

(c) Male ASD, using all HCs as references.

(d) Female ASD, using all HCs as references.

Fig. 2. Shapley values and their corresponding uncertainty (normalized to range [0, 1]).

(CPAC), band-pass filtering (0.01–0.1 Hz), no global signal regression, parcellated by Harvard-Oxford (HO) atlas. Skipping subjects with missing files, we downloaded 106, 175, 72, 71 subjects from UM1, NYU, USM, UCLA1 separately. For data augmentation, we used sliding window with length 32 and stride 1 to truncate fMRI series.

Implementation Details. The task we performed on the ABIDE datasets was to identify autism spectrum disorders (ASD) or healthy control (HC). We used the mean time sequences of ROIs to compute the correlation matrix as functional connectivity. The functional connectivity provided an index of the level of co-activation of brain regions based on the time series of rs-fMRI brain imaging data. Each element of the correlation matrix was calculated using Pearson correlation coefficient and Fisher transformation. As the correlation matrices were symmetric, we only kept the upper-triangle of the matrices and flattened the triangle values to vectors. The number of resultant features was defined by $N(N-1)/2$, where N was the number of ROIs. Under the HO atlas (110 ROIs), the procedure resulted in 5995 features. We designed an MLP (Input(5995) $\rightarrow$ Drop(Relu(FC(8))) $\rightarrow$ FC(2)) to classify the inputs. We randomly split 80% of the data as training data and the remaining 20% as testing data based on subjects. Adam optimization was applied with initial learning rate 1e−5 and reduced by 1/2 for every 20 epochs and stopped at the 50th epoch. We achieved 69.04% accuracy on the testing set.

Feature Importance Estimation Results. Given the pre-trained ASD vs. HC classifier, we showed the ROI importance results on a male ASD and a female ASD instance with different reference distributions ($\alpha = 0.1$ and $R = 100$) in Fig. 2. We interpreted the ROIs with high Shapley values as biomarkers for a certain group. Specifically, Fig. 2 showed the biomarkers to distinguish ASD in a certain sex group from different reference population. The results indicate that an ROI may have different importance scores when compared to different references. The uncertainty results presented on the right column of Fig. 2 gave us the confidence of trusting the biomarker detection results. For example, both Fig. 2(a) and Fig. 2(c) indicated high Shapley values and low uncertainty on frontal cortex. Existing clinical studies [15–17] have shown that frontal cortex is a salient biomarker in ASD reflected by reduced levels of GABA and reduced MeCP2 expression. In neurological biomarker detection studies, investigators could take the uncertainty information and assign higher priority to the trustable important ROIs (with large Shapley score and low uncertainty) in clinical investigation, as conducting clinical biomarkers evaluation without prior knowledge is costly and time consuming. With the flexibility of choosing subgroup references, our proposed DistDeepSHAP could explore the fine-grained neurological biomarkers of subgroups. The almost opposite importance scores for the male and the female instances showed the heterogeneous ASD neurological patterns in different genders [18,19].

4.3 Evaluation of Uncertainty on the Two Experiments

As we described in Sect. 3.3, a reasonable uncertainty measurement has calibration property. Namely, the uncertainty estimation can imply how the other instances agree on the importance score calculated from x, and how certain the feature importance estimates are on previously unseen sample images. Hence uncertainty can be used as out of distribution data importance score calibration, and an accurate uncertainty estimate of a given testing sample should be correlated with the Shapley value estimates on the **held-out** testing set. For example, the best uncertainty estimates with given significance level $\alpha = 0.1$ will include $1 - \alpha = 90\%$ of the testing samples whose Shapley values fall into this $1 - \alpha$ confidence interval. Given the estimation on a digit '8' input and a male ASD input, we sampled another 100 digit '8's and 10 male ASDs from the testing set of the two experiments separately. We used digit '3's as the references for digit '8's and used male HCs as the references for male ASDs. We applied the uncertainty evaluation method proposed in Sect. 3.3 for the given testing samples. The top 10% uncertain features were selected.

For a held-out testing sample, only if its Shapley value of the top 10% uncertain features fall in the estimated confidence interval of the previous given testing sample, we regarded the held-out testing sample as success. We varied $1 - \alpha \in \{0.7, 0.75, 0.8, 0.85, 0.9, 0.95\}$. The relationship between $1 - \alpha$ and the percentage of the held-out testing samples that are located in the confidence interval, was shown in Fig. 3. The correlations of $1 - \alpha$ and this percentage were 0.981 for MNIST and 0.916 for ABIDE respectively. The uncertainty estimated for the classification model on the MNIST dataset had higher correlation and closer percentage to $1 - \alpha$. The reason could be that the classification model on the MNIST dataset achieved higher accuracy and we had more held-out testing samples than those of ABIDE. Overall, the calibration patterns could validate the reliability of our proposed uncertainty estimates.

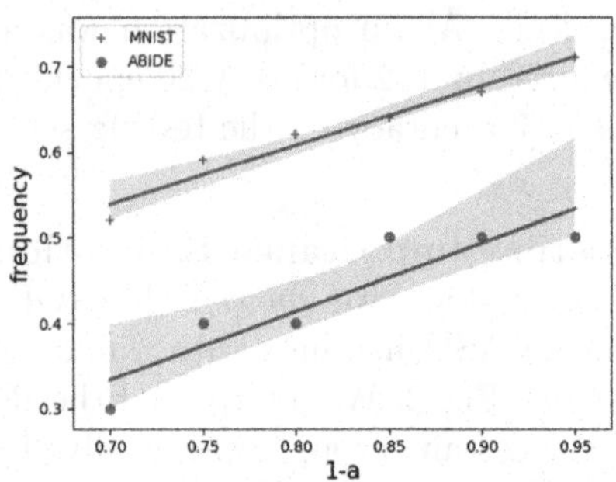

Fig. 3. Relationship between $1 - \alpha$ and percentage of held-out testing samples falling into the confidence interval for uncertainty evaluation.

5 Conclusion

In this work, we propose DistDeepSHAP, a post-hoc feature importance estimation method with uncertainty evaluation for deep learning models. DistDeepSHAP is based on the idea of DeepSHAP, but improves DeepSHAP by sampling the references from a distribution and calculating Shapley values for these references. Our proposed DistDeepSHAP has several advantages over DeepSHAP. First, it can obtain uncertainty estimates for the provided feature importance scores. Second, it can better utilize the empirical reference distribution and has the potential for better feature importance score estimation. Last but not least, it can be calculated with arbitrary subgroup references and interpret salient features with respect to a subgroup, which is crucial for neuroscience study.

References

1. Simonyan, K., et al.: Deep inside convolutional networks: visualising image classification models and saliency maps. arXiv preprint arXiv:1312.6034 (2013)
2. Sundararajan, M., et al.: Axiomatic attribution for deep networks. In: Proceedings of the 34th International Conference on Machine Learning, vol. 70, pp. 3319–3328. JMLR. org (2017)
3. Montavon, G., et al.: Methods for interpreting and understanding deep neural networks. Digit. Signal Proc. **73**, 1–15 (2018)

4. Shrikumar, A., et al.: Learning important features through propagating activation differences. In: Proceedings of the 34th International Conference on Machine Learning, vol. 70, pp. 3145–3153. JMLR. org (2017)
5. Ribeiro, M.T., et al.: "why should i trust you?" Explaining the predictions of any classifier. In: Proceedings of the 22nd ACM SIGKDD International Conference on Knowledge Discovery and Data Mining, pp. 1135–1144 (2016)
6. Lundberg, S.M., Lee, S.-I.: A unified approach to interpreting model predictions. In: Advances in Neural Information Processing Systems, pp. 4765–4774 (2017)
7. Schwab, P., Hlavacs, H.: Capturing the essence: towards the automated generation of transparent behavior models. In: Eleventh Artificial Intelligence and Interactive Digital Entertainment Conference (2015)
8. Shapley, L.S.: A value for N-person games. Contrib. Theory Games **2**(28), 307–317 (1953)
9. Kindermans, P.-J., et al.: The (un)reliability of saliency methods. In: Samek, W., Montavon, G., Vedaldi, A., Hansen, L.K., Müller, K.-R. (eds.) Explainable AI: Interpreting, Explaining and Visualizing Deep Learning. LNCS (LNAI), vol. 11700, pp. 267–280. Springer, Cham (2019). https://doi.org/10.1007/978-3-030-28954-6_14
10. Fen, H., et al.: Why should you trust my interpretation? Understanding uncertainty in lime predictions. arXiv preprint arXiv:1904.12991 (2019)
11. Adebayo, J., et al.: Sanity checks for saliency maps. In: Advances in Neural Information Processing Systems, pp. 9505–9515 (2018)
12. LeCun, Y., et al.: Gradient-based learning applied to document recognition. Proc. IEEE **86**(11), 2278–2324 (1998)
13. Springenberg, J.T., et al.: Striving for simplicity: the all convolutional net. arXiv preprint arXiv:1412.6806 (2014)
14. Di Martino, A., et al.: The autism brain imaging data exchange: towards a large-scale evaluation of the intrinsic brain architecture in autism. Mol. Psychiatry **19**(6), 659 (2014)
15. Goldani, A.A., et al.: Biomarkers in autism. Front. Psychiatry **5**, 100 (2014)
16. Nagarajan, R., et al.: Reduced MeCP2 expression is frequent in autism frontal cortex and correlates with aberrant MECP2 promoter methylation. Epigenetics **1**(4), 172–182 (2006)
17. Watanabe, T., et al.: Mitigation of sociocommunicational deficits of autism through oxytocin-induced recovery of medial prefrontal activity: a randomized trial. JAMA Psychiatry **71**(2), 166–175 (2014)
18. Rivet, T.T., Matson, J.L.: Review of gender differences in core symptomatology in autism spectrum disorders. Res. Autism Spect. Disord. **5**(3), 957–976 (2011)
19. Halladay, A.K., et al.: Sex and gender differences in autism spectrum disorder: summarizing evidence gaps and identifying emerging areas of priority. Mol. Autism **6**(1), 1–5 (2015)

Cartilage Segmentation in High-Resolution 3D Micro-CT Images via Uncertainty-Guided Self-training with Very Sparse Annotation

Hao Zheng[1(✉)], Susan M. Motch Perrine[2], M. Kathleen Pitirri[2], Kazuhiko Kawasaki[2], Chaoli Wang[1], Joan T. Richtsmeier[2], and Danny Z. Chen[1]

[1] Department of Computer Science and Engineering, University of Notre Dame, Notre Dame, IN 46556, USA
hzheng3@nd.edu

[2] Department of Anthropology, Pennsylvania State University, University Park, PA 16802, USA

Abstract. Craniofacial syndromes often involve skeletal defects of the head. Studying the development of the *chondrocranium* (the part of the endoskeleton that protects the brain and other sense organs) is crucial to understanding genotype-phenotype relationships and early detection of skeletal malformation. Our goal is to segment craniofacial cartilages in 3D micro-CT images of embryonic mice stained with phosphotungstic acid. However, due to high image resolution, complex object structures, and low contrast, delineating fine-grained structures in these images is very challenging, even manually. Specifically, only experts can differentiate cartilages, and it is unrealistic to manually label whole volumes for deep learning model training. We propose a new framework to progressively segment cartilages in high-resolution 3D micro-CT images using extremely sparse annotation (e.g., annotating only a few selected slices in a volume). Our model consists of a lightweight fully convolutional network (FCN) to accelerate the training speed and generate pseudo labels (PLs) for unlabeled slices. Meanwhile, we take into account the reliability of PLs using a bootstrap ensemble based uncertainty quantification method. Further, our framework gradually learns from the PLs with the guidance of the uncertainty estimation via self-training. Experiments show that our method achieves high segmentation accuracy compared to prior arts and obtains performance gains by iterative self-training.

Keywords: Cartilage segmentation · Uncertainty · Sparse annotation

1 Introduction

Approximately 1% of babies born with congenital anomalies have syndromes including skull abnormalities [13]. Anomalies of the skull invariably require treatments and care, imposing high financial and emotional burdens on patients

A. L. Martel et al. (Eds.): MICCAI 2020, LNCS 12261, pp. 802–812, 2020.
https://doi.org/10.1007/978-3-030-59710-8_78

and their families. Although prenatal development data are not available for study in humans, the deep conservation of mammalian developmental systems in evolution means that laboratory mice give access to embryonic tissues that can reveal critical molecular and structural components of early skull development [3,18]. The precise delineation of 3D chondrocranial anatomy is fundamental to understanding dermatocranium development, provides important information to the pathophysiology of numerous craniofacial anomalies, and reveals potential avenues for developing novel therapeutics. An embryonic mouse is tiny ($\sim$2 cm^3), and thus we dissect and reconstruct the *chondrocranium* from 3D micro-computed tomography (micro-CT) images of specially stained mice. However, delineating fine-grained cartilaginous structures in these images is very challenging, even manually (e.g., see Fig. 1).

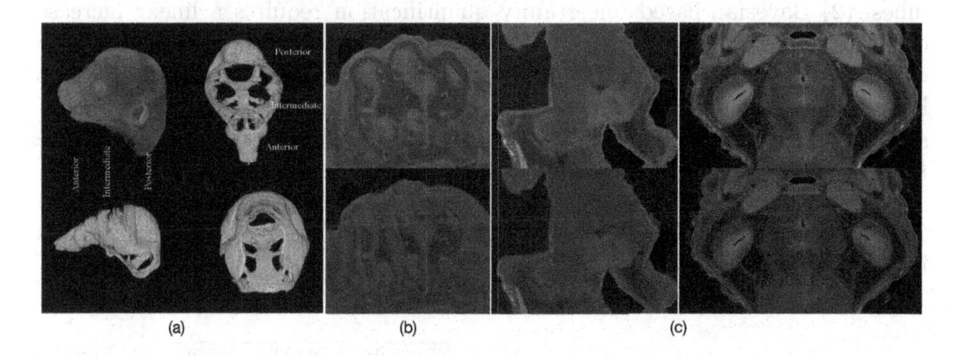

Fig. 1. Examples of micro-CT images of stained mice. (a) A raw 3D image and its manual annotation. The shape variations are large: the front nasal cartilage is relatively small (i.e., 300^2); the cranial vault is very big (i.e., 900×500) but extremely thin like a half-ellipsoid surface. (b) A 2D slice from the nasal cartilage (top) and its associated label (bottom); the image contrast is low and there are many hard mimics in surrounding areas. (c) Two 2D slices from the cranial vault (top) and their associated labels (bottom); the cartilage is very thin. Best viewed in color. (color figure online)

Although deep learning has achieved great success in biomedical image segmentation [11,12,19,20,22], there are three main challenges when applying existing methods to cartilage segmentation in our high-resolution micro-CT images. (1) The topology variations of craniofacial cartilages are very large in the anterior, intermediate, and posterior of the skull (as shown in Fig. 1(a)). Known methods for segmenting articular cartilages in knees [2,17] only deal with relatively homogeneous structures. (2) Such methods deal with images of much lower resolutions (e.g., 200×512^2), and simple scaling-up would precipitate huge computation requirements. Micro-CT scanners work at the level of one micron (i.e., $1 \mu m$), and a typical scan of ours is of size 1500×2000^2. In Fig. 1(c), the cropped sub-region is of size 400^2, and the region-of-interest (ROI) is only 5 pixels thick. (3) More importantly, only experts can differentiate cartilages, and it is unrealistic to manually label whole volumes for training fully convolution networks

(FCNs) [12]. While some semi-supervised methods [21,23] were studied very recently, how to acquire and make the most out of very sparse annotation is seldom explored, especially for real-world complex cartilage segmentation tasks.

To address these challenges, we propose a new framework that utilizes FCNs and uncertainty-guided self-training to gradually boost the segmentation accuracy. We start with extremely sparsely annotated 2D slices and train an FCN to predict pseudo labels (PLs) for unseen slices in the training volumes and the associated uncertainty map, which quantifies pixelwise prediction confidence. Guided by the uncertainty, we iteratively train the FCN with PLs and improve the generalization ability of FCN in unseen volumes. Although the above process seems straightforward, we must overcome three difficulties. (1) The FCN should have a sufficiently large receptive field to accommodate such high-resolution images yet needs to be lightweight for efficient training and inference due to the large volumes. (2) Bayesian-based uncertainty quantification requires a linear increase of either space or time during inference. We integrate FCNs into a bootstrap ensemble based uncertainty quantification scheme and devise a K-head FCN to balance efficiency and efficacy. (3) The generated PLs contain noises. We consider the quality of PLs and propose an uncertainty-guided self-training scheme to further refine segmentation results.

Experiments show that our proposed framework achieves an average Dice of 78.98% in segmentation compared to prior arts and obtains performance gains by iterative self-training (from 78.98% to 83.16%).

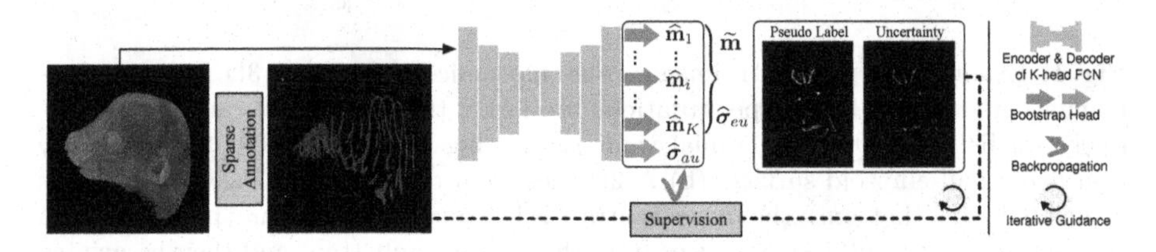

Fig. 2. An overview of our proposed framework.

2 Method

As shown in Fig. 2, our proposed framework contains a new FCN, which can generate PLs and uncertainty estimation at the same time, and an iterative uncertainty-guided self-training strategy to boost the segmentation results.

2.1 K-Head FCN

Initial Labeling and PL Generation. We consider two sets of 3D data, $\mathcal{A} = \{\mathcal{A}_i\}_{i=1}^{L}$ and $\mathcal{B} = \{\mathcal{B}_i\}_{i=1}^{U}$, for training and testing respectively, where each

$\mathcal{A}_i$ (or $\mathcal{B}_i$) is a 3D volume and L (or U) is the number of volumes in $\mathcal{A}$ (or $\mathcal{B}$). Each 3D volume can be viewed as a series of 2D slices, i.e., $\mathcal{A}_i = \{\mathbf{A}_i^j\}_{j=1}^{i_Q}$, where i_Q is the number of slices in $\mathcal{A}_i$. To begin with, experts chose representative slices in each $\mathcal{A}_i$ from the anterior, intermediate, and posterior of the skull and annotated them at the pixel level. Due to the high resolution of our micro-CT images, the annotation ratio is rather sparse (e.g., 25 out of 1600 slices). Thus, each $\mathcal{A}_i$ can be divided into two subsets $\mathcal{A}l_i = \{\mathbf{l}_i^j\}_{j=1}^{i_P}$ and $\mathcal{A}u_i = \{\mathbf{u}_i^j\}_{j=1}^{i_R}$, where each slice $\mathbf{l}_i^j$ has its associate label $\mathbf{m}_i^j$, and $i_Q > i_R \gg i_P$. Conventionally, using such sparse annotation, a trained FCN lacks generalization ability to the unseen volumes $\mathcal{B}$. Hence, a key challenge is how to make the most out of the labeled slices. We will show that an FCN can delineate ROIs in unseen slices of the training volumes (i.e., $\mathcal{A}u_i$) with very sparsely labeled slices. For this, we propose to utilize these true labels (TLs) and generate PLs to expand the training data.

Uncertainty Quantification. Since FCN here is not trained by standard protocol, its predictions may be unreliable and noisy. Thus, we need to consider the reliability of the PLs (which may otherwise lead to meaningless guidance). Bayesian methods [7] provided a straightforward way to measure uncertainty quantitatively by utilizing Monte Carlo sampling in forward propagation to generate multiple predictions. Prohibitively, the computational cost grows linearly (either time or space). Since our data are large volumes, such cost is unbearable. To avoid this issue, we need to design a method that is both time- and space-efficient. Below we illustrate how to design a new FCN for this purpose.

There are two main types of uncertainty in Bayesian modelling [8,16]: *epistemic uncertainty* captures uncertainty in the model (i.e., the model parameters are poorly determined due to the lack of data/knowledge); *aleatoric uncertainty* captures genuine stochasticity in the data (e.g., inherent noises). Without loss of generality, let $f_\theta(x)$ be the output of a neural network, where θ is the parameters and x is the input. For segmentation tasks, following the practice in [8], we define pixelwise likelihood by squashing the model output through a softmax function $\mathcal{S}$: $p(y|f_\theta(x), \sigma^2) = \mathcal{S}(\frac{1}{\sigma^2} f_\theta(x))$. The magnitude of σ determines how 'uniform' (flat) the discrete distribution is. The log likelihood for the output is: $\log(p(y = c|f_\theta(x), \sigma^2)) = \frac{1}{\sigma^2} f_\theta^c(x) - \log \sum_{c'} \exp(\frac{1}{\sigma^2} f_\theta^{c'}(x)) = \frac{1}{\sigma^2}\log\frac{\exp(f_\theta^c(x))}{\sum_{c'}\exp(f_\theta^{c'}(x))} - \log\frac{\sum_{c'}\exp(\frac{1}{\sigma^2}f_\theta^{c'}(x))}{\left(\sum_{c'}\exp(f_\theta^{c'}(x))\right)^{\frac{1}{\sigma^2}}} \approx \frac{1}{\sigma^2}\log\mathcal{S}(f_\theta(x))^c - \frac{1}{2}\log\sigma^2$, where $f_\theta^c(x)$ is the c-th class of output $f_\theta(x)$, and we use the explicit simplifying assumption $\left(\sum_{c'} \exp(f_\theta^{c'}(x))\right)^{\frac{1}{\sigma^2}} \approx \frac{1}{\sigma} \sum_{c'} \exp(\frac{1}{\sigma^2} f_\theta^{c'}(x))$. The objective is to minimize the loss given by the negative log likelihood:

$$\mathcal{L}_{UC}(\theta, \sigma^2) = -\frac{1}{N}\sum_i^N\sum_m^M \mathbb{1}_{m=c}\log(p(y_i = c|f_\theta(x_i), \sigma^2)), \tag{1}$$

where N is the number of training samples and $\mathbb{1}_{m=c}$ is the one-hot vector of class c. In practice, we make the network predict the log variance $s := \log\sigma^2$ for numerical stability. Now, the aleatoric uncertainty is estimated by e^{-s}, and we can quantify the epistemic uncertainty by the predictive variance by $\frac{1}{K}\sum_k^K \hat{y}_k^2 - \left(\frac{1}{K}\sum_k^K \hat{y}_k\right)^2$, where $\hat{y}_k = f_\theta(x)$ is the k-th sample from the output distribution.

K-Head FCN. To sample K samples from the output distribution, we adopt the bootstrap method into the FCN design. A naïve way would be to maintain a set of K networks $\{f_{\theta_k}\}_{k=1}^K$ independently on K different bootstrapped subsets (i.e., $\{D_k\}_{k=1}^K$) of the whole dataset D and treat each network f_{θ_k} as independent samples from the weight distribution. However, it is computationally expensive, especially when each neural net is large and deep. Hence, we propose a single network that consists of a shared backbone architecture with K lightweight bootstrapped heads branching on/off independently. The shared network learns a joint feature representation across all the data, while each head is trained only on its bootstrapped sub-sample of the data. The training and inference of this type of bootstrap can be conducted in a single forward/backward pass, thus saving both time and space. Besides, in contrast to previous methods where σ^2 is assumed to be constant for all inputs, we estimate it directly as an output of the network [7,16]. Thus, our proposed network consists of a total of $K+1$ branches—K heads corresponding to the segmentation prediction map and an extra head corresponding to σ^2. In all the experiments, K is set as 5, and the input image size is 512 × 512.

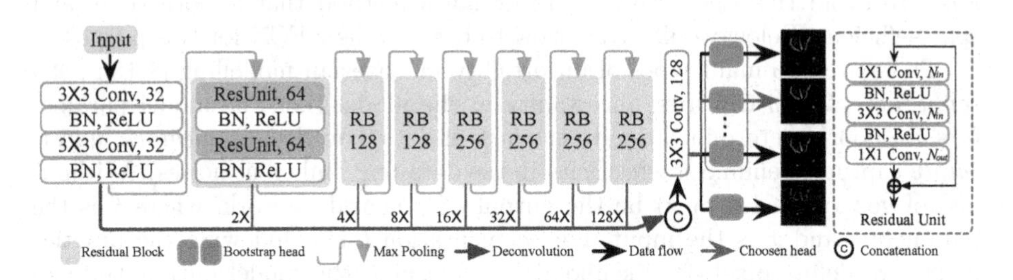

Fig. 3. The network architecture of our proposed method, K-head FCN. The output layer branches out to K bootstrap heads and an extra log-variance output.

Figure 3 shows the detailed structure of our new K-head FCN. There are 7 residual blocks (RBs) and max-pooling operations in the encoding-path to deliver larger reception fields, each RB containing 2 cascaded residual units as in ResNet [6]. To save parameters, we maintain the number of channels in each residual unit and a similar number of feature channels at the last 4 scales. Rich contextual and semantic information is extracted in shallower and deeper scales in the encoding-path and is up-sampled to maintain the same size for the input and output and then concatenated to generate the final prediction. The output layer splits near the end of the model for two reasons: (1) ease the training difficulty and improve the convergence speed; (2) incur minimal computation resource increases (both time and space) in training and inference. To train the network, we randomly choose one head in each iteration and compute the cross-entropy loss $\mathcal{L}_{CE}$. It is combined with the uncertainty loss $\mathcal{L}_{UC}$ to update the parameters in the chosen head branch and the shared backbone only (i.e., freezing the other $K-1$ head branches). Specifically, $\mathcal{L} = \mathcal{L}_{CE} + 0.04\mathcal{L}_{UC}$.

2.2 Iterative Uncertainty-Guided Self-Training

Since both $\mathcal{A}l_i$ and $\mathcal{A}u_i$ come from the same volume $\mathcal{A}_i$ and are based on the assumption that the manifolds of the seen/unseen slices (of $\mathcal{A}_i$) are smooth in high dimensions [15], our generated PLs bridge the annotation gap. However, the K predictions, $\{\widehat{\mathbf{m}}_i^{j,k}\}_{k=1}^K$, obtained from the output distribution for each $\mathbf{u}_i^j \in \mathcal{A}u_i$ could be unreliable and noisy. Thus, we propose an uncertainty-guided scheme to reweight PLs and rule out unreliable (highly uncertain) pixels in subsequent training. Specifically, we calculate the voxel-level cross-entropy loss weighted by the epistemic uncertainty $\boldsymbol{\sigma}_i^j$ for $\mathbf{u}_i^j$: $\mathcal{L}_{CE}(\overline{\mathbf{m}}_i^j, \widetilde{\mathbf{m}}_i^j) = \frac{\sum_v e^{-\sigma_v} \mathcal{L}_{ce}(\overline{m}_v, \widetilde{m}_v)}{\sum_v e^{-\sigma_v}}$, where $\overline{\mathbf{m}}_i^j$ is the prediction at the current iteration and $\widetilde{\mathbf{m}}_i^j = \sum_{k=1}^K \widehat{\mathbf{m}}_i^{j,k}$; $\overline{m}_v$ and $\widetilde{m}_v$ are the values of the v-th pixel (for simplicity, we omit i and j); σ_v is the sum of normalized epistemic and aleatoric uncertainties at the v-th pixel; $\mathcal{L}_{ce}$ is the cross-entropy error at each pixel. Note that we do not choose a hard threshold to convert the average probability map $\widetilde{\mathbf{m}}_i^j$ to a binary mask, as inspired by the "label smoothing" technique [14] which may help prevent the network from becoming over-confident and improve generalization ability.

With the expansion of the training set (TLs $\cup$ PLs), our FCN can distill more knowledge about the data (e.g., topological structure, intensity variances), thus becoming more robust and generalizing better to unseen data $\mathcal{B}$. However, due to the extreme sparsity of annotation at the very beginning, not all the generated PLs are evenly used (i.e., highly uncertain and assigned with low weights). Hence, we propose to conduct this process iteratively.

Overall, with our iterative uncertainty-guided self-training scheme, we can further refine the PLs and FCN at the same time. In practice, it needs 2 or 3 rounds, but we do not have to train from scratch, incurring not too much cost.

3 Experiments and Results

Data Acquisition. Mice were produced, sacrificed, and processed in compliance with animal welfare guidelines approved by the Pennsylvania State University (PSU). Embryos were stained with phosphotungstic acid (PTA), as described in [10]. Data were acquired by the PSU Center for Quantitative Imaging using the General Electric v|tom|x L300 nano/micro-CT system with a 180-kV nanofocus tube and were then reconstructed into micro-CT volumes with a resulting average voxel size of $5\mu m$ and volume size of 1500×2000^2. Seven volumes are divided into the training set $\mathcal{A} = \{\mathcal{A}_i\}_{i=1}^4$ and test set $\mathcal{B} = \{\mathcal{B}_i\}_{i=1}^3$. Only a very small subset of slices in each $\mathcal{A}_i$ is labeled for training (denoted as $\mathcal{A}l_i$) and the rest unseen slices $\mathcal{A}u_i$ and $\mathcal{B}$ are used for the test. Four scientists with extensive experience in the study of embryonic bones/cartilages were involved in image annotations. They first annotated slices in the 2D plane and then refined the whole annotation by considering 3D information of the neighboring slices.

Evaluation. In the 3D image regions not considered by the experts, we select 11 3D subregions (7 from $\mathcal{B}$ and 4 from $\mathcal{A}u_i$), each of an average size 30×300^2 and containing at least one piece of cartilages. These subregions are chosen for their representativeness, i.e., they cover all the typical types of cartilages (e.g., nasal capsule, Meckel's cartilage, lateral wall, braincase floor, etc). Each subregion is manually labeled by experts as ground truth. The segmentation accuracy is measured by Dice-Sørensen Coefficient (DSC).

Implementation Details. All our networks are implemented with TensorFlow [1], initialized by the strategy in [5], and trained with the Adam optimizer [9] (with $\beta_1 = 0.9$, $\beta_2 = 0.999$, and $\epsilon =$ 1e-10). We adopt the "poly" learning rate policy, $L_r \times \left(1 - \frac{iter}{\#iter}\right)^{0.9}$, where the initial rate $L_r =$ 5e-4 and the max iteration number is set as 60k. To leverage the limited training data and reduce over-fitting, we augment the training data with standard operations (e.g., random crop, flip, rotation in 90°, 180°, and 270°). Due to large intensity variance among different images, all images are normalized to have zero mean and unit variance.

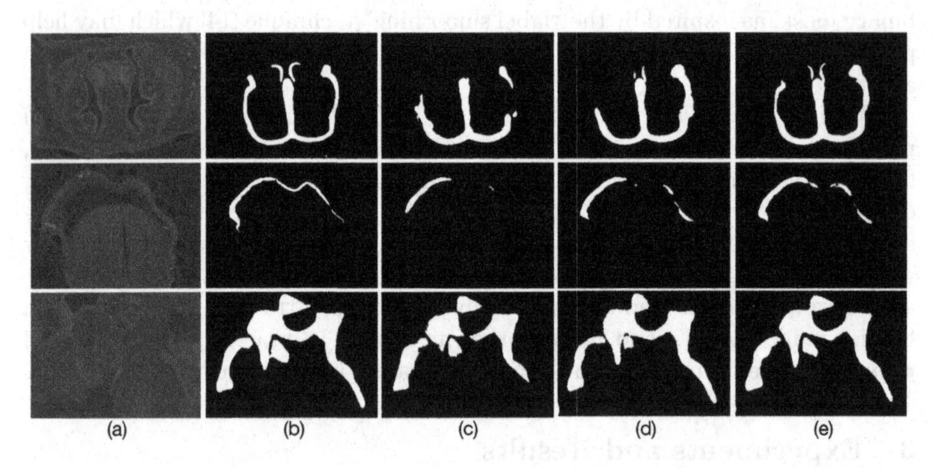

Fig. 4. Qualitative examples: (a) Raw subregions; (b) ground truth; (c) U-Net* (TL); (d) K-head FCN (TL); (e) K-head FCN-R3-U (TL∪PL). (XX) = (trained using XX).

Main Results. The results are summarized in Table 1. To our best knowledge, there is no directly related work on cartilage segmentation from embryonic tissues. We compare our new framework with the following methods. (1) A previous work which utilizes U-Net [19] to automatically segment knee cartilages [2]. We also try another robust FCN model DCN [4]. For a fair comparison, we scale up U-Net [19] and DCN [4] to accommodate images of size 512^2 as input and match with the number of parameters of our K-head FCN (denoted as U-Net* and DCN*). (2) A semi-supervised method that generates PLs and conducts self-training (i.e., 1-head FCN-R3).

Table 1. Segmentation results. Top: DSC (%) comparison of cartilages in the anterior, intermediate, and posterior skull, w/annotation ratio of 3.0%. TL: true labels; PL: pseudo labels. Bottom-left: "K-head FCN-R3-U (TL∪PL)" w/annotation ratio of 3.0%. Bottom-right: "K-head FCN-R3-U (TL∪PL)" w/different annotation ratios.

Method	Anterior	Intermediate	Posterior	Overall
U-Net* [19] (TL)	80.03	81.19	64.39	76.06
DCN* [4] (TL)	80.87	81.68	64.07	76.42
K-head FCN (TL)	82.23	84.46	67.52	78.98
1-head FCN-R3 (TL∪PL)	85.15	87.53	69.46	81.69
K-head FCN-R3 (TL∪PL)	85.77	88.34	70.30	82.45
K-head FCN-R3-U (TL∪PL)	86.31	89.17	70.98	83.16

Data	Iteration		
	1	2	3
$\{\mathcal{A}u_i\}_{i=1}^{L}$	83.19	86.39	87.08
$\{\mathcal{B}_i\}_{i=1}^{U}$	78.98	82.70	83.16

Data	Annotation Ratio		
	1.5%	3.0%	12.0%
$\{\mathcal{A}u_i\}_{i=1}^{L}$	80.12	87.08	89.20
$\{\mathcal{B}_i\}_{i=1}^{U}$	75.73	83.16	85.65

First, compared with known FCN-based methods, our K-head FCN yields better performance for cartilages in different positions. We attribute this to its deeper structures and multi-scale extracted feature fusion design, which leads to larger receptive fields and richer spatial and semantic features. Hence, our backbone model can capture significant topology variances in skull cartilages (e.g., relatively small but thick nasal parts, and large but thin shell-like cranial base and vault). Second, to show that our K-head FCN is comparable with Monte Carlo sampling based Bayesian methods, we implement 1-head FCN and conduct sampling K times to obtain PLs. Repeating the training process 3 times (denoted as '-R3'), we observe that using PLs, K-head FCN-R3 achieves similar performance as 1-head FCN-R3. However, in each forward pass, we obtain K predictions at once, thus saving $\sim K\times$ the time/space costs. Qualitative results are shown in Fig. 4. Third, we further show that under the guidance of uncertainty, our new method (K-head FCN-R3-U) attains performance gain (from 82.45% to 83.16%). We attribute this to that unreliable PLs are ruled out, and the model optimizes under cleaner supervisions.

Discussions. **(1)** *Iteration Numbers.* We measure DSC scores on both unseen slices in the training volumes ($\{\mathcal{A}u_i\}_{i=1}^{L}$) and unseen slices in the test volumes ($\{\mathcal{B}_i\}_{i=1}^{U}$) during the training of "K-head FCN-R3-U" (see Table 1 bottom-left). We notice significant performance gain after expanding the training set (i.e., TLs → TLs ∪ PLs, as Iter-1 → Iter-2). Meanwhile, because the uncertainty of only a small amount of pixels changes during the whole process, the performance gain is not substantial from Iter-2 to Iter-3. **(2)** *Annotation Ratios.* As shown in Table 1 bottom-right, the final segmentation results can be improved using

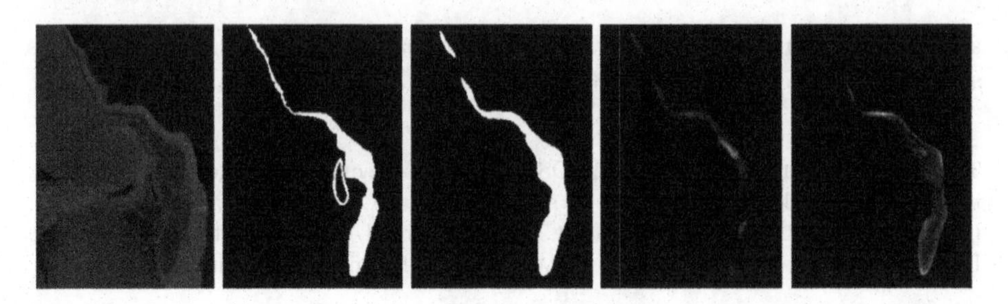

Fig. 5. Visualization of uncertainty. From left to right: a raw image region, ground truth, prediction result, estimated epistemic uncertainty, and estimated aleatoric uncertainty. Brighter white color means higher uncertainty.

more annotation, but the improvement rate decreases when labeling more slices. **(3)** *Uncertainty Estimation.* We visualize the samples along with estimated segmentation results and the corresponding epistemic and aleatoric uncertainties from the test data in Fig. 5. It is shown that the model is less confident (i.e., with a higher uncertainty) on the boundaries and hard mimic regions where the epistemic and aleatoric uncertainties are prominent.

4 Conclusions

We presented a new framework for cartilage segmentation in high-resolution 3D micro-CT images with very sparse annotation. Our K-head FCN produces segmentation predictions and uncertainty estimation simultaneously, and the iterative uncertainty-guided self-training strategy gradually refines the segmentation results. Comprehensive experiments showed the efficacy of our new method.

Acknowledgement. This research was supported in part by the US National Science Foundation through grants CNS-1629914, CCF-1617735, IIS-1455886, DUE-1833129, and IIS-1955395, and the National Institute of Dental and Craniofacial Research through grant R01 DE027677.

References

1. Abadi, M., et al.: TensorFlow: a system for large-scale machine learning. In: OSDI, vol. 16, pp. 265–283 (2016)
2. Ambellan, F., Tack, A., Ehlke, M., Zachow, S.: Automated segmentation of knee bone and cartilage combining statistical shape knowledge and convolutional neural networks: data from the osteoarthritis initiative. Med. Image Anal. **52**, 109–118 (2019)
3. Brinkley, J.F., et al.: The facebase consortium: a comprehensive resource for craniofacial researchers. Development **143**(14), 2677–2688 (2016)
4. Chen, H., Qi, X.J., Cheng, J.Z., Heng, P.A.: Deep contextual networks for neuronal structure segmentation. In: Thirtieth AAAI Conference on Artificial Intelligence, pp. 1167–1173 (2016)

5. He, K., Zhang, X., Ren, S., Sun, J.: Delving deep into rectifiers: surpassing human-level performance on imagenet classification. In: Proceedings of the IEEE International Conference on Computer Vision, pp. 1026–1034 (2015)
6. He, K., Zhang, X., Ren, S., Sun, J.: Deep residual learning for image recognition. In: Proceedings of the IEEE Conference on Computer Vision and Pattern Recognition, pp. 770–778 (2016)
7. Kendall, A., Gal, Y.: What uncertainties do we need in Bayesian deep learning for computer vision? In: Advances in Neural Information Processing Systems, pp. 5574–5584 (2017)
8. Kendall, A., Gal, Y., Cipolla, R.: Multi-task learning using uncertainty to weigh losses for scene geometry and semantics. In: Proceedings of the IEEE Conference on Computer Vision and Pattern Recognition, pp. 7482–7491 (2018)
9. Kingma, D.P., Ba, J.: Adam: a method for stochastic optimization. In: Third International Conference on Learning Representations (2015)
10. Lesciotto, K.M., et al.: Phosphotungstic acid-enhanced microCT: optimized protocols for embryonic and early postnatal mice. Dev. Dyn. **249**, 573–585 (2020). https://doi.org/10.1002/dvdy.136
11. Liang, P., Chen, J., Zheng, H., Yang, L., Zhang, Y., Chen, D.Z.: Cascade decoder: a universal decoding method for biomedical image segmentation. In: IEEE 16th International Symposium on Biomedical Imaging (ISBI), pp. 339–342 (2019)
12. Long, J., Shelhamer, E., Darrell, T.: Fully convolutional networks for semantic segmentation. In: Proceedings of the IEEE Conference on Computer Vision and Pattern Recognition, pp. 3431–3440 (2015)
13. Mossey, P.A., Catilla, E.E., et al.: Global registry and database on craniofacial anomalies: report of a WHO registry meeting on craniofacial anomalies (2003)
14. Müller, R., Kornblith, S., Hinton, G.E.: When does label smoothing help? In: Advances in Neural Information Processing Systems, pp. 4696–4705 (2019)
15. Niyogi, P.: Manifold regularization and semi-supervised learning: some theoretical analyses. J. Mach. Learn. Res. **14**(1), 1229–1250 (2013)
16. Oh, M.h., Olsen, P.A., Ramamurthy, K.N.: Crowd counting with decomposed uncertainty. In: Thirty-Fourth AAAI Conference on Artificial Intelligence, pp. 11799–11806 (2020)
17. Prasoon, A., Petersen, K., Igel, C., Lauze, F., Dam, E., Nielsen, M.: Deep feature learning for knee cartilage segmentation using a triplanar convolutional neural network. In: Mori, K., Sakuma, I., Sato, Y., Barillot, C., Navab, N. (eds.) MICCAI 2013. LNCS, vol. 8150, pp. 246–253. Springer, Heidelberg (2013). https://doi.org/10.1007/978-3-642-40763-5_31
18. Richtsmeier, J.T., Baxter, L.L., Reeves, R.H.: Parallels of craniofacial maldevelopment in Down syndrome and Ts65Dn mice. Dev. Dyn. **217**(2), 137–145 (2000)
19. Ronneberger, O., Fischer, P., Brox, T.: U-Net: convolutional networks for biomedical image segmentation. In: Navab, N., Hornegger, J., Wells, W.M., Frangi, A.F. (eds.) MICCAI 2015. LNCS, vol. 9351, pp. 234–241. Springer, Cham (2015). https://doi.org/10.1007/978-3-319-24574-4_28
20. Wang, Y., et al.: Deep attentional features for prostate segmentation in ultrasound. In: Frangi, A.F., Schnabel, J.A., Davatzikos, C., Alberola-López, C., Fichtinger, G. (eds.) MICCAI 2018. LNCS, vol. 11073, pp. 523–530. Springer, Cham (2018). https://doi.org/10.1007/978-3-030-00937-3_60

21. Yu, L., Wang, S., Li, X., Fu, C.-W., Heng, P.-A.: Uncertainty-aware self-ensembling model for semi-supervised 3D left atrium segmentation. In: Shen, D., Liu, T., Peters, T.M., Staib, L.H., Essert, C., Zhou, S., Yap, P.-T., Khan, A. (eds.) MICCAI 2019. LNCS, vol. 11765, pp. 605–613. Springer, Cham (2019). https://doi.org/10.1007/978-3-030-32245-8_67
22. Zheng, H., et al.: HFA-Net: 3D cardiovascular image segmentation with asymmetrical pooling and content-aware fusion. In: Shen, D., et al. (eds.) MICCAI 2019. LNCS, vol. 11765, pp. 759–767. Springer, Cham (2019). https://doi.org/10.1007/978-3-030-32245-8_84
23. Zheng, H., Zhang, Y., Yang, L., Wang, C., Chen, D.Z.: An annotation sparsification strategy for 3D medical image segmentation via representative selection and self-training. In: Thirty-Fourth AAAI Conference on Artificial Intelligence, pp. 6925–6932 (2020)

Probabilistic 3D Surface Reconstruction from Sparse MRI Information

Katarína Tóthová[1(✉)], Sarah Parisot[2], Matthew Lee[3], Esther Puyol-Antón[4], Andrew King[4], Marc Pollefeys[1,5], and Ender Konukoglu[1]

[1] ETH Zurich, Zurich, Switzerland
katarina.tothova@inf.ethz.ch
[2] Huawei Noah's Ark Lab, London, UK
[3] Imperial College London, London, UK
[4] King's College London, London, UK
[5] Microsoft Mixed Reality and AI lab, Zurich, Switzerland

Abstract. Surface reconstruction from magnetic resonance (MR) imaging data is indispensable in medical image analysis and clinical research. A reliable and effective reconstruction tool should: be fast in prediction of accurate well localised and high resolution models, evaluate prediction uncertainty, work with as little input data as possible. Current deep learning state of the art (SOTA) 3D reconstruction methods, however, often only produce shapes of limited variability positioned in a canonical position or lack uncertainty evaluation. In this paper, we present a novel probabilistic deep learning approach for concurrent 3D surface reconstruction from sparse 2D MR image data and aleatoric uncertainty prediction. Our method is capable of reconstructing large surface meshes from three quasi-orthogonal MR imaging slices from limited training sets whilst modelling the location of each mesh vertex through a Gaussian distribution. Prior shape information is encoded using a built-in linear principal component analysis (PCA) model. Extensive experiments on cardiac MR data show that our probabilistic approach successfully assesses prediction uncertainty while at the same time qualitatively and quantitatively outperforms SOTA methods in shape prediction. Compared to SOTA, we are capable of properly localising and orientating the prediction via the use of a spatially aware neural network.

Keywords: Uncertainty quantification · 3D reconstruction · Shape modelling · Deep learning

1 Introduction

High quality 3D surface models of internal organs constructed from MR imaging data are vital for diagnosis, disease tracking, surgical planning or interpretation

Electronic supplementary material The online version of this chapter (https://doi.org/10.1007/978-3-030-59710-8_79) contains supplementary material, which is available to authorized users.

A. L. Martel et al. (Eds.): MICCAI 2020, LNCS 12261, pp. 813–823, 2020.
https://doi.org/10.1007/978-3-030-59710-8_79

of functional data in clinical and research practice [3,21,22]. In cardiac imaging, for example, virtual ventricle surface meshes enable the use of patient-specific 3D models for investigation of valve and vessel function, or surgical and catheter-based procedural planning [28].

The problem of surface reconstruction has been widely studied in medical imaging research. Traditional approaches usually take advantage of parametric models through atlas or statistical shape model registration [2,10,19,21,22] or use predefined forces to evolve a deformable shape into the final surface [9,12,13,23]. In contrast to the complex frameworks that might be associated with such methods such as in [9], the advent of machine learning has led to the possibility of training deep neural networks in an end-to-end manner: from images to parametrised shapes. In their work on 2D shape reconstruction [5,20,27] employ convolutional neural networks (CNN) to modulate shapes generated from in-built principal component analysis (PCA) shape priors. The situation in *3D* machine learning surface reconstruction from medical images is slightly more complex. Besides the usual requirements on accuracy and availability of sufficiently large training sets, the 3D methods often have to deal with sparsity of the input imaging information—as is the case in 3D reconstruction of organs from a few imaging slices. Support for sparse input data is essential in the medical domain. It leads to faster acquisition times, fewer motion artifacts, less radiation exposure for a patient (e.g. for CT), and ultimately a cheaper and more accessible imaging method. Cerrolaza et al. [7] solve the problem by relating it to the one of a single-view 3D reconstruction in general computer vision. When constructing volumetric predictions of 3D fetal skulls from 2D ultrasound images, they utilise a conditional variational autoencoder (CVAE) via a TL-net inspired architecture [11]. The downsides of such encoder-decoder setups are, however, computational and memory demands associated with computing 3D convolutions on volumetric data and limited resolution of predicted surfaces. Moreover, to build an implicit shape prior they require volumetric training data to be pre-aligned and hence at test time predict 3D shapes in a canonical position and orientation. The topology of predicted shapes is not constrained, which may lead to undesirable artefacts. Finally, Tatarchenko et al. [26] suggest there is in general little statistical difference between the reconstructions of encoder-decoder methods and shapes obtained as nearest-neighbours or local cluster means in the training set.

A prediction system that is to be useful in practice should be not only highly accurate and precise, it should also indicate how confident it is about the results it provides. This is of utmost interest in medical imaging due to the nature of the task at hand, its application and the input data, which is often times sparse and comprised of noisy images of coarse resolution riddled with imaging artifacts. The ambiguity ingrained in the data—*aleatoric* uncertainty—is modelled via a probability distribution over the model output, which can be integrated into the optimisation process itself [4,15]. In [27], Tóthová et al. take this route by formulating 2D surface reconstruction as a conditional probability estimation problem. Another popular approach to assess the variability of the output of a method is the use of statistics aggregated on a set of plausible predictions sampled from a probabilistic model. Methods such as a conditional variational autoencoders [7], MCMC [8,19] or Gaussian processes [17] have been used to this end.

In this paper, we propose a novel probabilistic model for 3D surface reconstruction from MRI data that jointly addresses high resolution accurate reconstruction, sparse input data constraints and aleatoric uncertainty estimation within a single framework. We learn to reconstruct high resolution surface meshes from three quasi-orthogonal MR imaging slices through the object using a small sized training dataset. Our method extends the PCA shape prior based surface prediction of [20,27] into the probabilistic 3D setting by formulating the problem as a conditional probability estimation. Using a shape prior allows us to be robust in the absence of input information whilst our probabilistic formulation leads to an analytical expression capturing uncertainty. In contrast to autoencoder reconstruction methods [7,11], no training data pre-alignment is needed and model predictions are not only of the right shape, they are also correctly orientated, as shape orientation is predicted as a mode of the PCA model, and localised in the common world space. This is possible by augmenting the input features using coordinate maps relating the image and world space coordinate systems, a novel feature inspired by [18]. This approach also provides a simple yet efficient solution for dealing with possible heterogeneity in the input image acquisition setups, such as varying acquisition angles or positioning of patients in the scanner. The proposed approach was evaluated on 3D cardiac reconstruction using UK BioBank [24] MRI data. Our approach successfully assesses prediction uncertainty, outperforming SOTA methods [7,11,20] in terms of quantitative and qualitative evaluation. To our knowledge, we are the first to tackle the combined problem of reconstructing high resolution meshes in common world coordinates while preserving shape volume and topology and at the same time providing a principled quantification of uncertainty.

2 Method

Our goal is to devise a probabilistic 3D surface reconstruction model predicting organ surface meshes from a sparse set of 2D MRI input images. Specifically, we consider an input set of three MRI quasi-orthogonal slices across an organ denoted as $\boldsymbol{x} = \{\boldsymbol{X_1}, \boldsymbol{X_2}, \boldsymbol{X_3}\}$. Based on principles of probabilistic PCA [6], our framework addresses the inherent challenges linked with the sparse and heterogeneous input data via the use of a spatially aware deep CNN computing distributions over principal component scores of a PCA shape prior as in [27].

Probabilistic Model. We formulate surface prediction as a probability estimation problem where we aim to compute a probability distribution $p(y|\boldsymbol{x})$ over the coordinates of surface mesh vertices y.

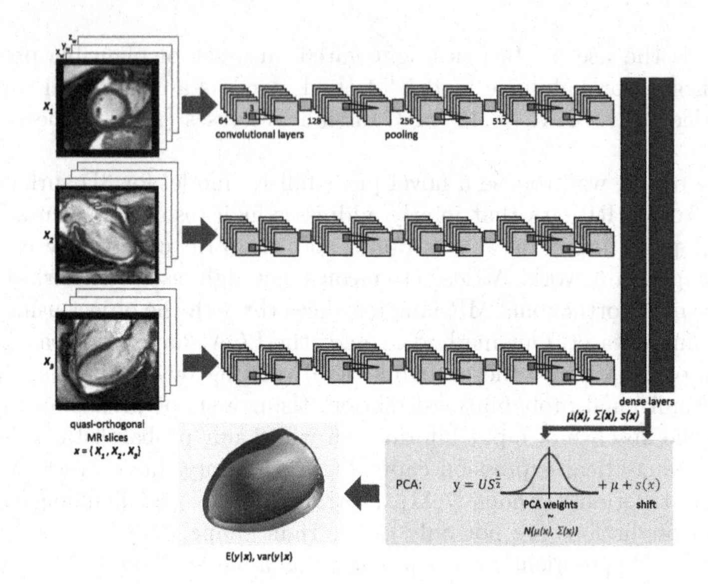

Fig. 1. Schematic setup of the surface and uncertainty prediction. The input to each of the three convolutional branches of the network is a single MRI slice (here pictured cardiac MRI) concatenated with the coordinate maps computed from the metadata and relating the pixel and world space. The output of the framework is a probabilistic mesh where the location of each vertex is modelled by a Gaussian distribution.

Probabilities are conditioned on the input MRI imaging stack $\boldsymbol{x}$ through a latent variable model

$$p(y|\boldsymbol{x}) = \int p(y|z, \boldsymbol{x})p(z|\boldsymbol{x})dz, \tag{1}$$

with latent variable z. For a given surface y, z are scores in the PCA shape prior which we define through $p(y|z, \boldsymbol{x})$ as

$$p(y|z, \boldsymbol{x}) = \mathcal{N}(y|US^{\frac{1}{2}}z + \mu + s(\boldsymbol{x}), \sigma^2 I), \tag{2}$$

where U is a matrix of principal vectors (columns of U), S represents the principal component diagonal covariance matrix and μ the data mean, all three precomputed using the surfaces in the training set, and s is a global spatial shift dependent on the input image stack $\boldsymbol{x}$. Variance σ^2 reflects the noise level in the data. The latent space is conditioned on the input and structured according to the Gaussian distribution with

$$p(z|\boldsymbol{x}) = \mathcal{N}(z|\mu(\boldsymbol{x}), \Sigma(\boldsymbol{x})), \tag{3}$$

where the mean $\mu(\boldsymbol{x})$ and covariance matrix $\Sigma(\boldsymbol{x})$ are estimated jointly from our input image stack $\boldsymbol{x}$ using a deep CNN, see Fig. 1. Note that in practice, we constrain the network to predict Cholesky factor $\Sigma^{\frac{1}{2}}(\boldsymbol{x})$ to ensure the positive definiteness of the estimated covariance matrix. Finally, a prior on the latent variable assumes $z \sim \mathcal{N}(0, I)$ such as in probabilistic PCA [6].

Inference. At test time, given a test image stack $\boldsymbol{x}$, we follow Eq. 1 and generate the target mesh predictions y by sampling from a Gaussian distribution $p(y|\boldsymbol{x}) = \mathcal{N}(\mathbb{E}(y|\boldsymbol{x}), \text{var}(y|\boldsymbol{x}))$ with mean and variance expressed by

$$\mathbb{E}(y|\boldsymbol{x}) = US^{\frac{1}{2}}\mu(\boldsymbol{x}) + \mu + s, \tag{4}$$

$$\text{var}(y|\boldsymbol{x}) = \sigma^2 I + US^{\frac{1}{2}}\Sigma(\boldsymbol{x})(US^{\frac{1}{2}})^T, \tag{5}$$

where σ^2 denotes noise in the data and constitutes a hyperparameter. Equation 4 and Eq. 5 follow directly from Eq. 2 and Eq. 3 and associate coordinates of each vertex in the predicted surface mesh with an explicitly expressed uncertainty. Full derivation can be found in Appendix A.

Training. In order to predict $\Sigma(\boldsymbol{x})$, $\mu(\boldsymbol{x})$ and $s(\boldsymbol{x})$, we train a deep CNN by minimising an objective function L consisting of two components: a surface prediction fidelity data term L_{data} aiming to maximise the conditional probability $p(y|\boldsymbol{x})$, and a regularisation term L_{reg} minimising the distance between the prior and observed distribution over z by means of Kullback-Leibler divergence (KLD)

$$L = \lambda L_{reg} - L_{data}, \tag{6}$$

where $\lambda > 0$ is a regularisation parameter.

Whilst Eq. 1 allows for direct maximisation of $\ln p(y|\boldsymbol{x})$ by marginalisation of the latent variable, this approach is not computationally feasible, as outlined in [27]. Even though the marginal distribution is Gaussian with closed form mean and variance, direct optimisation would require inversions of a large and often poorly conditioned covariance matrix (5) which could lead to numerical instabilities as was empirically observed. We apply Jensen's inequality to Eq. 1 and derive a lower bound for $\ln p(y|\boldsymbol{x})$, which will constitute our prediction fidelity data term as follows

$$\ln p(y|\boldsymbol{x}) \geq \mathbb{E}_{z|\boldsymbol{x}}\left[\ln p(y|z, \boldsymbol{x})\right] \cong \frac{1}{L}\sum_{l=1}^{L} \ln p(y|z_l, \boldsymbol{x}), \tag{7}$$

where z_l is sampled from $p(z|\boldsymbol{x})$ defined in Eq. 3. Sampling is done by means of a "reparametrisation trick" [16] through $z_l = \mu(x) + \Sigma_x^{1/2} * \epsilon$, $\epsilon \sim \mathcal{N}(0, I)$. L refers to the number of samples, in our case $L = 1$. Hence

$$L_{data} = \sum_{n=1}^{N} \frac{1}{L} \sum_{l=1}^{L} \ln p(y|z_l, \boldsymbol{x}_n). \tag{8}$$

Maximisation of the lower bound in Eq. 7 might not satisfy the prior set on the latent variable z in our probabilistic PCA model ($z \sim \mathcal{N}(0, I)$). To align the distribution of z observed in the training data $p(z) \cong \sum p(z|\boldsymbol{x}_n)$ obtained from the input image stack $\boldsymbol{x}_n \sim p(\boldsymbol{x})$ with the prior, we employ the KLD regularisation and minimise

$$L_{reg} = \text{KLD}\left(\sum_{n=1}^{N} p(z|\boldsymbol{x}_n), p(z)\right). \tag{9}$$

For further details on the derivations please refer to Appendix B.

Table 1. Results of surface prediction: average statistics collected over the test set. Results of our method were computed using mean predictions. Average DICE_{pair} score were computed over all pairs of predicted shapes across all subjects in the aligned test set and reflects an inter-subject shape variability.

Method	DICE	HD	ASSD	RAVD	DICE_{pair}
det PCA_O 12	0.15 ± 0.12	19.1 ± 10.3	7.38 ± 5.40	−15.6%	–
det PCA_O 16	0.14 ± 0.11	19.7 ± 10.4	7.50 ± 5.52	−14.6%	–
det PCA_O 20	0.15 ± 0.11	19.2 ± 10.0	7.35 ± 5.31	−12.6%	–
det PCA 12	0.57 ± 0.15	5.04 ± 1.90	$\mathbf{1.02 \pm 0.54}$	6.09 %	–
det PCA 16	0.56 ± 0.15	5.43 ± 2.38	1.08 ± 0.70	3.12%	–
det PCA 20	0.58 ± 0.14	5.09 ± 1.87	1.02 ± 0.58	0.97%	0.49 ± 0.21
TL-net	0.59 ± 0.19	7.75 ± 6.55	1.37 ± 0.76	−23.8%	0.52 ± 0.22
REC-CVAE	$\mathbf{0.61 \pm 0.13}$	8.02 ± 1.24	1.18 ± 0.31	−5.62%	0.97 ± 0.02
Ours prob PCA 12	0.54 ± 0.15	5.60 ± 1.92	1.15 ± 0.62	0.44%	–
Ours prob PCA 16	0.55 ± 0.13	5.37 ± 2.02	1.06 ± 0.52	0.63%	–
Ours prob PCA 20	0.57 ± 0.13	$\mathbf{4.98 \pm 1.66}$	$\mathbf{1.02 \pm 0.54}$	**0.29%**	0.70 ± 0.20

Spatially Aware Deep Network Architecture. To predict the distribution of the latent space along with the global shift of the surface, we build a 3-branch deep CNN where each branch takes one MR imaging slice as input, see Fig. 1. In each branch, 9 convolutional layers (stride = 1) are intertwined with 3 max pooling layers (stride = 2). Last features of the convolutional branches are concatenated and passed to two dense layers. All convolutional and the first dense layers are followed by ReLU activations.

An important challenge to be addressed is the fact that all three available input slices pertaining to a given organ are related in the real 3D world. Each slice is associated with a separate transformation matrix between the pixel and world coordinate system. The reference and reconstruction meshes are both defined in absolute world space coordinates. Hence, working solely in pixel space and applying the CNN to 2D images directly would lead to inconsistencies and poor performance as the accurate spatial localisation and orientation of image slices with respect to each other as well as to the reference meshes would be lost. Inspired by [18], we propose a simple yet effective solution to mitigate this issue by taking advantage of the known transformation matrices between pixel and world coordinates. These are available for every MRI volume. More precisely, the input to every network branch will consist of intensities of each MR imaging slice $\boldsymbol{X}_i$ $(i = 1, 2, 3)$ concatenated with the coordinates of each pixel in the 3D world space.

Alternative complex solutions such as spatial transformers [14] are infeasible in this situation as they would require converting the input 2D imaging plane into positions in 3D space and hence a construction of a full imaging 3D volume. This is however impracticable due to missing imaging information outside of the original imaging plane, since all we have are 2D slices.

3 Results

The proposed method was evaluated on the task of the left ventricle myocardium surface reconstruction using cardiac imaging volumes from UK BioBank [24]. We predict 3D coordinates of 22043 surface mesh vertices from 3 quasi-orthogonal MR image slices: one acquired along the short axis and 2 along the long axis. All images were resampled to isotropic pixel size of 1.8269 mm and cropped/padded to the size of 80×80 pixels. The data set used consists of 529 training, 178 testing and 178 validation examples. Reference meshes used for training and evaluation were constructed using an atlas based method described in [22] from the segmentations prepared automatically using expert-segmentations and combination of learning and registration methods from [1,25]. Mesh connectivity was fixed throughout the data set. Training was done via an RMS-Prop optimiser with a constant learning rate of 10^{-6} for batches of size 5. Hyperparameters were optimised on the validation set. Noise level in the data σ^2 and regularisation parameter λ were empirically set to $\sigma^2 = 5 \times 10^{-2}$ and $\lambda = 10^3$ respectively.

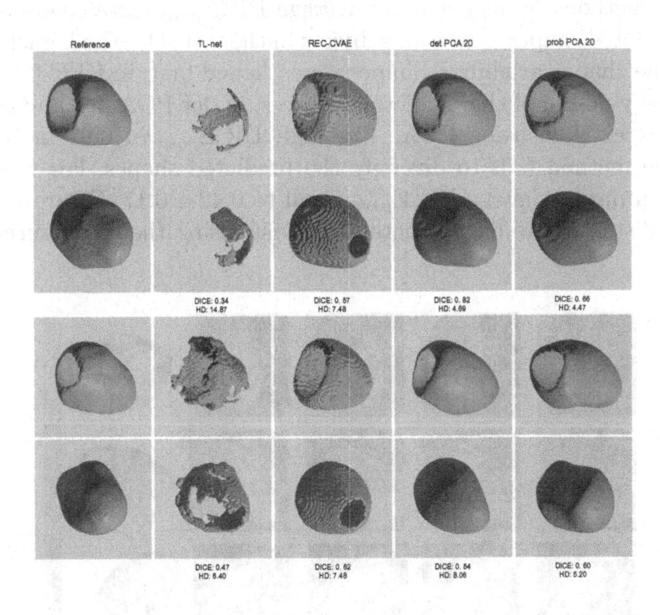

Fig. 2. Example of shape predictions for two different subjects visualised from two distinct angles.

We compared our method (prob PCA n, $n =$ number of principal components used) with four baselines: 1) deterministic PCA (det PCA$_O$ n) as proposed in [20] without the spatial transformer refinement and extended to 3D using an architecture analogous to ours, 2) det PCA enriched with input coordinate maps as in our method, 3) TL-net architecture [11] implemented as in [7] and 4) conditional variational autoencoder method REC-CVAE from [7]. The last two methods were trained and evaluated on a pre-aligned set of reference volumes.

Quantitative results can be seen in Table 1. The models were evaluated using a mixture of volumetric and surface measures: DICE score, Hausdorff distance (HD), Average symmetric surface distance (ASSD), and Relative absolute volume difference (RAVD). Where applicable, volumetric measures were computed using voxelisations of reference and predicted meshes. Surface measures were then aggregated on surfaces extracted from volumetric shapes. Results of our probabilistic model were evaluated using the mean prediction defined in Eq. 4. We can see that our approach outperforms all competing methods in terms of surface criteria and RAVD. Figure 2 shows that our results are qualitatively superior to coarse volumetric shapes of unrestricted topology obtained from volumetric methods. Our predicted meshes are genus 0 and respect the volume of the reconstructed organ.

We observe poorer performance in terms of DICE score, which can be explained through its definition. Methods that predict empty space in uncertain areas garner superior scores as they minimise the number of false positives in the prediction. We further quantified predicted shape variability for the top performing methods by means of an average DICE_{pair} score computed over all pairs of predicted shapes across all subjects in the test set—the higher the DICE, the lower the shape variability. Shapes were aligned first: in REC-CVAE [7] and TL-net [11] by design of the prediction process, in det PCA [20] and our method were aligned ex post using PCA. The high DICE_{pair} of 0.97 in REC-CVAE suggests the method fails to diversify the predicted shapes. For reference, the ground truth meshes have a DICE_{pair} equal to 0.42 ± 0.17. Figure 3 exemplifies the probabilistic nature of our method by visualising the mean surface predic-

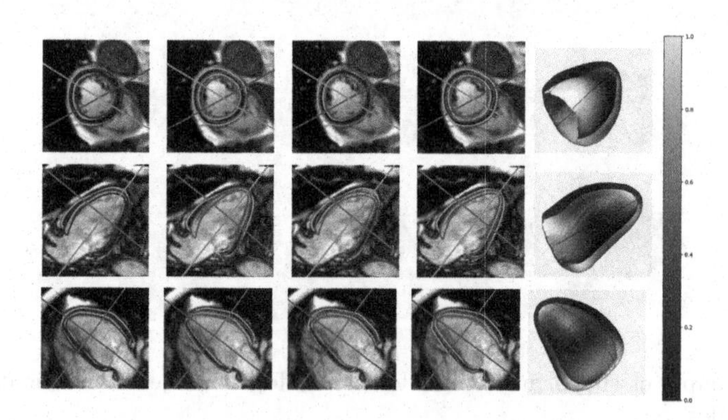

Fig. 3. Surface reconstruction results for a single subject. Column 1 on the left: intersection of the 3D surface with the input MR imaging slices - ground truth surface (yellow), mean surface prediction (blue). Straight lines represent the intersection of the other two input MR slices with the current imaging plane. Columns 2–4: example surface samples (magenta). Column 5: Prediction uncertainty visualisation - colour of the heatmap at each vertex of the mean shape mesh corresponds to the $\log\det\Sigma_i$, $\Sigma_i \in \mathbb{R}^{3\times 3}$ is the covariance matrix at the given vertex, rescaled onto interval $[0, 1]$. (Color figure online)

tion and sampled surface meshes by means of their intersections with the input imaging slices. Prediction uncertainty at surface vertices can be assessed via a heatmap - vertex colours correspond to scaled $\log \det \Sigma_i$, where $\Sigma_i \in \mathbb{R}^{3\times3}$ is the covariance matrix at the given vertex extracted from $\mathrm{var}(y|\boldsymbol{x})$. The larger the variance, the bigger the uncertainty over the position of the vertex that can be sampled from this distribution. Notice how uncertainty grows with the increasing distance from the center of the organ. Additional qualitative examples are available in Appendix C.

4 Conclusion

In this paper, we presented a novel probabilistic deep learning approach for concurrent 3D surface reconstruction from sparse 2D MR image data and aleatoric uncertainty prediction. Our method is capable of reconstructing high resolution large surface meshes from three quasi-orthogonal MR imaging slices from limited training data whilst modelling the location of each mesh vertex through a Gaussian distribution. We build on the principles of the probabilistic PCA [6] and 2D surface prediction methods from [20,27] and incorporate prior shape information via a linear PCA model. Experiments using cardiac MRI data from UK BioBank [24] show that our method qualitatively and quantitatively outperforms the deterministic and autoencoder baselines in shape reconstruction while correctly localising and orientating the prediction. Moreover, it enables generation of plausible surface reconstructions through sampling from the predicted model and evaluation of the prediction uncertainty. Future work will concentrate on generalisation of the method to surface reconstruction in the presence of pathologies and reconstruction of other types of organs of more variable shapes, which may lead to adaptation of the used shape model.

Acknowledgements. This research has been conducted using the UK Biobank Resource under Application Number 17806.

References

1. Bai, W., et al.: Semi-supervised learning for network-based cardiac MR image segmentation. In: Descoteaux, M., Maier-Hein, L., Franz, A., Jannin, P., Collins, D.L., Duchesne, S. (eds.) MICCAI 2017. LNCS, vol. 10434, pp. 253–260. Springer, Cham (2017). https://doi.org/10.1007/978-3-319-66185-8_29
2. Bai, W., Shi, W., O'Regan, D.P., Tong, T., Wang, H., Jamil-Copley, S., et al.: A probabilistic patch-based label fusion model for multi-atlas segmentation with registration refinement: application to cardiac MR images. IEEE Trans. Med. Imaging **32**(7), 1302–1315 (2013)
3. Bai, W., et al.: Automated cardiovascular magnetic resonance image analysis with fully convolutional networks. J. Cardiovas. Magn. Reson. **20**, 65 (2018)
4. Baumgartner, C.F., et al.: PHiSeg: capturing uncertainty in medical image segmentation. In: Shen, D., et al. (eds.) MICCAI 2019. LNCS, vol. 11765, pp. 119–127. Springer, Cham (2019). https://doi.org/10.1007/978-3-030-32245-8_14

5. Bhalodia, R., Elhabian, S.Y., Kavan, L., Whitaker, R.T.: DeepSSM: a deep learning framework for statistical shape modeling from raw images. In: Reuter, M., Wachinger, C., et al. (eds.) ShapeMI 2018. LNCS, vol. 11167, pp. 244–257. Springer, Heidelberg (2018). https://doi.org/10.1007/978-3-030-04747-4_23
6. Bishop, C.M.: Pattern Recognition and Machine Learning. Information Science and Statistics, 5th edn. Springer, Heidelberg (2007)
7. Cerrolaza, J.J., et al.: 3D fetal skull reconstruction from 2DUS via deep conditional generative networks. In: Frangi, A.F., Schnabel, J.A., Davatzikos, C., Alberola-López, C., Fichtinger, G. (eds.) MICCAI 2018. LNCS, vol. 11070, pp. 383–391. Springer, Cham (2018). https://doi.org/10.1007/978-3-030-00928-1_44
8. Chang, J., III, J.W.F.: Efficient MCMC sampling with implicit shape representations. In: The 24th IEEE Conference on Computer Vision and Pattern Recognition, CVPR 2011, Colorado Springs, CO, USA, 20–25 June 2011, pp. 2081–2088. IEEE Computer Society (2011)
9. Fischl, B.: Freesurfer. NeuroImage **62**(2), 774–781 (2012)
10. Garcia-Barnes, J., et al.: A normalized framework for the design of feature spaces assessing the left ventricular function. IEEE Trans. Med. Imaging **29**(3), 733–745 (2010)
11. Girdhar, R., Fouhey, D.F., Rodriguez, M., Gupta, A.: Learning a predictable and generative vector representation for objects. In: Leibe, B., Matas, J., Sebe, N., Welling, M. (eds.) ECCV 2016. LNCS, vol. 9910, pp. 484–499. Springer, Cham (2016). https://doi.org/10.1007/978-3-319-46466-4_29
12. Han, X., Pham, D.L., Tosun, D., Rettmann, M.E., Xu, C., Prince, J.L.: CRUISE: cortical reconstruction using implicit surface evolution. NeuroImage **23**(3), 997–1012 (2004)
13. Huo, Y., et al.: Consistent cortical reconstruction and multi-atlas brain segmentation. NeuroImage **138**, 197–210 (2016)
14. Jaderberg, M., Simonyan, K., Zisserman, A., Kavukcuoglu, K.: Spatial transformer networks. In: Cortes, C., Lawrence, N.D., et al. (eds.) Advances in Neural Information Processing Systems 28: Annual Conference on Neural Information Processing Systems 2015, 7–12 December 2015, Montreal, Quebec, Canada, pp. 2017–2025 (2015)
15. Kendall, A., Gal, Y.: What uncertainties do we need in Bayesian deep learning for computer vision? In: Guyon, I., von Luxburg, U., et al. (eds.) Advances in Neural Information Processing Systems 30: Annual Conference on Neural Information Processing Systems 2017, 4–9 December 2017, Long Beach, CA, USA, pp. 5574–5584 (2017)
16. Kingma, D.P., Welling, M.: Auto-encoding variational bayes. In: Bengio, Y., LeCun, Y. (eds.) 2nd International Conference on Learning Representations, ICLR 2014, Banff, AB, Canada, 14–16 April 2014, Conference Track Proceedings (2014)
17. Lê, M., Unkelbach, J., Ayache, N., Delingette, H.: Sampling image segmentations for uncertainty quantification. Med. Image Anal. **34**, 42–51 (2016)
18. Liu, R., et al.: An intriguing failing of convolutional neural networks and the coordconv solution. In: Bengio, S., Wallach, H.M., et al. (eds.) Advances in Neural Information Processing Systems 31: Annual Conference on Neural Information Processing Systems 2018, NeurIPS 2018, 3–8 December 2018, Montréal, Canada, pp. 9628–9639 (2018)
19. Madsen, D., Vetter, T., Lüthi, M.: Probabilistic surface reconstruction with unknown correspondence. In: Greenspan, H., et al. (eds.) CLIP/UNSURE -2019. LNCS, vol. 11840, pp. 3–11. Springer, Cham (2019). https://doi.org/10.1007/978-3-030-32689-0_1

20. Milletari, F., Rothberg, A., Jia, J., Sofka, M.: Integrating statistical prior knowledge into convolutional neural networks. In: Descoteaux, M., Maier-Hein, L., et al. (eds.) MICCAI 2017, Part I. LNCS, vol. 10433, pp. 161–168. Springer, Heidelberg (2017). https://doi.org/10.1007/978-3-319-66182-7_19
21. Peressutti, D., et al.: A framework for combining a motion atlas with non-motion information to learn clinically useful biomarkers: application to cardiac resynchronisation therapy response prediction. Med. Image Anal. **35**, 669–684 (2017)
22. Puyol-Antón, E., et al.: A multimodal spatiotemporal cardiac motion atlas from MR and ultrasound data. Med. Image Anal. **40**, 96–110 (2017)
23. Schuh, A., et al.: A deformable model for the reconstruction of the neonatal cortex. In: 2017 IEEE 14th International Symposium on Biomedical Imaging (ISBI 2017), pp. 800–803 (2017)
24. Senn, M.: UK BioBank Homepage. https://www.ukbiobank.ac.uk/about-biobank-uk. Accessed 24 June 2018
25. Sinclair, M., Bai, W., Puyol-Antón, E., Oktay, O., Rueckert, D., King, A.P.: Fully automated segmentation-based respiratory motion correction of multiplanar cardiac magnetic resonance images for large-scale datasets. In: Descoteaux, M., Maier-Hein, L., Franz, A., Jannin, P., Collins, D.L., Duchesne, S. (eds.) MICCAI 2017. LNCS, vol. 10434, pp. 332–340. Springer, Cham (2017). https://doi.org/10.1007/978-3-319-66185-8_38
26. Tatarchenko, M., Richter, S.R., Ranftl, R., Li, Z., Koltun, V., Brox, T.: What do single-view 3D reconstruction networks learn? In: IEEE Conference on Computer Vision and Pattern Recognition, CVPR 2019, Long Beach, CA, USA, 16–20 June 2019. pp. 3405–3414. Computer Vision Foundation/IEEE (2019)
27. Tóthová, K., et al.: Uncertainty quantification in CNN-based surface prediction using shape priors. In: Reuter, M., et al. (eds.) ShapeMI 2018. LNCS, vol. 11167, pp. 300–310. Springer, Heidelberg (2018). https://doi.org/10.1007/978-3-030-04747-4_28
28. Vukicevic, M., Mosadegh, B., Min, J.K., Little, S.H.: Cardiac 3D printing and its future directions. JACC: Cardiovas. Imaging **10**(2), 171–184 (2017)

Can You Trust Predictive Uncertainty Under Real Dataset Shifts in Digital Pathology?

Jeppe Thagaard[1,2](✉), Søren Hauberg[1], Bert van der Vegt[3], Thomas Ebstrup[2], Johan D. Hansen[2], and Anders B. Dahl[1]

[1] Technical University of Denmark, Lyngby, Denmark
[2] Visiopharm A/S, Hørsholm, Denmark
jept@dtu.dk, jth@visiopharm.com
[3] University Medical Center Groningen, Groningen, The Netherlands

Abstract. Deep learning-based algorithms have shown great promise for assisting pathologists in detecting lymph node metastases when evaluated based on their predictive accuracy. However, for clinical adoption, we need to know what happens when the test set dramatically changes from the training distribution. In such settings, we should estimate the uncertainty of the predictions, so we know when to trust the model (and when not to). Here, we i) investigate current popular methods for improving the calibration of predictive uncertainty, and ii) compare the performance and calibration of the methods under clinically relevant in-distribution dataset shifts. Furthermore, we iii) evaluate their performance on the task of out-of-distribution detection of a different histological cancer type not seen during training. Of the investigated methods, we show that deep ensembles are more robust in respect of both performance and calibration for in-distribution dataset shifts and allows us to better detect incorrect predictions. Our results also demonstrate that current methods for uncertainty quantification are not necessarily able to detect all dataset shifts, and we emphasize the importance of monitoring and controlling the input distribution when deploying deep learning for digital pathology.

Keywords: Deep learning · Digital pathology · Predictive uncertainty

1 Introduction

Motivated by the predictive performance of deep learning (DL) in research [3,21] and grand challenges [2], clinical-grade DL-tools for assisting pathologists in detection of lymph node metastases are now being developed. In clinical settings where algorithms can potentially affect medical decisions, it is crucial to know how well-calibrated the underlying model is, such that the model gives a reliable estimate of the quality of the predictions. However, there exists only limited research [4,20,22] on how different distributional shifts in pathology affect

A. L. Martel et al. (Eds.): MICCAI 2020, LNCS 12261, pp. 824–833, 2020.
https://doi.org/10.1007/978-3-030-59710-8_80

the accuracy of DL-based algorithms, and these do not consider predictive uncertainty. Dataset shifts are especially relevant in pathology as pre-analytical steps can introduce large variability, and the spectrum of the target indication of an algorithm can also be broad. This makes it difficult to include the whole spectrum within the training set. Rare incidental findings, which are clinically relevant, may also be missed by an algorithm because they are outside the distribution of the training set (Fig. 1).

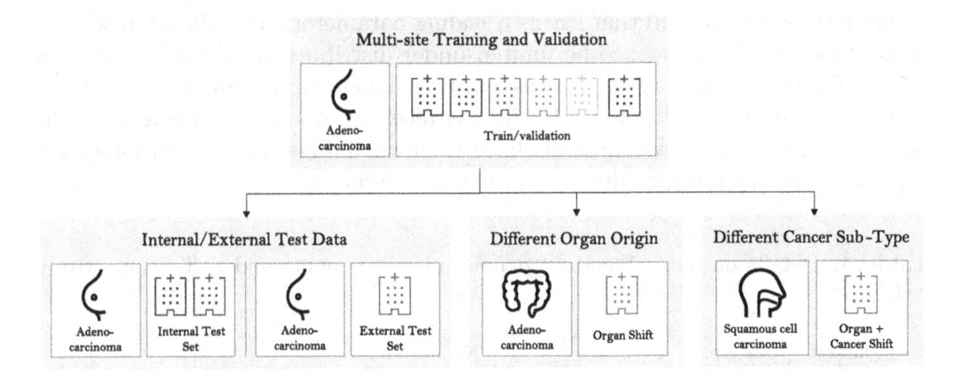

Fig. 1. Overview of experimental setup. Slides from 6 different sites are used as development data ($\mathcal{D}_{train}$ and $\mathcal{D}_{val}$), where blue (5 sites) represents CAM16-train and CAM17-train and red (one site) is DATASET2. CAM16-test defines the internal test set ($\mathcal{D}_{test,int.}$) as the 2 sites are also used as development data. DATASET3 (green) is denoted as the external test set ($\mathcal{D}_{test,ext.}$) as this site is not included in the development data. Slides from DATASET4 and DATASET5 (orange) with colon adenocarcinoma ($\mathcal{D}_{colon}$) and head and neck squamous cell carcinoma ($\mathcal{D}_{SCC}$) are used to test on different organ origin and different cancer sub-type than the original target task of detecting adenocarcinoma from breast cancer.

Our contribution is a thorough investigation of several state-of-the-art methods' ability to quantify uncertainty while keeping high accuracy. We focus on the problem of detecting cancerous tissue in digital pathology, specifically for the task of detecting lymph node metastases. This has not been covered in previous investigations such as [9,17], because the appearance and variation resulting from distributional shifts of histopathology images is very different from that of natural images. Therefore, we i) extend our evaluation to a unique real-world pathology setting with a multi-hospital single indication training set and perform an extensive evaluation on both internal and external test sets and clinically plausible distributional shifts. We ii) compare the methods in terms of performance and calibration in addition to iii) how accurate their predictive uncertainty can detect both incorrect predictions and out-of-distribution (OOD) inputs.

1.1 Related Work

Multiple popular methods have been proposed for quantifying predictive uncertainty for better calibration and robustness under distributional shifts and OOD inputs in deep neural networks (DNNs). Deep ensemble [13] is arguably the simplest method where multiple networks are trained individually and their predictions are averaged during inference. Monte Carlo Dropout (MC-Dropout) [6] is an approximate Bayesian method that uses dropout [19] during multiple forward passes during inference. Temperature scaling [7] is different as it serves as a post-processing method that learns a scaling parameter on a validation set but its performance has shown to be limited under distributional shifts [17]. Mixup [25] combines random pairs of images and their labels during training, originally aimed at increased performance but it has recently shown to improve the calibration of DNNs [23]. All methods have their advantages and limitations with regard to their complexity during training or inference.

Table 1. Details on data. * and ** denote adenocarcinoma and SCC, respectively † [14], ‡ [3].

Dataset	Purpose	No. of slides	Site
CAM16-train	Development ($\mathcal{D}_{train}$, $\mathcal{D}_{val}$)	270 (160 normal, 110 tumor*)	2 hospitals†
CAM16-test	Evaluation ($\mathcal{D}_{test,int.}$)	129 (80 normal, 49 tumor*)	2 hospitals†
CAM17-train	Development ($\mathcal{D}_{train}$, $\mathcal{D}_{val}$)	46 (0 normal, 46 tumor*)	5 hospitals‡
DATASET2	Development ($\mathcal{D}_{train}$, $\mathcal{D}_{val}$)	56 (41 normal, 15 tumor*)	Hospital-A
DATASET3	Evaluation ($\mathcal{D}_{test,\ ext.}$)	135 (67 normal, 68 tumor*)	Hospital-B
DATASET4	Evaluation ($\mathcal{D}_{colon}$)	81 (43 normal, 38 tumor*)	Hospital-C
DATASET5	Evaluation ($\mathcal{D}_{SCC}$)	60 (40 normal, 20 tumor**)	Hospital-C

2 Methods

2.1 Experimental Setup

To study a relevant application in pathology, we define the primary target task as detection of adenocarcinoma in hematoxylin and eosin (H&E) lymph node sections from breast cancer. To enable the development, we obtain datasets from public [2,3,14] and non-public sources (see details in Table 1) and evaluate both predictive accuracy and uncertainty using relevant metrics (see below).

In-distribution Shift. To evaluate whether we can trust the predictions on images not derived from the hospitals used in the development, we use DATASET3 as an external test set ($\mathcal{D}_{test,ext.}$) and CAM16$_{test}$ internal test set ($\mathcal{D}_{test,int.}$). The methods are evaluated based on their ability to generalize in terms of predictive accuracy and uncertainty.

As the same cancer sub-type can originate from different organs and metastasize to lymph nodes regardless of origin, we investigate the methods' ability to generalize to other organs than included in the training set. To enable this, we collect lymph node sections with adenocarcinoma from colon cancer ($\mathcal{D}_{colon}$).

Misclassification Detection. The ability to indicate incorrect classifications is attractive from a clinical automation perspective, so pathologists can better interfere and assess results when needed, especially when the input distribution change from the intended indication. It is easy to formulate as a binary classification problem using only the uncertainty as the prediction score, hence it is a popular downstream task to evaluate predictive uncertainty [10]. We hypothesize that current methods are better at detecting incorrect predictions when the dataset is more similar to the training distribution. To test the hypothesis, we use $\mathcal{D}_{test,int.}$, $\mathcal{D}_{test,ext.}$ and $\mathcal{D}_{colon}$ to assess the performance of the binary classification (correct vs. incorrect) on each dataset.

Out-Out-Distribution Shift. When pathologists assess lymph node sections for metastases, they are also aware of other clinically relevant abnormalities than the primary task. To mimic this setting, we collect slides that contain another histology sub-type (squamous cell carcinoma (SCC)) from head and neck cancer ($\mathcal{D}_{SCC}$), which includes both well- and un-differentiated SCCs. Since SCCs, especially well-differentiated cases, are morphological different than adenocarcinoma, we consider $\mathcal{D}_{SCC}$ a realistic out-of-distribution dataset because it contains unseen abnormalities from the same domain as the training set.

Here, our evaluation is two-fold: generalization to another cancer sub-type and the ability to detect novel classes using its predictive uncertainty. To achieve the latter, we denote all tumor regions from $\mathcal{D}_{SCC}$ as $\mathcal{D}_{out}$ and the in-distribution $\mathcal{D}_{test,ext.}$ as $\mathcal{D}_{in}$. We then compare each method to discriminate between $\mathcal{D}_{out}$ and $\mathcal{D}_{in}$.

Since poorly differentiated SCC can look morphologically similar to adenocarcinoma, we also take a subset of $\mathcal{D}_{SCC}$ diagnosed as well-differentiated SCC ($N = 5$) and treat only samples from these as OOD inputs in a final experiment.

Reference Standard. Similar to the Camelyon dataset, all ground truth annotations on the non-public datasets were carefully prepared under the supervision of expert pathologists with additional slides stained with cytokeratin immunohistochemistry (IHC). All work related to the non-public datasets was approved by their institutional review board.

2.2 Evaluation Metrics

We employ *Accuracy, Area Under the Receiver Operating Characteristics curve* (AUROC) and *Precision-Recall curve* (AUPR) to report classification performance (normal vs. tumor). As suggested by Guo et al. [7], we use the *Expected*

Calibration Error ECE [16] to measure the calibration for each model. First, we compute the confidence of each of N observation denoted $p(\hat{y}_n)$, and bin these into H bins. We then calculate the ECE by comparing the content of each bin to its average accuracy. Let B_h be the set of indices for bin h. We calculate the bin accuracy

$$\mathrm{acc}(B_h) = |B_h|^{-1} \sum_{n \in B_h} \delta(\hat{y}_n - y_n^*) \tag{1}$$

and the bin confidence

$$\mathrm{conf}(B_h) = |B_h|^{-1} \sum_{n \in B_h} p_n(\hat{y}). \tag{2}$$

Then we get

$$\mathrm{ECE} = \frac{1}{N} \sum_{h=1}^{H} |B_h| \cdot |\mathrm{acc}(B_h) - \mathrm{conf}(B_h)| \tag{3}$$

$$= \frac{1}{N} \sum_{h=1}^{H} \left| \sum_{n \in B_h} p_n(y) - \sum_{n \in B_h} \delta(\hat{y}_n - y_n^*) \right| \tag{4}$$

where $\delta(x) = 1$ if $x = 0$ or $\delta(x) = 0$ if $x \neq 0$, and y_n^* is the true label.

For misclassification and OOD detection, we use also AUROC and AUPR but on the classification performance of correct vs. incorrect and in- vs. out-of-distribution, respectively. We use *False Positive Rate at 95% True Positive Rate* (FPR95) to compare method at a certain operating point. As noted by [1], these metrics are more reliable to compare for OOD detection as the task remains the same regardless of method.

2.3 Overview of Methods

We focus on methods that model $p(y|x)$ as these are the most popular in medical image analysis [3,15] and are known to scale well [12,13]. As a baseline, we use the softmax of a standard DNN to obtain posterior probabilities. For all methods, we obtain the prediction as $\hat{y} = \arg\max_y p(y|x, \theta)$ and the confidence as the maximum softmax probability $p(\hat{y}) = \max_y p(y|x, \theta)$.

MC-Dropout. We train using dropout [19] with rate p and apply L forward passes during inference with dropout enabled as described in Gal et al. [6].

Deep Ensemble. We train M standard DNNs independently of each other following [13] and combine the predictions as

$$p(y = k|x, \theta) = \frac{1}{M} \sum_{m=1}^{M} p_m(y = k|x, \theta_m) \tag{5}$$

Mixup. Recently proposed as a simple method by [25] for training better DNNs where two random input samples (x_i, x_j) and their corresponding labels (y_i, y_j) are combined using:

$$\begin{aligned} \tilde{x} &= \lambda x_i + (1 - \lambda) x_j \\ \tilde{y} &= \lambda y_i + (1 - \lambda) y_j \end{aligned} \tag{6}$$

where $\lambda \in [0, 1]$ determines the mixing ratio of the linear interpolation. λ is drawn from a symmetric Beta distribution Beta(α, α), where α controls the strength of the input interpolation and the label smoothing. We train a DNN with mixup using standard cross-entropy calculated on the soft-labels instead of the hard labels. We refer to [25] for the full details on mixup.

Table 2. Evaluation of predictive performance. $^*\alpha = 0.3$

	$\mathcal{D}_{test,int.}$			$\mathcal{D}_{test,ext.}$			$\mathcal{D}_{colon}$		
	Acc	AUROC	AUPR	Acc	AUROC	AUPR	Acc	AUROC	AUPR
Baseline	90.5	96.5	95.1	**94.3**	97.9	94.3	**79.0**	90.7	92.8
Ensemble	90.1	**97.3**	**95.9**	**94.3**	**98.1**	**96.8**	78.1	**92.3**	**94.2**
MC-Dropout	**91.0**	97.0	95.7	93.8	97.7	96.2	78.0	90.9	93.4
Mixup*	86.5	95.6	94.2	93.4	97.1	94.6	75.8	91.0	92.6

2.4 Implementation and Training Details

We perform a train/validation split on the development dataset and use these to train and select hyper-parameters for all methods. All datasets are sampled in patches (512 × 512 pixels) at 20× magnification with 50% (strided) and 150% (overlapping) sampling fraction for normal and tumor, respectively. We employ a ResNet-50 [8] architecture as the backbone for all methods because there are negligible changes between different image classifiers [9]. We use $M = 5$ to create the ensemble as reported by [17] to be sufficient. For MC-dropout, initial experiments of different implementation variations showed no performance differences. Hence, we add a dropout before the logit layer similar to [12] with $p = 0.5$ and use $L = 50$. All models are trained for 15 epochs with ADAM [11] ($\beta = (0.9, 0.999)$) with weight decay (0.0005) using a mini-batch size of 16. We use an initial learning rate of 0.01 and drop it with factor 10 every 5th epoch for all methods except mixup which required a lower initial learning rate of 0.001 to converge. For mixup, we experimented with $\alpha \in [0.1, 0.3, 0.5, 1.0]$ and we report results with $\alpha = 0.3$ as this performed best on $\mathcal{D}_{val}$. In all experiments, we apply data augmentation similar to [15] and use Pytorch [18] and Pytorch-Lightning [5].

3 Results

3.1 Evaluating Predictive Performance Under Dataset Shifts

First, we evaluate the predictive performance on the primary task of detecting adenocarcinoma in lymph node sections. We summarize the results in Table 2, and the ROC-curves for all methods and dataset shifts are shown in Fig. 2. The results show that all methods can archive high predictive performance on both the internal and external test sets. All methods perform significantly worse when evaluated on the colon dataset $\mathcal{D}_{colon}$ with mixup performing worst. Interestingly, all methods have higher AUROC on $\mathcal{D}_{SCC}$ (see Table 4) compared to $\mathcal{D}_{colon}$ even though the cancer sub-type is histological different, especially in the well-differentiated cases. In general, deep ensemble slightly outperforms all other methods on threshold independent metrics like AUROC and AUPR.

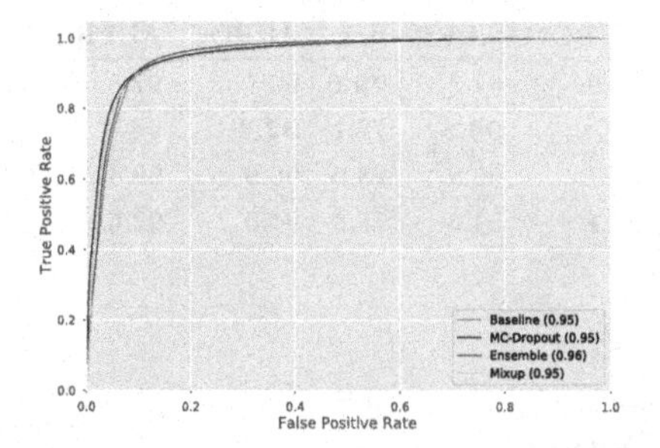

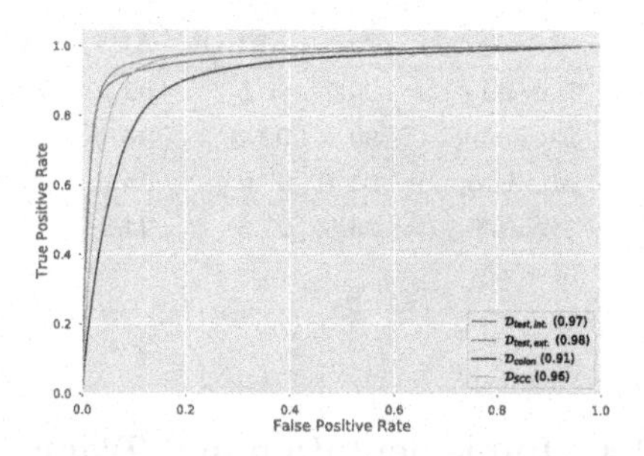

Fig. 2. ROC-curves for predictive performance. Left shows each methods with ROC curves averaged across all datasets. Right shows each dataset with ROC curves averaged across all methods.

3.2 Evaluating Predictive Uncertainty Under Dataset Shifts

We present results of calibration and detection of incorrect classified examples together in Table 3. In terms of ECE, deep ensemble and mixup improve calibration compared to the baseline method, whereas MC-dropout performs worse for the external and colon dataset. When using each method's predictive uncertainty to detect misclassifications on the test set, deep ensemble and MC-dropout have higher AUROC and AUPR on all three datasets than baseline and mixup. However, the quality of the predictive uncertainty for decreases slightly when dataset shift increases.

Table 3. Evaluation of calibration and misclassification detection. $^*\alpha = 0.3$

	$\mathcal{D}_{test,int.}$			$\mathcal{D}_{test,ext.}$			$\mathcal{D}_{colon}$		
	ECE	AUROC	AUPR	ECE	AUROC	AUPR	ECE	AUROC	AUPR
Baseline	4.9	82.6	35.7	2.1	77.7	28.6	11.8	76.7	42.0
Ensemble	**2.1**	83.9	35.6	**0.6**	**82.3**	**30.2**	**7.5**	**78.6**	**44.5**
MC-Dropout	4.6	**84.0**	35.3	2.6	79.8	29.7	13.3	77.2	43.5
Mixup*	4.2	79.1	**36.5**	0.9	80.9	29.3	9.7	71.5	41.4

3.3 Evaluating on Different Cancer Sub-type

The left part of Table 4 shows the performance on $\mathcal{D}_{SCC}$, while the right side summarizes the result of the OOD experiment. All methods show strong predictive accuracy, but fail to recognize SCC as an unseen class. Here, both ensemble and mixup outperform the baseline and MC-dropout methods.

Table 4. Evaluation of performance and OOD detection on $\mathcal{D}_{SCC}$. $^*\alpha = 0.3$

	Performance			OOD			ODD (only well-diff.)		
	Acc	AUROC	AUPR	AUROC	AUPR	FPR95	AUROC	AUPR	FPR95
Baseline	89.3	95.4	88.4	64.1	37.3	97.6	70.6	5.2	90.9
Ensemble	**89.7**	**96.3**	**91.8**	73.2	46.2	92.6	81.6	7.4	71.1
MC-Dropout	89.0	95.9	91.5	59.8	35.6	99.3	67.5	4.7	84.8
Mixup*	87.5	95.8	89.2	**86.3**	**53.6**	**47.5**	**86.5**	**8.1**	**44.6**

4 Discussion and Conclusion

We have evaluated current popular methods for predictive uncertainty on clinically relevant dataset shifts for the detection of lymph node metastases in pathology slides. All methods can generalize predictive accuracy from the internal test set to the external dataset while maintaining the quality of the predictive uncertainty. When applied to another organ, all investigated methods show both decreased performance and increased overconfidence. We have shown similar behavior when evaluated on the different cancer sub-type even-though the performance decrease was smaller than under organ shift.

As site-specific variations such as sectioning, staining and scanning variability are present in the experimental internal and external setup, we have shown that current methods are able to generalize to these sources of variability. We leave it to future work to quantify how site-specific pre-analytical variations affect the current methods as it requires a more controlled data acquisition scheme.

Our experiments show minimal benefits of MC-Dropout compared to the baseline method, and it can hurt the calibration performance on all dataset shifts. We contribute this to MC-Dropout being a too weak ensemble to achieve the same effect as a true ensemble. In general, deep ensemble increases predictive performance but also shows robustness in calibration under distributional shifts. It also displays decent capability in detecting incorrect predictions, but none of the methods are sufficient on this task. Based on the results and its simplicity, deep ensemble is an attractive method for predictive uncertainty but it comes with a computational overhead during both training and inference. Here, mixup might seem to be a cheaper alternative as our results show better calibration than baseline and MC-Dropout with a slight decrease in performance. We leave it to future work to investigate effects of different implementation of MC-Dropout and mixup extensions such as [24].

The ODD experiments indicate that adenocarcinoma and SCC, especially moderate and undifferentiated, are too similar in their morphological patterns to be treated as OOD. However, when we only assume well-differentiated SCC as an unseen class, ensemble and mixup are better to indicate the dataset shift without being sufficient for ODD detection.

Based on our results, we recommend that deep learning-based algorithms are ready for clinical implementation with reliable uncertainty estimates if used within the indication and organ included in the training set, but one should not expect current methods to alarm novel abnormalities.

Acknowledgement. The work was mainly supported by Innovation Fund Denmark (8053-00008B). Furthermore, it was partly supported by a research grant (15334) from VILLUM FONDEN, by the European Research Council (ERC) under the European Union's Horizon 2020 research and innovation programme (grant agreement n° 757360) and by The Center for Quantification of Imaging Data from MAX IV (QIM) funded by The Capital Region of Denmark.

References

1. Ashukha, A., Lyzhov, A., Molchanov, D., Vetrov, D.: Pitfalls of in-domain uncertainty estimation and ensembling in deep learning. arXiv preprint arXiv:2002.06470 (2020)
2. Bandi, P., et al.: From detection of individual metastases to classification of lymph node status at the patient level: the CAMELYON17 challenge. IEEE Trans. Med. Imaging **38**(2), 550–560 (2019). https://doi.org/10.1109/TMI.2018.2867350
3. Bejnordi, B.E., et al.: Diagnostic assessment of deep learning algorithms for detection of lymph node metastases in women with breast cancer. JAMA - J. Am. Med. Assoc. **318**(22), 2199–2210 (2017). https://doi.org/10.1001/jama.2017.14585
4. Ciompi, F., et al.: The importance of stain normalization in colorectal tissue classification with convolutional networks. In: ISBI, pp. 160–163 (2017)
5. Falcon, W.: Pytorch lightning. GitHub. Note. https://github.com/PyTorchLightning/pytorch-lightning (2019)
6. Gal, Y., Ghahramani, Z.: Dropout as a Bayesian approximation: representing model uncertainty in deep learning. In: ICML, pp. 1050–1059 (2016)
7. Guo, C., Pleiss, G., Sun, Y., Weinberger, K.Q.: On calibration of modern neural networks. In: ICML, pp. 1321–1330 (2017)
8. He, K., Zhang, X., Ren, S., Sun, J.: Deep residual learning for image recognition. In: CVPR, pp. 770–778 (2016)
9. Hendrycks, D., Dietterich, T.: Benchmarking neural network robustness to common corruptions and perturbations. In: ICLR (2019)
10. Hendrycks, D., Gimpel, K.: A baseline for detecting misclassified and out-of-distribution examples in neural networks. In: ICLR (2017)
11. Kingma, D., Ba, J.: Adam: a method for stochastic optimization. In: ICLR, pp. 1–15 (2014)
12. Kirsch, A., van Amersfoort, J., Gal, Y.: BatchBALD: efficient and diverse batch acquisition for deep Bayesian active learning. In: NeurIPS, pp. 7026–7037 (2019)
13. Lakshminarayanan, B., Pritzel, A., Blundell, C.: Simple and scalable predictive uncertainty estimation using deep ensembles. In: NeurIPS, pp. 6402–6413 (2017)

14. Litjens, G., et al.: 1399 H&E-stained sentinel lymph node sections of breast cancer patients: the CAMELYON dataset. GigaScience **7**(6), giy065 (2018). https://doi.org/10.1093/gigascience/giy065
15. Liu, Y., et al.: Artificial intelligence based breast cancer nodal metastasis detection: insights into the black box for pathologists. Arch. Pathol. Lab. Med. **143**(7), 859–868 (2018)
16. Naeini, M.P., Cooper, G., Hauskrecht, M.: Obtaining well calibrated probabilities using Bayesian binning. In: AAAI (2015)
17. Ovadia, Y., et al.: Can you trust your model's uncertainty? Evaluating predictive uncertainty under dataset shift. In: NeurIPS, pp. 13991–14002 (2019)
18. Paszke, A., et al.: PyTorch: an imperative style, high-performance deep learning library. In: NeurIPS, pp. 8024–8035 (2019)
19. Srivastava, N., Hinton, G., Krizhevsky, A., Sutskever, I., Salakhutdinov, R.: Dropout: a simple way to prevent neural networks from overfitting. J. Mach. Learn. Res. **15**(1), 1929–1958 (2014)
20. Stacke, K., Eilertsen, G., Unger, J., Lundström, C.: A closer look at domain shift for deep learning in histopathology. arXiv preprint arXiv:1909.11575 (2019)
21. Steiner, D.F., et al.: Impact of deep learning assistance on the histopathologic review of lymph nodes for metastatic breast cancer. Am. J. Surg. Pathol. **42**(12), 1636 (2018)
22. Tellez, D., et al.: Quantifying the effects of data augmentation and stain color normalization in convolutional neural networks for computational pathology. Med. Image Anal. **58**, 101544 (2019)
23. Thulasidasan, S., Chennupati, G., Bilmes, J.A., Bhattacharya, T., Michalak, S.: On mixup training: improved calibration and predictive uncertainty for deep neural networks. In: NeurIPS, pp. 13888–13899 (2019)
24. Verma, V., et al.: Manifold mixup: Better representations by interpolating hidden states. In: ICLR, pp. 6438–6447 (2019)
25. Zhang, H., Cisse, M., Dauphin, Y.N., Lopez-Paz, D.: mixup: beyond empirical risk minimization. In: ICLR (2018)

Deep Generative Model for Synthetic-CT Generation with Uncertainty Predictions

Matt Hemsley[1,2(✉)], Brige Chugh[3], Mark Ruschin[3], Young Lee[3], Chia-Lin Tseng[4], Greg Stanisz[1,2], and Angus Lau[1,2]

[1] Medical Biophysics, University of Toronto, Toronto, ON, Canada
matt.hemsley@sri.utoronto.ca
[2] Physical Sciences Platform, Sunnybrook Research Institute, Toronto, ON, Canada
[3] Medical Physics, Sunnybrook Research Institute, Toronto, ON, Canada
[4] Radiation Oncology, Sunnybrook Health Sciences Centre, Toronto, ON, Canada

Abstract. MR-only radiation treatment planning is attractive due to the superior soft tissue definition of MRI as compared to CT, and the elimination of the uncertainty introduced by CT-MRI registration. To facilitate MR-only radiation therapy planning, synthetic-CT (sCT) algorithms (for electron density correction) are required for dose calculation. Deep neural networks for sCT generation are useful due to their predictive power, but lack of uncertainty information is a concern for clinical implementation. The feasibility of using a conditional generative adversarial model (cGAN) to generate sCTs with accompanying uncertainty maps was investigated. Dropout-based variational inference was used to account for uncertainty in the trained model. The cGAN loss function was also combined with an additional term such that the network learns which regions of input data are associated with highly variable outputs. On a dataset of 105 brain cancer patients, our results demonstrate that the network generates well-calibrated uncertainty predictions and produces sCTs with equivalent accuracy as previously reported deterministic models.

Keywords: Generative adversarial networks · MR-only radiation therapy · Uncertainty

1 Introduction

Computed tomography (CT) is the standard imaging modality for radiation treatment (RT) simulation, planning, and image guidance [1]. Compared to CT, magnetic resonance imaging (MRI) provides improved soft-tissue contrast, which enables accurate delineation of target volumes and organs at risk [2], as well as additional information which can potentially guide biologically-based RT adaptation [3]. The emergence of MR-Linacs [4] for MR-based treatment delivery provides significant motivation for an MR-only RT treatment workflow, which would eliminate CT to MRI registration error, provide a more efficient and cost-effective workflow, and reduce patient exposure to ionizing radiation [5]. However, unlike CT images, in which the voxel intensities are

A. L. Martel et al. (Eds.): MICCAI 2020, LNCS 12261, pp. 834–844, 2020.
https://doi.org/10.1007/978-3-030-59710-8_81

measured in Hounsfield units, MR signal intensities are not directly dependent on the electron density of the imaged material. This limitation makes dose calculation based on MR images infeasible without correction for electron density [6]. Implementation of MR-only RT is crucially dependent on the generation of synthetic-CT (sCT) images with signal intensities that reflect electron density.

Recently, deep learning models have been reported for sCT generation [7–11]. Deep learning is an attractive option due to the high predictive accuracy and the computational efficiency for clinical applications. However, prediction errors can occur, posing potential risk to patients. A desirable feature for clinical implementation is an additional prediction of uncertainty. In this context, uncertainty, which can be understood as an estimate of prediction error, can be classified into two types [12]. The first type is aleatoric, or data dependent (DD) uncertainty, which is caused by incompleteness in input data, such as noisy inputs or the absence of visual features. The second type is epistemic, or model dependent (MD) uncertainty, which is caused by uncertainty in the model parameters. In this work, we investigate using a neural network model to place uncertainty predictions on conditional adversarial network (cGAN) generated synthetic-CTs. The standard cGAN model was modified so that for a given input MRI, the network learns (1) the corresponding sCT and (2) a heatmap of the magnitude and spatial regions of high predicted uncertainty in the sCT.

2 Related Work

Electron density assignment to MR images has previously been reported for MR-only RT [13], as well as for attenuation correction in PET/MR systems [14]. Deep learning solutions are attractive due to their prediction accuracy, ability to generalize to abnormal pathology, and computational efficiency. Studies have been conducted using various network architectures, including the U-Net [15], 3D-FCN [16], and cGANs [8, 9], and on various anatomical sites, including the brain [7], prostate [8], rectum [11], and lung [10]). Mean absolute error (MAE) between the ground truth and the sCTs is the most reported evaluation metric. Results vary with the anatomical site and exclusion criteria for patients, but generally methods for sCT generation report MAE in the range of 80 to 200 Hounsfield units (HU) [18] leading to a mean dosimetric agreement of $<2\%$, which is acceptable dosimetric agreement for RT planning [19].

Nix and Weigland [20] introduced the "heteroscedastic noise model" to estimate aleatoric uncertainty. They proposed training to minimize a Gaussian log-likelihood loss function, $\frac{1}{2}\frac{||y-\hat{y}(x)||^2}{2\hat{\sigma}(x)^2} + \frac{1}{2}\log[\hat{\sigma}(x)^2]$, returning predictions of variance, $\hat{\sigma}(x)$ in addition to the target, $\hat{y}(x)$. Gal et al. proposed using dropout to approximate Bayesian variation inference to determine epistemic uncertainty. Kendall and Gal combined these two techniques in a single multitask learning model [21]. Hu et al. [22] reported a similar technique wherein the epistemic and aleatoric uncertainty estimation is split into a two-step process, using distinct networks in sequence.

Neural networks with uncertainty predictions have been previously applied in biomedical imaging. Hu et al. validated their two-step uncertainty model on sparse MRI reconstruction [22]. Glang et al. [23] used a technique similar to [20] for chemical exchange saturation transfer (CEST) MRI reconstruction. Bragman et al. [24] performed sCT generation and target volume segmentation using a method similar to [21]. Klages et al. [25] used a cGAN for generating sCTs with aleatoric uncertainty by generating the sCT in overlapping patches. To the best of our knowledge, this is the first time that a cGAN has been used with both the heteroscedastic noise model and dropout for aleatoric and epistemic uncertainty prediction.

3 Methods

A cGAN based on the 'pix2pix' architecture [26] was implemented and modified to produce uncertainty predictions in a similar fashion to Kendall et al. [21] (Fig. 1). The dataset consisted of 2D slices of MRI and CT volumes of 105 patients. The network was trained to produce both target sCT predictions and uncertainty heatmaps.

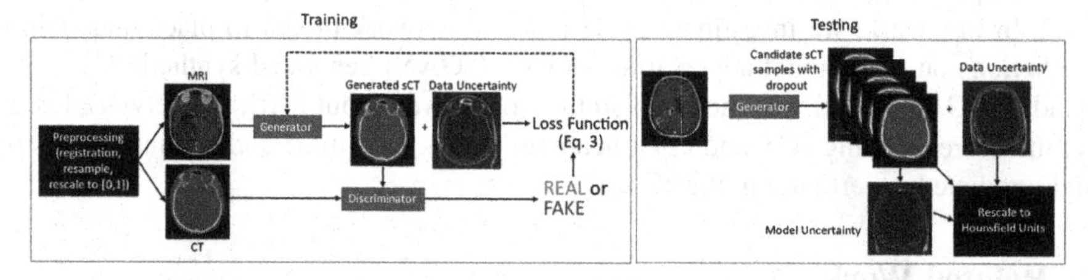

Fig. 1. cGAN architecture for uncertainty prediction. MRIs are input to the cGAN which predicts target sCTs and uncertainties dependent on the input data and error in model parameters.

3.1 Conditional Generative Adversarial Networks

The cGAN model consists of two distinct, competing convolutional deep neural networks, (1) a generator that produces candidate images based on a trained model distribution and an input condition, and (2) a discriminator which discriminates between candidate and ground truth images [26, 27]. The cGAN is trained to learn the transformation from conditioned sample $x \in X$ and random noise vector z to the desired output sample $y \in Y$, $G : \{x, z\} \rightarrow \{y\}$. The cGAN loss function can be written as

$$\mathcal{L}_{cGAN} = \mathbb{E}_{x,y}[\log D(x, y)] + \mathbb{E}_{x,z}\left[\log(1 - D(x, G(x, z))\right] \quad (1)$$

[26] where the functions G and D are the outputs of the generator and discriminator respectively, and $\mathbb{E}_{a,b}[f(x)]$ is the expectation value of $f(x)$ over the distributions of a, b. In this case, x is a 2D MR slice, y is the matching CT image, and the output of G is the 2D sCT. In the context of image-to-image translation, low-frequency information is better captured when an L_1 penalty is added to the loss function [26]. The L_1 term can be written as $|y - G(x, z)|_1$ and added to Eq. (1). The loss becomes

$$\mathcal{L}_{p2p} = \mathcal{L}_{cGAN} + \lambda \mathbb{E}_{x,y,z}\left[|y - G(x, z)|_1\right] \quad (2)$$

where λ is a regularization parameter.

Model Dependent Uncertainty. A Bayesian neural network replaces deterministic weights with probability distributions, which allows for uncertainty estimation [28]. Given a dataset $X = \{x_1, \ldots x_N\}$, $Y = \{y_1, \ldots, y_N\}$, computation of the posterior over the weights, $p(W|X, Y)$ requires the marginal probability $p(Y|X)$. In practice, $p(Y|X)$ is intractable, so dropout-based variational inference was used [29]. During testing, T stochastic forward-passes were made through the network with dropout. Each forward pass is considered a sample from the posterior distribution, $\left\{\hat{y}^{\hat{w}_t}(x)\right\}_{t=1}^{T}$ where the subset of weights used, $\left\{\hat{w}_t\right\}_{t=1}^{T} \sim q\left(\hat{W}|\theta\right)$, where $q(W|\theta)$ is the dropout distribution [21]. The prediction of the network is $\mathbb{E}\left[\hat{y}^{\hat{w}_t}(x)\right]$ and MD uncertainty is $Var\left[\hat{y}^{\hat{w}_t}(x)\right]$.

Data Dependent Uncertainty. Data dependent (DD) uncertainty is a measure of error on network output which propagates from noisy input. We replace the L_1 term in the cGAN loss function (1) with, $\mathcal{L}_{Unc} = \left[\frac{||y-G(x,z)||_1}{\sigma(x)}\right] + \log(\sigma(x))$ where $\sigma(x)$ is the standard deviation predicted by the network due to noise in the input data, x. This term is derived from the log-likelihood of the Laplace distribution. Unlike previous work, which used the log-likelihood of the Gaussian distribution [21, 23, 24], the Laplace distribution was used due to the analog to the L_1 penalty from Eq. (2). The loss function encourages assigning high $\sigma(x)$ to regions where the L_1 difference between the sCT and CT is large, and low $\sigma(x)$ when the L_1 is small. Replacing the L_1 term in Eq. (2) yields the loss used to capture DD uncertainty

$$\mathcal{L} = \mathcal{L}_{cGAN} + \lambda[\mathbb{E}_{x,y,z}\left[\frac{||y - G(x,z)||_1}{\hat{\sigma}(x)}\right] + \log(\hat{\sigma}(x))] \quad (3)$$

Following [21], the DD and MD uncertainties were added in quadrature, and the total uncertainty prediction is $\sqrt{\hat{\sigma}(x)^2 + Var\left[\hat{y}^{\hat{w}_t}(x)\right]}$.

Uncertainty Calibration. Calibrated uncertainty predictions represent the statistical distribution of outputs. Dropout-based uncertainty predictions are often not well-calibrated [30]. One method for calibration is to linearly rescale uncertainty predictions to match the true error between sCT and CT volumes using least squares [31]. A held-out training set (N = 2 volumes, 10% of the testing set) were used to determine the rescaling factors. 5-fold cross validation was used to prevent overfitting. Bland-Altman (BA) analysis was used to evaluate bias in the calibrated uncertainties [32].

3.2 Data Acquisition and Implementation Details

The dataset consisted of 105 patients with brain metastases or glioblastoma multiform scheduled for radiation treatment. Images were retrospectively analyzed in accordance with institutional ethics approval. Each patient received MRI and CT scans for radiation treatment planning (MRI: Philips Ingenia MR-RT 1.5 T, 3D T1w pre-Gd, T1w post-Gd, T2 FLAIR - matrix size 480 × 480, FOV = 240 mm, 215–250 slices 1 mm thickness;

CT: Philips Brilliance CT - matrix size 512 × 512, FOV = 450 mm, 215–250 slices 1 mm thickness). A thermoplastic mask was used for patient immobilization during the CT scans, but not for the MRI scans due to interference with the coil. No data was excluded based on the presence of imaging artifacts, air pockets, or abnormal pathology such as implants. The distribution of data containing these regions is described in Table 1. The MRI and CTs were rigidly registered, resampled to the same matrix size (256 × 256), 0.8 mm isotropic voxel size, and randomly grouped for training (n = 85) and testing (n = 20). For each patient, the CT intensities were linearly scaled from [−2000, 3000] HU to [0, 1]. The MR intensities were linearly scaled to [0, 1] using the full intensity range. The dimensionless network output was rescaled back to HU using the inverse of the previously applied scaling.

Table 1. The number patients and in brackets the average number of slices per patient containing metal artifacts, abnormal anatomy (implants, and post-surgery patients), bone/air interfaces.

	Anatomical abnormality	Air/Bone interface	Metal artifact
Testing n = 85	65 (21 ± 10)	85 (84 ± 9)	68 (33 ± 11)
Training n = 20	14 (32 ± 16)	20 (79 ± 7)	16 (25 ± 6)

Images were processed slice-by-slice in PyTorch [33], using a version of the publicly available 'pix2pix' cGAN code [26] modified for uncertainty prediction detailed in Sect. 3.1. The generator was a U-Net [34] and the discriminator was a patchGAN [26], with patch size 70 × 70. The model was trained with a batch size of 1 for 400 epochs. Adam [35] was used as the optimizer ($\beta_1 = 0.5, \beta_2 = 0.999$) with a learning rate of 0.0002 for the first 200 epochs and linearly decreasing to 0 for the final 200 epochs. Dropout with a probability of $p = 0.5$ was applied at each layer. 50 samples for each set of input MRIs were generated for inference. As suggested in the original GAN paper [27], and 'pix2pix' paper [26] the generator was trained to maximize $\log(D(x, G(x, z)))$ rather than $\log(1 - D(x, G(x, z)))$ and the network was trained to predict the log standard deviation, $s_i = \log \sigma$ rather than σ to increase stability. The normalization parameter λ from Eq. (2) was set to 100.

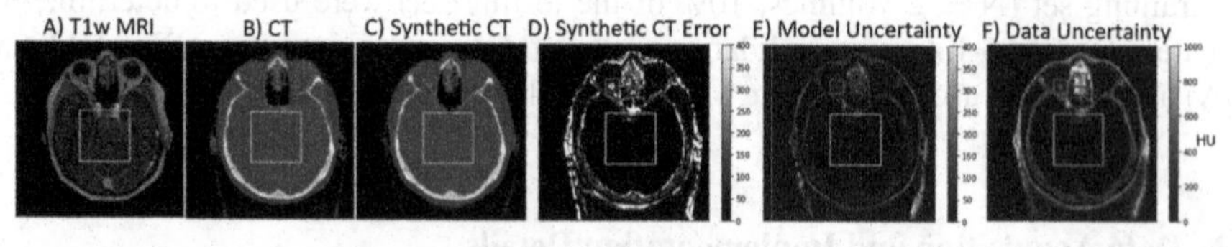

Fig. 2. sCT model output. The nasal cavity (within the green box) is incorrectly represented and the optic nerve (within the red box) was classified as bone. The immobilization mask found in the CT but not MRI was also generated. These regions where the sCT is qualitatively incorrect as well as tissue interfaces is represented as highly uncertain.

4 Experiments and Results

4.1 Evaluation

The trained model was evaluated on sCT generation by mean absolute error (MAE) $MAE = \frac{\sum_{i=1}^{n} sCT - CT}{n}$ where sCT and CT are the voxel values of the images, and n is the number of voxels. An cGAN was training with Eq. (2) as the loss function as a baseline comparison. Rectangular regions of interest containing artifacts, air pockets, and abnormal patient anatomy including implants, fiducials positioning, and areas affected by surgery were identified for each slice of the test set. The MAE and reported uncertainty of these regions were compared to identically sized regions of normal appearing soft tissue in the brain and presented in Table 2. We performed Bland-Altman (BA) analysis [32] between the predicted uncertainty and difference between the generated sCT and ground truth CT, The percentage of points for which predicted uncertainty is greater than the true error is reported for various regions of sample slices.

Table 2. MAE and uncertainty in different anatomical regions. After calibration, each region's predicted uncertainty better matches the MAE.

	MAE (HU)	Total Unc (HU)	Unc/MAE (no units)	Calibrated Unc (HU)	CalUnc/MAE (no units)
Normal soft tissue	6 ± 3	27 ± 6	4.21 ± 2.3	10 ±5	1.59 ± 1.2
Abnormal anatomy	16 ± 13	43 ± 21	2.71 ± 2.6	18 ± 14	1.13 ± 1.1
Air/Bone interface	237 ± 31	544 ± 110	2.29 ± 0.6	220 ± 46	0.93 ± 0.2
Metal artifact	246 ± 58	597 ± 250	2.42 ± 1.2	278 ± 96	1.12 ± 0.5

4.2 Model Performance

Training took approximately 12 h on two Nvidia Titan XP GPUs. Once trained a sCT brain volume could be generated with DD uncertainty in 7.5 s, and 150 s for both DD and MD uncertainty. The MAE per volume of the test set was found to be 93 ± 10 HU and the mean predicted total uncertainty was 158 ± 23 HU. The unmodified pix2pix network yielded a MAE of 89 ± 8 HU, indicating the uncertainty modifications slightly reduced voxel accuracy (two sample t-test p value: 0.012). Figure 2 presents a sample model output consisting of the conditioned T1w MRI, ground truth CT, generated sCT, and associated heatmaps of MD and DD uncertainty. Figure 3 shows the BA analysis for this slice, with and without calibration. The bias per volume of the uncalibrated uncertainty predictions over the test set was found to be −71 ± 18 HU, indicating an overpredicted error. The bias using the calibrated uncertainty was computed 5 times for each cross-validated fold. Using the parameters from the median performing fold, calibration reduced the bias to 6 ± 16 HU. In all cases calibration reduced the bias towards overpredicted error. Figure 4 shows the regression of three volumes, representative of the improved accuracy of uncertainty prediction after calibration. Figures 3 and 4 support that uncertainty prediction is correlated with sCT error. The Pearson correlation coefficients ranged between 0.59 and 0.81, indicating good agreement. Figures 3 and 4 also show uncalibrated uncertainty predictions are greater than true error for approximately 90% of voxels over the volumes and may be interpreted as an upper bound on error. Figure 5 shows an example of a failed sCT generation due to the presence of a metal artifact. The incorrect voxels were predicted as highly uncertain. The Pearson correlation coefficient between MD and DD uncertainty over the whole test dataset was 0.78.

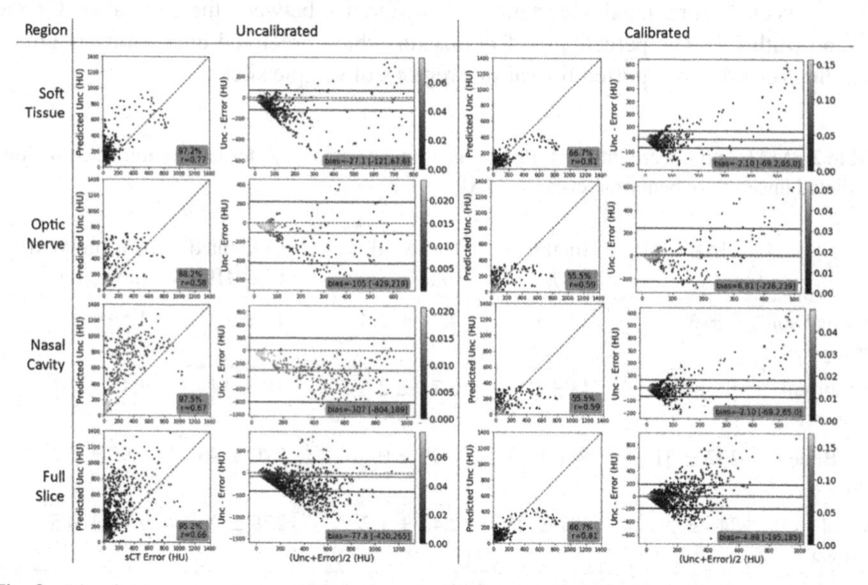

Fig. 3. Bland-Altman analysis of the regions highlighted in Fig. 2 with and without uncertainty calibration. Color bars correspond to the relative density of points, found using kernel density estimation [36]. Scatter plots show predicted uncertainty against error between the sCT and ground truth CT, report the Pearson correlation coefficient, and the percentage of points where uncertainty was greater than sCT error. The difference plots show that the uncalibrated uncertainty estimates have a bias towards over prediction, which is reduced after calibration.

5 Discussion

We presented a deep generative model for sCTs generation and accompanying uncertainty heatmaps. The trained networks were able to generate sCTs with both MD and DD uncertainty in a clinically suitable timeframe. No extra computational time was required to determine DD uncertainty. However, MD estimation increased computation time as multiple candidate samples needed to be generated for each input. In this study,

50 dropout samples were used so the inference time matched a commercial sCT solution [37]. The commercial solution does not provide uncertainty estimates. The benefit of generating more samples would be a better calibrated uncertainty prediction, however we explicitly calibrate the total uncertainty. Previous studies excluded volumes with image artifacts [7] and implants [11], or corrected for inconsistent air pockets between the MRI and CT [8]. No such measures were taken to see if these areas were predicted as uncertain. In this study, uncertainty predictions were positively correlated with sCT error and associated with spatial regions of failure, as demonstrated in Figs. 2 and 5. Table 2 shows that areas of high uncertainty are predominately areas where MRI offers incomplete information, such as regions where bone is adjacent to air deposits as well as artifact contaminated regions. The linear calibration is shown to be effective in these regions, demonstrated by the ratio of uncertainty and error approaching 1. Accurate sCT prediction in these regions is important as dose increases at tissue interfaces [38]. A non-dosimetrically relevant increase in accuracy was found when uncertainty modifications to the network were not included. MD and DD uncertainty predictions were found to be correlated, therefore adding a covariance term when determining total uncertainty may produce better calibrated initial predictions. While MAE is a commonly reported metric for sCT evaluation, it is highly susceptible to misalignment between the MRI and CT. In addition to the metal artifacts due to tooth implants, Fig. 5 shows an example where the tongue is in different positions in the MRI and CT. This registration error leads to a discrepancy between uncertainty and sCT error. Future work will focus on developing criteria for the automatic identification of failure cases using uncertainty prediction, such that patients with inaccurate synthetic CTs will be flagged for re-acquisition of images, or manual adjustment of the synthetic CT images.

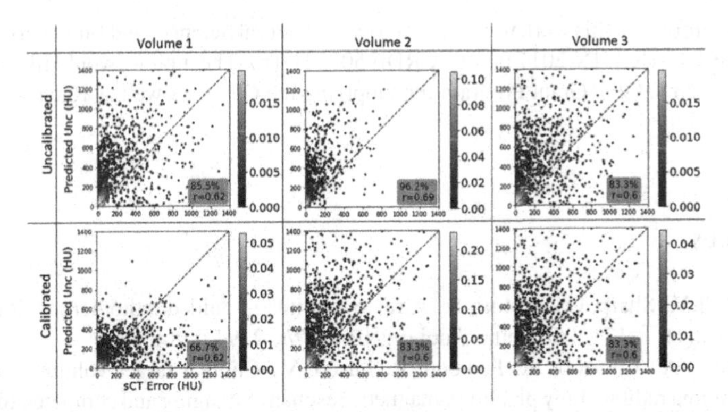

Fig. 4. Regression plots of three sample 3D patient volumes. The Pearson correlation coefficient and the percentage of points where uncertainty was greater than sCT error are reported. Color bars correspond to the relative density of points, found by kernel density estimation [36].

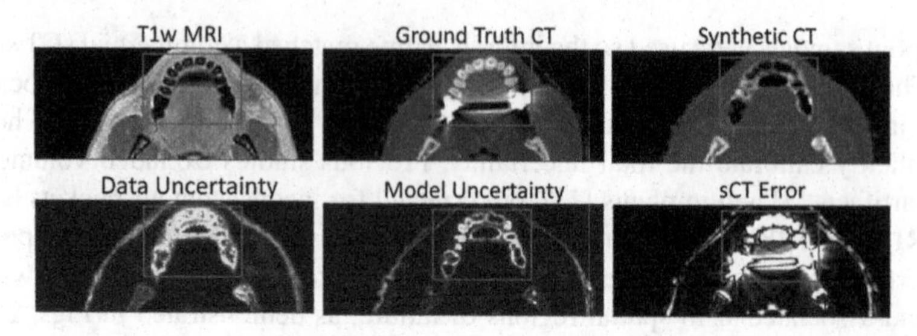

Fig. 5. Example of an sCT failure mode. The model incorrectly predicts several teeth and was affected by the metallic artifact caused by a dental implant. The regions of spatial failure are regarded by both MD and DD uncertainty as highly uncertain. The patient's tongue is also held in a different position in the MRI and CT slices, resulting in the unfair penalization of the network as measured by MAE.

6 Conclusion

MR-only radiation treatment planning requires accurate synthetic CT prediction. In this study, we used a modified cGAN to generate synthetic CT images from MRI, and to estimate spatial regions where the synthetic CT were incorrect. Synthetic CT images along with data dependent uncertainty were produced in 7.5 s, and 150 s for both data and model dependent uncertainties on two consumer grade GPUs. Due to the safety critical nature of using sCTs for radiation treatment planning, we anticipate that the methods developed in our study will become an important addition to MR-only radiation treatments.

Acknowledgements. This work was supported by the Natural Sciences and Engineering Research Council (NSERC) (RGPIN-2017 06596, CRDE 507521-16). The authors would like to thank the NVIDIA Academic GPU Grant Program for supplying the GPUs, as well as Rachel Chan for the helpful discussion.

References

1. Chen, G.T.Y., Sharp, G.C., Mori, S.: A review of image-guided radiotherapy. Radiol. Phys. Technol. **2**(1), 1–12 (2008). https://doi.org/10.1007/s12194-008-0045-y
2. Fiorentino, A., Caivano, R., Pedicini, P., Fusco, V.: Clinical target volume definition for glioblastoma radiotherapy planning: magnetic resonance imaging and computed tomography. Clin. Transl. Oncol. **15**(9), 754–758 (2013)
3. Bhatnagar, P., et al.: Functional imaging for radiation treatment planning, response assessment, and adaptive therapy in head and neck cancer. RadioGraphics **33**(7), 1909–1929 (2013)
4. Rai, R., et al.: The integration of MRI in radiation therapy: collaboration of radiographers and radiation therapists. J. Med. Radiat. Sci. **64**, 61–68 (2017)
5. Jonsson, J., Nyholm, T., Söderkvist, K.: The rationale for MR-only treatment planning for external radiotherapy. Clin. Transl. Radiat. Oncol. **18**, 60–65 (2019)

6. Johnstone, E., et al.: Systematic review of synthetic computed tomography generation methodologies for use in magnetic resonance imaging–only radiation therapy. Int. J. Radiat. Oncol. Biol. Phys. **100**, 199–217 (2018)
7. Dinkla, A.M., et al.: MR-only brain radiation therapy: dosimetric evaluation of synthetic CTs generated by a dilated convolutional neural network. Int. J. Radiat. Oncol. Biol. Phys. **102**, 801–812 (2018)
8. Maspero, M., et al.: Dose evaluation of fast synthetic-CT generation using a generative adversarial network for general pelvis MR-only radiotherapy. arXiv:1802.06468[physics.med-ph] (2018)
9. Emami, H., Dong, M., Nejad-Davarani, S.P., Glide-Hurst, C.K.: Generating synthetic CTs from magnetic resonance images using generative adversarial networks. Med. Phys. **45**(8), 3627–3636 (2018)
10. Wang, H., Chandarana, H., Block, K.T., Vahle, T., Fenchel, M., Das, I.J.: Dosimetric evaluation of synthetic CT for magnetic resonance-only based radiotherapy planning of lung cancer. Radiat. Oncol. **12**(1), 1–9 (2017)
11. Maspero, M., et al.: Feasibility of magnetic resonance imaging-only rectum radiotherapy with a commercial synthetic computed tomography generation solution. Phys. Imaging Radiat. Oncol. **7**, 58–64 (2018)
12. Der Kiureghian, A., Ditlevsen, O.: Aleatory or epistemic? Does it matter? Struct. Saf. **31**(2), 105–112 (2009)
13. Chen, L., et al.: MRI-based treatment planning for radiotherapy: dosimetric verification for prostate IMRT. Int. J. Radiat. Oncol. Biol. Phys. **60**(2), 636–647 (2004)
14. Delso, G., et al.: Performance measurements of the siemens mMR integrated whole-body PET/MR scanner. J. Nucl. Med. **52**(12), 1914–1922 (2011)
15. Chen, S., Qin, A., Zhou, D., Yan, D.: Technical note: U-net-generated synthetic CT images for magnetic resonance imaging-only prostate intensity-modulated radiation therapy treatment planning. Med. Phys. **45**(12), 5659–5665 (2018)
16. Nie, D., Trullo, R., Lian, J., Petitjean, C., Ruan, S., Wang, Q., Shen, D.: Medical image synthesis with context-aware generative adversarial networks. In: Descoteaux, M., Maier-Hein, L., Franz, A., Jannin, P., Collins, D.L., Duchesne, S. (eds.) MICCAI 2017. LNCS, vol. 10435, pp. 417–425. Springer, Cham (2017). https://doi.org/10.1007/978-3-319-66179-7_48
17. Wolterink, J.M., Dinkla, A.M., Savenije, M.H.F., Seevinck, P.R., van den Berg, C.A.T., Išgum, I.: Deep MR to CT synthesis using unpaired data. In: Tsaftaris, S.A., Gooya, A., Frangi, A.F., Prince, J.L. (eds.) SASHIMI 2017. LNCS, vol. 10557, pp. 14–23. Springer, Cham (2017). https://doi.org/10.1007/978-3-319-68127-6_2
18. Edmund, J.M., Nyholm, T.: A review of substitute CT generation for MRI-only radiation therapy. Radiat. Oncol. **12**(1), 28 (2017)
19. Korsholm, M.E., Waring, L.W., Edmund, J.M.: A criterion for the reliable use of MRI-only radiotherapy. Radiat. Oncol. **9**(1), 16 (2014)
20. Nix, D.A., Weigend, A.S.: Estimating the mean and variance of the target probability distribution. In: IEEE International Conference on Neural Networks - Conference Proceedings, vol. 1, pp. 55–60 (1994)
21. Kendall, A., Gal, Y.: What uncertainties do we need in Bayesian deep learning for computer vision? In: Advances in Neural Information Processing Systems (2017)
22. Hu, S., Pezzotti, N., Mavroeidis, D., Welling, M.: Simple and accurate uncertainty quantification from bias-variance decomposition. arXiv:2002.05582 (2020)
23. F. Glang et al.: DeepCEST 3T: robust MRI parameter determination and uncertainty quantification with neural networks—application to CEST imaging of the human brain at 3T. Magn. Reson. Med. mrm28117 (2019)
24. Bragman, F.J.S., et al.: Uncertainty in multitask learning: joint representations for probabilistic MR-only radiotherapy planning. arXiv: 1806.06595[cs.CV] (2018)

25. Klages, P., et al.: Patch-based generative adversarial neural network models for head and neck MR-only planning. Med. Phys. **47**(2), 626–642 (2020)
26. Isola, P., et al.: Image-to-image translation with conditional adversarial networks. arXiv:1611.07004 [cs.CV] (2018)
27. Goodfellow, I.J., et al.: Generative adversarial nets. Adv. Neural. Inf. Process. Syst. **3**(January), 2672–2680 (2014)
28. Kwisthout, J.: Most probable explanations in Bayesian networks: complexity and tractability. Int. J. Approx. Reason. **52**, 1452–1469 (2011)
29. Gal, Y., Ghahramani, Z.: Dropout as a Bayesian approximation: representing model uncertainty in deep learning. arXiv:1506.02142 [stat.ML] (2016)
30. Maddox, W., Garipov, T., Izmailov, P., Vetrov, D., Wilson, A.G.: A simple baseline for Bayesian uncertainty in deep learning. arXiv:1902.02476 [cs.LG] (2019)
31. Kuleshov, V., Fenner, N., Ermon, S.: Accurate uncertainties for deep learning using calibrated regression. arXiv:1807.00263 [cs.LG] (2018)
32. Bland, J.M., Altman, D.G.: Measuring agreement in method comparison studies. Stat. Methods Med. Res. **8**(2), 135–160 (1999)
33. Paszke, A., et al.: PyTorch: an imperative style, high-performance deep learning library. arXiv: 1912.01703 [cs.LG] (2019)
34. Ronneberger, O., Fischer, P., Brox, T.: U-Net: convolutional networks for biomedical image segmentation. In: Navab, N., Hornegger, J., Wells, W.M., Frangi, A.F. (eds.) MICCAI 2015. LNCS, vol. 9351, pp. 234–241. Springer, Cham (2015). https://doi.org/10.1007/978-3-319-24574-4_28
35. Kingma, D.P., Ba, J.L.: Adam: a method for stochastic optimization. In: 3rd International Conference on Learning Representations, ICLR 2015 - Conference Track Proceedings (2015)
36. Rosenblatt, M.: Remarks on some nonparametric estimates of a density function. Ann. Math. Stat. **27** (1956)
37. Tyagi, N., et al.: Clinical workflow for MR-only simulation and planning in prostate. Radiat. Oncol. **12**(1), 119 (2017)
38. Raaijmakers, A.J.E., Raaymakers, B.W., Lagendijk, J.J.W.: Integrating a MRI scanner with a 6 MV radiotherapy accelerator: dose increase at tissue-air interfaces in a lateral magnetic field due to returning electrons. Phys. Med. Biol. **50**(7), 1363–1376 (2005)

Author Index

Printed in the United States
By Bookmasters

Printed in the United States
By Bookmasters